Handbook of Road Technology

Handbook of Road Technology

M. G. Lay

4th edition

CRC Press
Taylor & Francis Group
Boca Raton London New York

CRC Press is an imprint of the
Taylor & Francis Group, an **informa** business
A SPON PRESS BOOK

CRC Press
Taylor & Francis Group
6000 Broken Sound Parkway NW, Suite 300
Boca Raton, FL 33487-2742

First issued in paperback 2019

ISBN-13: 978-0-415-47265-4 (hbk)
ISBN-13: 978-0-367-86487-3 (pbk)

This work has been produced in Times New Roman
from typeset copy supplied by the Author

British Library Cataloguing in Publication Data
A catalogue record for this book is available
from the British Library

Library of Congress Cataloging-in-Publication Data
Lay, M. G. (Maxwell G.)
Handbook of road technology / M.G.Lay – 4th ed.
p. cm.
Includes bibliographical references and index.
1. Roads—Handbooks, manuals, etc. 2. Highway engineering—Handbooks, manuals, etc. 3. Pavements—Handbooks, manuals, etc. I. Title.
TE145.L38 2009
625.7--dc22
2008049305

Visit the Taylor & Francis Web site at
http://www.taylorandfrancis.com

and the CRC Press Web site at
http://www.crcpress.com

Contents

Acknowledgments for the Fourth Edition

The first edition of this book was based in significant part on my earlier *Source Book for Australian Roads* (Lay, 1985a), with the text edited and expanded to suit an international audience. Subsequent editions have been revised in the light of new published work, reader feedback, my continued experience as a road practitioner and my own improved understandings. All this work has been done in my own time, but I acknowledge the ongoing support of my various employers, colleagues and clients over the last 23 fascinating years.

This fourth edition has been partly prepared during my periods as a principal engineer at Sinclair Knight Merz responsible for reviewing all aspects of the mammoth City Link build-own-operate-transfer (BOOT) toll-road project, as an advisor to the construction firm Thiess preparing bids for toll road projects, and finally as a Director of the Macquarie Bank company, ConnectEast responsible for the design, construction and operation of the EastLink toll road. All these experiences, which have added many practical and commercial facets to my existing theoretical and design knowledge, have greatly influenced this fourth edition.

Many of the changes also arise from reader feedback on the update pages published in *Highway Engineering in Australia* until the end of 2007.

Given my objectives, the book draws heavily on the work of others. Much of the material is not my own. To avoid presentation difficulties, it has not always been possible to acknowledge this in detail. The aim is to present available data in collected, collated and readable form, rather than to create new knowledge. Any perceived inadequacies in the referencing may thus be a result of deliberate policy, rather than laziness on my part. I particularly apologise if this has occasionally led to some imprecision or undesirable omission in the referencing of the work of others. It was either unintentional or unavoidable.

I do wish to acknowledge all those whose work I have drawn on so extensively to write this book. To you, my reader of this new edition, may I acknowledge in advance the comments that you send me on how future editions might be improved and enhanced.

However, my prime thanks must continue to go to my wife Margaret for her patience and understanding. Robbie and Cosette, with restraint, have kept their feet on the ground and off my papers. Grandchildren have provided many unexpected but essential distractions from the underlying tedium.

I have found writing and updating the book through all its versions to this fourth edition to be a rewarding and pleasant experience. I hope that in using it you may share some of my satisfaction.

M. G. Lay, AM

CHAPTER ONE

Introduction

The purpose of this book is to cover all aspects of the technology of roads and road transport. It is aimed at an audience which is able to lead itself, or to be led by others, through technically-oriented material. That is, the book is not an introduction to roads for the untrained. In writing it, I have aimed at:

* later-year undergraduates who have a lecturer to guide them, some background in independently using the technical literature, and an understanding of basic technical concepts;
* qualified professionals who have not been directly working in a particular sub-set of road technology but who wish to acquaint themselves quickly with that technology. The book is intended to give a background briefing and sometimes to point the reader to sources of deeper understanding should this be desired;
* informed laymen wishing to gain a quick understanding of the technical implications of a particular road-related technology; and
* librarians and information officers conducting information searches in the roads area.

The book begins by providing some core definitions in Chapter 2. In addition, in the body of the book, I have been careful to define all specific terms as they arise. Chapter 3 provides a history of roads, as the present can only make sense if seen in the context of the past. Chapter 4 discusses the management of today's roads at a political level.

To understand the layout and contents of the remainder of the book, it is necessary to recognise that the purpose of the road transport system is to aid the movement of goods and people. The system consists of three major interacting components:

(a) the drivers (or controllers),
(b) the vehicles (or moving parts), and
(c) the road network (or stationary parts).

Because it is interactive, there is no unique beginning or end to a logical sequential discussion on the subject. I have chosen in this book to break into the road system loop at the point at which the need for a particular road is being established.

Figure 1.1 shows the associated road system loop, which illustrates the consequential pattern of the sections of the book. We enter the loop at Chapter 5 by discussing road needs and the planning response to such needs. Once this response has been established, the road must then be designed in more detail and Chapters 5 to 24 deal with all the subsets of this design stage. These are followed by Chapters 25 to 32 on the construction, maintenance and operation of the road system. This then leads back to the point of entry into the loop as the specific need for a particular road arises only because of some more general set of transport and land use activities. Hence, the tour of the road system loop ends with the development and understanding of those causal activities which are the *raison d'être* of the road system, as met in Chapter 5. Incidental final chapters deal with essential peripheral technologies and information sources. 'Peripheral technologies' covers those related concepts — ranging from stress analysis to transport economics —

that are usually assumed to be common knowledge but are often discovered to be misunderstood or misconstrued.

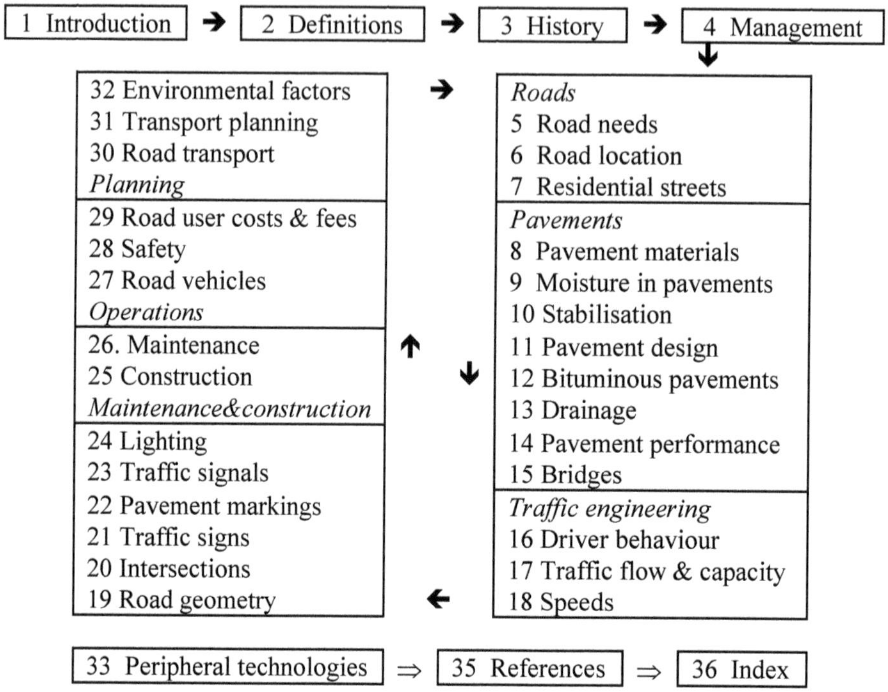

Figure 1.1 The organisation of the chapters of this book (➔'s indicate the road loop used in the book).

Because it is an international text, the book concentrates on principles and background rather than on particular local warrants for action, administrative policy, 'conventional' procedures and 'established' practice. These will all vary from locality to locality, often with neither rhyme nor obvious reason. The reader is, therefore, warned that it will usually be necessary independently to determine and understand these local matters. One aim of this book is to provide the basis for that understanding and to direct and guide readers to the next stage of their inquiry. Particular attention is therefore given to explaining the principles and defining the terms and jargon that the reader needs in order to enter the 'middle' stage of local manuals, engineering procedures and operational warrants.

In some areas, particularly where the book impinges on other disciplines, some effort has been made to provide checklists of things the reader should consider and pursue in order to take a particular issue further.

The book also acts as an information source by guiding the reader to more detailed or specialised material. In some instances it expands or cross-links these primary information sources.

The book is extensively indexed and cross-referenced to facilitate the search for answers to specific problems and queries. Such searches will sometimes then require reference to some other document. Internal cross-referencing is based on the terminology of 'section' for the parts of a chapter (e.g. Section 6.2 is the second part of Chapter 6).

CHAPTER TWO

Definitions

2.1 ROUTES AND ROADS

This chapter defines the common words, terms and jargon to be used throughout the book. *Travel* occurs when an object at a point A moves to its desired location at a point B (Chapter 31). In the context of this book, the basic terms to consider are *route* and *way*.

* A *route* is the course taken in travelling from starting point A to destination B and will be comprised of a series of ways.
* A *way* is a provision for travelling and the term *right of way* defines the traveller's right to move along that way (Section 3.3.3).

There is an hierarchical structure to the key words used to describe routes, ways and the road function and this is illustrated in Figure 2.1.

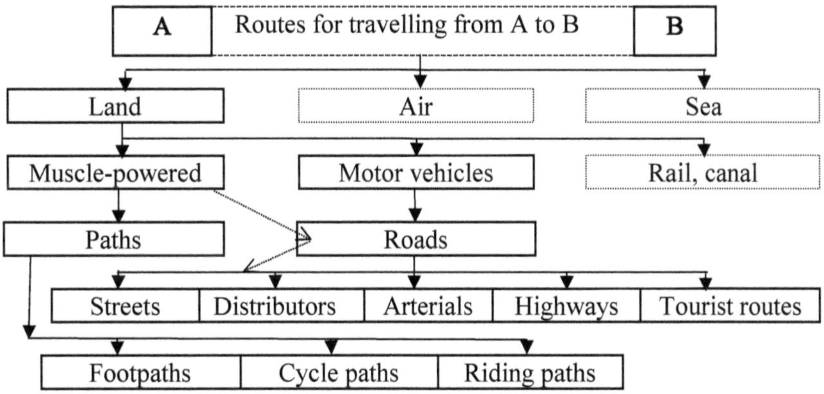

Figure 2.1 Linked definitions of routes, roads, streets, and paths. Refer also to Table 2.1.

A *road* is a ground-based route and so these definitions of way and right of way lead to a common formal definition of a road being:

> The way devoted to ground-based public travel, by foot or by wheel, and using a publicly-reserved (Section 6.3) strip of land lying between otherwise adjoining properties.

The Shorter Oxford English Dictionary now defines a road as (Oxford, 1993 and 2007):

> A path or way between different places, usually one wide enough for vehicles as well as for horses and travellers on foot.

These two definitions include routes comprised solely of cycle paths or footpaths. A more restrictive definition of a road is (Logie, 1980):

> A route trafficable by motor vehicles.

which more realistically restricts roads to routes that can at least accommodate a motor vehicle within their width. It is this definition which will be used in this book. A fuller version would be:

> The way devoted to effective travel by motor vehicles, and sometimes by foot and by cycle, and using a reserved strip of land lying between otherwise adjoining properties.

The term *path* is used to describe ways solely for pedestrians, cyclists, horse riders, and the like. Within the road, the operational surface actually used by vehicles is called the *carriageway.* The terms *roadway* and *travelled way* are alternatives which are not used in this book.

A road may serve for one or more of the following six major usages. Criteria to ensure that these usages are provided will be discussed in the indicated Sections.

(a) moving vehicles (see below and Chapter 17),
(b) parked vehicles (Section 30.6),
(c) pedestrians and other non-vehicular traffic (Section 30.5.3),
(d) people and freight accessing abutting property (Section 6.2.4),
(e) people involved in social intercourse (Section 7.4.1), and
(f) non-transport utilities (e.g. power, water, telecommunications) using easements within the road reservation (Section 7.4.3).

Usages (a), (b) and (c) follow directly from the above definitions. Usage (d) occurs if some origins A and destinations B are located within a length of road. Usages (e) and (f) are indirect consequences of the presence of a road.

Following on from the earlier definitions, usage (a) is the prime one and, in order to fulfil it, a road must be:

(1) wide enough to permit the passage of a vehicle (Section 7.4.2),
(2) smooth enough to give adequate riding quality (Section 14.4.1),
(3) rough enough to provide motive, braking and steering friction (Section 12.5.2),
(4) not so rough as to prevent the passage of a vehicle (Table 14.2),
(5) free of inhibiting dust and water (Section 12.1.2), and
(6) not so winding and/or steep as to prevent the ready passage of the vehicle (Sections 19.2&3).

A *street* is a special class of road and the current definition in Oxford (2007) is:

> A street is a public road in a city, town or village usually running between two lines of houses or other buildings on either side.

The term road denotes travel from place to place, but a street does not have to go anywhere and might only provide local access. In this context the definition of street is better expanded to (as in SA, 2002):

> A street is a public road in a city, town or village that has, and provides access to, (mainly) continuous buildings on one side or both. It does not make specific provision for through traffic.

Farm roads are countryside equivalents to streets. Some older, quite different, definitions of street are given in Section 3.3.3 and more specific sub-definitions of street are given in Section 7.2.1. The particular cases in Chapter 7 where the street services shops, offices and light industry require careful attention.

While streets provide access to contiguous properties, most roads predominantly serve a traffic function. As such, they fall into three broad groups called *distributors*, *arterials*, and *highways*. The distinction between the three is that:

* distributors do not service traffic from outside the residential or industrial precinct in which they are located (three classes of distributor are shown in Table 2.1), whereas
* arterials and highways also service traffic from outside their own area, as the following definitions illustrate:

Arterials are the major roads within a region or population centre and directly link its key centres of activity. They are major traffic ways intermediate between distributors and highways and are part of a regional road network.
Highways are roads linking major regions, activity centres and/or population centres. These major traffic ways are usually part of a national or international road network. *Bypasses* are highways that circumvent or bypass towns en route between the two ends of a highway.

Both arterial roads and highways are sometimes known as *main roads* or *primary roads*. These terms are not used in this book. Note that the above definitions relate to the function of the road, rather than to its physical characteristics.

All these definitions, classifications, various alternative terminologies, and some additional concepts from Section 7.2.1, are summarised in Table 2.1. Of course, the specific definitions and names used for the various road classes will vary between jurisdictional authorities.

Table 2.1 Summary of the terminology used to describe routes.
See also Section 7.2.1 and Figure 7.1. Alternative names are given in brackets.

Category	Common name	Short definitions
Car-free	Path	– a way without car access.
	Mall	– a provision for pedestrians only, and which is not primarily a path way.
Local access	Street, farm road	– a way which only provides access to adjoining properties.
	Loop	– a curved street with two entry points.
	Cul-de-sac, dead-end (or court)	– a street with one entry point. The cul-de-sac provides at its end for turning vehicles, whereas a dead-end may not.
Distributor	Local distributor (or street, secondary road, collector, or connector)	– a traffic way that only serves local access roads and other local distributors.
	Precinct distributor (or district distributor, limited distributor, or local crossing road)	– a traffic way which only services its own precinct.
	Primary distributor (or sub-arterial)	– a major traffic way linking precincts but not part of a regional network.
Arterial	Arterial (or regional distributor, bypass or main road)	– a major traffic way within a region and which is part of a regional network (see Section 2.1).
Highway	Highway (or primary, main, or strategic road)	– a major traffic way linking regions or key activity centres (see Section 2.1).

2.2 SPECIFIC TERMS

Specific terms used in this book are defined in the text where they first occur and are also listed in the Index. To aid the reader with terms and spellings that vary from one English-speaking country to another, the Index also provides cross-references to the preferred term used in the text [e.g. Tire, see TYRE]. Non-preferred terms are shown in (brackets) in the text.

The common terms used consistently in the book to define the road and its structure are illustrated in Figure 2.2. Most of the terms are either self-evident from the diagram or are defined elsewhere (see Index). There is a wide range of terms and definitions in use around the world and their misapplication can readily lead to confusion. For this reason, only the terms in Figure 2.2 will be used in the rest of the book and some alternative terms are either bracketed or discussed below. Specific technical definitions of these terms will be found in the listed cross-references.

The seven pavement terms that require further explanation are, working vertically down the pavement:

(1) *Surface course* (or *wearing course* or *running course*). This is the horizontal layer (course) that directly accommodates the traffic. Chapter 12 describes the types of surface course in common use.

(2) *Intermediate course* (or *binder course* in U.S.). The concept of a separate intermediate course was common in early asphalts (Lay, 1992) and in Britain was sometimes called a *basecourse*. It has recently returned to popularity with the availability of stiff asphalts (e.g. SMA, Section 12.2.2(d)). The term should be used with care, as it is sometimes also used to describe a form of seal coat (Section 12.1.3).

(3) *Basecourse* (or *base course*). This is the structural course that resists the traffic load. It was often called a *base, road base* or *roadbase* in Britain. Some European practice — which is not followed in this book — is to call this course the *binder* course if it is a bound course and the *base* if it is unbound (Section 11.1.4). The basecourse may actually be a composite structure composed of a number of individual structural courses, each possessing its own distinct structural characteristics. Basecourse properties are discussed in Sections 11.1–6.

(4) *Sub-base* (or *subbase*). This course separates the basecourse from the subgrade. It can serve a number of specific purposes, as listed in Section 11.1.2. It is sometimes called *lower base* in Britain, particularly when it is unbound.

(5) *Capping course*. This course protects the subgrade from penetration by the upper courses and is assumed to be part of the subgrade.

(6) *Subgrade* (or *basement*, *roadbed*, natural ground, or *substrate*). The subgrade is the non-pavement material below the pavement structure. It may be in situ natural material or compacted fill (Section 11.8). Subgrade properties are discussed in Sections 8.1–4. It is an integral part of the road formation (see 7 below).

(7) *Formation*. The formation includes the original ground below the road, as modified by the replacement of unsuitable material (Sections 8.1–4), the addition of drainage provisions (Chapter 13) and by any embankments and cuttings (Section 11.8). The term is usually also applied to the longitudinal section of the road and thus, in three dimensions, refers to the roadway after it has been brought to the shape required by the design drawings but before placement of the actual pavement structure. Note that only the two-dimensional definition of pavement formation is shown in Figure 2.2. A common progression over time from original ground to subgrade is described in Section 12.1.1.

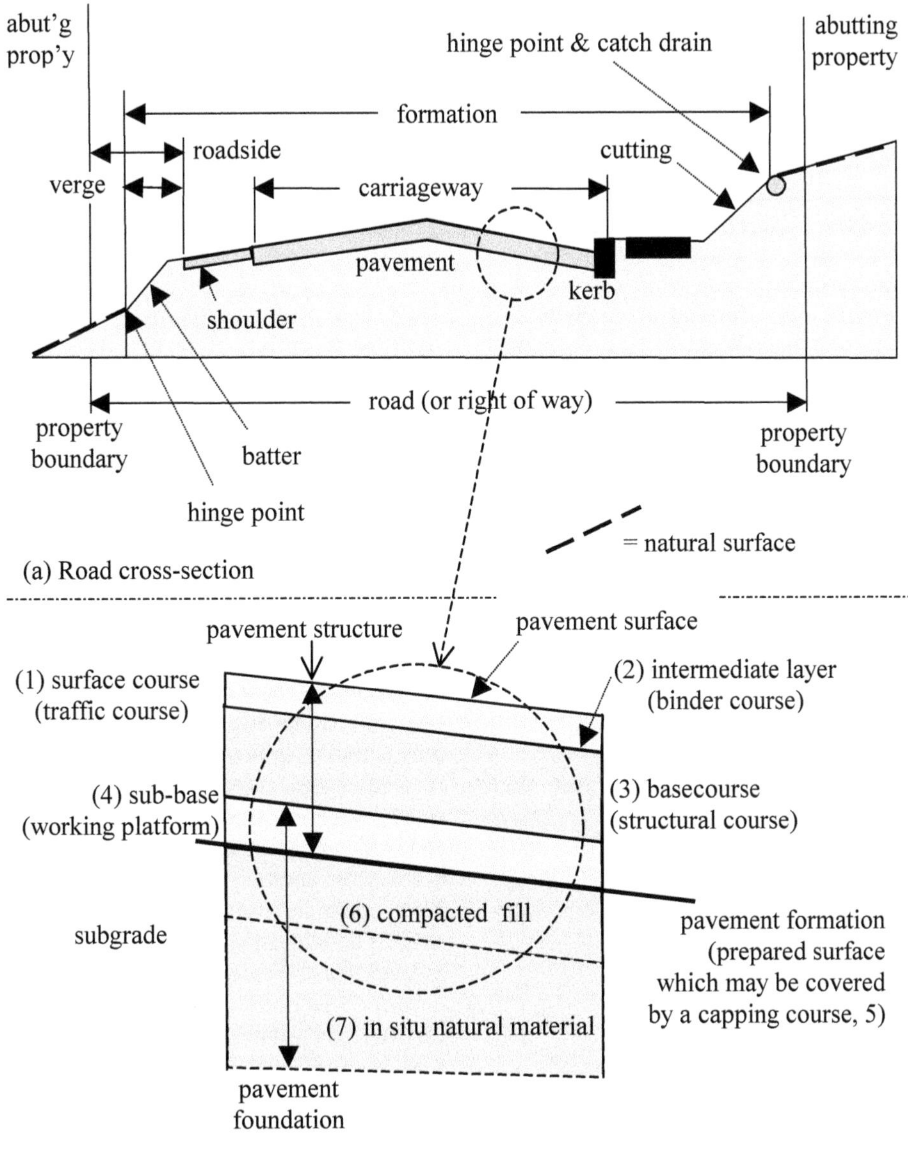

Figure 2.2 Commonly used road terms. See the text in Section 2.2 for various alternative names that have been used for all the pavement components shown in the lower half of the figure.

The Permanent International Association of Road Congresses (PIARC) has added its *Technical Directory of Road Terms* to its website (www.piarc.org) where it can be searched in twenty languages. An international set of road terms in English, French and German is contained in the ITRD Thesaurus (ITRD, 2005), which is commonly used to supply keywords for indexing road-related publications. It contains no definitions. A glossary of transport terms in English, French, Italian, Dutch, German and Swedish is given in Logie (1980), and Table 2.2 provides a multilingual link between some of the commonest terms used.

There is obviously some variation between the basic terms used in roads and the reader should therefore exercise care in comparing publications from different sources.

Table 2.2 Comparison of road terms in different languages. Notes: 'Thesaurus' = ITRD (2005), * = from Logie (1980) & ** = from PIARC Dictionaries.

English	German	French	Notes
highway	Strasse	route	
road	Strasse or Weg*	route**, roie* or chemin*	The Thesaurus uses *highway* rather than *road* for searching. However, *road* is now the common generic term (Section 2.1) and so this book uses *road* and gives *highway* the restricted definition of Table 2.1.
arterial road	Fernverkehr-strasse	route à grande circulation	This book uses *arterial* in the manner described in Table 2.1. It is not used in the Thesaurus, which prefers *main road.*
main road			This book uses *arterial road* in the manner described in Table 2.1. The Thesaurus uses *main road.*
expressway		route express**	An *expressway* is a road with dual carriageways separating oncoming traffic, but with some at-grade intersections and only partial or limited access control.
motorway	Autobahn	Autoroute**	A *motorway* is a road with dual carriageways separating oncoming traffic, and with no at-grade intersections and no access from abutting properties.
freeway	Autobahn*	autoroute	This common usage for the motorway defined above can confuse laymen who can assume that a *freeway* is the opposite of a *tollway* (Section 29.3.1b). This book, PIARC and the Thesaurus prefer *motorway.*
street	Strasse*	rue*	See Section 2.1. The Thesaurus prefers *highway* plus area descriptor.
local street	Ortstrasse*	route locale*	Not used in the Thesaurus, which prefers *highway* and *urban area.* This book uses *street.*
pavement	Oberbau	chaussée	This is the 'pavement structure' shown in Figure 2.2.

2.3 UNITS

The book is entirely in *SI* (Système International) units. SI is the internationally accepted current version of the *metric system.* It was developed by the International Bureau of Weights and Measures and codified in ISO Standard 1000 (ISO, 1992) and forms the basis for engineering units in most English-speaking countries, with the notable exception of the U.S. Dimensions on drawings are in metres, unless otherwise noted. A list of SI prefixes is given in Table 2.3. It is the practice in most English-speaking SI countries to use the *point* rather than the comma as the decimal symbol (i.e. 12.58 rather than 12,58) and this convention is followed in this book.

Table 2.3 SI prefixes (from ISO, 1992). Note: The unit is multiplied by the factor, 10^n.

Name	Symbol	n	Notes
zetta	Z	21	
exa	E	18	
peta	P	15	
tera	T	12	
giga	G	9	
mega	M	6	
kilo	k	3	
hecto	h	2	
deka	da	1	The multiples (1, −1, −2), their names and their symbols, are not preferred and their use should be limited as much as possible
–	–	0	
deci	d	−1	
centi	c	−2	
milli	m	−3	
micro	μ	−6	A *micron* is an old term for a μm.
nano	n	−9	
pico	p	−12	
femto	f	−15	
atto	a	−18	
zepto	z	−21	

CHAPTER THREE

History of Roads

[This chapter is a condensation of the author's *Ways of the World*, a history of the world's roads and the vehicles that used them (Lay, 1992).]

3.1 FOOTPATHS

To our present day minds a road is a black-topped carriageway for the use of cars and trucks. However, in order to trace the history of roads we must put this perception aside and recall the broader definitions in Section 2.1 where a road was defined as 'The way devoted to effective travel by motor vehicles, and sometimes by foot and by cycle'. Indeed, the term *way* will often be more useful in this short history than will *road*. Linguistically the word *way* derives from moving and travelling and has the same roots as the words *wagon* and *vehicle*.

Roads clearly began as pathways and were often created by the larger beasts that were more able than man to push aside vegetation and pound paths into existence. By about 30 000 BC convenient routes were being well used by human travellers. The first human pathways would have developed for very specific purposes leading to campsites, food, water, fords across streams, passes through mountains and routes through swamps and past dangerous areas. The earliest extant manufactured paths are lengths of corduroy (Section 3.6.1) built in a swamp in Glastonbury, England, in about 2500 BC.

Many of the early pathways rose quickly out of water-logged valleys to follow higher contours and ridges which were drier, less densely vegetated and safer. These high paths were called *ridgeways* and the term is still used to describe a number of English paths. The ridgeways probably date from at least 4000 BC and by 3000 BC they were being used as trade routes for salt, tin and rushes. Early U.S. highways often followed the ridgeways, making 'the ridgeway an American institution'. Where the ridgeway did descend to the valley to ford a river it was called a *harrow way*. As foot and hoof traffic over the millennia wore away the path surfaces, many became deep trenches known as *hollow ways*. Some in the Devon limestone are now 3 m deep trenches suggesting use over six millennia.

A number of the Roman roads, such as the Fosse Way in Britain, followed pre-existing pathways rather than one of their more famous straight alignments. Until well into the 18th century various ways were often the dominant transport route, with the packways of the pack-horse strings being the most recent example. Most of the new roads of the 18th and 19th centuries followed old pathways whilst many other ways remain today as independent foot and bridle paths.

Roads which developed from such interconnected local pathways could not be expected to produce a rational regional road network. What did arise was described by Belloc as 'haphazardly established roads, long neglecting opportunities that would have been obvious to the eye of the most cursory and moderately intelligent survey'. This seeming haphazardness is still observable in many of the inner streets of Old World towns and in local roads connecting adjacent small villages.

3.2 CARRIAGEWAYS

The next major transport development was the use of animals as beasts of burden and as haulers of sleds and, subsequently, of carts and carriages. Cattle, asses, mules, donkeys, dogs, goats, horses, camels, elephants, llamas and, of course, humans are some of the better known species which have found useful transport employment. Indeed, for most of its history, the world's road system has operated with animals as the principal motive force. Oxen were the original work-horses of the road, with small horses and donkeys arriving about 3000 BC. Modern horses and harnesses started appearing about 2000 BC.

The use of the horse for riding and the ox for hauling would have given mankind its first quantum jump in travel times and capacities. Indeed, four millennia were to pass before man found a faster means of travelling than on the back of a horse. Because the ridden horse, the loaded pack-horse, and the harnessed, hauling ox would have required larger horizontal and vertical clearances than the walking man, some path development in the form of bridle paths and trackways would have been necessary in order to utilise this new development.

The next step was the invention of the wheel. It was first developed in Mesopotamia, the Assyrian birthplace of civilisation, in about 5000 BC and a wheel dated at 3000 BC has been found in southern Russia. From a transport viewpoint, however, an important associated step was the invention of the axle joining two wheels and hence giving the system stability and the capacity to carry a useful payload.

After 3000 BC a variety of vehicles began to make practical use of the wheel, initially as two-wheeled carts and carriages and cumbersome four-wheeled wagons which first appeared in Mesopotamia. The development of useful four-wheeled vehicles required a third step in the invention sequence — an axle capable of swivelling to permit easy curve negotiation.

These wheeled vehicles gave rise to a whole new set of pathway needs as they were wider and heavier than man or ridden horse or beast of burden. To provide adequate strength to carry the wheels, the new ways tended to follow the sunny drier side of a ridge.

A fairly standard vehicle width of a little under 2 m arose and still exists in today's motor vehicles. Neolithic Cretan and Roman roads indicate a wheel gauge of 1.4 m, which is consistent with this vehicle width. Flagstones found at Maiden Castle near Dorchester in England and dated at 300–200 BC also show 1.4 m wheel ruts. A typical gauge for 19th century animal-drawn vehicles was 1.4 to 1.5 m and a common railway gauge was and still is 1.4 m. This historical liking for 1.4 to 1.5 m as a wheel gauge and under 2 m for vehicle widths led to common widths of 2.5 to 3 m for one-way and 4 to 7 m for two-way roads. Such widths would have been much more demanding to develop or construct than the 800 mm needed for footpaths or the 1.5 m needed for bridle paths.

The joint questions of the load carried by wheels and the width of tyres are also critical and have been key elements in road operations over the millennia, for they determine the contact pressures between the wheel and the surface of the path and the stresses on the pavement basecourse and subgrade (Figure 2.2). To reduce the weight of the wheel and its rolling resistance over good surfaces, the transport operator usually tries to minimise the width of its rolling surface. On the other hand, those charged with maintaining the pavement running surface have long fought to minimise pavement damage by requiring wider wheels and lower wheel loads. Such limits existed at least as early as Roman times.

From 1622 to 1661 a statute of James I of England attempted to control vehicle loads by only permitting two-wheeled vehicles to use English roads. In addition, loads greater than one tonne were prohibited on any vehicle. Scales for weighing vehicles were introduced in Dublin in 1555 'to eschew the loss to excessive and untrue tolls'. Such scales were legalised in Britain in 1741 to help protect the roads from damage. Their operation involved winching the vehicle to be weighed off the ground.

Up until 1830 tyres were made of heavy iron segments bolted or riveted to the wooden wheel. The Celts developed the hoop tyre, made as a hoop smaller than the wheel, expanded by heat, and then shrunk into place. One consequence of the use of iron tyres was that wheels could be made narrower. Even in medieval times, iron wheels were sufficiently common and damaging for a number of towns to have prohibited their use within the town limits.

Whereas man and beast of burden could manage steep slopes and tight curves and often preferred the ridgeways, the carts and carriages were far less adaptable, needing not only firm surfaces but also extra width, flatter grades, and wider curves. The trackways that developed to meet this need often sought the easy grades of the river valley, although the need for a firm smooth surface made many valley paths unsuitable. One of man's early construction achievements was the use of rush and corduroy (Section 3.6.1) roads to provide passage over swampy ground. At the other extreme, trackway surfaces could not be too smooth as some roughness was needed to provide traction for the feet of the animals hauling the carts.

3.3 PAVEMENTS

3.3.1 In the beginning

Natural material which is soft enough to be formed into a smooth well-graded surface is rarely strong enough to bear the weight of a solid-wheeled vehicle, particularly when the material is wet. On the other hand, rock that is strong enough under all moisture conditions is rarely able to be easily formed into a good running surface. Nevertheless, by the time of the wheel it would have been possible for man to produce an adequate running surface, given the time and resources. The first record is of cobblestone paving (Section 3.6.6) in Assyria in about 4000 BC.

The oldest extant road was constructed through the mountains of Crete, from Knossus to Leben in about 2000 BC. It could have been man's first effort at major road building as it was no mean structure with elaborate longitudinal drains, a 200 mm basecourse of sandstone in a clay-gypsum mortar, and a 4 m running surface of basalt blocks.

The first major arterial road was probably a 2.5 Mm route built by Assyria from its capital of Susa in eastern Iran, via Nineveh to the Mediterranean ports of western Turkey. The importance of the route was that it linked Assyria with the trading activities which had, by 2000 BC, begun to develop around the Mediterranean.

A number of other roads were also to radiate out from the Assyrian administrative and trading hub in the Persian Gulf. Perhaps the most famous was the Silk Road, a caravan route bringing both jade and silk from China. Again for trade purposes, by at least 500 BC another caravan road linked north-west India to the Assyrian hub. Indian road building, incidentally, was quite advanced by 1000 BC, with wide use of brick-paved roads and sub-surface drainage, and the suggestion that paving began there in 3000 BC.

The first recorded road builders were the engineer–pioneer corps (ummani) of the Assyrian kings. Their work in 1100 BC in constructing a well-aligned mountain road for their king was well documented for posterity.

Roadmaking using bricks was common in Mesopotamia. Excavations at Khafaje dated about 2000 BC uncovered pavements that used layers of brick bound by a bituminous mortar. An Assyrian temple in Assur contains a one kilometre processional road made of three or more layers of bricks in a bituminous mortar, all covered by a layer of breccia and limestone slabs. This 700 BC road is the oldest extant use of bituminous material for roadmaking. The Greeks and Romans used bitumen as a building mortar. However, the practice of using bitumen for roadmaking does not appear to have been adopted by other contemporary civilisations, which usually preferred simple clay mortars.

Naturally-occurring cements (or pozzolans) could well have been used in earlier road practice but the earliest records are of Indian roads built with cement-bound surfaces. The Greeks were also using cements for building by about 300 BC and their knowledge was certainly passed on to the Romans.

In about 500 BC the Persians under King Darius I built their famous Royal Road, duplicating the well-used Susa-to-the-Mediterranean way, although following a more time-efficient route and by-passing many significant towns. The road was restricted to Royal messengers in order to prevent them being impeded by common travellers and to protect the relatively poor quality surface.

3.3.2 Roman roads

All the above efforts pale into insignificance beside the Roman roads. They were a remarkable achievement and provided travel times across Europe, Asia Minor and north Africa that were not to be appreciably bettered until the coming of the train two millennia later.

The Romans learnt a little of their roadmaking from the Greeks, Etruscans and Carthaginians. They recognised that the fundamentals of good road construction were to provide good drainage, good material and good workmanship — fundamentals that were substantially aided by the fact that they also introduced cement and concrete to roadmaking. After the Romans left the scene in about AD 500 their knowledge took over a thousand years to rediscover.

The Greeks had used lime mortars for roadmaking, but the Romans took the lime and added volcanic pozzolans to produce a stronger and more durable mortar. They went a stage further by adding gravel to the mortar to make a concrete. Concrete was a major Roman roadmaking innovation which, after their departure, was to be lost for two millennia.

The first of the major Roman roads was built over 150 km between Rome and Capua, with construction commencing in 312 BC This was the Via Appia (or Appian Way, named after Censor Appius Claudius) which was eventually to extend south to the port of Hydruntum (Brundisium) in 244 BC.

The system steadily developed and at its peak in AD 200 involved 80 Mm of first class-road which completely ringed the Mediterranean and covered Asia Minor and Europe south-west of a line joining Istanbul, Vienna, Cologne and Edinburgh. Perhaps the greatest of the roads in terms of construction achievement was the 2 Mm Via Nerva that crossed the north African desert from the straits of Gibraltar to Alexandria via Carthage.

Compared to the pre-existing pathways and trackways, the straight and smoothly-graded Roman roads were more akin to a modern rail network than to the contemporary perceptions of a road system. However, this analogy ignores the somewhat steep gradients found on many Roman roads. If possible the Romans avoided cuttings. They preferred sidehill rather than valley floor routes. The alignment commonly consisted of long straight lengths with alignment changes at saddles or sighting points in ridges.

Roman roads were always constructed on a firm and formed subgrade, strengthened where necessary with wooden piles. An essential feature of the roads was that they were bordered on both sides by carefully constructed longitudinal drains. The next and key step was the construction of the *agger*, which was a raised formation up to a metre high and 15 m wide which was usually built from material made available from the excavation of the side drains and topped with a sand levelling course. The agger contributed greatly to moisture control in the pavement and not accidentally made the roadway a readily defended position.

The pavement structure on top of the agger varied greatly. In the more elaborate constructions the pavement could be another metre thick, leading to a surface up to 2 m above the surrounding countryside and said to have given rise to the term *highway*. One of their major types of pavement structure began with a course composed of stone at least 50 mm in size. Depending on soil conditions, the course could be up to 500 mm thick. This was followed by a cement-stabilised (Section 10.2) or mortared course up to 250 mm thick composed of smaller stone. Next came another mortared course up to 250 mm thick, this time of broken brick or similar material well compacted into place. Finally, to accommodate heavy traffic, there would be a surface course of large and neatly fitted hexagonal fitted flagstones, each about 250 mm thick (Figure 3.1).

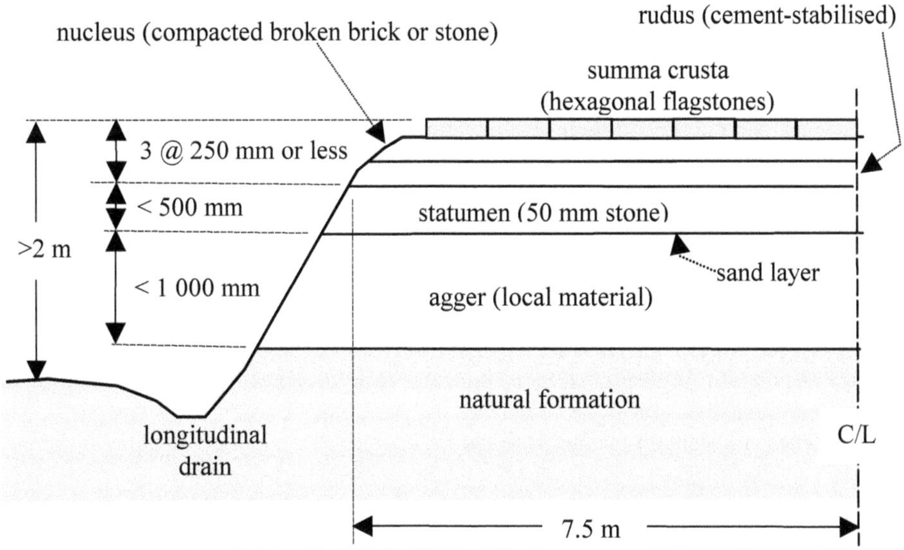

Figure 3.1 Relatively elaborate version of a Roman road.

Formation widths varied from 4 to 10 m, with an average of 5.5 m, thus providing shoulders on the agger very similar to modern road practice. Where the roads served a military function the roadside was cleared of vegetation for an arrow's flight on either side.

3.3.3 After the Romans

After the collapse of the Roman Empire in about in AD 400, both its road systems and the associated human skills decayed and then disappeared for over a millennium. Indeed the few on-going factors that kept many roads open were travel to markets and to the annual round of country fairs, and the needs of religious administration and pilgrimage.

The only glimmering of European road development during the millennium after the Romans was an attempt by Charlemagne in France to construct roads and place them under central control. Cordova, the wealthy showpiece of Saracen Spain, had its streets paved in 850 by its Moorish Caliph. Another brief flicker was the first paving of a street in Paris in 1184. The work, which utilised large flagstones, was ordered by Phillip II to reduce the stench in the street outside his castle at the Louvre.

With Norse dominance in Europe during the Dark Ages, the most significant new road was their 9th century Varangian (Viking) Road that ran south from the Baltic to Constantinople. The key towns on this road were Kiev and Novograd. The Silk Road trade routes to the east also had a major impact.

A number of monks routinely collected alms for road maintenance, some parishes offered indulgences to those who contributed to it, and some guilds dedicated themselves to the task. When monasteries declined in the fifteenth and sixteenth centuries, even that source disappeared in many areas. Nevertheless trade, the development of urban life and, finally, the Industrial Revolution were slowly to create a new set of needs.

Roads in medieval times would not fit our current perceptions as they were not so much useable pavements as legal rights of way (or passage). To protect their fields from being tramped upon, adjoining farmers therefore had an incentive to keep the original road clear. This philosophy still held in 1675 with one writer describing how much ground 'is now spoiled and trampled down in all wide roads, where coaches and carts take liberty to pick and choose for their best advantages ... (they) utterly confound the road in all wide places ... '.

As travel became more commonplace, an interesting 16th century development was the use of the word *road* in place of terms like *way* and *path. Road* came from the verb *ride* and implied a route along which riding, or at least reasonable progress, was possible. *Street* came from a word meaning constructed and was originally applied to a number of Roman roads (e.g. Watling Street). As paved roads in the Middle Ages were only to be found in towns, *street* came to be used for a town road and no longer has its original meaning. *The Oxford English Dictionary* indicates that the wider use of *street* occurred up to about 1550 and the term *road* came into the literature between 1590 and 1610. The word *pavement*, incidentally, came from the Latin word *pavimentum*, meaning a rammed floor.

There are graphic descriptions of just how bad were the roads of those times. Many reports tell of people drowning in potholes and roadside drains and of horses sinking to their bellies in mud. Abutting residents regarded the road as a convenient rubbish dump, a source of good garden soil and rammed clay for building. In the cities it was normal to empty the contents of chamber pots into the streets. Parisian practice was to require the emptier to call 'Watch out for the water' three times before tossing. However, in Edinburgh the obligation was on the traveller to shout 'Don't throw' to warn of his presence. Urban paving was generally drained towards a spoon drain down the centre of the street.

When pavement repairs were made, they were usually effected by filling holes and ruts with brush, branches, and other rottable organic material. Formal road maintenance was little better. High crowns were placed on the road surface in the unachievable hope of persuading surface water to run off the road and onto the roadside. The steep profile worsened rather than improved the situation. The most popular 'advanced' form of maintenance was the road plough, drawn by a team of eight or more horses and used each spring to plough the road and throw furrows towards the centre. The furrows were then flattened by harrowing. It was a process more attuned to agriculture than to road mending. In the 18th century some of these roads came to be known as *roof roads* because of their cross-section, which seriously endangered carriages and coaches (Figure 3.2).

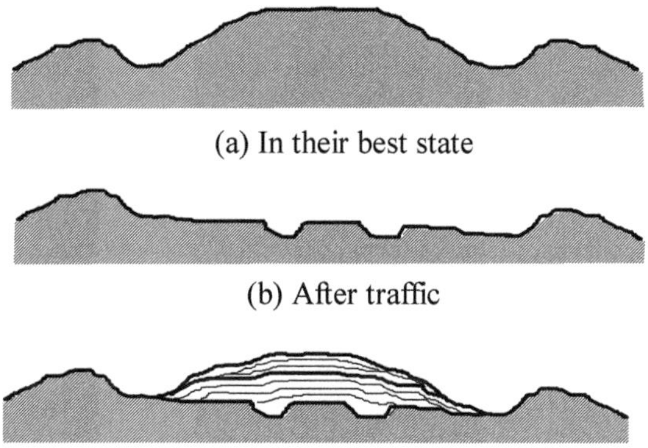

(a) In their best state

(b) After traffic

(c) After maintenance

Figure 3.2 British roads, 1500–1800

3.3.4 Pressures for change

It is estimated that by 1750 some 2000 cattle a day were being herded down the roads towards the London markets. They served well to keep those roads in 'a perpetual slough of mud'. Packhorse transport reached its zenith about this time, after which the canal system began to eat into its trade.

The need for good roads and road surfaces was reaching a crescendo. Fortunately, necessity had remained the mother of invention and so in the 18th century a number of talented people were beginning to make significant contributions to road technology — particularly Trésaguet in France and Telford and McAdam in the U.K.

Trésaguet lived from 1716 to 1796. From 1764 to 1775, he developed a cheaper method of road construction than the lavish and locally unsuccessful revival of Roman practice that occurred in France following the publication of Bergier's 1622 book on the History of the Great Highways of the Roman Empire.

Trésaguet's new pavement used large (*c*.200 mm) pieces of quarried stone of a more compact form than in the previous method and shaped to have at least one flat side which was placed on a cambered formation. Smaller pieces of broken stone were then compacted into the spaces between the larger stones to produce a level surface. Finally the running surface was made with a layer of 25 mm sized broken stone. All this structure was placed in a trench in order to keep the running surface level with the surrounding countryside. This last decision created major drainage problems which were counteracted by making the surface as impervious as possible, cambering the subgrade and providing deep side ditches (Figure 3.3). Trésaguet was certainly well aware of the need to keep the subgrade dry.

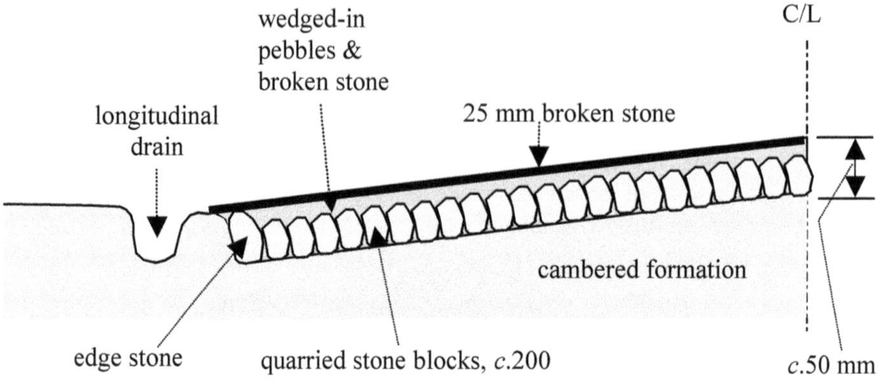

Figure 3.3 The Trésaguet pavement.

The random use of large stones was a problem. Not only did they provide a bumpy surface but they also caused high contact stresses at local points in the subgrade, resulting in large differential settlements. The French avoided this by placing their large stones over the entire formation, thus ensuring that the subgrade was subjected to a reasonably uniform and low stress. The system still had its problems. The large stones concentrated the load, mud worked up between them, and they tended to separate laterally.

Telford became involved in roads through the efforts of the British government in 1801 to halt the depopulation of the Scottish Highlands. Telford was himself a Scot, an ex-stonemason with an admirable record in building bridges, harbours, canals and buildings as well as founding the Institution of Civil Engineers.

He drew significantly from and then extended the practice of Trésaguet, using *c*.300 x 250 x 150 mm partially shaped stones, still with a flat face on the subgrade but with the other faces closer to vertical than in Trésaguet's method. Broken stone was wedged into the spaces between the tapering near-vertical faces to provide the layer with effective lateral restraint. Telford kept the formation level and cambered the upper surface of the pitchers, often by masons knapping the tops of the large stones. Indeed ex-stonemason Telford's pavement required more masonry work than the Trésaguet system but provided a stronger and more coherent basecourse. On top of this course Telford placed a further 150 mm basecourse layer composed of 50 mm maximum sized stone covered with a 20 mm gravel surface course (Figure 3.4).

Telford's flat subgrade relied on an impervious pavement structure to give maximum strength. To avoid drainage problems he would raise the pavement structure above ground level wherever possible. Where this could not be done he drained the area surrounding the roadside. The need to manage water flows and provide good drainage had been widely ignored by previous British road builders. Of course, the Romans had recognised those principles but their insights had been long lost and Telford's rediscovery of them was a major contribution.

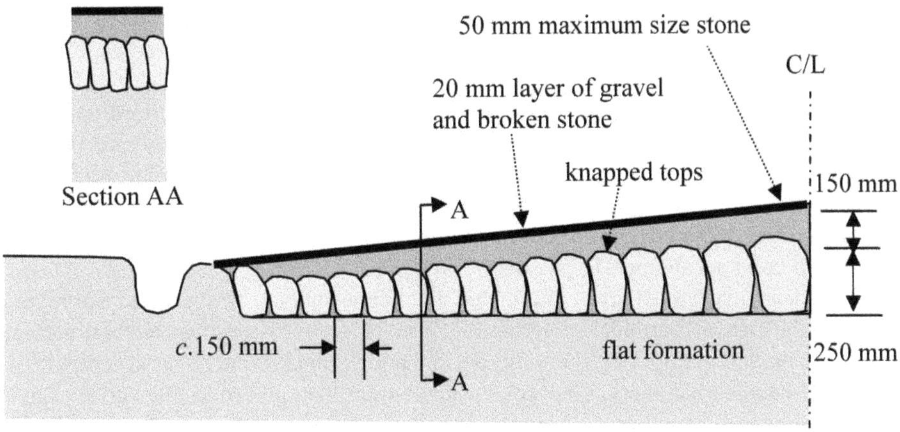

Figure 3.4 The Telford pavement.

3.3.5 McAdam

The next major technical breakthrough came with the work of J.L. McAdam who lived from 1756 to 1836. Concerned by the general low level of roadmaking knowledge and the degradation caused to roads by the coaches of the time with their narrow iron-tyred wheels and relatively high speeds, McAdam looked for alternatives to the then current methods of road construction and maintenance. By empirical observation of many roads, McAdam came to realise that 250 mm layers of well compacted broken angular small stone (Figure 3.5) would provide the same strength and stiffness as, and a better running surface than, a more expensive pavement based on a foundation of carefully made and placed large stone blocks (Section 3.6.6). Further, this course could reduce the stresses on the subgrade to an acceptable level, provided the subgrade was kept relatively dry and drained.

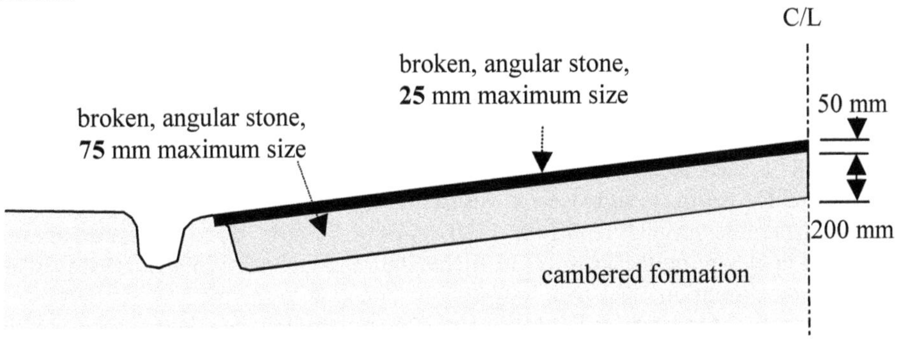

Figure 3.5 The McAdam pavement.

Stone size was an important element in the McAdam recipe. For the lower 200 mm of the pavement it was commonly 75 mm. However, for the upper 50 mm surface course the stone had to be small enough to fit into the stone-breaker's mouth and was checked by supervisors who carried in their pockets a set of scales and a stone of the correct mass.

The strength and stiffness of the course of compacted angular stone came from the mechanical interlock that developed between individual pieces of stone. The principle is still used in modern road construction and since 1820 McAdam's name has been remembered by the term *macadam* used to describe the courses of unbound angular stone that he introduced.

McAdam had thus broken away completely from the Trésaguet/Telford pattern. He observed that a layer of broken angular stone would behave as a coherent mass. By keeping the stone smaller than the tyre width, a good running surface could also be made. The layer could both carry the surface loads and spread them to an acceptably low subgrade stress. Moreover, the method could often make use of the large stones that beleaguered previous attempts at passage.

McAdam first put his ideas into major practice when he was appointed Surveyor of Roads for the Bristol Turnpike (Section 3.4.2) in 1816. In his original method men and women sat on small stools by the wayside, breaking sound stone to an acceptable size using small-handled hammers. The broken stones were then placed in piles prior to being spread on the road formation. A key aspect of the process was that no external smaller-sized binder material, particularly no soil, was added during the process. This led to macadam being an open-graded single-sized course. Presumably McAdam's strong views on the addition of extra material as binder arose from observing how soils and high plasticity materials had been unsuccessfully used as binders.

As experience with macadam increased, significant interparticle friction was observed to abrade the sharp interlocking faces and partly destroy the effectiveness of the course. This effect was overcome by introducing good quality interstitial finer material to produce what would now be called a well-graded mix. Such mixes also proved less permeable and easier to compact. The name macadam stayed with the new mixes and so, ironically, McAdam's name is now often used to describe a product outside his original specification.

3.3.6 Construction

In McAdam's time the broken stone was hand-rammed into place. Extra ramming produced a hard, smooth running surface and by blinding the surface gaps (i.e. filling the open 'eyes') with sand and stone sweepings from the stone-breaking operation. The crushing action of horseshoes and iron tyres produced further fine material to continue to bind the surface. The finer the material the more effective the binding when wet but the dustier the surface when dry.

By 1830 McAdam's method had been rapidly transferred to France, Russia, Australia and North America. Indeed it was first used in France in 1817 in trials and was given a favourable official report in 1830. Australia's first piece of macadam was constructed in N.S.W on the Prospect to Richmond road in 1822 although a letter-writer to the Sydney Gazette in 1821 had drawn public attention to the method. Its first use in the U.S. was in Maryland in 1823 between Boonesborough and Hagerstown on the Boonesborough Turnpike, which is now U.S. 40. It was then adopted as the standard for the National Road in 1825.

Despite macadam's cost and construction advantages, Telford's method was still used for soft ground construction until well into this century when a new generation of construction equipment made it easier to obtain, place and compact large thicknesses of crushed rock.

As the macadamising process spread, compaction with 1.5 m wooden, stumplike, hand-rammers became a trade art, as it was a key to a successful road surface. In 1846 a foreman introduced batteries of tradesman (paviors) working in unison to him beating time with a rod. The increase in productivity caused a public sensation. However, Telford and McAdam had both regarded compaction by traffic as adequate and this view prevailed until the 1860s. Horse-drawn rollers of wood, gaining their weight from carriages filled with roadmaking gravel, were first used in Ireland in the 1820s by Burgoyne and in England and France in the 1830s.

With such rollers it was difficult to achieve high contact pressures as heavier rollers required tractive efforts which were not available until the development by Ballaison in France of a practical steam-powered 17 t roller in 1860. *Steam roller* is a term which, like macadam, has now passed into the language. Steam power also led to the first use of mechanical plant (a jaw rock crusher) for stone crushing in 1858, producing stone for roads in Central Park in New York City. This innovation was accepted with some reluctance as:

(a) it was widely believed that hand-broken stone developed better interlock,
(b) hand-braking of stone was often a major source of local employment.

Compaction had traditionally been by hand or animal feet. Up till at least 1893 goats had been used for dam compaction and the observation in 1906 of the effectiveness of a stray flock of sheep running across a scarified, oiled Californian road is said to have led to the development of the sheep's-foot roller (Section 25.5).

Leaning-wheel graders, drawn by horses or steam-powered traction engines, were introduced in 1885. Self-powered devices arrived in the early 1920s although tractor-drawn machines were common into the 1930s. The petrol-powered crawler tractor was introduced in 1904 with the work of American Ben Holt, whose efforts also led to the founding of the Caterpillar Tractor Company.

Gunpowder (Section 25.4) was invented in 1856 and was being used for quarrying and road clearing in 1858. Other common road building tools of the time were ox-drawn ploughs, horse-drawn scoops, horse-drawn spreading and levelling drags and agricultural hand tools.

3.4 PURPOSE

3.4.1 Military motives

Some of the major uses of roads have already been implied in Section 3.1 — they provided local access to food and shelter, migratory routes to cope with climatic variations, paths for religious pilgrimages and avenues for interregional trade in such commodities as flint, tin and salt. Then the Romans showed that roads also served valuable administrative and control purposes. As society developed, roads also provided access to employment, education, entertainment, and culture.

Perhaps the most critical point to notice is that traders and migrants opened up most major routes, but that the development of these routes into manufactured roads has usually required military motives. The first wheeled military-vehicle was developed about 2500 BC and from that time roads facilitated efficient transport for attack and defence and thus a reason for rulers devoting significant resources to their construction and maintenance. In times of peace, such resources were rarely forthcoming, despite a variety of administrative techniques tried over the millennia and which will be reviewed in the next two sub-sections.

The Persian Royal Road (Section 3.3.1) was the first major example of a road built solely for militaristic or administrative reasons. However, Roman roads are the most obvious and lasting example of the effectiveness of military intervention in roadmaking.

Road building in North America was not a matter of high priority for the early colonisers. The first major manufactured road in the U.S. was of a corduroy road of military origin although it followed a route established in 1743 by settlers heading for the Ohio River. The road was built in 1775 by the British Army under General Braddock. It ran 170 km from Cumberland in Maryland to Pittsburgh and its purpose was to aid the capture from the French of that settlement's Fort Duquesne. The road followed an alignment laid out in 1753 by a 21-year-old surveyor, the well-known U.S. civil engineer, general and president, George Washington. While the road was a success, the mission failed and Braddock was killed in an ambush a few kilometres from the Fort. In an interesting link, Benjamin Franklin had supplied the horses and wagons for the expedition.

More recently, the construction of the grand boulevards of Paris between 1850 and 1870 was dictated in purpose and geometry by the potential requirements of a defending army and the need to permit the free movement of troops into areas of Paris that had previously been sources of insurrection.

The first motorways also had strong military links. A great deal of support for the Italian autostrade development came from Benito Mussolini who had come to power in 1922 with the intent of restoring Italy to the glories of the Roman Empire. The first autostrada was a single-carriageway controlled-access road built in 1922–24 by private enterprise as a tollway over 80 km from Milan to Varese en route to Lake Como.

The German autobahns were originally planned in the late 1920s to alleviate massive unemployment. The first was a 20 km stretch from Bonn to Cologne that was built in 1929–32 under the control of that city's Mayor, Konrad Adenauer, who was later to lead post-war Germany with great distinction. Hitler accelerated autobahn construction, despite his party's prior objection to the concept, soon after he came into power in 1933. The autobahn in the 1930s rapidly became a key part of the infrastructure of the German war machine with Hitler building 4 Mm of autobahn by 1942 with another 2.5 Mm under construction.

The world's greatest contemporary road system – the U.S. Interstate – was initially justified to U.S. Congress after the First World War and Second World War as a national defence system and this justification brought with it the requirement that its geometry and structure should be able to accommodate and aid the movement of large pieces of military equipment. Indeed the formal name of the system was the National System of Interstate and Defense Highways.

Although the autobahn at Cologne was the world's first freeway, it was not the first controlled-access road. The Italian autostrade of the 1920s have already been mentioned but they were preceded by the Long Island Motor Parkway in the U.S. The 10 km track with a 10 m single asphalt carriageway had originally been built on private land and intended as a racecourse for motor vehicles. It was opened to the public as a toll road in 1908 and operated as such until 1937. It was also the first public road to eliminate at-grade intersections and to employ superelevation on curves (Section 19.2.2), a frill that had not been needed in the low-speed days before the car.

3.4.2 Alternatives to the military

It is appropriate to now look at the peacetime alternatives for the funding and management of road construction and maintenance. The alternatives tried over a millennium or so can be put into three main categories that will now be discussed in turn, never forgetting that they were always subsidiary to the military alternative already discussed.

The first solution was to require adjoining land owners to fund or undertake the necessary roadworks, usually by providing a certain number of man-weeks of labour each year. Not surprisingly, adjoining landowners usually objected to working on roads intended primarily to serve the needs of passers-by and passers-through. Although this 'statute labour' system was at its best inefficient and ineffective, it remained in operation in England until 1835.

The second solution, therefore, was to pass the responsibilities for roads over to the local administrations through which the roads passed. Not surprisingly, these unpaid groups had little interest in the needs of through traffic and, with the advent of railways and canals, often actively oversaw the demise of the arterial road system.

A third solution was therefore to pass the road responsibility onto the road user, via tolls for using roads and bridges. This technique became popular in the Middle Ages where the toll bridge found initial favour as it was easier to manage and was harder to evade. As a result of petitions from local parishes and vestries north of London who were unable to maintain their roads to York and the north, the toll road became legal in Britain in 1663. In that year a toll road was established in Hertfordshire on the Great Road North from London towards Scotland. The first toll house was at Wadesmill. Such roads were commonly called turnpikes, as the common barrier was a pike (or horizontal bar) attached to a vertical post in the manner of military barriers of the time.

Nearly all the turnpikes were to be based on the routes of earlier roads. As these had had to have gentle grades for the low-powered animal-drawn vehicles of the time, they were frequently devious and tortuously located.

The toll-collecting agency was required to maintain the road from the tolls collected. As private enterprises they were also intended to return a profit to their owners. The fees charged were commonly proportional to the number of haulage animals being used. Wealthy landowners often regarded the toll they levied as merely a form of tax carrying no obligation to maintain the road. As poor citizens were often unable to afford the tolls, evasion levels were high. The turnpikes were often poorly received in local communities.

Nevertheless, the turnpike system became increasingly extensive, effective and significant. Telford and McAdam had major impacts. As a result, by the 1820s England possessed some 40 Mm of turnpike, covering virtually all main roads and many by-roads. The largest turnpike was the 300 km Bristol Turnpike Trust with which McAdam was associated, although the average Trust managed only about 40 km of road.

The first U.S. turnpike opened under government control in Virginia in 1785. Running from Snigger's Gap to Alexandria, it was probably the nation's first stone-surfaced road. By 1794 private enterprise had constructed the 7 m wide and 100 km long Lancaster Turnpike between Philadelphia and Lancaster following the 1683 Conestoga Road. It was the nation's first major paved arterial road.

But the turnpike era effectively ended throughout the world in the 1840s with the advent of workable and highly competitive steam-powered railways and canals. The turnpike trusts ended slowly in financial crisis and bankruptcy. When London's last turnpike closed in 1871, there were still 854 of the peak of 1100 English trusts still in some semblance of operation.

3.4.3 Central control

An obvious fourth solution to the problem of road administration and finance that confronted many central governments in the mid-19th century was for they themselves to supply the resources and skills. They had, after all, unsuccessfully tried to obtain them from everyone else. The Scandinavian countries had already headed in this direction, beginning with Norway in 1274.

In Britain, Telford and McAdam had seen the problems of the turnpike system and were aware of the success of the centralised system in Scandinavia and pre-Revolutionary France. With others they argued strongly but without effect for its adoption in Britain. Another century was to pass before English-speaking countries were to adopt central road administrations.

What happened instead between about 1840 and 1860 was a reversion to the previously discredited second solution – control by local government. Not surprisingly, the internal road systems under this 'solution' deteriorated rapidly during the latter half of the 19th century and in many places were effectively non-existent.

In the U.S. the Federal government in 1838 had relinquished its road responsibility for its National Pike or National Road. This road had been conceived in 1802 by a group including Thomas Jefferson and George Washington to aid westward expansion. It had a total length of 1350 km but was then abandoned well short of its goal. U.S. 40 now parallels much of its route. The move to abandon was initiated by constitutional arguments and sealed by the advent of far better expansionary forces – the railroads and the canals.

Federal systems of government complicate the sequence of this development a little. In the U.S. and in Australia, State Road Authorities were initiated; leading the way were Virginia in the U.S. in 1816 and Tasmania in Australia in 1827. However, by the 1850s these State administrations were to suffer the same fate, albeit temporarily, as the turnpikes.

A further distortion, but demonstrating the same effect, occurred in Germany. Until 1870 the German government exerted central control over its roads, motivated by military considerations. However, the new Imperial government of 1871 considered rail of prime military importance and roads of secondary importance and so gave road control to the largely uninterested States.

The move to central road administrations was to be successfully led by the cycle clubs and motor clubs at the turn of the century, often under the banner of 'Nationalise the Roads'. The move began with the U.S. State of New Jersey in 1891. The U.S. led the way with Federal funding introduced in 1916 and all States having a highway organisation by 1920.

3.5 POWER

3.5.1 Before steam

In most cases travel in the millennium after the Romans was a local activity based around one or two villages or towns. We saw earlier how longer distance travel slowly began to increase, but up to the 18th century such trips were predominantly made by human and animal foot traffic.

Humans on foot could manage between 15 and 40 km a day, depending on need and burden. A number of towns grew around these day's walk distances. The speed record for long-distance foot-travel was probably held by the Incas who developed a system which was able to transmit messages at 400 km/day by using fast runners over 2.5 km stages, suggesting that the runners were achieving about 15 km/h running stages by day and night.

A single horse and rider could manage about 60 km/day. To extend this meant using horses in stages and thus establishing 'post houses' where tired horses could be rested and fresh horses obtained. Horses could make 30 km trips each day but oxen (or bullocks) could only travel half that distance and were more difficult to organise. However, they had the advantage that they could haul larger loads, could keep going over more days, required less water and were easier to feed.

Pack horses (or donkeys or mules) could each carry about 120 kg shared between two pannier baskets for up to 25 km. Long strings of nine or more pack horses, called drifts, worked many regular freight routes. The horses were tied together and the lead animal had bells attached to its saddle. In the 18th century there were special pack horse paths (or causeways or causeys), sometimes stone-paved, with their own bridges and guarded with posts to deter use by carts.

3.5.2 Steam

When Roger Bacon predicted in the 13th century that, *inter alia*, 'one day we shall endow chariots with incredible speed without the aid of animals' he was jailed for being in league with the Devil. For 500 years, the decision must have seemed eminently wise.

The first doubts occurred when Newcomen completed his steam engine in 1712, after years of experimenting. However, its application to transport occurred very slowly. When it happened about a century later, man at last had a source of transport power much greater and easier to manage than the horse and the ox. But the power did not come without a price. The steam engines were heavy and large. Whilst the iron wheel could manage the new loads, few road pavements could carry the iron wheel.

To circumvent this problem inventors first tried bigger wheels. Such solutions were not effective and the inventors soon turned to placing their iron wheels on iron rails, carefully distributing and spreading the high loads via the rails, sleepers and macadam-like ballast. Flanges were placed on the wheels to stop them coming off the rails. The fact that this obviated the need for steering was incidental. Rail rapidly became the dominant land transport mode. Roads became no more than feeder roads to railway stations and even many of these roads were in such bad condition that they had begun affecting railway business.

Steam road-vehicles were far less successful. They suffered from a lack of mechanical reliability, commercial opposition from the stage coaches, railways, and turnpikes, and a range of only 20 km before a 15 minute rewatering stop. Indeed, they required about 10 L of water per kilometre. Clearly, the time was not yet technologically ripe for self-powered road vehicles.

The steam-powered vehicles were the cause of the notorious U.K. Red Flag Act (properly the Locomotive Act) which restricted the speeds of road locomotive to 7 km/h in the country and 3 km/h in the city. It also required the vehicle to be preceded by a man walking with a red flag by day or a lantern by night. Each vehicle had to have at least three operators. The Act was introduced in 1865 and not significantly relaxed until 1896, an event celebrated by the first of the famous London to Brighton car rallies. No other European country had had anything like the Act and it was a major restraint on British innovation in car technology during that time.

3.5.3 Bicycles

The great impact that the invention of a useable bicycle had in the closing third of the 19th century is often forgotten. Macmillan in Scotland introduced the first lever-powered bicycle in 1839. Despite the potential of these 'velocipedes', inventors were drawn away from them by the then-current railway mania. The first commercial bike was built in Paris in 1861 by Michaux, using foot pedals and a front-wheel drive. Attempts to improve the gearing then led to the 'ordinary' or penny-farthing bicycle. The 'safety' bicycle with pedal chain drive to the rear axle was introduced in 1877 and the first recognisable modern bicycle appeared in 1885.

The bicycle became popular and widespread through another great forgotten invention that changed the face of our transport world. This was the Scottish veterinarian Dr John Dunlop's invention in Northern Ireland in 1888 of the pneumatic tyre, an invention made specifically for the bicycle. The pneumatic tyre made the bicycle a useable and useful tool. Its key long term-effects were that:

* it overcame the millennia-old narrow wheel/high contact pressure problem, and
* it allowed high loads to be applied to wheels in the knowledge that the tyre would spread the load out over an area such that the contact pressure would, approximately, never exceed the tyre inflation pressure.

It is strange now to learn of the alarm with which many greeted the bicycle. It was pointedly noted that the cyclist, although using the road, 'was usually not even a ratepayer'. Cyclists were damned as 'cads on castors' and were said to have scared horses and pedestrians, raised dust, scattered mud, and travelled at excessive speeds of up to 20 km/h. An 1888 British regulation required all cyclists to carry a bell which would tinkle continuously whilst the cycle was in motion.

Whereas steam was providing the world — or at least the prosperous world — with long-distance travel, the bicycle was offering efficient and effective short-distance travel to all, in a way that the horse and the steam train never could. The bicycle initiated humanity to that 20th century right — the freedom of every citizen to travel when and where he wishes.

Thus it is not surprising that the bicycle clubs produced the first useful road maps, were the first group tourists, were the founders of many of today's automobile clubs and were the first organised protagonists for better roads via the Good Roads Association which arose in most developed countries.

3.5.4 Cars

Gottfried Daimler and Karl Benz in Germany developed the first useable petrol-powered internal combustion (IC) engines between 1882 and 1885. A key advantage of the new IC engines was that they easily utilised petroleum distillates, liquid energy that was readily dispensed, transported and consumed and which had a very high energy to volume ratio. Although they never met, the firms established by the two men finally merged in 1926. Benz's first successful vehicle (1885–86) was a tricycle powered by a one-cylinder gas motor that Daimler followed three months later with a powered bicycle.

The first practical trucks were steam powered and were in use on the roads from their introduction in France in 1892 until about 1930, although Britain, with an abundance of coal, did not stop building steam trucks until 1950. However, trucks powered by IC engines arrived in 1894 and had captured the market by 1900. The end of the horse era was surely the 1911 announcement by the British War Office that the motor truck was to replace the horse on a large scale in the British Army.

By the end of the 19th century there had thus been a congruence of events seemingly stage-managed to produce the modern motor vehicle. The new universal desire for private travel was supplied per medium of the pneumatic tyre and the IC engine. No flanged wheels restricted the driver's choice of route. The one missing factor was a good road system. Indeed, the roads of the day were such that it was wise to use large-diameter, wooden-spoked wheels to provide sufficient clearance over rocks, stumps, and other obstacles.

Although the pneumatic tyre brought the advantage of applying lower vertical contact pressures and better stress distributions to the road surface, it also delivered other bounties. In particular, the enhanced surface-friction — coupled with the greater power now available — allowed markedly greater acceleration and deceleration performance. These speed changes applied much greater horizontal loads to the pavement surfaces than had previously been encountered. Worse was to come — as braking capabilities increased, so did travel speeds, and high-speed cornering on curves produced yet another set of new forces. In addition the tyres of the speeding vehicles began to create significant uplift suctions on the pavement surfaces, leading to clouds of dust. Thus the new vehicles with their speeding, braking, accelerating, skidding and cornering began to seriously abrade the old road surfaces. The most obviously annoying characteristics of the car at the turn of the century were that it was dangerous and caused dust. The second loomed larger in the eyes of the public, the dust covering everything within 20 m of the roadways.

The enthusiastic adoption of the car meant that the expanding push for good roads that the cyclists had begun gained even more momentum as those roads that did exist began to deteriorate under the new traffic. The legacy of sixty or so years of inattention had left no effective infrastructure for the dazzling new invention.

3.5.5 Traffic engineering

Before the appearance of the bicycle and the car, vehicle speeds seldom exceeded 15 km/h. Thus the horizontal curvature of the road was only important in so far as it was necessary to enable long animal-teams hauling carts to negotiate tight bends. On minor roads a 15 m radius was adequate and 25 m was ample for major highways. However, by the turn of the century the self-powered vehicles had raised that minimum to 50 m. This led to road curves being frequently no more than kinks between pieces of straight alignment. Sight distance was an irrelevancy at the low operating speeds and, with the coming of the bicycle, was solved more by bell-ringing than by realignment.

Section 3.3.4 described how the older roads had often employed continuous and steep cross-slopes on both straight and curved sections in an often futile attempt to drain water from the road surface. However, banking (i.e. superelevation, Section 19.2.2) on curves was uncalled for at 15 km/h. The car soon changed the situation. Steep cross-slopes had to be removed from the straight lengths of road and superelevation provided at curves to encourage drivers to stay on the correct side of the road as they cornered. The provision of superelevation began in 1908 and met stiff opposition from the owners of horse-drawn vehicles whose lack of centrifugal force when cornering led to them slewing down the banking.

With all vehicles travelling at relatively low speeds and possessing short stopping distances, priority was usually by might rather than by decree. The few measures that were introduced, like the U.K. Red Flag Act discussed in Section 3.5.2, were largely aimed at preventing a few being a nuisance to many. The car changed all this and brought some new traffic disadvantages as well. Consequently, the practice of traffic engineering was born.

Traffic signals were introduced at a pedestrian crossing in London in 1868 and modern-style traffic signals in Cleveland, Ohio in 1914. The first roundabout was built in 1903 in France. Detroit claimed the title ‘Motor City’ in 1909 as the heartland of U.S. motor-vehicle manufacture. It obviously saw this as carrying traffic safety obligations as well and formed a Traffic Division of 13 policemen in that same year to control traffic via hand signals. In 1911 the surrounding Wayne County introduced centreline marking, initially on narrow bridges, and the city introduced pedestrian crossings. In 1915 the Traffic Division produced the first stop sign and in 1918 the first three-colour traffic light.

Route marking began with a 1704 Maryland law requiring trees beside a route to be marked with an elaborate system of notches, letters and/or colours. One South Maryland road has retained its name as ‘Three Notch Road’. More conventional numbering began in Wisconsin in 1918 but major routes were more commonly distinguished by coded bands of colour painted on roadside poles and trees.

One measure of some historical significance concerns driving on one side of the road or the other. During most of transport history such a requirement was irrelevant. Few roads could take wheeled vehicles and those that could were rarely wide enough for two vehicles to pass with ease. One vehicle, under the right-and-might rule, would pull aside for the other. Some Roman bridges did permit two-way traffic, with each bridge having its own priority arrangements. Of course, vehicle speeds on such bridges were so low that drivers did not require any advance knowledge of right of way.

The need for a clear rule grew as wheeled vehicles became more common in the 17th century. On vehicles with a seated driver and one line of horses, the use of the right hand for the reins and the whip led to the driver favouring the right and therefore passing to the left so that he could judge the passing manoeuvre. This was very much the case with the smaller English vehicles. However, with more than one line of horses the driver tended to the left so that his right hand could manage the full set of reins. Giving good centreline sight to the driver then favoured driving on the right. Such larger vehicles were more common in continental Europe. Similarly, those in charge of a string of pack-horses would hold the lead horse by the right hand and thus found right-hand passing easier.

Where firearms were needed, the right to left direction of the barrel across the body also favoured moving to the right. A major impetus for right-hand driving in the U.S.

came from the design of the Conestoga wagon, made famous as the 'Prairie schooner' that led to the winning of the West. The wagon was operated by the driver walking or riding a 'lazy board' alongside the vehicle and using his right hand to manage the horses and operate the brake lever. Thus passing required moving to the right to give the driver forward vision.

The first national decree was introduced in Saxony in 1736, requiring travellers to keep to the right when crossing a new bridge over the Elbe, whereas English legislation in 1756 required traffic on London Bridge to keep to the left. The rule became widespread in Britain after an Act of 1835. Right-side driving was encouraged by turnpike regulation in Pennsylvania in 1792, with the first general laws occurring in New York in 1804 and in Canada in 1812. Left-hand driving persisted in the U.S. on the National Pike (Section 3.4.3) until the 1850s.

3.6 SURFACES

3.6.1 The problem with macadam

A macadam surface of compacted broken stone would not have made an ideal pavement surface, even for the solid iron-tyred wheels then in vogue. Indeed, although the surface was *blinded* by working in pieces of fine stone and sand, these materials had little or no long-term cementing effect. The blinding process was thus far from effective, particularly when McAdam's strong advice was not followed (as was often the case) and organic material was used for blinding. Furthermore, the surfacing was regularly dislodged by the hooves of passing traffic and replaced by their excreta.

Thus urban macadam surfaces, although a great improvement on their predecessors, were frequently abraded, dusty in summer, muddy in winter, slippery, slimy and malodorous. In dry weather regular watering was required to allay the dust and in winter regular shovelling was needed to remove the mud. In urban areas people made a living as 'sweepers', cleaning paths for pedestrians wishing to cross the road without soiling their shoes and clothes.

This was not due solely to problems with macadam. There was also the logistics of the horse, for as its use peaked at the end of the 19th century, the problems of horse excreta and horse carcasses in large cities were becoming almost insurmountable. In 1900 it has been estimated that each day horse transport in New York created 1.1 kt of manure, 270 kL of urine and 20 carcasses. The motor vehicle might have alleviated this particular problem but the pneumatic tyre soon sounded the final death knell for unbound urban surfaces.

In the countryside macadam was still greeted with great joy as it provided passable all-weather roads. Its main problem was with the dust it created. Coach travellers and motorists often wore goggles and linen dustcatchers. The only serious alternatives to macadam in rural road construction were various uses of timber. The corduroy road (Figure 3.6a) was built of logs placed transverse to the direction of travel. The technique had been in use for millennia. It was demanding of timber, uneven and often short-lived. Nevertheless, the practice was widespread in forested regions. The corduroy surface was improved somewhat by adding sand or longitudinal running planks to the surface of the logs. One advantage of the technique in soft ground was that as the timbers settled more were stacked on top. An alternative was the plank road developed in Toronto, Ontario in 1835. It used longitudinal timbers with a flat upper surface placed slightly below ground level. Transverse planks were then pounded into place until they bore on the longitudinal timber (Figure 3.6b). The surface was then covered with sand. The technique became quite popular in North America and Australia but was far from durable.

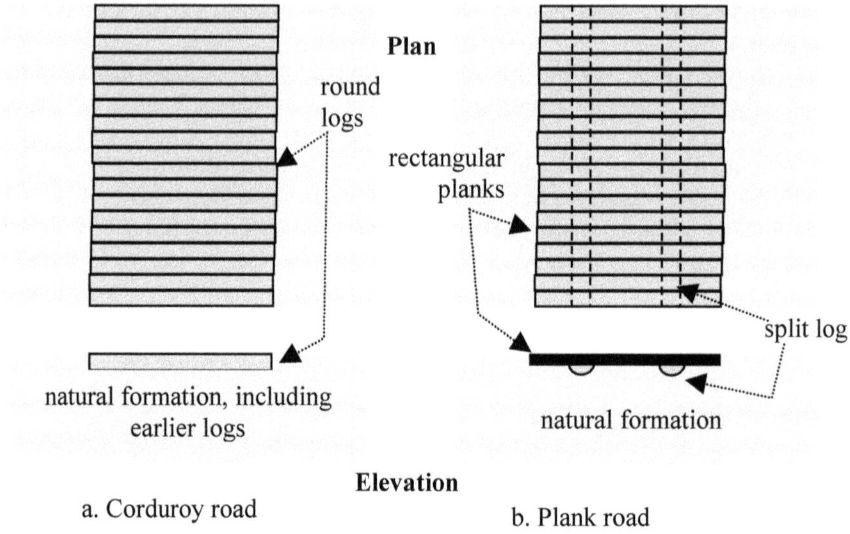

Figure 3.6 Timber roads.

3.6.2 Asphalt

The technological push for better road surfaces was not to come from the low volume rural roads, but from the demands of city life and city traffic. Bitumen is a component of petroleum. It was relatively plentiful in the Middle East and there are records of its use before 3000 BC, either directly or mixed with fine mineral particles, first for cementing inlaid jewellery and then for grouting bricks. Section 3.3.1 showed that by 3000 BC it was being used as a pavement mortar. Tar is similar to bitumen but is somewhat harder and less durable. It is obtained as a by-product of the processing of timber and coal. The major source of tar in McAdam's time was from Norwegian timber and was produced primarily to provide pitch for caulking ships. The first asphalts were naturally occurring mixtures of bitumen, mineral matter (commonly pieces of limestone) and occasionally organic matter. Asphalt deposits existed throughout the Middle East, including sources at Ur, Hit, Nineveh and the Dead Sea region.

After an initial discovery by Columbus in 1498, Raleigh re-discovered the bitumen-rich (40 percent) Trinidad Lake asphalt in 1595 and used it for caulking his ships, as it was less temperature-sensitive than Norwegian wood tar previously used. The mineral matter in Trinidad Lake asphalt is fine sand and colloidal clay. In 1712 a Russian physician and professor of Greek named d'Eyrinys found a major asphalt deposit at Val de Travers in Neuchâtel in Switzerland. The material had about 10 percent bitumen by mass and was usually viscous enough to be mined as a solid and then crushed or sawn into cubes. This led d'Eyrinys to invent a bituminous mastic, which he made by mixing the powdered asphalt with extra fluxed bitumen, to give about 15 percent bitumen by mass, thus giving a product intermediate between asphalt and bitumen in viscosity (Section 33.3.3).

In 1747 the Earl of Leinster in Ireland is reported to have copied Prussian methods in producing a road on his Dublin estate, mixing ships' bitumen with lime and gravel. The

Prussian methods would probably have used asphalt from the Braunschweig–Hannover deposits. The asphalt deposits at Seyssel in France began to be exploited commercially in 1797 as a mastic, mixing tar with the asphalt. Soon mastics were being widely used to waterproof bridge decks. Later, the addition of sand to the warm mastic surface was found to overcome problems of slipperiness under traffic. This led to the use of mastic for footpath construction in Paris and Lyon in 1810.

3.6.3 Tar

Meanwhile, a very significant alternative binder source arose as tar became increasingly available as a by-product of the growing 19th century production of 'town gas' from coal. Given the concurrent increase in the use of macadam, it was a natural but still innovative step to try the surplus tar and pitch as replacements for the sand, stone, dust and soil then in use as blinders for macadam surfaces. Tar was first used in this manner in 1822 at Margate Pier in England.

A problem with many of the tars tried was that they set very slowly. This perhaps aided the serendipitous use of the tar, not only as a blinder on the surface, but also as a binder for the entire macadam stone-course. The effectiveness of this process depended on the extent to which the tar penetrated into the interstices of the in-place open-graded macadam. Even when used accidentally it was clearly effective and by the 1840s the new material, known as tarmacadam, came to be widely accepted in the U.K. The abbreviation tarmac is still colloquially used to describe black topped pavements, particularly on airfields. As these in situ processes depended on the penetration of the tar or other binder into the open-graded macadam course, they also came to be known as penetration macadams. A subsequent major step was to precoat each particle with a thin film of binder before placing it in the pavement. Such products are called precoated (or coated) macadams. Tarred heated gravel was first used in this manner in a trial in Nottingham in England in 1840.

3.6.4 Concrete

During this same period cementitious binders — discovered and used by the Greeks and Romans (Section 3.3.2) but since forgotten — were having a rebirth. Natural cement was discovered on the Isle of Sheppey in England in 1796 and was initially called Sheppey Stone or Roman cement. Artificial cement was invented in 1824 by a Leeds bricklayer called Aspidin who named it Portland cement because it resembled in colour and texture the building limestone quarried on the Isle of Portland. It did not take over from natural cement until about 1890.

There was a rapid increase in the use of concrete made from cement, with Telford being a leading advocate of its use for basecourses. Indeed, concretes became expensive but effective bases for urban roads. There are reports in 1846 of existing lime concrete basecourses for wood blocks (Section 3.6.6) being removed from the Strand in London and replaced by Portland cement concrete carrying stone blocks. When mix design came to be well understood in the 1860s, concrete became a strong and reliable product. The development of continuous concrete mixers in 1875 aided this process. The first full concrete road was built at Inverness in Scotland in 1865, probably as a penetration macadam with the cement mortar forced (or grouted) and rolled into pre-placed macadam.

Cement mortars were tried with some effect to blind macadam surfaces but the treatment was expensive and produced a very slippery surface. Concrete had to await the advent of the pneumatic tyre and mechanical construction equipment before it could be effectively used for road surfacing.

3.6.5 Natural asphalts

Generally, the roadmaking problem lay not in the basecourse but in the surfacing. Not only was there a need for a surface that could be kept intact, clean, free of dust and mud, and waterproof but there also was a major need for a surface which would provide adequate friction for the tractive effort of iron-shod horses' hooves.

Asphalts were now in plentiful supply from Switzerland, Germany, France, Sicily and Trinidad. One technique was to cut or mould the asphalt into brick-like cubical blocks. The more common moulding method followed the mastic technology, grinding the native asphalt rock into a powder and pressing it into blocks in a factory. Blocks were first used for paving in Paris in 1824. A major trial was conducted in London in 1838-39 in Oxford Street near Soho Square in what was probably the world's first major road experiment. The tests favoured wood blocks (Section 3.6.6) and a number of the asphalt test strips performed poorly. It was not until 1880 that the development of a powerful compressing machine in the U.S. led to better asphalt blocks.

The second technique ground the asphalt to a powder, heated it to reduce its viscosity, and then pressed, stamped, rolled and/or hot-ironed it into place. The use of powdered asphalt in roadmaking is said to have originated about 1849 when an engineer called Merian noted that pieces of native asphalt rock falling from trucks taking it to the mastic factory were powdered by the truck wheels and then rolled by later wheels into a useful road surface. Merian conducted successful trials in the Swiss village of Travers in 1849.

Major applications of compressed powdered asphalt were undertaken on the Rue Bergère in 1854 in Paris. This product was commonly called sheet asphalt. Unfortunately, most of the old problems remained. The asphalt was just as slippery and wore just as quickly as its alternatives. However, it was claimed as a result of detailed observations that, when horses did fall on the new natural asphalts, they suffered less damage. It was also believed that the asphalt was more resilient under hoof impact and therefore generally easier on the horse.

3.6.6 Paving blocks

Major urban surfacing alternatives to macadam and natural asphalt were bricks, masonry cubes of natural stone known as setts, large paving slabs known as flagstones (or flags), wood blocks and cobblestones. Cobblestones were the traditional method of stone paving, using uncut and often waterworn stones or large pebbles of about 150 mm size. They produced a surface that was usually uneven and slippery. Setts were used to produce a better surface than the cobblestones. Belgian granite became famous for this purpose but many port cities had plentiful sett supplies based on discarded ships' ballast. Concrete blocks were used in Edinburgh in 1866. None of these segmental block methods had much inherent strength and commonly resulted in uneven and slippery surfaces. They were particularly slippery and noisy under the action of iron-shod hooves and wheels. Many also abraded easily.

Wood blocks had first been used in Russia in the 14th century. They were applied with some skill by Gourieff in Leningrad in 1820 when he set hexagonal blocks on a basecourse of crushed rock and sand. He filled the block interstices with sand, poured boiling tar on the surface and sprinkled sand on the cooling tar. The tar-impregnated blocks were still far from the perfect solution as they wore quickly into wheel ruts, and the whole pavement smelt when damp. Two factors helped overcome these difficulties. The first was the use of the tar overlay to reduce their slipperiness and seal them against moisture. The second was that Australian hardwood was imported from 1888 and this led to block pavements being far more durable and weather-resistant.

3.6.7 Manufactured asphalt

The next stage in the development of road surfacings was the production of manufactured rather than natural asphalts. Selected sand and stone were mixed with bitumen or tar and offered the advantage that the stone grading did not depend on either chance or crusher size, as was the case with powdered natural asphalt.

de Smedt was a Belgian chemist who emigrated to the U.S. in 1861. After trials beginning in 1869, he conducted the first successful application of manufactured asphalt in William Street, Newark, New Jersey. He developed an asphaltic concrete using crushed rock and sand and used it in applications in Battery Park and Fifth Avenue, New York, in 1872–73. He designed his mix to produce minimum air voids (a well-graded mix), using limestone, sand and fillers and used bitumen as his binder. The mix bitumen content was 10 percent by mass, with the bitumen fluxed from West Virginian native asphalt.

The suitability of the bitumen was determined by the tester chewing a sample as vigorously as possible. The oral tradition was to draw to a close as, in 1888 Bowen of Barber Asphalt introduced the first penetration tester for determining bitumen stiffness cum viscosity. The device employed a 'No. 2 sewing needle' and its descendants, little altered, still control the selection of most of the world's bitumen. The surface slipperiness problem was overcome in manufactured asphalt by following the technique used for mastic and wood blocks in which a covering of coarse sand or grit was incorporated into the still-sticky asphalt surface. Manufactured asphalt could thus provide a level surface with adequate skid resistance — it was just what was needed for the motor vehicle and its pneumatic tyres.

In 1887 Richardson succeeded de Smedt as Inspector of Pavements in Washington DC. His 1905 book probably contains the world's first reliable specification for manufactured asphalt. His technique for obtaining the specification was an admirable piece of empirical science. Over 20 years he visited 3 Mm of asphalt road in the U.S., Britain, and France, observing performance under traffic and taking samples to determine the composition of the asphalt. From this database he chose a specification which was to stand the test of time remarkably well.

Asphaltic concrete using bitumen found ready acceptance in New York and Washington and by 1910 it was the most commonly used paving material in U.S. cities. It was not used in Britain until 1894–95 when Richardson advised on the use of U.S. techniques in trials in London. The better grading of the new mixes made them stronger and stiffer, and yet they were easier to work into place as a coherent mixture than were the harsh open-graded macadams produced by the penetration methods. Hence placement by mixing, spreading, and rolling became possible. Mechanical spreading of these asphalt mixes did not occur until 1928. The development of asphalt is pursued further in Chapter 12.

Bitumen, which is a 'heavy end' by-product of oil refining (Section 8.7.2) and therefore sometimes referred to as a residual bitumen, had been produced in California in 1865, but was first used for roadmaking in the U.S. in 1900 — not coincidentally at about the time when internal combustion engines (Section 3.5.4) were consuming more and more of the light end of the oil refinery outputs.

The main roadmaking competitor for bitumen in this century has been Portland cement, whose development was discussed in Section 3.6.4. Concrete had rarely been used for surfacing in the 19th century as its smooth impervious finish made it unsuitable for use with iron-shod, excreting animal traffic. The disappearance of animal traffic and the appearance of rubber tyres were dual boons to the cement industry, prompting a vigorous post-First World War advertising campaign. The rivalry between the two materials continues to this day, with economics and local custom usually determining which is used.

3.6.8 Sprayed surfaces

Bitumen and tar had road uses other than as an asphalt binder. They were often emulsified or heated to a fluid condition and then brushed, painted, sprayed or squeezed onto urban macadam surfaces to suppress dust and to provide water (or slime) proofing. The process initially had a relatively short life of a few months. The techniques were gradually improved to a useful and reliable stage, particularly by the Swiss engineer Guglielminetti in experiments in 1901 at Monte Carlo on the French Riviera and at Geneva, in which tar was sprayed or brushed onto macadam and then covered with coarse sand. Its successor is used to this day as a common surface maintenance technique known as surface dressing.

Attempts to improve the poor performance of the first surface-coats led to the use of heavier and more viscous bitumens. When sprayed or brushed onto the pavement these were certainly stickier and stronger but produced a slick and very slippery surface. This problem was then overcome — as with asphalt — by applying coarse sand and small pieces of angular, broken stone to the still sticky surface. Priming of the surface before spraying proved a key ingredient in the process, as evidenced by successful U.S. trials in about 1910. Between 1910 and 1930 the process was taken a stage further in New Zealand and Australia where it was realised that the flexible but waterproof nature of the coating meant that it could be applied to unsurfaced roads that had developed via stage construction to traffic levels requiring a surface, but not needing a significant increase in structural strength. Such surface-dressed roads on good compacted natural material or on unbound crushed rock basecourses are still widely and effectively used in Australia and New Zealand as spray and chip seals.

3.7 BRIDGES

3.7.1 Early bridges

The first river crossings would have been by ford rather than bridge and many town names — Oxford, Stratford, Frankfurt — are indications of the importance of fords in early man's communication links. Some fords would have been natural but many were deliberately constructed and may have prompted the later introduction of bridge piers. The first bridges would undoubtedly have been natural arches (Section 15.1.3) and stepping stones and logs placed or fallen across relatively narrow streams.

The first practical extension of the stepping stone and natural arch was the use of vertical piers of stone or wood to allow multiple short spans to increase the distance bridged. An alternative approach to piers was the pontoon bridge, which used boats to support the spans. Inventive man also extended the stepping stone by carefully stacking stone upon stone in order to produce corbel arches, inspired perhaps by various natural arches. In this technique, the stone piers are made progressively wider, each layer of stone slabs cantilevering a little further out than the previous layer. Finally, the two piers are connected by a horizontal corbel slab or wooden beam. The technique could span up to 10 m (Figure 3.7).

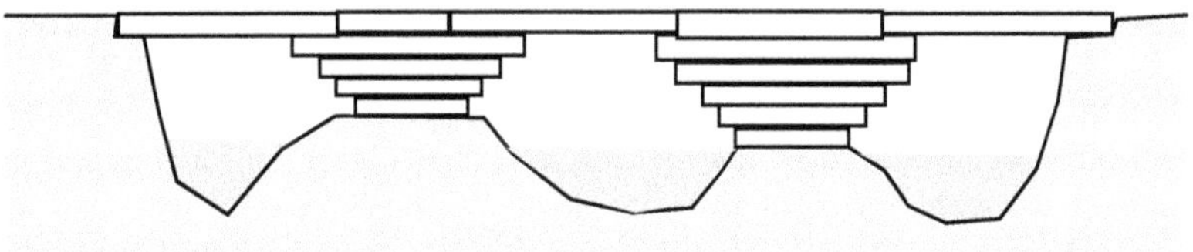

Figure 3.7 Corbel arch.

Greater spans required the use of the subtler keystone (voussoir) arch (Figure 3.8). The first keystone arch is believed to have been built in Mesopotamia about 4000 BC. As with roads, the Romans built on the knowledge of the Greeks and Etruscans to raise bridge building to a noble art. The semi-circular, keystone, masonry arch was their primary structural form for permanent bridgeworks. For prestigious sites the arch was infilled and the roadway brought to level; elsewhere they used the muleback arch with the roadway following the vertical curve of the arch. In difficult foundation conditions they used both piles and cofferdams.

Their oldest bridges still standing — although continuously repaired over the centuries — are the 178 BC Ponte Rotto in Rome which has three semi-circular arches and the 219 BC Martorelli 40 m span arch bridge in Spain. While the greatest span achieved was close to 50 m, possibly the most famous Roman bridges are the Puente Alcantara in Spain towering 30 m over the Tagus and the imposing Ponte du Garde aqueduct at Nîmes in south-eastern France, crossing 265 m at a peak height of 46 m. These used fitted stones at the lower levels and mortared joints for the higher tiers.

Vines have also provided an alternative crossing technique with the potential to span much larger distances than the other natural alternatives. The initial suspension bridges would have been of the simple hanging type, with the traveller either walking over or hanging or sliding from the rope or vine.

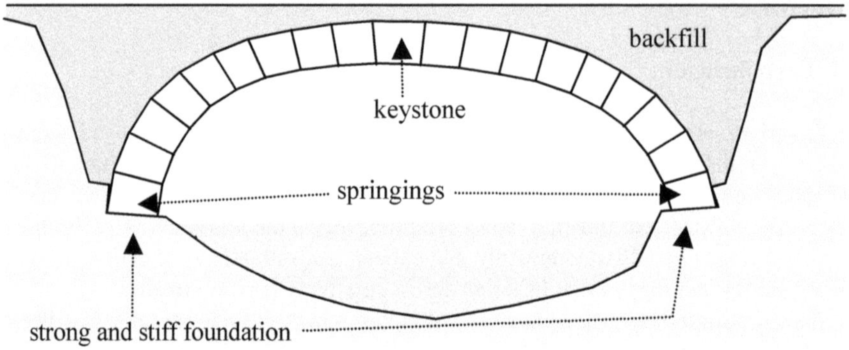

Figure 3.8 Keystone arch (form shown is elliptical).

Given the difficulty of bridgemaking and the relative ease of ford crossing, the major bridge building challenges arose on unfordable rivers. Such bridges were usually enormously expensive. Once such a bridge was established at one of these sites it became a foci for all the roads of the region. It was thus a major military feature and was usually built with fortified towers and a narrow carriageway to ease the task of defenders.

The art of bridge building died with the Roman Empire. One of the few extant bridges built in the millennia after the Romans is the Pont St Bénézet over the Rhône at Avignon in France, built between 1177 and 1187. It had twenty elliptical arches, each spanning 30 m. Some are still standing.

London had been located on the first fordable crossing of the Thames and in 994 the Saxons built a wooden bridge at this site. In 1176 the Normans, who had arrived in England from France in 1066, started building a 360 m total length stone bridge across the Thames. The bridge had 19 arches, each spanning 8.4 m. The work on London Bridge was completed in 1209 and it lasted 600 years until demolished in 1832. It was the only bridge over the Thames in London until 1750.

The next major step in bridge design occurred in 1569 when Ammanati in Italy introduced the ogive arch that permitted span to rise ratios as high as seven to one. His Ponte della Trinita in Florence, completed in 1569, was a radical departure from the old high pointed arches with their small spans and large piers. A restored version of the original bridge stands today.

The first major U.S. bridge was built in 1662 over the Charles River in Boston. Called the 'Great Bridge', it used thirteen piers and had a total length of 80 m, or about 6 m per span and so was possibly of timber beam construction.

3.7.2 Modern bridges

The father of modern bridge building was the Frenchman, Perronet, the first Director of the School of Bridges and Roads (l'École des Ponts et Chaussées). His major contribution was to realise that the thrusts on the piers in multiple-arch bridges cancelled each other out, thus permitting thinner river piers. Perronet's last and greatest existing bridge is the Pont de la Concorde over the Seine in Paris. Probably his most beautiful bridge was also over the Seine in Paris, this time at Neuilly. Although described as 'the most beautiful and graceful stone bridge ever built', it was demolished in 1956. Rennie made the next major breakthrough when he developed the semi-elliptical arch for London's Waterloo Bridge, opened in 1817 and replaced in 1942.

The development of steam-power had made it possible to make iron in large quantities and in 1783 Cort cast structural shapes in iron at Coalbrookdale, birthplace of the Industrial Revolution. The town was also the site of the first iron bridge which was built over the Severn a few years later in 1781. The 30 m span semi-circular arch Coalbrookdale Iron Bridge is still standing and has been described as Britain's best known industrial monument. Although the bridge broke new grounds in material use, its design is understandably a copy of a masonry arch with joints made in accordance with timber mortice carpentry procedures.

The next step was the beam-truss whose great advantage was that it exerted much lower horizontal thrusts on the piers. Although drawings of Trojan's AD 104 Roman bridge over the Danube show a number of truss-like features, the modern truss bridge probably developed from the example set by the use of trusses for roof construction. The oldest extant truss is the Kapellbrucke at Lucerne in Switzerland, built in 1333. Practical beam-trusses began in the U.S. in 1804 with Burr's bridge over the Hudson at Waterford, which utilised parallel chord beam-trusses outside the planes of its major arches. The major impetus came in 1820 with Town's lattice beam-trusses and Howe's 1840 use of iron and steel for the vertical tension members. The patented Howe and Pratt trusses soon became the popular favourites. The first iron truss bridge was built in the U.S. in 1851.

Scientific bridge analysis began with the Dutchman Stevin (1548–1620) who introduced graphical statics and Galileo (1564–1642) who explored the strength of rectangular beams in bending. However, bridge analysis did not flower until the widespread availability of wrought iron in the second half of the 19th century. Whipple built the first iron bridge of modern design with a 43 m span constructed in 1853.

The prototype of the modern bridge was Telford's Menai Straights Bridge, an essential part of the London–Holyhead road. This suspension bridge used wrought iron chains and spanned 174 m. Completed in 1826, it was then the world's longest bridge. Robert Stephenson introduced the wrought iron plate girder in 1841. The first all-*steel* bridge was built in 1876 over the Missouri River in Montana, employing five 150 m truss spans.

Many great bridges were built in the 19th century:

* Stephenson's three-span, 450 m long, hollow-tube Britannia rail bridge with a maximum span of 138 m, which opened in 1850 only a few kilometres north of Telford's Menai Straights Bridge;
* Ead's 1874 St Louis bridge, which was the first bridge to use structural steel and high-strength steel in its 156 m tubular arch ribs;
* Roeblings' Brooklyn suspension bridge of 1883, spanning 478 m and introducing the use of steel wire for suspension bridge cables (Roebling had previously used wrought-iron wire); and
* Fowler and Peter's Firth of Forth cantilever bridge of 1889 with its two 510 m cantilever spans.

The first use of cement in bridges since the Romans occurred in 1796 on the Isle of Sheppey bridge in England. Although reinforced concrete was being used for minor domestic items in 1849 it was first applied to bridges via the Monier method in the 1880s. However, Maillart, Hennebique and Freyssinet brought the material to perfection. Unreinforced concrete was used in an arch bridge at Grissoles, France, in 1840. Wayss built some reinforced arches in Switzerland in 1890. Probably the first modern concrete-arch bridge was a 50 m span designed by Hennebique and built in 1899 at Chatellerault in

France. Hollow section ribs were first used at Plougastel in France in 1930. Prestressed concrete (Section 15.2.1) is a relatively new invention that owes its development to Freyssinet. However, its popularity arose from the cement shortages after the Second World War. The first major post-war prestressed bridge was a 42 m span built in Hamburg in 1950.

CHAPTER FOUR

The Management of Roads

The various aspects of the management of the road itself and its traffic are discussed in all the following chapters of this book. This chapter explores the organisations responsible for that management.

4.1 INTERNATIONAL ORGANISATIONS

There are two major international road organisations — the International Road Federation (IRF) and the Permanent International Association of Road Congresses (PIARC). PIARC (AIPCR in French) is by far the older of the two, having been established in 1909. Its public name is now the World Road Association.

PIARC owes its existence to a decision of the French government in 1907 to hold an international congress in Paris during 1908 to study 'methods of improving roads to make them adaptable to the new forms of locomotion'. The Congress resolved that there was an ongoing need for its activities and a resolution led to the establishment of PIARC in April 1909. The Congress was, and still is, governed by a Commission with delegates from member countries and with offices and a small permanent staff in Paris. Although primarily an association of government road agencies, the World Road Association also accepts individual and company membership.

Except during periods of war, world Congresses have been held regularly ever since, using English and French as official languages. The 2007 Congress was the 23rd in the series. At each Congress a working program of preselected questions is considered, with additional items contributed by individual countries at the session. The Association began publishing its quarterly Bulletin in 1911. Many countries have national committees to foster their own involvement with World Road Association. In addition, the Association has many major Technical Committees of its own which keep their topics under review and report to each Congress.

The World Road Association publishes the Reports and Proceedings of each Congress, the Reports of its Technical Committees, Conference Discussions, a Technical Dictionary and a quarterly journal called *Routes/Roads*. It currently has about seventy-five member countries and each four-yearly congress attracts an audience of many thousands. The address of the World Road Association is AIPCR-PIARC, La Grande Arche, Paroi Nord, Niveau 5, 92055 La Défense Cedex, (Paris) France and www.piarc.org.

Whereas the World Road Association is primarily a meeting of the staff of government road agencies involved in administering, planning, designing, constructing and managing roads, the International Road Federation (IRF) could be said to predominantly represent the private enterprise component of roads. Thus each serves different market segments. Although it has a 'twin' office in Geneva, IRF was predominantly a U.S. influenced — as opposed to PIARC as a European influenced —

organisation. There is some significance in this division as the U.S. was absent from PIARC for much of the 20th century.

IRF was established in 1948 with the aim of promoting the development of road systems and road transport. Given the above, it is not surprising that it was a more effective public lobby group than PIARC, but a less effective technical organisation. IRF is funded by private sector groups and actively promotes the formation of similar bodies at a national level. It works closely with international road development agencies.

Like the World Road Association, IRF holds world conferences every four years but staggers them to avoid a clash with the World Road Association. The 2009 World meeting in Lisbon, Portugal, will be its 16th. The monthly World Highways magazine carries a dedicated section on IRF and its activities. IRF publishes an annual collection of world road statistics (IRF – & Section 4.3). It also provides a regular series of training courses enabling overseas road technologists to spend time studying U.S. practice. The U.S. address of IRF is 5th Floor, 500 Montgomery St, Madison Place, Alexandria, Virginia, 22314 and www.irfnet.org.

The Joint OECD–ECMT Transport Research Centre (JTRC) operates a collaborative research program. It was established in 2004 by merging the OECD Road Transport Research Programme and the European Conference of Ministers of Transport (ECMT) Economic Research activities. The Centre promotes co-operative transport research programs. Another significant international activity — begun by OECD — is the ITRRD database, described in Section 35.1.

4.2 NATIONAL ORGANISATIONS

The organisation of roads and road transport around the world follows many similar trends, whilst at the same time displaying marked and seemingly random differences. The common factors are important to pursue.

For instance, road planning is, with a few exceptions, a government task. Even when a non-government group might propose the construction of a road — typically a toll road — the road itself would be assessed as fitting within the overall transport and development system planned by government for the relevant region. A lesson which appears to have been almost universally learnt (Section 3.4.3) is that the control of such road issues must lie with government groups having a wider power base than those immediately affected, e.g. local government does not have the deciding role in arterial road issues which transcend local boundaries.

At another level, planning is recognised as an issue wider than roads and so interactions and decision roles commonly lie, at least in part, with planning, transport, developmental, economic, and environmental authorities. In many countries the Road Authority is part of a larger transport Authority, in order to ensure that roads are seen as part of a multimodal transport operation. The principle often works more in a titular than an effective sense.

The management, operation, and maintenance of roads are, with a few notable exceptions, government tasks, at least at the administrative level. The agencies responsible for these tasks, and for the planning role discussed above, are therefore responsible to elected officials and controlled by government ministries or departments. The major exceptions to government management and operations are the extensive systems of independent toll roads in operation in France, Italy, and Japan. In the Republic of South Africa a separate national agency — South African National Road

Agency — was established to provide overall management of that country's toll roads. Another common exception is found in the access roads to specific facilities such as mines and ports. Even these are usually under overall central government control, to ensure consistency and compatibility of any joint operations. For example, insurance risks are minimised if common operating procedures are followed.

It is common for the road system to fall into three categories (Section 2.1):
(a) national and strategic roads (or highways in the terms of Table 2.1);
(b) inter-urban and urban arterial roads;
(c) local sub-arterials, distributors, streets and farm roads.
A common pattern is for the major Road Authority (or Authorities in a regional or federal context) to be responsible for managing roads in categories (a) & (b) in fairly total manner, possibly receiving external funding and direction for category (a) roads; and, in the opposite way, to pass on some responsibility and financial assistance for roads in category (c) to local government bodies. This last issue is discussed in Chapter 7.

In most federal systems of government — such as Australia, Canada and the United States of America — the national government takes only a broad, policy-oriented interest in roads in category (a) and, although they may be national roads, the routine and operational decisions occur at a State level. The process is coloured by the fact that in most cases the major taxing and fund-raising powers lie with the national government.

Road Authority administrations follow a common overall pattern which is to a large extent reflected in the structure of this book (Figure 1.1). This pattern is illustrated in Figure 4.1. There have been occasional examples where police have been given traffic management powers, and/or where Road Authorities have been given police powers. These have rarely been successful. The main successful exception has been in giving the Road Authority the power to police truck operations (Chapter 27).

The internal structure within an Authority is usually modelled along the lines of box I in Figure 4.1, although not all the facets need necessarily be present. For example, the extent to which planning and financing roads and the management of traffic, vehicles, and drivers is part of the role of the Road Authority varies from region to region.
Three key support activities are:

* data management/computing/information supply/library,
* quality management/performance measurement/standards setting, and
* setting forward vision and policies.

In making this distinction, it is assumed that all the other groupings, e.g. surveyors and accountants, are covered within the key parts of box I. The execution of the individual tasks in box I may be done either in-house or contracted out to external groups. For example, there are many consulting engineers who undertake road and/or bridge design. There is wide variation in the degree to which tasks are contracted outside the Road Authority and this often reflects political rather than technical attitudes.

Within a Road Authority there is considerable divergence of practice with respect to regional authorities (sometimes called Divisions, Regions or Districts) and how much independence they are given. Administrations appear to go through cyclic patterns of favouring regionalisation or centralisation of administration, although the trend towards information networks would seem to tilt the balance towards centralised recording and decentralised decision-making. Some authorities further subdivide their regions into sub-regions (or residencies).

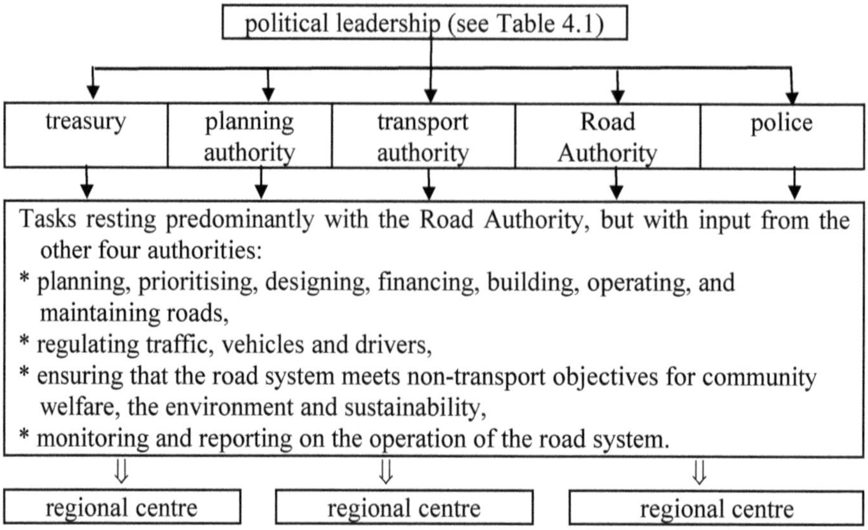

Figure 4.1 Road Authority contextual organisation.

The pattern of road administration in different countries can be grouped as shown in Table 4.1. To some extent grouping represents the overall political pattern in the country. Another dimension to the pattern occurs when the Road Authority is deliberately separated from Government. Such a 'separated' Authority might only manage part of the entire road network in an area. This separation might include giving the Authority its own fund-raising powers via hypothecation of road-related revenue (e.g. fuel taxes) either directly or via:

- a dedicated road fund,
- loans,
- tolls and other charges for road use,
- shadow tolls (the Government pays the operator on the basis of the number of vehicles using the road),
- rewards for service availability and/or quality, and/or
- road pricing.

Such charging and revenue-raising methods for roads are discussed further in Sections 29.3&4.

There are a number of ways in which the Authority can be structured, other than as a conventional government department. The possible models depend on the way in which the Authority is linked to the Government by its:

- sources of finance,
- political exposure and involvement,
- managerial control,
- adherence to Government policy,
- degree of underwriting of risk, and
- independence as a decision-maker.

Table 4.1 Pattern of road administration in various countries.

Pattern	Countries
1. Some national government funding and standard-setting; 2. State, provincial, or county Road Authorities; 3. Local government manages streets.	Australia, Austria, Brazil, Denmark, Germany, Greece, Netherlands, U.S.
1. No national government role; 2. State or provincial Road Authorities; 3. Local government manages streets.	Belgium, Canada, Switzerland
1. National Road Authority; 2. County, region, and/or local government manages streets and some highways.	Finland, France, Luxembourg, Spain, Sweden, U.K., New Zealand
2. County and/or local government manages streets and some highways.	Ireland

For instance, whilst formally and legally separated from Government, the Authority might still be totally owned by the State. Independence would then depend in part on the independence of the Authority's directors. It might also borrow money independently of Government, although in this case lenders would be particularly interested in the extent to which the Government guarantees loan repayments. Two key risks are:

1. the accuracy of any predictions of future traffic flows (Section 31.3), and
2. the funds invested are 'sunk' in the construction of the project, and cannot later be withdrawn or re-allocated.

Risk allocation is thus a critical component of all links between Government and the private sector.

Another variable in the separated Authority model is the degree to which ownership of the road asset, particularly the land involved in the right-of-way (Section 6.3), is handed over to the Authority. For example, what does the Authority own and for how long? In the completely privatised model, the Authority would own and operate all parts and components of a road network.

When the separated Authority has some level of private ownership, it is rare for Governments to grant the Authority full ownership of the road assets. For example, a toll road company may be required to give the entire road asset to the Government after a time (the concession period). This period is typically many decades and is set to at least allow investors to gain sufficient income from their investment.

4.3 NATIONAL ROAD TRANSPORT DATA

Table 4.2 is a collection of basic road-related numerical data from a range of countries which illustrates the role of roads and road transport using a range of indicators. The basic data are drawn from websites and the International Road Federation's *World Road Statistics* (IRF 2007), which is an annual volume giving data for the preceding five years for most of the world's 'road' nations. Other annual international transport data sources are the United Nations' *Statistical Yearbook* and the International Road Transport Union's *World Transport Statistics*.

Table 4.2 shows that expenditure on road construction and maintenance accounts for about 1 percent of most gross domestic products. As vehicle operating costs may be ten times construction and maintenance costs (Section 5.1), road transport as a whole can be seen to be a significant part of the gross domestic product of most countries. Typically, road transport expenditure is about 10 percent of gross national expenditure and road capital expenditure is about 20 percent of all public capital expenditure. A study by the World Bank in 1994 showed that a nation's stock of paved roads increased by 0.8 percent for every 1.0 percent increase in GDP. Factors other than finance also come into play. Demand for roads is also reflected by the need for those roads. As would be expected, road length per capita decreases as population density (per area) increases. The effect is less than linear. Studies of the effect of road investment on national economies (e.g. Aschauer, 1989) indicate that every 1 percent increase in spending on roads produces, approximately, a 0.25 percent increase in national productivity.

These effects tend to be self-perpetuating, as the data for developed countries suggests that road usage and per capita income tend to grow in tandem (Schafer and Victor, 1997). Specific details of personal expenditures on transport are explored further in Section 32.5.1. In particular, it is shown that households spend about 14 percent of their income on transport. Section 31.2.1 shows that people devote about an hour of each day to travel. Transport issues thus loom large in the lives of most individuals.

Table 4.2 Basic road data. From www and IRF (2007). The [n] refers to the year of the data, i.e. to year (2000+n). GDP = gross domestic product.

Item	Nation								
	Australia	Brazil	Canada	Germany	India	Japan	New Zealand	U.K.	U.S.
(GDP/person)/(GDP/person for U.S.)	0.82 [7]	0.21 [7]	0.83 [7]	0.75 [7]	0.03 [7]	0.73 [7]	0.59 [7]	0.77 [7]	1.00 [7]
Total road length (Mm)	813 [3]	1 750 [4]	1 410 [4]	231 [5]	3380 [2]	1 180 [2]	93 [5]	388 [4]	6 540 [5]
Motorway length (Mm)	1.82 [3]	–	16.9 [2]	12.40 [5]	0 [3]	6.92 [2]	0.17 [5]	3.48 [2]	75.40 [5]
Percentage of road network paved	41 [3]	10 [-10]	40 [4]	99 [-12]	47 [2]	78 [2]	65 [5]	100 [4]	65 [5]
Cars in use (M)	10.6 [4]	24.9 [4]	17.9 [4]	45.4 [5]	8.6 [3]	56.3 [4]	2.5 [5]	27.8 [4]	222.7 [3]
Cars/person	0.54 [5]	0.14 [4]	0.56[3]	0.55 [5]	0.01 [3]	0.44 [4]	0.61 [5]	0.45 [4]	0.46 [5]
Median car life (y)	10 [-3]	–	12 [-20]	10 [-20]	–	9 [-20]	18 [-20]	11 [-20]	9 [6]
Freight vehicles in use (M)	2.53 [5]	5.80 [4]	0.59 [3]	2.75 [5]	3.49 [3]	18.40 [4]	0.44 [5]	3.31 [4]	62.60 [5]
Total car usage (Tm/y)	171 [5]	2 [-14]	136 [-13]	578 [5]	204 [2]	526 [4]	37 [5]	398 [4]	2 720 [5]
Average travel distance per car (Mm/y)	14.7 [4]	–	17.2 [-13]	12.8 [3]	10 [-13]	9.7 [2]	12.7 [0]	16.6 [-1]	29.9 [1]
Total freight carried by road (Mt/y)	1 040 [-4]	–	137 [-1]	3790 [4]	5200 [?]	5900 [?]	–	1860 [4]	–
Road expenditure (G$/y) (U.S. $)	8.53 [5]	0.74 [4]	8.61 [1]	87.2 [4]	0.2 [3]	23.0 [1]	1.27 [5]	14.7 [4]	153.0 [5]
Road expenditure as a percentage of GDP	1.1 [7]	0.1 [4]	0.7 [7]	3.0 [4]	0.0 [3]	0.5 [3]	1.2 [5]	0.6 [4]	1.2 [5]

CHAPTER FIVE

Road Needs

5.1 THE DEVELOPMENT OF ROAD PROPOSALS

The sequence on which the arrangement of this book is based is explained in the continuous loop in Figure 1.1. This chapter discusses how the need for a road is assessed. This must be done in the context of the planning of the larger transport system, of which the road-based route is just one part. Transport planning is examined in Chapter 31 which covers (Sections 31.1–3) the full range of transport facilities — systems and networks, corridors and sectors, and individual routes. Section 31.4 shows that one of the key outputs of transport planning is the production of specific proposals for the expenditure of funds to bring about changes to a transport system, raising its existing condition to some improved level of service. Such studies can have time horizons of ten years or more. The resulting proposed facilities must satisfy overall and specific community and industry needs and constraints.

The study will occur at three levels. At a broad regional level it will assess whether a transport network or system improvement is justified. Section 4.3 indicates that road investments usually have a very positive impact on national productivity. The next level of analysis will consider the transport corridors which make up the network. These corridors represent the areas directly affected by those routes in the network which need improvement and are typically about a kilometre wide in an urban area and ten kilometres wide in a rural area. Corridor studies usually have a five to ten year time horizon, with the emphasis on the analysis of alternative sets of proposals for the corridor.

Finally, specific preferred routes within a corridor will be selected as road proposals in accordance with the principles in Sections 31.3.6–7 and 32.1. This chapter, which begins the road-oriented part of the book, discusses methods for the selection of a preferred proposal from a range of options and for prioritising those proposals into approved lists of road projects, meeting previously defined objectives and constraints. Each project will usually constitute a convenient construction package (Section 25.1) aimed at delivering a part of the approved proposal.

In beginning the assessment process it is firstly necessary to define the objectives of and constraints to any proposed change and then to determine how effectively the proposal meets those objectives and constraints. The decision as to whether resources should be used for a transport proposal is then based on:

* the objectives met and constraints satisfied,
* the resources consumed, and
* the merits of the competing needs, as any expenditure will require the diversion of resources from other competing needs. The decision-maker must balance the benefits and net value of an improved transport system against the net advantages to be gained by using the resources elsewhere.

Methods for doing this are given in the following section.

The objectives, consequences, and constraints which must be taken into account in the development of a new road proposal are listed in Table 5.1 where they are related to

Table 5.1 Objectives and consequences of road projects.

Item number	Objectives and direct benefits	Directly influence	Indirectly influence	Magnitude of benefits[a]	Costs	Cross-reference to Section
Part T, Travellers as the stakeholders						
T1	Minimise vehicle operating costs and cargo damage	freight haulers, travellers	public transport, regional[b] economy	15% of urban, 30% of rural	remaining operating and freight costs	29.1 30.2
T2	Minimise distance, time, and time variability of private vehicle trips	car travellers	regional economy, land use, urban growth	10% of benefits, 15% of costs	remaining travel costs, takes traffic from other routes	31.2
T3	Minimise distance, time, and time variability of commercial vehicle trips	freight hauliers and logistics, bus operators	local and regional economy, land use	30% of urban[c], 20% of rural benefits	creates extra demand	
T4	Minimise cost of vehicular and pedestrian crashes	all road users, travellers	regional economy, local groups	10% of urban, 5% of rural	remaining crash costs (Table 28.8)	28.8
T5	Maximise travel comfort and convenience, reduce driver stress	improved level of service, travellers	travel by road, and on other modes	indicated by patronage	higher operating costs	17.4 31.3 31.5
Part R, the Road Authority as the stakeholder						
R1	Minimise property acquisition, relocation of occupants	property owners, tenants, local politics	public attitudes	reduced property costs	property cost	6.3
R2	Minimise construction costs	Road Authority	regional economy	reduced costs	construction cost	25.2
R3	Minimise maintenance costs				maintenance cost	26.2
R4	Minimise traffic operations costs	Traffic Authority			operating cost	26.4
R5	Provide facility to meet future needs	land use, Road Authority		stable land use	planning cost	33.5
R6	Maintain road conditions	Road Authority	traveller	typically BCR>2.5	reconstruction	26.1

Table 5.1 (continued)

Item number ⇓⇓	Objectives and direct benefits	Directly influence	Indirectly influence	Magnitude of benefits	Costs	Cross-reference to Section ⇓⇓
Part C, the community as the stakeholder						
C1	Maximise economic investment	investors	everyone	exceed costs	Section 29.1.1	29.1
C2	Create construction activity	local workers	local business	construct'n dollars	Table 32.1 (1.1)	25 30.1
		have a *multiplier* of about 3 (industry, retailing) and 0.8 (employment)				
C3	Minimise adverse effects of planning blight and property acquisition	nearby land users	local growth, job creation	reduced compen-sation	measured losses, Table 32.1 (1.2,2.1–2)	4, 6.3, 32.1
C4	Minimise impact of construction on congestion, local access, and the environment	road user, local people, environ-ment	regional economy		disruption cost, Table 32.1 (1.4)	25.932 .1
C5	Provide a new traffic facility, thus raising accessibility	new and existing industry, tourism, retailing, transport demand	jobs, market options, regional economy, taxes, land use	land values and economic activity (30% of benefits)	reduced travel on other routes and modes, new travel demands	30.1–2 31.3
C6	Improve or extend transport network and operations					32.1 33.5
C7	Minimise personal disturbance to citizens, particularly loss of jobs, housing and parks	nearby land users	local groups		(1.4,2.3–5 & 2.6–8)	32.5
C8	Minimise social and cultural disturbances and community severance	lifestyle of local people		reduced compensation	cost of effects, Table 32.1	32.1 32.5
C9	Minimise disruption to public services and utilities	govern-ment	utilities		losses, Table 32.1 (1.4)	7.2.3,3 2.1
C10	Minimise permanent physical disturbances and environmental degradation (e.g. ground-water changes) and achieve sustainability[d]	local people	tourists	land values	visual losses, Table 32.1 (1.2, 1.6)	6.3–4, 32.1

Table 5.1 (continued)

Item number					Cross-reference to Section	
⇓ ⇓	Objectives and direct benefits	Directly influence	Indirectly influence	Magnitude of benefits	Costs	⇓ ⇓
C 11	Improve visual appearance					
C 12	Improve and add recreational facilities			local well-being	n.a.	6.3
C13 & C14: Minimise effects and nuisances caused by traffic effects and nuisances (e. g. community severance, residential intrusion, noise, air pollution, vibrations, crashes due to speed, crossing delays) due to:						
C 13	Existing traffic	nearby land users	land use	10% of urban and 15% of rural benefits[e]	social and human costs effects, Table 32.1 (3.1–7)	25.83 2.1–4
C 14	Increased traffic and traffic diverted by the final Project	abutting land owners, current road users	local travellers		decrease in land values, increase in local travel costs	

Notes
a. Projects justified by traveller benefits usually have BCRs>2 (Section 5.2.3).
b. Effects both industry activity and incomes.
c. The total time savings are shown as 60 percent of urban benefits. In heavily-developed networks, this can rise to 80 percent.
d. This is usually via the environmental impact process (Section 32.1).
e. Includes benefits from R3

the interests of the various stakeholders in the process. In addition to the break-down given in Table 5.1, in some instances it may be useful to categorise any project impacts as 'short term' or 'long term'.

Many of the consequences and constraints in Table 5.1 can be quantified in terms of the benefits and costs which would flow if the proposal became a completed set of road projects. Thus, the table gives the magnitude of the individual benefits flowing from a road project as a percentage of the total benefit to be derived from the project. An assumption that underlies the Table is that the prior planning and analyses that have occurred will ensure that any constructed scheme is economically warranted in that:

* it would return net benefits to the community (BCR>1, Sections 5.2 and 33.5.8), and
* delaying the expenditure would not produce a greater net benefit.

This emphasis is important as roads are a relatively unique commodity in that, except for some cases of tolling and road pricing discussed in Section 29.4, there are none of the normal pay-as-you-use consumer preference signals to ensure efficient investment and usage.

For context, a typical capital cost of a road might be about 14 c/passenger-km or 35 c/tonne-km, or about 27 k$/m for a heavily-used 6-lane road. About 60 percent of the road authority's cost will be the construction cost, 20 percent will be the land costs, and 20 percent the net present value (NPV, Section 33.5.7) of the operating and maintenance costs.

However, the major cost component of a road over its life will usually come from the operation of the road, rather than from its construction cost. For instance, the annual total vehicle operating costs (Section 29.1.2) for a piece of road with over 500 vehicles per day (veh/d) will typically be four times — and can be up to ten times — the cost of maintenance and the annualised cost of construction (Section 33.5.6). Systems to explore this are discussed in Section 26.5.2. As a result of this multiplier effect, the benefit–cost ratios for road construction schemes frequently exceed ten or more.

The benefits listed under 'Objectives' in Table 5.1 (e.g. reductions in the costs of vehicle operations, trip time, and crashes) are direct benefits. The effect on the overall economy of the expenditure of capital on a road project will be through its impact on such factors as incomes, employment, production, tax revenues, resource consumption, and pollution (Chapters 32 and 34). These are covered in the 'Indirectly influences' column in the Table.

One feature that may not be directly evident from Table 5.1 is that the actual impacts of a road project may often occur on a project-by-project rather than a regional basis. Generally, any unfavourable impacts will occur close to the project and be evident early in its life, whereas benefits will be more widespread and slower to occur. Indeed, it has been suggested that there may be a five-year lag before the full benefits of a road investment begin to flow.

5.2 ASSESSMENT OF PROPOSALS

5.2.1 Proposal review

To be acceptable, a proposal to invest in a road system must deliver the following outcomes in that it will:

1 Basic objectives (Chapter 31):
- 1.1 overcome an accepted deficiency,
- 1.2 provide a necessary facility,
- 1.3 be better than doing nothing (e.g. using a consumer-surplus analysis, Section 33.5.5)

2 Constraints:
- 2.1 be sound with respect to social factors (Section 31.1),
- 2.2 be sound with respect to environmental factors (Chapter 32),
- 2.3 meet sustainability criteria (Chapter 32),

3 Suitability:
- 3.1 be efficient (Section 30.2.2),
- 3.2 be effective (Section 30.2.3),

4 Feasibility:
- 4.1 be appropriate for the local situation (Sections 5.2.2–3), and
- 4.2 be technically (Chapter 6), socially, financially, and legally achievable.
- 4.3 be affordable.

Criteria based on achieving these outcomes can be combined with the objectives and influences in Table 5.1 to give the composite Table 5.2. The criteria can each be regarded as separate filters that a project must pass through in order to proceed to construction and operation. Another concept is that of 'cascading' in which projects that pass one criterion then cascade onto the next, until all are passed.

Table 5.2 Steps and stages in the review of a road proposal.

	1 Basic objectives (Chapter 31)	2 Constraints (Chapters 31 & 32)	3 Suitability (Chapter 30)	4 Feasibility (Chapter 6)
Acceptance criteria	**1.1** overcome a deficiency (T1–5)	**2.1** meet social needs (R1, C3–9,11–13)	**3.1** is effective (Section 30.2.2)	**4.1** is relevant (C2,5&6)
	1.2 provide a new facility (T1–5, C1)	**2.2** respect the environment (C10–14)	**3.2** is efficient (Section 30.2.3)	**4.2** is achievable (R1–5)
	1.3 maintain a facility (R6)	**2.3** meet sustainability targets (C10)		**4.3** is affordable (R1–3)
Stage in project	**A** State task or problem			
	B List options and select best			
	C Design chosen option	Check each constraint	Verify and test design	Confirm feasibility
	D Revise project	If a check is not passed, a new option must be selected or the project abandoned.		
	E Build chosen design	Check performance of completed project		

For T1 etc., refer to Table 5.1.

Stage B in Table 5.2 implies that the full range of reasonable options, or courses of action, will be examined. The general assessment of proposals will first be conducted using the broad techniques discussed in Chapters 31 and 32. In particular, a transport systems analysis in the terms of Section 31.3, and the subsequent application of Criteria 1 and 2 to its results, will mean that the problem or task to be addressed and its proposed solution are both clearly defined. These cover the engineering, economic, social, environmental, and planning aspects of the proposal. Criteria 3 and 4 then determine whether the proposed solution is appropriate and rely on procedures described in Chapters 6 and 30.

When a set of acceptable proposals for a suite of future projects has been established, the next road planning requirements are to determine the priorities for these projects and the funds to be allocated to them. This is done in terms of the overall objectives described in Table 5.1 and the specific needs of the local road network. Next, a benefit–cost analysis is commonly used to make monetary comparisons between specific proposed projects. This process is described in Section 5.2.3 below. If only incremental changes to the road network are being considered, Section 33.5.8 will show how their economic aspects can be explored by a comparison of marginal benefit–cost ratios.

Whilst all these processes are in train, every effort should be made to reduce the period of community uncertainty, concern, and conflict, and to minimise the effects of planning blight (Section 6.3).

5.2.2 The assessment system

Having tentatively selected a proposed road scheme, its merits can be assessed by modelling its operations over its intended life (Sections 14.6.2 & 26.5.2) and predicting the cash flows needed to maintain and operate it over that life. The question that can then be addressed is:

'During its intended life, will the proposed road and its associated regime of operations, maintenance, rehabilitation, and improvement:

* deliver its intended traffic and economic benefits and stay within the given social and environmental constraints, as measured by pre-defined evaluation criteria (or performance standards),
* satisfy the Road Administration's policy directives,
* be within budget,
* be insensitive to possible changed circumstances (e.g. how would the outputs vary if the underlying assumptions varied within feasible limits), and
* address new issues raised by external parties during the proposal stage, particularly via the environment impacts process (Section 32.1)?'

This is a broad scope and to be useful, assessments need to focus on the key issues and objectives involved (Section 31.1) and on the specific decisions that will face the relevant decision-maker. In addition, the evaluation must somehow draw together all the disparate issues discussed above — methods for doing this are outlined in Sections 5.2.3, 32.1.2 and 32.1.3.

A flow chart for resolving these questions and processes within a governmental structure is shown in Figure 5.1. The chart provides a logical path for policy decisions and, importantly, includes a mechanism for the adjustment of the performance standards. Note that the chart is for a specific road proposal and presupposes that the system, network, corridor, and route decisions discussed in Section 5.1 have already been made.

In particular, Figure 5.1 shows how the basic decisions stem from Government setting policy directives and funding levels for its Road Authority. An annual suite of road proposals will spring from these directives and funding levels. A model corporate structure would allow the Road Authority to proceed with a proposal that meets directives and funding levels. The Authority would only return to Government if it found that the funds allocated did not match the objectives and policy directives given. Government may also act at a project level as well as at a policy level. Many governments are more interventionist than the above model. For example, for pragmatic electoral reasons many governments will be concerned to see, and perhaps influence, the projects in the annual road program. It is essential that such Governments be presented with all the relevant factors, the viable options, and the implications of action and inaction. The next Section describes a framework for doing this.

5.2.3 Assessment methods

Assessments must address the issues raised in Section 5.2.2. Many of these are not quantifiable and decisions must be based on judgement rather than calculation. However, whether a road proposal is economically warranted can be assessed in a numerate manner by a benefit–cost analysis (BCA), usually performed by the Road Authority in selecting projects (Figure 5.1). The theory behind BCA is discussed in detail in Section 33.5.8. Basically, it compares the benefits and costs of a project over its life, calculated using the

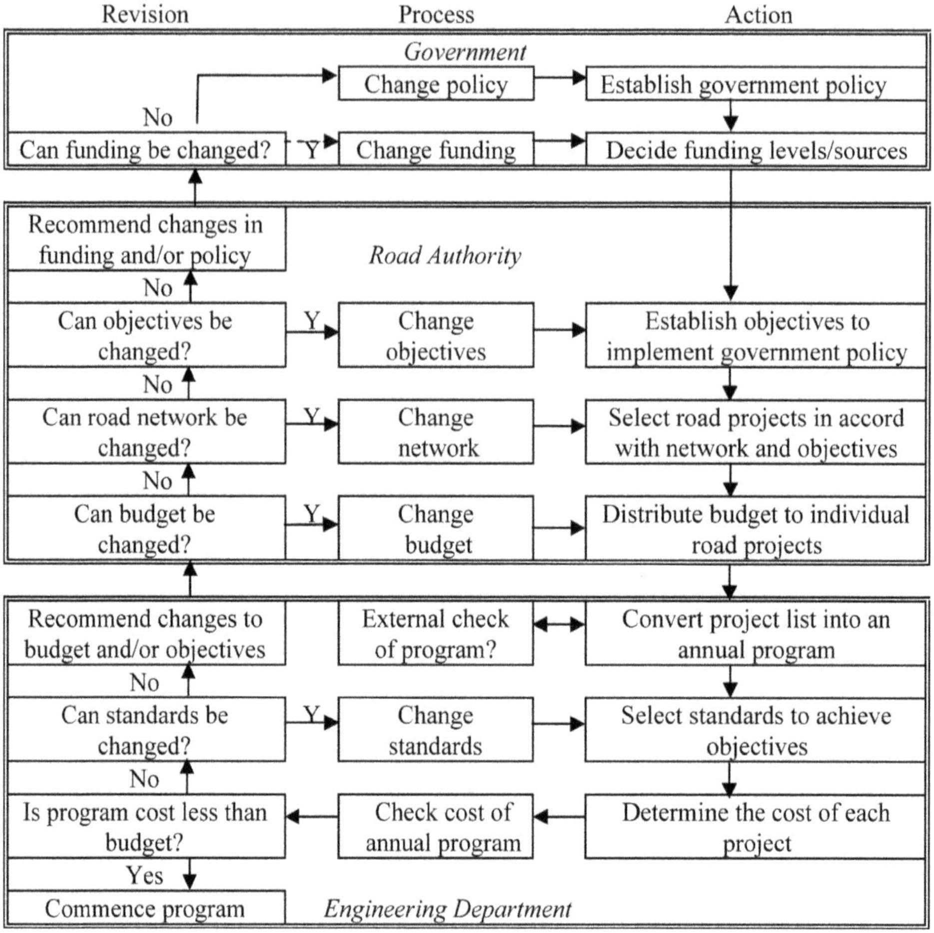

Figure 5.1 Government / Road Authority / Engineering decision-making process.

net present value (NPV) method. Benefit–cost analysis provides a powerful method for ranking and prioritising proposals in order to ensure the optimum economic allocation of the available resources. Because BCA focuses on monetary factors, it is a single criterion method using the benefit/cost ratio (BCR) as the criterion. Following Figure 5.1, the standards adopted can also be set via a benefit–cost analysis.

The first task in applying a BCA is to identify the appropriate benefits and costs. For a typical road project, these are given in Table 5.1. The next major task is to measure the identified benefits and costs, particularly the more intangible ones, and assign money values to them. The listing of benefits and costs in Table 5.1 reveals the major distributional problem associated with BCA in that those who gain the benefits (e.g. passing motorists) are usually a different group from those who pay the costs (e.g. local residents). Distributional factors cannot be properly included in a benefit–cost analysis and must therefore be separately assessed for each project. Another more general distributional problem is that the benefits may not be distributed in a manner consistent with the principles of welfare economics — that is, the benefits may go to people considered relatively undeserving in the eyes of the community. This can partly be overcome by weighting the benefits and costs to give constant marginal utility (Section 33.5.1) to all recipients. The weights would be determined in accordance with income utilities and the community's views of equity (Stanley *et al.*, 1973).

If a benefit–cost analysis is either not done (typically because of either the assessment, quantification or distributional issues mentioned above), or considered inadequate (typically because it struggles to cover social and environmental factors), then the ranking and prioritising process must rely on the decision-maker's judgement alone or on that judgement aided by defined processes known as multi-criteria (or sufficiency or deficiency) analyses (MCA). Sufficient information must be presented to allow the selection of a proposal to be based on adequately informed decision making. Nevertheless, the MCA process will still usually require the decision-maker to exercise considerable judgement and perception (see also Section 32.1.3).

An MCA begins by tabulating all the proposals and all their associated objectives. The objectives will include positive performance targets, assessment criteria, and the degree of minimisation of issues, impacts, and consequences (Table 5.1). The goals achievement analysis version of an MCA then lists how well each proposal meets the various objectives. A second MCA method is the elimination procedure which uses the tabulated objectives to compare individual proposals progressively and in pairs, to determine which projects tend to naturally dominate the others. Projects which are clearly out-performed by other projects are eliminated, leaving a smaller list for final selection.

One sub-method then gives a score between 0 and 10 to indicate how well each performance target is met. The planning balance sheet (or decision framework, Section 32.1.3) method then assigns relative weights to each objective. The scores are multiplied by these weights to give an overall rating of sufficiency. A pavement example of the method is given in Section 14.5.1. A social audit (or social impact study) can be conducted to check how well the project has met non-quantifiable targets such as:

* social, distributional, and equity issues,
* environmental issues,
* wider policy issues (e.g. defence and national development),
* externalities such as pollution, congestion, and subsidies (Section 33.5.2), and
* local issues such as significant changes in a community's well-being or concern (Section 32.5).

An approach for remote regions accepts that roads in these areas provide a social service and assumes that a certain amount of money is available to meet this social obligation. It then ranks the road schemes in order of their priority for the use of that money. The ranking can be based on such objective criteria as the resident and tourist populations served, the perishables saved and the freight tonnage moved. For example, the technique could positively weight the accessibility that the road provides for social, recreational and medical services and deliveries, the benefits from tourism, and the value of the export of perishable produce.

5.3 BUDGETING AND PROGRAMMING

Following the assessment of road needs and priorities, long-term budgets must be prepared for periods of up to ten years. This advance programming is necessary to:

* indicate the timing and extent of the funds needed to achieve the program objectives, and

* provide the means of integrating programs prepared by various branches within an organisation (e.g. planning, design, land acquisition, construction, operation).

This long-term programming period divides into seven stages:

(a) a conceptual and planning phase of the type discussed in Sections 5.1 and 5.2,
(b) the preliminary design stage (Section 6.2),
(c) an environmental studies, statutory procedures, and assessment stage (Section 32.1),
(d) a land acquisition stage (Section 6.3),
(e) the detailed design stage (Sections 11, 19 etc.),
(f) the construction stage (Chapter 25), and
(g) the maintenance phase (Chapter 26).

It is normal for the process through the first six stages to take at least six years. This can readily be lengthened to the ten years mentioned above by the demands of community involvement (Section 32.1.4) and of drawn-out statutory procedures.

Once a decision has been made during or after stage (c) to make a work commitment, then short-term programming and budgeting of the type described in Section 25.2 begins. For detailed actions in this area, it is useful to work from a standard list of jobs and activities. For construction, the broad headings in Table 25.1 may prove useful. Similar lists for maintenance purposes are developed in Section 26.1.

CHAPTER SIX

Road Location

6.1 RURAL AREAS

6.1.1 General

Chapter 5 discussed how the need for a road or a road improvement is established. The need commonly relates to various unfulfilled road transport demands within a planning corridor linking two transport nodes. Indeed, Section 5.1 described a form of geographic scaling wherein proposals to meet that need are progressively considered at regional, corridor, and route-related scales:

(a) at the regional level, alternative corridors are considered,
(b) at the corridor level, alternative alignments are considered, and
(c) at the route or detail level, alignment variations are considered.

The interaction between the three scales is shown in Table 6.1, which is a development of Table 5.2.

Table 6.1 Road location matrix

Stages⇒	Planning	Conceptual and functional design	Detailed design
Processes⇒ Scales⇓	Assess needs, establish purpose	Consider constraints, assess all options and select the best	Provide drawings, check suitability
a. Region	yes	–	–
b. Corridor	yes	yes	–
c. Route	–	yes	yes

Level (a) was discussed in Section 5.1. The investigatory and approval process outlined in the remainder of Chapter 5 tackles Level (b) and will usually nominate a preferred route within a corridor. It will also determine the extent to which the initiating transport need is to be satisfied; for example, by specifying such key factors as the traffic flows to be used in design. The next step is to locate the specific road within the broad zone embraced by the preferred route. This is Level (c) and is the topic of this chapter. As Level (c) relates to detailed construction drawings it is also discussed in Section 25.3.1 and in various specialist chapters such as Chapter 19.

The fundamental objective of locating a new road within a broadly specified route is to find a road alignment that will satisfy the design specification — particularly traffic flow and speed which respectively determine the carriageway geometry (Section 6.2.3) and road alignment (Chapter 19) — within the constraints of:

* staying within the specified route (typically a planning reservation),
* minimising project and whole-of-life costs (Chapters 31 and 34),
* maximising travel safety (Chapter 28), and
* protecting and enhancing the environment (Chapter 32).

This chapter concentrates on the location of rural roads, as in this instance most of the design variables are free and unconstrained. In urban design there are many additional constraints, usually in the form of community aspirations, the high cost of land, and the need to interlock with existing developments such as local levels, drainage facilities and underground services. These urban aspects are discussed in Sections 7.2.1 & 7.3.2. Road location in rural areas proceeds in four stages:

(1) reconnaissance of the corridor to disclose potential locations,
(2) preparation of alternative design solutions,
(3) assessment of the alternatives, and
(4) selection of the best road location.

In stages (2)–(4) the candidate locations must pass through a series of sieves based on various criteria, culling out particular locations for reasons such as the requirements of planning laws, environmental constraints and the presence of insurmountable physical conditions. In these stages, the best road location will follow the shortest path between its two ends.

However, there are many real constraints on this measure. First, the road cannot be too steep nor its curves too sharp for motor traffic (Chapter 19). Second, the cost of earthworks, drainage and bridges must be less than the benefits that they deliver (Chapters 8 & 15). Third, community and environmental impacts must be at acceptable levels (Chapter 32). This makes road location a complex exercise in optimisation which frequently involves considering many alternatives and iterations. There are some proprietary programs such as Align3D (Gipps, 1992) that assist in this process. They typically require ground terrain data in digital form.

The type of information needed to permit the alternative solutions developed in Stage (2) to be considered must include both current and forecast future conditions in the following four categories:

(A) *Physical conditions*: Information is obtained from maps (Section 6.1.2), photographs, site visits, drill cores, records of soil types and profiles (Section 8.4.1), agricultural data, hydrological data (stream flow and drainage, Sections 13.2 & 15.3.3) and inventories of available construction materials. Data must also be obtained on such factors as restrictions on access and working times, and the supply of water, drainage, sewerage, electricity, fuel, and communications. Climatic records must be scanned for flooding data (Section 15.3.3), high temperatures (Section 12.2), and low temperatures and freeze–thaw cycles (Section 9.5).

(B) *Demographic conditions*: Data is needed on population levels; demographic distributions, local employment generation; housing numbers; socio-economic status and car ownership of the residents (Chapter 31); journeys to work and their modal split; and land use. This data must include forecasts of future conditions.

(C) *Traffic*: Data is needed on trip origins and destinations; average and peak traffic flows, composition and growth rates; crash data; and vehicle speeds (Section 6.2.3 & Chapter 31).

(D) *Environmental and planning issues*: There is a need for catalogues of flora and fauna, landscape data, significant sites, maps of existing and permitted land uses, and lists of susceptibilities to the effects of the new road and its traffic (Section 6.4 and Chapter 32).

Data related to the factors related to these four categories can be obtained from the following sources:

* the relevant Road Authority,

* various government departments. For example, climatic data are obtainable from meteorological bodies and hydrological data from river and water Authorities.
* local government,
* special interest groups and other stakeholders, and
* community groups.

As items (B), (C) and (D) above are discussed in other chapters, most of the remainder of this chapter will concentrate on the evaluation of relevant physical data referred to in (A).

6.1.2 Maps

Maps come in a vast range of scales and types. The accuracy of the maps needed increases from (approximately) 1:1 000 000 for regional studies, to 1:100 000 for broad location studies, to 1:50 000 for corridor studies, to 1:20 000 for detailed location work, to 1:5000 for preliminary design, to 1:1000 for construction drawings. Similarly, whilst a location drawing might only show a centreline, a preliminary design might give alignments to ± 5 m and a construction drawing must specify all details to within a site tolerance (perhaps ± 5 mm).

Road technology typically uses geodetic maps, satellite imagery, geographic maps, planimetric maps, aerial photographs and photomosaics made from them, geological and agricultural maps, topographic, contour, and military maps and cadastral maps. Each of these various map types will now be discussed.

Geodetic maps cover such large areas that they take the spherical shape of the earth into account.

The original *Landsat satellite images* are based on U.S. satellites that remotely sense radiation from the earth's surface. The images come in various scales but are commonly at 1:1 000 000 covering about 33 000 km^2 with a minimum resolution of around 80 x 60 m. They are thus sufficiently detailed for gross road location purposes. There resolution is such that they do not show individual houses. They have been largely replaced commercially by images from other satellite, typically with 3 600 km^2 images and a resolution of around 10 x 10 m. More recent products such as Spot Image from commercial satellites provide a resolution down to 2.5 m which is suitable for many planning purposes.

Planimetric maps cover much smaller areas and provide a simple two-dimensional representation. *Geographic maps* are enhanced planimetric maps used in atlases and for road maps and are usually at about 1:250 000.

Photogrammetry is a means for obtaining ground measurements by means of photography and stereo photogrammetry means using two photographs of the same surface taken from different camera positions (see Lawrance *et al.*, 1993). The technique is therefore very similar to the manner in which human binocular-vision operates. Photogrammetric mapping relies on the stereoscopic examination of pairs of aerial photographs in the manner illustrated in Figure 6.1. The photographs are taken at a known altitude and camera separation. The figure shows how similar triangles can then be used to calculate contours. The technique is necessarily coupled with control ground surveys of specific locations. Prediction of horizontal distances requires the knowledge of some horizontal distance on the ground in order to establish the map scale. In practice,

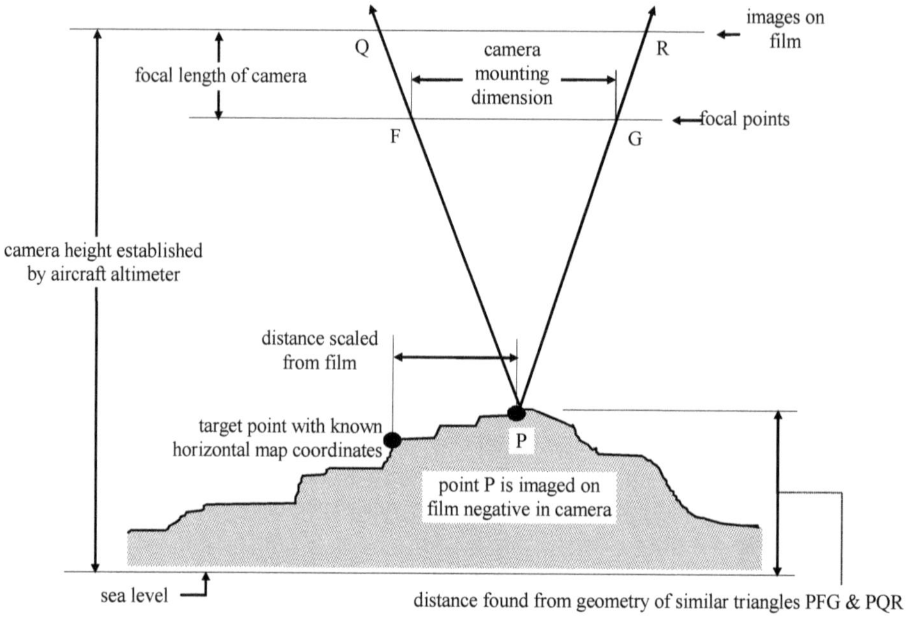

Figure 6.1 Basis of photogrammetric mapping.

other requirements often result in the need for further ground control data with respect to horizontal and vertical co-ordinates. The vertical accuracy is about ±10 m.

As well as providing contour information, the aerial photographs can be used to locate land features such as slopes, and stream beds, and to aid the recognition of geological jointing, fault zones, folds, rock outcrops, landslides, and old flood plains. Photographs for such purposes may be in black and white or colour and are commonly at scales between 1:10 000 to 1:100 000, covering about 400 km^2.

The most widespread map system for supplying basic information on the road network is the *topographical map* that adds vertical (hypsographic) data to the planimetric map. They thus give the position and shape of natural features and may also show many man-made features.

The commonest topographical map is a contour map which shows lines of equal altitude (contours) and thus gives the information needed for geometric design of a road in the vertical and horizontal planes (Section 6.2). It can also provide a first indication of likely sources of materials and of such problem areas as swamps and streams. Contour maps can be produced by conventional ground surveys or by photogrammetric mapping. Lines called hachuring can also be used as symbols to show the steeper side-slopes of hills or mountains. The military map is usually a contour map enhanced by the addition of most other ground features. It can thus be of great value.

A *geological map* indicates the physical features on the land surface and may suggest their relation to the underlying geology (geomorphology) of an area. It will indicate the presence of hard or soft rock, but will rarely give data about such unconsolidated material as alluvium. As certain rocks on weathering produce characteristic soils (Section 8.5.10), the geological map may give indications of these. For example, heavy clays frequently relate to basalts and light soils to sandstones. Lithology, the science of the composition of rocks (Section 8.5), can be of help in making such deductions. The geological map will also indicate sources of construction material. Existing site investigation reports (Section 6.2.2) are valuable adjuncts to a geological map. These may be obtained from construction and design agencies that have operated in an area.

Geological data is also gathered by such geophysical methods as seismic refraction and electrical resistivity measurements. Both can be used to indicate rock depth and strength and resistivity can also indicate soil type and the presence of gravel deposits. Seismic refraction is based on the travel times of seismic waves created by an explosive charge. The electrical resistance method measures the apparent resistance of the material when subjected to an electric potential. These two methods are only fully effective where significant contrasts exist between physical characteristics and where large areas need to be examined rapidly. They can be made more reliable when calibrated against drill cores or excavation records.

Data for inaccessible areas can be gathered by thermal, infrared sensing from an aircraft or satellite. This method measures the radiometric temperature of the ground to a resolution of about one square metre. Rock near the surface, groundwater, and vegetation can be detected.

Agricultural soil maps provide formal (pedological) classifications (Section 8.4.1) of the surface soil. Such maps must be used with care as soils with different classifications may be similar from an engineering viewpoint.

Land acquisition and management requires *cadastral maps* which are planimetric maps showing property boundaries and land ownership (title) details and any existing man-made services and facilities. They are typically very large-scale at 1:200. Although they often also show planning zones, these are usually more usefully consulted on smaller-scale planimetric maps.

Section 26.5.1 discusses map-like reference systems now widely used for managing data associated with specific locations on existing roads. On occasion, these can also be of value when new roads are being located near existing roads. Maps are increasingly being stored and manipulated using geographic information systems (GIS). GIS is a tool that also provides such database management tools as storage, querying, and reporting. As with CADD (Section 6.2.2), GIS stores its map data in a series of layers, each covering one particular class of information. The topological data within GIS includes intelligence about the spatial relationship between adjacent objects (e.g. where a line begins and ends). Lower levels of GIS are usually either automated mapping related to CADD or thematic mapping, in which attributes are added to map features.

6.1.3 Terrain

Terrain is defined as the topography, hydrology, drainage, geology, pedology, and vegetation of an area. It thus covers all the properties in (A) in Section 6.1.1 and embraces all the physical data related to road location. Areas of land occur which are of relatively uniform terrain and which often form part of a recurring pattern. Thus, terrain maps can be produced delineating areas with common identifiable parameters. This is most useful for large projects, giving a consistent basis for the evaluation of alternative routes and sources of roadmaking materials, prior to undertaking site investigations. It can also be of value for assessments in categories (B), (C) and (D) in Section 6.1.1.

In road layout, terrain particularly refers to the vertical alignment of the route. Typically, terrain is considered to be level (or flat) if the route encounters less than ten 5 m contours every kilometre or gradients of 3 percent or less, rolling when between 10 and 25 are encountered or gradients between 3 and 12 percent, and mountainous when more than 25 or gradients of over 12 percent are encountered. Upper limits on gradients for roads are given in Section 19.3.

A terrain classification system can be used to identify and delineate the similar repetitive terrain types, collecting all available information on a logical basis. The four common classification terms are components, units, patterns and provinces. A component is a local facet that is unique in terms of slope, geology, soil, and vegetation. It gives data on local accessibility, flooding and the suitability of materials for roadmaking.

A terrain unit (or facet) is an association of components forming a recognisable and distinct landform and which all have characteristic physiography or soil and vegetation associations. Road alignment, cuttings and embankments (Section 11.8.1), and roadmaking materials would be consistent within a terrain unit. A typical terrain unit might be 'ridge'. A simple classification for roads would involve, for instance, ignoring the 'ridge' units for layout but considering them for quarrying purposes. Indeed, terrain units are usually classified in terms of topography only, as this factor has the dominant highway impact. Mapping would be at 1:50 000. Terrain units are also recognisable on aerial photographs with scales ranging from 1:8 000 to 1:25 000 (Davies and Eades, 1980). Units identified from maps or aerial photographs should be checked by sample ground reconnaissance.

A pattern is an association of units. It can be based on geomorphology, drainage patterns, or common characteristics of the soil, rocks, or vegetation, quantified by such items as contours or stream flows. It will usually give data on the extent of earthworks needed and on numbers of bridges and culverts. Mapping would be at 1:250 000. A province is an association of patterns underlain by a constant suite of rock of a defined kind (Section 8.5). Contour, geological, and pedological maps are usually sufficient to allow province boundaries to be defined. The latter two require extensive pre-surveys of the area. A photomosaic of aerial photographs can also be examined to find areas of similarity with respect to surface form, geological features, topography, drainage, tone, and texture of the land surface (Davies and Eades, 1980).

Studies of the type described above allow preliminary road designs to be developed, major engineering problems to be noted, and preliminary economic evaluations to be undertaken. They are also used to check interpretations of aerial photographs, particularly at delineated boundaries, and to record geological exposures providing information on the underlying rock. A useful manual on terrain evaluation is available (Lawrance *et al.*, 1993).

Similar classifications can be produced from the viewpoint of landscape ecology. In this case the land mosaic viewed, as in an aerial photograph, can be seen as composed of a background matrix in which might be found various landscape patches, corridors (or strips) and barriers. The patches will have their own characteristics and will include specialised species living in their interior and species able to survive in the diversified zones at the edges of the patch. From a habitat viewpoint, the corridors might merely be 'stepping stones' through which animals may move between the patches.

6.2 GEOMETRIC DESIGN

6.2.1 General

Geometric design is the dimensional design of a road within the surrounding terrain, including the determination of all surficial dimensions but excluding dimensions related to pavement thickness and structure. These exclusions are covered in Chapter 11. The constraints on geometric design must include the fact that the road is only one interacting part of the total road transport system. The other two parts are the vehicles (Chapter 27) and the drivers (Chapter 16). The design will only prove successful if this total system performs satisfactorily. The art of geometric road design is the three-dimensional combination of the elements that establish the road layout:

* horizontal and vertical alignment,
* pavement and median width, cross-fall and camber (Section 6.2.5 & Figure 2.2),
* pavement type (Chapter 11), and
* intersection form

in such a way as to ensure that the road is an appropriate, efficient and effective part of the transport system.

An all-embracing concept that has arisen to ensure that a design is appropriate, efficient and effective is *contextual design*. This implies a collaborative, interdisciplinary approach to road design that ensures that, whilst all the technical needs described in this book are met, the design is also adapted to the overall physical and social context (or setting) in which the road is found, and at least preserves and preferably enhances the surrounding aesthetic, environmental and cultural resources. Much assistance in achieving this goal is given in Section 6.4 below and in Chapter 32 on environmental factors.

6.2.2 Preliminary alignment

Once the mapping, terrain evaluation and initial reconnaissance surveys discussed in Section 6.1 have been completed, a number of preliminary road alignments are developed, based on centrelines and carriageway widths. A ground-based feasibility study is then undertaken along these alignments. Ones leading into areas with obviously-difficult conditions should be immediately reconsidered and steps taken to find alternative alignments.

When a feasibility study has confirmed the potential of a particular proposal, detailed site investigations should be undertaken. These will systematically determine all the conditions relevant to the design, construction, and performance of a road based on the proposed alignment. A detailed flow chart for site investigation is given in Figure 6.2. Care is needed at this stage as a review showed that half the extra costs in British road contracts between 1957 and 1977 were due to inadequacies in site investigation and planning (Tyrrell *et al.*, 1983). Typical items of geotechnical concern that might be raised by a site investigation are:

* potential areas of slope instability (Section 11.8.1),
* areas of hard rock (Section 25.4),
* high water-tables (Section 9.2.3), or
* flood-prone areas (Section 13.2.2),
* areas of poor material which would need to be removed from the site.

The next step is to determine which of the proposed alignments minimises the sum of the whole-of-life construction (Chapter 25), maintenance (Chapter 26) and road-user (Chapter 29) costs of the proposed road, whilst satisfying the various planning and environmental constraints appropriate to the area. Firstly, broad costings are used to narrow the choice down to a specific ribbon of interest. Then a detailed costing will allow a specific alignment to be chosen within this ribbon.

At the other extreme, the traditional concept of earthworks balance in which the amount of excavation is made to equal the amount of fill still produces a good first design — although large cuttings balanced by large embankments are rarely economical. This restraint is usually demonstrated by a mass-haul diagram that plots at each location the product of the amount of material to be taken from the location by the distance it must be moved. The designer then attempts to minimise the sum of all these products, which is a surrogate for the cost of the earthworks and thus a better indicator than the simple volume of earthworks.

As a result of these reviews, any alternative proposals should be re-examined to ensure that the best alignment has been selected. For example, as the cost of bridges becomes clearer, consideration may be given to significant alignment changes in order to reduce their cost. Similarly, an assessment of pavement materials and of drainage proposals might indicate the need for some relocations. The costs of adjustments to various public utilities such as power, water, telephone, and gas may also lead to alignment reconsiderations.

Finally, any road design must accord with driver expectancies and not present the driver with unexpected difficulties. To a large extent, this is managed by the speed environment and design speed concept (Sections 18.2.4&6), but care must be taken with such factors as carriageway width which are not directly linked to a chosen design speed. For example, a wide carriageway often suggests to drivers that high speeds are appropriate (Section 18.2.1).

Once a favoured horizontal and vertical alignment and carriageway width have been selected, a set of computer-based digital mapping, drafting, and design techniques known as *CADD* (computer aided design and drafting) are commonly used to produce 'drawings' showing perspectives, longitudinal sections, cross-sections, local levels, grades of the centreline and kerbs, drainage systems, and earthwork quantities. CADD drawings flow logically from and back to the automated mapping and GIS systems discussed in Section 6.1.2.

When the CADD drawings are available, most of the economic, environmental, aesthetic, and safety features of the proposed road can be assessed in significant detail. In particular, the appearance of the road should clearly indicate to the driver the speed and path to be adopted in order to proceed with comfort, economy, and safety. If this is not possible, additional devices such as signing (Chapter 21), pavement marking (Chapter 22), signals (Chapter 23), and lighting (Chapter 24) should be used. The drawings also permit the assessment of the visual impacts of the proposal and the definition of any land acquisition requirements (Section 6.3).

The form of the individual elements — such as lane width — that make up the road is largely dictated by the pre-determined design standards adopted for each element. The principal factors influencing these individual elements are discussed in the following sections.

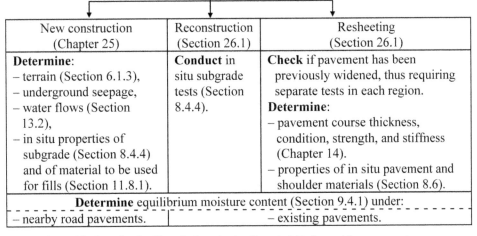

Obtain:
- any available construction plans (Chapter 25),
- locations of underground utility services to allow test holes to be located clear of these services (Section 7.4.3) and to plan future approaches to construction,
- results of any previous nearby investigations (Section 6.1.2).

⇓

Determine:
- testing procedure relevant to proposed type of construction (Chapter 8),
- location of test sites:
  # usually at 100 to 500 m intervals,
  # covering all combinations of terrain type (Section 6.1.3), soil type, and drainage, and
  # where the formation intersects the natural surface.

⇓

Determine the depth to subgrade (Section 2.2) at each test site.
- testing is usually only required more than 500 mm below subgrade when previous in situ strength tests (Section 8.4.4) indicate possible soft areas.

New construction (Chapter 25)	Reconstruction (Section 26.1)	Resheeting (Section 26.1)
Determine: – terrain (Section 6.1.3), – underground seepage, – water flows (Section 13.2), – in situ properties of subgrade (Section 8.4.4) and of material to be used for fills (Section 11.8.1).	**Conduct** in situ subgrade tests (Section 8.4.4).	**Check** if pavement has been previously widened, thus requiring separate tests in each region. **Determine**: – pavement course thickness, condition, strength, and stiffness (Chapter 14). – properties of in situ pavement and shoulder materials (Section 8.6).
Determine equilibrium moisture content (Section 9.4.1) under:		
– nearby road pavements.	– existing pavements.	

Figure 6.2 Steps involved in a site investigation for material to be used for a pavement course (based on Country Roads Board 1980a).

6.2.3 The influence of traffic factors

Geometric design must consider the type of road (Section 2.1) and the flow, composition, and speed of its traffic. Road type (or classification) is usually decided when the road is being planned and influences such items as access control and intersection type. The design traffic flow is commonly taken as the lower of:

* the forecast traffic flow at the end of the design life of the road (Section 31.5), on the assumption that some initial over-capacity to cater for future growth will be economically justifiable, or
* the traffic flow that occurs when all feeder facilities planned to be in operation during the design life of the road, are operating at capacity (Section 17.4.1).

If this choice provides only the annual average daily traffic flow in veh/d (AADT, Section 17.1), the design hourly flow (DHF) is derived using the Nth highest hourly flow concept (N HHF, Section 17.1), commonly selecting N = 30. As will be seen below, the DHF is the main determinant of carriageway cross-section. For congested urban roads and intersections, it is more common to use the peak weekday traffic flows — these will usually be very close to 30 HHF values.

Traffic composition relates to the mix of cars, trucks and other commercial vehicles (Section 27.1.1), motorcycles, bicycles, pedestrians, etc. which will use the road. Commercial vehicles are frequently slower and larger than passenger cars and so can have a significant influence on traffic behaviour. A specific discussion of the influence of traffic composition on capacity — and hence on carriageway width — is given in Section 17.4.3, on speed in Section 18.2.10, on alignment in Sections 19.2–3, and on intersection layout in Section 20.3.1.

Design speeds are discussed in Section 18.2.6 and the alignment requirements resulting from them are described in Sections 18.2.5 & 19.2&3. Alignment standards depend largely on the chosen design speed, as illustrated by the typical speed–radius relationship shown in Table 6.2 and used for determining horizontal alignment.

Table 6.2 Typical link between design speed and curve radius

design speed km/h	minimum horizontal curve radius m
130	1000
100	500
80	300
60	150

6.2.4 The influence of physical factors

Terrain particularly influences the road alignment and is discussed in Sections 6.1.3 & 18.2.5 and in Chapter 19. Specific alignment requirements will also arise from such restrictions as:

* a lower limit on horizontal curve radius (Section 19.2.6) and sight distance (Section 19.4),
* an upper limit on vertical gradient (Section 19.3), and
* the need to accommodate existing natural and man-made features within the terrain.

Environmental issues can significantly influence geometric design via such factors as landscaping and appearance (Section 6.4), minimising air and water pollution (Section 32.4), and reducing the impact of community severance (Section 32.5).

Abutting land ownership is also important, as roadside access to a road can greatly affect its capacity (Section 17.4.3), operating speed (Section 18.3), and safety. This is most dramatically seen in the better crash rate of motorways over other road types (Section 28.5). In law, the owner of a property abutting a roadway usually has access to the road immediately adjacent to the property, but not to every part of the roadway. In addition, the public's right of way (Section 3.3.3), i.e. of public passage, usually dominates over the rights of the abutting property owner. These two principles allow the effective control of access to public roads.

The location of access points from property or from the existing road system will often be a problem and will depend partly on the road's classification. The situation may range from a motorway with full access control, to an arterial road with service streets, through to a residential street with no access limitations (Section 7.2.1). Access control is one of the most important planning and design features affecting a road's safety performance.

As the design advances, the various requirements are progressively refined, leading to a consideration of such factors as the speed at approaches to steep grades, the length of the steep grades, the speed at the end of long grades, sight distance, and the elimination of long stretches of straight alignment. This process is aided by the production of engineering drawings of longitudinal sections (grade-lines), cross-sections, and any necessary earthworks. But to produce such drawings, it is first necessary to decide on the cross-section of the road.

6.2.5 Cross-sections

Important geometrical requirements also apply to the carriageway cross-section (Figure 2.2). A lane is a portion of carriageway allotted to the use of a single line of vehicles and so carriageway width is determined by the number and the width of the traffic lanes, and whether or not the shoulders are paved (Sections 11.7 & 14.2), i.e.:

carriageway width = (width of two shoulders) + (number of lanes)(lane width)

Most 'official' lane widths in current use derive originally from U.S. work in the 1930s and 1940s. Observations of driver behaviour on horizontal curves led to a recommended lane width of 3.6 m and this became standardised in the U.S. in the 1950s. However, lane widths are better based on studies of driver behaviour over a range of circumstances. Such studies show that very wide and very narrow lanes both demand excessive driving effort and cause a decline in driver performance (Section 16.5.1) and that, for each driver–vehicle combination, some optimum range of lane widths exists at around 3.0 to 3.7 m.

Table 7.3 will show that a typical maximum car width is 2.0 m and that clearance requirements mean that the minimum practical width is about 2.7 m. The use of narrow 2.8 m lane widths at intersections is discussed in Section 20.3.2. For two-way operations, lane widths as low as 3.0 m can be used on grounds of economy and where the AADT is below about 250 veh/d.

However, large vehicles and the demands of turning radii will modify the above conclusions. For example, trucks and buses in lanes less than 3.0 m will be a problem to other drivers, not only because of their size, but also because they often have low steering-sensitivity and slow steering-response. A small percentage of drivers also tend to shy away laterally from large trucks in adjacent lanes.

There is little safety or capacity benefit in increasing lane widths over 3.2 m (Sections 28.5 & 17.4.1), and this width is sometimes used when operating speeds are below 100 km/h. However, it is common practice to use lane widths of 3.5 m on rural roads and 3.7 m on motorways for AADTs of over 200 veh/d, as they allow large vehicles to pass without instinctive lateral movement. Internationally, motorway lane widths range between 3.5 m and 4.0 m. Furthermore, research has shown that most heavy vehicles could travel comfortably in a 3.5 m lane, with a few extreme cases requiring 3.7 m (Prem *et al.*, 1999). Lanes over 3.7 m in width can create problems if drivers are tempted to overtake within lanes.

Typical lane widths usually vary with traffic flow, in the manner suggested in Table 6.3.

Table 6.3 Link between AADT and lane width

AADT (veh/d)	Typical suggested lane width (m)	Notes
	2.7	Car clearance
0 – 150	See Tables 6.4 & 7.4	
150 – 250	3.0	Low speed traffic
250 – 800	3.3	
800 – 300	3.5	For trucks and all
1300 – 800	3.7	high speed traffic

Most traffic on two-way roads with a carriageway width of 5.5 m or less will behave as if it is on a single-lane road and vehicles will need to partially leave the pavement in order to accommodate occasional oncoming traffic. Pavements commonly need to be at least 5.5 m wide to allow oncoming vehicles to pass each other at low travel speeds without leaving the pavement (Table 7.3). Not surprisingly, crash studies reported in Section 28.5 suggest that crash rates decrease as carriageway widths increase. Pavements less than 6.0 m wide typically have very high maintenance costs associated with the pavement edges. This data is summarised in Table 6.4. Useful additional data is given in Section 25.1 of Lamm *et al.*, 1999. Thus, the use of two-way roads with a carriageway width of 5.5 m or less is usually discouraged in developed countries.

However, there is a growing trend to reduce these widths for lightly trafficked two-way roads. The minimum carriageway width suggested by Tables 6.3 & 7.3 is 3.0 m for low traffic flows travelling at low speed and 3.5 m for low flows of high-speed traffic. These narrow widths can cause problems with safety (Section 28.5b), operations (Section 17.4.2,3&5), and pavement maintenance (Section 14.1A10).

Table 6.4 Typical links between road type and carriageway width

Road type	Traffic flow (veh/d)	Typical carriageway width (m)
No passing, few trucks, restricted speeds	< 15	3.0
No passing, restricted speeds	< 30	3.5
Difficult passing, restricted speed	< 100	5.0
Some passing, moderate speeds	< 150	5.5
Speeds unrestricted, centrelines	> 150	6.0
Two-lane road, few trucks	> 500	6.5
Two-lane road	> 1000	7.0
Two-lane road	> 5000	9.5
Multi-lane road	> 10 000	≥ 12.0
Rural freeway	> 12 000	≥ 2 x 11.0
Urban freeway	> 30 000	≥ 2 x 12.0

Swedish and Australian road design standards permit the use of single-lane two-way roads for AADTs of up to 100 veh/d, but require the provision of passing opportunities (Section 18.4.4) when long lengths of such road are used. These single-lane roads are unacceptable in many developed countries where carriageway widths as low as 5 m are only tolerated for two-way traffic if:

* seen as a merged combination of overlapping shoulders and operating lanes,
* there is minimal truck traffic,
* traffic flows are below 100 veh/d, and
* speeds are low.

When two-way flows exceed about 150 veh/d, it is usually desirable to employ lane marking in the form of a separation line between the two flow directions (Section 22.2.1). This suggests a pavement width of at least 6.0 m. An AADT flow in excess of 5000 veh/d suggests two lanes within a 7.0 m pavement. For higher flows, the number of lanes depends on the predicted AADT and the assumed capacity of each lane (Sections 6.2.3 & 17.4.3). Typically, two lanes in flat terrain could manage an AADT of about 10 000 veh/d.

Carriageway widths based on Table 6.3 and the above discussion are shown in Figure 6.3. Street carriageway widths are summarised in Table 7.3. The associated formation widths are usually more generous in jurisdictions that regard single-lane roads as part of a process of stage construction and hence allow for future development to two lanes.

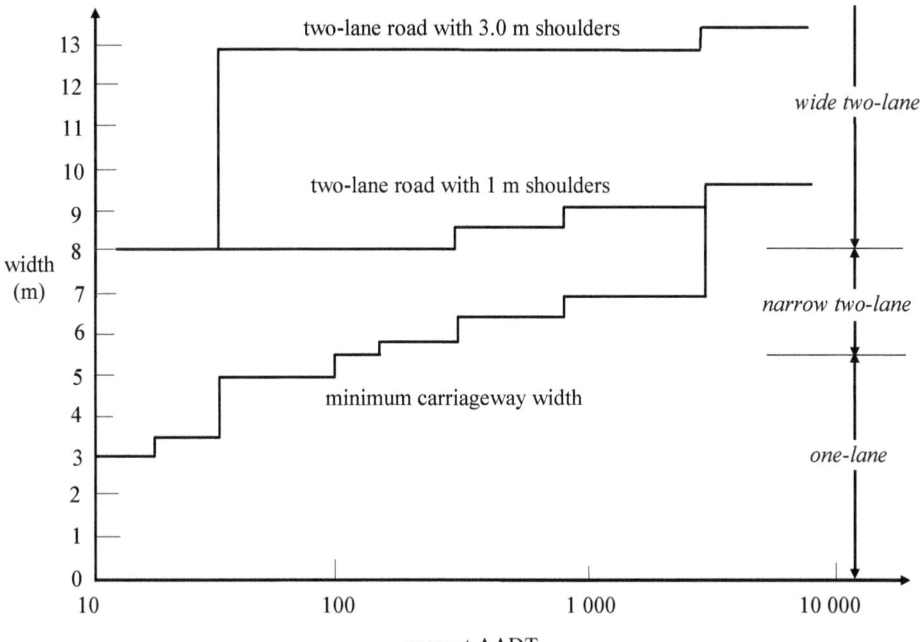

Figure 6.3 Typical carriageway widths, based on the discussion in Section 6.2.5.

The standards of such jurisdictions also require the provision of full shoulders (Section 11.7) at much lower traffic flows — typically, at AADTs of about 300 veh/d compared with 3000 veh/d. Even for traffic flows below 200 veh/d, the need to control moisture in the pavement structure suggests that shoulder widths should be at least 600 mm (Section 11.7).

Safety requirements and cycling needs demand a shoulder width of at least 1.0 m (Section 28.5). Based on data in Table 7.3, minimum shoulder widths for high standard roads are usually:

* on the kerb side, 3.0 m to permit emergency parking and travel by trucks or cars (some jurisdictions require slow vehicles to use the outer shoulder, Section 17.4.3i). Shoulders can also be used to enhance sight distance (Section 19.4.4). Typically, the emergency shoulder use is about one stoppage per 50 veh-Mm with the most common use being to change a flat tyre. International usage on major roads ranges from 1.2 m (urban Japan) to 3.3 m (U.K.), and
* on the median (or off) side, 1.2 m. For a maximum truck width of 2.5 m (Table 7.3), the combination of a 1.2 m shoulder and a 3.7 m lane provides a width of 5 m which would permit other trucks to at least slowly pass a truck stopped on the shoulder. Shoulders less than 300 mm in width have no beneficial effect and significant effects only begin when their width reaches 1.0 m. International practice uses shoulder widths from 0 to 3.0 m.

Assuming a minimum shoulder width of 0.5 m and a maximum of 3.0 m, typical carriageway widths are:

Table 6.5 Carriageway widths

Road type	Carriageway width (m)
one-lane	3 to 5.5
narrow two-lane	5.5 to 8
wide two-lane	8 to 13
three-lane (but see Section 28.5f)	13 to 17
four-lane	> 17

Dual carriageway (or divided) roads are roads where a separate carriageway is used for each traffic direction. Dual carriageways are usually considered when the traffic flows require two lanes to be provided in each direction — typically, when AADTs exceed 10 000 veh/d. For lower-speed suburban roads, this limit can rise to 20 000 veh/d, if turning movements are low. The number of lanes for a dual carriageway road is usually based on a design hourly flow (DHF, Section 17.1) rather than AADT.

Dual carriageways are produced by inserting a median (or central reserve) between the lanes serving each direction of travel. Medians markedly improve safety (Section 28.5), particularly for oncoming traffic, turning traffic (Section 20.3.5), and pedestrians (Section 20.4). Median widths between carriageways are usually at least 15 m. Detailed median design is discussed in Section 22.4.2.

If at this stage of design the traffic flow levels only justify a two-way road with no central median, the alignment must be rechecked to ensure that adequate overtaking opportunities are provided (Section 17.4.5). For example, it may be necessary to introduce either sections of straight road to provide adequate overtaking sight distance or specific lengths of overtaking lane (Section 19.4.4&5).

It is desirable to maintain a consistent minimum number of lanes along a major road, with auxiliary lanes used to provide for weaving between adjacent entry and exit ramps or to manage local peaks in the DHF. Lanes are usually added via dual lane entry ramps and deleted (dropped) by taking the lane into an exit ramp (see also the discussion of lane balance in Section 17.3.6). Such lanes must be well signed (e.g. EXIT ONLY) to prevent them capturing through traffic. Lanes are also often dropped when a carriageway splits into two once a median is introduced. Inconsistent practice with respect to lane drops can create considerable driver confusion.

Freeways are commonly considered when flows exceed 30 000 veh/d in the city or 12 000 veh/d in rural areas. A minimal cross-section freeway would have two lanes for through traffic and provision for merge and diverge lanes and/or for adding an additional lane on each carriageway. In this case, the lanes will take 6 x 3.7 = 22 m, the shoulders 2 x (3 + 1.2) = 9 m, the median 15 m, and safe clear distances (Section 28.6.1) on either side 2 x 10 m. This all leads to a minimum formation width of 46 m and a minimum right-of-way width of 66 m, although 100 m is commonly recommended for freeways. Section 29.3.1 indicates that *toll roads* with manual toll collection may require roads up to five times the above widths at toll collection plazas. At the other extreme, a separate review of the locally complex aspects of residential street width is given in Section 7.4.2 and Table 7.3.

Pavements are sloped transversely to produce a cross-fall to permit water to drain rapidly from the pavement surface (Section 13.3.1). This is usually achieved by creating a central longitudinal saddle — called the crown — by making the centreline of the carriageway higher than its edges, commonly by about 100 mm. Typical pavement cross-falls are shown in Table 6.6:

Table 6.6 Typical pavement cross-falls.

Situation	Slope	
	N to 1	%
grassed shoulders	12	8
cleared shoulders	18	6
natural soil pavement surface	18	6
gravel pavement surface	24	4
sealed pavement surface	32	3
asphalt or concrete pavement	48	2
minimum to avoid ponding	100	1
minimum for drainage	500	0.2

Note: N is the horizontal distance. It is common practice to refer to slopes as either 'N to 1' or '1 in N' where N is the ratio of horizontal to vertical slope distance, i.e. the horizontal distance to achieve a unit drop. The 'percentage slope' is 100(1/N).

Drivers prefer cross-falls to be as low as possible. In addition, high cross-falls will produce deleterious scouring on unsurfaced roads (Section 13.3.1). The continuous sloping of the full width of the pavement on curves is known as superelevation (or camber) and is discussed in Section 19.2.2.

6.3 LAND ACQUISITION

Once the alignment and cross-section of a new road has been established, it is possible to define the specific strip of land needed for building the road and the subsequent provision of public right of way. Commonly, this strip is then reserved for road purposes via a defined public planning and environmental impact process (e.g. Section 32.1.2). Consequently, the strip of land is often called either the right of way or the road reservation (Chapter 2).

When approval is received to go ahead with construction, a next step is to acquire any land not already owned by the Road Authority. Land may need to be acquired to not only provide for roadbuilding, but also to control access (Section 6.2.4), ensure compatible abutting land uses, and manage roadsides and landscapes (Section 6.4). Such action may well lead to significant property disturbance, strong protests, and high costs. Early acquisition of land can minimise these effects, prevent inappropriate development occurring and avoid the social disruption caused by unexpected land acquisition. Early purchase is only a sound economic investment, however, if the rate of increase in land prices exceeds the interest being paid on the money invested in the acquired land. Early acquisition can also cause planning blight in an area, i.e. a condition of zero or negative growth caused by uncertainty and community awareness of plans for future property acquisition and road development. Planning proposals can also raise land values as a result of property speculation.

Each region will have its own legal provisions for acquiring land for public purposes. Differences arise in such matters as methods of notifying the landowner, land valuation, negotiation, compensation, the settlement of disputed claims, and the powers of compulsory possession. The two common ways of acquiring land are:

(a) purchase by mutual agreement between the owner and the acquirer; and
(b) compulsory acquisition (or resumption or appropriation). In such cases, the acquiring Authority must often seek approval at a higher level than for purchase by agreement.

Measures (a) and (b) are often preceded by a preliminary publication of acquisition proposals, formal publication of final plans, and then a notice-of-intent served on affected owners. There are usually defined periods in which formal objections to proposals may be publicly lodged and proposals confirmed or modified.

When a final notice to acquire land is issued, it is also usual for there to be a defined period in which the recipient can accept or otherwise react to the notice. There is normally enough flexibility to allow mutual discussions to be held on the value of the land to be acquired and for revised offers made. It is common for the parties involved to be able to call for an additional valuation. It is normal for the Authority to pay for both the value of the land acquired and the effects of any severance of adjoining properties. However, if the owner has other land that is bettered or enhanced by the new road, it is uncommon for a payment to be required from the owner.

Finally, the various procedures normally allow reparation to be made for cases of special hardship. Disputed claims are usually settled by appeal, arbitration or legal action. If an owner refuses to accept a final decision against him, the Authority may need to enter the property, take compulsory possession and evict the former owner. Such processes should be rare. Many jurisdictions permit a Road Authority to take possession, even though price to be paid may still be a matter of legal argument.

Following land acquisition, the project can then move into the pre-construction planning stage and thence into construction. These aspects are described in Chapter 25. However, it is useful to conclude this part of the discussion of road location with a few comments on construction surveys.

It is normal to base the survey for a road on its centreline and on running distances along that centreline. However, to enable survey pegs to be placed away from construction work, reliance is often placed on pegs offset from the centreline by some defined amount. Pegs are placed at about 50 m intervals on long, flat straights and, at the other extreme, at about 10 m intervals on sharp curves or in steep terrain. The increasing reliability of GPS systems is making some pegging practices redundant. Curves, transition curves and superelevation transitions (Section 19.2.2) require special attention.

6.4 LANDSCAPING

6.4.1 Issues

The landscape is the viewed external environment and includes such elements as terrain (Section 6.1.3), water, flora, fauna, farms, towns, and roads. Landscape issues enter into the planning of a road (Chapters 5 and 31), the assessment of its environmental consequences (Chapter 32), alignment design (Section 6.2), construction (Chapter 25), maintenance (Chapter 26), and operations (Chapter 18). In considering landscaping issues, it is necessary to take into account its influence on:

* the road users (it is important to realise that the road user will be viewing the landscape whilst in motion),
* the occupants of the surrounding land,
* the intrinsic natural, social, cultural, and historic characteristics of the area, and
* items in the existing landscape which are unique and/or of high visual value.

The landscape resulting from road construction must therefore satisfy the following five objectives of helping the road to:

(a) fit harmoniously into the external visual environment, creating an interesting and pleasing scene for both travellers and non-road users, e.g. by linking components of the landscape and screening unsightly views,
(b) function sensitively within the processes that occur in the external environment, (e.g. by preserving local movement systems and by conserving existing topsoil, controlling water run-off, and preventing soil erosion, Section 13.3.1),
(c) perform its intended traffic function by:
  - leading the driver's eye,
  - signaling changes in alignment,
  - delineating the alignment,
  - screening the glare from oncoming headlights (Section 24.4.2),
  - providing forgiving traffic barriers to minimise damage to errant vehicles (Section 28.6.2),
  - reducing side winds,
  - relieving driver boredom, and
  - providing a reference point for speed indication.
(d) improve travel amenity by providing areas for rest and shelter, and
(e) respect local environmental, cultural, and conservation values (e.g. by preserving remnant vegetation).

Usually, the road builder can only influence the landscape in the verge beside the road, although the planners may have had an opportunity to locate the road within the overall landscape. Landscaping also covers such construction and maintenance matters as:

(1) site clearing, and methods of retaining as much as possible of the existing ground cover,
(2) disposing of unwanted, existing soil and vegetation and removing construction debris,
(3) locating borrow-pits to supply roadmaking materials,

(4) reducing maintenance costs by selecting plantings requiring low maintenance and controlling soil erosion, rockslides, and snow slides.
(5) selecting the plant stock, seeds, and fertiliser to be used,
(6) planning future plant maintenance,
(7) providing for the movement of local wildlife, and
(8) selecting the best way of utilising small parcels of land left over from land resumption activities (Section 6.3).

Useful guidance on road ecology is given in Forman *et al.* (2003).

6.4.2 Design

It will usually be found that the longitudinal vertical alignment will have greater visual impact on motorists than the horizontal alignment or cross-section, particularly when cuttings and embankments are involved. However, long straight sections, monotonous horizontal alignments, and isolated sharp curves can have a major influence on driver behaviour (Sections 16.5.2 and 18.2.1).

Cost conflicts will often arise. For example, flat rather than steep slopes on the sides and faces of cuttings and embankments will usually prove better at blending into the landscape and supporting vegetation. However, they will be more expensive than steep slopes. The correct balance between the two is not easy to achieve, as it is difficult to put a value on aesthetics.

Consideration should be given to wayside stops, rest areas, truck-parking areas, and scenic lookouts. Information bays or off-road facilities on the approaches to large towns and tourist facilities can be useful. Wayside stops are commonly provided at 50 km intervals, reducing to 20 km when AADT exceeds 2500 or to 10 km when tourist activity is high. Rest (or service) areas that additionally provide water, toilet and shelter, are recommended when AADTs exceed 1000 and comparable town facilities are more than 50 km apart. Drivers prefer rest areas that are large and shady (Pearce and Promnitz, 1982). From an engineering viewpoint, the sites should be at road level with good entry and exit sight distances and adequate drainage. Ideally, they should be able to be developed within the strip of land already reserved for the road.

Roads can be a barrier to the movement of many animal species. Where this will be a threat to the local species, movement can be catered for by well-placed overpasses, tunnels, culverts and pipes. The solution adopted must be tailored to suit the specific species under threat (Maningian, 1996).

6.4.3 Plantings

It is usually desirable to begin selecting the landscape planting by conducting a survey of the pre-existing flora and fauna (Section 6.1.1), as part of an assessment of the whole environmental framework of the area (Section 32.1.1). This can vary from comprehensive catalogues, to data on significant single specimens, to surveys of complex ecological areas. The stability, interdependency, sensitivity, vulnerability, and resistance to change of each type needs to be understood. Plantings (and future maintenance) should encourage the development over time of natural plant communities. Specific reference should be made to protected plants, pest plants, and historic plants. Possible surgery or maintenance of existing vegetation also needs consideration. Local seedstock can be collected for planting after construction is finished.

Natural features that require evaluation include environmentally-sensitive areas such as an unusual stand of trees or a creek line, the character of the landscape, and the existing land usage. Other desirable data relate to the quality, quantity and permanence of surface and subterranean water. Photographic records are useful. It is important to delineate those features that should be preserved. For example, stands of trees can be included in the road reservation and utilised as a roadside amenity or used as part of the overall planting pattern.

Construction often leaves the land in a highly disturbed state that may initially be unable to support plantings of the final mature species. Thus some patience and staged planting may need to be employed. For instance, many maturing plants favourably alter local site conditions and permit a 'succession' of plant species to take hold until, finally, a mature self-regenerating set of species arises. The resulting strata of grass, shrubs, and trees will often also provide a well-used animal habitat.

Shrubs can serve most of the same purposes as trees and also reduce mown areas. Evergreen shrubs with a dense, compact habit and growing up to 2 m high are particularly effective as anti-dazzle screens. To be safe as traffic barriers, shrubs should have resilient stems rather than single trunks of over 100 mm in diameter.

In planting trees, it is necessary to minimise collisions by cars (Section 28.6.1) and the effect of falling branches. The former factor usually controls tree placement and the common recommended 9 to 10 m clearance from traffic is often reduced to 2 m for shrubs. Trees are rarely planted on steep slopes, or where they would obstruct drains or reduce sight distance. Tree planting issues are discussed further in Section 7.4.2. Not every area needs to be intensively planted, and the roadside verge can also contain wildflower patches, sandy and rocky areas and wetlands.

Slope protection on embankments and cuttings (Section 11.8.2) is an important part of landscaping works. If preventative measures are not taken, such areas can be readily scoured by surface water flows (Section 13.3.1). The two keys are good surface cover and adequate drainage to prevent surface flows. Surface cover may be provided by a man-made cover (e.g. stones or concrete) or by planting. Plant growth may be made difficult because the construction process will usually leave the slope devoid of topsoil. Thus material may need to be imported and then held in place until leaf and root growth are adequate. This can be done by using soil sods containing young grass or by stabilising the slope with bitumen sprays, mulch mixtures, netting, brush, mesh, or various combinations of these.

6.5 URBAN AREAS

Many aspects of road location in urban areas will be dealt with in Section 7.2.2, where the emphasis is on the role of streets. This section concentrates on the major urban arterials (see Figure 7.1). The basic principles outlined in Chapter 5 with respect to the selection of schemes will still apply, as will many of the geometric requirements discussed earlier in Section 6.2. Section 18.3 describes how the features of the urban road determine its operating characteristics. Key additional features arising in urban area road locations are:

(a) the need for the new street to interlock with existing street geometry and traffic,
(b) a strong tendency to use land already reserved for road purposes, due to relatively high land costs and local opposition to land acquisition,

(c) very detailed drainage needs (Chapter 13),
(d) extensive provisions for utilities and services (Section 7.4.3),
(e) widespread interaction with community aspirations (Section 32.1.4),
(f) compulsory town planning requirements (Section 7.2.2),
(g) the strong role of commercial interests,
(h) the need to accommodate public transport (Sections 30.4),
(i) the importance of pedestrian needs (Section 30.5.3),
(j) provision for recreational areas,
(k) the presence of many historical sites,
(l) the great impact of the road on land values,
(m) high costs for the supply and delivery of roadmaking materials,
(n) a greater emphasis on the construction sequence (Section 25.2), and on minimising traffic diversion and such construction nuisances as noise, dust and vibration (Section 25.8), and
(o) sight distance requirements at each at-grade intersection (Section 20.3.2).

In planning to overcome the constraints provided by the items in this list, the urban road designer often works as a member of one of the inter-disciplinary teams discussed in Section 32.1.1.

The location process is usually an interactive one with routes being generally located in accordance with the principles laid down in earlier sections and then adjusted in accordance with the above list of additional constraints. As a result of the application of these additional considerations, it is common to find that urban roads are not ideally located from a purely engineering viewpoint. Urban design speeds will depend on local conditions and may range from 50 to 80 km/h for non-freeway designs (see also Table 18.2). A detailed discussion of urban road traffic capacity is given in Section 17.4.3.

CHAPTER SEVEN

Residential Streets

7.1 INTRODUCTION

This chapter is devoted to the special issues associated with the arrangement and design of residential streets, as these follow different principles and satisfy different objectives than do the major traffic routes discussed in the previous chapter. The chapter begins by examining the residential street as part of a local development and then explores the associated needs of the local community in terms of transport, amenity, and safety. Consideration is given to such specific matters as the road hierarchy, pavement surface, pavement width, drainage, kerbs, landscaping (Section 6.4), and street furniture (Section 7.4.3).

Following Section 2.1, the term residential street is used to emphasise the human orientation of the street as opposed to the vehicle orientation of the road or highway. The reader is also referred to the explanations of road terminology in Table 2.1, and particularly to the listed alternatives, as definitions vary from region to region. Many residential streets result from land subdivision in which large pieces of land are subdivided in order to permit more intensive development.

7.2 THE STREET AS PART OF THE LOCALITY

7.2.1 Definitions

As discussed in Chapter 2, the classification of streets is based on a traffic hierarchy (Figure 2.1) rising from culs-de-sac, loops, various grades of distributor (or collector) roads, through to arterial roads, although such definitions are not always clear-cut (Section 7.2.5) . The associated street types are shown in Figures 7.1 & 7.2. Culs-de-sac and loops service only abutting properties; local distributors only service culs-de-sac, loops, and abutting properties; and precinct distributors only service traffic from the adjoining precinct. As indicated in Figure 7.1, the ends of adjacent culs-de-sac can be usefully connected together by footpaths.

The prime purpose of a residential street is to provide access to contiguous residential properties along its periphery (Section 2.1). It may also serve other properties within the same residential precinct (or local area), which is typically the area bounded by the distributor roads (Figure 7.1). A pedestrian precinct is somewhat different as it is an area devoted solely to pedestrian traffic (Section 7.2.3). These residential precincts may be quite large — as in a typical suburb from the second half of the 20th century — or quite small and bordering other land uses — as in towns and cities developed before the car.

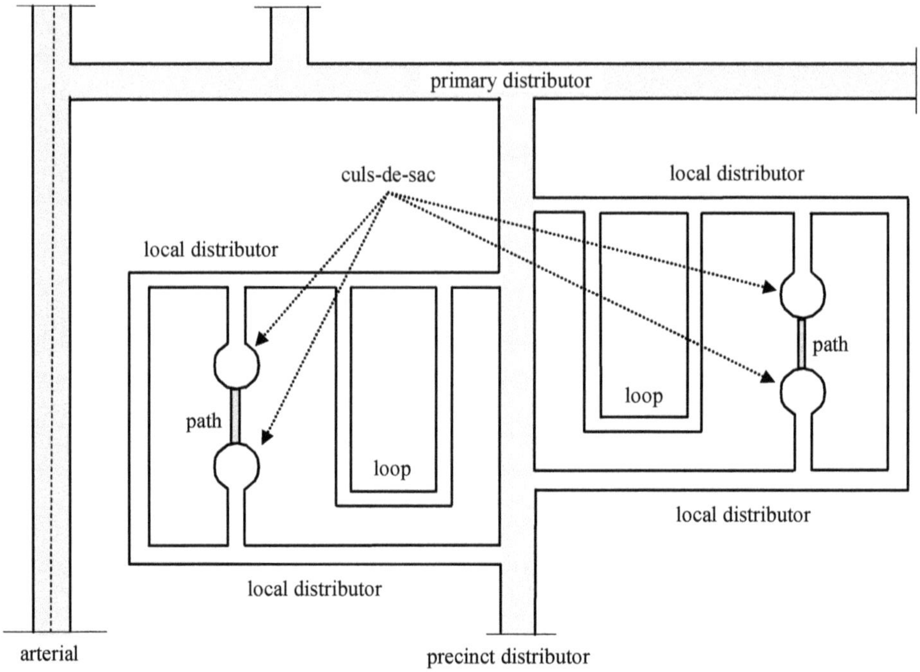

Figure 7.1 Definition of road and street terms in a residential or commercial precinct. Refer also to Figure 2.1.

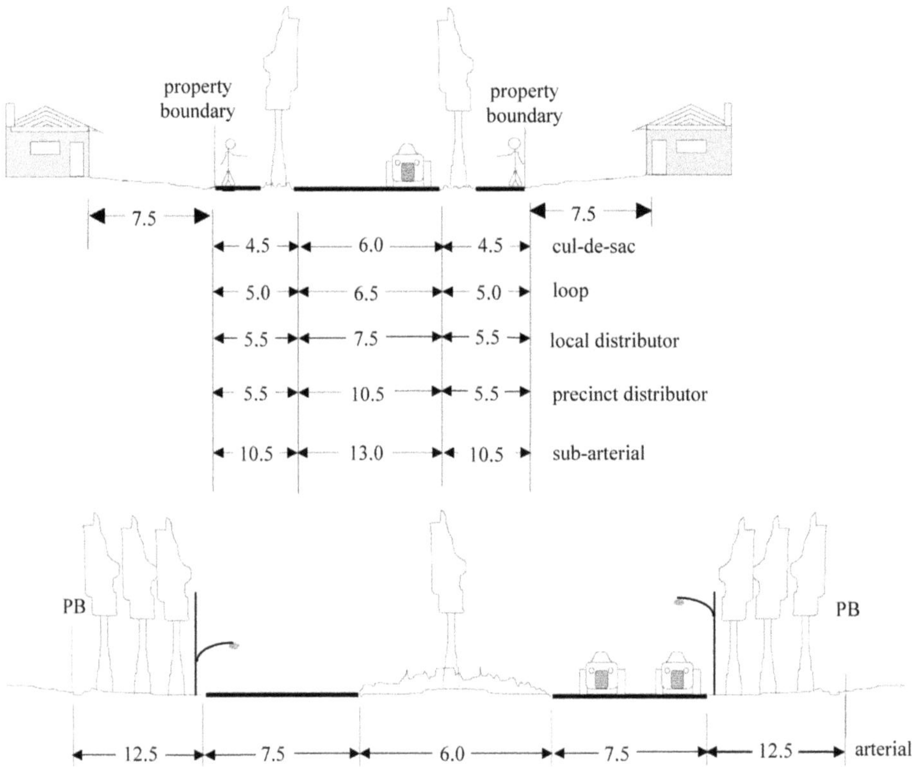

Figure 7.2 Typical road and street dimensions (metre). PB = property boundary.

7.2.2 Objectives

The planning of the streets and paths in a residential precinct should accord with the overall community objectives for the area. These commonly are considered in the three categories of efficiency, safety, and liveability discussed below.

1. *Efficiency* and adequacy in the movement of people and goods is achieved by a street design which provides:
   (a) access to the arterial road network, whilst avoiding inefficient access,
   (b) access to public transport, and easy interchange between travel modes,
   (c) access to private property, relying on streets but not on arterial roads, and including the movement of required large vehicles (e.g. garbage trucks),
   (d) local parking,
   (e) footpaths and bikeways (Section 19.6) for pedestrians and cyclists as separate elements in the system and not as optional appendages. Footpaths should accommodate the needs of children and of adults unable to drive. For instance, the visually-impaired can be helped by the use of a continuous line of guide blocks with dimples on their surface.
   (f) incentives to minimise vehicle trips through the area,
   (g) facilities for emergency, utility, drainage and statutory services, and
   (h) infrastructure which minimise its lifetime cost and which is economical in its use of land (Chapters 11 and 25).
2. *Safety* for all street users is achieved by a street design which provides (see also Section 7.2.6 and Chapter 28):
   (a) safe access to the arterial road network, whilst preventing unsafe access,
   (b) safe access to public transport and safe interchange between travel modes,
   (c) safe access to private property, relying on streets but not on arterial roads,
   (d) recognition that the street will frequently be used by children, pedestrians, and cyclists, thus ensuring their safety,
   (e) facilities which discourage speeding (Section 18.1.4, speeding cars are a prime source of resident distress),
   (f) minimal facilities — such as centreline markings (Section 22.2.2) and medians (Section 22.4.2) — which encourage speeding,
   (g) measures which encourage obedience of the traffic laws, and
   (h) a system which is inherently safe for all users, both in fact and in the perception of the residents.

   The reduced traffic flows in 1(f) are also very beneficial. Residents commonly place higher priority over (d) to (g) than over (a) to (c).
3. *Liveability* (and amenity) requires a street design which:
   (a) discourages through traffic and large vehicles,
   (b) encourages walking and cycling and safely provides for leisure activities such as strolling and jogging,
   (c) creates a good neighbourhood in which to live, with a sense of place rather than of passage,
   (d) is of pleasant appearance and scale, and is in harmony with the local natural environment,
   (e) includes adequate parks and open spaces to provide the local community — both children and adults — with spaces for socialising and shared activities,
   (f) is an extension of each abutting house and considers the link between each house and the street,

(g) protects residents from undue noise from vehicles (Section 32.2) and neighbours, dust (Section 12.1.2) and air pollution (Section 32.3) from vehicles, and obstruction of natural light, ventilation, and vistas by nearby houses.
(h) provides adequate privacy,
(i) creates a feeling of local pride, social cohesion and identity, and
(j) will be adaptable, in the face of future changes.

Amenity is somewhat less tangible and is defined as reflecting the convenient, pleasing, and agreeable features that raise being at a location above the level of social subsistence.

7.2.3 Design factors

The objectives of street design were listed in Section 7.2.2. The major road and street design factors which determine the details of the movement system in a residential precinct are related to land use, route layout, visual appearance, operating speed and the final design will depend on the relative weighting given to each of these factors at precinct and at suburb level.

(1) *Land use* within and adjacent to the precinct will mainly concentrate on the control of traffic generators (Section 31.3.4). They should be located with five key points in mind:
   * Trip origins and destinations should be kept as close together as possible to minimise trip lengths (ideally within 400 m, see Section 30.2.2a), and to allow as many trips as possible to be made without recourse to the private car. Separating trips is less important than minimising them.
   * Activities which generate high levels of pedestrian traffic should be located on only one side of an arterial street or, if at a major intersection, in only one quadrant of that intersection, unless all pedestrian movements across the arterial street can be controlled.
   * Neighbourhood shopping-areas and car parks should be located on distributors or arterials.
   * The provision of off-street parking (Sections 30.6.2) should be an integral part of the design of any major traffic generator. Parking exits and entries should be carefully located.
   * Major generators creating continuous ribbons (or strips) of development should not be placed along arterial streets as they create continual traffic interruptions and a poor visual environment.

(2) *Route layout* (or network design) particularly involves defining the relationship between a residential precinct's streets and the surrounding town or region's arterial roads (Figure 7.1), in terms of the objectives in Section 7.2.2. At this interface, the layout should control and minimise connections between roads and streets at different levels in the road hierarchy (Section 7.2.5), particularly intersections between minor and major streets. The arrangement of the roads within the precinct is discussed in Section 7.2.4.

(3) The *visual perception* of a street will be largely determined by the chosen landscaping and alignment treatment (Sections 6.4 and 7.2.3). The need for the street's traffic function to be unambiguous in terms of the various street classes shown in Figure 7.1 requires that each class of road should be sufficiently different in visual treatment to allow immediate identification of its position in the total network hierarchy. Vegetation and alignment both can dramatically influence visual perception.

(4) *Road design* is largely determined by the assumed vehicle *operating speed* (Section 18.2.10). This speed depends on whether the street is in:
 * a new area, in which case the chosen design speed (Section 18.2.6) will determine such factors as pavement width and street alignment (Figure 7.2 and Sections 6.5 & 18.3),
 * an existing area, in which case operating speeds may need to be controlled or manipulated by traffic devices in the manner described in Section 7.3 and 18.1.4).

 Control of vehicle speed is the greatest single factor influencing the liveability of a street system. Residents perceive a greater degree of safety on the street and noise levels are kept at acceptable levels.

(5) *Intersections* can also be used to control vehicle behaviour. The devices listed in Section 7.3 can be used at intersections to provide pseudo-gateways into residential precincts or small towns to announce the changed environment to drivers. Typical gateways are distinctive entrance structures, speed signs, sharp bends, road humps and plateaus (Section 18.1.4), rough paving, road narrowings (Section 7.3), roundabouts (Section 20.3.6), channelisation, and pedestrian crossings. They may not so much reduce the speed by physical means as create a driver perception of a low-speed area.

7.2.4 Street systems

Within a residential precinct, cars, pedestrians and cyclists are separated wherever possible. Where this is not possible, the design should ensure that the behaviour of each road user is apparent to and predictable by other users.

The traditional street layout utilises the rectangular grid. However, it can create long through streets with frequent intersections. Two of the more adventurous layouts that aim to alleviate the effect of through traffic are:

* *Radburn*, which segregates vehicle and pedestrian traffic at the local level by giving houses separate access to networks of roads and paths, and
* *Cluster* housing in which the dwellings are clustered around culs-de-sac and landscaped spaces, with the rest of the development being left for communal use. This system provides all residents with freedom from through traffic and an enhanced outside living area.

Pedestrian-only areas (or pedestrian precincts), such as shopping malls, should be developed where possible (Section 30.5.3). However, thought must be given to the servicing of any adjacent premises — particularly shops. Part-time malls are one way around this problem, with servicing restricted to non-pedestrian hours.

The development of a residential street system is usually controlled by local government land-use (or subdivision) codes, building and planning ordinances, and market forces. With the best of intentions, all of these usually restrict design innovation. In addition, such past influences have often left many planning issues unresolved. For instance, many people in conventional residential precincts live in houses fronting onto arterials, sub-arterials, or precinct distributors, and thus with through traffic passing their door.

7.2.5 Street hierarchy

The common approach to road hierarchy within a residential precinct has already been illustrated in Figure 7.1 where arterial and sub-arterial roads are the traffic carriers, with the distributor roads handling decreasingly lower traffic flows. The pattern outlined represents a tributary system characterised by a gradation of road types, with each lower road acting as a traffic tributary delivering traffic to roads higher in the hierarchy. The road hierarchy concept is used to:

(a) identify the organisation responsible for each road and street,
(b) allocate funding priorities,
(c) set road and street design standards (Section 7.2.3),
(d) provide a rational basis for locating land uses that generate significant traffic (Section 7.2.3),
(e) allocate access control, traffic priority, and routing,
(f) plan and provide traffic-restraint measures,
(g) provide drivers with assurance as to the traffic-related expectations of them and of other drivers, and
(h) operate measures to prevent driver behaviour that is inappropriate for the road being used, given its position in the hierarchy.

Unfortunately, road hierarchies and the associated road classifications are not always a useful concept in existing residential areas. This is because a large proportion of the existing road system services a mix of purposes, with many streets serving abutting houses also acting as distributors, taking traffic to other streets and with many precinct distributors and arterial roads having full roadside access from each abutting property. For example, the local distributor uneasily services its dual functions of traffic carrier and house-frontage function, a problem which persists in many current layouts.

Thus, despite the many words written over the years about road hierarchies, it is not surprising that this lack of real application exists, as the philosophy as originally expounded has a number of serious flaws. For example, the ‘classical’ concept of a gradual gradation of road types cannot be sustained within a residential precinct. Instead, new residential streets should be based on the principle that a road is either totally for local access or totally for traffic movement (Figure 7.3). This is quite feasible, as minor distributors can have total access control by careful street layouts and house sitings. The above criticism of road hierarchies applies only to existing residential precincts; a more positive view of their usefulness for considerations of wider areas and for major roads is given in Section 30.7.

In a residential precinct the through-traffic function is subordinate to its distributional, pedestrian, and social functions, and design speeds are kept low. This requires a sufficiently small spacing of distributor and arterial roads to provide adequate overall mobility. For example, for a low-density suburban housing development, arterial and sub-arterial spacings of the order of 1.5 km or grids of about 2.5 km^2 in area have been found to be effective, although intermediate distributors are also needed to service this system. These distributors should carry less than about 500 veh/d. Suburban population densities range from about 500 to 25 000 people/km^2 with 3 500 as a common average. At about 2.5 trips/day (Section 31.3.4.1), this gives between 1 000 and 60 000 veh/d, but less vehicle trips are usually generated by high-density areas. Each square kilometre enclosed can therefore be assumed to generate about 10 000 veh/d trips.

Section 7.4.1 shows that the environmental capacity of a street is about 3000 veh/d and so the above data imply that about three distributors (or arterials) are needed for each

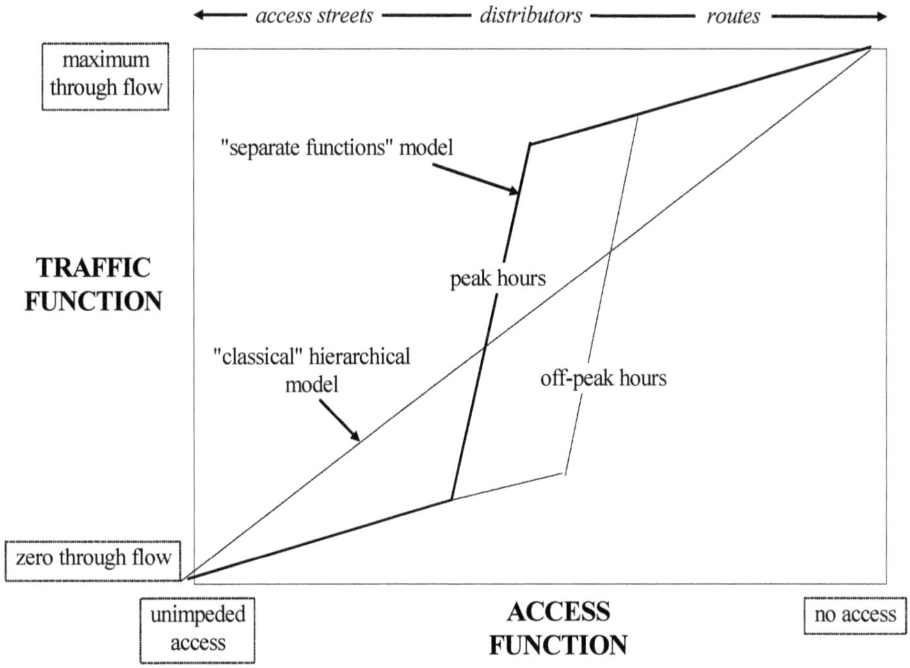

Figure 7.3 Functional mix of roads and streets according to both 'classical hierarchical' and 'separate functions' models (due to Brindle).

square kilometre. This leads to the diagrammatic representation in Figure 7.4. The precinct area of 2.25 km^2 will generate about 2.25 x 10 000 = 22 000 veh/d and this will require at least 22 000/3 000 = 7 distributors. The 500 m grid provides eight distributors, indicating that local distributors should be at about 500 m spacing. Practice suggests compromise spacings of between 500 and 800 m as this permits drivers joining from a previous intersection to reach peak speed prior to the next intersection (from Section 27.2.2, this distance is only 200 m for 60 km/h). In some jurisdictions it is somewhat obscurely argued that the spacing should be at least the stopping distance on the road in question (from Table 21.1, 100 m for 60 km/h). However, minor street intersections can safely be at 25 to 50 m spacing, as this gives a driver travelling at around 60 km/h about 2.5 seconds of travel between intersections, which accords with the driver response times in Table 16.1.

Another basis for checking a proposed street arrangement is to assume that no access street should serve more than 30 or 40 dwellings. About 6 to 10 trip/day/house, with at least one trip in each peak hour (see Section 31.3.4), will ensure traffic flows of below 300 veh/d. For a cul-de-sac, it is desirable to restrict flows to about 150 veh/d, which implies less than about 15 to 20 houses and lengths of under 200 m.

An extension of the residential precinct concept is the use of 30 or 40 km/h speed limit across an entire precinct in order to create low-speed areas, or *stillevej*. A useful rule of thumb for the extent of these precincts is that a driver should not have to travel more than 2 minutes at the low speed, which implies that the low-speed zone has a maximum radius of about one kilometre.

Perhaps the final stage in the evolution of the residential precinct is where the road space is formally and legally shared by cars and pedestrians on an equal basis. Such an area is known as a shared zone, or Woonerf. Woonerf is the Dutch word for residential

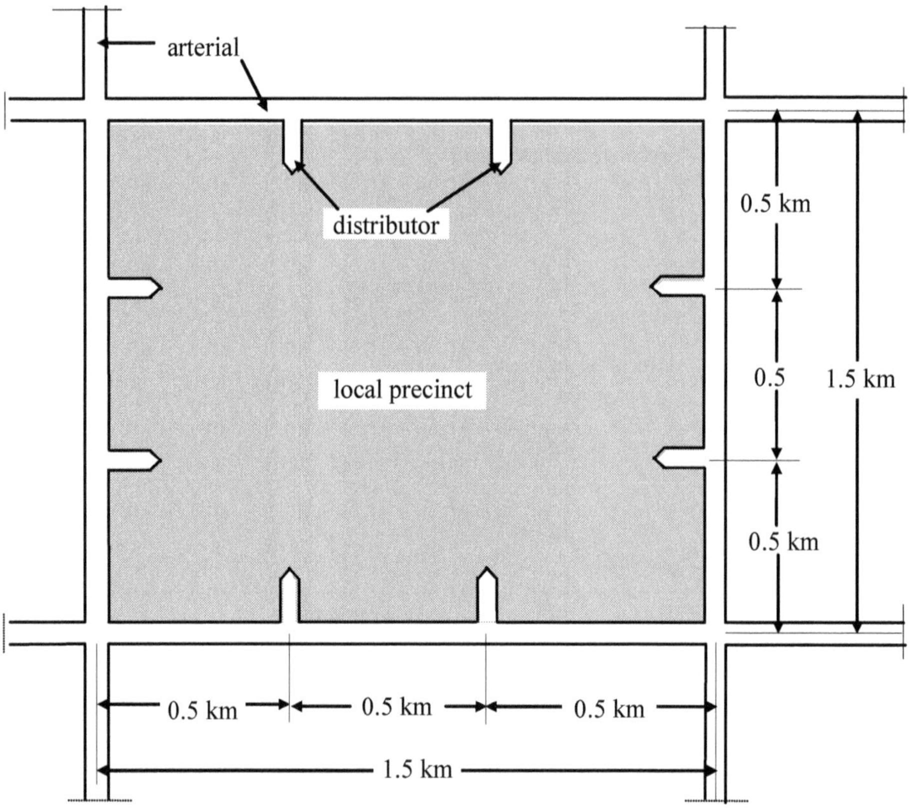

Figure 7.4 Simple model for calculating spacing of access-distributor points for a residential precinct.

precinct. Cars are reduced to pedestrian speeds and the residential function of the area predominates. This is enhanced by making the streetscape attractive and compatible with the above low-speed concept. A typical approach is to remove the distinction between footpath and roadway. Naturally, a shared zone can only be used where traffic flows are low.

7.2.6 Design for safety

Within the arterial road system, safety is controlled by the conventional road-related measures discussed in Section 28.5. Ideally, kerb-side parking and direct access from private property should not be permitted and positive traffic control should be provided at any arterial-to-arterial intersection.

Safety is also important within the residential precinct and is listed as a key planning objective in Section 7.2.2. For instance, on an exposure basis (Section 28.1.2), residential streets can be much more hazardous than arterial roads. In addition, the majority of the pedestrian accidents occurring within a residential precinct are typically to children under 12 years of age. Thus the designer of pedestrian facilities in residential precincts should not assume adult standards of behaviour.

Safety in new and existing residential precincts is affected by all the design factors in Section 7.2.3. Such planning and engineering treatments can reduce crash levels by up to 50 percent. Safety can be further improved by considering the following additional principles:

(a) Reduce traffic flows. This is a key point as the best predictor of traffic crashes in residential streets is the traffic flow in the street.
(b) Reduce traffic speeds. Measures in (c), (d) and (e) below will help achieve this. In addition, street trees and roadside vegetation that give an enclosed feeling will reduce the speeds that drivers adopt (Section 7.4.3).
(c) Avoid long (> 50 m), straight streets which encourage speeding (Section 18.3), and short, sharp curves (Section 28.5).
(d) Provide adequate sight distance at all times and minimise on-street parking. Section 19.4 shows that sight distance requirements drop as speed drops and Section 18.3 shows that reducing excessive sight distances can reduce vehicle speeds.
(e) Use T's and staggered intersections (Section 20.3.2), or roundabouts (Section 20.3.6), rather than cross-intersections. For uncontrolled intersections (Section 20.3) on lightly trafficked streets the use of two T-intersections is about seven times as safe as one cross-intersection. However, if one of the cross-streets is more heavily trafficked, then the two T's are only 20 percent safer. The use of cross-intersections where distributors and arterials meet is recommended as these are more amenable to the use of traffic control devices. Roundabouts may be more advantageous at intersections between distributors.
(f) Remove fixed objects from hazardous roadside locations (Section 28.6.1).

Traffic management measures to help achieve increased safety and amenity are described in Section 7.3. European experience is that targeted traffic management measures can reduce crashes by between 15 and 80 percent (Griebe *et al.*, 2000). One overall problem is that it is difficult to quantify the improvements resulting from such schemes, whereas the 'inconvenience' they cause is perceived by each motorist who uses a treated street and may outweigh the safety and amenity considerations in residents' collective minds. A related problem is that those who gain from such measures are usually a separate group from those who are disadvantaged. Trade-offs are therefore difficult. Successful schemes frequently involve including the affected residents through local traffic committees.

7.3 LOCAL AREA TRAFFIC MANAGEMENT

The use of local area traffic management (LATM) measures to produce safety and amenity improvements may occur at four levels: malls, residential precincts, corridors, and town-wide. This section deals with the use of LATM measures within the existing street system; i.e. at the first two levels. The other two levels are discussed in Chapters 30 & 31. The traffic aspects of LATM are examined further in Section 30.7.

LATM is sometimes also called traffic calming, but that term is best used to describe the creation of 'peaceful coexistence' between the mix of transport modes that the community wishes to see operating in a particular area. Traffic calming may therefore include both physical and non-physical measures such as road pricing (Section 29.4.6), the alleviation of the adverse effects of car and truck, and favoured treatment for public transport.

Many LATM measures are directly concerned with reducing vehicle speed and systematic, well-planned treatments can reduce local crashes by as much as 40 percent. However, LATM measures also have their disadvantages:

* it is important to consider the area-wide effect of the installation of each device, as they will rarely eliminate traffic. Instead, they will commonly divert traffic to other routes and the consequences of this diversion on those routes must be considered.

* there will often be an increase in average trip times, although this should not be of concern. Given an average residential precinct trip of about 750 m, a drop in speed from 60 to 50 km/h would add only 9 s to a trip within the precinct. Similarly, most devices would delay a worst-case large vehicle by 10 s or less.

* there may be inconvenience to large vehicles and care must be taken to continue to provide access for:

– emergency services,
– delivery vehicles,
– garbage collection,
– street-cleaning devices,
– public transport vehicles on defined routes.

Some *precinct-wide* LATM measures are:

(a) One-way streets. Whilst these can reduce through traffic, they usually have no net effect on safety. They are appropriate for areas having narrow or parallel streets that are frequently congested. In these situations, a change to a one-way system can increase both capacity and safety. However, such schemes usually inconvenience local traffic and create strong local opposition.
(b) Use of T- rather than X-intersections, to reduce through movements (Section 20.3.2).
(c) Bans on heavy vehicles, however some heavy vehicles will be found to be servicing facilities within the precinct and alternative delivery procedures will need to be provided.
(d) Parking controls (Section 30.6.2).

Localised LATM measures can be categorised as street closures, intersection treatments, and mid-block treatments.

(1) *Street closures* are LATM measures aimed at deliberately reducing street connectivity. The common forms are:
 (a) Half closures (or chokers) at one end of a street. These are rarely effective for streets over 300 m in length.
 (b) Complete closures of one arm of an intersection.
 (c) Complete closures at one end of a street, creating a cul-de-sac.
 (d) Diagonal closures at intersections. These can reduce speeds at the intersection to 25 km/h (Section 18.2.5) but are not very effective in reducing midblock speeds.
 (e) Midblock closures. These serve only to reduce local connectivity of the street network, they will have a lesser effect on intersection safety.

 Closure of an existing street will initially cause local traffic disruption, but common experience is that traffic soon readjusts and there is a net reduction in traffic as some drivers seek travel alternatives (the reverse of traffic induction discussed in Section 31.3.3).
(2) *Intersection design* is discussed in Section 20.3. The common forms of intersection treatment used in LATM are:
 (a) Partitioning of intersections to only permit turning movements – these may be the same as a diagonal street-closure or they may use a star-shaped diverter at the centre of the intersection.
 (b) Median strips (Section 6.2.5) to prevent turns into minor, side streets.

(c) Small roundabouts. By their geometry, these force drivers to deviate from a straight line, and therefore to slow down, when passing through an intersection. Casualty plus reported damage accidents at all classes of road are reported to have dropped by about 50 percent and at minor intersections by 90 percent with the installation of such roundabouts (Ashton and Brindle, 1982). Large roundabouts are discussed in Section 20.3.6.
(d) Signing measures at intersections, particularly Give Way and Stop signs (Section 20.2.2) and turn bans, and reduced green times at traffic signals (Section 23.2.3).
(e) Intersection channelisation – islands at intersections within the street length often operate almost as small roundabouts (McKelvey and Thomas, 1984).
(f) Plateaus (or speed tables) of raised pavement (Section 18.1.4) at an intersection, to both slow traffic and to also highlight the intersection and its pedestrian movements. As the intersection already requires drivers to slow down, a plateau can have minimal influence on capacity.
(g) Pavement narrowings at intersections, which in the extreme reduce to driveway or threshold entries to the street. When all four approaches to an intersection are narrowed, the result is a Catherine wheel intersection.
(h) Divider (or splitter) islands placed along the centreline to denote the entry to a residential precinct (Section 7.2.3).
(i) Small kerb radii at the corners to prevent high-speed turns from one street to the other and to aid pedestrians. Speeds are reduced to about 25 km/h (Section 18.2.5).

(3) Common *midblock* LATM treatments are:
(a) Short lengths of pavement narrowing can reduce speeds by up to 10 percent and can also prove effective in most aspects of traffic management (Daff and Siggins, 1982).
(b) Techniques which cause the horizontal deflection of the traffic path are particularly effective. The pavement may be narrowed on either one or both sides of the street. Such treatments are also called chicanes, midblock restrictions, pinch points, throttles, angled slow points, or neckings and can reduce speeds by up to 30 percent locally and 15 percent in the local road network. Long lengths of pavement narrowing can also be used to produce a meandering or wandering route — this may additionally be used for aesthetic reasons.
(c) Staggered or meandering traffic lanes within a wide pavement, to produce an effect similar to (b) at lower cost.
(d) Pavement narrowing to increase footpath width and thus reduce the potential exposure of pedestrians (particularly children) to traffic.
(e) Narrowed traffic lanes to reduce traffic speed (Section 16.5.1 & 18.2.1), and possibly provide extra space for cyclists (Section 19.6).
(f) Divider islands along the centreline at midblock to slow traffic.
(g) Median strips to narrow pavements and provide pedestrian refuges (Section 20.4).
(h) Kerbline or median fencing to discourage unsafe pedestrian movements.
(i) Rough surfaces or block paving to slow traffic (Section 18.1.4).
(j) Road humps, plateaux, ramps, and rumble areas to slow traffic, make the street context very apparent to the driver, deter through traffic, and improve safety

(Section 18.1.4). Spoon drains (Section 13.3.2) have also been used but are not recommended.

(k) Tree planting, which is an effective speed-reducer through its effect on drivers' perceptions of an area (Section 18.3).

7.4 THE STREET ITSELF

7.4.1 Street environment

A street in a residential precinct serves a number of distinct activities that are categorised in Table 7.1. Clearly, it should be a *place* rather than a *route* and seen as a transitional zone between households and the larger community.

Table 7.1 Street usage, from Gehl (1980).

Activity type	Example	Percent of all activities	Typical duration (min)	Percent of total use of street
Staying	Standing about	15	10	30
Doing	Working on car	10	110	25
Strolling	Visiting neighbourhood	5	5	10
Interacting	Chatting	15	2	10
Playing	Children's games	5	15	10
Moving	Walk to and from bus	30	2	10
Driving	Drive to and from work	20	1	5

In the example in the table, almost half (45 percent) of the activities on the street do not involve travelling and, moreover, in terms of time, these activities represent 75 percent of the usage of the street. Streets therefore need to be designed to permit their use for purposes other than travel. The data also indicate that children are greater street users than adults and that the edges of streets adjacent to the property boundaries are key activity areas. Design for children and design attention to the edges of the street space is therefore of importance.

Typically, the residents of a street place priority on traffic safety, aesthetics, travel comfort, sense of community, and the absence of traffic nuisance (such as noise). The operation of their street as a transport link is not high on their priorities. Traffic speed is used by residents to assess the safety of their street. Street width is inversely related to their perceptions of aesthetics and comfort.

The term environmental capacity (Section 17.4.1) is used to define the impact of road traffic on residential streets. It is convenient to specify this capacity in terms of the offending level of traffic flow in the street. This flow will depend on a number of factors that can be conveniently reduced to:

* *traffic noise*. Noise measurements and acceptable noise levels for residential areas are discussed in Section 32.2 and provide a readily quantifiable traffic measure. Section 32.2 indicates that such measurements frequently do not relate to householder noise annoyance and so limits on traffic flow deduced from noise level criteria should be used only as introductory guides.
* *visual intrusion due to traffic*. This is impossible to quantify, but can be based on such indicators as signs of personal care for the abutting properties.

* *delays to people* at pedestrian crossings (measured in time units, Section 17.3), a measure introduced by Colin Buchanan in 1963.

As the first two measures are difficult to quantify, environmental capacity estimates are often derived from delays to people at pedestrian crossings. If based on critical crossing-gaps and random vehicle arrivals (Section 17.3.4), such delays suggest, for example, that a 7.4 m two-way distributor (Figure 7.2) would delay more than 60 percent of pedestrians wishing to cross it, if the total traffic flows exceed about 300 veh/h or around 1800 veh/d (Holdsworth and Singleton, 1980). Data in Section 20.4 suggest that flows of over 600 veh/h indicate the need for a formal pedestrian-crossing. In a survey in a Melbourne suburb, more than 50 percent of the residents accepted traffic flows in their street of 1800 veh/d or less. Most residents considered 4000 veh/d as a traffic nuisance. A general summary of the influence of vehicle flows on pedestrians is given in Table 7.2.

The environmental capacity of a street therefore probably ranges between 300 and 500 veh/h, or about 1800 to 3000 veh/d. Hindrance effects due to parked cars can set the level as low as 1000 veh/d. On the other hand, sub-arterials are usually tolerable with flows up to 10 000 veh/d (Section 17.4.3).

Table 7.2 Influence of vehicle flows on pedestrian behaviour.

Traffic (veh/h)	Comment
100	Suggested limit for ways providing local access
200	Suggested limit for street in 40 km/h zone
300	Noticeable pedestrian delays, but usually acceptable (Section 20.4)
400	Near environmental capacity, possible danger to children, definite restraint on movement
500	Environmental capacity, see above text
700	Nuisance obvious and a cause of annoyance
1000	Serious noise annoyance (Section 32.2.3)
1300	Significant danger to pedestrians
1500	Fumes may be a health hazard (Section 32.3)
1800	Near traffic capacity (Section 17.4), delays are obvious

7.4.2 Street geometry

The broad aspects of road geometry are discussed in Section 6.5 and Chapter 19. This Section concentrates on street geometry. The key minimum widths needed for street design are given in Table 7.3.

These can be used to give the road and carriageway widths shown in Figure 7.2. Some small differences exist. For example, use of the data in Table 7.3 for a cul-de-sac would permit a minimum carriageway width of 5.5 m in conjunction with mountable kerbs used to permit parking and emergency passing off the paved surface (Section 7.4.3). However, Figure 7.2 conservatively suggests a minimum of 6.0 m for the carriageway and 15 m for the entire road reserve.

As another example, the carriageway width for two vehicles to pass is often assumed to be 7.4 m rather than the 7.0 m that could be deduced from Table 7.3. This last assumption determines the minimum width for roads with a distributor role or carrying over 200 veh/d, although it is based on no vehicles being parked in lengths where passing is to occur. In this context, note that it is unnecessary to design for two cars parked opposite each other within a residential precinct. Minimum widths are often prescribed by local regulation. A common minimum width between opposite property boundaries is 15 m, which is seen to correspond to the cul-de-sac width in Figure 7.2. This 'traditional' value originated over a century ago to allow for adequate light and air space between buildings.

Table 7.3 Possible basic minimum widths, in metres.

Item	Possible basic minimum widths, metre
1. **Paths**	
1.1 clearance from slow vehicles	0.5
1.2 bicycle width (Section 19.6.3)	0.7
1.3 operating width for pedestrians	0.8
1.4 clearance from fast vehicles	1.0
1.5 distance to kerb	1.0
1.6 footpaths, one-way	1.0
1.7 bicycle operating width (Section 19.6.3)	1.1
1.8 footpaths, two-way	1.3
1.9 paths for prams and wheelchairs	1.8
1.10 bicycle paths,	see Section 19.6.3
2. **Lanes**	
2.1 minimum spacing between vehicles travelling in same direction	1.1
2.2 bicycle lane, next to car parking lane	1.5
2.3 maximum car width	2.0
2.4 maximum truck width (this width varies between regions, Section 27.3.2)	2.5
2.5 lane devoted to kerbside car parking	2.5
2.6 maximum width for car with door open	2.7
2.7 generous parking bay width	2.7
2.8 lane at intersection (Section 20.3.3)	2.8
2.9 lane for cars only	2.8
2.10 lane able to accommodate parked trucks	3.0
2.11 normal lane	3.5
2.12 lane for high speeds (Section 6.2.5)	3.7
2.13 lane providing for kerbside parking and cycling	4.0
2.14 lane to permit disabled vehicles to be passed at reduced speed within lane	6.0
3. **Carriageways** (see Section 6.2.5),	
3.1 private drive	2.7
3.2 access for utility services	2.8
3.3 slow (< 20 km/h), isolated, access way	3.0
3.4 cars passing cars, low flows and speeds	4.0
3.5 cars passing trucks, low flows and speeds	4.8
3.6 cars passing cars, low flows	5.0
3.7 cars passing cars, moderate flows or some trucks	5.5
3.8 cars passing cars, low flows, high speeds	6.0
3.9 cars passing cars, moderate flows, high speeds	6.5
3.10 corner sight distance	90.0

Within a residential precinct, the factors in Table 7.4 can be used to assess the impacts of choosing particular *street widths*. Minimum widths should be used wherever possible to reduce:

* construction costs,

* pedestrian crossing distances,
* vehicle speeds (Sections 6.2.2 & 18.2.1(1)), and
* visual intrusion.

Table 7.4 Street width factors.

	Pavement width					
	Normal		Reduced	Narrow		
pavement type	normal paving	low noise*	normal paving	mountable kerbs	unpaved, no kerb or gutter	
Functions:						
Access rating	good	good	good	reasonable	adequate	weather sensitive
Traffic service	high	good	reduced	low	low	low
Vehicle speed	normal	normal	lower	low	reduced	low
Passing options	normal	normal	abnormal	use verge	passing bays	poor
Visual options	low	low	high	high	high	moderate
Mainten-ance	low	medium	verge costs up	low	medium	high
Conflicts:						
Vehicle / pedestrian	high	high	low	low	low	high
Vehicle / vehicle	low	low	increased	increased	increased	high
Parking	adequate	adequate	good on verge	good in bays	adequate	good on verge
Visual impact	high	high	low	low	low	dusty and muddy
Drainage impact	high	slightly reduced	low	minimal	marked reduction	reduced
Resident preference	high, for neatness	high, for quiet	moderate	moderate	moderate	very low
Costs	high	reduced	low	low	low	very low

* see Section 12.2.2f

With respect to *street grade:*

* Vehicles can ascend slopes of up to 30 percent at crawl speeds and for short distances; however,
  # 15 percent is generally considered to be the maximum practical grade, particularly for safety reasons.
  # furthermore, grades of more than 12 percent are usually considered unacceptable and a source of drainage problems — e.g. most of the water in the gutter bypasses the entry pits (Section 13.4.1).
* As Section 19.3 indicates, 8 percent is considered a limiting grade for streets carrying through traffic.

* The limiting grade for paths catering for people with disabilities should be 5 percent. There should be occasional flat sections at least 1.5 m long to accommodate people in wheelchairs.
* At the other extreme, grades of less than 0.3 percent can create drainage problems and noticeable local crests and sags due to construction tolerances.

The *superelevation* requirements discussed in Section 19.2.2 are rarely needed for the low speeds encountered on streets. Pavement surface drainage *cross-falls* are listed in Section 6.2.5; values of more than 5 percent are not recommended for residential streets as vehicles begin scraping driveways leading into private property.

With respect to *horizontal curves* in residential streets, inside kerb radii can range from 7 m when the 50 percentile speed is as low as 30 km/h, to 80 m at the common legal limit of 60 km/h. However, if the carriageway width is 3.5 m or less, the minimum kerb radius will need to be 9 m if articulated vehicles are to be accommodated. When the inside kerb radius is less than 40 m and the nominal carriageway width is under 5.5 m, it will need to be increased by about 1 m to accommodate the path that vehicles actually take on the curves (Section 19.2.2). Curved streets having a radius of less than 120 m may require a greater width to accommodate utility services.

7.4.3 Street appearance

The appearance of a street can be enhanced by its geometric layout, landscaping, and tree planting. Trees and other vegetation can be used to determine the user's perception of an area (Section 7.2.3), to create visual interest and variety, and to unify the visual elements within the street. They also help manage local humidity, act as windbreaks, provide shade, control glare and reflection, and reduce noise. On the other hand, they can obstruct light, drop leaves and branches, be an unyielding object when struck, impede overhead cables, be a traffic hazard (Section 28.6.1), cause root damage to underground services, cause soil shrinkage (Section 9.3.3), and require frequent pruning. The location, size and type of trees must therefore be carefully considered. In the absence of a tree plantation, trees are normally planted in the same alignments as utility poles and fire hydrants. The following points should be kept in mind when producing designs intended to enhance the appearance of a street:

(a) design in harmony with the landscape (Section 6.4),
(b) utilise existing stands of trees,
(c) aim for co-ordinated diversity,
(d) keep streets as narrow as possible, e.g. occasional parking-bays can be used to avoid the need for wide pavements (Section 7.4.2),
(e) curve roads through straight reservations (Section 7.2.3),
(f) consider omitting one footpath and meandering the one retained,
(g) consider pavement alternatives to asphalt (Section 7.4.3),
(h) consider mountable kerbs (Section 7.4.3),
(i) treat trees as an integral part of street design and restrict the varieties used in a street,
(j) consider gently-sloped and grassed table drains as alternatives to kerbs, gutters, and underground drains (Table 7.4 & Section 7.4.5),
(k) wherever possible, underground all utility services,
(l) design street furniture to co-ordinate with and enhance the street environment,

(m) site houses to suit the landscape. Houses set-back a common distance from the street and/or at a common angle to the street can provide a monotonous effect.

7.4.4 Construction

The various standards used in the construction of a residential street are usually the province of the local government agency who, as well as controlling the construction of the pavement, may also have requirements for kerbs, drainage, footpaths, driveways, and underground utility services.

Pavement surface types are compared in Table 7.4. Of the favoured conventional types, experience indicates that many residents prefer streets with asphalt surfacings (Section 12.2.6) over alternatives such as concrete blocks (Section 11.6), concrete (Section 11.5.1), and brick pavers (Section 11.6). Whilst porous paving may reduce roadside drainage needs, it should be used with great care as very few subgrades function effectively when saturated (Chapter 9). Pavement skid resistance is discussed in Section 12.5.4 where a minimum skid resistance of SFC = 55 is recommended. This can be relaxed a little on low speed local streets, but slippery paving finishes such as slate, cobblestone or stencilled concrete should not be used.

Kerbs and gutters are used at the edge of most urban streets and roads and are typically at least 150 mm high. They serve the following six main purposes:

(1) prevent lateral spread of the pavement and provide strength at the edge of the pavement (Section 11.7),
(2) act as part of the street's drainage system (Section 13.3),
(3) manage any difference in level between footpath and carriageway,
(4) act as a longitudinal edge marking for drivers (Sections 22.2 and 24.4.3),
(5) act as a traffic barrier for unprotected footpaths, and
(6) act as a cue for visually impaired pedestrians.

The four types of kerb and gutter are:

(a) *barrier*; the kerb face has a gradient of 4 vertical to 1 horizontal. These kerbs are always used for roads of the local access or higher class (see Figure 7.1) in urban areas.
(b) *semi-barrier*; has a vertical kerb face for its lower half, and a 1 to 1 slope of the upper half.
(c) *semi-mountable*; has a vertical kerb face for lower third, and a 1 vertical to 1.3 horizontal slope for the upper two-thirds.
(d) *mountable* (or layback or roll-over); has a kerb slope of 1 vertical to 12 horizontal.

Kerbs are usually constructed from 30 MPa concrete, but may also be built from asphalt.

Kerbs can pose a major barrier to the disabled, the visually impaired, and the pram-pusher. Thus, at potential crossings it is essential that there be gently-sloped ramps leading to smoothly-dished gutter crossings (or drop kerbs). If at all possible, the ramps must be placed across the full width of the walkway. If not, the ramp should be at least a metre wide and centred in the walkway. In addition, the ramps should be located so that they are on the same centreline as other relevant walkways and cross-overs (Bail, 1981). Typically, the ramp, gutter, and crossing pavement should present a continuous profile for wheelchairs and prams and use grades of 8 percent or less. It is also desirable to provide detectable advance warning of the ramp by both colour contrast and a tactile surface treatment. The street kerbs should also be sloped gradually at ramps, to provide another cue for visually impaired walkers. Pedestrian crossings are discussed in Section 20.4.

Footpaths are preferably independent constructions separated from the carriageway by at least a kerb, a change of level, and a planted strip (e.g. Figure 7.5). They must provide their users with safety, a good walking surface, and adequate width. Appropriate widths are suggested in Table 7.3 and Section 19.6.3. Cross-falls of more than 1 in 40 make usage difficult for disabled pedestrians. Disabled pedestrians and wheelchair users all require a smooth, even surface. Tactile pavement elements can be used to warn blind people of approaching hazards.

A driveway provides vehicular property access between the property boundary and the edge of the carriageway. Driveway requirements are influenced by the needs and rights of users, owners and occupiers of adjacent land, the carriageway width, the type, flow, and speed of traffic, and the characteristics of the area. A key criterion is that a driveway should be located at points with adequate sight distance. Driveways should discourage vehicles from reversing onto public streets and attempt to avoid vehicles queuing on the street for access into a driveway. It is also poor practice to locate them near traffic signals or intersections or adjacent to barrier lines (Section 22.2.2). The minimum recommended width for a driveway is given in Table 7.3, with a minimum turnout radius of 1 m. Vertical clearances from vehicle underbodies also requires attention (Section 19.3), particularly at the sags caused by gutter crossings.

7.4.5 Drainage

Drainage is needed to avoid flooding streets or property, endangering public health, and saturating the subgrade and basecourse (Section 9.3). It is discussed in detail in Section 13.4. In a residential street, kerb and gutter systems (Section 13.3.2) can increase rain runoff rates and therefore increase downstream flow and water pollution, whilst being of higher cost and poorer visual appeal than available alternatives (Table 7.4).

The common alternative is the table drain (or swale), that is formed by gently sloping the soil surface into a low micro-valley. The rough surface of the table drain advantageously lowers the speed of the water flow and allows the water to percolate into the soil. The process is aided if carefully selected grasses are planted on the swale surface. The underground drain may commonly be a French drain (Section 13.4.2). The main problem with these techniques is that they will increase the risk of local flooding. A development of the swale concept is the combination drain that includes an underground drain below the table drain and is designed to carry excess flows. Precast systems include a subsurface drain directly below and linked to a continuous, 'grated' drainage channel.

7.4.6 Public utilities

Public utilities frequently require accommodation within the road right-of-way and may possess the legal authority to achieve their requirements. In fact, most arrangements between road and utility agencies are of a voluntary nature. Underground facilities placed in the road reserve include cables, wires, ducts, and pipes carrying electricity, telecommunications, water, gas, sewerage, stormwater, oil, and various other public utilities. Their initial installation usually creates localised disturbance and underground access is also required for both regular and emergency maintenance.

There are usually local guides to the co-ordination of the location of utilities within the road right-of-way. They commonly contain such general principles as the placing of water pipes on the high side of an urban street to minimise flooding if the pipe bursts (and on the low side of a rural road to minimise scouring). Figure 7.5 is a typical set of recommendations for local distributors. It will be seen that the suggested minimum requirement of 4.25 m could be accommodated within any of the cross-sections in Figure 7.2. Nevertheless, for constrained conditions, the right-hand side of Figure 7.5 indicates that all facilities can be fitted into 1.8 m, which is the maximum footpath width recommended in Table 7.3.

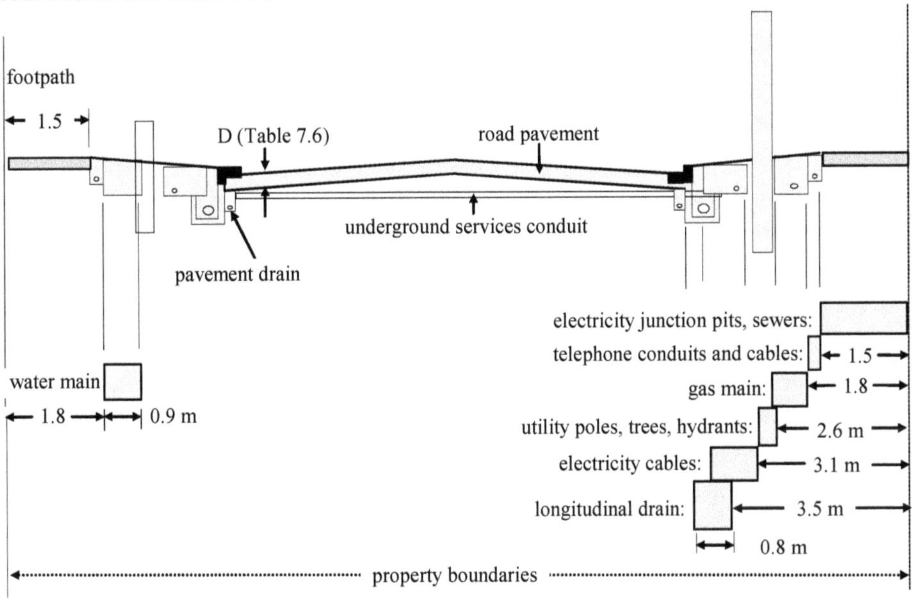

Figure 7.5 Typical allocation of underground space in a local distributor.

The use of common trenching for various utilities is often proposed but rarely adopted. One problem relates to the need to maintain minimum separation distances between some utility services, e.g. between electrical wiring and other potential conductors. Another problem relates to the large separations needed to provide for maintenance access.

It is essential that adequate depth of cover is provided over the underground utility services located in these trenches, to protect them from accidental damage and from damage caused by wheel loads on the surface. Minimum cover requirements vary from location to location: typical values are shown in Table 7.5 and design data are given in Section 13.4.2.

Street maintenance is discussed in Section 26.3.

Table 7.5 Typical protective cover, D (mm), above an underground utility service.

Service	Network service		Supply to consumer	
	verge	pavement	verge	pavement
Gas	600	600	450	450
Electricity	450	750	450	750
Telephone	300	450	300	450
Water	450	675	450	450
Sewerage	900	900	900	900
Drainage	750	900	600	670
Traffic signal cables	450	600	–	–
Combustible	750	1200	–	–
or flammable liquids	300 below subgrade			

CHAPTER EIGHT

Pavement Materials

8.1 INTRODUCTION

8.1.1 Fundamental definitions

The preceding chapters have discussed various aspects of road location and geometry. Having thus established where the road will be and what proportions it will take, attention is now turned to a discussion of the materials to be used in its construction.

The seven basic requirements for pavement material are shown in Table 8.1. Criteria developed to detect whether or not a material meets these requirements must be:

(a) sound theoretically,
(b) validated experimentally,
(c) obviously related to the requirement in question, and
(d) capable of reasonable testing and comparison.

In order to satisfy these requirements criteria must be well known and clearly expressed. This section therefore firstly discusses the basis on which the associated physical properties of paving materials are established.

Table 8.1 Pavement material requirements.

Property	Definition:	Section
1. **Construction**	Must have the ability to be placed, compacted, and formed:	
1.1 workability	– to the required condition and shape.	8.1
1.1 economy	– at an acceptable cost.	8.5
2. **In-service**	Must have the ability to resist unacceptable:	
2.1 strength & stiffness	– damage and deformation under load.	8.4.3&4
2.2 durability	– deterioration in desired characteristics over time.	8.5.10–11
2.3 volume stability	– changes in volume as conditions, such as moisture content, change.	8.4.2
2.4 wear resistance	– scour, abrasion, polishing, and loss of frictional resistance of its surface.	12.5.2
2.5 impermeability	– penetration by water, particularly at its surface.	9.2.2

Generally, pavement materials are composed of solid, liquid and gaseous phases and can be represented by a phase diagram (Figure 8.1). The total mass and volume of the original sample are given by the quantities M and V defined in Figure 8.1, and can be determined by simple direct measurements on the initial sample. The conventional density, M/V, is called the bulk density.

The mass of the solids in the sample, M_s, is determined by drying the sample in an oven until its mass is constant. This gives the mass and volume of water, M_w and V_w, as:

$$M_w = M - M_s = \gamma_1 V_w$$

where γ_1 is the density of water. Note that in the SI units used in this book:

(a) γ_1 = 1000 kg/m^3 = 1 t/m^3 = 1 kg/litre, and need not be included in any formulae using the [tonne, metre] or [kilogram, litre] combinations; and

(b) the old term specific gravity is not used in SI as it is numerically equal to density.

As water can be present as free liquid, vapour, adsorbed onto surfaces (e.g. with clays), or absorbed into solid particles, it is necessary to follow a standard drying procedure, in order to avoid debate over which water forms to include. The procedure commonly involves drying the sample to a constant mass at 105 C for at least 30 minutes. Various techniques are reviewed in test 8.6u below.

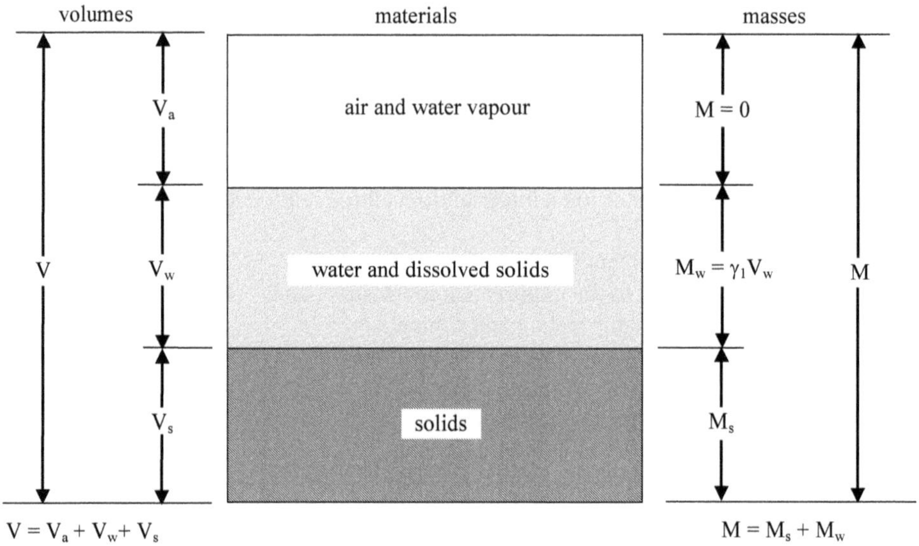

Figure 8.1 Phase diagram showing the three phases that constitute a roadmaking material.

The terms derived from the simple quantities in Figure 8.1 are:

bulk density, t/m^3 $\quad \gamma = M/V$ (8.1a)

dry density, t/m^3 $\quad \gamma_d = M_s/V$ (8.1b)
(or dry (bulk) density)

solid density, t/m^3 $\quad \gamma_s = M_s/V_s$ (8.1c)
(or soil particle density)

wet density, t/m^3 $\quad \gamma_w = M/(V_s + V_w)$ (8.1d)
(or saturated density)

moisture content $\quad w = M_w/M_s$ (8.1e)
(or water content or gravimetric water content) (often quoted in percent)

volumetric moisture content $\quad w_V = V_w/V_s$ (8.1f)

total void ratio $\quad e = (V_a + V_w)/V_s$ (8.1g)

air void ratio, $\quad a = V_a/V$ (8.1h)
(often quoted in percent)

degree of saturation $\quad S_r = V_w/(V_a + V_w)$ (8.1i)

porosity, $\quad n = (V_a + V_w)/V$ (8.1j)
(often quoted in percent)

specific volume $\quad v = V/V_s$ (8.1k)

The determination of the volume of solids, V_s, takes much care and hence the dry density, defined by Equation 8.1b, is commonly used in preference to γ_s, the solid density given by Equation 8.1c. In the field, even the determination of dry density is a complex matter, as is illustrated in Section 8.2.2. In the various definitions given below it is therefore important to note the special definition of dry density as it differs from the 'layman's' definition in which the volume of the dried material would be used. Dry density is best thought of as a measure of the amount of solid matter present at the current level of compaction and a better term for it might be dry (bulk) density. On the other hand, the solid density is a function only of the solid component and so is independent of the compaction level.

The total void ratio, e, is equivalent to the term VMA (voids in mineral aggregate) which is sometimes used in discussions of asphalt (Section 12.2.3). Bulking is the proportional increase in volume caused by the addition of water to a dry sample. If the water initially simply displaces the air in the voids in the sample, then bulking would not occur until $V_w > V_a$. When all air voids are full of water ($V_w = V_a$), the material is said to be saturated.

The moisture content, w, is defined in Equation 8.1e as the ratio of the mass of water to the mass of solids. It is discussed further in Sections 9.2 and 9.3. It is obviously not a continuous property within a mixture of solids, water, and air as major variations can occur over microscopic dimensions. The determination of moisture contents in the field has a coefficient of variation of about 25 percent, although this can be reduced to 5 percent in well-controlled operations (e.g. in placing an upper basecourse). In addition, repeatabilities can be 10 percent and reproducabilities 20 percent (Section 33.4.1), so precise results cannot be expected. The variations are caused by variations in the amount of water added, in the pre-existing moisture content, in the absorptivity of the material, and in the grading of the material (most of the water fills the voids within the mix, Section 8.3.1).

In-service pavement performance is closely related moisture content (e.g. Section 14.1) and so the accurate determination of moisture content is a critical part of pavement construction. As defined in Equation 8.1f, the volumetric moisture content, w_v, is the ratio of the volume of water to the volume of solids and is related to w by Equation 8.2h.

8.1.2 Relationships between fundamental variables

The common relationships between the above terms can be derived from the following equations:

$$n = e(1 + e) = ev \qquad (8.2a)$$

$$\gamma_d = \gamma/(1 + w) \qquad (8.2b)$$

$$\gamma_d = (1 - a)/([1/\gamma_s] + [w/\gamma_1]) \qquad (8.2c)$$

$$\gamma = \gamma_d + \gamma_1 S_r n \qquad (8.2d)$$

$$\gamma_w = \gamma/(1 - a) \qquad (8.2e)$$

$$e = (\gamma_s/\gamma_1)w + (\gamma_s/\gamma_d)a \qquad (8.2f)$$

$$\gamma/\gamma_s = (1 + w)/(1 + e) = (1 + w)/v \qquad (8.2g)$$

$$w_v = M_w/(\gamma_1 V_s) = w(\gamma_s/\gamma_1) = eS_r \qquad (8.2h)$$

$$S_r = 1/(1 + [\{a/w\}\{\gamma_1/\gamma_d\}]) \qquad (8.2i)$$

$$v = 1 + e \qquad (8.2j)$$

Equation 8.2c is plotted in Figure 8.2, which uses the two readily measured quantities γ_d and w as variables. The resulting line is called the air voids line.

Some specifications are based on wet density (Equation 8.1d) as it does not require drying of the specimen and thus can be done much more quickly than a dry-density test by using Equation 8.2e and a knowledge of air voids. Placing a = 0 in Equation 8.2c gives the zero air voids (or saturation) line:

$$1/\gamma_d = 1/\gamma_s + w/\gamma_l$$

The presence of entrapped air means that it is usually practically impossible to reach the zero air voids line (i.e. to achieve a = 0 and $S_r = 1$). Similarly, the saturation moisture content for a particular dry-density, γ_d, is given by Equation 8.2c with a = 0 as:

$$w_{sat} = [(1/\gamma_d) - (1/\gamma_s)]\gamma_l$$

Using this in Equation 8.2f with a = 0 shows that at saturation:

$$e_{sat} = w_{sat}(\gamma_s/\gamma_w)$$
$$= (\gamma_s/\gamma_d) - 1$$

Compacted density and moisture content are clearly critical quantities and are discussed further in Section 8.2.

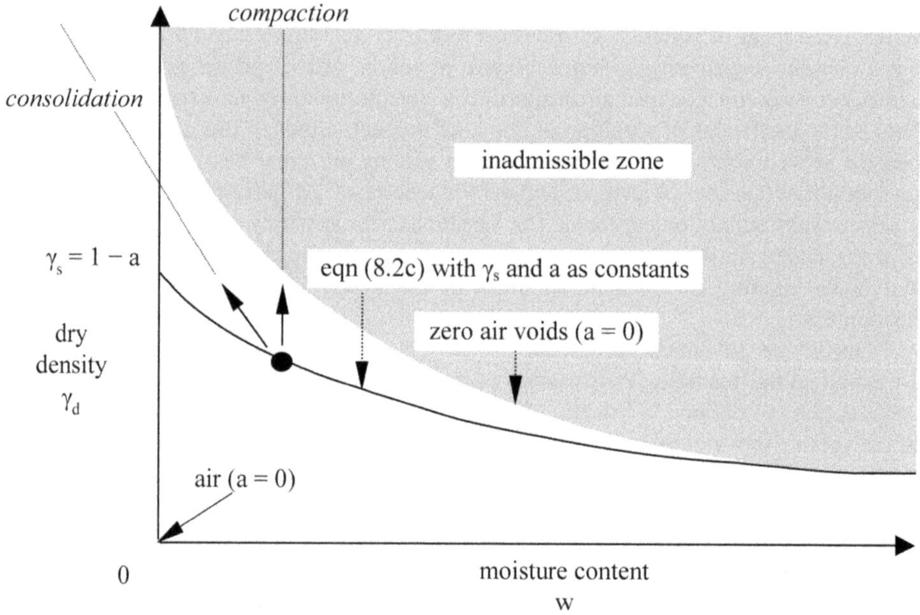

Figure 8.2 Basic dry-density vs moisture content curve.

It will be seen later that the nature of the individual particles of which the soil (Section 8.4.1) is composed is also critical. The relevant particle factors and the associated subsequent sections are:

* size distribution (Section 8.3.1),
* proportion of fine particles (Sections 8.3.1 & 8.4.1),
* maximum particle size (Section 11.8.2),
* surface texture (Section 12.5.2),
* particle shape (Sections 8.3.2, 8.6a & 11.1.3), and
* physical and chemical stability (Section 8.5.10).

8.2 COMPACTION

8.2.1 Background

Compaction is the process by which the air void ratio, a, of a soil — and hence its volume V — are both reduced. This is usually done by mechanical means, e.g. by rolling (Section 25.5), and takes place with insufficient time for any incidental change in moisture content, w. Consolidation is the process by which both the volume and the moisture content are reduced under long-term sustained load. From Equation 8.2c, consolidation will therefore cause the dry density, γ_d, to increase. This explains the orientation of the consolidation arrow shown in Figure 8.2. The simple device used to measure consolidation is called a consolidometer. It is similar to the oedometer, which is used to measure the swelling of soil (Section 8.4.2).

A loosely-compacted material will deform under load as its constituent particles adopt a more closely packed arrangement and as air (and possibly water) is expelled. On the other hand, a well-compacted material will mainly deform by deformation of the particles themselves. This will clearly result in a much stiffer response and the well-compacted layer of material will suffer much less permanent and elastic deformation under subsequent traffic, and hence have a longer service life (Section 14.3). In addition, compaction raises the shear strength of the material by increasing particle interlock (Section 11.1.2), and decreases permeability by decreasing available air voids (Section 9.2.2).

Finally, for unpaved roads (Section 12.1.1), compaction produces a more coherent structure that reduces surface erosion and gravel loss. At the other extreme, over-compaction will cost extra money with little return on the investment, and may cause high interparticle stresses, leading to the breakdown of individual particles, and thus to premature failure (see test 8.6(b) for instance).

Thus, once a pavement material has been selected, the prime need is to compact it at reasonable cost to a density at which it will behave as a stiff and strong structural course. This explains much of the pavement engineering concentration on in situ density.

8.2.2 Density measurement

Given the importance of achieving acceptable densities, the first issue is to be able to measure it reliably. The practical implications of this are discussed in Section 25.5. The more common field methods are listed and described below. They have a coefficient of variation of about 4 percent:

(a) *sand replacement*. In this method a cavity is made in the material and its volume measured by filling it with a known volume of sand (or linseed). The IPCAD (in-place coarse-aggregate density) device allows the amount of sand required to be reduced by also inserting a cylinder of known volume into the cavity. The material removed is weighed before and after drying. Density and moisture content can then be determined from Equations 8.2b&e. IPCAD gives densities about two percent higher than the all-sand method (Jameson, 1985).

(b) *balloon densometer*. The volume of the cavity is measured by inserting a balloon and filling it with a measured volume of water.

(c) *core cutters*. A coring device removes a fixed volume of material.

(d) *nuclear density meters*. The nuclear density meter can be used for the in situ determination of both moisture content and wet or bulk density (Section 8.1), with a maximum penetration depth of 300 mm. Its basic components are a radioactive source, a radiation detector, a counter and a display unit. It utilises the fact that the attenuation of gamma rays passing through a material is density dependent. The meters emit radiation and operate on either backscatter mode in which the emitter is located directly above the surface, or in direct transmission mode in which they are at the far lower boundary. Most experts recommend use of the direct transmission mode. The detectors are at the surface.

Current commercial devices use either radium 226 or caesium 137 as X particle emitting sources for density determination and americum 241-beryllium as a neutron source for the determination of moisture content. The devices should meet local safety regulations. Although most commercial meters are robust and safe, care is still needed in their handling as they are potentially hazardous as radiation levels from the unshielded ray source could be as high as 500 mR/h (Hamory, 1980). In the field, the devices are often seated on a layer of 425 μm bedding sand.

Calibration curves are needed to relate measured X or neutron counts to density or moisture content. It is a difficult process. Calibration using standard blocks of known density is probably the most widely used method for obtaining such curves (Hamory 1980). However, this method relies on the assumption that the calibration is material-independent. Tests have shown that the use of blocks supplied with an instrument can result in substantially biased estimates (Jameson, 1985). Thus calibration of each material against sand replacement or IPCAD data is probably desirable. A final problem is that the density and moisture content calibrations are inter-related.

One disadvantage of the nuclear meter is that it is not obvious which particular sample of material is being tested. On the other hand, the device does not cause any significant damage to the material being tested whereas the alternatives are all destructive tests. Another concern is that the device can be affected by surface irregularities.

8.2.3 Theory

Compaction is not a trivial process and requires considerable constructive effort and energy. The resistance of a material to compaction and in-service deformation depends on the following effects:

(a) *Mechanical effects* (*internal friction* and *particle interlock*, Section 11.1.2. A graphical, stress–strain definition of internal friction is given later in Figure 8.9. It is also called friction angle or angle of shearing resistance.)

Mechanical effects are influenced by the shape (Sections 8.3.2, 8.6a & 11.1.3) and surface texture (Section 12.5.1) of the coarser particles in the material being compacted. They increase markedly as density and normal compressive stress increase and are little influenced by moisture content. They are the dominant factors with granular soils such as gravels, sands, and coarse silts (Sections 8.4.1&3).

(b) *Chemico-electrical effects* (*cohesion*). A graphical, stress–strain definition of cohesion is given later in Figure 8.9. Cohesion can be considered to be the shear strength of a material at zero normal stress ($\sigma_l = 0$ in Figure 8.9). A deceptive term is unconfined cohesion which usually means the maximum compressive strength of an unconfined cube of soil ($\sigma_c = 0$ in Figure 8.9). Cohesive effects arise from the properties of the fine particles in the soil giving rise to molecular attractive forces between the particles (Section 8.4.2), and to tension in the soil water (Section 9.2.3). Cohesion dominates compaction behaviour in clays, which are the typical cohesive soil. Silts, with particle sizes between 2 and 60 µm, usually gain their cohesion solely from surface tension and hence are only cohesive when moist. Granular soils — see (a) above — are usually cohesionless (or non-cohesive).

Cohesion is greatly influenced by moisture content, peaking when molecular bonding or surface tension is maximised, and then decreasing as the moisture content increases, to reach zero when the material becomes liquid (Section 8.4.2). Cohesion usually increases with density, typically because the higher density indicates a less granular material.

Cohesiveness is a useful property for material used in unsurfaced roads as it creates inherent surface tightness (Section 12.1.2). However, Section 8.4.2 shows how the cohesiveness of clays can add to the difficulties associated with their compaction.

(c) *Drainage effects.*

Pavement materials can also be divided into two additional categories, viz.:

(1) *those from which water can be readily squeezed during construction.*

These are typically well-graded mixes (Section 8.3.2) governed by factor (a) above. The moisture content at which they are placed is not critical, although the highest densities are obtained with the materials either saturated or completely dry. Although the densification process here is commonly called 'compaction', theoretically it is short-term consolidation.

(2) *those from which water is not readily squeezed during construction.*

These are typically either cohesive materials governed by factor (b) above or poorly-graded mixes. The compaction process may realign and physically breakdown some of the particles and reduce porosity by producing a more closely packed particle arrangement. If this means that the contained water and air cannot readily leave the mix, it will take some of the compaction pressure and inhibit the physical processes. The water pressure produced is called pore pressure (Section 9.2.3&5). The water usually contains the last 2 or 3 percent of air entrapped as small bubbles.

Methods of compacting soils are discussed in Section 25.4. As stiffness and strength increase with density, the energy used during compaction — the compactive effort — increases as density increases until no further compaction is possible. For a fixed moisture content and fixed amount of compactive effort, one particular dry-density will be achieved for each soil type. A plot of such data for one soil on a graph such as Figure 8.2 gives curves such as 'cc' and 'bb' in Figure 8.3, traditionally called Proctor curves. These may be obtained by laboratory procedures, as described in Section 8.2.5, or by field measurements of the effect of compaction plant.

Each particular laboratory or construction procedure will have its own characteristic curve. These curves will each have an obvious maximum dry density, $\gamma_{d,max}$, which will commonly range from about 1.3 t/m^3 for some clays to about 2.1 t/m^3 for some rocks. The coefficient of variation for maximum dry density tests is about 1 percent (Smith and Crews, 1983).

In addition, when water is initially added during the compaction of a dry material, the density also increases. This is because the added water is denser than the air it displaces. However, as the material becomes damper, the incompressible water begins displacing denser solids rather than air and overall densities begin to drop. Thus the addition of water during compaction must be carefully controlled, as will be seen in the next sub-section.

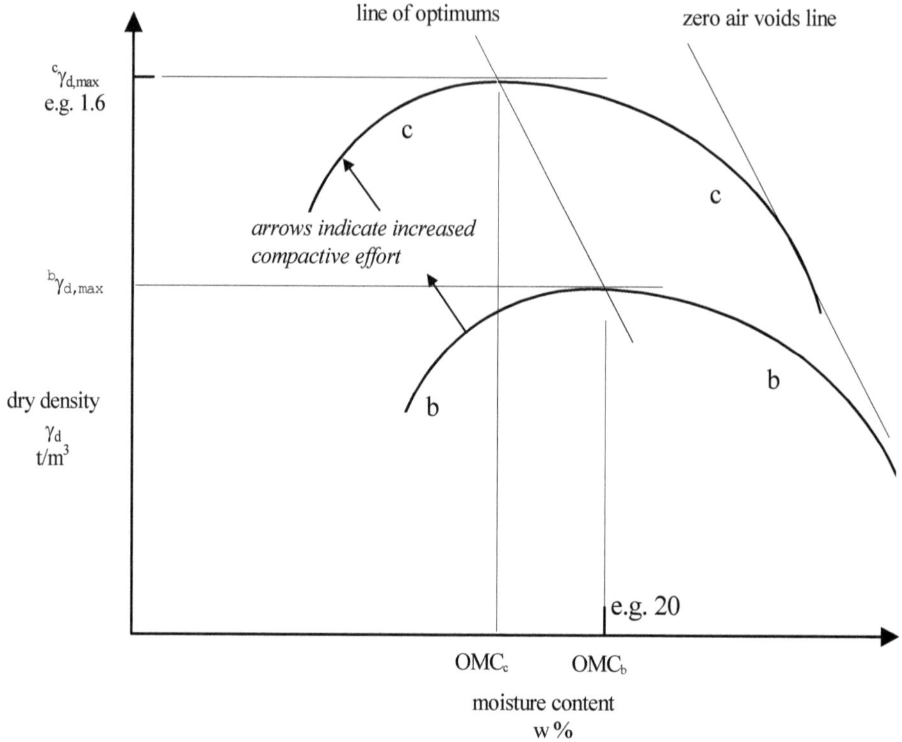

Figure 8.3 Classical dry-density vs moisture content curve for impermeable soils comprised only of fine particles (from Figure 8.7, this is a soil where the majority of particles are below 60 μm). See Figure 9.3 for experimental data.

It is not possible for compaction to fill all the voids in a soil with water, realigned particles, or newly-ground finer solids. For example, at maximum dry density even clays will have an air void content of about 5 percent. For this reason, some specifications use air void content as an alternative to density for checking compaction.

8.2.4 Optimum moisture content

It will be noted from Figure 8.3 that there is an optimum moisture content (OMC, also referred to as w_o) at which the maximum density will be achieved for a particular compactive effort. OMCs range from under 10 percent with some granular soils to over 20 percent with some clays. The OMC is not easy to determine accurately and has a coefficient of variation of about 20 percent, although it has been suggested that the value can be as high as 45 percent or as low as 10 percent (Smith and Crews, 1983). For moisture contents below (or 'dry of') OMC, cohesive forces and particle contact pressures both resist attempts at compaction; thus the resulting orientation of individual particles is usually random. Density begins to increase again when the material becomes extremely dry.

When material is compacted at moisture contents above (or 'wet of') OMC, resistance comes from increasing pore pressure (Section 9.2.3) and the particle arrangement is usually non-random and highly oriented. Compaction plant usually have difficulty operating on such material due to rutting and bogging, and the compactive effort used will be thwarted. For such reasons, moisture content is sometimes referred to the OMC as a reference level.

By definition, field compaction should take place at OMC for the particular compactive effort in order to achieve maximum density. Amongst other things, this will decrease permeability and reduce the potential for further compaction and rutting. However, Sections 8.4.2 and 9.4.2 show that, for expansive clays, another option the builder must consider is to compact at the long-term moisture content when moisture conditions have reached equilibrium. This will minimise volume change effects, but at the price of a lower strength, lower stiffness, and more permeable pavement. As low permeability will limit the entry of water, the balance for expansive clays is in favour of compaction at OMC at any interface areas such as pavement edges (Section 9.3) and at the equilibrium moisture content (Section 9.4.1) for the remainder of the pavement.

Another variation of the above direction towards the use of OMC arises in areas where water is scarce. In these circumstances, compaction at dry of OMC must be accompanied by greater compactive effort in order to achieve the intended dry density. Compaction below OMC can be difficult, due to the absence of water to act as a lubricant for particle rearrangement. Compaction may thus cause the soil particles to breakdown into finer sizes. This will reduce soil strength and increase moisture susceptibility, due to enhanced plasticity (Section 8.3.3). To avoid this effect, the particles may need to be tested for strength and abradability, e.g. by tests 8.6b,i&n in Section 8.6. Another concern is that quite high strengths can be achieved at high air void contents (e.g. the bottom left hand segment in Figure 9.3). Such strengths can be misleading as closer packing over time can lead to long-term settlement problems.

A line drawn in Figure 8.3 joining the peaks of the various compactive effort curves (such as bb and cc) is called the line of optimums and often tends to parallel the air void lines. Attempts to compact above the line of optimums often result in lower than expected pavement stiffness and in-service problems.

In comparing test results from basically the same materials taken from different sites, the simplest approach is the single-point (or one-point) method. This only requires the determination of the moisture content and density of a field sample and the assumption that the field dry-density vs moisture content curve is identical to that determined in the laboratory. Separate field maximum dry density and OMC values are not evaluated.

If dry-density vs moisture content curves are obtained for two 'sites' (e.g. for laboratory and field), a useful technique is to adjust the moisture content results for the second site by multiplying them by the ratio $OMC_{first\ site}/OMC_{second\ site}$. For comparing densities, any dry density at the second site is then adjusted by multiplying it by the ratio of the dry density of the first site at the adjusted moisture content to the dry density at the second site and at the unadjusted moisture content.

8.2.5 Achieved density

It is usual to specify the density required in the field in relation to densities achieved in the laboratory, although a few specifications define density indirectly in terms of the air void content. Samples are prepared at a number of moisture contents to determine the maximum dry density and the corresponding OMC, which are known as standard (or Proctor) compaction values. For the common laboratory test, the material is compacted in a cylindrical one-litre mould. Conventionally, 25 blows of a 2.7 kg hammer dropped through 300 mm have been used for each of the three equal layers. The compactive effort is about 600 kJ/m and the procedure is about equivalent to the use of a medium vibrating-roller (Section 25.8). The technique is also used as an impact compaction test. A vibratory test is sometimes used for cohesionless soil (Section 8.2.3).

Another version of the impact compaction test uses much greater compactive effort with a 4.9 kg hammer, 450 mm drops and five layers. This produces higher densities at lower OMCs, which are known as modified compaction values. The compactive effort is about 2700 kJ/m. For material compacted below the line of optimums, this density is about equal to that at which the maximum stiffness of the compacted layer is achieved (Shackel, 1973). This confirms general roadmaking experience that modified values better represent successful field conditions for pavements (Metcalf, 1978), although standard values are still used, particularly for underlying subgrades and earthwork formations (Section 2.2). When performed carefully, these two versions of the impact compaction test have a repeatability of about 2 percent and a reproducibility of about 4 percent.

If the material being tested includes particles whose size is too large (e.g. 20 mm) for the test cylinder, it is necessary to remove the large material and empirically adjust the test results (e.g. AASHTO T224). If the proportion of removed rock is p, the adjustments are typically of the form:

$\gamma = (1 - p)$(measured density of compacted, testable material) + αp(dry density of removed rock)

where α is an empirical multiplier.

The effect of increased compactive effort on the compaction curves is illustrated in Figure 8.3 by the shift from curve b to curve c. Figure 8.4 shows how the stage is reached, however, where further compactive effort produces little additional densification. This is called compaction to 'refusal' and occurs when the resistance of the material exceeds the stresses applied by the compaction process. For cohesionless material, this resistance is best estimated by the bearing capacity (Section 8.4.5) of the compacted layer. For cohesive, impermeable materials, the maximum resistance is usually greater and is best estimated by their compacted shear-strength (Section 8.4.3). In addition, the achieved density increment is approximately proportional to the logarithm of the energy input (HRB, 1962).

In the field, compaction is achieved by rolling the placed material, as described in Section 25.5. It may take up to sixteen passes of a roller to achieve the field equivalent of points 3 and 4 in Figure 8.4 (Morris, 1975). It has been suggested that the maximum dry density achieved can be related to moisture content for a remarkable range of materials by the relationship (Neeson, 1973):

$$w\,(\%) = 30(2.4 - \gamma_{d,max})$$

Density in the field as a percentage of either the standard or modified maximum dry density values is called relative compaction. The ratio is known as the dry density ratio, R_D, and a typical value for roadmaking would be between 95 and 101 percent modified maximum dry density. R_D has a coefficient of variation of 2 to 10 percent (Smith and Crews, 1983), however values below one percent are feasible for good quality work on upper basecourses.

A related term is relative density, which is the state of compaction of a soil with respect to its loosest and densest states (i.e. from 0 to 100 percent). The loosest state is defined by gently pouring the soil into a mould and the densest by either the standard or modified maximum density (BS 812).

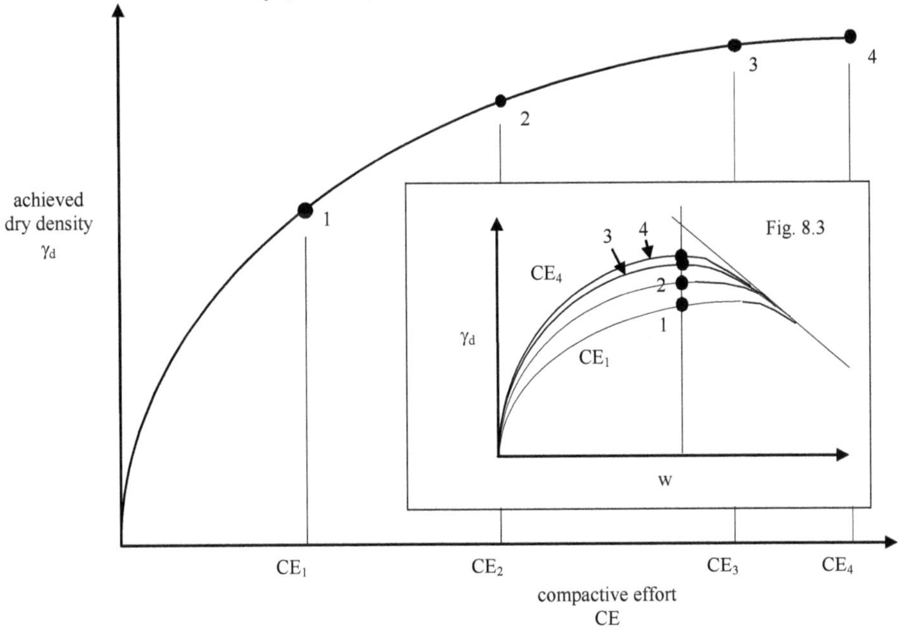

Figure 8.4 Effect of continued compactive effort. Compactive effort could be measured by the number of hammer drops in the laboratory test or by the number of roller passes on the job site.

There is often a need to rapidly determine the relation between field conditions and the desired optimum conditions. One technique is the Hilf method. Its main advantage is that it does not require moisture content determinations in order to determine the in situ dry density ratio and the deviation from OMC with reasonable precision. It works best for fine-grained material dry of optimum. The desired test points are obtained by simple compaction tests on the sample as removed, and then on the sample with the addition of 2 percent and 4 percent more water. Occasionally it is also necessary to dry a specimen back to obtain a peak in the data points. The adjustments to the data to give OMC and maximum dry density are complex, and the reader is referred to Hilf (1957) for more details.

In recent times the importance of reproducing field conditions has led to the wide use of gyratory compaction, using devices such as the electro-pneumatic Gyropac, for laboratory compaction, particularly for asphalt.

8.3 GRADING

8.3.1 Particle size

In assessing the grading of a material, the first step is to determine particle size. This is established by sieving a soil sample through standard sieves or by using sedimentation methods and an hydrometer if the soil particles are too small to sieve. The coefficient of variation is usually below 5 percent and test results should be within 5 percentage points of the specification (Jameson, 1984).

Standard sieves are usually defined by their aperture size, which is the side dimension of the square space between each pair of equally-spaced precisely-woven wires. Common sieve sizes are based on the British (BS 410), U.S. (ASTM D422), or Australian (AS 1152) standards with openings typically ranging from 22 μm to 125 mm, over a series of some 60 or so sizes. The series is commonly geometric, providing an approximately constant ratio between successive sizes.

The test results from sieving are commonly plotted as the percentage of the soil sample passing a given size of sieve (related to the mass of material smaller than that diameter). This sieve size is the maximum particle diameter in millimetres and is plotted on a log scale to provide a particle-size distribution (or grain-size distribution or grading) curve. Typical curves are shown in Figure 8.5. Particle size distribution is often represented by the exponent, n, in the expression:

$$(\text{\% passing sieve size d})/(\text{\% passing sieve size D}) = (d/D)^n \qquad (8.3)$$

which is also known as Talbot's grading curve. N is usually found to be between 0.3 and 0.5.

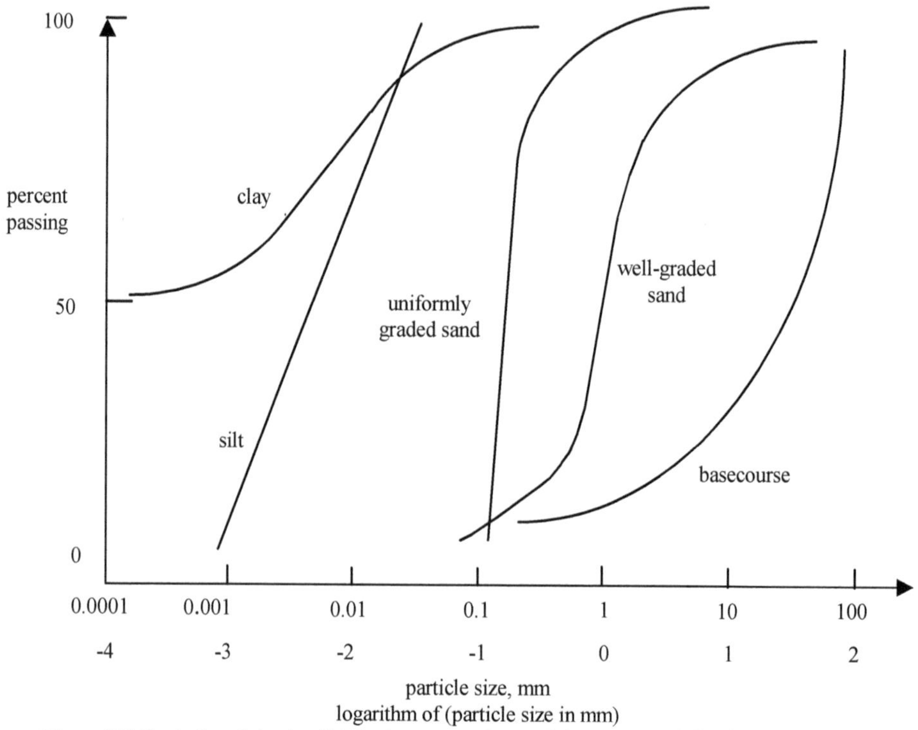

Figure 8.5 Typical particle-size distribution curves. Some of the terms are defined in Section 8.3.2.

Gradation is a term used to describe the particle size distribution of a soil. The region into which the data for a new material must fit when plotted on a particle size distribution curve is called a grading envelope. In a soil context, the word *grain* is often used as synonymous with 'particle'. A further definition of grain for rock is given in Section 8.4.1.

In the sedimentation/hydrometer test, the rate at which the particles settle gives an indication of their respective sizes. Stokes' Law is used to calculate the size of particle that has settled a particular distance in a particular time and the hydrometer measures the density of the water at these various times in order to determine the amount of material in that size range.

Roadmaking material is often referred to by its particle size as determined by sieving and the common terminology is shown in Table 8.2. The proportion of smaller sized particles can have major influences on performance, as the discussion on clays in Sections 8.3.3 & Section 8.4.2 will show. Whereas the properties of coarse-grained soils are largely influenced by their grading and particle shape (Section 8.3.2), and indirectly by gravity, the fine-grained soils are more influenced by their surface properties and, indirectly, by the electrical forces generated by charges on those surfaces (Section 8.4.2). In addition, the presence of a myriad of small interconnected voids, rather than a few large disconnected ones, will allow positive pore pressure to develop more readily and thus reduce shear strength (Section 9.2.3).

Table 8.2 Particle size terminology.

Term	Definition
Coarse fraction:	Retained on a 4.75 mm sieve
Fine fraction:	Passing a 4.75 mm sieve
Coarse aggregate:	Retained on a 2.36 mm sieve
Fine aggregate:	Passing a 2.36 mm sieve, retained on a 425 μm sieve Note: some specifications have different sizes for the coarse and fine aggregate boundary.
Binder:	Passing a 425 μm sieve Note: material is called a *binder* when it forms a soil mortar that embeds the coarse aggregate and prevents its movement. Thus it should be cohesive, impermeable, non-swelling, and – as a test – able to be formed into a shape with the hand and to retain that shape upon drying.
Filler:	Passing a 75 μm sieve (see Section 12.2.2) Fillers are thus fine particles. They were defined in the context of Section 8.1 as < 60 μm, rather than < 75 μm.

8.3.2 Grading types

Classification of particulate material by grading is important in pavement engineering as the value of many relevant properties such as internal friction (Section 8.4.3), voids content, wear resistance, and permeability depend on the distribution of particle sizes. In addition, pavement materials can be in either of density categories (1) or (2) in Section 8.2.3, depending on their grading and the characteristics of their finer particles.

A *well-graded* (or dense-graded) particle size distribution is one which will permit each particle to fit into the voids created by inter-particle contact of the larger sizes, thus producing close-packing and maximum mix density. Grading changes can thus directly change the density of the placed material.

Mixes which are not well-graded have either an excess of one size in an otherwise well-graded mix, or are *gap-graded.* A gap-(or skip-)graded mix has at least one size range of particles missing. One form of gap-graded mix is the *uniformly-graded* mix that has a preponderance of particles of a single size (Figure 8.5). Another gap-graded mix is the *open-graded* mix that lacks particles of one intermediate size. Although the term open-graded conventionally also excludes particles below 1 mm in size, some fine particles may be needed to make handling practical. Open-gradings are not frost-susceptible and have advantages for surfacings where noise reduction and permeability are required (Section 12.2.2). Gap-graded mixes — containing mainly coarse particles — will be below their potential strength, permeable, and prone to raveling (Section 12.1.2).

Mixes that are not well-graded are limited in their compaction potential as it is impossible to produce a close-packed geometric arrangement. During construction, such mixes will be difficult to work, shape, and compact and are often described as *harsh*. The finished layer will therefore be permeable and prone to excessively high interparticle-stresses, leading to crushing. Densities will be below that of the well-graded mix. As some of the sizes needed for close-packing will inevitably be missing, gaps and voids will always exist within the pavement layer.

If the deviation from a well-graded mix is caused by an excess of fine material, this will diminish the potential for direct contact and mechanical interlock between the larger particles (Sections 11.1.1&2), and hence reduce the strength and stiffness of any placed course. Such mixes will also be of low density due to a lack of close packing at a microscale and will have a surface that will be slippery when wet and dusty when dry. Section 8.4.2 shows that these mixes can also be very sensitive to the presence of water — becoming weak and unstable — particularly when the fine material is composed of clay particles. As little as 3 mm of rain can render unsurfaced clay roads impassable.

In terms of Equation 8.3, values of n between 0.45 to 0.50 indicate well-graded mixes. Open-graded mixes have a low n and mixes with an excessive proportion of fine material have a high n. A grading with n = 0.5 is known as Fuller's maximum density curve and stems from the 1907 work of Fuller and Thompson. It is a useful approximation to the maximum density (i.e. minimum porosity) grading for a large range of materials although other workers have advocated that exponents as low as 0.4 must be used to produce a maximum density, and 0.6 is often used in asphalt (Section 12.2.3).

As discussed in Section 8.2, strength and stiffness will decrease as density decreases. Thus lower density mixes, such as open-graded and macadam (Section 11.1.2) mixes, are less stiff than well-graded mixes and are often termed unstable (thus perpetuating the incorrect use of stability for 'stiffness' which pervades much pavement technology (Section 10.1.1).

In order to achieve a desired distribution of particle sizes, it may be necessary to mix various particle fractions together. The crushing process (Section 8.5.9) usually produces some fine material automatically and this crusher-run mix will usually be the cheapest of the available gradings. Nevertheless, it is quite common to supplement the crushed rock with fine material from earlier crushings, or from other rock sources, in order to meet particular grading requirements. *Fine crushed rock* is a term used to describe many such (usually well-graded) mixtures. In addition, some breakdown of particle size will occur during compaction (Section 8.2.1).

The *coefficient of uniformity* of a grading is defined as d_{60}/d_{30} where d_n is the sieve dimension at which n percent by mass of the soil is passed during sieving. A coefficient of uniformity of about 5 is the border between good and poor grading from the viewpoint of a well-graded mix. However, a gap-graded mix can have a deceptively high coefficient of uniformity. The d_{30} diameter fraction has a major influence on permeability (Section 9.2.2) and cohesion (Section 8.2.3) and is known as the *effective size* of the soil. Nevertheless, some definitions of the coefficient of uniformity are based on d_{60}/d_{10}. The coefficient is a good measure of the permeability of a mix.

Segregation is a problem that occurs with a uniform mix when various particle sizes segregate together during the handling process, thus destroying the intent of the grading selection.

Some soil components such as mica (Section 8.4.2) meet nominal grading requirements but, because of their flat, plate-like (lamella) shape, interfere with the compaction process and achieve little interlock (Section 11.1.3). They thus deleteriously effect the in situ properties achieved. Angular, cubical shapes are to be preferred and some control on particle shape as well as on particle size is therefore necessary. This is provided by test *a* in Section 8.6.

8.3.3 The Atterberg Limits

Section 8.1 mentioned that pavement materials consisted of three phases: gaseous, liquid, and solid. As water is added to a particulate material, an increase in water pressure was seen to diminish both cohesion and interparticle interlock (Section 8.2.3). Eventually the mixture practically becomes a liquid with the fine material particles in suspension and no effective interlock. It can then flow under its own weight. The moisture content at which this extreme condition occurs is known as the *liquid limit*, LL (percent). From the reverse viewpoint, the LL can also be defined by removing water from the liquid mix. The LL is then the moisture content at which the mix changes form a liquid to a plastic condition. In this condition a mixture can be permanently deformed under load without losing its strength; e.g. it can be moulded and rolled into threads. The effect can be easily observed in clays which can readily be made plastic by moistening and kneading.

Note that *plastic* describes that property of a material that allows it to deform in a ductile, non-brittle, and permanent fashion under steady load. Plasticity is formally defined in Section 33.3.3. Commonly, soils are made more plastic by adding clay and less plastic by adding sand. A soil exhibits *dilatancy* when hand-shaking a sample horizontally brings water to the surface and when the water recedes again after the sample is pressed with the fingers.

The LL is determined by a standard test which has a coefficient of variation of about 6 percent (Jameson, 1984). It involves making a groove in a sample and observing whether the groove closes when the base of the sample container is struck 25 times. A maximum LL value of 25–35 is a typical specification limit to control the presence of particles that would deleteriously affect compaction and cohesion (e.g. see items 2 & 10 in Table 8.3). The test is only applied to fine particles likely to be suspended in water. Its application to layers of stones is thus almost irrelevant.

The moisture content at which a material becomes too dry and 'solid' to be plastic is called its *plastic limit* and is determined by a standard rolling test. It has a coefficient of variation of about 10 percent (Jameson, 1984). In many fine-grained soils the plastic limit is a little below OMC. Soils at the plastic limit are typically about a hundred times stronger than when near the wetter liquid limit.

For very fine-grained soils, the plastic limit can provide a more appropriate moisture content than OMC for reference purposes. In this case, a *moisture ratio* is defined as:

(moisture ratio) = (moisture content)/[(plastic limit)(proportion finer than 425 μm)]

A moisture ratio of one represents an upper field limit for well-compacted, well-drained soils (Wallace, 1981).

The further removal of water will cause volume changes within the soil until a moisture content — called the *shrinkage limit* (or *linear shrinkage*) — is reached, below which no volume changes occur with the further removal of water. It has a coefficient of variation of about 15 percent. The *shrinkage product* is the product of the shrinkage limit and the proportion of the material finer than 425 μm (Section 8.3.1). A low shrinkage product indicates a material that may ravel and corrugate if used on unsurfaced roads, whereas a high product indicates material that could be slippery in service.

These three limits — shrinkage, plastic and liquid — are called *Atterberg* (or Consistency) *Limits*. They must be used with caution when the particles involved are porous. In many ways the limits define the clayiness of the fine fraction of a soil.

The *plasticity index* (PI) — also referred to as I_p, which is the international symbol — is the difference between the plastic and liquid limits. It is thus a measure of the range of moisture contents over which the soil will remain plastic and derives mainly from the surface activity of the clay component of the soil. A PI above ten usually indicates a clayey soil; the higher the PI, the 'heavier' the clay (Section 8.4.2). For many clays:

$$PI = 0.7(LL - 20)$$

However, silts (Section 8.4.1) have low PIs, passing rapidly into the liquid stage once they have reached the plastic limit.

The plasticity index is basically a control on the moisture susceptibility of the material, with a high value indicating susceptibility. Empirical evidence indicates that materials with low plasticity indices make the best subgrades and basecourses and a maximum value of six is a typical specification limit for high quality material.

However, in an examination on Australian pavement materials, Jewell (1969) found that only six of his 20 samples met the Atterberg specification limits but 15 of the 20 were performing satisfactorily in pavements with an impervious surface. Indeed, only one of the 20 complied strictly with all the grading and Atterberg limits of a typical specification. In a detailed study of a further 102 arid Australian road sites, Brodie (1970) found no correlation between plasticity index and pavement performance.

Although the index is widely used, in addition to the above caution, Ingles and Noble (1975) note that it can also be one of the least useful of tests because of both its high variability, poor reproducibility and relatively high cost and time to complete. Although the within-sample coefficient of variation is about 20 percent (Jameson, 1984), Ingles and Noble suggest that it has a field coefficient of variation of about 75 percent. It is better to try to understand the material, than to rely on PI values.

8.4 SOILS

8.4.1 Soil types

The term *soil* is used in preference to *earth* or *dirt*. It refers to naturally occurring, uncemented, unconsolidated, and/or loose material found above bedrock. Soils composed of particles formed by the physical disintegration of rock (Section 8.5.10) are usually called granular (or mineral or inorganic) soils. Soils formed from the chemical decomposition of rock are usually clays (Section 8.5.10). Soils formed from living material are called organic. Most soils include particles of both rock and organic material.

When soils are being considered for roadmaking, the following key steps should be followed. The soils should be:

1. classified so that the characteristics and past performance of other similar soils can be considered. Classification processes are discussed below.
2. tested for compactibility (Section 8.2).

3. assessed for strength, stiffness and swell potential (Section 8.4).
4. tested for permeability (Section 9.2.2).

Soils can be classified by their method of formation as:

(a) residual (or eluvial) — formed in-place from the weathering of parent rock (Section 8.5.10),
(b) alluvial — deposited from running water,
(c) lacustrine — deposited from lake water,
(d) marine — deposited from sea water,
(e) glacial — the remains of glacial action,
(f) aeolian — deposited by the wind,
(g) colluvial — deposited by gravity (e.g. land slides),
(h) cumulose (or organic or histostol) — formed from decaying vegetable or other organic or fibrous matter (or humus). Typical examples are peats, bogs, marshes, moors, and muskegs. They occur predominantly in thick layers in wet areas. They often emit an odour and retain large amounts of water, with moisture contents often exceeding 100 percent. Organic soils are usually unsuitable under load and are rarely used in roadmaking, unless contained by geofabrics (Section 10.7).
(i) leached — deposited from natural salts in solution in ground water. The presence of such salts can cause problems for spray and chip seal surfaces (Section 12.1.4). Sodium chloride (NaCl), sodium carbonate ($NaCO_3$) from limestone (Section 8.5.4), sodium sulfate (Na_2SO_4), and gypsum ($CaSO_4$) are the most commonly encountered and may occur in either the soil itself or the groundwater (although gypsum is relatively insoluble). The problems associated with their use can be reduced by using an impermeable layer to prevent the migration of salts to the surface via the evaporation of groundwater at the surface.
(j) scalpic — previous human activity has removed the original soil and left the bedrock exposed, possibly covered by recent landfills or refuse disposal.

Another common classification for soils is based on the *unified soil classification system*, which is often used in terrain classification studies (Section 6.1.3). The unified system builds on the above grading distinctions, from sand to heavy clay. Key overall characteristics of a soil zone are its soil components and colours, bedding characteristics, uniformity, geological history, and degree of weathering (Section 8.5.10). General variations in soil type across the land surface were discussed in Section 6.1.3.

Patterns reflecting variations of soil with depth are termed *soil profiles*. The following associated terms are used (Figure 8.6):

A horizon (or topsoil); the surface layer of residual, leached and/or cumulose (organic) soil,

B horizon; an underlying transition layer, usually alluvial and/or clayey,

C horizon; residual soil immediately above the parent rock and formed from that rock,

D horizon; the parent bedrock,

G horizon; a layer mottled by water logging.

The A and B horizons are not normally used for subgrades and are often very erodable. They are sometimes described as solodic (alkaline, $pH>7$) or podzolic (acidic, $pH<7$). Podzols usually have an ashy feel and appearance and occur in a range of colour pairs. A pH difference between the A and B horizons usually implies that much soluble material has been leached from the A to the B horizon, leaving it silica-rich.

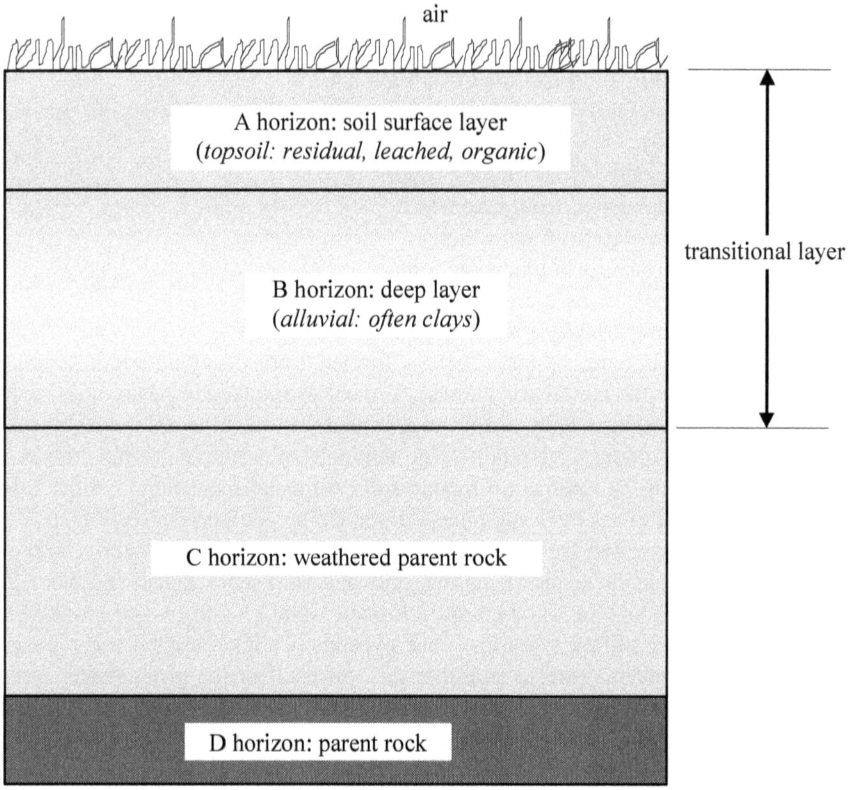

Figure 8.6 Soil profile definitions. The G layer is omitted, see text.

Major soil groups are also defined by their grading variation as:

U (uniform) Soils in this group are composed of particles of the same size; they are often sands or heavy, expansive clays (Section 8.4.2).

G (gradational) Soils in this group contain a well-graded range of particle sizes, which makes them potentially useful for roadmaking.

D (duplex) Soils in this group contain both the A and B horizons, sharply contrasted with respect to the size, shape, and arrangement of the component particles. The horizon boundaries occur over 100 mm or less, except in the possible presence of an intervening laterite layer (Section 8.5.4). A typical duplex soil is sand over clay. Duplex soils often have salinity present when a heavy clay (Section 8.4.2) is below a silty layer.

The common engineering subdivision of soils depends on grading (Section 8.3) and particle size and is illustrated in Figure 8.7. The soil 'names' in this figure range from boulders (or floaters) to the fine-grained clays. Actual soils are, of course, usually mixtures of sizes. Sand-clays are one such obviously named mixture in which neither of the two components constitutes more than about two-thirds of the mixture. Roadmaking loams are usually mixtures of fine sand and clay and are thus both friable and coherent. They are usually distinguished from sand-clays which have a coarser sand fraction. The term loam does not imply the richness in humus associated with gardening terminology.

Gravel is a potentially confusing term. This text uses the civil engineering usage which applies it only to naturally-occurring rock particles (as defined in Section 8.3.7).

However, the soils literature uses it for all rock particles, which this text calls *stones* (Section 8.4.1). Gravel is sometimes also taken to include sand and/or clay particles.

The word *rock* is used (Section 8.5) to refer to a large body of relatively-cohesive, naturally-occurring minerals. The word *stone* is used to describe particles of rock. One set of definitions for fine- and coarse-grained materials is given in Figure 8.7. For a soil mixture to be termed fine, it is usually required that it contains more than 50 percent of fine material. For very fine-grained material at the right-hand end of Figure 8.7, use is often made of the more detailed classification scheme in Table 8.3, which is based on ten fine-grained soil components, the most critical of which are the clays.

8.4.2 Clays

The commonest and most demanding fine-grained soils are those with a high clay content (Groups 6, 7 and 8 in Table 8.3) and so clays deserve special attention.

(a) Formation

Clays are commonly formed from the by-products of rock-weathering (called secondary minerals, Section 8.5.10), whose further breakdown is prevented by the presence of cations and ionised water. A large variety of geological depositional processes have then led to the creation of a clay layer. Other geological layers then frequently cover the layer, and a degree of *preconsolidation* of the clay layer occurs. Preconsolidation increases the inherent stiffness of the clay and is usually associated with a fissured and heterogeneous layer.

(b) Clay components and particles

Clay components are usually chemically-complex hydrated alumino-silicates. They differ markedly from the other soil components in that the resulting particles (*grains*, in the context of other soils) are very small, plate-like, and carry ionic electrical charges which are negative over the plate surfaces and positive around its edges. Their surface chemistry means that the resulting materials are inherently unstable. The final, stable, crystalline products are described in Section 8.5.4.

The particles attract soil cations and come together edge-to-side, as opposite charges attract through ionic bonding. An open, flocculent, loosely-packed structure is produced. In many circumstances, these electrical charges produce inter-particle forces that are far more significant than any gravitational forces. This electrical surface-activity explains the role of clays in the Atterberg limits discussed in Section 8.3.3. The inter-particle attachment also leads to clays being plastic, sticky, smooth to touch, cohesive (Section 8.2.3), strong when dry, weak when very wet, and sometimes prone to major volume changes. The plasticity arises because the edge-to-side clay structure can be forced into new, but still strong, interparticle arrangements.

The total ion exchange capacity of a clay is an indicator of how reactive it is. The capacity is sometimes measured by a methylene blue test in which exchangeable cations on the clay surfaces are replaced by methylene blue cations. The test is sometimes used as an indirect measure of the presence of clays.

If water is present, the water ions can preferentially attach to the plates and thus weaken the clay by preventing interparticle linking. Thus, wet clays have a low resistance to deformation, are difficult to compact, and are almost impermeable. The low permeability means that the other changes may not directly coincide with climatic conditions. Similarly, laboratory measures of the swell potential of clays (see below) can

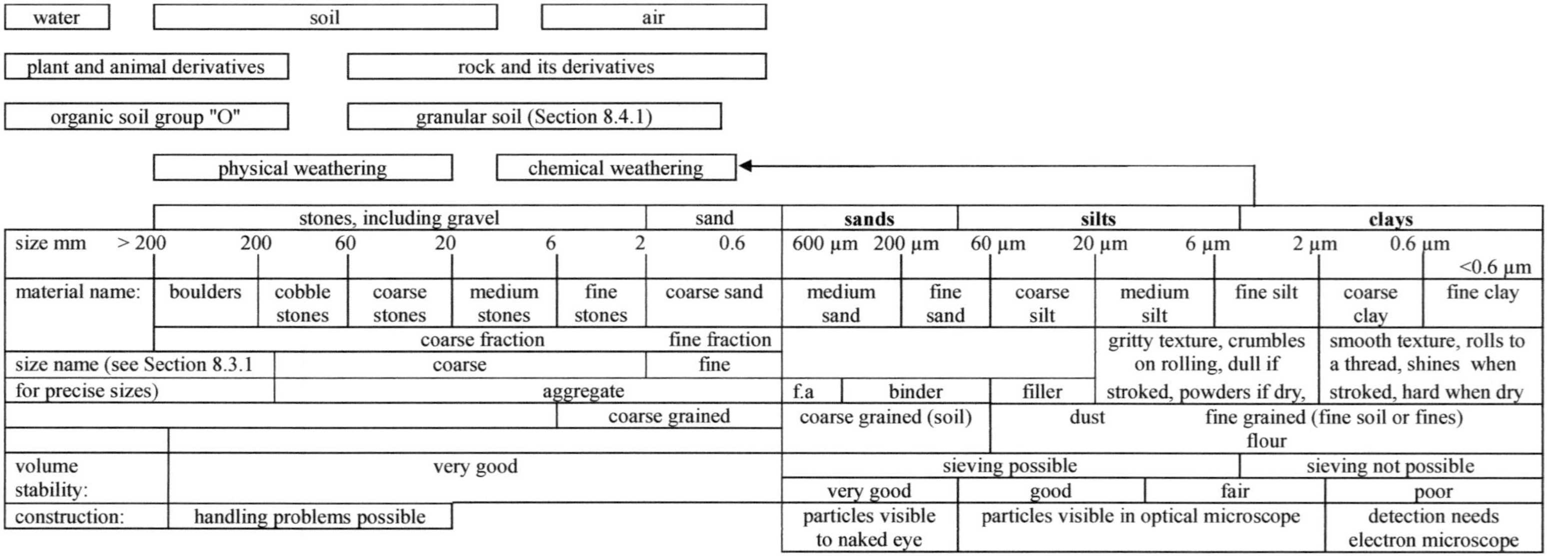

water
soil
air
plant and animal derivatives
rock and its derivatives
organic soil group "O"
granular soil (Section 8.4.1)
physical weathering
chemical weathering
stones, including gravel
sand
sands
silts
clays
size mm
> 200
200
60
20
6
2
0.6
600 µm
200 µm
60 µm
20 µm
6 µm
2 µm
0.6 µm
<0.6 µm
material name:
boulders
cobble stones
coarse stones
medium stones
fine stones
coarse sand
medium sand
fine sand
coarse silt
medium silt
fine silt
coarse clay
fine clay
coarse fraction
fine fraction
gritty texture, crumbles on rolling, dull if stroked, powders if dry,
smooth texture, rolls to a thread, shines when stroked, hard when dry
size name (see Section 8.3.1 for precise sizes)
coarse
fine
aggregate
f.a
binder
filler
coarse grained
coarse grained (soil)
dust
fine grained (fine soil or fines)
flour
volume stability:
very good
sieving possible
sieving not possible
very good
good
fair
poor
construction:
handling problems possible
particles visible to naked eye
particles visible in optical microscope
detection needs electron microscope

Figure 8.7 Classification of soils by size. Note that the particle size range may vary slightly between classifications and is more closely defined in Section 8.3.1.

be misleading if the impermeability of a clay shields it from the effect of changes in moisture content. Some clays can accept large amounts of water (see below).

The presence of external cations (e.g. from stabilisation, Chapter 10, or a change in the pH of the water) can alter many of the above properties by also adhering to the charges on the clay plates. In addition, the cations can leave the balance of the interparticle forces repulsive, with the particle sides repelling each and the plates becoming parallel. The clay is then in a dispersed state. The presence of an electrolyte can counteract this repulsive effect and lead to flocculation.

Behaviour varies from clay to clay and so it is necessary to study Groups 6, 7 and 8 in Table 8.3 separately.

(c) Kaolinite

Kaolinite (number 6 in Table 8.3) is formed when sodium, potassium, calcium, magnesium, and iron — commonly present as a feldspar (Section 8.5.1) — are leached away in an acidic environment typical of many tropical areas. It has a relatively inactive surface and is composed of single sheets of gibbsite and sub-sheets of silica. Gibbsite is a sheet of hydroxyl ions surrounding aluminium (and occasionally iron or magnesium) atoms in an octahedral pattern. The silica sheets have a silicon atom at the centre of a tetrahedron of oxygen atoms with an hydroxyl layer balancing the oxygen layer formed by the tetrapod bases. Each sheet is about 300 pm thick and the actual particle consists of many such layers. The sheets are joined together by strong hydrogen bonds so the structure is stable and non-expansive and water is unable to penetrate. Halloysite is similar to kaolinite except that a layer of water molecules separates each sub-sheet.

(d) Illite and montmorillonite

Illite (number 7 in Table 8.3) and montmorillonite (number 8) develop in an alkaline, poorly drained environment where sodium, potassium, calcium, magnesium and iron are not leached away and stay to be part of the crystal structure. The structure of illite and montmorillonite is a plate composed of a gibbsite sheet between two silica sheets. The plates can have thicknesses from 1 nm to 10 nm and lengths about 100 times their thickness. The silicon cations sometimes (more commonly in illite) exchange with aluminium of lesser valency. This exchange results in a net negative charge on the plate surface, which attracts cations such as Na^+, Ca^{++}, Mg^{++}, and (in illite) K^+ in the soil water and leaves the water highly ionised. Na^+ causes the most swelling. The montmorillonite plates have no K^+ to bond them together, being dependent on the lesser cation exchanges. They are therefore only weakly bonded together and can be easily separated by ionised water.

(e) Swelling and expansion

The attachment of ionised water discussed in the preceding sub-Sections can result in the volume of absorbed water being many times the volume of the actual clay particle. This volume change can cause significant swelling (or expansion) of the piece of clay. When the water disappears there is consequential shrinkage and cracking. Clays can also swell when water entry reduces high internal suction stresses (Section 9.2.4).

Thus, clays containing illite and/or montmorillonite (sub-Section d above) are called expansive (or active — referring to their micro-surface activity — cracking or swelling) clays. Illite does not swell as much as montmorillonite as it attracts K^+ ions in a non-exchange mode that links the plates together. In practice, both illite and montmorillonite are prone to swell and shrink significantly with changes in moisture content.

Table 8.3 Ten fine-grained soil components (from Ingles and Metcalf, 1972). Note that specific area = (surface area)/mass.

Group	Minerals present	Mean size	Chief physical properties
1. silica	quartz (Sections 8.5.1&3)	> 2 μm	cohesionless, very fine sand, abrasive.
2. mica	muscovite (iron and magnesium silicates), hydrous aluminosilicates. Muscovite (white) is the common form but it is also found as biotite (dark) and chlorite (#9 below).	> 1 μm	cohesionless, flat plate-like shape, weathers easily, resists compaction, white in colour. Micas have perfect cleavage patterns and exist as sheet-like crystals.
3. carbonate	calcite, dolomite	any	pulverises easily
4. sulfate	gypsum	> 1 μm	can disrupt concrete
5. allophane	amorphous aluminosilicates, etc.	any	high void ratio, high plasticity, air drying, degrades permanently
6. kaolin	kaolinite, halloysite (silica-poor)	1 μm	low cohesion, often red-brown in colour, non-expansive, low plasticity, less surface-active, specific area = 10 to 20 m^2/g, friable, permeable
7. illite	illite, degraded micas	100 nm	expansive, medium plasticity, moderately surface-active, specific area = 100 m^2/g, low permeability
8. smectite	montmorillonite, bentonite, etc; silica rich, (Section 8.5.10)	> 10 nm	highly expansive, very plastic, very surface-active, specific area = 400 to 1000 m^2/g, impermeable
9. olivine	chlorite, vermiculite	100 nm	green in colour, slightly expansive, low shear strength
10. organic matter	humic acid, humates	any	degrades in oxygen, permeable, resists compaction

Swelling is typically measured by an *oedometer*, or consolidation machine. A disc of soil is placed in a mould between two discs of porous stone. The sample is saturated via the porous stones and its expansion normal to the stones is measured. The swell potential is this expansion divided by the original distance between the two stone discs and is usually quoted as a percentage.

The swelling/contraction occurring in expansive clays will depend on the clay type and the size of the change in moisture content. Materials that swell by more than 2 percent when saturated can cause problems within a pavement. Soils are usually called expansive if they swell by more than 2.5 percent. However, the swelling of some clays can be of the order of 20 percent and can lead to fissures opening in the soil in dry weather. If resisted, the expansion can create large swelling pressures (e.g. 100 kPa) which can disrupt road surfaces, tilt poles, and break utility pipes. Seasonal vertical movements (heaves) can be up to 65 mm, with diminishing effects to a depth of 2 m.

Such volume changes can be overcome by preventing changes in moisture content by using a number of techniques:

* *material placement*. Material is usually best placed at OMC to minimise air voids (Section 8.2.1). However, Section 8.2.4 shows that volume changes can be minimised

by compacting expansive clays in critical areas at their equilibrium moisture content (Section 9.4.1), rather than at OMC. In practice, systematic testing will usually show that there is a combination of density and moisture content that minimises the swell potential of a clay in a particular environment.

* *moisture control.* The expansive material can be protected from moisture by techniques discussed in Sections 9.3&4 and 13.3&4. A specific method is to cover the expansive material with an impermeable layer of non-swelling material. This capping layer (Sections 9.4.2 & 11.1.2) should be at least 150 mm thick. Roots of roadside trees can be a particular problem by drying out a soil (Section 9.3.3).
* *material removal.* If the above measures are unsuitable, or fail, it may be necessary to replace the expansive clays to depths of about 2 m with an alternative material having greater volume stability and to ensure that the remaining expansive material is protected from moisture change.

(f) Clay classification

Clays can be classified by their particle/component type (sub-Sections b to d above) or by their expansion potential (sub-Section e). Another common clay classification is into:

* stiff (cannot be hand-moulded, Section 8.3.3),
* firm (can be hand-moulded with difficulty), and
* soft (easily hand-moulded).

Table 8.4 quantifies the effects discussed above in more detail.

Table 8.4 Expansive potential of a clay (from Holland and Richards, 1982)

Potential for expansion	Arid to semi-arid and cold climatic areas		Humid climatic areas	
	shrinkage limit (%)	plasticity index (%)	shrinkage limit (%)	plasticity index (%)
low	0 to 5	0 to 15	0 to 12	0 to 30
moderate	5 to 12	15 to 30	12 to 18	30 to 50
high	> 12	> 30	> 18	> 50

Clays are also sometimes classified as:

* heavy (or fat) clays. The term *heavy* originally referred, not to density, but to the difficulty encountered when digging the clay. Heavy clays can be readily rolled into thin strings and are highly compressible when moist. They are now taken as clays having plasticity indices over 20 percent and liquid limits over 50 percent (Section 8.3.3) and are from groups 7 and 8 in Table 8.3. They are thus also likely to be expansive clays. Heavy clays often make poor subgrade material, but unfortunately commonly occur in areas where useable natural rock for roadmaking is scarce.
* lean clays.

High plasticity indexes in any soil indicate the presence of clays, as noted empirically in Section 8.3.3. The presence of some clay in a material is desirable, as small amounts of clay will increase strength and wear resistance. Further, as the clay content of a soil increases, its surface activity increases, and hence more and more water can be added to a material with poor cohesive strength without destroying its (low) shear strength. A clay content of over 20 percent will mean that clay properties will dominate.

8.4.3 Stress–strain properties

(a) Required properties

Having discussed fundamental components of a soil, it is now appropriate to examine the physical properties of that soil which are relevant to pavement analysis, design and performance (Chapters 11 & 14). These are particularly its *strength* and its *stiffness* (see Figure 8.8 and Section 11.3).

Strength and stiffness depend on the soil's composition, structure, and density and, for fine-grained soils, moisture content and so a soil should be tested in as close to its in-service conditions as possible, although these service conditions are often difficult to determine (Section 9.4.1). In particular, strength and stiffness properties depend very much on moisture contents. For instance, the effect on clays was discussed in Section 8.4.2 and Section 9.2.5 shows how:

(a) in dry soils, soil suction can increase normal stress and hence increase shear strength, and
(b) in wet soils, pore pressure can reduce normal stress and hence reduce shear strength.

Once they are known, composition, density and moisture content can be fairly readily replicated in testing. However, *soil structure* can arise from:

* the manner in which the material was formed,
* preferred particle orientations created by the compaction process, or
* the effects of traffic loading.

Care must therefore be taken to ensure that the testing process and any remoulding or kneading that accompanies it does not significantly alter the structural characteristics of the material. The important role that soil structure plays in determining soil behaviour can be appreciated by considering the influence of fissures and shrinkage cracks and from the discussions of clay properties in Section 8.4.2 and of the effect of loading on granular soils in Section 11.2.2.

The prior loading history of the material — particularly its preconsolidation (Section 8.4.2) — will have a significant effect on how any test data is interpreted.

(b) Shear strength

The three key strength measures are the compressive strength, tensile strength and shear strength (Section 33.3 and Figure 8.9). Many pavement failures (Section 14.1) exhibit shear rather than compressive failure. In particular, tests for shear strength will often be the most appropriate tests for fine-grained soils affected by moisture content.

Shear strength can be measured directly by placing the soil in a box that can be sheared along a plane parallel to and midway between two of the faces of the box. The field (or in situ) vane shear test (ASTM D2573) can also be used to determine the in situ, undrained shear strength of soft to medium clays and coarse silts. The vanes are commonly square ended with a height-to-diameter ratio of two and are turned on a shaft. The torque at failure is simply related to the shear stress over the cylindrical failure surface. Typical shear strength (SS) indicators are:

SS < 25 kPa,	soft enough to squeeze between the fingers of a closed fist.
25 < SS < 50 kPa,	easily moulded by the fingers. Termed a *soft* material.
50 < SS < 100 kPa,	moulded with strong finger pressure. Termed a *firm* material.
100 < SS < 150 kPa,	dented by strong finger pressure. Termed a *stiff* material.
150 < SS < 200 kPa,	only slightly dented by strong finger pressure.
200 < SS	only slightly dented by pencil point. Termed a *hard* material.

With poorly-graded materials, a bearing test (Sections 8.4.4&5) is often most appropriate. Unbound materials such as gravels have low shear strength.

(c) Stress–strain response

The stress–strain response of a soil was once primarily established by using a shear box. Today, it is much more common to use a form of the triaxial test (Section 33.3.4). The resulting stress–strain curve (Section 33.3.1) for a fine-grained soil is as shown in Figure 8.8. The loading stress is the stress σ_1 in Figure 33.2 and this implies that the confining stress σ_c is defined, perhaps by the ratio σ_c/σ_1. When $\sigma_c = 0$, the peak stress shown in Figure 8.8 is called the *unconfined compressive strength* (ucs). In this case, it follows from equilibrium considerations (Figure 33.2) that:

$$SS = (ucs - \sigma_c)/2$$

The determination of the unconfined compressive strength of field specimens has a coefficient of variation of about 40 percent.

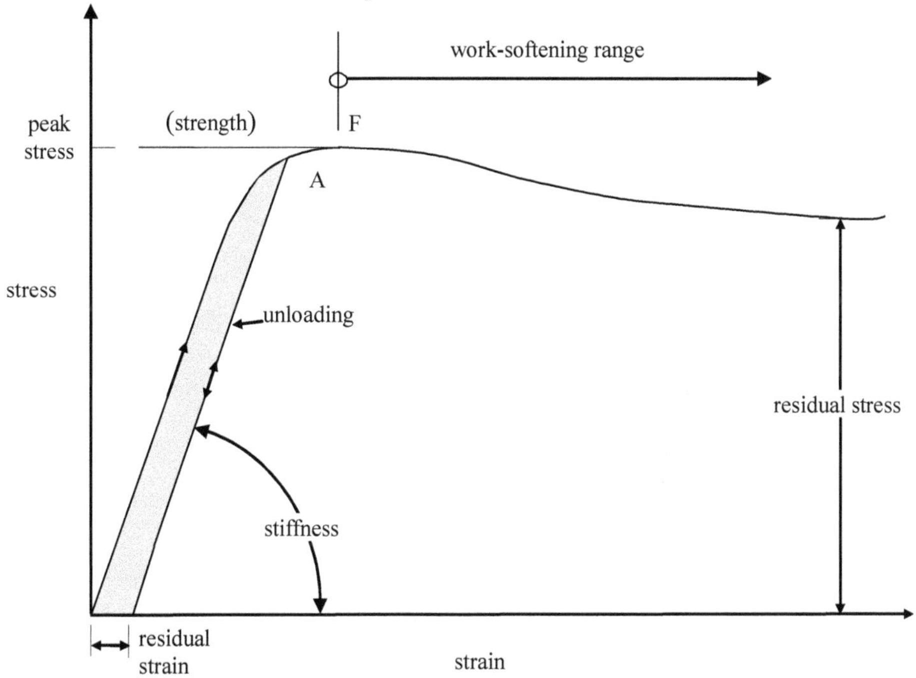

Figure 8.8 Stress–strain curve for soil. The shaded area represents hysteresis and indicates the energy absorbed in loading to A and then removing all the load.

As noted in Section 11.2.2, *stiffness* properties are very dependent on load path. Stiffness leads to the resilient modulus of pavement courses discussed in Section 11.3.1.

The *elastic limit* for many cohesive soils occurs at a strain between 0.001 and 0.005, and for non-cohesive soils at about 0.0001, whereas applied strains can often exceed 0.001 (Section 11.2.3) and can be as large as 0.1. A key characteristic is the drop in strength after the peak strength has been reached at point F. The remaining strength at large strains is called the residual strength and the overall post-F behaviour is called *work-softening*. The strains that contribute to this behaviour are localised in thin zones called slip zones.

Typical unloading behaviour is also illustrated in Figure 8.8 and this pattern is later repeated in Figures 14.2a & b. Energy is absorbed in each loading cycle and permanent strains occur in a process called *hysteresis* (Section 33.3.3). Their effect is explored further in Section 14.3.2. In particular, permanent strains result in permanent deformation of a pavement.

The Mohr–Coulomb failure criterion for strength is the most widely used in studies of the structural mechanics of soils (often called soil mechanics or geomechanics). In this approach, triaxial test strength results for a range of vertical and horizontal stress conditions are plotted as *Mohr's circles* (Sections 8.6r & 33.3.1). The envelope of these circles provides the failure criterion — it is often called the *Mohr–Coulomb failure envelope*. The process is illustrated in Figure 8.9, which shows how the *cohesion* and *internal friction* of the soil can be obtained from the envelope. The failure envelope is thus given by:

SS = (cohesion) + (SS).tan(internal friction angle)

From Mohr's circle:

$SS = (\sigma_1 + \sigma_2)/2$

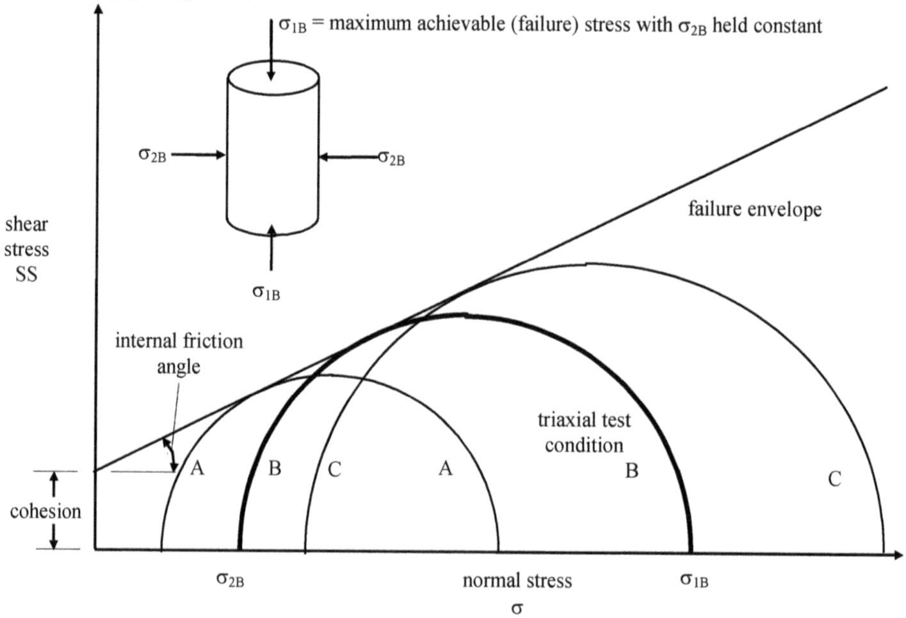

Figure 8.9 Mohr–Coulomb failure criterion.

Figure 8.9 can thus be used to divide soils into those with cohesion and those that rely purely on internal friction for shear strength. The position of the failure envelope in Figure 8.9 also forms the basis of the Texas method for classifying pavement materials (Section 8.6s).

For materials that are not water-prone, measures of brittle behaviour such as repeated-load triaxial tests (Section 8.6r) could be more appropriate. Most soils in service are subjected to repeated loading and so such testing has considerable relevance and the use of these tests in pavement design is described in Section 11.4.2.

8.4.4 California Bearing Ratio

The vertical stress that a material can carry without excess permanent deformation is referred to as its *bearing capacity*. One of the most common bearing capacity tests for soil for roadmaking is the *California Bearing Ratio* (CBR) test that was developed in California in the late 1920s.

The CBR test is essentially a small-scale loading test which measures the force needed to cause a 50 mm diameter plunger to penetrate 2.5 mm in about 2.5 minutes into either an in situ site which has been levelled over a 50 mm diameter, or into a soil sample compacted into a 150 mm diameter, 200 mm high cylindrical mould and subjected to a surcharge to reproduce field conditions. This deformation is usually predominantly permanent and is much greater than, and the strain rate is much less than, those that would be produced by traffic. To eliminate misleading size effects, with large isolated particles providing all the resistance, the soil used must all pass a 19 mm sieve. To eliminate zero effects, the actual penetration range used is usually about 5 mm, rather than the nominal 2.5 mm.

The measured penetrating force is compared as a percentage of a force of 13.2 kN, the force originally obtained with a particular Californian crushed rock which required no cover other than a wearing course. It must be said that the only subsequent justification for comparing the CBR of a compacted clay to that of crushed rock is that the test has a record of relative reliability and usefulness (Metcalf, 1979b). The test has an overall coefficient of variation of about 20 percent. However, at one test location the coefficient can be as high as 60 percent (Smith and Pratt, 1983).

Two of the main problems with the CBR test are reproducing the correct moisture content and density. As would be expected from the discussion of compaction in Section 8.2.3, the CBR of a clay drops as the moisture content increases or density decreases (Figure 8.10). A comparison of Figures 8.3 and 8.10a shows that the CBR of fine-grained soils can be increased by applying further compactive effort.

One of the problems with the test is deciding what moisture content to use. The solution often adopted is to assume the worst case and use a specimen that is saturated with water. This leads to the soaked CBR that is obtained by immersing the sample in water for four days. A surcharge is applied during soaking to reproduce field conditions. The soaked CBR test is the appropriate test for a material likely to be saturated in service and remains the basic test condition in many countries. However, the soaked test would be very conservative in semi-arid areas. The unquestioning use of the soaked CBR thus tends to sidestep the more desirable determination of in-service equilibrium moisture contents (Sections 8.2.4 & 9.4.1).

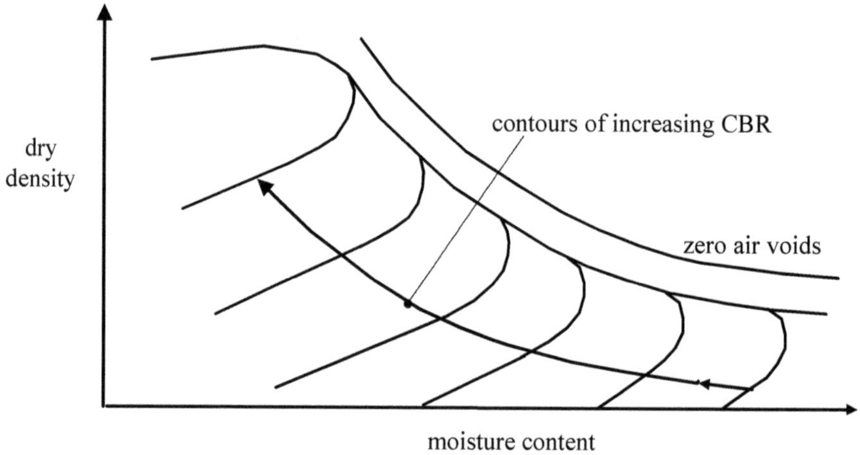

Figure 8.10a Variation of CBR with density and moisture content for a clay. A typical experimental curve is shown in Figure 9.3.

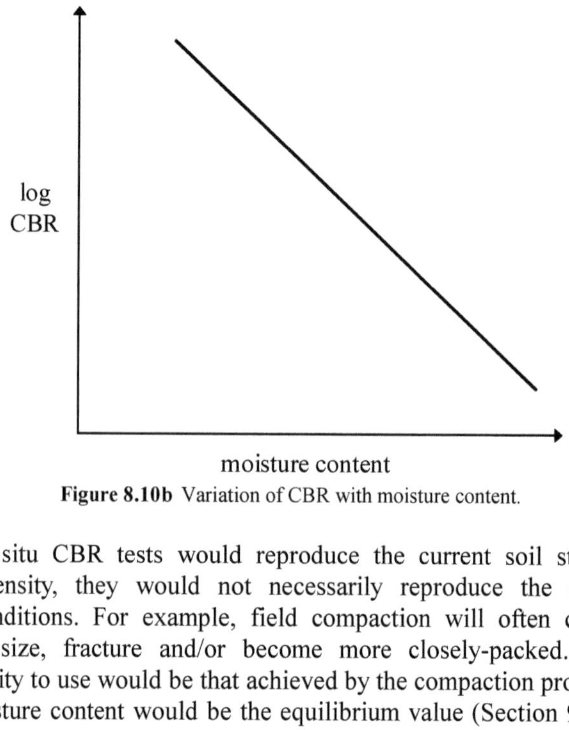

Figure 8.10b Variation of CBR with moisture content.

Whilst in situ CBR tests would reproduce the current soil structure, moisture content, and density, they would not necessarily reproduce the appropriate post-construction conditions. For example, field compaction will often cause particles to break-down in size, fracture and/or become more closely-packed. Thus, the most appropriate density to use would be that achieved by the compaction process and the most appropriate moisture content would be the equilibrium value (Section 9.4.1). To achieve these conditions, a soil sample will usually need to be remoulded and perhaps pre-conditioned by repeated compaction in the laboratory. If the test is to be used to predict in situ conditions, this remoulding may well destroy the existing soil structure, giving a result quite different to the in situ value.

The CBR test has received most validation as a subgrade assessment test. As the top layer of an in situ subgrade will be worked by construction traffic, the preconditioned CBR may also be appropriate for this layer. An alternative to pre-conditioning is to take the test specimen from a compacted layer.

As will be seen in Section 11.4.6, the CBR test is the basis for major empirical methods of pavement design. A flow chart for the selection of a CBR for use in design is given in Figure 8.11. Typical subgrade CBRs for moist conditions are:

* graded granular material	80+
* natural gravels	30+
* gravelly soil	15+
* sand	8 to 30+
* sandy-clay with a PI < 10.	8 to 30
* sandy-clay with a high PI	3 to 15
* silty clay	3 to 7
* heavy clay	1 to 4

Traffic could be expected to damage the surface if the CBR is less than 15. However, a CBR of over 10 represents a strong subgrade.

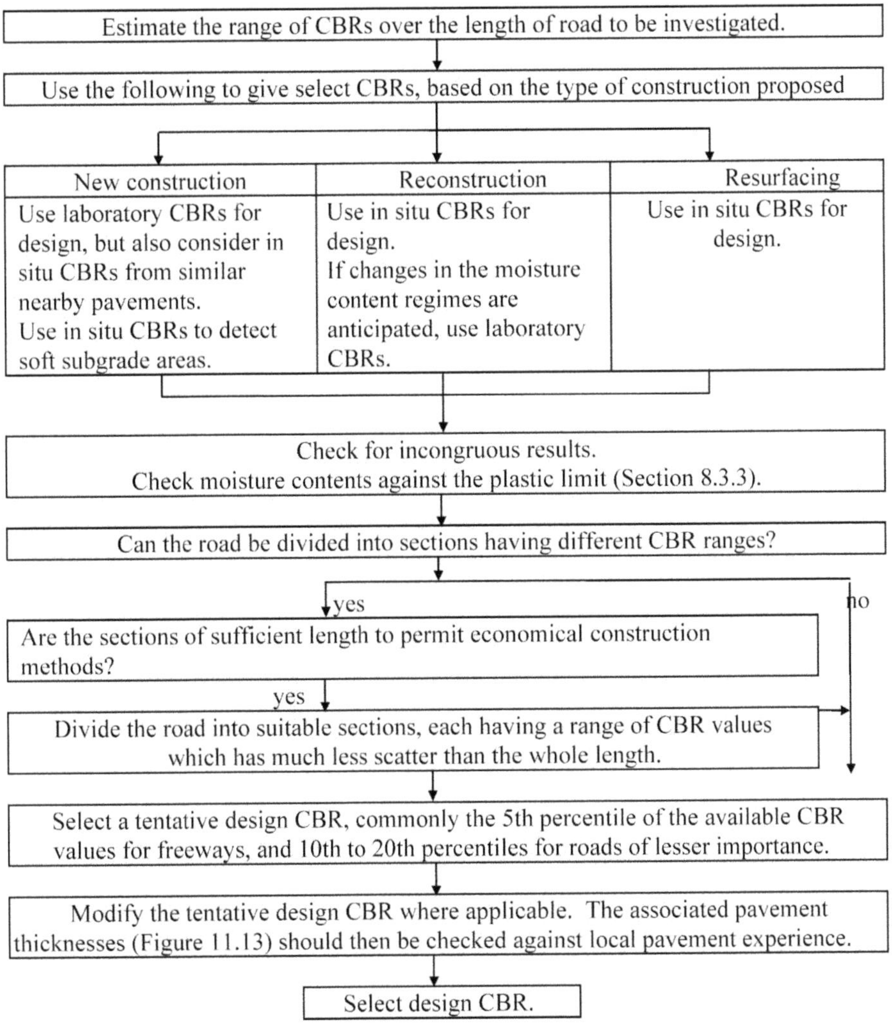

Figure 8.11 Selection of a design CBR (CRB, 1980a)

More generally, the test is a measure of both the vertical stiffness and the shear strength of the material being tested. For stiff materials, the latter property will predominate and the test will therefore be a measure of bearing strength. It can also be used to empirically estimate other properties. For instance:

* the unconfined compressive strength of a soil (Section 8.4.3c) in kPa is about 46 times the CBR in percent (Freeman *et al.*, 2003),
* the repeated-load shear strength (Section 8.6r) of a soil in kPa is about six times the CBR in percent (Uzan *et al.*, 1980),
* the single-load cohesion (Figure 8.9) of a soil in kPa is about 25 times the CBR in percent (Uzan *et al.*, 1980),
* the elastic stiffness modulus, E, in MPa is about ten times the CBR in percent (Figure 8.8 and Section 11.3.4), and
* a link has also been suggested between the CBR test and the single-load Texas triaxial test (Section 8.6s & Clegg, 1983).

Another term sometimes used for resistance to permanent deformation is *support value* or soil support value. It is closely related to the resilient modulus discussed in Section 8.4.3(3) and Section 11.3.1. This measure was used in the AASHTO pavement design guides from the 1960s to 1980s (Section 11.3.1).

8.4.5 Alternatives to the CBR

The alternatives to the CBR test are soil classification tests, miniaturised strength and bearing tests, and full scale strength and bearing tests.

Soil classification tests include grading analyses (Section 8.3.2), plasticity indices (Sections 8.3.3), and the sand-equivalent test (Section 8.6h). They are usually simple, quick, and cheap and so are often used in preference to the CBR. However, they are even further removed from reality than the CBR test and their application to pavement design requires the use of local empirical relationships between their results and either pavement performance or CBR results.

Miniaturised strength and bearing tests are typically the Proctor needle and various penetrometers, which measure in situ penetration resistance. Needle tests are widely used in many process industries and the Proctor needle was used to determine the moisture content of fine-grained cohesive soils.

Penetrometers are a larger development of the needle test and utilise the point resistance and side (or sleeve) friction of a rod pushed into the ground. The more sophisticated units can separately measure these two resisting forces. Penetrometers perform best in cohesive, homogeneous soils such as clays and silts, and become unreliable when penetration resistance is high. Their output may not reflect the contribution of thin layers of problematic material.

The *static cone (Dutch) penetrometer* test is the commonly-preferred penetrometer because of its greater reliability and small disturbance of the surrounding soil. Typically, it uses a hardened steel rod with a 1000 mm^2 area and a 60° conical point angle at its tip. The rod is forced into the soil at a rate of about 1 mm/min and the resisting force is measured and used to estimate soil strength and stiffness. The test commonly requires a vehicle and jack to apply the load and a proving ring to measure it. Testing is usually done at spacings of 15 m or more.

The *dynamic cone penetrometer* (dct) is often used for sites with difficult access, to check variability, or for stiff materials. It can be used to assess the properties of the subgrade of a completed road. The test typically uses a hardened steel rod with a 320 mm^2 area, a 30° conical point angle at its tip, and a 10 kg hammer dropped through 500 mm. Some larger penetrometers use masses of 40, 60, and 80 kg. The test measures penetration in millimetres per blow. Broadly, course stiffness and CBR are inversely related to the measured penetration. A number of different dynamic cones are in use around the world and the dimensions given above are only illustrative.

At a single location the dct has a coefficient of variation of about 40 percent, which is less than that of the in situ CBR (Smith and Pratt, 1983). It is suspect for material with CBR > 20 as this commonly represents good material (Section 8.4.4) which may have particle and void sizes as large as the cone, and as the penetration per blow may be only a few — difficult to measure — millimetres. Impact acceleration devices such as the Clegg impact tester may be better tools in such cases (Section 25.5). Nevertheless the dct is a useful quality control device and can give useful estimates of CBR and thickness without the need to dig pits in the pavement.

Penetrometers are simple, quick, and cheap, but their empirical basis means that they are more useful for control than for design — typically to estimate in situ CBRs (Section 8.4.4). Generally, both types assume that the CBR is proportional to the penetration resistance and therefore, approximately, inversely proportional to the penetration achieved. Frequently, the product of CBR and penetration per blow is a constant, thus giving an easy first estimate of CBR once the constant is established. For the static test, the equivalent CBR is commonly estimated as 4.5 times the penetration force in kN. A typical dynamic test relationship is:

$$\text{CBR} = (200 \text{ to } 300)/(\text{penetration in mm/blow})$$

and, with some non-proportionality,

$$\log(\text{CBR}) = 2.56 - 1.16\log(\text{mm/blow})$$

or

$$\log(\text{CBR}) = 1.90 - 0.61[\log(\text{mm/blow})]^{1.5}$$

Piezocones are penetrometers that also measure pore-water pressure (Section 9.2.3–5). This allows them to differentiate between loose sands (low pressure) and soft silts (high pressure) with the same penetration resistance.

Full-scale performance tests include the field CBR, plate bearing, Benkelman beam (below, & Section 14.5.2), falling weight deflectometer (FWD, Section 14.5.2), and dynamic modulus (Section 11.3.4) tests. They are not easy to conduct, delay traffic, and are relatively cumbersome.

A static cone test requires a force of only 9 kN, whereas a field CBR test would require a force of up to 14 kN. A Benkelman beam test needs 80 kN, and a Westergaard plate-bearing test (based on the load to cause a 750 mm diameter plate to 'penetrate' 2.5 mm) up to 300 kN. Therefore, the load is usually applied via a loaded truck and applied in four increments that may need to be repeated to remove bedding errors. A single test might take about 20 minutes to conduct. Plate-bearing tests are often based around ASTM D1195.

Field CBR and plate bearing tests can give 'correct' in situ soil strength and stiffness at the current moisture content, but their results are specific to one location and circumstance. Consequently, they are rarely used (CRB, 1980a). Occasional use has been made of laboratory plate-bearing tests.

The *Benkelman beam* is described in detail in Section 14.5.2. It is relatively good as a variability test, provided the material under test is able to support the test truck. CBR values and Benkelman beam readings are obviously related, with CBR proportional to beam deflection, d_0, to a power of about −1.5 (Project 010, Lay, 1978), i. e.

$$\text{CBR} = c\, d_0^{-1.5} \tag{8.4}$$

A typical conversion graph between Benkelman beam test data and design CBR is given in Figure 8.12. It uses the actual shape of the *deflection bowl* caused by the load via the spreadability term defined in the figure. If used alone, the graph's predictions will probably be conservative (Rufford, 1977). Section 11.3.4 discusses the use of the shape of the deflection bowl to predict pavement stiffness levels.

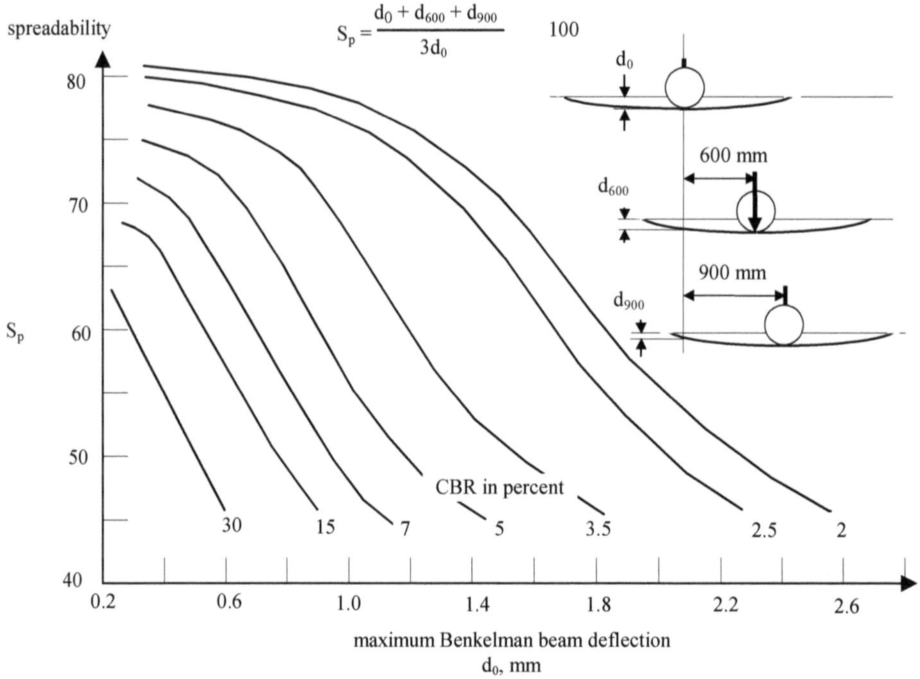

Figure 8.12 Conversion of Benkelman beam readings to CBR predictions (DMR–NSW, 1980). Note that d is the *rebound deflection*; that is, the deflection change when the load is removed (Section 11.3.1).

Sections 11.3.4 & 14.5.2 respectively discuss the more sophisticated FWD and dynamic modulus tests which circumvent the need for CBR data. Theoretically, the most attractive of the miniaturised strength and bearing capacity tests is the design-oriented triaxial test to be discussed in Section 8.6r&s. Its output and that of the FWD can feed directly into the mechanistic methods of pavement design to be described in Section 11.4.2 and their use is therefore becoming increasingly more widespread.

8.5 AGGREGATES

Aggregate is the term used to describe the relatively small pieces of rock — as defined in Section 8.4.1 & Figure 8.7 — used for roadmaking. It may occur naturally — as with gravels (Section 8.5.2) — or require the crushing of large deposits of rock (Section 8.5.9). Section 8.3.1 and Figure 8.7 suggest that coarse aggregate is between 2 mm and 60 mm in size, and fine aggregate between 400 μm and 2 mm. Some definitions put the division between the two at 5 mm, rather than 2 mm.

In searching for new aggregate sources, the first step is to consult existing geological maps and descriptive reports and bulletins available from the relevant government agencies. Once a likely area is determined, the physical condition of the subsurface material can then be ascertained from surface outcrops, existing excavations, past experience, geophysical methods, and drilling (Section 6.1.4). The principal aggregate sources are discussed in Sections 8.5.1–8 below. The generic rock name is given in each heading and the specific associated rock types in the following italicised paragraph.

8.5.1 Igneous rocks

adamellite, andesite, basalt, breccia, diorite, dolerite (diabase), granite, granodiorite, granophyre, picrite, porphyry, rhyolite, scoria, serpentine, toscanite, tuff.

The components of igneous rocks usually formed as crystals during the cooling of molten rock. The resulting interlocking crystal structure produces hard, strong rock, at least in the 'fresh' state. This latter qualification is necessary as it will be seen in Section 8.5.10 that the crystal structure is sometimes prone to attack and the rock may therefore degrade with time. Nevertheless, most hard-rock quarries are located in igneous rock.

The major components of igneous rocks are the *feldspars* (or *felspars*) which are silicates of aluminium with potassium, sodium and calcium. The potassium alumino-silicates are called *orthoclase* (*potash*) and the sodium and calcium ones are called *paglioclase* (*soda lime*). Feldspar is the most abundant of the rock-forming minerals and is also found in sedimentary rock (Section 8.5.2) and metamorphic rock (Section 8.5.3). Another important component of igneous rocks is *quartz*, which is a form of silica dioxide. It is slightly superior to feldspar in terms of hardness and weather and abrasion resistance. However, it may be alkali-reactive and thus lead to alkali-aggregate problems (Section 8.9.1). As a rock form, quartz is discussed in Section 8.5.3.

Basic igneous rock forms at high temperatures, has less than about 55 percent silica, and is relatively dark and dense compared with a*cidic igneous rock*, which forms at lower temperatures and has more than 55 percent silica, often as free quartz.

Basalt, rhyolite, and andesite are formed by the cooling of volcanic lava flows on the earth's surface — they are thus called 'surficial (or effusive) igneous (*volcanic*) rocks' — and are typically found in flows, dykes, or sills. Shrinkage cracking during cooling from the molten state often produces columnar jointing. These lava-based rocks are fine-grained as a consequence of rapid cooling. This produces high-quality rock that may, however, polish readily under traffic and hence will be poor surfacing material. This can be detected by the polished-stone-value test (Section 8.6o).

Note that a *grain* is a coherent portion of rock containing an unvaried crystalline structure. Igneous rock grain sizes are categorised as:

coarse > 2 mm > medium > 100 μm > fine
(naked eye) (lens) (microscope)

Basalt is a basic rock that is commonly black in colour when fresh, but can weather to reddish or greenish hues. It usually contains either pyroxene or olivine. *Pyroxene* is a silicon trioxide chain that is normally dark green or black. *Olivine* is a magnesium-iron silicate that forms a vitreous, relatively hard, olive-green mineral that can decompose to clay, causing major durability problems (Sections 8.4.2 & 8.5.10). Basalts can also appear *green* due to the presence of glass or palogonite, which have no such predisposition. Basalt is characteristically microcrystalline in structure with a grain size of less than 50 μm. Durable basalts are often the best roadmaking material available and are a good source of crushed rock for heavily trafficked roads, although their hardness (Section 33.3.4) and lack of planes of weakness often means that an inadequate proportion of fine particles is produced during crushing. This may necessitate the addition of imported fine particles to produce a well-graded product (Section 8.3.2). Basalt has an elastic modulus (Section 11.3.1) of 80–100 GPa.

In its unaltered form *rhyolite* is an acidic igneous rock, rich in obsidian and a mixture of glassy and crystalline forms, with the former predominating. *Andesite* is a term that covers a wide range of fine-grained crystalline porphyritic acidic igneous rocks containing a range of feldspars and some ferro-manganese minerals and may include breccias and tuffs. Andesites often occur as surface deposits. *Toscanite* is a form of andesite that contains quartz and both orthoclase and plagioclase feldspars.

Dolerites are basic rocks that contain up to 10 percent olivine, the similar but weaker *picrites* contain at least 25 percent. These two rock types are effectively basalts formed in small intrusions, rather than from lava flows, with a grain size of between 50 and 500 μm.

Granite, *porphyry*, and *serpentine* are formed by cooling of large intrusions within an existing rock mass. They are thus called *intrusive rocks* and usually have a visibly coarse-grained, crystalline texture. This can imply internal fracture planes and thus poor crushing resistance.

Granite is an acidic rock containing orthoclase feldspar, mica (Section 8.4.2), and more than 10 percent free quartz. Granite thus has the same composition as basalt, but a much larger grain size (indeed, the word 'granite' comes from the Italian for 'grainy'). Most granites are of only average strength, although a few are very strong and some very weak. Granite is less widely used for roadmaking than basalt as it typically crushes and weathers more readily, particularly when the feldspar respectively cleaves or turns to kaolin (Section 8.5.10). Crushed granites are usually deficient in their proportion of fine particles.

Granodiorite is a granite rock-type containing paglioclase rather than orthoclase feldspar. *Adamellite* is a granite containing equal portions of orthoclase and paglioclase feldspars. *Diorites* are also coarse-grained and contain less than 10 percent free quartz. *Granophyre* is another granite, having a smaller grain size and internal quartz and feldspar grains which produce a 'graphic' texture. *Porphyry* commonly describes a rock where large crystals — usually feldspar — are set in a fine-grained matrix.

Scoria results from frothy lava flows and therefore has a relatively weak skeletal structure. It is occasionally used for sub-bases. *Breccias* are igneous rocks formed by rock collapse or bursting, magna surges, gas fluxing, or explosive actions. Rock particles of various sizes are usually randomly embedded in a basaltic or carbonate matrix. The particles are commonly fragments of basalt, sandstone, shale (Section 8.5.2), chert (Section 8.5.4) or quartzite (Section 8.5.3) and are 30 mm or less in size. Breccia is distinguished by uneven or angular particles in a matrix of finer particles. Material containing larger rock particles is usually known as *agglomerate*. Particles below 5 mm usually give rise to *tuff*, which may be very strong in situ but is very degradable and therefore possess a low service life. Tuff may also form the breccia matrix. Most breccia has been subjected to alteration and weathering (see Section 8.5.10).

8.5.2 Sedimentary rocks

arkose, claystone, conglomerates, greywacke (or greywakke or graywacke — grey refers to the colour of the rock), mudstone, sand, sandstone, shale, siltstone.

Sedimentary rocks are formed from:

(a) water deposition of granular (insoluble) soil;

(b) air (or wind, or aeolian) deposition of fine granular soil. A typical example is silt deposited by the wind – *loess* – which can be very strong and stable when dry but which collapses when wet. These materials can also suffer from piping due to subsurface water flows

(c) deposition of organic remains of plants and animals; and

(d) crystallisation of soluble materials from solution.

The granular material in (a) and (b) is usually the result of weathering (Section 8.5.10) of other rocks.

Although an essential part of the formation process is a hardening and, possibly, a cementing together of the constituent particles, sedimentary rocks are usually mechanically weak and can be ripped and grid-rolled (Section 25.5). Because of the formation process, they will possess bedding planes, often with distinctly anisotropic properties relative to these planes. They are not widely used in roadmaking as they are typically:

* prone to abrasion and erosion,
* water-sensitive,
* of variable quality,
* likely to absorb large amounts of bitumen, and
* of suspect durability.

Conglomerates are natural concretes containing rounded rock-particles of relatively large (gravel) size in a cementitious mortar.

Sandstones are compacted sedimentary layers, consisting mainly of quartz grains. The binders between the grains (Section 8.3.1) can include:

* clay, producing *argillaceous* sandstone (argillaceous means clayey!). Theses are the commonest sedimentary rocks.
* ferrous oxides, producing *ferruginous* sandstones,
* limestone-derivatives (Section 8.5.4), producing *calcareous* (i. e. calcium carbonate-based) sandstones,
* sand and other silica forms, producing *arenaceous* and *siliceous* sandstones.
* other natural cements.

Sandstones such as *arkose* and *greywacke* contain admixtures of feldspar, quartz, mica, and lithic fragments. Greywacke occurs in thin, tough, layers. Dry sandstones often gain their strength from high negative pore pressures (Section 9.2.3). *Ragstone* is a vague British term for hard and/or rubble-like rock. It sometimes refers to coarse-grained sandstones.

Mudstones are comprised of particles that are silt-sized or smaller. However, they are not plastic (Section 8.3.3). *Claystones* and *siltstones* are specifically composed of clay or silt particles. Many mudstones and siltstones and all claystones are argillaceous. *Shale* commonly derives from claystone and is composed of laminated, compacted, and cemented layers that can be slightly metamorphosed (Section 8.5.3). Its specific structure depends on its composition and degree of cementation. The size and composition of the clay fraction in a shale are the key factors determining its roadmaking performance. Shales usually fracture easily along their bedding planes — this is called *fissile* behaviour. Some shales are alkali-reactive (Section 8.9.1).

Sands are the more resistant residue of rock weathering. Calcareous sands are discussed in Section 8.5.4. Granitic sand can be either good or poor as a roadmaking material, with excess mica being an indication of problems (Section 8.3.2). Limestone sands are detrital in that they are formed from the sedimentation of pre-existing (Section

8.5.4) limestone fragments. Quartz sands (Section 8.5.3) are usually of poor quality. There are also sands based on bauxite (Section 8.5.4) and iron oxides.

Many very fine sands prove unable to resist traffic loads. Cement stabilisation (Section 10.2) can be effective in alleviating some of the problems encountered with sands. However, two special problems are:

* *Particle shape*. Desirable sands are sharp-edged (often called *sharp* sands). However, natural sands are frequently comprised of well-rounded particles that therefore achieve very little interparticle interlock (Section 11.1.3). Some pavement specifications therefore restrict their use.
* *Collapsible sands* exist in an open structure with the individual grains arching over openings and held together by colloidal bonds at the contact points. This structure is quite strong at low moisture contents and they are therefore found in arid and semi-arid areas. However, the open structure collapses as the moisture content increases and the bonds are broken. This will result in major surficial settlement. A typical collapsible sand is Pindan, a red-clayey sand found in north-west Western Australia.

8.5.3 Metamorphosed rocks

amphibolite, argillite, gneiss, greenstone, hornfels, phyllite, quartz, quartzite, schist, slate.

These are rocks in Sections 8.5.1 or 2 that have been subjected to heat and/or pressure of such magnitude that new minerals and textures are formed. *Argillaceous* metamorphic rock is derived from argillaceous sandstones, such as shale, and *argillite* is metamorphosed shale. *Hornfels* — which has a high silica content — and *slate* are other examples of argillaceous rocks. They can provide good rock for roadmaking but usually polish easily and produce poorly-shaped aggregate. *Schist* is predominantly mica (Section 8.4.2). *Phyllite*, which may be alkali-reactive (Section 8.9.1), is intermediate between slate and schist.

Arenaceous metamorphic rock is derived from sand, siliceous sandstone, or argillaceous sedimentary rocks. *Quartz* (SiO_2, etc.) is a typical arenaceous metamorphic rock and is often alkali-reactive (quartz as a component of igneous rocks is discussed in Section 8.5.1). *Quartzite* is derived from an arenaceous sandstone that has metamorphosed into a solid quartz rock. It usually contains at least 90 percent quartz and is thus high in silica and snowy white in colour. Quartzite can be used for roadmaking and has good polishing resistance, but can suffer from poor bitumen adhesion and may be frost-susceptible.

Gneiss is derived from granite and hence is an acidic, crystalline rock. Its granite minerals are arranged in distinct bands. *Amphibolite* contains amphibole minerals, which are metasilicates. *Greenstone* is a broad term generally embracing metamorphosed rocks containing the green chlorite mineral (Section 8.4.2).

Although metamorphic rocks are widespread, they are little used in roadmaking. One reason for this is that the rocks may often be closely fractured or jointed, permitting water to cause weathering within the rock mass (Section 8.5.10). For example, argillite degrades in this manner to produce fine particles that are water-sensitive. Other reasons are that the rocks can exhibit relatively low stiffness under load and fail under repeated loading.

8.5.4 Chemically-formed (or duricrust) rocks

bauxite, calcareous soils and rocks, calcrete, chalk, chert, dolomite, ferricrete, flint, gypcrete, ironstone, laterites (but see Section 8.5.7), limestone (but see Section 8.5.5), marl, silcrete, whinstone.

(a) Secondary minerals

The intense leaching of some rocks in hot, humid conditions removes almost all silicates and bases. This leaves secondary oxides, hydroxides of aluminium and iron, and remnant kaolinite and quartz. This produces new chemical compounds commonly called *secondary minerals* (Section 8.5.10). The presence of these minerals in rocks appears to play a role in causing bitumen to strip from aggregates in wet conditions (see Section 12.3.2). This particularly applies to some limestones.

In the longer term, if the natural breakdown process is stalled at an intermediate stage by the presence of cations and ionised water, the secondary minerals will form *clays*, which congregate in a distinct soil horizon (Section 8.4.2).

Deposits where a host material has been modified by introduced minerals, usually via groundwater, are called *pedogenic* and the resulting products are *pedocretes*. The consequent (or secondary) cementing effect is usually quite variable and can range from total to non-existent.

Aggregates made from chemically-formed rocks may polish readily under traffic and hence be a poor surfacing material. This can be detected by the polished-stone-value test (Section 8.6o). Such tests are worth doing, as there are some notable exceptions (e.g. gritty limestones) to the above rule.

(b) Laterites

If the free ions are not present or are removed, the secondary mineralisation continues and produces a more stable crystalline residue. For instance, *laterite* occurs when leaching in a well-drained environment results in sesquioxidic secondary compounds (R_2O_3). They then form in a concretionary manner and will harden (*indurate*) on exposure to air. Indeed, the term laterite comes from the latin word for brick, *later*. The final lateritic product usually has a hard vesicular structure, is nodular and/or slag-like (Section 8.5.6) in appearance, and can resist subsequent erosion. *Vesicular* means that the rock contains many small internal cavities caused by gas escaping from the molten rock. The actual cavities (or vesicles) are sometimes too small to see with the naked eye. Laterites are prone to irreversible changes when exposed to air, or dried, or remoulded.

Laterites are often referred to as *ferruginous*, *aluminous, titaniferous*, or *siliceous*. Their colours reflect their composition.

* *Ferruginous laterites* (iron oxides, typically FeO & Fe_2O_3 – haematite), are red and are commonly called *ironstone* and *ferricrete*.
* *Aluminous laterites* (or bauxitic, Al_2O_3 etc.) are yellow and are known as *bauxite*. *Calcined bauxite* is a manufactured product that is very hard and wear resistant and has a PSV (Section 8.6o) of 75, compared with 50 for many natural aggregates. It is therefore sometimes used to provide surfaces with high skid resistance.
* *Sliceous laterites* (silica oxides) are purple and are commonly called *silcretes.* They often represent an alternative to clay as a leaching end-product.
* *Gypcrete,* which is a salty crust based on gypsum ($CaSO_4$, Section 8.4.1).

Some laterites are self-cementing when used in pavements, but this property is rarely relied upon in a formal or specified manner.

Most laterites do not have the expansive characteristics of some of their clay cousins and those that do are often of volcanic origin (as are most expansive clays, Section 8.4.2).

Laterites can exist in a range of forms ranging from rock layers, massive boulders, stones through to gravel. Lateritic gravels are discussed in Section 8.5.7.

(c) Limestone and similar rock

Many rocks are formed by the deposition of *carbonates*, which are relatively soft. For example, at least 50 percent of the components of most *limestones* are carbonates, predominantly a crystalline *calcium carbonate* encountered as calcite (see Section 8.5.5), aragonite and dolomite, in order of common occurrence. Calcite and dolomite come from algae and aragonite comes from polyps, particularly from prehistoric coral beds. The calcium carbonate often performs as a natural cement. Materials associated with calcium carbonate are called *calcareous*.

There are six main types of limestone: *marble* (a coarse-grained calcite), *calcrete*, *calcareous siltstone*, *dune limestone*, *chalk* (a very loose, degradable form of limestone), and *marl* (soil containing a high proportion of limestone, such as *calcareous shale* or *calcareous clay*). Limestone has an elastic modulus of about 40–90 GPa. The older the limestone geologically, the stiffer it is and the more useful it will usually be for road making. Limestone is the most common rock used for road making in the United Kingdom.

Dolomitic rock contains at least 50 percent calcium- and magnesium-carbonate and is a replaced limestone. It may be alkali-reactive and susceptible to water — particularly if there is clay present (Section 8.9.1). It is usually preferable to reserve the term *dolomite* for the crystalline mineral listed in Table 8.2 and discussed further in Section 8.5.5.

Cherts are non-crystalline silica minerals found in limestone as concretions or as independent, bedded masses. *Flint* is a form of chert. Because of their hardness, and usually-inert nature, cherts may be found as components of alluvial gravel and other residual deposits (Section 8.5.7). Chert should be used with care as it is often alkali-reactive, particularly if fine-grained. It has an elastic modulus of 40–100 GPa. *Whinstone* is a British term for any hard, dense and dark-coloured rock. In the south the term usually refers to cherts, in southern Scotland to greywacke, and further north it can refer to dolerites and basalts.

(d) Calcretes

Calcretes are rock-like materials resulting from the secondary cementation of soils (particularly sands) by calcium- or magnesium-carbonate. Material where the cementation is too weak to be useful is usually called *calcareous soil,* whereas the strong calcretes of use in engineering are classified as calcareous rock. This may exist as nodules, boulders, or large, flat, slab-like areas. These slabs are sometimes called *hardpan*, *duricrust*, or — if extensive — *kunkar* (or *kankar*). Deposits are usually worked to a depth of about one metre, with plasticity increasing rapidly with depth. Grading is often poor and breakdown can occur during testing. The basic reason why calcareous materials can be unstable is that the calcium carbonate itself is only metastable and is therefore prone to alteration. For example, depending on conditions, it will either dissolve in, or precipitate from, water.

(e) Pozzolans

Pozzolan is a natural material that is inert in itself but which will react with lime and water to form cemented products (Section 3.3.1).

8.5.5 Organic rocks

coral, limestone (but see Section 8.5.4), seashell.

Coral is composed of crystalline calcium carbonate (Section 8.5.4). Despite a relatively low strength, coral spreads easily, compacts well and sets quickly, with good limey qualities of adhesion and some self-cementing. It commonly crushes if used as a basecourse but the resulting fine particles are usually non-plastic (Bullen, 1984). Coral has been used successfully for roadmaking and was widely employed for pavement construction in the Pacific during the First World War and Second World War (Vines and Falconer, 1980). *Seashell* has also been used for roadmaking. Clearly, both coral and seashell are closely related chemically and geologically to *limestone* (Section 8.5.4).

8.5.6 Artificial rocks

ash, flue dust, slag, colliery wastes, glass.

Slags are mainly non-metallic by-products from various metallurgical processes, particularly from iron- and steel-making. *Blast-furnace slag* is an iron making by-product formed from a combination of a limestone flux with ash from coke, silica, and earthy alumina components of the iron ore. Slow cooling (e.g. in air) produces a crystalline product resembling igneous rock. Typically it contains about 40 percent CaO, 35 percent SiO_2 and 15 percent Al_2O_3 and its main mineral component is melelite. It is a hard, coarse, and stable material with structural properties appropriate for a graded basecourse material and it is thus the most commonly used slag for roadmaking (Heaton and Bullen, 1980). There is only a slight cementing action, with the in-place material predominantly acting as an unbound layer (Section 11.1.2). However, over time some increase in strength does occur due to a slow cementitious reaction. Nevertheless, it is not a suitable cement extender.

A specific roadmaking product called *granulated slag* (or *slag sand*, *grave laitier* or *gravel slag*) is produced by rapidly chilling (or quenching) blast-furnace slag with water sprays. The resulting sand-sized granules are sharp-edged, glassy, and non-crystalline. Gravier latier is typically a blend of granulated blast furnace slags and fly ash. Granulated slag, when finely ground, can hydrate in the presence of water and is used in the production of blended cement but, unlike the fly ashes discussed below, it is not pozzolanic. However, the relatively weak air-cooled slag has some self-cementing properties. Slag has low heat of hydration but is slow to gain strength. As lime (or cement) is a good catalyst for the hydration reaction (Heaton and Bullen, 1982); granulated slag containing small amounts of lime has potential as a binder in soil stabilisation (Chapter 10).

An aerated, vesicular (Section 8.5.4) form of blast-furnace slag called *foamed slag* has been used as a skid-resistant aggregate. However, it is relatively weak. Pelletised slags are also used.

Steelmaking slag comes from either the BOS or electric arc process and contains unhydrated, 'dead', burnt lime. It is therefore cementitious and expands on contact with water. If used for roadmaking or as fill, it should be finely ground, well-weathered for many months, and well compacted to avoid future expansion. Steelmaking slag can also be used to provide the lime catalyst for granulated slag, but is not itself a suitable cement extender. If used in asphalt, it will absorb large amounts of bitumen. Table 8.5 gives some interesting comparative chemical compositions (percent):

Table 8.5 Slag composition (Heaton *et al.*, 1996, and elsewhere).

Material/ Component	Blast furnace slag	BOS slag	Cement	Fly ash	Bottom ash	Basalt
SiO_2	35	15	20	60	50	45
CaO	40	40	65	5	5	10
MgO	5	10	–	–	–	10
Al_2O_3	15	5	5	20	30	15
Fe_2O_3	–	25	5	5	5	5
Other	5	5	5	10	10	15

Slag is used in both the raw condition or crushed to meet a particular grading specification. To obtain a maximum-density grading, fine particles of shale or stone dust from quarrying and crushing (Section 8.5.9) are usually added at up to about 4 percent by mass. Slag without fine particles has been found to provide a good free-draining layer.

Flue dust comes from cement manufacture and is used as a fine aggregate. However, care must be taken to avoid problems with the cementing action that might occur. This may make its application in the upper layers of a basecourse undesirable.

Power-station ash comes from the burning of coal. About 85 percent of the burnt coal is carried out by the flues and is usually collected on precipitators as *fly ash.* The remaining 15 percent drops back to the furnace floor as *furnace-bottom ash* or *boiler slag*, depending on the boiler type. Typical fly and bottom ash compositions are given in the above table. Fly ash has a low density of only about 2.4 t/m^3 and represents about 5 percent by mass of the coal used. The particles are usually spherical and usually consist of finely divided siliceous and/or aluminous crystalline products, much of it in the glassy phase. It is also often rich in iron. The material is pozzolanic and becomes cementitious after lengthy contact with water. The coarser furnace-bottom ash is a sinter product of low density, uniform grading and vesicular structure, and may resist polishing under traffic (Section 8.6o).

Bottom ash also arises from the incineration of domestic waste. For health reasons, such ash is usually only used when encapsulated in a bound course (Section 11.1.4). It has been used as a substitute for up to 30 percent of conventional aggregate.

Colliery wastes are usually quite variable and either very abradable or susceptible to the weather.

Glass has been used as a partial (< 10 percent) substitute for natural aggregate in asphalt.

8.5.7 Gravels

alluvial, buckshot, colluvial, glacial, granitic, hill, ironstone (but see Section 8.5.4), lateritic, pea, pedogenic, quartzitic, ridge, river, sandstone, till, and volcanic gravels

Gravels are naturally-occurring particulate rocks, as defined in Figure 8.7, and are often excellent roadmaking material. In addition, relatively poor gravels are the only roadmaking materials available in some areas and therefore find use despite such problems as poor grading and shape and a tendency to lose strength when wet. Gravels fall into the following four categories.

(a) *Ridge* (or *hill*) *gravel*: angular rock particles found on higher ground, often in association with clay binders. *Colluvial gravels* are gravels formed by gravity as aprons or fans on slopes. All are usually the result of in situ weathering of sedimentary or metamorphic rocks and are technically a residual rock. They can be very variable and may require ripping (Section 25.4), screening (Section 8.5.9), and grading adjustment. If the PI and grading are under control, ridge gravels can be satisfactory basecourse materials.

(b) *Alluvial* (or *river*) *gravel*: water-worn, smooth, rounded particles. They are often very strong, but poorly graded. Their strength made them much sought after for early roads carrying steel-tyred traffic. However, their rounded shape makes it difficult to achieve the inter-particle interlock (Section 11.1.3) needed for bulk coherence and compaction and they must often be crushed (Section 8.5.9) before use. *River shingles* are poorly-graded mixtures of alluvial gravel and coarse sand that are strong, but difficult to work. Alluvial gravel may be found:
   * in river channels,
   * as terrace deposits that represent the remains of old flood plains,
   * in alluvial fans, that represent a form of inland delta,
   * at water lines, as coastal gravel due to wave and tidal action,
   * in solid formations, possibly formed from old coastal gravel, and
   * as long, low, smooth profile ridges, which are often evidence of old alluvial gravel deposits.

(c) *Glacial gravel* (or *till*): this material has many of the characteristics of alluvial gravel, but contains a wide range of particle sizes and is far less predictable than alluvial gravel. Clay balls in the till can have a particularly deleterious effect.

(d) *Lateritic* (or *pedogenic*) *gravel*: these surficial deposits come from the in situ weathering of laterites (Section 8.5.4). They therefore are often collocated with laterites and found on ridges and slopes. *Ironstone* is one laterite that typically occurs as gravel. *Buckshot* (or *pea*) *gravel* is one of these iron-rich gravels. It will be noted from Section 8.5.4 that most of these gravels are also concretionary. By their nature, they are often gap-graded, being deficient in the coarse sand fraction. If the grading is under control, pedogenic gravel can be useful shoulder material.

Gravels can also be known by their rock of origin. Common examples are granitic, quartzitic, sandstone, and volcanic gravels.

8.5.8 Recycled material

Recycled road material comes from existing road pavements which, in most cases, have deteriorated to the stage of needing rehabilitation (Section 26.1). The material is usually removed from the existing pavement by either ripping or cold milling. *Ripping* is typically done with tines operating to the full depth of the course. The material is subsequently broken down to size in a rock crusher. *Cold milling* is commonly employed when only a portion of the course must be removed. The next stage in the process may occur in a central plant or on the road surface, with the salvaged old material mixed with up to 50 percent new material, particularly binders, and then replaced in the pavement. With due care, it is possible for recycled material to have properties on a par with new material. A key requirement is for the recycled material to possess reasonable uniformity. The advantages of recycling are:

* preservation of new sources,
* reduced disposal needs,
* reduced fuel and transport costs, and
* lower net emissions.

Recycled material is commonly bitumen-bound or cement-bound. It is necessary to check that the original *bitumen* has not hardened excessively by measuring penetration or viscosity (Sections 8.7.4 and 33.3.3) and ageing (Section 8.7.7). It is also necessary to check the actual bitumen content and the grading of the new material. If these checks prove inadequate, fluxing oils and soft bitumens can be added to supplement the original bitumen (Section 8.7.8). Care must be taken in the selection of additives, as some can lead to premature ageing of the bitumen. Cement or lime stabilisation (Chapter 10) may also be used to improve the properties of the recycled mixture. Grading deficiencies can be rectified by blending, scalping, or crushing (Section 8.5.9). The resulting asphalt is tested in the same manner as new material.

Material recycled from other human activities can also be used for roadmaking. Section 8.5.6 discusses the use of blast furnace, power station and colliery wastes and glass, and Section 8.7.8 describes the use of plastic scrap rubber and waste oil.

Less-tightly specified materials can be used for embankments (Section 11.8.1). Typical examples would be crushed concrete, crushed brick and demolition wastes.

8.5.9 Quarrying

Aggregate is sometimes obtained from deposits of naturally occurring and adequately sized and shaped gravels (Section 8.5.7). For instance, *pit-run* aggregate is material taken *directly* from a gravel pit with no further processing. On the other hand, a quarry is an excavation in a rock deposit. The rock at a quarry face is converted into pieces of stone by either conventional earthmoving equipment (e.g. by ripping the surface with tines), or by using explosives to blast the in situ rock (or *fragmentation*). These two technologies are described in Section 25.4. A large quarry can produce 1 Mt/y of rock. The next step is the *crushing* (or breaking) of large pieces of rock into a more satisfactory particle size distribution (Section 8.3.1). This is usually done by either *impact crushers* using mechanically-driven hammers or jaws, *gyratory crushers* based on an inverted pestle and mortar, or *cone crushers*. Crushers may be fixed or mobile. On major jobs, an impact crusher may feed into gyratory and/or cone crushers. The product from a crusher is called *crushed rock* and is distinguished by both its size and the very useful angularity of the individual particles. The importance of the latter is discussed in Section 11.1.2. The grading, separating, and blending of aggregate is discussed further in Section 25.4. In situ crushing refers to the use of grid rollers (Section 25.5) or mobile *hammer-mills* to crush rock already placed on the road formation. It will be appreciated that many of these processes can deleteriously induce microfractures in the aggregate.

Screening is a sieving process used to adjust size gradings by passing the material through a screen, thus removing oversized material. *Dry screening* is a very dusty sieving and winnowing operation used to remove very fine, undersized material (called *stone dust*), although *washing* is used to remove undersized particles from sand. *Screenings* are under- or over-sized pieces of aggregate rejected by the screening process. *Scalping* is a process by which undesirable material is removed from the aggregate, usually by visual selection of soft, weathered or oversize material. Subsequent processes are discussed in Section 25.4.

8.5.10 Durability

Durability is a measure of a rock's ability to resist repeated loading, wearing, and weathering. Tests for resistance to repeated loading and wearing are discussed in Sections 8.4.3 & 8.6r. This section discusses weathering.

Weathering (or *degradation*) refers to physical changes (*disintegration*) and chemical changes (*decomposition*) at the rock surface, brought about by reaction with the adjacent environment. Weathering is most commonly observed in freshly quarried rock that has not had an opportunity to degrade naturally. Staining or discolouration are early and common signs of weathering. However, it may also have been proceeding for long periods of geological time and can result in weathering to a considerable depth.

Disintegration is particularly associated with splitting or fracturing along joint planes in the rock mass. It is aided by:

* *thermal expansion and contraction* during rapid temperature changes, causing cracks to open,
* contained *water freezing* and hence swelling (Section 9.5) and thus causing cracks to open,
* growing *salt precipitation* from ground water in the rock capillary system forcing cracks to open,
* the consequences of chemical changes such as the growth of oxidation products,
* surface abrasion by water or sand, or
* interparticle contact between pieces of aggregate in the basecourse of a trafficked road (Section 11.1.3 & Wylde, 1982). The redistribution of such fine particles under traffic is a primary factor in the in-service deterioration of basecourses as it reduces the local strength and stiffness of the course (Wylde, 1982).

Disintegration is more common than decomposition in cold, dry regions. It results in a reduction in particle size and the production of relatively coarse and inert material.

Decomposition, which causes a change in the type of material present, is more common in warm, moist regions. It is usually an accelerating process as it exposes progressively more area to further attack. Decomposition usually occurs as a result of the action of water, oxygen or carbon dioxide on the rock. The processes involved are usually:

* *solution*, where water dissolves the more soluble constituents of the rock,
* *hydration*, where water is absorbed (e.g., to turn anhydrite into gypsum),
* *dehydration*,
* *hydrolysis*, is an acid-base reaction in which the water ionises to hydrogen or hydroxyl ions, which then combine with other ions. Hydrogen ions replace the cations of silicon and other elements on the microsurface layers of the primary rock and leave a surfeit of hydroxyl ions in the residue. Thus, the pH of the liquid phase rises. Atomic polyhedra are released from the crystal structure into the water. The cations become soluble hydroxides and insoluble residues such as silicates. These fine and potentially chemically-active units eventually settle in an oriented fashion and, typically, clay is formed (Section 8.4.2). The process is mainly controlled by the quantity of water passing over the exposed micro-surfaces.
* *carbonation*, where cementation occurs, typically via calcium carbonate,
* *decarbonisation*, where cementation is destroyed,
* *oxidation*, typically turning ferrous iron to ferric iron and thus destroying the crystal structure,
* *reduction*.

In situ weathering in a rock mass commonly proceeds from the surface downwards. If in situ weathering is not present, it can mean either that the rock was not accessible to weathering agents, or that it is resistant to decomposition. As an example of the former, some rocks that can be extremely difficult to quarry will decompose rapidly after exposure to air and water. The process is accelerated if shrinkage cracks or vesicular structures (Section 8.5.4) render the internal rock structure permeable (as in columnar basalt, Section 8.5.1).

The original rock (which is usually hard and strong) is referred to as the *primary mineral.* As discussed in Sections 8.4.2 & 8.5.4, the decomposition products are called *secondary (or alteration) minerals* and are often soft and hydrophilic. A typical alteration process is the change of the primary mineral, *olivine* (Section 8.5.1), into secondary minerals in the chlorite and smectite groups (Table 8.2) and thence into clay. Minerals formed when the rock has almost solidified are called *deuteric minerals* and, for pavement purposes, are usually also classified as secondary minerals. Given the above process, secondary minerals will often be found in the vesicles of the rock.

It has been found that an initial secondary mineral content in excess of 25 percent will usually lead to decomposition problems in paving aggregate (Section 8.6p). The most deleterious of all the common secondary minerals is expansive *smectite* (Table 8.2; Cole and Sandy, 1980). This forms montmorillonitic clay, which was shown in Section 8.4.2 to be the most difficult of all the clays. Other problem secondary minerals include *iddingsite* and other granite-to-feldspar derivatives such as *kaolinite* (Sections 8.5.1–2), chalcedony (quartz), mica, kaolinite, vermiculite, and zeolite (many of these are discussed in Section 8.4.2). *Iddingsite* derives from olivine and thus from basalt.

Rock types having known durability problems include:

* basic igneous rocks such as *basalt* and *granite*, which derive their strength from the strong interlock of their constituent crystals. Even a small amount of decomposition can seriously weaken this interlock. Some *green basalts* have a poor record for durability. Granites tend to show kaolinisation of feldspars, with a loss of dry strength (Section 8.5.1).
* poorly-consolidated sedimentary rocks (especially argillaceous types),
* many low-density rocks, such as volcanic breccias, and
* rocks containing some *sulfides.* Sulfide-bearing rocks from mine dumps are particularly prone to deterioration. The sulfides oxidise when exposed to the atmosphere and form soluble and destructive acidic salts. These salts can, for example, breakdown and debond thin asphalt surfacings. *Hydrated lime* is sometimes used to counteract this acidity.

8.5.11 Predicting aggregate performance

Usually, a prediction of the performance of a rock first requires knowledge of its mineralogy, petrological texture, and geological history. The items to look for have been recounted in the preceding text. Note particularly that initial appearances can be deceptive in classifying the usefulness of rock. For example, Section 8.5.10 has shown how material can present considerable resistance to quarrying, but then rapidly deteriorate when exposed to air and weather, or wetting and drying.

Tests for aggregates must relate to the source rock, the ready-to-use quarried aggregate, and the intended service conditions, typically in a compacted basecourse or on a road surface. These tests will involve both current and long-term properties (e.g. both crushing strength and durability). The final material used on a job should preferably contain less than 5 percent of uniformly distributed material which fails to pass the specified tests.

To derive in-service data from laboratory tests will often require special methods of preparation, or pretreatment, to ensure that the test specimens are in a condition equivalent to that achieved in service. Any test procedure must then attempt to reproduce the load and the environment to which the rock will be exposed:

(1) during production (blasting, loading, crushing, washing, screening, weathering),
(2) during construction (loading, spreading, additives, moisture content, heating during asphalt manufacture, compaction, weathering),
(3) under traffic (impact, abrasion, fatigue), and
(4) under weather (temperature changes, freezing and thawing, changes in moisture content, leaching, chemical attack).

Before testing aggregate for stages (3) and (4), it is often advisable to use pretreatment procedures to reproduce the effects of stages (1) and (2) (ARRB, 1982). For example, it may be desirable to use repeated compaction to imitate the effect of field-compaction equipment and the associated wetting and drying cycles (e.g. test 8.6f & Smith, 1980).

The eight key characteristics of an aggregate type are its:

* strength (e.g. Section 8.6j),
* stiffness (e.g. Section 8.6r),
* shape (Section 8.6a),
* hardness (Sections 8.6b & 33.3.4),
* durability (Sections 8.5.10 & 8.6c&d),
* surface texture (Sections 12.3.2 & 12.5.1),
* wear resistance (Section 8.6o), and
* hydrophobia (i.e. are not hydrophylic, Section 12.3.2).

Surrogate measures sometimes used to predict these basic properties are its integrity (e.g. whether planes of separation are present) and its density (Section 8.1). Integrity arises because rock strength will depend on both the inherent strength of the grains of which it is composed and the adhesion between the individual grains. The search for high-density rock usually arises from empirical observation and the fact a dense rock will probably have few of the vesicles that aid the decomposition process.

The more relevant of the specific tests used to determine aggregate suitability are given in Section 8.6, with the exception of tests for density and grading, which have been discussed in Sections 8.2.2 and 8.3.2.

8.6 TESTS FOR AGGREGATES

The basis for aggregate testing was outlined in Section 8.5.11. Most of the following specific tests give useful comparative measures but do not give absolute measures. They therefore need to be interpreted in the light of local service conditions and engineering practice.

(a) Flakiness Index

This test eliminates aggregates that — on account of their shape — would be unsuitable for bituminous spray and chip seals (Section 12.1.4) and would have a low potential for developing inter-particle interlock (Section 11.1.2). The Index is the percentage by mass of stones having a least dimension less than 0.6 times their average dimension (this least-to-average quotient is called the *flatness ratio*). It is determined by direct measurement using callipers or sensors or, less commonly, from the ratio of the mass of material passing specially slotted sieves to the total mass of the size fraction. For example, for material passing a 19.0 mm sieve but retained on a 13.2 mm one, the slots are about 10 mm wide by 50 mm long. A typical specification value is to limit the Index to an upper value of 35.

The test has a coefficient of variation of about 20 percent (Ingles and Noble, 1975). It has a repeatability of only 6 percentage points and a reproducibility of only 10 percentage points.

ASTM versions of the test are D3398 based on gradings and void levels giving a particle index, and D4791 based on actual measurements of particle shape. Another version is given in BS 812, Part 105.

Recently, imaging methods have been used to quantify the shape, angularity and texture of aggregates (Papagiannakis and Masad 2008).

(b) Los Angeles Abrasion test

This is one of the oldest and commonest of the aggregate tests (Orchard, 1964). It is a relatively severe, dry-condition, abrasion test that evaluates the hardness of the source rock (Section 33.3.4), particularly during crushing and compaction, in terms of a *Los Angeles Value (LAV)*. It is also useful as a durability indicator and as a measure of the abrasion resistance of sealing aggregate (Section 12.1.4). However, it does not provide a good representation of service under traffic. The ASTM version of the test is C131–06 (small aggregate) and C535 (large aggregate). It is listed as AASHTO T96.

To conduct the test, conditions during crushing and compaction are simulated by, typically, placing a 10 kg washed, dry sample of the test rock and twelve 47 mm diameter steel balls (weighing 5 kg in total) into a drum rotating at about 0.5 Hz for 1000 revolutions. The process abrades the rock and the percentage of particles produced that pass a 1.70 mm sieve is taken as the Los Angeles Value. The test has a coefficient of variation of about 30 percent (Ingles and Noble, 1975). It can be unreliable if used to compare dissimilar types of aggregate.

Broadly speaking, rocks having an LAV of over 40 are considered soft and with an LAV of under 25 are considered hard. An LAV between 15 and 25 suggests that the rock source should be retested annually, at least. The actual LAV specified will depend on the circumstance (e.g. sealing, asphalt, cement concrete, basecourse or sub-base), the traffic volume and the rock type. Table 8.6 shows typical limits on the size of the LAV:

Table 8.6 Typical limits on the size of the Los Angeles Value (LAV).

Situation	Normal sources	Granite
Sealing course, < 250 veh/d	30	40
Sealing course, 250–10,000 veh/d	25	35
Sealing course, > 10,000 veh/d	20	not permitted
Asphalt surface course,	25	30
Asphalt basecourse	30	40
Concrete basecourse	30	45
Unbound basecourse	30	40

Dense basalts typically test at about 12 to 15, whereas vesicular (Section 8.5.4) ones may have values between 20 and 30. Otherwise satisfactory rocks can show high LAVs if they are porous, have planes of weakness, produce flaky particles, or have excessive clay minerals.

The results of the test correlate:

* strongly and positively with those of test j (Minty *et al.*, 1980),
* weakly and negatively with the results of test g, which was developed to overcome deficiencies in this test, and
* strongly and negatively with the solid density (Section 8.1) of the test material (Orchard, 1964).

Test e is, in some ways, a field version of this test.

(b1) Micro-Deval abrasion test

This is a French abrasion-resistance test that is a modern version of the Deval test (AASHTO T327), first introduced in the 1870s. It uses a similar test method to test b, but with soaked aggregate and a much smaller testing arrangement (relying on a 5 L steel jar). It is thus a wet abrasion and grinding test and its results are closely linked to test c by:

percent loss = 12 + 0.62[test c percent loss]

(b2) British abrasion test

This test is specified in BS 812 and is more like test o than either tests b or b1. The sample is tested for surface wear by pressing circular abrasion wheels against the test aggregate. The percent loss due to abrasion is measured. High-quality aggregate would have a loss of under 16 percent.

(c) Sulfate soundness test

This test measures the resistance of aggregate to disintegration. It can also detect such flaws as fine cracks and weak internal (shear) planes. The ASTM version of the test is C88; the BSI version is in BS 812, Part 121; the AASHTO version is T104.

The test was developed in 1818 to detect the effect of freezing water on building stone. It is now used for assessing both:

* weathering susceptibility (Section 8.5.10), including resistance to freezing and thawing (although there is considerable question as to whether it does reproduce freeze–thaw conditions), and
* whether the aggregate could be expected to perform well under a new set of environmental conditions for which no service data is available.

The test has a coefficient of variation of about 85 percent and is rated as one of the poorest of the aggregate tests (Orchard, 1964; Ingles and Noble, 1975). It should not be allowed to over-ride the evidence of successful field experience or to independently accept or reject aggregate.

The test is favoured because it requires little equipment and is easy and quick to perform. To conduct the test, a sample of aggregate in the 10 to 14 mm size range is immersed in a 20 percent solution of sodium or magnesium sulfate in water, and then dried. The solution penetrates into fine cracks in the aggregate, depositing salts on the crack interfaces during wetting and drying. The volumetric growth of these salts causes splitting of the aggregate. The percentage loss of weight from the sample is then measured and is called the magnesium sulfate soundness value (MSSV). An MSSV of over 12 is a cause for concern. The results of this test are correlated with those of test k (Minty *et al.*, 1980).

(d) Freeze–thaw soundness test
This variation to test c uses freezing and thawing rather than chemical deposition. It is AASHTO test T103. It is not very reproducible, due to the lack of control over temperatures, time, and moisture contents.

(e) Aggregate impact test
This test is similar in object and style to test b except that it uses hammer blows to produce impact loading and is thus more suitable for field conditions. The percentage of the sample that passes a 2.36 mm sieve after 15 blows of a 14 kg hammer dropped 380 mm is the aggregate impact value. Good aggregate has values under 30. The test is described in BS 812.

(f) Repeated compaction test
This test is used to bring source samples for other tests to the condition they would reach in the field after placement in their final location. It also provides evidence as to how a material will degrade during field compaction, and thus influence final grading (e.g. Section 8.2.5). Although pretreatment of aggregate prior to testing is often essential if service conditions are to be reproduced, it is time-consuming and difficult and therefore sometimes neglected.

To conduct the test on aggregate, the sample is commonly compacted in a 150 mm diameter (standard) cylindrical mould in three layers of equal thickness. Each layer is subjected to 56 uniformly distributed blows of a 2.7 kg hammer falling 300 mm. Alternatively, gyratory (Section 8.2.5) or rolling-wheel compaction (Section 12.2.4) may be used. The rolling-wheel method best reproduces in-service conditions. These compaction methods are also used to prepare (or precondition) samples for other performance tests listed in this Section.

(g) Washington degradation test
This test is used to assess the decomposition of coarse aggregate as a result of contact with air and water (Fielding, 1980a). It is used for assessing the degradation potential of igneous rocks by giving a measure of the quantity and activity of clay-type particles produced by attrition and dispersion from the surface of the aggregate. The test was developed to overcome deficiencies in test b and is closely linked with test p. Generally, tests p and x are used for basic igneous rocks and this test for other igneous rocks and for metamorphic and sedimentary rocks (although it is still widely used for basic igneous rocks).

The sample is crushed to pass a 19 mm sieve and then is shaken in water for 20 minutes using a sieve shaker. Mutual attrition of the particles occurs and some of the secondary minerals present leach out as clay. The test can therefore supplement test p. The sample is then washed with a flocculating agent comprising calcium chloride, water, glycerine, and formaldehyde; the amount of sediment obtained is measured; and test values are calculated from a complex scale. Good basalts typically test at above 70. Values below 60 indicate a need for concern, 50 is a typical specification limit, although values as low as 40 are permitted for sub-bases and below 30 usually indicates an unsound aggregate. The crushing technique used, and various other factors, can exert a major influence on the results obtained (Moors, 1972). In some standards, the result is called the *degradation factor*. An alternative version of the method is used in ASTM D3744 to determine an *aggregate durability index*.

The test gives a measure of the quantity, quality and accessibility of secondary minerals in basalts and of soft micaceous and claylike material in granites and hornfels. For the latter materials, the test should always be conducted in association with tests b or l as poor results may not necessarily mean that the material lacks durability, however, for hornfels the test is probably a better guide than test b. There is a relationship with test b such that the results of this test are predicted very approximately by (Fielding, 1980a):

5(70 – [test b result])/3

(h) Sand equivalent test

This test measures the quantity and quality of the fine particles in a mixture and provides an indication of the fine grading and plasticity index of a sample. The original form was developed by Hveem of the California Division of Highways in 1953. However, the common current version is based largely on a French variant. It is a useful and relatively quick quality control test but is generally considered to have insufficient resolution to be used as a unique specification requirement. The ASTM version of the test is D2419.

To conduct the test, a sample of the material is shaken with a flocculating agent (as in test g) for 45 s. Mutual attrition of the material occurs and the amount of clay that flocculates out (above the sand) is measured. Because the shaking time is less than in test g, less attrition material is produced. The sand equivalent is the ratio of sediment height to total flocculent height in the tested samples and specification values range between 30 and 80 percent. It has a coefficient of variation of about 20 percent (Ingles and Noble, 1975). A value over 65 indicates that clay is present.

(i) Extended sand equivalent test

This test assesses the durability of fine aggregate by attempting to reproduce the effect of weathering. It is suitable only for igneous and metamorphic rock. The test measures the quantity and quality of fine particles produced by abrading the aggregate in water. It extends the shaking time in test h from 45 s to 10 or 20 minutes. The extended sand equivalent value is the ratio of the square of the 10 minute sand equivalent to the 45 s (test h) value.

Poor results may indicate:

(1) the presence of highly expansive clays (Section 8.4.2),

(2) excessive mica-like minerals,

(3) weak intergranular-bonding (Section 8.4.1), or

(4) excessively soft rock.

The acceptable values are determined by comparison with results on the same source material subjected to test g. Indeed, a linear relationship has been established between the two tests (Fielding, 1980b).

(j) Aggregate crushing value test

This test is performed on an oven-dried sample of aggregate particles tamped into a testing cylinder. A compression-testing machine is used to crush the sample with a force rising from zero to (usually) 400 kN at about 670 Ns^{-1}, i.e. it takes about 10 minutes to reach the peak force. The percentage of particles produced when the aggregate is crushed under this load and which pass a 2.36 mm sieve is called the *Aggregate Crushing Value*. The test is thus a measure of aggregate crushing-resistance. It has a coefficient of variation of about 10 percent (Ingles and Noble, 1975) and a repeatability of about 3 percent (Shipway, 1964). It is described in BS 812. Typical specification values range between 15 and 35 percent.

The test has been found to be very sensitive to small changes in aggregate quality (Shipway, 1964). There is a strong correlation between its results and those of test b (Minty *et al.*, 1980). Very approximately, the Aggregate Crushing Value obtained in this test is about 60 percent of the Los Angeles value, plus 5. Refer also to the discussion of tests k, l and m.

(k) Ten percent fines test

This test — sometimes called the TFV — is a variation of test j in which the load in kN to produce 10 percent of 2.36 mm particles is measured. Given that test j typically produced at least 15 percent below 2.36 mm, this test will usually require a much lower crushing force than the 400 kN of test j. Indeed, it is usually used for softer aggregates such as mudstones (Section 8.5.2) with a test j Aggregate Crushing Value exceeding 30 percent. A typical minimum crushing load is 140 kN, but 110 kN is used in less demanding circumstances. The test is described in BS 812.

Many now place emphasis on the wet/dry variation of this test (test m) as the prime means of determining aggregate durability. There is a strong correlation between the results of this test and tests b and j (Minty *et al.*, 1980). Very approximately, the results of this test are predicted by:

$500(1 - [\{\text{test b or j result}\}/35])$

The approximate nature of such relationships is stressed and they are given only to indicate trends and highlight inter-relationships between tests. As an illustration of the approximations, note that if zero force is needed to produce 10 percent of 2.36 mm particles, giving a value of zero in this test, one would expect tests b or j to give values of 100 (rather than the 500 predicted above). A more general relation would be:

$([\text{test b or j result}]/100)^{0.4} + ([\text{test k or l result}]/800)^{0.4} = 1$

There is also a link with test g (Fielding, 1980a).

(l) Soaked ten percent fines test

This is a variation of test k, using a soaked aggregate that is saturated but surface dry. The load to produce 10 percent of 2.36 mm particles is determined as for test k. The test results are strongly correlated with those of test j (Minty *et al.*, 1980), with the approximate relationship being that its results are predicted by:

$500(1 - [\{\text{test j result}\}/30])$

The test is coupled with test k to give test m.

(m) Wet/dry strength variation test

This test uses the variation in strength of an aggregate tested after drying in an oven (test k), and when saturated but with a dry surface (test l). It is useful for softer rocks. The wet/dry strength variation is defined as:

$100([\text{test k result}] - [\text{test l result}])/[\text{test k result}]$

Typical results range from 50 percent for breccia and 40 percent for dolerite to 10 percent for diorite. Values of 35 percent or less are taken to indicate a durable material and values as high as 60 percent are used in undemanding circumstances. From the earlier discussion, a noticeable link with the test j results can be expected.

(n) Texas ball mill test

This test is only suitable for sedimentary and similar soft rock, and is widely used for these materials. Soaked aggregate is abraded in a watertight cylindrical drum containing six steel balls. The percentage of fine particles produced by this milling is measured as the percent passing a fine sieve (e.g. see Figure 8.7). The test thus measures the ability of the material to withstand breakdown in the presence of water. Values of between 30 (high quality) and 55 (sub-bases) are commonly specified.

(o) Polished stone value (PSV)

This test assesses whether an aggregate will polish under traffic, as this will be a major determinant of the pavement's skid resistance (Section 12.5.2). The British PSV test involves the accelerated polishing of aggregate specimens with a pneumatic tyre coated with an abrasive paste. The procedure is defined in BS 812, Part 114. The ASTM version is D3319. The aggregate particles used are sieved to be between 10 and 14 mm. Thirty-five to fifty of these particles are mounted in mortar in a single layer in a curved-tile mould. Their exposed surfaces are kept proud of the mortar backing. Fourteen such tiles are made and then mounted around the periphery of a steel wheel. The stones on the tiles form a continuous circumferential surface 45 mm wide and 406 mm in diameter. This wheel is held in contact with a second, pneumatic-tyred wheel allowed to rotate freely at about 320 Hz on an axis parallel to that of the steel wheel. The second wheel is 203 mm diameter and 50 mm wide and is applied to the stone periphery with a force of 390 N. Its tyres are smooth, two-ply, truck pneumatics, inflated to 310 kPa. Fresh natural corn-emery and water are continuously and uniformly spread over the tyre–tile contact area. The specimens are tested on the wheel for 3 h at 20 C. Some U.S. versions of the test extend this period to 8 h.

The aggregate is assumed to then be in the condition it would attain after being in service on the road for a lengthy period. In the second stage of the test the skid resistance of the curved specimens of polished aggregate is measured with the associated *British Pendulum Tester* developed by TRRL in 1952. The ASTM version of this part of the test is E303. The Pendulum Tester uses a pendulum with a spring-loaded rubber slider at its end. The total mass of the swinging arm is 1.50 kg located 410 mm from the centre of suspension. The pendulum is released from the horizontal and swings across the wet surface of the specimen. The device measures the energy lost during this process in terms of scale readings (called SRV) which are reasonably close to the percentage value of the coefficient of friction (Section 33.3.5) of the surface. The scale-values are multiplied by 1.6 in order to compensate for the curved surface of the specimen, which presents a diminished surface to the rubber slider. The resulting quantity is called the polished-stone value and is usually quoted as a percentage (PSV). The test has a repeatability of about 4 and the variation between stones from the same sample can exceed 5.

The Pendulum Tester is portable and is also known as the Portable Skid Resistance Tester. Thus, it is also sometimes used to measure the low-speed skid resistance of a localised area of an actual pavement surface, Section 12.5.3. Pendulum Testers should be regularly calibrated (Oliver, 1978). The reliability of the test diminishes as the surface texture becomes coarser, with the effect becoming noticeable when the stone size at the surface exceeds about 12 mm.

A typical specification might limit the PSV of aggregate used in wearing courses to values of 45 or more. Aggregate with $45 < PSV < 50$ and an $LAV < 25$ (see test b) should be rechecked very six months, at least. High quality stones would have a PSV of over 60.

(p) Secondary mineral content test

This soundness and durability test for basalt and other basic igneous rocks requires a competent investigator to examine thin sections of the aggregate under a petrological microscope, in order to identify the secondary mineral content (Section 8.5.10). The tests usually use a point-counting device on a microscope able to do about 500 counts for each thin-section slide. Methylene-blue organic dye is often used to stain the clay minerals for ease in point counting. The secondary-mineral content is the percentage of the secondary-mineral count to the total count and maximum values of 20 to 30 percent are commonly specified (Nyoeger, 1964).

The test generally discriminates against altered basalt containing abundant green clay-minerals — the so-called *green* basalts (Sections 8.5.1&10). It is sometimes conducted in association with test g, but should take precedence over that test. The results of test g can be approximately predicted from those of this test by (Cole and Ceram, 1981):

test g result = 110 – 2.6(test p result)

The test is often conducted in conjunction with test x.

(q) Brazil (or Brazilian splitting or indirect tension) test

This test measures the tensile strength of a rock by testing a cylindrical specimen (e.g. a drill core) loaded in compression along two opposite surface generators of the cylinder. The compressive load produces tensile strains within the specimen as the Poisson effect (Section 33.3.2) causes it to expand transversely. The test thus avoids the experimental complications of producing specimens that can be clamped and then pulled in uniaxial tension. The MATTA device is a typical machine produced for performing this test (AS 2891.13.1) and an application is discussed in Section 12.2.4. The test is also covered by ASTM D–3967.

The maximum tensile stress produced is (load)/πlr, where l and r are the length and radius of the cylinder. The test is also used to measure the stiffness of the material. It can give misleading results due to premature compressive or shear failures occurring under the loads. The test is sometimes used for soils and is increasingly used for asphalt (e.g. for fatigue in Section 12.2.4, for mix design — particularly stiffness (resilient modulus) — in Section 12.2.5, and for adhesion in Section 12.3.2).

(r) Repeated-load triaxial test

The simple triaxial test is described in test s1. In the repeated-load version of the test the load on the cylinder ends fluctuates and load on the cylinder surface varies in phase with this end load. It is usually necessary to measure the longitudinal and transverse strains in the sample, and these data are often used to control the amount of stress applied in each load cycle.

The basic role that this test plays in characterising pavement material behaviour is discussed in Sections 8.4.3, 11.1.3, 11.2.4, 11.3.1, & 11.4.3. The prime basis of the test is that the loading used closely resembles the loading experienced in a basecourse or subgrade under traffic loading. It also allows good control of specimen density and moisture content and provides easy measurement of stress and strain. It is the only test that reasonably characterises materials with regard to deformation under load. It provides useful values for stiffness moduli, permanent deformation (rut resistance) data, and shear strength. It is probably the best test for an 'unknown' material being considered for use in an unbound pavement course. Note that it would normally be expected that materials loaded dynamically, as in this test or in service, would appear stiffer, stronger, and less ductile than if slowly loaded (Lay, 1982).

(s1) The triaxial test

This fundamental test is described in Section 33.3.4. It is usually conducted on a cylindrical specimen whose length is about twice its diameter. The manner in which the specimen is prepared is critically important. This is usually done by machine-controlled dynamic compaction. The prime load, which represents the vertical traffic load, is applied to the flat ends of the cylinder. A uniformly distributed pressure is applied to the curved surface of the cylinder. It represents the horizontal compressive stresses induced by the restraint that the remainder of the pavement structure applies to the attempted horizontal Poisson expansion of the cylinder. From Figure 33.3 it can be seen that this horizontal stress will be somewhere between zero (no restraint) and $\nu/(1 - \nu)$ times the loading stress, where ν is Poisson's ratio.

It is a good test for determining shear strength. Considering the Mohr–Coulomb failure criterion in Figure 8.9, a typical test uses three loading configurations (A, B & C in Figure 8.9).

(s2) Texas triaxial (or accelerated triaxial compression) test

This test is a version of the triaxial test (s1). It is used particularly to quantify the moisture susceptibility and the deformation resistance of basecourse materials. It is a good test for determining the shear strength of a material. The ASTM version of the test is D3397.

In conducting the test, material that passes a 20 mm sieve is used to form a standard 150 x 100 mm cylindrical test specimen. It is compacted with a drop hammer in a steel mould to mimic field compaction. The static peak (or failure) stress is then determined for five different confining pressures. The results are usually dependent on the sample moisture content. The Mohr–Coulomb failure envelope of the five resulting Mohr's circles is used as the basis for classifying the material (Section 8.4.3). The classes range from:

* very good (0); a very steep failure-envelope, with shear stress at least 3.3 times normal stress,
* satisfactory (3); shear stress at least 1.5 times normal stress,
* very poor (6); a very flat failure-envelope with shear stress no more than 0.25 times the normal stress.

These categories were established by comparing laboratory results with satisfactory service performance (Hamory and McInnes, 1972; Gerrard *et al.*, 1972). For instance, a material is usually acceptable for basecourses on sealed, lightly trafficked (< 2000 veh/d) rural roads, if it has a Texas Classification number of 3 or less and an average measured compression modulus of 35 MPa or more. For roads carrying up to 2000 heavy vehicle/lane/day, the acceptable Classification number drops to 2 (Giffen, Youdale and Walter, 1978).

The test results can usefully be coupled with a consideration of the internal friction and cohesion of the material (Figure 8.9), which can both be obtained from the same testing arrangement. The test can also be linked to the CBR test of Section 8.4.4 (Clegg, 1983). Clearly, the test is a measure of the strength of a material, i.e. of its load capacity prior to gross permanent deformation. It is not a measure of the stiffness of the material, i.e. of its resistance to deflection under working loads. However, it is sometimes argued with some reason that the two (strength and stiffness) are related for common basecourse materials. Nevertheless, the Classification number cannot be directly related to the fundamental performance measures needed for a structural analysis of the pavement (Chapter 11).

(t) Full-scale testing

Full-scale testing falls into one of four categories:

(t1) *Circular test tracks*, in which loaded wheels on the ends of long arms rotate about a central vertical pin and run over an annular test track. They are relatively widely used, with over 20 having been in existence around the world. A typical device is the TRRL 'Road Machine' which has a 34 m diameter, 3 m wide track. Its single test wheel is electrically driven and loaded pneumatically. The main disadvantages of circular test tracks are the difficulty of placing a realistic, circular piece of pavement, using conventional equipment, and their inability to reproduce the field environment.

(t2) *Linear test tracks*, in which a loaded wheel is moved up and down (or up and over in one-way loading) a straight test track. It is mechanically less efficient than the circular track but has the advantage in allowing the construction of more realistic pavements. It is otherwise fairly similar to the circular track. Linear tracks are also fairly widely used, with at least 11 having been in use around the world. Small versions of this test are used to determine the rutting characteristics of asphalt (Section 12.2.4).

(t3) *Test loading of real pavements* in situ. This is usually done using a portable version of (t2). The two major instances of this approach are the South African Heavy Vehicle Simulator and the Australian Accelerated Loading Facility (ALF, ARRB, 1985). They overcome a number of the disadvantages of (t1) and (t2), but experimental control is more difficult and loading is quicker than would be encountered in practice. The CBR test (Section 8.4.4) should also be mentioned in this context.

(t4) *Test pavements under traffic*. In this technique, either an actual pavement length subjected to real traffic is monitored or, as in the AASHO Road Test (Section 11.4.3), test pavements are set aside solely for the purpose and loaded with specially selected and controlled trucks. Experimental control is even more difficult than in (t3) and the monitoring of the axle loads in the traffic stream may well present difficulties — both practically and technically.

Another distinction between test tracks is based on loading method. There are three main types:

* rolling wheel,
* miniaturised rolling wheels, and
* simulated rolling wheel, e.g. by vibrating or pulsed loads applied by jacks.

(u) Water-absorption test

A basic, practical, preliminary test for aggregate is to measure the amount of water a dried sample will absorb (Section 8.1 & BS 812). If it absorbs between 2 percent and 4 percent of its mass, it should be carefully examined by other tests. If it absorbs in excess of 4 percent of its mass, it will rarely prove to be an adequate roadmaking material.

(v) Scraping test

A practical preliminary test is to hold a piece of proposed aggregate in the hand and firmly scrape it along a clean, smooth, steel surface. If a residue is left on the steel, the aggregate should be treated with caution.

(w) Wheel tracking tests

Wheel tracking tests have developed as laboratory tests from the larger wheeled tests t1 and t2. They are increasingly used for 'proving' asphalt mix design (Section 12.2.4). In a typical test a loaded wheel rests on the test slab which is carried on an oscillating table. The horizontal table movements are of the order of 250 mm. The progression of rutting with load cycles is measured.

(x) Accelerated soundness index by reflux

This test estimates the soundness of basic (basaltic) igneous rocks which contain smectite/chlorite clays (Table 8.2) as a result of alteration and/or weathering (Section 8.5.10). These are typically some of the 'green' basalts (Sections 8.5.1&10). The test sample largely passes a 19.0 sieve and is retained on a 13.2 mm sieve. The sample is immersed in boiling ethylene glycol and subjected to five boiling and soaking cycles over 40 h. 'Reflux' relates to the reflux condenser used during the boiling process. The amount of physical degradation is observed. Numerically, the soundness index relates to the proportion of the sample after testing which is retained on a 6.70 mm sieve. The test is usually conducted in conjunction with test p. It is covered in Australian Standard AS 1141.29.

8.7 BITUMEN

It is necessary to begin with two sets of definitions. This text uses the word *bitumen* as a technically explicit term for the highest boiling point fraction of crude petroleum, and uses *asphalt* for a mixture of bitumen and aggregate (Section 12.2). Note that in North America, *bitumen* used for roadmaking is usually called *asphaltic cement* and sometimes *asphalt* or *petroleum asphalt.* This confusing practice arose from the range of binder types originally marketed in North America (Lay, 1992).

It is also useful to recall that *hydrocarbons* (HC) are compounds composed only of hydrogen and carbon atoms (C_nH_m). The simplest hydrocarbon is methane, CH_4, which can be represented as:

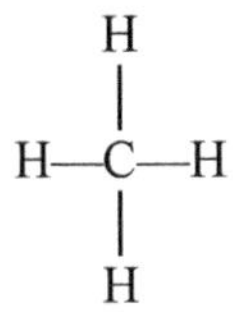

The carbon–hydrogen bond can be replaced by carbon–carbon to form straight (n-alkanes) or branched (iso-alkanes) chains. For example:

alkane-pentane, C_5H_{12}

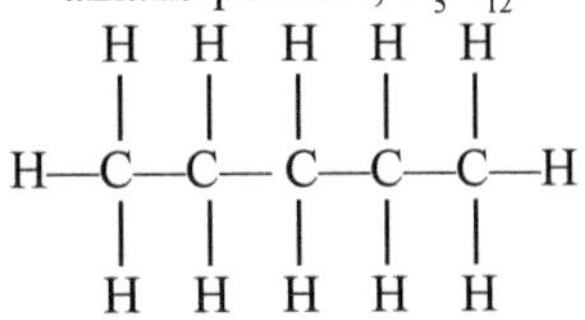

iso-alkane-isopentane, C_6H_{14}

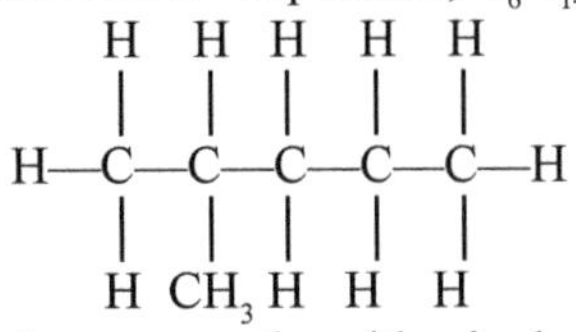

These examples both illustrate saturated compounds with single carbon bonds. Compounds such as ethylene have double carbon bonds and ones such as acetylene have triple bonds. Bituminous hydrocarbons can contain from 50 to 1 000 carbon atoms.

8.7.1 Composition

(a) Initial conditions

Bitumen is a mixture whose parts are composed mainly of hydrocarbons (Section 33.1) and their derivatives. By weight, it is about 85 percent carbon. The precise composition of bitumen depends on its source, and can vary from oil well to oil well and from time to time at a particular oil well.

The four main parts of bitumen are:

(1) *Maltene* (or malthene or petrolene). This part is the main component of the mix — at about 55 percent — and supplies the mix with its visco-elasticity (Section 33.3.3). Maltene is an aliphatic non-polar, hydrocarbon oil composed of products such as acidaffins and paraffins. In the laboratory, it can be subjected to further fractionation. It is a solvent and is soluble in petroleum ether. Maltene has an atomic carbon-to-hydrogen ratio of less than 0.8 and a relatively low molecular weight. The molecular structure tends to be chain-like.

(2) *Asphaltenes.* This part is probably a hydrocarbon oxidation product and represents about 20 percent of bitumen by weight. It supplies the hard 'body' of a bitumen. Asphaltenes are in colloidal suspension in maltene, usually as a sol (particles dispersed) rather than a gel (particles in contact). They can be separated by precipitation in low molecular weight hydrocarbon solvents. They are polar and this means that they control the adhesion properties of the bitumen (Section 12.3.2). Due to their polarity, they form into hydrocarbon micelles (elementary lamella a few molecules thick) held together loosely by hydrogen bonding, creating a series of polynuclear sheets which are brown-black in colour, aromatic, amorphous, hard, and relatively inert. On their own, they are solid but powdery materials. Heating bitumen breaks down the asphaltene adhesion and therefore reduces the viscosity of the bitumen.

Asphaltenes are soluble in carbon disulfide but not in petroleum and are thus easily distinguished from maltene. The molecules are relatively large and have an atomic carbon-to-hydrogen ratio of more than 0.8, a high molecular weight (*c.* 1000-20 000), and are more aromatic than maltene. The molecular structure tends to be ring-like.

(3) In between the maltene and asphaltenes phases are resin-like intermediate molecular-weight hydrocarbons. These comprise about 20 percent of the mixture, and usually exist as a sheath of adsorbed material covering the asphaltene molecules. They assist in maintaining the colloidal stability of the sol suspension of the asphaltenes in maltene and provide ductility and adhesion.

(4) The remainder of the mixture — usually about 5 percent — consists of atoms of sulfur, nitrogen, and oxygen that are mainly attached to the various hydrocarbons and give them a polar character.

The resulting mixture of the four parts — bitumen — is consequently a black or brown viscous liquid, although at its lower operating temperatures it is almost a solid. It is soluble in carbon disulfide, benzene, and trichloroethylene. It is largely non-volatile and is resistant to most acids, alkalis, and salts. Its deformation response is dependent on both its temperature and the rate at which it is loaded. Its relevant roadmaking properties are its adhesion, cheapness, workability, strength, durability, and imperviousness. Indeed its main virtue is that it is currently the cheapest durable glue (or binder) available and in asphalt is primarily used as a glue to hold particles of aggregate together and thus form a stiff and impervious composite material (Section 12.2).

A comprehensive review of bitumen properties and applications is given in Dickinson (1984). A formal description and definition of bitumen is given in WHO (2004).

(b) Changes with time

Over time, bitumens:

* harden, stiffen and become more viscous,
* become brittle and lose their ductility, and
* lose their adhesiveness.

The assessment and management of bitumen durability is discussed in Section 8.7.7. The ageing process that causes these effects results from a combination of:

volatilisation,
oxidation,
steric hardening (molecular restructuring over time and related thixotropic effects), and
actinic light (ultraviolet light effects).

Although the maltene and other volatile, 'lighter' components of bitumen provide much of its plasticity and fluidity, hardening with age due to the direct loss of such components by *volatilisation* is not as common a problem as is oxidation.

Oxidation occurs when hydrogen in the maltene combines with oxygen and is removed via water molecules. It therefore requires the presence of oxygen, which must be able to diffuse into the bitumen for this effect to occur (Dickinson, 1982). The presence of light speeds up this oxidation reaction but, because bitumen is a good light-absorber, this is confined to the top 5 μm of the exposed bitumen. Oxidation in the absence of light is much slower and is temperature-dependent, with the rate doubling for every 10 C temperature rise. A similar oxidation process occurs with air blowing during the manufacture of bitumen (see Section 8.7.2b). The speed of reaction is thus partly controlled by temperature, partly by light, and partly by the rate at which the oxygen can diffuse into the bitumen. As oxidation progresses, the asphaltene proportion in the bitumen increases and it plays a more dominant role; and so the bitumen itself hardens and becomes less durable. A colour change from black to grey is also common.

Asphalt production is discussed in Sections 12.2.1 and 25.6. It will be evident from comparing those descriptions with the above causes of bitumen hardening, that a considerable amount of bitumen hardening occurs during asphalt production.

8.7.2 Manufacture

Bitumen exists in solution as a natural constituent of most crude petroleums, and may constitute from zero to 60 percent of the crude. The higher percentages are usually associated with crudes having a relatively high specific gravity. The bitumen is usually obtained by treating the *heavy-end* residue from the distillation process. This product is therefore known as *residual bitumen.* The other product obtained from the heavy-end residue is residual (or heavy) furnace oil. Section 8.7.5 shows how bitumen is sometimes recombined with the lighter distillation products to produce cutback bitumens. In all practical cases, bitumen is produced to meet a performance specification, rather than to achieve a specified chemical composition.

Bitumen manufacture is usually a multi-stage process (Figure 8.13). Three main production processes are used:

(a) *distillation.* Distillation of crude oil at atmospheric pressure removes lighter fractionation products such as petrol and kerosene, leaving a 'long', higher boiling point, residue. Atmospheric distillation of a crude oil must occur at restricted temperatures to prevent *cracking*, which is a decomposition of the chemical structure of the bitumen during which the higher molecular-weight hydrocarbons are split into lighter oils and free carbon. Thus the long residue is usually treated by vacuum distillation, reduced pressure, or steam distillation, to produce a 'short' residue that provides both fuel oils and the various bitumen grades. These processes permit temperatures as high as 400 C to be attained during processing without causing cracking. However, the bitumen produced becomes harder as either the vacuum or the temperature is increased. The resulting products are called *straight-run bitumens*. As the above processes are physical rather than chemical, the products can be later recombined to produce a homogeneous material.

(b) *distillation plus air blowing.* Blowing air into the hot, liquid distillation residue causes some oxidation to occur and higher molecular-weight hydrocarbons to form. These effects raise the viscosity of the straight-run bitumen and reduce its temperature susceptibility (Sections 8.7.4 and 33.3.3). However, air blowing also increases hardness (Sections 8.7.1 & 33.3.4) and colloidal instability (the tendency to change from a sol to a gel, Section 8.7.1(a)), and lowers ductility. The amount of air blowing is therefore a compromise between these two sets of effects.

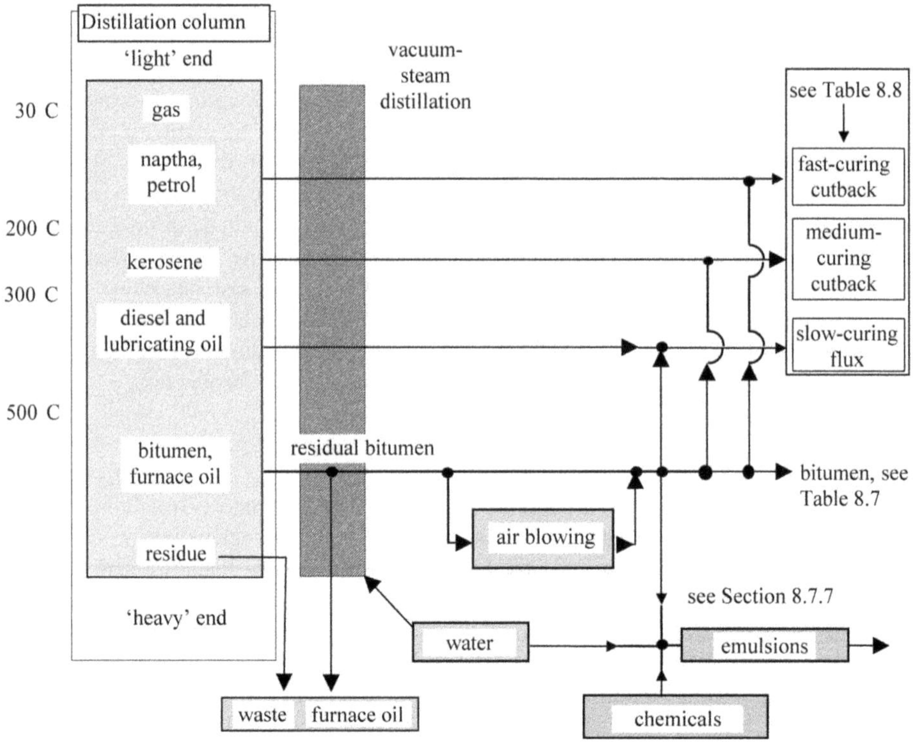

Figure 8.13 Bitumen production as a by-product of the distillation of crude oil.

(c) *distillation followed by blending with solvent-precipitated residue.* The solvent is usually propane and the residue is usually a pitch. The resulting harder bitumen is therefore called a *propane precipitated (PPA) bitumen.*

Historically, a significant quantity of bitumen was also obtained from the naturally-occurring *native asphalts* (see Sections 3.6.2 and 12.2.1), particularly Trinidad Lake asphalt. The Lake asphalt contains about 55 percent soluble bitumen that was often fluxed with oil to raise the bitumen percentage to about 65 percent. More recently, the natural product was mixed with residual bitumen to raise the percentage to 75 percent. Some native asphalt is still used in the U.K. for surface courses, due to its good weathering properties (Section 12.5.2).

Very hard bitumens cannot be obtained from the oil-refining process, but may be found in naturally-occurring deposits. Some, like *Gilsabind/Gilsonite*, are useful in treating existing bitumen surfaces to increase their viscosity.

8.7.3 Performance

Temperature plays a major role in determining bitumen performance. The highest handling temperature for bitumen is about 150 C and the highest pavement temperature in service is almost 70 C. At temperatures between 70 and 150 C bitumen is predominantly *viscous* and can be modelled as a viscous fluid (Section 33.3.3). Bitumen is therefore made workable by heating. This also means that the transport and storage of bitumen requires sophisticated equipment and a great deal of energy.

When bitumen is heated, some fumes are produced. Their organic component comprises benzene-soluble matter that in turn contains *polycyclic aromatic hydrocarbons*. These are potential *carcinogens* (WHO, 2004) and so breathing the fumes from heated bitumen should be avoided. At higher temperatures, the fumes increase and may also contain semi-volatiles that may irritate the skin and respiratory system. Thus temperatures should be kept as low as possible, in the context of proper handling and application procedures. It has been suggested that an acceptable level of total particulate matter might be 5 mg/m^3.

At pavement temperatures below 70 C, the temperature and loading rate effects become important and behaviour can be described as linearly visco-elastic or pseudo-plastic. Thus the bitumen behaves in a predominantly elastic manner when cold or subjected to rapid loading, and tends towards viscous flow when hot or subjected to long-term loading. True viscosity measurements are no longer possible. At temperatures of around 0 C or less, most bitumens begin to fracture without any prior plastic deformation. At the bottom of the range of service temperatures, bitumen reaches the *glass transition temperature* and becomes a weak and brittle elastic solid. This is thus an important temperature for roads in cold climates and its value for many bitumens is between 0 and –40 C (Dickinson, 1984). Generally, the viscous properties of a bitumen should be such that it is:

(a) sufficiently fluid at high temperatures to permit it to be handled during construction and to coat the entire surface of pieces of aggregate;

(b) sufficiently viscous at high pavement temperatures to ensure that it will not permanently deform under traffic; and

(c) sufficiently plastic at low temperatures to avoid fracture and cracking.

Clearly, there is conflict between these requirements and some compromise must be reached. A performance specification for bitumen would control its:

(1) deformation response over the whole service range of loading rates and temperatures, and
(2) tendency to harden during handling and service (Section 8.7.1).

The popular spray and chip seal process (Section 12.1) relies on thin bituminous films. Two types of loading are of particular importance in relation to their performance (see Section 12.1.4). The first is the result of vehicles passing over the surfacing and is periodic and approximately sinusoidal with a duration of about 40 ms. This traffic loading can be critical at both ends of the temperature range. The second is due to *thermal expansion and contraction* and has a duration of about 10 ks. Thermal contraction will be critical at low temperatures when ductility is low. For a fresh bitumen, the deformation response to traffic loading at 60 C is rheologically equivalent to thermal contraction at 5 C.

8.7.4 Properties

As with most materials, the static resistance of a bitumen increases as its loading rate increases. At the other extreme, viscosity (Section 33.3.3) measures the ability of a material to resist flow at low loading rates. The key issue in determining fundamental bitumen properties is to establish its viscous properties as a function of loading rate and temperature. The sensitivity of a bitumen to loading rate is known as its *shear susceptibility*.

Figure 8.14 shows the deformation response of a bitumen loaded sinusoidally in shear (Dickinson, 1981 and Section 33.3.3). The absolute value of the complex shear modulus (G*, a stiffness measure) increases, at an increasing rate, with loading frequency (i.e. with strain rate) and decreases with temperature. The dynamic shear rheometer developed as part of the U.S. SHRP program measures this modulus by applying shear stress at different frequencies to a bitumen sample held between two parallel plates (as in the sliding plate viscometer described below).

Curves of the complex shear modulus against loading frequency for various temperatures can be superimposed by a loading-rate shift to give a master curve (Section 33.3.3) at a reference temperature usually taken as 25 C (Figure 8.15). The temperature susceptibility can be estimated from the shift needed. A high value means that the effect of temperature on shear modulus is high. That is, the bitumen can be readily deformed when hot and fractured when cold. Temperature susceptibility is relatively independent of bitumen composition.

The slope of the master shear-modulus vs loading-frequency curve in Figure 8.15 is used to define another shear-susceptibility parameter, ß. A high ß value means that, with increasing loading frequency, there is a slow transition from viscous to elastic behaviour. The value of ß after a Rolling Thin Film Oven (RTFO) treatment should be as high as possible, up to a maximum of 2.25 to prevent the acceptance of colloidally unstable material (Sections 8.7.2&7). The limiting (or maximum) viscosity is obtained from the reference master curve by extrapolation to the estimate of the viscous constant at zero loading rate.

The *sliding plate viscometer* measures viscous response by shearing a thin film of bitumen between two plates at a defined strain rate. Very large strains can be applied, but the method lacks the precision of the rotational and capillary viscometers. A typical device is the (Shell sliding plate) microviscometer (AS 2341.5). It gives an apparent (steady-state) viscosity that is calculated as the ratio of shear stress to shear strain rate at a

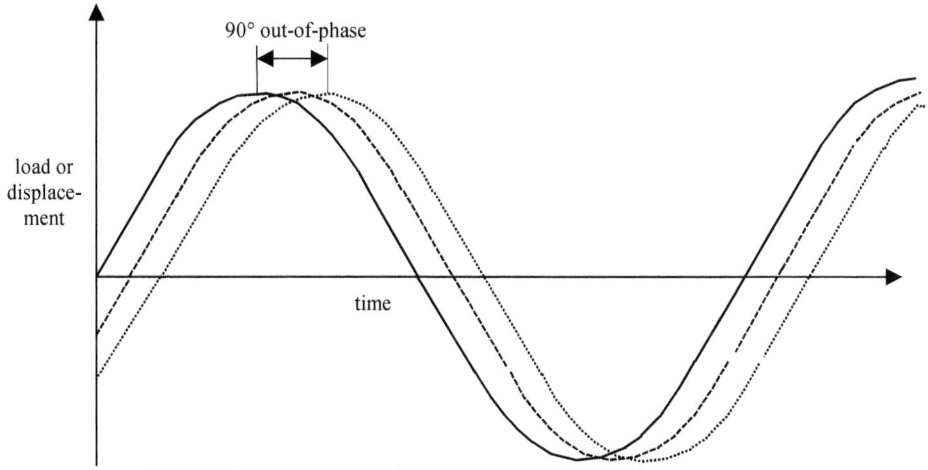

key:	deformation response due to:
————————	purely elastic behaviour
- - - - - - - - - - - -	visco-elastic behaviour
··························	purely viscous behaviour
————————	this is also the load–time curve

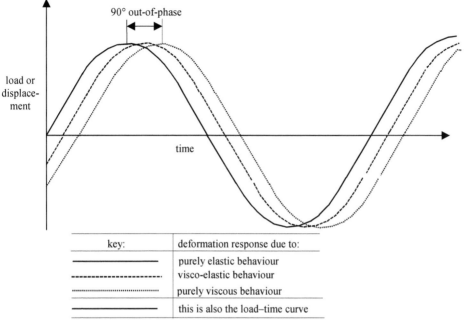

Figure 8.14 Deformation response of bitumen.

defined strain rate and temperature. This quantity is sometimes termed the *consistency* of the bitumen. For much of its range of application, it is an 'invention' (hence, the use of 'apparent') as true bitumen viscosity is not measurable at temperatures below about 60 C. Apparent viscosity drops as the strain rate increases. The temperature susceptibility parameter increases as the ratio of the 25 C apparent viscosity to the 70 C viscosity

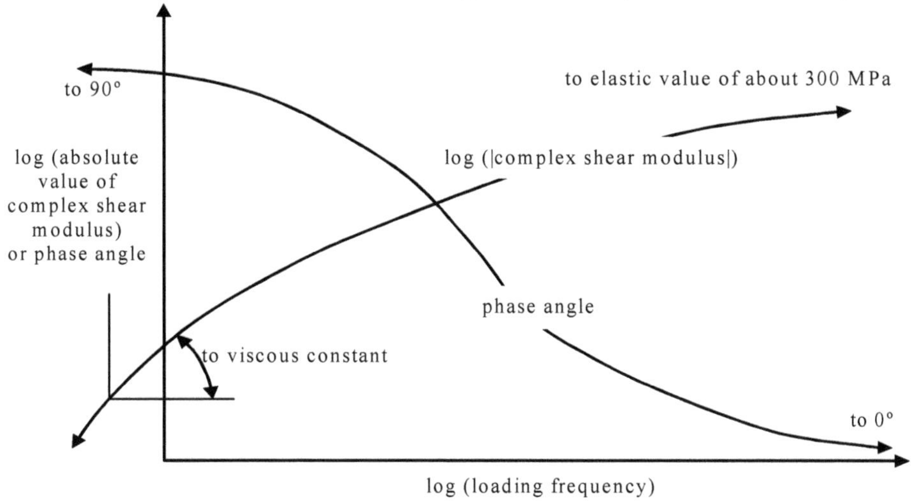

Figure 8.15 Master curve showing the influence of strain rate on the load response of bitumen (from Dickinson).

increases. The *fracture stress* of unconfined bitumen is about 5 MPa at a strain of between 0.01and 0.10. The value of G* also drops once strains exceed 0.01.

The *softening point* of a bitumen is a simple measure of its transition from a viscous solid to a liquid. Its measurement is specified in ASTM D36 & AS 2341.9. It is conventionally set at least 10 C above the maximum anticipated air temperature. Even inherently soft bitumens would usually have a softening point of at least 30 C. The *flash point* of a bitumen is the temperature at which its vapour can momentarily take fire. It is therefore the maximum temperature to which bitumen can be heated safely without the danger of an instantaneous flash in the presence of an open flame. A typical value might be 250 C measured by the Pensky Martens Open Cup test. The *fire point*, at which the bitumen will actually burn, is usually well above the flash point. A minimum flash point is used also to safeguard against contamination by cutting and fluxing materials.

8.7.5 Classification and specification

Kinematic viscosity is used as a classification measure in some specifications. It is a function of the dynamic viscosity, test temperature, and material density and its assessment usually involves measuring the time for a defined volume of bitumen to flow through a capillary under a defined pressure head. Multiplying this time by a calibration constant for the instrument and dividing by the density of the material gives the kinematic viscosity. The test is specified in ASTM D2170.

Bitumens have traditionally been specified on the basis of their *penetration* at 25 C, measured by the distance (in mm) that a standard 1 mm diameter sharp needle will penetrate a specimen under a force of either:

* 2 N (supplied by a 200 g mass) applied for 60 s at 15 C, or
* 1 N applied for 5 s at 25 C (e.g. AS 2341.12).

The longer loads and times quoted above (2 N, 60 s) were introduced to improve the sensitivity of the test at 15 C. The penetration value is given in millimetres or tenths of a millimetre, respectively. The range of this value is used to control the consistency of a grade. A low-penetration bitumen is a hard bitumen and a high-penetration bitumen is a soft one. Penetration testing down to 5 C is used in colder climates to help control low-temperature cracking. The penetration test can only be used for comparative rather than absolute purposes as it is conducted at one, fairly irrelevant, loading rate. Penetration increases with temperature, as represented by:

$$\log(\text{penetration}) = f + h(\text{temperature})$$

where f and h are positive constants. When combined with Equation 8.5 below, this accords with the common observation that:

$$\log\log(\text{viscosity}) = j - k(\text{temperature})$$

where j and k are positive constants. The *penetration index* (PI) is an empirical measure of the rate at which the penetration value changes with temperature and is thus another measure of the temperature susceptibility of a bitumen. A low PI bitumen has a low temperature susceptibility.

In older specifications, it is common to see reference to 85–100 Pen bitumens and the like. This usually refers to the penetration range in 0.1 mm units, for instance, 85–100 refers to a bitumen with a penetration of between 8.5 and 10 mm. There are typically five penetration ranges specified, 40–50, 60–70, 85–100, 120–150, and 200–300. However, European standards are moving towards nine ranges between 20–30 and 250–330. Penetration test data can be related to viscosity by an empirical equation of the form:

$$\log(\text{viscosity}) = m - n(\text{penetration}) \tag{8.5}$$

where m and n are positive constants. McLeod developed a *pen-vis number (PVN)* that relates directly to n and which was used as a measure of temperature susceptibility. PVN decreases as n increases with a PVN of 0 indicating low temperature susceptibility and of –1.5 (a higher n) indicating increased susceptibility. The link between penetration and viscosity values is illustrated in Table 8.7.

Table 8.7 Typical bitumen classes, based on ASTM D3381 and SAA, 1980

ASTM AC grade	SAA class	Viscosity range (Pa.s)		Penetration (mm) at:	
		60°C	135°C	5 s, 25 C, with 100 g (standard)	60 s, 15 C, with 200 g
2.5		20–30		20–	
5	50	40–60	0.17–0.30	18–20	17
10		80–120		7–	
	170	140–200	0.25–0.45	8.5–12	8
20	160–240			4–	
	320	260–380	0.60–0.65	5–6.5	6
40		320–480		2–	
	600	500–700	0.60–0.85	3.0–4.0	3

The viscosities in this table can be related to the following practical requirements:

coating aggregate	0 to 0.2 Pa.s
steel wheel rolling	0.3 to 10 Pa.s
tyred rolling	2 to 100 Pa.s

Another traditional empirical test of bitumen is the *ring and ball test* that gives the temperature at which a disc of bitumen contained in a ring, sags 25 mm under the weight of a ball.

The specification requirements for bitumen frequently involve limits on viscosity at 60 C and 135 C, on penetration at 15 C or 25 C and on ductility at 15 C, after RTFO treatment (Section 8.7.7). The viscosity temperatures of 60 C and 135 C represent, respectively, a typical maximum pavement temperature and a temperature in the construction handling range. The specification of viscosities at the two temperature limits controls temperature susceptibility. Viscosities above about 20 Pa.s make an asphalt difficult to compact. In some specifications, the old penetration test at 25 C is replaced by penetration tests at 15 or even 5 C , which serve to prevent the use of bitumens prone to low temperature brittleness. The low-temperature ductility limit is used to exclude bitumens prone to colloidal instability (Section 8.7.2) in service.

Bitumens may be classified according to their 'average' viscosity at 60 C in Pa.s units. Four common classes are shown in Table 8.4 and range from the relatively fluid Class 50, to the soft Class 170, to the relatively stiff Class 600.

* Class 50 bitumens are used for emulsions (see Section 8.7.6 & Table 8.5).
* Class 170 is the bitumen grade often used in warm climates.
* Class 320 is used where stiffer mixes are required, e.g. in high temperature regions and in thick asphalt layers resisting heavy traffic. It is thus the common grade for major roads. However, its high viscosity is of little value for loads applied at slow strain rates — as with slow-moving and stationary traffic.
* The stiffer Class 600 is rarely used, although it would still be adequately fluid to be able to be placed in other than low temperature circumstances.

8.7.6 Bitumen types

(a) Conventional bitumen

The refinery production of undiluted straight-run bitumen is discussed in Section 8.7.2. This material must be heated to become of low enough viscosity to be workable and to properly wet aggregate surfaces.

(b) Cutback bitumen

As an alternative to heating bitumen, its viscosity can be temporarily reduced by the addition of a suitable volatile diluent: the diluent is called a *cutter* and the new compound is referred to as a *cutback bitumen*. Typical cutters are:

* the relatively volatile petrol and naphtha, used to produce very short-term effects;
* the lighter petroleum oils such as kerosene and turbine fuel which are the most common materials used and provide short- to medium-term cutting. They are often called *cutter oils*.
* furnace and diesel (or automotive gas) oils which are used as long-term cutters and create a fluxed (i.e. 'permanently' softened) bitumen. They are sometimes called *flux oil*.
* most adhesion additives (Section 12.3.2) are also diluents.

These materials may cause skin and respiratory irritation (Section 8.7.3), pollution and fire-risk problems during preparation and placement and while the cutter is evaporating.

Cutback bitumens retain their low viscosity for a much longer period and at lower temperatures than is practical with a heated straight-run bitumen. The effects of the diluents diminish with time. Relatively rapid evaporation is usually required to permit the bitumen to become stiff enough to carry construction traffic. This stiffening process caused by cutter evaporation is a form of curing.

As with conventional bitumen, cutback bitumens can be classified according to their viscosity at 60 C — as shown in Table 8.8.

Table 8.8 Cutback bitumen types.

Surface treatment application (Section 12.1.4)	60°C viscosity range, (Pa.s)	Application temperature, (°C)
Priming	0.008–0.016	10–30
	0.025–0.050	35–55
	0.060–0.120	60–80
Primer-seal and premix	0.22–0.44	75–100
	0.55–1.1	95–110
	2.0–4.0	110–135
Sealing	5.5–11	120–150
	3–26	135–160
	43–86	150–175

(c) Emulsified bitumen

A second alternative to heating bitumen to reduce its viscosity is to emulsify it. An emulsion is the intimate dispersion of one immiscible fluid into another. A bitumen emulsion is a finely divided dispersion of minute (1–10 μm) droplets of bitumen in water. Preparation of the emulsion requires the use of a high-speed, high-shear mechanical device such as a colloid mill. The emulsified state is then preserved by the use of a water-soluble emulsifying agent — or emulsifier — which creates surface charges on the droplets and thus causes mutual repulsion between them.

A range of emulsifiers is available; some can be used cold but others must be heated. They must be water-soluble and are commonly surfactant molecules (such as proteins and soaps) with a long hydrophobic hydrocarbon tail that seeks a bitumen droplet, and an ionised hydrophilic head that seeks water. They thus collect on droplet surfaces in an oriented fashion. Emulsifiers may be either anionic or cationic:

(i) An *anionic emulsifier* places negatively charged anions on the surfaces of the bitumen droplets. The 'water' is usually alkaline. The emulsion breaks as the water evaporates and the emulsion changes in colour from black to brown. This is usually a very slow process and anionic emulsions remain fluid for long periods. They are therefore liable to flow if placed on slopes or to be affected by traffic using the road within a few hours of application. They can assist with adhesion (Section 12.3.2) when aggregate surfaces are positively charged, as is commonly the case with basalt, dolomite, and limestone. Anionic emulsions were first developed in the 1920s.

(ii) A *cationic emulsifier* places positively charged cations on the surfaces of the bitumen droplets. The 'water' is usually acid. Cationic emulsions do not rely significantly on water evaporation in order to break and so are not as fluid as anionic ones. The break is irreversible and shorter than the break with anionic emulsions. Nevertheless, cationic emulsions can still take a significant time to break and become stiff enough to carry traffic. They can assist with adhesion (Section 12.3.2) when aggregate surfaces are negatively charged, as is commonly the case with quartz, quartzite, granite, and sandstone gravel. Cationic emulsions were first developed in the 1950s.

Typically, the emulsion will be 60 percent bitumen, 40 percent water, and a small amount of emulsifier. Emulsions therefore avoid the pollution and safety problems caused by the use of volatile and flammable hydrocarbons in cutback bitumen. They are also often cheaper than cutback bitumens. Care must be taken to ensure that the additives used do not become active carcinogens.

When the repulsion disappears the emulsion is said to *break*. The separation of the bitumen particles can then no longer be sustained and the bitumen droplets recombine or coalesce. This can occur when the water is evaporated or absorbed by another agent, when certain ions or salts are introduced into the water, or when agitation in the presence of rock particles occurs. The first stage of breaking is called the cheesy or curing state. Breaking is called *instability* when it occurs earlier than intended and *setting* when it occurs as intended. The control of breaking is critical to emulsion use and can be managed through either chemical or physical means. Common means are increasing the pH, increasing bitumen content, and water evaporation.

The common emulsion grades are:

(a) ARS — anionic rapid-setting. This emulsion breaks rapidly (in less than 3 minutes) and so in warm climates it cannot be mixed with aggregate. It is useful for tack and seal coats (Sections 12.1.3&4).

(b) ASS — anionic slow-setting. This emulsion has sufficient stability to last for at least 8 minutes and therefore can be used for all operations involving mixing.

(c) HF — high float. This is a medium-setting anionic emulsion that is useful in low temperature applications.

(d) CSS — cationic slow-setting. This is used for soil stabilisation (Section 10.4), surface enrichment (Section 12.1.5), and laying dust (Section 12.1.2).

(e) CAM — cationic special-grade. This is used for making asphalt mixes intended to be stockpiled at ambient temperatures (coldmixes, Section 12.2.7).

(f) CRS — cationic rapid-setting. This is the cationic form of ARS.

They are designated by the relevant acronym, followed by the name of the bitumen grade used.

Cutters and emulsions can be more expensive than bitumen but their ability to be used when cold makes them particularly appropriate for maintenance operations. Emulsions are also used for filling cracks in pavements.

8.7.7 Durability

The durability of bitumen is mainly limited by oxidation in the presence of sunlight. The process is described in Section 8.7.1(b) where it is seen that bitumen is most susceptible to chemical decomposition in hot, exposed conditions, in porous mixes, and/or when used in thin films to coat aggregate. The thin film condition is also the most critical condition for the physical properties of bitumen. However, in all applications an ageing bitumen becomes more brittle and crack-susceptible.

To imitate the stone-coating situation and to simulate and accelerate the bitumen hardening to be expected during the manufacture and placement of asphalt (Section 8.7.1), the Californian State Highways Department in 1963 developed the Rolling Thin Film Oven (RTFO) test (ASTM D2872, AS 2341.10 and Dickinson, 1984). 35 g of bitumen is placed in a bottle that is then rotated in a convection oven. The sample viscosity is then measured. The RFTO test has been modified to also represent long-term hardening of asphalt in the field, as in the SHRP-PAV test.

To reproduce the demanding spray and chip seal conditions, the RFTO was modified to further test thin films of bitumen in order to assess their relative hardening rates under long-term exposure at the pavement surface and hence permit the selection of bitumens least likely to lose durability in service. This test is called the *ARRB Durability test* (Witt 1976, SAA 1980). Its object is to find how long a bitumen must be exposed to a temperature of 100 C in order to reach an apparent viscosity of 5.67 log Pa.s which is the viscosity that had been found to be associated with the onset of cracking in spray and chip seals in the winter in the southern portion of Australia. The data suggest that the hardening viscosity for distress is lower in cold conditions whereas the rate of hardening is higher in hot conditions. To find this exposure time, the test is done at a range of times and interpolation used to give the appropriate one. A test result of at least ten days is considered desirable (Dickinson, 1984). For service conditions, the following empirical relationship has been established:

$$\log(\text{viscosity, Pa.s}) = 3.59 + 0.0476Y\sqrt{t} - 0.0227D\sqrt{t}$$

where Y is the yearly mean of the maximum daily air temperature (C) at the site, t is the seal service life in years, and D is the Durability test result in days. The equation indicates the improvements likely from selecting high-durability bitumens (Oliver, 1984). The use of the 100 C test temperature can be questioned as it represents only one set of reaction conditions and processes, and as it is well above most service temperatures.

There is disputed evidence that the addition of some hydrocarbons can reverse the oxidation of bitumen. Further, the addition of oils to oxidised bitumen may lower its viscosity and hence extend the life of a surfacing. However, rapid penetration of oil into non-porous surfacings is not easily achieved. Thus diffusion times are very long and the process may only be practical for porous surfacings.

8.7.8 Rubber and other additives

As seen in the preceding sections, bitumen has a number of mechanical disadvantages – such as poor creep resistance as temperatures increase and brittleness as temperatures decrease. A number of additives have been introduced to minimise these deficiencies. There are five key requirements for any bitumen additive. It must:

(a) significantly improve important properties,
(b) maintain the properties in service,
(c) be able to be processed by the range of manufacturing and construction equipment,
(d) not detrimentally affect basic bitumen properties, and
(e) be cost-effective. Additives are typically about ten times the cost of bitumen and so must lead to significant improvements.

Additives which have been dispersed in minor (e.g. 3 percent) concentrations in bitumen have predominately been polymers (a chain of individual monomers or molecules). The product is called *polymer-modified bitumen (PMB)*. Its behaviour depends on both the physical system of the polymer and the chemical compatibility of the polymer with bitumen. Polymers used include:

1 Elastomeric polymers.

Elastomers are defined in Section 33.3.2. These bitumen enhancers are typically rubber. Natural rubber is a polymer class distinguished by the fact that it possesses molecular chains that are very long, flexible, without weak links, and cross-linked in the common vulcanised form. Its chemical formula is $(C_5H_8)_n$ where n is about 10 000. C_5H_8 is the monomer isoprene:

```
link ← CH2                 CH2 → link
          \               /
           C=============C
          /               \
       CH3                 H
```

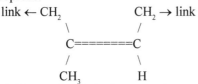

The additive works via its long, tangled, cross-linked, and flexible molecular chains conferring rubber-like properties on the bitumen. This enhances elasticity, crack-resistance (and thus improves an asphalt's resistance to reflection cracking, Section 14.1B), fatigue strength, ductility at high strain rates, viscosity at low strain rates and service temperatures, and adhesion. Rutting resistance is limited by the fact that these elastic materials require large strains to develop significant resistance. One subsidiary benefit is that higher bitumen application rates are possible in spray processes (Section 12.1.5).

Rubber (usually scrap rubber broken down — comminuted — to small particles) can be added to the bitumen at the refinery, prior to mixing with aggregate (the wet process), or as a solid during the final mix (the dry process). Careful testing should be done before selecting a process as heat and oxidation can adversely affect polymers. Properties are better controlled in the wet process.

In the wet process, the rubber is digested in bitumen by heating the stirred mixture at 200 C for about 45 minutes; kerosene is often added to aid the process. The rubber remains in the mixture as a series of discrete porridge-like particles. Rubber in the dry process usually performs more like a piece of aggregate than a binder.

For concentrations of 5 percent or less, the properties of the mix of bitumen and rubber differ little from those of bitumen alone. Concentrations of up to 25 percent of scrap rubber or up to 15 percent of polymer have been used in sprayed applications.

Many pavement crack fillers and sealants are mixtures of bitumen and about 20 percent rubber.

One disadvantage of polymer additives is that the resulting polymer-modified bitumens cool at a faster rate than unmodified bitumens and therefore have a faster increase in viscosity. This makes their field application more difficult (Section 12.4.2) and they should not be used if the air temperature is above 20 C or if there is a chance of wind chill. Another disadvantage is that they have poorer adhesion characteristics (Section 12.3.2) and so they should not be used if moisture is present.

The elastomeric polymers come in three main forms:

1.1 Natural rubber and scrap rubber.

In terms of Figure 8.15 and compared with bitumen, rubber's modulus master-curve is more of a logistic curve (Section 33.2) and its phase angle is low except for a peak over one frequency range. Most of the rubber particles used are produced as buffings (or crumbs) during the preparation of tyres for retreading. Some particles are also made by the cryogenic shattering of tyres. Particles of natural rubber come from truck tyres, whereas car tyres produce particles of synthetic rubber. Natural rubbers are usually superior to synthetic ones. For spray and chip seals, the morphology of the rubber particles is the most important factor affecting the elastic properties of the resulting digestion. This morphology is determined by the manner in which the particles are manufactured. Fluffy particles are desirable. The morphology can be measured by a bulk density test (Section 8.1).

The useful elastic-type properties of the product are usually measured by recording the elastic strain recovery of a specimen loaded at the intended service temperature to well past the elastic limit. The recovery is linearly related to its rubber concentration.

The process does not work optimally if there is cutter (Section 8.7.6) in the bitumen. In addition, the cross-linking makes the mixture more difficult than bitumen to handle or spray during construction.

1.2 Thermoplastic synthetic rubbers such as:

1.2.1 Polybutadiene (PBD),

1.2.2 Styrene-butadiene rubber (SBR) which exists as random-copolymer (a polymer containing two or more kinds of polymer molecule) lattices and is usually added to bitumen as a water-based emulsion of polymer droplets (or latex), and

1.2.3 Styrene-butadiene-styrene (SBS). This contains about 30 percent hard polystyrene and 70 percent rubbery polybutadiene. Styrene and butadiene are both monomers which are compatible with the components of bitumen and this produces strong interactions between SBS and bitumen. SBS is an unusual block copolymer in that the two components are themselves incompatible and fight to keep apart, with the polystyrene settling as blocks at the tips of the polybutadiene. This structure aids the formation of cross-links. These links breakdown at high temperatures, thus avoiding the handling problem associated with natural rubber. Indeed, SBS is a rubbery elastomer (Section 33.3.2) at service temperatures, and only becomes a non-rubbery thermoplastic (Section 33.3.3) with bitumen-like properties at construction temperatures. A drawback of the tendency of SBS to breakdown and separate when hot, is that the material can become irretrievably dispersed in the bitumen prior to cooling. A petroleum-based combining agent is commonly used to aid the dispersion process.

SBS *fumes* may also cause eye, ear and respiratory infection and nausea.

SBS reduces permanent deformation in bitumen, and improves its temperature susceptibility (and thus reduces low-temperature cracking in surface seals and high-temperature rutting in asphalt, Chapter 12). SBS can treble the fatigue life of a bitumen. This is because the styrene domains of the polymer take-over the bitumen as the dominant phase in the mixture for breaking behaviour. For such reasons SBS is one of the most common of the bitumen additives.

1.3 Neoprene.

Neoprene was one of the first additives used, with applications in the 1950s aimed at 'rubberising' bitumen. It was then replaced by the use of natural rubber (see 1.1 above).

There are six other classes of bitumen additive:

2 *Fibres* made from polymers such as *polypropylene*. These act as reinforcement (see also Section 8.9.2) to improve the ductility of the material. They are added as either as random fibres or a preformed mesh.

3 *Thermo-setting epoxy resins* (a disadvantage with these is that it is difficult to bond future bitumen layers to an existing layer containing an epoxy additive).

4 *Thermoplastics:*

4.1 Polyvinyl chloride (PVC),

4.2 Polystyrene (PS, a styrene polymer which is stiff but brittle). See 1.2.3 above.

4.3 Plastomeric polymers such as:

4.3.1 Polyethylene (e.g. polyolefins). Polyethylene can be made from recycled plastic.

4.3.2 Ethyl-vinyl acetate (EVA, Elvax) copolymers. EVA is produced by copolymerisation of ethyl and vinyl acetate. The vinyl acetate content varies between zero and over 50 percent and is adjusted to suit the properties of the bitumen. EVA is relatively non-elastic, as the polymer is not cross-linked, working solely through the elasticity of each strand. Thus it lacks the elasticity of rubber and SBS at high service temperatures. It is most effective at increasing workability and raising stiffness and toughness at low temperatures. It is thus useful in a bitumen that is to be used in asphalt (Section 12.3). EVA tends to disperse and can suffer from a temperature-dependent phase change.

4.3.3 Ethylene methacrylate (EMA) copolymers.

5 *Antioxidant and oxidant additives* (direct chemical processes are used to increase bitumen strength and stiffness).

6 *Extenders* such as waste oil, lignin and sulfur, which replace the more expensive bitumen.

7 *Fillers* such as carbon black, fly ash and lime (Section 12.2.2).

8.8 TAR

Tars are produced commercially from the following sources:

(a) carbonisation of coal at temperatures over 600 C,

(b) residue from coke ovens, usually at steelworks,

(c) residue from plants for making gas from coal, using either horizontal or vertical retorts. The latter operate at relatively low temperatures and produce tars with a high phenolic content and which are susceptible to oxidation (McGovern and Alderton, 1972).
(d) cracking petroleum at high temperatures, or
(e) the distillation of wood and the fluxing of associated pitch. These traditional processes are now rarely encountered (Section 3.6.3).

Compared with bitumen (Section 8.7.4), tars are composed of organic materials with a:

* lower molecular weight,
* lower percentage of carbon in paraffins,
* higher percentage of carbon in aromatic ring structures, and
* higher percentage of polynuclear aromatic hydrocarbons with three to seven fused rings.

This last category is important for worker health, as some of its members are *carcinogens*.

As with bitumens, tars are best characterised by their viscosity. Although tar has excellent adhesive properties, it is more susceptible to temperature than is bitumen. It is also less durable because its plasticising oils are more volatile, an effect which is heightened at high temperatures. The durability can be improved by removing low boiling-point compounds from the tar. On the other hand, compared with bitumen, chemical changes due to the effects of air and light are relatively minor. Maximum safe-handling temperatures are relatively low at about 120 C. Tars have a temperature range between brittle and soft behaviour of only 30 C, which is half the value for bitumen. This means that tars must be carefully selected.

Tar is now little used in roadmaking, with Britain one of the last countries to be major users.

8.9 CONCRETE

8.9.1 Material description

Concrete is a composite mixture of Portland cement, water and aggregate. It is a well-known and well-described material used in many fields other than roads. Hence no comprehensive discussion of concrete properties will be given here.

Portland cement is basically a burnt limestone (Section 8.5.4) and clay mixture that is subsequently ground to a fine powder. Gypsum (hydrated calcium sulfate) is added at this final stage to retard the subsequent hydration process. Cement's chief chemical constituents are lime (CaO) and silica (SiO_2), with the former comprising about two-thirds of most cements. Cement is sometimes extended by the addition of cheaper cementitious materials, such as fly ash and granulated blast-furnace slag (Section 8.5.6). Indeed, the lime in the cement encourages pozzolanic action in the fly ash. Slag cements also have low heat of hydration and high chemical resistance.

Cement is mixed with water to form a *mortar*. This sets into a useful product as a result of the hydration reaction between the cement and water which produces calcium silicates and a lesser amount calcium aluminates. The full hydration process occurs over some months and requires a continuing supply of water in order to complete the processes and achieve maximum strength. This is known as *curing*. It particularly affects the 100 mm outer layer of a concrete member, and hence has a major influence on the member's durability. It is important to keep the concrete moist and surface evaporation rates low whilst curing is occurring.

Seventy-five percent of the final rock-like mortar is tri-calcium silicate (3 $CaO.SiO_2$) and dicalcium silicate (2 $CaO.SiO_2$), which are commonly called C_3S and C_2S. They are stable, strong, and resistant to chemical attack. C_3S supplies the early strength, whereas C_2S provides the long term strength. A rapid-hardening cement will therefore contain more C_3S, as well as being more finely ground. The presence of tricalcium aluminate ($3CaO.Al_2O_3$), commonly called C_3A, also contributes to early strength. C_3S, C_2S, and C_3A are all light in colour.

Concrete is formed by mixing the still-fluid mortar and the aggregates together. To take maximum advantage of the mortar, the aggregates should at least match its strength. Normally, no chemical reaction occurs between them. Indeed, in selecting concrete aggregates it is important to choose those which will not react with the alkaline cement. The two main reactions are alkali-silica and alkali-carbonate. Many siliceous aggregates or their secondary minerals (Section 8.5.10) are alkali-reactive (or alkali-silica-reactive (ASR) or alkali-aggregate-reactive) due to the presence of reactive silica, the chief offenders being opal, chalcedony, some phyllites, some volcanic rocks (Section 8.5.1), and dolomitic limestones (prone to alkali-carbonate problems). Small proportions of reactive aggregate are enough to cause serious problems. The problem is accentuated if the aggregate is of fine size.

The expansive by-products of ASR can cause internal tensile cracking, surface pop-outs and surficial alligator cracking (Section 14.1A) a year or more after the concrete has been cast. If water enters these cracks, the effect is often exacerbated by freeze–thaw cycles. When corrosion-inducers enter the cracks, the expansive corrosion of the reinforcing steel will seriously compromise the durability of the concrete.

Quality control (Section 25.7.3) during concrete production and placement is crucial, particularly with respect to the potential for:

(a) aggregates to segregate during placement (Section 8.3.2),

(b) inadequate compaction, as it impacts compressive strength, flexural strength, and fatigue strength. The plate-like structure of pavements makes them very susceptible to poor compaction.

Vibrators are normally used in the placement of concrete, particularly in complex areas around reinforcing and prestressing (Section 15.2.1). However care must be taken as excessive vibration can lead to segregation (see a above). The first 5 s or so of the compaction process allows the mix to consolidate and fill the forms. The next 15 s or so allows any entrapped air (which can initially be as high as 20 percent by volume) to escape.

(c) large air voids to be trapped within the concrete.

(d) low strength, due to insufficient water to service the slow hydration reaction (see above).

The use of concrete in roadmaking is described in Sections 11.1.4 & 11.5.1 and in bridge building in Section 15.2.1.

8.9.2 Physical properties

The physical properties of concrete are dominated by its *unconfined compressive strength* after 28 days, F'_c which is usually discussed in terms of a specified minimum value, F_c. This latter value is a 95th percentile characteristic resistance (Sections 15.3.2 & 33.4.1).

It is normally established from unconfined axial compression tests (Section 33.3.4) on concrete cylinders cast at the same time as the concrete component, although in emergencies it can be evaluated from cores cut from the component itself.

The most important variable controlling concrete properties is the initial *water content*, which affects nearly all the engineering properties of the material. For instance, a reduction of the initial water content increases:

* compressive strength, F'_c
* shear, tensile and bond strengths,
* elastic stiffness,
* impermeability and durability,
* shrinkage and creep resistance (see below) and
* entrained air content.

The disadvantages of a dry concrete are that it is stiff and difficult to place. Super-plasticisers are additives used to help overcome this problem.

An accurate value of the stiffness modulus of concrete is needed in order to calculate deflections and the initial prestressing changes caused by member shortening under prestressing forces. Concrete has a non-linear stress–strain relationship, even at low strains and so two stiffness moduli are used at the stress–strain point in question:

(a) the tangent (or gradient) modulus at the point, and

(b) the secant (or chord) modulus, E_{cT}, which is given by total stress divided by total strain at the point. This is usually preferred as it fits simply into the grosser calculations needed to calculate prestressing losses. It is commonly given as a form of Pauw's formula which is for a stress of $0.40F'_{cT}$:

$$E_{cT} = 6200\sqrt{F'_{cT}}$$

where E_c and F'_c are both in MPa, the c subscript means compression, and the T subscript highlights the fact that both E_c and F'_c vary with time. The appropriate time in calculations of prestressing loss, for instance, is the time when the prestressing load is applied. It has been suggested that the 6200 can be replaced by $k\gamma^{1.5}$ where k depends on the concrete type and varies between 0.040 and 0.060, and γ is the density of the concrete in kg/m^3 (Carse and Behan, 1980).

Whereas the need with properties such as F'_c is to obtain 95 percentile values to ensure little chance of sub-strength material, the need with modulus values used in the prediction of deformation is to obtain average values.

Shrinkage and creep properties comprise a second set of relevant concrete properties. *Shrinkage* is the change (shortening) in length of an unloaded member due to changing internal and external environmental conditions. The amount of shrinkage depends on the:

* cross-sectional shape,
* cement content of the mix,
* amount of reinforcement,
* environmental conditions, and
* time since casting.

Generally, the amount of shrinkage is linearly related to log time.

Creep is deformation that occurs under load over a long period and after the instantaneous elastic strains have occurred. It is composed of recoverable and non-recoverable parts, with respect to the removal of the applied load. Whilst shrinkage eventually effectively stops, creep can be open-ended — although occurring at a diminishing rate. Ultimate values are assumed to occur over 30 years. As discussed in Section 15.2.1, when a prestressed concrete member shortens due to shrinkage and creep, the prestressing force is relaxed and the structural capacity of the member is reduced. The accurate estimation of shrinkage and creep is therefore important. Appropriate formulas are given in design codes, but there is still some doubt as to their validity.

Indiscriminate cracking and surface abrasion in concrete can be reduced by the use of fibre reinforcement, usually involving the use of many small (20 x 0.55 mm) pieces of randomly distributed steel (or plastic) wire, at about 75 kg of wire per cubic metre of concrete (see also Section 8.7.8).

For conventional concretes, *durability* problems are usually due to either:

* porous concrete, possibly due to
  # a harshly graded mix, or
  # inadequate placement and compaction of the concrete,
* cracked concrete, possibly due to poor curing (Section 8.9.1),
* insufficient cover for the reinforcement, or
* the use of the wrong concrete — for example, too little cement (under 250 kg/m^3) or too much water (water/cement ratio > 0.60) in the mix. Density or unconfined compressive strength can be guides here as durability increases as each of these increase.

CHAPTER NINE

Water in Pavements

9.1 INTRODUCTION

The deleterious effects that water can have on pavement performance are explored in Chapters 8, 11 and 14. Water movement within a pavement can be considered as occurring in three phases:

(a) an entry phase (Figure 9.1), which occurs quite rapidly;

(b) a redistribution phase (Figure 9.1) when water moves within the pavement in response to soil suction (Section 9.2.4) and gravity; and

(c) an evaporative phase when water, as vapour, leaves a material or moves to other layers. Water vapour movements can occur under temperature gradients, with the vapour travelling from a warm to a cool area where it then condenses.

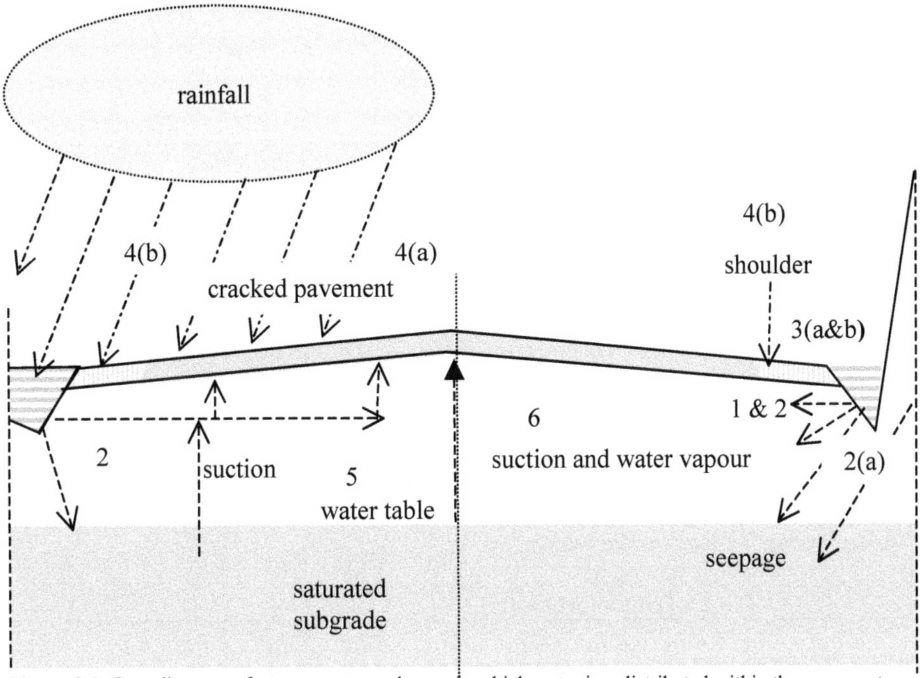

Figure 9.1 Overall sources for water entry and ways in which water is redistributed within the pavement. See Table 9.1 for the key to the numbers. The water table is defined in Section 9.2.3.

The means by which water can enter a pavement are shown in Figure 9.1. The indicated sources are listed in Table 9.1 and are fairly self-evident. However, the rate of water entry from the sources in Figure 9.1 may vary by several orders of magnitude as a result of barely perceptible changes. The 'associations' given in the table permit the design engineer to take the necessary preventative action in each particular situation.

Table 9.1 Causes of water-related effects and their associations. The table also provides the key to the numbers in Figure 9.1.

Causes	Associations
1. *Aquifers* carrying water from higher ground or saturated zones; transversely and longitudinally:	
	1.1 cuttings in undulating or gently sloping terrain
	1.2 permeable (e.g. sand or loam) subgrades
	1.3 shallow side-drains
	1.4 surface formations in duplex soil profiles (Section 8.4.1) with perched water-tables (Section 9.2.3)
	1.5 undrained cross-sections with shoulders of relatively low permeability (Section 13.4)
2. *Ponding* at the edge of the road at (a) cuttings and (b) embankments;	
	2.1 fissured or sandy subgrades
	2.2 inadequate or poorly maintained table-drains (Section 7.4.3)
	2.3 low embankments on flood plains
	2.4 undrained excavations close to embankments
3. *Flooding* of the road surface at (a) pavements and (b) shoulders:	
	3.1 floodways
	3.2 poorly maintained drains and culverts
	3.3 full-width permeable sub-bases
	3.4 cracked pavements
	3.5 permeable shoulder material
4. *Rainwater* on the road surface at (a) pavements and (b) shoulders:	
	4.1 full-width permeable sub-bases (Section 13.4)
	4.2 cracked and/or pot-holed seals and pavement courses
	4.3 rutted pavements (Section 14.1)
	4.4 shoulder material deficient in fine particles or poorly compacted or segregated
	4.5 undrained cross-sections with shoulders of relatively low permeability (Section 13.4)
	4.6 poorly maintained shoulders
5. Rise in *water-table*:	
	5.1 subgrades composed of fine sand
	5.2 flat littoral terrain
6. Water moving from *water-table*:	
	6.1 suction
	6.2 vapour movement

Apart from these local causes, the climate and surrounding topography must be considered in estimating water entry.

Water entry from the surrounding formation (Causes 1, 2, 5 & 6 in the table) is discussed in Section 9.2. Water can also enter via the pavement surface (Causes 3 & 4) as these surfacings can be quite permeable (Sections 9.2.2 & 12.5.4). In particular, infiltration through *shoulders* is a common problem (Section 11.7) and will depend on the permeability of the shoulder material, on its surface tightness, and on its surface slope and shape. The last two factors encourage the ponding of water on the shoulder surface. See also Sections 11.7, 13.3, and 14.2.

9.2 FLOW OF WATER IN SOIL

9.2.1 Flow laws

Phase (b) in the list in Section 9.1 concerned the movement of water within the pavement structure. The 'unrestricted' flow of water can be usefully understood by the flow net approach. Flow nets are based on the concept of a streamline, which is the track that a drop of water will take in flowing through the soil. This path is governed by Bernoulli's theorem that states that the total energy will not change from point to point, i.e. the sum of the potential, pressure, and speed energies will be constant. Flow nets for water in open bridge-waterways are discussed in Section 15.4.5.

In soils, flow speeds are usually quite low and flow nets are based on Laplace's equation, combining the incompressibility of water with the equations of motion. Flow nets are also usually coupled with the fact that Darcy's underground flow law governs the laminar flow of water in saturated soils (Section 8.1):

flow velocity = (permeability)(potential-head gradient) (9.1)

or volumetric flow = (permeability)(potential-head gradient)(cross-sectional area)

Darcy's law is the most appropriate model for water in porous materials subject to a relatively low, positive potential-head. Non-laminar (i.e. turbulent) flows result in lower flows than predicted by Darcy's Law. The law must also be used with caution for coarse-grained materials where the permeability depends on the potential-head gradient. The potential-head gradient is also known as the *hydraulic gradient*. The *potential* (or pressure) *head* is the summation of the hydrostatic pressure at the water surface and gravity pressures due to differences in water elevation. Considerable potential heads can arise in confined regions where the water enters at a relatively high elevation.

High water-flows can occur in large pores, cracks, and fissures in a soil. An *aquifer* is a permeable soil layer. However unfissured clay only contains very small, interlinked pores through which water flow can be as low as 10 fm/s, making drainage virtually impossible.

Furthermore, only the water that is not electrically bound to the clay particles (Section 8.4.2) is able to flow. This is called *free water* (or *groundwater* or *phreatic water*) whereas the bound water is called *soil water*. In a fissured clay, nearly all the water flow takes place around the clay lumps, rather than through them.

9.2.2 Permeability

Permeability is a measure of the ease with which water passes through a material and is defined by Equation 9.1. The opposite of permeability is impermeability. Permeability is usually highly anisotropic and difficult to measure. Vertical pipes called *open standpipe piezometers* can be used to estimate in situ permeability. The water head in the standpipe can also be used to estimate free water levels and pore pressure (Section 9.2.5) and to provide free water samples for chemical analysis. Water enters the piezometer via a porous tip located at the level under study. It is then bailed or pumped from the piezometer and the rate at which the water level then recovers its original position is measured. In the pumping version of the test, the equilibrium level of the achieved water-table drawdown is also measured (Section 9.2.3). The pumping test usually uses one bore for pumping and an adjacent vertical bore as the piezometer tube. Its data are usually applicable over a larger area than those of the bailing test (Evans and Haustorfer, 1982).

Permeability is most influenced by the size, shape, and connectivity of the water passages that are themselves largely predicted by the soil's:

* *particle grading.* As a first approximation, permeability increases as the square of the d_{10} particle size (Section 8.3.2) and as the sixth power of porosity (Equation 8.1j). For materials with a grading of $n < 0.5$ (Equation 8.2), permeability is particularly sensitive to the fine particles present, decreasing as plasticity increases (Section 8.3.2). Specifically, a well-graded material with less than 5 percent of fine particles (i.e. passing a 75 μm sieve) will be relatively permeable. As the percentage of fine particles increases, the permeability decreases until about 20 percent, after which no further effect occurs. However, use of more than 10 percent of fine particles creates problems due to plasticity and positive pore pressure (Section 9.2.5). The coefficient of uniformity (Section 8.3.2) is also a useful measure of permeability, permeability decreasing as the coefficient increases. The d_{30} fraction also has a major influence on permeability (Section 9.2.2).
* *size segregation.* Segregation of grading sizes increases permeability.
* *degree of compaction* (or *density*). Permeability is usually at a minimum for material compacted at OMC — because of the associated optimal packing configuration — and highest for material with the open structure typical of compaction dry of OMC (Section 8.2.4), due mainly to the effects of size segregation. Compaction of clays wet of OMC can increase permeability, due to clay dispersion leading to an open structure (Section 8.4.2). Despite this caution, compaction is often the best means of reducing permeability.
* *crack propensity.* Section 9.2.1 showed that fissures in dry clay greatly increase permeability. Likewise, cement stabilisation (Section 10.2) will not reduce permeability if cracking occurs and permits easy water passage.
* *degree of saturation.* Permeability is usually taken as applying to a saturated soil and any trapped air in an unsaturated soil may significantly lower permeability, as discussed in Section 9.2.4.

The difference of over a million-fold between the upper and lower permeability values in the list in Table 9.2 should be noted. To some extent this reflects differences in material structure — e.g. between sand and clay — as well as in grading.

Table 9.2 Typical permeability values for saturated soils.

Course	Permeability	Description
gap-graded crushed rock (Section 8.3.2)	> 30 mm/s	
gravel	> 10 mm/s	free draining
coarse sand	> 1 mm/s	
medium sand	1 mm/s	permeable
fine sand	10 μm/s	
sandy loam	1 μm/s	practically
silt	100 nm/s	impermeable
clay (homogeneous)	< 10 nm/s	impermeable
bituminous surfacing (homogeneous)	1 nm/s	Section 12.5.4

Permeability can also be readily altered by adding cement slurries, sealants, or *deflocculants* (e.g. polyphosphates; see Section 8.4.2) to the soil, or by the use of moisture barriers. Slurries and sealants cannot fill all the fine pores, so they cannot totally prevent water flow whereas deflocculants can seal a soil completely. On the other hand, *flocculants* (e.g. lime and gypsum) and *wetting agents* can be used to increase permeability. Care must be taken as some agents that work by producing a hydrophobic reaction in a soil can also lower its cohesion (Ingles and Metcalf, 1972). Moisture barriers are discussed in Section 9.4.2.

Section 12.5.4 discusses the factors that cause a pavement surface to be permeable and the implications of the subsequent water infiltration into the pavement structure.

9.2.3 Water-tables

Water will percolate downwards under gravity. The water-table is the highest level at which free water can be found in the soil (Figure 9.1) and can be detected from water levels in wells or piezometers (Section 9.2.2). Saturated conditions (Section 8.1) will obviously exist below the water-table. A perched water-table is an unexpectedly high water-table, due to an intermediate impermeable layer preventing water from draining further downwards.

However, even in permeable soils, not all water will be free to drain downwards. Some will rise up from the water-table due to the attractive forces between each water molecule and:

(a) other water molecules (*surface tension*), and

(b) most internal surfaces within the soil (*wetting*).

The amount by which water can rise by this *capillary action*, and thus saturate soils above the water-table, can range from 20 mm in coarse sands to over 4 m in fine clays, although the effect in clays is countered by their relative impermeability. Silts often exhibit extreme effects. If there is a water-table within 6 m of the surface in clay soils, or 3 m in sandy clays or silts, or 1 m in sands, it will generally exert a controlling influence on subgrade moisture contents. In such cases, water movements due to capillary action can be reduced by the introduction of a very permeable layer — the net effect can actually reduce total permeability.

Water also tends to move from warm to cool zones, and so is drawn upward in cold weather and downward in hot weather. The upward movement can be halted by a descending frost line (Section 9.5). Water also tends to condense out of humid air onto the surface of granular material. However, a relatively stable temperature regime exists in most subgrades and so movements of water vapour are not usually significant.

9.2.4 Soil suction

The water pressure at the surface of a water-table is atmospheric. As water rises above the surface by capillary action, the pressure in the capillaries drops away linearly with the height above the water-table. This creates a balancing tension, called *matrix suction*, in the water. This tension is resisted by the molecular attractions and a counter-balancing compression in the soil. In more general terms, matrix suction is defined as the suction relative to atmospheric pressure to which the soil water would need to be subjected in order to keep it in equilibrium with an identical solution on the other side of a permeable membrane.

The presence of dissolved salts will cause additional *solute tensions* due to *osmosis*. [Osmotic suction is that which would need to be applied to pure water to keep it in equilibrium with the soil water, with the two fluids separated by a membrane permeable to water molecules only.] In clays, the presence of highly-charged surfaces (Section 8.4.2) means that there are also internal osmotic and *surface-adsorption* tensions at work. The combined effect of all these tensions on the water is to produce a sub-atmospheric pressure called *soil suction*, or *negative pore pressure*. Soil suction is therefore the sum of matrix suction, solute tension and surface-adsorption tension. Soil suction can also be defined directly from the partial pressure of the water vapour in equilibrium with the soil water.

It follows from the above that soil suction is zero at the surface of a water-table; it increases above it such that a soil which is partially saturated is in a suction condition, and decreases below it as conventional potential water heads take over. The flow of water through an unsaturated soil thus depends, via a form of Darcy's underground flow law (Equation 9.1), on a permeability and a suction gradient, which both themselves depend on moisture content. Consequently, permeabilities increase from near-zero values at low moisture contents and high suctions, to the conventional saturated values as the moisture content increases. Furthermore, Section 9.2.5 confirms the intuitive view that a dry soil with soil suction will experience a shear strength increase.

A counteracting effect is that the tensions due to soil suction will create shrinkage cracks in many soils. These cracks will increase the overall permeability of the soil as they can be equivalent to a strain of 1000 $\mu\varepsilon$.

The height that water will rise due to suction is inversely proportional to the capilliary radius, i.e. the smaller the capilliary, the higher the rise. A low air void ratio, 'a' (Section 8.1.1), indicates a small radius. Typical pore radii vary from 1 μm in clay to 100 μm in sand. Approximately, the height rise in mm can be estimated as 100/(pore radius in mm). The capilliary rise will also be aided by the ability of water to wet the surface of the voids (this hydrophilic property is discussed in Section 12.3.2).

Suction is commonly measured in pF, which is the logarithm of the height, in centimetres, that water would have to rise in a capillary tube to attain an equivalent internal pressure at the top of the column of water. pF = 2 would thus correspond to capillary water which had risen 1 m above the water-table. A suction with pF > 2 will have removed the bulk of the water from all but the finest-grained soils (clays). With pF > 4, only molecular layers of surface-absorbed water (Section 8.4.2) will remain. Suction values for a fine-grained grey brown clay (33 percent below 2 μm) are shown in Figure 9.2.

Methods for determining soil suction essentially employ specialist psychrometers or less accurate filter paper methods. In the latter method, filter paper is left in contact with the soil sample until it reaches suction equilibrium. Its moisture content is then measured and related to a previously-established correlation between the moisture content of the filter paper and its suction.

Soil suction levels can be predicted using average annual rainfall values or the *Thornthwaite Moisture Index*. This index measures the overall availability of water during a year and uses three basic climatic categories:

(a) Wet (approximate pF, 2 to 3): water drains from the soil almost continually and little change of moisture content or strength occurs with season. Moisture contents under the pavement are similar to those in exposed soil.

(b) Intermediate (approximate pF, 3 to 4): either a shallow water-table or shoulder and edge conditions dominate.

(c) Dry (approximate pF > 4): desert to semi-desert with small seasonal changes of moisture content. Moisture content is governed by atmospheric humidity.

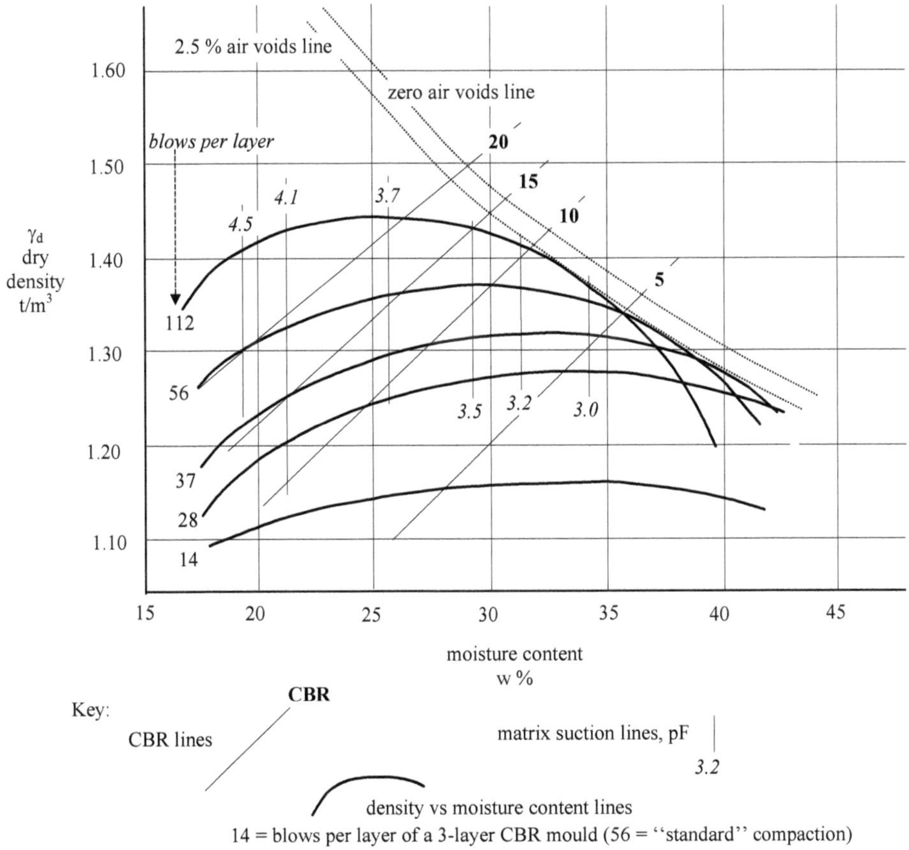

Figure 9.2 Dry density, moisture content, CBR and suction relationships for a grey-brown soil as its moisture content is increased from 15 to 35 percent. The moisture content was determined after CBR testing and taking of samples (from Morris *et al.*, 1968). The dry density was $\gamma_s = 2.67$ t/m^3.

Although soil suction is a more meaningful measure than moisture content in many arid circumstances, it is less commonly used in road practice. This is mainly a consequence of the greater experimental and conceptual difficulties associated with the measure.

9.2.5 Positive pore pressure

The previous section showed that soil suction can be considered to be a negative pressure increment within the soil pores. Positive pore pressure occurs due to *applied stresses* caused by external loading due to traffic or compaction. In a saturated soil, such applied (or *total*) stresses are transmitted by being shared between pressure stresses in any entrapped water and interparticle stresses in the network of soil particles (the mineral skeleton). These interparticle soil contact forces create an *intergranular* (or *effective*) *stress* in the soil particle structure. That is:

applied stress = (intergranular stress) + (pore pressure) (9.2)

For dry soils, the previous section showed that:

pore pressure = – (soil suction)

hence for dry soils:

intergranular stress = (applied stress) + (soil suction)

Intergranular stress is the normal stress in Figure 8.9. It is therefore a major determinant of a soil's response to the load. Equation 9.2 shows that the increase in pore pressure as moisture content increases, reduces intergranular stress and thus interparticle normal stress. Figure 8.9 then shows that this reduces the shear strength due to interlock as the presence of pore pressure simply moves a material to the left on the graph, indicating a lower intergranular shear strength, i.e.:

(shear strength) = (cohesion) + [(normal stress) – (pore pressure)].tan(internal friction)

The three related factors that:

* shear strength is reduced by positive pore pressure,
* pressured water alone has no shear strength, and
* cohesion is reduced by rises in moisture content (Section 8.3.2)

explain many of the engineering problems encountered in moist soils, some of which will be described below. Indeed, applied loads vary little from road to road and frequently the main variable is the positive pore pressure, with its debilitating effect on shear strength. Note also how, conversely, soil suction increases the shear strength of dry soils.

The pattern of pore pressure effects is illustrated in Figure 9.3. It is the resulting pore pressure level that is critical, rather than the actual volume of water present. As will be seen later in Figure 9.4 quite small quantities of water can create high positive pore pressures. Stiffness is similarly affected, being approximately proportional to the square root of the intergranular stress.

The range of applicability of Equation 9.2 depends on the sharing of the load between the liquid and solid phases. For saturated cohesive soils, Equation 9.2 applies fairly totally. However, granular non-saturated soils carry their load largely by interparticle contact. The spectrum of effects between the two is handled by a *compressibility factor* which ranges from 0 for non-load-sharing granular soils to 1 for full load-sharing cohesive soils.

A build-up of positive pore pressure during compaction can arise in impermeable materials, such as clays and granular materials with a high content of fine particles (Section 9.2.2), where there is no means by which the pressure in the water can be quickly dissipated. Of course, this build-up deleteriously resists any densification (Section 8.2.1), as there is no stress available to 'work' the soil.

9.3 EFFECTS OF WATER

9.3.1 General

As seen in Sections 8.2.4 and 8.4.2, water entry will reduce the strength and stiffness of subgrades and pavement material. It can also lead to frost damage (Section 9.5), slope instability in cuttings and embankments due to loss of shear strength (Section 9.2.4), volume changes in expansive clays (Sections 8.4.2 & 9.4.2), bitumen stripping (Section 12.3.2), and potholing (Section 9.3.2). The effects are particularly marked for moisture contents above the plastic limit (Section 8.3.3) or close to saturation. Thus many pavement failures are the direct result of water entry. For instance, the loss of strength and stiffness can lead to rutting and other forms of pavement deformation or, more commonly,

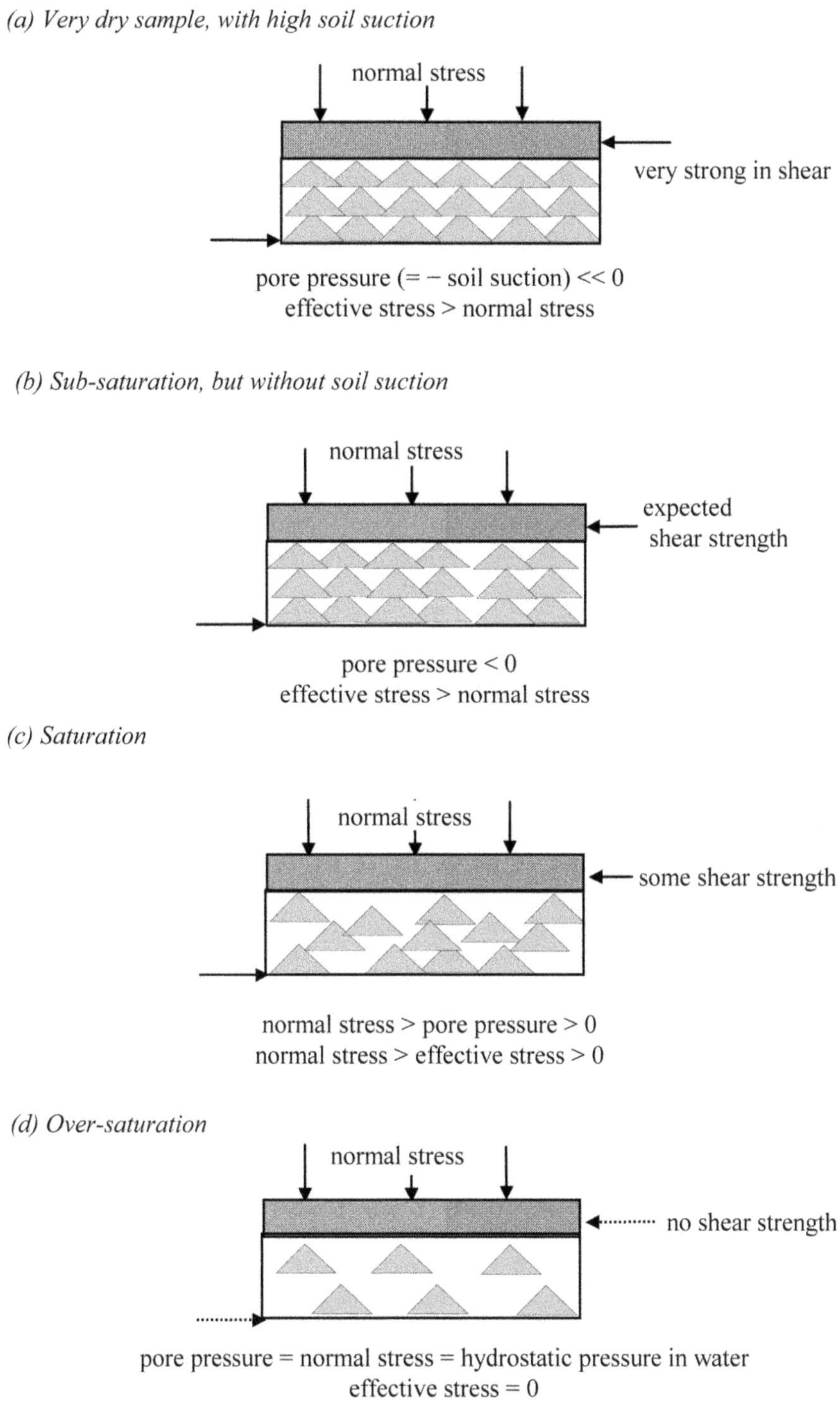

Figure 9.3 Illustration of pore pressure effects as sample moisture content increases from totally dry (a) to totally fluid (d). Note that the shear stress shown is proportional to the intergranular stress.

to pavement edge failures (see Section 14.1A). Certainly, the rate of traffic- induced deterioration of a road will increase noticeably when water is present.

The following general comments can be made with respect to the influence of small changes in moisture content on the following soil types:

(a) *Sandy or granular soils*: little change in volume or strength, some potential for positive pore pressure effects. The worst situations occur with poorly-compacted material where applied wheel loads induce pore pressures that effectively eliminate the shear strength of the material (Section 9.2.5). Excessive deformation then occurs under load.

(b) *Silty soils*: little change in volume but a possibility of large shear strength changes, as the rapid development of pore pressure can lead to very low intergranular stresses.

(c) *Clays*: little change in ultimate strength, but effects on soil electro-chemistry (Section 8.4.2) will cause a drop in effective strength and stiffness and the potential for large volume changes. With cohesive, expansive soils, moisture content changes may cause damaging volume changes and property decrements, particularly in subgrades (Section 8.4.2). This problem is worsened by the fact that the volume changes are rarely uniform.

9.3.2 Potholes and pumping

Water can also directly affect the performance of the surface course by causing:

* the bitumen to strip from the aggregate (Section 12.3.2),
* separation of horizontal pavement layers or courses (Section 11.1.2), and
* pothole formation.

The mechanisms are similar in the last two cases in which the incompressibility of water permits an hydraulic pressure system to operate in any permeable and saturated sub-surface layer. Thus, any applied pressure from tyres is transmitted as pore pressure and will need instant relief. Often the lowest resistance is encountered from the surrounding surface course, which is then 'blown out' and upwards. The upward pressures will be of the same order as the tyre contact-pressure and, without a sub-layer to help resist them, it is not surprising that blow-out occurs. The mechanism is illustrated in Figure 9.4. A typical case would be a layer of water in the interface between a wearing course and a stiff and impermeable basecourse.

The effect also occurs more generally as the phenomenon known as *pumping*, which involves free water held below the pavement surface, typically in voids between delaminated pavement layers (Section 14.1C.13) or left by curling or warping of concrete slabs (Section 11.5.2). Vertical movements in the pavement structure induced by the traffic create water pressures in this free water. This pumps the water to the pavement surface through cracks or joints. The water flow commonly leaches or erodes pavement fine particles out of the basecourse — the process is called *migration of fines* (Section 14.1C11) — and can thus cause structural deterioration of the pavement and staining of the pavement surface. In addition, the hydraulic pressures generated in a saturated basecourse can result in the subgrade stresses being concentrated far more than is assumed in flexible pavement design (Figure 11.5a), leading to early subgrade failure. Pumping occurs in:

* rigid pavements (Section 11.1.5), where the pumped water and fine particles are initially trapped between pavement and subgrade,

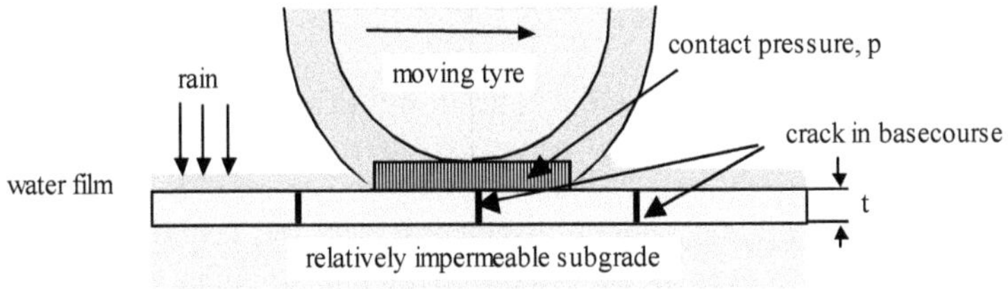

STAGE 1: Tyre contact pressure, p, forces water downwards through the cracked, bound basecourse.

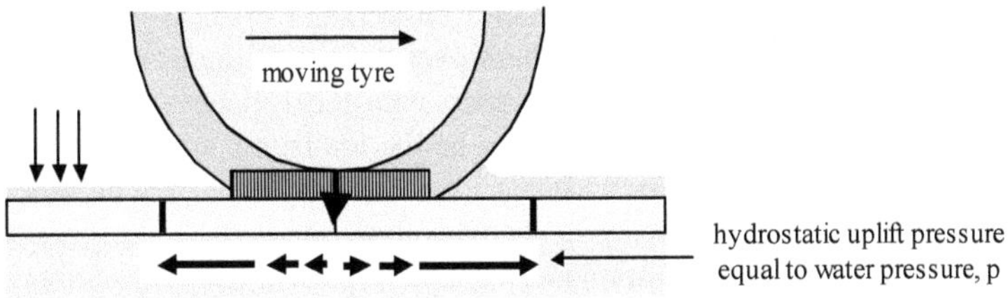

STAGE 2: Water under pressure travels along interface between basecourse and an impermeable subgrade, seeking relief via existing cracks.

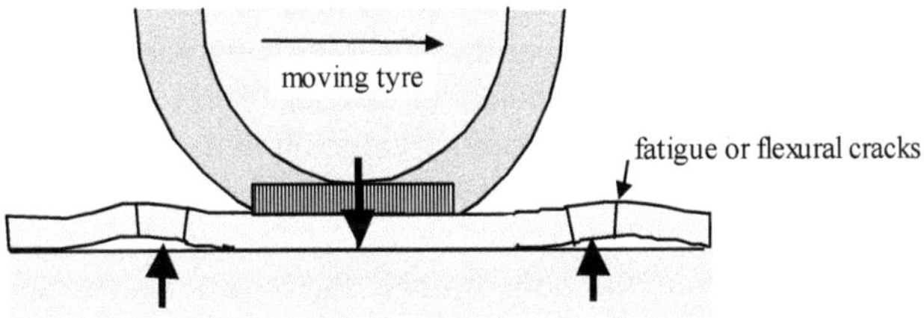

STAGE 3: Bound course lifts to allow water pressure to dissipate, allowing more water to be pumped in.

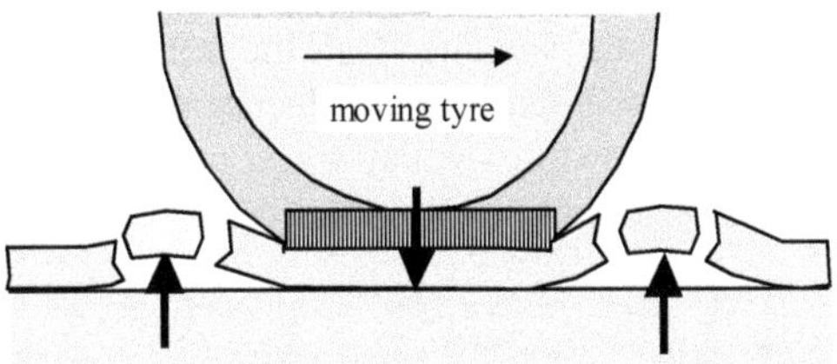

STAGE 4: Pieces of the bound course dislodge due to traffic action and water pressure, forming potholes in the basecourse.

Figure 9.4 Layer separation and pothole formation (after Gerke, 1987).

* asphalt pavements, if trapped water is not drained from the pavement structure (Section 13.4.2), and
* between layers in a stabilised pavement, and
* in gap-graded unbound courses (Section 8.3.2), with slurries of water and fine particles being sent to the surface. In this case the inherent permeability of the gap-graded course is increased by the tendency of the particles to spread out laterally under vertical load.

9.3.3 Variations in moisture content

Significant changes in the moisture content beneath the surface of a pavement will only occur immediately after rainfall, if very permeable layers exist. Otherwise changes in moisture content will be very slow. The initial moisture contents in the pavement and subgrade will tend to persist in most regions, whether wet or dry. The main variations will be seasonal, annual, and over even longer periods. Seasonal variations are common in the upper 1 to 2 m.

With low-volume rural roads, where the use of imported or treated material cannot be justified, it is particularly important to establish low, near-OMC, moisture contents at construction, and then to maintain them over the life of the pavement. Spray and chip seal coats (Section 12.1.4) used in association with local material also place heavy emphasis on the need to prevent any increase in the moisture content of the basecourse and subgrade. The stabilisation process discussed in Chapter 10 can be used as a low-cost way of reducing the moisture sensitivity of many materials.

Most of the change in moisture content occurs in the metre or so of pavement width at the edge of the pavement surface, as illustrated theoretically in Figure 9.5 and experimentally in Figure 9.6. In this region, moisture contents tend to increase in the long term (Beavis, 1984). The change is usually due to gravity rather than to soil suction, particularly with an impermeable subgrade. The consequence of the moisture content increase is that the pavement edges and their associated outer wheel-paths are the most failure-prone regions of a pavement, particularly if the subgrade is moisture-sensitive.

The paving of shoulders (as noted in Section 11.7) will therefore have major performance benefits in moving the region of variation in moisture content away from the outer wheel-path. There is relatively little change in moisture content in the rest of the pavement structure, even in the long term.

Trees draw water from the soil through their root system and can remove all available water from the soil to depths of 4 m or more. This can cause major drying of the soil during a hot, dry period and lead to above-average shrinkage and cracking of the root-affected area. The lateral spread of the roots of a tree will be about equal to: (a) its height, or (b) its above-ground radius, i.e. to the *drip* line. Thus the influence of trees on subgrade moisture contents can be considerable. Factors that influence such pavement damage due to trees include tree species, rapid growth of young trees, and the removal of old trees. Clays are associated with most cases of pavements suffering from tree damage.

9.4 DESIGN FOR WATER

9.4.1 Equilibrium moisture contents

From the discussion in Section 9.3.3, it can be seen that pavement designers — if they prevent subsequent water entry — can rely on most of their pavement being at an unchanged moisture content over its entire design life. Two consequences therefore arise:

(a) most material placed at or near OMC can be expected to maintain that desirable state; and

(b) if a material is placed in an excessively wet condition, and the pavement is then surfaced, the wet area probably will not dry out. Instead, it may remain a problem over the life of the pavement.

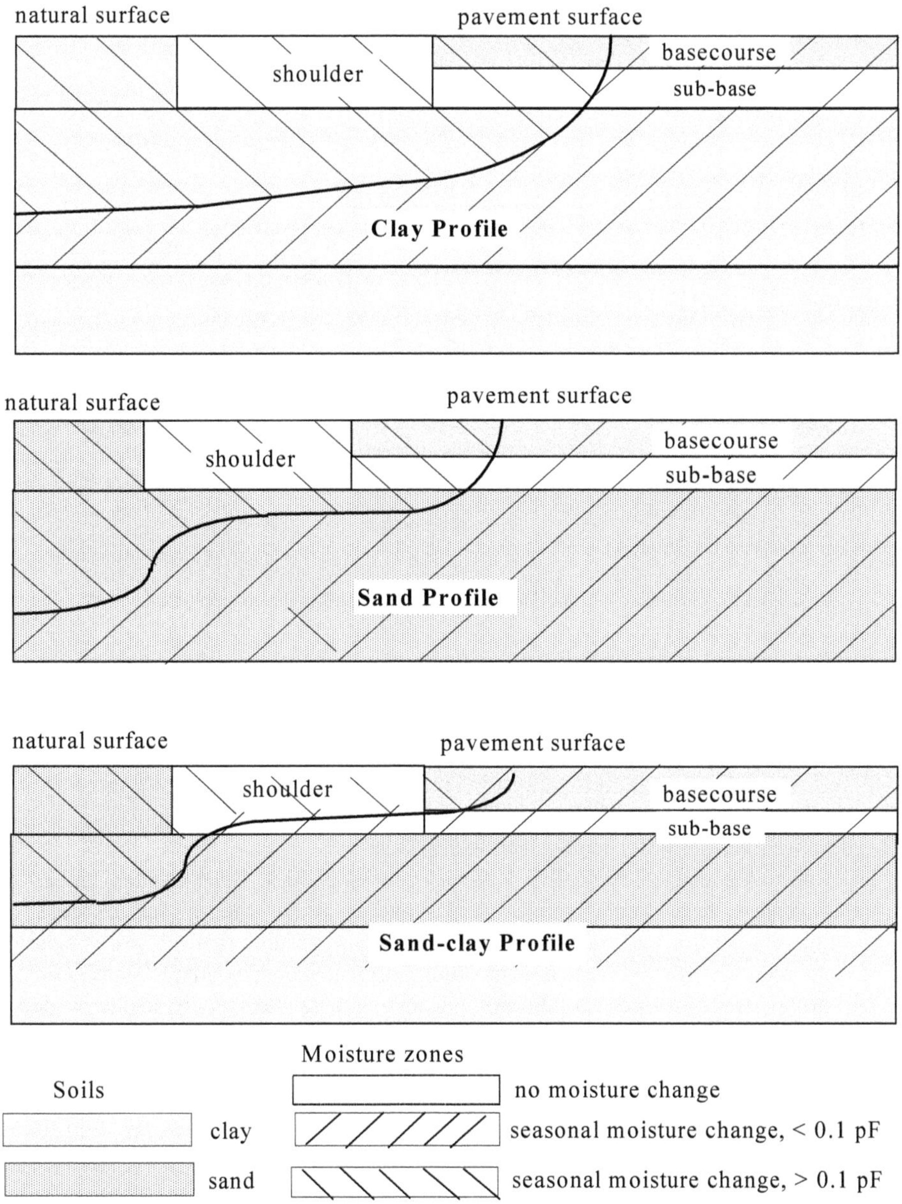

Figure 9.5 Zones of predicted variation for three soil profiles in a semi-arid area (from Richards and Chan, 1971).

An extreme moisture content is the *saturation* value that occurs when all voids are full of water. It would be the conservative value to use in design and it is so used in many jurisdictions. However, the estimation of a moisture content for design purposes is often more realistically based on the *equilibrium moisture content*. This is the stable (i.e. not subject to seasonal change) moisture content found under the pavement surface but away from the pavement edge some time (usually two or more years) after construction. It occurs when the moisture content of the pavement structure reaches an *equilibrium* with

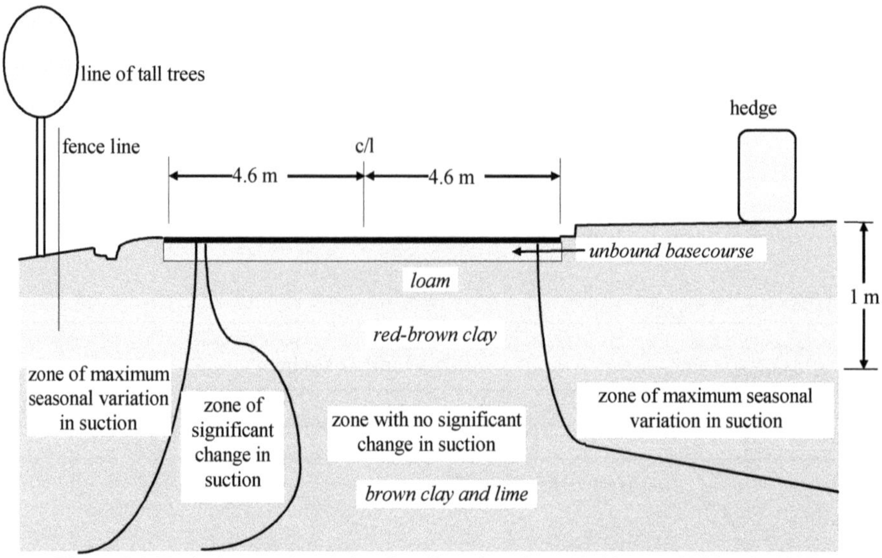

Figure 9.6 Typical seasonal variations in soil suction (from Aitchison and Richards, 1965).

the moisture content of the underlying subgrade, via the processes of soil suction (Section 9.2.4). As a matter of physical course, suctions tend towards values in equilibrium with those of the surrounding soil zones. The unique relation between suction and moisture content then leads to an equilibrium moisture content (Morris *et al.*, 1968). In most cases it will be well below the saturation moisture content and closer to OMC (Figure 9.2).

The three broad categories to consider in determining equilibrium moisture contents depend on whether the water-table (Section 9.2.3) is:

(a) close to the surface: typically, within 4 m for fine clays, 1 m for sands, and 20 mm for coarse aggregate (Section 9.2.3), and/or where the rainfall exceeds 600 mm/year. In this category it will be necessary to assume that the soil is saturated and that water will exert a controlling influence on pavement behaviour. This applies whether such conditions are either persistent or intermittent due to flooding, watering, or irrigation of adjacent land. In such cases the design approach in Section 11.4.6 would be based on the soaked CBR value (Section 8.4.4).

(b) deeper than in category (a), and the rainfall exceeds 250 mm/year. Equilibrium moisture contents will exist outside the seasonal fluctuation zone discussed in Section 9.2.3.

(c) deeper than in category (a), and the rainfall is less than 250 mm/year. Equilibrium moisture contents can be assumed throughout the subgrade and in the basecourse below the pavement surface.

The equilibrium moisture content will depend on soil type and soil-water composition. For a given situation, it will then depend on climate and the depth of the water-table depth. The size of the equilibrium moisture content can be predicted:

(1) from first principles;

(2) by sampling existing roads and using the 90th percentile value of moisture contents measured in the field;

(3) from climatic data, e.g. using the Thornthwaite Moisture Index (Section 9.2.4);

(4) from empirical, equilibrium moisture content versus OMC, plasticity index, or liquid limit ratios; or

(5) by such approximations as the use of 90 percent of the OMC where the rainfall is under 500 mm/year (Metcalf, 1978).

For design purposes, the equilibrium moisture content used must be:

* a characteristic value, to account for normal variability (Section 34.4.2), and
* based on the most adverse equilibrium moisture conditions that could occur under the intact pavement surface.

The equilibrium moisture content is thus an in-service moisture content and is the appropriate value to use in the design of much of the roadway. In addition, volume changes during the life of the pavement (Section 8.4.2) are minimised if the material is compacted at the equilibrium moisture content. One qualification is the observation in Section 9.3.3 that usually a 1 m outer strip of pavement will be affected by seasonal changes in moisture content. Thus the outer wheel-path is usually the critical design region. This is often best handled by ensuring that this outer metre will not be subjected to as high a traffic loading as the rest of the pavement (e.g. it is a shoulder).

If the equilibrium moisture content is less than OMC, the design CBR value may be obtained from classification tests. Otherwise, a CBR test should be done on the subgrade at its equilibrium moisture content. Variations of CBR with moisture content are illustrated in Figure 9.2.

Using the equilibrium moisture content wherever possible will usually produce a less expensive design than one based on assuming that the materials used in basecourse and subgrade are saturated.

9.4.2 Water management

The following five steps for water control will minimise moisture contents and hence lead to minimum cost designs (Wallace and Leonardi, 1978b).

(a) *Determine* whether the terrain (Section 6.1.3) is of a type with a high probability of water entering the basecourse or subgrade, in the manner of Cause 1 in Figure 9.1. Preventing this water flow from reaching the pavement is a key to the design of efficient low-cost roads (Chapter 13).

(b) *Design* the pavement cross-section to minimise the probability of water entry and penetration, following the guidelines set out in Section 9.1. There are a number of processes:

(b1) A design recommendation introduced in Section 9.3.3, and which is pursued further in Section 11.7, is to use wide, impermeable surfaces and *shoulders* to prevent water entry and place the moisture-affected edges outside the outer wheel-path (Metcalf, 1978). In residential streets, infiltration through shrinkage cracks in the soil at the rear of kerbs can be minimised by extending the basecourse beyond the kerb line.

(b2) A *capping* course (Section 11.1.2) may be placed immediately above any water-sensitive courses to prevent the vertical entry of water into those courses. To be effectively impermeable, a capping course should:

* be compacted to maximum density at OMC (to minimise voids, Section 8.2.1),
* have a permeability of under 10 nm/s (Section 9.2.2) although under 5 nm/s is often required as a conservative measure,

* a thickness of over 150 mm, and
* not be water-sensitive.

The capping layer should extend laterally to the edge of the formation.

Lime stabilisation (Section 10.3) can often be used to produce a suitable capping layer.

(b3) Another effective technique is to use vertical *cut-off trenches* and underground drains along the edge of the pavement structure (Figure 13.1a).

When the pavement contains expansive clays (Section 8.4.2), this may need to be supplemented by installing *vertical moisture barriers* up to 3 m deep along the pavement edge. The depth will usually exceed 2 m and must exceed the crack depth seasonally encountered in the soil. Whilst many early barriers were only a metre deep, experience has shown that water will sometimes find its way around such shallow barriers. The barrier may be:

* an open-graded drainage-course (Figure 13.1a & Section 11.1.2),
* cement-stabilised (Chapter 10), although these barriers tend to crack with time, or
* an impermeable geomembrane in deep, narrow slit trench (Section 10.7).

Care must be taken to prevent penetration by tree roots (Section 9.3.3). An excellent review of vertical barriers is given in Evans *et al.,* (1996).

Of course, the preferable solution is to keep expansive clays out of the pavement structure entirely. As a minimum, such clays should *not* be placed within 400 mm of the surface. For heavily trafficked roads, the restriction could be at least 1000 mm.

(c) *Design the pavement cross-section* to promote internal drainage and reduce water retention, including capillary water. Ideally, the permeability of the pavement layer should increase with depth. This encourages gravity drainage of any water that does enter, whilst preventing capillary movement upwards (Section 9.2.4). If no permeable layers exist within the pavement structure, then the removal of any water that does enter will be very slow, unless subsurface drainage is provided (Section 13.4.2).

Many subgrades are impermeable and a special permeable *drainage course* (as in b above) may be needed to permit drainage and act as a capillary cut-off. Its thickness will need to exceed the soil-suction height (Section 9.2.4). On the other hand, a minimal thickness should be used to limit the drainage course's own water retention and restrict lateral infiltration. The less permeable material immediately below a permeable layer must be either moisture-insensitive (e.g. a capping course, Section 11.1.2) or designed to act in the soaked state. As a part-way measure, the permeable layer can be restricted to the shoulders to at least provide for water egress. This aspect is discussed further in Section 13.4.3.

An impermeable layer is capable of creating a *perched water-table* (Section 9.2.3). It is important to avoid this by ensuring that any exposed edges at the upper surface of the impermeable layer will provide horizontal drainage outlets.

Boxed construction (Section 11.7 & Figure 11.17b) places the pavement in a trench and should be avoided wherever possible. Where it is unavoidable, it is necessary to employ special drainage provisions, or use permeable shoulders, to permit trapped water to flow out of the pavement. Design would then be based on soaked CBR values (Section 8.4.4).

(d) *Construct* the pavement materials and the upper subgrade layers to a state at which their strength and stiffness is least affected by water. The ideal way to achieve this is to use good quality, well-compacted granular material. If the exposed subgrade has recently been moistened by rain, it should be allowed to dry to normal equilibrium conditions (Section 9.4.1), or better, before proceeding with construction.

(e) *Maintain* the pavement surface to prevent ponding and reduce permeability. Items such as surface patching, crack-filling, shoulder grading, drain maintenance, and resurfacing can significantly reduce infiltration (Section 12.5.4).

9.5 FROST DAMAGE

The *dew point* is the temperature at and below which moist air is saturated and water then condenses on available surfaces. When temperatures drop below freezing, the dew point becomes the *frost point*.

As atmospheric temperatures drop to freezing (0 C) and below, there will be a drop in the temperature of the pavement surface and, subsequently, of the sub-surface components of the pavement. Frost first occurs when the surface temperature drops to 0 C and the dew point is above 0 C. Frost and ice can produce tricky driving conditions due to the dramatic lowering of skid resistance levels.

The lowest depth at which freezing occurs during the course of the winter is called the *frost line*. Because soil and snow are effective insulators, the frost line will rarely be more than a metre below the surface. The amount that the frost line does descend will be related to the length of time during which sub-freezing temperatures occur and the extent of those temperatures. This is measured by the *Freezing Index*, which is the (negative) area of the time-temperature graph at a site, from the beginning of freezing to the beginning of thawing.

Section 9.2.4 discussed how soil suction can raise water through a pavement structure and thus saturate some of the pores above the water-table. Near-freezing temperatures also create a suction effect, raising the free water in the soil towards the frost line. As water in very small pores has a lower than normal freezing point, the cold-suction effect also causes this small pore-water to be drawn towards any already-frozen larger pores. The soil below the frost line from which the water has been drawn will now be dry and contain shrinkage cracks, giving it a characteristically rough appearance.

As a consequence of this migration of water to frozen sites, lenses of ice can form and these will subsequently draw even more water towards themselves. Water expands about 9 percent upon freezing and large lenses can cause the soil surface to heave by amounts of up to 300 mm. The process is called *frost heave*. By opening up surface cracks, cold air can then enter the basecourse and cause small subsidiary ice lenses to form, thus creating further expansion. Significant pavement damage in the form of differential frost-heave can occur as a result of these factors.

For dry soils and in less severe climatic conditions, ice lenses will not form and the soil may sink rather than heave as a result of normal thermal contraction.

Pavement thawing can occur either downwards, upwards, or in both directions. At the beginning of the thaw, the pavement surface will usually be clear of snow, whilst the verges will probably have a higher-than-average covering of snow. This will lead to more rapid thawing and settlement under the pavement, particularly with sudden thaws. The melt water will fill shrinkage cracks as well as the original interstices and will have little chance of lateral or vertical drainage. Pore pressures will develop and, following Sections 9.2.4–5, the material will exhibit low strength and stiffness. The effect will be exacerbated by the constraint of the surrounding frozen soil.

Thus, although a frozen pavement will be quite strong, thawing will completely reverse the effect to the extent that traffic at the early stages of thawing will therefore cause a new cycle of major deformation, usually called *spring break-up*. It is therefore necessary to exercise some restraint in permitting traffic to use pavements after the spring thaw has left the upper courses saturated whilst the lower drainage provisions are still frozen and inoperative. The length of the freeze–thaw cycle is therefore of critical importance.

Not surprisingly, frost damage is usually the major cause of pavement damage in freezing areas. Susceptible soils for frost heave are clearly those that permit easy water movement, through permeability or suction. Thus, silts, seamed clays, and poorly-graded mixtures are susceptible to frost heave; whereas well-graded sand and heavy clays are not, unless drainage is poor — e.g. if there is a perched water-table able to feed water to the lenses. Susceptibility tests for frost heave have been developed (Roe and Webster, 1984). Typically, susceptible material to about 70 percent of the depth to the frost line is replaced. Given common limits for the penetration of the frost line, this implies soils within 700 mm of the surface in normal, and 1000 mm in extreme conditions. The thermal conductivity of the pavement and subgrade will also be a relevant property.

In summary, design should concentrate on:

* avoiding frost-susceptible material,
* designing for saturated conditions and/or limiting traffic during spring break-up,
* avoiding the entry of water into the pavement structure, and
* using insulation to manage the frost line.

Anti-icing operations are pro-active treatments conducted at the pavement surface. Commonly, a chemical such as salt brine (liquid NaCl) is applied to lower the freezing point of the surface water. This prevents any snow or frost from bonding to the pavement surface and thus improves driving conditions and simplifies snow removal. *De-icing* is a reactive treatment that breaks the existing bond between the snow and the pavement surface. Salt (NaCl) is commonly used for anti-icing and de-icing. It is usually used as a brine to reduce losses. However, it increases vehicle corrosion, damages concrete and steel structures, produces surface grime, and adversely affects the environment. One commonly used replacement for salt is calcium magnesium acetate.

The frequency with which snow is removed depends largely on the levels of traffic flow. Typically, if:

* AADT > 10 000, all lanes are kept clear;
* 10 000 > AADT > 2 000, wheel-paths are kept clear for two lanes; and
* 2 000 > AADT > 800, intermittent wheel-paths are kept clear.

Dry snow generally needs to be over 30 mm in depth to begin to affect traffic performance in communities accustomed to snow conditions.

CHAPTER TEN

Stabilisation

10.1 INTRODUCTION

10.1 1 Principles

When confronted at a job site with local (in situ) material, a pavement engineer has three choices:

(a) produce a design using the inherent properties of the in situ material,
(b) remove the in situ material and use better, imported material, or
(c) treat the in situ material on site to enhance its ability to perform its pavement function.

This chapter discusses this last alternative. It is called *stabilisation* — a term which possibly arose from the common misuse by pavement engineers of 'stability' for 'stiffness' (see Section 12.2.2). Alternatively, it might have implied preventing pavement properties from deteriorating. The simplest stabilisation processes are:

* compaction (Sections 8.2 & 25.5),
* drainage (Chapter 13),
* improved grading (Section 8.3.1, or *mechanical stabilisation*),
* heating the soil,
* dust palliatives (Section 12.1.2), and
* the use of additives (Section 10.1.2).

Using the nomenclature developed in Figure 8.6, Figure 10.1 gives recommendations for which of these stabilisation processes to use for a particular material type.

10.1.2 Additive stabilisation

The first three processes in the list of stabilisation processes in Section 10.1.1 — compaction, drainage and grading — are discussed elsewhere in the book, this chapter concentrates on the use of additives. The fourth process — thermal treatment — is an old method which involved heating in situ clay in a manner similar to primitive brick-making (Lay, 1992). The remainder of this chapter discusses the fifth process — additive stabilisation — in which the addition to some soils of a few percent of compounds such as lime, Portland cement, calcium chloride, bitumen, or some polymers produces disproportionately large permanent increases in the properties of the soil.

There are two main mechanisms at work in lime and cement stabilisation. The first, slower, mechanism is more dominant with lime than cement. It occurs as a result of the soil water (and any clay minerals) reacting with the additive to produce a calcium silicate gel which acts as a conventional mortar within the soil structure (Section 8.9.1). It is thus a cementitious reaction.

The second mechanism depends on the surface reactivity of the material particles, and is pozzolanic rather than cementitious. It is thus most relevant to the addition of lime

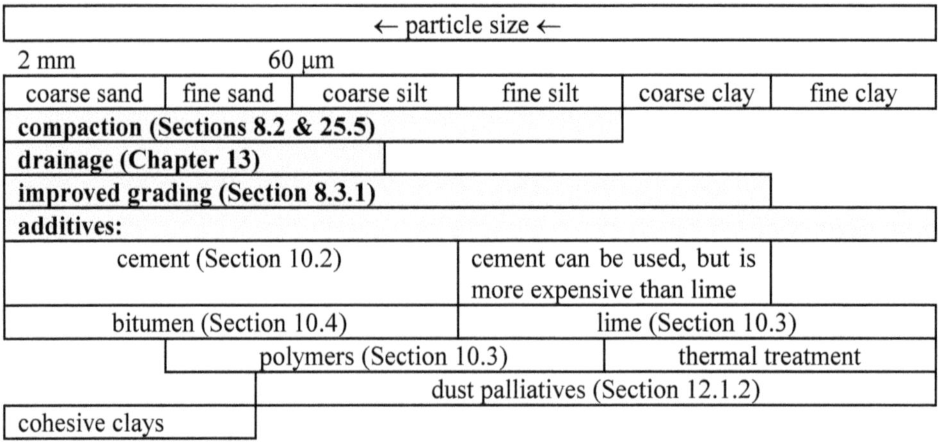

← particle size ←
2 mm
60 μm
coarse sand
fine sand
coarse silt
fine silt
coarse clay
fine clay
compaction (Sections 8.2 & 25.5)
drainage (Chapter 13)
improved grading (Section 8.3.1)
additives:
cement (Section 10.2)
cement can be used, but is more expensive than lime
bitumen (Section 10.4)
lime (Section 10.3)
polymers (Section 10.3)
thermal treatment
dust palliatives (Section 12.1.2)
cohesive clays

Figure 10.1 Zones of maximum efficiency for various stabilisation processes used with a range of material types (after Ingles and Metcalf, 1972). Particle size data based on Figure 8.7.

or cement to materials with a high surface reactivity, such as swelling clays. As discussed in Section 8.4.2, clay consists of plate-like particles with electrically active surfaces. The cement, lime and calcium chloride stabilisation mechanisms act by additive ions, such as hydroxyl and calcium cations, attaching themselves to the clay particle edges, displacing cations such as sodium and hydrogen, which are naturally present in soil.

This surface electro-chemistry permits much greater linking together of the individual clay particles. The water ions are displaced and the clay plates lose their slipperiness. As a consequence of the linking, the fine particles of clay form into coarser, more friable particles. The immediate effect of the process is often seen as flocculation or agglomeration. When the effect is due to natural additives, this cementitious form of stabilisation is colloquially known as *setting up*. In fact, what occurs is that, in the presence of free ions, the secondary minerals that formed the clay are now encouraged to continue their natural breakdown to a stable crystalline state (Sections 8.4.2 & 8.5.4).

The additive stabilisation processes thus improve the workability, compressive strength, stiffness, moisture susceptibility (plasticity index, Section 8.3.3), and volumetric stability of the material. Permeability is reduced by a pore-blocking process. At high additive contents, some tensile strength is added to previously-unbound layers. Because of the variability of natural materials, the correct proportions of the stabilisation additive to use will depend on the specific application and are best determined by testing specimens before construction (Dunlop, 1980). This issue is pursued further in Section 10.5.

Additive stabilisation is not particularly effective when more than 2 percent of organic matter is present in the material being stabilised. Stabilisation can be destroyed by some soil chemicals: for example, the presence of more than 1 percent of soil sulfates removes the effects of cement and lime stabilisation. The sulfate can be water-borne to the site, or it can arise internally — usually by the production of highly-expansive calcium sulfate byproducts such as *ettringite*, $Ca_6\{Al(OH)_6\}_2(SO_4)_3 26H_2O$. Using sulfate-resistant cement is rarely an effective solution. At the other extreme, some degradation-prone (Section 8.5.10) aggregates such as laterites and basalts can be successfully stabilised with cement and lime.

Given the above mechanisms, the main purposes of additive stabilisation are to:

(1) provide a working platform for construction equipment operating on soft material;
(2) enhance subsequent workability;
(3) improve the strength and stiffness properties of sub-specification material;
(4) reduce moisture susceptibility;
(5) improve volumetric stability;
(6) decrease permeability (provided no cracking occurs, Section 9.2.2);
(7) reduce the required thickness of a pavement course by enhancing subgrade properties (e.g. Section 11.4.6); and
(8) possibly produce a bound course (Section 11.1.4).

As a rough rule of thumb, stabilisation to improve material properties is usually justified when the alternative would require hauling imported materials for over fifteen kilometre.

10.2 CEMENT STABILISATION

The properties of cement were discussed in Section 8.9.1 and the principles of cement stabilisation were described in Section 10.1.2. Figure 10.1 suggests that cement stabilisation is best used for granular materials, particularly if they are well-graded. It is thus a useful process to use when recycling a pavement. It can also be used for sand-clays and non-expansive clays with low plasticity, and *with caution* for heavy clays. The cement contents used cover a wide range of circumstances:

A. When *less than 3 percent of cement* is added, the process is called *cement modification* and the material produced has little tensile strength and is essentially unbound (Section 11.1.3). Cement modification is often employed:
   * to meet specification requirements for unbound materials (Section 8.6),
   * to reduce moisture sensitivity,
   * to enhance an in situ subgrade, or
   * as a construction expedient. With well-graded unbound mixes, cement modification will aid compaction interlock (or 'set up'), and prevent ravelling under construction traffic.

B. *Cement contents of 3–4 percent* are commonly used:
   * in rehabilitating existing unbound basecourses; e.g. by restoring grading (Section 8.3.2) — and hence strength and stiffness — to courses containing particles broken-down under traffic, or
   * enhancing the properties of well-graded sands.

C. *Material with 3–5 percent cement* is often called *soil-cement*. It has some tensile strength but is still only partly bound and far from elastic. Soil-cement may deteriorate under traffic and, if used beneath a spray and chip seal (Section 12.1.4), may absorb excessive bitumen. It may fail due to water erosion of material at the upper or lower faces of the soil–cement layer.

D. *Cement contents of up to 6 percent* are used to stabilise fine materials, whilst avoiding cracking. However, with poorly-graded sand, compaction may be very difficult and segregation of the cement may occur.

E. Stabilisation with *cement contents at around 6 percent* will produce a stiff, brittle material. In flexible basecourses, drying shrinkage, thermal movement, or traffic loading could then lead to both transverse and longitudinal cracks (Metcalf, 1979a). Such cracks frequently reflect through to the pavement surface (Section 14.1B7). They have a major effect on the stiffness of the cement-treated layer, and thus on its structural performance (Section 11.3).

F. As the *cement content increases above 6 percent*, the crack size and spacing increases. Rather than reduce the cement content to control cracking, a number of other measures may be used:

* the cement-treated layer can be shielded from shrinkage or thermal effects by controlled curing (Section 8.9.1),
* the cement-treated layer can be covered with a relatively ductile surface course,
* the effect of load-induced cracking is most directly reduced by increasing the thickness of the stabilised layer and thus reducing the flexural strains induced in it (Section 11.2.3). Even for light traffic, this may require thicknesses of over 300 mm and may not be practical.
* a smooth-wheeled roller (Section 25.5.) can be used to cause controlled cracking during construction, or
* the cement-treated layer can be overlain with an unbound course unable to propagate cracks. This is called an upside-down pavement (Section 11.1.5).

G. *Material with 6–15 percent cement* and rolled into place for use in basecourses and sub-bases is known as *dry, lean concrete* (or *roller-compacted concrete* or *econocrete*). [In this context, conventional concrete is called 'internally vibrated' concrete.] The need to apply heavy compactive effort means that dry, lean concrete must usually be placed on a good granular basecourse. It has much less cement than conventional concrete (Section 8.9.1), is far less workable; cannot be vibrated, and its placement is more analogous to that of asphaltic concrete (Section 12.2.2) than to slip-formed concrete. The water content is kept sufficiently low to just support the heavy compaction roller required for placement, but this means that the surface cannot be finished or textured.

The economic advantage of dry, lean concrete lies in the cheapness of the construction method, rather than in the cheapness of the materials used. Its low water content suggests that it should be stronger than conventional concrete, however it is usually unwise to rely on this potential strength gain. Dry, lean concrete can produce a good basecourse, but may crack over time, leading to subsequent reflection cracking in the surfacing. This is best controlled by using an adequate thickness, usually in excess of 120 mm for routes carrying significant traffic, and by placing joints at about 4 m centres.

H. *Cement added at around 10 percent* to wet sandy soils will produce a plastic soil-cement which can be placed like lean concrete.

I. The addition of *more than 15 percent cement* usually results in a conventional concrete (Section 11.5.1).

The above gradation of the types of cement stabilised material is illustrated in Figure 10.2. Properties of typical mixes are given in Table 10.1. In general, the strength of the soil–cement mix steadily increases with an increase in cement content, giving an extra unconfined compressive strength of about 500 kPa to 1000 kPa for each 1 percent of cement added. Poisson's ratio ranges from 0.1 to 0.3, with values at the higher end predominating. Swelling is usually about 2 percent. Satisfactory durability results are commonly obtained from the wet/dry test (test 8.6m).

Cement stabilisation can also be used for sand-clays and non-expansive clays with low plasticity. Some typical cement contents for this situation are shown in Table 10.2. Shrinkage, and thus cracking, is particularly evident in swelling clays (Section 8.4.2) and a swell and shrinkage test for cement-stabilised material is given in ASTM D559.

<table>
<tr><td>soil</td><td>modified soil</td><td>soil-cement</td><td>lean concrete</td><td>concrete</td></tr>
<tr><td colspan="5">→ increasing cement content →</td></tr>
<tr><td colspan="2"></td><td colspan="3">range of cement use</td></tr>
<tr><td></td><td colspan="2">range of lime use</td><td colspan="2"></td></tr>
<tr><td colspan="5">→ increasing strength and stiffness →</td></tr>
<tr><td colspan="2">tensile strength below 80 kPa</td><td colspan="3">tensile strength above 80 kPa</td></tr>
</table>

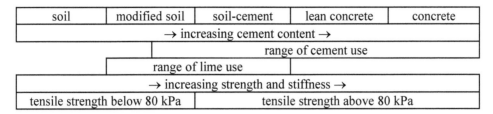

Figure 10.2 Gradation of properties of cement- and lime-stabilised materials (from Dunlop, 1980).

Table 10.2 Cement contents for various clay types (from Metcalf, 1979a). Note that Table 10.1 is on the following pages.

Clay type		Cement requirement (percent)
Well-graded sandy clay-gravels		2–5
Sandy clay		4–6
Silty clay	cracking much	6–8
Heavy clay	↓ becomes	8–12
Very heavy clay*	more likely	12–15

* Mixing may be very difficult — pretreatment with lime may help.

10.3 LIME STABILISATION

Lime is usually made by burning limestone or dolomite (Section 8.5.4) to convert their calcium (or magnesium) carbonate to calcium oxide.

$$CaCO_3 + \text{heat} = CaO + CO_2$$

The CaO is known commercially as *quicklime* (or *agricultural* or *unslaked lime*) and is occasionally used for the short-term drying of wet soils or for deep stabilisation (Metcalf, 1977). However, its corrosive properties mean that it requires special handling and the pH of the soil-lime slurry can be as high as 12.5. Adding water to quicklime — a dangerous process — produces *hydrated* (or *slaked*) *lime*, which is calcium hydroxide ($Ca[OH]_2$).

$$CaO + H_2O = Ca[OH]_2 + \text{heat}$$

Whilst less corrosive than quicklime, even hydrated lime needs to be handled with care. Nevertheless, as quicklime lime can be obtained in a larger particle size than hydrated lime, it is easier to manage and this tends to cancel out the negative effects of its extra corrosivity.

Thus, stabilisation with lime usually refers to the use of hydrated lime and not to unslaked lime. The hydroxides provide the calcium (and magnesium) cations referred to in Section 10.1, and which are not nearly so freely available from the original carbonates. Indeed, the term *available lime* refers essentially to cation availability. However, it is measured as the percentage of pure calcium oxide present in quicklime or the percentage of pure calcium hydroxide present in hydrated lime (an earlier, similar, term was equivalent calcium oxide content). Whether the lime is calcitic or dolomitic (measured by its relative proportions of Ca and Mg) will affect its performance as a stabiliser with respect to individual soils.

Following the discussion in Section 10.1, the increases in strength due to lime stabilisation can be attributed to:

Table 10.1 Typical properties of cement-stabilised materials (based on Metcalf, 1977). B = before, A = after, w/ds = wet/dry strength. Continued on facing page.

#	Material type	Strength range[a] (MPa)	Stiffness range[b] (GPa)	Soaked CBR[c]	Permeability (μm/s)
1	Well-graded sand, gravel-sand-clay, or gravel	1.5–10+ w/ds = 2/3	5–30	>> 200	High B = 150 A = 0.18
2	Silty sand, sandy clay, poorly-graded gravel	1.5–3.5	5–7	> 200	High
3	Poorly-graded sands, or silty-sandy clays	0.7–1.7	2–5	100–200	B = 0.05 A = 0.001
4	Very poorly graded sand, silts, or silty clays	0.3–1.0	< 3.5	20–100	Low
5	Heavy clay, organic and sulfate-rich soil	< 0.7 w/ds = 1/3	< 1.4	< 50	Very low μB = 0.1 μA = 0.1

Notes:

a. Strength (UCS) is for 7 days curing (Section 8.9.1) at constant temperature and moisture content. The ratio of UCS to flexural strength is about 4:1 in sand and 3:1 in kaolin clay. The ratio of UCS to indirect tensile strength is about 10:1. The loss of strength on soaking should not be more than 20 percent.

* in the short term (or modification phase), surface reactivity and the associated ion exchanges leading to flocculation of clay particles, rapid drying of the soil and improvements in such properties as plasticity, volumetric stability and workability,
* in the medium [1 to 3 days] term (or stabilisation stage), the amorphous products of the lime/clay chemical reaction — composed largely of calcium hydrates of silica or aluminium — surround, and then extend from, any clay particles to form a continuous cementitious phase that increases the strength and durability of the material, and
* in the long term, a gradual crystallisation of the amorphous products. The lime also gradually reacts with pozzolanic material in the soil (Section 8.5.6), producing additional gradual gains in strength, particularly in hot weather.

Many of these processes result in strong chemical bonding and so the treatment is effectively permanent once the minimum lime content needed to service the process has been exceeded. Thus, lime rapidly produces a stabilised layer that, for a given compactive effort:

* is stronger,
* is more granular,
* is less plastic,
* is less moisture-sensitive (typically, soaking a lime-stabilised soil in water does not reduce its strength by more than about 30 percent),
* has a lower maximum dry density,
* has a higher optimum moisture content, and
* is less prone to volume change,

Table 10.1 continued from facing page.

#	Thermal expansion (με/C) [d]	Volume change [e]	Comments	Uses
1		Very small, < 1%	Brittle, leading to widely-spaced wide cracks	Base for heavy traffic, embankment protection
2		Small	Good material	Base for heavy traffic
3	B = 100 A = 70		Compaction is a problem. Needs good drainage.	Sub-base, base for light traffic and/or with good drainage in a warm region
4		Moderate		Low-grade sub-base, trench backfill
5	10	High, > 4%, increases with cement content	Extreme problems in mixing	Possibly for ungraded subgrade or backfill

b. Data is from flexural tests. Static and dynamic values are similar. Typically, for uncracked samples, $E = 2[(UCS)^{0.88} + 1]$ GPa. Alternatively, $E = 1.5(UCS)$ GPa

c. Approximate figures for mixes with a 7 days UCS of 1.7 MPa.

d. The value for concrete is from 3 to 8, for bitumen about 600.

e. Based on very limited data.

than the original material. Although the final setting of lime-stabilised soils is slower than with cement — and so compaction can be delayed longer — it is usually convenient to gain advantage of the lower OMC value associated with the pre-set material.

Cement disperses poorly in fine material and so lime stabilisation is widely used as a construction expedient with heavy clays and material with more than 10 percent clay and with a plasticity index (Section 8.3.3) in excess of 10. It has an almost instant effect on strength and plasticity, hence its common use in construction. Lime is also used to stabilise low-grade breccias, reacting with any clay minerals present in the breccia (Sections 8.5.1&10) and has been successful with some laterites (Section 8.5.4).

Wet, heavy clays present mixing problems, and a staged approach to the addition of lime may be needed, with the first stage being used only to provide a working platform. The conventional mixing process (Section 10.6) is often followed by a further pulverisation hours or days later.

The amount of lime to be used can be established by tests on specimens to determine when the requisite physical properties (e.g. Table 10.1) have been achieved. Alternatively, a lime-demand test can be used to determine the amount of lime necessary to satisfy the cation exchange requirements (Section 10.1.2).

Mixed lime and cement stabilisation is effective as a construction expedient with very wet subgrades, producing higher unconfined compressive strengths than an equivalent pure lime addition, and providing a better immediate working platform than the equivalent pure cement addition (Grahame and Goldsborough, 1980).

There is a long record of stabilising with fly ash (Section 8.5.6) and related industrial by-products such as rice-husk ash, blast-furnace slag, and silica fume, frequently blended with cement (Section 8.5.6) or lime. Some fly ashes will produce higher strengths and more rapid strength gains than lime alone (Linn and Symons, 1981). Many cements blended with fly ash or blast-furnace slag offer the advantage of retarding the setting process. They therefore allow construction to continue over a longer period.

A recent development uses dry powdered polymers (DPP) to stabilise moisture-sensitive materials. DPP is typically bound thermally at a particulate level to a fine-particle carrying-medium such as fly ash (Section 8.5.6). When mixed with hydrated lime, it coats clay particles (Section 8.4.2) within the pavement. This both reduces permeability (Section 9.2.2) and lowers the moisture-susceptibility of the pavement.

10.4 BITUMEN STABILISATION

Various bituminous materials can be used to stabilise a wide range of soil types (Metcalf, 1979a). However, the process is usually used with the more granular materials where mixing is easier and the required strength gains are not as great. The bitumen acts predominantly as a glue-like cohesion enhancer, increases the impermeability of the course, and can aid compaction. Bitumen stabilisation can also be used to water-proof clays.

Conventional (straight) bitumen is rarely used; instead cutback bitumen (Section 8.7.6), foamed bitumen (see below), road oil (Section 12.1.2), low-viscosity bitumen (Section 8.7.5), and bitumen emulsion (Section 8.7.6) are preferred as they are easier to disperse into the soil than is straight bitumen.

In the stabilisation process, bitumen droplets coalesce and are then preferentially attracted to the high surface-area/mass ratios of the finer aggregate particles. There is therefore a quite non-uniform dispersion of the bitumen. The process enhances the stiffness of the course via the cohesive inter-particle contact with the larger particles, which improves the internal friction in the course and thus produces a stiffer and stronger material. To avoid flooding of the mix (Section 12.2.3), the bitumen percentage is kept well below the air voids percentage. For strength gains, it is common to use about 1.5 percent low-viscosity bitumen (Shackel *et al.*, 1974). A higher percentage is used to reduce permeability, but will lead to a weaker material.

The addition of a small amount of cement or lime to the bitumen will help coalescence of the bitumen, improve binder viscosity by acting as an inert filler, and remove excess water by hydration of the cement. Thus this step can further increase the strength and durability of the course.

The bituminous material is added either on the road, in a pugmill, or mixed with the water being added to aid compaction.

Bitumen is also used to stabilise limestone (Section 8.5.2) which, if used in a basecourse in its natural state, is likely to be uncohesive, difficult to compact, and a dust nuisance. The use of about 2–3 percent bitumen emulsion produces a marked improvement in handleability and deformation resistance (Sparks and Hamory, 1980).

Hot bitumen naturally foams when it comes into contact with water. This can be prevented by the use of silicone-based antifoaming agents. On the other hand, in *foamed bitumen* about 3 to 4 percent by mass of water or steam is added to hot bitumen to produce an inverted emulsion which — due to bubbles of steam — occupies about 12 times the volume of normal bitumen (Bowering, 1970). The product is a sticky foam with a low viscosity. In its short life span of just a few minutes before 'breaking' (bubbles collapsing), the foam is able to coat most surfaces, even if they are cold and moist. Fine particles are preferentially coated and large particles may be only partially coated. The mix can still be compacted for about an hour after the foam has collapsed. Once the mix cools, the coated fine particles form a binding mortar. The prime value of foamed bitumen is that it avoids the use of the cutters or emulsions discussed in Section 8.7.6.

Foamed bitumen is also useful for spray and chip sealing (Section 12.1.4), where expansion ratios of about 8 and half-lives of about 15 minutes are used. It is not useful with materials with a plasticity index of over 12, or when there are insufficient fines to form a well-graded end product. When used on an unbound surface course, the product will have a life of three years or less.

10.5 DESIGN PROCEDURES

Design procedures for stabilised pavements fall into one of three categories, in which the pavement is designed:

(1) *empirically*, as a *flexible* pavement (Section 11.1.5), with perhaps an increase in the assumed stiffness of the stabilised layer. A material is considered unbound if it has an unconfined compressive strength (UCS, F'_c) below 700 kPa. This level permits little advantage to be gained from stabilisation, other than to bring initially sub-specification material to within the specification requirements for an unbound material. In this case, the following practical criteria can be used in addition to those flowing from the normal design process (Dunlop, 1980).
   (a) Subgrades are only stabilised if they have a soaked CBR of 5 or less, and the amount of cement used is limited to keep the final CBR below 50.
   (b) Sub-bases after treatment should have a plasticity index of 5 or less.
   (c) Basecourses after treatment should have plasticity indexes of 2 or less and a CBR in excess of 100. Granular or modified courses will have a final CBR above 40 and a UCS between 700 kPa and 1.5 MPa

(2) *mechanistically* as a *flexible* pavement (Section 11.4.2) utilising the new characteristics of the stabilised layer, particularly its increased stiffness and its tensile strength as a bound layer. The enhanced properties used in design are established by prior testing of samples and must recognise the consequences of any subsequent shrinkage cracking. In the past, stiffness was sometimes determined from the UCS using Pauw's formula (Section 8.9.2). However, the increased emphasis on stiffness in the mechanistic methods means that it is better determined from a repeated-load triaxial test (test 8.6(r)). Work of this type relates stiffness to UCS using formulas such as those given in Note b of Table 10.1. A material is commonly taken as bound if it has a UCS of at least 1.5 MPa.

(3) as a *rigid* concrete pavement (Section 11.5.2).

The correct proportion of additive to use in any particular situation is best determined by prior testing of the material (Dunlop, 1980). It will often be possible to detect an optimum additive level based on measures such as the Atterberg Limits (Section 8.3.2), or tests for permeability, shrinkage, stiffness, or unconfined compressive strength (Section 8.4.3).

10.6 CONSTRUCTION

To convert the required additive percentage to an actual volume to be added at site requires knowledge of the thickness of the layer to be treated. This will depend on the capabilities of the plant to be used and on the design needs. Layers of under 150 mm in thickness are often ineffective due to high roughness and excessive cracking, and stabilised thicknesses commonly range between 150 and 300 mm with 400 mm a practical maximum set by equipment capacity and the setting time of the additive used.

When clayey soils are to be stabilised, they often only need to be coarsely pulverised as the lime or cement additives diffuse naturally into the soil lumps and affect every clay particle in the soil structure. However, the finer the pulverisation, the greater the strength achieved.

The stabilisation effect is almost immediate, with expansive clays responding more rapidly than non-expansive ones. General-purpose cements will set (due to the hydration and bonding processes discussed earlier) in 2 or 3 hours, or less in warm climates. Once cement has been added to a layer, it is necessary to compact the mixture to maximum density and trim the surface to the required level before setting occurs. It is common to specify this 'working time', which is the time between mixing and the completion of compaction and trimming. Thus, the earlier that compaction and the later that setting can occur, the better the outcome will be and slow-setting stabilisers are often preferred. This can range between 2 h for rapid setting general cement in the summer to 24 h for lime in the winter. Thus, if the construction process requires a longer time, the alternative of blended cements or lime (Section 10.3) should be considered. Some blended cements (e.g. with blast-furnace slag) can be reworked a day after placement. Similarly, for deep-lift pavements it may be necessary to use slow-setting additives to ensure that the full layer can be properly compacted before the binder has set.

Immediately after construction, drying will increase the strength and propensity to cracking of a stabilised layer, whereas wetting will reduce its strength. Subsequent curing of the layer produces a strength that is mainly dependent on time, moisture availability and temperature. Significant curing can occur for up to two months after placement and minor increases can still be occurring after six months. As with conventional concrete (Section 8.9.1), curing is mainly a continuation of the hydration process. Subsequent removal of moisture from the soil can lead to noticeable shrinkage cracking.

In cold climates it is also necessary to check for the effect of freezing and thawing, which can cause disintegration of the stabilised structure (Section 9.5). ASTM D560 suggests cement contents to avoid this problem.

The construction process requires both pulverisation of the existing soil and mixing of the additive. A key requirement is that the mixing spreads the stabiliser uniformly throughout the material, avoiding a natural tendency to produce lumps of stabiliser. The mixing of stabilisation additives into the in situ material is usually carried out by either:

* agricultural rotary hoes and/or disc ploughs,
* graders, or
* purpose-built single-rotor or multi-rotor equipment.

Agricultural equipment and graders can produce work to reasonable standards. However, where the stabilised layer is to be included in the pavement design, following design category (2) in Section 10.5, additive tolerances of about 0.6 percent (absolute) are required and purpose-built rotary equipment is needed (Grahame and Goldsborough, 1980).

The single-rotor equipment is mounted on a tractor with the rotor either between the front or back wheels or at the rear of the tractor. Mixing efficiency depends on both speed of travel and rotor speed. It is common for more than one pass to be required. Multi-rotor devices usually have each rotor designed for a different function and are able to complete most jobs in a single pass. A further development is a self-powered machine with its rotor fitted with hooks rather than blades and capable of cutting into the in situ material to depths of up to 500 mm. It operates counter to the travel direction, cutting to a free face and lifting the soil-additive mixture up for recycling (Grahame and Goldsborough, 1980). These devices are sometimes called 'stabilisers'. Their noise levels may be too high for urban areas.

The additive is carried in a storage hopper on the rotor vehicle and fed to a transversely-mounted rotor fitted with a number of rotor blades. The material is then mixed dry with the pulverised soil. The amount of additive placed is a function of both the mechanical operation of the device and the speed at which it travels. A water-delivery system is also essential in order to achieve optimum moisture contents.

The outcome may be very dependent on the moisture content used – this is best established by prior laboratory testing.

The layer may then need to be remixed before compaction. After the first mixing, some time must be allowed for the setting process. In stabilisation this is sometimes referred to as the *ageing period*, or *mellowing period*. Recommended mellowing periods commonly range between 1 and 2 days. Holt and Freer-Hewish (2000) suggest that modification improves for up to 3 days but stabilisation effects reduce with prolonged modification. It may be necessary to keep the material moist during this period. It is essential that any testing for mix design (Section 10.5) replicate this mellowing period.

Contraction joints are not commonly used, as they cannot control the majority of the natural cracking patterns that occur in a stabilised pavement. An excellent review of the construction aspects of lime stabilisation is given in Holt and Freer-Hewish (2000).

10.7 FABRICS AND MEMBRANES

The use of *fabrics* (or *textiles* or *geotextiles*) and *membranes* (or *geomembranes*) within a pavement structure can also be considered as a form of stabilisation. Fabrics and membranes usually operate by either:

* preventing the mixing of the subgrade and the pavement material, particularly with poor subgrades (CBR < 3, swampy material, or organic material) or in bad weather,
* protecting a subgrade from the effects of construction traffic,
* reinforcing weak, deformable, brittle or cracked material. The relatively stiff and strong fabric or membrane acts as a layer of tensile reinforcement, raising the tensile strength of the course and preventing existing cracks from widening.
* reinforcing a spray and chip seal (Section 12.1.4) placed over expansive-clay basecourses (Section 8.4.2),
* providing a tensile membrane to carry courses across voids,
* controlling the movement of water by acting either as a filter, water barrier or drainage layer (Section 13.4.3), or
* reducing the possibility of slippage by containing material in embankments.

There are a number of basic types of fabric and membrane in use:

(a) *Woven fabric* consists of lengthwise threads (4 mm x 0.1 mm) interlaced with orthogonal threads. The tightness of the weaving is varied to give different permeabilities.

(b) *Knitted fabric* consists of a single thread systematically intertwined with itself.

(c) *Non-woven fabrics* have threads randomly placed on a low strength base and attached by either chemical, thermal or mechanical bonding. Most geotechnical fabrics are in the thermally or mechanically bonded non-woven category, although the non-woven fabrics are not as strong or as stiff as the woven ones. Mechanical bonding usually uses barbed needles to give a felt-like product. Such material is usually needle-punched to provide the required transverse permeability.

(d) *Geomembranes* consist of impermeable plastic sheets or non-woven fabrics impregnated with bitumen or plastic.

(e) *Geostrips* are polymer strips used as reinforcement. They can range from rope to kevlar.

(f) *Geogrids* (or *geonets, plastic grids,* or *plastic meshes*). Some geogrids consist of open weaves of geostrips bonded at their intersections, some are extruded on dies and others are formed by punching. They are used as a form of tensile reinforcement, for crack control for pavement courses and for embankments, and can play a secondary role separating material at layer interfaces. These materials are all characterised by having large apertures in them. Uniaxial grids have strength in only one direction. Because of their structural role as a reinforcing agent, they must maintain their strength, ductility and stiffness throughout the expected range of service temperatures and over their expected design life. They are usually visco-elastic (Section 8.7.4).

(g) *Geocomposites* consist of combinations of the above systems, as with two-layer non-woven fabric containing a permeable natural fabric core.

Fabrics are usually permeable and membranes are usually impermeable. The materials used may be natural — as with *coconut* fibre and *hessian* — or man-made synthetics such as glass fibres, plastics, and polymers, which are called *geosynthetics* when used in this application. The common plastics used are *polypropylene*, *polyester*, *polyamide* (e.g. *nylon*), and *polyaramid* (e.g. *kevlar*). Polypropylene burns easily so must be stored carefully. Polyaramids are expensive, but match the properties of steel. Of the others, only polyesters can produce good strength at low strains. The threads (or fibres) used in fabrics are usually non-natural, about 50 mm in diameter and light-sensitive. Tests for fabrics and membranes fall into two main categories:

(1) *Classification or index tests*, which establish basic properties used for classification, quality control, and constructability. They are usually fairly simple to perform. Typical important classification properties are stiffness, tensile strength, creep, tear propagation, burst strength, puncture resistance, surface friction, permeability, filter size, durability and resistance to common materials such as soil and diesel fuel. A typical test is pushing a pyramidal tip against a geotextile to measure its resistance to tearing by aggregate penetration.

(2) *Design tests*, e.g. for structural performance, soil-interaction and management of water flows. These tests attempt to relate to field conditions and are usually relatively complex. They usually need to be performed over a considerable length of time under field conditions of deformation, temperature and moisture.

CHAPTER ELEVEN

Pavement Design

11.1 PAVEMENT COMPONENTS

11.1.1 Pavement structure

A *pavement* is a longitudinal structure whose primary purpose is to provide a traffic surface of acceptable riding quality (Section 14.4.1), adequate skid resistance (Section 12.5), favourable light-reflectance (Chapter 24), and low noise-generation (Section 32.2). The latter three constraints are handled solely by the pavement surface and are discussed at length in Section 12.5. This chapter concentrates on the structural strength and stiffness needed to support the pavement surface and maintain riding quality, remembering that much of this strength and stiffness will depend on the control of moisture entry into the pavement (Chapter 9, particularly Section 9.4). The pavement — with suitable maintenance (Chapter 26) — must fulfil the above purpose over its entire intended life whilst subjected to the continual influence of weather and traffic loading.

The pavement structure is composed of a number of horizontal *courses* (Figure 2.2), each comprised of a different set of materials. The use of different courses allows the most appropriate material to be used for the varying conditions that exist through the depth of the pavement structure. Chapters 8 and 10 discuss the individual materials used. Pavement courses are rarely less than 75 mm in thickness, particularly given the size of common construction tolerances (Section 25.7). As a consequence of limits on the ability of the construction process to place and compact large quantities of material in a single pass (Section 25.5), courses are often built in a series of independent *layers* of otherwise identical material. For a pavement to act as a coherent structure and deliver its maximum potential strength and stiffness it is also necessary for the horizontal interfaces between layers and courses to be able to transfer all the required shear flows without separation or sliding occurring at the interface.

As the pavement structure is built from the bottom up, an important purpose of each pavement construction stage is to provide adequate support for the next round of construction activities, ranging from compaction equipment (Section 25.5) to surfacing processes (Sections 12.1.4 & 25.6). This intermediate purpose may be a controlling factor for lower levels of the pavement structure and Sections 10.1.2 & 10.3 discuss how stabilisation is often used to assist in providing a working platform for construction on soft material.

11.1.2 Pavement courses

Pavement courses are known by their location and function within the pavement structure (Figure 2.2). Each will now be described in turn, working downward from the surface used by the passing traffic.

The uppermost course is called the *surface course* (or *wearing course*) and its prime objectives are to (Section 12.5.1):

(a) carry the vertical and horizontal wheel-loads resulting from the traffic using the road, without either impeding that traffic or being appreciably damaged by it,
(b) provide adequate skid resistance (Section 12.5.2), favourable light-reflectance (Section 24.1), and low noise-generation (Section 32.2),
(c) withstand the atmospheric environment, and, usually,
(d) keep water from entering the underlying pavement structure (the major exception is the porous friction course, Section 12.2.2).

The surface course is not a significant direct contributor to the overall strength of the pavement and is not included in most conventional pavement design processes.

Most vertical deformation in an asphalt pavement occurs in the highly stressed upper layers of the pavement (Figure 11.2 & Section 11.2.3). To improve the deformation resistance of heavy-duty pavements there is, therefore, a growing tendency to use an upper course of very stiff asphalt (e.g. SMA, Section 12.2.2(d)) known as an *intermediate course*.

The *basecourse* (Figure 2.2) directly supports the surface and intermediate courses and is the main load-carrying (or structural) course within the pavement. It is the subject of most of this chapter. The basecourse may itself be composed of different courses, each chosen to reflect the structural requirements for its level of the basecourse.

The *sub-base* is a course that is sometimes placed below the basecourse. Its prime purposes are to provide:

(1) a *working platform* over weak (CBR < 6) subgrades, with sufficient:
 * strength, for construction traffic to pass without damaging the sub-grade,
 * stiffness, for the basecourse and sub-grade to be adequately compacted, and
 * coherence, to protect and separate the basecourse from the subgrade.

(2) a *physical barrier*, with sufficient:
 * insulating characteristics, to prevent the subgrade from suffering from the effects of low temperatures (Section 9.5),
 * impermeability; to prevent undesirable moisture movement within the pavement, and protect a moisture-sensitive subgrade from water infiltration. The impermeability may need to be supplied by:
 # a separate, impermeable *capping course* (Section 9.4.2);
 # stabilising the subgrade surface layer (Sections 10.1.2 & 10.3); or
 # the use of impermeable membranes (Section 13.4.3).
 Impermeability may create a perched water-table (Section 9.2.3) and require at least the lower layer of any overlying basecourse to be moisture-insensitive.
 * coherence, to prevent the *pumping* of subgrade fines by traffic action on the basecourse slab in concrete pavements (Section 9.3.2). Such sub-bases may be built of lean concrete (Section 10.2).

(3) a *physical path*, using a *drainage* (or *filter) course* to provide a permeable path for water to drain horizontally from the pavement structure (Section 9.4). As a result of their open structure, permeable layers will inherently be of relatively low stiffness.

The *subgrade* is the lowest part of the pavement system and is the foundation on which the pavement structure is constructed. It comprises either undisturbed soil or rock or compacted material within an embankment (Section 11.8.1) and its surface is called the *pavement formation* (Section 2.2).

11.1.3 Unbound courses

Courses are termed bound or unbound. An unbound course is formed of independent granular particles and behaves under load as if its component particles are not inherently linked together. However, the use of angular particles of stone known as *crushed rock* (Section 8.3.1) and similarly sharp-edged sand (Section 8.5.2) will ensure that significant mechanical interlock can occur between particles and thus that:

* the course can be compacted into a relatively dense layer (Section 8.2),
* significant shear strength can be provided by interlock of the angular faces, and
* compressive strength can be provided by high particle-to-particle, intergranular contact-stresses (these are the *effective stresses* discussed in Section 9.2.5).

This gives unbound courses made of these materials a major advantage over those using smooth, rounded particles such as river gravel. Even with interlock, horizontal tensile strengths can be as low as 30 kPa under vertical load and 10 kPa without load. The repeated-load triaxial test (Section 8.6r) is widely used to predict subsequent course performance. To gain maximum strength and stiffness, the courses should be restrained against kerbs or shoulders (Section 11.7) to prevent them from moving laterally due to the Poisson's ratio effect (Section 33.3.2).

Unbound courses are occasionally treated with additives to improve properties other than strength (e.g. grading and plasticity). Common size gradings for the granular particles of aggregate (Section 8.5.9) are discussed in Section 8.3.1.

Macadam is a term properly used in an historical sense to describe a uniformly-graded, unbound pavement-course, constructed using single-sized (uniformly-graded) angular aggregate (Section 3.3.5). It was named after its inventor, whose advance at the beginning of the 19th century was to replace the large, rounded natural stones which were then in common use by smaller broken angular particles, thus taking advantage of the natural interlock that develops between such particles. As seen in Section 9.2.2, uniformly-graded courses are quite permeable. They thus provide a good drainage course (Section 11.1.2) and prevent the build-up of pore pressure. On the other hand, such courses can lead to problems, viz.:

(a) they can permit water to enter a moisture-sensitive course.
(b) stones at the surface can be pushed around by construction traffic,
(c) their 'harshness' makes them very difficult to compact to an adequate density.
(d) the subgrade tends to penetrate their lower layers. Conversely, the aggregate of the unbound course can penetrate the subgrade, thus rapidly losing the beneficial and essential effects of lateral confinement and high compaction levels. These serious consequences can be avoided by the use of sub-bases (Section 11.1.2) or imported membranes (Section 10.7).
(e) the high intergranular stresses (Section 11.2.2) from compaction and traffic can cause local crushing leading gradually to an increase in fines and an increase in permanent deformation. A number of the aggregate tests in Section 8.6 are designed to detect stones that might be prone to this effect. This issue is discussed further in Section 11.2.2.
(f) if required to carry traffic directly, smaller fine aggregate must be rolled or vibrated into surface interstices to produce a fine-textured finish. If water is not used in this surfacing stage, the process is called *drystone* and the resulting product is *drybound macadam*.

In *waterbound* (or *wetbound*) *macadam* courses, water is added before, during or after surface rolling, depending on local construction practice. Initially the wet, fine aggregate acts like a cementing medium. The material should be used with care over a sub-base or subgrade that is susceptible to moisture, due to the effects of excess construction water. Historically, waterbound macadam was a natural predecessor to asphalt and concrete (Sections 3.6.3–4).

The present-day addition of fine material to produce a more continuous grading than in macadam is used mainly to reduce permeability and to improve construction workability, leading to higher achieved densities and less segregation. It does not automatically increase strength and, indeed, may sometimes cause some strength reduction. *Wetmix* is one such well-graded mix of crushed rock, to which water is added at the crushing/mixing plant in order to further aid compaction and avoid segregation.

The selection of the required thickness of the unbound course is discussed in Section 11.4.6.

11.1.4 Bound courses

A bound course is one in which the particles are positively linked together by glues (or binders) such as cement or bitumen. The course behaves under load as a continuous system able to develop tensile stresses. Bound courses are discussed in some detail in Sections 11.5.1 (Portland cement concrete) and 12.2 (asphalt).

The choice between bound and unbound courses is usually on the basis of cost, construction constraints, and traffic levels. Bound courses will have a higher volumetric cost but will require less material. Hence, their total cost may be lower. Construction constraints generally favour bound courses. For example, bound courses are often the only alternative where:

(a) the water-table is high,
(b) drainage is poor,
(c) a working platform is needed,
(d) pavement thickness or construction times must be minimised (common in urban areas), or
(e) traffic flows are high.

11.1.5 Pavement types

Pavements are classified as either *rigid* or *flexible*, depending on their stiffness relative to that of the subgrade. A *rigid* pavement is not, of course, actually rigid and the terminology is merely an arbitrary attempt to distinguish between pavement types. In particular, it has become common to speak of Portland cement concrete pavements as rigid and of all other (e.g. bound asphalt or unbound) as flexible. The relatively-stiff *rigid pavement* produces a relatively uniform distribution of stress on the subgrade, whereas the flexible pavement deforms and concentrates its subgrade effect over a smaller area (Figure 11.1). The difference between the two pavement types is one of degree rather than of fundamental mechanism and there are many gradations between these modes. Rigid pavements are discussed further in Section 11.5.1 and flexible pavements in Sections 11.3–4.

The two modes may also be mixed together in one application. This is sometimes called a *composite* pavement. For example, flexible sub-bases are frequently used beneath rigid basecourses to control pumping (Section 11.1.2), provide a drainage course (Section 11.1.2), absorb subgrade volume changes (Section 8.4.2) and provide a working platform during construction. Another composite pavement example is the use of an asphalt overlay (Section 12.1.6) over a Portland cement concrete slab.

The illogical term *semi-rigid* is sometimes used to describe a pavement with a bituminous surface course and containing a course that has been cement- or lime-stabilised (Chapter 10).

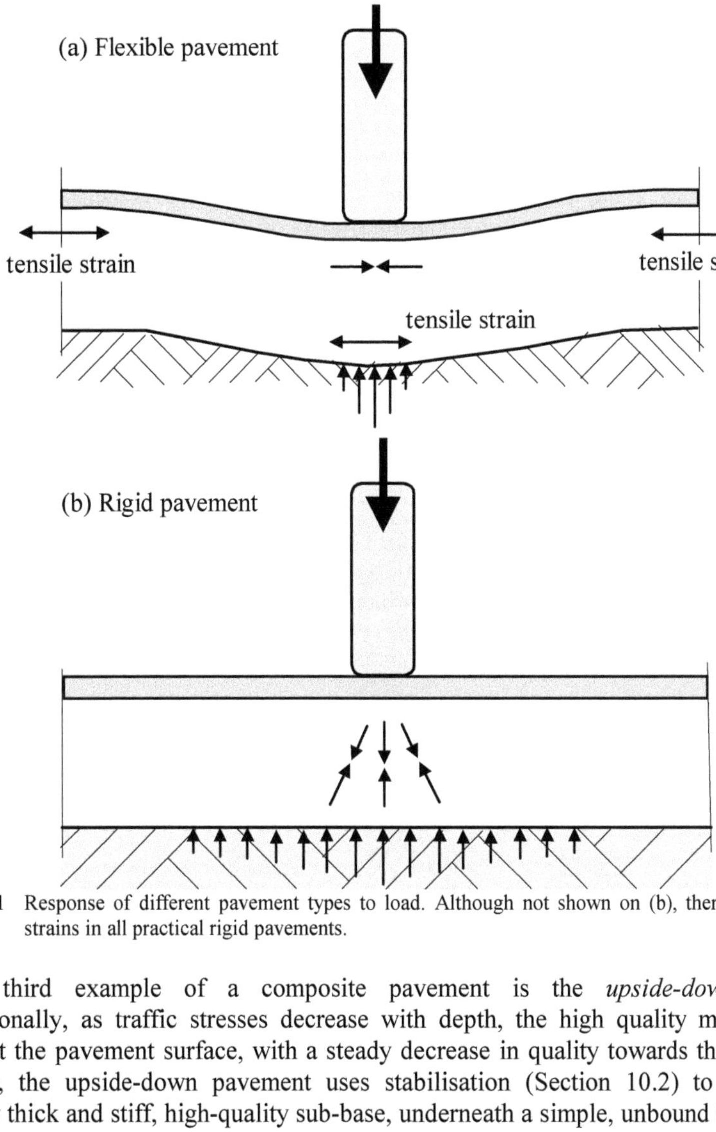

Figure 11.1 Response of different pavement types to load. Although not shown on (b), there are bending strains in all practical rigid pavements.

A third example of a composite pavement is the *upside-down* system. Conventionally, as traffic stresses decrease with depth, the high quality materials are located at the pavement surface, with a steady decrease in quality towards the subgrade. However, the upside-down pavement uses stabilisation (Section 10.2) to produce a relatively thick and stiff, high-quality sub-base, underneath a simple, unbound basecourse. Such a pavement may be used to control cracking, which cannot reflect through the unbound course.

Flexible pavements lend themselves to stage construction and ease of subsequent access to buried services. Rigid pavements come into their own in poor conditions or when the traffic loading is very high. Otherwise, the choice of pavement type and material will be determined by:

* construction constraints,
* material availability,
* local costs, and

* a need to minimise the total cost of the overall pavement system. This requires an estimation of the maintenance costs of the pavement over its design life. Techniques for doing this are given in Section 14.6.

11.2 PAVEMENT BEHAVIOUR UNDER LOAD

11.2.1 Basic functions

The purpose of a pavement is given in Section 11.1.1. To achieve this purpose a pavement must, under peak and cumulated traffic loads,

(a) reduce subgrade stresses to such a level that the subgrade does not deform excessively;

(b) reduce pavement stresses to such a level that the pavement courses do not crack or deform excessively;

(c) protect the pavement structure and the subgrade from the effects of the environment, particularly moisture (Section 9.4) — given the major role that moisture plays in determining pavement behaviour, it is not surprising to find that a significant proportion of pavement design effort is related to this requirement; and

(d) provide an acceptable running surface (Section 12.5.2).

Requirements (a) and (b) are achieved by using the thickness and stiffness of the pavement courses to disperse the concentrated surface load to stress levels acceptable to the various materials encountered (Section 11.4). The stiffer the courses, the lower will be the peak subgrade stresses. Thickness-related dispersion is illustrated in Figure 11.2.

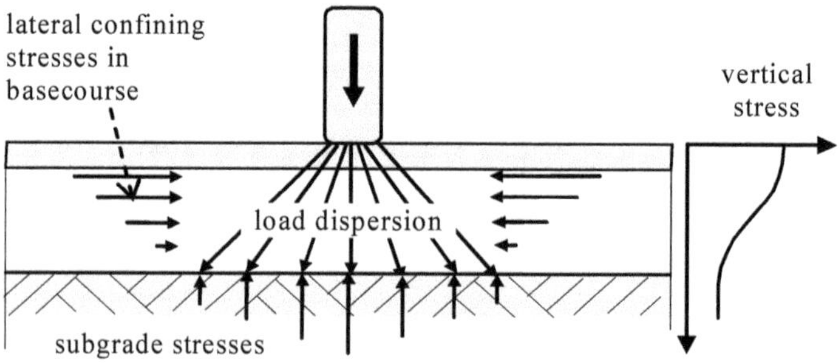

Figure 11.2 Thickness-related load dispersion.

The pavement structure does not only behave as a very stiff bearing-pad, dispersing load through its thickness. Its finite bending-stiffness means that it also acts as a flexible slab operating across a large area of subgrade. This flexing of the pavement course under load means that horizontal bending stresses are produced in each course, with their magnitude and sign depending on their location and the amount and direction of the flexing in the course (Figure 11.1a). The resulting horizontal stresses create cracking in bound courses and horizontal deformation in unbound ones. Requirements (c) and (d) are the subject of Chapters 12 and 13.

11.2.2 Behaviour of real pavements

Before discussing stress analysis (Section 11.3) and pavement design (Sections 11.4–5), it is worth noting seven examples of the actual — rather than assumed — behaviour of a real pavement, less the subsequent development of mathematical techniques based on assumed structural models should get out of perspective.

First, as a result of placement and compaction effects, even the properties of a pavement course composed of some uniform material such as homogeneous sand would not be uniform in each direction — i.e. the uniform material, when placed, would not be isotropic. For instance, there is a marked tendency for compacted sand grains to have their maximum dimensions in the horizontal plane (similar preferred particle orientations occur with clays). Consequently, the horizontal elastic modulus (Section 11.3.1) can be 60 percent of the vertical value, leading to a simple isotropic elastic stress analysis underestimating vertical stresses near the surface and horizontal stresses at depth and overestimating horizontal stresses near the surface. This form of anisotropy, which is associated with a common horizontal stiffness, is called *cross-anisotropy*. Anisotropic pavement structures usually must be analysed by the finite element method of stress analysis.

Second, the stiffness properties of even a homogeneous mass of uniform sand are load-dependent, as a result of improved interlock between loaded sand particles. Consequently, a relatively stiff column of sand forms under an applied load, raising stresses near the load axis. This phenomenon varies with the applied stress level, the number of cycles of load that have been applied, and on whether the current loading is increasing or decreasing. Two consequences are that linear behaviour does not occur (see also Figure 8.8) and Poisson's ratio varies with stress level (Section 11.3.4). On the other hand, for cohesive materials, stiffness tends to drop as shear stresses increase and local shear failures accumulate. Behaviour is thus more plastic than elastic. Methods for allowing for non-elastic behaviour of both granular and cohesive materials are discussed Section 11.3.1.

Third, the properties of the course change with traffic usage. In unbound courses the load-dispersion path illustrated in Figure11.2 is achieved by the load being transferred from one piece of aggregate to another by stone-to-stone (or intergranular) contact over quite small areas. The resulting high, localised stresses can produce cracking or local crushing, which both add to the rearrangement/packing effects noted above, and increase the fines content of the basecourse. This can lead to increases in the short-term in the rate of *permanent deformation* and the plasticity of the material, and in the long-term to rutting (Section 14.1A8) and increased moisture-sensitivity (Section 8.3.2). There is thus a need for basecourse material to be:

(a) well-compacted prior to loading (Section 8.2) to reduce rearrangement effects and increase the number of contact points, and

(b) composed of aggregate able to resist high local stresses under various environmental conditions (Section 8.6).

Fourth, the results will depend on the speed (or rate) at which the loads are applied. For example, the behaviour of bitumen is visco-elastic (Sections 8.7.1&3 and 33.3.3) and so its response to the load caused by a rolling wheel is time-dependent. This means that *creep* becomes significant at low speeds. At high speeds, rapidly-applied loads give no time for particle rearrangement and so many unbound pavements are effectively elastic. Further, the dynamic loads caused by trucks operating on rough roads can be substantial (Section 27.3.5).

Fifth, a wheel rolling along a pavement produces a bow-wave effect as the pavement surface in front and beside the wheel rises to relax the compression under the wheel. Horizontal lateral expansion of the pavement due to the Poisson's ratio effect (Section

33.3.2) also occurs under and away from the wheel, and any permanent deformation which occurs will also cause some permanent horizontal movement away from the wheel-path. This effect is clear for outward movement transverse to the wheel-path but, in the line of the wheel-path, it will tend to be in the rolling direction due to the asymmetry of the loading/unloading curves. This tendency for movement in the rolling direction will usually be overwhelmed by effects due to the horizontal braking, accelerating, and cornering forces at the tyre–pavement interface. Thus, a one-dimensional load soon begins to have three-dimensional effects.

Sixth, tensile stresses can create significant internal discontinuities as many pavement courses have relatively little tensile strength and may already be cracked.

Seventh, Section 25.7.4 discusses the variabilities in actual pavement thickness and should caution designers against excessive analytical accuracy with respect to pavement thickness assumptions or design predictions.

Further aspects of pavement behaviour under load are discussed later in the chapter, as the discussion of pavement properties unfolds.

11.2.3 Theoretical stress analysis

It has just been shown that a real pavement is neither single-layered, elastic, isotropic, homogeneous, nor statically loaded. Nevertheless, the mathematical analysis of stresses and displacements in flexible pavements is founded on the basic stress–strain equations are given as Equations 33.2 and subsequently on Boussinesq's (1885) prediction of the stress effects produced in a semi-infinite (i.e. infinitely deep), elastic, isotropic and homogeneous course loaded on its upper surface by a uniform static pressure, p, over a circular area of radius, r. The results directly under the load, at a depth of z are:

vertical stress $= p(1 - [z^3/\{r^2 + z^2\}^{1.5}])$ (11.1)

horizontal stress $= (p/2)([1 + 2\nu] - [2\{1 + \nu\}z/\sqrt{\{r^2 + z^2\}}] + [z^3/\{r^2 + z^2\}^{1.5}])$ (11.2)

where ν is Poisson's ratio. These equations result in stress distributions of the type illustrated in Figure 11.3. Note that the elastic modulus, E, does not affect either prediction, and that the horizontal stress depends on Poisson's ratio.

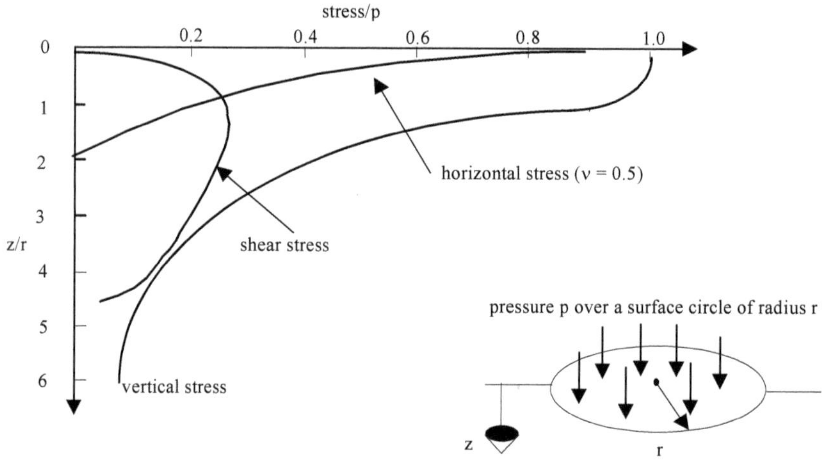

Figure 11.3 Theoretical (Boussinesq) stress distributions for a uniform pressure over a circular area.

The maximum local stress is compression at the loaded surface (p), although this peak value can rise to 1.5p as a result of localised elastic body effects around the load (Timoshenko, 1941, p. 356).

Shear stress governs the failure of ductile materials (Lay, 1982). The maximum shear stress is given by a Mohr's circle analysis as half the difference between the vertical and horizontal normal stresses (Section 33.3.1). In Figure 11.3 it is seen to peak at a distance r below the surface, where r is the radius of the contact area (e.g. a tyre imprint). Actual tyre contact areas and local, peak stresses are discussed in Section 27.7.3.

Closed-form stress predictions for concrete pavements are often based on Westergaard's equations for a slab resting on a Winkler (dense fluid) foundation, although the assumption that no shear transfer occurs between slab and subgrade can make the results questionable. Further developments are discussed in Section 11.5.2.

The surface *deflection*, d, at the centre of the loaded area is given by:

$$d = 2pr(1 - \nu^2)/E \qquad (11.3)$$

Deflections at other surface locations are close to inversely related to their horizontal distance from the load centre. The deflection at a distance R from a concentrated load, P, is:

$$d = P(1 - \nu^2)/\pi ER$$

Burmister produced elastic solutions for the multi-course case in the 1940s, thus greatly extending the Boussinesq solution. Subsequently, tables became available giving theoretical solutions for the stresses, strains and elastic deformations in three-layer, isotropic, elastic systems, assuming full shear and tension bonding between courses — i.e. no sliding or separation at horizontal interfaces (Gerrard and Wardle, 1976). The surface loads considered were (Figure 11.4a) vertical normal, inward shear and uni-directional shear, all applied over circular areas. These loads due to a vehicle wheel are respectively intended to represent the vertical forces, shear forces due to tyre contact effects, and vehicle braking and accelerating forces. The three courses are usually taken to be the surfacing, basecourse, and subgrade.

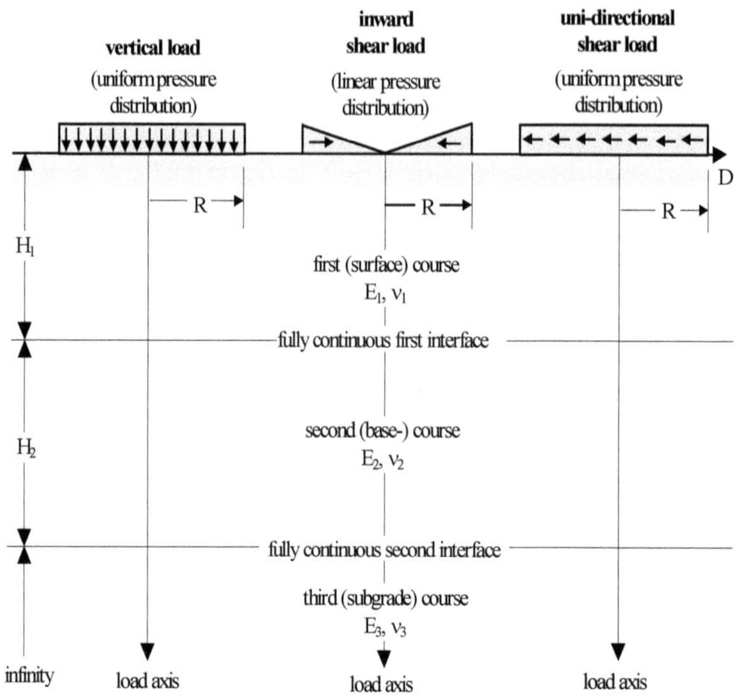

Figure 11.4a Analysis of three-course pavements – load types acting on a three-layer system.

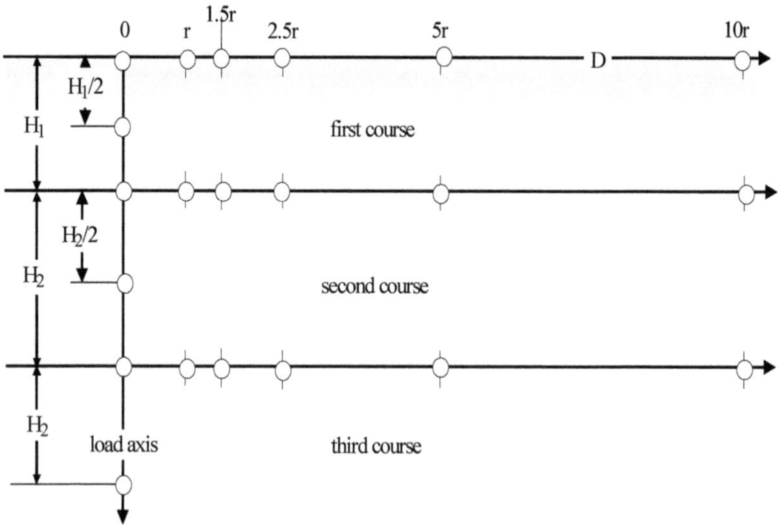

Figure 11.4b Analysis of three-course pavements – showing the points O at which elastic solutions were applied in Gerrard and Wardle (1976). The solutions cover the intercourse elastic modular ratios E_1/E_2 of between 1 and 25.

Solutions available are shown in Figure 11.4b and allowed consideration of most of the primary variables in flexible pavement design. For a three-layer system — it is necessary to know:

E_1/E_2 = surface to basecourse elastic modular ratio,

E_2/E_3 = basecourse to subgrade elastic modular ratio,
ν_1, ν_2, ν_3 = Poisson's ratio for surface, basecourse and subgrade (Section 11.3.4).
H_1, H_2 = surface and basecourse thickness. The depth of the bottom course, H_3, is infinite.
p = load/area is the contact pressure over the loaded area.

A sample set of solutions is given in Table 11.1. The Tables are for sets of *modular ratio* (E_i/E_j), relative course-thicknesses, and Poisson's ratio illustrated below and for three values of H_1/r (0.5, 1.0 & 2.0) where r is the radius of the circular loaded area.

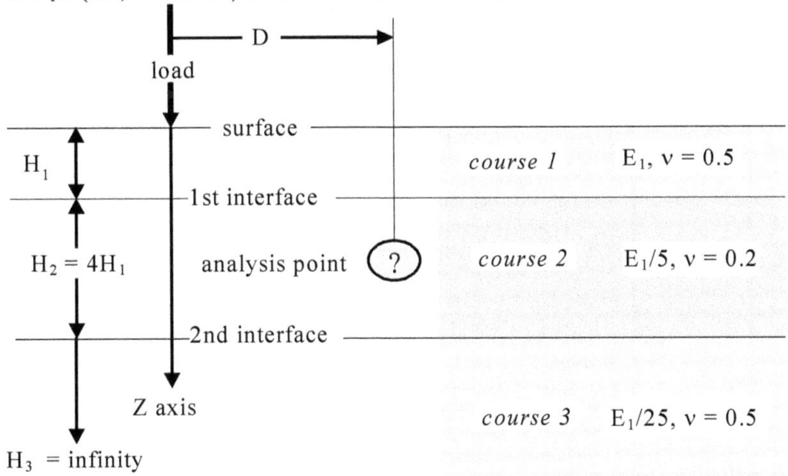

Figure 11.4c Cases illustrated in Table 11.1.

The outputs are:
vertical deflection under the wheel load: = WRp/E_3
vertical stress = pσ_{z1}
horizontal stress at an interface = pσ_{r1}
vertical strain at an interface = pε_{z1}/E_3
radial strain = pε_r/E_3

The subscript b indicates an effect in the bottom course at an interface, otherwise the effect is at the upper course. The deflections are in the same units as R, the stresses are in the same units as p, and p and E_3 are in the same units. Compressive stresses are positive. Note that in Gerrard and Wardle (1976), (Table III), ε_{z1} = EZZ, ε_r = ERR, σ_{z1} = ZZ, σ_{r1} = Rr (RR for bottom course).

As an example of the use of the tables, consider the three-layer pavement shown in Figure. 11.5a. The property values selected are typical of a thin asphalt pavement supported on an unbound basecourse and a clayey subgrade. The basecourse and subgrade stiffnesses used indicate the presence of water. This relatively weak pavement is subjected to a 40 kN wheel load with a circular contact area of 100 mm radius. A single wheel is used to simplify the analysis; normally such a load would be carried on a set of dual tyres (Sections 27.3.2 & 27.7.3). The contact pressure, p, is thus:

$$p = (\text{load})/(\text{area}) = (40 \times 1000)/(\pi[100/1000]^2)\ \text{Pa}$$
$$= 1.27\ \text{MPa}$$

Table 11.1 Sample table of elastic solutions, from Gerrard and Wardle (1976). The sample covers vertical loading, $E_1/E_2 = 5$, $E_2/E_3 = 5$, $\nu_1 = 0.5$, $\nu_2 = 0.2$, $\nu_3 = 0.5$, and $H_2/H_1 = 0.4$.

	Variable	$H_1/r = 0.5$					
Location	Z/H_1	0.00	0.50	1.00	3.00	5.00	9.00
Load axis	W	−0.4714	−0.4797	−0.4725	−0.3775	−0.3384	−0.2421
	σ_{z1}	1.0000	0.9005	0.7203	0.2497	0.0809	0.0380
	σ_{r1b}	2.7147	0.9235	−0.0027	−0.0818	0.0113	0.0043
	ε_{z1}	0	−0.0008	0.0584	0.0565	0.0319	0.0337
	ε_{z1b}	−0.0684					
	D/r	0.00	1.00	1.50	2.50	5.00	10.00
Surface	W	−0.4714	−0.3954	−0.3346	−0.2622	−0.1673	−0.0836
	σ_{r1b}	2.7147	1.2727	0.2001	0.0447	−0.0201	−0.0292
	ε_r	0.0344	0.0102	−0.0031	−0.0033	−0.0023	−0.0013
1st inter–face	W	−0.4725	−0.3988	−0.3403	−0.2653	−0.1680	−0.0835
	σ_{z1}	0.7203	0.3543	0.1021	0.0066	0.0015	0.0002
	σ_{r1}	−0.7415	0.0445	0.4885	0.2359	0.0316	−0.0061
	ε_{z1b}	0.1443	0.0687	0.0167	−0.0004	−0.0001	0.000
	ε_r	−0.0292	0.0012	0.0166	0.0074	0.0005	−0.0003
2nd inter–face	W	−0.3384	−0.3188	−0.2985	−0.2528	−0.1653	−0.0835
	σ_{z1}	0.0809	0.0679	0.0558	0.0342	0.0108	0.0013
	σ_{r1}	−0.1968	−0.1404	−0.0911	−0.0159	0.0248	0.0134
	ε_{z1b}	0.0696	0.0551	0.0420	0.0201	0.0018	−0.0013
	ε_r	−0.0347	−0.0240	−0.0148	−0.0009	0.0057	0.0028
		$H_1/r = 1.0$					
	Z/H_1	0	0.50	1.00	3.00	5.00	9.00
Load axis	W	−0.2902	−0.2932	−0.2785	−0.1972	−0.1739	−0.1217
	σ_{z1}	1.0000	0.7660	0.3937	0.0814	0.0221	0.0099
	σ_{r1b}	1.8295	0.4018	−0.0479	−0.0265	0.0026	0.0009
	ε_{z1}	0	0.0146	0.0467	0.0184	0.0088	0.0090
	ε_{z1b}	−0.0330					
	D/r	0.00	1.00	1.50	2.50	5.00	10.00
Surface	W	−0.2902	−0.2514	−0.2188	−0.1806	−0.1311	−0.0833
	σ_{r1b}	1.8295	0.9919	0.2547	0.0398	0.0121	−0.0033
	ε_r	0.0167	0.0081	0.0023	−0.0019	−0.0008	−0.0005
1st inter–face	W	−0.2785	−0.2485	−0.2232	−0.1840	−0.1327	−0.0836
	σ_{z1}	0.3937	0.2358	0.1244	0.0328	0.0009	0.0004
	σ_{r1}	−0.7764	−0.2352	0.0897	0.1759	0.0573	0.0087
	ε_{z1b}	0.0826	0.0481	0.0241	0.0053	−0.0002	0.0000
	ε_r	−0.0234	−0.0056	0.0046	0.0064	0.0018	0.0001
2nd inter–face	W	−0.1739	−0.1708	−0.1672	−0.1571	−0.1269	−0.0821
	σ_{z1}	0.0221	0.0209	0.0196	0.0162	0.0086	0.0029
	σ_{r1}	−0.0553	−0.0449	−0.0440	−0.0295	−0.0030	0.0098
	ε_{z1b}	0.0195	0.0182	0.0166	0.0128	0.0050	0.0086
	ε_r	−0.0097	−0.0087	−0.0076	−0.0049	0.0000	0.0014
		$H_1/r = 2.0$					
	Z/H_1	0	0.50	1.00	3.00	5.00	9.00
	W	−0.1752	−0.1637	−0.1464	−0.0950	−0.0827	−0.0551

Load axis	σ_{z1}	1.0000	0.5720	0.1510	0.0219	0.0055	0.0023
	σ_{r1b}	1.2699	0.1161	−0.0283	−0.0074	0.0003	−0.0001
	ε_{z1}	0	0.0182	0.0211	0.0050	0.0023	0.0023
	ε_{z1b}	−0.0107					
	D/r	0.00	1.00	1.50	2.50	5.00	10.00
Surface	W	−0.1752	−0.1500	−0.1295	−0.1114	−0.0852	−0.0599
	σ_{r1b}	1.2699	0.7204	0.1733	0.0851	0.0078	0.0047
	ε_r	0.0055	0.0042	0.0029	0.0008	−0.0006	−0.0002
	W	−0.1464	−0.1381	−0.1300	−0.1140	−0.0869	−0.0608
1st	σ_{z1}	0.1510	0.1133	0.0826	0.0397	0.0074	0.0002
inter–	σ_{r1}	−0.3766	−0.2182	−0.0999	0.0271	0.0471	0.0142
face	ε_{z1b}	0.0325	0.0240	0.0171	0.0078	0.0012	−0.0001
	ε_r	−0.0106	−0.0056	−0.0019	0.0017	0.0017	0.0004
	W	−0.0827	−0.0823	−0.0817	−0.0802	−0.0758	−0.0578
2nd	σ_{z1}	0.0055	0.0055	0.0054	0.0051	0.0040	0.0020
inter–	σ_{r1}	−0.0149	−0.0145	−0.0140	−0.0127	−0.0080	−0.0013
face	ε_{z1b}	0.0052	0.0051	0.0050	0.0047	0.0034	0.0014
	ε_r	−0.0026	−0.0025	−0.0024	−0.0022	−0.0013	−0.0001

This pressure is relatively high (Section 27.7.3) and stems from the simplifying assumption of a single tyre, rather than a set of dual tyres. A more common design value is 700 kPa. For these conditions Table 11.1 gives (sourced data in *italics*) the following deflections and stresses on the load axis:

vertical deflection under the wheel load: = WRp/E_3

W = *0.2902*

= 0.2902 x 100 x 1.27/50 = 740 μm

horizontal stresses at the surface = pσ_{r1b}

$\sigma_{r1b(surface)}$ = *1.8295*

= 1.27 x 1.8295 = 2.3 MPa (c)

horizontal stresses above subgrade = pσ_{r1}

$\sigma_{r1(subgrade)}$ = *−0.0553*

= −1.27 x 0.0553 = 70 kPa (t)

vertical subgrade stress = pσ_{z1}

$\sigma_{z1(subgrade)}$ = *0.0221*

= 1.27 x 0.0221 = 28 kPa (c)

The 28 kPa vertical stress indicates an effective dispersion of the original 1.270 MPa contact stress. The subgrade contributes 60 percent of the 740 μm deflection (i.e. [−0.1739]/[−0.2902]).

The results are illustrated in Figure 11.5c and show how, in design, the 740 μm deflection and 70 kPa subgrade tensile stress calculated above can be compared with some criterion surface deflection and subgrade course tensile stress. This process is discussed in more detail in Section 11.4.3. Of course, stress–strain analyses are now usually conducted using one of a number of commonly available computer programs. The best known of these are:

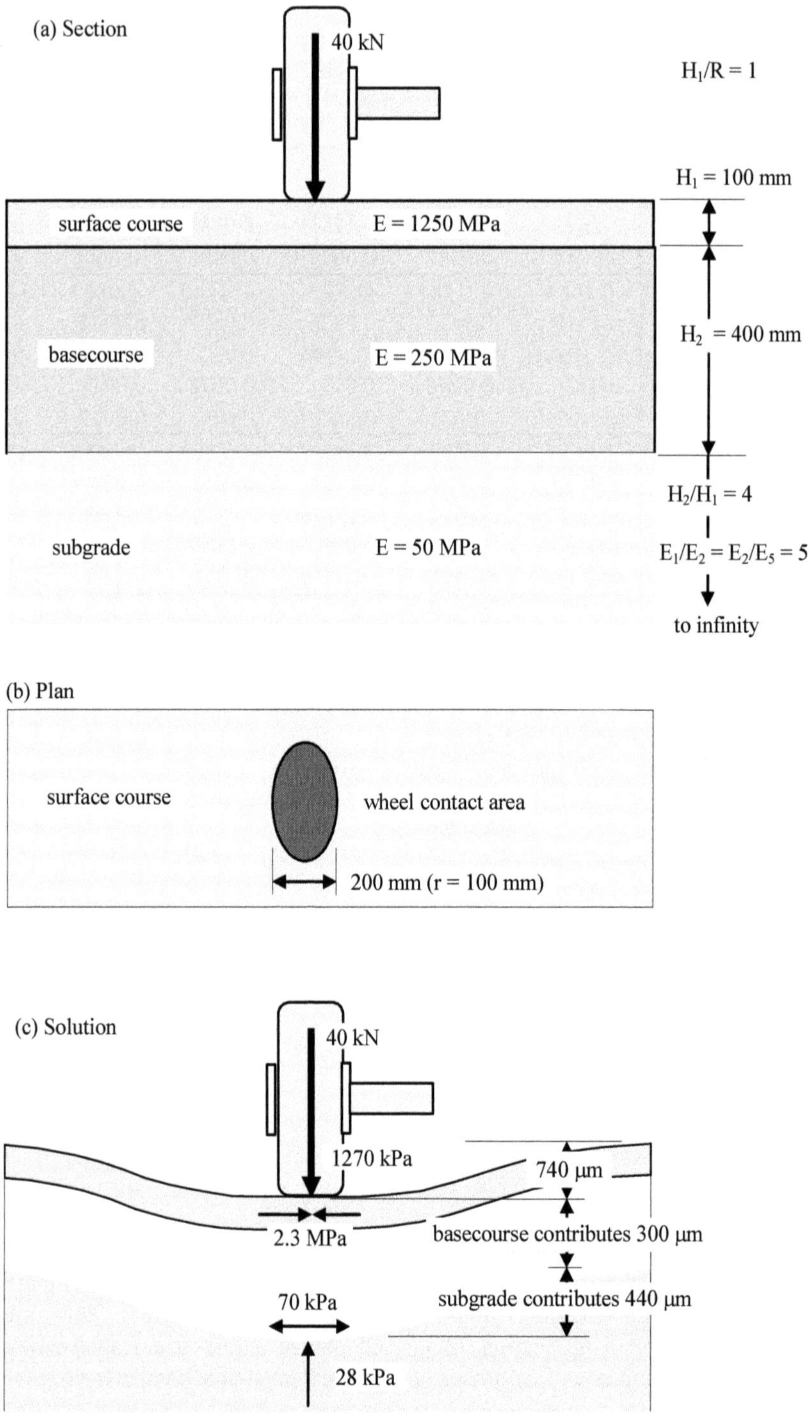

Figure 11.5 Stress analysis example.

* *CIRCLY*, which is an extension of the above tabular solutions. It allows the material to have anisotropic horizontal and vertical properties. CIRCLY 5 was released in 2004 and is supported by Minicad.
* *CHEVRON* & *CHEV5L* (Haas *et al.*, 1994, Croneys, 1991 and Section 11.4.2). This is a five-layer elastic solution developed by the CHEVRON OIL Research Company. A 15 layer version is available.
* *SHELL* [based on *BISAR*, a ten-layer elastic solution developed by Shell] (Shell 1978, Haas *et al.*, 1994, Croneys, 1991, and Section 11.4.2), and
* *ELSYM5* (ITTE, University of California, Berkeley & McTRANS). This is a five-layer elastic solution developed by Ahlborn, and utilising principles developed in CHEVRON. It is similar to CIRCLY but does not handle anisotropy as well.
* *VESYS* (Kenis *et al.*, 1982). This is a five-layer system developed by Kenis in 1978. VESYS is particularly interesting as it accepts linearly visco-elastic material response (Sections 8.7.4 & 33.3.3) and can therefore account for loading speed. It also employs reliability concepts and assumes that material variability contributes to longitudinal roughness (Section 14.4). VESYS 5 is now a well-developed probabilistic and mechanistic model.

11.2.4 Analysis for real conditions

Seven aspects of the behaviour of real pavements that are beyond simple isotropic elastic analyses were discussed in Section 11.2.2. Despite the negative views expressed there with respect to the elastic analyses methods discussed in Section 11.2.3, they form a useful basis for rational pavement design in that they permit general stress distribution patterns to be established and limiting conditions to be defined (Section 11.4.2). However, following the time- and stress-dependent effects and other non-linearities discussed in Section 11.2.2, elastic models and a few tests on material components will not be capable of adequately predicting stresses, strains and deflections over the full range of pavement behaviour.

The designer is therefore faced with the daunting task of handling a non-linear, non-isotropic, time-dependent, environment-sensitive material — often without a set of clearly-defined material strengths and stiffnesses. In such circumstances, it is necessary to use the more elaborate material response models such as VESYS (Section 11.2.3), iterative non-linear *finite element* solutions such as ILLI-PAVE and MICH-PAVE, and more sophisticated material testing. Non-linear finite-element stress analyses based on repeated-load triaxial test data (Section 8.6r) have been successfully used in the laboratory to predict pavement performance (Vuong 2008).

Despite the above misgivings and the trend towards analytical complexity made possible by better computing tools, practical experience does fortunately suggest that designs based on a multi-layer elastic analysis are adequate for many purposes. For example, experience has indicated that materials with a stiffness of over 1.5 GMPa can be taken to be isotropic. An appropriate multi-layer elastic analysis method will be explored in Section 11.4.2 below.

11.3 STIFFNESS PROPERTIES

11.3.1 Representations of E

The preceding section showed that the elastic stiffness modulus, E, was the key material parameter determining the response of a pavement to load. A high E implies good load-spreading ability (Figure 11.2b) and a low E implies that loads will be concentrated on the subgrade and that high flexural strains will occur (Figure 11.2a). The correct determination of E is therefore an important aspect of pavement analysis.

However, the real-life situation is somewhat more complex than implied by the elastic analysis in Section 11.2.3. As shown in Section 11.2.2, roadmaking materials will have a non-linear load-deformation response and so even the estimation of the appropriate values of E will be difficult.

To partly overcome this, the unloading load-deformation curve is used to define stiffness as it is more closely linear (and more reproducible) than the loading curve. Thus, pavement design often uses an elastic modulus known as the *resilient modulus*, M_R, that is defined in terms of the recoverable strain upon unloading after a sequence of repeated loads (Section 11.4.2). Methods for determining the resilient modulus are discussed in Section 11.3.4.

A related subgrade property is the *modulus of subgrade reaction*, which is the applied pressure divided by the deflection and has units of MPa/m. In effect, it is a stiffness modulus integrated over all stressed areas but depends on the shape of the loaded area as well as on the soil stiffness properties.

The associated deflection that occurs when a load is removed from a pavement is called the *rebound* (or *resilient*) *deflection.* In bound materials and in well-placed unbound granular materials, the tendency towards closer packing under cyclic load means that over time the load deflection and the rebound deflection tend to equalise. In this sense, therefore, the rebound deflection and associated resilient modulus are *shakedown* values of the deflection and stiffness modulus. In shakedown the build-up of internal residual stresses means that the total deflection stabilises after a certain number of load passes, rather than continues to increment plastically (Lay, 1982). Thus loads below the shakedown load do not result in incremental failure. This is discussed further in Section 14.3.2.

A typical stress situation in the field is given in Figure 11.6. σ_1 is the applied compressive stress due to traffic, and σ_3 is the consequential orthogonal restraining stress. As a vehicle passes by, σ_1 and σ_3 will rise to their peak loaded values from their base levels, which are usually zero, although σ_3 may sometimes be above zero, even in the unloaded state. In a repeated-load triaxial test, σ_3 is kept as a constant (surrounding) hydrostatic stress and so the vertical stress varies from σ_3 to σ_1. Once the vehicle passes, the stresses will subside to their base levels. The stress-time curves will be approximately sinusoidal. Between passes, there will often be a rest period at the base stress level. This period can significantly influence the response of many materials to repeated loads.

The ratio σ_3/σ_1 is called the *inverse stress ratio*. The difference $(\sigma_1 - \sigma_3)$ is sometimes incorrectly called the *deviator stress*, although from elasticity theory (Section 33.3.2) the real deviator stress is $2(\sigma_1 - \sigma_3)/3$. This slip in terminology then leads to the common definition of the *resilient modulus* obtained from the repeated-load triaxial test as:

$$M_R = \text{(modified, repeated deviator-stress)/(resilient strain)}$$
$$= (\sigma_1 - \sigma_{3u})/(\varepsilon_1 - \varepsilon_{3u})$$

where the primary stress and strain are in the same '1' direction, and u denotes unloaded rather than initial values (i.e. the strain path is ε_3, $\varepsilon_3 + \varepsilon_1$, ε_{3u}). Note that some definitions use the axial strain, ε_1, rather than the deviator strain, $\varepsilon_1 - \varepsilon_{3u}$. Similarly, M_R, is somewhat different to E which is conventionally taken as σ_1/ε_1, although the two are identical for a triaxial test on an elastic material.

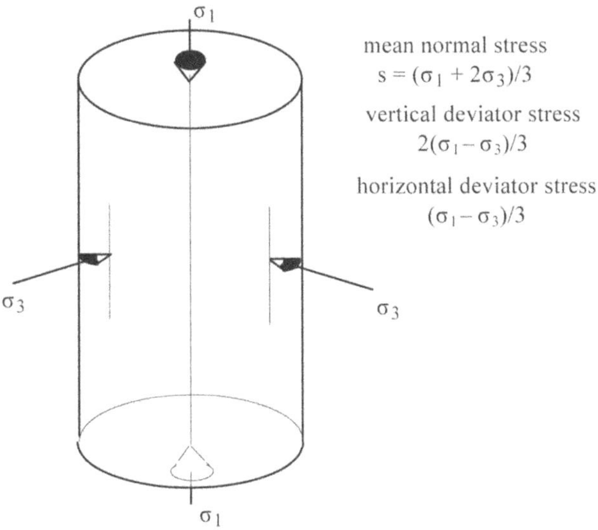

Figure 11.6 Stress definitions. The stresses are assumed to be effective rather than total (Section 9.2.5). For this stress case, the deviator stress also equals the basic J_2 stress invariant (also called the octahedral shear stress. Lay (1982)).

The use of the deviator stress stems from the fact that the shear stress is given by $(\sigma_1 - \sigma_3)/2$ and the failure of cohesive material is related directly to shear stress levels (Section 8.4.3). The modulus obtained in this way (AASHTO T274) forms the basis of the subgrade aspects of the 1986 AASHTO pavement design method (Section 11.4.6). The 1993 method weighted the measured modulus to account for seasonal effects.

As the stress–strain response is nonlinear, M_R will depend on the stress level. A good review is given in Vuong 2008. In strain-softening materials such as most soils, increasing stress will lower the stiffness measured in the stress direction (σ_1). On the other hand, increasing the restraining stress (σ_3) will raise the stiffness. In dry cohesive materials negative stresses due to soil suction (Section 9.2.4) will also raise stiffness.

Typical empirical relationships used for estimating the variation of M_R with stress for *cohesive materials* are of the form (Lay, 1993):

$$M_R/M_{Ri} = [1 - (\sigma_1 - \sigma_3)/(\sigma_1 - \sigma_3)_{failure}]^2 \quad (11.4a)$$

or $$= 1 - [(\sigma_1 - \sigma_3)/(\sigma_1 - \sigma_3)_{failure}]^2 \quad (11.4b)$$

or $$= 1 - c_1[\sigma_1/\sigma_3] \quad (11.4c)$$

or $$= 1 - c_2\log[\sigma_1/\sigma_3] \quad (11.4d)$$

or $$= c_3(\sigma_1 - \sigma_3)^{-n} \quad (11.4e)$$

or $$= 1/[1 + c_4(\sigma_1 - \sigma_3)^{-n}] \quad (11.4f)$$

where c is a constant which can be estimated from triaxial test data (Section 33.3.1) and M_{Ri} is the value of M_R at zero applied stress. Equations 11.4a&b calibrate against $M_R = M_{Ri}$ at $\sigma_1 - \sigma_3 = 0$ and $M_R = 0$ at $\sigma_1 - \sigma_3 = (\sigma_1 - \sigma_3)_{failure}$. Note that these equations assume isotropic behaviour. A typical n in Equation 11.4e for a cohesive material would be 0.3.

Non-cohesive materials — such as unbound granular sand and coarse aggregate (Section 11.1.3) — rely generally for their stiffness on the integrity of the whole course and, particularly, on the particle interlock resulting from compaction and confinement (Section 11.2.2). In stress terms, that integrity and interlock is provided by the confining effect of the *mean normal stress*, s (Section 33.3.1), where (Figure 11.6):

$$s = (\sigma_1 + 2\sigma_3)/3$$

An increase in s will clearly raise the values of M_R and E. In dry conditions s can be favourably increased by the effect of soil suction (Section 9.2.4). Repeated-load triaxial tests indicate that the modulus can be represented by:

$$M_R = ks^m \qquad (11.5)$$

where k and m are 'constants' which depend on the number of load repetitions (and on the units used, as s^m will have 'unique' units). The constants are commonly obtained from repeated-load triaxial tests. The range $0.275 < m < 0.325$ includes many common mixes made from angular, crushed rock. m can rise to 0.9 for a well-compacted mix of good quality, angular aggregate. A negative m would indicate a *strain-softening* material (Section 11.3.3) which was probably cohesive. Values of k and m for a typical crushed rock are shown in Figure 11.7. The stiffening under trafficking indicated by the increase in k with load repetitions is consistent with Section 11.2.2 and is a result of increased granular interlock.

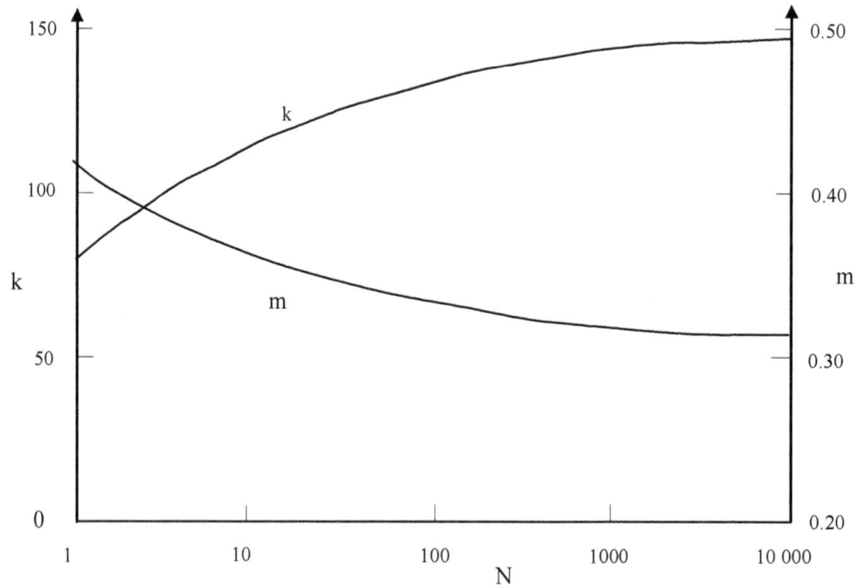

Figure 11.7 Variation in stiffness parameters in Equation 11.5 with load repetitions, N, for crushed dolerite with a clay binder. s is in kPa and E in MPa (Barrett and Smith, 1976). See Kennedy (1985) for values for other materials.

Similar equations have been reported by other workers, e.g. Richards (1980) and Heaton and Bullen (1980) for blast-furnace slag. As Equation 11.5 is sometimes written:

$$M_R = K_1\sigma^{K_2}$$

where σ (= $3s = \sigma_1 + 2\sigma_3$) and K_n are constants, the relationship is often referred to as the *K-σ* or *K-θ model*. Some analytical procedures only consider conditions after 1000 load cycles, and call the earlier cycles 'conditioning' cycles. Note that k and m in Figure 11.7 do tend to stabilise after 1000 cycles.

Equation 11.5 may be written more generally as *Janbu's equation*:

$$E = E_0(s/s_0)^m \quad (11.6)$$

where E_0 is the conventional elastic modulus at some basic level of confining stress, s_0. To avoid using s_0, Equation 11.6 can be generalised to:

$$E = E_i(1 + gs^m)$$

where g is a constant. However, this equation cannot be integrated to give a stress–strain equation (Section 11.3.2).

From the above analyses for cohesive and granular materials (e.g. Equations 11.4b & 6), useful general expressions for E covering the effects of both mean normal and deviator stress are seen to be of the form:

$$E/E_i = f[s^m,(\sigma_1 - \sigma_3)^n]$$

or

$$= [1 + gs^m][1 - c(\sigma_1 - \sigma_3)^n] \quad (11.7)$$

The form using Equation 11.4e is sometimes called *Uzan's equation*:

$$E/E_i = s^m (\sigma_1 - \sigma_3)^n$$

Alternative general equations use Equations 11.4c or d and thus the stress ratio σ_1/σ_3 rather than $(\sigma_1 - \sigma_3)$. However, analogous to the run-out situation in metal fatigue, there are 'shakedown' stress levels below which no amount of cyclic loading will effect the material response.

The stress-dependence of E reflected in these equations will cause E to vary continuously with depth. This analytical complication can be avoided by applying the elastic model to a series of sub-courses of the pavement such that the modulus change between sub-courses is insignificant.

11.3.2 Asphalt stiffness

Asphalt stiffness from the viewpoint of mix design is discussed in Section 12.2.4. Typical stiffness moduli for asphaltic pavements are between 500 MPa and 10 GPa (Wallace and Monismith, 1980). These stiffnesses have two major components:

(a) *friction and interlocking* of the aggregate pieces, as in unbound courses (Section 11.1.3). These effects depend on aggregate contact pressures transmitted through the skeletal structure formed by the aggregate particles. This in turn depends on aggregate type, size, shape, grading, and roughness but is not dependent on loading rate or temperature. Friction and interlocking is the more dominant mode in asphaltic concrete (AC, Section 12.2.2); and

(b) *the stiffness and cohesion of the matrix* of bitumen and fine filler.

These properties increase, and hence asphalt stiffness *increases*:

* in proportion to the bitumen stiffness to a power of about 0.4 (bitumen stiffnesses vary between 1 and 100 MPa)
* with bitumen viscosity (Section 8.7.4),
* as the level of compaction increases (Section 12.2.3), and
* with loading rate.

and asphalt stiffness *decreases*:

* dramatically with temperature, as Figure 11.8 shows for asphaltic concrete. To a first approximation, the stiffness is estimated by:

  $E = a - b.\log T$, or

$\log E = c - dT^{1.7}$

where a to d are positive constants, typically about 6, 2, 4.2 & 0.0023 for E in MPa and temperature T in C.

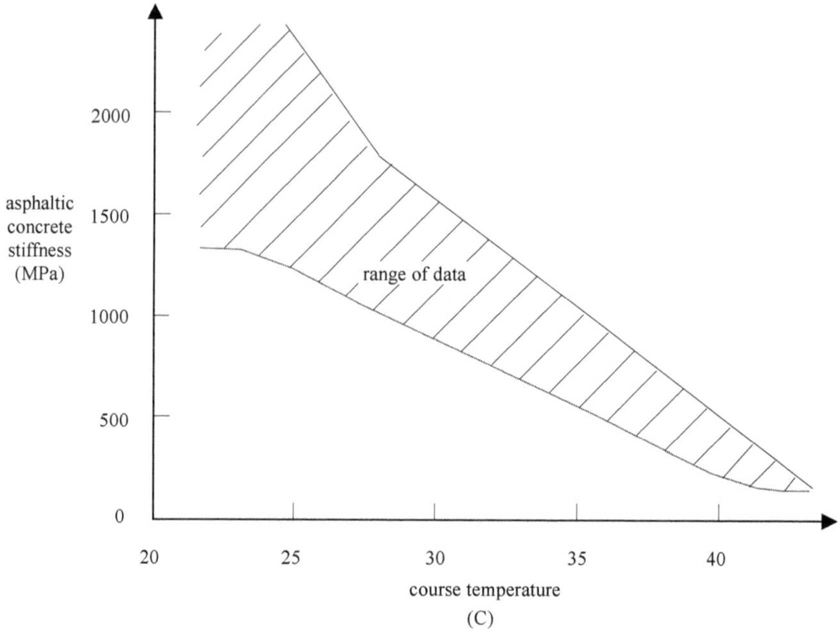

Figure 11.8 Effect of temperature on the stiffness modulus of asphaltic concrete, after Gray and Baran (1979).

* with the air void content of the mix (Section 12.2.3), as Figure 11.9 illustrates, also for asphaltic concrete, and
* as the pavement cracks in fatigue under repeated loading (Section 11.4.4).

Stiffness and cohesion of the matrix is the more dominant mode in hot rolled asphalt (HRA, Section 12.2.2).

Historically, asphalt stiffness was 'measured' in terms of its Marshall Stability (Section 12.2.4), although the newer mechanistic methods (Section 11.4.2) use the E values discussed in this section and determined by either the indirect tension test or the repaeated load triaxial tests (tests q and r in Section 8.6).

For in-situ asphalt the modulus is dramatically reduced by high temperatures and by the amount of cracking in the asphalt. For example, at 30 C or 15 m of cracks per m^2, few asphalts will have a stiffness of over 1 GPa.

11.3.3 Stress–strain implications

If it is assumed that E is a tangent ($d\sigma/d\varepsilon$) rather than a secant (σ/ε) value, the equations in Section 11.3.1 can be integrated to produce stress–strain curves. For instance, integrating Equation 11.4a gives:

$$c_1(\sigma_1 - \sigma_3)/(\sigma_1 - \sigma_3)_{failure} = 1 - [(\sigma_1 - \sigma_3)/\{E_i(\varepsilon_1 - \varepsilon_3)\}]$$

or, more simply,

$$\sigma/\sigma_{failure} = 1 - (\sigma/E_i\varepsilon)$$

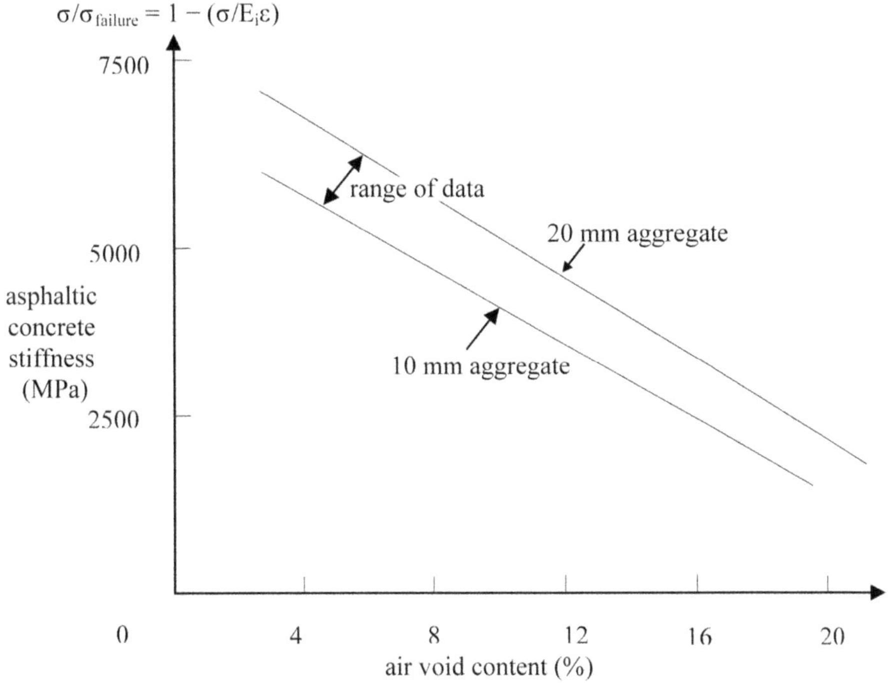

Figure 11.9 Influence of air void content, as measured by the air void ratio, on stiffness of asphaltic concrete, after Mackenzie and Fletcher (1980).

which is the widely used hyperbolic equation. It is termed *hyperbolic* because it comes from the 'reasonable' mathematical assumption that the stress–strain curve for a plastic material will be of the form first seen in Equations 11.4a&b, i.e.

$$\varepsilon = (\sigma/E_i)/(1 - \sigma/\sigma_{failure}) \quad (11.8)$$

Integrating Equation 11.6 gives an equivalent stress–strain curve of the form:

$$\sigma = E_q\varepsilon^{1/(1 - m)} \quad (11.9)$$

where E_q is a constant. A key difference between the two equations is that:

* Equation 11.8 is used for plastic, cohesive materials. It represents a *strain-softening* response (Lay, 1993 & Figure 11.10) in which the strain increment for a given stress application increases with the number of stress applications, whereas
* Equation 11.9 is used for angular, granular materials with a typical $1/(1 - m) = 1.5$ from Figure 11.7. It represents a *strain-hardening* response, which is a consequence of the granular interlock that occurs. Strain-hardening is one reason why in pavements granular materials are greatly preferred, compared to the plastic cohesive materials. It can be destroyed by the powdering of granular aggregate at inter-particle contact points (Section 11.1.3). It is strain-hardening that allows shakedown (Section 11.3.1) to occur.

The equations include the effect of overload and plastic strains and so they can be combined with Equation 33.6 to give the total *deformation,* Δ, as:

(total elastic and plastic Δ) =	(Δ based on E)	+	(creep Δ)
	e.g. on Equation 11.7		e.g. on Equation 33.6

A particular case where the plastic strain is linked directly to the number of load applications at a constant stress will be discussed later in Equation 14.1. Section 11.2.2

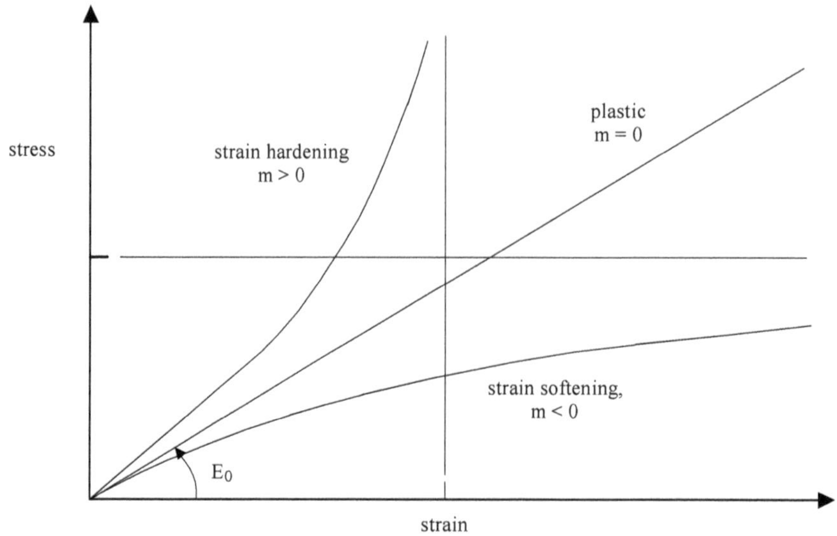

Figure 11.10 Forms of stress–strain response (from Lay, 1993).

showed that the stiffness modulus will also depend on the number and rate of prior load applications, increasing as the rate increases.

11.3.4 The determination of E

In the laboratory, the elastic stiffness of a material, E, and particularly its *resilient modulus*, M_R, can be determined from a repeated-load triaxial test (Section 8.6r), an indirect tension test (Section 8.6q), a shear test (Section 8.4.3) or a bending test (Section 12.2.4C). Standards for these tests usually specify a variety of restraining stresses, σ_3, somewhat in excess of those that might be encountered in the field. To ensure that testing to establish the resilient modulus will reproduce real conditions, the sample preconditioning and loading/unloading sequence should be associated with the expected loading history.

In the field, the modulus can also be determined from plate-bearing tests (Section 8.4.4–5), particularly CBR tests, and from deflection and stiffness tests (see below). The determination of E and Poisons ratio, ν, from field data is not easy, with each having a coefficient of variation of about 35 percent.

In addition to the dependence of the modulus on the number and rate of prior load applications, Chapters 8 and 9 suggest that it will also depend on material type, moisture content, density, soil suction, and grading.

Section 8.2 demonstrated how stiffness will generally drop as moisture content rises. However, if a saturated condition exists in some confined layer well below the surface, and if the water is also confined, it may happen that the incompressibility of water may lead to compressive stiffness increasing as moisture contents rise. This has relevance in back-calculation (see the following text). For non-cohesive materials, the modulus is usually unaffected by moisture contents below 85 percent saturation (Equation 8.2h).

Stiffness falls away from some peak value located on the dry density/moisture content graph (Figure 8.3) below the line of optimums and between the standard and modified maximum dry densities. Below OMC, E is often found to increase linearly with dry density.

Soil suction and grading will have more influence on subgrades than on basecourses. Finally, subgrade E values are temperature-dependent in cold climates, rising as freezing occurs (Section 9.5) and falling with the thaw to a value well below the normal level.

Some typical E values are given in Table 11.2 below.

Table 11.2 Typical E values for pavement components

	E in MPa	
Material	Average (MPa)	Range (± MPa)
clay or silt subgrade	100	± 60
sand (Shackel 1973)	150	−100/+800
unbound aggregate sub-base	350	± 200
unbound aggregate basecourse	350(c)/1(t)	± 100
minimum acceptable design value	50 (upper course) to 100 (lower course)	
cement-stabilised basecourse	see Table 10.1	
asphalt	see Section 11.3.2	
cement concrete	see Section 8.9.2	
steel	200 000	± 0

A common design approach is to estimate E for subgrades from measured subgrade CBRs (Section 11.4.6). Given the fact that the CBR test result does depend partly on the deformation of the sample, it is not surprising that some relationship exists between the two. The discussion in Section 8.4.4 and Uzan *et al.* (1980) suggest that E in MPa is somewhere between 10 and 16 times the CBR in percent. U.K. practice is to take E as 18 times CBR to the power of 0.64 (Powell *et al.*, 1984). This suggests subgrade E's of between 10 MPa for clays and 300 MPa for rock. For basecourses, E in MPa approximately equals the CBR in percent (Clegg, 1983).

Deflection testing is described in Section 14.5.2. A *deflection bowl* resulting from such a test is defined in Section 14.5.2 and shown in Figure 14.9 and in the inset to Figure 8.12. An analysis of the bowl shape can be used to estimate in situ E values. First, define d_0 and d_m as the deflections at the point under consideration when the load is respectively at the point and m mm away from it. The ratio d_m/d_0 is linked to the curvature of the loaded pavement surface and depends almost entirely on the modular ratio E_l/E_s, where E_l is the surface-course elastic stiffness and E_s the subgrade stiffness for practical multi-layer pavements (Scala, 1979). Table 11.3 gives some actual stiffness estimates based on m = 250 and thus on d_{250}/d_0. Design values for unbound sealed pavements typically ranged from $d_0 - d_{200} = 300$ μm at 10^5 load applications (N_{esa}, see Sections 11.4.1 & 27.3.2) to 100 μm at 10^8 load applications.

The following equation has been used for the estimation of E in MPa for a two-layer pavement (Scala, 1978), with r and d in mm:

$$E_l = 185([d_{250}/d_0] - 1)^{-1.75} \text{ MPa} \qquad (11.10)$$

More recently, design methods have been based on d_{200} rather than d_{250} and on $d_0 - d_{200}$ rather than d_{250}/d_0 as a bowl curvature measure. In particular, it was found that $d_0 - d_{200}$ better correlated with the tensile strains at the base of the asphalt layer that led to fatigue (see Section 11.2.3).

Table 11.3 Deflection bowl stiffness estimates, from Scala (1979). The evidence to support these values is not extensive.

d_{250}/d_0 (single tyre)	d_{250}/d_0 (dual tyres)	modular ratio, E_1/E_s	Notes
0.6	0.50	1	light, unbound pavement
0.7	0.60	2	
0.8	0.75	5	
0.8	0.80	10	
	0.85	20	heavy, rigid pavement

Following the observation that surface deflections well away from the load point depend largely on subgrade stiffness, it has also been tentatively suggested that:

$$\log E_s = -([\log d_{900} - 1.166691]/0.92497)$$

or $$E_s = 11.4/d_{900}$$

Alternatively, d_0, d_{600} and d_{900} can also be used with Figure 8.12 to obtain CBR. Knowing CBR makes it possible to then use one of the E vs CBR relationships at the beginning of this Section to estimate E_s.

A related and increasingly popular method for determining E, and other material response factors such as Poisson's ratio, ν, is by *back-calculation*. In this method, measurements are made of pavement deflection response under known loads. Analytical stress–strain models such as CIRCLY and ELSYM (Section 11.2.3) are used to provide a set of deflection equations that can be solved for all the E and ν in the pavement, usually by a lengthy iterative process. An n-layer pavement will require (n + 1) sets (Figure 11.4a) and the total number of unknowns will depend on the degree of anisotropy in each course (Sections 11.2.2 & 11.2.3). To reduce computing effort, many methods assume values for ν (see below).

One such back-calculation program is Texas Transportation Institute's MODULUS, which uses BISAR and WESLEA (or BISDEF & WESLEF) for its stress analysis (Section 11.2.3). In operation MODULUS uses a database of precalculated deflection bowls that it matches with the actual deflection bowl. Another method is Cornell's MODCOMP which uses CHEVRON. Two methods associated with the Falling Weight Deflectometer (FWD, Section 14.5.2) are ELMOD (Ullidtz, 1987) and ARRB's CIRCLY-based EFROMD. ELMOD is relatively rapid. The main problems with back-calculation are that:

* the deflections must usually be known to a high degree of accuracy,
* the approximations made may lead to non-unique solutions (e.g. a number of layer combinations can give the same surface deflection),
* the results may depend on the seed values used for the iterations,
* the results can seriously mislead a user who does not cross-check their reasonableness (e.g. with respect to data 'outliers'),
* the method does not work:
  # for thin courses,
  # when adjacent courses have very similar or very different moduli,
  # when the pavement response is not elastic (Section 11.2.4), and
  # when the shear bonding between courses is indeterminate.

However, proven back-calculation methods give estimates of E far more quickly and cheaply than do laboratory tests.

Even FWD estimates of E have a coefficient of variation of about 30 percent for basecourses and 10 percent for subgrades. A major advantage of the FWD is that it estimates E at strain levels likely to be encountered in service — an important factor given the variability of E with strain. The FWD can also be used to determine dynamic moduli (Section 14.5.2).

E can also be estimated in the laboratory from triaxial tests (test 8.6(r)).

Poisson's ratio, ν, can be quite variable and dependent on the stress ratio (σ_1/σ_3). Theoretically, it must lie between 0 (no lateral deformation under normal stress) and 0.5 (no volume change under stress). Typical ranges are 0.10 to 0.20 for stabilised soil, 0.2 for concrete, 0.25 to 0.35 for AC (the higher the strength the lower the ratio), 0.30 to 0.40 for unbound material, 0.35 to 0.45 for silt, and 0.40 to 0.45 for clay. Values in excess of 0.5 are measured in some unbound courses in a phenomenon known as elastic dilation; it is a result of the aggregate particles moving out, under shear, from a close-packed arrangement (Lay, 1982). Some values assumed in design are given in Section 11.2.3.

E can also be determined non-destructively and dynamically using *seismic wave propagation* (or *spectral analysis of surface waves*) methods (Potter, 1977). The pavement surface is subjected to vertical vibrations of a known frequency. The energy is mainly transferred from the vibrator by *surface* (or *stress* or *Rayleigh*) *waves*. The speed of the waves radiating horizontally is determined for a number of different frequencies. By comparing these measurements with theoretical predictions, the elastic moduli for the various pavement courses can be deduced. Use of the method requires a knowledge of the number and thickness of any pavement courses present. The values calculated represent averages over about a metre of pavement. The test has an accuracy of about ± 30 percent.

In the Shell Method (Section 11.2.3) it is assumed that E for an unbound granular basecourse of thickness T mm can be predicted from:

$$E_{basecourse} = 0.20E_{subgrade}T^{0.45}$$

This relationship was empirically based on deflection measurements and therefore relates to a horizontal modulus (Section 14.3.1). It is difficult to see the rationale behind the approach, except that the role of the subgrade is diminished as the basecourse becomes thicker. Note also that the vertical stiffness modulus may be up to three times higher than the horizontal one (Section 11.2.2).

11.4 PAVEMENT DESIGN

In addition to the theoretical constraints mentioned in earlier sections, pavement design deals with a relatively thin, imperfect structure, fabricated from cheap local materials with difficult quality control, subjected to a wide variety of conditions, and often designed with no margin of safety. With these cautions, the various steps involved in the design process will now be examined.

The behaviour of each pavement type under traffic loads will depend on the properties of each of its components. As the subgrade properties and drainage conditions will vary along the road in a manner beyond the control of the designer, the road will need to be divided into design lengths within which the subgrade and drainage conditions are essentially constant. As in any design procedure, the next step is for the designer to select a tentative (or trial) design and check its performance in the given conditions. In the case of a pavement, this means selecting how many pavement courses will be used and deciding on their thickness and composition. If this selection is shown by analysis to be unsatisfactory, a revised design must be selected. If it proves satisfactory, further design trials are still required to ensure that a minimum-cost solution has been obtained. These whole-of-life cost-minimisation aspects of pavement design are discussed in Section 14.6.2, after pavement performance data has been described.

11.4.1 Loading

A key next step in pavement design is to determine the design loadings, which come largely from the *heavy-vehicle* wheel loads discussed in Section 27.3.4. The stresses and strains produced by a wheel were discussed in Section 11.2.3. Some allowance must be made for the transverse distribution of a series of wheel loads passing longitudinally down the road. The distribution will probably be normal with a standard deviation of about 300 mm, but will be site specific (e.g. wheel tracking on curves).

As the wheels are invariably attached to axles, pavement life studies are usually in terms of axle loads and their magnitudes and configurations. These are so diverse that a particular loaded configuration is commonly characterised by its number of *equivalent standard axles* (*ESA*s or *ESAL*s, Section 27.3.2&4), N_{esa}. This is defined as the number of passes (or repetitions) of the standard axle load for the local jurisdiction which, when carried on an axle with pair of dual tyre sets (Figure 27.2), would cause the same pavement damage as a single pass of the axle configuration in question. The method brings all vehicle loading configurations to a common denominator. Methods for calculating N_{esa} for a particular axle configuration are discussed in Section 27.3.4. So the first stage of design is to assess N_{esa} from the estimated traffic and freight flows.

The next stage of the design requires the designer to select an intended life, or *design life* (e.g. 20 years) for the pavement, a process discussed in Section 14.6. An estimate of the traffic and freight carried over that design life, will lead to the prediction that, over the design life, there will be n_c passes of each vehicle configuration c with its ESA of N_{esac}. A method for doing this is given in Section 31.5. Thus the design 'load', N_{esad}, is:

$$N_{esad} = \sum_{\text{for all configurations, c}} (n_c N_{esac})$$

In structural engineering terms, pavement design is therefore predominantly a form of fatigue design.

Alternatively, N_{esad} may be specified directly as a key design variable (e.g. $N_{esad} = 10^8$), as in many national approaches (e.g. Powell *et al.*, 1984). Attention must also be given to:

(a) occasional overloads (for $N_{esad} < 1000$, design will usually be governed by the magnitude of the individual wheel overload, rather than by the size of N_{esad}),
(b) turning, cornering, braking, and accelerating loads, which produce horizontal loads (particularly at points of heavy braking and accelerating, as at bus stops),
(c) areas where heavy vehicles congregate or where their wheels all follow the same paths (e.g. as occurs in climbing lanes for heavy vehicles, Section 18.4.4), and
(d) environmental loads such as those due to volumetric changes in the surrounding soil.

11.4.2 Mechanistic design

The design methods alluded to in Section 11.2.3 are called *mechanistic* as they are based on the methods of mechanics. Alternative terms are 'analytical', 'theoretical' or 'rational'. In a mechanistic approach to pavement design the next step, after design loads and the design life have been established (Section 11.4.1), is to use techniques such as those described in Section 11.2.3 to calculate the stresses and strains caused when these load repetitions are applied to the selected pavement. It is then necessary to use this data to predict the condition of the pavement at the end of the design life and to assess whether this condition is acceptable. This is best tackled by first establishing a set of defined *serviceability* and/or *failure criteria*. This is a difficult task that will be discussed further in Section 14.3.1. If the assessed condition is unacceptable, the pavement design is then revised and a further iteration of the above cycle is conducted.

From the viewpoint of stress analysis, and following Figure 11.5, the following intermediate acceptability criteria provide relatively simple checks of pavement condition by considering:

(a) the passage of traffic causing permanent deformations in the wheel tracks. Note that the shakedown load (Section 11.3.1) may be very small as the typical non-linear stress–strain response of many basecourses and subgrades (Section 11.3.3) means that even low stresses may result in some permanent deformation during each passage of a loaded vehicle. For example, silty soils can be very stiff but still rut. The acceptability criteria are limiting values of:
   (a1) the accumulation of permanent vertical compressive strains in the pavement,
   (a2) the accumulation of permanent vertical compressive strains in the subgrade,
   (a3) the permanent deformation (rutting, Section 14.1A8) caused by the sum of (a1) and (a2).

(b) a single vehicle permanently dishing the portions of pavement directly supporting its heavily-loaded wheels. The criteria are limiting values of:
   (b1) the vertical compressive stress in the subgrade,
   (b2) the vertical compressive stress in the basecourses, and
   (b3) rutting due to the sum of (a1) and (a2).

(c) a single vehicle causing flexural plate failure of the basecourses directly beneath its heavily-loaded wheels. The criteria are:
   (c1) a limiting value of the tensile horizontal bending stress in the basecourses, to prevent flexural cracking of bound courses, and
   (c2) the peak value of the compressive horizontal bending stress in the basecourses, to prevent flexural crushing. This criterion is usually supplanted by criterion (c1).

(d) the passage of traffic causing fatigue failure due to accumulating flexural tensile stresses in the bound basecourses, managed by:
   (d1) a limiting value of the accumulation of permanent horizontal tensile strains in the basecourses, predominantly at the bottom of the basecourse.
   (d2) fatigue in old asphalt can also progress from the surface down.
   (d3) for spray and chip seals (Section 12.1.2) fatigue due to tensile strains may occur in the overlying bound surface course rather than the basecourse. This form of fatigue is hastened by bitumen ageing (Section 8.7.7).

(e) the effects of moisture, shrinkage, creep, thermally-induced strains and frost heave on pavement and subgrade. These can lead to pavement failure, even in the absence of any significant traffic loads. They must be managed on a broader basis using techniques discussed in Chapters 8 and 9.

The pavement distress types associated with these criteria are described in Section 14.1 and the particular limiting criteria in common use in pavement design are discussed in Sections 11.4.3&4.

A common design approach is to select a pavement arrangement based on criteria (e) and pavement thicknesses based on criterion (b), followed sequentially by checks on criteria (a), (c1), and (d1). These guide the selection of the basecourse thickness and material properties prior to the design check of the entire pavement.

Some conflict may arise in satisfying all the criteria simultaneously. For example, in already-thick courses, it may not be possible to control rutting by further increasing the course thickness as there may be sufficient deformation occurring within the course. As an alternative, it may be necessary to use stiffer materials. However, a higher stiffness modulus may decrease fatigue life.

Modern pavements frequently employ a stabilised sub-base (Section 11.1.2). In this case, the fatigue design can proceed in two time phases. The first utilises the sub-base as a flexural layer and continues until that layer cracks in fatigue. The second phase only utilises the flexural properties of the basecourse and surface course.

To avoid the need for elaborate testing programs for each design, a number of semi-empirical limiting criteria have been developed, based on particular pavement failure modes. These are described below.

11.4.3 Failure criteria for overload and rutting

The overload and rutting failure mode covers limiting criteria (a) and (b) in Section 11.4.2, as both are related to permanent deformation. Criterion (a) for overload can be estimated from plastic deformation in the triaxial test (Section 33.3.1), or even from plate bearing tests (Section 8.4.3 & 4).

Rutting is an accumulation of vertical permanent strains leading to a longitudinal surface depression in a wheel-path (Sections 14.1A8 & 14.5.3). It may arise from either the plastic and visco-elastic creep strains resulting from the passage of a single high load (criterion b), or from the repetitive nature of ordinary traffic loading (criterion a), or a combination of the two. Rutting may be due to basecourse, and/or subgrade deficiencies, and Equations 14.1 to 4 respectively give the accumulated strains and deformations in the pavement courses. The final rut depth is the sum of subgrade and basecourse contributions.

In a layered pavement for heavy traffic it may well be that the upper part of the basecourse is a bound layer and the lower part in contact with the subgrade is unbound. This unbound layer will also need to be designed for the effect of construction traffic and compaction of the bound layer. Its traffic stresses will be lower and — under traffic — its stiffness (Section 11.3.4) will often be as important as its strength.

To specify a limiting criterion, use is sometimes made of the terminal rut depth levels of 10 to 13 mm for good quality roads suggested in Section 14.1A8. These deformation levels are translated into strains accumulated over the design life by dividing them by the thickness of the contributory course. Therefore, any single strain criterion will necessarily be an empirical selection. Nevertheless, single limiting-strain criteria concentrating on the dominant contribution have been widely and successfully used, beginning with the 1962 Shell method (see below) which limited the vertical compressive strain at the top of the subgrade.

Acceptable stress and strain levels in the subgrade are usually associated with the surrounding pavement material providing (favourable) horizontal confining stresses (Sections 11.2.3 & 11.3.1 and Figure 33.3). These horizontal stresses are relatively low at about 150 kPa, compared with a vertical pressure from truck tyres of about 800 kPa (Section 27.7.3).

Figure 11.11a shows how data from repeated-load triaxial tests (test 8.6r) can be represented on a Mohr's circle. The approach uses permanent strain as the distress criterion (Section 33.3.1, Barrett and Smith, 1976; Gerrard and Wardle, 1980) with a

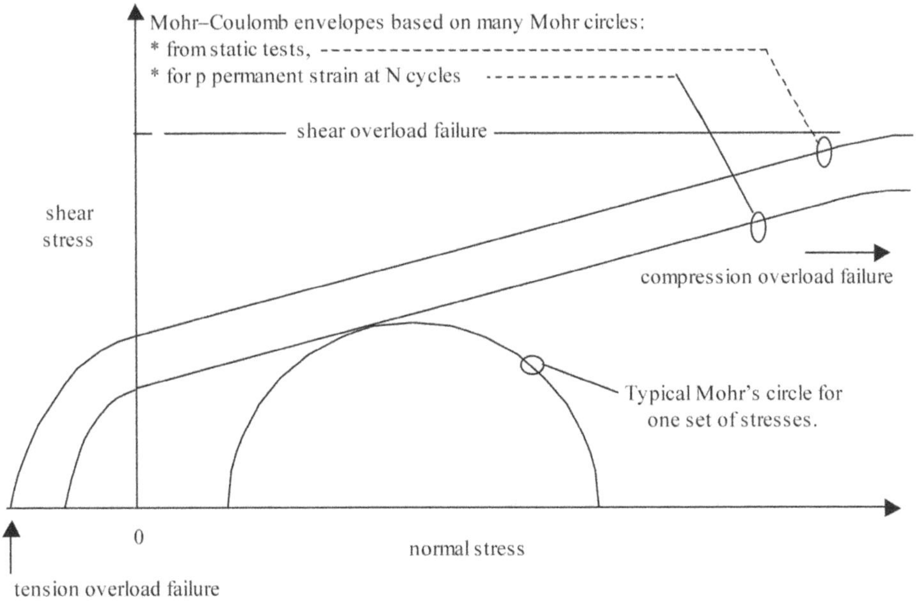

(a) Establishment of permanent strain contours and Mohr–Coulomb envelopes

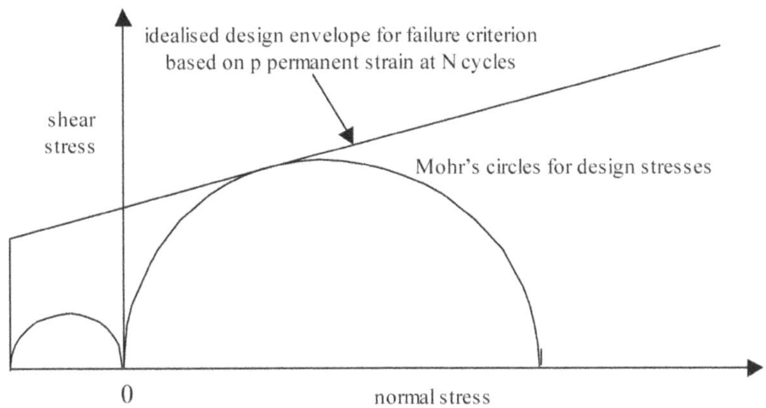

(b) Design curves based on work in (a)

Figure 11.11 Mohr–Coulomb representation of limiting deformation criterion calibrated against the repeated-load triaxial test.

permanent strain well in excess of 0.01 as the criterion for static conditions. σ_{c1p} and σ_{t1p} are the maximum uniaxial compression and tension stresses at a strain of p and after N cycles of load. They can thus be readily determined. The Mohr–Coulomb envelope (Section 8.4.3) of a set of test results is shown in Figure 11.11b for limiting conditions defined by the chosen permanent strain criterion. For a particular permanent strain criterion and number of load cycles, the limiting envelope is then idealised as shown by the solid line whose slope is independent of the stress levels used. It is therefore possible to use a straightforward series of tests to define the limiting levels of load and load repetitions.

For example, criterion (a2) — subgrade vertical permanent deformation leading to rutting — is a common cause of pavement failure. A typical limiting criterion for this failure condition was first suggested by Dorman and Metcalf (1964) after analysing data from the AASHO Road Test in the late 1950s, and further developed by TRL (Powell *et al.*, 1984). It links the vertical strain due to loading, ε_v, at the top of the subgrade to the number of ESAs, N_{esa}:

$$\log N_{esa} + 3.95\log\varepsilon_v \leq -7.21 \qquad (11.11)$$

or $$\log N_{esa} + 3.95\log(\mu\varepsilon_v) \leq 16.5$$

where $\mu\varepsilon_v = \varepsilon_v/1{,}000{,}000$. Data from other sources suggests that the critical 3.95 value should be replaced by 4.48. Austroads (2004) suggests:

$$\log N_{esa} + 7.00\log(\mu\varepsilon_v) \leq 28$$

Experimental data is inherently variable and Equation 11.11 is based on an 85 percent probability of survival, i.e. 15 percent of subgrades would have failed at these N_{esa}, ε_v limits. Applying this criterion to the example in Figure 11.5, shows that ε_v at the top of the subgrade is 0.0195 x 1.27/50 = 0.000495 and so Equation 11.11 gives N_{esaf} = 700,000 for subgrade compressive failure of this relatively weak pavement.

Equation 11.11 and similar criteria used in other approaches to flexible pavement design — such as Shell and CHEVRON — are plotted in Figure 11.12. The limiting criterion used in the original 1962 Shell design method was also derived from AASHO Road Test. The AASHO team had used data from their full-scale tests to correlate the number of axle passes to 'failure', N_f, with the axle (or wheel) load, P, used in each test. The result was the now-notorious *fourth power law* (see also Section 14.3.3 & Equation 14.2):

$$N_f P^4 = \text{constant} \qquad (11.12)$$

An identical $N_f P^g$ relationship arises in classical metal fatigue, with g = 3.1 for steel. In metal the accumulation of small individual permanent strains also leads to an eventual failure-level permanent strain (Lay, 1982).

Failure of the AASHO test roads was defined as equivalent to the condition of a road with a need for major maintenance. Considering its dependence on U.S. freeze–thaw failure conditions, it is surprising to see how widely and confidently the fourth power law is utilised in temperate countries. The form of the equation is not in serious doubt; the main question should be directed to the value of the exponent used in it. Shell also assumed that ε_v was linearly related to P.

The new criterion curve in the 1978 Shell method resulted from a drop in the value assumed for Poisson's ratio from 0.50 in the 1960s edition to 0.35, and the direct introduction of dual wheels into the calculations (Claessen *et al.*, 1977). The Wallace and Monismith (1980) curve in Figure 11.12 was suggested by trial and error and after comparison with CBR methods (Section 11.4.6). This relationship is further discussed in Section 14.3.3.

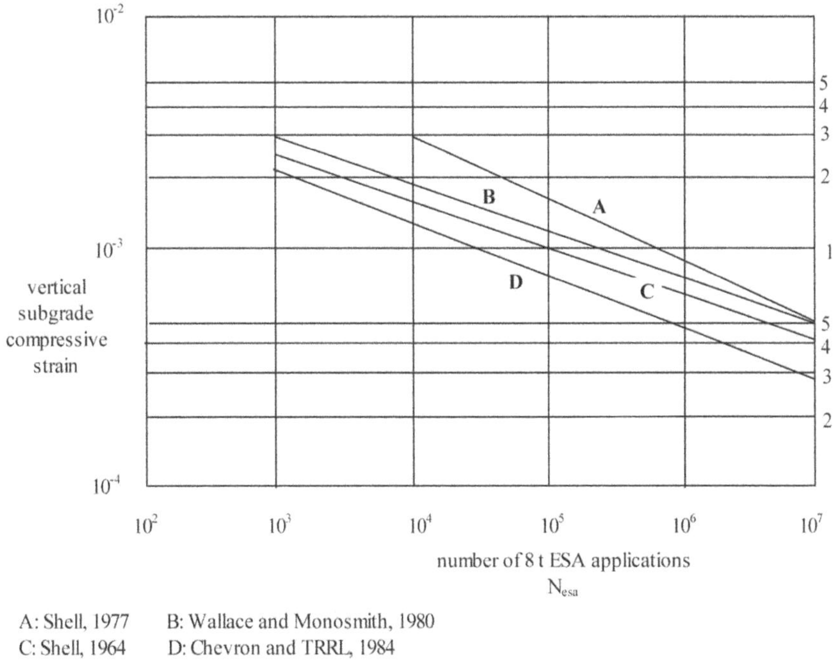

Figure 11.12 Subgrade strain criteria (Wallace and Monismith, 1980). Note that the graphs suggest, reasonably, that the limiting strain for failure under a single load pass is about 0.01.

In comparing the various curves, the slopes and intercepts are seen to vary. For example, alternate values to the TRL intercept (–7.21) in other common codes range from –5.80 to –10.3. Some of the variation arises from the definitions used. To illustrate this, the Australian code uses an adjustment to the Shell criterion to give (Austroads, 2004):

$$\log N_{esa} + 7\log \varepsilon_v \leq -27.6 \tag{11.13}$$

This permits strains at $N_{esa} = 10^6$ which are many times as large as Equation 11.11. The main cause for this difference is that Equation 11.13 is closer to a mean (50 percentile curve) than to the 85 percentile curve of Equation 11.11. The difference between these two percentiles is explained further in Section 15.3.2.

Equation 11.11 may be rewritten as (TRL 85th percentile):

$$N_{esa}(\varepsilon_v)^{3.95} \leq 10^{-7.21} \tag{11.14}$$

$$(N_{esa})^{0.253}\mu\varepsilon_v \leq 15\,000 \tag{11.15}$$

or (Australian 50th percentile):

$$N_{esa}(\varepsilon_v)^{7.14} \leq 10^{-14.77}$$

$$(N_{esa})^{3.95}\mu\varepsilon_v \leq 8\,500$$

Generally, the repeated-load triaxial test (Section 8.6r) or full-scale testing (Section 8.6t) can be used to establish the form of Equation 11.11 for a range of materials. Test data suggests that 3.95 in Equation 11.11 is also a reasonable value for the exponent for rutting of asphalt, with values in the literature ranging from 3.57 to 4.48. Smaller values — typically, 2.5 — apply for crushed rock when the densities or moisture contents are sub-optimal. The exponent can be as low as 0.1 for unbound materials, indicating that in this material, the permanent vertical strain does not increase greatly with load cycles.

It is important to conclude this section with the reminder that these methods remain predominantly empirical and should therefore be applied to non-typical cases with considerable caution. In addition, a poorly constructed pavement will suffer rapid and deleterious changes in pavement properties in the early stages of its life — thus invalidating any predictions of performance based on tests at the end of construction.

11.4.4 Failure criteria for fatigue

Fatigue failure is a result of tensile strains causing a small amount of cracking to occur on each load pass; these crack lengths accumulate during repeated loading, the crack grows and gradually reaches a critical condition. It is notoriously hard to define the advent of 'fatigue failure' experimentally. Some take it as when the flexural stiffness is half its initial uncracked value, but this implies that cracks have initiated and actively grown (Lay, 1982) and is probably a little too late in the fatigue life. In a pavement the major tensile strains occur:

(a) at the bottom of a bound course and immediately under the vertical wheel-load (Figures 11.1 & 5): these are horizontal flexural strains, as in limiting criterion d1 in Section 11.4.2, and

(b) in surface courses: these are due to reversed flexure away from the load axis and/or horizontal traction forces caused by braking, accelerating, and turning (Section 11.4.1).

A typical fatigue criterion equation is the TRL relation for dense bituminous macadam asphalt (Section 12.2.2):

$$\log N_{esa} + 4.16 \log \varepsilon_h \leq -9.38 \qquad (11.16)$$

or, $$N_{esa}(\varepsilon_h)^{4.16} \leq 10^{-9.38} \qquad (11.17)$$

or, $$(N_{esa})^{0.240} \mu\varepsilon_h \leq 5\,560$$

where ε_h is the calculated horizontal strain that occurs at the base of the bound course (Powell *et al.*, 1984). Equation 11.17 is a classic relation in material fatigue and is known as *Wöhler's* equation and produces the S–N curves commonly used in metal fatigue (S is the stress range). Different load levels can be combined using Miner's rule (Lay, 1982):

$$\Sigma(N/N_{esa})_i = 1$$

where i is a load producing a strain of ε_{hi} and Σ covers all loads. Substituting into Equation 11.17 gives:

$$\Sigma[N_{esa}(\varepsilon_h)^{4.16}]_i \leq 10^{-9.38}$$

Applying this criterion to the example in Figure 11.5, shows that ε_h at the bottom of the basecourse is 0.0097 x 1.27/50 = 0.000246 and so Equation 11.17 gives N_{esaf} = 430 000 for basecourse fatigue failure of this relatively weak pavement. From Section 11.4.3, in this case fatigue occurs before subgrade compressive failure.

The coefficients in Equation 11.16 and similar equations are established from laboratory tests on small beams or indirect tension specimens (test 8.6q). The coefficient of variation of $\log N_{esa}$ from field samples is usually around 40 percent. Test results suggest that the coefficient of $\log \varepsilon_h$ (4.16 in Equation 11.16) commonly lies between 2.6 and 6.4, and increases as the bitumen content of the mix increases. U.S. studies based on the AASHO Road Test suggested 3.3 for asphalt, Shell (1978) uses 5 for asphalt and the Australian code uses 5 for asphalt and 18 for cement-stabilised material. The limiting strain level in fatigue, ε_f, usually lies between 0.000050 and 0.001.

The Shell (1978) version of Equation 11.16 — sometimes called the *fifth power law* — also determines the RHS of the equation from an estimate of mix stiffness and bitumen content to give:

$$\log N_{esa} + 5\log\varepsilon_h \leq -10.81 + \log(0.856b_{ve} + 1.08) - 1.8\log E \qquad (11.18)$$

where b_{ve} is the bitumen content in percent by mass (Section 12.2.3(3)) and E is in MPa. Thus, fatigue life increases as the bitumen content of the asphalt increases (see also Section 12.2.2). In a related effect, fatigue life also increases with an increase in relative compaction and with a decrease in void content. However, these factors also increase E. This can result in higher stresses and strains occurring in the course, and thus reduce pavement life. It has also been argued that fatigue life depends on the visco-elastic properties of the bitumen, as these indicate the energy absorbed (Section 8.7.4 & van Dijk *et al.*, 1972).

Many modern local pavement design codes use minor variations of Equation 11.18, which thus represents the core of current best practice. Care should be taken to use the definitions and units specified in the local codes.

An alternative method, which avoids dealing with individual strains, is based on the energy dissipated (or absorbed) in the process. Energy being force by distance, this energy is proportional to stress by strain. Indeed, energy per unit volume is given by stress by strain and is equivalent to the shaded area in Figure 8.8. It is derived from the visco-elastic contribution discussed in Section 8.7.4. Such energy-balance models have been well used in fracture mechanics since the 1960s. The more energy that is dissipated, the longer the fatigue life. A noticeable change in the energy dissipated per cycle indicates that significant crack growth has occurred.

Methods for determining the fatigue strength typically involve testing cantilevers or beams in bending, special specimens in tension, or cylinders in torsion or indirect tension (test 8.6q). Fracture mechanics tests (Lay, 1982) at one extreme and full-scale pavement tests (test 8.6t) at the other can also yield valuable data.

11.4.5 Design criteria for temperature

Temperature plays a critical role in all the limiting criteria. Section 11.2.3 and Figure 11.8 showed that lowering the temperature increases the asphalt stiffness. Hence strains and deflections will be reduced, leading to reduced rutting levels. However, crack susceptibility will increase and the net effect of a drop in temperature is usually to increase cracking rates. One approach to the design of pavements in cold climates where cracking can be a major problem is to limit the stiffness levels of any bound courses. Limits due to McLeod in Canada are given in Table 11.4 (Haas, 1973). Although specific temperatures are shown, the cracking is caused by a temperature differential between the pavement surface and the rest of the pavement or by frost heave (Section 9.5).

Table 11.4 Variation of cracking stiffness with temperature (Haas, 1973).

Minimum temperature (C) at 50 mm asphalt depth	Maximum stiffness (MPa) for cracking: expected	Maximum stiffness (MPa) for cracking: eliminated
– 40	7000	3500
– 30	5000	2000
– 20	3000	1500
– 10	700	350

A similar approach compares the thermal and shrinkage strains in a cooling, restrained asphalt pavement with the fracture strain at the appropriate temperature. In both approaches the key properties are based on the behaviour of the bitumen at low temperatures (Section 8.7.4). Hence the problem worsens as the temperature drops and as the bitumen ages.

Higher temperatures will lead to decreased stiffness, increased strains, and therefore increased rutting. These temperature effects are partly accounted for in the Shell and CHEVRON methods by techniques such as altering the coefficients in Equation 11.10. Usually, thin (< 25 mm) pavements are most susceptible to temperature effects and thick (> 100 mm) pavements have minor temperature susceptibility (Youdale, 1984). A positive aspect of high temperatures is that many bitumen cracks will close due to autogenous healing.

11.4.6 Catalogue and simple empirical design methods

Generally, design methods can be placed into three main categories, although some are a conglomeration of all three. The categories are:

(a) *mechanistic* methods, as described in Section 11.4.3;

(b) *catalogue* (or *prescription*) methods which use past successful practice to provide a list of standard configurations of cross-section, thickness, and materials. They are usually unique to the area in which they were developed.

(c) *empirical* methods where design is based on some measured or estimated property, such as CBR (Section 8.4.4 and Figure 8.10), resilient modulus (Section 11.3.1), or deflection, and where the design limits applied to these properties are related to past successful practice.

In an international text such as this it is impractical to give a comprehensive discussion of design methods in categories (b) and (c), as these will vary from region to region. The reader is advised to consult the design guides issued by the local Road Authority to obtain specific guidance. Typical such guides are AASHTO (1993) for the U.S., Highways Agency (2008) for Britain, AUSTROADS (2004) for Australia, and RTAC (1977) for Canada.

Category (b) methods usually require the designer to have a knowledge of subgrade conditions, basecourse material type, and expected traffic. A typical method in this category is the specification of minimum pavement thicknesses for particular pavement and subgrade types and traffic loadings. For instance, for a subgrade with a CBR of 3 and traffic of 10^6 ESA, a catalogue design might suggest 100 mm of asphaltic concrete on a 400 mm granular unbound basecourse. Naturally, such prescriptions are very location-specific and must be used with caution.

The TRL method (Powell *et al.,* 1984) — particularly for heavy traffic — falls into category (b) and is based on the assumption that most subgrades in moist areas will have a CBR of 5 or less (but see Section 9.4). It assumes that measures will be taken to raise the CBR to 5; CBR can thus be eliminated as a design variable and catalogues produced for CBR = 5 situations, although provision is made for higher CBRs. There are two problems with this approach. First, it will often prove more economical to upgrade weak subgrades — e.g. by stabilising (Chapter 10), or using a sub-base (Section 11.1.2) — than to use pavements able to accommodate low subgrade CBRs. Second, it is often difficult to achieve adequate compaction of basecourse layers placed on top of a low CBR subgrade.

The *structural number* (Section 11.4.7) and *layer equivalency* (Section 11.4.8) *methods* can also be placed in category (b) and are discussed separately below.

Category (c) methods were once very common — particularly given the earlier discussion about uncertainties in analysis and materials characterisation — and remain so for unbound courses. The CBR test was based on the work of O. Porter at the California State Highways Department from 1928 onwards and formally linked to a pavement design method by the U.S. Army Corps of Engineers (USCE) at the beginning of the First World War and Second World War. An underlying assumption was that there was a linear relation between the CBR result and the maximum shear stress possible in the subgrade without causing failure under a defined number of load applications. The maximum shear stress is in turn linearly related to the maximum vertical compressive strain in the subgrade. Equation 11.14 has shown how this strain can be linked to the number of load applications. This suggests that $N_{esa}(CBR)^{3.95}$ is a constant. The USCE pavement-thickness prediction curves are based on an equation of the form:

$$\text{thickness} = f_1(\log CBR)f_2(N_{esad})$$

or, more specifically:

$$\text{thickness} = c_3(CBR)^{-0.5}(1 + c_4\log N_{esad}) \quad (11.19)$$

where c_3 and c_4 are constants and N_{esad} is an aggregate of different N_{esa} (Section 11.4.1). Thus, pavement construction costs increase with $\log N_{esad}$ (i.e. less than linearly with N_{esad}). Typically, costs double for each million N_{esad}, and doubling N_{esad} adds 24 mm to the pavement thickness (OECD, 1988a).

An example of an empirical category (c) method based on the USCE–CBR approach is as follows (Mulholland, 1989):

(1) An assessment is made of subgrade strength in terms of its CBR. Because of the variability of CBR values, it is necessary to base the design on some sub-mean characteristic value (Section 15.3). CBR depends on moisture content (Section 8.4.4) and the appropriate moisture content to use is discussed in Section 9.4.1.

(2) As subgrades with a CBR of 5 or less can be easily damaged by construction traffic, it is usual to first increase their strength by stabilisation (Chapter 10) or the addition of a sub-base of low-cost local material.

(3) The thickness of the basecourse is determined using procedures based mainly on an assumed relationship, such as Equation 11.19, between subgrade CBR, basecourse thickness, and traffic in terms of N_{esad} (Section 11.4.1). A typical set of curves is shown in Figure 11.13. Similar curves are used as one basis for the British design guide (Powell *et al.*, 1984) and in a number of other countries and the reader is advised to use the appropriate local graph. In particular, the minimum thickness levels are based on local practice.

(4) Steps are taken to ensure that the CBR of the basecourse is at least 80.

11.4.7 Structural number methods

The *structural number* method was common in the U.S. (AASHTO) and in U.K. Road Note No. 31. The method provides simple techniques for the:

* design of multi-course pavements composed of different materials, and
* comparison of different pavements, as it is assumed that pavements with the same structural number or layer equivalency will behave identically.

The value of the structural number, SN, is given by

$$SN = \Sigma a_i H_i$$

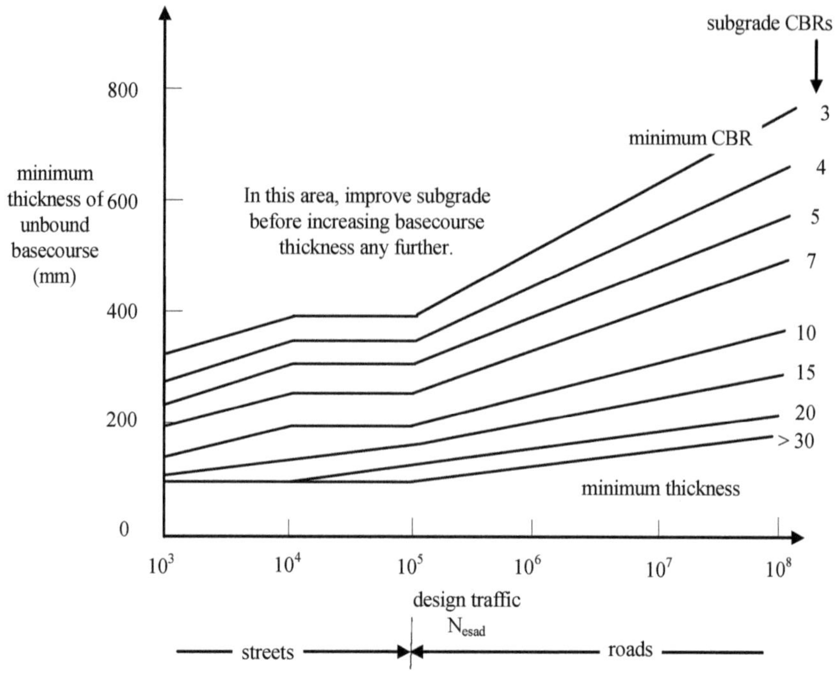

Figure 11.13 Typical CBR pavement design chart for N_{esad} — the design value of N_{esa} (Section 11.4.1) — based on an 8.2 t standard axle and, for less than 10^5 axles, a 5 percent probability of failure during the design life (Mulholland, 1989). For the 'roads' part of the graph, refer — for example — to AustRoads (2004).

where H_i is the course thickness. The coefficients, a_i, in this equation were determined by regression analysis of the AASHO Road Test (Section 11.4.3) serviceability data and were biased towards cracked pavements (Yoder and Witczak, 1975). Corrections were introduced to account for local regional conditions such as freezing and thawing and the efficacy of the drainage system. For an asphalt pavement, typically:

$$SN = 0.44\,(\text{thickness of stiff asphalt}) + 0.14\,(\text{thickness of unbound bases}) + 0.11(\text{thickness of unbound sub-base}) \quad (11.20)$$

More generally, AASHTO in 1986 suggested that:

$$a_i = a_s\,(E_i/E_s)^{1/3} \quad (11.21)$$

where s relates to the standard layer [e.g. from Equation 11.20 $a_s = 0.44$ if E_s is the value for stiff asphalt]. The layer stiffness basis for this Odemark approach is explained in Section 11.4.8.

To allow SN to take account of the subgrade strength, as well as the properties of the imported basecourses, TRL (Hodges *et al.*, 1975) introduced the concept of the *modified SN*. It is based on the SN that would be needed if the pavement structure were to carry the same traffic but on a CBR of 3 percent (the CBR in the AASHO Road Test). It is given by (mm units)

$$SN_{modified} = SN + SN_{subgrade}$$
$$= SN + 90\log CBR - 22(\log CBR)^2 - 36$$

In the AASHTO (1986) method a minimum acceptable pavement structural number (or Thickness Index), *SN* was calculated on the basis of N_{esad}, the desired pavement condition (PSI, Section 14.4.1), and the resilient modulus of the subgrade. The resulting AASHTO equation for flexible pavements is:

$$\log N_{esad} = 9.36\log(SN + 1) - 0.20 +$$
$$\{[\log[0.37(4.2 - PSI)]/[0.40 + \{1094(SN + 1)\text{-}5.19\}]\} + 0.38(SSV - 3.0) - 0.97\log R$$

where SSV is a soil parameter linked to resilient modulus (and thus indirectly to CBR) and R is a regional factor. A basic advance introduced in the method was the use of a reliability factor to allow the designer to vary the risk of failure and thus use cheaper designs in non-critical areas.

For an existing pavement, SN has only a weak link to pavement deflections, thus implying that it covers strength as well as stiffness. Nevertheless, equations of the log form (Section 8.4.5):

$$SN = SN_0 - 4\log d_0 + \log d_{900}$$

or power form, using Equations 8.4 and 11.21 (d_0 is in mm),

$$SN = cd_0^{-0.5}$$

or, empirically,

$$SN = 3d_0^{-0.63}$$

were often used to estimate SN. The first of the three equations is counter-intuitive and, at best, implies some form of relationship between d0 and d900.

The modified structural number, now called SNC, was used in HDM-III (Section 26.5.2). To account for changing pavement conditions over time, HDM-IV adjusted SNC to a new term, SNP, which has a weighting factor that reduces the contributions from the deeper pavement courses. SNC and SNP are similar for pavements less than 700 mm in thickness.

A typical modern sealed granular pavement might have an SNP of between 3 and 7 with a median value of 4. Within a given uniform section of such a road, SNP could have a coefficient of variation of about 5 percent. Pavements with an SNP of over 5 will have a noticeably better rutting performance (Section 14.1.A8) under traffic.

To cover the drop in SNC with time, SNCt, the ratio SNCt/SNC0 has been empirically linked for Australian uncracked sealed granular pavements (Section 12.1.4) to age, y, divided by design life, Y, by the following equation (Martin, 2005):

$$SNC_t/SNC_0 = 0.96(2 - e^{y/3Y})$$

11.4.8 Layer equivalency methods

In view of the definition of layer and course, one should more properly speak of *course equivalency*. However, common practice is to use *layer* in this context. Note also that *load equivalency* is discussed in Section 27.3.2.

A variety of empirical design methods use a layer equivalency approach to undertake the design of courses of different materials without using computer programs for multilayer systems. In the approach all the diverse courses in the pavement to be designed have their thicknesses adjusted to convert them to courses of the 'equivalent' material. Equivalencies often are not applied to the top 50 mm or so of the pavement as this upper course will be in compression (Figure 11.2a) and there will be little difference between the behaviour of bound and unbound courses (see below). The approach is a design simplification and should not be used where exceptional or unusual circumstances are encountered.

In using the method it is necessary firstly to clearly define the 'equivalent' layer. It is common to use asphaltic concrete (Section 12.2.3) for this purpose and it therefore has an *equivalent thickness* (or *equivalence*) factor, EF, of 1.00. EF for another material is then the amount that its course thickness would need to be multiplied by in order that an asphaltic concrete course of this new thickness would perform identically to the real thickness of the material in question. For example, if a granular base has EF = 0.50, then 400 mm of the granular base could be replaced in calculations by 0.50 x 400 = 200 mm of asphaltic concrete.

There is a need to be particularly careful with this definition as the term *layer equivalency factor* — defined as:

> the thickness of a material which would be equivalent in performance to a given thickness of the standard material, divided by that given thickness

is also in use and is thus the reciprocal of EF. In summary, if a thickness T_n of a new material performs in an equivalent fashion to a thickness T_s of a standard material, then:

$$\text{equivalent thickness factor, EF} = T_s/T_n$$
$$\text{layer equivalency factor} = T_n/T_s = 1/(EF)$$

Typical EF values are given in Table 11.5 which illustrates that the 'natural' progression in basecourse usage as N_{esa} increases is to move from unbound aggregate to a weakly-bound cement-stabilised course (Chapter 10) to a bound course.

The above equivalencies are based primarily on load-spreading behaviour (i.e. on course stiffness), rather than on flexural strength, and are estimated by $(E/E_s)^{0.33}$, following the work of *Odemark* in Sweden in the late 1940s (and therefore prior to the availability of computer programs for analysing multicourse systems) and Equation 11.23 below. The 0.33 (= 1/3) power comes from the fact that elastic plate stiffness is proportional to thickness to the power of 3. If EF is largely dependent on the ratio of the effective stiffness modulus of the two courses, then care needs to be taken to manage the factors which affect these two moduli, i.e. temperature, load level, number of load repetitions, underlying course stiffness and course thickness (Section 11.2.3). The last two do not directly affect E but, by altering the boundary conditions, do alter the effective E, e.g. by a change from plane stress towards plane strain (Lay, 1982). As mentioned in Section 11.4.7, a similar approach is sometimes used to determine structural numbers, SN.

The AASHTO approach (Equation 11.20) implies an EF of 0.14/0.44 = 0.32 for unbound bases and 0.11/0.44 = 0.25 for unbound sub-bases. Equation 11.17 can be rewritten as:

$$SN = 0.44(T_a + \Sigma[EF_iT_i])$$

where EF_i is the equivalent thickness factor for course i of thickness T_i and T_a is the asphalt thickness. Note that SN values are usually quoted in inch units and so, if T_i are in mm, comparison with U.S. data may require division by 25.4. The specific EF coefficient for unbound bases is often estimated by:

$$EF_{ub} = 0.66(CBR) - 0.45(CBR)^2 + 0.11(CBR)^3 \qquad (11.22)$$

It can be seen from Section 11.2.3 that equivalent thicknesses could also be calculated from an elastic stress analysis, by picking some criterion value such as vertical subgrade stress beneath the load and maintaining it constant for various combinations of stiffness modulus and course thicknesses. Thus it is argued that conditions below the changed course are unaltered if the stiffness of the changed course is unaltered. This leads to the equivalent thickness for two courses being given by (Kennedy, 1985):

$$(T_n/T_s)^3 = (E_n/E_s)/([1 - \nu_n^2]/[1 - \nu_s^2]) \qquad (11.23)$$

and by a series expression for multilayer equivalencies:

$$T_n = \Sigma T_i\{(E_n/E_s)/([1 - \nu_n^2]/[1 - \nu_s^2])\}^{1/3}$$

Table 11.5 Typical equivalent thickness factors, EF, taking asphaltic concrete in good condition as the standard material. (CBR) = [CBR in percent]/100 and UCS is the MPa value given in Table 10.2.

Material	Description of material	Equivalent thickness factor, EF
1. asphaltic concrete	1a. In good condition with a resilient modulus > 4 GPa	1.0
	1b. Some cracking and deformation	0.80
	1c. Large cracks and noticeable deformation	0.50
2. asphalt surface course		0.45
3. stabilised soil	3a. uncracked	0.60
	3b. cracked	0.16 + 0.07(UCS)
4. unbound bases	4a. (CBR) > 80	0.50 – 0.30
	4b. 80 > (CBR) > 20	0.40 – 0.10
	4c. From Paterson (1985)	0.02 + 0.15log(CBR)
	4d. From AASHTO (Equation 11.20)	0.32
	4e. Common procedure	Equation 11.22
5. unbound sub-base	from AASHTO, see text	0.25 or 0.02 + 0.15log(CBR)
6. spray and chip surface		0.20

where T^3 comes from the second moment of area of the course and E_s and ν_s are commonly taken as subgrade properties.

11.5 CONCRETE PAVEMENTS

11.5.1 Concrete pavement types

Because of a general tendency in the preceding text to emphasise flexible pavements, this section collects together some special aspects of concrete pavements, which are the most common form of rigid pavement. Concrete as a material has been discussed in Section 8.9, the general distinction between rigid (concrete) and flexible (bituminous) pavements was examined in Section 11.1.5, the pavement design methods in Section 11.4 included techniques for the design of concrete pavements, and Section 11.6 describes concrete blocks. Concrete pavements offer advantages for:

* residential streets, due to their construction simplicity and low maintenance costs. However, the problem of access to under-street services remains a major concern, with concrete pavements being difficult to break through and to reinstate. On the other hand, as the pavements are usually thinner than flexible ones, they offer benefits in less construction disturbance to services close to the surface and less excavation.
* heavily-loaded roads on poor subgrades, due to the ability of the stiff pavement to disperse the loads widely (Figure 11.1b).

The five main types of concrete pavement are as follows:

(1) *Jointed unreinforced (or plain) concrete pavement* (*PCP*). This pavement contains no reinforcement. It is the lowest initial cost concrete pavement, although its maintenance costs can be high as cracking progresses.

(1a) *Short-slab* paving. In the common application of the PCP method, narrow contraction joints are placed across the traffic direction (often at a skew) — typically by sawing — at intervals of 5 m or less. Steel dowels are usually employed to prevent vertical movement across the joints and thus maintain rideability (Section 14.4). It is important to ensure that the concrete is adequately consolidated around the dowels. The paving strips are rarely more than a lane in width. The method permits large-scale machine paving. For heavy traffic, this basecourse is placed on a well-bound sub-base.

(2) *Jointed reinforced concrete pavement* (*JRCP*) (Section 15.2.1) in which steel reinforcing – typically mesh – is provided to control cracking. Note that (a) cracking is not prevented and (b) the reinforcing does not increase the load carrying capacity of the slab relative to a PCP slab. The concrete is placed in continuous strips, with dowelled joints constructed afterwards, rather than in alternate bays between joints. The joints are made as saw cuts about 15 mm wide are used at about a 12 m spacing. The elimination of two-thirds of the contraction joints used in PCP leads to greatly improved rideability. However, JRCP requires higher construction standards and better sub-bases. Transverse cracks typically develop in the initial years, when concrete shrinkage (Section 8.9.2) is restricted by subgrade friction forces. Vertical loading shear is carried across the cracks by aggregate interlock. This effect diminishes in effectiveness over time and slab deflections will gradually increase.

(3) *Continuously reinforced concrete pavement* (*CRCP*), which has no transverse joints other than construction joints and expansion joints at bridge structures (Section 15.2.5). The longitudinal reinforcing steel is continuous for the entire pavement length and, at about 0.6 percent, is much greater than for (2). There is no technical limit to CRCP lengths. Terminal anchorage is provided at the ends to reduce length changes due to temperature and shrinkage. Fine random cracks about 300 μm wide will develop at about 2 m spacing. These cracks and a filigree of even finer cracks at closer spacing allow the CRCP to accommodate time-related volume changes. The reinforcing must be adequate to overcome the effects of this cracking. The more steel that is used, the closer the crack spacing.

(4) *Prestressed concrete* (Section 15.2.1) in which there are few joints, and the reduced tensile stresses make lower thicknesses possible.

(5) *Fibre-reinforced concrete* (with dowelled joints at about 10 m) in which small quantities of short steel or plastic fibres are distributed randomly through the concrete (Section 8.9.2).

In these concrete slabs, joints and reinforcing steel are used to:

(a) prevent or control cracking due to concrete shrinkage;
(b) prevent and control expansion, contraction and warping due to temperature and moisture changes;
(c) accommodate subgrade movement; and
(d) provide construction convenience, e.g. the centreline longitudinal construction joint in a two-lane carriageway constructed one lane at a time or the transverse construction joint at the end of a day's pour.

Joints for (a) to (c) are called *control joints* and joints for (d) are called *construction joints*. Early shrinkage (a) is usually managed using transverse contraction joints formed or sawn to at least a third of the slab depth. They are typically 15 mm wide and are usually filled with a sealant. It is necessary to pay considerable attention to joint design as it has proved a critical feature determining concrete pavement performance (Section 14.1A). For instance, joints must not permit the entry of water (which will damage the subgrade or sub-base and/or cause pumping, Section 9.3.2) or of grains of dirt and rubble (which will then keep the joint permanently open). This may require joints to be sealed with preformed or poured sealants (Sections 14.1B1&C18). Poured sealants are commonly cold-applied elastomers (Section 33.3.2).

Particular emphasis is thus directed at joint detailing, and also to construction procedure and reinforcement placement. Sections 9.3.2 & 11.1.2 described the need to employ a non-erodable sub-base and a drainage layer under a concrete slab in order to avoid pumping. This provision must be coupled with longitudinal drainage at the edge of the slab (Section 13.4.2).

Construction is discussed in Section 25.4.

11.5.2 Concrete pavement design

Concrete pavement design is based on elastic analysis of the pavement slab supported on a sub-base. Initially, techniques were largely based on Westergaard's method (Section 11.2.3), but today more reliance is placed on finite element methods such as the University of Illinois' *ILLI-SLAB*.

The sub-base usually comprises at least 100 mm of a bound material. If the sub-base is not bound, its structural contribution should be ignored. The calculated stresses in the concrete will thus depend on the stiffness of sub-base and subgrade and on the flexural properties of the concrete slab. These flexural properties will themselves depend on the compressive strength (Section 8.9.2) of the concrete and the thickness of the slab. The stresses will vary inversely with the square of the pavement thickness, but are only weakly dependent on the subgrade properties. The most critical stress situation usually arises from loads at the corner of the pavement slab, although all edge loadings are relatively severe.

Design formulas often resemble Equation 11.16, with thickness a function of CBR and N_{esa}. A common design method is given in AASHTO (1993). The addition of up to 0.6 percent reinforcement is assumed not to make any contribution to the slab strength. It will be initially unstressed, but will develop stress as the shrinkage of the concrete is restrained by the surrounding subgrade and shoulders. The role of the reinforcement is to control the concrete shrinkage and then to manage the inevitable cracking by limiting the amount by which the cracks open. Thus the reinforcement content is most closely related to the spacing and design of joint details, rather than to any flexural stresses. The performance criteria (Section 11.4.2) recommended are:

- * the level of permanent deformation,
- * the subgrade stress,
- * the horizontal tensile stresses at the base of the concrete slab causing fracture or fatigue (Figure 11.2). These stresses may be caused by:
  - # traffic (Section 11.4.1–2),
  - # overall thermal effects. Overall temperature changes will create horizontal stresses in the slab as its thermal movement is restricted by adjacent slabs and by friction between the slab and the sub-base; or
  - # differential temperature changes through the thickness of the slab will cause it to curl and warp. At night, the corners of a slab will curl upwards and, during the day, the centre of the slab will tend to rise. Bending stresses will be caused by the inevitable restraints to this movement.

* joint behaviour, particularly subgrade or sub-base erosion due to moisture movement caused by repeated movement at the joints (Sections 9.3 and 14.1A13).

Typical minimum concrete thicknesses are:

* culs-de-sac, 100 mm
* streets, 120 mm
* distributors, 140 mm
* truck routes 160 mm
* arterials, 180 mm
* highways, 200 mm.

11.6 CONCRETE BLOCK PAVEMENTS

11.6.1 Block types

This section discusses concrete blocks for road paving. Stone blocks (or setts), wooden blocks and bricks (or clay pavers) have been used in pavement construction for many years (Section 3.6.6) and the modern concrete and brick blocks can be regarded as their successors. Although written in terms of the concrete block, much of this Section also applies to brick blocks.

Concrete blocks are produced from high strength concrete in automatic machines to high ± 2 mm dimensional interlock tolerances. They are about the size of an ordinary house brick with specifications commonly limiting their plan area to 60 000 mm^2. The top edges of the blocks are chamfered to reduce tyre suction, provide a path for water under tyre pressure, hide minor vertical discrepancies between blocks, ease handling, and reduce edge spalling. Blocks are commonly manufactured to 60, 80 and 100 mm thicknesses, with the 80 mm block being the most widely used. The 60 mm blocks are just able to carry traffic and the 100 millimetre blocks are able to withstand the very large vehicle loads encountered in shipping container terminals. If blocks are too large horizontally, there is a high risk of tilting under wheel loads. An increase in block thickness is significantly more effective for load resistance than an increase in basecourse thicknesses. Somewhat larger and thinner concrete slabs known as *flagstones* (Section 3.6.6) are also used occasionally for footpath construction. Typical flagstone dimensions are 400 x 400 x 40 mm.

The blocks need good abrasion resistance and durability, but their main direct requirement is for compressive strength during the manufacturing process. The common compressive strength values are 35 MPa, 45 MPa, 50 MPa and 60 MPa. Cement contents are always above 250 kg/m^3 (Section 8.9.2), are typically in the range 250 to 300 kg/m^3, but may be raised to 380 kg/m^3 or more if resistance to frost damage is needed.

Paving behaviour depends greatly on geometric interlock between the blocks. Block shapes fall into three shape categories, depending on whether they develop (Figure 11.14):

(a) full wedge interlock,
(b) some wedge interlock, and
(c) no wedge interlock.

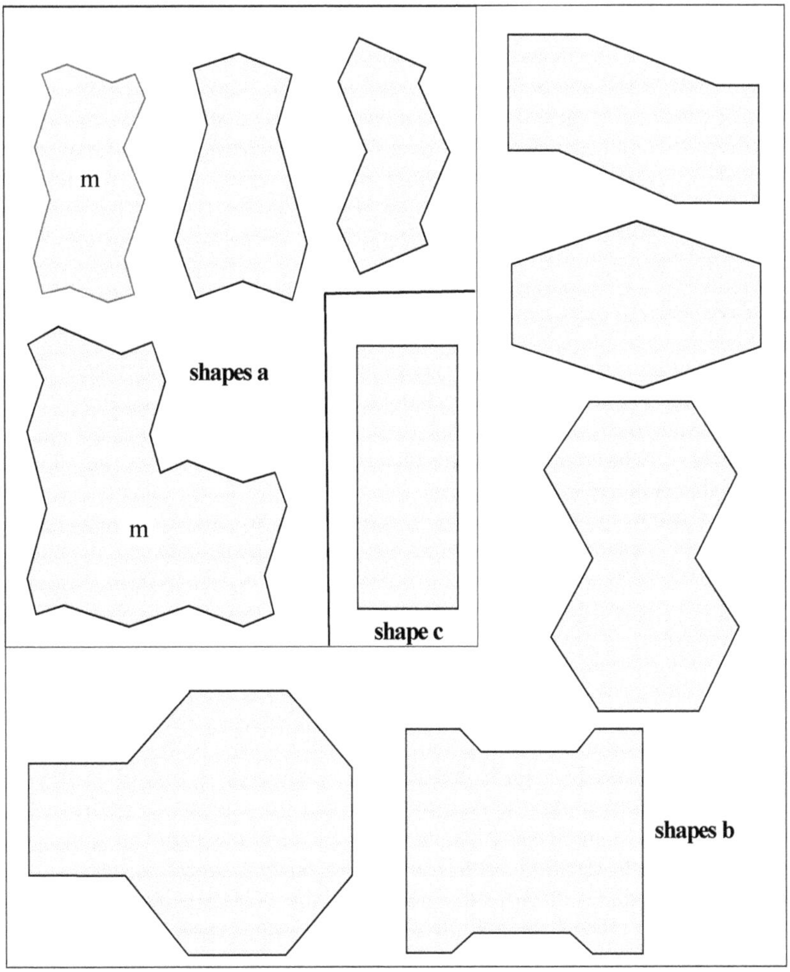

Figure 11.14 Concrete block shapes for horizontal interlock.

Although the wedge interlock provided by shapes a and b adds to pavement strength and makes it possible for the blocks to be well laid by relatively unskilled labour, rectangular blocks (shape c) are widely and successfully used in a number of countries, particularly Holland and the United Kingdom. The block labelled 'm' in Figure 11.14 can be placed by mechanical devices that take a pre-laid square metre of block from a pallet and set it in place in the pavement.

Blocks are occasionally used which are shaped to provide vertical interlock. These ensure that a smooth surface is maintained in service across the inter-block joints. Another form of block contains large vertical perforations that are filled with soil and then planted with grass. These provide a landscaping alternative in car parks.

11.6.2 Block paving

The advantages of concrete block pavements are that they:

(a) require simple equipment and relatively unskilled labour for placement in the field;
(b) have a long life and are easy to maintain, modify and relevel; and
(c) can produce a variety of textures, colour and appearances. In dry weather a driver can detect different block colours and distinguish concrete blocks from asphalt. However, in wet weather only the joints in the block surface will provide a visible cue (Jenkins and Sharp, 1985).

One disadvantage of block paving is that its noise and riding quality may restrict operating speeds to less than 60 km/h.

Block paving is usually laid on a prepared sand *bedding* (or *layer*) *course* about 25 mm in thickness. The blocks are considered to be the basecourse. It is normal to also use a sub-base below the bedding sand to help meet the design requirements for unbound pavements. During compaction of the blocks, the bedding sand infills the lower gaps between the blocks and jointing sand is swept and vibrated into the upper portion of the gaps. The choice of the grading and thickness of the bedding sand is therefore crucial to achieving good performance.

It is necessary to use an edge-restraint strip such as a kerb and channel at the edges of the paved areas to prevent blocks moving sideways and becoming loose. Within the kerbed area, horizontal movement is resisted by contact between the faces and so it is essential to ensure that no unfilled gaps or compressible, permeable material is present between the vertical faces. Such gaps would also permit moisture to enter the lower pavement courses.

As no mortar is used between the blocks, the permeability of the pavement may require special attention (Sharp and Armstrong, 1985). As water may enter through the joints, the bedding sand must have sufficient permeability to remove that water as a drainage course (Section 13.4.3). If the subgrade is moisture-sensitive, it may be wise to apply a polymeric sealant to the pavement surface to prevent moisture ingress. This treatment will also help prevent the loss of sand into the basecourse (Section 11.1.2).

Block pavements respond to traffic loading in a manner similar to an unbound granular course (Seddon, 1980). Pavement design therefore follows the principles outlined in Section 11.4 for flexible pavements. The equivalent to granular interlock in unbound courses arises from the horizontal interlock that develops between the mating vertical faces of the blocks. It is probable that some arching and precompression of the block course also occurs. These three factors allow the block course to take some part in flexural behaviour.

The principle mechanisms by which interlock develops are shown in Figure 11.15 and are:

(a) friction between adjacent block faces transmitted by sand in the joints (Figure 11.15a);
(b) wedging interlock between adjacent blocks (Figure 11.15b); and
(c) geometrical interlock (Figure 11.15c).

Mechanisms (a) and (b) have been discussed above and in Section 11.6.1 and mechanism (c) will now be examined.

To utilise mechanism (c), blocks are commonly laid in either *herringbone* or *stretcher bond* patterns (Figure 11.16) . The herringbone pattern (Figure 11.16a) is ideal for geometric interlock development with the longitudinal stretcher bond (Figure 11.16c) of intermediate value. Another advantage of the herringbone pattern is that it can easily accommodate curved alignments. In the stretcher, the long side of the brick is in the exposed face, in the header the short side is in the face (and so the brick is transverse to

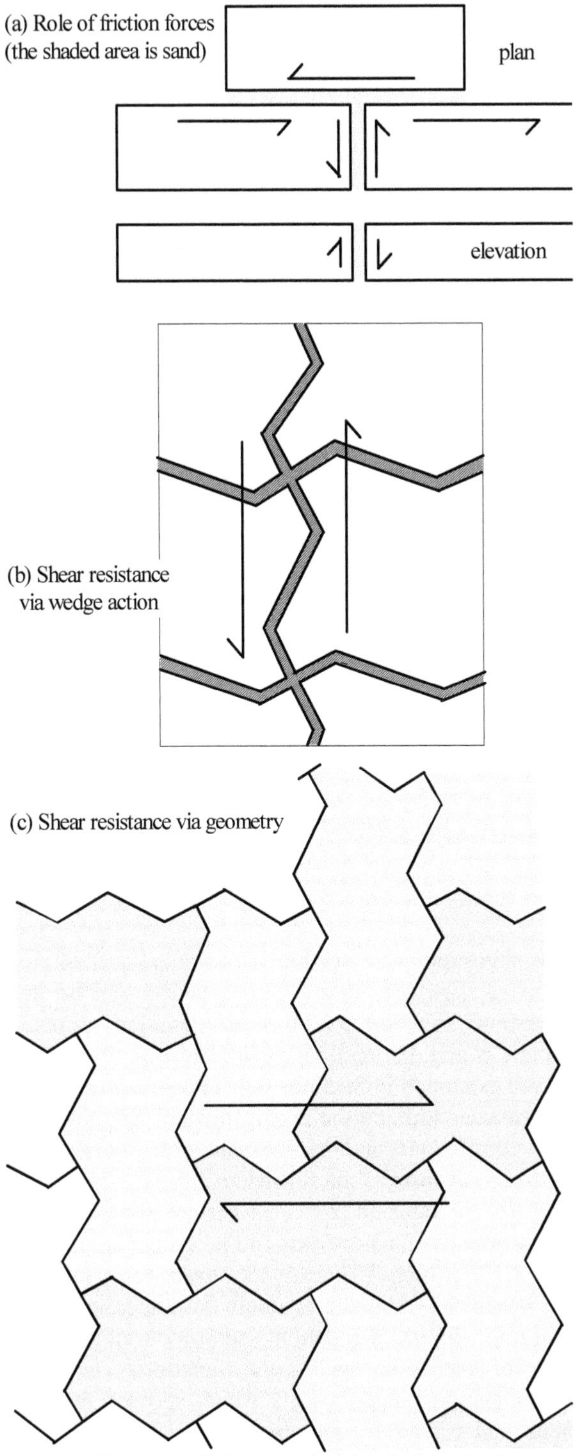

Figure 11.15 Concrete block interlock mechanisms.

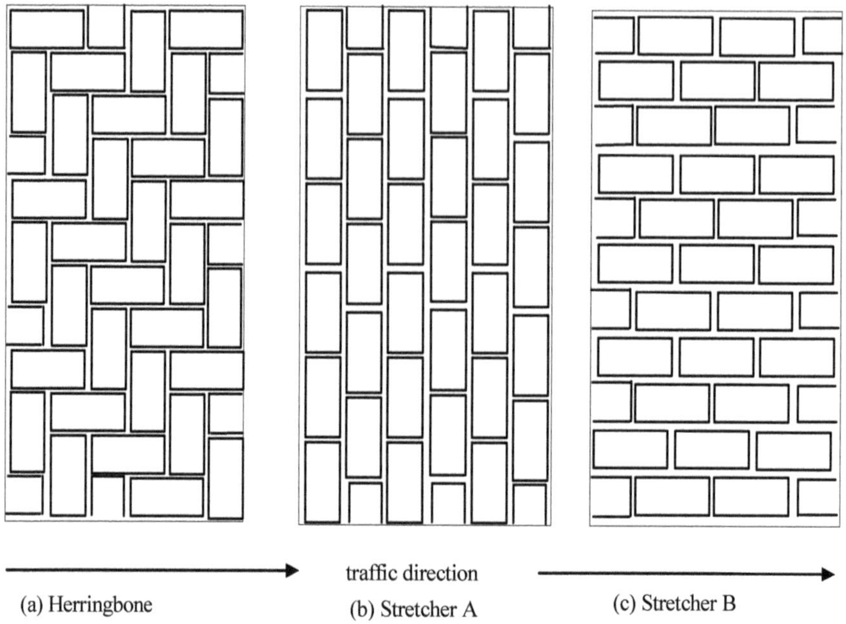

Figure 11.16 Types of block pattern.

the face). Stretcher bond patterns transverse to the traffic (Figure 11.16b) are not recommended as poor geometric interlock develops.

11.7 PAVEMENT CROSS-SECTION

11.7.1 Cross-section types

The selection of the pavement cross-section is a further facet of pavement design. The number of lanes and carriageways that will be needed will be determined by traffic engineering considerations (Section 6.2.5); super-elevation, camber and grade by the discussions in Chapters 6 and 19; and pavement thickness by Section 11.4. Associated drainage provisions are examined in Section 13.3. The remaining major aspects to resolve are the form of the cross-section and the extent to which the pavement structure extends beyond the edge of the traffic lane, and the associated role of the pavement shoulder.

The ideal cross-section form is the full-width arrangement shown in Figure 11.17a, as it minimises the effect of water on pavement performance. However, it is not feasible for most urban roads where pavement widths and levels are constrained. This leads to the situation in Figure 11.17b which is known as *boxed construction* and which has no effective shoulders. When the subgrade is relatively impermeable, such a construction forms a natural water trap with the attendant moisture-induced problems (Section 9.4.2).

For pavements of the same width, the superiority of full-width over boxed construction has been demonstrated in tests which showed that the largest measured pavement deflections under load in both boxed and full-width trials were, under all conditions, at the edges of the box sections. This was due to (Metcalf, 1978):

* water penetration between the pavement basecourse and the shoulder,
* lack of support from the shoulder, and
* the difficulty usually experienced when compacting a basecourse against an edge of the formation.

In many cases underground drains (Section 13.4.2) will be required in order to alleviate the problems of boxed construction. Another ameliorative measure is to use an extended box construction (Figure 11.17c) which takes the boxed region to at least the extremities of the shoulders and thus moves the most serious moisture problems away from the pavement running surface.

11.7.2 Shoulders

Shoulders are strips of constructed material on either side of the pavement, not intended to regularly carry traffic but still considered part of the carriageway (Figure 2.1). However, in some countries the roles of the shoulder and the verge (Figure 2.2) are combined and the pavement may appear to lack a shoulder.

Two related and contentious issues with respect to shoulders are the extent to which they should be surfaced and the extent to which their use by traffic should be encouraged. These issues are best seen in relation to an examination of the role of the shoulder, which is to (Armour and McLean, 1983):

1. *pavement-related*
   1a. laterally restrain the pavement courses (Section 11.1.1);
   1b. separate construction-related pavement edge weaknesses from the influence of traffic;
   1c. drain water away from the pavement and prevent ponding (Sections 9.1, 9 4 and 13.3.1);
   1d. remove the zone of sub-surface moisture change from the outer wheel-path area (Section 9.4.2);
2. *traffic-related*
   2a. increase lateral clearance between vehicles by allowing them to drive closer to the outside of their lane (Section 16.5.1);
   2b. make overtaking easier by allowing overtaken vehicles to move to the side (Section 18.4.2);
   2c. provide a recovery area for errant vehicles (Section 28.5);
   2d. provide safer and better parking for vehicles stopped by breakdown or emergency;
   2e. provide an emergency traffic lane in cases of lane blockage,
   2f. provide a lane for use by slow vehicles and bicycles (Section 17.4.3i).

Following these varying different objectives, shoulder construction practice falls into three categories:

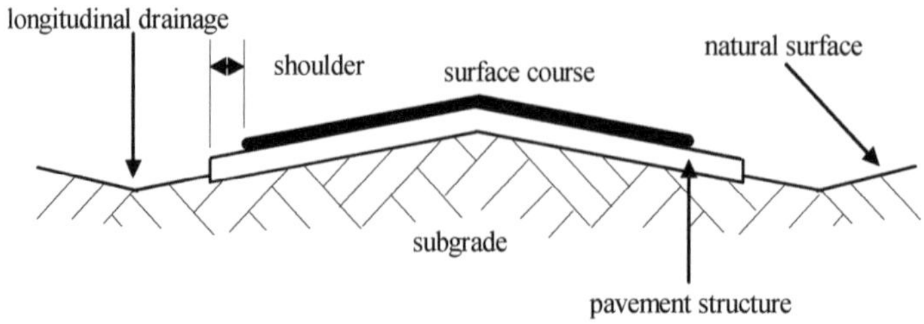

(a) Preferred full width construction, minimising the influence of moisture by using minimal "boxing" and by keeping the pavement structure above the natural surface level.

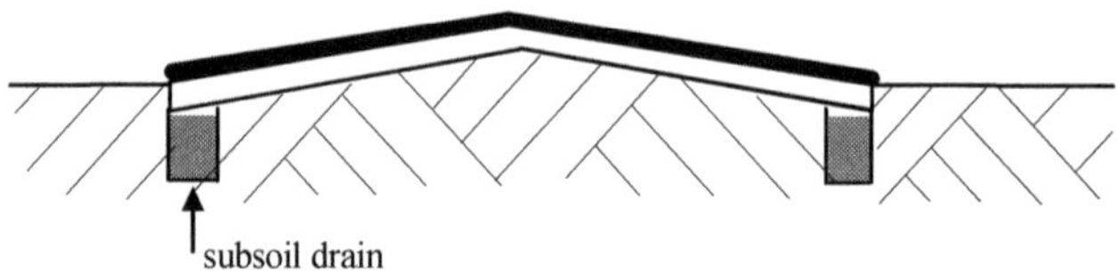

(b) Boxed construction, used where pavement width and surface level are both constrained.

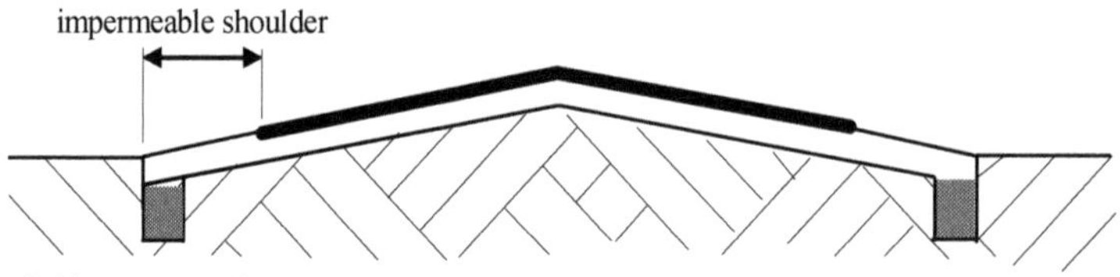

(c) Extended box construction

Figure 11.17 Common forms of road cross-section.

(i) *Complete surfacing of the entire shoulder.* A surfaced shoulder is sometimes called a *hard shoulder*. It is the most expensive of the categories and meets all the above requirements. However, it may still not be completely effective unless traffic can be persuaded not to regard the shoulder as another traffic and/or parking lane. Nevertheless, it has significant driver behaviour (Section 16.3.4), maintenance (as an alternative traffic-lane), and safety (Section 28.5) benefits.

(ii) *Surfacing of the metre or so closest to the traffic lane.* This method discourages use as a traffic lane, keeps the outer wheel-path away from the moisture-affected zone (objective 1d) and still has significant driver behaviour and safety benefits. It meets all the above needs except (2d) and (2e). It has a quite high benefit–cost ratio (often about 10) in terms of crash reduction and a ratio of about two for edge maintenance costs. An additional advantage on low volume roads is that the shoulder may not need full pavement strength as it may rarely take truck loads. High volume roads (> 3×10^6 N_{esa}), on the other hand, will require the shoulder to have full structural strength and be at least 1.2 m wide.

(iii) *Unsurfaced shoulders.* These should be kept impermeable and capable of emergency use. In climates where grassing is feasible, grassed shoulders may keep vehicles away from the pavement edge, but will seriously inhibit the drainage of surface water away from the pavement.

Shoulder width should be at least:

* 500 mm to provide structural strength — a load dispersion angle of 30 degrees to the horizontal would indicate a minimum extension of only twice the basecourse thickness.
* 600 mm if the traffic exceeds 200 veh/d, to provide safe operating conditions (Section 28.5f),
* 1.0 m, in order to move moisture-affected edges from the outer wheel-path (Section 9.4.2), and
* 3.5 m (a lane width, Section 6.2.5), to provide ideal options for emergencies.

Shoulders should drain water away from the pavement to a verge extending down to side drains (Section 13.3.2), but to avoid a traffic hazard should have a slope (Section 6.2.5) of no more than 1 in 20 away from the travelled way. The slope of the verge is discussed in Section 28.5. The distinction between shoulder and verge is that the shoulder consists of basecourse material whereas the verge is part of the road formation.

It is important to ensure that the vertical step between the shoulder and the verge is minimised (e.g. less than 50 mm), otherwise a driver unintentionally leaving the shoulder at speed might lose control and attempt to regain the pavement at an angle which would risk re-entry impact with the normal through traffic.

11.8 EMBANKMENTS AND CUTTINGS

11.8.1 Theory

Embankments (or *fills*) and cuttings are a consequence of earthworks required by the chosen road alignment and its associated formation (Section 2.2). The material requirements are usually not very severe as the internal pressures are low. The main need is to use material with ready compactibility and low compressibility to ensure that the traffic surfaces remain smooth. Some material would clearly be inappropriate: peat, material from marshes, swamps and bogs, timber, rottable material, combustible material, frozen material, hazardous material and clays with a high sensitivity to water. The slope face of the formation is also called a *batter*.

The key design criteria for both embankments and cuttings is usually that their slopes should be stable. As the general issues associated with embankments and cuttings are not unique to roads and are well covered in standard texts on soil mechanics, they will be given little attention in this book. Nevertheless, they are particularly important as an unstable slope — above or below the roadway — will usually cause major traffic disruption and may cause loss of life and major property damage. The first sign of an unstable embankment is often longitudinal cracking on the road surface (Section 14.1(B.5)).

Slope instability usually takes the form of *landslips* (or *slides)* of material, in one of five mechanisms:

(a) Translational or planar slipping of one stratum on another along a shear surface approximately parallel to the ground surface. When this occurs in rock, it is termed a *rockslide* and is associated with weak bedding and jointing planes in the rock. It is probably the commonest form of slope instability associated with roads and is usually associated with the road cutting exposing a lower plane that then becomes a shear plane.

(b) Rotational movement of a segment of the soil mass (or *slump*) along a curved shear plane (or *slip circle*). The moving mass tends to rapidly deteriorate into a debris slip. It is less frequent than (a).

(c) *Slips* of thin surface layers, commonly *earthflows* (or *mudflows*). These are often associated with poor or altered drainage (Sections 9.4 and 13.4.2). They thus arise in saturated, weak and unconsolidated soils. *Terracettes* (or *sheep tracks*) are a common small-scale, slow-acting form of this type of slip and are often observed on steep hillsides.

(d) *Falls* of rock from exposed rock faces in mountainous regions. They may be either the result of one large rock-fall, or may have grown progressively (a *talus*) from the accumulation of many small rock-falls. They frequently relate to jointing within the rock mass, particularly when the jointing (or bedding) plane aligns to a potential slip direction (e.g. is parallel to the face of the cutting). Rock-falls can be a serious driving hazard and their chance of occurring increase as the slope height increases. They can be managed by
   * minimising blasting intensity when the cutting is being constructed,
   * constructing mid-slope benches,
   * using energy-attenuators and nets on the slope, and
   * catching the rocks in wide ditches and wire mesh fences at the base of the slope.

(e) Major *avalanches*, often of rock debris.

A good warning of the risk of future landslips is evidence of past slips and of recent clearing of vegetation, earthworks, high rainfall, and/or altered drainage. It is wise to install monitoring devices in locations where the risk of slope instability is high.

In many cases, heavy rainfall will be the trigger which initiates the actual landslip. Generally, water must be carefully managed as it has six major deleterious effects on embankments in that it:

(1) causes surface scour,
(2) encourages weathering of the material,
(3) increases the mass of the material,
(4) produces pore pressure that reduces shear strength (Section 9.2.5), thus increasing the risk of slope instability,
(5) causes volumetric expansion, and
(6) can cause special problems during any rapid draw-down of water levels after flooding.

Factors (3) to (5) are very relevant for clays. The expansive clay problem is particularly severe with high embankments, where the movement can accumulate over a considerable depth and where slopes with large surface areas can promote moisture changes.

Cuttings with slopes (Table 6.6) as steep as 0.25 to 1 may be used in rocky material on minor roads whilst urban freeways frequently use slopes of 2 to 1 with horizontal ledges (*berms or benches*) at 8 to 10 m vertical intervals to enhance slope stability. Embankment slopes may range from 6 to 1 to 1.5 to 1 on rural freeways. Flat slopes are desirable for vehicle safety reasons (Section 28.6.2). Slopes steeper than 1.5 to 1 will usually need to be faced to prevent scour during rain, slopes steeper than 3 to 1 are difficult to mow, and slopes steeper than 4 to 1 in cohesive plastic soils will be prone to slope instability. Slope protection against scour and other forms of surface erosion is critical and is discussed in Section 6.4.2.

An embankment may be required to operate as a flood-control levee, which will require design and construction considerations beyond those discussed in this book.

11.8.2 Settlement

In practice settlement is a major cause of road deficiency (Section 14.1(A.1)). Thus, an important requirement for embankments is that any vertical settlement should not compromise the pavement structure or its rideability. Settlement may result from a range of causes, primarily from compaction and consolidation (Section 8.2.1), both of the embankment itself and of the underlying subgrade. It can occur quite slowly, reaching a critical level over a number of years. All settlements must be carefully examined as they may be signs of imminent slip failure. The first signs of settlement may be longitudinal pavement cracks (Section 14.1(B.2)) and this may misleadingly suggest a problem with the pavement rather than the embankment.

The mass of the embankment, and hence the consequential settlement, can be reduced by using light-weight material, such as blocks of expanded polystyrene (EPS, or rigid cellular PS). EPS has a density of about 20 kg/m^3 whereas most construction materials are over 1000 kg/m^3 (Section 8.1.1). EPS has a compressive strength of about 100 kPa at 10 percent deformation, which is comparable to low CBR soils (Section 8.4.3 & Table 10.1). EPS must be laid flat to avoid local crushing and protected from solvents (e.g. most petroleum products), sunlight and abrasion. One of its associated advantages is that it can be placed very quickly.

In a cutting, the removal of overburden — and a subsequent reduction of gravity-induced stresses — may actually result in reverse compaction and consolidation (i.e. soil expansion), with surface levels rising with time in response to diminished vertical stresses. In addition, as the subgrade material in the cutting will have been preconsolidated by the overburden, it is reasonable construction practice not to attempt to further compact or otherwise work such preconsolidated material.

A particular problem occurs at the approach to a *bridge* where the settlement of subgrade and possible embankment forming the approach to the bridge will be naturally greater than that of the bridge deck, supported on relatively stiff bridge foundations. This difference may lead to a noticeable depression in the pavement adjacent to the bridge deck. The difference may be increased by the consequences of:

* poor compaction of the embankment adjacent to the abutment wall,
* the impact of truck wheels dropping into the depression from the stiff deck,
* erosion of fill material,
* poor drainage at the face of the abutment, and/or
* settlement of poor natural material underlying the embankment at the riverside.

The first two are usually the major contributors.

The problem is best avoided by careful attention to compaction of the embankment near the deck (compaction theory and measurement is discussed in Section 8.2 and compaction methods in Section 25.5), or by the use of a special approach slab (Section 15.2.4 & Figure 15.2). A typical slab would span about 8 m and be about 400 mm thick. It would keep the change in slope to under 1 in 200. Settlement of the approach pavement and any associated embankment can also cause large extra downward forces on adjacent bridge abutments. A related problem is the effect on the bridge abutments of horizontal confinement pressures within the approach structure.

11.8.3 Construction techniques

Embankments may be categorised as either:
* free-standing, gravity structures,
* internally restrained or reinforced, or
* externally restrained.

Likewise, cuttings are either:
* free-standing,
* restrained by the surrounding material, or
* externally restrained.

Free-standing embankments are usually built from material removed from cuttings elsewhere along the alignment. Their shape is determined by slope-stability requirements. They are sometimes constructed from boulder-sized (Figure 8.6) material obtained from excavations in hard rock. This material presents compaction and density testing problems (Section 8.2.2). To ensure good compaction and uniform deformation, the layer thickness must be at least 50 percent more than the maximum boulder size and this can require very large compaction plant. In ideal circumstances, layers up to 2 m can be compacted (Sparks, 1984). As most test methods are limited to 20 mm particles, it is often necessary to check achieved densities by calibrated proof-rolling, rather than by attempting to recover and use very large specimens.

A typical internally-restrained embankment or cutting is one faced with *retaining walls* in order to achieve steeper side slopes than would be possible with natural slopes. The retained material must be well-drained, in order to prevent hydrostatic pressures building up at the back of the wall. It is also important to avoid making the fills of potentially-expansive material (Sections 8.4.2, 8.5.4&6, 8.9.1 & 10.1.2) — for example, material with a sulfate content of over 1 percent is a significant risk. Retaining walls may be:

* tied back to some fixed point or heavy weight (called a *deadman*), or
* vertical cantilevers of reinforced concrete or steel, or
* rely on their own mass. A *crib wall* is one such gravity system and is built of interlocking precast or timber pieces (cribs) stacked on top of each other in a pattern, with the intervening spaces filled with soil. Typical crib walls are formed of precast reinforced concrete units about 100 x 200 x 100 mm. As with brickwork, the longitudinal pieces on the faces are called stretchers (Section 11.6) and the transverse pieces are called headers. The walls are generally built on a compacted basecourse with a 1/4 to 1 slope.

More sophisticated, internally-restrained systems rely on the mass of the whole fill to provide lateral restraint and enhance slope stability. *Reinforced earth* is one such system. Layers of fill are placed between layers of horizontal steel (or suitable plastic) reinforcing strips attached at each end to the segments of two confining retaining walls. Friction develops between the fill and the reinforcing and is enhanced by the normal forces created by the self-weight of the embankment. The reinforcing effectively adds cohesion to the material and, at high concentrations, can also increase internal friction (Figure 8.9). In addition, because the soil is not permitted to expand horizontally, it is also possible to design on the 'at-rest' horizontal restraining pressure, rather than the lower 'active' pressure associated with the horizontal expansion possible in more conventional embankments. Important design variables in reinforced earth are the steel–soil coefficient of friction, the reinforcing spacing and the reinforcing strength and stiffness.

The technique permits the use of relatively large free-standing vertical walls. It also has the advantages of being simple, rapid, uncomplicated to construct, relatively cheap, and tolerant of foundation movement. It works best with soils with high internal friction, such as sand and gravel, and requires caution with clays where creep or saturation may occur. Most construction follows one of a variety of commercial procedures.

The *soil nailing* method enhances the slope stability of cuttings, typically by drilling a series of closely-spaced (1 to 1.5 m) near-horizontal holes into the face of the cutting. The hole length is typically equal to the depth of the excavation at the hole. Steel anchor rods are then inserted and grouted into place. Alternative methods place the rods by percussion or compressed-air methods. The rods are passive in that, unlike tiebacks, they are not pretensioned when installed. They only come into tension when the soil moves horizontally. The rods are placed in lifts, typically of 2 m or less, as the excavation progresses. Indeed, the excavation usually cannot proceed until the previous lift of nails is completed. Before the nails are installed, a layer of wire mesh is placed in front of the face of the excavation, and the entire face and mesh are covered with shotcrete. Some tensioning is conducted to test the efficacy of the grouting.

CHAPTER TWELVE

Bituminous Pavements

12.1 SURFACE TREATMENTS

12.1.1 Unpaved roads

Unpaved roads may be either:

* *natural* (or unformed),
* *improved* by shaping to be *formed* roads, or
* *improved* by shaping and use of imported gravel to be *formed and gravelled* roads.

Each of these categories will now be discussed in turn.

Many rural roads began life as natural, unpaved (dirt) tracks for horse-drawn or off-road vehicles. The main forms of damage that such roads suffer are wheel rutting and erosion due to wind and water. With occasional attention to drainage, unpaved tracks are able to carry up to about 10 veh/d. Over the years, they may be progressively up-graded by a process known as *stage construction.* Typical stages will now be described.

The *first* stage in the upgrading of an unpaved track is to level the natural material to provide a satisfactory running surface. This treatment usually only lasts until the first heavy rains. However, it is the first step in the production of a *formed road.* If a firm crust does form on such a road, it should be protected from damage by a thin layer of loose stone typically gravel (Sections 8.4.1 & 8.5).

The *second* stage is to enhance the transverse drainage of the road by using natural material taken from pits along the edge of the road to provide a cross-fall of at least 4 percent. The result of this shaping and forming is called the *road formation* and its prime purpose is to control moisture entry and thus inhibit the occurrence of ruts and potholes. Longitudinal drainage is also provided, often by utilising the same pits. The drains are frequently diverted into the surrounding land to prevent the build-up of water flows large enough to cause flooding and scouring of unlined drains (see also Section 13.3.2). The provision of *culverts* (Section 15.1.2) to take water from adjacent land under the road formation and away from the pavement structure represents an important subsequent stage.

The formation is generally about 200 mm above the natural surface and 500 mm above the invert of the longitudinal drains. Its raised surface is maintained with *graders* or, in less advanced technologies, with *drags* (Section 25.4) and similar planing devices. Up to this stage, most of the basecourse (Section 11.1.2) compaction is a consequence of traffic; weak spots are remedied by maintenance grading. Such formed, unpaved roads made from in situ material with a soaked CBR of at least 5 (Section 8.4.4) are usually suitable for up to about 120 veh/d, but local economic stringency sometimes sees them carrying up to 400 veh/d. Note that Section 12.1.2 suggests that imported material for the surface should have a soaked CBR of at least 12, and for the basecourse at least 45.

12.1.2 Use of imported material

(a) Material selection

With the use of specially-selected material imported to the site, the natural formation can be protected from traffic damage. This may require removing surficial material which would be unsuitable for carrying traffic. At this stage, the natural formation has, technically, become the subgrade (Section 2.2) and the imported material is the basecourse. The acceptable traffic capacity of these unpaved but gravelled roads can be raised to as high as 400 veh/d.

Thus an improved unpaved road is the result of engineering interventions to remove unsuitable in situ material and replace it with better imported material. There is often a gradual progression from a natural to an improved surface and some definitions require the latter to contain at least 25 mm of imported material. Ideally, soil–aggregate mixtures imported for use in unpaved roads should, when placed, be able to:

(a) provide a good riding surface,
(b) resist permanent deformation due to traffic in wet and dry conditions,
(c) resist the horizontal tractive forces due to vehicles,
(d) resist the formation of corrugations (Section 14.1A5),
(e) provide good skid resistance,
(f) resist water and wind scour at the surface,
(g) produce minimal dust (effects both adjacent properties and vehicle handling and braking),
(h) resist water penetration,
(i) require low maintenance, and
(j) be of low cost.

Due to the need to use low-cost material, the aggregates used will commonly come from local sources. Igneous rocks (Section 8.5.1) and many gravels (Section 8.5.7) can usually be successfully employed on unpaved roads. Typical measures to achieve a suitable roadmaking mix include requiring the basecourse to have a:

(1) mixture of wear-resistant aggregate, sand, and natural binders such as clay (Sections 8.3.1 & 8.4.2) (the material from some gravel pits may naturally include some clay),
(2) clay content to be less than the void content of the sand component of the mix (Section 8.1), (the clay acts as a pseudo-cement, binding the sand particles together),
(3) well-graded particle size distribution (Section 8.3.2) to reduce permeability (Section 9.2.2) and allow adequate compaction to occur (Section 8.3.1),
(4) maximum stone size of less than 40 mm and preferably less than 20 mm, and less than two-thirds of the course thickness (larger stones are displaced by traffic, give a poorer ride and interfere with maintenance of the surface),
(5) soaked CBR of at least 60, although this can be relaxed to 30 in arid conditions. For the surface course, this can be further relaxed to 12, to retain some clay binder (Section 8.4.4 and Paterson, 1985),
(6) dry compressive strength of at least 2.8 MPa (Section 8.4.3), and/or
(7) PI of below 4 (Section 8.3.2). For the surface course, this can be relaxed to 14, to retain some clay binder. The clay makes it possible to grade the road to a good shape but results in a surface which is slick when the road is wet and dusty when it is dry.

A balance must be maintained in selecting the clay content. Naturally well-bonded, fine materials such as sand-clays and granite sands can be maintained at traffic volumes and speeds which would cause coarse and poorly-bonded gravels and many sandstones to give poor service.

As mentioned in Section 11.1.3, *waterbound macadam* courses are occasionally used for unpaved roads, with mixed in situ sand–clay mixtures being common. However, they are costly to maintain under high-speed traffic and their use for this purpose is now minimal.

The thickness of the imported material course must be at least sufficient to reduce the stresses on the weaker subgrade to an acceptable level. It will typically be between 100 and 200 mm, and will depend primarily on prior local experience with the material being used.

(b) Maintenance

Pavement surfacing may be lost through the action of traffic, rain, wind, and encroaching vegetation. For example, dust will be lifted from the road surface by the action of vehicle aerodynamics, tyre suction (Section 27.7.1), and local winds. Approximately a millimetre of surfacing is lost for every 2000 vehicles using the road. Even without traffic, about 7 mm of surfacing will be lost each year. The amount lost increases linearly with the quantity of fine particles present (Figure 8.5) and with the square of the traffic speed. High wind speeds also increase the rate at which dust is generated. This will result in:

* a loss of pavement material and thus a deterioration of the road surface, and
* pollution of nearby water and air, and thus creating:
  - an environmental nuisance, and possibly reducing crop yield,
  - a traffic hazard, due to reduced visibility.

A large part of the maintenance task with unpaved roads involves replacing this lost material — typically by replenishment in 30 to 50 mm layers. The process is called regravelling and may be needed every couple of years. The replacement material is usually spread with a grader, but rarely compacted. This process is sometimes called dry maintenance. Another major task is replacement of material lost due to water erosion. Other aspects of the maintenance of unpaved roads are discussed in Section 26.2.3. A good summary of unpaved road practice is given in ARRB (1993).

(c) Dust suppression

Common dust palliatives are:

* *water and wetting agents*
  Dust can be diminished for very short periods by spraying the surface with water, which binds the fine particles together, increasing the effective mass of the dust.
* *low-viscosity, slow-curing oil or modified bitumen* (Section 8.7.6)
  Dust can be diminished for short periods by spraying the surface with these materials, which bind the fine particles together, increasing their effective mass. As the amount of bitumen (or 'oil') used is increased to around 5 percent and the depth of penetration is increased, these *oiled roads* come closer to being bitumen-stabilised roads, as discussed in Section 10.4, and can have lives of two to three years.
* *calcium chloride* ($CaCl_2$)
  This is sometimes used to control dust in areas that are not very dry (the relative humidity must be over 50 percent). Calcium chloride's hygroscopic and deliquescent properties enable it to absorb more than twice its mass in water and its high surface tension further helps to reduce evaporation. This keeps the pavement surface in a

desirably moist condition and increases the period over which the basecourse is at a desired moisture level (such as OMC). In dry conditions the material can crystallise and form a hard surface crust that resists traffic abrasion. Note that hygroscopic means absorbs moisture from the air; deliquescence means dissolving in the presence of moisture to form a liquid.

Similar comments apply to magnesium chloride (bischofite).

* *sodium chloride*

  Sodium chloride works in a similar fashion to calcium chloride by inhibiting the evaporation of surface moisture but — as it permits higher evaporation rates — it requires humidity levels of 75 percent or more to be effective.
* *lignin sulfonate* (or *sulfite lye* or *lignosulfonate*)

  This natural polymer is a gummy, malodorous by-product of the paper-industry which has been used as a successful short-term dust suppressant.
* *molasses and tannin extracts*

  These natural polymers have been used as successful short-term dust suppressants.
* *polymer emulsions*

  These are typically vinyl acetates or acrylic-based co-polymers, are suspended in water by surfactants and about half solid particles by weight. As the water evaporates the polymer particles start to coalesce. The resulting soil-polymer matrix prevents the escape of soil particles as dust.

Each of the above products rapidly leaches away and is rarely effective for more than a year.

12.1.3 Surface priming

An unbound surface is sometimes sprayed with bitumen derivatives of suitable viscosity in a process known as *priming*. The functions of this process are to:

(a) coat and thus increase the mass of any fine particles in the surface and hence reduce its propensity to become dust;
(b) strengthen the surface;
(c) provide a uniform surface;
(d) seal surface pores to help waterproof the pavement; and
(e) aid the development of bond with a subsequent spray and chip seal (Section 12.1.4) or asphalt course.

To achieve these functions, the primer used must be able to penetrate the finely divided dust film covering each particle and provide a strongly-adhering bituminous film. Vertical penetration of a primer into the surface should average about 5 mm (Dickinson, 1984). Primers must therefore be quite fluid with viscosities between 10 and 20 mPa.s at 50 C (Section 8.7.5). The main primer used is *cutback bitumen* (Section 8.7.6), although there is a diminishing and small use of tar (Section 8.8). Viscous (heavy) primers that leave significant bitumen on the surface are sometimes called *primer binders*.

A surface to be primed should be free of moisture as this will both inhibit bonding of bitumen and stone (Section 12.3.2) and physically block the penetration of bitumen.

Before priming, the basecourse must be rolled to provide a tight, firm surface. The main pre-priming types of surface finish are:

(1) Tightly-bonded surfaces. These ring when struck with a heavy handle. Such hard, dense surfaces may be created by *armouring*, i.e. rolling into the top of the

basecourse a one-stone-thick layer of good quality smaller stones, as with drybound macadam (Section 11.1.3).

(2) Medium-porosity surfaces. Although they look tightly bonded, they do not ring when struck with a heavy handle.

(3) Porous surfaces. These have an open texture.

Care must be taken when the basecourse to be primed contains a noticeable clay proportion, as good bonding may not be achieved.

The amount of primer needed depends on how the basecourse surface ranks in the above categorisation. For instance, priming can be omitted if the pavement surface is in category (1) above. Otherwise, a typical primer would be applied at about 800 mL/m^2 (i.e. a nominal thickness of 800 μm), with a range of 500 to 1400 mL/m^2, depending on surface porosity. Low rates and low viscosities are used with tightly-bonded surfaces and high rates and high viscosities with porous surfaces.

After priming the coat should be allowed to cure for at least a day (depending, for example, on the emulsifiers or cutters used, Section 8.7.6) and any excess primer removed, e.g. by blotting it with sand. This sand must then be removed.

A primer may only last a week or two under traffic. Where longer service is needed before the work is completed, a *primer seal* (or *primerseal*) is used. This is still of a temporary nature and is used:

(a) to allow a subsequent spray and chip seal surface (Section 12.1.4) to be applied in favourable weather, after traffic compaction or in an appropriate construction sequence;

(b) to rectify local deficiencies in an existing bound surface; or

(c) when a conventional primer would be slow drying and hence subject to traffic damage.

A primer seal is produced by applying a relatively viscous primer that is then covered with a surface of fine aggregate or sand in order to extend its traffic life. The use of a more viscous binder aids the retention of the fine material in the surface cover but basecourse penetration is reduced to about 3 mm. A typical primer seal would be applied at a rate of about 1.0 L/m^2, although rates as high as 1.7 have been successfully used. If well maintained, a primer seal will last up to six months. Increasing the stone size and the primer viscosity produces a *heavy primer seal* with a life of up to 12 months.

The fine-textured surface of a primer seal provides lower skid resistance than conventional surfaces (Section 12.5). Thus, if used as part of a high-speed curve it could lead indirectly to a driver's loss of control as the vehicle enters the primer-seal. Such a friction change within a curve should be either avoided or appropriately signed.

A *tack coat* is a primer applied to substrates, concrete slabs, or stabilised basecourses in order to provide good adhesion for the next layer. To achieve the requisite fluidity, it is usually either a cutback bitumen or a bitumen emulsion (Section 8.7.6).

12.1.4 Spray and chip seal surfaces

A number of new techniques were developed around 1900 to overcome the problems of the fast new pneumatic-tyred cars using unpaved roads (Section 3.6.8). In particular, the practice of using surface priming (Section 12.1.3) to 'seal' the pavement surface with a thin bituminous film was developed to protect the basecourse from traffic and moisture entry. This *seal coat* also proved able to provide a pavement of sufficient structural strength for the new traffic. A rapid development was to protect the seal coat by embedding angular stone particles into its surface (Dickinson, 1984). This form is now commonly known as *spray and chip sealing* (or *sprayed seal* or *bituminous surface treatment, BST* or *bituminous seal* or *chipseal*) where the word *chip* refers to the use of angular particles of broken stone. A spray and chip seal surface can typically carry 500 veh/d, but it is often found to be cost-effective to apply when traffic is as high as 600 veh/d. In good conditions, this is a conservative estimate. There are many instances where such surfaces, when placed over a good basecourse and with good maintenance, have carried 1,200 veh/d, and occasionally 2 000 veh/d. The purposes of spray and chip sealing are to:

(a) waterproof porous surfaces, in order to protect the underlying pavement and subgrade from water damage (Section 9.4). The flexible, inert nature of the seal allows it to maintain its impermeability for many years.
(b) transmit traffic forces to the basecourse;
(c) provide a surface that resists wear and erosion due to traffic, wind and water;
(d) protect the pavement against suction forces from the tyres;
(e) provide a smooth riding surface;
(f) provide adequate skid resistance;
(g) provide adequate light reflectance;
(h) provide a dust- and mud-free surface;
(i) bind surfaces at the onset of cracking-related deterioration, thus increasing maintenance intervals;
(j) provide a low cost surface — commonly about 40 percent of the cost of the equivalent 25 mm of asphalt.
(k) repair a pavement, as discussed in Section 12.1.6(4).

Purposes (b), (d) and (e) are achieved by the bitumen locking the surface layer of stones into position. Note that the seal itself does not strengthen the pavement or correct its shape.

From a user's viewpoint, the spray and chip sealing of an unsealed road provides a dramatically better driving surface (higher skid resistance) and reduces roadside dust and mud. There are also suggestions that the crash rate per kilometre of travel is at least halved. These benefits must be weighed against the net costs of providing and maintaining the sealed surface. The surface is relatively noisy and so is often unsuitable in residential areas. Another disadvantage is that it lacks sufficient shear strength to last in areas where braking or turning produce high shears at the tyre–road interface.

12.1.5 Spray and chip seal construction

(a) The spray and chip process

The basecourse for a spray and chip seal may be either an existing and adequately-strongbound course, a sound unsurfaced pavement (Section 12.1.1), or a newly-placed and well-compacted unbound course. The basecourse is firstly coated with a primer (Section 12.1.2). The stiffness of the basecourse and the firmness and immobility of the primer are key factors defining the future life of the spray and chip seal.

After the primer has set, the surface is sprayed with bitumen to produce a seal coat (hence the term *spray seal*, occasionally called a *flush seal*). The bitumen may be straight-run, cutback, emulsified (Section 8.7.6), or foamed (Section 10.4). Cutback and emulsified bitumens are used to extend the sealing season into colder weather, when straight-run bitumen would become too viscous and solidify too quickly.

Before spraying, the bitumen is kept in an insulated tank where it is continuously heated and stirred at temperatures of up to 220 C. The tank feeds a sprayer system consisting of a spray bar distributing the bitumen to individual slotted spray nozzles. Each nozzle typically sprays about 300 mL/s, so the nozzle spacing determines the spray rate for the machine. A typical 100 mm spacing along a transverse spray bar gives 3 L/s/m. For a sprayer travelling at 20 km/h (5.5 m/s), the spray rate becomes 3/5.5 = 0.55 L/m^2. The key performance determinants for the sprayer thus relate to the width of spray and the sprayer speed. In practice, the rate at which bitumen is applied is usually between 0.3 and 4.0 L/m^2 for sprayer trucks travelling at 20 km/h or less. The higher rates require multiple spray bars. The spraying equipment must be capable of providing a uniform spray, and nozzle cleanliness and adjustment and spray bar height are key factors. Spraying is best undertaken in warm, dry weather.

Single-sized crushed-stone is then spread onto the still-sticky surface (Figure 12.1), usually from the rear of a tip truck reversing over the placed surface. The flakiness index (Section 8.6a) is commonly employed to ensure that the stone particles used are reasonably cubical. They also need to be clean, strong, tough, and durable. Their abrasion resistance and hardness are often measured by the Los Angeles test (Section 8.6b) and tests c, e, j, k and/or m in Section 8.6 are also often used as guidance for stone selection. The stone must also be able to adhere to the bitumen (Section 12.3.2). This is often aided by pre-coating the stones with bitumen or kerosene to overcome the adverse effects of dust and water.

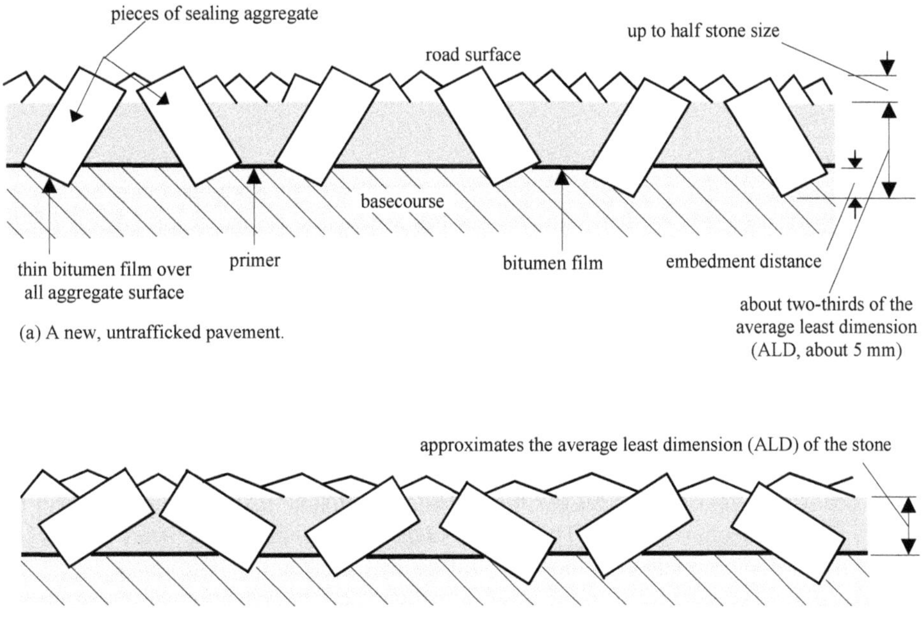

Figure. 12.1 Spray and chip seal. The basecourse shown is idealised — it, of course, is also composed of interlocked stones.

Next, the stone cover is rolled with self-propelled pneumatic-tyred multi-wheeled roller-compactors to:

* press the aggregate partly into the bitumen,
* give the surface a protective mosaic finish,
* cover most of the layer of bitumen, and

* place the stones in shoulder-to-shoulder interlocking contact (Figure. 12.1b) with their average least dimension (ALD) close to vertical. *ALD* is the average of the minimum cross-section dimension of each piece of aggregate (Dickinson 1984), often weighted in proportion to the surface area covered by each stone when the stones are all lying with their least dimension vertical. Typically (Section 8.3.2):

$$ALD = d_{50}/(1.09 + [0.0118 \times \text{flakiness index}])$$

The final thickness of the surfacing is typically 25 percent greater than the ALD. The rollers usually do not need ballasting to provide the required kneading action. Steel rollers are not preferred as they can fracture the aggregate. Rollers are discussed further in Section 25.5. The rolling process commonly becomes ineffective after six to nine roller passes, but the stones are also rearranged by subsequent traffic action.

This process produces a flat, stable configuration of stones, which assists the durability of the surface. The stones generally protrude both (Figure 12.1b):

* above the bitumen film, providing a good surface texture and protecting the bituminous film from traffic damage, and
* below the film into the basecourse, transmitting traffic load through the bitumen.

It is common for the surface to be rolled again prior to opening to traffic. Vehicle speeds must be controlled during the first hour or so of traffic operations to ensure that the stones in the surface layer are rearranged rather than removed. Usually there will still be an excess of unbonded stones and/or surplus aggregate which has not become part of the stable configuration. These loose stones should be removed from the surface as they present a hazard to traffic (e.g. may cause skidding) and may cause damage to vehicles (e.g. broken windscreens and headlights) and to adjacent facilities (e.g. lawns and drainage pits).

(b) Process design

It is common to design stone and bitumen application rates for spray and chip sealing by F. M. Hanson's (1935) method, which uses the average least dimension (ALD) of the stone particles in its calculations. The volume of stone used must be just sufficient to spread the required layer of stone over the primed surface, allowing for a wastage of about 10 percent.

The bitumen volume used must be sufficient to retain the stone surface layer in the seal coat, but must finish well short of the top of the final location of the stones, in order to avoid bleeding (Section 14.1C8) and thus lower skid resistance. Typically, the bitumen might rise half way up the aggregate layer. From Figure 12.1a it can be seen that the amount of bitumen needed is given by the air-void content (Section 8.1) of the rolled layer of surface stones, less allowances for the embedment volume and the volume within the surface texture (Section 12.5.1). The void content will typically range from 0.2 to 0.4, depending on the amount of rolling and trafficking. Applying an allowance of 0.5 for partially filling the voids leads to the bitumen volume needed being typically between 0.1 and 0.2 times the volume given by the surface area times the ALD. This corresponds to a bitumen application rate of 1 to 2 L/m^2 for ALD = 10 mm.

The type and quantity of bitumen needed will also depend on the:

* absorptivity of the basecourse surface and the effectiveness of the earlier priming;
* texture of the basecourse surface;
* absorptivity of the stones used in the surface;
* adhesion properties of the stones used in the surface layer (Section 12.3.2);
* shape of the stones used in the surface layer (Section 8.6a); and
* local weather conditions.

Common practice is for the bitumen to fill about 50 percent of the air voids in the surface layer of stones. During rolling the percentage usually increases to about 80 percent and the thickness of the finished layer approximates the average least dimension of the stones. The use of too much stone will prevent the reorientation shown in Figure 12.1b from occurring and may lead to stones being plucked from the surface. The use of too soft a basecourse upper surface will stop the stones from realigning themselves with their average least dimension vertical. Void contents will therefore increase. The use of too rough a basecourse upper surface will stop the stones from repositioning themselves to form a continuous horizontal mosaic. The use of too much bitumen may lead to bleeding (see above and Section 14.1C8).

(c) Treatment types

The application of primer followed by a surface seal coat covered with stones produces the spray and chip seal. As described in the next section, the term *surface dressing* describes the same process when used to improve the surface properties (e.g. skid resistance, impermeability, cohesion, or appearance) of an existing bound course. The five types of spray and chip seal are defined by the number and sequence of construction events used and are:

(1) The most common case is one application of bituminous binder and one layer of stones, as described above (called a *single coat* or *a single/single seal,*).

(2) One application of bituminous binder and two layers of stones. The second layer of stones is of smaller size to permit interlocking into the first layer. A major reason for using this method is that the smaller stones reduce movement of the larger aggregate during rolling or under initial traffic. This process is variously called *single coat, double chip; single-spray, double chip*; *racked-in; single coat with dry locking coat, added scatter coat* or *single/double seal.* Such a surface has about twice the traffic carrying capacity of a single-coat.

(3) Two applications of bituminous binder and two layers of stones (*double seal, double spray or double/double seal*). The second application may be delayed until the first seal has been tested under traffic. Binder rates are lower and stone sizes smaller in the second seal, to permit interlocking into the first coat. Triple applications are occasionally used.

(4) Two applications of bituminous binder and one intermediate application of stones. The second binder application is more of a slurry seal (Section 12.1.6.2). The resulting surfacing is called a *Cape seal.* This treatment is used when there is a need to ensure retention of the surface aggregate.

(5) Two applications of bituminous binder and two layers of stones (*two-coat*). The second layer uses larger stones.

Compared with asphalt (Section 12.2), spray and chip seals are relatively difficult to construct, requiring skilled operators and responsive equipment. They are also more likely to be affected by weather during construction. Spray and chip seals are widely used, however, because of their relatively low cost, greater ductility, and lower permeability.

12.1.6 Seal management

Spray and chip seal surfaces have good fatigue and fracture resistance and most will last between 5 and 15 years. As noted above, they can carry quite high traffic volumes, if constructed on a good basecourse and well-maintained (see below). The effects of traffic will usually relate to wear of the surfacing and rutting of the unbound basecourse (Section 11.4.3). Indeed, if the seal is kept intact, rutting that will determine the service life of the pavement.

If the surface has been designed and constructed satisfactorily using a durable stone, the onset of distress will often be decided by the rate at which the bitumen hardens (Section 8.7.1&7), rather than by the traffic volumes. The main factors determining the rate of hardening of the bitumen will be:

(a) service temperature levels;

(b) exposure of the bitumen to oxygen and temperature, and

(c) the inherent durability of the bitumen.

The distress will be in the form of surface cracking (Section 14.1B), loss of surface stone (Section 12.3.2), or edge cracking of the pavement (Dickinson, 1982).

In order to minimise whole-of-life costs, it is usually best to repair a distressed spray and chip seal as soon as cracking occurs in the seal — typically after about seven years — and thus delay the need to add a new seal. There are four main techniques used to do this. They are:

(1) *Surface enrichment.* This is a light application of a fluid bitumen, such as an emulsion (Section 8.7.6), to an existing sealed or asphalt surface. It can thus be considered as an extension of the primer seal technique. Its purpose is to enter and seal minor cracks and to recoat surface stones to delay them being stripped from the surface. When a very light application is used, the technique is referred to as a *fog coat* (or *fog seal* or *overspray*). Heavier applications are sometimes called *wash coats.*

(2) *Slurry seal.* A slurry seal typically uses a mixture of filler (e.g. cement, Section 8.3.1), fine aggregate, and 5 to 20 percent by mass of bitumen emulsion. It is applied cold by screeding the slurry onto the surface in layers less than 5 mm thick. It may take up to 4 hours for the emulsion to break and before traffic can use the road. The mix design is a compromise between being sufficiently fluid to fill cracks and being sufficiently stiff to attain a good microtexture (Section 12.5.1). The slurry seal is used to treat residential street pavements to:

* fill cracks,
* provide a good surface appearance,
* provide a more shear-resistant surface at intersection approaches than is available from spray and chip seals,
* protect and waterproof an old pavement, mainly by filling cracks, surface voids, and other minor surface deficiencies,
* raise the microtexture component of skid resistance (Section 12.5.1),
* fill shallow ruts, and
* minimise any increases in pavement surface level.

However, a slurry seal:

* has no influence on structural strength,
* will only slightly improve riding qualities,
* is often too smooth to use on streets carrying fast traffic,
* is an ineffective sealant,
* will permit cracks to reflect through to the surface (Section 14.1B7), if there is any movement, and
* has limited flexibility and poor fatigue and fracture resistance.

The method is also sometimes referred to as a surface enrichment seal, but surface enrichment (1 above) is better defined as excluding the use of filler and stones.

Microasphalt is a cross between a slurry seal and a thin overlay and is discussed in Section 12.1.7.

(3) *Resealing or surface dressing.* This is the application of a spray and chip seal (usually of type 1 in Section 12.1.5) to an existing sealed or asphalt surface. It is used to:
* seal a badly cracked spray and chip seal (a cost-minimisation study has indicated that it is an appropriate treatment when the cracked area exceeds about 50 percent of the surface area, Potter and Hudson, 1981),
* regain the advantages of the initial spray and chip seal,
* counteract the effects of bleeding (Section 14.1C8), and/or
* raise skid resistance levels.

If applied to an asphalt surface, it is sometimes called a *sandwich seal*, as the new layer of aggregate is sandwiched between two bituminous layers.

Before resealing, local defects such as cracks more than 2 mm wide, potholes, and non-structural ruts should be filled, edge breaks should be repaired, and structural defects should be reconstructed. Indeed, even if wide (> 2 mm) cracks are filled, they will usually reflect through (Section 14.1B7) at a later date if the underlying cause of the cracking is not removed. It is also important to ensure that the underlying bitumen has hardened — otherwise it may bleed through the reseal.

Bitumen application rates may need to be reduced across wheel-paths where the voids are full and the surface is flushed due to bitumen bleeding.

Most roads can accommodate one or two reseals before either a thick overlay (Section 12.2.6) or structural reconstruction of the basecourse is needed (Mulholland, 1989).

(4) *Use of geofabrics to produce geotextile reinforced seals (GRS).* Geotextiles are used to:
* lengthen seal life,
* reduce crack propagation, and
* reduce reflective cracking (usually effective for between 1 and 2 years).

The normal surface dressing in (3) is reinforced by including a layer of geofabric (Section 10.7) which typically is non-woven, needle-punched, and about 160 gm/m^2 in weight. To increase adhesion, the fabric is placed directly on top of a sticky primer applied to the original surface. The new seal coat (or spray and chip seal or asphalt) is then sprayed on top of the fabric.

(5) *Modified bitumens.* Bitumen can have its ductility, fatigue resistance, elasticity, adhesion, workability, temperature susceptibility, and strain rate susceptibility (Section 8.7.4) enhanced by the addition of rubber, other polymers, or geotextile reinforcement. The range of these additives is reviewed in Sections 8.7.8 and 10.7.

For example, the improvements in ductility, fatigue resistance, and waterproofing caused by the addition of rubber are of particular relevance when reseals are being placed over cracked surfaces, as reflection cracking will be reduced or eliminated (Section 14.1B7). Typical rubber-in-bitumen proportions by mass are:
* for stone retention in heavy traffic, 5 percent,
* to cover fine cracking, or for very heavy traffic, 15 percent,
* to cover severe cracking in structurally sound pavements, 20 percent.

Rubber-in-bitumen seals used in this way are sometimes (rather oddly) referred to as *stress-absorbing membranes* (*SAM*). This presumably refers to the ductility of the layer — i.e. to its ability to deform without cracking — rather than to any absorbing of stress. A material with low stiffness and high ductility will not transfer all stresses through to the new surface layer. The product is therefore more appropriately described as a *strain-absorbing membrane* (or *strain alleviating seal*), which fortunately has the same initials as SAM. SAMs can be quite thick and — in the manner of an asphalt, Section 12.2 — can include aggregate to help carry the traffic.

Stress- or strain-absorbing membrane interlayers (*SAMI*) are thin SAM layers which prevent cracks in the original pavement reflecting through to new pavement layers placed on top of them. SAMIs do not use aggregate.

SAMs can also be made of ductile unbound material and/or may contain geotextile fabric. In the latter case, the first step is to spray a bitumen tack coat at about 0.6 l/m^2. The geotextile is then immediately placed on top of the tack coat and rolled with a multi-tyred roller. The remainder of the bitumen required to complete the layer is then sprayed on to the fabric.

12.1.7 Thin overlays

The previous section discussed the application of seals of negligible thickness to an existing pavement. *Resurfacing* is the application of a new surface (or wearing) course of significant thickness to an existing pavement. The courses are called *overlays* and they are categorised as either thin or thick depending on whether or not they add structural strength and stiffness to the pavement. Thick overlays add strength and stiffness and are discussed in Section 12.2.6.

Thin overlays are usually classed as maintenance rather than rehabilitation (Section 26.1) and can last up to 15 years (using polymer-modified binder). They do not add strength or stiffness to a pavement and can only be used in cases where the pavement strength is satisfactory. They have two main purposes:

* to prevent structural deterioration. Cost-minimisation studies usually indicate that frequent maintenance using either thin overlays or surface dressing (Section 12.1.7) to control pavement deterioration at an early stage, and thus prevent major structural damage, is preferable to less frequent major rehabilitation (Potter and Hudson, 1981).
* to improve surface properties. In this respect, they are usually more effective and efficient than extensive local patching.

Thin overlays can be quite stiff, but their lack of strength means that they can only be placed over sound pavements. The use of thin overlays on cracked pavements is risky unless the cause of the cracking is removed, as the basecourse will continue to move and the cracks may reflect through the overlay (Section 14.1.7). This problem can be managed by the use of:

(a) binders which are mixtures of rubber-in-bitumen (Section 12.1.6),
(b) the use of somewhat thicker layers, which are more resistant to reflection cracking and can allow cracks to heal in warm weather,
(c) fabric reinforcement (Section 10.7),
(d) strain-absorbing membranes and/or crack-arresting layers (Section 12.1.4), and
(e) intentional joints in the overlay (Section 12.4.1).

There are three main types of *thin overlay*:

(1) A thin course of conventional asphalt (Section 12.2), usually between 25 and 40 mm thick, placed on a surface primed with bitumen. It is used where it is necessary to:
   * correct shape deficiencies, e.g. as a levelling course, regulating the surface and filling ruts and depressions,

* improve surface properties, e.g. by enhancing surface texture and skid resistance (Section 12.5) and reducing ravelling, permeability and patching,
* maintain surface integrity, and/or
* replace a failed surface course.

To achieve workability and durability the asphalt generally has either:

* for lightly trafficked roads, a high bitumen content, or
* for heavily trafficked roads, a harsher, lower bitumen-content mix, with workability improved by using a polymer-modified bitumen (Sections 8.7.8 & 12.1.6(4)).

(2) A very thin surface course (or *ultra thin open-graded asphalts, UTA*) which is usually 7 to 20 mm thick. It thus minimises any loss of vertical clearance, but can only correct minor shape deficiencies. Its main purpose is to improve surface properties.

First, a heavy tack coat (Section 12.1.3) is applied to the receiving surface to ensure impermeability and adhesion. The resulting strong bond to the basecourse is a feature of this surface type. Sufficient time must be allowed to permit the volatiles in the coat (Section 8.7.6) to evaporate, or else fatty areas (Section 14.1C8) will develop.

The course placed on the tack coat is a gap-graded (Section 8.3.2) asphalt comprised of 6 or 10 mm aggregate and a polymer-modified bitumen. The gap grading means that the running surface has an open texture which can produce good noise and spray suppression.

The *NOVACHIP* version of the process uses a polymer-modified bituminous tack coat that can be placed at the same time as the asphalt. The use of 10 mm aggregate results in a minimum course thickness of about 12 mm. The product has many of the good properties of a spray and chip seal (Section 12.1.5).

(3) A *slurry surface* (or *microasphalt* or *veneer surface*) overlay utilising the slurry technology described in Section 12.1.4(2). Slurry surfacing is used primarily to correct shape deficiencies.

Eight percent or more of bitumen emulsion or polymer binder is added to fine, sandy aggregate (below 10 mm). This produces a highly viscous material that is placed cold in layers from 7 to 30 mm thick. It does not need rolling and sets within 30 minutes.

12.2 ASPHALT

12.2.1 Terminology

An overview of asphalt terminology and processes is given in Tables 12.1 & 2. Asphalt has many similarities with Portland-cement concrete (Section 8.9), but the binders used in asphalt are bitumen (or asphaltic cement in the U.S., see Section 8.7.1) and tar (Section 8.8), rather than cement. The use of tar is now mainly of historical interest (Section 3.6.3) and in the remainder of this section it will be assumed that the binder is bitumen. Thus the text will generally use the word *bitumen* rather than *binder*, even when the material is probably a modified bitumen (Sections 8.7.6&8).

The stone used in asphalt can comprise coarse or fine aggregate (Section 8.3.1) or a mixture of the two. The coarse aggregate must meet:

* most of the requirements given in Section 12.1.3 for spray and chip seal aggregate,
* the adhesion requirements given in Section 12.3.2, and
* many of the requirements for an unbound aggregate discussed in Section 12.1.2. In the case of asphaltic concrete (Section 12.2.2), all these requirements must be met.

Table 12.1 Asphalt overview; giving asphalt components, their roles, and the resulting products. The shading indicates gaps in the grading.

<table>
<tr><td colspan="6">Asphalt components:</td></tr>
<tr><td colspan="2">coarse aggregate
Sections 8.3,5&6 & 12.3.2</td><td colspan="2">fine aggregate
Sections 8.3,5&6</td><td colspan="2">bitumen
Section 8.7</td></tr>
<tr><td colspan="6">→decreasing component size→</td></tr>
<tr><td colspan="6">Useful component properties:</td></tr>
<tr><td rowspan="2">provides interlock shear Strength</td><td rowspan="2">provides compressive strength</td><td rowspan="2">stiffens the bitumen</td><td>fills voids, displacing bitumen</td><td>fills voids, displacing air</td><td>glue providing tensile strength</td></tr>
<tr><td colspan="3">lowers permeability</td></tr>
<tr><td colspan="6">Products formed from these components:</td></tr>
<tr><td colspan="6">well-graded mix, typically asphaltic concrete (AC), Section 12.2.2a</td></tr>
<tr><td></td><td colspan="5">gap-graded mix, typically hot-rolled asphalt (HRA), Section 12.2.2b</td></tr>
<tr><td colspan="2"></td><td colspan="4">typically mastic, Section 12.2.2c</td></tr>
<tr><td colspan="2">stone mastic</td><td></td><td colspan="3">asphalt, Section 12.2.2d</td></tr>
<tr><td colspan="2">bituminous macadam,
porous friction course</td><td colspan="2"></td><td colspan="2">Section 12.2.2e,
Section 12.2.2f</td></tr>
<tr><td colspan="4">unbound course, Section 11.1.3</td><td colspan="2"></td></tr>
</table>

Table 12.2 Asphalt overview, giving asphalt types.

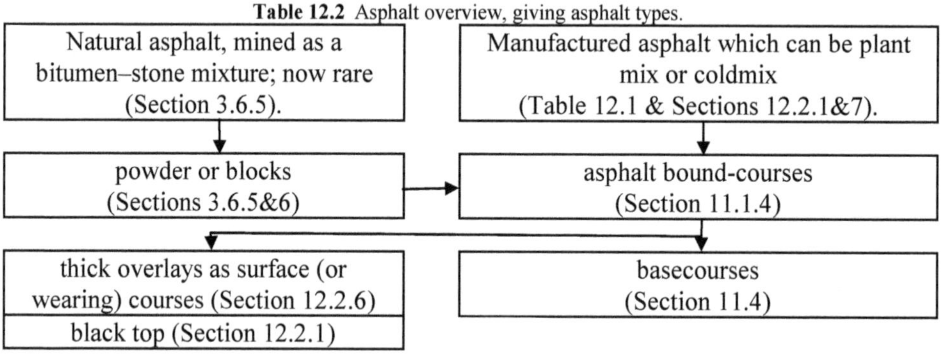

A large range of products is produced from the aggregate–bitumen mixture, permitting asphalts to be used for many purposes. A small quantity is mixed by hand or in small mixers and used for repair-work and other minor applications (Section 12.2.7). However, most asphalts are manufactured by a major piece of mechanical plant of the type described in Section 25.6 and are thus called *plant mixes* (or *hotmix*, *hot-mix asphalt* (*HMA*), or *hot pre-mix*). These terms originally applied to mixes where the aggregate was pre-heated to remove moisture and the bitumen pre-heated to lower its viscosity. The terminology has tended to drift to now infer that the mix is not only prepared hot at about 150 C (Section 25.6), but then maintained at that temperature and spread and compacted whilst still hot. Therefore, for practical purposes, plant mix and hotmix are usually synonymous and cover the bulk of the product used. *Black top* is a layman's term for all bituminous surfacings. There are some specific terms used to define particular asphalt usages:

* *deep lift asphalt* refers to courses at least 75 mm thick.

* *full depth asphalt* is a term used to describe a pavement in which all the courses are comprised of asphalt, and will usually be over 200 mm in total thickness and placed on a subgrade with a CBR of at least 10 (Sections 8.4.4 & 11.1.2).
* *deep strength asphalt* is a full depth asphalt placed on a cement-stabilised sub-base (Sections 10.1 & 11.1.2).
* *thick or long-life asphalt* is an asphalt course designed for a long, indeterminate life — e.g. to eliminate the future need for major rehabilitation works under heavy traffic. The asphalt will usually be at least 250 mm thick. It may, nevertheless, suffer from cracking and rutting of the surface layers.

Thin asphalt overlays and other surface courses can have thicknesses down to 25 mm (Section 12.1.7), but bound structural courses of asphalt are rarely less than 75 mm thick. An exception is the Canadian use of thin (20 mm) asphalt courses for low-cost road construction.

It is necessary to ensure that plant mix does not segregate into its components during transport and application. A common problem relates to the bitumen dripping away from the aggregate. The compactibility of asphalt is discussed in Section 12.4.

When plant mix is used as a basecourse it is in competition with Portland-cement concrete and unbound material. In the latter case it must rely on its lower equivalent thicknesses (Section 11.4.6) to compensate for its higher volumetric cost. When used as a surface course it is in competition with the spray and chip seal surfaces discussed in Section 12.1.4. In this instance its greater cost is compensated for by its contribution to structural performance and its ability to produce a more regular and immediately useable running surface. It is also often preferred in urban areas because of its speed of construction, the relative ease with which it can be swept, and its smoother surface. The last attribute aids non-vehicle uses and reduces noise, whilst not seriously compromising skid resistance (Section 12.5.2).

The early history of asphalt is discussed in Sections 3.6.2,5&7. Today, the term *asphalt* is commonly used for all mixtures of bitumen and aggregate that are placed hot. However in British practice the term sometimes implies the use of hard bituminous binders (e.g. Croneys, 1991). This usage probably arose from the fact that many binders were originally made from powdered *native asphalt* (Section 3.6.2) and hence incorporated some fine aggregate; this would have produced a stiffer binder than the alternative use of tar from coal works (Section 3.6.3).

12.2.2 Mix types

There are a number of different types of asphalt in common use. These are discussed below.

(a) Asphaltic concrete

Asphaltic concrete (*AC*) — or *bituminous concrete* — is a term best used to describe asphalts where the aggregate sizes used are well-graded and thus in particle-to-particle, intergranular, contact in the final product. This means that the coarse aggregate is the primary load-carrying medium, forming a structural skeleton within the course. As well as meeting the aggregate requirements in Section 12.2.1, AC aggregate must therefore also meet the appropriate requirements for aggregate used for unbound courses, i.e. it should be hard, relatively cubic (test 8.6a), and have sharp, angular corners to ensure good interparticle contact and interlock (Section 11.1.3). These attributes provide AC with good compressive and shear strength and stiffness (Figures 8.7&8). However, AC can be limited in tension and fatigue due to its reliance for tension on the thin bitumen films formed within the small natural voids available in a well-graded mix. The high stiffness also makes AC relatively harder to work and compact than the more bitumen-rich mixes. In addition, when any voids do occur — usually the result of poor placement — they tend to be large and interconnected.

The fine aggregate in AC acts as both a stiffener for the bitumen and as an inert material, bulking (or 'extending', i.e. cheapening) the mixture. It is usually angular (sharp) sand, which develops good interparticle interlock (Section 11.1.3) and hence strengthens and stiffens the mix in shear. The sand is sometimes supplemented by finer material (*fillers*) such as fly ash (Section 8.5.6), stone dust (Section 8.5.9), ground limestone (Section 8.5.4), cement (Section 8.9.1), carbon black, and sulfur. In most cases, the finer the filler and the higher its inherent void content, the better it performs.

AC aggregate size is typically between 20 and 40 mm in size. A misleading British practice is to refer to a well-graded mix where the aggregate size is limited to 20 mm as *dense bituminous* (or *bitumen*) *macadam*. It is thus a form of AC using more smaller-sized coarse aggregate. Wearing course for lightly trafficked roads can use aggregate sizes as low as 7 mm.

A heavy duty asphalt pavement might typically comprise 200 mm of AC placed over a sub-base of 150 mm of cement treated crushed rock. A medium duty pavement might comprise only 50 mm of AC over a crushed rock sub-base.

(b) Gap-graded asphalt

Gap-graded asphalt (Section 8.3.2) (or *hot rolled asphalt* (*HRA*), *rolled asphalt* or *low stone-content mixes)* usually has less aggregate than needed for a well-graded mix. Thus the initial void content is relatively high, but good mixing ensures that the voids are well distributed and not inter-connected. The bitumen fills all these voids and this leads to higher bitumen contents than with AC, producing mixes that are dense and impermeable and, at low temperatures, relatively stiff. However, unless specific measures are taken, the high bitumen content will lead to poor surface-friction for traffic.

Mixes of this type are common in Britain where they are referred to as *hot rolled asphalt*. The gap in the grading is between the 2.36 mm and 10 mm sieves — so the coarse aggregate is suspended in a mastic of bitumen and sand. A characteristic of HRA relative to AC is that it relies more on a good, stiff bitumen, rather than on the coarse aggregate, to provide mix stiffness and strength. In particular, HRA has relatively good tensile and fatigue strength, reflecting the tensile attributes of the extensive bituminous network. However, HRA must be used with some caution where high temperatures would soften the bitumen and lead to rutting and bleeding (Section 14.1C8). This effect is counteracted by using a stiffer bitumen further stiffened by the addition of angular sand or other fine aggregate. To achieve good surface friction, pieces of angular 14 mm broken stone are rolled into the surface during field compaction.

Greater workability, particularly ease of compaction (Section 12.4.1), means that HRA is often more suited to residential streets than is AC.

(c) Mastic

A straight mix of bitumen and sand produces *mastic* (or *sand asphalt*), that may be considered an extreme form of gap-graded mix. It usually contains about 30 percent bitumen, 40 percent filler, 30 percent fine aggregate, and virtually no air voids. Mastic cools to a stiff material but is quite fluid when hot. Thus it that can be poured or hand-floated into place and can provide a smooth surface finish. It gains its fluidity from the fact that, as with HRA, the aggregate particles are not in contact. The filler content is kept high to enhance the stiffness of the bitumen, but must be carefully proportioned or else voids will occur in the mix. In cold climates mastics can be exceptionally hard.

Mastic is used for footpath construction, surfacing low-speed residential streets, and as a coldmix (Sections 3.6.2 & 12.2.7). A coarse surface texture is rarely required in residential streets and the sandpaper texture produced by the fine-aggregate content of HRA or mastic is usually adequate. Mastic asphalts require lower temperatures than AC or HRA to achieve workability and so can be used for repair work.

(d) Stone mastic asphalt (SMA)

Stone mastic asphalt (or *split mastic*) is an open-graded mix (Section 8.3.2) in which the larger aggregate pieces — usually restricted to a maximum aggregate size of 20 mm — are in contact and able to interlock. Thus the behaviour in compression is similar to that of AC and the aggregate requirements are similar to those for AC. However, the voids in the open-grading are just filled with a stiff bituminous mastic, gaining some of the tensile and fatigue strength of HRA.

Cellulose or polymer fibres are often added to the mastic in order to achieve adequate mastic stiffness during high temperature conditions and thus prevent the binder flowing out of some of the voids in the aggregate mix during transport and placement. They also act a micro-reinforcement during subsequent service conditions. The fibres used are typically made of polyacrylonitrile, about 5 mm long, and represent about 1.5 percent of the mix by weight. This gives about 15 M fibre/m^2. However, the fibres can increase mix porosity and therefore reduce long-term stiffness.

SMA was originally developed in Germany in the 1960s to provide good wearing resistance to studded tyres. Today it is used to achieve:

* good rut resistance based on the compressive strength and shear stiffness of AC.
* good fatigue strength due to the high bitumen content of about 7 percent by mass. This means that it is suitable for heavily trafficked roads and cold-weather conditions.
* a rough, wear-resistant surface texture that provides good skid resistance, reduced water spray and low noise generation.

SMA is relatively expensive compared with other asphalts and so it is usually used as a thin surface course or as an intermediate course (Section 11.1.2).

(e) Bituminous macadam

Macadam was defined in Section 11.1.3 as a single-sized (or *uniformly-graded*, Section 8.3.2) mix. The mixture of macadam and bitumen is commonly called *bituminous macadam* or, if tar is used, *tarmacadam*. *Tarmac* is a British trade name for a tarmacadam mix using slag (Section 8.5.6) as the aggregate. The binder content of these mixes is usually kept relatively low to preserve stiffness and to keep costs low. This often results in a quite stiff mix with a high air-void content. In order to make the mix manageable, it is common practice to apply some bitumen to pre-coat the aggregate and aid compaction. The resulting product is called *coated macadam* or *pre-coat* and is frequently used for basecourses. Modern mixes usually use thicker than normal pre-coating films of bitumen to prevent unacceptably rapid hardening of the bitumen under exposure to air and light (Section 8.7.1).

(f) Porous friction courses
On roads carrying medium- to high-speed traffic, a coarse surface texture is often desirable in order to provide skid resistance and to reduce spray, noise, and specular light-reflection under wet conditions (Section 24.2.1). To meet these objectives, bituminous macadam mixes (e above) with conventional aggregate sizes (up to 20 mm) can be used to produce *porous friction courses* (or *open-graded friction courses [OGFC]*, *plant-mix seals*, *popcorn mix*, *pervious asphalt*, *porous asphalt*, *very open asphaltic concrete (VOAC)*, or *open-graded asphalt*). Typically, in a mix with a maximum stone-size of 14 mm, the 2 to 8 mm sizes are omitted and most of the stones would be above 8 mm. Some fine filler is also added. The low bitumen content ensures that most of the voids in the open grading remain unfilled. Stiff or rubberised bitumens are often used to prevent the hot bitumen draining into the open pores.

The open surface of a porous friction course inherently enhances skid resistance, and the inter-connected voids allow:

* noise to dissipate via alternative noise paths to such an extent that the courses are up to 3 dB(A) quieter than conventional surfaces (Section 32.2.3), and
* surface water to drain rapidly away, thus both improving visibility by reducing spraying and splashing and increasing wet-weather skid resistance.

To ensure that the course is self-draining at its lower longitudinal edge, it must be placed above the vertical face of any abutting kerb or channel. It must also be placed on an impervious underlying course as it will not protect those courses from water.

Porous friction courses have the following drawbacks relative to conventional asphalt in that they:

* are normally more expensive, requiring higher-quality aggregate and bitumen,
* are of lower strength, as there is little interlock between the aggregate particles — any strength comes from the bitumen, which is thinner and more flexible than in conventional hot rolled asphalt. Porous friction courses are therefore inappropriate at intersections.
* may lose their permeability, spray suppression and good frictional properties after a relatively short period of time (typically, after about 7 year), particularly as a result of continued compaction under traffic,
* polish easily, unless the coarse aggregate is particularly suited to providing frictional resistance, i.e. is angular rather than round and has good polishing resistance (Section 8.6o),
* cannot retain their good friction properties by attrition of the stone surface, as there will be no alternative new stones to take the place of worn ones, and
* may freeze more readily, require more salt during the snow season, and hinder snow removal.

12.2.3 Density, air voids, and bitumen content

1. Density
Following Section 8.1, there are some specific volume related terms used in asphalt:

* *bulk volume* is the total volume, V, of an asphalt sample,
* *apparent volume* is the volume of stone used in the asphalt, and
* *volume concentration of aggregates* (*VCA*) is the ratio of apparent volume to bulk volume.

As will be seen in Section 12.2.4, asphalt design places a strong emphasis on *density* — and thus on the compaction of the in situ mix. This is similar to the emphasis on density encountered for unbound courses in Chapter 8 and elsewhere. For instance, increasing density:

(a) increases stone interlock and therefore produces a stronger and stiffer mix, and
(b) minimises air-filled voids, which in turn:
  * increases the effective bitumen stiffness, thus increasing mix stiffness, and
  * decreases permeability, thus raising durability by preventing the entry of:
    – air, leading to the oxidation of bitumen (Section 8.7.7).
    – water, leading to bitumen stripping (Section 12.3.2).

To achieve high densities, mix proportioning for AC is commonly based on the use of Fuller's maximum density curves for the grading of the aggregate (Section 8.3.2). These curves originally used n = 0.5 in Equation 8.3, but values of n = 0.6 are common in current procedures. The proportions of each constituent in a typical AC mix are shown in Figure 12.2.

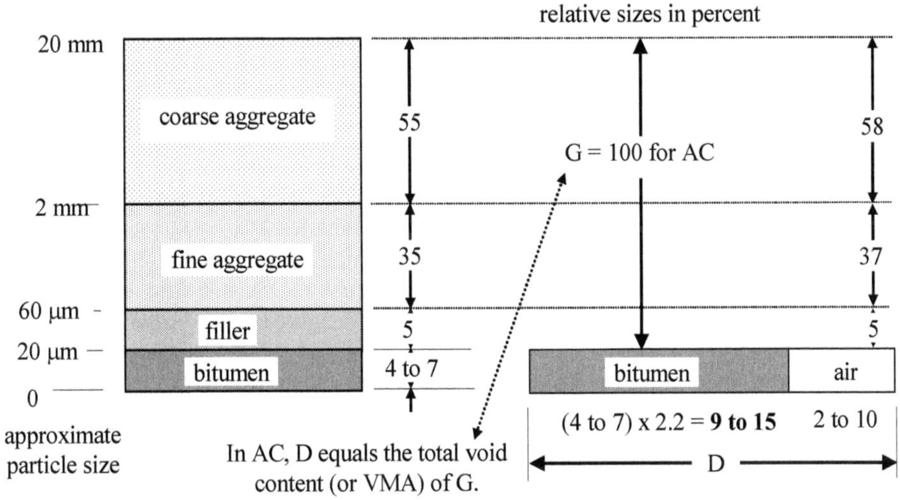

Fig. 12.2 Asphalt terms and proportions. The typical percentages shown are for AC. The 2.2 in the bitumen volume calculation is its density relative to the aggregate density, i.e. (γ_b/γ_{agg}, see Equation 8.1f.) Particle size definitions are from Figure 8.7.

2. Aggregate air-voids

The volume of the voids in the batch of aggregates to be used to make an asphalt will depend on the aggregate grading and will directly influence the amount of bitumen needed. This initial void volume is precisely defined as the volume of water that the batch could absorb in 24 h, excluding any absorbed air. In asphalt technology, the *total void ratio* (e, Section 8.1) of the batch of aggregate is usually called the *voids in mineral aggregate* ratio (VMA, Figure 12.3a), i.e.

$$VMA = e_{agg.}$$

VMA ranges from around 10 percent for mixes using well-graded stones, to 17 percent for conventional AC mixes, to almost 30 percent (typically 18 to 23 percent) for open-graded mixes. Similarly, if

$$VCA + VMA = 1$$

then VCA is given by

$$VCA = 1 - e_{agg}$$

These approximations do not apply to mixes where the aggregates are separated as a consequence of:

* the stiffness of the bitumen,
* the bitumen more than filling the voids, or
* the bitumen overflowing the original aggregate voids, even though some air voids remain. Thus the two approximations will not apply to many gap-graded mixes, even though they contain air voids.

These effects result in the final asphalt volume exceeding the original aggregate volume.

3. Bitumen content

When bitumen is added to the aggregate, some is absorbed into the aggregate, but most at least partially fills the VMA. No satisfactory method is available for measuring the amount of bitumen absorbed into a piece of aggregate, but the issue arises if surface pore diameter exceeds 50 nm. AC and asphalt for porous friction courses normally contain 4–7 percent by mass of bitumen (Figure 12.2). HRA typically contains 6–8 percent. Increasing the bitumen content of a mix:

* decreases mix stiffness and can lead to rutting (Sections 11.3.2 and 12.2.4). Increasing the air void content also decreases stiffness.
* increases mix fatigue resistance (Section 11.4.4). Increasing the air void content will increase fatigue resistance,
* reduces the loss of surface aggregate (Section 14.1C3), and
* improves workability, until the bitumen volume becomes excessive.

The percentage of bitumen by mass in a mix can be estimated by:

* extracting the bitumen with a chlorinated solvent, and then measuring its mass,
* for AC, assuming the bitumen volume is equal to the air void content of the aggregate mix, and then dividing by the ratio of aggregate to bitumen densities (see Equation 8.1f),
* measuring the masses of bitumen and aggregate used in making the mix, and
* using a nuclear gauge in accordance with ASTM D4125 (Section 8.2.2d).

The determination of bitumen content commonly has a coefficient of variation of about 5 percent. The accuracy is important, as the bitumen content needs to be kept within ± 0.5 percentage points of the specification value to avoid ravelling due to insufficient bitumen or rutting due to excess bitumen.

The volume of bitumen which is not absorbed by the aggregate is called the *effective bitumen volumetric content* and is given in consistent nomenclature by $b_{ve}V$. For AC, the bitumen only operates as a thin film (perhaps 10 μm thick) confined between the relatively rigid aggregate particles and the voids are not filled with bitumen (Figure 12.2). The minimum bitumen content relates as much to surface area as to initial void content. Thus, for AC and other mixes where the air voids are not flooded with bitumen, the link between bitumen content and the final air void ratio, a_f, is closely given by:

$$b_{ve} = e_{agg} - a_f \qquad (12.1)$$

If the densities of bitumen and aggregate are similar, b_{ve} will also be the bitumen content by mass.

The overflow of bitumen beyond the initial aggregate voids ($a_f = 0$) is sometimes called *flooding* (Figure 12.3c below). The excess bitumen permits positive pressures to build up in the bitumen and the mix loses interparticle stiffness and interlock shear strength in a manner analogous to the effect of positive pore pressure in saturated soils (Section 9.2.5). The mix must therefore rely on the bitumen matrix to replace these losses with its own inherent stiffness and strength. However, a side effect may be plastic permanent deformation of the bitumen matrix in hot weather, leading to rutting and corrugating (Section 14.1A5). Flooding may occur due to:

* a high initial bitumen content. This is the case with HRA where stiffer bitumens are used to obviate flooding problems, while still taking advantage of the improved tensile strength and fatigue resistance provided by a continuous matrix of bitumen.
* hot weather, leading to the bitumen softening and thermally expanding out of the voids, or
* the course compacting and the void content diminishing with age.

A further term sometimes used is the ratio *voids bitumen filled* (*VABF*), which is the ratio of the (unabsorbed and unflooded) bitumen volumetric content to the void volume $e_{agg}V$. Thus, for AC only, VABF — which is typically about 0.75 — is given closely by:

$$VABF = b_{ve}/e_{agg}$$

and using Equation 12.1 leads to:

$$= 1 - a_f/e_{agg} \qquad (12.2)$$

Measurements of VABF have a coefficient of variation of about 5 percent (Auff, 1984). A linked term is the *effective volume*, which is the total volume less the volume of bitumen; i.e.:

$$(\text{effective volume})/V = 1 - b_{ve}$$

For AC and using Equations 12.1&2, this becomes:

$$(\text{effective volume})/V = 1 - e_{agg} + a_f$$

$$= 1 - a_f(VABF)/[1 - (VABF)]$$

The density of a mix reaches a maximum at a particular bitumen content. This is usually slightly higher than the bitumen content that produces maximum stiffness. The decrease happens as the bitumen effectively fills the available voids and begins to separate the higher-density aggregate particles.

Densities can be predicted from the known properties of the mix constituents (Sections 8.1 and 8.2.2), from volume and mass measurements on cores, or from nuclear density meters (Section 8.2.2d). The nuclear density approach is covered by ASTM D2950. The method can be extended by mounting the meter on the roller and continuously monitoring density during compaction. As density is therefore relatively easy to measure, it is often used as a primary control test for asphalt production. For example, as discussed in Section 8.2.1, the degree of compaction and air-voids content of a mix may be estimated by measuring its density. Mix testing is discussed in the next Section.

4. Mix air-voids

The voids which are not filled with bitumen become the air-void content, *a*, of the mix (Figures 12.2&3). The coefficient of variation for the determination of *a* is over 20 percent (Auff, 1984). As the air-void content will drop due to traffic compaction, surface courses can be placed at relatively high initial air-void levels. (Of course, the bitumen mass content, b, will not change during traffic compaction.) The final air-void ratio, a_f (Equation 8.1h) will range from about 2 percent for mastics, to 2 to 9 percent for AC, to 15 percent for other asphalts, to 15 to 22 percent for porous friction courses (Section 12.2.2). Of course, for a lightly trafficked road traffic compaction cannot be relied on and the initial air voids must be kept low.

Bitumen hardening due to oxidation (Section 8.7.1) is usually negligible for air-void contents of 2 percent or less. However, with such low air voids, there is a risk of rutting and shoving problems from soft bitumen flooding the mix, as discussed below. Hence, it is common to require that the final air-void ratio be at least 3 percent, or to use a very stiff

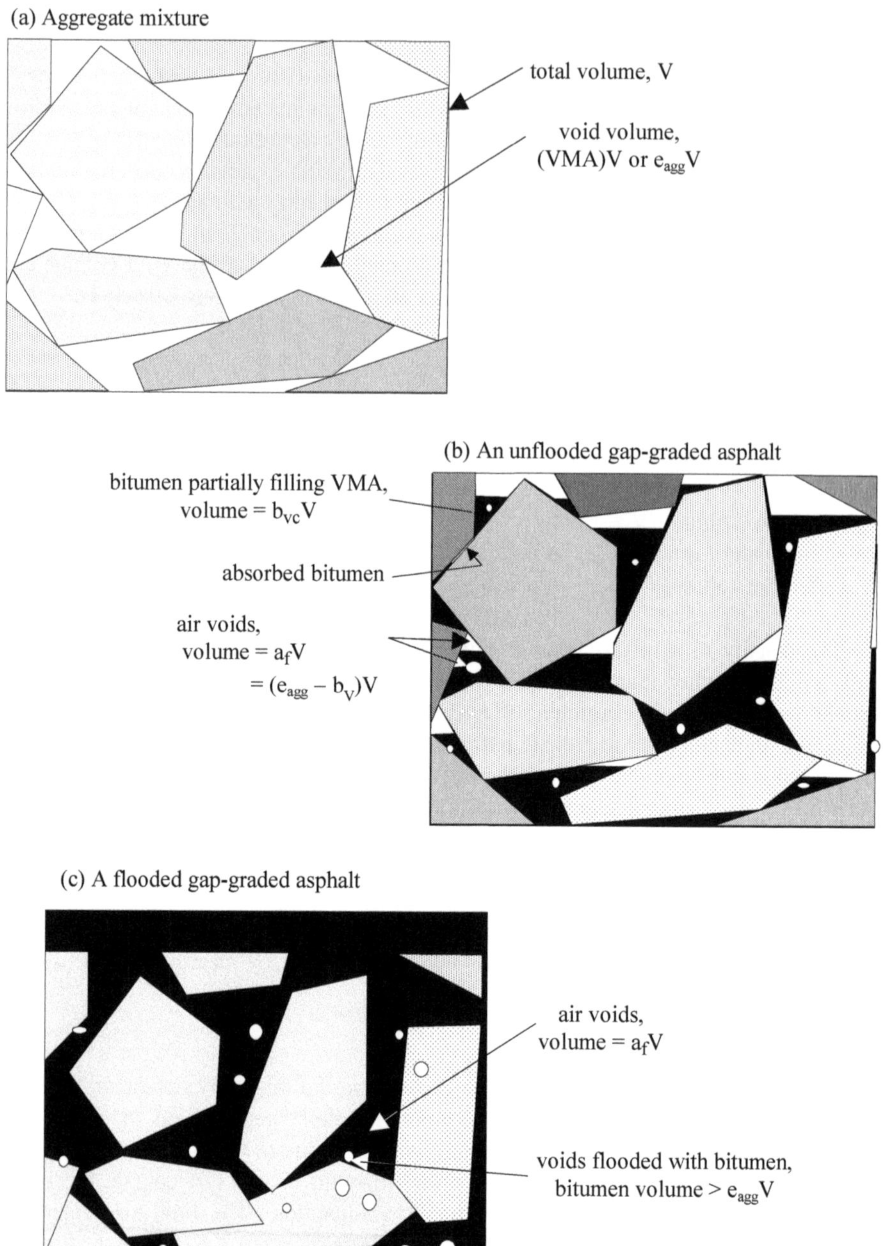

Fig. 12.3 The geometry of air voids and bitumen in an asphalt mix.

bitumen (Section 8.7.5). Many mixes exhibit maximum deformation resistance when their air-void content lies between 3 and 7 percent:

* 4 percent is a typical recommended level for lightly trafficked roads (< 500 kESA)
* 5 percent for medium to heavily trafficked roads, and
* 6 percent for heavily trafficked sites

For polymer-modified mixes (Section 8.7.8) these values are usually reduced by 1 percent to cover increased binder contents.

Asphalt will become permeable if the air-voids content exceeds about 6 percent. If the air voids then fill with water, positive pore pressures can develop (Section 9.2.5) and lead to aggregate stripping (Section 12.3.2), bleeding (Section 14.1C8) or rutting of the stripped bitumen, and/or surface pocking as the voids explode to the surface. Some specifications therefore define maximum values for a_f to maximise asphalt durability.

Desirably, air voids should be small, dispersed and occluded as their deleterious effect will be worse if they are:

* interconnected, thus permitting water and oxygen to enter, or
* large and infrequent, thus permitting large local deformations.

12.2.4 Mix design

The primary aim of the design of asphalt mixes is to determine the proportions of bitumen, filler, and aggregate that will produce a mix that will:

* be spreadable and compactable,
* be adequately stiff in hot weather and crack-resistant in cold weather,
* be adequately strong, durable, and impermeable, and
* where necessary, have a suitable surface texture with respect to skid resistance, light reflectance, and wear resistance (Section 12.5.1).

A recipe method specifies the components to be used in the mix, and their proportions. From a *recipe* viewpoint, Section 12.2.3 demonstrated that mix density is a suitable surrogate for many of these properties and can therefore be used to test or even certify a proposed recipe. Secondary aims then relate to providing the chosen mix proportions with special properties, such as rut resistance under heavy traffic.

However, a *performance* view shows that, in particular, an asphalt course must resist rutting and fatigue cracking under traffic, and thermal cracking over time. The key structural design properties of asphalt are therefore:

a. stiffness in the short and long term (creep),
b. static and dynamic strength in compression and tension,
c. fatigue strength,
d. ductility, and
e. durability.

These mirror the general pavement failure criteria listed in Sections 11.4.3–5 and are now discussed in turn.

a. Stiffness

Stiffness is a fundamental component of pavement design (Section 11.3). The key aspects of stiffness in the short term are discussed in Section 11.3.2 and in the long term in Section 11.4.3. To meet the associated pavement demands, asphalt mix design must provide products with adequate stiffness and so this section concentrates on the links between mix design and asphalt stiffness.

Generally, mix stiffness will increase with increasing bitumen content up to some maximum value. This initial increase is due to the greater triaxial cohesion of the bound course. However, as the bitumen content continues to increase, aggregate interlock tends to be lost and more reliance is placed on bitumen stiffness. Stiffness then begins to fall. A good design must therefore utilise the binding and viscous properties of the bitumen without compromising the strength and stiffness of the mixture.

As discussed in Section 12.2.2a, for stiffness and strength in AC, it is only necessary to have enough bitumen to coat the aggregate particles. Stiffness then comes from particle interlock and empirical studies indicate that AC mix stiffness decreases linearly with increasing VMA.

On the other hand, interparticle contact may not occur in HRA (Section 12.2.2b) and bitumen viscosity will therefore play a greater role in determining HRA stiffness. Bitumen stiffness is temperature-dependent and, as discussed in Section 11.3.2, the temperature in an HRA course therefore has a major influence on its stiffness. Recall that in HRA the deleterious effect of bitumen on the strength and stiffness of a mix can be controlled by adding filler or polymers (Sections 8.7.8 and 12.2.2a) to the bitumen and hence increasing its inherent stiffness.

Stiffness is commonly measured by an indirect tension test (Section 8.6q), using equipment such as the British NAT or the Australian MATTA devices.

b. Strength

The static strength of an asphalt is normally considered in relation to its resistance to permanent deformation, which guards against rut formation (Section 11.4.3). The deformation observed will be partly due to compaction (Section 12.2.3) and partly to shear failure under load. A lack of shear strength is a common cause of so-called *tender* mixes (Section 12.4.1).

Deformation can be measured using strength tests under repeated-load situations (Section 8.6r and Equation 14.7), from wheel tracking tests (test w, a small versions of tests t1 and t2, Section 8.6), or from indirect tests of mix viscosity (or creep, Section 33.3.3). The tracking test has been well correlated with full-scale accelerated loading tests (test ts, Section 8.6o (Paul, 2003)). In recent times, wheel tracking tests (test 8.6t2) have been developed to such a stage that in many countries they are now the prime tool used to 'prove' a mix design.

c. Fatigue

Criteria for fatigue strength are of the form shown in Equations 11.13 and 14 relating the log of the fatigue life ($\log N_f$) to the loading strain. The data are usually established directly from fatigue tests on rectangular beams, rotating cantilevers, tension members, triaxial specimens (test 8.6r), or diametrically-loaded cylinders in the indirect tension test (test 8.6q). Wheel-tracking tests (test 8.6t2) have also been used to imply fatigue performance. As with stiffness, $\log N_f$ decreases linearly with air void content. Field data current gives a life about one-tenth of that predicted from laboratory data (Maccarrone *et al.*, 1996).

d. Ductility

If an asphalt is too stiff and/or lacks ductility, it may be susceptible to low-temperature cracking, particularly at the pavement surface. The basic mechanism is that large strains are induced in the asphalt by either thermal shrinkage or by cracking in associated underlying layers. In a stiff asphalt, this will result in high stresses and heighten the chance of tensile cracking strains being achieved. Values to control this are given in Section 11.4.4.

e. Durability
As shown in Section 12.2.3, bitumen durability is not a problem with a well-compacted mix with few air voids, as only the bitumen at the traffic surface is exposed to weathering. This can be controlled by managing the bitumen-related factors discussed in Section 8.7.7. Bitumen loses durability at high temperatures as the reaction rate at which it is attacked by atmospheric oxygen doubles for every 10 C rise, thus preparing the way for subsequent low-temperature cracking (Section 8.7.3). This critical combination of high and low temperatures can be found daily in many dry inland regions.

f. Test conditions
Within porous structures, water vapour movements are affected by temperature change and can cause cracking. As a consequence of this, the load-spreading capabilities, crack-resistance and strength of an asphalt course will be quite temperature-dependent. Designers must therefore determine appropriate maximum and minimum service temperatures for a pavement.

It is common to base the associated pavement design (Section 11.4.2) on values of some or all of the above properties obtained from tests on the proposed mix. If this route is followed, experience indicates that sample preparation is of major importance. For consistent results, the component material and mixing equipment should be brought to a uniform temperature (typically 150 C) and held there for a considerable time (perhaps 4 hours) before mixing takes place. The test mix should then be preconditioned by compaction aimed at simulating field conditions. It is essential that this laboratory compaction process represents field conditions, as test results will be very sensitive to the method of compaction. There are two main types of appropriate test compaction method:

* *rolling-wheel compaction* (Section 8.6t2) is often preferred as it best reproduces field loading.
* *gyratory compaction* (Section 8.6f, using devices such as Gyropac or the French PCG) in which the specimen gyrates during kneading compaction can be made — with careful calibration — to closely reproduce field conditions.

On the other hand, for diagnostic work, the asphalt samples are usually based on coring or results from nuclear density meters (Section 8.2.2).

12.2.5 Empirical design methods

Asphalt mix design has traditionally used either direct recipes or recipes modified by empirical methods. As an example of a modified recipe method, the determination of mix proportions — particularly bitumen content — often follows the empirical *Marshall, Hubbard Field*, or *Hveem procedures* (Sections 25.3 and 25.7.2). These methods were developed for aggregates with a maximum size of 25 mm.

The *Marshall test* (ASTM D1559) is the common method used for both approaches. As a first step, a laboratory trial mix is produced. Cylindrical specimens commonly 64 mm long and 100 mm in diameter are prepared and compacted following standard practice. Compaction is based on the Proctor hammer from soil testing (Section 8.2.5) and the hammer is in contact with the entire end-face of the cylinder. The number of blows is chosen to represent expected traffic levels — commonly 75 for heavy traffic. In examining an existing pavement, the test can use cores removed from the pavement — which will also help establish the pavement thickness. A similar but less dynamic method is followed in the *Hubbard Field* procedure which was originally developed for mastics (Section 12.2.2C).

The density and air voids are then measured (Section 12.2.3) before the specimen is heated to 60 C (a maximum service temperature). It is then loaded in compression on two diametrically opposite surface generators. The load is applied at a deformation rate of 830 μm/s and the maximum load in kN is recorded as the *Marshall Stability*. The *Flow Value* is the deformation at this peak load. The test is similar to the Brazil Test (Section 8.6q) for indirectly measuring tensile strength, although its test temperature is usually too high to lead to tensile failure. The test loses its effectiveness with asphalts using poorly-graded aggregate.

The Stability and Flow Value are thus particular measures of the dynamic strength and stiffness of the mix and do not accord with the conventional technical meanings of the terms stability and flow. The terminology arose because the test originally detected mixes with high void contents that tended to deform plastically when loaded. Typical values for these two quantities are 4–7 kN for Stability and 2–5 mm for Flow. They have coefficients of variation of about 10 and 15 percent respectively (Auff, 1984). The Marshall properties have proved useful for quality control and producing constructible mixes, but offer little ability to predict field performance directly and — if used for this purpose — can be misleading.

The Marshall Stability of AC increases with increasing bitumen content to a maximum of about 6 percent by mass. It increases more than linearly with increases in mix density. The Flow Value increases steadily as the bitumen content increases. The two properties for tests on specimens with a range of bitumen contents are then used in empirical procedures which allow the appropriate bitumen content for a particular mix to be evaluated.

Some design methods (Section 11.3.1) analyse the bound asphalt layer using soil mechanics principles, relying on knowledge of its cohesion and internal friction (Figure 8.9). Two special mix tests give semi-empirical estimates of these properties. They use the *Hveem stabilometer*, which is a form of triaxial test, and *Hveem cohesiometer*, which is a form of bending test (ASTM D1560). Hveem specimens are compacted by kneading, with the hammer in contact with only a small proportion of the end-face of the cylinder.

Related volumetric design methods determine the mix proportions based on a recipe modified by laboratory compaction tests.

12.2.6 Thick overlays

Following the cost-minimisation remarks made in Section 12.1.7 for thin overlays, the pattern of events in the life of the pavement is that deflections will tend to increase at a low, steady rate until the occurrence of such deterioration causes as surface cracking and water entry. These create snowballing effects and accelerate the deterioration process. Ideally, overlaying should occur before this snowballing stage is reached (Section 14.5.3).

Section 12.1.7 discussed the use of thin overlays and noted that they cannot alter a pavement's shape (e.g. to improve its ride quality), reduce its roughness (Section 14.4), or add to its strength and stiffness. The term *reconditioning* refers to work that does reduce the roughness or increase the strength and stiffness of an existing pavement. The reconditioning technique used will depend on the current condition of the existing pavement, and the expected future traffic and moisture environment.

The most elaborate reconditioning method is to replace the whole pavement, a process known as *reconstruction* (Section 26.1). The simplest reconditioning method is to place a new course of appreciable thickness on top of the existing surface. This process is called *resheeting*; the added courses are called *thick overlays* and are usually composed of asphalt, plain concrete, or reinforced concrete. The concrete overlays may or may not be bonded to the underlying pavement. Resheeting is usually classed as rehabilitation (Section 26.1) rather than as maintenance.

A thick overlay is over 40 mm thick and is used primarily to stiffen and strengthen a pavement. If the overlay is under 70 mm thick, it will still be significantly influenced by the structural performance of the overlain pavement. Thick overlays can improve roughness; typically each 15 mm of thickness reduces roughness by an IRI (Section 14.4.5). However, the effect is indirect and care must be taken to avoid the overlay simply repeating the poor longitudinal shape of the existing pavement.

An overlay in which the thickness is deliberately varied to correct or adjust shape is called a *levelling course*. A similar course applied during construction is called a *regulating course*. The aggregate size rarely exceeds 14 mm and needs to be less than the planned minimum thickness.

In urban areas the need to maintain surface levels may require the existing surface to be removed by milling, scarifying, or planing before the new overlay is placed. The material removed may be suitable for *recycling* (Section 8.5.8), particularly if the existing pavement is being replaced as a consequence of age-related cracking.

Hot-in-place asphalt recycling (HIPAR) involves a single pass of the recycling train heating the asphalt surface to between 140 C and 170 C to soften the bitumen without degrading it, scarifying the pavement to a depth of up to 60 mm, blending the scarified material with fresh bitumen, asphalt and aggregate, spreading the new mix, and then compacting it to the required density. The process is capable of re-establishing layer strength and stiffness. Thermoplastic line markings (Section 22.1) must be removed before recycling begins. The addition of fresh bitumen may not be necessary on lightly trafficked roads.

Design methods for determining overlay thickness are usually based on a deflection or FWD assessment of the stiffness of the existing road (Section 14.5.2). This practice is followed because the moisture and density conditions under the pavement will have reached equilibrium (Section 9.4) and the in situ deflection measurements will allow the design to be based on actual rather than assumed conditions. It is necessary to use a deflection that represents a reasonable upper bound to the values measured along the length of the road to be overlain. Deflections taken at temperatures other than 27 C should be modified using the procedure given in Section 14.5.2. The deflection that accounts for all these corrections is called the *adjusted measured deflection*.

However, it is necessary to reconcile any high deflection readings on the existing pavement with observed conditions, and to then rectify any causative malfunctions — such as poor drainage — before placing the overlay. The deflection to use should be that associated with the corrected condition. Similarly, a badly cracked pavement may have allowed considerable water entry (Section 9.1), raising deflection readings above a reasonable equilibrium value. An overlay that successfully sealed such cracks over a free-draining basecourse and subgrade could be designed for drier conditions and lower measured deflections.

Most thick overlay design systems for placement over asphalt are based on the assumption that the required overlay thickness is proportional to:

log([adjusted measured deflection prior to overlay]/[desired deflection after overlay])

For thickness in millimetres, the proportionality constant ranges from 300 to 500. However, overlay thicknesses in excess of 150 mm should be designed by one of the mechanistic methods (Section 11.4.2). Many design codes also include provisions for estimating the amount of additional unbound material required to reduce measured deflections to an acceptable level. For example, RTAC (1977) suggested in its Figure 5.4 that reducing deflections from 2 to 1 mm would require an extra 250 mm of unbound material.

For overlays over cracked concrete pavements, any cracks moving vertically under traffic by more than 50 μm before overlay, if not repaired by dowelling, will probably reflect through the overlay; 200 μm movements will certainly reflect through. These reflection effects usually occur very early in the life of the overlay. The problem can be treated by:

* the use of fabric reinforcement (Section 10.7) in the overlay,
* the use of strain-absorbing layers (Section 12.1.7) under it,
* treatment of the original crack by dowelling, etc,
* placing joints in the overlay above the original cracks, and/or
* breaking the underlying slab into small pieces.

12.2.7 Maintenance

Asphalt maintenance is commonly conducted with plant mix kept warm in insulated or heated boxes. As this creates severe practical problems in many areas, *coldmixes* are also used. These may be manufactured either hot or at the relatively cold temperature of 100 C, and then stored for weeks or months. They must, of course, be heavily treated with up to 20 percent cutter or emulsifier (Section 8.7.6) or foamed (Section 10.4) to retain their workability at low temperatures. Coldmixes are usually open-graded, to ensure that the bitumen is able to cure by evaporation of the cutter. Therefore, a subsequent spray-seal waterproofing coat is usually essential to avoid porosity leading to permeability and bitumen oxidation.

Well-graded coldmixes producing dense mixes (Section 8.3.2) are possible if a suitable cutter or emulsifier is used and these are often easier to place. As emulsion technology improves (Section 8.7.6), coldmixes are increasingly being used for thin overlays as well as for spot maintenance (Section 12.1.7). However, problems can arise in using a coldmix based on cutbacks, as it is difficult to arrange for it to set quickly when placed in the road, e.g. in a pothole. To avoid this, only relatively thin layers are placed to ensure fairly rapid volatilisation of the cutter. Another problem with coldmix is that the cutter or emulsifier can soften the bitumen in the covering seal or in the adjacent asphalt, leading to fatty spots, or bleeding (Section 14.1C8). Coldmixes operate at between 25 and 60 C and therefore are far less demanding on heating fuel than are conventional mixes.

Similar processes have recently seen the application of *warm-mix asphalts,* produced at temperatures between 60 and 140 C.

Maintenance techniques for cracked asphalt will depend on the cause of the cracking (Section 14.1B). The appropriate overlay techniques are described in Sections 12.1.7 and 12.2.6. Where it is desired to prevent cracking reflecting through, the strain-absorbing membranes described in Section 12.1.7 are often effective. The repair of rutted asphalt follows a similar theme. Thick overlays are appropriate, provided the cause of the rutting (or large deflections) is also controlled.

Methods for addressing low skid resistance are discussed in Section 12.5.2. Surface deficiencies such as bleeding and excessively shiny asphalt (Section 14.1C8) can be treated by surface dressing (Section 12.1.4). Open cracks and potholes (Section 9.3.2) should be filled as soon as possible to prevent water entering the pavement structure. Cracks are often filled with rubberised bitumen (Section 8.7.7).

12.3 AGGREGATES

Aggregate properties for spray and chip seals were discussed in Section 12.1.5, for asphalt generally in Section 12.2.1, and for AC in Section 12.2.2. The ranges of aggregate and grading were discussed throughout Section 12.2. This section covers additional, specific aspects of aggregate properties.

12.3.1 Aggregate size and shape

The nominal size of a mix refers to the largest aggregate size used in it and this rarely exceeds half the layer thickness (a course is constructed as a series of layers, Sections 11.1.1 and 12.4.1). Increasing aggregate size reduces the cost and increases the strength and stiffness of a mix, but also decreases workability and wear resistance.

The design of bituminous mixes with a large aggregate can present a problem as common design procedures (Section 12.2.3) are unsuitable for mixes with a maximum particle size exceeding 25 mm. Modified procedures permit sizes of up to 40 mm, but large-sized test equipment is not commonly available. The problem is important because, with larger particles, the maximum-density gradings used for conventional mixes show segregation during handling and laying (Section 8.3.2).

The use of angular aggregate will markedly improve the deformation- and rutting-resistance of an asphalt, particularly an AC-type mix (Section 11.1.3). On the other hand, rounded aggregate is sometimes used to help achieve workability in non-structural courses.

12.3.2 Aggregate adhesion

In spray and chip seals and in asphalt mixes, the bitumen must first wet and then adhere to the aggregate surface. The success of this process is a key to the performance of the mix. Failure to continue to adhere is called *stripping* (or *disbonding*). A spectrum of adhesion mechanisms — from hydrogen bonds to salt formation — occurs at the bitumen-aggregate surfaces. The basic form is molecular *bonding* between the polar components of the bitumen (Section 8.7.1) and electrochemically-active sites on the aggregate surface, although the bonding may include the gamut of electrostatic, chemical, hydrogen, and Van der Waals effects. Bitumen as a whole has little surface-charge bias (or polarity), and any bonding predominantly involves the oxygen-containing groups in the polar asphaltenes (Section 8.7.1) within the bitumen. Basalts, dolomites, and limestones have positively-charged surfaces (i.e. are basic) however many aggregate surfaces are weakly negative, as a result of silicon ions. Quartz, quartzite, granite, and sandstone gravels have relatively strongly negative surfaces (i.e. are acidic) and may have problems with bitumen bonding.

All the bonding mechanisms are susceptible to change as a result of either the amount or the pH of the water present. Water is strongly polarised, and is drawn to the interface and then bonds to the aggregate surface in preference to bitumen. This process causes some stripping of the bitumen from the aggregate. If water is able to form microdrops under the bituminous layer, the aggregate surface will raise the local pH and the previously-bonded bitumen molecules may be made to ionise. Disbonded bitumen surfaces are usually negatively charged. Hence, if the aggregates also have negatively-charged surfaces, mutual repulsion will encourage a snowballing effect once stripping has begun. There is thus a need to avoid moisture in a mix. Adhesion between aggregate and bitumen thus depends on:

* avoiding aggregates whose surfaces have a high negative surface-charge (see the above listing),
* avoiding aggregates with a marked affinity for water (such aggregates are called *hydrophilic* and include some shales, limestones, stones with a high SiO_2 content, and many water-worn stones),
* restricting the presence of water (consequently, it is important to reduce the freewater present, dry aggregates before mixing, and work in dry weather to avoid water on the aggregate surface),
* reducing the presence of dust and dirt on the surfaces, as these physically interfere with the bonding process. This can be done by using clean (e.g. no adherent silt or clay) and dust-free (typically, less than 0.5 percent dust) aggregate and can be done by screening and/or washing the aggregate. Note that it is relatively rare to encounter aggregates that are neither wet nor dusty,
* increasing the texture of the aggregate surface,
* keeping the bitumen viscosity from being too high, making it unable to coat all the aggregate. This can be achieved by avoiding low temperatures and/or 'cutting back' the bitumen (Section 8.7.6),
* ensuring that mixing conditions provide adequate initial wetting of the aggregate by bitumen,
* pre-coating the aggregate with bitumen or kerosene,
* using a *wetting agent*. These operate by:
  (a) improving the spreading of the bitumen over the aggregate surface, filling up potential water-containing cavities, and
  (b) providing strong interfacial chemical bonds;
* adding an *adhesion additive* (see below). It is more economical to introduce the additive into the pre-coat rather than into the binder.

Adhesion (or *antistripping*) *additives* (or *agents*) take a number of forms. Bitumens with anionic emulsifiers are useful with basic aggregates and cationic emulsifiers can help with acidic rocks (Section 8.7.6). Anionic and inorganic salt additives will only be successful if salt surface bonds can form (Scott, 1978). Cationic agents are commonly used. They are typically organic ('fatty') amines or imidazolines of high molecular weight. They orient themselves such that the hydrocarbon tail is in the bitumen and the amine attached to the aggregate surface, providing a tie between aggregate and bitumen (Dickinson, 1984). Many of these cationic adhesion additives will delay but not prevent stripping, as the surface environment changes with time. Many also become unstable at temperatures above 120 C. Most are diluents (Section 8.7.6).

The presence of alkali salts (e.g. from the addition of about 1 to 2 percent of hydrated lime, Portland cement or fly ash) will raise the pH at which stripping will occur and hence protect or even enhance adhesion in the presence of water. Hydrated lime also enhances calcium-silicate bonds and works on any clay coatings on the aggregate. Such alkali agents should not be used in conjunction with cationic agents.

Care must be taken lest such additives adversely affect other bitumen properties. All adhesion agents should first be laboratory tested for compatibility with the intended aggregate type. More generally, it is necessary to consider the chemistry of the local environment and the surface chemistry of the bitumen. However, the aggregate properties are usually of more importance than the bitumen properties in determining adhesion.

Tests used to assess the propensity of an aggregate surface to encourage water to compete with bitumen to adhere to its surface include:

* setting stones in a thin (2 mm) bitumen layer and then standing the product in a water bath for 24 hours. The stones are then plucked from the bitumen and the average percentage of the contacted area still coated with bitumen is calculated.
* setting stones in a thin bitumen layer applied to a steel plate. The plate is then hit on the reverse side by a falling ball. The amount of aggregate dislodged is measured. This is called the *Vialit test*. Another form of the test uses a pendulum for impact.
* conducting the indirect tensile test (Section 8.6q) on wet and dry specimens. This is sometimes called the *Lottman test.*
* observation of coated stone specimens that have been in boiling water for about 10 minutes (U.S.) or about 24 h (Europe).

12.4 ASPHALT COMPACTION

12.4.1 Mix and placement practice

The production of asphalt is discussed in Section 25.6 and compaction plant in Section 25.5. Compaction plays a major role in determining the performance of asphalt. Its main benefits are to:

(a) increase particle interlock and reduce the thickness of the bitumen film between aggregate particles, thus increasing overall stiffness and strength, particularly in AC, and
(b) reduce air voids and thus minimise permeability and maximise durability (Section 12.2.3).

Asphalt acquires significant compaction during placement, just as unbound courses do, and often achieves further major compaction under traffic. Compaction is more critical, but harder to achieve, for AC than it is for HRA.

Compactibility is influenced by mix proportions, aggregate shape (Section 8.6a), bitumen type and content, and mix temperature. The role of temperature is discussed in Section 12.4.2. Excess bitumen in the mix, or poorly-shaped aggregate may result in a *tender mix* (Section 12.2.4b) which exhibits low shear stiffness when being compacted. In this context the need for a stiff in-service asphalt competes with the need for a mix that can be readily placed and compacted.

Asphalt is usually placed in layers of at least twice the aggregate size and about a lane in width. Sometimes thick layers are preferred in order to reduce heat loss during compaction. Asphalt is then compacted with a static or vibrating steel-wheeled roller in a process known as *primary* (or *initial* or *breakdown*) compaction. This stage achieves nearly all the increase in density over that present at placement. It is followed by a pneumatic-tyred roller for *secondary* compaction, which is mainly aimed at closing surface cracks produced by the primary compaction process. To prevent asphalt pick-up by the tyres, they should be warmed by rolling an existing pavement for 10 minutes and, possibly, by using a coating agent on the tyres. *Finish rolling* is the third and final stage and is used to remove rolling marks and other surface imperfections. A light covering of sand or grit is often used after compaction to prevent the initial pick up of bitumen by tyres.

Care must be taken with the longitudinal joints between layers as they can be regions of low density and high porosity. This may permit excessive moisture entry that will reduce pavement life by causing layer separation, ravelling, or subgrade failure. The low density is usually in the first layer placed or near an edge. It is difficult to compact against an edge which is unrestrained and subject to relatively high heat loss. Any level difference between the two layers may also result in water forming ponds over this relatively porous area.

Mixes for residential streets and low-volume roads are sometimes placed in thin (25–35 mm) layers under conditions where satisfactory compaction is not always possible. This often results in larger than desirable air voids and subsequent early hardening of the bitumen (Section 8.7.1), which leads to cracking and loss of surface material (Section 12.3.2). One of the other causes of poor compaction in residential streets is the lack of subsequent heavy truck traffic to cause traffic-induced compaction. Many hundreds of load cycles are needed to bring a conventional asphalt close to maximum compaction. In circumstances where all compaction must take place at construction, the mixes should be easily worked and compacted. This suggests the use of gap-graded mixes such as HRA for residential streets rather than the well-graded ACs.

12.4.2 Influence of temperature on compaction

For normal mixes, the main factor influencing compactibility during placement is the initial temperature and subsequent rate of cooling of the placed layer. This is because the viscosity (Sections 8.7.3 and 33.3.3) and brittleness of the bitumen (Sections 8.7.3&4) increase dramatically as the temperature falls. A bitumen with a viscosity above 20 Pa.s is difficult to compact. The effect can be severe with some polymer-modified bitumens (Section 8.7.8). It is therefore essential that asphalt be spread and compacted to its final shape whilst the mix still retains adequate heat.

There are some useful clues as to the temperature of an asphalt mix delivered to the site. A dull, bluish glow suggests that the mix has been overheated, whereas the presence of an outer crust indicates that the mix is too cold. A typical procedure for calculating cooling rates gives the time for a placed layer to cool to a given temperature in terms of layer thickness, placement and receiving layer thickness, layer thermal conductivity and wind speed (Dickinson, 1978b). The estimates require Mean Monthly Air Temperature (*MMAT*) data, although at any particular pavement site the primary climatic factor is the daily *solar radiation* (or *insolation*). Layer *thermal conductivity* usually lies within the range of 1 (basalt aggregate) to 3 (quartz aggregate) W/m.K.

Compaction temperatures commonly range from 75 C to 145 C, extending to 65 to 165 C, depending on mix type. Primary compaction usually takes two to four and secondary compaction eight to twelve passes and should be completed as quickly as practical. On cooler days, primary compaction should be completed within 10 minutes and secondary compaction 25 minutes of laydown. In addition, layers less than 50 mm thick may cool too quickly to be compacted.

12.5 PAVEMENT SURFACE PROPERTIES

12.5.1 Surface characteristics

A comprehensive review of road surface characteristics is given in OECD (1984). The key service properties of a road surface are its frictional characteristics and its impermeability. Both can be measured directly, as discussed below. However, they can also be deduced by smaller-scale measures. For instance, air voids and their connectivity give a permeability measure (Section 12.5.5) and, as will be shown below, aggregate properties and surface texture will give a measure of frictional attributes.

The surface of a road can be considered from a number of foci, or scales, as illustrated in Figure 12.4. At one extreme, roughness (Section 14.4) provides a broad description of the longitudinal profile of the road and is the facet of the surface felt by the passing motorist. It has a wavelength of at least 500 mm. At shorter wavelengths are a series of effects such as bumps, corrugations, and faulting (Section 14.1A). These are sometimes called *megatexture* and have a 'wavelength' between 5 and 500 mm. This wavelength is about the same size as the contact area of a tyre (Section 27.7.3) and can therefore directly interact with a vehicle suspension.

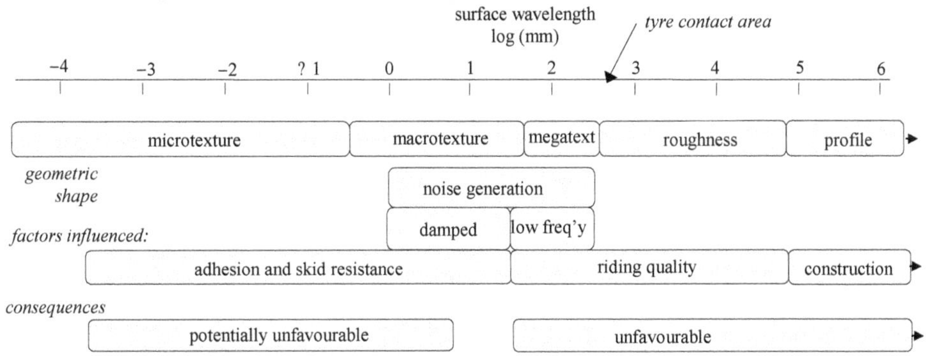

Fig. 12.4 Road surface wavelength spectrum, from Huschek (1990).

The focus on the next level is from the viewpoint of someone standing or squatting on the road surface. The picture of the road surface presented to the naked eye now includes the individual pieces of aggregate in the surface and is also the texture that can be felt by the human hand. It is called the *macrotexture* of the road and relates to the way in which the pieces of aggregate are arranged. Macrotexture has a wavelength between 0.5 and 50 mm and a texture depth range commonly between 0.2 and 3 mm. Texture depth increases as the size of the coarse aggregate increases and as the proportion of fine aggregate decreases. One factor that reduces this effect is the alignment of the coarse aggregate under traffic, as illustrated in Figure 12.1b. Texture depths less than 1.2 mm can lead to problems in attaining a suitable skid resistance. Texture depths below 0.4 mm over more than 5 percent of a braking or cornering area deserve immediate attention.

Care must be taken when placing asphalts to ensure that a good surface macrotexture is achieved. Even then, it may take a couple of weeks of traffic before the bituminous coating is removed from the exposed aggregate surfaces and the maximum friction potential is achieved. For concrete, an adequate macrotexture will not automatically occur and common practice is to lightly brush or bag the surface before it has set. Techniques such as grooving (Section 12.5.5) or the addition of a surface layer of stone chips are used where friction demands are likely to be high.

Macrotexture is measured by the *sand patch method* (ASTM E965) in which a defined volume of fine sand — typically 24 mL — is used to fill a measured area of the surface to the top of the aggregate pieces. The area so covered is an inverse measure of the mean depth of the texture. Macrotexture can also be measured by laser-based methods. Macrotexture plays a number of roles, influencing:

* *skid resistance,* by providing both (Section 12.5.2):
  # strong but hysteretic keys into the tyre tread (Section 33.3.3), and
  # drainage channels for the rapid egress of any surface film of water from beneath a vehicle tyre.

  Typically, about 75 percent of skid resistance comes from macrotexture.
* *noise generation* (Sections 12.5.2 and 32.2), with the noise levels produced related to the mean texture depth as measured by the sand patch method, and
* *wet weather visibility* characteristics of the surface, by reducing both water spray and night-time specular reflection (Section 24.2.1).

Finally, someone lying on the road and using a magnifying glass or their fingertips could detect the roughness of the surface of the exposed pieces of aggregate. This texture is called the *microtexture* of the surface and is sometimes described as *harshness*. The dividing line between macro- and microtexture is at a wavelength of about 250 μm. It is important at a wavelength down to about 25 μm, which means that it may need to be assessed by electron microscopy. Microtexture acts like sand paper when providing some skid resistance (typically about 25 percent) and is a major contributor to tyre wear. Adhesion (Section 12.3.2) also plays a role at this level.

12.5.2 Principles of skid resistance

The frictional properties of the pavement surface allow frictional forces to be developed at the tyre–pavement interface and these allow vehicles to change momentum via braking, accelerating, and cornering. When the braking or cornering forces exceed the frictional resistance, the vehicle is said to *skid* and it is common to describe the frictional properties of a pavement–tyre interface as its *skid resistance*. This phenomenon is explored in more detail in Section 27.7.2.

The role of *macrotexture* in improving skid resistance was introduced in Section 12.5.1 and is discussed further with respect to tyre behaviour in Section 27.2.2. Its effect in reducing aquaplaning during wet weather is described in Section 13.3.1. Mean texture depths between 200 and 900 μm will provide adequate wet friction without excessive tyre abrasion (Sections 14.1A8 and 27.7.1).

For high speed travel, levels of at least 750 μm are desirable although typical mean texture depths measured by the sand patch method on safe urban roads range from 300 μm upwards. However, British data suggests that crash rates for all classes of road begin to increase as the texture depth drops below 700 μm. In the absence of any macrotexture, the necessary wet-weather friction is supplied solely by the tyre tread and the sand-papery nature of the microtexture. Macrotexture becomes increasingly important in skid resistance as vehicle speeds increase. Skid resistance typically drops with increasing speed.

Unless special measures are taken (e.g. porous friction courses, Section 12.2.2), noise levels will increase if skid resistance is raised by increasing the macrotexture. This is not a problem with low-speed residential streets, which can provide adequate skid resistance through good microtexture alone. However, effective skid resistance drops exponentially as speed increases and, at speeds over about 75 km/h, a good microtexture and macrotexture are both required.

Skid resistance also drops due to polishing of the protruding stones. Tests for measuring the polishing propensity of an aggregate are described in Section 8.6o. The two main causes of polishing in service are vehicle tyres and the weather. The effect of an average truck is equivalent to about 18 passenger cars. As well as individual pieces of aggregate polishing and hence changing the microtexture of the surface, the macrotexture of the surface can also be reduced by traffic action either tearing protruding aggregate from the bitumen or pressing it down into the bitumen (Figure 12.1b).

A good stone for skid resistance purposes would be either hard with sharp angular faces or composed of hard particles in a softer matrix. The stones should perform well in the PSV test (Sections 8.6o and 12.5.3). Similarly, the skid resistance of asphalt composed of aggregate with a low polishing resistance can be improved by the addition of harder, fine particles.

In asphalts such as HRA (Section 12.2.2) the bitumen present may produce a slippery surface. For this reason it is common to:

(a) roll a layer of (usually) pre-coated, polish-resistant aggregate chips into the new surface in the manner of surface dressing (Section 12.1.4) — this is sometimes called a *sprinkle treatment*, and

(b) select a bitumen for such applications that will weather back and leave the coarse aggregate as the dominant exposed material.

Of course, this approach requires the use of aggregate with a high polishing resistance.

Porous friction courses (Section 12.2.2) are far more effective than conventional asphalts in providing good friction.

The skid resistance of a new road usually falls for the first two years of its life — the amount depends on the stone used and the traffic volume. Seasonal variations then occur and can be as much as ±20 percent. The summer often gives low values as the surfaces polish under the action of tyres on the dry, hot surface. The winter gives high values as the texture is restored by a weathering effect involving the action of rain and low temperatures on the stone surface, and thus on the microtexture of the road. The longer the surface is in contact with rainwater, the higher is the recovery in skid resistance values (Dickinson, 1989). In-service polishing predominates in areas of severe braking, cornering, or accelerating. Spillages (e.g. of oil) on the road can also lower skid resistance.

12.5.3 Test methods for skid resistance

The *polished stone value* (PSV) test for determining the polishing characteristics of aggregates and of local areas of pavement is discussed in Section 8.6o. The associated *British Pendulum test* is an effective and simple control measure and gives local pavement skid resistance in terms of a defined skid resistance value (SRV, Section 8.6o). SRV is representative of the friction coefficient generated by tyres with good tread patterns at speeds of about 50 km/h in wet weather (Oliver, 1978). However, it covers only a very small area of pavement, requires traffic to be stopped, and contacts the surface at a fixed and relatively low speed equivalent to about 7 km/h. It therefore cannot be expected to provide direct guidance on normal road operations. In particular, the effective skid resistance of a relatively smooth (low macrotexture) road declines significantly with speed.

Thus, a number of on-road devices have been developed to measure the in situ skid resistance of road surfaces, usually at some standardised speed chosen to represent traffic speed. These involve the measurement in the critical wet road condition (typically, using a 500 μm water film) of:

(a) *the braking torque or force needed to*:

(a1) stop a running wheel from rotating, thus producing a *locked wheel* or 100 percent *slip*. 'Locked wheel' and 'slip' are defined in Section 27.7.4.

The method is exemplified by U.S. tests specified in ASTM E274. These use a *skid trailer* with either a ribbed or a smooth tyre skidding on a wet pavement. The smooth tyre allows more emphasis on the role of pavement surface properties. The retarding force is measured over the one second after the wheel is fully locked and so the method usually detects the peak braking force.

A longitudinal friction coefficient called the *skid number* (*SN*) is calculated as the retarding force divided by the vertical force, expressed as a percentage. In terms of the PSV test, a change of 1 SRV causes a change of 0.17 in the 100 km/h SN value (Heaton *et al.*, 1978).

The SN data is standardised to a common speed — say, 70 km/h. If the test is conducted at some other speed, v km/h, then the measured SN_v is converted to the standardised value by:

$$SN_{70} = SN_v + (v - 70)/35$$

(a2) permit a fixed slip percentage. For instance, the *Swedish Skiddometer* controls wheel slip to about 15 percent.

(a3) produce the peak friction, often by using a variable slip. (The peak force is usually 50 percent greater than the locked wheel force.) Typical devices are the Norsemeter and Reibungsmesser.

(b) *the cornering (or side) force needed to* hold a wheel allowed to rotate freely whilst placed at an angle to the direction of travel. This force is measured perpendicular to the plane of the wheel and hence the category relates more to cornering than to braking. It includes such common on-road devices as the μ Meter (or Mu Meter) and the SCRIM.

(b1)The μ (Mu) Meter (ASTM E670) consists of a trailer with two wheels each toed-out at 7.5° to the direction of travel. It is towed by a truck delivering water to the road surface ahead of the test tyres. The force developed by the two smooth-tyred wheels is measured and is linearly related to the effective coefficient of friction.

(b2)SCRIM (Sideways-force Coefficient Routine Investigation Machine) was developed by TRRL (Salt, 1977). The device provides a useful and efficient tool for skid resistance measurement. Two smooth-tyred test wheels are free-rolling parts of the test truck, which is basically a 10 t water tanker commonly operating at 80 km/h or less. The wheels carry a vertical load of 2 kN (about a carload) and run in each wheelpath at an angle of 20° to the travel direction. The road in front of the wheel is kept wet by a water tank in the vehicle and the test wheel has its own deadweight and suspension. Load cells measure the sideways force produced, which is divided by the vertical force to give a coefficient of friction known as the *sideforce coefficient* (*SFC*). Thus, to a good approximation:

$$SFC = f_{ls}$$

where f_{ls} is the longitudinal coefficient of friction (Section 19.4.2). The test output is the lowest 20 m of SFC in each 100 m. The wheels are made of low hysteresis rubber (Section 33.3.3) and so may under-estimate SFC for modern tyres.

12.5.4 Skid resistance applications

Skid resistance measurements were categorised in Section 12.5.3. It would appear that category (b2) SFC values are about 45 percent higher than category (a1) locked-wheel SN-type values (OECD, 1984). The British divide the SFC by 1.28 to correlate to their earlier data. The SRV of a stone is measured on specimens polished in the laboratory (Section 8.6o). A road pavement may take about four years to reach the same level of polishing. After that has been taken into account, and considering that SRV is in percentage units (Salt, 1977):

SRV = 105[SFC]

The SFC and SRV/100 values are both a form of coefficient of friction for the pavement surface, with values ranging from, perhaps, 0.25 up to about 1.2 on dry surfaces. A further discussion of the role of the coefficient in curve negotiation is given in Section 19.2.3. The role of the tyre in this friction process is discussed in Section 27.7.1. There, it is pointed out that tyre design usually places a maximum value of about 0.8 on the coefficient of friction, with Figure 27.8 illustrating its specific components.

On a busy road, SFC may drop by 0.02 units a year. The rate will be higher in locations subjected to high tyre friction. However, as was the case with SRV, the SFC of a pavement surface may not continue to deteriorate with traffic, but may reach some equilibrium level that depends on the rate of use (i.e. on the 'daily' flow rather than absolute traffic volume).

Many studies show a decline in *crashes* as a result of skid resistance improvements. For example, the risk of crashes due to skidding is measurably influenced by SFC values of 0.55 or less. The determination of desirable skid resistance levels is a thorny issue. The problems relate to measurement difficulties, variations with season and recent rainfall, and the fact that skid resistant levels are rarely the only causative factor in a crash. Thus skid resistance levels are less important on dual carriageway roads away from intersections, and very important at intersections. This aspect is usefully explored in Viner *et al.* (2005) and incorporated into Table 12.3.

Typical SFC levels at which remedial measures should be investigated are shown in Table 12.3.

Table 12.3 Investigatory skid resistance levels (Austroads 2000 & Viner *et al.*, 2005)

Site	Investigatory level of SFC at 50 km/h, deduct 0.05 if flow below 2.5 kveh/lane/day
traffic lights, crossings etc	0.50 – 0.55
curves and ramps	0.45 – 0.55
intersections	0.45 – 0.55
undivided roads, smooth flow	0.40 – 0.45
divided roads, smooth flow	0.35 – 0.40
freeways	0.30 – 0.35

Note: For high speed roads, the texture depth should be at least 750 μm to ensure adequate macrotexture (Section 12.5.2).

Skid resistance levels also fall with time. This is illustrated in Table 12.4.

Table 12.4 Indicative friction life (from VicRoads, 2002). N = not expected to have a suitable friction life.

Stone PSV	Traffic volume (kveh/lane/–day)	Indicative friction life in years			
		Curves with R<100 m	Traffic control unit	R<250 m curves	Uncontrolled intersections
48	5	N	N	3 to 6	9 to 12
	10	N	N	1 to 2	4 to 6
52	5	N	N	6 to 8	15 to 18
	10	N	N	3 to 6	8 to 10
56	5	N	3 to 7	14 to 16	> 20
	10	N	3 to 5	7 to 10	12 to 14
60	5	4 to 6	12 to 15	> 20	> 20
	10	2 to 4	6 to 8	11 to 14	16 to 17

The skid resistance of an unpaved road will usually lie between 0.15 and 0.25 and so will be significantly below the above 'desirable' levels. If coarse ice or snow is present, skid resistance values for paved roads will drop to as low as 0.25, conventional dry ice or snow will see values drop as low as 0.15, and in very icy conditions skid resistance may be zero.

Care should also be taken in cases where there are major differences in SFC values between wheelpaths.

12.5.5 Permeability

Well-maintained spray and chip seals (Section 12.1.4) can be very water-tight. However, thicker asphalt surfacings can exhibit significant permeability (Section 9.2.2) as a result of:

(a) a linking of air voids (Section 12.2.3). Construction practice may have left the asphalt poorly compacted, leaving a high void content. In addition, because of the role of traffic compaction (Section 12.4.1), newly-laid asphalt can be expected to be more permeable than the same material after some weeks in service. Permeability increases as the void content of the mix increases, with typical values ranging from 300 pm/s at 2 percent air voids to 30 μm/s at 12 percent air voids. The general relation is given by:

$$\log(\text{permeability in nm/s}) = 0.55\ (\text{air voids in percent}) - 2.25$$

Typically, a 1 percent increase in air-void content will result in a three-fold increase in permeability.

(b) poor construction practice leading to large open passages (e.g. due to poor compaction) or to gaps at poorly-made construction joints between layers;

(c) passages created by cracking in service (Section 14.1B). These cracks can be sealed during maintenance. However, attempts to waterproof and seal the permanently-moist surfaces of inherently permeable materials will be only partially effective and relatively short-lived. The permeability of the surface will clearly depend on the length, width, and depth of the cracks.

Surface permeability is sometimes measured by an *infiltration factor* that relates to the proportion of the water which falls on the surface and which then infiltrates the surface.

For in-service pavements, the factor ranges from less than 0.2 for spray and chip seals, 0.3 for asphalt, 0.4 for concrete and 0.5 for unsealed shoulders. Because of surface tension effects, a depth of some 50 mm of surface water is usually required to begin water penetration of a surface that is reasonably intact but untrafficked. However, the local pressure-raising effects of passing tyres will readily overcome the surface tension barrier and cause penetration under much thinner water layers.

Another occasional measure of in situ permeability is obtained by placing a container of water on the pavement surface, sealing the edges and recording the rate at which the water permeates into the pavement. Field permeabilities can be expected to be in the range 1 to 2000 nm/s with 200 nm/s being a typical value. The lower rate would be achieved with homogeneous impermeable clays and the upper rate with open-graded sands. At the upper rate, moisture would penetrate 7 mm of asphalt in about one hour if no lateral infiltration occurred. Lateral infiltration reduces this penetration depth and would be relevant if the water was applied at only one location on the pavement surface, or if there was a free-draining asphalt edge. Lateral infiltration may therefore distort such in situ permeability measures, by suggesting that a pavement is more permeable than it actually is.

Unless the underlying course is impermeable, a permeable pavement surface course is to be avoided as it will reduce the structural serviceability of the pavement by the mechanisms described in Section 9.3.1.

12.5.6 Rehabilitation

If a pavement's surface properties must be upgraded, the commonest techniques are:

(a) surface dressing (Section 12.1.4);
(b) the application of a thin overlay (Section 12.1.7);
(c) machine grooving of the pavement surface by a cold-planer or custom-made saw;
(d) improving the pavement profile (Section 14.4) by cold-milling the surface (this technique can also be used to prepare the surface for an overlay). Cutting is done by teeth attached to a hollow steel drum. Cutting depths range from 5 to 300 mm.
(e) using epoxy to bond gritty aggregate to the surface.

CHAPTER THIRTEEN

Drainage

13.1 INTRODUCTION

This chapter discusses drainage in three stages:

(a) rainfall and water flow within the catchment area (Section 13.2),

(b) water flow within the roadway (Section 13.3), and

(c) the removal of water from the roadway (Section 13.4).

Following Chapter 9, the chapter concentrates on drains beside or under the carriageway and which have been designed to control the entry of water into and facilitate its removal from the pavement structure.

An important concept in drainage design is that of the *recurrence interval* (Section 33.4.6) which is the time between an event and its recurrence. For example, a flow with a ten-year recurrence interval is one likely to occur once every ten years, or with a 10 percent chance of occurring in any one year. If an event has a recurrence interval of T years, then the probability, p, that it will occur in any L year period (say the design life of the structure) is given by:

$$1 - p = (1 - T^{-1})^L$$

or:

$$1 - p = e^{-L/T}$$

Design is typically based on an average recurrence interval, T_{av}. Thus the average exceedence probability is $1/T_{av}$.

A comprehensive review of road drainage for Australian conditions is given in Alderson (2006).

13.2 WATER FLOW WITHIN THE CATCHMENT

13.2.1 Storms and rainfall

Estimates of design flows within a watercourse can be based upon either *stream-flow* or *rainfall* records. The former is preferable but, if it is not available for the 15 or more years usually needed, recourse must be made to rainfall records.

(a) Stream-flow records

Stream-flow records are usually held by River Authorities and given in terms of the largest flow in each year. If these flows are ranked in order of size over the T years of record, then the probability of the ath highest flow being exceeded in any future year can be estimated as $a/(T + 1)$. The discharge with a particular recurrence interval can be best estimated from the data, using a Pearson Type III distribution to represent the logarithms of the largest annual flows. However, a graphical plot of the annual flows against their probability, $a/(T + 1)$, will often give useful first estimates.

The importance that should be attached to the selection of the recurrence intervals for flows is seen in the following example of some typical urban average flow ratios (Argue, 1979):

two year flow = 0.70 (ten year flow)
five year flow = 0.90 (ten year flow)
twenty year flow = 1.20 (ten year flow)
fifty year flow = 1.40 (ten year flow)

(b) Rainfall records

The calculation of flows using rainfall records is based on storm data, particularly data relating to the frequency of storms of a given magnitude. The magnitude is measured by the storm's intensity and duration, which are inter-related. The method is therefore often called the intensity–frequency–duration (IFD) method (e.g. Pilgrim, 2001).

Typical storm recurrence intervals for road and street design are given in Table 13.1.

Table 13.1 Typical recurrence intervals for storms. After Argue, 1986.

Item	Years
Road drainage, where overflow would not cause property damage	1
Nuisance flooding of non-freeway pavements	1 – 2
Road drainage in residential areas and open space	2 – 5
Road drainage in business and commercial areas	5 – 10
Nuisance flooding of freeway pavements	10
Traffic disruption on minor roads	5 – 20
Flood flow in minor culverts, drainage inlets, and table drains (Section 7.4.3&5) in areas of intense development	10 – 20
Flood flow in bridges and major culverts	20 – 50
Traffic disruption on arterials and freeways	20 – 50
Traffic disruption on critical highway links	50 – 100
Rare floods (Section 13.2.2)	50 – 100
Flooding of major structures	100
Design floods for forces on bridges	2000*

* The 2000 y stems from the need to ensure a low 5 percent probability of failure of the bridge during its design life and is thus 20 times the typical design life of 100 y (Section 15.4.2).

For a given storm, it is then necessary to estimate the amount of water deposited. *Rainfall intensity*, I, is the rate at which rainfall is deposited over an area and can be approximated by expressions of the type:

$$I = \text{constant}/(\text{rainfall duration})^{0.5} \qquad (13.1)$$

The constant depends on local conditions and on the chosen recurrence interval. For catchments under 25 km^2 in area, it is assumed that the rainfall is of uniform intensity over the whole catchment. It can be seen from Equation 13.1 that rainfall (or storm) intensity increases as rainfall duration decreases.

The minimum storm duration in Equation 13.1 is calculated by first estimating the *flow time* within the catchment from the most distant point to the point of discharge or, alternatively, the shortest time necessary for all points in a catchment to contribute simultaneously to run-off past a specified point (Alderson, 2006). This is commonly called the *time of concentration*.

For example, for a building, its components from roof through the property to street gutter may be about 5 minutes and 15 minutes, respectively. Gutter speeds are of the order of 1 m/s and can be calculated from Section 13.3.2 and Equation 13.2 below. For larger catchments, a variety of locally derived formulae are in use (e.g. Alderson, 2006, Table 3.3).

The storm duration used is that which produces the worst design conditions, given that intensity drops with duration. Experience indicates that the critical storms for small catchments (area < 10 ha) often have durations of 30 min or less, whereas for moderate urban basins (10 ha < area < 25 km^2) the relevant duration may be 2 h.

The quantity of water discharged due to a storm may be calculated by the so-called *rational method* (Pilgrim, 2001). This assumes that:

* there is a known catchment area,
* the maximum discharge rate is proportional to the chosen rainfall intensity (see above),
* there is no retention of water in the catchment by soaking, etc.,
* there is a known *run-off coefficient* converting the rainfall volume into a flow volume. The coefficient can be as high as 0.95 for a surfaced road and so the presence of a road will alter local water flow patterns.

The method has an accuracy of about ±30 percent and is not recommended for catchments over 25 km^2 in area. Some of the method's weaknesses are that it gives no data on the total volume of run-off or on the flow-time characteristics of the run-off.

Flows from a catchment tend to increase as the 0.7 power of the catchment area — i.e. less than linearly — but as the 2.3 power of the annual precipitation (Boyd, 1979). The much greater than linear effect in the latter case probably reflects the loss of soaking-in capacity in high rainfall areas.

13.2.2 Floods

A flood is a temporary inundation of useable land resulting from a period in which the quantity of water discharged from the area is less than the quantity entering it. Thus, neither storm run-off over the land surface nor natural swamps represent flood conditions. The maximum flood level reached during a flood is the most common descriptor of the event and clearly indicates how much of the road system is inundated. In addition, flood levels are also sometimes defined relative to the elevation reached in some previous epoch flood. The duration of the flood is also a key road parameter as it determines the length of time during which the roads will be inoperative. Flood hazard is the risk of a flood of some defined severity occurring and is generally given in terms of recurrence intervals for floods of a particular elevation and duration. This in turn leads to predictions of how many days per year a road can be expected to be closed by flooding.

In considering urban road networks, floods can be divided into four main types depending on the flood consequences (Argue, 1986):

(a) *minor* floods: excess stormwater flows causing only nuisance to the public. Such floods are discussed in Section 13.3.2 and their recurrence interval is given in Table 13.1.

(b) floods: these cause some damage to outdoor property and some traffic disruption.

(c) *major* floods: these cause some damage to indoor property, major traffic disruption, and public distress.

(d) *extreme* floods: these cause severe financial losses, complete traffic disruption, and deaths in the community.

In bridge design it is common to speak in terms of a *design* flood (Section 15.4.3). Floods greater than this will stop the structure performing its intended function.

On roads carrying infrequent traffic, many pavements and water crossings will have a relatively high probability of flooding. The designer's job is to balance the cost consequences of that risk against the cost of avoiding it. In simple terms the design recurrence interval for the flood is selected to minimise the following whole-of-life cost:

(initial cost of structure) + $\Sigma_{\text{life of structure}}$ [(cost of flooding)(probability of flooding)]

In remote areas, there may be a requirement to provide all-weather accessibility (Section 5.2) which may over-ride the economic factors in the above equation.

When a flood does occur, the following data can be usefully collected:

* rainfall (see Section 13.2.1),
* maximum water levels,
* maximum water levels upstream and downstream of structures,
* rates of rise and fall of water levels,
* flow patterns and speeds,
* stream gradients,
* locations of river-bank overflows, and
* photographs.

13.3 WATER FLOW WITHIN THE ROADWAY

13.3.1 Surface drainage

Drainage is important during both the construction and operational phases of a road. Surface drainage at *construction* sites can cause surface scouring, erosion, sedimentation, and flooding on site and water pollution off the site (Section 32.4). The situation is exacerbated by such activities as clearing of the right-of-way, topsoil stripping, haul-road construction, temporary drains, borrow pits, equipment-tracks on slopes, and stockpiles of soil. Areas particularly prone to trouble are the batters of cuts and fills, partly completed drainage facilities, and temporary stream diversions and crossings. Long-term batter protection measures are discussed in Section 6.4.

When rainfall occurs on a *completed* road, the water will initially fill any shallow depressions in the pavement surface. Further rain will lead to surficial water draining as sheet flow under the influence of gravity. For low flows and smooth surfaces, the initial flow would be laminar, however most pavement flows are turbulent. The eight main reasons for providing adequate pavement surface drainage are to:

(a) maintain adequate skid resistance, as surface water depths of over 5 mm can cause tyres to aquaplane (Sections 12.5.2 and 27.7.4),

(b) improve the effectiveness of lighting by reducing specular reflection (Chapter 24),

(c) increase visibility by reducing spray,

(d) avoid hiding pavement markings (Chapter 22),

(e) avoid causing unexpected behaviour when drivers are confronted with a large sheet of water,

(f) minimise the infiltration of surface water into the basecourse (Figure 13.2 and Sections 14.1 and 12.5.5). (The longer the water lies on the pavement surface, the more chance it has of penetrating that surface.),
(g) prevent excessive scouring of the shoulders, and
(h) diminish the risk of flooding adjacent properties.

With respect to (a) to (e), Section 12.5.3 will show that water films on a road surface can significantly raise the risk of crashes.

Every effort is therefore made to prevent films of water from forming on the pavement surface by providing:

* good macrotexture (Section 12.5.1),
* (possibly) porous friction courses placed as a surfacing on top of an impermeable pavement course (Section 12.2.2), and
* transverse cross-falls on the pavement to remove the water quickly to the pavement edge (Section 6.2.5). This is particularly important with flat, longitudinal grades.

The water at the pavement edge is then taken away using longitudinal gutters or well-maintained shoulders. Problems can arise on steep longitudinal grades when the water tends to flow longitudinally down the pavement. The pavement surface water depth in millimetres in this case is approximated by $0.045S^{-0.2}[LI]^{0.5}$ where L is the flow length in metre, I is the rainfall intensity in mm/h, and S is as defined in Section 13.2.1.

Unpaved roads usually have a greater cross-fall to minimise water penetration, but this will also encourage rainwater to scour the surface. When the local materials used are of poor quality, rutting can occur under traffic and the ruts form natural channels and ponds to retain water and further exacerbate the rutting problem. Similarly, any potholes or depressions will fill with water and worsen during rain. Regular grading is the only solution.

13.3.2 Gutter flows

In rural areas, surface water running from the pavement is usually collected in longitudinal table-drains laterally beyond the road shoulder (Section 7.4.3). Paved channels are used if scour is a problem; for instance, rainfall intensities of over 25 mm/h will usually cause destructive scouring of unpaved drains. The drain capacity will usually be determined by the consequences of overflow.

In residential areas, surface water is usually removed by longitudinal gutters adjacent to the carriageway and formed as part of the kerb, although a number of alternative systems are described in Section 7.4.3. For maximum capacity, the *wetted perimeter* of the gutter should be kept to a minimum and this requires a full-flowing semi-circular gutter. Steep-sided drains approaching this shape are known as *spoon drains*. Such a cross-section is rarely possible in street design because of unacceptable common flow depths and the awkwardness of the resulting gutter shape for both construction and domestic traffic. Thus most gutters are of a simple, but hydraulically inefficient, triangular cross-section.

In order to use the methods in Section 13.2.1 to determine the magnitude of the gutter flow, it is first necessary to establish a few parameters. The catchment area is usually taken as the area of pavement, gutters and verges served by the gutter. Gutters are usually designed to cope with catchment flows of very short duration — typically 5 to 30 minute storm peaks. The run-off coefficient is usually 1.0. Table 13.1 suggests one year as a typical value for the recurrence interval.

Once flow magnitudes are established, it is necessary to check flow speeds and capacities for the selected gutter cross-section. Because gutter flow is usually shallow and of triangular cross-section, the flow speed will vary from a peak near the kerb face to zero at its outermost lateral extremity. Thus, it is common to talk in terms of mean speed, which is total flow divided by the cross-sectional area of the flow (flow in mL/s divided by area in mm^2 gives speed in m/s). Flow speed is then calculated using *Manning's formula* that gives:

$$\text{speed (m/s)} = r^{2/3}S^{1/2}/n \qquad (13.2)$$

where n is the roughness coefficient (or *Manning's n*), S the slope, and r the *hydraulic radius* (the flow cross-sectional area divided by the wetted perimeter) in m. Typical values of Manning's n for roads are 0.010 for smooth, 0.012 for asphalt and 0.035 for rough surfaces (Dowd *et al.*, 1980). It has been suggested that:

$$n = 0.01(1 + d)$$

where d is the sand patch texture depth in mm (Section 12.5.1). Manning's n also varies with flow depth and gutter slope and only gives consistent values when the flow is non-laminar.

For shallow, wide flows of the type that occur in gutter and table drains (Section 7.4.5), the hydraulic radius is approximated by the depth of flow, d, leading to *Izzard's flow formula*:

$$\text{flow } (m^3/s) = 0.375Zd^{8/3}S^{1/2}/n$$

where Z is the reciprocal of the cross-slope. This allows the flow depth d to be calculated for a given flow and gutter geometry. Gutter flow widths can also be deduced geometrically from the above data. The above formulas indicate that both speed and depth will increase at a less than linear rate with flow. Typical design criteria for selecting gutter capacity are then based on preventing:

* gutter flow *depth* from exceeding 100 mm, and thus being a pedestrian hazard,
* gutter flow *width* intruding (or 'spreading') by more than 1 m into a through traffic lane and thus forcing traffic out of that lane,
* gutter *overflow* causing damage to adjacent properties,
* excessive flow *speeds*. For roads having a steep grade, flow speed is the key variable as this will determine the effectiveness of the removal systems, such as the side-entry pits to be discussed in Section 13.4.1. In addition, flows of the order of 4 m/s can cause injury to frail or weak people in their path. Such speeds are normally only encountered with longitudinal slopes of 10 percent or more.

The flow will change from streamlined to non-laminar at very low speeds. There are then four major types of gutter flow (Dowd *et al.*, 1980):

(a) *Slug*, containing surges superimposed on an otherwise steady and uniform flow. This occurs with shallow (< 2 percent) slopes.

(b) *Smooth surface.*

(c) *Rough surface.* This occurs when the gutter has surface protrusions of 20 mm or more.

(d) *Undular.* This occurs with steep slopes and very smooth surfaces. Flows on grades over 2 percent are likely to be super-critical.

Large, flat areas are frequently drained by precast concrete surface drainage channels, incorporating both the subsurface channel and a continuous surface opening. They usually operate on grades of about 5 percent to ensure some self-cleansing and operate at around 0.75 m/s when flowing at capacity.

Table drains are used in unpaved areas and are discussed in Section 7.4.5.

13.4 THE REMOVAL OF WATER

Extensive general discussions of water removal are given in Alderson (2006). The need to manage the pollutants in the removed water is discussed in Section 32.4.

13.4.1 Pits

Water flow is carried away from the roadside gutters and into the underground drainage system by *kerb pits* in the gutter's horizontal surface or by *side-entry pits* in the kerb's vertical face. The pits connect directly to the underground drains. Their location, capacity, and spacing are therefore key design variables. However, some pit locations will be governed by external constraints or by road design requirements.

A kerb pit has its opening covered by a grating that is usually depressed below kerb level by 25 to 50 mm. Its design must be such as to avoid problems with items such as bicycle wheels. Clearly, longitudinal grating bars are the best for water entry and the worst for wheels, whereas transverse bars are good for bike wheels but poor for water entry. Some compromise solution is therefore needed. A side-entry pit has an opening in the kerb face about 125 mm high and between 1 and 2 m in length.

A major design problem with pits is that the speed and volume of the water in the gutter (Section 13.3.2) may cause a substantial portion of the flow to wash past the pit and continue down the gutter. The percentage of water captured by a pit is called its efficiency, with 80 percent being a common design goal. Side-entry pits can be excessively inefficient on steep grades, although their efficiency can be improved by forming deflectors in the gutter surface to direct water towards the pit opening.

The pit itself represents a tricky design problem as the water entering will do so in free-fall, losing much of its energy, and must then be taken away by underground pipes. There is little guidance available to allow this device to be designed for maximum flow, although a basic design principle is to encourage streamline flow (Section 15.4.5). Pits are commonly fitted with special traps to collect debris and pollutants before they enter the overall water system (see Section 13.4.2).

13.4.2 Underground stormwater drains

The discussion to date has focussed on the management of surface water, arising predominantly from recent rainfall. This water is collected into the pits described in Section 13.4.1 and then removed by a system of underground drains.

The pipes used to form the drains will need to have a smooth bore in order to operate at small pavement grades. Indeed, the fact that drainage grades must usually exceed 1 percent, whereas many carriageways have a much flatter grade, can make it expensive to drain long, flat lengths of road. Traditionally, pipes have had a circular cross-section, but geo-composite panel drains are proving to be efficient, effective and price-competitive.

Water flow in pipes is commonly based on *Darcy's turbulent-flow equation* that predicts that the square of the water speed is proportional to the change in head per length and the pipe diameter and inversely proportional to the pipe friction. As standard charts and tables are widely available for pipe design — particularly pipe diameter selection, this aspect will not be pursued further here.

It is usual for pipes in a load-carrying region to be placed on a well-compacted granular bedding in order to ensure uniform support of the pipe as most failures of small pipes are due to overstressing due to localised bending. Pipes over 3 m deep will be little affected by traffic load; pipes within 300 mm of the surface will require special structural protection. Minimum pipe cover is given in Table 7.6.

Fine granular back-fill is often used over pipes to ensure a uniform distribution of any traffic loads. In many cases, loads from construction traffic prove to have a more severe effect on pipe systems than do traffic loads. Backfilling is a common cause of pavement and drainage problems and particular attention is drawn to the remarks in Section 26.3.

Underground pipes must be used with care in expansive clays (Section 8.4.2) where the soil movements can upset grades and joints.

If spills of hazardous materials are possible, it is desirable to have isolatable holding-reservoirs in the drainage system and readily available plans showing all drainage outlets.

In recent years there has been a major move towards the installation of devices in drainage inlets and/or outlets that remove pollutants such as solid waste, hydrocarbons, stormwater sediment and oil from the drainage flows (Section 32.4). These can dramatically improve the quality of the water flowing into streams, lakes and wetlands. The simplest solution is to use some form of settlement pond, large sump or gully pit. There are also proprietary devices that utilise centrifugal water flows to separate out the solids more quickly. Most solutions require regular removal of the accumulated waste.

13.4.3 Underground freewater drains

As was demonstrated in Section 9.3, the performance of most pavement materials is deleteriously affected by the presence of water and so considerable attention must also be paid to the movement of water within the soil. This water is known as *freewater* (Section 9.2.3) and it is controlled by a further set of underground drains sometimes called *subsoil* (or *subsurface*) *drains*.

These underground drains prevent the entry of water into and remove water from the pavement structure (Figure 9.2) and so they are intended to either (Figure 13.1):

(a) stop freewater from entering the pavement,
(b) intercept water flowing along permeable courses (Section 9.2.2),
(c) remove collected freewater, or
(d) remove water from the pavement structure.

Somewhat shallower underground drains running longitudinally along each edge of a carriageway (Figure 13.1–2) are used to continuously capture water moving towards the pavement edge, with an emphasis on applications (c) and (d) above. This can usually be achieved by a variety of methods:

1. *Trench methods:*
   * *French* (or *rubble*) *drains*. These are used for low-cost, low-flow applications. An excavated trench is back-filled with a porous material such as open-graded crushed rock which serves as both a filter and a drain.
   * *Wrapped drains*. Before back-filling with crushed rock, the trench is lined with a geofabric which helps maintain the porosity of the filler.

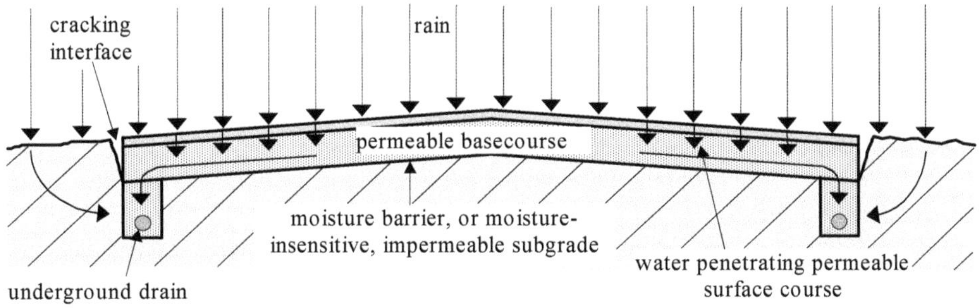

Figure 13.1 Water infiltration removed by underground drains. Arrows indicate water movements.

The first two applications are sometimes called *cut-off* drains (Figure 13.2). Cut-off drains that are intended to protect the entire road formation are usually relatively deep and often called *formation* (or *groundwater*) *drains*. All the other cut-off drains are sometimes referred to as *pavement* (or *structural*) drains.

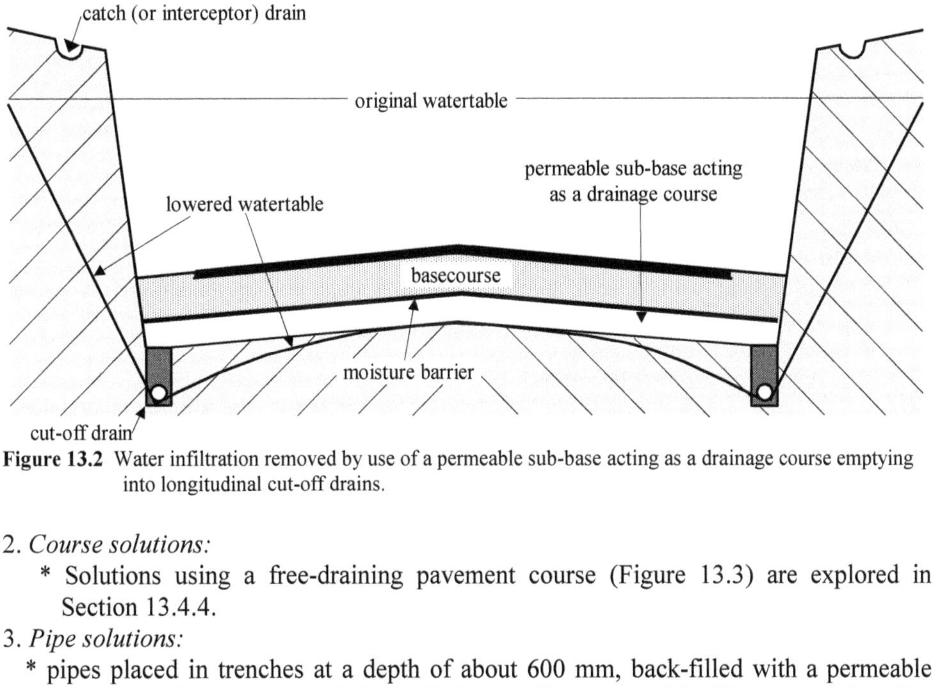

Figure 13.2 Water infiltration removed by use of a permeable sub-base acting as a drainage course emptying into longitudinal cut-off drains.

2. *Course solutions:*
   * Solutions using a free-draining pavement course (Figure 13.3) are explored in Section 13.4.4.
3. *Pipe solutions:*
   * pipes placed in trenches at a depth of about 600 mm, back-filled with a permeable material and then surrounded by higher quality permeable filter material and/or permeable geofabric (Section 13.4.4). There are two main types:
     – *Agricultural drains* – the pipes are made of porous material.
     – *Perforated pipes* – the pipes are perforated with thin (e.g. 1 mm or less) longitudinal slots.
   * pipes placed within a pavement drainage course. Apart from economic advantages, this also avoids the ponding of water within the trench. The pipes used in such a system need to be of relatively high strength, as they will be closer to the trafficked surface than conventional pipes (Section 13.4.4).

4. *Geofabric solutions:*
Systems using geofabrics have proved to be particularly effective, provided they are carefully chosen and installed to accommodate the actual conditions encountered. The main two types are:
* *Preformed geofabric composite drains.* The internal core drain may be a rigid pipe or a flexible material (as in the fin drain below). The rigid core alternative is often easier to install.
* *Fin drains.* A filter fabric (Section 10.7) is wrapped around a permeable core, typically 600 × 30 mm in cross-section, to provide both a filter and a permeable layer in the plane of the fabric. It is also used at interfaces between vertical and horizontal layers, as well as for pavement edge drains (Figure 13.2 and Section 13.4.2). Fin drains are very narrow and so the trench width need only be about 125 mm, resulting in considerable construction savings. Generally, a fin drain collects the water more quickly and discharges it more slowly than a conventional drain.

5. *Prefabricated solutions:*
Stiff polymer box-like structures are wrapped in geofabric and placed in a narrow trench. They may buckle under traffic load.

Applications 3 and 4 above are particularly relevant in the case of boxed construction described in Figure 11.13b, where the use of longitudinal subsoil drains to remove pavement water is essential if water is present and the surrounding material is impermeable. Even if the pavement material is not moisture-sensitive, trapped water will rapidly lead to pumping (Section 9.3), and hence to such failure modes as surface scouring, erosion, bleeding and potholing (Section 14.1A12, A13 and C10). Care must be paid to the design of the drain outlets to:
* ensure good flow speeds in the drain and thus aid the flushing out of any pavement fines that pass the filter barriers,
* prevent rodent entry, and
* prevent scour below the outlet.

Figure 13.2 also shows a *catch* (or *interceptor*) drain which is used to prevent water entering the road reservation. Such drains are particularly useful on the top of cuttings to prevent scour of the cutting face.

Drainage systems require regular inspection and maintenance (Table 26.1). Common problems are inaccessible inlets and outlets, lack of provisions for cleaning, blocked inlets and outlets, blocked pipes, and clogged filters. Only the first pair of these can be readily detected and corrected; hence strong attention is required at the design and construction stages to prevent the later occurrence of blocking and clogging.

13.4.4 Drainage courses

Permeability is discussed in Section 9.2.2 and Section 13.4.2 showed how permeable courses can be used as *drainage courses* (or *horizontal drainage blankets*) within the pavement structure. Each of the cross-sections in Figures 13.1–3 contains a permeable course. The application shown in Figure 13.2 illustrates the use of a drainage course to help lower a surrounding water-table. The discussion of water tables and soil suctions in Section 9.2.3 showed that suction can cause water to rise some distance above a water table. This upward movement of water is common in fine-grained materials and can be intercepted by using a drainage course (Figure 13.2).

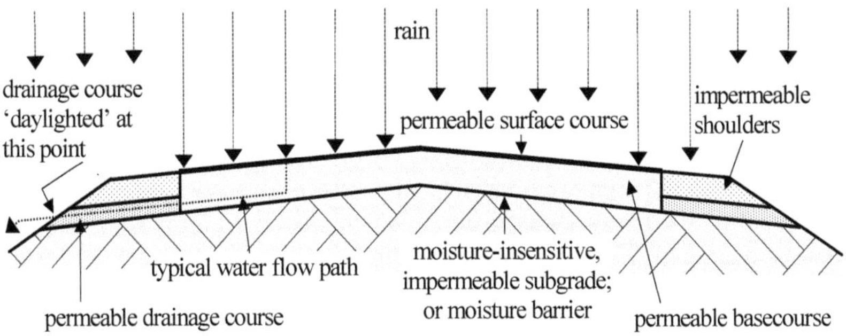

Figure 13.3 Water infiltration removed by permeable course in shoulders.

Obviously drainage courses should only be used in appropriate circumstances, e.g. when the pavement is on impermeable fill, a cement-stabilised sub-base or other impermeable layer (Section 10.2), and/or when the underlying course is not moisture-sensitive.

Material selected for a drainage course usually requires a permeability of above 1 mm/s. However, the variable nature of permeability values must be recognised. For example, the values quoted in Section 9.2.2 are very dependent on the total void ratio, e, and thus on the material grading and degree of compaction. As the void ratio drops and the degree of compaction increases, the permeability will decrease. The material used must be well chosen as it is almost impossible to rectify a clogged course without overall reconstruction of the pavement. Common design solutions are:

* gap-graded coarse sand,
* 20 mm crushed rock, typically with no more than 3 percent of material under 75 μm (Sections 9.2.2 and 11.1.2),
* filter fabrics of the type discussed in Sections 10.7 and 13.4.2), or
* porous (open-graded) asphalt (Section 12.2.2f).

Three important design points with drainage courses are to ensure that:

* water always travels along a path with progressively greater permeability and without flow constrictions, and
* any water movement in a course does not cause water to be trapped in the pavement structure at vertical sags or changes in course thickness. It is wise to provide direct transverse drainage at such locations,
* there is at least 40 mm of basecourse above the drainage course to protect it from traffic damage and surface intrusion.

The water from the permeable basecourse can drain transversely through a drainage course to exit via the shoulder (Figure 13.3) or flow directly into an underground drain (Section 13.4.3 and Figure 13.2).

13.4.5 Filter material and pipes

The permeable filter material used in drainage systems (Sections 13.4.2 and 3) must, of course, be in contact with the permeable courses of the pavement. To keep the filter material in place and to prevent the entry of fines from the pavement, the filter material is often wrapped in a permeable geofabric (Section 10.7).

Following Section 13.4.4, the permeable filter material is commonly a granular open-graded aggregate mix (Section 8.3.2) that is porous enough (Section 9.2.2) to permit water to pass through it and enter the pipe via the perforations. If the filter material is not wrapped in a permeable geofabric, the particle size distribution must be sufficiently dense to inhibit the movement of fines through the filter material. This conflicting requirement is to be avoided, as it reduces the efficiency of the whole system.

A tubular *fabric filter sock* (Section 10.7) is often fitted over the perforated drainage pipe to prevent material from entering the pipe. If this is not used, more care must be taken to prevent fines from the filter material which must be comprised of stones that are individually too large to fit through the pipe perforations.

13.5 Beyond the roadway

The presence of a road and its associated drainage provisions will inevitably alter natural drainage patterns in the associated region. The principles of good drainage practice within a road catchment are that:

(a) natural watercourses and existing drainage patterns should be interfered with as little as possible;
(b) water courses intercepted by a road formation should be allowed to return to their natural course as soon as possible; and
(c) it is usually unwise to merge a number of watercourses together in order to reduce construction costs.

Conditions during construction were discussed in Section 13.3.1 and will usually be governed by the requirements of appropriate drainage and environmental Authorities. For an operating road, once water has been collected by the above systems, it is delivered to the surrounding drainage catchment. This must also be done:

* to avoid nuisance to users of the adjacent land,
* with proper consideration of the catchment characteristics, and
* in association with whatever plans and requirements have been imposed by the appropriate drainage, catchment and environmental Authorities.

Treating water from the road at the roadside has many environmental benefits. A major technique involving streetside drains is discussed in Section 7.4.5. More elaborate systems may involve the retention of some drainage water within the road reserve. The process, depending on local conditions, will probably involve some combination of pollution traps (Section 13.4.2), sediment and erosion control measures, sedimentation ponds, wetlands and other pollution management measures in accordance with conditions imposed by other Authorities. Some discussion of sediment control was given in Section 13.4.2 and erosion control in Section 13.4.3. Water pollution is discussed in Section 32.4. Performance monitoring and ongoing maintenance of these systems is critical.

The management of the drainage flows will commonly be a dual system with a 'minor' component designed to manage the frequent storm events and a 'major' component to manage the rarer high-intensity storms (Alderson, 2006).

CHAPTER FOURTEEN

Pavement Performance

14.1 PAVEMENT CONDITION

The behaviour of a pavement during its life cycle passes through a number of stages:

1. After construction has been completed (Chapter 25) and the road is in operation, it enters a settling-down stage, which may last for one or two years (Section 12.1.1). During this time, construction imperfections become evident, particularly after a wet season. Such faults can usually be corrected by routine maintenance (Chapter 26).
2. As the pavement is subjected to continual traffic, it is progressively damaged by the effects of the more structurally-severe portion of that traffic. Its condition in this second stage therefore changes with and reflects its usage. The symptoms of the accumulated damage and the evidence of current condition are observed as tolerable or easily repairable pavement surface deficiencies. During this stage many deficiencies — e.g. rutting (see A.8 below) — will reach a level where maintenance to *prevent* further deterioration is desirable.
3. If preventative action does not occur the road enters the third stage in its life cycle when the surface deficiencies have worsened to such an extent that the pavement needs major repair, for instance by the application of a thick overlay (Section 12.2.5). This *restores* the pavement to its previous condition.
4. If restorative action does not occur, the fourth and pseudo-final stage in the life of the pavement occurs when the deficiencies represent terminal conditions in the pavement — for example, severe permanent-deflections related to structural deficiencies occur. These can only be rectified by some form of *major rehabilitation* of the pavement structure, such as ripping up the existing basecourse and either replacing it or adding additional material (Section 26.1). The conditions at this stage may also be at *maximum acceptable levels* in terms of their impacts on users.

The various deficiencies and defects which can occur in a pavement throughout these stages can take a large number of forms, the most important of which are listed in Sub-sections A, B, and C below. The terms used are from various sources. Some causes have been assigned to each definition to give strength to it. Whilst these are common causes, such lists cannot be comprehensive and each pavement failure deserves an independent and unbiased assessment.

A. Deformation-related deficiencies and defects

A.0 *Distress*: pavement deterioration visible at the surface. Distress is a broad term that embraces many of the subsequent deficiency modes.

A.1 *Settlement*: a general lowering of the road surface. Settlement may be due to such factors as a change in the water content of the subgrade, compressible layers in the subgrade, inadequate compaction during construction (Section 25.5), long-term consolidation (Section 8.2.1) or embankment problems (Section 11.8.1).

A.2 *Subsidence* (or *sags*): a localised or rather abrupt lowering of the road surface. Subsidence may result from poorly compacted backfill, poor local drainage (Section

13.4.2), the collapse of underground cavities, or heavy or frequent axle loads (Section 27.3.2). Compaction under traffic is usually called *densification.*

A.3 *Bumps* (or *heaving*): localised upward displacements of the pavement. Bumps may be caused by:
* clay soils expanding as a consequence of absorbing moisture (Section 8.4.2) which may have entered through cracks in the pavement (Section 14.1B) or via the shoulders (Section 14.2),
* load-induced plastic deformation in the basecourse or subgrade, leading to material not directly under the load being pushed outwards and then upwards (Section 11.2.2), or
* frost heave (Section 9.5).

A.4 *Waves*: wave-like vertical displacements of the pavement surface, with crests at least 500 mm apart. May be due to deep differential deformation of the subgrade, expansive clays or frost heave (see defect A.3). Although not traffic-induced, waves will normally be classed as roughness (see defect A.11).

A.5 *Corrugations* (or *rippling* or *washboarding*): closely spaced waves (crests less than 500 mm apart) on the road surface and transverse to the traffic direction. They are part of the megatexture of the pavement (Figure 12.4). Corrugations are important when they are visually noticeable, and moderately severe when they produce a rough ride.

On *paved roads*, corrugations are relatively rare and may be the result of:
* a poorly-laid surface course,
* a low-stiffness asphalt, or
* high horizontal wheel-forces due to braking, accelerating or turning.

On *unpaved roads* (Section 12.1.1), corrugations are commonplace and occur as the almost inevitable consequence of dynamic interaction between the road surface and the tyres of any braking, accelerating, or turning vehicles. Specifically, a spinning wheel tends to throw loose material backwards, troughs appear in areas of relative looseness, and subsequent wheels may bounce from trough to trough — both exacerbating the initial effect and creating new troughs wherever the bouncing continues. Thus, in addition to the conventional pavement factors, corrugations are also a function of the dynamic properties of the vehicle suspension and tyres (Sections 27.3.5 & 27.7). They arise on straight and curved roads, but are more common on curves. They occur most commonly where the surfacing is sandy and devoid of larger aggregate, or where the grading of the surfacing material lacks fines. Methods for successfully controlling the formation of corrugations on unpaved roads currently do not exist. However, they are reduced by making the road surface of well-graded cohesive material.

A.6 *Depressions* (or *birdbaths*): an area of dished subsidence resulting from either localised or variable failure of the subgrade or by the failure of an underground pipe (Section 13.4.2). It will probably lead to further pavement damage. See also defect A.8.

A.7 *Distortion*: irregular deformation of the pavement. Distortion may be the result of differential traffic-induced permanent deformation of a pavement layer which, in turn, may be due to either compaction or plastic shear failure of the material. One of the operational problems with any form of pavement distortion is that it can lead to poor drainage of surface water after rainfall.

A.8 *Rutting* (or *troughs* or *channels* or *instability*): longitudinal wheel-path depressions (defect A.6), usually accompanied by smaller longitudinal ridges on either side of the rut (defect A.3). Rutting is discussed analytically in Sections 11.4.3 & 14.5.3. Ruts may be due to:

(a) *abrasion* (Section 12.5.2), which is common with:
   * unpaved roads, and
   * the use of *studded tyres* to give better traction on ice and snow (causes about 3 mm of wear per million wheel passes).

(b) *traffic-induced compaction.* This may become evident very early in the life of the pavement and is caused by both:
   # the aggregation over time of routine incremental vertical deformations in a typical subgrade — and in some basecourses — with non-linear load-deformation responses (Section 11.2.2), *and either*
   * poor construction compaction (Section 25.5),
   * lack of lateral restraint at pavement edges, causing lateral shoving (Section 11.2.2), leading a corresponding vertical deformation,
   * moisture contents in excess of OMC (Section 8.2.3), or
   * high bitumen contents leading to flooding — and thus softening — of the asphalt (Section 12.2.3).

(c) *material* (or structural) *failure* within the pavement. This is more likely to occur in the longer term and in slow lanes where there is a high proportion of heavy traffic. Material failure is due to either:
   * high temperatures reducing asphalt stiffness (Section 11.3.2),
   * stone breakdown due to high inter-particle stresses (Section 11.1.3),
   * under-strength material,
   * overloading, or
   * long-term loading causing creep (Section 8.9.2), particularly at intersections where there are pseudo-stationary loads.

      Under-strength material, *overloading* and *creep* each lead to plastic failure. These shear failures (Section 8.4.3) occur without any volume change and so are often accompanied by noticeable lateral movement and/or vertical upheavals (defect A.3).

Although both subgrade and basecourse deformation can contribute to rutting, in a well-designed pavement the subgrade effect is usually insignificant, with the main contribution coming from incremental traffic-induced material failure in the upper basecourse layers. Increasing the basecourse thickness, by increasing the amount of deformable material, may increase rather than decrease rutting.

Ruts are of concern as they will usually:

* give drivers no advance warning of their presence,
* reduce riding quality (Section 14.4.1),
* lead to further pavement damage by causing the surfacing to crack (Section 14.1B). Even a uniform rut will create transverse surficial tension, leading to longitudinal cracking. The cracks allow water in the ruts to be pumped by tyre action into the pavement structure. Even shallow (< 10 mm) ruts are of structural concern if over half a rutted length is cracked.
* when over 10 mm deep, lead to water ponding in the wheel-paths and thus to:
   + braking problems due to hydroplaning or freezing of water (Croneys, 1991 and Section 27.7.4),
   + visibility in wet weather being reduced by spray from the ponded water.
* when over 20 mm deep, experience indicates that they will lead to rapid deterioration of the pavement (Croneys, 1991).

The correction of ruts of 10 mm or more in depth will usually involve major repair work. For such reasons many road authorities institute preventative intervention when ruts are only 5 mm deep. This matter is explored further in Figure 14.4 and Section 14.3.2. In terms of extent, rutting over more than:
5 percent of a road length deserves routine maintenance intervention,
20 percent of a road length is a cause for immediate concern, and
40 percent of a road length is critical.
The various intervention rutting levels given above can be relaxed for lower cost roads.

A.9 *Shoving (or creep)*: horizontal displacement of the surface material. *Shoving* is the preferred term as technically 'creep' in this instance may actually be plastic or overload deformation caused by horizontal wheel forces due to braking, accelerating, or cornering. If the resulting displacement is permanent it can result in a single ridge or, in the extreme, a series of shallow transverse depressions resembling corrugations (defect A.5). Permanent displacements generally result from a:
* low stiffness surface course (possibly due to poor compaction),
* low strength surface course, or
* lack of bond between the surface course and the underlying layers (Sections 12.1.3 and 12.3.2), usually due to dust, clay or water on the surface at the time of priming.

A.10 *Edge failures*:
There are two types of edge failure:
* Step-downs associated with fretted, scalloped, or broken edges of pavements. These are caused by:
  – traffic using narrow pavements travelling on the pavement edge (Section 6.2.5). Heavy and recreational vehicles are a particular problem.
  – pavement run-off water scouring the shoulders (Section 13.3.1), or
  – shoulder maintenance operations (e.g. by grader blades, Section 14.2).

  Such edge failures can be controlled by preventative maintenance (see Section 14.2). Any step-down of 50 mm or more will create a significant safety hazard (Section 28.5f).
* Step-ups which occur when long lengths of the pavement edge heave upwards. This is caused by water entry into expansive-clay subgrades.

A.11 *Roughness*: see Sections 14.3.2 & 14.4.

A.12 *Pumping (or blowing or bleed-out)*: See Section 9.3.2 and also defects B.11 & C.10.

A.13 *Faulting (or edge stepping)*: relative vertical movement of slabs in a Portland-cement concrete pavement, which can occur at either cracks or joints (Section 11.5.1). Faulting is usually the only deformation-related deficiency found in such pavements. It is caused by either:
* a lack of support under one slab due to subgrade failure or subgrade erosion due to pumping and/or poor drainage (Section 9.3.2),
* the build-up of fines under one slab, as a result of pumping (Section 9.3.2),
* a lack of intended load transfer between slabs due to joint failure (Section 11.5.1),
* frost heave under one of the slabs (Section 9.5), or
* overloading.

Faulting can be alleviated by grinding of the raised edge or by jacking the depressed edge.

B. Cracking-related deficiencies and defects

B.0 Introduction

From a fracture mechanics viewpoint, there are three types of crack:
– Mode 1 cracks open in tension in the plane of the pavement,
– Mode 2 cracks shear in the plane of the pavement, and
– Mode 3 cracks tear normal to the plane of the pavement.

Once a crack forms in one of these three modes, it may then propagate under further load applications. Cracking can also be defined with respect to its:

cause (see entries below),
consequences (see entries below),
repair. If cracks are not associated with surface deflections, they can usually be remedied by some form of surface treatment. Less extensive or lower severity cracking can be repaired or controlled by one of the surface enrichment or resealing methods described in Section 12.1.6. With the larger cracks, such as block cracks, it will first be necessary to seal the actual cracks before applying the surface treatment. Sealing of cracks over 2 mm in width involves filling and/or *overbanding* (or *striping*).
Cracks are filled by cleaning the crack and then injecting a bituminous or rubbery binder (Section 8.7.7). Overfilling must be avoided as it can lead to the filler bleeding across the surface.
In overbanding, an area 50 mm or so on either side of the crack is covered with a sealant or geotextile or bitumen sheet. This method is used where large relative movements are expected across the crack.
extent. Cracking over more than 10 percent of a length of road is cause for concern and is critical if over more than 40 percent.
severity. Cracks over 500 μm wide are a problem. Cracks up to 1 mm wide are called *hairline* (see defect B.9 below) and cracks over 3 mm are called *wide*. The width will depend on the temperature, with cracks tending to close in hot weather and open in cold weather. This crack opening movement can be up to 10 mm. Wide cracks should be sealed as soon as possible and are of major concern if they occur over more than 5 percent of a pavement.
intensity: crack length per area.
pattern. Different patterns are described under defects B.1,2,5,6&8 below.
location. Different locations are described under defects B.3,4,10&11 below.

Cracking causes the following problems in that it may:

* reduce tensile strength and thus limit the structural strength and stiffness of bound courses (Section 11.1.4),
* permit water to move from water-resistant to moisture-sensitive parts of the pavement structure (Section 9.3), thus producing deleterious effects. Indeed, any cracking which penetrates an impervious layer can allow moisture to accelerate the overall deterioration of the pavement.

Cracking which penetrates through the pavement surface may also:

* prove unsightly,
* deteriorate further under the localised effects of tyres, and
* permit surface water to enter into the pavement structure (Figure 9.1).

In pavements, cracks can propagate upwards or downwards. In the simple model in Section 11.2.3 and Figure 11.5d, cracking will begin in the tension layer at the base of the pavement and, under repeated traffic loading, propagate upwards to the surface as fatigue cracks. When the crack in a pavement course passes upward through a newer overlay (Sections 12.1.6 & 12.2.5), the effect is called *crack reflection*.

However, observation shows that many cracks begin at the surface and subsequently propagate downwards. The reasons for their occurrence in this manner are:

* lack of bonding between the surface layer and the underlying layers creating high local stresses in the surface layer (Section 11.1.4),
* rapid ageing of the exposed surface layers (Section 8.7.7), and
* local edge stresses at the tyre–pavement interface (Section 27.7.3).

The cracks propagate downwards under both thermal cycling and the tensile bending stresses that occur in upper pavement layers in the vicinity of (but not directly under) the wheel load (see Table 11.1).

Crack types

B.1 *Alligator* (or *block*, *chicken wire*, *crazing*, *crocodile*, *fishnet*, or *map*) *cracks*: interconnected cracks forming a series of approximately straight-sided polygons, or blocks. The cracking pattern is called:

crazing when the uncracked blocks are small (e.g. below 100 mm) and the cracks are hairline. *Chicken wire* cracking is crazing in association with rutting.
crocodile when the blocks are up to 300 mm in dimension and resemble a crocodile skin,
map when the blocks are up to 1 m in diameter, and
block when the blocks are larger than 300 mm.

Alligator cracks may be due to:

* shrinkage of age-hardened bitumen,
* exceptionally low temperatures,
* flexural fatigue of a surface course,
* the reflection through to the surface of basecourse cracking, itself due to inadequate pavement strength, and/or
* over-loaded vehicles.

Most alligator cracks will allow water to enter the pavement structure and therefore probably lead to further pavement damage, such as potholing (defect C.9). Any alligator cracking which is leading to the dislodgment of pieces of pavement should be considered to be severe. Cracking over more than 10 percent of the area suggests a need for treatment.

B.2 *Longitudinal* (or *line*) *cracks*: may be caused by:

* poor joint practice in a concrete pavement (Section 11.5.1); for example, longitudinal *warping cracks* often occur at the centreline of a pavement without a centreline joint.
* settlement of embankment fill, which usually leads to long, crescent-shaped cracks (defect A.1 & Section 11.8.1),
* a change in relative subgrade stiffness where a pavement has been widened,
* rutting of underlying courses (defect A.8),
* wheel-path factors (see defect B.4),
* moisture changes in basecourses or subgrades of expansive clay (Section 8.4.2),
* edge failure (defect A.10),
* pavement *shrinkage*, if near the pavement edge or if there is a set of parallel longitudinal cracks, or
* *thermal contraction* and expansion during diurnal temperature changes.

Longitudinal cracks will permit water to enter the pavement structure and therefore should be sealed in the manner described under defect B.1, especially if they occur over more than 5 percent of the pavement length.

B.3 *Edge cracks*: cracks near the pavement edge which may be caused by:
* settlement of embankment fill (Section 11.8.1) or shoulder movement (Section 14.2),
* inadequate pavement width for the traffic (Section 6.2.5),
* a very flexible surface course,
* shrinkage due to seasonal drying of shoulders and the outer metre of the pavement (Section 9.3.3),
* bitumen hardening (Section 8.7.7), or
* curling upwards of the corners of concrete slabs, due to climatic effects (Section 11.5.2).

Edge cracks may be accompanied by branching transverse cracks (defect B.6). They will probably lead to further pavement damage.

B.4 *Wheel-path cracks*: cracks along the wheelpath which may be caused by:
* fatigue (Section 11.4.4),
* local wheel-induced stresses in a wheel-path (Section 11.2.3),
* transverse tensile strains due to wheel-path rutting (defect A.8 & Section 11.2.2), and/or
* water infiltration from the shoulder into basecourse and/or subgrade (Section 13.4.2), if the cracking is in the outer wheel-path.

Such cracks can lead to surface attrition (defect C.18) and edge spalling (defect B.10). They are often a sign that critical structural conditions have been attained. In such cases, their cause should be diagnosed and early treatment prescribed.

B.5 *Meandering cracks*: caused by poor construction leading to settlement of embankment fill (Section 11.8.1).

B.6 *Transverse cracks* (or *contraction* or *shrinkage cracks*): particularly common in old, hardened pavements that are otherwise in good condition, in rigid pavements, and/or in low temperature regions. They may be caused by:
* temperature strains and shrinkage in rigid pavements or old asphalt pavements. In rigid pavements, these cracks usually begin near construction joints and grow towards a meandering polygonal pattern. Reinforcing usually does not inhibit such cracks.
* late saw-cutting of a construction joint in a rigid pavement. Secondary cracking can then occur between the joint and the transverse crack.
* an expansive-clay subgrade (Section 8.4.2). In this case, any subsequent moisture increase will cause bumps or ridges (defect A.3) to occur around the crack.

Transverse cracks may lead to edge spalling (defect B.10).

B.7 *Reflection cracks*: any form of crack in the basecourse or subgrade which reflects through to the surface layer and may thus fall into any of the defect categories B.1 to B.6 above or B.12 below. Reflection cracks will probably lead to further pavement damage by permitting water to enter the pavement structure and therefore should be treated in the manner described in Section 12.1.6 — e.g. by putting a ductile overlay over the cracked area.

B.8 *Crescent* (or *bearing*, *parabolic* or *slippage*) *cracks*: usually relate to the traffic direction and occur as a set of closely spaced parallel cracks. They may be caused by:
* slippage of the surface course under horizontal wheel loads, and/or
* inadequate surface course thickness.

B.9 *Hairline* (or *D-*) *cracks*: small, fine, closely spaced, irregular cracks. For above-zero temperatures, cracks are defined as hairline if they are less than 1 mm wide. They may be caused by:
* a low bitumen content,
* poor compaction (see also defect B.13),
* normal drying processes in a rigid pavement, although they can lead to slab breakage near corners and edges (see defect B.11), or
* the deterioration of moisture-sensitive aggregate, particularly when subjected to repeated freezing and thawing (Section 9.5). Porous aggregates are most sensitive to the effect.

Hairline cracks may develop into craze cracking (see defect B.1).

B.10 *Edge spalling* (or *local cupping* or *lipping*): pieces of concrete spall off the pavement at joint or crack edges. Edge breaks over 25 mm in depth are described as *deep spalls*. Spalling may be caused by:
* the stresses due to wheel loads applied to a free edge,
* relative movement at cracks,
* in unreinforced concrete pavements, the crack widening rapidly and trapping rubbish which causes edge stresses at the crack face when contraction occurs subsequently, and
* in sub-zero temperatures, the presence of freezing water swelling in cracks (Section 9.5) and leading to edge stresses at the crack face.

Edge spalling at the surface will increase pavement roughness (defect A.11) and give the road a poor appearance. It can also reduce load transfer across a joint.

B.11 *Corner cracking*: occurs at the corners of slabs in rigid pavements. It is usually associated with:
* overloading (Section 11.5.2),
* subgrade failure (due to pumping, etc. — see defect A.12),
* curling of slab corners upwards due to climatic effects, or
* poor joint design.

B.12 *Isolated cracks*: usually associated with depressions (defect A.6) or reflection cracks (defect B.7).

B.13 *Checking*: cracks which occur during the construction of an asphalt pavement, particularly as a consequence of rolling. They are often hairline (defect B.9) and sealed at the surface by the use of rubber-tyred rollers (Section 25.5c).

C. Surfacing deficiencies and defects

See also A.9 (creep) and the cracking modes in B, which are also in this category when they occur in the surface course.

C.1 *Scoring, indentations, and wheel imprints*: localised marks which may be caused by car underbodies striking sharp changes in vertical alignment (Section 19.3, scoring) or prolonged parking (wheel imprints), possibly coupled with the use of a low-stiffness asphalt mix.

C.2 *Stripping (or disbonding or scabbing)*: loss of smaller pieces of surface aggregate from bituminous surfaces. Stripping is discussed in detail in Section 12.3.2.

(a) For unpaved roads, aggregate loss is discussed further in Section 12.1.2 which indicates that it is due to:
* wind,
* water,
* traffic volumes and speeds,

* the fines content of the surfacing material.

(b) For bound bituminous surfaces, it may be caused by:
* insufficient binder,
* binder which is too hard (Section 8.7.5),
* binder which has lost its ductility with age (Section 8.7.7),
* use of weak or friable aggregate,
* use of aggregate coated with dirt or with certain surface characteristics (Section 12.3.2),
* poor mix design,
* the presence of water,
* poor compaction, or

Stripping will probably lead to further pavement damage, but can be controlled by surface enrichment (Section 12.1.4).

C.3 *Ravelling* (or *fretting*): removal of the larger pieces of surface than in defect C.2, leaving pits (or craters) in the surface. In many ways this defect is simply a progression from defect C.2 and is largely caused by the same possible factors listed under defect C.2. It may also be due to the use of over-sized aggregate.

Ravelling is considered of slight severity when there is noticeable loss of material, moderate severity when the pitting is noticeable across the surface, and severe when the pitting is extensive and loose material is in evidence.

Ravelling should be attended to rapidly when it covers more than about 4 m^2.

Ravelling can lead to further pavement damage, such as potholing (defect C.9). In minor cases it can be repaired by the addition of extra binder via a reseal or similar treatment (Section 12.1.5). Otherwise, a thin overlay (Section 12.1.6) will be needed.

C.3A *Spalling* (or *pop outs*) *of rigid pavements*: small pieces of concrete detach from the pavement surface. This effect is usually caused by concrete which has been weakened by poor construction practice (e.g. excess surface fines) or chemical attack, or has succumbed to tyre suction.

C.4 *Weathering* (or *long-term ravelling* or *scaling*): the pavement surface initially pits and then becomes rough in texture over time, often as a progression from defect C.3. It can be initiated by the factors listed in defects C.2&3 or by:
* bitumen which has lost its ductility with age (Section 8.7.7),
* weathering of the aggregate (Section 8.5.10), with pockmarks left by the removal of the clay products of weathering,
* the abrasive action of traffic,
* aggregate fracture,
* overloading of asphalt causing shear failures which, in turn, pull aggregate from the mix along the failure plane. Any moisture entry then causes further stripping of the aggregate.

Weathering can lead to further pavement damage, such as potholing (defect C.9). In minor cases it can be repaired by the addition of extra binder via a reseal or similar treatment (Section 12.1.5). Otherwise, a thin overlay (Section 12.1.6) will be needed.

C.5 *Protrusion of aggregates* (*stoniness*): caused by the loss of binder on surfaced roads and the loss of fines on unpaved roads. It may be caused by ageing of the binder or the flow of the binder in high temperatures.

C.6 *Glazed* (or *slick*) *surface*: the surface becomes smooth, hard, shiny, and slippery. This results in low skid resistance (Section 12.4) and poor light reflectance (Section 24.2.2). It may be associated with defects C.2, C.3, C.7 or C.8, or with poor aggregate grading or distribution.

C.7 *Polished aggregates (or abrasion)*: the skid resistance of exposed pieces of aggregate is lowered by the polishing action of heavy vehicular traffic. Even a rough-looking surface may have a low skid resistance. See the discussion in Sections 8.6o & 12.5.2.

C.8 *Bleeding of binder (or fatty surface or flushing or black spot)*: the road surface has a sleek, smooth appearance because of excess binder on the surface. It becomes soft in hot weather — leading to loss of shape, bitumen pick-up by tyres, and aggregate stripping, and slippery in wet or cold weather — leading to low skid resistance. It will also increase water spray in wet weather and provide poor bond for future overlays. Bleeding may be caused by:

* too heavy a prime coat,
* excessive bitumen content, which may be a result of calculations for spray rates (Section 12.1.6) not allowing for lower voids in the wheel-path,
* high seasonal temperatures softening the bitumen,

These three factors will cause the bitumen to flood any available air voids (Section 12.2.3) and rise to the surface. Other causes are:

* flaky aggregate,
* excess aggregate embedment during construction (Section 12.1.5),
* a loss of cover aggregate due to stripping (defect C.2 above), or
* water in the voids in the asphalt (Section 12.2.3). This causes small pockmarks on the surface as the trapped water-vapour creates bubbles in the binder and then breaks through to the surface, allowing the bubble to collapse.
* overloaded wheels causing surface aggregate to be pushed below the surface.
* aggregate being forced into an under-compacted sub-base.

A flushed appearance may also be caused by an accumulation of oil droppings and road refuse.

As a first stage remedy, sand, grit or a fine-grained hard bituminous product (e.g. gilsonite) is applied to, and the rolled into, the 'bleeding' surface. It may be necessary for the bleeding areas to be sprayed with about 100 mL/m^2 of an aromatic kerosene to soften the surface prior to rolling — this is usually only effective for bitumen which is less than two or three years old.

If this method is not appropriate, or where continuous fatty strips of over 10 m in length occur, more positive remedial action must be taken, particularly on curves. This usually involves some form of surface dressing (Section 12.1.3) and may require the excess bitumen to be burnt off before any dressing is undertaken. Some proprietary products are also offered to harden bitumen. These should be tested before field application as it will be difficult for many materials added at the surface to achieve adequate penetration into the soft bitumen.

Bleeding should be attended to rapidly when it is in a critical location for braking or cornering or covers more than 4 m^2.

C.9 *Potholing*: a steep-sided bowl-shaped cavity, usually at least 150 mm in diameter and 25 mm deep. A pothole meeting both these criteria is considered to be severe. Potholes may be caused by:

* loss of surfacing,
* basecourse erosion,
* advanced cracking under traffic or severe weather, or
* water-induced blow-outs (see Section 9.3.2 and Figure 9.5).

Potholes can be controlled by preventative or early maintenance. They are usually repaired by thoroughly cleaning, drying, and priming the cavity and then filling it with asphalt. Any bituminous surfacing (Section 12.1.3) should not be placed over the filling for at least two months, to reduce the risk of bleeding (defect C.8).

Potholes create a poor ride, worsen under traffic, and permit water to readily enter the pavement structure. Once they have formed, their growth is accelerated in wet weather by the removal of fine material in suspension in the water being splashed out of the hole.

C.10 *Bleeding of water* (or *wet spots*): excess water is pumped through surface cracks under the action of traffic (see defect A.12 & Section 9.3).

C.11 *Migration of fines*: as for defect C.10, but with the water containing fine particles (e.g. clay) from the pavement structure.

C.12 *Inadequate surface drainage*: see Section 13.3.1.

C.13 *Peeling* (or *delamination* or *surface lifting* or *seal break*): a pavement layer debonds away from its underlying layer. It usually occurs between the surface course and the underlying basecourse, with large pieces of surface course breaking away.

Peeling may be caused by:

* no bond being present, due to poor construction practice,
* an adhesion failure between layers,
* inadequate surface thickness causing excessive shear strains between the surface course and the basecourse,
* pumping (Section 9.3.2), or
* chemical attack on the bond.

C.14 *Blistering*: localised areas of the surfacing develop dome-like ruptures, which usually then crack and expose the underlying basecourse. It may be caused by chemical reactions within the pavement.

C.15 *Blow-ups*: localised buckling or shattering of a rigid pavement, usually at a transverse crack or joint. It is caused by excess longitudinal movement or by dirt accumulating in a contraction joint during cold weather and subsequently preventing the joint from closing. Its cure will usually include the provision of an expansion joint.

C.16 *Streaking*: alternative lean and heavy lines of bitumen running longitudinally. It may be caused by non-uniform spraying of bitumen (Section 12.1.4).

C.17 *Erosion gullies*: created by surface water run-off from an unpaved road.

C.18 *Joint sealant distress*: elastic joint-sealants used in manufactured joints in rigid pavements (Section 11.5.1) can deteriorate or be dislodged. This leaves an ineffectively sealed joint which will permit moisture to enter the sub-base and allow pieces of rubbish to lodge in the joint, keeping it permanently open. Water trapped in poorly sealed joints can cause localised frost heave (Section 9.5).

C.19 *Rough surface*: A rough pavement macrotexture is desirable in that it provides good skid resistance (Section 12.5.1) and low spray and aquaplaning risks in wet weather (Section 27.7.4). However, it does increase traffic noise generation (Section 32.2.3). In the case of a concrete slab, a poor surface usually indicates poor overall construction practice.

D. Diagnosis

The surface deficiencies and defects listed are all visible on the surface. However, the equivalent subgrade and intermediate course deterioration will never be directly evident although the consequent differential or excessive deflection will become apparent in various pavement surface defects, such as alligator cracking (B.1), rutting (A.8), or roughness (A.11). This illustrates the importance of determining whether or not the surface deterioration is caused by deterioration in basecourses and subgrades. If it is, surface repairs will not be of great value. However, if the pavement structure is sound, then the repair of surface deficiencies will not only restrict further surface deterioration but also will prevent moisture-induced damage to the pavement (Potter and Hudson, 1981). However, it will probably increase perceived roughness levels.

The data given above can be recast in terms of 'causes' to give Table 14.1, as an aid to avoidable causes of future deterioration.

Table 14.1 Causes and effects of pavement deterioration.

Cause	Defects listed in Sections 14.1A–C
1 Subgrades, fills and natural formations	
1.1 wide change in subgrade water content	A.1, A.7, A.8, A.12, B.2, B.3
1.2 local change in subgrade water content	A.2 – subsidence
1.3 general subgrade overload	A.1, A.4
1.4 local subgrade overload	A.6, A.13, B.10, B.11
1.5 collapse of underground cavities	A.2
1.6 collapse of underground pipes	A.6
1.7 poorly compacted fill	A.1, A.2, B.2, B.3, B.5
1.8 structural failure of fill	A.2, B.2, B.3
1.9 expansive clays (Section 8.4.2)	A.3, A.4, B.2, B.6 – a seasonal effect
1.10 frost heave (Section 9.5)	A.3, A.4
2 Pavement courses	
2.1 sub-standard pavement material (Section 8.6)	A.7, A.8, A.11, B.3, B.9, C.2, C.4, C.7, C.8, C.14
2.2 inadequate pavement thickness	A.7, A.8, A.11, B, B.3, B.8
2.3 widespread inadequate pavement compaction	A.1 – soon after construction
2.4 local inadequate pavement compaction	A.7, A.8, B – soon after construction
2.5 pavement compacted above OMC	A.8, B – soon after construction
2.6 water trapped in pavement structure	C.9, C.10, C.11
2.7 wheel overloads	A.2, A.3, A.7, A.8, A.13, B.1, B.3, B.4, B.11, C.2, C.8
2.8 excessive wheel passes (Section 27.3.2)	A.8, A.11, B.1, B.4, C.3
2.9 compressive stress internal overload	B.2
2.10 tension or fatigue stress internal overload	B.1, B.7
2.11 pavement shrinkage	B.2, B.6, B.9
2.12 pavement expansion	C.15
2.13 joint failure in a rigid pavement	A.13, B.2, B.6, B.11, C.18
3 Surfacings	
3.1 an unbound surface	A.5, C.3, C.17
3.2 poor surface for priming	A.9, C.13; soon after construction
3.3 poor sealing technique (Section 12.1.3)	C.4, C.13; soon after construction
3.4 high bitumen content in surface course	C.8
3.5 low bitumen content in surface course	B.9,C.3, C.4
3.6 inadequate thickness of surface course	B.8, C.13
3.7 inappropriate asphalt in surface course	C.2, C.3, C.6
3.8 poor concrete in surface course	C.3
3.9 low stiffness asphalt surface course	A.5, A.9, B.3, C.1

3.10 poorly-laid asphalt surface course	A.5, C.2, C.3, C.13, C.16
3.11 poor bond between surface course and basecourse	A.9, B.8, C.13
3.12 low temperatures	B.1
3.13 high temperatures	C.8
3.14 aged or hardened bitumen in surfacing	B, B.1, B.3, C.3, C.4
3.15 chemically-attacked surface	C.3, C.13
3.16 cracked pavement surface	C.9
3.17 repaired pavement surface	A.11
4 Cross-section	
4.1 inadequate pavement width	A.10, B.3,B.10
4.2 poor pavement edges	A.10, B.2,B.3
4.3 moisture changes in shoulders	B.3, B.4
4.4 poor shoulders	A.10, B.3

14.2 SHOULDER CONDITION

The major issues with shoulder condition must be seen in the context of the role of the shoulder, as outlined in Section 11.7. These issues are as follows:

(a) Pavement deterioration in the outer wheel-path. Entry of water into the pavement will cause or accelerate this process. The critical role of the shoulder in determining the amount of moisture which does enter is discussed in Sections 9.1&4 and 11.7. For this reason, shoulder condition is of vital importance for the entire pavement.
An impermeable shoulder is very desirable and the universal use of some form of surface treatment (Section 12.1.3–5) is increasingly favoured. Impermeability can be achieved in unpaved shoulders by the use of well-graded material, regular surface grading, and avoiding the use of material over 25 mm in size (Section 9.2.2).

(b) Pavement edge failures caused by loss of shoulder support (14.1A.10). The shoulder adjacent to the pavement edge is exposed to traffic abrasion and water and wind erosion. This creates a step down from the surface which is a traffic hazard and source of further deterioration. It can be temporarily alleviated by grading but this must be done carefully to ensure that cross-fall drainage is not destroyed and the grader does not cause further edge deterioration.

(c) Fretting of the pavement edge. This is a consequence of (b) and occurs when the basecourse at the pavement edge weakens and wheel loads break off pieces of the pavement surface (see also 14.1A.10 above).

(d) Poor shoulder surface, including potholes and corrugations. An unpaved shoulder behaves and deteriorates as an unpaved road and can therefore become unsuitable for emergency use by vehicles. Shoulder material should be selected with this in mind (Section 12.1.2). Regular grading can also assist.

(e) Surface scour. This is due to high water flows and high cross-falls, and is exacerbated when large vehicles travel on the shoulder. Remedial action includes rerouting the water and surface protection of the shoulder and batter.

14.3 MODELLING PAVEMENT DETERIORATION

14.3.1 Serviceability and useability

All roads will deteriorate to some extent under most commercial vehicle traffic, due to the irreversible components of the load-deformation response of pavement materials (e.g. Figure 8.8), caused largely by continued compaction, aggregate crushing at local contact points, and material plastic deformation. Pavements can also deteriorate in the absence of traffic, as illustrated by the discussions of aggregate degradation in Section 8.5.10, bitumen oxidation in Section 8.7.7, and moisture-related volume changes in Section 9.3. Thus, deterioration is most commonly a combination of traffic and environmental effects. The various forms of pavement deterioration were discussed in Sections 14.1 & 2 and were seen to occur within both the surface course and the pavement structure. In most cases the deterioration process is incremental and cumulative, as illustrated in Figure 14.1.

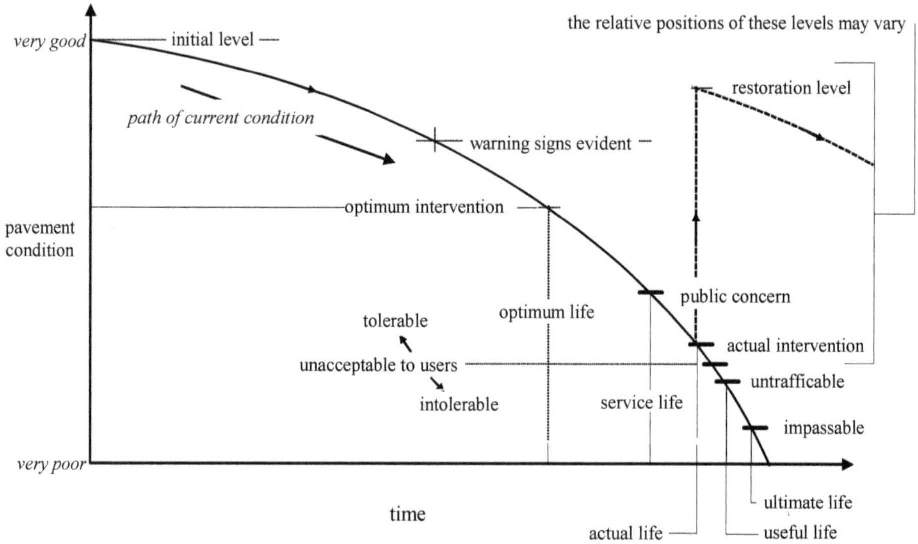

Figure 14.1 Relationship between different pavement conditions levels (Widdup, 1980). Most of the terms used are self-explanatory or defined in the associated text.

The rate of deterioration is determined by five factors:

(a) the design of the road (Chapter 11);
(b) the construction of the road and thus its initial condition (Chapter 25);
(c) the maintenance of the road (Chapter 26);
(d) the type and volume of traffic imposed on the road (Chapter 17); and
(e) the environment in which the road operates (Chapters 9 and 13).

Pursuing factor (c), as the road deteriorates, various maintenance procedures of relatively moderate cost can be used to prolong its useful life. For instance, repairing cracks and potholes will retard structural deterioration and prevent moisture entry. Non-structural surface treatments (Section 12.1.7) can improve skid resistance and riding quality. However, a stage will be reached when pavement strengthening will be needed in order to counteract the irreversible structural deterioration that has occurred. Practical implications of this need for major intervention are discussed in Section 26.1.

The *serviceability* of a road at any point in its life is defined as the extent to which it meets the requirements of its users. The *present serviceability index* (*PSI*) was developed during the AASHO Road Test (Section 11.4.3) and is probably the best known method for evaluating and quantifying pavement serviceability. The PSI of a piece of road is typically obtained from a panel of engineer-drivers traversing the road, which is usually of at least an arterial road (Table 2.1). The panel rates the current ability of the road to provide a smooth and comfortable ride for high-volume, mixed traffic at speeds of 50 or 80 km/h. The panel assumes four hours of driving for the 80 km/h road and one hour for the 50 km/h road. It ignores the formal classification, alignment, shoulders, any settlement of embankments, and skid resistance of the road. To obtain a numerical value for PSI, the panel uses a linear scale from 0 ('extremely poor') to 5 ('extremely good'), with 2.5 being rated as 'fair'. The quantity determined is called the *Present Serviceability Rating* (*PSR*). In Canada, a *riding comfort index* or *rating* is used with a rating scale from 0 ('very poor') to 10 ('excellent'), which is double the PSI value, and a rating panel which includes lay-people as well as experts. Results are influenced by the age of the panel and the type of vehicle used.

Because of the obvious practical difficulties with the routine use of this panel approach, PSI results have more often been related to the more readily-measured road *roughness*. Roughness measurement and equations linking such data with PSI are given in Section 14.4.4.

The 'warning level' shown in Figure 14.1 signals a need to monitor pavement conditions more closely. This situation is returned to in Section 14.5.3.

The optimum intervention and rehabilitation levels are discussed in Sections 14.5 and 14.6 and the *optimum life* of a road is defined in Section 14.6.3 and Figure 14.13.

If suitable intervention does not occur, a road will continue to deteriorate until it becomes unserviceable and no longer fulfils its intended purpose. As with PSI, this definition of road *failure* also depends on road-user expectations (Figure 14.1). As the road deteriorates beyond most users' perceptions of failure, travel becomes increasingly more unpleasant and difficult, until the route is completely untraffickable for normal traffic. Further deterioration makes the road *impassable* for all traffic. This is the *ultimate failure state* and is consistent with the definition of a road in Section 2.1.

14.3.2 Loading effects

The causes of traffic-related pavement deterioration will now be examined. First, as can be seen from Section 11.4.3 and Figure 11.10, a single overload of sufficient size can produce immediate structural failure. Vehicle mass regulations attempt to prevent this mode from occurring (Section 27.3.2).

The second mode of deterioration even occurs with well-regulated traffic. As illustrated in Chapter 11, each pass of a wheel load produces stresses and strains in a pavement. Vehicle regulations deliberately keep individual axles far enough apart to avoid significant simultaneous cumulations of the stresses and strains from two adjacent axles (Section 27.3.1). The influence of traffic on pavement deterioration therefore depends largely on the size and number of the individual wheel loads. With respect to size, each wheel-induced stress above some threshold level causes a small increment of permanent strain (or damage) within the subgrade and/or pavement structure (Section 11.3). Truck wheels loaded to the legal limit produce stresses in an average pavement which are above that damage threshold. From Section 11.3 and typical cyclic stress–strain behaviour, this deformation increment is proportional to 1/E and remains a monotonic function of the size of the wheel load.

The resulting damage from the passage of each axle therefore accumulates with the number of axle passes. Accumulated tensile strain damage eventually results in fatigue cracking (Section 11.4.4) and accumulated compressive strains in rutting (Sections 11.4.3 & 14.1A8). In both cases the total pavement damage — as typically measured by permanent deformation, rut depth or crack length — is a monotonic function of the number of axle passes. This behaviour, basically for the compressive case, is illustrated in Figure 14.2a which reflects the hysteretic material response shown in Figures 8.8 & 11.10 and assumes that all axles are loaded to the same level. For a perfectly plastic material there would be no further increase in permanent deformation after the first loading cycle, however the behaviour in Figure 14.2a represents a more realistic situation (Lay, 1982). It is often implied that the pavement damage increases linearly with passes of identical axles. However, as Equations 11.4 & 6 illustrate, the monotonic effect is less than linear with *strain-hardening* stone courses and more than linear with *strain-softening* plastic materials. When the loading varies from axle to axle, the size of the permanent deformation per cycle also varies (Figure 14.2b).

The accumulated permanent damage cannot be calculated from the stress–strain curve for a single load-cycle, but can be estimated from empirical stress–strain equations such as 11.5 and 11.9 which estimate the total strain as:

$$\varepsilon_p = \sigma^{1-m}/(1 - m)E_p \qquad (14.1)$$

where the factor E_p will also be a function of the total number of loads, N, applied up to the point being considered. The resulting stress–strain responses are illustrated in Figure 14.3. Rather than introduce complex functions mimicking Figure 11.7, it is useful to take the general power form of Equation 14.1 and to then consider Sections 11.4.3&4 and the fourth power law, particularly Equations 11.11–15, as these give the conditions at failure and are of the form (Equation 11.12 with $P \propto \sigma$):

$$N_f\sigma^c = \text{constant} \qquad (14.2)$$

where N_f is the number of load passes to failure, usually in equivalent standard axles (N_{esa}, Section 27.3.4), and c is a constant, commonly $2 < c < 8$.

Making the reasonable assumption that the actual strain conditions at failure are related to material properties more than to loading sequences, suggests that the failure value of ε_p from Equation 14.1, ε_f, should be stress-independent of σ. This independence can be achieved by using Equation 14.2 to add N to the form suggested by Equation 14.1 in such a way that ε_f will be a constant. This leads to:

$$\varepsilon_p = E_k(N\sigma^c)^{(1-m)/c} \qquad (14.3)$$

where E_k is a calibration constant. For typical data and for σ and E in MPa units, Equation 14.3 reduces to:

$$\varepsilon_p = \sigma^{0.8}N^{0.2}/12E_0 \qquad (14.4)$$

which reflects Figure 14.3 and eventual strain-softening, leads to a common failure criterion, and introduces N into Equation 14.1. Other empirical relations have suggested that the key variable is not $\sigma^{0.8}N^{0.2}$ but $\sigma^{0.3}N^{0.3}$.

σ can be linearly related to the wheel load P, and thus to the elastic loading strains ε_v and ε_h in Section 11.4.3&4, by the elastic analysis methods in Section 11.2.3. In these cases it is useful to see ε_v or ε_h as the above-damage-threshold elastic strain caused by P. This strain is the run-out strain, in fatigue terminology.

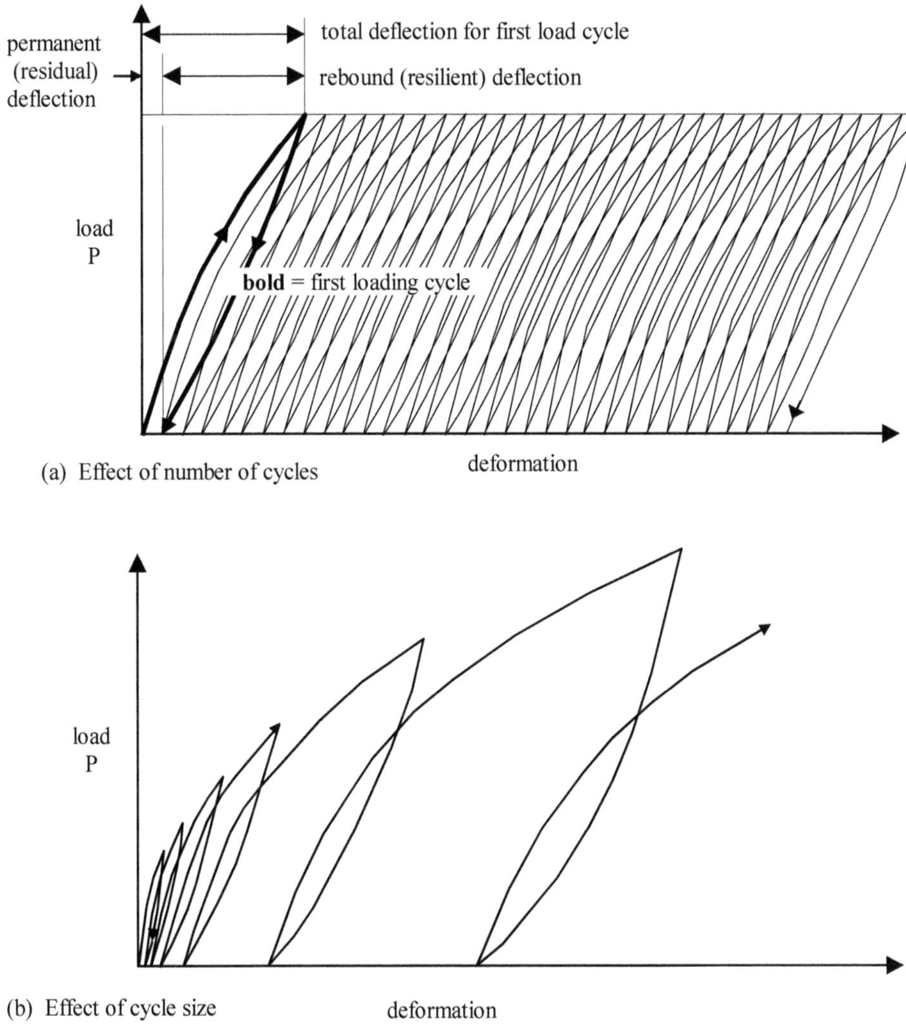

Figure 14.2 Cyclic load patterns for a strain-softening material – such materials are defined in Figure 11.10.

With strain-softening materials, such as a typical subgrade, Equation 11.8 can be used but with σ/σ_f replaced by the ratio of ε_v^4N to its limiting-criterion level for compressive failure, raised to some power. The compressive failure limiting-criterion is given by Equation 11.14 as $10^{-7.21}$, which allows N_f to be calculated. Given the form of Equation 14.4 and the use of σ/σ_f as the dominant first fourier-expansion term in the similar *magnification* (or amplification) factors used in structural engineering, it is assumed that the power is 0.2 in order to preserve σ/σ_f and the link with Equation 14.4. If the stress cycle remains constant, Equation 11.8 becomes, typically:

$$\varepsilon_p = \sigma/E_0(1 - [N/N_f]^{0.2}) \qquad (14.5)$$

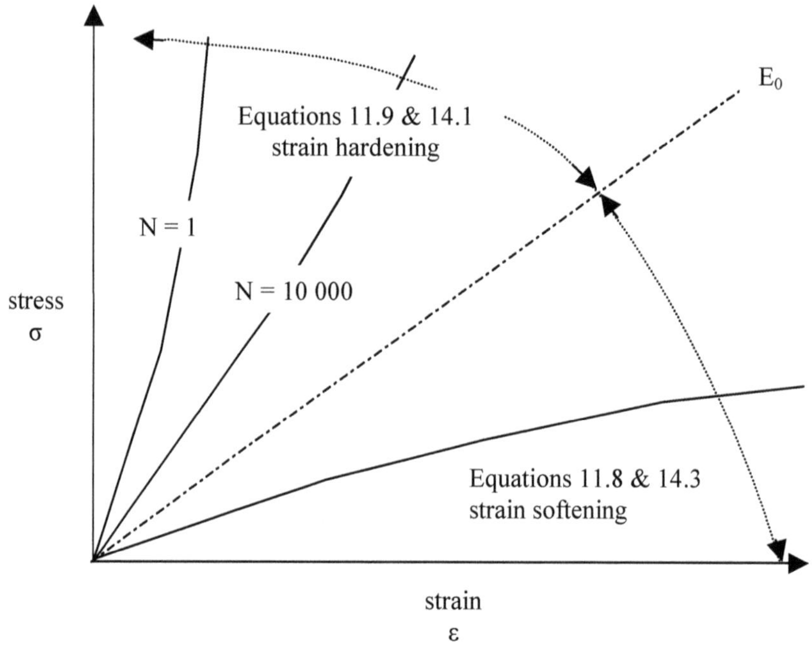

Figure 14.3 Effect of load cycles on stress–strain response.

For the example, if the stress cycle is constant and N_f = 700 000, then the estimate of the total strain equivalent to the strain-hardening Equation 14.5 is:

$$\varepsilon_p = 495/(1 - [N/700\ 000]^{0.2})\ \mu\varepsilon \quad (14.6)$$

Recall from the above that the 0.2 power might be 0.3.

To illustrate the use of Equations 14.4–6 to estimate of ε_p, consider the stress analysis example in Section 11.2.3. Sections 11.4.3&4 showed that basecourse fatigue failure occurs at N = 430 000 and that fatigue rather than rutting controls in this relatively weak pavement. Similarly, the elastic deflection of the pavement can be calculated from Table 11.1 as 740 μm. The permanent deformation can be estimated by applying Equation 14.4 to the upper two strain-hardening courses and Equation 14.6 to the subgrade, using the values of σ_v and ε_v derived from Table 11.1 for each course.

The permanent deformation can be estimated by applying Equation 14.4 to the upper two strain-hardening courses and Equation 14.6 to the strain-softening subgrade, using values of σ_v and elastic deformation, d, derived from CIRCLY for each course. Deformation rather than strain is used to take advantage of the integration already performed by CIRCLY to convert strains to deflections. The deformation-load passes relationship is therefore:

$$d_{total} = 0.295 + (0.929^{0.8}100/1250 + 0.261^{0.8}200/250 + 0.025^{0.8}200/250)(N^{0.2}/12) + 0.442/(1 - [N/700\ 000]^{0.2})$$

$$= 0.295 + 0.0325N^{0.2} + 0.442/(1 - 0.0678N^{0.2})$$

The resultant load-passes versus deflection plot is shown in Figure 14.4. It has the same form as measured data (Lay, 1993). The graph also shows the effect of assuming N/N_f rather than $(N/N_f)^{0.2}$ in the strain-softening Equation 14.5 and indicates that, in terms of absolute deflections, this is the most sensitive of the assumptions made.

Permanent deformation under wheel loads leads to ruts (Section 14.1A8). Hence, rut depths are seen initially to grow increasingly slowly (with $N^{0.2}$) while strain-hardening

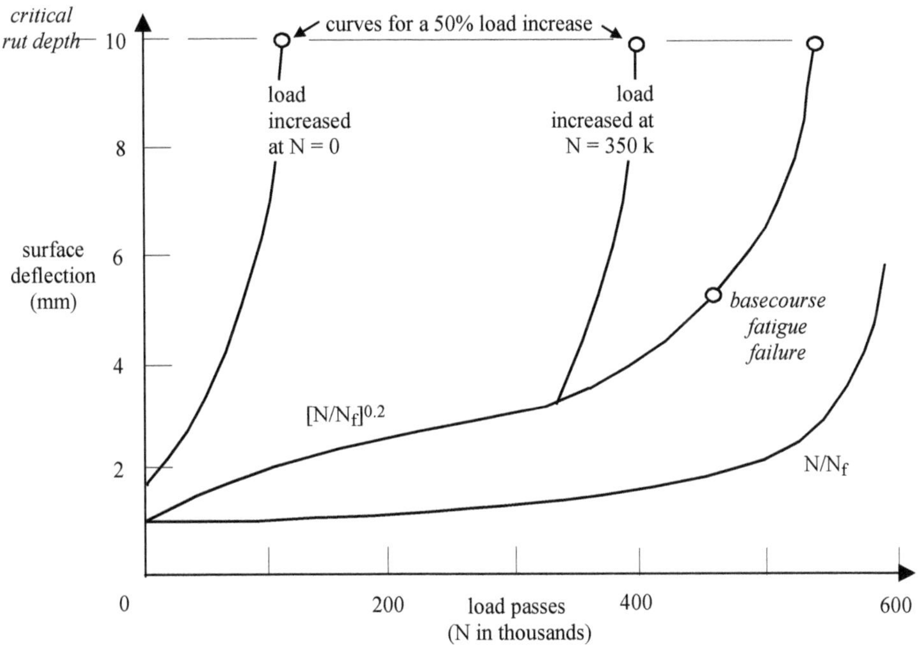

Figure 14.4 Strain-softening load–deflection curve with increasing axle passes for pavement used in example in Section 11.2.3 and Figure 11.5.

dominates, but to then begin to snowball as the effects of the strain-softening subgrade begin to dominate. In the relatively weak pavement used in the example, the rut was over a millimetre deep at 100 000 passes. For a heavier-duty pavement, rut depth would not reach 1.0 mm until about 3 000 000 passes. Note that some empirical studies have suggested that rut depth increases with $N^{0.5}$.

Clearly, the total strain ε in Equations 14.3–5 is closely linked to the damage the road has suffered. Thus, it can be reasonably assumed that initially the damage is proportional to $N^{(1-m)} = N^{0.2}$. If it is further assumed that the serviceability of a road (Section 14.3.1) is linearly related to the damage it has suffered, then, using equivalent standard axles (Section 27.3.4) it can be suggested that:

$$PSI = 5(1 - [N_{esa}/N_f]^{0.2}) \qquad (14.7)$$

where N_f is the number of equivalent standard axle passes to unserviceability. This does give a PSI vs N (effectively, time) curve of the type shown in Figure 14.5. Equation 14.7 is based on asphalt behaviour. Rigid pavements behave in a different way, with less initial deterioration but a more rapid deterioration as failure approaches. Even asphalt finally deteriorates more quickly than the above model would suggest.

14.3.3 The fourth power law

The *fourth power law* was first discussed in Section 11.4.3 and Equation 11.12, and then used in Section 14.3.2 and Equation 14.2 to estimate pavement damage. This empirical law covers repeated loading and suggests that the number of passes, N_f, of a wheel load, P, needed to cause failure is approximately inversely related to the fourth power of P. For different combinations of N_f and P, Equation 11.12 leads to:

$$N_{f1}/N_{f2} = (P_2/P_1)^4 \quad (14.8)$$

This ratio is sometimes called the *load equivalency factor (*LEF*)*. The related concept of equivalent standard axles is discussed in Section 27.3.4.

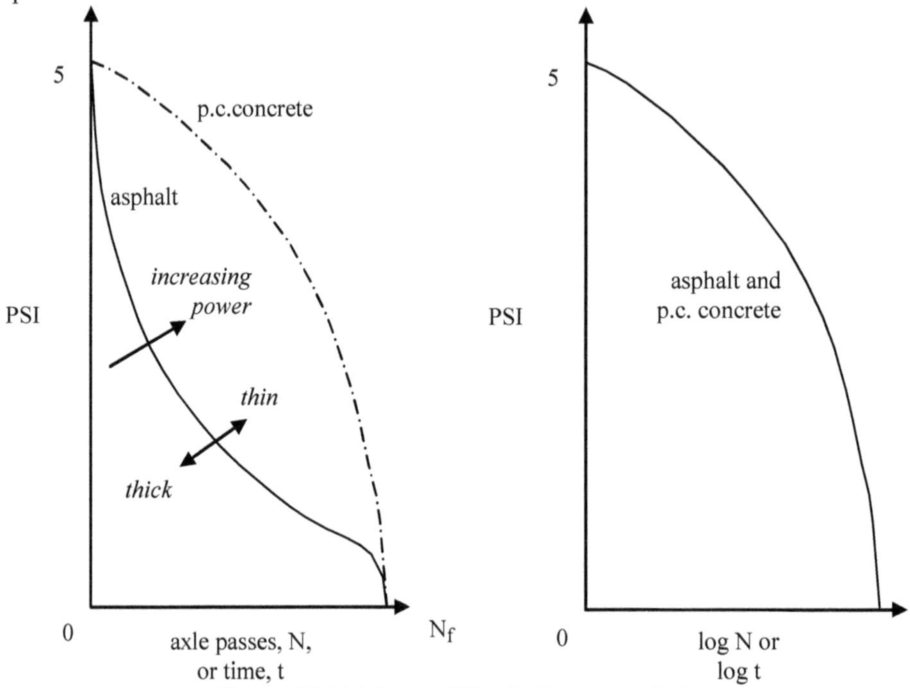

Figure 14.5 Link between PSI and axle passes (or time).

In place of the original 4, values of power law exponents ranging from 2 to 8 have been proposed. Lower powers (e.g. 2) are usually associated with surface cracking, the power of 4 is used for rutting and roughness, and higher powers (e.g. 6) are probably more appropriate for weaker, heavily-loaded, pavements (Paterson, 1985 and Kennedy, 1985). OECD (1991) suggests that the power might vary from 2 for 20 percent cracking to 6 for 60 percent cracking.

The effect of increased wheel loads on a pavement can be examined using Equation 11.12. If the wheel loads on two identical pavements are P and (1 + ß)P, then the load passes to failure of the second pavement will be $1/(1 + ß)^4$ times the number needed to terminate the life of the first pavement. However, it is important to note from Section 11.3 that the cost of new construction to cater for the increased wheel loads will increase by approximately (1 + ß) times greater and not by the more severe $(1 + ß)^4$ times greater that might be implied from the fourth power law. From Equation 14.4, the increased basecourse damage caused by a pass of a load of (1 + ß)P is $(1 + ß)^{0.8}$ times the damage caused by a pass of P.

There is, of course, no difference between the application of the fourth power law and the incremental assessments given in Section 14.3.2 and Equation 14.4, although the latter would perhaps imply that load increases have a lesser role. For the example pavement, Figure 14.4 shows the influence of a 50 percent load increase on the deformation vs number-of-load-passes curve taken over the entire life of the pavement. Figure 14.6 shows the effect of a single overload applied about halfway through the fatigue life at 200 000 passes.

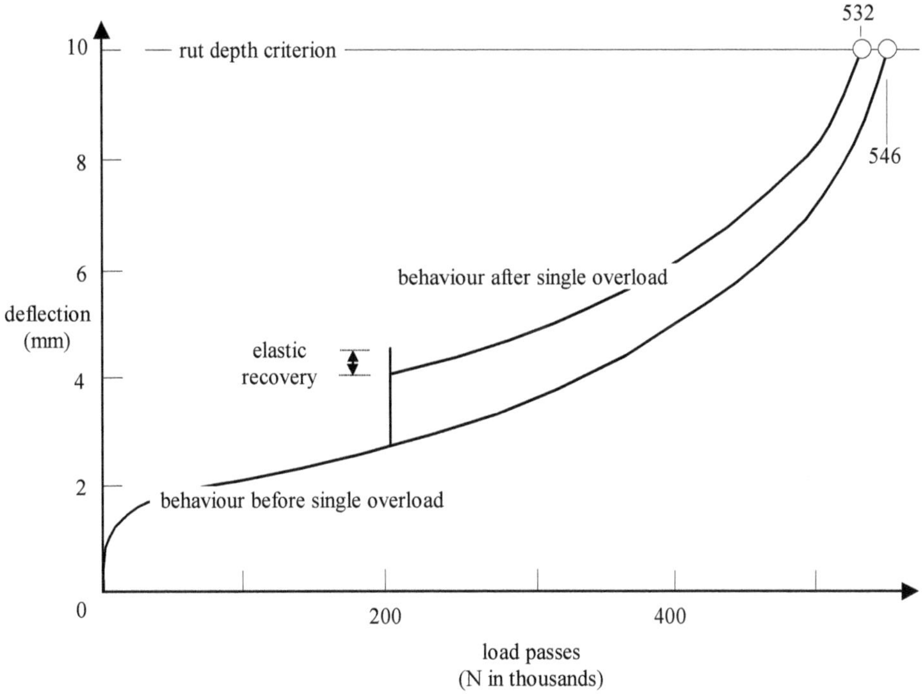

Figure 14.6 The effect of a single 50 percent overload at 200 000 load passes on pavement performance.

Thus, if fleet wheel loads increase, it is not possible to neatly predict the effect of that increase if it occurs part way through the life of the pavement. Instead, the specific case must be examined using the method shown in Figure 14.4. The results obtained will depend on the relative role of subgrade and basecourse and on the failure criterion being applied.

The effect of a single truck with a wheel load of (1 + ß)P amongst a fleet with loads of P is to cause an extra amount of damage, ε_p, which can be estimated from Equations 14.4&5. Assuming a 10 mm critical rut depth, one passage of a legally-loaded vehicle would reduce the pavement life by 100/546 000 = 0.0002 percent. However, one passage of an overloaded vehicle would reduce the pavement life by 2.6 percent (Figure 14.6). This is 14 000 times as big, and much larger than predicted by $(1 + \beta)^4$, as is often proposed.

The massive difference comes from the fact that it is the initial passage of the high load that causes most of the damage. Put another way, Figure 14.6 illustrates how overload can produce rutting far more effectively than can load passes. Pavement managers must therefore give priority to strategies for overload prevention, rather than for overload detection.

14.4 PAVEMENT ROUGHNESS

14.4.1 Development of roughness

Roughness is a measure of aspects of the longitudinal profile of a pavement. European practice is to use the term *evenness*, the inverse of roughness. In some countries the term *smoothness* is used. Roughness measures were originally developed to indicate the quality of the ride, as perceived by a road user, and is now also seen to indicate the structural deterioration of the pavement. In summarising the discussion in Section 14.1 and elsewhere, roughness depends mainly on:

(a) *inherent factors*
- (a1) the initial, as-built longitudinal profile of the road surface. This profile variation may have arisen at construction (Section 25.7.3), or as a consequence of factors ranging from design variations in the vertical alignment of the road (Section 6.2.2) to variations in the pavement macrotexture (Section 12.5.1).
- (a2) the texture of the road surface (Section 12.5.1),

(b) *traffic factors*
- (b1) the presence of heavy traffic wheel loads likely to cause permanent deformation (Section 27.3.4 & 5),
- (b2) random differences (Section 25.7.3) along the road in the permanent deformation response of the pavement structure to normal traffic wheel loads (Section 14.3.2),
- (b3) differences in loading along the road, due mainly to vehicle suspensions interacting with existing roughness (Section 27.3.3&5). This creates dynamic forces within the vehicle so that the peak loads applied to the pavement at some longitudinal locations will exceed the static load and encourage further roughness. Roughness therefore tends to be self-enhancing.

(c) *environmental factors* (for a well-maintained pavement, this will contribute about half the roughness)
- (c1) the environment to which the road is subjected,
- (c2) random differences in the environment along the road (Section 9.3.3),
- (c3) random differences along the road in the pavement's response to the environment.

(d) *age factors*
- (d1) the effluxion of time, as most roads steadily deteriorate — whether or not they are used (Section 14.4.5),
- (d2) progressive cracking of the road surface (Section 14.1B),
- (d3) generally inadequate maintenance (Chapter 26). If maintenance is not adequate, as the road further ages and degrades, the property variations needed to create roughness become smaller and smaller.

(e) *pavement factors*
- (e1) major surface defects in the wheel-path such as ruts, potholes and localised repairs (Section 14.1A8, C9),
- (e2) for concrete roads — the presence of stepped joints (Sections 11.5.1 & 14.1A13),
- (e3) for unpaved roads — the presence of corrugations (Section 14.1A5) and large stones.

The list in (a) to (d) indicates that developing roughness represents a normal pattern of pavement deterioration. However, the occurrence of factors in the final category, (e), is usually a sign of a rapid increase in roughness levels. It is sometimes postulated that roughness is a linear summation of:

* a road's initial roughness (R_0, see a above),
* an increment due to load-related factors (see b, c, and d),
* an increment due to non-load but time-related factors (see e, f, and g above), and
* an increment due to maintenance practice (see h, i, j, k and l above).

Roughness is not directly affected by rut depth (Section 14.1A8) as ruts are the result of traffic causing uniform permanent deformation along a road, whereas roughness arises more from factors that vary longitudinally. However, as with ruts, roughness changes will occur predominantly in wheel-paths.

Thus the longitudinal profile of a road, particularly in a wheel-path, will contain both random and regular variations in elevation, occurring over various wavelengths. Theoretically, such fluctuations are usefully handled by the *power spectral density* approach (Section 33.2), but simpler concepts are used in this text. By definition (Figure 12.4), roughness has a wavelength band which fits between megatexture (< 500 mm) and longitudinal profile (> 50 m). This profile variation induces vertical motion in a travelling vehicle and thus undesirably influences ride comfort (via the PSI, Section 14.3.1) and safety (Section 28.5j). The roughness-induced loads may also damage the pavement and the vehicle and, possibly, its contents (Section 29.2.7). A good historical review of this topic is given in Sayers, 1995.

14.4.2 Profile measurement

The longitudinal profile of relevance to road roughness can be established by a number of methods:

(a) conventional straight-edge or survey-based levelling techniques. These pose a number of practical problems, such as:
   (1) the difficulties associated with operating survey devices on roads,
   (2) allowing for the surface texture, and
   (3) the slowness and costliness of the process.
   Problem (3) has led to the development of profile-measuring devices relying on either the direct measurement of vertical distance or the double integration of vertical acceleration. They are usually either:

(b) *rolling straight-edges* (or *profilographs*) which are usually pushed along by hand and used on new works. Transducers measure the vertical distance from straight-edge to pavement. The straight-edge is either supported directly on the pavement or carried on sets of wheels spring-loaded to keep them in contact with the pavement. Microcomputers in the device give direct outputs of profile and its derivatives. Accelerometers can be used to monitor the absolute vertical location of the device's reference plane, via double integration. This approach permits a small device to produce a full profile.

(c) geometric devices such as *CHLOE*, the *General Motors Research (GMR) profilometer,* first produced in 1966. It uses sensor wheels in contact with the road and accelerometers on the supporting frame to measure vertical movement of the sensors and provide a reference frame. The use of contact wheels means that it is restricted to a maximum speed of 30 km/h. One recent version is the Surface Dynamic Profilometer.

(d) the French *APL* machine, which is a single-wheel trailer using a pendulum to maintain a reference plane whilst travelling at 70 km/h. It produces a signal representing the wave number of the profile and which is computer-processed to provide relevant output. The machine measures wavelengths from 1 to 40 m.

(e) relatively sophisticated direct-measuring devices such as the TRL laser-based *High-Speed Road Monitor* (*HRM*) which consists of a trailer supporting a rigid 4.2 m beam carrying four laser-based vertical distance sensors. It can operate at speeds of up to 80 km/h whilst measuring wavelengths from 0.5 to 100 m and can also record rutting and texture depth. Other similar profilometers are those produced by VTI in Sweden,

ARRB in Australia and K. J. Law in the U.S. Movement of the frame carrying the sensors is calculated from accelerometer readings. The devices effectively sample every 50 mm longitudinally and have a vertical resolution of 0.2 mm. They can capture profile wavelengths from 0.2 m to 33 m.

Experience indicates that profiles should be measured at least every 300 mm (Sayers, 1995).

14.4.3 Profilometry

Once the longitudinal profile has been measured, it must be converted to some meaningful indicators, such as a prediction of its effect on the level of service or quality of ride offered to travellers (Section 14.3.1). This will mean accommodating the characteristics of the vehicle suspension (Section 27.3.5), the driver and any associated passengers, and any freight being hauled. There are a number of ways of doing this.

The *first* method is to use the RMS (root mean square) of the measured profile heights within a specified frequency band. The RMS squaring process is used so that upwards and downward displacements are added, rather than cancelling each other out. This produces the *Profile Index* which has been subjectively calibrated against PSI (Section 14.3.1).

A *second* method is to use the second derivatives of the profile heights. These will be proportional to the vertical accelerations of any object passing along the profile, and in continued intimate contact with it, and hence proportional to the motion forces that it experiences. People respond particularly to accelerations in excess of 0.04g RMS and to vibrations in the 3 to 8 Hz range (Jordan, 1984). An aggregate measure along the profile is obtained by calculating the RMS sum of the vertical accelerations (RMSVA). By calculus, the vertical acceleration, a_v, at horizontal distance x, vertical distance z, and time t, can be calculated as:

$$a_v = d^2z/dt^2 = d/dt(dz/dt)$$

but:

$$dz/dt = (dz/dx)(dx/dt) = Sv$$

where S is the instantaneous longitudinal slope of the pavement, dz/dx, and v is the speed, dx/dt, of the wheels along the road. Thus:

$$a_v = (dS/dt)v + S(dv/dt)$$

if $dv/dt = a_h = 0$:

$$a_v = (dS/dx)(dx/dt)v = (dS/dx)v^2 \qquad (14.9)$$

The RMSVA must therefore be related to the base-length over which it is measured (Gillespie and Sayers, 1983).

A *third*, more realistic measure, is obtained by using a computer to run a hypothetical car over the road profile measurements, calculating and recording the dynamic vertical motion of a car with a defined sprung mass, unsprung mass, suspension stiffness, suspension damping, and tyre stiffness. Because only one wheel of the vehicle is simulated, the method has come to be known as the *quarter car* method. An advantage of the quarter car is that it can be applied to data from all common methods of profile measurement. The vertical motion (commonly the vertical displacement between the sprung and the unsprung masses) per distance travelled, calculated in terms of mm/km or instrument-counts/km, is usually called *QI* (Sayers *et al*., 1986).

Such calculations only consider the wave band seen by the passing vehicle at its given speed. For example, pavement wavelengths greater than 30 m do not normally excite vehicle suspensions into resonant vibration. With current vehicles, people in cars appear sensitive to wavelengths between 1.5 and 20 m and in trucks to wavelengths between 5 and 20 m (see also Section 15.4.7)

A variation is to calculate the *reference average rectified slope (RARS)*, where the use of *rectified* rather than RMS means that the plus/minus problem is handled by using the absolute value of each quantity, rather than by squaring. RARS is usually measured in mm/m and values range from 0 for a smooth surface to 16 mm/m for a road with gullies and deep depressions.

The World Bank's *International Roughness Index (IRI)* uses the RARS of a specified car travelling at 80 km/h, as at this speed it appears that the IRI is sensitive to the wavelengths associated with vehicle vibrations on real roads. Its quarter car is mostly influenced by wavelengths between 1.2 and 30 m. Wavelengths of 2.4 m and 15.4 m have a major excitation effect on the model suspension. If perchance these are present in a road, they can distort IRI readings upwards. The IRI is calculated from a single longitudinal profile. IRI and QI are linearly related.

The profile is commonly measured from a longitudinal profile in a wheel-path. This is called a single track profile and leads to a *single track IRI*. The average of the single track IRIs for a pair of wheel-paths is called a *lane IRI*. For a car width of 2 m (Table 7.4) wheel-paths are typically about 1.5 m apart.

The related *rectified average velocity* method uses the same general definition, but is based on the vertical speed of the vehicle axle and so is more vehicle-dependent.

14.4.4 Response-type meters for roughness measurement

Response-type meters (or *road meters*) are mechanically reliable, low-cost devices which record the vertical movement of the vehicle body relative to the vehicle axle per distance travelled, as the vehicle travels along the road at a constant speed. The output is therefore directly related to the QI, RARS and IRI measures defined in Section 14.4.3. It is not possible to calibrate response-type meters in an absolute sense as their output data are not independent of the vehicle properties and there are usually relatively high gains at vehicle-body resonance (usually about 1 Hz) and axle resonance (usually about 10 Hz).

The devices can be mounted in a towed trailer — as with the *BPR roughometer*, or installed in a car — as with the Mays *ridemeter*, the *PCA meter* and *NAASRA roughometer*. The BPR machine is the oldest of the response-type meters, dating from 1941 and based on an earlier 1925 device (Lay, 1992). It is mounted on a special single-wheeled trailer employing a leaf-spring suspension (Section 27.3.5). It measures the vertical movement of the trailer body in one direction and the usual units are inch/mile.

The quarter car described in Section 14.4.3 effectively models a response-type meter and also is sensitive to excitation frequencies of 1 and 10 Hz. There is thus a close link between QI and response-type meter output. However,

* a major difference between the BPR roughometer and the commonly-used quarter car model is that the QI model vehicle has a suspension stiffness (Section 27.3.5) to sprung mass ratio that is about half that found in the old BPR vehicle, and
* QI does not model the interaction across a vehicle travelling in two wheel-paths of differing roughness. This is a significant problem as the moisture-related effects discussed in Section 14.2a usually mean that the wheel-path nearer to the pavement edge is relatively rough. Roughnesses measured by a device running in

two associated wheel-paths are typically about 80 percent of the roughness measured in the rougher of the two wheel-paths.

The TRRL *Bump Integrator* (BI) is a development of the BPR Roughometer and operates at 32 km/h. In mm/km (or μm/m) units, its scale ranges from 0 (perfect) to 16 000 (very bad) and is converted to IRI (mm/m) by:

$$IRI = 0.0032(BI)^{0.89}$$

Many trailers do not run in a commonly-used wheel-path and so can give deceptively optimistic results. Even if this is overcome, the trailer data will only apply to one of the two wheel-paths. Thus, many other authorities (and the author) favour the car-based meters that operate in both wheel-paths simultaneously. Nevertheless, Sayers *et al.* (1986) favoured the trailer-running-in-a-wheel-path approach, as it gives more specific information.

In a comparison of roughness machines in Brazil, the NAASRA machine gave the most consistent results of all the devices tested (Sayers *et al.*, 1986). For the NAASRA meter, 'movement' is recorded in 'counts' which are 15.2 mm relative movements in one vertical direction. The unit is thus count/km and the result is called NRM (NAASRA roughness number). The regression relation between IRI and NRM is:

$$IRI \text{ (average of two wheel-paths)} = 0.0479 + 0.0378(NRM)$$

A key point to determine in making a comparison between standard data and the numerical output from a particular device is whether the device measures in one or two directions.

Assuming a linear response of the devices, the following comparisons with an IRI of 5 mm/m provide a simple cross-check:

IRI = 5 mm/m
QI = 68 count/km
BI = 4000 μm/m
NRM = 130 count/km
PSI = 2.4.

This issue is pursued further in the next section.

The coefficient of variation of the devices is about 5 percent. Roughness readings are also affected by vertical steps in the pavement with each 10 mm step locally increasing the IRI by about 0.33 mm/m.

14.4.5 Roughness relationships

The calibration of profile (Section 14.4.3) and response-type (Section 14.4.4) measures to drivers' perceptions (PSI) requires either:

(a) the use of a PSI rating panel to directly link profile shape or response readings to PSI (Section 14.3.1),
(b) operation of the response meter over a profile of known PSI, vertical profile, and/or 'roughness' value, or
(c) comparison with results from a calibrated vehicle.

Such techniques are extensively outlined in Sayers *et al.* (1986).

Most profile measures and response-meter counts by their physical nature will relate linearly to the profile slope, and Equation 14.8 shows that they are therefore closely related to the vertical accelerations experienced by the vehicle. Thus it is not surprising that surveys frequently show a simple relationship between driver perceptions (PSI) and response-type measures of roughness or RARS. Studies by the World Bank and others suggest the following links between PSI and IRI levels.

Table 14.2 Links between road condition, present serviceability index (PSI), and two roughness measures.

Road condition – the description given is for asphalt, unless in ( )	PSI	IRI	NRM
extremely good condition	5	0	0
very good condition		0.0–1.6	
new freeway		1.4	
(new spray and chip seal)	4.5	1.6	40
good condition		1.6–2.5	
new main road (speed > 80 km/h)		1.7	
new main road (speed < 80 km/h)		2.0	
(best unsurfaced road)	4.2	2.5	65
some signs of deterioration, surface cracking		2.5–4.0	
unacceptable for some trips	3.6	4	110
signs of deterioration		4–5	
roughness rate accelerating	3.4	5	130
rapid deterioration	3	6	160
visible irregularities: unacceptable for a major road in a developed country	3.4–2.5	5–8	130–200
fair condition	2.5	8	200
unavoidable potholes, poor condition		8–10	
bad condition	1.9	10	250
just tolerable	0.6	13	350
equivalent to an unpaved surface		10–16	
extremely bad condition	0	16	400
equivalent to a poor unpaved surface		16–20	
impassable		20	500

AASHO Road Test data (Section 11.4.3) showed that roughness measurements explained 85 to 95 percent of a pavement's PSI. A linear link is sometimes noted, i.e.

$$\mathrm{PSI} = a(\mathrm{R} - \mathrm{R}_0)$$

where R is roughness, R_0 is R for a new or high quality road, and a is a constant. Calibrating this function for a PSI of 5 at R_0 and a PSI of 0 for an unacceptable roughness of R_u gives:

$$\mathrm{PSI} = 5[\mathrm{R}_u - \mathrm{R}]/[\mathrm{R}_u - \mathrm{R}_0] \tag{14.10}$$

$$= 5(1 - \mathrm{r}) \tag{14.11}$$

where:

$$\mathrm{r} = [\mathrm{R} - \mathrm{R}_0]/[\mathrm{R}_u - \mathrm{R}_0]$$

is a normalised measure of roughness (r = 0 is perfect, r = 1 is unacceptable). The data in Table 14.2 leads to:

$$\mathrm{r} = \mathrm{IRI}/16 \text{ or } \mathrm{NRM}/400$$

A major reason for introducing r is that R_0 and R_u will vary with both measuring device and community expectations (see Table 14.2).

Despite Equation 14.10, the link between PSI and roughness is extremely variable and is probably far from linear (Sayers *et al.*, 1986). To illustrate the variability, many drivers would consider a road to be 'unacceptable' if it had a roughness of IRI = 4 or more and they were travelling on it for many hours, but would consider it acceptable for trips of an hour or less.

The alternative assumption that log PSI is linearly related to roughness gives (calibrating at a PSI of 1 at a lesser $R_{u'}$ to avoid problems with PSI < 1:

$$\log PSI/\log 5 = 1 - r'$$

or

$$PSI = 5^{1-r'}$$

A common field expression is:

$$PSI = 5e^{cIRI}$$

where c = –0.23 for asphalt and –0.29 for concrete.

Another assumption is that PSI is linearly related to log roughness. This gives:

$$PSI = 5\{1 - [\log(R/R_0)/\log(R_u/R_0)]\}$$

and so for the above conditions:

$$PSI = (75/16)(1 - [\log R/\log 16])$$

Measurement of the roughness of a road over a period of time gives graphs of the form shown in Figure 14.7. Broadly, an annual rate of roughness deterioration over time of:

* about 0.5 NRM or 0.05 IRI represents background conditions,
* less than 2 NRM or 0.15 IRI represents good pavement performance, and
* over 4 NRM or 0.20 IRI represents poor pavement performance.

Explicit algebraic forms of the graph in Figure 14.7 are needed for various purposes, such as strategic pavement management models of the type discussed in Sections 14.6 & 26.5,

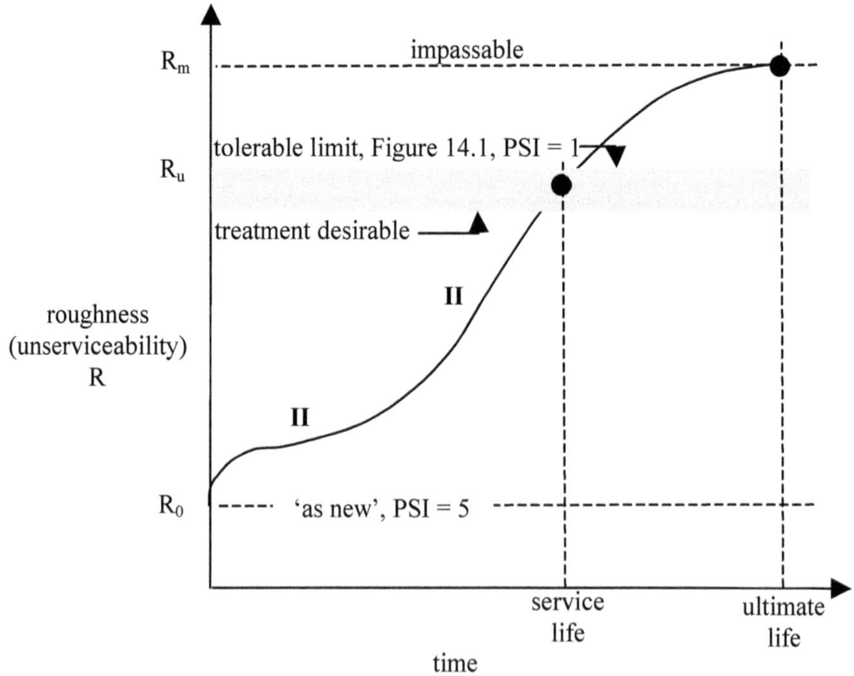

Figure 14.7 Change in roughness with time. Note that point A is not necessarily an inflection point but could be anywhere on the roughness curve. Stages I, II and III will be explained in Section 14.5.3.

and these will now be developed. The link between PSI and traffic use can be predicted simply by combining Equations 14.7&11 to give:

$$IRI = 16(N_{esa}/N_f)^{0.2}$$

which fits stages I and II in Figure 14.7, and stage III as well if the unmodelled downturn near $N_{esa} = N_f$ in Figure 14.5 is considered. Others have suggested using variables between $N_{esa}^{0.7}$ and $N_{esa}^{0.02}$.

The form used in the TRRL RTIM model has roughness increasing linearly with N_{esa}, the rate depending on structural number (Section 11.4.6 & Parsley and Robinson, 1982). The Brazil Study data (Geipot, 1982b) has roughness increasing linearly with $\log N_{esa}$, with age, and with the square of deflection. Alternative forms of the Brazil model are available, depending on the variables regressed, and include a linear link between roughness and N_{esa}.

As N_{esa} can be estimated on a time basis, it is common to plot roughness as a function of time, y, in years. Roughness also deteriorates solely with time, due to ageing of the pavement materials. Thus, a simple linear model can be used (e.g. Martin, 1994):

$$r = y/y_u \tag{14.12}$$

where y_u is the typical life at which R_u occurs. Some have suggested other powers ($r \propto e^y$, $y^{1.5}$, or y^2), with the roughness rate increasing with time, as in the early parts of Figure 14.7. Others use the prediction that roughness might be exponentially given by:

$$r = 1 - e^{-3y/y_u} \tag{14.13}$$

The equation is asymptotic to $r = 1$ ($R = R_u$) and bears some resemblance to the latter part of Figure 14.7. It can be substituted in Equation 14.10 to give a PSI vs time function of:

$$PSI = 5e^{-3y/y_u}$$

Of course, roughness will depend both on N_{esa} and on y, as it is partly due to traffic and partly to age, and when expansive clays are present (Section 8.4.2), roughness may also vary with the time of year. Thus more general expressions are of the form:

$$r = c_1(N_{esa})^{c_2} + c_3(1 - e^{c_4 y})$$

or

$$r = c_5(N_{esa})^{c_6}(1 - e^{c_7 y})$$

or the ARRB model (Martin, 1996) which is of the form:

$$r = c_8(N_{esa})^{c_9} y^{c_{10}}$$

where c_i are constants. It has been suggested that $c_6 = 1$ and c_7 ranges from 2 to 5.

Following the definitions in Section 14.3.2, it is important to determine whether an increase in roughness is associated with structural deterioration. If it is not, it may be improved by the application of a thin overlay (Section 12.1.6). If it is, it can only be remedied by major reconditioning — using at least a thick overlay (Section 12.2.6). Deflection tests (Section 14.5.2) provide a means of making this distinction. Evidence from Brazil is that most initial increases in roughness are not due to structural deterioration.

14.4.6 Applications

Roughness measurement is widely used in pavement management as it provides:

(a) a field indicator of the quality of new construction,

(b) a regional indicator of changes in the condition of an existing road (Section 14.5.1), and

(c) a policy indicator of the overall condition of a road network (Sections 5.1 and 26.2).

In each case it will only provide part of the necessary input to decision-making. A typical specification is that 95 percent of a new road should have an IRI less than 1.6 (or 2.0) and that 95 percent of an operating highway should have an IRI less than 4.0.

Although roughness was initially introduced as a criterion for riding quality, it provides data having wider application, as an increase in roughness will also mean that:

(1) water will accumulate in local areas of the road causing spray and lowering skid resistance, and thus decreasing road safety,
(2) vehicle speeds will reduce slightly,
(3) vehicle control will become more difficult,
(4) vehicle operating costs will rise (Section 29.1.2),
(5) damage to fragile freight will increase, and
(6) dynamic truck–road forces will vary from static levels, with the lower forces reducing skid resistance (Section 27.3.5) and the higher forces causing increased pavement damage (Section 14.4.1). Indeed, roughness is by far the major cause of dynamic pavement loading and is thus a self-aggravating effect. Pavement loading via roughness is discussed further in Section 27.3.5

14.5 PAVEMENT EVALUATION

14.5.1 Procedures for rating pavement condition

Given the roadmaking need to use local materials in natural environments, the role of experience and judgement in pavement design, construction, and maintenance is so great that good engineering practice demands that the performance of all roads be continually and consistently evaluated. The basic aims of this evaluation are to:

(a) provide guidance for planning future maintenance and rehabilitation (Section 26.2);
(b) check if the intended pavement functional and performance objectives are being achieved;
(c) provide feedback for improvements to existing design, construction, and maintenance procedures;
(d) establish a reservoir of road performance data for use by future designers and economic analysts; and
(e) detect condition changes from one year to the next.

Most evaluation (or rating) systems are forms of multi-criteria analysis (Section 5.2.3). Their four main components are typically:

(1) intrinsic data for that piece of road (Table 14.3) collected from a database for the network (Section 26.5.1); e.g. pavement type,
(2) current road attributes which will give guidance on the condition of the road (Table 14.4: leftside); e.g. roughness,
(3) the rating of the present condition of each attribute in terms of both its severity and extent: e.g. 5 percent of the road length is repaired to a fair quality (Table 14.4: rightside),
(4) the empirical combination of the rating of each individual attribute item into a single total rating, or *Pavement Condition Index*, which is intended to represent how well the pavement meets its service objectives (e.g. ride quality, load capacity, safety, appearance). Such indices can be convenient but misleading indicators and their empirical basis is not examined further in this text. Constructing such indices is essentially a matter of combining apples and oranges and the weighting adopted to add the different values into a single

number will therefore reflect local priorities and have little or no absolute technical justification.

Table 14.3 Intrinsic road data. The numbers in brackets link the items to the item numbers in Tables 14.4, 25.1 & 26.1.

General:
* Topography (Section 6.1), climate (Section 9.2), road alignment (Section 19.2.4)
Road-related:
* Road identifier, Road name, road type (Section 2.1) * Road section, chainage from, chainage to, section length
Pavement-related:
* Surface drainage (310), subsurface drainage (340), watercourses (620): * Formation type (Ch 11), shoulder type (140), shoulder width (140) * Surface type (Ch 12), pavement type (Ch 11), pavement age * Carriageway, lane number, wheel path (150)
Traffic-related:
* Traffic flow in AADT (Section 6.2.3), traffic composition (Section 6.2.3), speed limit, load limit.

Typically, most of the items in Table 14.4 are collected by a trained inspector accompanied by a local foreman (Porter *et al.*, 1980). About sixteen kilometre lengths could be rated in one day. For urban roads, inspection rates might be only one quarter of this amount. To ensure a uniform rating, the pair would annually inspect the same selected uniform length of about one kilometre of road. British practice is to sample annually randomly selected 100 m (rather than 1000 m) road lengths.

Routine maintenance matters such as the need for minor repairs are not covered in these procedures as it is assumed that they would be undertaken without recourse to a formal condition survey (Section 26.1). At the other extreme, the system does not demand the use of deflection data (Section 14.5.2), on the assumption that it would not be widely available. However, if deflection results are available they should be used as a prime input as deflection is a valuable indicator of pavement structural condition. Other factors listed in Section 14.1 should be included, if warranted by local conditions.

For part (4) above, the system in Table 14.4 produces pavement, surface and shoulder condition indices (they are sometimes called *distress indices*). Each index begins with a value of 100 and points are then deducted if the pavement is less than perfect. The methods are therefore sometimes called deduct value methods. The maximum numbers of deduct points for the items are also shown. These are usually of a matrix form and the independent variables in the matrix are shown in Table 14.4. Specific points are found in locally produced sub-tables. For example, some skid resistance and surface texture matrices are given as Tables 14.5–6. Table 14.7 shows the recommended action for a particular Pavement Condition Index. Overall, 100 represents the best possible conditions and 0 represents an unacceptable condition.

Table 14.4 Pavement evaluation for items routinely collected. Based on Tables 5.3 and 26.1. Continued on page 339.

Pavement item categorisation	Location of item in Book	Table 26.1 ID
Pavement	Chapter 11	110,200
deflection survey	Sections 12.2.5 & 14.5.2	110
settlement	Sections 14.1A1,B2,3&5	110
roughness survey	Section 14.4.5	110
pavement failures	Sections 14.1A&B	110
rutting or pumping	Sections 11.5.1 & 14.1A8&12	110
edge failures	Sections 14.1A10,B3,B10	130
Surface course	Chapter 12	120
repairs	Section 14.4	122
alligator cracking	Section 14.1B1	122.1
longitudinal cracking	Sections 14.1B2&4	122.1
transverse cracking	Section 14.1B6	122.1
glazed or bleeding surface	Sections 14.1C6&8	122.2
scoring, spalling	Sections 14.1C1&3A	122.3
loss of aggregate, ageing, and weathering	Sections 14.1A0,C2–5 & Chapter 12	123
skid resistance survey	Sections 12.5, 14C6–7	123
surface texture	Section 12.5	123
Shoulders	Section 14.2	110,140
shoulder width	Section 11.4	140
shoulder type	Section 11.4	140
shoulder condition	Section 11.4	140
Footpaths	Sections 19.5 & 30.6	150
Drainage	Chapter 13	300
surface drain condition	Section 13.3	310
kerb and gutter condition	Sections 7.4 & 13.3	315
culvert condition	Section 15.4	330
subsurface drain condition	Section 13.4	340
Roadside	Chapters 6 and 19	400
trees, grass, sight distance	Sections 6.4 & 19.2	413
service areas	Section 6.4	420
fixed roadside hazards	Section 28.6	432
batter stability and scour	Section 11.7	440
fences	Chapter 6	450
Traffic devices	Chapters 21, 22, 24 and 28	500
signs	Chapter 21	510
guard fences	Section 28.6	520
guide posts and delineators	Section 24.4	520
pavement markings	Chapter 22	530

Table 14.4 (rightside)

Independent variables used to determine points	Scoring
Pavement condition index	= 100 – A – B – C – D – E – G – H – J
Measured deflection levels	Action specified directly
Percent of pavement area settled and its severity	Action specified directly
Measured roughness levels	A = 0 to 60
Failures per km, wheel-path location and severity. Indications of excess moisture in pavement or layer delamination	B = 0 to 25
Number of rutted wheel-paths, percent of length, depth, severity; signs of loss of support under concrete slabs	C = 0 to 30
Percent of length with failures, extent from edge, severity	D = 0 to 25
Surface condition index	= 100 – E – F – G – H – J – K – L – M – N
Number per km, percent of area, condition, roughness	E = 0 to 10
Percent of area, severity	F = 0 to 25
Percent of area, severity	G = 0 to 20
Percent of length, crack width, severity, wheel-path or general	H = 0 to 20
Percent of area, severity	J = 0 to 30
Number per km	K = 0 to 30
Percent of area, severity	L = 0 to 30
Skid resistance, importance of location, Table 14.5	M = 0 to 50
Texture depth, percent of area, importance of location, Table 14.6	N = 0 to 20
Shoulder condition index	= 100 – 2D – P – Q
Width	P = 0 to 10
Visual score, shoulder type	Q = 0 to 40
Usability	Action specified directly
Footpaths	
Drainage	
Section 13.3 Clear channel, pits operational Unobstructed, inlets and outlets Flow at outlet	Action specified directly
Roadside	
Appearance, fire hazard, sight distance Tidiness, appearance Clearance, protection, safety audit results Falls, slips, scouring Length in effective service	Action specified directly
Traffic devices	
Maintained, in-place In-service reflectivity In-service reflectivity In-service conspicuity	Action specified directly

Table 14.5 Deduct points, M, for skid resistance (Table 14.4). After Porter *et al.*, 1980.

	Skid resistance value			
Importance of site	60	50	40	30
Easy – straight flat road with no intersections	0	10	20	30
Average – high design speed and high AADT	5	20	30	40
Difficult – sharp, steep curves; frequent stops	10	30	40	50

Table 14.6 Deduct points, N, for texture depth (Table 14.4). After Porter *et al.*, 1980.

	Texture Depth (mm)			
Speed limit (km/h)	>= 1.0	1.0–0.6	0.6–0.3	<= 0.3
≤ 60	0	0	0	5
60 – 100	0	5	10	15
≥ 100	0	10	15	20

Table 14.7 Index scale and associated actions for Pavement Condition Indices. Note that the expected variance of computed condition indices is 13 points. After Porter *et al.*, 1980.

Score range	Pavement management consideration	Treatment method (Section 26.1)
0<	Signposting to tell users that the road fails to provide the expected level of safety and comfort. If the road is needed, work should have funding priority.	Reconstruction, with minor salvage.
0 to 20	Signposting (as above) to be considered, depending on alignment and weather. Work a priority as funds permit.	Reconstruction. Weekly inspections until work begins.
20 to 40	Work placed on current program, to avoid loss of value of existing asset.	Rehabilitation.
40 to 60	Work placed on forward program, using the deterioration rate to assign priority and preserve the existing asset.	Rehabilitation or major maintenance.
60 to 80	Work determined by deterioration rate.	Preventative maintenance.
80 to 100	Routine monitoring of condition.	None.

14.5.2 Deflection testing

Deflections are an important indicator of pavement condition. They have the potential to deliver much more information about the pavement structure than the other measures listed in Table 14.4, but are usually about an order of magnitude more expensive to obtain.

The calculation of elastic and permanent deflection was illustrated in Section 14.3.2. Elastic deflections predominantly reflect pavement geometry and material stiffness properties. Permanent deflections relate directly to deterioration. Ideally the complete elastic and non-linear load-deflection response should be obtained by measurement or back-calculation (Section 11.3.4). This is usually impractical: indeed, even the increment in permanent deflection per pass of a legal wheel-load (an ESA, see Section 27.3.4) is too small to measure easily. For instance, for the example in Figure 11.5 and Section 14.3.2, a permanent deflection of (5.52 – 0.74) = 4.8 mm occurs over 430 000 passes, which is only 11 nm/pass. Consequently, the much larger *elastic rebound deflection* (Section 11.3.1) is commonly used, as it is much more readily measured and as:

* the cyclic stress–strain curves (Figure 14.2b) and Equation 14.4 show that it is monotonically related to the permanent deflection and thus changes in it will also reflect changes in pavement structural condition,
* Section 11.3.4 shows that it is a practical non-destructive measure of pavement stiffness, and
* Equations 11.1&2 show that it is proportional to the elastic vertical subgrade stress and the horizontal tensile strain.

Thus the elastic rebound deflection should give a useful measure of pavement deformation performance under a single wheel-load, but will be less effective in warning of imminent fatigue and overload failures and will not give an absolute measure of structural condition. Elastic rebound deflections have now gained almost universal acceptance for use in structural evaluation. The smaller the deflection, the sounder the pavement. The major variable influencing such deflection levels is moisture content.

Field methods for determining pavement deflection can be represented by Table 14.8.

Table 14.8 Deflection testing methods

Loading method	Static	Slow (quasi-static)	Highway speed
Loaded plate	Plate bearing (Section 8.4.5)	–	Falling weight deflectometer
Loaded wheel	–	Benkelman Beam, Deflectograph	Transducers in the pavement (expensive and usually impractical)

The *Benkelman beam* (or *Deflection beam*) was developed by A. C. Benkelman for the U.S. *WASHO Road Test* in the early 1950s. The simple hand-operated device consists of a base beam sitting on two supports 1.53 m apart (Figure 14.8). It was hoped that, at 2.69 m, the supports would be well away from the influence of wheel-load deflection, but this is not the case for weak subgrades. The measuring beam is in contact with the road surface at the point under test. At the other end, a dial gauge measures movement relative to the base beam with a 2 to 1 reduction. Readings are normally taken to 10 μm accuracy and have a coefficient of variation of about 20 percent. The beam is sufficiently narrow to pass between dual wheels of a truck suspension and the truck thus travels over the beam at about 0.5 km/h. The axle load used is commonly the legal limit (Section 27.3.2). The change in deflection recorded when the loaded wheel is moved from the test region to a 'no influence' zone located beyond the beam supports is the elastic rebound deflection.

As the truck travels along the road, the graph of the deflections measured at the beam contact point and plotted at the wheel location, is known as the *deflection bowl* for the contact point (Figure 14.9). It is thus a form of *influence line* and is *not* the deflected shape of the pavement for a load at the contact point. A deflection is represented by d_n where n is the distance in mm between the location of the load and the beam point at which the deflection is measured. For instance, d_0 is the pavement deflection at the load point and d_{250} is the deflection when the load is 250 mm away. For ease of measurement, the deflection is usually measured under dual-tyre loading with d_0 taken between tyres and d_{250} on the line of travel.

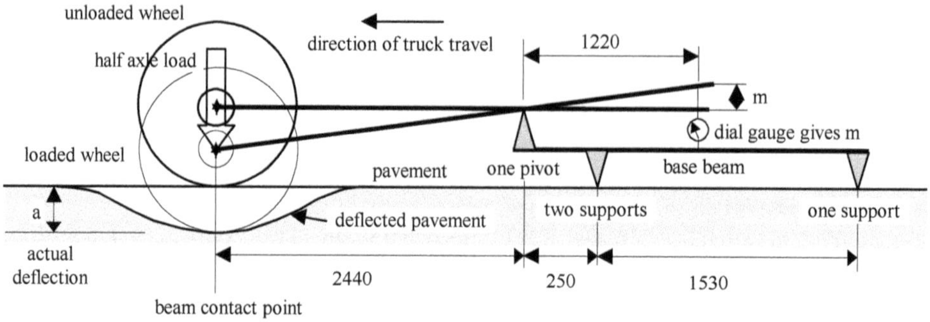

Figure 14.8 Benkelman beam operation. Note that the measured movement, m, is half the actual deflection, a. In the case shown, the beam contact point is at the point of maximum deflection.

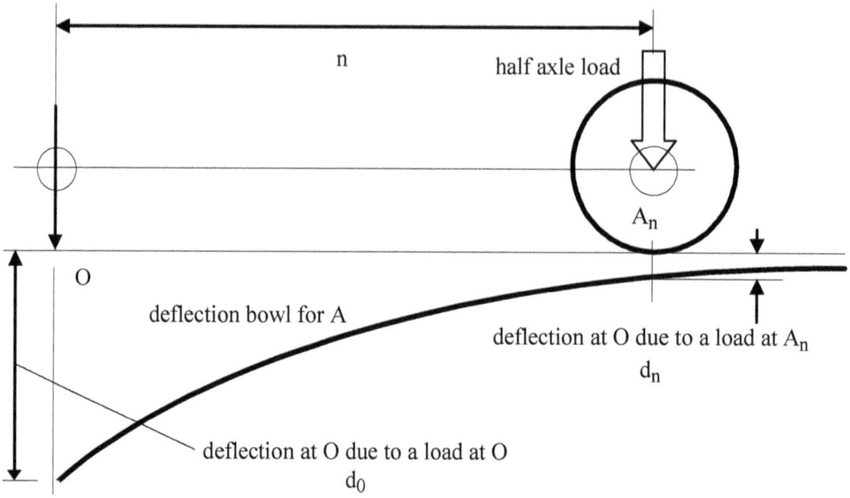

Figure 14.9 Deflection bowl (or influence line) for Figure 14.8. Normally, d_n is quoted, where n is in mm.

Section 11.3.4 describes how the bowl is used to determine pavement stiffness. A key parameter is the *curvature function* of the bowl, which is usually measured by ($d_0 - d_n$) and typically n = 200 or 250 (mm). For a given subgrade or lower layer, Equation 11.10 indicates that E will depend on this curvature function to the power of −1.75. Thus, the flatter the deflection bowl and the lower the curvature function, i.e. $d_n/d_0 \to 1$, the stiffer the basecourse and subgrade. At the other extreme, if d_{1000} is large (> 200 μm), the subgrade is probably unusually weak. If the region near the contact point is steep the curvature function will be high. This usually indicates that the basecourse lacks stiffness and strains will be high, probably leading to cracking in any bound material. A related term is the *deflection ratio*, d_n/d_0. Thus,

$$\text{curvature function} = d_0 - d_n = d_0(1 - [\text{deflection ratio}])$$

Deflection ratios are typically above 0.8 for bound materials, between 0.7 and 0.6 for good unbound pavements, and below 0.6 for weak pavements.

As the demand for deflection readings increased, a variety of travelling, partially-automated *deflectometers* were developed, based on extending the Benkelman beam concept. For example, the *Lacroix deflectograph* is a truck with a rear axle that produces 1 ESA, and a deflection-measuring undercarriage. It is capable of operating at 3 km/h whilst measuring deflections about every 4 m. It is not suitable for unbound materials. An equivalent U.S. device is the Californian travelling deflectometer.

Measured deflections have a coefficient of variation of at least 4 percent. Deflection readings are typically taken *along the road* at intervals of 30 m or less. This requires a *characteristic deflection* (Section 15.3.2) to be calculated to represent the road length and it is common to use the mean of the readings plus one standard deviation (i.e. the 85th percentile) for this purpose. The coefficient of variation of deflection readings along a road can be interpreted as shown in Table 14.9.

Table 14.9 Coefficient of variation of deflection readings

Coefficient of variation of deflection readings along a road (percent)	Interpretation
< 15	very uniform construction
20–30	fair construction
40–50	non-uniform construction
> 50	consider remedial action

Deflections will depend on the *applied load* (Equation 11.3) and the *loading rate* (Section 11.3.2). The load level is usually taken as the legal wheel load. The loading rate will alter both stresses and material response and therefore wheel speeds should be kept constant to provide comparable results. Low speeds will give higher readings on smooth surfaces.

In addition, deflections will obviously reflect any variations in *moisture content* (e.g. between inner and outer wheel-paths, Section 9.2.3), or from one season to another. For example, the variation between beam readings in rural areas in wet and dry periods can be about 30 percent in the outer wheel-path and 5 percent in the inner wheel-path. For similar reasons, outer wheel-path readings will average about 20 percent above inner wheel-path ones, if the shoulder is unpaved (Section 11.7).

As deflections depend on material properties they will also depend on the *temperature* of any bituminous layer. In this context, it is necessary to know both the surface temperature and the vertical temperature profile within the pavement. The temperature input will mainly come from solar radiation, possibly mitigated by the effects of wind (Section 12.4). Deflection readings for asphalt will be very temperature-dependent, reflecting the variation of E with temperature (Section 11.3.2). One practice is to divide measured readings by (0.29 + [T/38]) where T is the air temperature in degrees C, to bring them to a common value at 27 C. This is much less than the effect predicted from Figure 11.8 and reflects the fact that the lower asphalt layers will not have directly followed the surface temperatures.

Deflection tests can also be used to give guidance on subgrade CBR during construction. Using Equation 11.3 in Section 11.4.1, adjusted for a value between two dual tyres, typical tyre contact-pressures from Section 27.7.5, a tyre contact-radius of about 100 mm from the same section, and the relations between E and CBR in Section 11.3.3, leads to equations of the type:

subgrade CBR = 6/(subgrade deflection in mm)

An alternative method of pavement evaluation is provided by devices that determine the stiffness and deflection of a pavement under a known dynamic load. One such machine is the *Falling Weight Deflectometer* (*FWD*) which is mounted on a trailer that can be towed behind a small vehicle (Ullidtz, 1987). The test is specified in ASTM D4694. To conduct a FWD test, the trailer is halted and seven seismic detectors (or speed detectors) are placed on the pavement surface. A 30 ms, roughly triangular, load-pulse is applied by a raised mass falling vertically onto a set of springs that are mounted on a relatively rigid 300 mm diameter circular plate resting on a 5 mm thick ribbed-rubber base placed on the pavement surface. The resulting load vs time pulse is intended to reproduce the load from a truck travelling at about 60 km/h, although it is a little narrower than the average tyre. Peak loads of between 5 and 100 kN are produced. The force is proportional to the square root of the drop height. Section 27.3.2 shows that the available force range will cover most truck half-axle loads, plus impact, with 40–50 kN being a typical truck wheel load.

The springs ensure that there is only one impact. The seismic detectors measure the maximum deflection at each detector, which occurs after about 20 ms. This deflection is used to give the shape of the resulting deflection bowl. The FWD gives more reliable deflection estimates than the Benkelman beam and the Deflectographs as it integrates effects over a larger area and as it reduces the problem of having its supports within the affected area.

Using elastic pavement theory, the deflection output of the device may be converted to predict such factors as the stiffness, thickness and strain regime of each pavement layer. This conversion process is discussed in Section 11.3.4.

The *Loadman* is a portable version of the FWD, with many similarities to the Clegg Impact tester (Section 25.3). It has been found to be a good predictor of basecourse stiffness within about 150 mm of the surface (Pidwerbesky, 1997). A similar device is the Light Drop Weight Tester developed in Germany, although it is more an alternative to the plate bearing tests described in Section 8.4.5 than to the FWD described above. It produces peak loads of only 7 kN. It is easy to use and gives stiffness values, but these require considerable calibration.

Alternatives to the FWD are the *Dynaflect* that uses counter-rotating masses producing a maximum force of 4.5 kN via two 72 mm diameter wheels; and the *Road-Rater*, which uses electro-hydraulic systems and produces a maximum force of 36 kN. Both apply a static preload followed by the alternating load and record deflections in a similar manner to the FWD.

14.5.3 Pavement assessment

As was the case with roughness, deflection is a surrogate for pavement condition. However, deflections are more sensitive to local conditions than is roughness. For pavement assessment purposes it may be desirable to measure deflections under the worst conditions; viz. high moisture levels, outer wheel-path, high temperatures, and critical loading rate. The critical loading rate will be at zero speed for a smooth pavement — to allow for creep to occur — but may be at some higher speed, as roughness levels increase and raise impact effects (Section 14.4.2).

As suggested in Section 14.5.1, deflection readings can be used for pavement assessment and condition rating and for construction control. Section 8.4.5 showed how the deflection bowl can be used to estimate subgrade CBR and Section 11.4.2 described its use for estimating pavement stiffness. Section 12.2.5 discussed how deflection readings are used for overlay design.

For pavement assessment purposes, if a pavement is tested after N_{esa}, then the measured deflection, d, will typically relate to N in the manner shown by the solid line in Figure 14.10, which is modelled on Figure 14.4. The line is composed of three stages linked directly to the three stages in the life of a pavement, as described in the opening to Section 14.1. Nevertheless, they may not all occur in any one pavement. In Stage I, the pavement with an initial deflection of d_o continues to deform steadily under traffic. This stage will be accentuated if initial *compaction* levels (Section 8.2) were not adequate. In Stage II, the deflection rate steadily decreases as the pavement reaches a stable, optimally-compacted state and begins to strain-harden. In Stage III, the pavement structure has begun to deteriorate under load and environment (Section 14.3.2), strain-softening has commenced and is reflected by an increasing deflection rate. A common initiation of Stage III is when surface cracking permits water to enter the basecourse and subgrade.

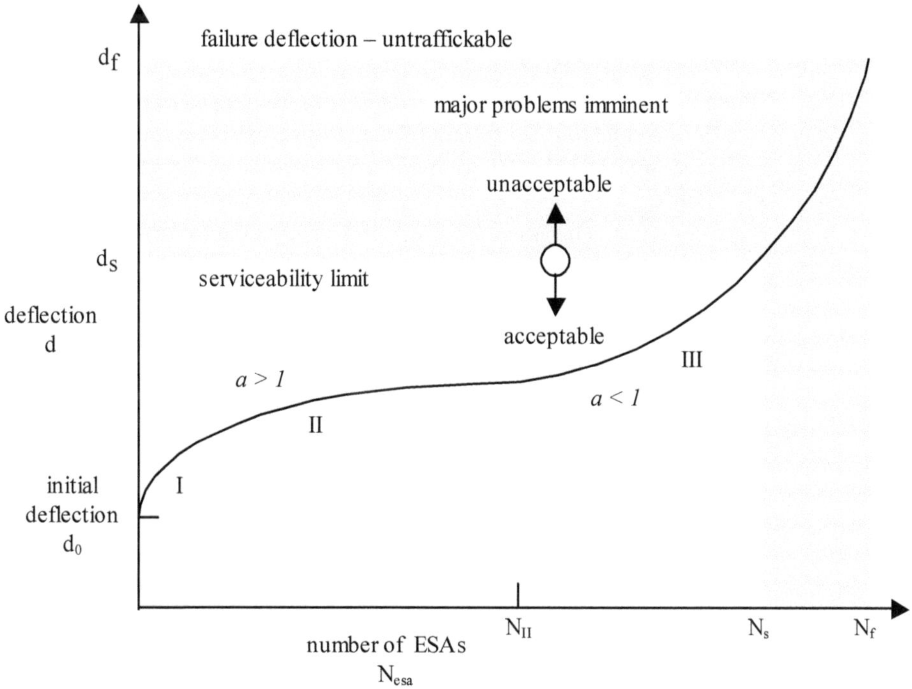

Figure 14.10 Pavement deterioration in deflection terms for an individual pavement. The figure is based on Figure 14.4. Compare it with the roughness curves in Figures 14.5 & 7. *a* is the power in Equation 14.14.

A number of researchers, with an eye on Equations 11.11–15, have linked N and d by a power relationship of the form:

$$N = C(d - d_o)^a \qquad (14.14)$$

where $2 < a < 6$ in Stage II and $-6 < a < -4$ in the critical Stage III. The value of −4 flows directly from the *fourth power* Equations 11.12 & 14.2. From Section 14.3.2 and Equation 14.4 it could be hypothesised that for Stage II:

$$a = c/(1 - m)$$

Thus the apparent major change in value in a (e.g. from 6 to −6) actually represents a smooth change in m moving on either side of unity. The assumption that the deflection, d, is proportional to the accumulated permanent strain ε_p, gives a = 5 for the example analysed in Section 14.3.2. c is a constant that could vary between stages.

In terms of Section 14.3.2, Stage III is caused by the advent of significant strain-softening. From Figure 14.1, further limits occur at the end of its service life when the pavement condition is no longer acceptable to the users (N_s), and when the road is untrafficable and then impassable and will have failed by all definitions (N_f). Another critical point is when the pavement reaches a stage at which further deterioration would require pavement reconstruction rather than strengthening. This is not easy to define, but is probably between N_{III} and N_s.

The assessment of N_s is critical. For a number of similarly-constructed pavements it has been found, for a given d_s, that N_s is governed largely by the value of d_o. The relationship depends on the definition of *serviceable* (Section 14.3.1), and a typical pair of functions is shown in Figure 14.11 for asphalt rutting (curve 1) and for cracking of stabilised pavements (curve 2). Figure 14.11 can also be used to:

* assess the remaining life of a pavement of known d_s and d_o, and
* provide a design criterion by giving the d_o required if a pavement is to achieve a given N_{esa} (Section 11.4.3).

The form of Figure 14.11 is not surprising, for it can be reproduced by taking the empirically-based design equations for rutting (Equations 11.13–15) and assuming that $d_o \alpha E \alpha \varepsilon_v$, as introduced in Section 14.3.2, giving:

$$\log d_o = A - B\log N_{esa}$$

where A and B are constants. The essential $d_o \alpha E$ link reflects the fact that both d_o and E characterise the initial condition of the pavement.

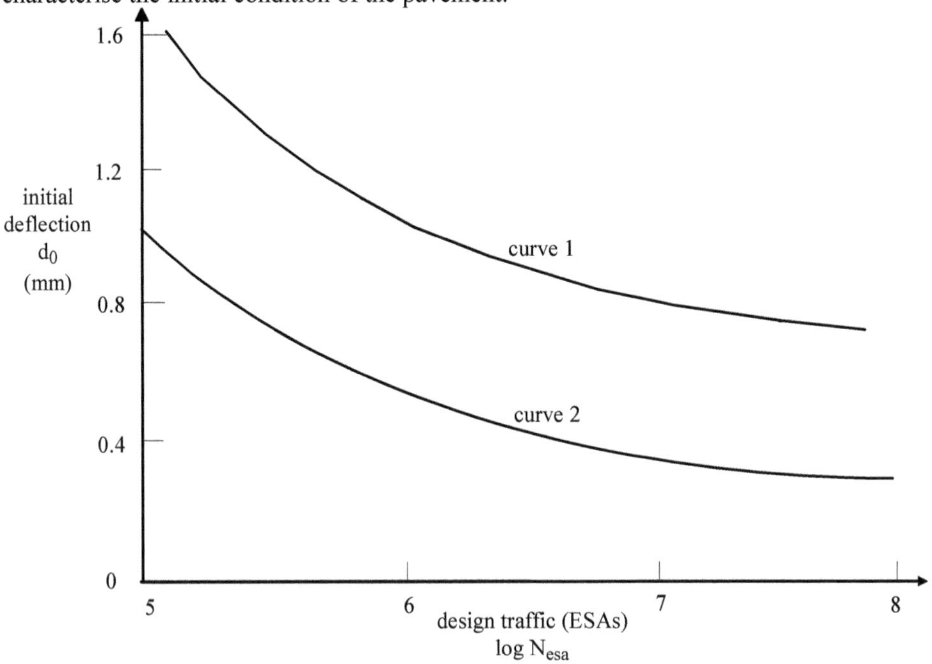

Figure 14.11 Typical graphs of the number of ESAs, N_{esa}, needed to cause just-acceptable service conditions, for a given initial deflection d_0 (from AustRoads 1992, Figure 10.3)). Curve 1 is for rutting of asphalt and curve 2 is for cracking of a pavement with a stabilised basecourse.

For more lightly trafficked streets with flexible pavements, the typical deflection values shown in Table 14.10 have been used to decide whether or not *maintenance* is needed. Note that these deflections are quite high and indicate a weak subgrade suitable only for light traffic.

Table 14.10 Deflection criteria for intervention.

Traffic on:		
local streets	major roads	
Commercial vehicles per day.	Total number of equivalent standard axles in one direction.	Intervention level for 95th percentile elastic deflection (mm). Subtract 0.1 mm for AC surface layers over 50 mm thick.
20		2.3
30		2.0
150		1.8
	10^5	1.7

Section 14.3.2 showed how permanent deformations, and hence rutting, can be expected to be monotonically related to measured deflections. Hence rutting can be expected to follow similar patterns to those illustrated in Figures 14.10 & 11. This is illustrated in Figure 11.12 and Equations 11.11–15, for example. Thus, from Equation 14.13, rut depth is probably proportional to:

$$d - d_o = (N/C)^{1/a}$$

14.6 PAVEMENT MANAGEMENT SYSTEMS

14.6.1 Objectives

In conducting a systems analysis of a pavement, the first need is to define the system objectives. These may include:

(a) minimising the total Road Authority capital and maintenance costs of the project over its life (probably using a net present value approach, Section 33.5.7),

(b) minimising total systems costs by also including external costs such as *vehicle operating costs* (Sections 5.1.1 & 29.1.2),

(c) improving such less quantifiable factors as safety (Chapter 28), comfort (Section 14.4.1), environmental impact (Section 32.1), and level of service (Section 17.4.2), and

(d) reducing decision-making time and allowing alternative management options to be explored.

The objective of a *pavement management system* (*PMS*) is to maximise the benefits listed in (c) whilst minimising the cost disbenefits in (a) or (b). A number of PMSs only minimise the costs in (a) above and are called *introverted* systems. They are particularly suited to the non-arterial streets managed by local government. A typical such system is *MicroPAVER* produced by the American Public Works Association. Systems that consider all the factors listed in alternative (b) above are called *extroverted* systems (Lay, 1985b). A typical extroverted PMS is the *Arizona PMS*. The consequences of a PMS analysis are to:

* alter the way in which road funds are spent by providing a logical structure for considering and comparing alternative propositions; e.g.
  + between alternative uses of public funds (Section 5.1).
  + between competing projects (Section 5.2),
  + between aspects of a project (e.g. between alignment and structure),
  + as capital within a project (e.g. between subgrade and basecourse), and

+ over its life (e.g. between first cost and maintenance),
* provide guidance on the programming of work, and
* give greater emphasis to the condition monitoring and evaluation methods discussed earlier in the chapter. As discussed in Section 5.2 (e.g. Figure 5.11), and Section 26.2 (e.g. Figure 26.1), this is an essential subsidiary part of PMS decision-making.

A PMS is thus a logical extension of the planning and programming (Chapter 5), design (Chapter 11), construction (Chapter 25), maintenance (Chapter 26) and evaluation (this chapter) phases of a road. The PMS operates by coordinating and controlling all relevant pavement activities through the sequence of eight activities or stages shown in Table 14.11.

Table 14.11 Stages in a pavement management system.

1	*List* the *external constraints* on the system and its operation. For instance, the pavement must be: # affordable (Section 25.2), # useable (Section 14.3.1), # safe (Section 28.5), # comfortable for people and suitable for freight (riding quality, Section 14.4.1), # maintainable (Chapter 26), # environmentally acceptable and sustainable (Section 32.1).
2.1	*Provide* an inventory of *design data*: # budget ranges and limits (Figure 5.1), # in situ material and its properties, e.g. subgrade type and CBR (Section 8.4), # imported material characteristics and costs (Sections 8.3–6), # pavement design standards (Chapter 11), # results from pavement trials (Section 8.4(t)), # weather, # design storms and floods (Section 13.2), # moisture regime (Chapter 9), # temperature regime (Chapter 12), # expected construction quality, # extent, condition, and usage of the current pavement, # restrictions on construction methods, # the overall geometry of the road, as determined from separate traffic and planning analyses, # the expected traffic, particularly in terms of truck ESAs, # locations requiring high skid resistance, # design life (Section 11.4.1) and thus design traffic ESAs (Section 33.5) including lane *ESA*s — see Section 14.6.2, # limiting criteria for serviceability (Section 14.5.3), # environmental and sustainability standards (Chapter 32).
2.2	*Provide* an inventory of *design choices*: # materials to be used (Section 8.6), # cross section (Sections 11.1&7), # subgrade type (Section 14.6.2), # overlay limits (Section 14.6.2), # pavement surface type (Chapter 12), # method of analysis (Chapter 11), # construction method (Chapter 25),

	# maintenance method (Chapter 26), # rehabilitation strategy, # reconstruction strategy.
2.3	*Provide* an inventory of: # direct detailed design outputs such as *course type and thickness*, # direct broad design outputs such as *expected life*, water flows and pollutants, # for extroverted systems, *user costs*.
3	*From* the inputs from Stages (1) and (2), and the *performance prediction models* developed in: # Chapter 11, # Sections 14.1–5 (e.g. Equation 14.14), and # sub-models within the network models described in Section 26.5.2, (see also Section 14.6.2), *predict* at some future time the: # extent and condition of the system, and # outcomes for each pavement condition.
 4.1 4.2	From *analysis* and design: # *catalogue* appropriate pavement treatment processes, # *calculate* the outcomes associated with various proposed strategies.
4.3	*Assess* benefits and costs of (Chapter 5): # construction, # maintenance, # rehabilitation, # reconstruction, # vehicle operations, # safety, # salvage value of the unwanted road.
4.4	*Assess* the benefits and costs of (Chapter 5): # environmental and sustainability effects, # social impacts.
4.5	*Iterate* Stages (2) to (4) to select the: # optimum design and pavement management strategy to give whole-of-life net benefits, # best affordable strategy, if optimum strategy is not affordable.
4.6	*Select the best strategy*, considering all the above factors and using a decision analysis.
5	*Provide planning and budgetary advice* (Figure 5.1) as to how the system can be brought to, and then maintained at, some desired extent and condition (including examining alternative strategies).
6	*Produce* short and long term *budgets and prioritising work schedules*.
 7.1 7.2	When in-service: # *undertake* the work to the chosen design, # *manage* the pavement according to the chosen strategy.
 8.1 8.2	After some service: # *monitor* pavement and user performance, # *modify* models in light of observations.

14.6.2 Pavement models

Pavement management systems (PMS) at a network level are described in Section 26.5.2. This section deals with PMS at a project level. At the next level down, the consideration of an individual pavement is best done on a one-off basis but still using the core modules of the project-level PMS model at stages (2) to (4) in Section 14.6.1. These cover the design component, cataloguing the details of pavement design covered in Section 11.4 and exploring their implications for future pavement performance.

Stage (2.1) requires a knowledge of the *design life* (Section 11.4.1) of the pavement. However, it is not necessary to use it as a fixed constraint. It is better regarded as a variable that is optimised with respect to whole-of-life costs by re-running the pavement model with different design lives. In the absence of such a value, it is common to use a life of 20 to 30 years. N_{esa} is then predicted from a knowledge of annual traffic. Either years or N_{esa} can be used in most models (Section 11.4.1). In the absence of such data, some have assumed that the number of repetitions, N_{esa}, of the design load would be between 10^3 and 10^8 ESA (Powell *et al.*, 1984).

Discussing life in years rather than in N_{esa} is appropriate for low-volume roads where much deterioration is time-dependent and traffic volumes predictable. Of course, pavement life will be extended by intermediate maintenance and there are many examples of low-volume roads having effective lives of over 30 years. On high traffic volume roads it is preferable to use N_{esa} rather than years and to use measured N_{esa} to then calculate remaining lives. The limits on pavement life will be:

(1) failure under traffic passes as measured by N_{esa};
(2) failure under an overloaded vehicle, as determined by the axle load level and the then-current pavement conditions; and
(3) time-dependent degradation (e.g. bitumen hardening, Section 8.7.7).

In practice, design lives rarely coincide with actual lives to failure due to:

* imperfect methods of pavement analysis,
* the lack of relevance of the failure mode chosen, and
* the real traffic, materials, and environmental conditions differing from the design assumptions.

The allocation of *subgrade* as a design input in Stage (2.2) is on the basis that the subgrade properties are a fixed set in an area. This is somewhat controversial as these properties are capable of modification by techniques such as drainage, stabilisation, the use of geofabrics, and capping layers.

The *overlay* limit in Stage (2.2) exists to impose a restriction on the minimum time between overlays and/or a constraint on overlay thickness due to kerb lines or height restrictions.

Stage (7.2) covers the deterioration of pavements and the remedying of this deterioration by reseals, overlays or rehabilitation is covered in Stage (2.2) and was discussed in Sections 12.1.4–5 & 12.2.5. Section 14.5 showed that the earlier such action occurred, the less expensive it would be. The pavement designer must therefore balance three main strategies:

(a) A *high initial-cost pavement* and reduced future strengthening costs. The main effect of high initial cost is to extend the time between major maintenance actions (Figure 14.12). High initial cost pavements tend to find favour:

* if funds are available, as it is relatively cheap to obtain extra design life by increasing thickness and quality at initial construction (Section 11.4), or

* where heavy traffic would result in a large increase in user costs and/or toll revenue during reconstruction.

(b) A *medium initial-cost pavement*. This is followed by frequent low-cost strengthening by thin overlays (< 50 mm, Section 12.1.6), rather than infrequent high-cost strengthening by thick overlays or reconstruction (Section 12.2.5). The approach has been supported by cost-minimisation studies (Potter and Hudson, 1981). Operating experience suggests at least restrengthening before significant damage has occurred, and Danish work (Schacke, 1984) showed benefit–cost ratios of two or more for keeping the pavement PSI (Section 14.4.1) above two.

(c) A *low initial-cost pavement*, followed by frequent low-cost strengthening by thick overlays. This approach finds favour for light traffic when initial funds are limited but a reasonable flow of maintenance funds is possible. It is the classic stage-construction route (Section 12.1.1).

Resealing to improve skid resistance, etc. (Section 12.1.5) is not part of this discussion.

One ready way of affecting the balance between strategies (a), (b) and (c) is by controlling the design life, with a long design life leading to approach (a) and a very short life to approach (c). Studies referred to in the following section suggest that strategy (a) results in the minimum total cost.

Stage (3) in Section 14.6.1 is difficult and is sometimes circumvented. The options are:

* model future behaviour using Sections 11.2 & 14.3,
* utilise past empirical evidence by using a Markov chain to predict probable future condition states, as in the Arizona PMS. (The Markov method assumes that the probability of a future event depends only on the present state and is independent of the past. Transitional probabilities are calculated for a move from one state to another.), or
* use expert systems and/or neural networks to capture previous successful decision-making.

The entire process is not completely accurate or reliable. The term *project reliability* indicates the probability that a pavement will outlast its design life. Austroads 2004 estimates that this ranges from 95 percent for freeways to 85 percent for lightly trafficked roads. In terms of probabilities of failure, this suggests that the probability of a pavement failing to achieve its design life is 0.05 for freeways and 0.15 for lightly trafficked roads.

14.6.3 Pavement management strategies

The deteriorated condition of the pavement and lane closures during reconstruction will both raise vehicle operating costs (Section 29.1.2), often to levels exceeding the costs of construction and reconstruction. Clearly, the selection of the best course of action will require an optimisation of the total costs and benefits of the pavement-vehicle system. This is sometimes called *pavement life-cycle costing* (*PLCC*) or *whole-of-life costing*.

It is usually found that the rate of pavement deterioration is the major factor in PLCC. Roughness is the best single indicator of the overall pavement performance and the riding quality of roads as it is an objective and low-cost measure (Section 14.4), gives a good indication of pavement performance (Section 14.4.1), and can be related to road user costs (Section 29.2.7).

Thus, by using roughness as the prime measure, Figure 14.7 can be extended in Figure 14.12 to examine strategy (b) in Section 14.6.2. Note that the graph does not show that higher initial cost also raises the initial serviceability of the road. Whether major maintenance is undertaken before the road reaches an unacceptable roughness level will depend on the optimisation process referred to above. Typical current construction practice means that all new roads are initially constructed to a fairly common, low roughness level. Thus the effect of more effort being placed in initial construction would usually be to extend the life of the road rather than improve its initial quality.

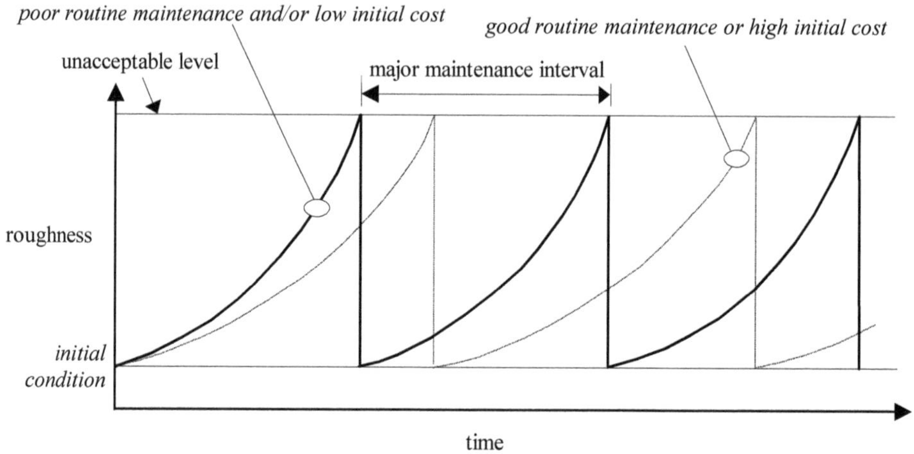

Figure 14.12 Comparison of alternative construction and maintenance strategies.

Specific approaches to pavement maintenance and its strategic application to a network of roads are discussed in Chapter 26. The discussion in this chapter concentrates on issues related to a single piece of road. As shown in Figure 14.12, routine maintenance (Section 26.2) can also influence the rate at which the road deteriorates and delay the need for major effort. If this major effort involves structural reconstruction of the road, then the major maintenance interval indicated in Figure 14.12 is probably close to the service life of the pavement defined in Section 14.3.1.

There is doubt as to the roughness levels indicating the need for structural reconstruction and inadequate field evidence to suggest whether or not roughness is a good indicator of an approaching terminal structural condition. On the other hand, there is no doubt that a road needing structural reconstruction will record high roughness levels; the doubt concerns whether those levels are reached gradually or dramatically over a short interval of time.

Strategy (a) in Section 14.6.2 can also be represented in graphical terms, as shown in Figure 14.13, which suggests that an optimum overall capital/maintenance strategy can be established by whole-of-life costing. The mathematical combination of initial cost and costs incurred at an interval of years is best done using the discounted cash flow techniques explained in Sections 5.2 and 33.5.7 with respect to benefit–cost analysis (BCA). Following Figure 14.12, the effect of improved routine maintenance on the predicted optimum can be treated by considering it as equivalent to an improved standard of initial construction. The major maintenance interval corresponding to the minimum total cost is sometimes called the *optimum life* of the pavement.

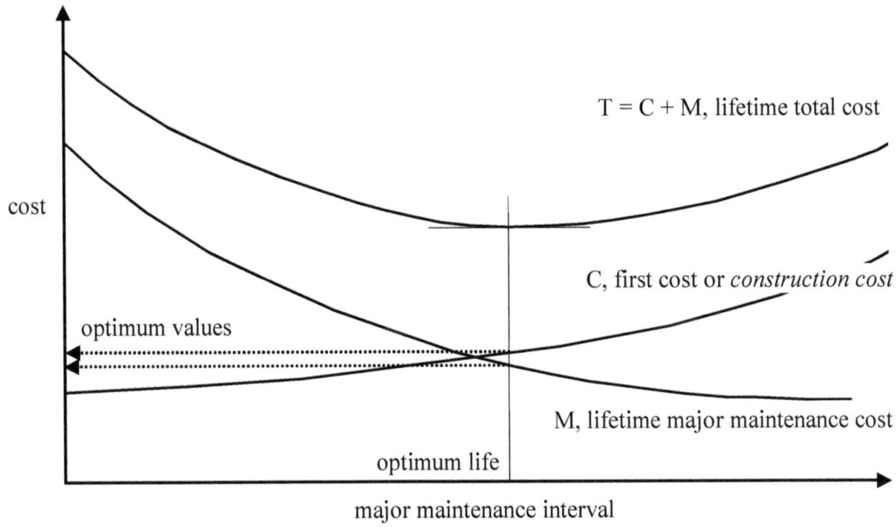

Figure 14.13 Overall maintenance strategy.

A few parametric studies of these issues have tended to confirm the author's experience that strategy (a) in Section 14.6.2 produces the best solution for minimising total (user plus road agency) costs, and that an optimum roughness level was an IRI of about 3 ± 0.4 (80 ± 10 NRM). The author is not aware of significant explorations of this issue.

A broad review of PMS is available in Haas *et al.* (1994).

CHAPTER FIFTEEN

Bridges

15.1 BRIDGE TYPES

The purpose of a bridge is to carry traffic over an obstacle. In doing so, it must also be affordable, buildable, safe, durable, and pleasing to the eye. The four main types of bridge are as follows.

15.1.1 Fords

A *ford* is a dip in a road which passes through (rather than over) a stream, with a submerged pavement able to carry traffic and a water level low enough not to impede traffic. As a ford will be overtopped in any flow, there is no value in raising the pavement level above the level of the stream-bed. All vehicles will travel slowly through fords. The critical water depth at which the road becomes impassable is commonly 250 mm for cars and 450 mm for trucks. Fords are common when the traffic and/or the stream flows are usually low. A *vented* ford is one in which routine, low stream-flows are carried under the pavement by small pipes. The crossing becomes a ford when an uncommon, high stream-flow exceeds the pipe capacity and overtops the pavement. Because it creates a greater obstruction to water flow, a vented ford is more likely than a conventional ford to cause local scouring of the streambed and banks and of the approach road.

A *floodway* is a piece of road deliberately designed to be overtopped during a flood. It is important to design any associated roadside structures to minimise the difficulties associated with clearing water-borne debris from the roadway. The downstream face of the embankment must also be well designed to prevent scour from both surface flow and channel eddies. In this respect, skewed crossings can create major turbulence.

15.1.2 Beam bridges

A *beam* (or *girder*) is a structural component that carries load by bending. The three major beam structural arrangements are shown in Figure 15.1. The *continuous beam* is stronger and stiffer than the *simply supported* one, but is sensitive to relative settlement of the foundations. The *cantilever bridge* allows longer spans to be achieved without a disproportionate increase in material requirements. A *culvert* is a minor structure somewhere between a large pipe and a small bridge. A further discussion of beam types is given in Section 15.2.

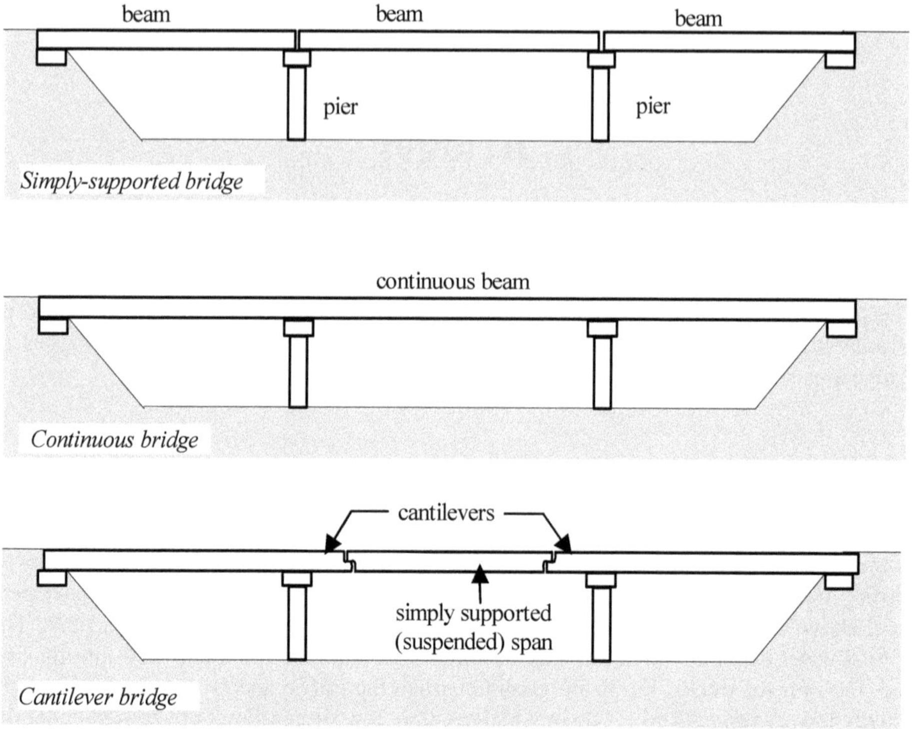

Figure 15.1 Types of beam bridge.

The following types of simple beam bridges may be used for spans of up to:

* 3 m, precast reinforced-concrete box culverts,
* 4 m, cast-in-place culverts,
* 6 m, reinforced concrete slabs,
* 7 m, good quality timber logs,
* 10 m, steel culverts,
* 10 m, precast *reinforced-concrete planks* of solid cross-section
* 15 m, precast *reinforced-concrete planks* with internal voids.
  These rectangular or inverted T or U planks are generally about 600 mm wide. They are laid side by side to form a multi-beam bridge and are often linked together with transverse prestressing (Section 15.2.1).
* 20 m, *composite beams* (Section 15.2.1),
* 30 m, reinforced concrete boxes,
* 40 m, prestressed concrete I's and T's,
* 60 m, steel plate girders (Section 15.2.2), and
* 80 m, prestressed or steel box-section beams.

Although most people think of beams as having solid webs, many *triangulated steel trusses* actually function as beams and are used for beam spans of over 30 m. Trusses are lighter, but more expensive than plate girders, and their span is limited only by the available depth (Section 3.7.2).

15.1.3 Arch bridges

The *masonry arch* bridge was discussed in Section 3.7.1 (Figure 3.8). The individual wedge-shaped stones in the arch ring (or barrel) are called *voussoirs* — the topmost voussoir is the *keystone* and is the last voussoir to be placed in position. Once is it is in place, the arch can begin to be self-supporting. It then carries the applied loads by compression in the ring and transfers this load to the ground at the *abutments* (or *springings*). Masonry arches usually require expensive temporary support during construction and are very sensitive to relative movement of the two foundations. However, modern concrete and steel arches can avoid most of these problems and possess greater internal continuity.

At the other extreme, many *culverts* also act as a form of arch and, in the larger corrugated-steel culverts, the soil and culvert work together to provide both the arch and abutment action. Conventional corrugated steel culverts can span up to 8 m. They usually have a concrete invert if required to act as a water-carrying structure. Corrosion is particularly important and even galvanised steel may corrode at 15 μm/y in an aggressive environment. Reinforced-concrete *pipes* are also a form of arch and are commonly used for smaller culverts. Cover requirements for pipes are given in Table 7.2 and design data in Section 13.4.2.

15.1.4 Suspension bridges

Early suspension bridges were discussed in Section 3.7. In this bridge form, the span of each longitudinal traffic-carrying beam is reduced by using cables draped down from high towers to carry hangers which support the longitudinal beam at intermediate points within its span. The steel cables carry their load in tension and are therefore structurally efficient. They may be curved, as in the suspension bridge, or straight, as in the *cable-stayed* bridge. The unbalanced horizontal load on the towers is usually carried by continuing the cables over the tower to either an adjacent span or to a sound anchorage. Anchorage is normally provided at the abutments, but one design technique uses the longitudinal compressive forces in the deck to 'anchor' the cables.

15.2 BRIDGE COMPONENTS

15.2.1 Concrete

In general, the materials used in bridges are well described elsewhere and only a few unique aspects are discussed in this chapter. For example, the properties of concrete are discussed in some detail in Section 8.9.2. *Reinforced concrete* now has a century of successful structural use behind it (Section 3.7). In a reinforced concrete member, steel rods are used to carry any tensile stresses developed in the member during bending or shear. Even at low load levels, these tensile stresses will be associated with tensile strains in the concrete that will exceed its fracture strain. Hence, reinforced concrete members will be cracked under design loads. One consequence of this fact is that reinforced concrete is not an efficient user of the properties of the concrete cross-section and its use in bridge design is relatively limited.

The development of *prestressed concrete* in the 1930s (Section 3.7.2) further extended the ductility and tensile strength of concrete members. It does this by using steel rods or cables to stress the beam before it is loaded, in order to apply high initial compression to the beam cross-section. It is necessary to use high-strength steel for this task as the creep and shrinkage strains that occur in concrete (Section 8.9.2) would be sufficient to relax all the prestressing force that could be applied using conventional steels. This is because concrete creep and shrinkage strains about equal the yield strain in mild steel. Relaxation of the steel is discussed in Section 15.2.2.

When any subsequent tensile service stresses are applied to the member, they are counterbalanced by the compressive stresses due to prestressing, which thus keep the combined stress out of the concrete's brittle tension regime. Thus, no cracking occurs in prestressed concrete. Of course, it is necessary to check that the combination of service compressive stress and prestress does not exceed the crushing strength of the concrete. Other design checks must be applied to long-term deflections, as influenced by concrete creep.

Partial prestressing lies between reinforced and prestressed concrete. Cracks are not permitted under dead load alone (Section 15.4.1), but are permissible under service loads. A partially prestressed member contains both reinforcing rods and prestressing cable. The reinforcing steel controls the crack width, provides better durability near ultimate load, controls deflections, and removes tensile strains as a design constraint.

There are two types of prestressed concrete.

(a) *Pretensioned*: high-strength steel wires are tensioned; concrete is then cast around them; when the concrete has set and there is longitudinal bond between concrete and steel, the external tensioning forces are released and transferred by the bond to the concrete, placing the concrete in compression (i.e. prestressing it). Some of the tension is lost due to elastic contraction (and subsequent creep and shrinkage) of the concrete, but the bond between the wire and the concrete maintains most of the tension in the steel.

(b) *Post-tensioned*: The concrete is cast with ducts left for the insertion of the wire cables through the set concrete. The cables are then tensioned against the concrete with the force transfer mainly at the ends of members. However, the ducts are usually grouted to control corrosion and so some bond transfer may occur also. Post-tensioning is the simpler technique when curved cables are needed.

Precast concrete has largely replaced cast-in-place construction for smaller structural members. Prestressed concrete beams are commonly I, inverted T or U in cross-section and require a concrete deck (Section 15.2.5) to be placed above and keyed to them, although the use of an inverted U avoids the need for a separate deck. I and T beam spacings are usually between 1.5 m and 2.0 m.

To some extent, the alkaline environment provided by concrete (Section 8.9.1) results in a protective covering of oxide forming on reinforcing steel and prestressing wire. However, some corrosion protection of the steel is often essential, particularly in environments — such as seawater and de-icing salts — with high chloride contents. The corrosion products are expansive and lead to cracking and then spalling of the concrete cover. The cracking exacerbates the problem by allowing further water to enter, causing more corrosion and possible freeze–thaw cycle expansion. Corrosion protection usually involves:

* using a sufficient cover of good quality (e.g. high cement content) concrete,
* preventing water and chloride entry,
* coating the steel, and/or
* cathodic protection.

15.2.2 Steel

Steel is a common and well-documented structural material (Lay, 1982) and only its behaviour as a prestressing wire is of particular relevance here. The main issue is the amount that the steel *relaxes* (i.e. relieves itself of load) under permanent tension as this will release some of the pre-compression in the concrete. The relaxation process that occurs is one of long-term plastic (*creep*) deformation within the wire. It is usual to prestress the wire cables initially to quite a high (*c*.80 percent) proportion of their failure or 'ultimate' tensile stress, as this will be the highest stress encountered during the life of the cable. The wire must be of high strength steel to ensure that the residual strains are still sufficiently high to provide adequate precompression.

Corrosion protection is critical. Without it, member cross-section is lost and severe notching may be introduced. Corrosion requires the presence of water and oxygen and can therefore be reduced by eliminating crevices and pockets likely to hold water and damp rubbish. Overall protection is provided by galvanising or by well-maintained coats of paint applied over a clean, scale-free steel surface.

Steel beams used in bridge construction are usually either:

* *sections* rolled to their final shape by the steelmaker, or
* *plate girders*, which are built up to shape by a steel fabricator. The plates are joined by welding.

Composite construction involves the interconnection of a concrete deck (Section 15.2.5) to the steel bridge beams by means of steel outstands known as *shear connectors*. This allows the steel to work in tension and the concrete in compression, thus gaining the best from both materials. Composite beams are not as widely used as prestressed concrete beams, but find favour for short spans, where beam depth is limited, where long transport distances are involved, or where access is difficult.

15.2.3 Timber

Timber is a traditional material in bridge construction (Section 3.7), but often suffers from poor durability, high variability, and relatively low mechanical properties. Hence, it is infrequently used in modern bridge practice. When timber is used, it can be employed for decking, structural members, and substructure.

Timbers are classified as *hardwood* or *softwood*, depending on their cell structure. However, the names are misleading and not a good guide to performance. Within each category, a timber is either sapwood or heartwood. The lighter-coloured sapwoods are the new growth and usually lack durability. The four main factors determining timber durability are the:

(a) type of timber,
(b) exposure conditions,
(c) bridge detailing, and
(d) maintenance practices adopted.

The first is usually the most important. The key aspect of factors (b) and (c) is to keep moisture away from the timber, as degradation is largely dependent on the time the timber stays moist. Timber preservatives will also assist. It may also be necessary to tighten timber connections as the timber shrinks and changes shape.

15.2.4 Foundations

Bridge foundations are usually either *spread footings*, *piles* or *caissons*. A spread footing is a large reinforced-concrete pad that distributes the bridge loads at an acceptably low bearing stress applied to the in situ material immediately beneath the footing. It is usually a relatively cheap foundation to a depth of about 4 m. When used in water, spread footings are usually placed at least 2 m below bed level to avoid the stream water *scouring* away the material under the footing (Section 15.3.3).

The *pile* is normally used in poor foundation areas where insufficient local bearing stress is available. It is a slender column of timber, concrete, or steel that gains its support from both side friction along its length and end bearing at its tip. Timber, steel, and prestressed-concrete piles are brought to the site as complete units. Piles are installed by either:

* driving. Timber piles are almost always driven into position with their leading end shoed with steel or cast iron, and the driven end (or pile head) fitted with steel rings to prevent splitting. For steel piles, special driving mandrills are usually placed over the driven end. Concrete piles usually do not need such extra provisions but must be lifted carefully to avoid cracking under bending stresses. Similarly, they must be driven in a controlled, concentric fashion to avoid moments due to eccentric forces. Although pile driving is very common, it can create undesirable effects on adjacent areas.
* screwing.
* boring and then casting reinforced concrete in situ,
* driving a hollow shaft and then casting reinforced concrete in situ. *Caissons* are a special version in which large diameter, hollow shafts are sunk into position, with material subsequently removed from inside the caisson. They are used where a larger bearing area is required than is available from piles and where conditions make it impossible to build a spread footing within a cofferdam.

Piles are frequently tested to determine their structural integrity and their load capacity. Structural integrity is usually determined by a form of sonic test in which the speed at which a stress wave travels down and then back up the pile is measured. The four main types of capacity test use:

(a) slowly applied, maintained load increments,
(b) a constant rate of penetration,
(c) cyclic loading, or
(d) measurements of dynamic load behaviour.

Load tests are usually carried to some multiple (e.g. 1.5 to 3.0) of the serviceability load (Section 15.4.2) rather than to failure.

Concrete and steel can be corroded by acidic sulfate soils, particularly where such soils have been disturbed or drained during construction. They can sometimes be detected by their sulfurous smell.

A *pile cap* is a beam cast on top of a group of piles and which then serves a similar function to the spread footing, sharing the bridge load amongst all the attached piles. Columns placed on the pile caps or footings are known as *piers* and are commonly built of reinforced concrete. There are four main types of pier:

(1) Frame piers. These are frame-shaped, as illustrated in Figure 15.2, and are the conventional form.

(2) Single-column piers. These are often used for:
* high piers,
* where streamlining is necessary for debris flow (Section 15.3.3),
* where the pile cap and roadway are skewed, or
* where the structure is torsionally stiff.

(3) Wall piers. These are continuous transverse walls acting as structural members. They are used for piers under about 3 m in height, or where thin piers are needed (e.g. between carriageways).

(4) Trestle piers. These are used when there is relatively low ground clearance. They are formed by extending the piles upward to directly connect to the crosshead (Section 15.2.5).

Bridge *abutments* are constructed at each end of the bridge to support the end spans and to retain the approach pavement. An approach slab may be used to eliminate the dip that often forms in the pavement adjacent to the bridge deck (see Briaud *et al.*, 1997 & Section 11.8.2 for a further review). To support their own weight and the vertical structural load, abutments are usually carried on piles (as in Figure 15.2) or strip footings.

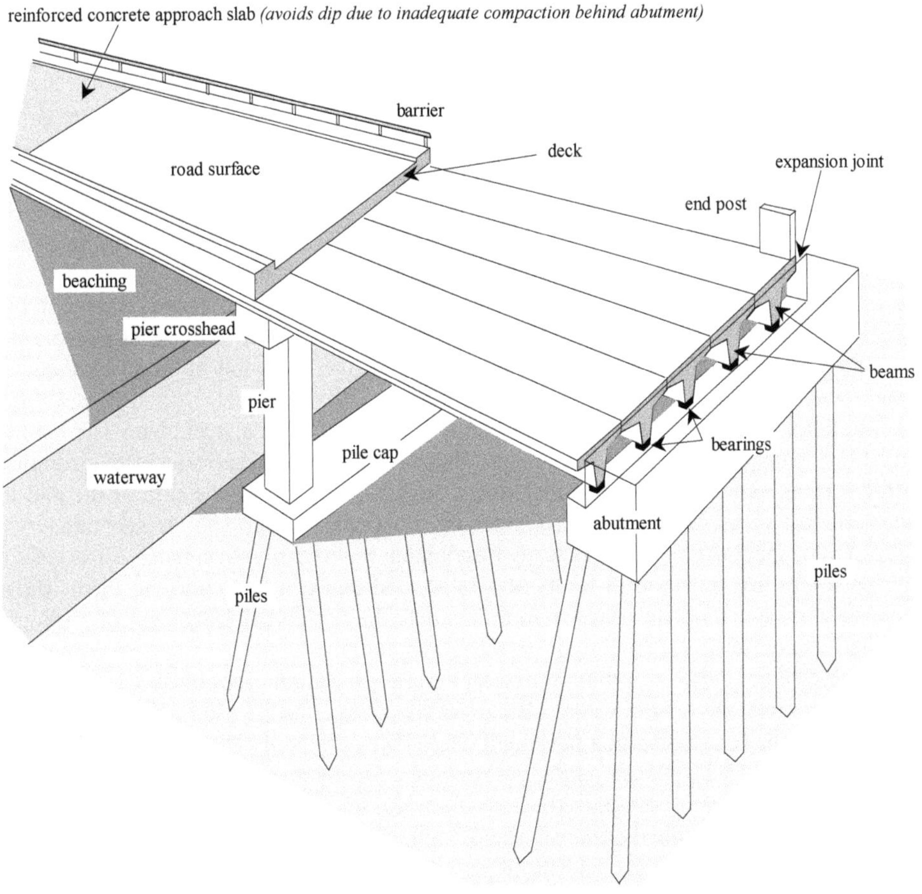

Figure 15.2 Some bridge components (based on CRB, 1980b).

To serve their retaining function, they may be either the spill-through type illustrated in Figure 15.2 or, less commonly, a complete retaining wall. This latter type is usually restricted to very short spans. Stone or concrete surfacing (called *beaching*) is often placed on the embankments under the bridge and leading up to the abutments to prevent surface scour and improve appearance. It is necessary to ensure adequate drainage behind any continuous retaining structure to prevent the build-up of full hydrostatic pressure. This may lead to either lateral sliding on the foundations, or bulging of the retaining structure.

15.2.5 Superstructure

Referring to Figure 15.2, the piers discussed in the previous section support a transverse horizontal beam known as the *pier crosshead.* The *superstructure* of a bridge is that part of the bridge above the pier crossheads and *substructure* refers to the pier crossheads, piers, and bridge foundations. In a typical bridge, the superstructure cost is about double the substructure cost.

The pier crossheads carry the *bearings* on which the longitudinal bridge structure is seated and supported. Bearings permit any necessary translation or rotation to occur at the load transfer point as the bridge superstructure bends, expands and contracts. They also spread the reaction force from the structural members so that the resulting bearing stresses on the supporting structure are acceptably low.

Most bearings are either natural or synthetic *rubber* or, occasionally on larger bridges, steel. The commonest bearings are *elastomeric,* composed of interleaved laminated plates of steel and of natural rubber. The rubber and steel are bonded together during vulcanisation. The rubber used has an elastic modulus of about 3 MPa, a bulk modulus of 2000 MPa and a breaking strain of about 6. Horizontal movement of the bridge superstructure is accommodated by shearing of the bearing.

In theory, one reaction point should be fixed to prevent total movement of the superstructure. This particularly applies when the superstructure is on a slope (e.g. to accommodate super-elevation (Section 19.2) or drainage cross-fall or in an earthquake zone.

In a *pot bearing*, a circular rubber pad is contained within a steel plate. The load is applied by a matching circular steel plate. There is little vertical movement, but rotation readily occurs as the contained rubber flows 'hydraulically' from one side of the pad to the other. In a *spherical bearing*, two matching spherical steel surfaces are separated by a Teflon (pfte) sheet. Again, there is ready rotation but no vertical movement.

Bearings are specified in terms of a rated load that can be applied to them whilst subjected to a specified shearing strain to simulate the horizontal movement of a bridge. They have a coefficient of friction of between 0.05 and 0.20.

The *bridge deck* is a flat, horizontal plate placed on top of the longitudinal structural components of the bridge. It carries the bridge traffic surface and is thus the equivalent to the basecourse in a pavement. Decks are commonly either in situ reinforced concrete (Figure 15.2), precast and/or prestressed concrete, or open-mesh steel. Metal or synthetic *expansion joints* are placed across the width of a bridge deck to:

* allow for longitudinal expansion and contraction, due to thermal effects (Section 15.4.6) or material creep and shrinkage (Section 15.2.1), and
* avoid the transfer of unwanted forces (e.g. Section 15.4).

Commonly, the joints are between two concrete deck-slabs and are recessed below the deck surface to allow easy sealing. They are usually either poured in situ, elastic (e.g. rubber), geometric (e.g. a concertina) or mechanical (e.g. interlocking metal fingers). Poured in situ joints can usually only handle movements of 10 mm or less. Joint performance can be noticeably affected by relative vertical movement on either side of the joint. The requirement that these be trouble-free, of long life, and waterproof under traffic has proved very difficult to meet. Thus there is a growing tendency to eliminate joints in small bridges and to make the bridge an integral structure restrained by the surrounding ground and the approach embankments. Such bridges must be able to safely accommodate any movements and resist the resulting restraining forces.

Guardrail is provided on the bridge approaches to protect errant vehicles from unintentional impact with the relatively-rigid bridge structure (Section 28.6.3). A number of studies have indicated the high relative danger caused by the close proximity of solid, unprotected bridge abutments and parapets to a carriageway.

A traffic surface may be placed on the deck to achieve desired cross-falls, profiles, and/or noise levels. It is common for safety reasons to make the deck sufficiently wide to continue the full pavement and shoulder width across the bridge, as the presence of shoulders can reduce the crash rate by about 25 percent for narrow shoulders and 85 percent for full shoulders.

Guardrail or other *bridge barriers* along the edge of the deck reduces the severe crash rate by about 80 percent and prevents errant users from falling from the bridge (Section 28.6.2). About 3 percent of impacts with bridge barriers result in a truck going through or over the barrier. Hence special barriers may be needed when the bridge is over a busy area.

15.3 BRIDGE DESIGN

15.3.1 Aesthetics

Bridges are very functional structures with a clearly understood role; they must be affordable, serviceable and safe. The pleasing appearance of a particular bridge usually results from the active expression of that role in harmony with the surrounding natural environment. That is, the bridge — which will frequently be visually prominent — must not only fit into its environment, but must add to it. Thus, the creation of a visually satisfying bridge is a task requiring a combination of structural skill, artistic judgement, and attention to both concepts and detail.

Excellent guidance on bridge aesthetics — covering such factors as function, proportion, order, and form — is given in Leonhardt (1991) and the accompanying papers in the same volume.

15.3.2 Structure

The analytical processes used in bridge design are usually precisely defined in the local bridge design code and so this section will concentrate on examining the basis for those processes.

The design criteria in use in bridge specifications evolved through experience and generally have led to safe and serviceable structures. However, the degree of conservatism involved can be unknown and variable. Furthermore, different design philosophies evolved within different sections of many codes (e.g. between reinforced and prestressed concrete). The method of *Limit States design* provides a mechanism by which the above defects can be safely and consistently remedied. The method is also sometimes called *Load Factor* design, *Ultimate Strength* design, *Load and Resistance Factor* design, or (ungrammatically) *Limit State* design.

A general objective of Limit State design is to provide a unified approach by requiring the designer and code-writer to define all the relevant objectives and design criteria for the bridge being designed. The points at which the bridge ceases to fulfil these intended functions are called the *limit states*. The design criteria must include a definition of each limit state and the acceptable probability for the attainment of that state during the *design life* of the bridge. Design life is typically taken as 100 y, although longer lives are sometimes chosen for major structures. Most limit states fall into one of two categories.

(a) *Serviceability limit states* cover the routine operation of the bridge and include such aspects as crack growth, deflections due to creep or foundation settlement, and unpleasant vibration under traffic (Section 15.4.2). Introducing an initial reverse deflection, called *hog* (or *camber*), into the structural members can reduce the effective deflection due to load and/or long-term creep.

(b) *Ultimate limit states* cover the total *failure* (or collapse) of the bridge. The three commonest forms of bridge failure in modern times are (OECD, 1979):
   * floods causing foundation movement, 60 percent,
   * material inadequacy, 20 percent,
   * overload or accident, 15 percent.

Historically, erection failures were also very common (Lay, 1992).

The structural analysis of a bridge calculates the manner in which the structure uses its strength and stiffness to resist the various actions imposed on it. These actions may be loads, imposed deformations, or temperature changes. The basic design requirement for each limit state is that:

design resistance, r > design action, a

The difference between the mean design resistance (r_{50}) and the mean design action (a_{50}) is ($r_{50} - a_{50}$) and is somewhat misleadingly called the *margin of safety*. Limit State design employs a more probabilistic approach based on the statistical description of the actions and the resistances via a probability density plot (Section 33.4.2) of the type shown in Figure 15.3. The *probability of failure* (or *risk*), p_f, can be expressed as the *convolution integral* (Section 33.2):

$$p_f = \int_0^\infty f_r\{\int_x^\infty f_a dx\}dx$$

where f_r and f_a are the probability density functions (Section 33.4.2) for the resistance and action, respectively, and x is a variable for action and resistance. Ideally, the designer would decide if the design was satisfactory by checking whether this failure probability was acceptable.

There is rarely adequate knowledge of the variability of most input parameters, and semi-probabilistic methods have evolved to overcome some of the hurdles. For instance, actions and resistances are not currently known to an accuracy which would permit p_r or p_a to be evaluated and the above convolution integral to be solved. It is therefore common to work in terms of *characteristic values* which are defined (Section 33.4.2) as values which have an F percent chance of being exceeded over the design life (Section 15.3.2).

For actions, F = 5 (a_5), and for resistances, F = 95 (r_{95}) are often adopted for the ultimate limit states (Figure 15.4). The ultimate design situation is then:

$$r_{95} \geq a_5$$

However, the serviceability limit state deals with more conventional and less extreme situations, and commonly uses either lifetime mean (i.e. F = 50) characteristic values, or values with a 95 percent or 5 percent probability of occurring within a year.

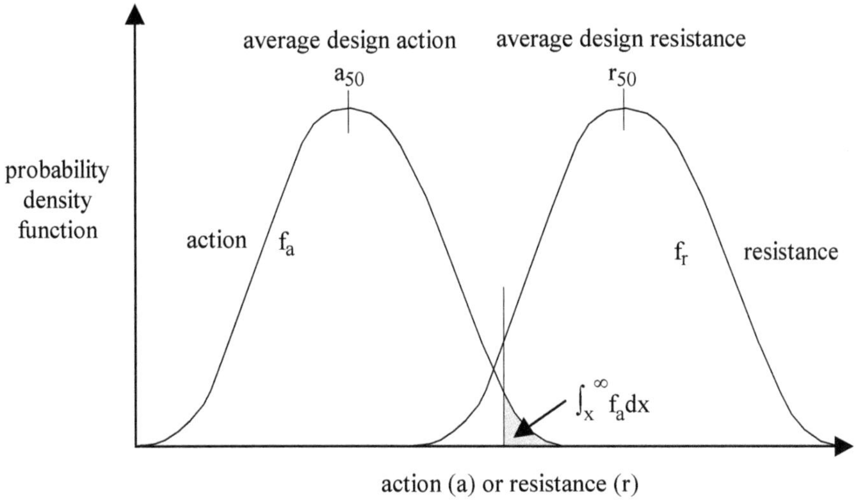

Figure 15.3 Distribution of actions and resistances.

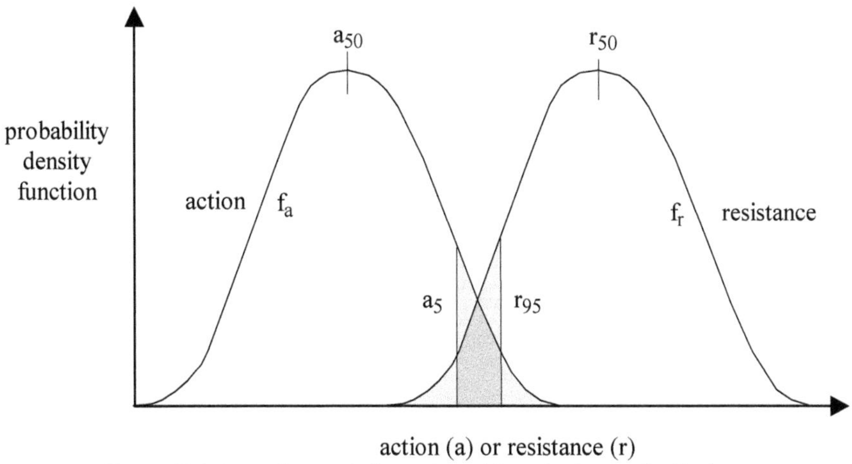

Figure 15.4 Approximate solution to the problem of unknown distributions.

The use of 5 percent characteristic values for traffic loads is qualified in the following discussion.

It will be clear from Figure 15.5 that there will still be some probability of failure, even if the design inequality is met. To control this, additional factors, ϕ (< 1) are applied to produce the final design inequality:

$$\phi_r r_{95} \geq a_5/\phi_a$$

Two ϕ factors are used to account for the possibility of quite different probability density functions applying to a and to r. The ϕ_r factor is called a *design coefficient* (or *load factor*) and ϕ_a is a resistance (or performance) factor. As shown in Section 15.4, bridges are subjected to a range of loading types. As a different ϕ applies to each, the inequality becomes:

$$\phi_r\, r_{95} > \Sigma_i(a_{5i}/\phi_{ai}) \qquad (15.1)$$

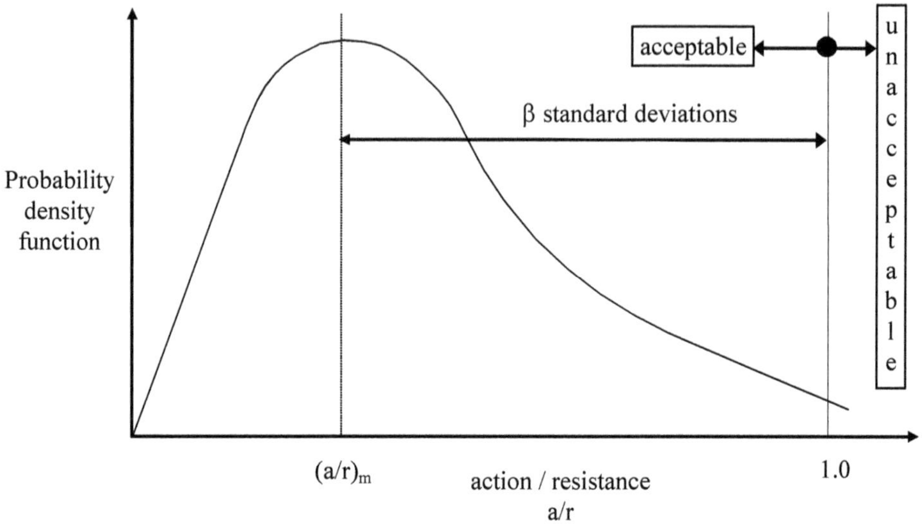

Figure 15.5 Representation of the safety index, β.

An index of the overall probability of failure is obtained by observing the probability density function of the ratio a/r (Figure 15.5). Failure will occur if a/r > 1. A *safety* (or *reliability*) *index*, ß, is defined as the number of standard deviations that the failure case is away from the mean, e.g. the difference between $(a/r)_{mean}$ and unity. It can also be calculated as $(r_{50} - a_{50})/([\text{variance of a}] + [\text{variance of r}])^{0.5}$. ß values of 3 to 4 are common for bridges, and are most sensitive to changes in the resistance parameters. For a normal distribution of a/r, the associated failure probabilities are given in Table 15.1.

Table 15.1 Failure probabilities.

	probability of failure	
β	–log	1 in
3	3	1,000
3.75	4	10,000
4	4.5	33,000
4.25	5	100,000
5	7	10,000,000

ß is used to calibrate values of ϕ_a and ϕ_r to achieve some parity with designs based on non-statistical approaches.

When a bridge is checked for its capacity to carry a specific overloaded vehicle, it is common to adjust the product $\phi_r\phi_a$ in inequality (15.1) to represent the actual and known conditions. This will take into account the greater degree of certainty to which these particular actions and resistances are known. Problems particularly arise when the bridge is old or in poor repair or made of timber. Test loading or strengthening should be adopted if there is uncertainty over the bridge's capacity. On the other hand, bridges will not always behave according to the original design assumptions. For instance, it is commonly observed that loaded bridges:

(a) act in a more composite manner than assumed,
(b) act with more continuity than assumed, and
(c) distribute the load more widely than assumed.

15.3.3 Water

An overall approach to *waterway* design is illustrated in Figure 15.6. The three main tasks are seen to be:

(a) the collection and assessment of information on primary water flows,
(b) the design of compatible bridge (or culvert) structures and road embankments,
(c) the assessment of the acceptability of the resultant flow conditions.

a. Collect and assess primary information.
a1 Collect hydrological (Section 13.2.1) and topographic data, including obstacles to flood flows (Section 6.1.2–3).
a2 Assess the flow-regimes upstream and downstream, with emphasis on erodability, debris deposition, and effects of future flooding.
a3 Determine the level of traffic serviceability (flood-free days) to be provided by the bridge.

⇓

b. Design the bridge, including any approach embankments.

⇓

c. Assess effect of bridge on water flows.
c1 Assess approach and exit conditions, and debris control
c2 Check bridge waterway conditions, particularly for energy drops and flow speeds

⇓

d. Assess the effect of the new waterway and bridge on the environment.

⇓

e. Reiterate to (b) if (c) or (d) are unsatisfactory.

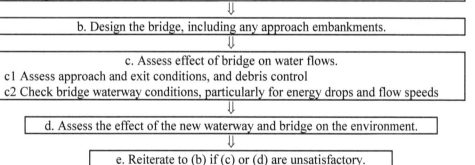

Figure 15.6 Decision flow diagram for waterway design.

The *hydrological* data required by the first box in Figure 15.6 can be obtained from the general information on flood flows and their estimation given in Section 13.2.2. Following Section 15.4.2, design floods could be defined with the same recurrence intervals given there for live loads. Design flood levels are currently based, however, on *recurrence intervals* of the type defined in Table 13.1.

Water flows, and particularly *floods*, are a key factor in Figure 15.6. They present a major problem to bridge designers as the flows associated with the recurrence interval used in design must be passed safely through the openings in the bridge substructure. This problem is often accentuated by the fact that considerations of cost and traffic flow mean that bridges are not necessarily located on the best possible site from an hydraulic viewpoint. For example, there is the common bridge practice of, for economic reasons, partially crossing the flood plain with an impenetrable embankment. The flood flow is thus constricted at the bridge site and water is impounded upstream (an *efflux*).

At the same time, *flow speeds* through the bridge opening will be increased, raising the possibility of erosion of the bed of the waterway. Such *scouring* takes three main forms:

* *general scouring* occurs when the whole streambed is lowered, usually as a result of increased flows (as discussed above), or the exposure of a weak layer at the streambed,
* *contraction scouring* occurs due to the increased flow speeds caused by the piers locally contracting the waterway area, and
* *local scouring* occurs as a result of the eddies and vortices caused by constricted water flow at the piers and abutments. This is managed by placing erosion-resistant material (riprap) around the obstruction to a distance of about twice the pier width, but level with the streambed.

At all levels of flow, water can act to undermine a bridge and its approaches. At high flood levels, water pressure exacerbated by debris trapped under the bridge can exert severe horizontal transverse and longitudinal forces on the bridge structure. Thus, when a bridge is required to pass a flood of specified recurrence interval, clearance must also be provided beneath the bridge for the passage of debris. For heavily-timbered catchments it may be necessary to allow 600 mm of freeboard.

If the roadway *embankment* is to be overtopped by the design flood, then some form of embankment protection will be needed, particularly on the downstream face, if the tail water on that side could be at least 350 mm below the edge of the formation. In such circumstances, protection work can be minimised by lowering pavement levels over a short length to localise overtopping.

Some consideration must also be given to what would happen if a flood exceeding the design flood occurred during the life of the structure. The immediate possibilities are that:

(1) high headwaters caused by the structure would create upstream flooding,
(2) high flows would cause scouring at the structure and downstream of it,
(3) the structure would be overtopped and traffic halted, and
(4) the road embankment approaches would be overtopped.

15.4 BRIDGE LOADS

The loads that a bridge must carry are mainly *dead, live* and *environmental loads*. Loads in the last two categories can be horizontal as well as vertical, and so it is necessary in bridge design to make specific provision for horizontal forces to be transmitted to the foundations or abutments. Similarly, *expansion joints* (Sections 15.2.5 & 15.4.4) are often used to avoid longitudinal loads being carried to inappropriate members. For instance, if a bridge is carried on tall piers, it is better for horizontal deck forces to be carried in compression through the deck to the abutments, rather than in bending via the tall piers. Hence, expansion joints and hinges may be used at the top and bottom of the piers.

15.4.1 Dead load

Dead load is a permanent vertical load caused by the mass of the bridge. It can be the major load applied to long-span bridges, as the bridge must support itself as well as the traffic using it. Indeed, long-span bridges are mostly devices to carry their own weight, with the traffic causing only minor loads in the major members. Because dead load is derived from the mass of the bridge itself and of all components permanently associated with its operation, it is fairly readily assessed and controlled and is subjected to relatively small variation.

15.4.2 Live load

Live loads are traffic-induced transient loads and have been discussed a little in Section 11.4.1 as they are caused by the same vehicles that load the pavement structure. The design values are always in excess of the current legal vehicle loads, in order to cover such contingencies as illegally overloaded vehicles, special or permit vehicles, and future increases in legal loads during the life of the bridge. With respect to the last point, some conservatism is required as there is an historical pattern of steadily increasing live loads, fuelled by the natural economies of scale associated with the haulage of larger loads. Thus, today's design loads must attempt to consider the future vehicle loads that might occur over the life of the bridge. Live loads are more fully specified in Section 27.3.2.

The author has advocated the use of the *recurrence interval* approach for live loads (Lay, 1980b & Section 33.4.6). In this case,

* the load which would cause structural failure (Section 15.3.2) and which would be used in design for the ultimate limit state is a 5 percentile *characteristic load.* This is the largest load with a 5 percent chance of occurring in the design life of the structure (Section 15.3.2).
* the load for operational measures is the *serviceability load* which is the largest load with either a 95 percent chance of occurring in the design life or a 5 percent chance of occurring in any one year. The '95 percent chance' definition has a 3 percent chance of occurring in any one year, so there is little difference between the two.

It has been shown (Lay, 1980b and Equation 13.2) that the recurrence intervals which define these ultimate and serviceability loads are 20 and 0.25 times the design life, respectively. An alternative is to view the loads as those which have a chance of occurring in any one year. This is the reciprocal of their recurrence interval, i.e. a 25 year truck has a 4 percent chance of occurring in any one year. Recurrence intervals are, of course, also used for wind loads and hydraulic design, as discussed in Sections 13.2.1 and 15.4.3&5.

Care is needed in using 5 percentile characteristic loads for bridges for, if all the cars and two-wheelers in the traffic flow are counted, the top 5 percent of these loads would embrace most truck loads and not just extreme values of these loads (Lay, 1980b). The 5 percent is better applied to loaded trucks. The other problem with the use of characteristic loads is that they are unrelated to the frequency with which the load occurs during the life of the structure.

Once a definition has been chosen, the key subsequent issues in determining live loads are the:

(a) number of trucks on the bridge,

(b) total mass of each truck,
(c) load on each tyre and axle, and
(d) distribution of those axle loads within the truck length.

Issue (a) can be estimated from a knowledge of the traffic and is usually specified by design codes. One problem with many such loads is that they are based on queues of heavy vehicles side by side in adjacent lanes. Whilst this represents the maximum possible loading, it is not always a very probable load. Issues (b) and (c) are covered by truck regulations, as discussed in Section 27.3.2. Issue (d) is often covered by the use of a *bridge formula* that controls the truck mass as a function of its length. The longer the truck, the more load it can carry. A typical bridge formula is:

(truck mass) = [a(truck length) + b] < c

where a, b, and c are constants.

The transport of heavy *indivisible loads* (often for power stations or mining companies) is a common source of exceptional bridge loadings. These loads are discussed in Section 27.4.1.

Live loads can also have dynamic effects and care must be taken to ensure that the *natural* (or *resonant*) *frequency* of a bridge does not coincide with any sensitive live-load frequencies. Footbridges are particularly susceptible to vibration problems (Section 15.4.7), but resonance analyses are desirable for all types of bridges.

15.4.3 Wind loads

Regional windstorms arise from the following effects.

(a) *Tropical cyclones* — caused by wind circulation about a vertical axis.
(b) *Extra-tropical cyclones* — the speeds are usually lower than for (a) but the affected area is larger.
(c) *Strong gradient winds* — these arise from pressure gradients between high and low pressure systems.
(d) *Severe local storms* — these usually result from local convection currents associated with cold fronts or depressions.
(e) *Thunderstorms* — these arise from particular synoptic conditions and develop cold downdraughts in the leading half of the storm.
(f) *Tornadoes* — these may be of short duration and narrow path width.
(g) *Local wind accelerations* — these result from peculiar features of the local topography.

The wind speeds arising from these storms will vary from location to location. For design purposes, the speeds are usually given as contours on a regional map, for a particular *recurrence interval.* These speeds are then converted into transient pressures on structural members, using standard structural formulas.

15.4.4 Earthquakes

Earthquake loads are usually specified in terms of *ground accelerations* at a particular site. It is usual to divide a region into about five recurrence-interval zones and to specify design provisions appropriate for structures in each zone.

Roads normally suffer from embankment failure, ground liquefaction, or gross vertical distortions. The commonest earthquake failure modes for bridges are:

* foundation failures,
* beams sliding over or lifting off bearings. Gross horizontal movements are common and so horizontal ties are critical, especially if expansion joints are used.
* inadequate pier connections to foundations and cross-beams,
* underestimation of the forces imposed on elevated structures (often leading to pier failures).

15.4.5 Water loads

Some discussion of the loads induced by water flows was given in Section 15.3.3. At a basic level, water flow is defined by the *streamline*, which is the track that a drop of water takes in flowing through the bridge. There is thus no water movement across a streamline. Its path is governed by *Bernoulli's theorem* that states that the total *energy* will not change from point to point, i.e. the sum of the potential, pressure, and speed energies will be constant. The speed energy is proportional to the square of the speed. Following the continuity condition, no streamline can begin or end within waterway lengths that have no points of entry for additional water.

Bundles of streamlines forming a small, closed passage carrying a defined, constant water flow are known as *streamtubes* and their walls are called *streamsurfaces*. As each streamtube contains the same discharge as any other streamtube, their spacing is inversely proportional to their speed. Surfaces drawn at right angles to the streamtubes are known as *equi-potential surfaces*. This orthogonality condition provides the first check on any visual sketching of flow nets. The spacing of the surfaces along the streamtubes is also inversely related to the speed in the streamtube. It is common to choose the spacing of the *equi-potential lines* to equal that of the streamtubes. The fact that these spacings are both inversely proportional to the same speed means that their relative proportions will be maintained throughout the closed system.

The two-dimension representation of streamtubes and equi-potential lines produces a *flow net,* which allows the movement of the water to be visualised. Each enclosed quadrilateral forming the net approaches a square as the grid spacing is reduced. This simple flow net treatment assumes that:

* no water enters or leaves the system,
* there are no large eddies, cross-currents, or flows separating from boundaries,
* the depth is relatively uniform, and
* the boundaries relatively smooth, and thus the Reynolds number is high.

Flow nets can be most useful in deciding on a design arrangement that will minimise the water loads on a bridge. For example, a key design understanding is obtained by considering the behaviour of water entering and then leaving a constriction in its path, forcing the streamtubes to converge and to subsequently diverge with the following effects:

Flow net characteristic	Event	
	Water enters constriction	Water leaves constriction
Streamtubes	converge (narrow)	diverge (widen)
Total energy	unchanged	unchanged
Speed	increases	decreases
Speed energy	increases	decreases
Pressure energy	decreases	increases

Steamlining is used to achieve a shape change such that the flow net changes are not so large as to create discernible energy losses through such causes as the development of eddies and boundary layer separation. Even in a streamlined practical structure there will, of course, be some energy loss due to friction at the boundary surfaces but, if it is minimised, a free-flowing condition will be produced.

15.4.6 Temperature

Temperature induces structural strains and stresses through the expansion and contraction that occurs with temperature changes. These are in the same category as the strain-induced effects caused by creep and shrinkage of concrete (Section 8.9.2). The temperature-induced stresses and strains in a bridge will thus depend on:

* the bridge structure,
* the material used, and
* the temperature variations which occur.

The role of the bridge structure relates to the degree of restraint imposed upon temperature-induced deformations. For example, uniform heating of the beams in Figure 15.1 will cause longitudinal expansion of the members. Any restraint on this expansion will create longitudinal compressive stresses in the beams. Movements in excess of about 5 mm are usually alleviated by the use of transverse *expansion joints* (Section 15.2.5). Likewise, temperature gradients caused by solar heating of the deck or cooling of the exposed undersides of the beams will cause the beams to bow upwards. Any rotational restraint due to end fixing will then cause moments in the bridge beams.

The role of the bridge material is that:

(a) its coefficient of thermal expansion will determine the amount of strain occurring, and

(b) its post-elastic behaviour will determine whether the bridge has any permanent memory of the restrained temperature deformation.

With respect to temperature variation, for most bridges vertical heat flow through the deck and beam depth will be of greater significance than horizontal heat flow, as the depths are small relative to the plan dimensions. At any given time, it can usually be assumed that the temperature in a bridge varies only with the depth of the bridge. A common approach is therefore based on two design variables; an average temperature and a temperature variation through the bridge cross-section. There is a need to compare the average temperature with a base temperature in the bridge, which can be taken as the temperature at which structural continuity was established. Variations in the average temperature from the base will result in the longitudinal strains discussed above.

Maximum bridge temperatures generally occur towards the end of a hot day and minimum values on a cold morning. Maximum temperature differentials are usually reached a little earlier in the afternoon than the maximum average temperature. Minimum temperature variations occur early in the morning.

15.4.7 Footbridge loads

The design of footbridges follows similar principles to the design of vehicular bridges except that:

(a) the loads are much smaller and hence the bridges are relatively lighter, and

(b) pedestrians are much more sensitive than motorists to the motion of the bridge. Site geometry usually dominates the design. The vibratory motion that disturbs the pedestrian is usually excited by their own passage, and the design criterion is their subjective reaction rather than any structural overstress. Usually, if a pedestrian can perceive a vibration, it will cause some negative personal reaction. The design solution is to reduce either:
(i) the magnitude of vibrations, by an increase in stiffness, or
(ii) the time over which they are significant, by the provision of *damping*.

The force which a pedestrian imposes on a bridge ranges from 120 percent to 270 percent of their static mass, over a spectrum from *slow walking* to *running* (Wheeler, 1980). The time over which this peak force occurs decreases from 600 to 100 ms (i. e. from about 2 Hz to 10 Hz), over the same spectrum. The travel speed, incidentally, increases from 0.75 to 8 m/s (or 5 footfall/s). The data are illustrated in Figures 15.7 & 8. There is also a lateral force of about a tenth of the vertical force applied at about half the footfall frequency. These vertical and horizontal forces may correlate, due to the synchronisation that sometimes occurs naturally with a large crowd.

Acceptable user limits are not easy to establish. Rapid damping has been found to render otherwise unacceptable vibrations acceptable. Vibrations are unacceptable if (Wheeler, 1980):

(vibration frequency in Hz) $\geq$ C/(maximum displacement in mm)

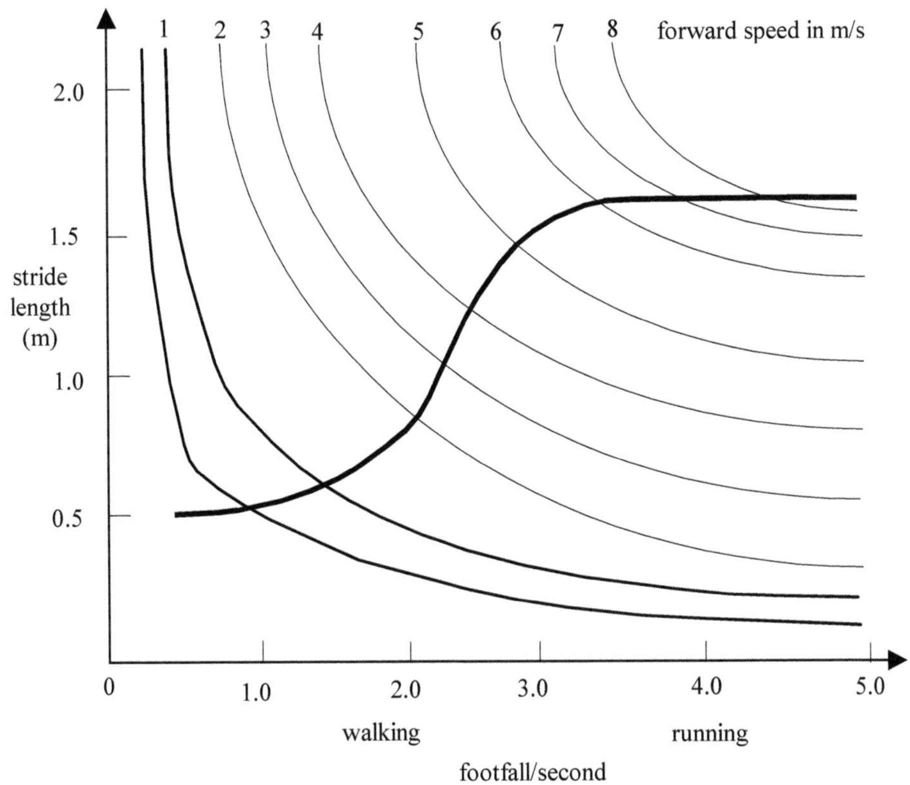

Figure 15.7 Characteristics of pedestrian movements.

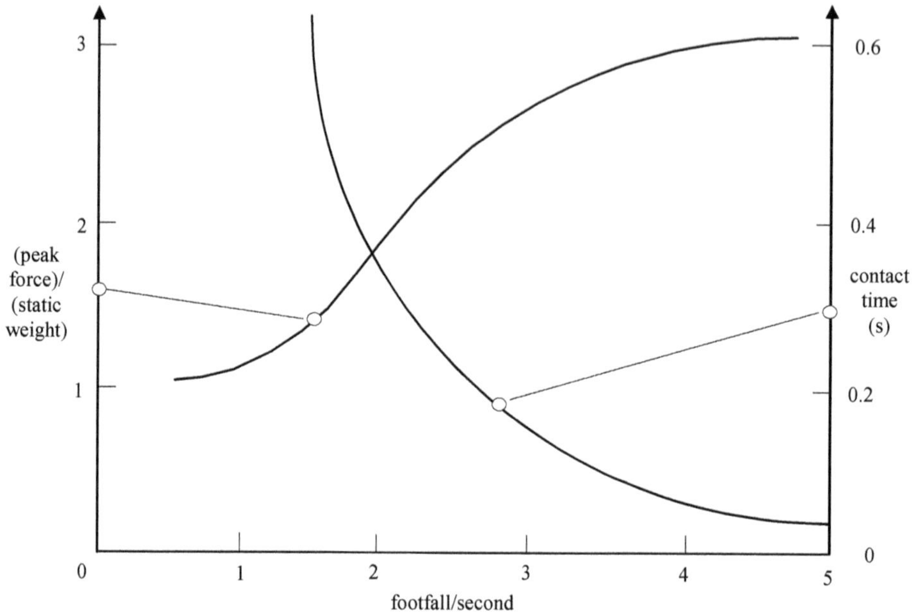

Figure 15.8 Design data for pedestrian characteristics (Wheeler, 1980).

where C = 12 for vibrations which are definitely perceptible and C = 34 for ones which are unpleasant. Thus, a 5 Hz vibration is unpleasant if its amplitude exceeds 7 mm. Furthermore, if the pedestrian footfalls occur at a frequency about equal to the natural frequency of the bridge (Section 15.4.2), their effect tends to be reinforced and disproportionately noticeable.

15.4.8 Load combinations

The main practical caution to the designer is to avoid combining loads which each have a low probability of occurrence (e.g. a flood and a temperature extreme) and whose joint probability would be insignificant (Lay, 1980b). Even many events with recurrence intervals less than the design life of the structure are unlikely to occur simultaneously, although any such assessment must take account of the actual duration of the events in question.

The commonsense approach embodied in *Turkstra's Rule* is to work through each transient load type in turn, using combinations of the ultimate load of that transient load type, the permanent loads, and the common levels of all the other transient loads, to find the combination giving the design ultimate load.

15.5 BRIDGE CONSTRUCTION AND MAINTENANCE

The general principles of construction management are outlined in Chapter 25. However, for bridgework special attention should be given to:

* the provisions for setting out the job,

* constructing falsework (experience suggests that construction staff may pay inadequate attention to falsework strength and stiffness),
* working in traffic,
* working in water, and
* supervision of prestressed concrete construction (Section 15.2.1).

Once a bridge is constructed, it must be entered into a bridge inventory system containing details of the bridge's:

(a) location,
(b) design calculations, as-constructed drawings, and construction diaries,
(c) traffic loading history,
(d) subsequent structural alterations,
(e) damaging extreme events, and
(f) administrative and management arrangements.

It is then necessary for the bridge to be routinely inspected in order to determine maintenance needs. The inspection reports are also added to the bridge inventory file. Inspection should follow carefully defined procedures and schedules. Typically, bridges are given a routine observational check annually by regular inspection staff, a general inspection every two years, and a major inspection every five years by technical experts. Bridges should also be inspected after any damaging event. Reports should cover the bridge's:

(1) condition,
(2) current capacity,
(3) need for immediate repairs,
(4) need for any usage limits, and
(5) need for future maintenance.

Inspectors should be trained in 'what to look for' and have some knowledge of materials and structural behaviour. Particular attention should be given to older bridges, timber bridges, and bridges in regions where de-icing salts are used. Timber may need to be regularly cored to monitor its internal soundness. U.S. data suggest that the percentage of deficient bridges in an age group is approximated by (OECD, 1983d):

percent deficient = 10 + (age in years)

As a result of inspection, it should be possible to prioritise maintenance work on the basis of the size of the product:

(probability of failure, p_f, Section 15.3.2) x
(cost of failure due to the work not being undertaken).

The probability of failure will depend on the design factors given in Section 15.3.2, modified by the inspectors' view of the bridge's (a) current condition, and (b) its propensity for further degradation. Factor (b) will relate to such factors as diverse as the bridge environment and the actual traffic loading on the bridge — as compared with the loading assumed in design.

Maintenance in general is discussed in Chapter 26. Bridge maintenance falls into three main categories:

(1) ordinary maintenance such as cleaning the bridge and its drainage systems, repairing the traffic surface, and repairing the effects of traffic damage to railings and similar devices;
(2) specialised maintenance such as painting steelwork, patching concrete, replacing joints and bearings (Section 15.2.5), and renewal of waterproofing and drainage systems; and

(3) exceptional maintenance such as correction of settlement, scouring of piers, or a weakening of structural members.

These three categories together will typically require an annual budget of about 1 percent of the initial cost of the bridge. Over the life of the bridge, this means that the present value of all the maintenance costs may be about 50 percent of the construction cost.

A review of methods for evaluating of the load-carrying capacity of bridges is given in OECD (1979) and techniques for rehabilitating and strengthening bridges are reviewed in OECD (1983d). Whilst rehabilitation can improve the condition of a bridge, field experience is that subsequent deterioration rates of bridges over 50 years of age will be greater than with a new bridge.

Bridges are typically designed for a life of 100 years (Section 15.3.2) and field evidence is that many bridges will have suffered severe deterioration by the end of this period.

CHAPTER SIXTEEN

Driver Behaviour

16.1 THE DRIVER–VEHICLE–ROAD SYSTEM

Understanding the behavioural characteristics of the driver is an essential component of an overall understanding of the road system. Indeed, the efficient operation of road traffic ultimately depends on the performance of the human users of the system. This chapter therefore draws together the common aspects of driver behaviour in order to lay a basis for the subsequent chapters which cover such specific aspects of driver behaviour as normal driving (Chapter 17, Traffic flow), speed selection (Chapter 18, Speeds), driving on curves (Chapter 19, Road geometry), driving through intersections (Chapter 20, Intersections and Chapter 23, Traffic signals), visual issues (Chapter 21, Traffic signs and Chapter 24, Lighting) and driver–vehicle interactions (Chapter 27, Vehicles). Moreover, a number of major aspects of driver behaviour such as crash causation, the influence of alcohol and social maladjustment, risk-taking, and the potential for behavioural modification are discussed in Chapter 28 on safety.

Overall, driving is part of the driver–vehicle–road system shown in demand and task terms in Figure 16.1. This system is:

* based on the basic psychological paradigm of:

  input→stimulus→organism→response→output

  or,

  information→decision→action→observation

  and often expressed by the equation:

  response = sensitivity x (stimulus)
* designed on a demand and task basis, although Section 16.3.2 will show that the driver behaviour component does have *cognitive-motivational* basis,
* driver-controlled,
* based on information inputs which are mostly visual,
* *closed-loop* and *cognitive*, in an overall sense in that the driver receives feedback by sampling the road and vehicle parts of the system.

  [A closed-loop system responds to feedback received during operation. An open-loop system, on the other hand, cannot be further controlled once it is activated. Open-loop systems are more like fired bullets than cars on the road.]
* *open-loop* and *precognitive* in between the driver's intermittent sampling of the external environment. Indeed, Section 16.4.5 shows that *sampling* rates will not exceed approximately 2 Hz, or once every 500 ms. The precognitive components cover a driver's 'automatic' conduct of practised, skilled manoeuvres appropriate to a particular set of circumstances. Any untoward event during open-loop control will see the driver revert back to the basic closed-loop model, and so the overall feedback system will continue to operate.
* based on negative, compensatory feedback, in that the information it provides about errors is used during the control process to reduce those errors. Such systems without continuous error-correction are known as *supervisory systems*.

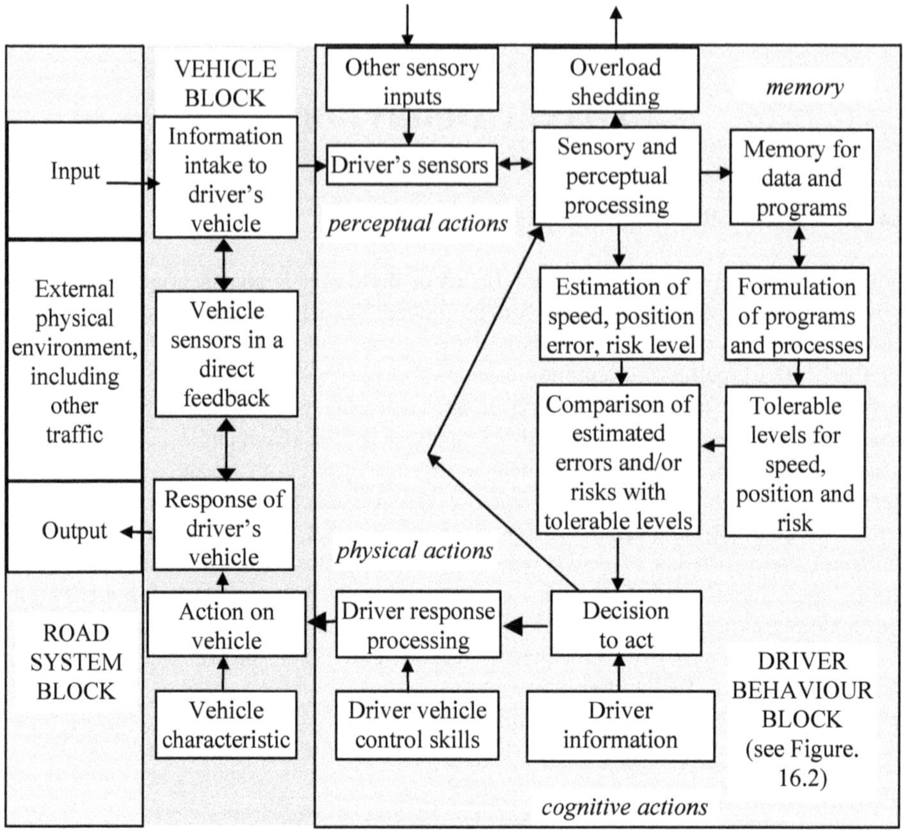

Figure 16.1 Demand and task model of the driver–vehicle–road system.

* capable of extension to a *pursuit* control-model (Section 16.4.6), in which information on the way ahead allows for forward feedback to produce short-term strategic actions,
* unable to routinely handle out-of-range inputs or uncommon events,
* relatively inflexible, and
* reliant on very interdependent physical parts.

16.2 DEFINING DRIVER BEHAVIOUR

The driver's information→decision→action→observation system discussed in Section 16.1 is shown as the right-hand block in Figure 16.1. The individual processes managed by that system can be subdivided into the following nine tasks:

(a) *strategy selection.* This determines the choice of departure time, transport mode, and overall route. Strategy is pursued further in Chapter 31.

(b) *navigation.* This determines which roads to select, and then ensures that the correct journey path is maintained. To some extent navigation is discussed in Section 21.3, but is not a major part of this book.

(c) *vehicle guidance.* This determines which piece of the selected road to use and is discussed in Sections 16.5.1 & 21.3.

(d) *vehicle manoeuvring*. This determines how to move from one spot on the road to another and is discussed in Sections 16.5.1 and 20.2.1.
(e) *vehicle control*. This determines the incremental changes in vehicle position, resulting in a vector rate of change of vehicle position, and is discussed in Sections 16.5 & 16.6.1.
(f) *compliance with traffic rules and laws*. This is an attitudinal issue determined following principles outlined in Sections 16.7 and 28.2.4.
(g) *interacting with other traffic*. This determines relationships with other vehicles and with the roadside and is discussed in Chapters 17 and 28. This process becomes quite slow in older drivers and pedestrians.
(h) *using peripheral communications*. This includes observing the view, and using car radios, navigation devices and telephones.
(i) *responding to unexpected emergencies*.

Initially, a driver learns task (d) followed by task (e). These skills then allow task (c) to be undertaken.

Driver behaviour within each of these tasks can occur at four levels:

(1) *skill-based tasks*, employing precognitive *psychomotor skills*. This supplies automatic or routine or habit-based responses in familiar environments, without conscious reflection on the part of the driver. These responses often do not require the full decision sequence in Figure 16.1 to be activated, as stereotype behaviours are called into play. This level includes actions in tasks (c) to (e), and the sensing component of all tasks. The performance of skill-based tasks will improve with training and practice. Performance deteriorates with age and the tasks are more difficult for older drivers to acquire.
(2) *rule-based tasks*, based on the selection and application of some behavioural rule. This level includes actions in tasks (f) to (g), and the perceptual component of all tasks. They are used when the skill-based tasks in (1) are completed and/or inappropriate.
(3) *knowledge-based tasks*, based on deductive reasoning. They are used where neither tasks (1) nor (2) apply. Their use is always a conscious decision. This level includes most actions in tasks (a) & (b), any situations which are new to the driver, and the decision component of all tasks.
(4) *style-based tasks*, where the driver has clear preferences as to how some tasks should be executed (Section 16.3.2). This level particularly covers the selection of acceptable risks and attitudes to other drivers. It relates to the driver's goals and objectives in a broader social context.

Clearly, the more experienced a driver is in a particular process, the more behaviour tends towards tasks in level (1).

Overall, driving is a relatively simple task. However, the driver behaviour model indicates that a driver is burdened with the following basic problems:

* There may be inadequate input available for the task at hand, e.g. while driving at night (Section 24.4.2), encountering hazards on the road, or negotiating complex intersections (Section 20.2).
* The driver may sample inappropriate input as a result of distraction or inexperience.
* The driver may process input too slowly to usefully act on it. For instance, items 1 to 4 in Table 16.1 below show that when a significant proportion of drivers are presented with a visual input, they will take about 2.7 s to reach the decision-making stage. Thus traffic engineers often allow a driver 3 s between key decisions.

* Input overload is dealt with by simply shedding some of the required tasks in order to deal with that part judged to be more important. This may result in inappropriate responses when inexperienced or older drivers encounter complex traffic situations. [To tackle this problem requires a knowledge of task demand but this has not proved easy to measure. Assessments of a driver's performance of secondary tasks have been tried fairly widely but with limited success.]
* When confronted with novel multiple tasks, the driver must often choose which task to give priority to, and decide how to tackle two tasks simultaneously.
* The processes of making comparisons and reaching decisions based on the input received are both affected by stress, arousal, conditioning, motivation, and type of input (Section 16.3.2).
* The driver may ignore messages and signals.
* The driver may make serious errors.

These eight problems all carry lessons for the road and vehicle engineer. Nevertheless, most driving is of a routine nature, with the driver operating at skill-based level (1), observing simple cues, and responding according to habit. In other cases, the processes involved are complex and it is not possible to precisely predict either the information a driver will capture or the form of any decisions based on that information. Even driving at level (1) is not quite 'routine', as most drivers experience a heart rate increase of at least 12 percent.

16.3 DRIVER CHARACTERISTICS

16.3.1 Introduction

Section 16.2 lists the nine key tasks associated with driving and the four levels at which they are performed. The manner in which they are performed depends largely on the characteristics of the individual driver. These characteristics must be taken into account in considering both normal traffic behaviour and the type of abnormal behaviour that leads to many traffic crashes. They can be considered under four major headings:

(a) *psychological traits* such as intelligence, learning ability, motivation, desires, temperament, emotional stability, and attitudes (Section 16.3.2).
(b) *sensory abilities* such as vision and hearing (Sections 16.3.3 & 16.4).
(c) *physical abilities* such as skills (Section 16.2), training (Section 16.6.1), response time (Section 16.5.3), and mechanical limitations on body movement.
(d) *medical factors* such as the influence of drugs, alcohol (Section 28.2.2), disease, fatigue (Section 16.5.4) and physical impairment.

16.3.2 Psychological traits

The role of *psychological traits* in driver behaviour is to exert a direct influence on the driver-behaviour block of the task-demand model shown in Figure 16.1. This is illustrated in Figure 16.2, which demonstrates how a consideration of psychological traits moves the task-demand model of Figure 16.1 to a cognitive-motivational sub-model.

In studying these influences, it cannot be assumed that driving is different from any other human task. Drivers bring to the driving task all the personal traits in Figure 16.2 and these will usually reflect not merely their personal views, but the views of their

Attributes	**Needs and expectations**
Physical * skills, experience, training; * visual and response capacity; * fatigue, influence of drugs; * health; * living conditions.	*Emotional* * personality and attitudes; * social values, e.g. age- and gender-related; * social norms, peer-group pressures; * lifestyle; * transient emotional state.
Cognitive * self-control; * response to distractions, within and outside the car; * response to overloads.	*Motivational* * momentary, e.g. hedonism; * short term, e.g. preoccupations; * long term, e.g. trip goals; * general, e.g. attitudes to risk.

Figure 16.2 Personal modulating factors in the cognitive-motivational model of driver behaviour. This is the right-hand block in Figure 16.1.

family, their peer group and, often, society as a whole. How are such traits acquired and can they be changed? Three that need particular discussion are:

* *Social values* which are typically influenced by family, friends and schooling. Once acquired, they are not easily changed.
* *Social norms* particularly involve the driver's perception of what the relevant peer group will think of an intended driving action. That is, drivers will usually behave in a manner which they assess to be favoured by those who are important to them, such as their friends and people they admire.
* *Attitudes* are internal states of mind, usually based on social values, and which predispose a person to respond in a particular way; thus, in driving attitudes relate to how drivers view themselves and others. Indeed, it is possible to discern five major sets of driver attitudes as defined by the following spectral extremes:
  1. law-abiding or law-breaking,
  2. socially-concerned or unconcerned for others,
  3. tolerant or intolerant of others
  4. cooperative or selfish, and
  5. strategic or interested only in immediate needs.

  In life, many attitudes are learnt 'across the kitchen table'. In driving they are often passed over the back seat of the family car. Because they are part of the driver's social values, they will be difficult to change.

All these personal traits will influence and often determine the driver's performance of each driving task, from the immediate traffic effects of a decision to the long-term social consequences of a planned driving manoeuvre. The important road safety implications of this reality are discussed further in Sections 16.6.1 & 28.2.3.

As illustrations of the broad significance of societal influences, consider that — on top of the simple task of driving from A to B — the driver self-imposes another set of motives. Examples are:

in the long term:

(a) *goal fulfilment* (e.g. minimise the journey time for a trip),

(b) *satisfaction* from driving (e.g. an English study of car drivers found that 63 percent thought that driving was pleasurable, and 7 percent often drove simply for the pleasure of driving, Hallett, 1990),

(c) *exhibitionism* (e.g. displays of social status, risky overtaking),
(d) *hedonism* (e.g. competing in traffic),
(e) *risk-taking* is discussed further in Section 28.2.4. The level of deliberate risk-taking — or, alternatively, the degree of prudence exhibited — is a personal characteristic.
(f) *need-fulfilment* (e.g. the car may be a compensation for inferiority, a substitute for other goods, a source of privacy, self-assertion, power or a virility symbol).
(g) *peer-group* pressure (e.g. travelling at a speed that permits subsequent boasting rather than apologising),
(h) *attitudes* (e.g. a general desire to show aggression or care towards other people),
(i) *emotions* external to the trip (e.g. pre-trip arguments, social stresses),
(j) *intellectual commitment* (e.g. the maximum level of mental effort that the driver is prepared to devote to driving).

and in the short term:
(k) *goal fulfilment* (e.g. pass the slow vehicle ahead),
(l) *emotions* internal to the trip (e.g. annoyance at another driver, impatience).

A key factor, which illustrates how much driving depends on all these motivations, is that much driving, particularly in rural areas, involves self-paced tasks. That is, the level of task-demand, measured by its complexity and extent, is essentially under the driver's control. For instance, it is the driver's choice of speed, headway, and lateral position that determines the difficulty and risk of the driving task. In particular, a driver's choice of speed is a powerful measure of driving attitudes and skills. The driver is thus an active, creative part of the traffic system and not merely a passive responder to external events. Consequently, many of the stresses that a driver experiences whilst driving stem from the driver's own self-determined actions.

Another example of the role of psychological traits is that drivers often do not use all the information carefully made available to them. For instance, Section 21.2.3 discusses how their awareness of a traffic sign depends on the relevance they subjectively assign to the sign.

Children have a different range of psychological traits:
* From 2 to 7 years of age their behaviour is characterised by an egocentric concentration on the immediate present. This makes them unreliable in traffic.
* From 7 to 11 they are able to take a more abstract view of traffic and anticipate future situations and risks. Nevertheless, they cannot be relied upon to base their behaviour on a process of logical reasoning from general (e.g. safety) principles.
* After 11 they begin exhibiting adult traits. However, their size limits their ability to see and be seen and their visual capabilities are still underdeveloped (see Sections 16.3.3 & 16.4.2&6). This means that they still cannot be treated as adults in traffic.

16.3.3 Acquiring information

The driver has various sensors for receiving information from both outside and inside the vehicle–driver part of the overall system. The *internal* information can be from both system feedback and from the driver's memory (Figures 16.1 & 2).

The *external* sensing is by vision, hearing, and touch. Of these senses, vision provides about 90 percent of a driver's information, and is the only channel for course-keeping, detecting obstacles, and obtaining information from signs (Chapter 21), pavement markings (Chapter 22), traffic signals (Chapter 23), and delineation (Section 24.4). It is the subject of Sections 16.4–5. Of the other sensory inputs, drivers have fairly accurate acceleration sensors and can detect changes of 0.01 g (0.1 ms^{-2}). However they are not inherently good judges of the speed of either their own vehicle or of any other vehicle. Speed is usually assessed by learnt visual cues (Section 16.4.6) or from sounds generated by the vehicle. Thus hearing does help in speed determination and also provides cues concerning unexpected events. In the future, it is probable that hearing will be further utilised for conveying system messages to drivers. Touch helps via the feel of the vehicle 'ride' (Section 14.4) and the feedback the driver receives through the steering wheel (Section 16.5.1).

A driver does not need to observe the road ahead continuously and Section 16.4.5 discusses minimum, feasible visual-sampling rates of about once every 500 ms (or every 14 m at 100 km/h). Examples of visual sampling leading to manageable interruptions to the scene ahead are eyeblinks, making visual observations well away from the road scene, viewing vehicle controls, turning to look at passengers, using the rear-vision mirror, and driving in heavy rain with the windscreen wipers operating.

This sampling approach to the acquisition of information is possible because the driver often has sufficient information to perform the vehicle control task without intermediate input or immediate output. Obviously, the more the driver is aware of the scene ahead, the less it will be necessary to sample it. Thus, local knowledge, traffic memory, the complexity of the scene ahead, perceptual ability (Section 16.4.2), ambient light level, and vehicle speed will be important factors in determining the sampling that does occur. Sampling will also be affected if a driver:

* is fatigued (Section 16.5.4) or alcohol-affected (Section 28.2.2),
* is insufficiently vigilant and fails to pay sufficient attention to the driving scene,
* pays too little attention to critical tasks,
* is unable to divide attention between two or more critical tasks,
* attempts a non-driving task such as looking for an object in the vehicle, or
* is distracted by less relevant events.

Inattention is a significant factor in road crashes (Section 28.2.1).

With respect to ambient light, as the light level falls, it will be shown in Section 16.4.3 that visual performance deteriorates and contrast sensitivity, acuity, distance judgement, speed of seeing, colour discrimination, and tolerance to glare (Section 24.2.2) are all impaired. Critical luminance levels are discussed in Section 21.2.2.

There will be times of high driving-demand when the limit of the driver's sampling ability is reached. These are obviously critical times for a driver. Their management is simply explained by a single-channel information processing model in which, in the context of Figure 16.1, all processing in any one link is done sequentially and the system components attend to only one input at a time. The single-channel links have limited capacity and so the driver can only deal with a limited quantity of information in a given time.

However, the single-channel model does not account for many observations of people undertaking simultaneous tasks with high information-inputs and high action-outputs. Examples range from simultaneously reading and eating, to complex multi-task actions in music and sport. These suggest that, with practice, people are able to receive data simultaneously from a number of channels and produce multiple outputs. This multitask model explains many of the rehearsed but spontaneous responses of experienced drivers. However, some functional limit to multitask performance will obviously exist. For example, difficulties often arise if:

* the tasks have not been rehearsed,
* one of the tasks is intellectually demanding,
* the two key tasks have a different rhythm or timing, or
* the stimulus for one of the tasks clearly precedes the others.

Driver workload is a vague term sometimes used to describe the number (or length?) of tasks a driver must accomplish in a given time. Experience suggests that driver performance suffers when the work load is relatively high (taking priority away from some driving tasks) or relatively low (causing inattention, Section 28.2.1).

When information is received, but not immediately required, it is held in short-term memory for about 30 s. After this time, it is forgotten unless reinforced or rehearsed.

16.4 DRIVER VISION

16.4.1 The eye

To explain vision, it is first necessary to describe a few features of the eye. The lens of the eye is called the cornea. It is made from a hard, transparent material and is set at a curve at the front of the eye. The curvature is changed by muscular action and this alters the angle at which light is refracted through the cornea. However, the bulk of the refraction occurs at the air/cornea interface and the muscular action only produces a fine-tuning effect. As the lens ages, it becomes stiffer and harder to focus.

The iris is a shutter in the lens which has an opening known as the pupil. The iris contracts and dilates, depending on the available light. The pupil reduces in size as ageing occurs until it reaches about 10 percent of its original area in extreme old age. It is also less able to dilate in low light conditions. In addition, the corneal lens 'browns' and also reduces light throughput. Hence, old people can suffer losses of 90 percent in available light — causing a similar effect to wearing very dark glasses, or two pairs of normal sunglasses. The phenomenon makes night driving particularly difficult for many old people. As poor vision does not appear to be a major factor in crashes involving elderly drivers, these drivers would appear to satisfactorily compensate for their loss.

The light which passes through the lens follows a fairly clear path to the retina, a relatively spherical backing to the eyeball with millions of small photosensitive receptors standing up from its surface. The receptors are divided into cones and rods which both respond to in-coming light by sending signals to the brain. The cones respond to ordinary light levels and to colour, and not to low light levels. They perceive electromagnetic wavelengths between 380 and 780 nm. The rods respond only to low light levels (< 1 cd/m^2, Section 24.2.2) and are insensitive to colour, perceiving only 505 nm radiation. Furthermore, the response of individual rods is summed before being sent to the brain, and so visual acuity is much poorer than for the cones. The retina thus operates as a duplex system of cones and rods.

Both visual acuity and colour sensitivity (Sections 16.4.2&4) are at their peak in the fovea, a dished central region of the retina which has an angular radius of only 1°. The eye normally uses the fovea and a driver's line-of-sight passes through its centre, called the *fixation point*. Visual acuity (Section 16.4.2) and colour sensitivity (Section 16.4.4) are still quite good within 3°, reasonable within 10°, and have some value within 35° of the line-of-sight. The region that can be usefully seen whilst looking straight ahead along the line of sight is called the *central visual field* (or *functional field of view*). A driver uses central, foveal vision for the tasks of steering, braking, accelerating, and navigating.

Eye movements represent the first step in examining areas initially outside the small (±1°) central visual field. The movements typically take about 50 ms (Section 16.5.3) and cover a region within ±15° of the line-of-sight. Larger scans are undertaken by head movements. These are much slower than eye movements (typically 700 ms) and drivers often depend more on eye movements and peripheral vision.

Peripheral vision allows the eye to detect objects within 95° of the line-of-sight. Detail discrimination in the peripheral visual field is poor (e.g. visual acuity at 50° is only 25 percent of the foveal value) and objects usually need to move to be detected. The extent of available peripheral vision is influenced by a number of factors:

* *gender*. Women have a larger peripheral field than men.
* *age*. Peripheral vision is not well developed in children and hampers their performance in traffic. At the other extreme, the adult peripheral field moves 10° closer to their line of sight in old age.
* *alcohol*. Impairment begins at a blood alcohol content of 2 gm/L (Section 28.2.2).
* *speed*. For a person travelling at 30 km/h, the peripheral field has dropped to within 50° of the line of sight, at 60 km/h to 40°, and at 100 km/h to 20°.
* *vehicle design*. Features such as window columns can restrict peripheral vision.
* *object being detected*. Response to movement in the peripheral field is enhanced when the objects being detected are large and/or highly contrasted.
* *light level*. Peripheral vision disappears at low light levels.
* *observer*. Response to movement in the peripheral visual field is reduced when complex visual decisions are being made in the central visual field.

Peripheral vision is used by drivers to detect potential hazards outside the central visual field where attention is focussed — such as pedestrians and traffic coming from cross streets — and assists in judging vehicle speed (via streaming, Section 16.4.6).

Given this general background, the following sections discuss the human factors that determine a driver's inherent visual capability.

16.4.2 Visual acuity

Visual acuity is the ability of the eye to see fine detail. It varies markedly between individuals, and with the level of illumination, the length of viewing, and the eccentricity of viewing (Section 16.4.1). It is measured by either (Figure 16.3a):

* the angle subtended at the eye by the smallest detail that the eye can see and resolve (called the *minimum angle of resolution*). People with normal vision are able to distinguish high-contrast detail with a minimum angle of resolution as low as 150 μrad, although the value covering 85 percent of people below age 45 — the so-called 'average young eye' — is closer to 300 μrad. [Luminance contrast is defined in (b) below.]
* the ratio of a standard distance to the furthest distance at which an average young eye can see the smallest detail that the eye under test can see at the standard distance. The distance components of this definition are the inverse of the relevant minimum angle of resolution.

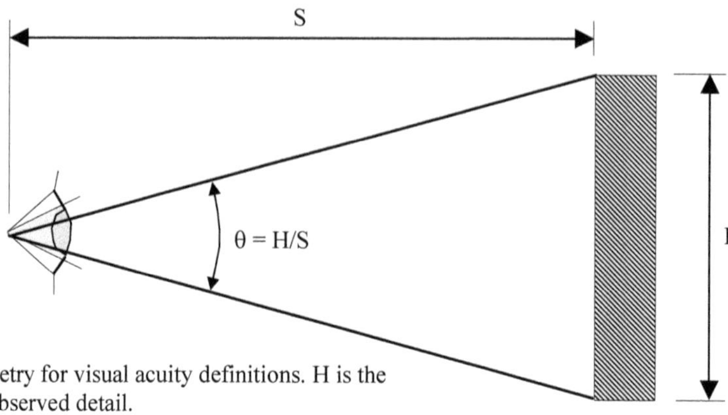

(a) Basic geometry for visual acuity definitions. H is the size of the observed detail.

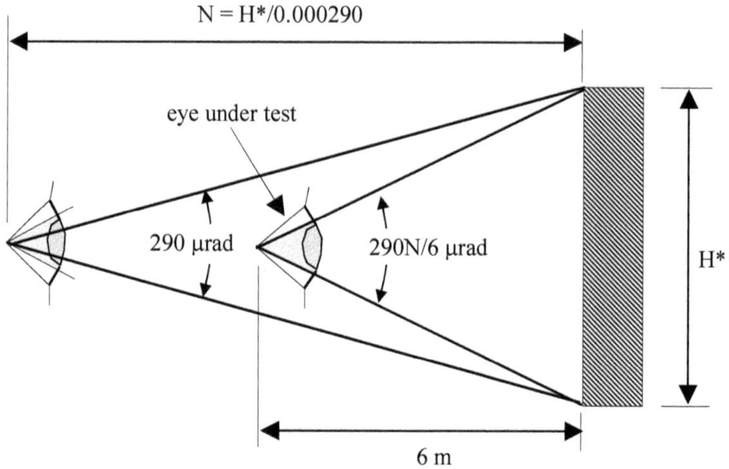

(b) Usual definition of visual acuity. H* is the smallest value of H that can be seen with the eye under test placed 6 m away.

Figure 16.3 Visual acuity definitions.

Visual acuity according to the second definition is commonly measured by the *Snellen visual acuity* test that determines a person's ability to see high-contrast black letters on a bright white screen. The results are given as 6/N where the 6 indicates that the letters are viewed by the eye under test from 6 m away — the 'standard distance' in the second definition above — and N is the distance at which the smallest detail thus resolved subtends an average-young-eye angle of 290 μrad. This is illustrated in Figure 16.3b where it is seen that high visual acuity is associated with a small N. To restate the definition in two other ways, a person with 6/N vision:

* must stand 6/N as far from a sign to resolve the same smallest detail as can a person with 6/6 vision, or
* can resolve small detail, which is N/6 as large as, can be resolved by a person with 6/6 vision.

Low visual acuity results in a blurred, unfocussed image (e.g. about 0.1 percent of the population cannot see a nearby pedestrian signal light). It arises from the following key factors:

* inherent ocular muscle defects
* diseases such as cataracts, glaucoma, diabetic retinopathy, field of view losses or strokes.
* age. The ability of the eye to focus universally declines from the mid-forties onwards. Thus the proportion of the population with 6/6 vision drops from 85 percent at age 40 to 70 percent at age 50, 55 percent at age 60, and 30 percent at age 70. A similar visual decline occurs in the ability to adjust for distance (presbyopia).
* alcohol. Visual acuity drops with blood alcohol contents above 0.2 gm/L (Section 28.2.2).

Poor visual acuity can be corrected by the use of spectacles supplying supplementary refraction. Overall, it has been estimated that:

* about 30 percent of the total population have visual acuity below 6/6, even when wearing their usual spectacles and
* 6/12 vision must be assumed, if the capabilities of 95 percent of the driver population are to be covered.

Thus, a common design assumption is that drivers — possibly wearing spectacles — have at least 6/12 vision in at least one eye. Licence tests generally require between 6/9 and 6/20, with 6/12 being common. Of course, drivers are able to use spectacles to attain the acuity levels. Indeed, visual impairment is only considered to exist when a person's corrected visual acuity is worse than 6/12.

Section 21.2.4 shows that low visual acuity will decrease a driver's ability to read traffic signs and will increase the time taken to respond to them

Visual impairment is considered to exist when a person's visual acuity — even when corrected with spectacles — is worse than 6/12, and *low vision* occurs when their corrected acuity is inadequate for them to perform essential social functions. Low vision occurs in about 1.5 percent of the population. The percentage rises to 5 percent in the 65–75 year group, 10 percent in 75- to 85-year-olds, and 55 percent in 80- to 90-year-olds. A decrease in effective visual acuity can also be caused by external factors such as:

* halation, which is the irradiation of light from a sign or signal producing a fuzziness around its edges, particularly at low levels of ambient light,
* glare from oncoming lights (Section 24.2.2), and
* low light levels, as at night and in rain, fog or snow. In low light, the eye becomes more sensitive to light but suffers a compensating loss of acuity.

As the Snellen test applies to high-contrast, well-lit conditions, it may not be a good guide to performance in the three conditions discussed above.

The effects of the reductions in effective visual acuity are discussed in Sections 21.2.2 & 24.2.1. In particular, Section 21.2.4 shows that low visual acuity will decrease a driver's ability to read traffic signs and will increase the time taken to respond to them.

16.4.3 Visual sensitivity

Visual sensitivity is the ability of the eye to detect differences in light levels. These levels are measured in terms of their *luminance* in cd/m^2 (Section 24.2.2). The eye is able to detect light stimuli over the entire range of luminance levels encountered between day and night, i.e. from −6 to +6 log cd/m^2. However it takes up to 30 minutes to achieve maximum night-time sensitivity, although the major effect occurs after only 200 ms. This adaptation to low light levels is known as *scotopic vision* and occurs when the luminance is below 10 mcd/m^2 (or −3 log cd/m^2). Normal day vision occurs when the luminance is over 10 cd/m^2 and is called *photopic.* Twilight vision occurs between 10 mcd/m^2 and 10 cd/m^2 and is called *mesopic.* Most night driving is also done in mesopic conditions.

The *threshold luminance contrast* is the minimum difference that the eye can detect in the luminances of an object and its background. This determines the eye's ability to draw useful messages from the light signals received. It is measured by contrast sensitivity tests. The *threshold luminance contrast ratio* is this luminance difference divided by the background luminance and the *threshold luminance contrast value* is the difference divided by the sum of the two luminances. Practically, the latter is half the former. A typical threshold luminance contrast ratio is about 2 (luminances increase exponentially, Section 24.2.2).

Section 16.4.1 noted that in conditions of low luminance — such as at night — the eye becomes more sensitive to light but suffers a compensating loss of resolving ability. This decrement occurs at luminances of 30 cd/m^2 or less and becomes dramatic below 3 cd/m^2. About 5 percent of the population are visually deficient with respect to detecting low luminance contrasts. The ability declines with age, particularly for people over 40 years, and deteriorates with alcohol intake, thus making the reading of signs more difficult for drivers who are old or alcohol-affected. For instance, for elderly drivers the threshold luminance contrast ratio is about double that of 'normal' drivers.

16.4.4 Colour vision

The eye detects colours mainly in terms of their:

* *hue*, which relates mainly to the electromagnetic wavelength of the radiation,
* *brightness* (Section 24.2.2), and
* *saturation* (or *purity*), which relates to how much white they contain.

Measures of perceived colour (or *chromaticity*) are given in Section 21.3.1. In addition, the eye's response is far from constant as a process of adaptation goes on in which the properties of the eye are modified according to the luminances of the colour stimuli presented to it. *Light adaptation* occurs above a few cd/m^2 and *dark adaptation* below about twenty cd/m^2. Colour vision is inoperative below 0.3 cd/m^2.

Colour vision abnormalities and anomalies (commonly called *colour blindness*) influence about 8 percent of males and 0.5 percent of females. About a third of these are *dichromates* who are likely to confuse red, yellow, and green if the colours are not carefully chosen (the problem is for wavelengths over 530 nm). Another third are *protans* who have difficulty seeing the long wavelengths present in red. Such groups are helped by a bluish tinge in the green and a red near the yellow chromaticity limit. A reduction in the luminosity of red light is aided by the use of red signals of greater intensity than required by normal drivers. Loss of colour vision becomes more prevalent with age and may accompany losses in acuity (Section 16.4.2) and light adaptation (Section 16.4.4). Loss of all colour discrimination is much rarer and can only be catered for by alternatives such as the careful use of pattern. The influence of colour on pattern is discussed in Section 21.3.1 and on traffic signal design in Section 23.2.2.

16.4.5 Visual recognition times

When a stimulus suggesting the need for visual attention occurs, a driver's basic *response time* (or *latency*) will be about 250 ms. This time will increase as the number of possible responses increases. The driver will then begin a visual search for an object related to the stimulus by using eye movements to scan the available visual field. This scanning process will take about 50 to 100 ms to complete. Section 16.4.1 showed that the need for a head movement would add another 700 ms. The size of the visual field that can be scanned by an eye movement is reduced by alcohol.

Thus, a driver will take 250 + 50 = 300 ms to have just one 'look' at the scene presented in even a centrally-located visual field. This time is much longer in children, who have poorly developed search strategies. In addition, Section 16.3.2 suggested that eye movements and the subsequent detection of objects in the visual scene would both depend upon the viewer's motivation, arousal, and experience, as well as upon more direct physical abilities. This behaviour is part of a person's *perceptual* (and *cognitive*) *style* and is measured by *field-dependence* tests such as the embedded-figures test common in psychological testing. Drivers who are field-dependent will find it relatively difficult to extract relevant information from the roadside scene.

After the visual field is scanned, the brain takes about 60 ms to merely perceive what the eye has seen. It will then usually allow the eye to dwell on a detected object (*eye fixation*) for at least 200 ms. Therefore, it is necessary to allow about 300 + 200 = 500 ms for a driver to visually locate one object.

Next, the driver may need to recognise the detected object and this typically may take another 600 ms. Thus, even a glance at some known location such as a convenient car mirror may usefully take about 200 + 600 = 800 ms. Table 16.1 in Section 16.5.3 takes all these times a stage further. Note that most of these required times are further increased by fatigue (Section 16.5.4) and by alcohol at BACs above 5 gm/L (Section 28.2.2).

These required times place major limits on the amount of new information that a driver can obtain. For instance, a driver travelling at 100 km/h will have moved 14 m in 500 ms. Required time is also linked to *Hick's law* which states that information is transferred at a constant rate. Hence the more time there is available, the greater the amount of information that can be extracted. There is an associated *economy law* that states that people only extract as much information as they believe is necessary. These laws suggest the use of simple and readily identifiable visual messages. This issue is taken much further in Chapter 21.

16.4.6 Moving objects

The visual system is also affected by the groups of factors intrinsic to the nature of the information. These predominantly relate to its form (e.g. symbolic or verbal), size, shape, luminance, colour contrast, detail, structure, and motion. With the exception of motion, each of these is discussed in Chapter 21 and is largely the province of the traffic engineer. Section 21.2 particularly provides a review of the driver's application of the visual sensory processes and emphasises that drivers must not only see, but also recognise. This section will therefore concentrate on objects moving in the driver's visual field.

Dynamic visual acuity is the detection of movement by the eye, often in the peripheral visual field. It is of relevance in detecting speed and speed changes (Section

16.3.3) and cross-traffic at intersections (Section 20.2). Section 16.4.2 indicated a typical static visual acuity of 300 μrad. Normal drivers can detect a movement of 400 μrad, which exceeds their static acuity level as movement tends to blur objects and reduce contrast. Dynamic visual acuity begins to decline noticeably in drivers over 50.

An important use of dynamic visual acuity occurs in the common driving situation of *car following*, which is discussed in a response sense in Section 16.5.3 and in a traffic sense in Section 17.2.3. When following another car, a driver must detect changes in the spacing and relative speed of the two vehicles. This is sometimes called the *pursuit* mode (Section 16.1), in distinction to the 'looking ahead' preview mode to be discussed in Section 16.5.2. A driver visually manages car following by firstly detecting a change in the size of the image of the rear of the lead car on the driver's retina. In this respect, the lack of need to detect speed changes in earlier times has meant that evolution has provided us with poor means for detecting speed changes.

Geometrically, the retinal width change process requires detecting changes in the visual angle, θ, that the rear of the lead car subtends in the driver's eye. This requires detecting changes in H in Figure 16.3, where H is now the width of the rear of the lead car and S is its distance from the driver's eye. After time δt and speed v, the angle changes to $\theta = H/(S + \Delta v\delta t)$ where Δv is the relative speed of the two cars. The rate of angular change at the eye is therefore given by (m, s units):

$$d\theta/dt = H\Delta v/S^2 \text{ rad/time} \qquad (16.1)$$

The detection threshold for this rate is about 3 ± 1 mrad/s. Below this value (S large; H and Δv small), relative speed changes will not be detected. For a typical car width of 2 m, this suggests a threshold speed change of $(S/13)^2$ km/h. Other studies suggest that:

* using learnt visual methods, speed changes can be detected to accuracies of about 8 km/h, (equivalent to S = 37 m in the model above),
* drivers can rarely detect changes until S changes by more than 10 percent.

Sedatives and tranquillisers affect a driver's ability to detect speed changes. Because drivers rely on detecting subtle changes, most car-followers operate by allowing their headways to vary above and below their desired mean, with a period of about 1 min and an amplitude of about 5 m. Using Equation 16. 1, the time to collision if the pursuit driver does not respond to a ΔV is $S/\Delta V = \theta/(d\theta/dt)$.

Although a vehicle's *speedometer* gives factual data on its *absolute speed*, drivers require cues to remind them to consult the speedometer and in some circumstances may not have time to glance at it. In these cases, a driver gains speed data from:

* changes in the subtended angle of centrally-placed, fixed objects in the forward visual field (e.g. an overhead sign), using the same processes discussed above for car-following,
* the relative movement of laterally-displaced, fixed objects in the forward visual field (e.g. roadside posts). The effect is called *optical flow* and is discussed below.
* binocular vision and changing perspectives of nearby objects,
* vehicle 'feel', aerodynamic and engine noise, and other background cues (Section 16.3.3), and
* lateral accelerations during cornering (Sections 19.2.3&4).

Optical flow (or *streaming* or *retinal cross-movement*) is often the most significant of these methods. As the vehicle moves, objects in the driver's visual field follow lines of flow radiating (or streaming) out from a point towards which the vehicle is heading. The point is called the *focus of expansion.* Objects in the foreground expand as the driver approaches them, and the move into peripheral vision. Usually, therefore, the greatest angular velocity due to vehicle motion will be in the peripheral field (see Section 16.4.1). Peripheral vision allows movement to be noticed but is not particularly useful in detecting the size of the change unless objects are within 100 m of the viewer. Thus roadside posts and trees help drivers assess their speed and, consequently, at night roadside lighting will be of more use in this respect than will reflective lighting (Section 24.2).

The same optical flow phenomenon used for speed cues also warns a driver of a likely collision with an object seen ahead. The vehicle usually must be within 100 m of the object before its detected optical flow indicates to the driver that it is not on the line of travel. For example, if the lateral displacement of an object from the line of sight is D and its distance ahead is S, then its subtended displacement angle is D/S and, following Equation 16.1, the rate of change of angle is $Dv\delta t/S^2$. A typical minimum safe lateral clearance for adjoining 3.5 m lanes would be D = 2 m, and using the above threshold rate of change of 3 mrad/s gives $S_{max} = 26v^{0.5}$, in m, s units. This gives a time before collision of $26v^{-0.5}$. For a speed of 100 km/h, this gives the critical time and spacing as about 5 s and 135 m. Table 16.1 will show that these are more than adequate, as driving experience would also indicate.

Adults predict absolute speeds fairly well, but tend to overestimate high speeds and underestimate low speeds. Children have much poorer speed estimation capabilities as it is a learnt ability.

To perform a safe overtaking manoeuvre (Section 18.4.1), a driver must also be able to estimate how far away other vehicles are. This requires a skill known as *depth perception* and is usually done by comparing the retinal size of the vehicle with data in the driver's memory. Subsequent driver predictions of vehicle spacing are very poor, with errors of up to 100 percent being common and worsening as the spacing increases.

16.5 PHYSICAL SKILLS

The role of physical ability in driver behaviour relates particularly to the driver–vehicle part of the driver–vehicle–road system. Visual ability was covered in Section 16.4 and this section covers other necessary physical abilities. The general issue of vehicle control (Section 16.2e) involving managing the steering, accelerating, and braking of a car is not discussed, as it not a significant issue with modern vehicles driven by other than inexperienced drivers (Section 16.5.5).

16.5.1 Tracking

Tracking (or *course-keeping* or *stabilisation*) is a task continually undertaken by all drivers. In the long term the issue is navigational. In the short term, tracking is a feedback response in which a driver maintains a required lateral position by moving the steering wheel in response to the effects of the previous movement of the steering wheel, often in the reverse direction. The amount of effort devoted to tracking can thus be measured by the frequency with which the driver performs a tracking manoeuvre, e.g. by the number of movements of the steering wheel that the driver makes whilst travelling down a length of traffic lane. Commonly measured quantities are the number of times the steering wheel crosses the zero angle position and the number of times the steering wheel direction reverses through a finite angle.

Typical movement frequencies for a steering wheel are around 200–400 mHz. The common 200 mHz frequency indicates that such drivers give themselves about a 5 s 'clearance' between navigation changes. Higher frequencies indicate:

* a poor driver, or
* a more difficult driving situation such as an increase in speed or a decrease in lane width.

On the other hand, a low steering frequency may indicate driver fatigue (Section 16.5.4).

Factors affecting tracking performance are fatigue, alcohol, visual deprivation, other driving task demands (such as reading signs), distractions, external disturbances (such as wind gusts), rough roads, and poor steering.

Drivers use lane markings, kerb lines, or their perceived equivalent as their cues for determining lateral position. Peripheral vision (Section 16.4.1) plays an important role, as a driver will primarily focus on the way ahead. Steering within a defined traffic lane is usually a closed-loop process in which the driver determines the vehicle's future path relative to its intended path, within loosely constrained physical and subjective bounds associated with the longitudinal lane markings. The driver then adjusts the vehicle's controls accordingly, without attempting to minimise the actual directional error. Lane boundary-crossings and lane position-keeping both deteriorate with driver age, mainly as a consequence of a deterioration in peripheral vision.

For moderate lane widths and speeds, drivers use a strategy of dominantly controlling the direction (or heading angle) of their vehicle, with steering interventions at about 200 mHz (as above). This manoeuvre is only indirectly linked to the lateral position of the vehicle. The situation is illustrated in Figure 16.4a.

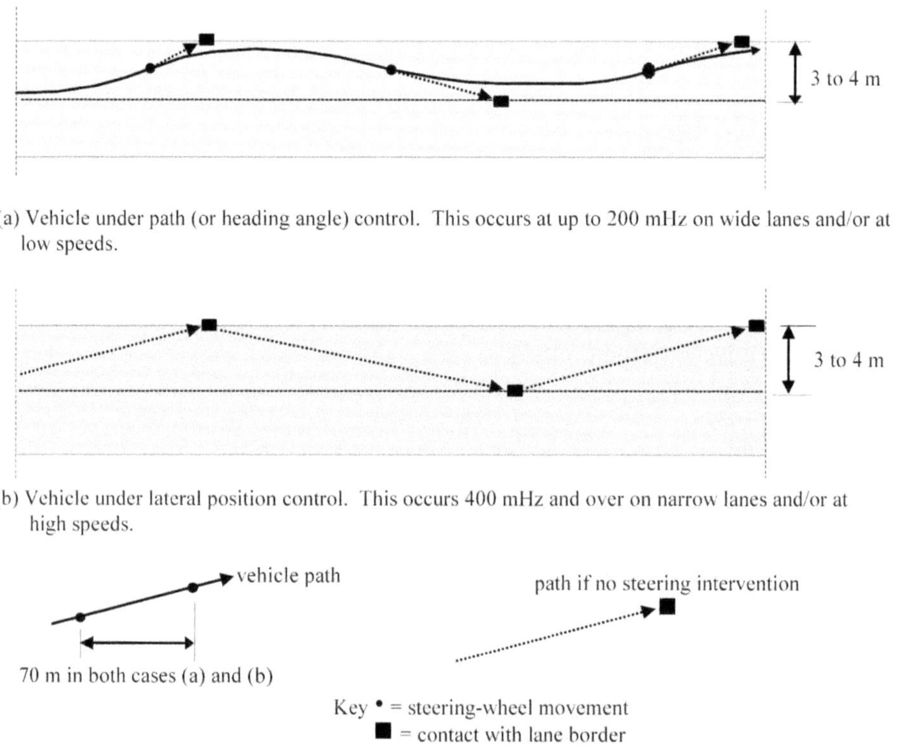

(a) Vehicle under path (or heading angle) control. This occurs at up to 200 mHz on wide lanes and/or at low speeds.

(b) Vehicle under lateral position control. This occurs 400 mHz and over on narrow lanes and/or at high speeds.

Figure 16.4 Vehicle tracking behaviour.

Wide lanes have effects other than easing the driver's tracking task. As would be expected, driver performance initially improves, with drivers showing shorter response times to other events. For reasons that will be apparent from Section 17.2.2, the traffic capacity of the lane increases at about 1 percent for every 50 mm of extra lane width. There is also an effect on mean position, with drivers travelling about 150 mm closer to the pavement edge as the lane width increases from 3.2 to 3.7 m (Armour, 1985). This last point demonstrates the dangers of making lanes too wide in urban areas — not only is the practice wasteful of road space, but it may also lead to poor lane discipline and risky attempts at multiple use of a single lane. However, wide 6.0 m lanes have been successfully used on arterial roads in Germany and Sweden, where they permit easy overtaking and result in higher travel speeds.

On the other hand, narrow lane widths and high speeds create inherently difficult driving conditions. Drivers change their strategy to the direct control of lateral position error (or tolerance) using high control frequencies above 400 mHz. Thus narrow lanes create an active steering strategy (Figure 16.4b) as the driver attempts to maintain an intended course. This results in a decline in driver performance. For instance, on two-way roads overtaking times and distances are increased with a subsequent loss in safety (Troutbeck, 1984). Crash rates for run-off-the-road and oncoming-traffic crashes decrease as the lane width increases from 2.5 to 3.5 m. Because the size of the acceptable error increases with speed, crashes with oncoming traffic also increase with speed. Thus the overall effect of lane widths on crash rates (Section 28.5f) is marginal, particularly for lane widths above 3.5 m.

Under- and *over-steer* refer to the tendency of a vehicle to under- or over-emphasise the turning movement imposed by the steering wheel. Tests conducted on cars with various degrees of under- and over-steer, steering ratios, and steering response times suggest significant driver difficulties with lanes under 2.5 m and, for a 1.9 m wide car, an optimum width of about 3.1 m. [Vehicle widths are usually tightly controlled (Tables 7.3 and Section 27.3.2)]. Allowances for conditions in adjacent lanes often raise this width to closer to 3.5 m. Lane width selection is also discussed in Section 6.2.5 and the implementation of the theory in Section 22.2.1.

Regardless of the lane width, the presence of any lane marking will improve a driver's lateral position-keeping. Generally, drivers stay closer to the centre of the road than they intend and/or perceive. On a two-way two-lane road, drivers respond to an oncoming vehicle by moving towards the kerbside about 3 s ahead of meeting the oncoming vehicle. A related effect is that drivers in outer lanes travel closer to wide shoulders and to surfaced shoulders than they do on narrow or unsurfaced ones (Armour, 1985).

Lane changing is often prompted by the need to either overtake after following behind a slower vehicle or to utilise a forthcoming exit ramp. Technically, lane changing relates closely to diverging, merging and overtaking, which are discussed in Sections 17.3.6 and 18.4. It increases as traffic density increases and is influenced by the degree of urgency of the manoeuvre.

16.5.2 Curve negotiation

A driver unaware of the road alignment will look ahead, seeking the first signs of an oncoming curve. The point at which the alignment first deviates from the straight is called the point of curvature (Figure 19.1) and once detected, many drivers tend to concentrate their attention on this point. When the driver is located on the inside of the curve, the point will be located at the inside edge of the pavement. However, for an outside curve on a two-way road, most drivers are reluctant to focus their attention further across than the centreline of the road. Successful negotiation of a road curve depends upon a driver's choice of:

* approach speed,
* curve entry-speed,
* speed profile through the curve, and
* lateral position through the curve.

The curve direction and radius perceived by a driver in advance of the curve determine the approach and entry speeds that the driver adopts (Sections 18.2.5 & 19.2.5) and are therefore key elements of the curve negotiation process. Traffic engineers must therefore take care to provide only correct cues. Drivers will also base their expectations of an oncoming curve on their immediate previous driving experience. This is the reason for the emphasis on design consistency in the discussion of alignment design in Section 18.2.6.

The basic need for a driver to be able to see well ahead is generally achieved by meeting the sight distance requirements discussed in Section 19.4. More specifically, because of the way the curve presents itself to the driver, the cues the driver uses will be more readily found for right-hand curves in the case of left-side driving, and for left-hand curves for right-side driving. For instance, many drivers base their visual estimation of road curvature on the visual shape of the curve formed by the inside edge of the road curve beyond the point of curvature. The sharper the initial curve seems, the greater the perceived curvature (Figure 16.5). The actual curve angle (Figure 19.1) is a more direct cue that is sometimes available to drivers on a descending grade. Gradually increasing curve radius via a transition curve (Section 19.2.7) can deceive a driver as to the true radius of the curve.

This open-loop pattern of strategically searching ahead to anticipate conditions — the *preview mode* — alternates with the closed-loop tracking process described in Section 16.5.1. A more intense 400 mHz search frequency typically begins at about 100 m (or 4 s) from the curve. A driver's perceptions of the characteristics of a curve will therefore be dominated by information received in this preview distance (or time). Decisions made too early may result in incorrect perceptions and decisions made too late may leave too little time for speed reductions (Fildes and Triggs, 1982). To quantify this in terms of the pre-curve demands that an alignment places on a driver, Shinar (1978) introduced the *preview index*, which is:

preview index = [(initial speed) – (entry speed)]/[approach length]

Despite receiving advance notice, drivers only begin slowing down from their current speed to their selected entry speed, as the curve becomes imminent. The entry speeds actually adopted are discussed in Section 18.2.5. To at least the curve entry stage, the driver continues to use any available longitudinal road lines (Section 22.2) as the dominant closed-loop tracking control on vehicle trajectory, scanning to left and to right to find the best cues.

In steering around a curve, a driver must anticipate the curvature of the remainder of the curve and adjust the angle of the steering wheel in order to attain the desired heading angle. Drivers are less able to anticipate the required steering-wheel angle for sharp curves and when travelling at high speeds, indeed their errors increase with speed squared.

For curves of constant curvature, the steering-wheel angle at curve entry is maintained at a fairly constant value throughout the curve. However, in the absence of

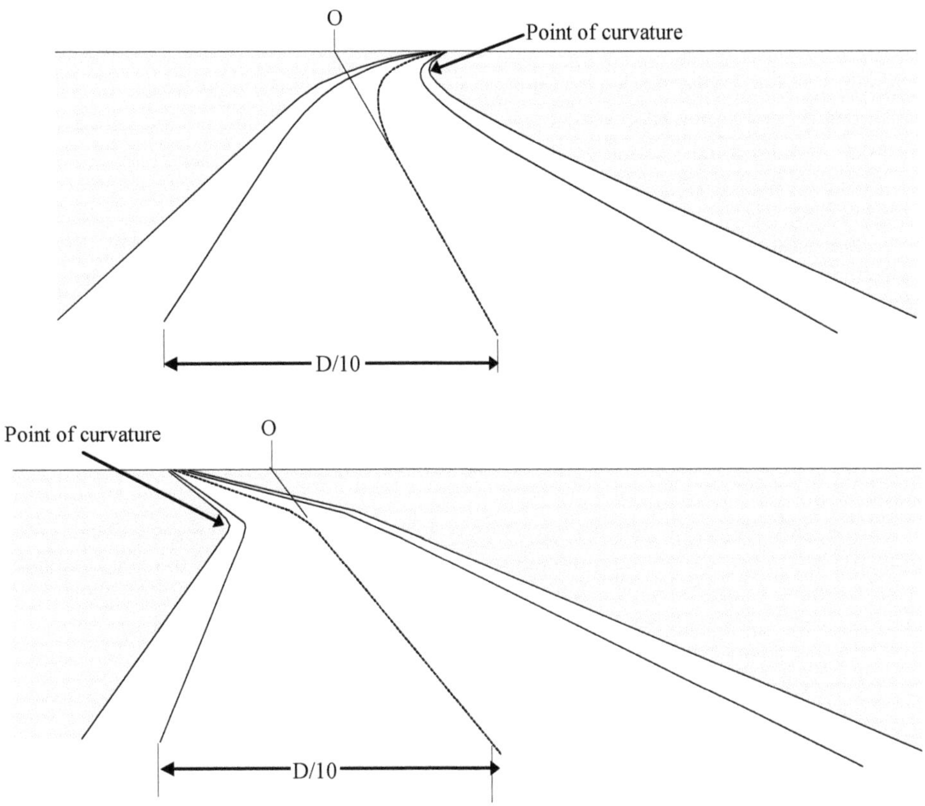

Figure 16.5 Curves and points of departure. From ten Brummelaar (1983). Observers should view the picture from a point perpendicularly above the origin O and at a distance D. Mirror-image for driving on the right.

good tracking delineation, drivers tend to wander and usually 'cut the corner' by decreasing the curvature of their path, and hence reducing the cornering side-forces that they experience within their vehicle (Equation 19.1). This effect is accentuated by alcohol (Johnston, 1983). Such vehicle encroachments across edge- or centrelines have been found to be indicators of a relatively high crash rate on a curve and thus can be used as advance warnings of possible problems.

Most drivers decelerate in the first half of a curve — typically in the first 80 m, with the highest decelerations being undertaken by those with the highest approach speeds. Of course, good driving practice is to accelerate out of the curve so that centrifugal forces can help bring the car to the correct final heading. Few drivers do this and over-braking is more the norm.

When a driver loses control on a curve, it is reasonable to infer that this inability to maintain an intended lateral position occurred because of excessive speed induced by either or both high prior speed and a poor perception of the curve geometry. That is, the cause of the problem usually arises before the driver enters the curve — hence the above emphasis on correct curve perception. There is a need therefore to ensure that drivers have accurate advance knowledge of the radius of an impending curve. Ways of ensuring this are discussed in Section 19.2.3. Curve design and delineation are discussed further in Sections 19.2.2 & 24.4.

16.5.3 Response (or reaction) time

Figure 16.1 shows that the human component of driving is the conversion of perceived information into some action by the vehicle. If the situation is known, practised, or familiar, then the time to respond to relevant received information can be very short and the action of high quality. The time to respond is known as the *response* (or *reaction* or *latency*) time, T_r.

The components of the response time for the average (50th percentile) driver are collected together in Table 16.1. The sequence of detect, identify, decide, and respond (or act) used in the table is from Figure 16.1 and is discussed again in the sub-sections in Section 21.2. The extent to which a driver draws on the full capability of the driver is often limited by the driver's physical capacity. For instance, the peak decelerations that a driver can tolerate and is therefore willing to apply are about 4 ms^{-2} longitudinally (Section 27.2.2) and 8 ms^{-2} laterally (Section 19.2.5), coupled with the above response times. This data permits a calculation of manoeuvre and avoidance distance from a given initial speed.

The table suggests that the range of response times for unexpected events range from 1.0 to 3.6 s with 1.9 s as a median value. The commonly used 2.5 s is a reasonable 'safe' estimate of the time for a vehicle to begin to respond to an unexpected event. In the *car-following* mode with the driver focussed on the tail lights of the car ahead:

* items 1 and 2 are zero,
* item 3 is 400 ms,
* item 4.2 is 200 ms, and
* item 5.1 remains at 300 ms.

Figure 17.11 suggests a common car-following headway of 1.5 s, which implies that the typical driver allows a further 600 ms to accommodate braking differences between the two vehicles.

Equation 19.3 with $a = 4$ ms^{-2} and Table 21.1, Column 6, suggests a total braking time of about 3 s by an 'experienced' driver in an emergency and 12 s at the uncomfortable level for normal drivers (Section 27.2.2). The level of deceleration actually used will depend on the driver's estimates of risks of:

* a collision with the lead vehicle,
* a loss of control of the driver's vehicle
* cargo damage in the driver's vehicle, and/or
* a rear-end collision with the driver's vehicle.

In these circumstances drivers tend to significantly underestimate the time to collision.

The variance between drivers is very large and the distribution is probably lognormal. Values of up to 7 s have been recorded at one extreme and, at the other extreme, 1 s times have been measured with forced stops. One reason for the large variability is that response time will depend on a driver's level of alertness at the time; e.g. distractions due to conversations can readily add 300 ms to the overall response time. Similarly, anticipation or presignalling of an event, the absence of uncertainty on multiple choices, and familiarity with the task can each lower response time.

Given the above, it has been suggested that most unalerted drivers in an urgent situation can respond simply to a clear stimulus in less than 2.5 s. In addition, 2.5 s represents an upper (possibly 85th percentile) value for normal drivers, and is close to the mean for degraded drivers (Triggs and Harris, 1982).

Table 16.1 Typical driver response times (milliseconds). The ± values are indicative variations and define a range that covers about 90 percent of the population.

Typical driving actions & vehicle responses	Response times mean	Response times variation
1. Detect (Section 16.4.5)		
1.1 Basic perceptual response to an intellectual or visual stimulus – can increase for complex tasks	250	
1.2 Eye movement locates the object in one look	50[a]	+50/–0
1.3 Eye dwells on the detected object	200	
Σ1 Total *detection* time[b]	**500**	**+300/–0**
2. Recognise, identify & interpret the detected object (see Section 16.4.5 & footnotes b & c)	600	+300/–400
3. Decide on a response to this identification (see also items 6.1&2 below & footnote d)	500	+500/–250
4. Driver responds; e.g. by lifting foot from or applying it to the accelerator	200	+350[e]/–100
Σ1–4 Total driver *response* time	**1800**	**+1800/–800**
5. Vehicle manoeuvre time		
5.1 Operate pedal	50	+1000/–100
5.2 Vehicle begins to respond	100	+1000[f]/–0
5.3 Specific braking aspects		
5.3.1 Brake lights illuminated	50[g]	
5.3.2 Braking system completes operation, Equation 19.4 & Table 19.1, Col 6		
5 Total vehicle response time	**150+**	**+1000/–100**
Σ1–5 Total driver and vehicle *response* time	**1950+**	**+?/–900**
6.1 Additional time if two responses are needed	600[h]	+/–200
6.2 Additional time if three responses are needed	1200[h]	+/–300

Notes:

a. If the object is outside the desired visual field (typically, ±15°, Section 16.4.2) and a head-movement is needed, this will add another 700 ms.

b. Increases in the presence of alcohol and become more variable in the presence of fatigue (Section 16.5.4).

c. The effect of age on this component is discussed in Section 16.5.5.

d. Section 16.5.5 shows that this time increases with age, by about 50 ms for every ten years.

e. Occurs if driver is not watching rear light of lead car. Thus, in close traffic, the time for a following car's rear light to illuminate after the lead vehicle's rear light illuminates is 250 (1.1) + 200 (1.3) + 200 (4) + 100 (5) + 50 (light latency) = 850, +350/–100 ms.

f. This may extend to 4000 ms for some large vehicles.

g. Standard brake lamps take 250 ms to reach 90 percent of full output.

h. Add 200 ms for older drivers.

16.5.4 Fatigue

The role of alcohol and drugs (prescription and illegal) has been described earlier in this chapter (e.g. Sections 16.4.3&5 & 16.5.2&3) and is discussed at some length in Section

28.2.2. The major current medical factor influencing driving and not dealt with elsewhere in this chapter is fatigue. Fatigue can be defined generally as a diminished capacity or inclination to perform, and may be both psychological and physical.
The three types of fatigue of relevance to driving are either:

(1) of an emotional nature (e.g. worry, conflict, or irritation), and usually overcome only by removal of the cause,
(2) of a permanent physical nature due to some organic cause such as narcolepsy, and requiring medical treatment, or
(3) of a temporary physical nature and predominating over the other two in its influence on driving. It lies between being wide-awake and being asleep, and is best described as a state of drowsiness. It can result from:
* sleep deprivation,
* monotony,
* an adverse environment (e.g. background noise),
* over-work and other over-extensions of physical capability, or
* physiological factors such as over-eating.

Fatigue that exists before the journey starts is usually chronic fatigue; fatigue caused by the journey is known as acute fatigue.

Fatigue is only overcome by rest and recuperation. However, drivers' self-awareness of their fatigue impairment commonly underestimates their actual impairment. Furthermore, they are poor at recognising the symptoms of fatigue and consequential impairment.

Fatigue results in a deterioration of the body's physical and mental effectiveness. Its immediate symptoms are loss of attention to a task, and/or a drop in task performance. In the context of driving, fatigue affects the driver's cognitive, sensory, and physical skills. It results in visual scanning decreasing or stopping entirely, visual detection times increasing, and response times increasing (Sections 16.4.5 & 16.5.3). These seriously degrade driver performance.

Fatigue can lead to the dangerous onset of sleepiness. Sleepiness is particularly dangerous during the driver's normal sleeping hours. In the common first signs of sleepiness, a driver completely loses visual awareness and begins eye rolling over a few 2 s periods. Other observable signs are a dull facial expression, beady eyes, and closed eyelids. If unaddressed, these microsleeps become progressively longer until the head drops, hopefully giving the driver one final warning. It is thus important for drivers to recognise the earliest signs of sleepiness. Short naps (15 min) and/or coffee can provide some respite, although coffee takes some 30 minutes to become effective. Short rest breaks are usually effective only if taken at least every three hours. If sleep does occur, the driver will usually make minimal steering adjustments and the vehicle will deviate only slightly from its previous path.

Fatigue effects are difficult to deduce from crash data, but estimates are that between 10 and 30 percent of serious crashes have a significant fatigue component. Crash studies have shown a significant increase in relative crash rate. after 4 hours driving and an increase in truck crashes between the 7th and 10th hours of a driving shift. Even after adjustments for exposure (Section 28.1.2), crash rates still peak late at night and in the early hours of the morning. Thus, there is little doubt that fatigue contributes significantly to road crashes.

16.5.5 Age

Age effects in driving occur mainly in the two extremes of the young and the elderly.

Youthfulness

The role of youthfulness in driving relates to:

(a) many of the psychological and attitudinal issues discussed in Section 16.3.2, and

(b) *inexperience*, which produces five main sets of problems. Note that a number of the following 'inexperience' factors, apply to all new drivers, and are independent of their age.

(1) A lack of well-rehearsed, basic *psychomotor skills* (Section 16.2) in:

* vehicle control (Section 16.2e), such as the ability simultaneously to observe the road, steer, use the foot-pedals, change gear, and activate the turn indicators. The lack of these skills produces jerky driving responses.
* interaction of the vehicle with the road and with other traffic, such as subconsciously keeping the vehicle within the lane markings and away from other vehicles. The lack of these skills produces poor lateral control and imprecise vehicle manoeuvring.

Once drivers are experienced, their level of psychomotor skills will have no significant bearing on their crash record.

(2) A lack of *strategic driving*, i.e. incorrectly anticipating and assessing the traffic ahead, and late detection of distant events and hazards. This is caused by:

* 'heads down' rather than 'heads up' driving, with the eyes focussed on the road immediately ahead of the vehicle and the mind focussed on performance of the tasks currently in hand [this behavioural pattern is also encountered in older pedestrians],
* not scanning the visual field efficiently, tending to fix on objects for relatively long periods (Section 16.4.5),
* making little use of peripheral vision (Section 16.4.1),
* insufficient use of rear vision mirrors (for example, looking more at the speedometer than at the rear-vision mirror),
* devoting too much attention to the immediate task of vehicle handling, and
* under-developed cognitive and perceptual skills in many in the 18-year-old and younger age group.

These all result in young drivers taking longer than they should to detect and identify hazards. Thus overall driver response times (Section 16.5.3) improve — i.e. decrease — with age, despite declining physical ability, as experienced drivers give themselves earlier warning of forthcoming events.

(3) Poor risk management (Section 28.2.4&6). There are two age-related effects:

* *risk assessment*. This falls into three categories:
  # Inexperienced drivers are likely to use inadequate information to assess risks.
  # Inexperienced drivers do not expect some dangerous movements by other road users (e.g. pedestrians from the side or close-following drivers at the rear, Catchpole, 2005). In this context, Catchpole has noted that for many new drivers the crash-free experience gained over the first few years of driving can lead to a growing complacency with respect to such unexpected movements, and to a tendency to adopt higher travel speeds.

Young drivers may take a relatively cavalier approach to risk-assessment (Section 28.2.4), sometimes leading to reckless behaviour (i.e. driving with low concern for consequences).
The accuracy of an inexperienced driver's perception of the actual risks associated with particular driving manoeuvres is relatively low and only improves with increased driving experience.

* *risk exposure*. The social habits of young drivers often lead them to drive at times of maximum risk (e.g. end-of-week evenings) – i.e. although they may drive no further than older drivers, their risk exposure (Section 28.1.2) is higher.

(4) Tending to drive more impulsively and therefore making their behaviour more difficult for other drivers to predict.

(5) Receiving little positive feedback with respect to their bad driving, in terms of crashes, near-misses, frights, and abuse from passengers and other road users. This issue is important as only a small proportion of unwise actions will actually lead to undesired consequences.

The very real consequences of this behaviour in terms of road crashes are explored in Section 28.2.6. The dramatic reduction in risk after a few years of driving suggests that for most new drivers, inexperience (b above) is probably more of a determining factor than is age (a above).

Therefore, there are good reasons why drivers should be given as much driving experience as possible at an early age and in a safe environment. Some of this experience can be achieved off the road, but much of it can only be gained on the road. At the moment, driving whilst inexperienced is a gamble many must take at the beginning of their driving career (Section 16.6.1).

Ageing

In distinction to the discussion of youthfulness, with older drivers the driving problem is very much one of declining physical abilities and slowing of performance. For most people, driving skills begin to decline at age 55 and decline rapidly after 75. Older drivers compensate for this decline by driving more slowly and braking more frequently and by limiting their driving to off-peak traffic and to short trips.

Eyesight begins to decline for most people after age 45 (Section 16.4.1) and, even with glasses, most people over 60 have a visual decrement (Section 16.4.2 & OECD, 1986b). The loss of light transmission through the pupil tends to make night driving increasingly difficult for people over 40 (Section 16.4.1). For such people, illumination levels must be doubled for each additional 13 years of age, in order to maintain equivalent visual performance. The size of the central and peripheral visual fields and contrast sensitivity also begin declining at about this age (Section 16.4.2&3). The decline often accelerates after age 55.

The older driver's ability to acquire information is also restricted by failing hearing and by stiffer joints and/or cardiac conditions preventing rapid head-movements. Joint problems can also hinder use of the vehicle control devices. Older drivers also have:

* longer recognising and decision-making times (Items 2 & 3, Table 16.1),
* decreased performance levels for skill-based tasks,
* poorer management of task-overload in complex situations, and
* increased error rates in such key driving tasks as gap selection (Section 17.3.1), and steering control.

As a consequence of these factors, older drivers particularly have trouble with:

* reading and then understanding signs,
* route selection,
* observing and then interacting with traffic, particularly when:
  # making centre turns, if unaided by green arrows (Section 23.2.5) or protected turn slots (Figure 20.10),
  # crossing through traffic, and
  # merging into another traffic stream.

This class of problem can be aided by making the associated signs larger, brighter, simpler and more legible. Finally, older people recover much more slowly from a crash, in part because of their more fragile bones.

16.6 DRIVING TRAINING AND LICENSING

16.6.1 Education and training

Educational and training courses for drivers and motorcycle riders aim at improving three different facets (or categories) of driving:

(a) *Psychomotor, perceptual and cognitive skills*
These skills are improved via training at either:
(a1) a basic level, as in pre-licence courses, or
(a2) an advanced level, as in post-licence courses.

At level (a1) an inexperienced driver must learn and rehearse the physical and mechanical psychomotor processes involved in vehicle control (Section 16.5.5). Perceptual and cognitive skills develop after vehicle control is mastered and take longer to learn and are associated with a move towards strategic driving.

Once level (a1) has been passed, further level (a) measures have little relevance as a modern car is a relatively simple device to operate and most potential drivers have the requisite knowledge and skills to do so successfully.

Only a very small percentage of crashes can be traced to a driver's lack of psychomotor, perceptual or cognitive skills. Most crashes are due to errant driver behaviour with errors in perception and attention predominating (Section 28.2.1). Such crashes could be avoided by alert and attentive (rather than skilful) driving (OECD, 1981b). The particular role of driver attitudes, which cannot be changed in a short skill-based course, was discussed in Section 16.3.2.

With respect to specific measures at level (a2), evaluations of post-licence driving schools catering for the general population of drivers commonly find a greater number of traffic violations among their graduates than among the general population. Part of this effect is probably due to the type of driver attracted to such a course and part to a tendency to raise a driver's self-confidence (Brown *et al.*, 1987). In addition, French studies have shown that drivers soon lose the useful emergency-procedure skills that they learn in such courses.

(b) *Knowledge of traffic law, road-use conventions, and traffic system characteristics*
This knowledge is provided by education — frequently with a strong safety bias — as with *defensive driving* tuition. Such courses in the short term can do little more than improve violation records and reinforce existing attitudes. Whilst effective for

professional drivers, for ordinary drivers they have no record of safety benefit or of altering driver attitudes (see Section 28.2.5).

People with reading difficulties, or who are illiterate, need special assistance and supervision during training and licensing. Drivers with a relatively low IQ have more total, repeat and serious traffic convictions, particularly offences associated with requirements needing some literacy such as driving licence, insurance and vehicle registration offences (Boyce and Dax, 1981).

(c) *Modified attitudes*
The task of modifying attitudes is mainly addressed by the use of publicity material and educational courses in schools. A basic aim of many school programs is to bring about a gradual change in community attitudes to driving over a comparatively long period of time, with specific effects not becoming apparent until the school population reaches adulthood. This approach underlines the importance of the role of the parents in the 'driver' education task (Catchpole, 2005). The role can be both positive and negative as often driving behaviour is learnt more through family than through institutionalised sources and positive correlations have been found between fathers' and sons' traffic conviction records.

It is often argued by laymen that any form of driver education must make better and safer drivers and be a ready and effective way of reducing road crashes. Surprisingly, perhaps, there is no scientific evidence to support such claims for courses in categories (a) and (b). The combination of (a), (b) and (c) is sometimes called *roadcraft* and it is this complete package which must be the basis of driver education.

One of the reasons for the poor effectiveness record of many driver-training courses is that they appear to have developed without any basis in either theory or experiment. Another is that in some countries driving instructors have themselves rarely had better than superficial training in their task. Probably the basic reason why driver courses fail to produce safety effects is that safe driving, as shown in Sections 16.2, 16.3.2 and (c) above, depends on attitudes and motivation rather than on the acquisition of additional skills. For example, drivers determine most of the difficulty of the driving task that they undertake.

(d) *Summary*
Most current driver training contributes little to reductions in crash involvement or crash risk among drivers of all age and experience groups. Improving driver knowledge and skill does not always lead to a change in on-road behaviour or reduced crash risk among trainees. Conventional driver training is unlikely to undo firmly established past learning, or to alter motivation or personal values. Conventional additional training often leads to increased crash risk among novice drivers. A better alternative for novice drivers is to promote extensive supervised driving experience and graduated licensing. This summary is largely taken from Christie (2001).

16.6.2 Licensing

The granting by an appropriate statutory Authority of a licence to drive is a common requirement before a person is permitted to drive. The process typically implies that the holder has:

(a) attained a minimum age,

(b) adequate knowledge of traffic laws,
(c) adequate physical ability (Section 16.5), and
(d) adequate driving skills.

There are usually intermediate stages involved in obtaining a full licence. Most jurisdictions have a *learner* stage to allow trainee drivers to obtain the on-road experience needed to meet requirement (d).

Structured (or *graduated* or *probationary*) *licensing* systems are used during the critical years of driving inexperience to:

* provide a training mechanism, allowing the new driver to learn on-road skills in a sequenced and controlled fashion (Section 16.5.5),
* minimise exposure to high risk situations (Section 28.1.2), and
* cover the driver's initial lack of experience and judgement.

Structured schemes involve a sequence of stages with the novice driver operating under restrictions and laws not applied to drivers on an unrestricted licence. Typically these are:

(1) an initial period driving under full 'parental' supervision,
(2) driving only in daylight hours,
(3) driving cars with a limit on engine power,
(4) driving with zero blood alcohol level (Section 28.2.2), and
(5) under laws which impose more severe penalties for traffic offences than those imposed on a driver on an unrestricted licence.

Other arrangements are possible. Despite their apparent commonsense, little evidence exists to support the effectiveness of the various schemes.

At the other extreme, special licences are required for people taking special driving responsibilities; for example, those driving buses, vehicles used for commercial purposes, unusual vehicles, or vehicles carrying hazardous cargoes. In these cases, it is appropriate to require applicants to have had past satisfactory driving experience and to test them for relevant special skills and knowledge.

It is normally argued that a driving licence represents the granting of a privilege rather than the affirmation of a right. The privilege, once granted, may thus be withdrawn after some offences against traffic laws and only regranted after a new demonstration of competence and awareness. However, as shown in Chapters 30 and 31, the availability of a motor car is a near necessity in many current societies. This set of circumstances has resulted in a de facto reluctance to withhold or withdraw licence privileges. Even when licences are withdrawn, there is strong evidence that many people continue to drive whilst unlicensed.

16.6.3 Licensing and road safety

Aspects of driver training and licensing related to road crashes are discussed further in Section 28.2.5. Driver licensing is often cast in a road safety context. Even with such a specific intent, it is difficult to set or review the standards for requirements (c) to (d) in Section 16.6.2. This is partly because crashes are too rare and random to be used as direct measure (Section 28.1). Indeed, it is sometimes argued in the popular press that licensing should be used as a crash countermeasure. However, no system of testing drivers exists which could deny licences to those highly likely to be involved in crashes without at the same time denying licences to a very large number of drivers who would subsequently be crash-free. The reason for this is that (Section 28.2.3):

* no measures have yet been developed of any physical, sociological, psychological, or personality traits that could explain more than a small proportion of the variation in crash risk from one person to another, and
* the risks of having a crash are so low that, although some groups of people are over-represented in crashes, their predicted crash rate is too low for them to be labelled as crash-prone. To give three illustrative examples:
  (1) The major group over-represented in crashes is young male drivers, but it would be patently unreasonable to debar all young males from driving.
  (2) The prime medical reason for denying a person a licence would be, on the basis of crash statistics, evidence of an alcohol problem. Yet many people with an alcohol problem do not have car crashes.
  (3) Even a relatively high-risk driver, with either frequent past crashes or traffic offences, has only a one in five chance of being involved in a significant crash in any one year.

It is therefore not surprising that a retrospective look at road crashes indicates that most crashes involved drivers with a satisfactory driving record at the time of the crash. On the other hand, drivers without valid licences are disproportionately represented in crash statistics, mainly because the attitudes that lead to unlicensed driving also lead to hazardous driving.

As young drivers figure so prominently in road crashes (Section 16.5.5), the greatest safety effect of the driver licence is usually via its role in controlling the minimum driving age and thus limiting driving by young people. There are two countervailing arguments with respect to the minimum driving age.

* Raising the minimum driving age reduces the absolute exposure that young people have to dangerous driving conditions. This trend is aided by the observation that in most jurisdictions the younger the driver, the more likely he or she is to pass the driving test.
* On the other hand, keeping the age low separates learning to drive from the age when young people are susceptible to peer pressure, unreceptive to parental advice, and able to legally drink alcohol. People then learn to drive when their minds are receptive to advice and caution.

The main demonstrated safety traffic merit in current driving licensing and training is the graduated approach outlined in Section 16.6.2. However, perhaps the greatest benefit of a driving licence is that it provides a means of driver identification.

16.7 THE DRIVER AND THE LAW

Traffic laws impose a duty to comply on a driver and are introduced to ensure safe, smooth, and efficient traffic flow and to prevent that flow from harming other activity or property. Traffic law is usually complex, extensive, and detailed — a situation exaggerated by the fact that many traffic requirements can be defined by *regulations* made under a law and therefore may not be embodied in the law itself.

Traffic law enforcement strives to control driver behaviour in accordance with the requirements of the local traffic law. It does this by deterring inappropriate behaviour by preventative, persuasive and/or punitive means. Specific deterrence happens when a driver is prevented from undertaking a particular traffic action. General deterrence happens when the driver acts as if the relevant traffic laws are being specifically enforced at that moment and in that area.

To be effective, the intent and reasonableness of most traffic laws should be obvious to the driver and therefore self-enforcing. This particularly applies to the set of laws designed to minimise conflict between drivers at intersections (Section 20.2.1). However, complying with a safety law will often impede a driver's basic purpose of travelling from A to B in minimum time. Nevertheless, the law is usually obeyed, even when enforcement levels are low, because the driver:

(a) wishes to be law-abiding;
(b) does not wish to risk any chance of prosecution;
(c) does so from habit;
(d) sees the safety and social benefits of the law as outweighing its disadvantages; and/or
(e) sees the societal benefits of the law.

Laws limiting speed are often an exception to this tendency to obey, mainly because in a particular circumstance their intent and reasonableness may not be apparent (Section 18.1.2).

A traffic violation (or offence) occurs when the driver fails to comply with a traffic law or regulation. However, this failure does not necessarily make the driver a criminal as traffic offences are not usually considered crimes or felonies. The driver in such a case is an offender rather than a criminal. Penalties for a traffic offence fall into five classes:

* fines,
* driver improvement processes,
* community service,
* vehicle impoundment, or
* disqualification from driving.

The offences might also be suspended or the offender placed on probation. The criteria for determining offences and their penalties are usually based directly on the law rather than on earlier court cases. Indeed, it is not common for traffic law cases to be published or reported.

Enforcement is a relatively expensive measure and Section 28.2.4 suggests that the risk of being apprehended whilst committing a serious traffic offence is about 1 in 1000. Indeed, actual enforcement levels will usually be very low, and so general deterrence is usually more effective than specific deterrence. This suggests a value in a visible police presence on the road. On the other hand, such a policy may soon lead drivers to realise how low actual enforcement levels are. Speed enforcement is discussed in Section 18.1.2.

It is appropriate to conclude this section by noting that there is only a weak statistical link between a typical driver's crash record and traffic-offence record.

CHAPTER SEVENTEEN

Traffic Flow and Capacity

17.1 GENERAL

Traffic is the movement that occurs in a particular transport mode (Section 30.1). This book deals solely with the traffic due to road transport. *Traffic flow* measures the rate at which movement occurs and therefore indicates the use that road transport is making of a piece of road. In some places the term *traffic volume* is used instead of *traffic flow*. This book uses *flow* rather than *volume*, as *volume* refers to a quantity rather than a rate. Indeed, *flow rate* would be a more precise term than the conventional *flow*. *Traffic capacity* is the maximum possible traffic flow on that road. The difference between flow and capacity is discussed further in Section 17.4.4. Although flow and capacity are thus dominant numerical indicators of traffic, it must be emphasised that they are not the only criteria for assessing road usage. Safety (Chapter 28), economics (Section 33.5), environmental impact (Chapter 32) and aesthetics (Chapter 6) are other major factors that must be considered.

In traffic engineering, *traffic flow* is specifically defined as the number of vehicles passing a point on a road per hour (veh/h). The integration of this flow over time gives the total number of vehicles to pass a point, which is often the more important measure in traffic signal work (e.g. Figures 23.9&10). Various flow measures are used, depending on the period over which the flow is measured. For instance, the *average daily traffic* (*ADT*) is the daily flow calculated as the average flow over many days. Indeed, the *annual average daily traffic* (*AADT*) specifies that period as a year and is specifically defined as the total number of vehicles passing the point in a year, divided by the number (365) of days in a year (thus ignoring weekend effects), i.e.:

AADT = (total vehicles in a year)/365 (17.1)

For a two-way road, AADT commonly uses the two-way traffic flow, however it is worth checking whether presented AADT values are actually two-way ones.

AADT is most useful as a broad planning measure and is not adequate for many traffic engineering needs. Thus a number of specific measures have arisen. For example, average weekday daily traffic (AWDT) is sometimes used when considering commuter flows.

More specifically, traffic engineering design requires design flows to be specified (Section 6.2.3) and these are commonly peak hourly flows. For instance, the *maximum annual hourly flow* is the *highest hourly flow* (*HHF*) in a given year and is often directly used in studying congested conditions. However, for lighter flows it may represent too extreme a condition. If the traffic is perfectly uniform over a year, then:

uniform HHF (uHHF) = AADT/24 = 0.042AADT

with increases above 0.042 indicating traffic non-uniformity. For real-world non-uniform traffic the Nth highest hourly flow, *N HHF*, is the flow that is exceeded in (N – 1) hours in the given year. For HHF, N = 1, but N is commonly between 10 and 100. There is no simple relation between N HHF and AADT. Some empirical relationships have been proposed but N HHF/AADT will vary with both seasonal and locational factors.

N HHF provides a defined 'characteristic' traffic flow for design purposes (Section 33.4.2), with an N/(365 x 24) probability of being exceeded in a given year. In 1950 the U.S. Highway Capacity Manual introduced 30 HHF as a 'reasonable' design flow and it became the *design hourly flow (DHF),*

DHF = 30 HHF

and its use for design is now common practice, although it is increasingly coupled with measures of *level of service* (Section 17.4.2). DHF has a 30/(365 x 24) = 0.003 probability of being exceeded in a given hour, which increases to about 0.03 if only peak hours are considered and to 0.06 if only local peak flows are considered. This suggests that DHF represents about a 5th percentile annual 'failure' level (Section 33.4.2), although it will be shown below that only 1.3 percent of drivers will be adversely affected by the decision.

On a comparative basis, DHF is less than the extreme HHF but more than the unrealistic u HHF:

u HHF < DHF < HHF

An alternative justification for using DHF for the design of traffic facilities, is the observation that if N is plotted against the associated traffic flow, the curve will commonly have a knee at about N = 30, however this latter argument is very judgmental and scale-dependent and has little to recommend it. Ideally, design flows should be selected by maximising the benefit–cost ratio of the scheme, including the annual costs of the occasional traffic failure implied by using N > 1. In practical terms, if design for N < 30 causes only marginal cost increases (commonplace in urban design), then DHF should be discarded. To estimate DHF directly from AADT figures, note that typically,

0.10 < DHF/AADT < 0.25

with the ratio depending significantly on road type, e.g.:

DHF/AADT = 0.25 for recreational roads, (17.2a)
= 0.15 for rural arterials, (17.2b)
= 0.12 for outer urban arterials, and (17.2c)
= 0.10 for inner urban arterials. (17.2d)

Another empirical relationship is:

(80 HHF)/DHF = (80 HHF)/(30 HHF) = 0.85

Estimates of this type allow the number of drivers adversely affected by a facility with a capacity of DHF to be estimated as 30 x 1.15 x DHF where the 1.15(DHF) represents the average flow in the 29 hourly flows between DHF and HHF. From Equation 17.1, the total flow for the year is 365(AADT) and using Equation 17.2b changes this to 365(DHF)/0.15. Thus the percentage of drivers adversely affected by a design at DHF is:

30 x 1.15 x DHF/(365 x DHF/0.15) = 1.3 percent

The field determination of flow levels and thus of AADT, ADT, N HHF, and DHF and their subsequent presentation as traffic flow maps, is discussed in Section 31.5.

17.2 TRAFFIC FLOW

17.2.1 Uniform flow

Traffic flow is a stochastic process, with random variations in vehicle and driver characteristics and in their interactions, and its full treatment clearly requires statistical and probabilistic methods. These are developed at some length in Section 17.3. However, it is firstly necessary to analyse flow when conditions are uniform.

A basic traffic element is the *traffic lane* carrying a single line of vehicles. Continuing the flow analogy, this is sometimes called a *traffic stream.* The simplest case occurs when the conditions are steady and all vehicles behave in a uniform manner. In particular, the uniform speed that the vehicles adopt is called the *stream speed*, V. [In Sections 18.2.14 & 15 this speed is more precisely defined and will be seen to be a *space mean speed* that for many purposes can be taken as the *running speed* (Section 18.2.11).]

Uniform flow also implies that the vehicle *spacing* (or *occupancy distance*) — measured from, say, the front of one vehicle to the front of an adjacent vehicle — will be a constant, L_s. L_s is the average spacing for all vehicles in a traffic stream and therefore is also given by the length of the traffic stream divided by the number of vehicles in it.

In a reciprocal fashion, *traffic density* (or *concentration*), K, is defined as the number of vehicles per length and for steady uniform flow conditions it is the exact reciprocal of the vehicle spacing, i.e.

$$K = 1/L_s \tag{17.3}$$

veh/km = 1000/m

The italicised equation gives the units typically used, where veh = vehicle.

From simple continuity considerations, the instantaneous vehicle flow, q, within such a lane is related to V and L_s by:

$$V = qL_s \tag{17.4}$$

km/h = kveh/h x m

Equation 17.4 is sometimes called the *continuity equation.* The continuity equation may also be written in terms of density as:

$$q = VK \tag{17.5}$$

veh/h = km/h x veh/km

This form is sometimes called the *fundamental equation* and attributed to Wardrop (1952).

The *occupancy ratio* for a lane is the proportion of the lane length actually occupied by vehicles and is given by:

(occupancy ratio) = L_v/L_s

(percent) = (m) / [(m) x (100)]

where L_v is the mean vehicle length (typically, about 5 m).

From Equations 17.4&5:

(occupancy ratio) = $L_vq/V = L_vK$

(percent) = (m) x (veh/km) / (10)

or = (mean vehicle length) (traffic density)

The time for a vehicle to pass a point is the *vehicle passage time*, s_v, which is given by:

$s_v = L_v/V$:

(second)= 3.6 (m) / (km/h)

With these quantities defined, by taking partial derivatives of Equation 17.5 with respect to distance along the traffic lane, x, the basic vehicle continuity equation (sometimes called Gazis' principle of the conservation of cars) for cars entering or leaving a traffic flow at uniform speed can be written as:

$$\partial q/\partial x = -V\partial K/\partial x = -\partial K/\partial t$$

where t is time.

17.2.2 Non-uniform flow

In non-uniform flow, the various vehicles are travelling at different speeds. So far the discussion has concentrated on a condition known as *unforced flow* (or *free* or *normal flow*), which occurs when vehicle speeds are not influenced by vehicle spacing. This happens when either:

* the flows are so low that any vehicle is too far ahead of the vehicle following it, to exert any influence on the following vehicle's speed, or
* overtaking is possible and vehicles are passing rather than queuing (Section 18.4).

Sections 17.3.9 & 17.4.2, and Figure 17.11 in particular, will show that unforced flow occurs at a vehicle time spacing (*headway*, Section 17.3.1) of about 4 s or more or a flow of less than half the maximum possible flow ($< 0.5q_c$). In such a situation the maximum speed that a driver will adopt is only influenced by considerations of inherent safety and comfort and not by the presence of other vehicles in the lane. In Section 18.2.2 this speed is called the *basic desired speed*, V_{bd}.

In unforced flow it is possible to extend Equation 17.5 to the case of a line of vehicles with different speeds by taking q, K, and V as average values for the vehicles in the line, as measured over a given length of road or over a given time interval. For calculating these means, suggested lengths are between 1 and 5 km, and time intervals between 30 s and 5 min. Applying this approach to a length of road leads to a mean V called the *space mean speed*, which is defined further in Section 18.2.14.

Actual speeds are handled by representing the distribution of the speeds by their probability density function, f (Section 33.4.2). The distribution of speeds over a length of road is given by f_s and over a time interval by f_t. The link between f_s and f_t is obtained by realising that they will both predict the number of vehicles passing a point over a time δT as:

$$qf_t\delta T = Kf_s\delta x = Kf_sV_a\delta T$$

where $V_a = \delta x/\delta T$ is the actual speed at time T and position x. Substituting this into Equation 17.5 gives:

$$f_t = (V_a/V)f_s \qquad (17.6)$$

A method for handling the case where the vehicles in the traffic stream impede each other is discussed in Section 18.4.3.

There are two main approaches to the useful further quantification of traffic flow to accommodate behavioural non-uniformity:

* a *macroscopic* approach which looks at the flow in an aggregate sense. For example, at one time hydrodynamic (fluid flow) and electrical (current) models of traffic were favoured. These approaches are often too general to be useful.
* a *microscopic* approach which examines the response of each individual vehicle in a disaggregate manner, and then integrates this behaviour to attempt to predict overall response. This is more possible with modern computing capabilities.

The first step under the microscopic approach is to continue to assume that a stream of traffic is composed of identical vehicles, but that they are now moving at various constant speeds. A practical condition introduced is that there is a minimum spacing that vehicles will adopt at a given speed. For a stream of vehicles travelling at uniform speed, there is thus one maximum possible vehicle flow, as defined by Equation 17.5. Driver expectations of abrupt external influences on the traffic stream will affect the level of this minimum spacing and lead to two main types of driver behaviour. These are:

* *confident*, which occurs when the driver expects to be uninterrupted by factors external to the traffic stream (e.g. a child running across the road);

* *cautious,* which occurs when the driver expects to be *interrupted* by external influences.

Drivers may change from confident to cautious mode during a journey — for example, after encountering a queue on a motorway. This approach is discussed further in Section 17.2.4.

17.2.3 The forced flow model

As the traffic flow increases in the non-uniform speed model, two new categories of flow are encountered:

* *forced* (or *impeded*) *flow,* which occurs when flows are sufficiently high or speeds are sufficiently low for a vehicle to be impeded by a slower vehicle ahead and forced to adopt its speed and the minimum spacing associated with that speed and the expected traffic conditions. Thus the vehicles form a moving *queue* (Section 17.3.9).

* *unstable flow* typically occurs when:

- drivers first move from free to forced flow,
- localised random events cause a change to cautious behaviour, or
- new vehicles are inserted into the traffic stream.

These can change vehicle spacing and/or flow from free to forced. New moving queues form and then discharge, giving an appearance of *instability.*

Section 17.2.1 showed that forced flow depends on a link between speed and spacing and therefore, from Equation 17.5, on a link between speed and flow. Thus, the curves that describe forced flow for a stream of uniform vehicles are typically of the curved speed–flow form shown in Figure 17.1. A key feature is that this curve reaches a point of maximum flow, q_c, known as the *capacity flow*. This occurs at a speed, V_c, which is called the *optimum speed.* The associated traffic density is called the *critical density*, K_c. The capacity flow, q_c, is a key traffic parameter and is determined by the speed-spacing relationship. From Equation 17.4,

$$dq/dV = [1 - (V/L_s)(dL_s/dV)]/L_s$$

so when $q = q_c$ then $dq/dV = 0$ and:

$$dL_s/dV = L_s/V$$

in the speed-spacing relationship. Above optimum speed the traffic flow drops in a relatively linear fashion and is increasingly less forced.

In the past, V_c was typically 80 km/h and q_c was 1800 veh/h. Recently, closer headways have raised these values to 100 and 2200. Refer also to the discussion in Sections 17.4.1&2.

In practice vehicle speed is limited by a number of factors discussed in Section 18.2, including the legal limit and the road alignment. This maximum practical speed is called the *free speed* (V_f, Section 18.2.5) and two cut-offs are shown in Figure 17.1. Free speed is a measure of the quality of the road but is not influenced by features associated with the alignment of high-speed roads (Section 18.2.6).

Practical measures of capacity are discussed further in Section 17.4 and particularly depend on those road features which will affect driver behaviour up to and at optimum speed. Empirically, the optimum speed and the capacity flow tend to increase as the road's free speed increases. The observations imply that the speed-spacing relationship also depends on free speed.

Conditions below the optimum speed (Figure 17.1) begin at the *stalled* (or *jammed* or *congested*) condition, which occurs when the traffic has completely stopped and each

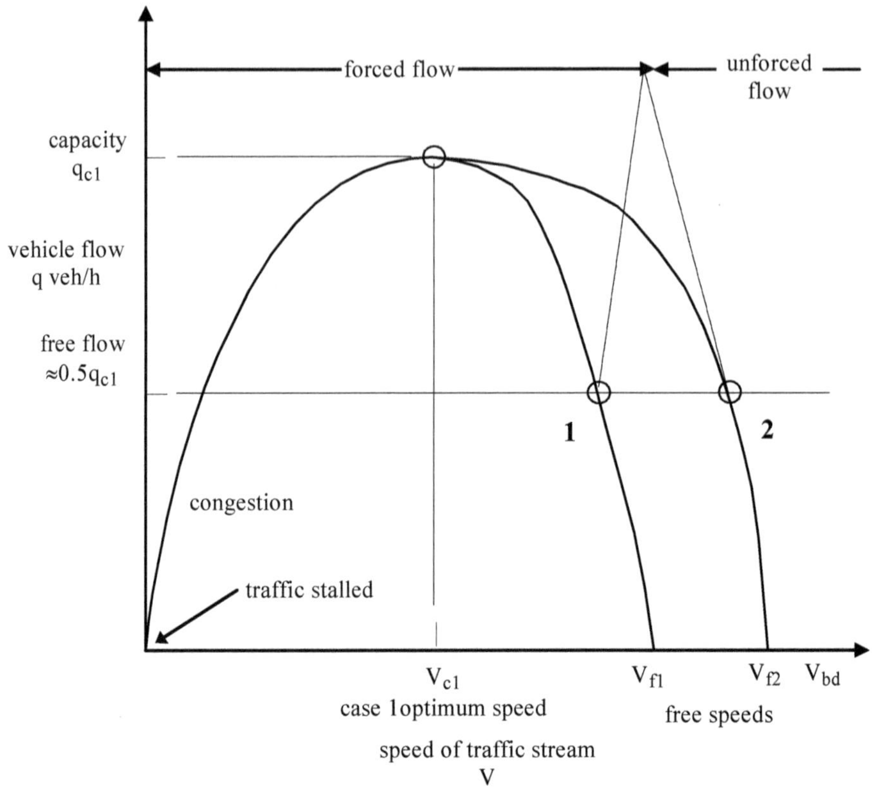

Figure 17.1 Speed–flow relationship for a uniform traffic stream. Specific data for a variety of conditions is given in TRB (2000). Curves 1 and 2 are for two different sets of local road design features.

vehicle has adopted its *minimum vehicle spacing* (or *jam spacing*). As shown in Figure 17.2 this length is equal to the actual vehicle length, L_v, plus some territorial surround:

$$L_{sm} = L_v + (\text{territorial surround, typically 2 m}) < L_s$$

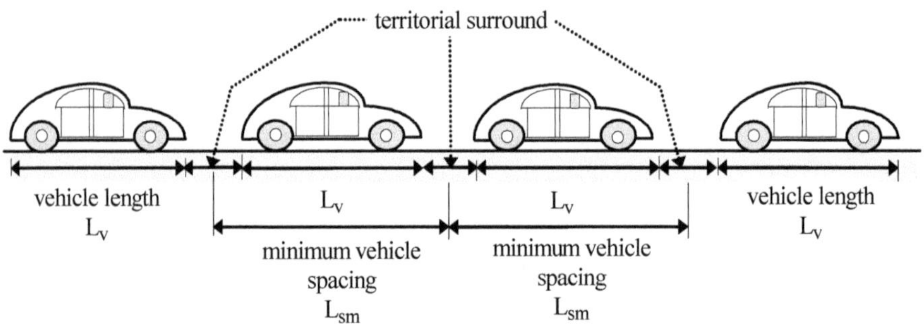

Figure 17.2 Traffic stopped in the stalled condition. Note that the minimum vehicle spacing, L_{sm}, which exceeds L_V, defines the maximum lane occupancy.

As a consequence, the traffic flow and speed are both zero, a stationary queue exists, and the traffic density is at a maximum value known as the *jam density*, K_j.

As the traffic begins to move from the stalled condition, vehicle spacing increases above L_s. Whether or not this represents congestion will also depend on conditions elsewhere in the network (Section 17.4.4).

The form of the graph in Figure 17.1 arises from the fact that drivers tend to increase their spacing as they increase their speed. It may be assumed that a driver will wish to be able to respond safely to changes in the speed of the vehicle ahead. That is, the driver will keep a safe distance from the vehicle ahead, given the driver's own finite response time (T_r, Table 16.1) and the braking capacity of the driver's vehicle (Section 27.2.2). By using data on a driver's allowance for stopping distance (D_s, Table 21.1) and overtaking (Section 18.4), it is possible to use Equation 17.5 to predict that speed–flow curves are of the form:

$$q = V/(L_{sm} + D_s + T_r V)$$

As D_s is a function of V, this can be generalised to:

$$q = V/(c_1 + \Sigma[c_{2n}V^n])$$

where c is a constant. Using just one term of the series gives:

$$q = V/(L_{sm} + c_3 V^n) \qquad (17.7)$$

which provides a reasonable approximation to Figure 17.1 for $n \geq 1$. The simple assumption that n = 1 leads to:

$$q = V/(L_{sm} + c_4 V)$$

European data suggests n = 0.5, although this alone will not produce curves like Figure 17.1. Other constraints are also required such as ones on maximum separation distance (also proportional to $L_{sm} + cV^{0.5}$) and to changes in relative speed.

Assumptions such as these which relate vehicle spacing to vehicle speed have been the basis of most predictions of uniform flow in traffic lanes, with spacing as a parabolic function of speed (n = 2) being very common (McLean, 1989). Another assumption relates the spacing to e^V rather than to V^n in Equation 17.7. Finding dq/dV = 0 on these curves provides a value for q_C.

The value of *n* will also depend on conditions. For instance, the spacings drivers adopt in uninterrupted conditions can be much less than for interrupted flow. This explains the relatively small spacings used by high-speed traffic on urban freeways.

To date the discussion has been based on a uniform traffic stream. Figure 17.1 is for a uniform set of overall conditions. Real traffic is far from uniform and shows a very considerable amount of scatter. Thus some imagination is often required to relate the real data to Figure 17.1. For real traffic with a range of characteristics, there is considerable empirical evidence that a distinct break occurs in traffic behaviour at speeds above and below optimum speed, i.e. that the traffic flow curve is not continuous through (V_c, q_c). Many studies have fitted two distinct sets of curves to empirical data above and below optimum speed, but have had difficulty modelling data at the discontinuity. An analysis based on cusp *catastrophes* has suggested that the major discontinuity that occurs is in speed, rather than density or flow (Hall, 1987). One other consequence is that rear-end collisions increase significantly when speeds are below V_c (Section 28.5j).

These factors have sometimes led to the behaviour of real traffic at about V_c being described as *unstable*. Certainly, for an individual driver, as a consequence of disturbances as minor as the presence of one relatively slow vehicle in the traffic stream (Banks, 1991), the flow can contain sharp changes in speed and sudden queuing. One reason for this is that the traffic densities are too great to permit easy lane-changing and overtaking (Section 18.4) and so traffic queues soon form behind the slow vehicle and shock waves (Section 17.2.5) travel upstream. As there is no slack in the system, these perturbations are not readily damped. Instead, the events distort the measured speed–flow relations and give the suggestion of instability.

The relations between the three pairs of variables that can be selected from the trio of speed, flow and density produce the *fundamental diagrams*. Figure 17.3 is such a set

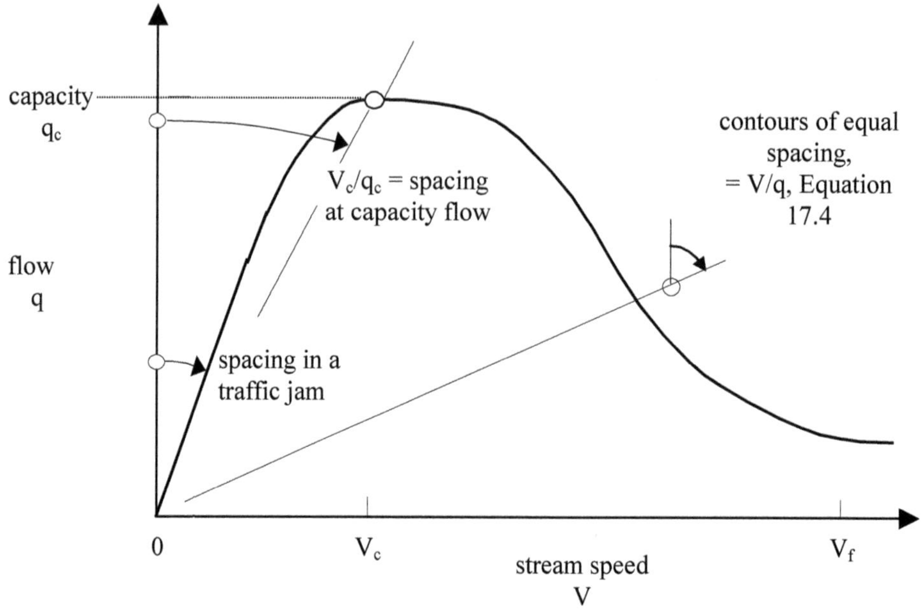

Figure 17.3a Speed–flow relationship.

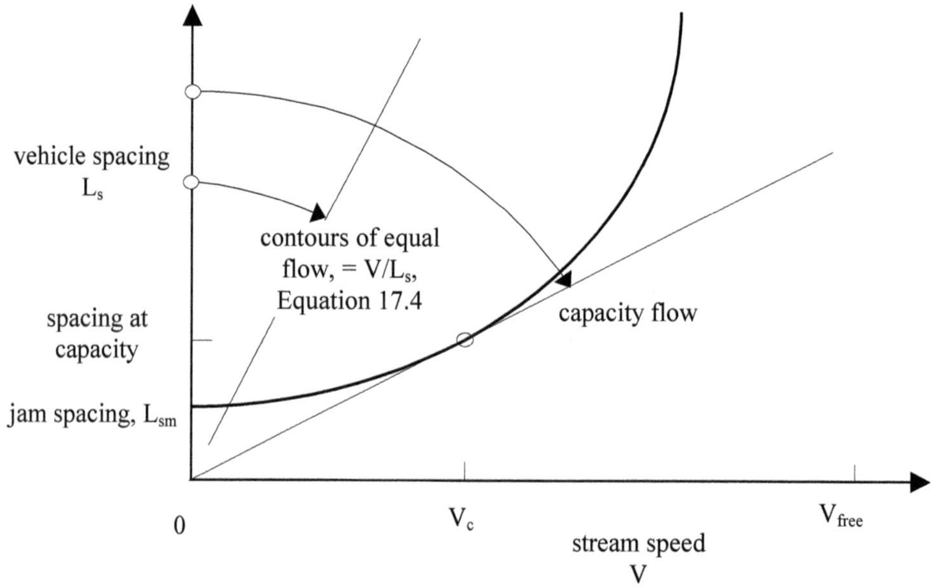

Figure 17.3b Speed–spacing relationship.

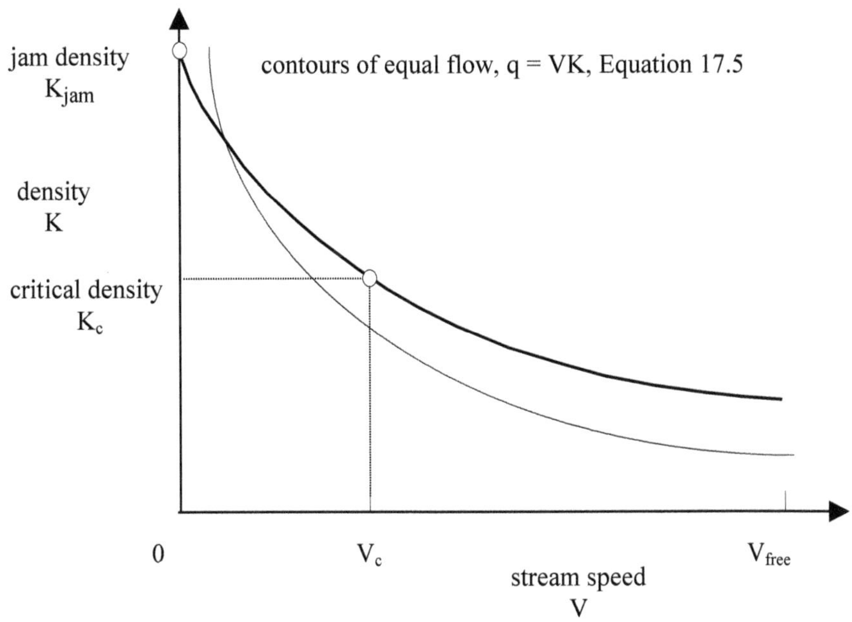

Figure 17.3c Speed–density relationship.

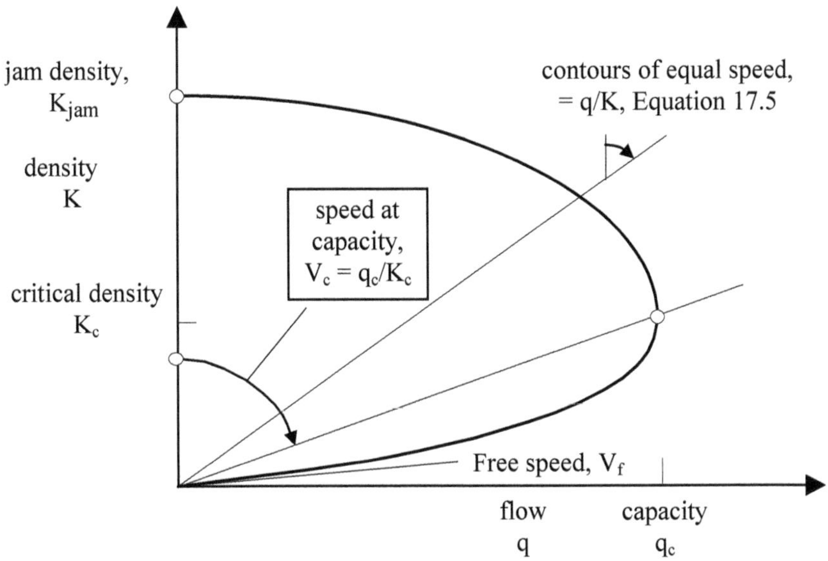

Figure 17.3d Density–flow relationship.

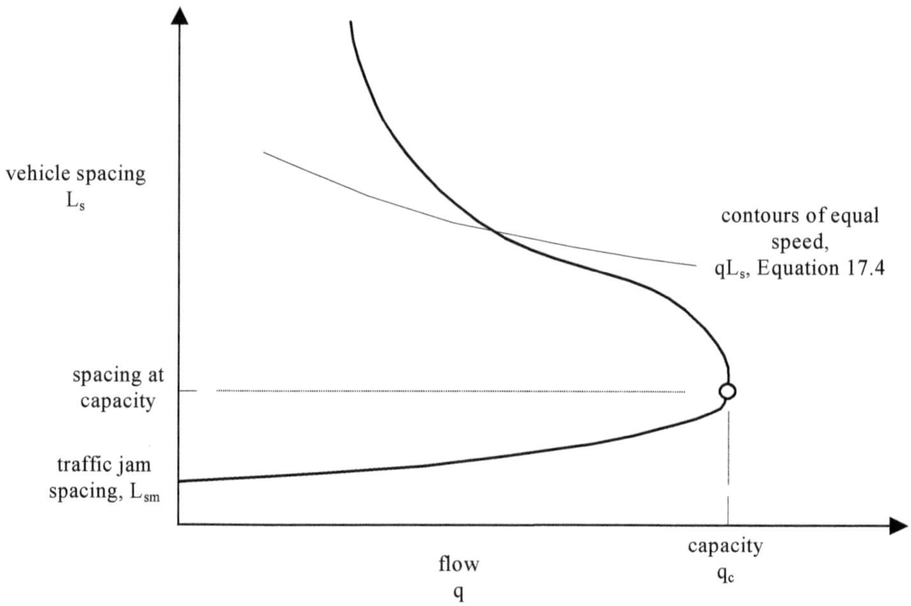

Figure 17.3e Spacing–flow relationship.

of diagrams. The advantages of Figures 17.3b&c over the other two forms is that they provide a single-valued relationship whereas, for example, Figure 17.3a gives two possible speeds for a given flow level.

It has been common to assume a linear relationship between density and speed:

$$K/K_{jam} = 1 - V/V_f$$

and the consequences of this model are shown in Figure 17.4. The assumption is due to empirical assessments made by Greenshields in the 1930s as shown in Figure 17.4c, although data often looks more like Figure 17.3c. When substituted into Equation 17.5 the model gives flow as a parabolic function of speed, the optimum speed as half the maximum speed and the optimum density as half the jam density.

Some have generalised Greenshield's density–speed equation to:

$$V/V_f = [1 - (K/K_{jam})^{d1}]^{d2}$$

where d_1 and d_2 are constants to be determined (for the original Greenshields case $d_1 = d_2 = 1$), which can be converted into an equation for flow q by using Equation 17.5. Other work suggests that the density–flow relationships in Figure 17.3d are composed of three linear curves, including a queue-discharge length during which density decreases whilst flow remains constant. In addition, the Greenshield's model assumes zero flow at maximum speed. However, it was shown in Figure 17.1 — and will be seen again in Figure 17.11 — that about 900 veh/h would pass by if that maximum speed was the free-flow speed and some lesser but non-zero flow would occur if the maximum speed was a free speed. The model is also unrealistic in that dK/dV should approach zero at maximum speed, as a small increment in traffic there would be expected to have a minor effect on traffic density. The reader is therefore recommended to concentrate on Figure 17.3 as a somewhat more realistic representation of traffic. Finally, the various curves in each of Figures 17.3 and 17.4 are merely planar representations of a three-dimensional curve, as shown in Figures 17.5.

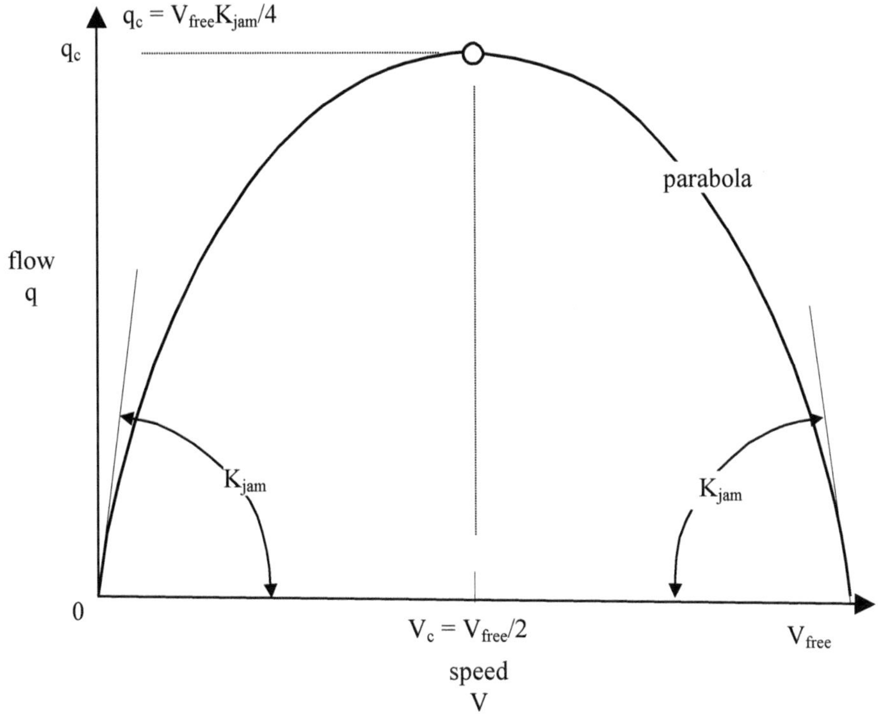

Figure 17.4a Flow–speed relationship for linear K–V assumption.

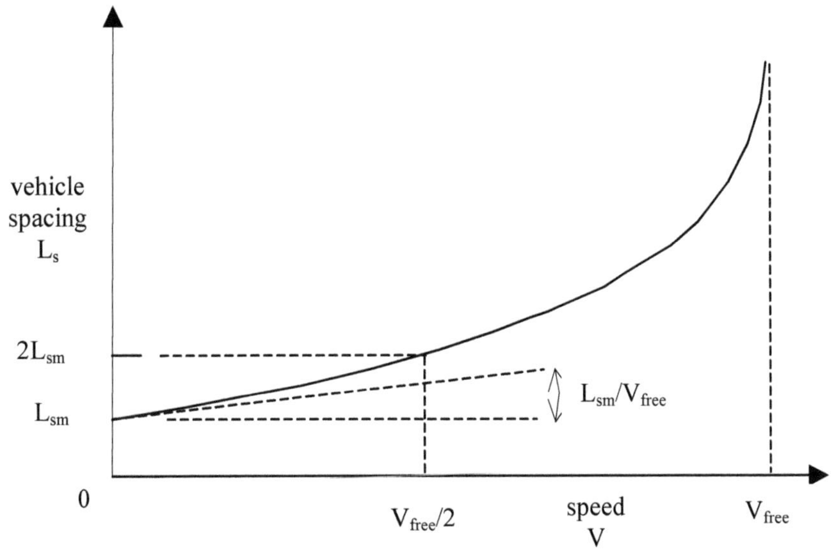

Figure 17.4b Spacing–speed relationship for linear K–V assumption.

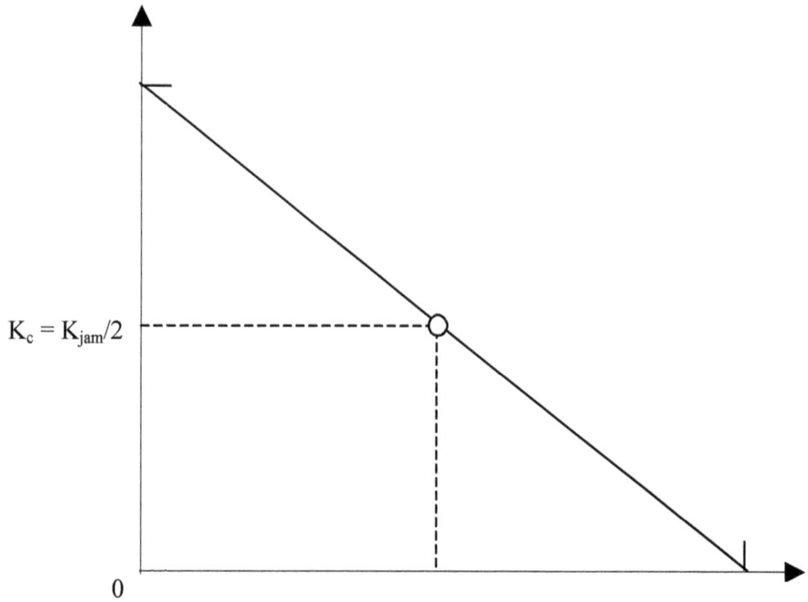

Figure 17.4c Density–speed relationship for linear K–V assumption.

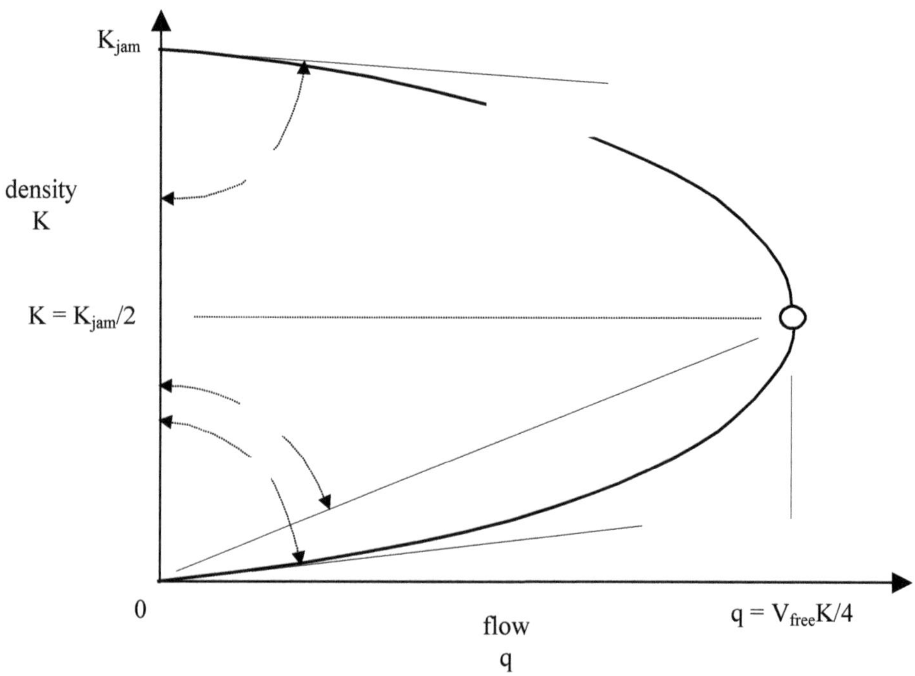

Figure 17.4d Density–flow relationship for linear K–V assumption.

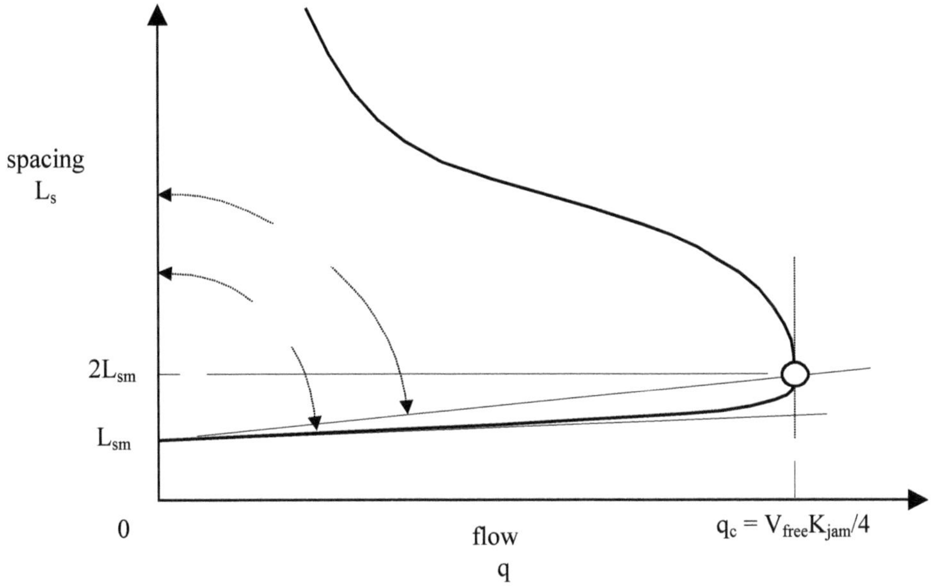

Figure 17.4e Spacing–flow relationship for linear K–V assumption.

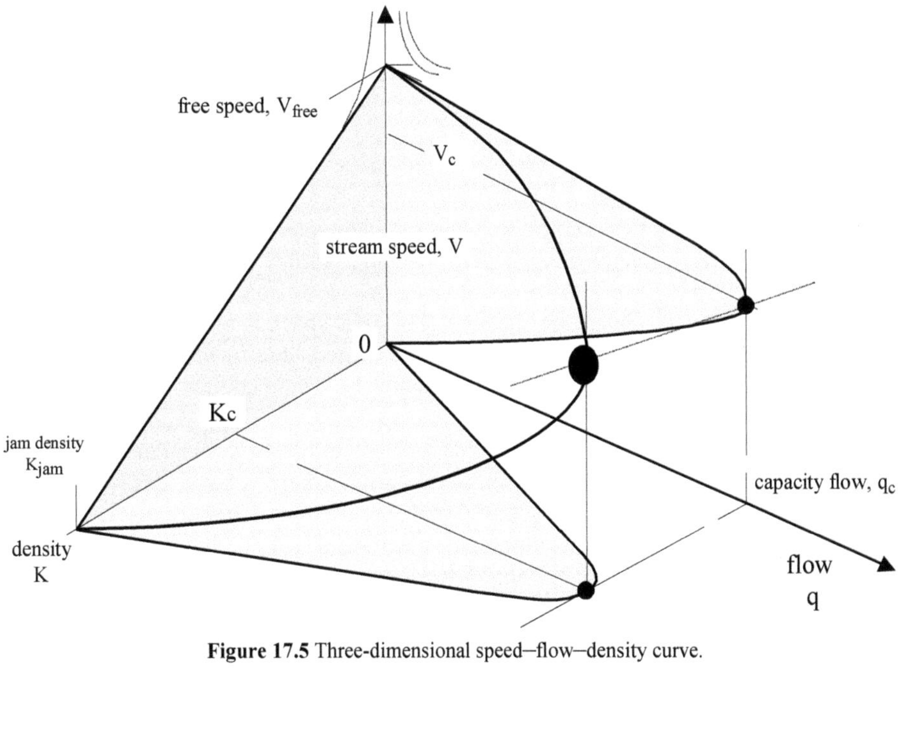

Figure 17.5 Three-dimensional speed–flow–density curve.

Note that the capacity flow, q_c, and associated optimum speed, V_c, may not represent the optimum situation from a transport systems viewpoint. For instance, we will see:

* in Section 29.2.2 that fuel consumption will be optimum at a speed unrelated to V_c,
* in Section 31.2.3 that the value of a traveller's time will force travellers to avoid congested conditions, and
* in Section 18.1.1 that the risk and consequences of a crash increase steadily with speed.

17.2.4 Car-following

The speed-density relationship in Figure 17.3c and the associated contours of equal flow, are based on the basic Equation 17.5 governing driver–vehicle behaviour. They are useful tools in considering a traffic stream in the uninterrupted model where the total flow cannot alter as a consequence of vehicles entering or leaving the system. Traffic behaviour must therefore lie on one of the contours of equal flow. If vehicles in the stream increase their speed, or slow down, then there must be a corresponding opposite change in traffic density in order to maintain constant flow. Taking partial derivatives of Equation 17.5 with q constant gives:

$$\partial V/\partial K = -q/K^2 = -V/K$$

or $$\partial V/V + \partial K/K = 0 \qquad (17.7)$$

Although car following was encountered in forced flow (Section 17.2.2), the term is usually restricted to cases where speeds vary and accelerations occur. Typical traffic changes that a driver in a single lane might encounter are:

* increased spacing to the vehicle ahead, as a vehicle immediately ahead leaves the traffic stream,
* decreased spacing to the vehicle ahead, as a vehicle joins the traffic stream immediately ahead, and
* catching up to a slower vehicle and thus reducing inter-vehicle spacing.

Each of these requires a change of speed and spacing to satisfy the driver's speed-spacing behaviour.

In making a speed change to achieve a desired spacing, the acceleration of the following car takes place after some delay has occurred, due to the following-driver's response time (Section 16.5.3). It can reasonably be assumed that a driver uses an acceleration or deceleration that is either:

* constant at, typically, $\pm$ 1 ms^{-2} in unconstrained circumstances or 2 ms^{-2} when there is some urgency (Section 27.2.2). In normal traffic there is some acceleration noise with a typical standard deviation of 0.2 ms^{-2}, or
* dependent on the difference between the intended speed and the current speed. The most common car-following approach assumes, fairly unsuccessfully from a modelling viewpoint, that:

  (acceleration of following car) =

  $$g_1(\text{speed differential between cars})^{h1}/L_s^{h2}$$

  or $$g_2(\text{speed differential between cars})^{h3} + L_s^{h4}$$

  or $$g_3(\text{speed of following car})^{h5}(\text{speed differential between the cars})/L_s^{h_6}$$

  where g_i and h_i are constants. It was often been assumed that $g_1 = h_2 = 1$ (Herman and Potts, 1961).

Mathematically, the general Equation 17.8 can be solved to give density as a negative-exponential function of speed, in the manner of Figure 17.3c. Using Equation 17.5 then gives:

$$q = K_jVe^{-kV}$$

where k is a constant. Other solutions give equations such as:

$$q = V/(k_1 + \Sigma[j_nV^n])^m \qquad (17.9)$$

and (May, 1990):

$$V = (1 + j_1q^{j2})^{j3}$$

where j_i are constants. Note that with m = 1 and the first term in the series, Equation 17.8 leads to Equation 17.7 and further putting n = 1 gives the case in Figure 17.4.

Extending the car-following assumptions to non-uniform conditions will usually require some form of car-by-car analysis, which is only practical via computer simulation (Section 17.3.10). For example, the *MULTSIM* model for simulating the explicit traffic behaviour of a variety of cars, and thus the speed–flow relationships, assumes that each vehicle in the traffic stream is either accelerating towards its desired speed or braking to maintain a safe distance behind the car ahead (Gipps, 1981). To do this, it assumes specific acceleration and deceleration characteristics for each vehicle. Using this method, the model is able to generate curves such as Figure 17.1 for realistic traffic conditions. The underlying Gipps model is that:

$$\text{(acceleration of the following car)} = k_2a_m(1 - V/V_m)(k_3 + V/V_m)^{0.5}$$

where a_m and V_m are the maximum desired acceleration and speed.

The individual vehicles travelling along a road are often shown on a distance–time trajectory. These graphs are discussed in Section 18.4.1.

17.2.5 Shock waves

Consider a steady, uniform stream of traffic vehicles adopting a speed, V_i, flow, q_i, and density, K_i, as in Figure 17.6. If for some reason — e.g. an incident (Section 17.4.4), a constriction, or a short length of steep grade — a lead vehicle changes its traffic speed from V_i to V_j, then the following vehicles will eventually need to alter their speed to adjust to the speed change of the leader. The local form of the adjustment was the subject of Section 17.2.4. In addition, the reaction to this event will flow like a wave backwards along the traffic stream, as all drivers sequentially respond to the change in their own lead vehicle. The wave is called a *shock wave*, and occurs at the time–space boundary between the two sets of conditions. Shock wave intensity is measured by the drop in speed across the shockwave.

Section 16.5.3 showed that drivers do not react instantaneously and gave estimates of the likely response time, T_r. One appropriate situation relates to a lead driver touching the brakes and thus activating the brake lamps. The following driver then reacts to the lamps and/or the visible deceleration of the lead vehicle by touching his/her brakes. In this case the relevant T_r is the time between the same initial response in the lead and following vehicles.

As the vehicles in the traffic stream are separated by $1/K_i$ (Equation 17.3), the shock wave will travel back down the traffic stream at a speed of:

$$V_s = 1/(K_iT_r) - V_i \qquad (17.10)$$

Sections 17.2.3–4 can be used to estimate the vehicle spacing as:

$$1/K_i = L_{sm} + T_rV_i + \text{(distance to brake from } V_i)$$

therefore,

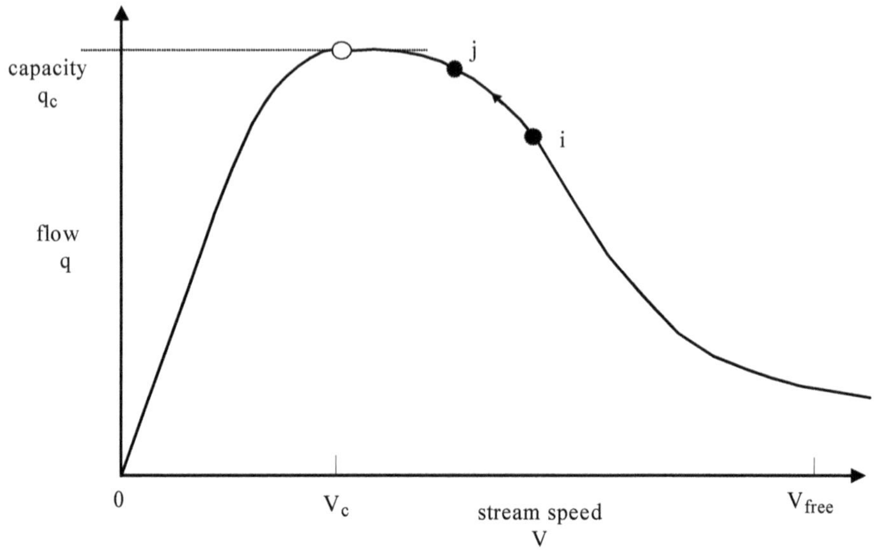

Figure 17.6 Shock-wave theory. The graph is based on Figure 17.3a.

$$V_s = \{L_{sm} + (\text{distance to brake from } V_i)\}/T_r \qquad (17.11)$$

which is always positive. The braking distance data in Table 21.1 and T_r data in Table 16.1 suggest that this shock wave will travel at speeds equivalent to the traffic speed.

The shock wave will occur at the completion points of the series of adjustments from condition i to condition j which take place between each subsequent pair of vehicles in the traffic stream. The lead vehicle decreases its speed from V_i to V_j. The following vehicle responds T_r later and the spacing between them closes by $(V_i - V_j)T$, where $T > T_r$, to accommodate the new acceptable separation distance of $1/K_j$. Hence T is given by:

$$1/K_i - 1/K_j = (V_i - V_j)T$$

so $$T = (1/K_i - 1/K_j)/(V_i - V_j) > T_r$$

and can be used with Equation 17.11 to estimate a shock-wave speed that will be much slower than the 'brake light' speed given by Equation 17.10.

If conditions before and after the shock wave are known, then the shock-wave speed can be calculated from a knowledge that the flow relative to the shock wave speed will be the same in each condition:

$$K_j(V_i - V_s) = K_i(V_j - V_s)$$

so: $$V_s = (q_i - q_j)/(K_j - K_i) = dq/dK$$

This leads to the observation that V_s is the slope of the line joining the two condition points i and j on the density–flow curve (Figure 17.3d). The shock wave speed is thus positive (forward) in light traffic and negative in heavy traffic. The typical case is where the traffic comes to a halt; the shock-wave speed is then $q/(K - K_{iam})$.

Applying this last expression to q vs K curves such as Figures 17.3d & 4d gives these second equilibrium shock-waves speeds as typically at about walking speed, e.g. 8 km/h, with the highest speeds in the relatively-meaningless jam condition. For example, if a crash or traffic signal change causes a sudden halt in a traffic stream with q_i = 1500 veh/h, then K correspondingly increases from 25 to 250 veh/km. The speed of the shock wave will travel back at about 7 km/h (rather than 70 km/h). Nevertheless, for each hour it takes to clear the incident, even this low-speed shock-wave will produce a queue some 7 km in length.

The straight line ij in Figure 17.6 is a speed vector for the shock wave. In real traffic, the non-uniform nature of driver behaviour usually means that drivers at the tail of long, moving queues will come to a complete halt.

17.2.6 Flow–time graphs

Most urban transport-planning models (Sections 31.3.7&9) use estimates of the speed–flow relationship for each link in the road network, although such relationships are usually expressed in terms of trip time on the link rather than vehicle speed and are called *trip time* (or *congestion*) *functions*. For a trip made at uniform speed, trip time will be proportional to the inverse of trip speed. Otherwise, the link between the two is more complex and is discussed in Section 18.2.12.

Section 17.2.3 defined free speed as the maximum practical speed on a link. The free speed for urban streets, V_{uf}, (sometimes misleadingly called the speed at capacity) is discussed in Section 18.3. It will usually be quite low, with values ranging from 40 km/h for an unpaved road to 70 km/h on a smooth surface, and may well be below the optimum speed, V_c, discussed in Figure 17.3. The traffic flow associated with this maximum speed will then represent the practical capacity, q_{pc}, of the urban link (Figure 17.7). Section 17.4.3 shows that many factors cause practical capacity to be below q_c. Practical capacity is rarely below 800 veh/h and is sometimes assumed to be $0.75q_c$ or $0.80q_c$. Greenshield's model (Figure 17.4) would lead to $V_{uf} = 0.5V_c$.

Figure 17.7 has a time–flow curve superimposed on it, as indicated by the darker-shaded area. If L_l is the length of the link, the theoretical minimum trip time assumes that the trip is not influenced by other traffic and therefore uses the local free speed, V_{uf}, to give a theoretical minimum, free-flow trip time of $T_{tm} = L_l/V_{uf}$. The practical minimum trip time, T_{pm}, is somewhat greater and, as with q_c and q_{pc}, the difference will depend on the traffic features of the link. For example, if the link includes traffic signals, T_{pm} will include the most likely delays caused by those signals in the absence of other traffic (Section 23.6.3). In an urban system, network effects (Section 23.6.3) may cause the incoming flow to rise above the practical capacity, q_{pc}, of a link. This will cause traffic delays to further increase as the traffic flow increases. The combination of these effects is shown by curve A and the associated shaded-region in Figure 17.7. The curve between B and A in Figure 17.7 becomes straighter as the traffic becomes more irregular, and tends to BDA' as it becomes more regular.

The average speed vs flow graph is as shown in Figure 17.8 and is obtained in part as the reciprocal of the shaded addition to Figure 17.7. Therefore, the separate flow–speed curves of traffic engineers and transport planners should not be confused, and can be distinguished as follows:

* *traffic models* (Figures 17.3&7): used for predicting:
  (a) flow on a link between nodes;
  (b) arrival flows upstream of an intersection, or
  (c) link demand (demand predictions are discussed in Section 23.6),
* *transport models* (Figure 17.8): used for predicting:
  (a) the flow on a link and its nodes,
  (b) departure flows downstream of an intersection, or
  (c) behaviour when demand exceeds link-node capacity. For example, flows above q_{pc} in Figures 17.7&8 represent the over-saturated situation where the arrival flow exceeds the maximum departure flow.

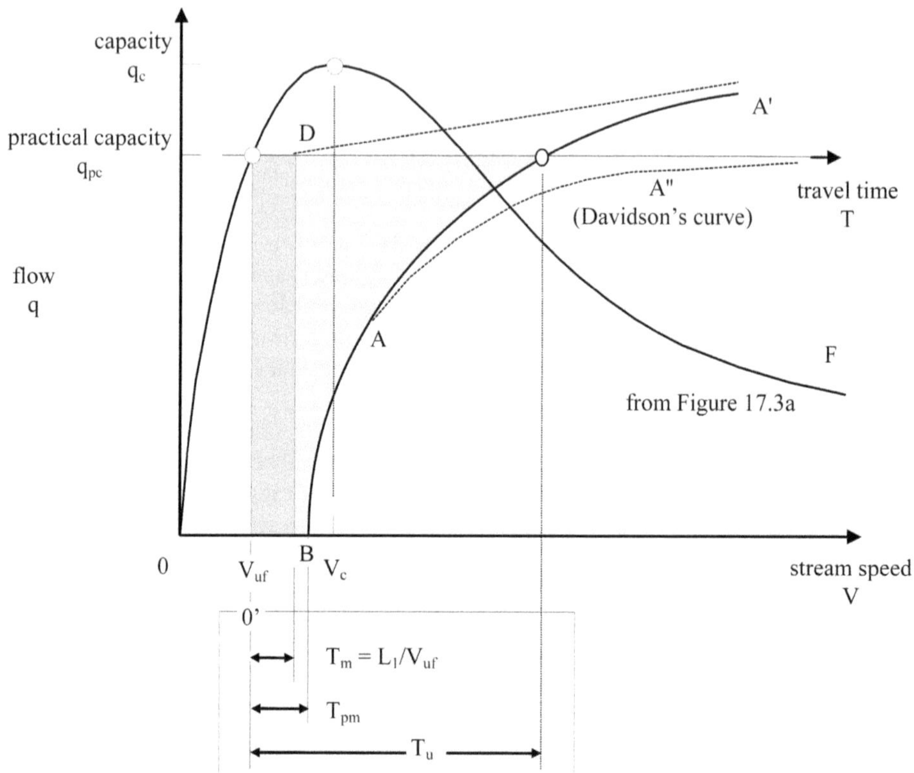

Figure 17.7 Time–flow for an urban road (traffic engineering). Note that V_{uf} is the maximum 'free' speed dictated by local conditions. The lightly-shaded area shows network-induced queuing and slowing time.

Typical algebraic representations of Figure 17.7 used in transport modelling are shown below. Note that the forced flow equations in Section 17.2.3 (including Greenshield's approximation) and most of the car-following models in Section 17.2.4 cannot be written explicitly in terms of V. All but Equations 17.12j&k combine the two conditions (unforced and congested) and their respective curves into one equation. The Normann equation (Equation 17.12b) does not permit flows above capacity, i.e. $x > 1$. The equations all use $x = q/q_{pc}$.

$V/V_{uf} = 1 + C_1x$ (17.12a)

$= 1 - x$ (Normann equation, 17.12b)

$= 1 + C_2x + C_3x^2 (x < 1)$ (17.12c)

$= (1 + C_4x^p)^r$ (for car-following, commonly p = 1, r = 0.5),(17.12d)

$= (1 + C_5x)^{0.33}$ (Smeed equation) (17.12e)

$= 1/(1 + C_6x)$ (produces a linear flow–time curve) (17.12f)

$= 1/(1 + C_7x^4)$ (BPR–FHWA equation, with $C_7 = 0.15$) (17.12g)

$= 1/(1 + C_7x^{10})$ (Dowling's equation, with $C_7 = 0.15$) (17.12h)

$= 1/(1 + C_8[1 - x + \{(1 - x)^2 + C_9x\}^{0.5}])$ (Akcelik's equation) (17.12i)

$= 1 + C_9x^{27}$ if $x < 1$ (Akcelik & Besley, 1996) (17.12j)

$= x + C_{10}x^{2.55}$ if $x > 1$ (Akcelik & Besley, 1996) (17.12k)

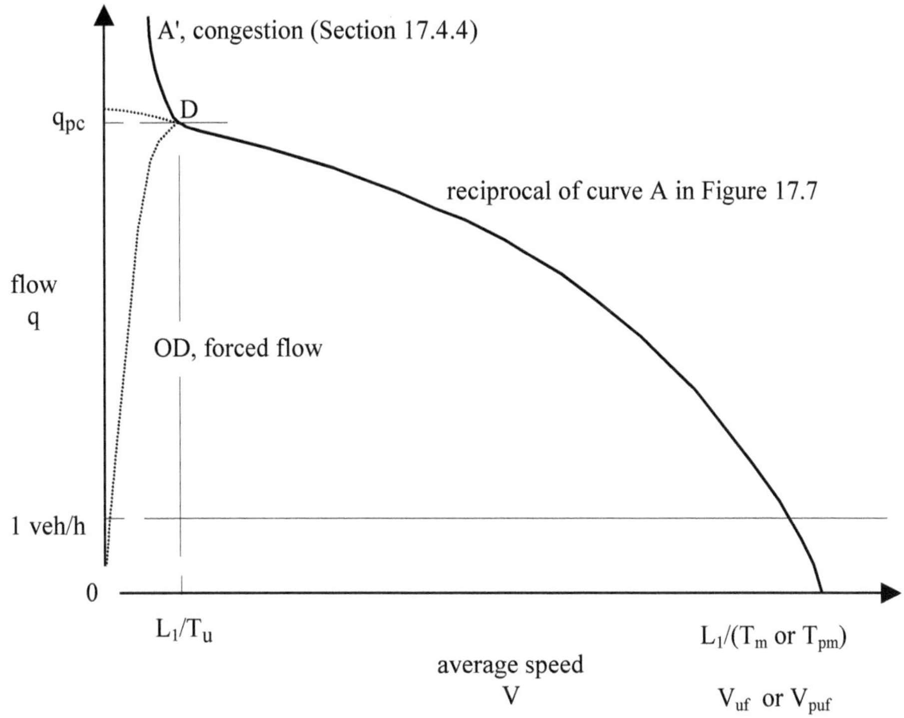

Figure 17.8 Speed–flow for one link in a road network, as used in transport planning. Note that T_u is the trip time for the link when it is at practical capacity (see Figure 17.7).

Akcelik's equation, Equation 17.12i, is discussed further in Section 23.6.5 as Equation 23.6b.

From Section 17.2.3, at $x = 1$, $V = V_c$, putting $s = V_c/V_{uf}$ allows some of the constants in Equations 17.12 to be given some physical significance; viz.:

C_1, C_9, C_{10}, and $(C_2 + C_3) = (1 - s)$

$$C_4 = -(1 - s^2)$$

$$C_5 = -(1 - s^3)$$

$$C_6 \text{ and } C_7 = -(1/s - 1)$$

$$C^2_8C_9 = (1/s - 1)^2$$

Figure 17.9 shows a piecewise-linear form of the relationship used in some computer models. The increase in flow above capacity at low travel speeds reflects the situation when demand exceeds supply (congestion) and acknowledges that this volume does eventually pass through the system. Using the Normann speed–flow Equation 17.12b gives trip time, T, as:

$$T = T_{pm}/(1 - x)$$

A popular modification of this model is Davidson's time–flow equation (Davidson, 1978):

$$T = T_{pm}\{1 - (1 - j)x\}/\{1 - x\} \qquad (17.13)$$

where j is a level of service factor (Section 17.4.2) which Davidson derived from queuing theory with specific values determined empirically from traffic flow measurements. Its significance can be seen by using Equation 17.13 to calculate the implied traffic delay $(T - T_{pm})$ as:

$$(T - T_{pm})/T_{pm} = j/[1/x - 1]$$

For a given q, j therefore determines how much delay will occur. The term (1 – j) is a flow efficiency factor going from 0 (worst case), through a typical value of 0.7, to 1 (best case).

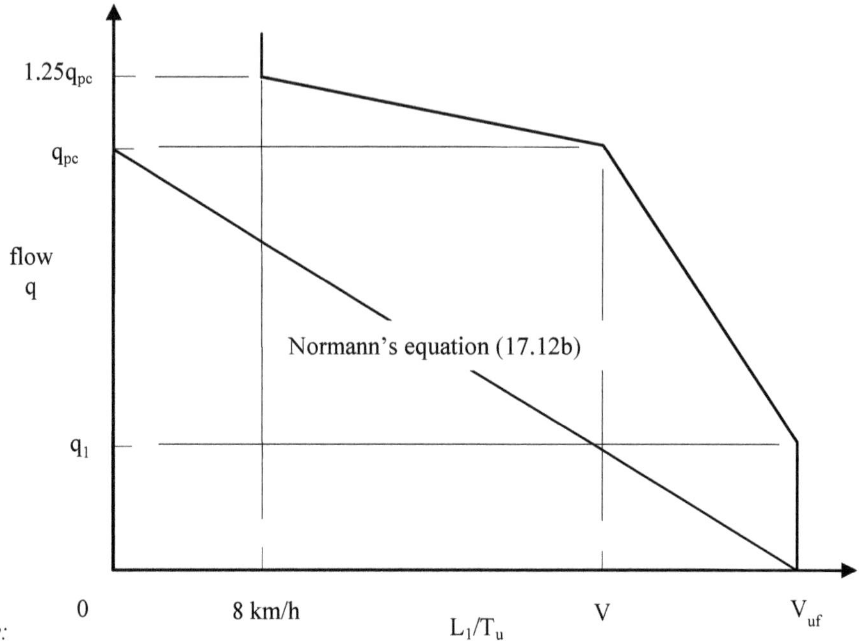

Key:

q_1/q_{pc} = 0.1 for one- and two-lane roads
= 0.2 for three- and four-lane undivided roads
= 0.3 for divided roads
= 0.4 for freeways

Figure 17.9 Speed–flow diagram for a link in a road network, as used in some computer-based models.

Equation 17.13 implies steady-state conditions and so is unrealistically asymptotic to the q_{pc} line (BAA") in Figure 17.7. In congested conditions there will not be a steady state, and conditions will depend on the relative magnitude of the processes of queue creation and extension, and queue discharge (see Figures 23.9&10). Modifications to BAA" produce the more realistic BAA' curve for flows above 85 percent of the capacity flow, q_{pc} (Akcelik, 1978). This new curve is asymptotic to a straight line based on a steadily increasing queue length during congestion (Section 17.4.4).

17.3 HEADWAYS AND GAPS

17.3.1 Definitions

The word *gap* in traffic engineering is usually reserved for time separations, T_g, between vehicles passing a defined point, measured from the rear of the lead vehicle to the front of the following vehicle. *Gap* is sometimes applied to space separations as well, but will not be so used in this book. The discussions in Sections 17.2.4 & 17.3.8 show that there are minimum gaps which vehicles will adopt at a particular speed.

A related term is headway, T_h, which is the time interval between vehicle arrivals at a fixed point based on the same sighting point on each vehicle (e.g. the front of each vehicle). The term *time headway*, T_h, is sometimes used, with *space headway,* L_s, being the same as *vehicle spacing* defined in Section 17.2.1. Given the above definitions, the relationship between gap and headway is (Figure 17.2):

$$T_h = T_g + \text{(time for the following vehicle to pass)} \quad (17.14)$$

However, it is often assumed that:

$$T_h = T_g \quad (17.15)$$

which is reasonably accurate at higher speeds.

For distance rather than time, the equivalent of Equation 17.14 is:

$$L_s = \text{(space gap)} + \text{(length } L_V \text{ of following vehicle)}$$

For uniform traffic, applying Equation 17.4 to Equation 17.14 leads to:

$$T_h = 1/q = L_s/V = 1/(KV) \quad (17.16)$$

$$(s) = 3600/(veh/h) = 3.6(m)/(km/h) = 3600/[(veh/km)(km/h)]$$

The headway given by Equation 17.16 is only a mean value when the traffic flow is not uniform. T_h is reasonably constant over speed so Equation 17.16 suggests that space headway increases linearly with speed.

Section 17.4.1 suggests that the maximum flow in a typical urban traffic lane, q_c, is about 1800 veh/h, which Equation 17.16 indicates to be equivalent to a minimum headway of T_h = 2.0 s. The speed, V_c, at this particular maximum flow ranges between about 40 to 60 km/h, depending on the quality of the road. From Equation 17.4, these speeds correspond to a vehicle spacing of 22 to 33 m. From Equation 17.14, at 60 km/h the T_h = 2.0 s minimum headway corresponds to a gap of about 1.8 s. Minimum headways are discussed further in Section 17.3.7 and measured headways in Section 17.3.9.

Knowledge of the statistical distribution of headways and gaps within a traffic stream is of relevance in:

* predicting basic traffic flow (Section 17.2.4), and
* calculating crossing (Section 17.3.5), merging (Section 17.3.6), and overtaking opportunities (Sections 18.4.2&3).

Distributions that focus on the number of arrivals are called *counting distributions* and those that consider the number and size of gaps are called *gap distributions*. This issue is pursued in Section 17.3.2 for low traffic flows and Section 17.3.7 for heavy traffic.

17.3.2 Theory for low traffic flows

Section 17.2.2 showed that a line of vehicles will operate independently of each other (i.e. flow will be *free*) if the traffic flow is below about 900 veh/h. Such low traffic flows may be almost random in that the probability of a vehicle passing a given point during some interval will be independent of the pass-by times of other vehicles. To derive the associated probability density function, (p(x), Section 33.4.1), for x vehicles passing by, it is assumed that:

(a) the traffic flow is q and so the mean number of vehicles arriving over time δt is $q\,\delta t$,

(b) the number of vehicles arriving is small ($q\delta t \ll 1$),

(c) the probability of two vehicles arriving in δt is zero, and

(d) the number of vehicles arriving is independent of the number that arrived in the previous time interval.

The probability of x vehicles arriving in a time $(t + \delta t)$ is thus the sum of the probabilities of:

[x vehicles in t and of none in δt] + [(x – 1) vehicles in t and one vehicle in δt].

Now the probability of one vehicle arriving over δt is $q\delta t$ ($\ll 1$) and the probability of no vehicles in δt is thus $(1 - q\delta t)$ and so:

$$p_{t+\delta t}(x) = p_t(x)[1 - q\delta t] + p_t(x - 1)q\delta t$$

[x veh in t, 0 veh in δt] [(x – 1) veh in t, 1 veh in δt]

In the limit, as $\delta t \to 0$:

$$dp(x)/dt = q[p(x - 1) - p(x)]$$

The solution of this differential equation leads to the well-known *Poisson distribution* (Section 33.4.5) for the number of vehicles in time t, i.e.

$$p_t(x) = e^{-qt}(qt)^x/x! \quad (17.17)$$

Defining the mean (small) number of vehicles expected to arrive in time t as m leads to:

$$m = qt \quad (17.18a)$$

and $$p_t(x) = e^{-m}m^x/x! \quad (17.18b)$$

Thus the Poisson distribution is the appropriate distribution for random, independent arrivals when the overall flow, q, remains a constant. As q^{-1} is the average headway in the traffic stream, some have taken t as the headway and interpreted Equation 17.18 as:

$$m = T_h/(\text{average } T_h) \quad (17.19)$$

The Poisson distribution does not apply when the traffic flow loses its random nature when:

* vehicles interact at high traffic flows. This typically occurs when $q > 900$ veh/h (forced flow, Section 17.2.3), but may occur for q as low as 400 veh/h (Figure 17.11). Alternate models for higher traffic flows are discussed in Section 17.3.9.
* traffic is congested, interrupted, or restricted.

Despite its widespread use in practice, the Poisson distribution thus properly has restricted application.

For low flows, the Poisson model can also be used to predict the distribution of the headways. Writing $P(T_h > t)$ for the probability that a headway T_h is greater than some interval t, then the probability that T_h will be greater than $(t + dt)$ is:

$$P[T_h > (t + dt)] = P[T_h > t] \, . \, [1 - qdt]$$

(prob. of the event in Th) (prob. of no event in dt)

hence $$P(T_h > t)/dt = - P(T_h > t)q \quad (17.20)$$

and $$P(T_h > t) = e^{-qt} \quad (17.21)$$

As the probability of any headway is 1.0, then

$$(T_h < t) = 1 - e^{-qt} \quad (17.22)$$

Hence, the actual number of headways per time greater than and less than t is:

number of headways $> t = qe^{-qt}$

number of headways $< t = q[1 - e^{-qt}]$

These equations show that vehicle headways in a random Poisson traffic flow follow a *negative-exponential distribution* (Section 33.4.5). This distribution has a mean of $1/q$ and a variance of $1/q^2$. Field data suggests that the log-normal distribution is also effective in representing these headways.

An alternative solution is to define a headway as a time t with no vehicles in it, i.e. with $x = 0$. Symbolically;

$$P(T_h > t) = p(x = 0)$$

Substituting into Equation 17.20 gives:

$$P(T_h > t) = e^{-qt}(qt)^0/0! = e^{-qt\,t}$$

which corresponds to Equation 17.21.

17.3.3 Applications of the Poisson model

For traffic applications, the *gaps* between vehicles in a traffic stream are far more important than the headways that they adopt. The relation between the two was described in Equation 17.14 and the following discussion makes the simplifying assumption that they are equal (Equation 17.15).

Individual drivers exhibit different vehicle-spacing and car-following characteristics (Section 17.2.4) and so gaps of various sizes will occur in a traffic stream, even when all drivers are travelling at constant speed.

Equation 17.21 is the core system descriptor. It can be used to predict that the gap which is 95 percent certain to be encountered in a flow of 900 veh/h is $\ln 0.95/(-900)$ h = 0.2 s and that there is a 50 percent chance of a 2.8 s gap. The equation may also be interpreted as giving the proportion of intervals greater than t, i.e.:

$$(\text{proportion of gaps} > t) = e^{-qt} \qquad (17.23)$$

The number of gaps in time T is qT, so the number of gaps greater than t is given by:

$$(\text{number of gaps} > t\text{, in } T) = qTe^{-qt} \qquad (17.24)$$

or,

$$(\text{frequency of gaps} > t) = qe^{-qt} \qquad (17.25)$$

These are all negative-exponential distributions (Section 33.4.5).

Similarly, a representative size of each gap may be obtained by multiplying it by its own probability dP given by Equation 17.18, viz.:

$$dP = -qe^{-qt}$$

The time occupied by gaps greater than t is:

$$(\text{time with gaps} > t) = \int_{t=t}^{\infty} t\,dP = (-1/q)\int_{t=t}^{\infty}(-qt)e^{-qt}\,d(-qt)$$

$$= [(1 + qt)/q]e^{-qt}$$

The equivalent total time is the same integral from 0 to ∞, which has the value of $1/q$ (i.e. an average gap) and so the time with gaps greater than t in any time T is:

$$(\text{time with gaps} > t \text{ in } T) = (1 + qt)e^{-qt}T \qquad (17.26)$$

and the proportion of time T with gaps greater than t is:

$$(\text{proportion of time with gaps} > t) = (1 + qt)e^{-qt} \qquad (17.27)$$

The use of Equations 17.24 and 17.27 is illustrated in Figure 17.10, which shows the number of gaps in excess of 10 s available in 5 minutes. Although the number of gaps peaks at 11 at 360 veh/h (from Equation 17.24), this is not an optimum flow as Equation 17.25 shows that the actual time available to people waiting for gaps over 10 s continues to increase as the flow drops to zero. The equations are clearly of considerable use in calculating prospects for crossing through a traffic stream. The importance of the 10 s gap used in the Figure 17.10 example is that it does appear that practically all drivers will accept gaps of 12 s or more.

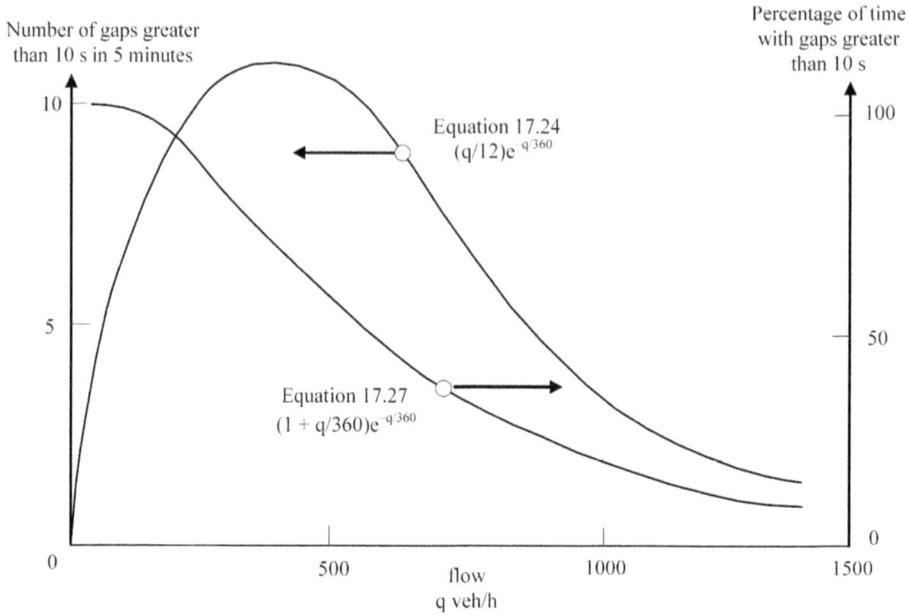

Figure 17.10 Example of a Poisson gap calculation, based on the number and percentage of gaps greater than 10 s.

The average duration of a gap is obtained by dividing Equation 17.27 by Equation 17.24 to obtain:

$$(\text{average duration of gaps} > t) = (1 + qt)e^{-qt}T/qe^{-qt}T$$

$$= (1 + qt)/q \quad (17.28)$$

$$= t + (1/q) \quad (17.29)$$

$$= (\text{overall average gap duration}) + t$$

Similarly, the average duration of gaps less than t is:

$$(\text{average duration of gaps} < t) = 1/q - [te^{-qt}/\{1 - e^{-qt}\}]$$

17.3.4 Critical gaps

Consider a driver in one traffic stream who is planning to cross (or pass through) another stream at an uncontrolled intersection (Section 20.1.1), or to enter (or merge into, Section 17.3.6) the other stream. In a process known as *gap acceptance*, the driver will wait until a gap of acceptable size occurs in the other stream. The driver assesses gap acceptability by intuitively establishing an acceptable gap size and then comparing this with the size of the gap the driver expects to encounter during the manoeuvre, based on the driver's current estimate of the separation and relative speed of approaching vehicles. Section 16.4.6 described how drivers are not skilled at estimating the separation and relative speed of approaching vehicles, and their meagre abilities decline with age.

Drivers will probably estimate the separation as a space separation, rather than as a time gap, as they make their decision on the basis of the spatial separation that they see at the decision time. Nevertheless, the theory has developed around the time gap as the observed safety of the manoeuvre depends directly on the time gap at the crossing point.

The *critical gap (*or acceptance gap), T_{gc}, is the smallest gap that a driver will accept. Gaps larger than the critical gap are called *super-critical.* The size of the critical gap will depend on:

* the characteristics of the vehicle, particularly vehicle length and acceleration capability,
* the characteristics of the driver. [Size depends on the mood of the driver and will vary between drivers: for example, males typically accept smaller gaps than females.]
* the type of trip. [Drivers require larger critical gaps when carrying passengers.]
* the driver's ability in the particular circumstance to estimate the size of the gaps in the approaching traffic,
* the type of traffic flow,
* the type of conflicting vehicle. [Drivers require larger gaps when a motorcycle is involved, possibly because its performance is difficult to judge. They require smaller gaps for bicycles, possibly because the risks are much lower.]
* the type of movement being attempted,
* how long the driver has been delayed. [Size is reduced by driver impatience caused by an absence of suitable gaps, particularly when the driver's total queuing delay has exceeded about 20 s. Beyond 20 s, the critical gap can drop by a second for every further 10 s of delay.]

Because the size of the critical gap is so variable, for design purposes it is often defined as the gap size which, when offered, is accepted as often as it is rejected, i.e. there is a 50 percent probability that a chosen driver will accept the offered gap. Critical gaps within the driving population appear to be log-normally distributed, although the Erlang, gamma, negative-exponential, and normal distributions have also been used (Section 33.4.5). Critical gap values for specific circumstances are:

* *high-speed merging* (Section 17.3.6). Critical gaps are usually in the 0.5 to 6 s range, averaging about 3 s. Means increase linearly from 0.7 s at zero relative speed to 4 s at a 50 km/h relative speed.
* *merging at a roundabout entry.* Critical gaps are typically 3.5 s (Section 20.3.6).
* *crossing a traffic stream* (Section 20.1). Critical gaps are commonly in the range 2 s to 10 s, averaging about 5 s to 6 s. They depend greatly on the circumstances and on the difficulty of the crossing manoeuvre and the high values apply to crossings of multilane highways. Nearly all gaps over 12 s will be accepted.
* *follow-up gap.* Applies when following other vehicles through a super-critical gap (Section 17.3.5). Typical values are in the range 1 to 5 s, with a mean near 2 s. It is equivalent to the headway in a saturation flow (Section 17.4.4).
* *low-speed turning into a major road.* Critical gaps for standing starts usually lie between 3 and 8 s, with a mean near 7 s. Drivers require smaller critical gaps when centre-turning, compared with their needs when kerb-turning.
* *overtaking.* Critical gaps are between 20 s and 30 s (Table 18.3). They are much larger than the above values as they must also allow time for one vehicle to pass another travelling in the same direction (Sections 18.4.2&3).

Further discussion of the role of critical gaps will be found in Sections 18.4.2 and 20.1.3.

A *lag* is the unexpired portion of a gap that remains when a driver arrives at a traffic stream, and a *critical lag* is the smallest accepted lag. It is closely related to, and usually larger than, the critical gap. Indeed, for convenience, gaps and lags are often taken to be equivalent.

A sequence of sub-critical gaps is called a *block* and the name effectively describes the effect that it has on drivers wishing to join or pass through it. An *anti-block* describes a sequence of super-critical gaps.

The number of gaps that will be accepted will depend on the distribution of gaps in the traffic stream being crossed or joined. Assuming that the Poisson distribution applies, Equation 17.24 gives the number of gaps of T_{gc} or greater (i.e. anti-blocks) in time t as:

$$(\text{number of gaps} > T_{gc} \text{ in } t) = qte^{-qT_{gc}} \quad (17.30)$$

which is another negative-exponential distribution (Section 33.4.5). Equation 17.29 gives average duration of such a super-critical gap as:

$$(\text{average super-critical gap duration}) = (1/q) + T_{gc}$$

As T_{gc} is unacceptable to anyone arriving within that time of the end of a gap (i.e. a sub-critical lag), the average single anti-block length is 1/q.

Using Equation 17.21, the total time available for joining or crossing a traffic stream in time T is $Te^{-qT_{gc}}$ and therefore:

$$(\text{proportion of time that a driver entering a stream is delayed}) = 1 - e^{-qT_{gc}} \quad (17.31)$$

From Equation 17.30 and Figure 17.10, it can be seen that the number of super-critical gaps in a busy traffic stream will be low in conditions of high flow. Generally, the greater the flow, the smaller the gaps and hence the probability of encountering a critical gap will diminish as the flow increases. Merging or crossing will be difficult. In practice, the situation will be improved by forming the traffic into platoons (Section 17.3.8) and, incidentally, destroying its randomness. Whilst the gaps within a platoon will be very small, there will be much larger — and hopefully super-critical — gaps between the platoons.

17.3.5 Traffic absorption

Unsignalised intersections and freeway entrance ramps are cases where other traffic uses some of the unutilised capacity (or *ullage*) of a traffic stream. The number of vehicles that can cross or join such a traffic stream, with flow q_m, is found by assuming that the gap in the stream sufficient to allow n or less vehicles to enter is t_n. The number of gaps per time unit exceeding t_n is given by Equation 17.30 as $q_m e^{-q_m t_n}$. The number of gaps of a particular size per time unit is therefore $q_m[e^{-q_m t_n} - e^{-q_m t_{n+1}}]$. Hence, the total number of vehicles that can be absorbed into the traffic stream per time unit, q_N, is given by:

$$q_N = \sum_{n=1}^{\infty} n q_m [e^{-q_m t_n} - e^{-q_m t_{n+1}}]$$

$$= q_m \sum_{n=1}^{\infty} e^{-q_m t_n}$$

The value of t_n is given by:

$$t_n = T_{gc} + (n-1)T_{gf}$$

where T_{gf} is the time for successive minor flow vehicles to move up and enter the stream after the first vehicle (sometimes called the *follow-up gap*). Typical values of T_{gf} are in the range 1 to 5 s, with a mean near 2 s.

Substituting this value for t_n into the q_N expression gives q_N as:

$$q_N = q_m \sum_{n=1}^{\infty} [e^{-q_m T_{gc}}][e^{-q_m \{n-1\} T_{gf}}]$$

$$= q_m e^{-q_m T_{gc}} \sum_{m=1}^{\infty} e^{-q_m \{n-1\} T_{gf}}$$

$$q_N = q_m e^{-q_m T_{gc}} / [1 - e^{-q_m T_{gf}}] \qquad (17.32)$$

Following Equation 17.30, the first part of the expression in Equation 17.32 — $q_m e^{-q_m T_{gc}}$ — gives the number of acceptable gaps in a time unit. Thus the second part — $\{1/[1 - e^{-q_m T_{gf}}]\}$ — covers the fact that more than one vehicle may use a gap.

Equation 17.32 gives q_N, the *absorption capacity* in vehicle/time units for vehicles crossing a major traffic stream with a flow of q_m. In particular, it may be used for unsignalised intersections and roundabouts. However,

* it is common practice to assume that the practical absorption capacity is about 80 percent of the value predicted by Equation 17.32,
* some design codes omit its denominator and account for T_{gf} by replacing T_{gc} with $[T_{gc} - (T_{gf}/2)]$.

Simpler empirical models have also been developed. For example, PICADY and TRANSYT basically assume that:

$$q_N/q_{Nc} = 1 - (q_m/q_{mc})$$

SATURN uses $[1 - (q_m/q_{mc})]^n$ where n is a function of the critical gap. A common approximation is, for q in veh/h:

$$q_N = 750 - 0.4q_m$$

When the minor stream flow is considerably below q_N, then the *average delay* to each minor stream value is calculated as the sum of the delays due to:

(1) finding that the first gap is below T_{gc},
(2) meeting subsequent gaps which are below T_{gc},
(3) waiting for the previous vehicle to proceed, and
(4) arriving too late to use the current super-critical gap.

The resulting equation, known as *Adam's delay*, is:

$$\text{average (or Adam's) delay} = [1/(q_m e^{-q_m T_{gc}})] - T_{gc} - [1/q_m]$$

and applies to minor-stream vehicles with a critical gap of T_{gc} waiting to cross a major stream flow of q_m. The average delay of delayed vehicles only is:

$$\text{(average delay of delayed vehicles only)} = [1/(q_m e^{-q_m T_{gc}}] - [T_{gc}/(1 - e^{-q_m T_{gc}})]$$

Models developed to handle these problems are commonly known as A/B/N models. *A* refers to the main flow distribution, *B* to the crossing or joining flow distribution, and *N* to the number of lanes. A and B describe the distributions assumed which can be either M (random, as in the Poisson-based distributions discussed in Section 17.3.2), G (undefined), D (regular), or E (Erlang, Section 33.4.4).

17.3.6 Merging

Sections 17.3.4–5 discussed vehicles merging in to a traffic stream in terms of the vehicles joining the stream by utilising acceptable gaps within it. The actual process is somewhat more complicated. Drivers in a lane attempting to merge into a primary lane have a difficult set of eight tasks to undertake. These are to (Figure 17.11):

(a) from an access ramp from a secondary road to a primary road, *enter* a *merge* (or *merging* or *acceleration*) *lane,* commonly at a lower speed than the traffic in the primary road [another use of the term acceleration lane is given in Section 20.3.6],
(b) steer and *maintain safe headways* in the merge lane, and in any speed-change lane provided at the end of the merge lane,

(c) if the merging carriageway has more than one lane, with some of the lane marking stopping before this carriageway ends, *watch* for other merging drivers,
(d) in the merge lane, *accelerate* towards the speed of the traffic in the adjacent lane of the primary road,
(e) *search* for a super-critical gap in this primary lane (this process often does not commence until (d) is well advanced),
(f) *steer* laterally into the primary lane when a super-critical gap passes (this will be at a relative speed equal to the speed difference between the traffic in the two lanes),
(g) *accelerate* up to the speed of other traffic in the primary lane. [Recall that mean critical gaps for merging increase linearly from 0.7 s at zero relative speed (Section 17.3.4). Clearly the optimum merging speed is as close as possible to the speed of the primary traffic lane. However, experience suggests that the merging speed for trucks can be about 15 km/h below that of the traffic stream.],
(h) if there is a failure in any of steps (e), (f) or (g), either:
   * *accept a smaller gap* than normally desired,
   * *force a merge* into the primary lane, with a subsequent increase in crash risk: primary lane drivers will often move into the next lane to avoid this, or
   * *stop or slow down*, which will greatly increase the speed difference between the two lanes and will mean that subsequently only much larger (and therefore less frequent) gaps will be able to be accepted.

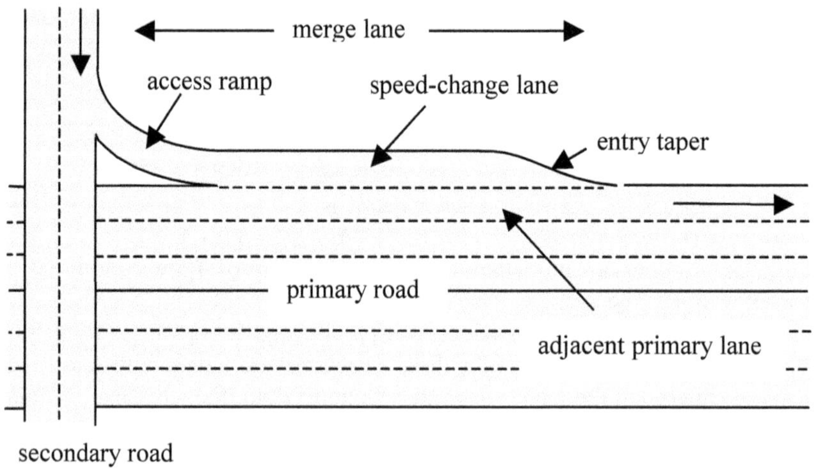

Figure 17.11 Merging geometry. Note: the horizontal scale is greatly foreshortened. Mirror image for right-hand driving.

A merge lane must therefore at least include an adequate length to allow a driver to:
   * accelerate from the expected speed on the access ramp to the speed of the traffic in the primary lane,
   * detect any super-critical gaps presented, and
   * accept the gap and merge into the primary stream.

This is achieved by either using a *speed-change lane* parallel to the primary lane or a large entry taper. Whilst the former is more common, some prefer the taper alternative as it requires the driver to make one less steering manoeuvre. Either of these facilities is of added assistance as they:
   * allow a driver to complete tasks (a) to (d) independently of the next four,

* remove angular velocity considerations from the speed-matching process in tasks (d) and (g), and
* heighten the chance of a super-critical gap being presented.

Speed change lanes will frequently be over 500 m in length. All ramps will eventually taper in width from the ramp width to zero. It is common to only consider the portion of the taper which is at least the width of a conventional lane as useful in the speed change process. Guidance on merging lane design is given in AASHTO (2004).

To avoid further complicating the merging/diverging process, care is taken in freeway design to ensure:

* adequate separation between the end of an entrance ramp and the start of the next exit ramp to permit effective *weaving* (at least 200 m and typically 500 m, Sections 6.2.5 and 20.2.1). It is desirable to have at least two through lanes in addition to the weaving lane, and
* *lane balance*, in order to accommodate demand changes after a merge and an exit. Ramp lanes will usually have a lower capacity than even a weaving lane. Thus, it is common to assume that if N lanes enter or leave at a ramp, then (N – 1) lanes should be added or subtracted on the freeway (see also the discussion of *dropped lanes* in Section 6.2.5).

From a driver behaviour viewpoint, it takes some courage and experience for a driver to approach freeway speed on an on-ramp. Many unskilled drivers nearing the end of an on-ramp decelerate rather than accelerate. Not only do they increase their relative speed — and thus the difficulty of the merge task — but they also make it much harder to finally bring their car up to the speed of the traffic on the primary road. We saw above that an acceptable gap could be 3 seconds. Whilst this can be easily selected at a relative speed of zero, few normal drivers are able to reach 100 km/h in 3 seconds from a standing start. This would require an average acceleration of about 9 m/s^2 whereas few drivers — particularly those that slow at the end of an on-ramp — are willing to exceed 2 m/s^2 (Section 27.2.2).

There is also a real safety hazard related to stopping or slowing at the end of an on-ramp. The driver behind the slowing vehicle may well be looking to the driver's side rear in order to detect an oncoming acceptable gap, assuming 'normal' accelerative driving from the driver ahead. This driver may not notice the vehicle ahead slowing down. This behaviour is a cause of some end-to-end crashes at the end of on-ramps.

This leads to the problem for the 'normal' merging driver who reaches the end of the merge lane at close to the freeway speed, without having found an acceptable gap. In a fall-back case, some drivers in this situation will force a merge into a less-than-acceptable gap, at least subconsciously relying on the fact that even a 2 s gap on a lane can usually be temporarily reduced without dire consequences — particularly when all involved drivers are aware of what is happening.

However, the far more desirable situation is that the drivers on the outer lane of the primary road are sufficiently competent and well-trained to permit the on-ramp drivers to slot into their outside lane, one merging vehicle for each gap. This very effective merge mode is called the 'zip fastener effect' and is a legal requirement imposed on drivers in many countries.

Another successful approach used in some countries is 'ramp metering' (Section 17.4.4). In this case, a signal system such as an intermittent red light restricts the number of vehicles entering the on-ramp to a number which can find acceptable gaps in the primary road traffic, usually based on upstream monitoring of that traffic.

Section 17.2.3 described the flow instabilities that occur near capacity due to the undamped consequences of gap acceptances. Therefore it is not surprising to observe that capacity breakdowns often occur just downstream of the merge point for an entry lane or after a through-lane has been dropped.

The *diverging* (or separation) process — commonly into a large exit taper leading into an *exit ramp* — also involves a number of steps. [Less commonly, the taper is preceded by a deceleration lane.] To diverge successfully the driver must:

(1) detect the oncoming need or opportunity (Section 16.5.3, Table 16.1),

(2) steer out of one direction (or lane) and into the new direction (or exit ramp). Experience indicates that drivers can readily achieve the required lateral movement of a lane width at about 1 ms^{-1} and so will require about 3.5 s or 100 m for this repositioning, and

(3) decelerate by gear and then by braking to the exit speed required by the ramp geometry and sign-posting. Deceleration distances can be calculated using Table 21.1 or Section 27.2.2.

Diverging can cause flow problems if low capacity at the end of the diverge (e.g. at a signalised intersection) causes traffic in the diverge lane to queue back onto the primary lane. Guidance on diverging lane design is given in AASHTO (2004).

Finally, two lines of traffic travelling in approximately the same direction may need to cross rather than merely merge. The process is called *weaving* (Section 20.2.1) and involves joining the second lane by merging and then leaving it by diverging. These processes are done without the aid of traffic control devices. The weaving length is usually slightly greater than the combined merging and diverging lengths. Weaving stops either lane operating at capacity and raises the risk of crashes. It is therefore essential to give drivers adequate advance warning of an oncoming need to weave. Weaving is common at freeway entry and exit ramps and when drivers position themselves for an oncoming turn on a multilane highway and these cases can be aided by advance signing.

17.3.7 Minimum headways

The Poisson random-traffic model developed in Section 17.3.2 (e.g. Equation 17.18) leads to a negative-exponential distribution of headways. This is not suitable for congested traffic, mainly because there is a minimum headway (Section 17.2.4 & Figure 17.2) that vehicles will accept, whereas Equations 17.18 to 17.22 of the negative-exponential distribution assume a continuous range of headways down to zero. If this minimum headway is T_{hm} then it can be argued that it is the component of headways in excess of T_{hm} that is subject to a negative-exponential distribution (Section 33.4.5). The mean of this component will be:

$$1/q_H = (1/q) - T_{hm}$$

$$q_H = q/(1 - qT_{hm})$$

Typical values of T_{hm} are between 0.6 and 2.5 s (Section 17.3.9). The value of qt for use in Equations 17.21, 22 & 25, for example, then becomes:

$$qt = q_H\,(t - T_{hm})$$

The resulting distribution is known as a *displaced* (or *shifted*) *negative-exponential distribution*. It is useful to then rewrite Equation 17.25 with $q = q_H$ and $t = t - T_{hm}$ as:

$$(\text{frequency of headways} > t) = 0, \quad \text{if } t < T_{hm} \qquad (17.33)$$

$$= q\{e^{-q[t - Thm]/[1 - qThm]}\}/\{1 - qT_{hm}\}, \text{ if } t > T_{hm} \qquad (17.34)$$

Following Equation 17.23:

(proportion of headways > t) = $e^{-q[t - T_{hm}]/[1 - qT_{hm}]}$ (17.35)

Similarly, Equation 17.32 becomes:

$$q_N = q_m e^{-q_H\{T_{gc} - T_{hm}\}}/(1 - e^{-q_H T_{hf}}) \quad (17.36)$$

The q in Equation 17.32 is q and not q_H as it relates to qT in Equation 17.24. It was shown (Tanner, 1962) that the consideration of T_{hm} imposes an effective *average delay* of $T_{hm}^2 q_H/2$ on all vehicles in the major traffic stream, relative to the situation where $T_{hm} = 0$. Equation 17.36 is a version of Siegloch's formula.

Alternative distributions that have been used for headways in heavy traffic are other forms of the negative-exponential, such as a combination of two displaced negative-exponential distributions. The *Erlang* (Section 33.4.5) distribution is effective for the very short headways encountered in signalised intersections.

Measured car-following headways are discussed in Section 17.3.9.

17.3.8 Platoons

In real traffic drivers all travel at different speeds, and the faster ones catch up to the slower vehicles. Consider one slow vehicle travelling at a speed V_s km/h in a stream of faster vehicles which have a speed of V km/h, density K veh/km (Figure 17.3c), and probability density function (Figure 33.4c) f_V h/km. The rate at which the faster vehicles catch up to the slow one is:

$$K\int_V^\infty f_V(V - V_s)dV \text{ veh/h} \quad (17.37)$$

If vehicles are to be unimpeded by other vehicles (i.e. their free speeds are maintained), then Equation 17.37 also gives the required *overtaking rate*, i.e. the rate at which fast vehicles must change lanes to prevent being impeded by the slow vehicles. Typically, Equation 17.35 predicts an overtaking rate that increases exponentially as the traffic flow increases. As there will usually be constraints on overtaking, the number of *queuing* vehicles will also increase exponentially as flow increases. Overtaking rates for vehicles travelling in the same direction — e.g. on one carriageway of a multilane freeway — depends on available gaps (Section 17.3.3) and thus on the flow in the adjacent lane. Overtaking will be virtually unimpeded for free flows, i.e. for adjacent flows below 900 veh/h (Section 17.2.2). The rates for vehicles on two-way roads are more complex and are discussed in Section 18.4.3.

Once a vehicle has caught up to a slow vehicle it will either overtake in a *flying overtaking* (Section 18.4.2) or decelerate to the speed of the slower vehicle and proceed to queue behind it until it has a chance to overtake. This will lead to *platoons* (or *bunches*) of cars, with a relatively slow vehicle as the platoon leader. Platoons will arise:

* in the absence of overtaking opportunities along a road length, and
* as a consequence of a set of vehicles being halted at a stop line by a stop sign or red traffic signal (Section 23.2.3). When an acceptable gap appears or the signal changes to green, the vehicles move off as a platoon.

Section 17.2.3 showed how the queuing caused by the slow vehicle will cause shock waves and instabilities in the traffic flow in the lane in question. The effect is still there, even if overtakings are easy, as drivers usually take some time to return to their optimal spacing. Furthermore, the movement of overtaking vehicles into the adjacent lanes may also cause flow instabilities in those lanes. Thus one slow vehicle can influence up to three upstream lanes. As discussed in Section 17.4.3, the platoons caused by the slow vehicles will also influence lane capacity, particularly if they are travelling below optimum speed (Section 17.2.3).

The general traffic condition in a platoon is known as the *following* state, as opposed to the free state described in Section 17.2.2. Each vehicle in a platoon will adopt the (slow) lead vehicle's speed but its own individual minimum headway (Section 17.3.7). Due to feedback effects, the forced speed adopted by following vehicles in the platoon drops as the vehicle's position down the queue increases.

Traffic is composed of a mix of diverse vehicle types (Section 17.4.3) and in a long road length, this mix reaches equilibrium conditions with respect to mean speed and intra-platoon spacing. These conditions will be maintained by a continuous process of vehicles catching up and then overtaking within the traffic mix. In this situation there are two flow rates of importance. The first is the rate at which vehicles catch up to the slow vehicle from behind (Section 17.3.7) and the second is the rate at which vehicles are able to overtake the lead vehicle and leave the platoon. Overtaking rate is discussed in Section 18.4.3 where it will be seen to depend on the lane provisions, traffic flows in each direction, and sight distance. Thus equilibrium conditions will also depend on these traffic and alignment factors. Platoon size is best fitted by the *Boral–Tanner distribution* that is derived from basic queuing theory as the distribution of busy periods for fixed service times and random arrivals.

Platooning will influence the capacity of a traffic stream to accept other vehicles. Clearly, such vehicles will find it relatively difficult to cross or merge into a platoon and so, if the proportion of free (non-platoon) vehicles is a, then Equation 17.36 must be modified to:

$$q_N = a q_m e^{-qH\{Tgc - Thm\}}/(1 - e^{-qHThf}) \qquad (17.38)$$

Section 17.3.7 showed that the negative-exponential distribution was the best form to use in heavy traffic. The above use of a to accommodate platoons produces the *bunched negative-exponential (or bunched exponential) distribution* with $(1 - a)$ of the vehicles having headways of T_{hm} and a having headways predicted by Equation 17.18 or its displaced variant. Their frequencies are thus as given by Equations 17.33&34 respectively and their proportions by Equations 17.33&35 as:

$$P(T_{hm} < t) = 1 - a e^{-\lambda(t - Thm)}$$

where $\lambda = aq/(1 - qT_{hm})$. This dichotomised distribution is also known as *Cowan's M3 distribution.*

The value of a in Equation 17.38 can be estimated by following Tanner and observing that $a = 1$ when $q = 0$ and $a = 0$ when $q = q_{max} = 1/T_{hm}$ and thus hypothesising that:

$$a = 1 - qT_{hm} \qquad (17.39)$$

This linearity can also be predicted from the basic model. Akcelik and Troutbeck have empirically suggested:

$$a = e^{-bq}$$

with values of b between 1.5 and 9 for q in veh/s, or alternatively,

$$a = e^{-cq}T_{hm}$$

Using Tanner's approximation (Equation 17.39), Equation 17.38 takes the commonly used form:

$$qN = a q_m\{1 - q_m T_{hm}\} e^{-qH\{Tgc - Thm\}}/(1 - e^{-qHThf}\} \qquad (17.40)$$

The platoons disperse once overtaking rates are appreciably in excess of catch-up rates. This can occur as a result of improved sight distances, the provision of overtaking lanes or an overall increase in road capacity by the addition of extra lanes.

In an urban area, platoons will affect the behaviour of co-ordinated signals (Section 23.5.1). Observations a kilometre downstream of a traffic signal indicate that dispersion effects may be inadequate to produce random flow and that platoons can still persist after 1.5 km (Smelt, 1984). This means that traffic on routes containing frequent traffic control devices will rarely follow the random flow patterns described earlier in this section and forming the basis of Equations 17.20 to 17.40.

17.3.9 Gaps and headways in heavy traffic

It was shown earlier in Equations 17.24&25 that the assumption of a random flow of vehicles led to a negative-exponential distribution of headways. Equations 17.33&34 have shown how it is probably better to use a displaced negative-exponential distribution based on ($t - T_{hm}$) rather than on t. Experimental data indicate that, whereas this distribution is a good representation of the gaps between platoons, the *log-normal distribution* better describes headways within a platoon, as it is biased against zero values (Section 33.4.5). These headways are relatively small and constant but vary between drivers, as shown by the distribution in Figure 17.12.

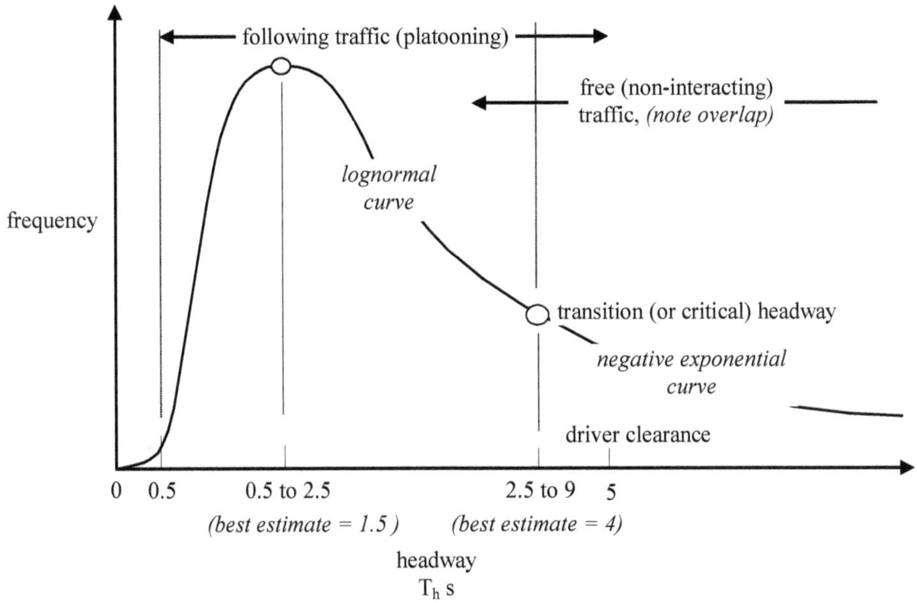

Figure 17.12 Car-following headways found within and between platoons.

The 1.5 s estimate of the mean headway includes the *response times* quoted in Table 16.1 for an alert driver looking at the right spot and aware of the required action when the brake light illuminates on the car ahead. It represents, for instance, a driver travelling in the median (fast) lane of a freeway. An 85th percentile value of T_{hm} used to cover all situations is typically taken as 2 s. The time headway adopted is relatively independent of absolute speed but does depend on the:

* relative speed between vehicles, increasing by about 1 s for each 1 km/h difference in speed, and
* driver's allowance for relative braking performance (Section 16.5.3).

The headways of 0.5 s shown as the lower limit in Figure 17.12 are observed in such circumstances (Postans and Wilson, 1983), but in other scenarios most drivers find them to be potentially dangerous and thwart with imminent disaster. This is consistent with the 0.4 s action time given in Item 4 in Table 16.1.

As suggested in Figure 17.12, drivers are affected by the presence of the vehicle ahead of them for headways of 9 s or less (McLean, 1989). Indeed, the transition headway between the following and free states — sometimes called the *critical headway* and the associated flow the *breakpoint flow* — is still a matter of some debate with values quoted between 2.5 and 9 s, with a best estimate of 4 s, and most in the range from 3.0 to 4.5 s. It is suggested that, as the transition between the free and the following state will be gradual, the reality is probably the composite picture shown in Figure 17.12. The best headway estimate of 4 s corresponds to 900 veh/h (Equation 17.16). This explains the dashed line in Figure 17.1. This 'best estimate' is close to the 5 s driver 'clearance' value noted in Section 16.5.1. Thus, only high speed flow would be unaffected by vehicle interactions

17.3.10 Traffic simulation

From the above discussion, it will be seen that the prediction of the behaviour of real traffic under real conditions is a difficult and algebraically intractable problem. Frequently, computer-based traffic simulation models are used to examine the problem, providing a comprehensive picture of traffic operations and thus giving much more than capacity estimates. For example, Robinson's TRARR model has been successfully used to provide estimates of trip time and vehicle operating costs (Section 29.1.2). Simulation models can also be used at a smaller scale to study such factors as the role of sight distance and alignment and the effect of passing lanes. One particular advantage of simulation models is that they permit 'experiments' to be conducted in which only one variable changes (such as the traffic composition or an alignment element).

As with the traffic flow studies discussed in Section 17.2, a model can either consider individual vehicles in a microscopic approach or treat traffic flow in an aggregate or macroscopic sense. The former is far more common for urban traffic, which requires greater attention to be given to close-following and turning movements. Microscopic models describe the detailed time trajectories of individual vehicles in the traffic stream (Section 18.4.1). They therefore need a car-following algorithm (Section 17.2.4) and data on the road, the traffic, the response characteristics of the various vehicles and their drivers, and the rules by which they interact with each other and with the road geometry. Macroscopic models concentrate on groupings of similar vehicles, as in a platoon.

There are many assumptions made in microsimulation. Thus the models require careful calibration, particularly with respect to lane changing in congested traffic. For example, some models destroy or suppress vehicles in congested conditions.

In addition to TRARR, other common microscopic (or microsimulation) models are the FHWA's NETSIM and FRESIM, MULTSIM (Section 17.2.4), INSECT, TWOPAS, AIMSUM (UPC, Barcelona), SITRA B+, VISSIM (from Germany) and PARAMICS. They all invoke a detailed car-following algorithm.

NETSIM applies to conventional road and street networks and simulates the acceleration and deceleration of individual vehicles, car-following, lane obstructions, lane-changing, traffic signals, Stop and Give way signs, and pedestrian traffic. It uses four vehicle types and sixteen vehicle categories and updates its vehicle position records every second. Lane-changing is instantaneous. Each network link can have a source and a sink to simulate traffic entering and leaving the system. NETSIM covers road networks, but MULTSIM and INSECT only apply to single roads. NETSIM is based on the earlier *UTCS* package reprogrammed for FHWA's *TRAF* suite of simulation programs. Two related FHWA programs are *FRESIM* for freeways and *ROADSIM* for two-way rural roads. FRESIM also simulates each vehicle, covers ten different driver types, lane changing, weaving, and ramp metering (Section 17.3.6). CORSIM combines the freeway and urban models.

Subsequently, PARAMICS (Parallel microscopic simulation) was developed by Quadstone and SIAS in Scotland using parallel processing. It has a programmer's interface which allows users to code their own algorithms and provides comprehensive 3-D visualisation. Paramics uses a car-following model based on British driver behaviour. PARAMICS and AIMSUN both use networks based on intersection nodes connected by multi-lane road links. PARAMICS can also accommodate roundabouts. VISSIM does not use nodes, instead it models each turning movement.

Today, PARAMICS, AIMSUN and VISSIM are probably the most widely used microsimulation models. They each will run under Windows and have no restraints on network size. They run at at least real-time speeds.

17.4 CAPACITY

17.4.1 Definitions

The term *capacity* has been frequently used in the preceding sections. It is now necessary to give it a more practical definition. The capacity of a traffic element is the maximum number of vehicles that could be reasonably expected to pass repeatedly through that element in a given time. It thus depends on prevailing road and traffic conditions. A mathematical definition of capacity has already been encountered in the maxima, q_m, observed on the flow–speed curves in Figures 17.1, 17.3a and 17.4a. From Equation 17.16 and Section 17.3.7, if headways are known, capacity is also given by:

$$q_m = 1/T_{hm}$$

For a link with signalised intersections, if the average delay time, d, is known (Section 23.5.7), then the capacity of the link is estimated by:

$$q_{m(sig)} = 1/(T_{hm} + d)$$

More generally, it is possible to consider the capacity of a traffic route in terms of the maximum number of vehicle-kilometres or vehicle-hours of travel that are accommodated within a given time. For vehicle-kilometres, if the given time is δT then the total distance travelled at speed V is $\sum_{1}^{q\delta T} V\delta T$ and the capacity measure is thus $\sum_{1}^{q\delta T} V$ for all the qδT vehicles, e.g. *veh-km* of travel per h. For vehicle-hours, the capacity measure becomes $\sum_{1}^{q\delta T}(1/V)$, e.g. *veh-h* of travel per km. For constant speed, these are equivalent to qδTV and qδT/V respectively. Hence vehicle-kilometres can be considered as a speed-weighted version of the original capacity measure.

The time interval, δT, can be an important parameter. As shown in Figure 17.12, traffic passes by at intervals of 1 s or more. Capacity and flow rates are generally quoted in units of vehicles per hour (veh/h) and counting lengths for long-term flow data are discussed in Section 31.5. For capacity flows, it is usual to consider flows in the peak 15 minute flow interval. During this interval, some 450 vehicles may have passed and, from Section 23.6.2, some 10 traffic signal cycles occurred. The δT value used is obviously a compromise and is best based on empirical evidence that smaller values of δT do not lead to statistically stable measures.

Traffic capacity can be seen as the supply side of the demand–supply traffic interaction. Generally, demand is represented by the desire to travel but more specifically it is represented by the traffic flow. Any imbalance in the demand–supply situation can be corrected by either altering the demand (a transport management task, see Section 30.3) or by altering the supply, i.e. by increasing capacity. It will be seen in Section 17.4.3 that capacity can be increased by either removing friction from the existing system (e.g. by banning kerbside parking or by using co-ordinated signal settings) or by building new lanes. A comprehensive review of traffic capacity on major routes is given in OECD (1983b).

Five important but somewhat subjective capacity terms are:

* *actual capacity*: Section 17.3.1 and Equation 17.6 show that maximum flow (capacity) occurs at minimum headways. The common assumption that the minimum headway is 2.0 s leads to a capacity flow of 1 800 veh/h/lane. The implication that the capacity of a modern traffic lane is 1800 veh/h/lane was first sceptically reviewed by Yagar (1983) who provided data on much larger measured flows and flows as high as 2900 veh/h/lane have been recorded — typically on a length of well-designed freeway without flow interruptions, merges, local platooning, flow instabilities or shock waves, Section 17.2.3. Figure 17.11 shows that the most common following headway is 1.5 s, which implies an average capacity figure of 2400 veh/h. However, Section 17.2.3 suggests that actual capacities are closer to 2200 on multilane rural highways and 2300 on freeways. For intersections, the range drops to 1500–1600 veh/h/lane.
* *design capacity*: The design of a road is based on the expected AADT and the design hourly traffic flow (DHF) discussed in Sections 6.2.3 & 17.1. However, various decisions during design may alter the actual capacity on each traffic element away from this initial DHF. For example, the decision as to how many lanes will be used is usually made in the context of providing a particular level of service (Section 17.4.2). The final capacity of a new traffic facility as predicted from its design features is its design capacity.
* *economic capacity*: This is the traffic flow needed to economically justify a particular action.
* *environmental capacity*: This is the capacity of a facility, considering the presence of parked vehicles and the effect of maintaining appropriate environmental standards. The environmental capacity indicates that the community's best interests may not be served by operating a piece of road at its design capacity. See also the discussion of the term in the context of residential streets in Section 7.4.1.
* *tolerable capacity*: This is the flow (typically 1800 veh/h/lane) above which traffic conditions are frustrating, and comfort and convenience are low. Flows at tolerable capacity will involve considerable periods (in space and time) of congested flow. Speeds are close to optimum speed (Section 17.2.3). Tolerable capacity drops by about 10 percent in poor weather.

17.4.2 Level of service

The concept of *level of service* relates to the operating conditions encountered by traffic. It is a qualitative measure of such factors as speed, trip time, interruptions, interference, freedom to overtake, ability to manoeuvre, safety, comfort, convenience, and vehicle operating costs. Nevertheless, the six common levels of service are defined in terms of traffic flows. They are:

A *Free flow* (Section 17.2.3) with high speeds and low flows (≤ 700 veh/h/lane). Drivers can drive at their own free speed with little interference. (Note that Section 17.2.2 would indicate 900 rather than the 700 commonly used).

B Appropriate to rural roads with moderate design flows (700–1000 veh/h/lane). Drivers have reasonable freedom to select their speed.

C Appropriate to design flows encountered on urban roads (1000–1500 veh/h/lane). Drivers are restricted in their freedom to select speed or manoeuvre, but speeds are still at or above optimum speed.

D Appropriate to flows near tolerable capacity, as defined in Section 17.4.1 (1500–1800 veh/h/lane). For intersections, this range drops to 1300–1500.

E At or near actual capacity (1800–2000 veh/h/lane). There may be momentary stoppages.

F Demand exceeds capacity, with queues and delays (0 to 2000 veh/h/lane). There is stop–start driving in congested conditions.

These six levels of service can be located on the speed–flow graph described in Figure 17.1 to produce the representation illustrated in Figure 17.13. The level-of-service approach does not justify a particularly close examination (Hoban, 1984). For example, at a given speed, a flow of, say, 20 percent of that given by the speed–flow relationship would clearly offer a better level of service than the maximum flow for that speed. Yet both are in the same level of service as it is assumed that drivers would immediately adopt a changed behaviour rather than conform to the local speed–flow relationship.

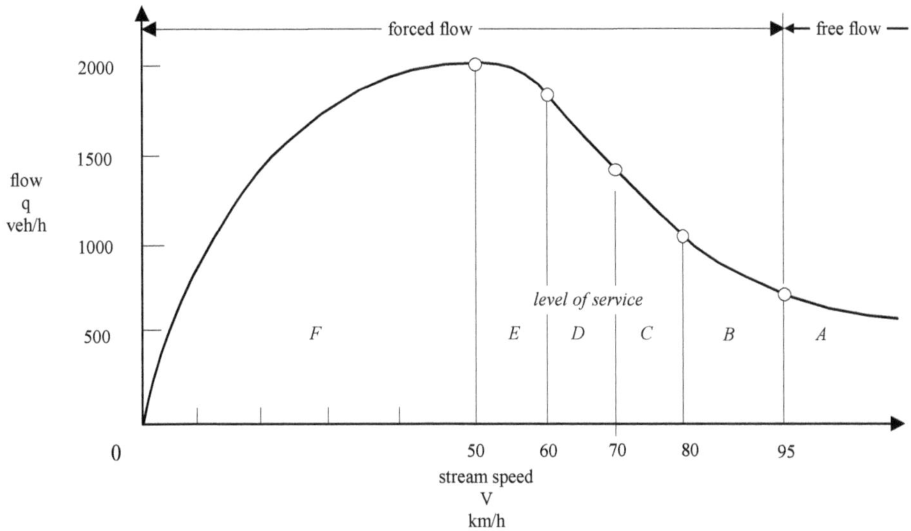

Figure 17.13 Level of service representation (refer to basic curve in Figure 17.3a).

The stream speeds (Section 17.2.1) used are trip speeds (Section 18.2.12) and, on urban roads, would include average delays at signalised intersections (TRB, 2000). The capacities are for the uninterrupted flow (Section 17.2.2) of cars on a two-way road in good conditions, but with some sight distance restrictions.

17.4.3 Capacity reduction factors

As discussed above in the previous section, it is common to assume that the capacity flow in a traffic lane is between 1800 and 2400 veh/h. However, practical capacity flows are often below this level as a result of a number of factors, many of which were described generally in Section 17.2.6 and are listed below. The urban factors in the list are usually of particular importance in defining the various saturation flows to be discussed in Sections 17.4.4 & 23.6.3.

(a) *surface conditions*: An unpaved gravel surface reduces capacity by about 50 percent and a natural earth surface by about 60 percent.

(b) *lateral clearances*: Clearances of under 2 m to obstacles beside the carriageway will affect capacity. This is one of the factors causing a reduction in capacity at construction sites (Section 25.9).

(c) *shoulders*: Increasing shoulder width increases capacity by easing overtaking and reducing the effect of incidents.

(d) *lane widths*: See Section 6.2.5. Generally, capacity increases by about 1 percent for each extra 50 mm of lane width, up to a lane width of about 3.2 m, there are lesser gains between 3.2 and 3.6 m, beyond which there is no further increase. (However, some U.S. data suggests this upper limit is 3.7 m, or even 4.2 m.) Thus with narrow 2.8 m urban lanes (Section 20.3.3), the capacity flows would be reduced by about 15 percent. The reduction is greater if there are significant roadside obstacles within 2 m of either side of the lane.

(e) *lane distribution*: Lane arrangements are important; for example, stopping a through lane can have an adverse effect (Section 6.2.5).

(f) *horizontal alignment*: Section 18.2.5 discusses how reducing the horizontal curvature lowers free speed, and thus at least causes flow instabilities.

(g) *vertical alignment*: See Section 19.3. For example, each 1 percent increase in grade decreases capacity by about 0.5 percent. Grade has an effect on the truck pcu's (defined in (i) below). Common values are 5 in rolling country and 10 in mountainous areas, with slightly lower values (4, 8) on multi-lane roads.

(h) *overtaking provisions*: The absence of overtaking opportunities has a major effect on the capacity of two-way arterial roads (Sections 17.3.8 & 18.4.4).

(i) *traffic composition*: Traffic is composed of a disparate mix of vehicles, each with their own separate capacity flows (Section 17.4.3): e.g. maximum capacity flows for trucks are about 400 truck/h. Thus overall capacity depends greatly on the composition of the traffic flow, particularly the presence of trucks. For example, the four main factors that determine the different effect of a truck compared with a car are the truck's:

(1) greater length causing it to occupy more space than a car,

(2) poorer acceleration characteristics,

(3) lower speeds on steep up-grades (in addition, some jurisdictions impose lower speed limits on trucks), and

(4) inhibiting effect on the behaviour of other drivers.

These factors in turn are influenced by terrain and traffic conditions.

It is therefore necessary to relate the capacity effect of various vehicle types to conventional cars and this is done by use of equivalent *passenger car units* (*pcu*), i.e. if a vehicle has a pcu of x, it has the same effect on capacity as x passenger cars. [pcu's are sometimes called passenger car equivalents (*pce*)].

Because headways are easier to measure than capacities and because they directly determine capacity (Equation 17.16), the most common method of assessing pcu is by comparing headways at capacity flow. That is, the pcu of a vehicle, pcu_v, is taken as its capacity headway, T_{hcv}, divided by the capacity headway for a passenger car (typically, $T_{hc} = 3600/1800 = 2.0$ s), i.e.:

$$pcu_v = T_{hcv}/T_{hc}$$

Typical pcu's are given in Table 17.1. As the proportion of the vehicle type, v, in the traffic stream increases the value of pcu_v will tend to q_{cv}/q_{cv}.

Table 17.1 Passenger car units (pcu's) for through traffic. See TRB (2000) and Patrick *et al.* (2006) for more extensive data.

Vehicle type	Intersections	Urban	Rural
Private car	1.0	1.0	1.0
Motorcycle, moped	0.5	0.75	1.0
Bicycle	0.2	0.3	0.5
Truck (see text)	2.0	2.0	3.0
Bus, tram	2.5	3.0	2.0

Trucks have a significant influence on capacity with an average pcu of 2, although values generally increase with truck size and frequency and range from 1 to 100. A operational solution to this problem is to create special lanes for large and slow-moving vehicles. Alternatively, some jurisdictions allow such vehicles to use the outer shoulder as a *slow lane*. Methods for analysing the issue are discussed in Patrick *et al.* (2006).

Another term is the *passenger car space equivalent* (pcse) which only considers the space occupied by particular vehicle, relative to the space occupied by a car. The concept is most useful when the space required is defined as, not the territorial surround (or jam space, Figure 17.2), but the space required to maintain unforced (or free) flow (Section 17.2.3). Possible pcse values would be: bicycle, 0.1; motorcycle, 0.4; bus, 1.5.

A further term is the *traffic composition factor*, f_c, which is the capacity flow of the traffic mix, $q_{c(all)}$, divided by the passenger car capacity flow, $q_{c(car)}$, i.e.

$$f_c = q_{c(all)}/q_{c(car)} = [\Sigma(q_{cv}pcu_{cv})]/[\Sigma q_{cv}]$$

where q_{cv} is the capacity flow for a particular vehicle type, and the sums are over the entire vehicle fleet found in the traffic flow.

(j) *traffic flow variations*: Flow variations can lead to flow instabilities (Section 17.2.2–3).

(k) *oncoming* (or *opposing*) *traffic*: The capacity of two-way roads depends on the ability of fast vehicles to overtake slow ones, and thus on the flow in the opposite direction as this will decide the size of the gaps presented to the driver wishing to overtake (Sections 17.4.5 & 18.4.1–2). For these reasons, the two-way capacity of a two-way road is commonly taken as 3200 veh/h (for a two-lane road) rather than 2x1800 = 3600 veh/h.

(l) *traffic interruptions*: See Section 17.2.2 generally and specific instances in (m) to (p) below.

(m) *adjacent land use*: See also Section 18.3. The effect mainly relates to a friction reaction between the traffic flow and the traffic activities generated by the adjacent land uses associated with the local urban environment.

(n) *cars parked or in the process of parking in a kerbside lane*: These can have a dramatic effect on capacity, effectively taking one lane permanently, and another lane partly, out of service.

(o) *intersections and turning vehicles*: The capacity of an urban road network is very largely determined by the number and capacity of the intersections. Traffic signals at intersections reduce the proportion of the time available for traffic flow. The major effect will simply be the time lost due to the red and yellow signals (Section 23.6.2). Two additional factors are the start-up and close-down effects. The start-up time between vehicles queued at signals receiving the green signal and actually moving off is typically between 4 s for the lead vehicle to 2 s for vehicles in the queue. This *entering (*or *discharge) headway* will affect capacity as it reduces the effective green time. Similarly, the flows decline near the end of green times if the green time exceeds a minute.

The pcu concept in (i) above is extended to the *through car unit* (tcu) which indicates the effect on capacity of a turning vehicle relative to a vehicle moving straight ahead in a lane. The smaller the turn radius, the larger the effect. Typically:

$$tcu = 1 + (1.5/r)$$

where r is the turn radius in metres.

The tcu increases by a further 10 percent for kerb-turn lanes. In this case, the small radius of curvature forces vehicles to turn at low speeds. This in turn forces the drivers to accelerate from an almost stationary position (rather than merge into the cross flow at speed, Section 17.3.6). They thus require much larger critical gaps. In addition, drivers in kerb-turn lanes must frequently give way to pedestrians. A typical kerb-turning vehicle has a tcu of 1.40 pcu. *Centre-turn* tcu's are a function of:

* the opposing flow [i.e. traffic coming in the opposite direction]
* the turning vehicle's ability to filter through both that flow and to turn through gaps provided by the signals (Section 23.2.4), and
* delays caused by centre-turners in the cross-traffic who are leftover at the end of their green phase.

[A centre turn is a right turn for left side driving and a left turn for right-side driving. A kerb turn is a left turn for left side driving and a right turn for right side driving.]

Capacity flows for filtering centre-turners can range between zero and 1200 veh/h (Pretty, 1980). The calculations are compounded by the fact that centre-turners are often mixed in lanes with through vehicles. The centre-turn tcu has typical values between 1.6 (green/cycle = 0.35, opposing flow = 200 veh/h) and 6.0 (green/cycle = 0.75, opposing flow = 1400 veh/h). The mean car value is 3 and the truck (or other commercial vehicle) value is 4. These considerations led to the values in Table 17.2. The very high tcu equivalent of centre-turners has led to the common strategy of banning these movements in peak periods.

Table 17.2 Turning vehicle tcu's.

Type of turn	Car	Commercial vehicle
Through vehicle and unopposed turns	1 (pcu)	2
Restricted kerb turns	1.3	2.6
Average value for opposed centre turns	3	4

(p) *pedestrian movements*: This factor occurs particularly in association with traffic signals and turning vehicles (Section 23.6.2). It will often arise at the beginning of the green time when a platoon of waiting pedestrians will move off with priority over any turning vehicles. A similar delay will occur subsequently when pedestrians from the other side of the road begin to inhibit the turning vehicles.
(q) *bus operations*: See Section 30.4.2. The deleterious effect that stopped buses have on other traffic can be alleviated by providing special bays at bus stops.
(r) *absorption of traffic into other traffic*: The ability of a traffic flow to be absorbed into another traffic stream is limited by the characteristics of that stream, as was discussed in Sections 17.3.5 and given by Equation 17.32.
(s) *traffic management measures*: Most of the traffic management measures discussed in Section 7.3 will tend to reduce effective capacity.
(t) *driver behaviour*: Capacities are particularly sensitive to the headways that drivers adopt (Section 17.3.9).
(u) *weather*: Routine wet weather will reduce flows by about 5 percent. However, Canadian experience indicates that the 1800 veh/h figure drops to about 1400 in a cold (Edmonton) winter to 1200 in a 'severe' winter. Cautious, conservative driver behaviour can also cause a drop of about 200.
(v) *incidents*: Incidents are caused by such factors as:
  * roadworks (Chapter 25),
  * maintenance operations (Chapter 26),
  * special events, such as entertainment and sporting functions,
  * severe weather conditions, or
  * disabled, illegally parked, or crashed stationary vehicles blocking a lane,
  * drivers slowing to observe incidents in other lanes (*gawking* or *rubbernecking*). Thus a blockage in one lane not only eliminates the flow in that lane, but also reduces — typically, halves — the flow in the adjacent lanes. The effect is worst at night and for incidents that create significant activity.

Given the influence of all the above factors, the capacity of a real road can only be fully described in terms that are well defined, probabilistic, and specific to a particular locality.

17.4.4 Saturation and congestion

(a) Saturation

Section 17.2.3 showed that capacity flows were the peak flows that can be carried on a road length. If the traffic flow wishing to use a link in the road network exceeds the capacity of the link, queues will form at the beginning of the constricting link and extend upstream (Section 17.2.5). This response may result from one of three main factors.

1. An *ongoing* or recurring general excess of demand over supply. To address this:
   1.1 *travel demand may be suppressed.* Travel demand management is discussed in Section 30.3. For normal commodities, recurrent excess demand would be relieved by raising the price of the product until demand is sufficiently diminished. For roads, this solution would require some form of road or congestion pricing, as discussed in Section 29.4.5. In the absence of such measures, the excess demand is only gradually suppressed by drivers' reactions in the medium term to the delays that they experienced during previous trips.
   1.2 the existing system may be made to *operate more efficiently* by such measures as:

* advance travel information,
* management of traffic within the travel corridor (transport systems management, Section 30.2.2),
* better traffic management (Section 30.7),
* improved signal timing (Chapter 23),
* better lane marking, (Chapter 22), and
* work-zone management during reconstruction (Section 25.9).

1.3 *extra capacity may be provided.* If extra road capacity is added and does not satisfy all the demand, then it will merely provide the same trip times as the existing road network. This is because, if the link offers any inherent speed advantages, vehicles will queue to use it until the sum of the new running time (Section 18.2.11) plus queuing time equals the time (or cost) to travel via an alternative route. For a congested road network, the evidence is that adding capacity that is less-than-demand will increase journey lengths rather than decrease trip times.

2. A *temporary* or nonrecurring loss of capacity. These are in two categories:

2.1 losses caused by *internal traffic incidents* (Section 17.4.3v). The cost of incidents can be estimated by multiplying the time delay caused to all traffic by the cost of travel (Section 31.2.2). Traffic incidents are the major problem in many traffic systems often causing 60 percent of congestion. For example, a disabled vehicle in one lane of a three-lane carriageway will reduce capacity by closer to 50, rather than 33 percent as other drivers slow down to 'inspect' the incident. This is not surprising as Section 17.2.5 shows how extensive queues can quickly arise as the result of an incident. Thus most Traffic Authorities operate traffic surveillance systems and incident-management teams to allow incidents to be detected and remedied as soon as possible. Incident management is discussed in Section 26.4.

2.2 losses caused by the range of *external factors* listed in Section 17.4.3. These include inadequate maintenance of the facility, poor parking control, extreme weather and/or a lack of weather alerts, and construction in the road space.

3. The operation of *traffic signals* (Chapter 23) and other traffic control devices (Chapter 21).

3.1 Even if demand does not generally exceed supply, as discussed in #1 & #2 above, *queues will form at traffic control devices* whilst traffic waits until it is appropriate to cross a conflicting traffic stream (as at a red signal from a traffic light).

3.2 *Local demand/supply mismatches*. The arrival flow upstream of a traffic signal is a measure of traffic demand. The departure flow is a measure of the traffic capacity (or supply). It is commonly called the *saturation flow*, q_s, and is the maximum departure rate possible when a waiting queue at a signalised intersection is provided with continuous green time (a fuller definition is given in Section 23.6.3). Following Sections 17.3.4 & 17.4.3, q_s will be between 0.20 and 1.0 veh/s with 0.50 a typical value. The capacity flow for the road link containing the traffic control device is thus [(green time)/(cycle time)]q_s. Refinements to this model are shown in Figure 23.5.

Intersection saturation occurs when the saturation flow during the green part of the signal cycle is not enough to handle the arrival flow that occurs over the full cycle. That is, when demand exceeds supply. If saturation occurs over more than one signal cycle, the queues at the traffic signal will increase from one cycle to the next. The resulting accumulating queues are routine parts of many traffic systems. However, much traffic design is for unsaturated conditions and so the overall system usually operates ineffectively when saturation occurs. This saturation is usually measured by the *degree of saturation* (Section 23.6.3). It can lead to congestion.

(b) Congestion

The previous discussion suggests a mechanism by which increasing local demand/supply failures leading to network congestion. This is explored further below. A more macroscopic equivalent definition would be that congestion arises when demand on a network exceeds supply, however it is difficult to formulate a practical measure that would indicate this event. In transport economics it is sometimes argued that congestion occurs when the marginal cost of a trip exceeds the marginal benefits of a trip (see Figure 33.6). Such a definition has some theoretical basis, but bears little relevance to the common use of the word *congestion.*

An operational definition is that *congestion* (or *oversaturation*) begins to occur when the accumulating queues resulting from local saturation interfere with other unrelated traffic movements within the network, in a process known as *spillback.* Indeed, some authors call saturation *primary congestion* and refer to congestion as defined in this book as *secondary congestion.*

Consider two intersections, A upstream and B downstream. In the first level, traffic will still stop and queue at many intersections but there will be no wider consequence. As traffic increases a second level is reached when the queue end in the link AB does not extend back as far as the upstream intersection, A, but its presence inhibits the ability of the link AB to receive the saturation flow $q_s g$ from A. The effect may be exacerbated by shock waves from the tail of the queue affecting the flow at A. At the third level, the queue may block crossing traffic at an upstream intersection or may cause slow departures of traffic from an upstream intersection when drivers see the queue ahead. Thus saturation at one intersection begins to have an influence on the capacity of a wider road network. As congestion worsens, more intersections become involved. This effect can only be dissipated when sufficient spare capacity appears in the system — it may therefore last much longer than the period during which it occurred. It can be modelled by models such as SATURN (Section 30.7). In the limit, the traffic comes to a complete stop, as in a *traffic jam* (Section 17.2.3), which is called *gridlock* when it extends over a large area.

Measures of congestion at this local level are the rate of queue formation (m/s), queue speed and queue discharge speed. The discharge speed will relate to the departure flow from the traffic control device. The fact that this flow cannot exceed the flow arrival capacity will often be a major factor governing queue discharge.

Two common wider measures of congestion are queue length (Section 23.6.6) and hours of congestion. Perhaps the best overall numerical measure of congestion is to compare travel times with the equivalent times for lightly trafficked, unsaturated conditions (with the signals still operating). However this measure is more difficult to calculate than are the first two.

More generally but less usefully, congestion is defined as occurring when the flows are such that a small increment of additional traffic or loss of capacity causes a major drop in the level of service (OECD, 1981a). At a systems level, it is necessary to look for broader causes of congestion. Table 17.3 (Lockwood, 2005) provides useful guidance.

Table 17.3 Percentage contribution to total delay (from Lockwood, 2005). Note that the percentages are very approximate.

Causes of delay		Urban area > 1 million	Urban area < 1 million	Rural area
recurring	demand > capacity	35	25	0
	poor signals	5	10	0
	total	**40**	**35**	**0**
non-recurring	crashes	35	25	25
	breakdowns	5	5	25
	work zones	15	25	40
	weather	5	10	10
	total	**60**	**65**	**100**

Congestion is difficult to tolerate as it leads to increased driver frustration, pollution, noise, residential intrusion, delay, fuel consumption and overall transport inefficiency. It has been argued that congestion can be used as a form of pricing and flow metering, to control traffic entering a system. As mentioned above, this can create severe problems for those obliged to suffer the congestion, not only as users but also as innocent bystanders. It also fails to distinguish between journeys of different length and importance and results in an inefficient use of resources if users are not fully informed of the extent of the congestion that they will encounter.

(c) Ramp metering

A related and more refined technique than the use of inherent congestion has been the use of *ramp metering* in which traffic signals are used to control external vehicle flows onto a facility (typically a freeway, see Section 17.3.6). To be fully effective, it requires a cordon of metered entry points. Whilst such systems can be made to operate effectively and increase freeway speeds and capacity, they can cause overall problems as their automatic tendency is to give priority to longer distance travellers over shorter distance travellers. This effect may be contrary to local transport policies. The ramp delays may also force queues, cause congestion on the adjacent street system, and result in undesirable trip diversions.

Two forms of traffic metering have been developed to overcome this problem. *Release metering* controls the release of vehicles from traffic generators within the system — typically from large car parks. *Internal metering* involves the traffic management philosophies discussed earlier and in Section 7.3 and in which various impediments to traffic flow are introduced in order to reduce flows and speeds and discourage road use.

17.4.5 Rural capacity

For two-lane, two-way rural roads, the presence of oncoming traffic limits overtaking opportunities (Section 18.4.2) and therefore has a marked effect on capacity. For example, whereas the capacity of a single lane with uninterrupted flow is given in Section 17.4.2 as about 2000 veh/h, the sum of the capacity in both directions of a two-lane two-way road is somewhere between 2000 and 3000 veh/h. Indeed, 2800 veh/h is a typical design assumption (TRB, 2000), and is often assumed to be independent of the proportion of traffic in each direction — the *directional split*. The corresponding two-way flow values for various levels of service is given in Table 17.4 (Yagar, 1983).

Table 17.4 Rural road level of service.

Level of service	One-way	Two-way	Percentage of vehicles platooning	Percentage of time delayed (Hoban, 1984)
	flow in veh/h (Section 17.4.1)			
A	700	400	10	30
B	1000	900	40	40
C	1500	1400	60	60
D	1800	1700	80	70
E	2000	2000	95	80
F	≤2000	≤2000	100	100

Most two-way rural roads will consistently operate below capacity as their throughput will be constrained by specific restrictions — such as lengths with poor sight distance — rather than by overall factors. At these sub-capacity flows, users will therefore place considerable importance on operating conditions and particularly on the presence of overtaking opportunities. The proportion of the journey time providing such opportunities (Sections 18.4 & 19.4) is thus a measure of service quality. In the absence of sight-distance restrictions, the overtaking rate is given by Equation 18.1. When combined with the no-overtaking lengths, it can be used to predict the proportion of following vehicles, which is a measure of platooning (Section 17.3.8). Some specific measures are (Hoban, 1984) *point bunching* — the percentage of all vehicles passing a point which are following a slow vehicle — and *time spent following* — the percentage of journey time spent following a slow vehicle (sometimes called *time delayed*).

In this case, level of service will depend not so much on flow but more on the demand for overtaking created by the presence of slow vehicles and on the supply of overtaking opportunities created by gaps in the opposing flow (Section 17.4.1) and the alignment of the road. A suggested link between level of service and percentage of vehicles platooning is summarised in Table 17.4. Further data is given in Section 18.4.4.

CHAPTER EIGHTEEN

Speeds

18.1 SPEED LIMITS

18.1.1 Speed and crashes

Following the general discussion in Section 16.3, the speed at which a driver travels depends on the driver's:

* vehicle's characteristics,
* travel motivations,
* personal characteristics, particularly with respect to risk-taking,
* perception of the safety, alignment, and comfort of the road,
* perception of the chance of an unexpected event occurring,
* perception of the level to which the legal limit is enforced,
* assessment of the legal speed limit, and
* self-imposed speed limit.

As the driver's selected speed increases, the following four major safety-related factors occur:

(a) The vehicle becomes less stable in some modes, e.g. cornering (Section 19.2.3). Consequently, crash investigations usually show that speeding is over-represented in crashes in which drivers lose control of their vehicles, particularly on curves.

(b) The driver has less time and distance available to respond to a potentially hazardous situation. A driver travelling at 100 km/h with a typical response time of 1.7 s (Table 16.1) will have travelled 100 x 1000 x 1.7/3600 = 47 m after noticing an event requiring a speed reduction, before even beginning to decelerate. Furthermore, the second power law relationship between distance travelled and speed (Equation 27.1d, constant deceleration) means that most of the speed drop does not occur until the length just prior to the stopping point.

(c) Other road users similarly have less time to react to the detected presence of the speeding vehicle.

(d) There are more task demands on the driver, the interactions are more complex, and the consequences of an error all place a greater emphasis on quick and accurate responses to external stimuli.

(e) The severity of any consequent crash increases. An extensive review of empirical data suggests that — using the ratio of before and after speeds in a collision as the variable — the risk increases with this speed ratio variable by (Elvik, 2005):

* a power of 2.2 for property damage crashes,
* a power of 3.2 for any injury,
* a power of 4.0 for a serious injury, and
* a power of 4.9 for a fatality.

(the empirical results should not be taken too far — e.g. by applying them to speeds outside the speeds at which the data was collected.)

In urban areas, the risk of death for a pedestrian in a car crash rises from 20 percent at 40 km/h to almost 100 percent at 80 km/h (Griebe *et al.*, 2000).

(f) There will be a distribution of speeds adopted by drivers on a particular road. As speeds increase, speed variances increase in absolute terms. There is some evidence that crash rates increase as speed variance increases and when a vehicle's speed exceeds the mean traffic speed by about 25 km/h. British studies suggest that a driver travelling 25 percent above average speed is six times more likely to be involved in a crash than a driver travelling at average speed. There is also evidence that crash rates are lowest for drivers adopting the 90th percentile speed. The rates then rise sharply as speeds approach the 100th percentile and are much higher than for those driving at the similar (0th percentile) extreme below the limit (Harkey *et al.*, 1990).

Thus, speed contributes to many road crashes, although it may be the sole cause of less than 20 percent. This is because the attitudes of the driver (Section 16.2) and the manner in which the vehicle is driven may be bigger contributors to crash risk than the speed adopted. Nevertheless, although speed may not be the prime cause of a crash, it will be a major determinant of the severity of that crash (see (e) above).

Although the link described above between speed and crash risk is very strong, there are two system-related counter factors that need to be considered.

* Because crashes are relatively rare, an area-wide economic analysis could show that the value of the time saved by increasing speed limits will exceed the cost of the extra crashes caused by the increased speeds.
* Raising speed limits on safe roads may move drivers away from more dangerous alternative roads. For example, it was suggested that there had been a reduction in fatalities per distance travelled of about 4 percent following the 1987 speed limit change in the U.S. from 55 to 65 mph (Lave and Elias, 1994).

18.1.2 Legal limits

Speed limits are imposed by law and regulation to curb a common tendency among drivers to travel above an appropriate maximum speed, particularly with respect to the safety of all road users. The speed limit may be chosen on the grounds of:

* safety (Section 18.1.1),
* fuel economy (Section 29.2.3),
* noise control (Section 32.2.3),
* load control (Section 27.3.2),
* neighbourhood amenity and anxiety (Sections 7.4.1 & 18.3), and/or
* mix of users (e.g. cyclists and pedestrians in urban areas, Section 18.3).

Some factors that influence safe speeds — such as extreme weather and heavy and/or unexpected traffic — will vary with time. It is rarely possible to accommodate these factors via changing speed signs and the general and somewhat unrealistic expectation is that drivers will self-impose lower speed limits on such occasions.

The five major types of speed limit are the:

(1) *blanket* (or *default* or *overall*) *speed limit*, which is a specific speed limit, applied over an entire area. Speed limits are usually indicated by displaying the maximum permissible speed in black within a red ring on a white sign (Section 21.3.1). Although simple in application, the blanket speed limit creates many anomalies due to local variations in conditions.

(2) To accommodate the problems with blanket speed limits, *speed zones* with lower speed limits are applied to specifically-signed lengths of road, with signs at the beginning and end of the zone. The exit sign may be either a special *derestriction sign,* directly give the blanket limit, or introduce a new speed zone.

The speed limit in a zone is usually determined by first considering the actual speeds that drivers have been adopting in good ambient conditions. This is called the *speed environment* and its measurement and statistical definition are given in Section 18.2.4. Its use as a base case ensures an obvious reasonableness in the selected speed zone limit. This is essential, due to the practical difficulties associated with the continuous enforcement of speed limits. The speed limit may be reduced below the speed environment following a consideration of the following factors:

* crash history (Section 28.3),
* road geometry (Section 18.2.5),
* likely traffic interruptions from roadside land usage (Section 17.4.2),
* roadside safety (Section 28.6),
* noise constraints (Section 32.2.3), and
* the presence of road construction or maintenance (Section 25.9).

(3) *prima facie speed limit,* which means that there is no nominated speed limit, but that drivers may be obliged to prove the safety of their chosen speeds. This approach has been largely superseded by the wide availability of speed-measuring devices and by many legal challenges by prosecuted drivers.

(4) *limits on specific drivers*. Special speed limits may be imposed on some drivers whilst their skill levels are low (e.g. learners and newly licensed drivers, Section 16.6.2).

(5) *limits on specific vehicles*. Differential speed limits between vehicle types using the same road may help allay a fear many have of large trucks, but have little technical value and some obvious disadvantages. The disadvantages can be minimised by making the limits lane-specific. Vehicle speed limits may also be imposed to control vehicle loads on bridges (Section 27.3.5), or to discourage the use of certain vehicle types in special areas such as residential streets or tourist sites.

(6) *limits on specific times*. For example, when children are approaching or leaving school, or where large groups of people congregate.

It is common layman's belief that the imposition of a speed limit will automatically reduce driver speed. Certainly, it is general experience that a lowering of the speed limit reduces the variance of the speeds between drivers, the number of vehicles travelling at peak speeds, and the severity of the average crash. However, if the new speed limit is unnatural and not vigorously enforced (Section 16.7), the effect on driver behaviour will often be less than expected. One reason for this is that drivers are aware of the chances of their illegal speeding being detected (Section 28.2.4). However, automatic methods for detecting speeding vehicles (e.g. speed cameras) have very effectively raised the risk of detection and hence altered driver speed-behaviour (Section 28.2.5). Studies of more conventional practices for enforcing speed limits indicate that visible forms of enforcement that drivers were able to relate to punishment produced a temporary modification of behaviour (Millar and Generowicz, 1980). The effect appeared to be to reduce violations rather than conflicts or crashes. Young drivers in particular see speed limits as unrelated to crash risk and rely heavily on their own (largely untested) assessment of risk levels (see also Sections 16.3.2 & 28.2.4).

As a lowered speed limit results in a drop in average speeds and speed variance, then — consistent with the data in Section 18.1.1 — crash numbers will drop and their consequences diminish. Lowering speed limits has also been strongly argued as an energy conservation measure. The U.S. drop in speed limit in the 1970s from 60 to 55 mph was seen to have reduced the U.S. petrol consumption by about 3.5 percent, which was about 1.5 percent of transport energy. European countries reported drops of about 8 percent in fuel consumption for speed limit drops similar to those in the U.S. Deductions from such calculations are complicated by the fact that distances travelled are usually reduced in conjunction with the speed limit reductions. There may also be changes in route choice and trip generation. Thus, studies of the incremental benefit/cost ratio of speed limit reductions, viz.:

(reduced fuel and crash costs)/(cost of increased travel time)

have indicated values of less than one, due to the relatively high cost of longer, slower trips.

The discussion in this section has concentrated on open road speeds where the safety concerns are with vehicle handling and inter-vehicle effects. These speed limits are typically 90 km/h, 100 km/h, 110 km/h or 130 km/h, depending mainly on the sight distance to potential hazards. Urban speed limits, which are typically 60 km/h, 50 km/h or 40 km/h are discussed further in Section 18.3.

18.1.3 Advisory speeds

Sign-posted *advisory speeds* should be viewed as a warning of the relative difficulty of the forthcoming length of road, rather than as a piece of absolute advice. The safe speed on a road clearly depends on the road condition, the characteristics of the drivers and their vehicles, the traffic conditions, and the weather and cannot be represented by a single, static advisory speed.

The advisory speed may be used to warn drivers of a particular hazard or to indicate that the comfortable speed may be less than the speed limit.

* *Hazard warnings*. These depend on the nature of the hazard (e.g. stray animals, Section 28.7.2) or the sight distance to the hazard (e.g. roadworks around a curve). The techniques to use in determining the speed and sign placement are found in Section 19.4 and Table 21.1.
* *Comfortable speeds on curves*. Section 19.2.4 shows that the *friction factor* assumptions used in curve design are largely based on criteria related to driver comfort, rather than to driver safety. They do not warn of the speed at which skidding (Section 19.2.3) or rollover (Section 19.2.2) will occur.

  The advisory values are usually determined by the ball-bank indicator described in Section 19.2.5. Typically, signs are then displayed if the advisory speed so determined is more than 10 km/h below the free speed (Section 18.2.5) adopted by drivers on the preceding length of road. If this is the case, the advisory speeds are usually displayed in 10 km/h multiples and accompanied by an appropriate warning sign (Section 21.3.1).

18.1.4 Physical controls

A number of measures other than advisory signs and the threat of prosecution can be used to lower driver speeds. They can be classified into measures which:

* warn of the need to change speed (e.g. kerb extensions into the traffic-way and road surface changes),
* warn of hazards ahead (e.g. transverse painted lines on the pavement),

* reduce a road's propensity to induce speeding (e.g. alignment tightening, Section 18.2.5, pavement narrowing, Section 7.3, or lane width narrowing, Sections 6.2.5 & 20.3.2),
* inhibit speed by the use of deliberately poor surface treatments,
* inhibit speeding by tactile speed-control devices (see below),
* cause intersection delay (e.g. Stop signs, Section 20.3.7),
* reduce network connectivity (e.g. diagonal closures at intersections, Section 7.3), or
* control or manage a vehicle's ability to speed (e.g. speed limiters on trucks and buses, external signs indicating that a vehicle is speeding, and cruise control permitting drivers to select a speed).

It is useful to see many of these measures in the context of the *self-explaining* road which will lead to drivers intuitively recognising the speed expected of them when using that piece of road. This approach mainly works by providing a range of cues which consciously or sub-consciously affect driver behaviour. The cues may come from the surrounding community, the roadside, the road itself, or signage. They must leave drivers in no doubt about the type of road they are using, and the speed expectations that that places on them.

Specific external physical controls on speeding — listed in terms of increasing directness — are either tactile or brutal:

(a) Tactile speed control devices:

These are often very effectively used to warn a driver to reduce speed due to an approaching visual signal or critical location, or because the vehicle has strayed into an area where it should not be. Their basic method of operating is via tactile (i.e. felt) and audible stimuli conveyed to the driver. To be effective, audible devices should raise noise levels within a vehicle by at least 6 dB(A) (Section 32.2). Spacings of about 250 mm maximise vehicle vibration and noise, and about 2000 mm maximises the jolt felt in the vehicle. However, due to cabin design, most truck drivers are relatively immune from the noise and vibration produced by these devices. Two of the reasons for their effectiveness are that they:

* do not interfere with the driver's visual functions, and
* produce quicker driver responses than do visual signs.

Tactile devices include rumble areas, jiggle bars, and road humps:

(a1) Rumble areas (or *rumble* or *chatter strips*)*:*

These are areas of coarse or grooved pavement surfacing, often intermittently spaced in the direction of vehicle travel. The texture is commonly achieved by:

* a spray and chip seal surface treatment (Section 12.1.3),
* treating an existing asphalt surface,
* surface grooving,
* low transverse bars,
* thermoplastic strips (Section 22.1.3), or
* replacing the existing pavement with concrete blocks (Section 11.6).

The upstands or grooves usually have a vertical projection of about 10 mm. Rumble strips as edgelines are discussed in Section 22.1.3.

(a2) Jiggle (or *rumble*) *bars:*

These are usually precast concrete blocks 50 to 100 mm wide and projecting about 15 mm above the road surface. Jiggle bars are spaced in series so that when travelled over by a car they produce a rumbling noise and a body jolt. They are smaller than the *safety bars* discussed in Section 22.4.2. Jiggle bars are thus more aggressive than are rumble areas, and so are used for more critical applications.

(a3) Road humps:

The discussion so far has been about relatively-passive, tactile speed-control devices. A far more active device is the *road hump*, which causes drivers to slow down by threatening them with severe discomfort, and even vehicular and personal damage, if they were to attempt a high-speed crossing. Because of the hump's potentially aggressive nature, it is essential that drivers:

* are made fully aware of its presence by advance signing and easy detection, and
* experience minimum discomfort if crossing it at an appropriately-low speed.

For any hump installation there will be a speed below which there will be no damage to crossing vehicles or discomfort to their occupants (Jarvis, 1980a). This is called the *design crossing speed* and it is typically between 15 and 25 km/h and produces peak vertical accelerations of the order of 0.7g. The *maximum safe speed* is the speed below which there is no risk of vehicular damage or of unsafe behaviour. It is typically above 80 km/h, which would also be above many urban speed limits and about four times the design crossing speed. It is important that it exceeds a reasonable upper limit of vehicular speeds across the hump. Between the design crossing speed and the maximum safe speed driving would therefore be safe, but there would be noticeable and increasing occupant discomfort.

The usual hump crosses the entire carriageway and in cross-section is a segment of a circle 3700 mm long and 100 mm high, following a profile developed by Watts (1973). Section 19.3 indicates that some special vehicles when laden can have ground clearances as low as 75 mm and so these vehicles will have difficulties traversing 100 mm high humps.

More recently, practice has tended towards a *plateau* (or *table* or *raised pavement*) profile in which the segment is split in two and separated by a flat top a couple of meters wide, sometimes in the form of a pedestrian crossing (Section 20.4). The aproach ramps are typically flatter than 1 in 15. To accommodate buses and their passengers comfortably, the flat top may need to be up to 8 m long.

A road hump is an effective speed-reducer but will only influence vehicle speeds within its immediate vicinity. Therefore, if low speeds are required over a greater road length, humps will be required about every 60 m.

Humps have a number of side-effects. The driving behaviour that a hump produces will increase trip times. However, if the humps also reduce traffic flows, then the net effect on travel time may be reduced. A British review also indicated a 70 percent reduction in crashes on treated streets and an 8 percent reduction on adjacent streets (Webster, 1993). If the traffic stream consists mainly of light vehicles, overall noise levels can reduce by about 7 dB(A), due mainly to the reduction in vehicle speed (Section 32.2.2). When there is about 10 percent heavy vehicles (Section 29.2) the noise effect is neutral, and gradually worsens as the percentage increases further (Abbott *et al.*, 1995). Other pollution effects will follow a similar pattern. The sharp acceleration and deceleration profiles produced by humps will usually cause a net increase in fuel consumption (Section 29.2.3).

A major problem associated with the use of road humps on public roads is the question of *legal liability* for the consequences of any deleterious effects of the humps on people or property. The general issue is discussed in Section 28.5. The author has not seen any evidence of crashes attributed to humps, whereas there are a number of indications of improved safety. With respect to *tort* claims for damages due to *negligence*, there are sufficient data available to permit humps to be designed with the care, skill and diligence needed to avoid the accusation of negligence (Jarvis, 1980b). Thus, it should be possible to demonstrate that the construction of a properly-designed road hump was not a negligent act, and thus counter the charge of *misfeasance*.

[A *tort* is a private or civil wrong or injury, other than a breach of contract, to a person or property. It is committed when a person breaches a duty fixed by statute or common law. A tort has the potential to be corrected by legal action providing damages to the victim to remedy the tort. *Negligence* is failing to exercise the care that a reasonable and prudent person would exercise in the same circumstances. It is a form of tort in which the wrong or injury was not caused intentionally. *Misfeasance* (i.e. 'wrong doing') is doing a lawful act in a negligent manner. It is contrasted with *malfeasance* (or *non-feasance*, i.e. 'not doing'), which is not performing a legal duty. A typical malfeasance would be doing inadequate maintenance. In some jurisdictions, road authorities are given immunity from malfeasance, usually by precedence rather than by written law (e. g. Sarre, 2003). However, some courts do not accept that immunity (Mihai, 2008).]

(b) Vehicle arrester beds:

These are at the extreme end of the spectrum of devices intended to slow vehicles down. They are lanes or entire carriageways used to halt runaway vehicles, particularly trucks, unable to brake on long, steep grades (Section 19.3). Occasionally, it is possible to provide a run-off track, which can divert the vehicle onto an upgrade or long, safe flat on which it can decelerate. Grades of up to 20 percent over 100 m lengths have been used.

More commonly, it is necessary to provide a positive means of arresting the out-of-control vehicle. This is usually done by using beds of soft earth, sand, or loose gravel placed on the run-off track. The vehicle is halted by the high rolling resistance and so it is essential that it not be able to plane over the material. The desirable properties of the bed material are that it be composed of round, smooth particles of a single size between 5 and 10 mm, giving a low angle of internal friction (Figure 8.8). Beds are typically 500 to 1000 mm deep. When the vehicle is intended to run straight into the bed, it must be at least a lane in width. When the bed is placed alongside a road, the bed need only be half a lane in width, thus accommodating only one wheel track. This half-width arrangement is only suitable when truck speeds are low. The length of a two-wheel track bed when sand is used is typically about $V^2/100$ m, where V is the entry speed in km/h. Mean deceleration rates of about 0.4g are possible at entry speeds of 50 km/h and over.

18.1.5 Speed measurement

Speed is measured either:

(a) directly, using external devices such as *radar* or in-vehicle equipment such as *speedometers* based on wheel or transmission revolutions. The speeds measured by such devices are effectively instantaneous and are called *spot speeds* (Section 18.2.13): or

(b) indirectly, by measuring the time the vehicle takes to travel a known distance. Typically, this involves measuring the time between a vehicle passing over two adjacent vehicle detectors, such as *loops* (Section 23.7), *pneumatic tubes* or electric *treadle switches* (Section 31.5). It is also possible to estimate speed from the time the vehicle occupies one detector. These are called *space speeds* (see Section 18.2.14).

Trip speeds (or *overall* or *journey* speeds, Section 18.2.12) are overall trip times divided by trip distance. In general traffic, these times are obtained from the *floating cars* (or *moving observers*) method in which special cars placed in the traffic flow. The drivers are instructed to either follow particular cars, attempt to 'float' in the traffic, overtake as many vehicles as overtake them, or drive 'naturally' and record overtaking events in which they are involved. Trip times derived from this method have a coefficient of variation of about 0.15. Sample sizes should be at least 20 to establish a trip time and 80 for before-and-after studies. Other applications of the technique are given in Section 31.3.7.

Trip speeds can also be deduced from a series of monitors of the spot speed of passing vehicles at regular locations along the trip path. Component trip times are then estimated from component distance divided by measured component speed. Component times can also be obtained from observers using stop watches on short routes, from video surveillance, or from cordon surveys (Section 31.3.7).

18.2 DRIVER SPEED BEHAVIOUR

18.2.1 Factors influencing speed

The factors that affect the speed that drivers adopt can be divided into physical factors, external factors, traffic factors, driver–vehicle factors, and factors peculiar to the trucking industry (Galin, 1981).

1. *The physical factors are*:
    1.1 carriageway and lane width (narrow widths slow traffic, see also Section 6.2.5),
    1.2 number of lanes (adding lanes increases speed),
    1.3 presence of a median (increases speed),
    1.4 horizontal curvature, sight distance, and grade (Section 18.2.5),
    1.5 speed-limit signing (Section 18.1.2),
    1.6 intersections (slow traffic),
    1.7 pedestrian crossings (slow traffic), and
    1.8 pavement quality (rough roads slow traffic, Section 18.2.5d).
2. *The external factors are*:
    2.1 socio-economic conditions (speed is higher in good times),
    2.2 weather (speed is higher in good weather and lower in rain, dropping by about 1 km/h for every 3 mm/h of rainfall intensity), and
    2.3 land use (Section 18.3).
3. *The traffic factors are*:
    3.1 traffic flow (Section 17.2),
    3.2 traffic composition (see 5.4 below & Section 17.4.3),
    3.3 parking (Section 18.3),
    3.4 cross traffic (Section 18.3),
    3.5 platooning (Section 17.3.8), and
    3.6 traffic complexity (drivers reduce speed as the cognitive demands from driving increase (Section 16.1).
4. *The driver–vehicle factors are*:
    4.1 trip purpose (leisure drivers drive more slowly),
    4.2 trip duration (Speed increases as trip time increases due to *speed adaptation* [or *velocitisation*], a process in which drivers increasingly underestimate their speed as their trip time increases. They also find speeds in speed-restriction zones to be inordinately slow),
    4.3 vehicle occupancy (speed drops as occupancy increases),

4.4 vehicle age (speed decreases as vehicle age increases),
4.5 vehicle type (vehicles towing trailers, vans and light commercial vehicles travel more slowly),
4.6 sex of driver (males drive faster than females, although female speeds tend to increase with age),
4.7 crash record of driver (drivers with a recent crash history travel faster),
4.8 cost of travel (Section 31.2.3, speed increases as the cost of trip time increases, e.g. as income level rises), and
4.9 influence of drugs, alcohol, and fatigue (Sections 16.5.4 & 28.2.2).

5. *The truck-related factors are* (Section 30.5.2):
5.1 pressures due to time lost seeking loads, loading, and unloading,
5.2 pressures from cargo owners,
5.3 pressures to increase driver's income, and
5.4 the different performance characteristics of a truck relative to a car (e.g. Section 27.2.2).

To discuss these factors the following carefully defined speeds are used. They are all summarised and interconnected in Table 18.2 at the end of Section 18.2.16. The definitions given are not universally accepted, and the ones put forward are those preferred by the Author and consistent with the logic of this book.

18.2.2 Basic desired speed

A driver's *basic desired speed* is the maximum safe and comfortable speed at which that driver would travel when *not* influenced by (McLean, 1989):
(a) the overall road standard,
(b) the surrounding terrain (or road environment),
(c) the speed limit,
(d) the traffic,
(e) local road design features, such as vertical grade, horizontal curvature, and sight distance restrictions, or
(f) the weather conditions.

The basic desired speed therefore depends only on the driver and vehicle, and will rarely be encountered.

18.2.3 Desired speed

A driver's *desired speed* (or *limiting speed*) occurs when the *basic desired speed* (Section 18.2.2) is constrained by that driver's perception of the:
(a) overall road standard,
(b) surrounding terrain, and
(c) speed limit.

Desired speed is thus a compromise drivers make between their basic desired speed and the general risks and discomfort associated with travelling at that speed on the actual road. It remains independent of the *traffic*, *local road design features*, and the *weather*.

Drivers usually consider themselves independent of other traffic when:

* vehicles are separated by at least 6 s, which approximates to a uniform flow of below 600 veh/h in the driver's lane. Estimates of this value vary greatly (Section 17.3.9); and
* the driver is not inhibited from overtaking. Table 18.3 will later suggest that on two-way roads this means a uniform opposing flow of under 100 veh/h. This can be taken as light traffic.

On a high standard road or any long stretch of straight good standard road, a driver's desired speed will only be influenced by the speed limit. On a good standard road where there are some restrictive localised design features, a driver on long straight or tangent (Section 19.2.1) sections of road will be uninfluenced by such a local feature until within about 100 m of that feature (Figure 18.4 & Table 21.1).

A distribution of desired speeds will occur within the driving population, even in light traffic. For the same piece of road, the distribution is commonly normal, with a coefficient of variation of about 0.14. Rural speed distributions collected at a number of sites on the same road class tend to converge to a single distribution, when standardised with respect to the mean speed at the site.

18.2.4 Speed environment

The *speed environment*, V_e, of a road is the 85th percentile of the desired speed (Section 18.2.3) adopted by all drivers using the road. Some approaches use this as their definition of operating speed, but this book defines operating speed (Section 18.2.10) more consistently as the speed environment lowered by the effect of other traffic.

18.2.5 Free speed

Free speed is defined as the *speed environment* (Section 18.2.4) influenced by *local road design features*, but uninfluenced by *weather* and *other traffic*. A general application of free speed was given in Section 17.2.3. Specifically, the following *local road design features* influence free speeds:

(a) *grade*: typically, a 1 percent increase in grade reduces free speed by 3 km/h for cars and 6 km/h for heavier vehicles. The reduction is greater for down-grades of over 8 percent, due to concerns over braking capacity, and for up-grades of over 5 percent, due to vehicle response. Practical effects are discussed in Section 19.3. The maximum speed on up-grades is primarily limited by the performance characteristics of the vehicle, whereas on down-grades the characteristics of the driver will also play a major role.

(b) *sight distance*: typically, a 100 m drop in sight distance (Section 19.4) drops free speed by 1 km/h. When sight distance drops to 100 m, speeds will typically be below 50 km/h.

(c) *road curvature*: typically, for a single circular curve of radius r and curvature 1/r, a 1 km^{-1} increase in curvature drops free speed by 4 km/h. For curves with a small radius, the speed V in km/h can be estimated by $10\sqrt{(r - 10)}$ where r is the radius in metres. However, once the curve radius becomes 15 m or less — effectively a right angle — drivers are able to negotiate such a minimal curve in a normal lane width at about 25 km/h. These factors and the definition of desired speed in Section 18.2.3 suggest an overall equation of the form:

$V/V_{dt} = 1 - [V_{dt}/21]^2/r$

where V_{dt} is the desired speed in the local terrain in km/h and r is in km. This gives a linear link between speed and curvature. Figure 18.1 gives a range of such typical V vs r predictions, which are each asymptotic to the relevant speed environment (Section 18.2.4). The British approach assumes a 1 km/h drop for every ten degrees of bend per kilometre of road (i.e. a curviness index of 10, Section 19.2.4). The behavioural basis for driver response to curves is discussed in Section 16.5.2.

Note that the 'speed environment' input applies to the road length as a whole and only becomes the curve approach speed for an isolated curve. Without this definition the model would predict that a series of curves would bring traffic to a halt. The 'free speed' output can be taken as the departure speed from the curve.

(d) *roughness*: free speed decreases linearly with road roughness (Section 14.4 and Geipot, 1982a), although the effect is minor,

(e) *carriageway and shoulder widths* have no effect. There is a minor effect for lane widths below 3.4 m.

Each of these factors, together with the desired speed, can be considered as applying its own upper limit to the free speed adopted.

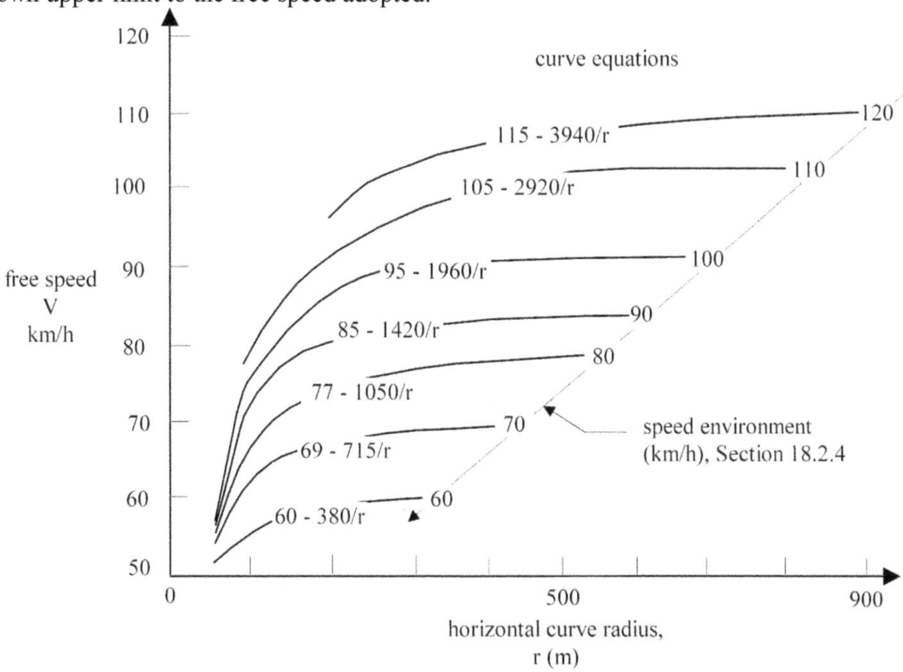

Figure 18.1 Typical graphs predicting free speeds on curves.

Free speeds for trucks are sometimes different to the above guides (Geipot, 1982a). In particular, trucks normally negotiate long up-grades and curves at lower speeds than do cars.

Note that a number of other quite different definitions are used in the literature. For instance, free speed is sometimes defined identically to our desired speed and the term *unimpeded speed* used where this book uses free speed (e.g. McLean, 1989). The reader is cautioned to check the particular definition in use in any other document.

18.2.6 Design speed: general concept

To avoid confusing or misleading drivers, roads must be designed to a safe, appropriate and consistent underlying philosophy. These design imperatives directly apply to features which influence vehicle operations such as horizontal and vertical alignment, sight distance, shoulder and lane width, lighting levels, braking performance, roughness and kerb provision (Sections 18.2.5 & 19.2.5). The design approach must avoid causing:

* dangerous or inappropriate driving within the road length being designed,
* driver expectations to be confounded by inconsistent design along the road length, and
* drivers to make rapid or substantial changes in speed or direction or preparedness when they leave a road length. Such changes can lead to poor outcomes, particularly as drivers will often carry behaviour adopted in a length with high design standards into a length with lower standards. This slow adjustment to poorer conditions is most pronounced for changes in alignment.

Within a length of road, road designers should therefore design the various road elements using a safe, appropriate and consistent *design speed.* From the definition in Section 18.2.5, the chosen design speed will therefore determine the free speed that drivers adopt within a newly constructed road length. That free speed will need to be checked for safety and appropriateness.

In a sense, the process works in reverse. From past observation, it is known that a particular road configuration will result in a particular free speed (e.g. Figure 18.1). To accommodate the variability between drivers, the speed used is that which experience has suggested will represent the safe and appropriate behaviour of 85 percent of drivers on that road configuration. This gives a link between 'speed' and road configuration. For a particular design speed it is therefore possible to use this link to work backwards to find an appropriate and safe road configuration for that speed. This issue is pursued further in Section 18.2.7 below.

The selected design speed — rather than the observed surrounding speed environment (Section 18.2.4) — is thus the major determinant of the standards adopted for the horizontal and vertical alignment of a road. Nevertheless, the chosen design speed must be compatible with and — as far as possible — equal to the speed environment. Furthermore, it is necessary to consider the wider implications of the choice, as the consequent road characteristics, together with the terrain and the applicable speed limit, will determine the subsequent speed environment of the completed road.

With respect to speed limits (Section 18.1.2), a design speed *above* the current maximum speed limit could be justified on the basis of:

* expected future changes in that limit,
* the need to provide drivers with a margin of safety,
* the prospect that many drivers will exceed the limit, and/or
* a desire to raise the overall standard of the road.

A design speed *below* the general speed limit could be justified where local features or cost constraints prevent the selection of a higher standard alignment. Such a road would need to carry advisory speed signs related to the design speed.

Problems in this area can arise in the following ways:

* The free speeds that drivers adopt on the completed road may not match the design speed used, if that speed was not consistently applied or lacked a sound behavioural basis.

* Some misguided design policies specify minimum standards but urge the use of higher standards where economically warranted. A set of disjointed free speeds then arises, creating the driver adjustment problems discussed above.
* Some stretches of road, such as long, flat straights, will naturally have a higher free speed than the specified design speed, as they will have no physical characteristics which limit safe and appropriate speeds and drivers will respond to their perceptions of the speed environment and not to the formal design speed.

Some further matters that require consideration are:

* It has been widely suggested that incremental changes in design speed along a length of road should not exceed 15 km/h.
* Not all road element characteristics directly relate to the design speed, and so designers must separately maintain consistency in such features as carriageway width (Section 6.2.5) and lighting levels (Section 24.3) to accord with driver expectancies.
* The influence of design speed on alignment economics is discussed in Section 19.5. The selection of a design speed in difficult terrain will therefore need to consider the available funding and the level of road usage (e.g. using the model in Figure 5.1). In urban areas and public parklands, the design speed may be limited by environmental factors (e.g. noise increases with traffic speed, Section 32.2.3). Design speeds below the optimum speed (Section 17.2.3) will reduce the traffic capacity of a road.

Beyond these general issues, there are three important specific design cases to consider, viz. long new lengths of road, short lengths of new road in an existing road, and a series of tight curves. These are covered in the next three sections.

18.2.7 Design speed for long lengths of new road

In adopting the design speed approach for long lengths of new road, many Road Authorities give broad guidelines for the selection of a design speed, as illustrated in Table 18.1. However, there is a considerable risk attached to the unthinking use of such tables. For instance, it has been suggested above that the design speed should be sensitive to the speed environment and that the final choice may often need to be a subjective one. More specific concerns are given below. These do not suggest that the design speed approach should not be used, but that it should be used with judgement. Individual design speeds should thus be selected on the basis of a specific engineering assessment of each situation, and not simply from a list such as Table 18.1.

The speeds in Table 18.1 fall into three bands (Austroads, 2003):

* *high speed roads*, design speed > 100 km/h,
  operating speed = desired speed > free speed
* *intermediate speed roads*, 80 km/h < design speed < 100 km/h,
  operating speed = free speed,
  drivers accelerate when feasible, and
* *low speed roads*, design speed < 80 km/h,
  operating speed = free speed,
  drivers are alert to low standard but still tend to drive faster than appropriate.

[Free speed is defined in Section 18.2.5 and operating speed in Section 18.2.10]. A basic problem with an over-reliance on design speeds is that drivers are largely unaware of the speeds that have been used by the designers. For instance, typical field measurements of the ratio of free speeds to design speeds are shown in Figure 18.2. Below 110 km/h,

Table 18.1 Typical design speeds in km/h for long lengths of road. Note that the text (Section 18.2.7) suggests that the direct and unquestioning use of the data in this table is an undesirable design procedure.

Type of road	Terrain			
	Flat rural	Rolling (hilly) rural	Mountainous rural	Urban
Low cost	80	60	40	50
Minor undivided	100	80	60	60
Major undivided	120	110	100	70
Divided	130	120	110	70–80
Freeway	130	130	120	120

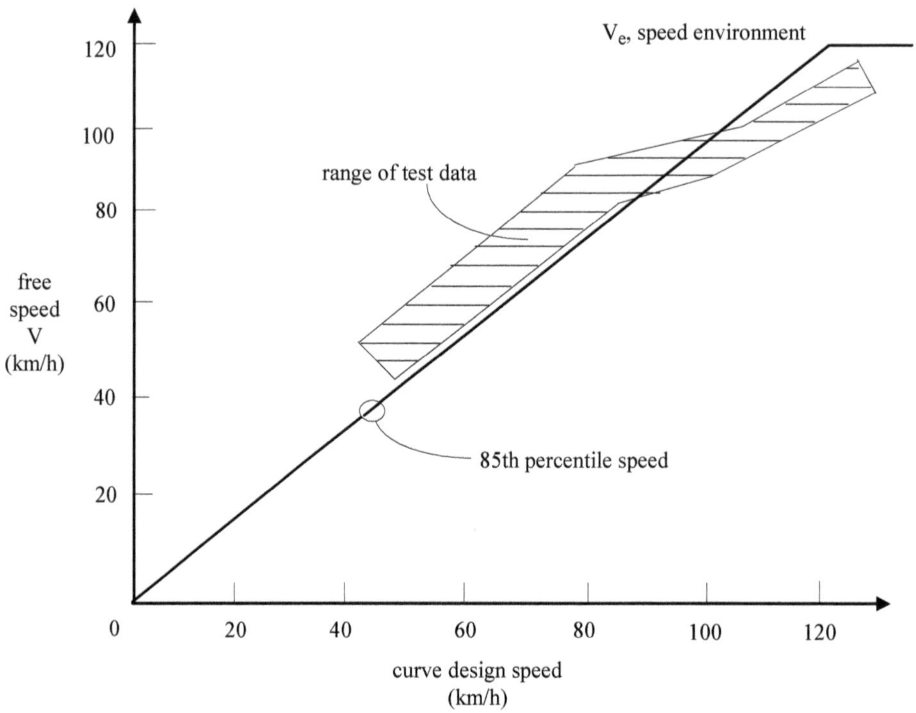

Figure 18.2 Relationship between the observed free speeds of cars on curves and the design speeds for those curves (McLean, 1979).

behaviour is influenced more by alignment and the ratio is approximately given by (McLean, 1979):

[free speed]/[design speed] = 2 – [design speed/120]

suggesting that the two increase proportionately but that drivers take more than is offered for design speeds below about 110 km/h.

Above this speed, free speeds depend less on the design speeds as drivers tend to be less affected by road alignment and more by external factors such as the speed limit. Consequently, design speed begins to exceed the speed environment. The benefits of increased design speeds would therefore not be in reduced travel times, but would be in increased safety resulting from the elimination of relatively low standard — and perhaps unexpected — road elements. Roads with design speeds of over 110 km/h could therefore be properly considered to be 'high-standard' roads.

The link between design speed and road feature is seen by equating the design speed to the free speed (Section 18.2.5) for that feature. Thus, for a curve, the free speed then gives curve radius via Figure 18.1a.

Recall that free speed only covers 85 percent of drivers. The other 15 percent are assumed to operate beyond the design limits — usually by exceeding their comfort limits as determined by the side-friction factor f (Section 19.2.3) or, less commonly, by making alternative safety assessments. They are commonly excluded from design consideration as a matter of economic constraint (Section 19.5) and also as it has been observed that there is always a percentage of the drivers who drive beyond whatever limits are formally provided.

18.2.8 Design speed for a short length of new road in an existing good-quality road

This design situation occurs when a single length of road with a tight alignment is being designed as part of a higher-quality existing road, or where one length of a new road must be designed for a lower speed than the rest. Here the speed environment is known and directly influences entry speeds to the new length. For example, drivers may adopt high speeds on a lightly trafficked length of straight, flat road and then enter a zone, such as a gorge, where topography will influence the alignment. Figure 18.1 can be used to give the amount by which a driver's speed will need to drop in order to safely and comfortably negotiate any such sub-normal curve.

There is increasing evidence that speed reductions from one piece of road to another are linked to crash rates (Polus and Mattar-Habib, 2004). Reductions of 20 km/h represent a potentially dangerous situation (Tate and Turner, 2007). The key design issues are to:

* recognise the features of each length of road which will determine the free speeds that drivers adopt,
* ensure that drivers do not encounter local conditions that require speeds markedly different from the speed environment, and
* ensure that design speeds follow a consistent pattern. Four ways of assisting this are to keep:
  (1) the driver's perceptions and estimates similar to the actual conditions.
  (2) design speed changes between successive road lengths to less than 10 km/h (used in the German road design guide, RAS-L-1). Some countries limit the radius change to 50 percent, which Figure 18.1 shows to be similar to the 10 km/h criterion.
  (3) design speeds to within 20 km/h of the speed environment, and
  (4) truck speeds to within 15 km/h of car speeds.

Further guidance in this matter can be obtained by modifying Figure 18.1, in the manner shown in Figure 18.3. The driver 'comfort' curve is obtained from Equation 19.2a with the typical design values for (e + f) ranging from 0.14 at 120 km/h, to 0.21 at 90 km/h, to 0.36 at 60 km/h. The (2) line is based on (2) above, which is always more severe than criterion (3). The two criterion curves are seen to be very close, and the shaded area between them can be taken as a band defining suitable conditions for drivers as being to the right of the band. The left-hand vertical segment of the band is based on the poor crash record of these tight curves, as discussed in Section 28.5(a).

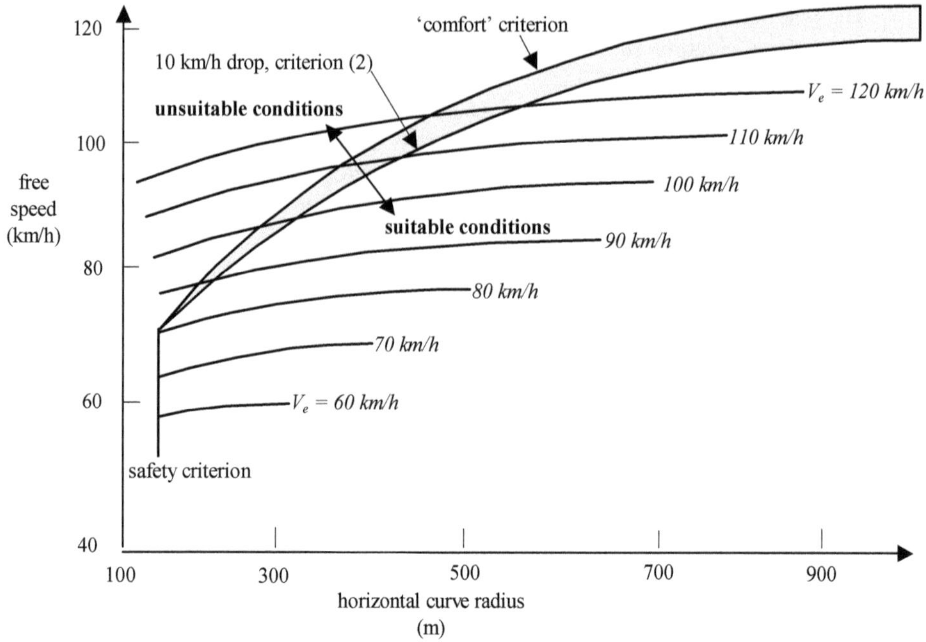

Figure 18.3 Design restraints added to Figure 18.1. This curve is typical only, actual designs should be based on local data.

If the required drop in speed is worrying, measures may need to be taken to either:
(a) improve the alignment,
(b) use perceptual cues to slow drivers down on the approaches (see Section 19.2.3), or
(c) use signs to warn of the need to slow down.

18.2.9 Design speed for a series of tight curves

This design case arises when a series of tight curves is being designed within a constrained alignment. In using Figure 18.1 in the conventional way, the speed environment input, V_e, is usually taken as the exit speed from the preceding curve and is thus the free speed of the preceding alignment. However, this tends to attenuate speeds rapidly and, if the exit speed predictions become unrealistically low, it is also necessary to estimate the speed environment that drivers will naturally seek over the particular stretch of winding road and use that value. This can be done by judgement or by observing exit speeds from particular curves and then back-calculating in Figure 18.1 to estimate the speed environment.

The reduction in speed that occurs due to a curve is plotted in Figure 18.4 for the data in Figures 18.1&3. Using the criterion contour in Figure 18.3 as a limit, only the speed reductions below the solid '10 km/h' contour line occur within the road curve; the rest must be achieved by the driver slowing before entering the curve. Ways of ensuring this were listed in Section 16.5.2.

It has been noted (Austroads, 2003) that in practice drivers entering such a situation tend to make most of their speed reduction over the first few curves. When they reach a speed they feel comfortable with, they tend to maintain that speed uninfluenced by other, relatively minor, alignment changes. For example, they will not increase their speed again

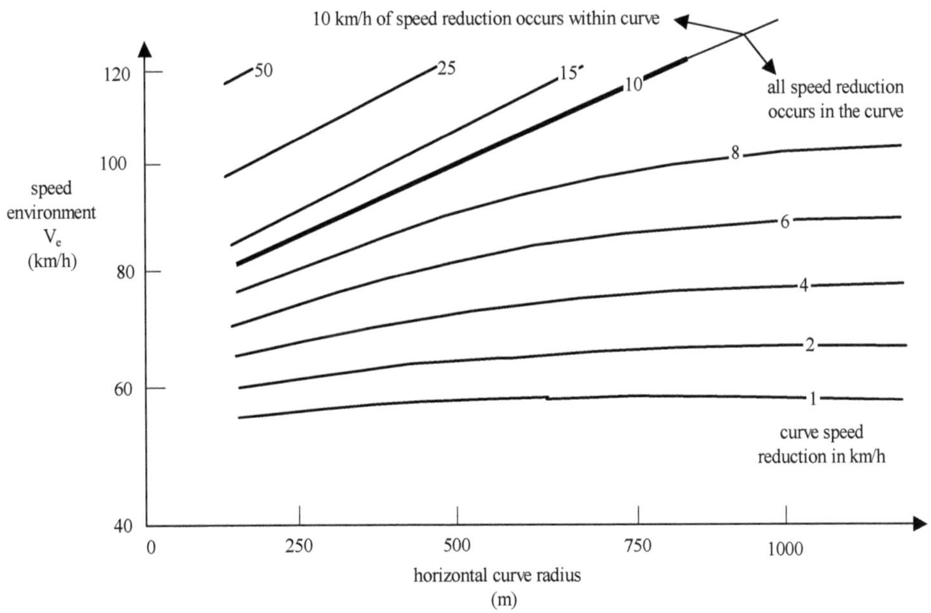

Figure 18.4 Speed reductions due to curves, based on Figure 18.3. Note that of the reductions above the (solid) 10 km/h contour line, only 10 can be assumed to occur within the curve. The rest must be achieved before the vehicle enters the curve.

unless they encounter a straight section at least 200 m long. Designs should be checked against this common perception.

18.2.10 Operating speed

The operating speed is the 85th percentile speed that drivers judge to be possible under prevailing traffic conditions on the road in question, but in the absence of cross-traffic. When alignments are a constraint, it would be expected that the operating speed would be close to the design speed. The traffic levels that cause operating speeds to drop below free speeds were defined in Section 18.2.3. On two-way roads, operating speed also drops by about 1 km/h for every 100 veh/h traffic flow in the opposing direction. On constrained alignments, commercial vehicles tend to operate at speeds that average 10 to 15 km/h below car speeds.

The speeds adopted by drivers on a piece of road are often normally distributed (and the operating speed is the 85th percentile of this distribution), although there will be considerable between-site variation in the distribution parameters. The 'pace' of the traffic is, typically, the 20 km/k speed range containing the greatest proportion of the total traffic.

18.2.11 Running speed

Running (or cruise) speed is the urban equivalent of operating speed. It is the average speed of a given vehicle travelling over a defined length, L_l, of road link, whilst the vehicle is in motion. In particular, it is the speed a vehicle adopts between traffic control devices and under the prevailing traffic and road conditions. Similarly, running time is the time during which the vehicle is in motion. Thus:

running speed = L_l/(running time)

18.2.12 Trip speed

Trip speed is the average speed of a vehicle during a journey, including any stops such as delays at traffic control devices. Thus:

trip speed = L_l/(link trip time) (18.1)

but

link trip time = (running time) + (stopped time)
= L_l/(running speed) + (stopped time)

Substituting into Equation 18.1 therefore gives:

1/(trip speed) = 1/(running speed) + (stopped time)/L_l

or

trip speed = (running speed)/(1 + X)

where

X = (running speed)(stopped time)/L_l

18.2.13 Spot speed

Spot speed is the instantaneous speed of a vehicle at a specific location (or spot) and at a point in time, as measured, for instance, by a radar speed meter (Section 18.1.5).

18.2.14 Space mean speed

Space (or distance) mean speed for a length of lane, L_l, can be defined as either:

* the average of the simultaneous spot speeds, V_i, of each vehicle in L. If speeds are constant, this is the same as the average speed of the vehicles passing some point in L_l.
* an estimate made from measurements of the time the vehicles take to traverse L_l, using:
  space mean speed = L_l/(the average time taken by all vehicles to traverse L_l)

or

1/(space mean speed) = (1/N)Σ(1/V_{ts})
= harmonic mean of the trip speeds, V_{ts}

where N is the number of vehicles in the count. The tendency of slow vehicles to be over-represented in such recordings will bias the estimates. If speeds are constant, the estimate equals the basic speed definition.

Space mean speed is often used for the speed variable in Equations 17.1 and 17.2 when vehicles in a real traffic stream are adopting different speeds. In this case the vehicle flow, q, and density, K, also become average values as the chance of a vehicle being in L_l in some time T is V_iT/L and thus the average flow for N vehicles in L in T is:

q = (1/T)Σ(V_iT/L_l) = (space mean speed)N/L_l
= (space mean speed)(average K)

which is the general form of Equation 17.2.

18.2.15 Time mean speed

Time mean speed is the arithmetic mean of the spot speeds of a line of vehicles measured over a given time interval, T_l, as they pass the same point. By analogy with space mean speed, it can also be estimated from 'aerial photographs', using the average of the distances travelled by all vehicles in the constant time interval between adjacent photos. Hence:

time mean speed = (average distance travelled in T_l)/T_l
= average speed

The tendency of fast vehicles to be over-represented in such samples will bias the estimates. It has been shown that, by multiplying both sides of Equation 17.3 by v and then integrating (Wardrop, 1952), time mean speed is related to the space mean speed by:

time mean speed = (space mean speed) + (variance of space mean speed)/(space mean speed)

or

(time mean speed)/(space mean speed) =
1 + (coefficient of variation of space mean speed)^2

From Section 18.2.4 the coefficient of variation for typical traffic can be put at about 0.14 which gives the difference between the two speeds at about 2 percent.

18.2.16 Stream speed

Stream speed is the speed of the traffic stream. As both space and time mean speed are average speeds, either can be used to give stream speed and Section 18.2.15 showed that the difference between the two estimates will usually be small.

Table 18.2 summarises the various speed definitions introduced in this section.

18.3 URBAN SPEEDS

Travel on urban roads is usually influenced more by the other vehicles and by non-geometric aspects of the road layout, than is rural road travel. Hence, the estimation of running speeds and times on urban roads is a far more complex issue than the rural road case discussed above. An introduction to the topic was given in Section 17.2.5. For urban transport planning purposes (Section 31.2.3), the basic need is often for a relationship between trip time and traffic flow (Section 17.2.6 & Figure 17.7). Trip time is the inverse of trip speed (Equation 18.1), and care is therefore needed to distinguish between running speed (Section 18.2.11) and trip speed (Section 18.2.12).

Urban trip speeds will depend on the:

(a) traffic flow in the direction of travel (Section 17.2.3),
(b) opposing traffic flow on a two-way road (Section 18.4.3),
(c) proportion of through traffic, which can raise speeds by up to about 10 km/h,
(d) presence of public transport vehicles, cyclists, and pedestrians,
(e) traffic composition, particularly the proportion of heavy vehicles (Section 17.4.3),
(f) extent of kerbside parking. [With high levels of parking activity, speeds can drop by 5 km/h, and frequent parking manoeuvres can halve speeds on local roads. The latter can thus be the largest single cause of local speed reductions.]
(g) presence of trucks loading and unloading,

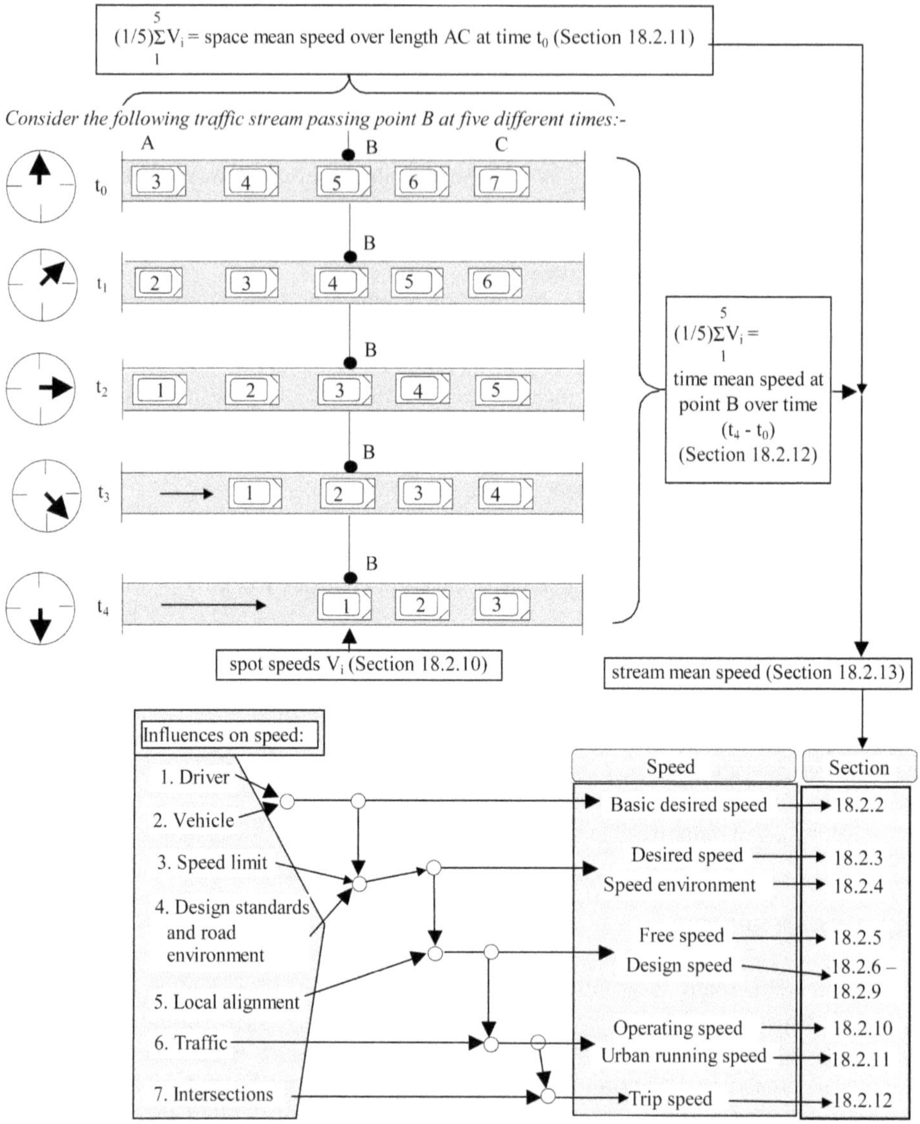

Table 18.2 Example giving a comparison of speed definitions in Sections 8.2.2 to 8.2.13.

(h) abutting land use, particularly the amount of traffic it produces. [In extreme cases speeds can drop by 10 km/h.]

(i) number of signalised intersections (Chapter 23). [Typically, each intersection per kilometre drops the speed by about 2 km/h.]

(j) number of unsignalised intersections (Chapter 20). [Typically, each intersection per kilometre drops the speed by about 0.75 km/h.]

(k) use of traffic management measures (Section 7.3 & 18.1.4),

(l) sight distance down the road. [In a residential street, every 70 m of apparent street length adds about 1 km/h (Armour, 1982). However, there is little or no evidence that restricting sight distance will lower vehicle speeds.]

(m) streetscape, as the presence of trees and shrubs reduces speed (Sections 7.2.3–5),

(n) pavement width, as increasing the width leads to increases in speed on residential streets, and

(o) tight urban curves ($r < 100$ m, $V < 60$ km/h)) reduce speeds to levels that can be estimated from (McDonald *et al.*, 1984):

$$V = 6\sqrt{r} \text{ km/h}$$

with r in metre. [This is associated with a lateral acceleration of about 3 ms^{-2} (Section 19.2.3).]

Factors of possible relevance that actually have little influence are lane widths, pavement quality, and pedestrian crossings.

Speeds in urban areas are often normally distributed with a coefficient of variation of about 0.30.

Section 18.1.2 suggested that urban speed limits typically lie between 40 km/h and 60 km/h. It also indicated that there are many safety benefits of travelling at lower speeds and that in urban areas crashes vary with $V^{1.5}$. Evidence from most countries is that a speed limit of 60 km/h is too high for urban areas. Dropping the speed limit to 55 km/h typically halves the crash rate.

18.4 GAPS AND OVERTAKING

18.4.1 Vehicle trajectories

The path of an individual vehicle along a road is defined by its distance–time trajectory. For example, Figure 18.5a shows a vehicle which stops for a time (B – A) at distance C.

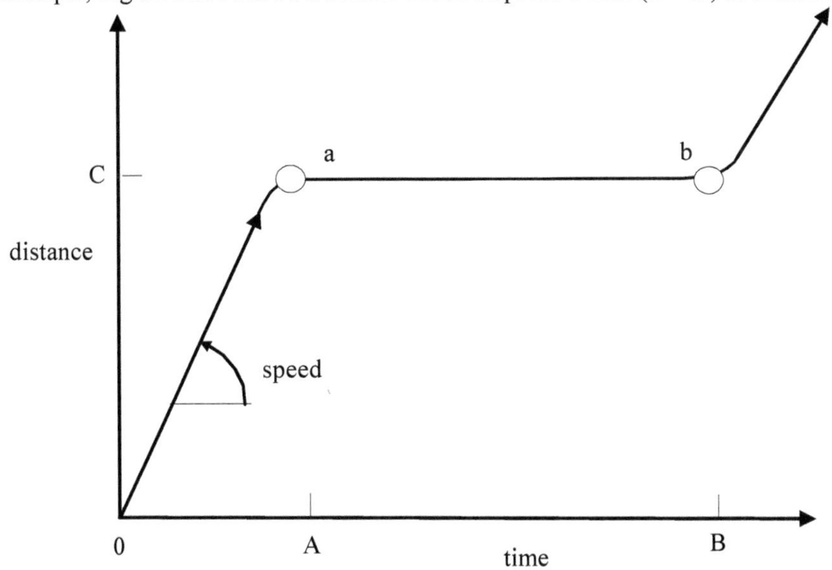

Figure 18.5a Trajectory for a vehicle travelling at constant speed, except for a stop at distance C for a time (B – A).

Typical trajectories for two vehicles in a traffic stream are shown in Figure 18.5b. Vehicle 2 begins at a time t_2 after vehicle 1 but travels faster and so catches up at point d. Vehicle 2 must then slow down to the speed of vehicle 1 and adopt the spacing, L_{s1}

(Figure 17.3b), appropriate for that speed. If the adjacent lane is fully available and thus imposes no constraints on vehicle 2, after some indecision, driver 2 may decide to overtake at point e by moving into the adjacent lane and accelerating back to vehicle 2's original speed. The point at which the two trajectories cross corresponds to the two vehicles being next to each other in adjacent lanes. Vehicle 2 returns to the original lane at point f once the vehicle spacing, L_{s2}, equals driver 2's accepted value for that speed.

The more difficult situations occur on two-way two-lane roads, as illustrated in Figure 18.5c. In some countries it is customary at point d for vehicle 1 to move to the side by using the road shoulder, allowing vehicle 2 to pass without delay or deviation. However, such courtesy does not occur in most countries and vehicle 2 must usually pass vehicle 1 by moving into the lane used by opposing traffic. This overtaking manoeuvre is usually done after some driver uncertainty.

When vehicle 2 reaches point d in Figure 18.5c, has adequate sight distance (Sections 19.4.5 & 22.5) and is in sight of oncoming vehicles 3 & 4, its driver must assess whether an overtaking of vehicle 1 could be completed well before vehicle 2 meets an oncoming vehicle. The 'well' is covered by the safety margin, which is defined below. The driver of vehicle 2 begins to overtake if the distance to the oncoming vehicles is considered to be more than an 'acceptable gap'. If sight distance is not adequate, drivers should — for safety — assume that the next oncoming vehicle is about to appear around the sight distance obstruction: high risk-takers may not make this assumption.

These difficult processes are the usual cause of the indecision referred to above and in Figure 18.5b. Indeed, the time between an acceptable gap in the oncoming traffic becoming available to a driver and the driver's acceptance of it is called the *indecision time* and is lognormally distributed with a modal value of between 0.6 and 1.0 s (Troutbeck, 1981). Indecision time decreases as the speed of the overtaken vehicle increases, although the reverse is true when the overtaken vehicle is a truck.

In Figure 18.5c, driver 2 decides that an overtaking from point d would be unsafe, and so lets vehicle 3 pass, but later decides that one from point e could be safely completed before vehicle 4 arrives at their meeting point, g.

Overtaking is thus quite a complex task for a driver and is made even more difficult by the driver's innate inability to estimate accurately the speed of an oncoming vehicle (Section 16.4.6). However, overtaking is not associated with a high crash-rate, typically representing about 1 percent of two-way crashes on rural roads in developed countries. Sideswipe crashes predominate in no-passing zones and head-on crashes in passing zones. The overtaking process is explored further in Section 19.4.5 and passing zones are described in Section 22.5.

If there are insufficient overtaking opportunities, a queue of vehicles will form behind the slow-moving vehicle, 1, as illustrated in Figure 18.5d.

Vehicle trajectories are discussed further in Section 23.6.6. The method is also useful in judging the behaviour of vehicles at traffic signals (e.g. Figures 23.9–11).

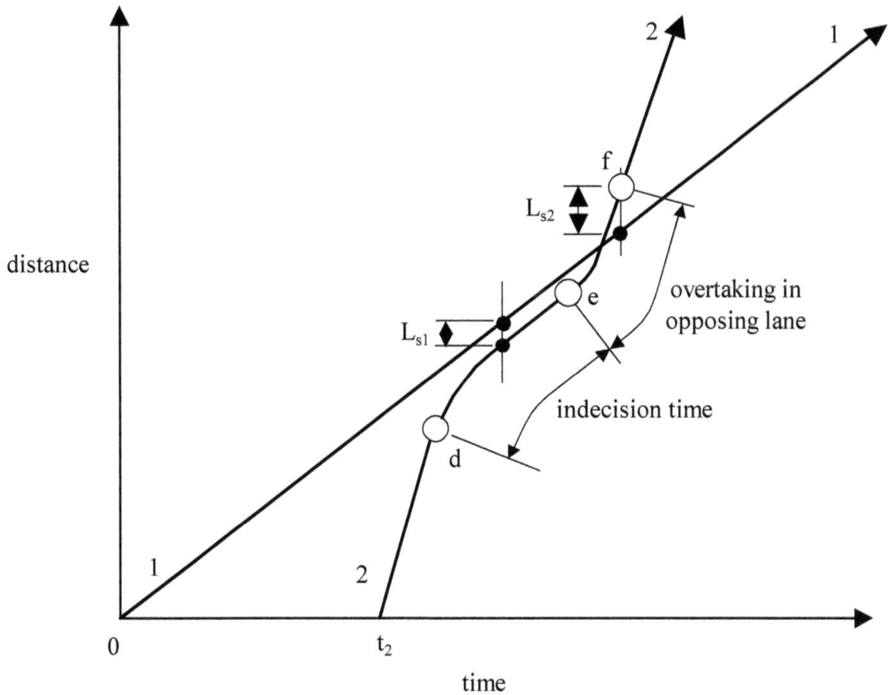

Figure 18.5b Vehicle 2 overtaking vehicle 1 without external constraints.

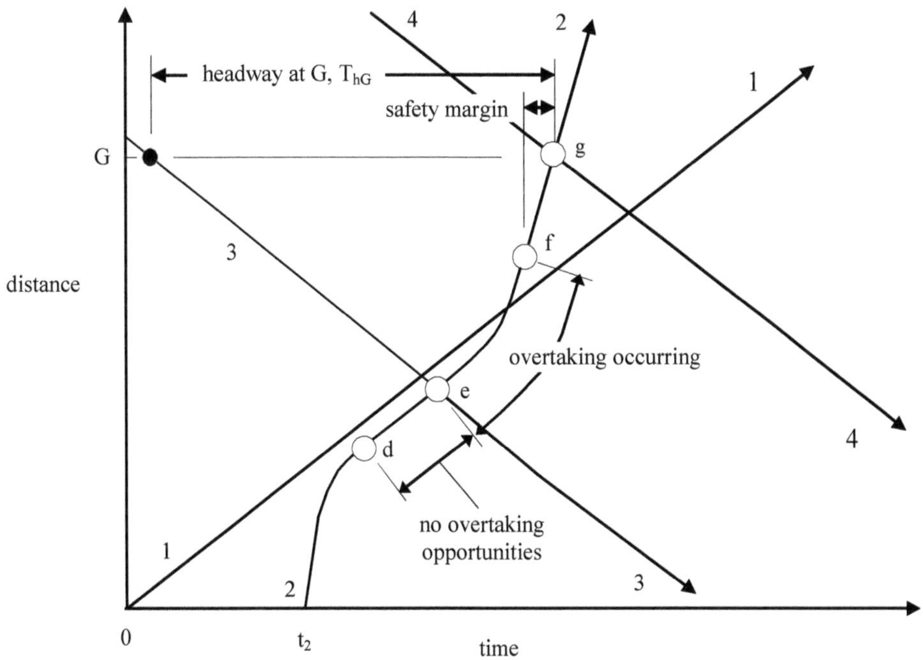

Figure 18.5c Vehicle 2 desires to overtake vehicle 1 on a two-way, two-lane road. It is restricted in its attempts to do so by oncoming vehicles 3 and 4. Note that the other parameters are as in Figure 18.5b. Headway is defined in Section 17.3.1.

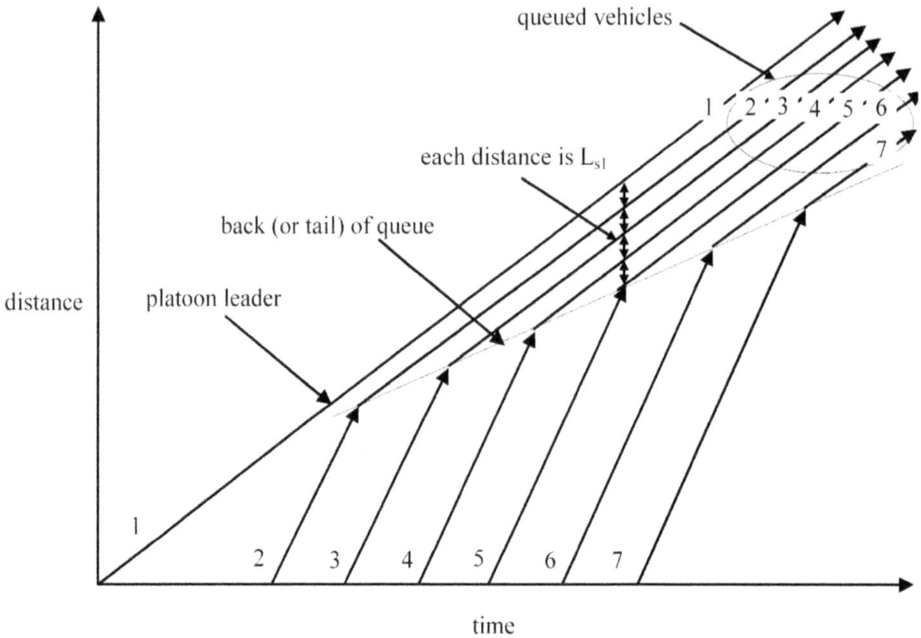

Figure 18.5d Queuing of vehicles 2 to 7 behind slow vehicle 1, in absence of overtaking opportunities.

18.4.2 Overtaking gaps

Gaps in traffic were discussed generally in Section 17.3. The events that indicate the start of an overtaking gap occur when:

(a) the vehicle intending to overtake catches up to the vehicle to be overtaken and encounters adequate forward visibility and no oncoming traffic (d in Figure 18.5b),

(b) a vehicle to be overtaken reaches the end of a sight distance restriction indicated by the end of a length of barrier line (Section 22.2.1) and there is no inhibiting oncoming traffic, or

(c) an oncoming vehicle passes the vehicle wishing to overtake (e in Figure 18.5c) and there are no further inhibiting oncoming traffic and sight distance restrictions.

The end of a gap occurs when either:

(d) the next oncoming vehicle is sighted and assessed to inhibit overtaking (similar to d in Figure 18.5c), or

(e) the vehicle to be overtaken reaches the beginning of a set of barrier lines.

The two main types of overtaking are:

(1) *flying*, when an overtaking vehicle does not need to slow down as it catches up to the vehicle to be overtaken, and

(2) *accelerative* (or *delayed*), when the overtaking vehicle must accelerate from the speed at which it was following the overtaken vehicle.

As a consequence of differences in driver behaviour, both types of overtaking, surprisingly, take about the same time (Troutbeck, 1981).

Actual overtaking times depend only marginally on the time to actually pass the slow vehicle and are largely dependent on the times required to move out of a lane and come alongside the slow vehicle, and later to move back into the lane in front of the slow vehicle. In turn, these times depend on the size of the acceptable gaps (Section 17.3.1) that the overtaking driver adopts and on whether the overtaking is flying or accelerative. Nevertheless, overtaking times and accepted gaps do increase as the speed of the overtaken vehicle increases.

The headway between vehicles 3 and 4 in Figure 18.5c for an observer at G is T_{hG}. This will be a constant for other locations such as d, e & f provided vehicles 3 and 4 are travelling at the same speed. The driver of vehicle 2 will perceive an available gap (assuming headways and gaps are identical, Section 17.3.1) given by:

$T_h/(1 + [V_2/V_4])$

due to the effect of the driver's own speed V_2. Perceived gaps can be measured as the time between an event in categories (a) to (c) above, and a second event in categories (d) and (e).

Section 17.3.4 defined the *critical gap* as the minimum gap that a driver will accept, and noted that impatience will often reduce the size of that gap. Such impatience would commonly develop from an absence of overtaking opportunities, e.g. in a driver confronted with a succession of sub-critical gaps. Critical gap estimates are shown in Figure 18.6 and are seen to depend on whether the overtaken vehicle was a car or a truck. The effect of narrow lanes on gap acceptance behaviour is discussed in Section 16.5.1. Given these factors, the distribution of these gaps and of most of the other time parameters describing overtaking can be represented by a log-normal distribution (Section 33.4.4 and Troutbeck, 1980). Typical design values for critical gaps are given in Table 18.3.

Table 18.3 Critical gaps for a driver overtaking a 5 m car.* = extrapolated values.

Overtaken vehicle speed (km/h)	85th percentile critical gap* (s)	Median accepted gap (s)	Associated flow (light traffic) (veh/h)	85th percentile accelerative overtaking time (s)
70	22	24	150	12
75	24	25	144	12
80	26	27	133	13
85	28	29	124	14
90	30	32	112	15

The fact that opposing flows of over about 100 veh/h can reduce overtaking opportunities in a lane will reduce the capacity of that lane and the effect is called *medial friction.*

The 85th percentile overtaking time is about half the 85th percentile critical gap, indicating that most drivers need a safety margin about equal to the overtaking time. The *safety margin* is defined precisely as the time by which an overtaking could be delayed and still completed before either jeopardising the safety of an oncoming vehicle or reaching barrier lines. Most negative safety margins are associated with the crossing of barrier lines, where drivers perceive the risk to be quite low. About 15 percent of the manoeuvres observed by Troutbeck were 'illegal' in this context, if not unsafe. The most common (modal) safety margin was 4.5 s. The calculation of the modal gap necessary to permit a driver to undertake an overtaking (the *establishment gap*, Troutbeck, 1980) can therefore be defined as the sum of the modal indecision time (0.9 s), the 85th percentile overtaking time and the modal safety margin (4.5 s).

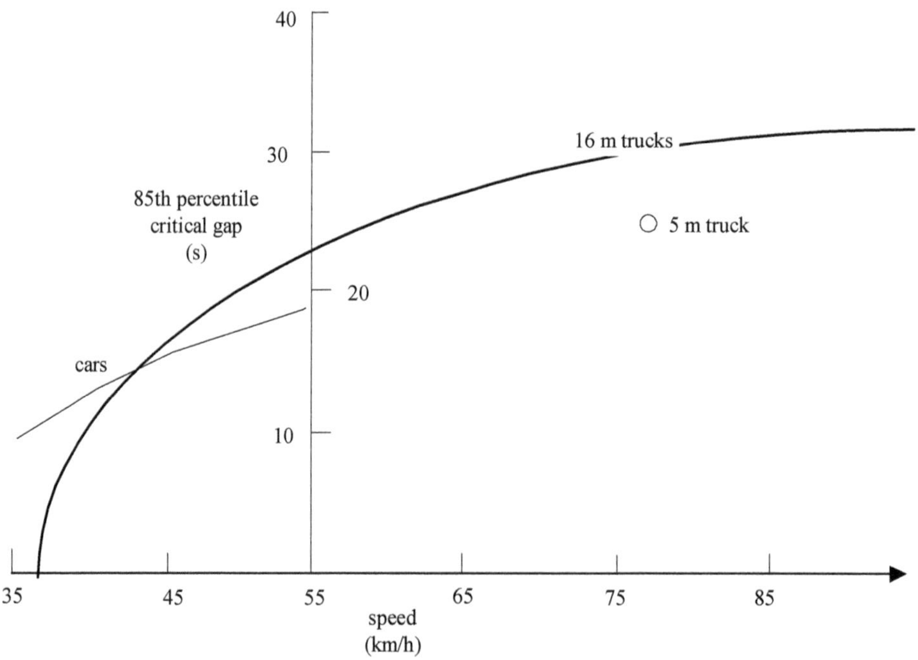

Figure 18.6 85th percentile critical gap for overtaking (Troutbeck, 1981). The car data is from Ashton *et al.* (1968) and the truck data from Troutbeck (1980).

18.4.3 Overtaking rates

Section 17.3.2 showed that the gaps in a traffic flow can be represented by the probability density function given by Equation 17.17. As individual drivers travel at various free speeds, the faster drivers will inevitably catch up to and then attempt to overtake the slower drivers. Catch-up rates were given in Equation 17.37.

If the speed of the overtaking flow is V, the mean speed and flow of the opposing flow is V_{op} and q_{op}, then the opposing flow measured relative to the vehicle to be overtaken, q_{or}, is given by:

$$q_{or} = (1 + V/V_{op})q_{op}$$

The equation also shows that the *passing rate* per km-h for two streams of vehicles travelling in the opposite direction on two-way roads is given by:

$$\text{passing rate} = qq_{or}/V = qq_{op}(1/V + 1/V_{op})$$

If it is assumed that the probability of a vehicle accepting a super-critical gap of T_g (= $1/q_{or}$) is linearly related to ($T_g - T_{gc}$) where T_{gc} is the critical gap and a is the proportionality coefficient, then the overtaking rate (veh/h) is given by:

$$\text{overtaking rate} = q_{or}\int_{T_{gc}}^{\infty} a(T_g - T_{gc})q_{or}e^{-q_{or}T_g}dT_g$$

$$= ae^{-q_{or}T_{gc}} \qquad (18.2)$$

Equation 18.2 shows that the overtaking rate is a negative exponential function of the critical gap size (Ashton *et al.*, 1968) and leads to the desired rate of overtakings per time per distance being given by:

$$\sigma K^2/\sqrt{\pi} = 0.56\sigma(q/V_{or})^2$$

where σ is the standard deviation of the vehicle speeds and K the traffic density (Section 17.2.1).

Typical catch-up rates for a slow vehicle vary from 32 percent of the rest of the flow if it is travelling at the 2 percentile speed of the traffic to 6 percent of the flow if it is travelling at the mean speed. These figures assume that on average vehicles can overtake as readily as they can catch up (Troutbeck, 1980). Overtaking demand is only satisfied at very low traffic volumes and once the easy-overtaking assumption becomes invalid, the overtaking rate drops below the catch-up rate and platoons form (Section 17.3.8).

Overtakings out of a platoon of impeded traffic where speed is steady will depend mainly on opposing traffic flow. The rate can be estimated in veh/h units as (McLean, 1978):

$$\text{overtaking rate from a platoon} = 2750q_{op}^{0.62}$$

Overtaking is obviously inhibited when the relative speed between following vehicles is small, the presence of oncoming traffic reduces the acceptable gaps in the opposing flow, and there is a lack of suitable sight distances (Section 19.4). Methods of increasing the supply of overtaking opportunities are discussed in the next section.

18.4.4 Overtaking lanes

An *auxiliary lane* is a lane provided on a road for a particular local purpose. For instance, it may be a speed-change lane for merging or diverging (Section 17.3.6), or it may be a lane to provide additional passing opportunities, as discussed below. Typically, an auxiliary lane will begin with a taper of about 150 m and end with one of about 250 m.

An overtaking (or *passing*) lane is a length of extra lane, typically of between 500 and 1000 m in length, which is provided in all terrain types in order to improve overtaking opportunities. It thus reduces both platooning and running time and shifts overtakings to safer sections of road. The overtaking lane therefore reduces the amount of time by which a vehicle is delayed, increases the driver's perceived level of service (Section 17.4.2), and reduces crash risks. The effects of the lane can be deduced from Table 17.2. Capacity will be largely unaffected. Overtaking lanes also have a strategic importance in that they can permit the staged upgrading of a section of two-way road.

The need for an overtaking lane is best determined from operating experience or from traffic simulation studies (Section 17.3.10). However, a typical European rule is to use overtaking lanes where the 15th percentile truck speed falls below half the 85th percentile car speed. If the indicated distance between overtaking lanes is less than 800 m, the two lengths of lane are made continuous.

Overtaking lanes must be clearly marked for use by only one direction of traffic. In almost all cases, it is preferably provided on the outside of the carriageway. The use of an unallocated third lane on two-way roads as an overtaking lane is relatively hazardous and usually inadvisable. Similarly, overtaking lanes are rarely advisable when there are more than two lanes of traffic in one direction, as the arrangement will require some slow-moving vehicles to cross two lanes of faster traffic. The hazards associated with this may exceed those avoided by using the overtaking lane.

Typical overtaking lane arrangements are shown in Figure 18.7. They are best located on straight stretches of road and away from vertical crests. There is no need to restrict overtaking lanes to steep grades; indeed, their location on flatter grades will minimise construction cost and make overtaking easier. Drivers usually find a 5 km spacing of overtaking lanes to be acceptable. Overtaking lanes can be highly cost-effective (Hoban, 1982). Three factors contribute to this:

(a) the benefits of breaking up bunches are carried over to downstream operations,
(b) the overtaking lanes are highly utilised, as demand has been concentrated, and
(c) the location can be chosen to minimise construction cost.

Advance notice of an overtaking lane will be beneficial in reducing driver frustration and risk-taking.

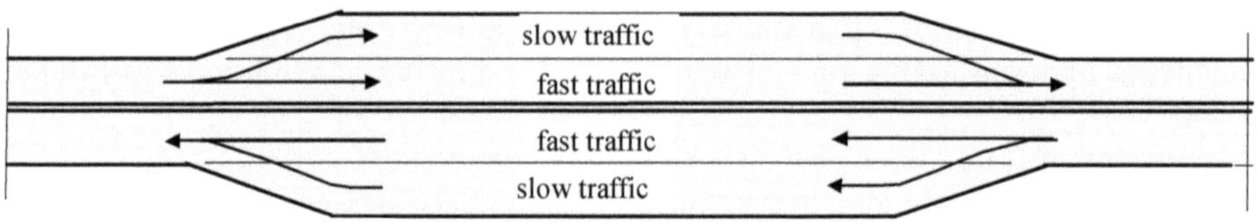

(a) Adjacent set of overtaking lanes, requiring a relatively-wide but short four-lane section.

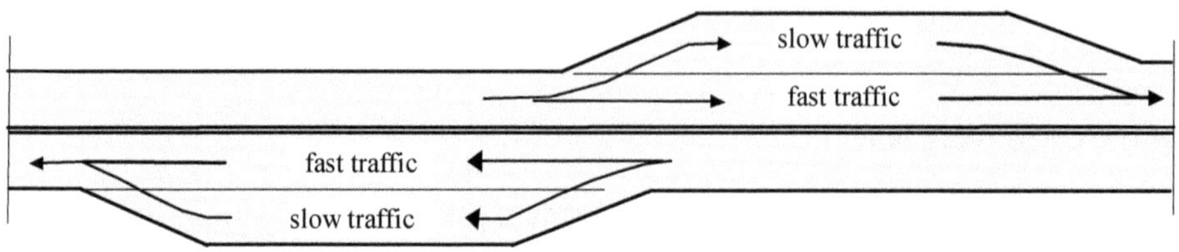

(b) Staggered set of overtaking lanes. The arrangement keeps maximum pavement width to three lanes, but affects a longer length of road.

Figure 18.7 Typical overtaking lane arrangements (drawn for leftside driving: reverse flow for rightside driving). ═══ indicates that crossing is prohibited.

It was noted above that it is best not to locate an overtaking lane on a hill. When it is not possible to avoid this — e.g. to overcome a local bottleneck created by slow vehicles on long upgrades — *climbing lanes* are used. When a climbing lane over the full length of steep grade is not practical, partial lanes or — in the extreme — very short *passing bays* (or *turnouts*) can be provided to allow very slow (20 km/h) vehicles to pull aside, effectively stop, and be overtaken. They are usually most effective when the local law requires their use, e.g. by slow leaders of long platoons of reluctant followers.

A key design input is an estimate of slow-vehicle speed. As noted in Section 18.2.5, trucks normally negotiate long up-grades at lower speeds than do cars. It is common to use the 15th percentile truck speed as the design value, whereas the 85th percentile for cars is used for features depending on the maximum speed adopted (Section 18.2.4). The difference between the two can lie between 40 and 80 km/h, depending on grade. This difference decreases as the curviness of the road increases. Section 27.2.2 shows that the mean acceleration of a truck is about 1.0 ms^{-2} compared with 2.0 ms^{-2} for a car.

CHAPTER NINETEEN

Road Geometry

19.1 GENERAL

The basic aim of the geometric design of roads is to provide operational efficiency, effectiveness, and safety. Considerations of traffic capacity (Section 17.4.3) will often dominate urban and arterial designs whereas safety (Section 28.5) and flow conditions (Section 17.4.5) will usually be the key factors in rural road design. In addition, road design should meet driver expectancies in such a way that drivers:

* can complete their current tasks in an unhurried manner,
* are mentally ready for the form of the oncoming road, and
* are not subjected to dramatic or unexpected changes in road conditions or in driver workload.

As discussed in Sections 18.2.6–10, a consistent design speed and a subsequent consistent operating speed are good measures of the way in which driver expectations have been handled. Section 18.2.8 included four specific guidelines related to design speed changes. This chapter will demonstrate that adequate sight distance is another measure. The main geometric elements to be considered in road design are:

(a) the cross-section, including widths (Sections 6.2.5 & 7.4.2), profile (Section 11.7), and cross-fall (Section 6.2.5). Width, in particular, depends largely on traffic flows. Cross-section issues are not pursued further in this chapter.

(b) the horizontal trace of the centreline, including curvature (Section 18.2.5), curve angle (Section 19.2.2), superelevation (Section 19.2.2), transition curves (Section 19.2.7) and localised pavement widening (Section 19.2.6), and

(c) the vertical trace of the centreline, including grades (Section 19.3).

The horizontal and vertical traces of the centreline collectively describe the road alignment. In a design sense, they depend largely on assumed speed levels (Section 18.2.6), however the underlying factors determining their main characteristics can best be considered under the following three headings:

(1) *Basic factors*: these come from the driver–vehicle–road–traffic system and include response time (Section 16.5.3), vehicle performance characteristics and dimensions (Chapter 27) and tyre–road friction (Section 12.5.2 & 27.7.4).

(2) *Intermediate factors*: these link the basic factors with the design elements and include sight distances (Section 19.4) and driver behaviour models (Section 16.2).

(3) *Operational factors*: these describe how the system is operating and include operating speed (Section 18.2.10), traffic capacity (Section 17.4) and safety (Section 28.5).

The specific interactions between these factors and curve design will be considered in the following section.

19.2 HORIZONTAL CURVES

19.2.1 Definitions

The horizontal alignment of a road consists of a series of straight and curved lengths. The (possibly extrapolated) intersection of two consecutive straight lengths is called an *intersection point* (Figure 19.1) and the angle so formed, θ_c, is called the *curve* (or *deflection*, *deviation* or *intersection*) *angle*. With respect to the design of the straight lengths, there is nothing to add to the discussion in Section 6.2.

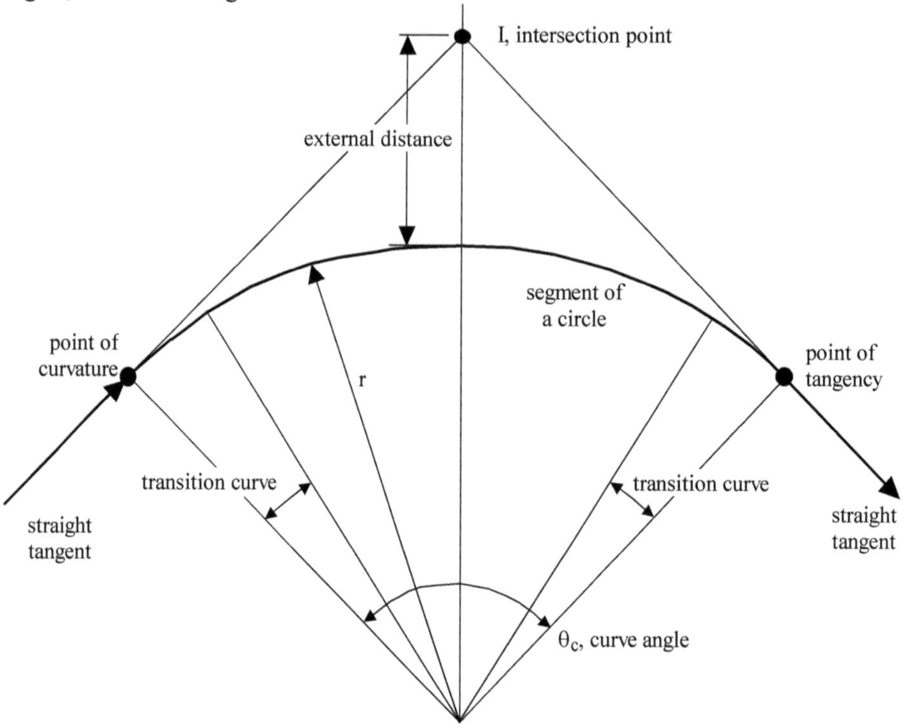

Figure 19.1 Road alignment components.

Curved lengths are used to accommodate the terrain or other physical features or whenever the curve angle between two consecutive straight segments exceeds about 1°. The point where a driver on a straight length first encounters a curve is called the *point of curvature* and the point where a further straight length is encountered is called the *point of tangency*. The distance between these two points needs to be at least 200 m to avoid the appearance of the curve being a mere kink. The curves are usually segments of circles as the constant radius ensures that drivers do not experience changing lateral forces.

A single circular segment directly and tangentially connected to two straight lengths is called a *simple curve*. The curve should be made as long as possible to avoid the appearance of a kink in the alignment. Frequently, the circular curves are connected to the straight *tangent* lengths by non-circular *transition curves*, as discussed in Section 19.2.7. A *compound curve* is formed from two circular segments of differing radius which are tangentially connected together and which curve in the same direction; a *multicompound curve* has a sequence of three or more circular segments curving in the same direction. If at all possible, compound curves should be replaced by single curves, as drivers find it difficult to detect changes in curve radius and therefore may choose inappropriate curve speeds (Section 18.2.5). Similarly, abrupt changes between the radii of adjacent segments in a compound curve can create a safety hazard, particularly for large trucks. Compound curves can also appear to be kinked if the joining tangent is less than about 500 m in length. A *mixed curve* is any sequence of three or more curves; the hairpin in Figure 19.4b below is one example of a mixed curve and many of the interchange ramps in Figure 20.1 are also mixed curves.

In a *reverse curve* the two segments curve in opposite directions. They can create a number of difficulties, viz.:

* it may not be possible to reverse the superelevation (Section 19.2.2) over a short distance,
* drivers may have inadequate time between turns of the steering wheel, and
* drivers may feel that their car is unstable.

Thus it is usually better to separate the two reverse curves by a transition length. In a *broken curve*, a straight segment joins two curves of the same curvature direction. The resulting alignment usually looks ugly and inappropriate.

On high quality, high speed roads, the large horizontal curve radii required (of the order of 1000 m, Figure 18.3), will usually be the major constraint on road improvements, quickly taking the road outside its original reservation (Section 6.2.3).

19.2.2 Background theory

Driver behaviour on approaching and negotiating a horizontal road curve is discussed in Sections 16.5.2, 18.2.5 & 18.3 (the last for curves with a radius of under 100 m). That work highlighted the dominance of the driver's forward perception of road curvature (i.e. 1/radius) in determining what speed to adopt on entering the curve. A misperception of that curvature can therefore lead to a dangerous driving situation, particularly as the curve characteristics of direction and curvature may need to be assessed typically at 4 s ahead and possibly up to 9 s ahead (Sections 16.5.2 & 17.3.9). Even detailed tracking data for actual curve negotiation may be required 70 m ahead of the curve, which amounts to 3 s ahead at 100 km/h (refer to the discussion of headlight sight distance in Section 24.4.1). Curve design is therefore of critical importance in both giving the correct cues to the approaching driver and in then providing a curve that can be negotiated on the basis of those cues and the driving actions resulting therefrom.

A pervasive concept underlying most current curve design philosophies is that of an assumed design speed. Section 18.2.6 showed how the selection of the design speed determined the minimum requirements for both horizontal and vertical alignment. Minimum horizontal curve design standards are based on the design speed for a new piece of road and the observed free speed for an existing road (Section 18.2.5). The two design criteria are then that the:

(a) available sight distance (Section 19.4) is adequate, and

(b) lateral accelerations, and hence lateral forces, are not excessive.

Criterion (a) is discussed in Section 19.4. Criterion (b) is applied by checking that the required *side friction* between tyres and pavement is not excessive at the design or operating speed. Traditionally, the check has been based on resolving lateral forces in the plane of the pavement surface to give (Figure 19.2a):

(centrifugal force) = (centripetal force)

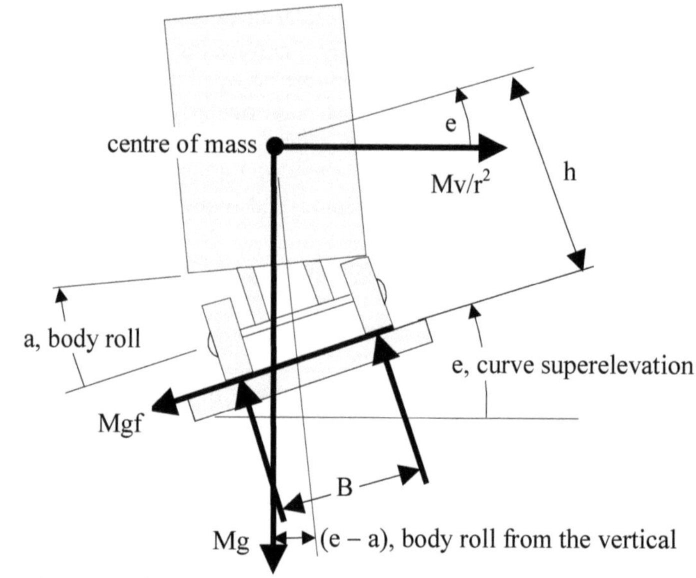

(a) Forces on vehicle on curve. M is the suspended mass of the vehicle g is gravity acceleration.

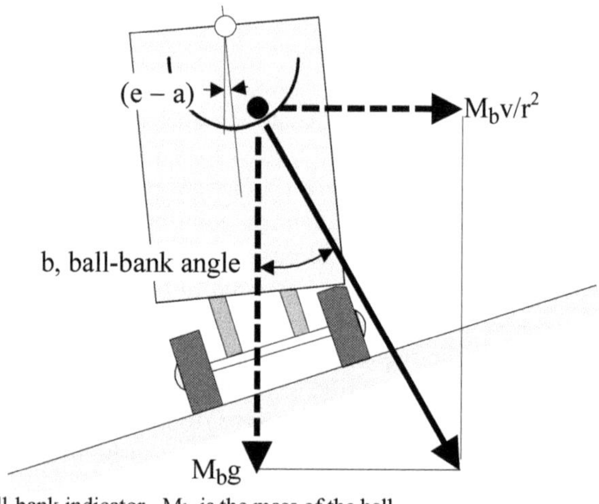

(b) Ball-bank indicator. M_b is the mass of the ball.

Figure 19.2 Mechanics of a vehicle on a curve (steady state). Both figures are drawn with the frame of reference in the moving vehicle. Hence, a centrifugal force is shown to maintain overall equilibrium.

$$Mv^2/r = \text{(friction force)} + \text{(gravity force)}$$
$$= \quad Mgf \quad + \quad Mge \qquad (19.1)$$

where v is the speed in m/s which is usually assumed to be constant throughout the curve, M is the mass of the vehicle in kg, g the gravity acceleration in m/s^2, and r is the curve radius in m. The non-dimensional term f is the *side friction factor* (or *sideways force coefficient*) and provides the indeterminate part of the equilibrium-of-horizontal forces statement given by Equation 19.1. It is only the *sideways coefficient of friction*, f_s, if the vehicle is on the point of sliding sideways. f is discussed in more detail in Section 19.2.3 where it is noted that it is only rarely equal to f_s.

The *superelevation* (or *camber*), e, is a transverse cross-slope extending across the entire carriageway width and across any unpaved shoulders. It is used on a curving road to accommodate cornering vehicles by countering the effects of centrifugal force on those vehicles. The development of superelevation along a longitudinal road curve is shown in Figure 19.3. Typically, two-thirds of the superelevation is developed on the tangent length (the *tangent run-out distance*) and one-third on the curve length (the *superelevation runout distance*).

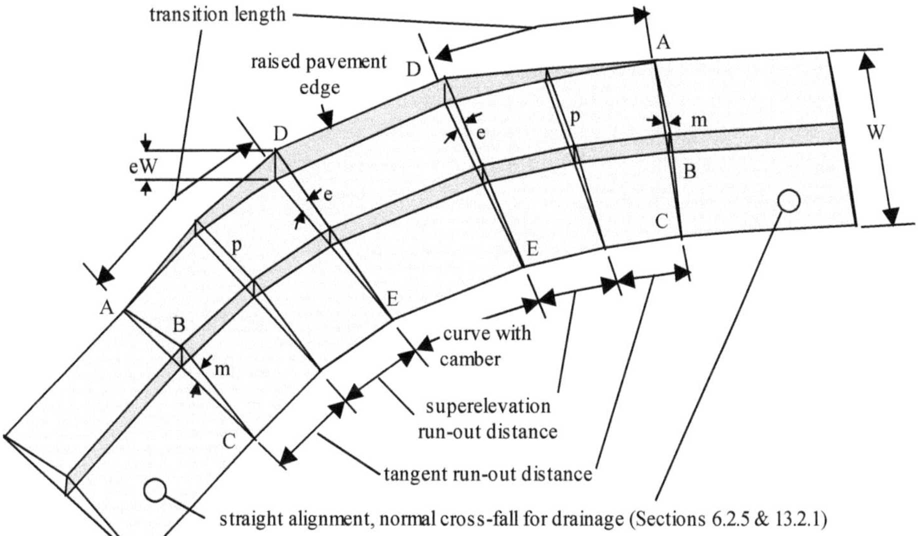

m = cross-fall slope, ABC is the shape of the crossfall.
e = superelevation slope, DE is the shape of the superelevation.
p = level

Figure 19.3 The development of superelevation.

Normal drainage cross-falls are 3 percent or less (Section 6.2.5), whereas superelevations can be much higher. Limits on superelevation of about 10 percent or less are imposed by:

(a) the need to reduce the cross-flow of rain-water, leading to aquaplaning (Section 27.7.4), particularly in cases where vertical and horizontal curves are combined with superelevation. A serious case arises on a superelevated downhill curve, where the water streams back across the road and, in some instances, across the curve where the friction demand is highest for cornering vehicles.

(b) a tendency for vehicles to track towards the inside of a superelevated curve,

(c) construction problems (e.g. matching with footpath levels),

(d) the longitudinal distances needed to develop large superelevations,

(e) from a behavioural viewpoint, drivers make their decision as to what speed to adopt in a corner before they reached the corner, basing their decision on curve curvature rather than on pavement camber,

(f) camber is optimised for one particular speed, and yet appropriate speeds will vary with conditions — e.g. between night and day, and

(g) the inwards roll-over (or sideways overturning or lateral stability) of slow-moving high vehicles (outwards roll-over is discussed below).

If snow or ice is present, superelevations are usually kept below 8 percent.

It is assumed that e is sufficiently small for cos(e) = 1 and sin(e) = e. Applying this process to Equation 19.1 gives:

$$e + f = v^2/rg \qquad (19.2)$$

or

$$= v_k^2/127r \qquad (19.2a)$$

where the symbol v_k is used for speed in km/h and r is in m. Equation 19.2 shows that a key role of superelevation is clearly to assist side friction in resisting the radially outward movement of the vehicle as it negotiates the curve. As v^2/r is lateral acceleration, Equation 19.2 can also be used to estimate lateral acceleration as (e + f)g.

Crash data is sometimes interpreted to suggest that inadequate superelevation does significantly increase crash risk on curves, particularly with respect to run-off-the-road crashes. (Crash data for curves is discussed in Section 28.5e).

A further limit on curve speed comes from the tendency of high vehicles to rollover outwards when negotiating curves at speed (see also Section 27.3.9). Figure 19.2a can be used to show that a vehicle will be on the point of rolling over in an outward direction on a curve when, ignoring the minor assistance of superelevation,

$$Mv^2h/r = MgB/2$$

or

$$v^2 > gBr/2h$$

where B is the lateral distance between tyres. Following Equation 19.2 this reduces to the rollover condition (Section 19.2.2):

$$e + f > B/2h$$

or, following Equation 19.2a, the condition becomes:

$$v_k^2/r > 63.5B/h$$

With a typical B/h of 0.75, the rollover conditions are:

1. $(e + f) > 0.375$,
2. a lateral acceleration of over 0.37g, and
3. $v_k^2/r > 48$.

The effect of superelevation, e, is to increase v_k^2 by the ratio:

$$([B/2] + he)/(h - [Be/2]).$$

19.2.3 The side friction factor, f

Given Equation 19.2, the choice of a design speed, V, allows the selection of a compatible minimum curve radius, r, and superelevation e, provided a value for f is known.

It is reiterated from Section 19.2.2 that f does not reflect what is available in terms of a pavement-type friction coefficient, f_s, but instead indicates what is actually used by the average driver. In the past there was assumed to be a link between this perceived 'safe and comfortable' value of f and the curve speed, V, and this was embodied into many design codes. Unfortunately, the link was not based on any justifiable data and had been mainly influenced by a 1936 survey of U.S. car passengers, who were asked to report when they felt a 'side pitch outward' when traversing a curve, sitting on bench seats and without seat belts. Thus f was more a measure of an unrestrained, cornering driver slipping laterally on the smooth bench seat of a car, than of the tyres slipping on the pavement. Of course, subsequent changes in vehicle type, seating, and restraints and in tyre and suspension design will have further reduced the relevance of the results.

One measure of f in a modern car is to leave a smooth, flat object like a large coin on a flat surface in the car (e.g. on the dashboard). This particular f is reached when the coin begins sliding sideways during cornering. At the other extreme, another f occurs when the tyres start squealing. As an intermediate measure between these two, Section 27.2.2 shows that drivers generally find accelerations above 2 m/s^2 to be disturbing. Hence, Equation 19.2 provides the following *discomfort* criterion:

v^2/r (m, s units) < 2

With this criterion v_k = 130 km/h (v = 36.1 m/s) requires a curve radius of at least 650 m. This is a very easy criterion to satisfy, as Section 19.2.6 will show that safety and sight distance will require far larger radius curves for travelling at 130 km/h. This illustrates another reason why drivers rarely exercise their full available f. The gamut of possible f criteria is therefore (in order of increasing size):

* objects in car slide sideways,
* passengers feel uneasy (safe and comfortable f),
* passengers feel discomfort (2 m/s^2),
* tyres squeal, and
* car slides sideways (f_s).

A second weakness in the original approach was the assumption that the side friction demand could be estimated from a vehicle's speed and the centreline radius of the curve. However, drivers rarely negotiate curves in this fashion, often cutting corners and thus following tighter radii (Section 16.5.2). Thirdly, Section 18.2.5 showed that drivers do not respond to superelevation when selecting the speed at which to negotiate a curve.

For these three reasons, assuming a value of f for Equation 19.2 has no behavioural basis. Nevertheless, the model has been empirically modified to provide a reasonable design tool. Typically, values of f specified in modern codes are based on local observation of conservative driver behaviour and range from 0.35 at 50 km/h to 0.11 at 130 km/h. The German design approach alternatively uses the need to leave some friction available for braking (see end of Section 19.4.2) to determine the appropriate value of f.

Measurements show that conventional drivers do not use values of f in excess of 0.30 (Rawlinson, 1987), although operating values of f measured when professional racing drivers were using dry public roads have averaged about 0.8 with a peak of 1.02. Even these were below the coefficient of friction, f_s and Figure 27.8 shows dry weather pavement values of f_s of up to 1.7 before slipping occurred between tyre and pavement. However, wet weather values can be very low, as also demonstrated in Figure 27.8. Furthermore, the predominantly-longitudinal pavement friction coefficients, f_{ls}, discussed in Section 12.5.3 are measured for skidding during braking in relatively-adverse wet weather conditions and usually exceed 0.35 but may often be below 0.70.

19.2.4 Curve indices

It is sometimes useful to have an index of the curviness or hilliness of a length of road. The *curviness index* is usually obtained for horizontal curves by dividing the sum of all the absolute values of the curve angles (Section 19.2.2) by the length of the road. A hilliness index is described in Section 19.3.

The *curve speed standard* is defined as the speed at which a driver makes full use of the available superelevation and the assumed side friction. It not a useful concept and it is suspected that the development and use of this definition arose in ignorance of some of the arguments in Section 19.2.3.

It is useful to examine Equation 19.1 when the friction force is zero; that is, when a vehicle can negotiate a curve without requiring any lateral force from the wheel–road interface. Putting Mgf = 0 in Equation 19.1 gives this *hands-off* (or *comfort*) *speed*, v_{ho}, as:

$$v_{ho} = \sqrt{(rge)}$$

At this speed, the vehicle steering would feel as if the vehicle were travelling on a straight, flat track. Indeed, it could negotiate a circular track without any torque being applied to the steering wheel. Put another way, a vehicle could negotiate a curve at v_{ho} even if the surface were frictionless. Thus superelevation will normally make steering easier and reduce side forces between the tyre and the road. It can be seen from Figure 19.2b that it will also reduce the likelihood of a cornering truck rolling over due to centrifugal forces (Section 19.2.5). A related term is the *comfort index*, which is given by:

$$\text{comfort index} = (v_{ho}/v)^2 = rge/v^2$$

Vehicles travelling at other than v_{ho} rely on the presence of some tyre–road friction (f) to stop them sliding outwards if $v > v_{ho}$, or down the track elevation if $v < v_{ho}$. Whilst the former requirement may require the driver to 'naturally' steer into the curve, the latter may require an unnatural outwards and upwards steering to maintain tracking. Following the above, a comfort index of one gives maximum steering comfort on a curve.

A problem with vehicles sliding laterally is that they can be readily rolled over if they encounter an obstacle as low as a kerb at lateral speeds as low as 10 km/h.

19.2.5 Advisory speeds

Advisory speed was defined in Section 18.1.3. The advisory speed to be signposted on curves can be determined from Equation 19.2 if e, f, and r are known. This is made difficult by the indeterminate nature of f.

A *ball-bank indicator* is often used as an alternative to measuring e and r. It utilises the force-equilibrium concept, as illustrated in Figure 19.2b, and measures the angle ϕ of a ball on a curved surface within the car. If the ball is only subjected to normal forces, then force equilibrium gives:

$$\tan(\beta + e - \phi) = v^2/gr$$

where β is the ball-bank angle. As $(e - \phi)$ is small (typically between −1° and +3°),

$$\tan \beta = (v^2/gr) - (e - \phi)$$

The body roll, ϕ, must be proportional to mgf (moments about c in Figure 19.2a) and assuming a linear body-roll stiffness, k, leads to:

$$\tan \beta = (v^2/gr) - e + kf$$

where k is a vehicle constant. Using Equation 19.2 then gives:

$$\tan \beta = (1 + k)f$$

Thus the utilised side friction, f (Section 19.2.3), can be predicted from the measured value of β in the car. For practical values, β is effectively a linear function of f.

Provided the ball is running on a smooth surface, its position on the surface will be that at which the normal to the surface is also at β from the at-rest position and so the actual shape of the surface is not important. However, it is common to use a circular tube in which a metal ball moves in a fluid such as alcohol or water to dampen vibrational movements. The original ball-bank indicators were indeed banking indicators manufactured for aircraft instrument panels. The tube is placed in front of an angular protractor scale which is set to read with the ball at zero when the loaded vehicle is at rest on a level surface.

In order to obtain the actual advisory speed, the indicator must be used in conjunction with a table relating the value of f calculated from the above equation to a safe speed, typically the curve design table discussed in Section 19.2.3. Cruder, unrecommended, methods based on 1930's conditions (Section 19.2.3) directly specified the maximum acceptable ball-bank angle for a given speed: e.g.:

ball angle (degrees), β	side friction factor, f	speed (km/h), v_k
14	0.21	30
12	0.18	40
10	0.15	> 60

Note that the advisory speed will not be an absolute measure of safe speed, as the value of f will not be f_s at sliding but a 'perceived' comfort value. As the f vs v link is not linear, the ball-bank test is best repeated at the predicted advisory speed, and the test further iterated until test speed and predicted speed are the same.

19.2.6 Sharp curves

The significant role of small-radius curves — commonly called *bends* — in road crashes is well known (Section 28.5e) and the problem is most noticeable on rural roads. Due to sight distance restrictions (Section 19.4.1), overtaking is rarely safe on curves with a radius of 400 m or less and severe crashes are disproportionately associated with curves with a radius of 600 m or less (Johnston, 1982). Thus, isolated curves of less than 400 m radius are usually avoided on roads where the likely free speeds (Section 18.2.5) on approach straights are in excess of 100 km/h. All isolated curves are best designed for the speed environment of the approach road, as many drivers will enter the curve at this speed (Sections 18.2.4 & 19.2.3). The sight distance provided (Section 19.4) should be matched to measured free speeds rather than to any assumed design speed. Fuel consumption also increases noticeably on curves under 600 m radius.

For curves of radius less than 600 m, line of sight clearances beyond 4.6 m from the roadway are needed to achieve safe stopping sight distances (sSD, Section 19.4.3). Pavements are sometimes widened on curves to permit the same lateral clearance between passing vehicles on curves as on straight sections of road. The technique also incidentally improves sight distance and permits safer cornering via managed driver curve-cutting. The extra tracking width required is a function of the swept path (Section 27.3.8) of the vehicle and allows for such features as the rear wheels taking a different path to the forward wheels (inside at low speeds and outside at high speeds), front overhang, and inaccurate steering. The widening can range from about 2 m on narrow 5.5 m pavements on sharp 60 m curves to zero on 7 m pavements with curves of over 150 m radius. Curve widening has been observed to produce a significant improvement in the frequency of curve crashes.

A longer term solution is to reconstruct the alignment to increase the radius of the bend. This is often difficult to achieve in hilly terrain and, by shortening the road length, may lead to unacceptable increases in the road gradient (Figure 19.4a). An alternative solution is to reconstruct the bend as a bulbous hairpin bend, which also increases the road length and may therefore permit the gradient to also be improved (Figure 19.4b).

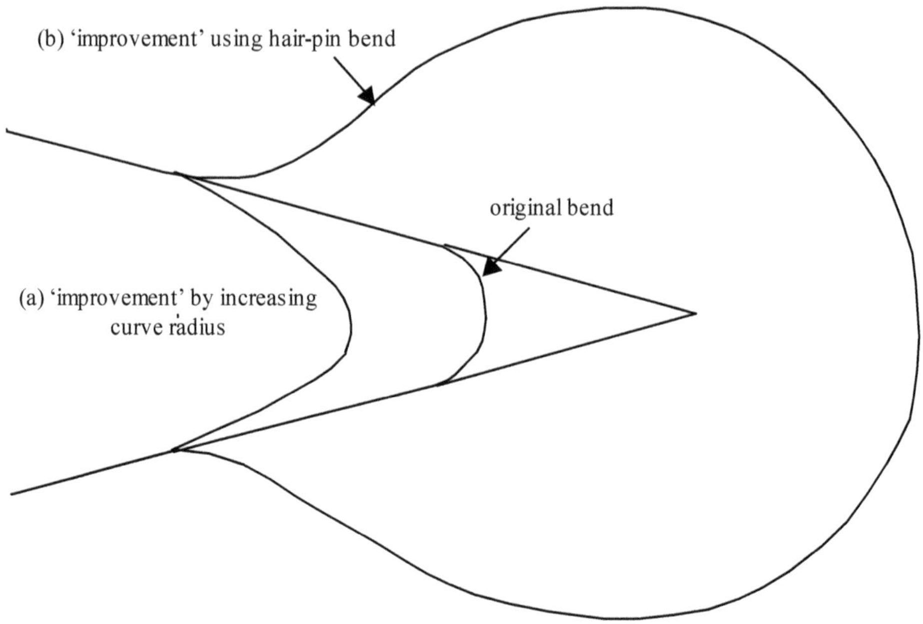

Figure 19.4 Methods for improving the alignment of a sharp bend.

It is also good practice to ensure that below-average vertical and horizontal curves do not occur at the same location.

19.2.7 Transition curves

Section 19.2.2 showed that the lateral acceleration of a vehicle in a curve is given by $v^2/r = v^2$(curvature); thus a gradual change in curvature produces the same gradual change in lateral acceleration. Sudden changes in horizontal road-curvature are sensed as a lateral *jerk* (measured as a change of acceleration). This produces occupant discomfort, and so drivers tend to cut corners in order to diminish their effective rate of curvature change (and also to increase their sight distance). Road designers try to minimise the problem by using *transition curves* to join two road lengths with different curvatures (Figure 19.1). Curves connecting two straight lengths or a straight length and a circular length are known as *simple transition curves* and those connecting two circular curves are known as *compound transition curves.*

Many now advocate that transition curves are unnecessary. They may even be undesirable in so far as they mask a driver's perception of the start of the curve and the true curve radius, and hence lead to an inappropriate choice of curve speed (Sections 16.5.2 & 18.2.3). A counter argument is that a transition curve that is at least equal to a driver's response distance ($T_r v$, Section 16.5.3) allows the driver time to respond smoothly to the new curvature.

The form of the transition curve is conventionally chosen to produce a gradual change from the zero curvature of the straight line to the peak curvature of r_m. A number of *geometric spirals* provide such gradual curvature changes. For example, if the curvature is taken as proportional to the length, L, along the curve, then rL will be a constant equal to the product of the curve length and the curvature change occurring over that length. Such a curve is known as the *transition spiral* (or *Euler's spiral*, *Cornu's spiral* or the *Clothoid*) and is represented algebraically by:

spiral angle = $\theta = L^2/2rL$

Alternatively, use can be made of cubic curves or the following *Fresnel integrals*:

$$x = \pi\int_0^L \cos(\pi\theta^2/2)d\theta$$

and

$$y = \pi\int_0^L \sin(\pi\theta^2/2)d\theta$$

Whatever transition curve is chosen, it must enhance rather than diminish the driver's perception of the sharpness of the oncoming curve and the future direction of the road. For this reason, long transition curves are usually avoided. Indeed, the transition length is often chosen as the length necessary to develop the required curve superelevation from either (a) its initial zero value (A in Figure 19.3), or (b), more safely, from the zero cross-fall value (B in Figure 19.3).

Transition curves can be omitted where there is no truck traffic and where the curve radius provides a free speed well above the operating speed. Research also suggests that the use of transitions raises crash rates for curves with a radius in excess of 400 m, possibly because the transition masks the driver's perception of the safe speed for the forthcoming piece of road. When transitions are not used, superelevation should be developed in the straight rather than in the curved portion of the road.

Criteria used by some past code-writers for the selection of transition curves were based on:

(a) unsophisticated vehicle dynamics,
(b) the search for a single criterion, and
(c) a deference to railway engineers.

Indeed, the main design criterion still used was proposed in 1909 to limit the amount that a railway carriage would swing during the transient roll-oscillation caused by a lateral jerk. The derivation was in error, nevertheless in 1932 an Englishman, Royal-Dawson, misapplied it to road curves. For many years it was used for road design in both the U.K. and U.S.

19.3 VERTICAL CURVES

Vertical curves are used to accommodate vertical changes in the topography through which a road is passing. They usually consist of straight lengths of vertical alignment linked by a parabolic transition curve. A useful geometric feature of the parabola is that the tangents at each end always meet midway between the two tangent points (Figure 19.5). The coefficient of the square term in the parabola is sometimes called the *road constant*, K. If y is the vertical axis, then the parabola in Figure 19.5 can be represented by:

$$y = gx - Kx^2$$

and

$$dy/dx = g - 2Kx$$

where:

K = (change in grade across the transition)/2(length of transition)

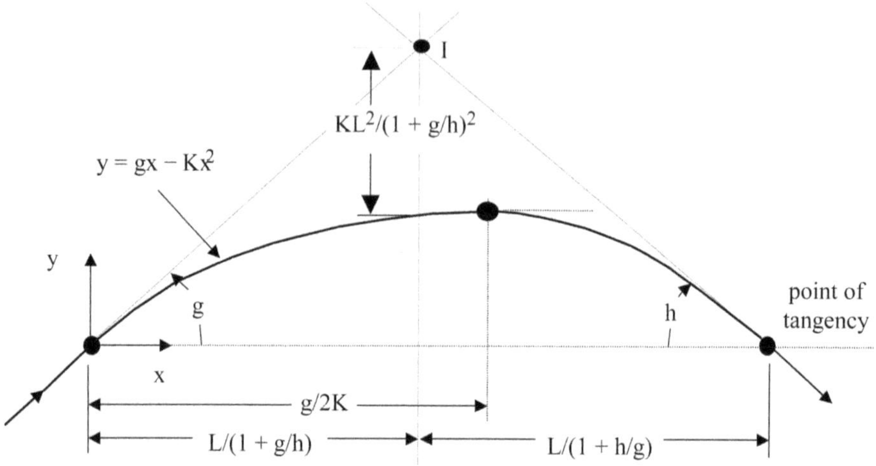

Figure 19.5 Geometry of a vertical transition curve, using two parabolic curves.

Only one tangency point may be freely selected if a single parabola is used. If the two tangency points are given, then normal practice is to use a pair of parabolas with a common tangent point at the same horizontal location as the intersection point.

Sight distance (Section 19.4) is usually the key factor with vertical curves, particularly on crests. On lower quality roads, the need to avoid earthworks will often lead to vertical curvature being a significant constraint on sight distance. On two-way roads, a short length of vertical curve in a valley or trough (a *sagging curve*) can be dangerous as it can hide an oncoming vehicle in the sag from the view of potential overtakers outside the sag. This *hidden dip* problem is readily checked using straight edges, an accurate longitudinal profile, and a knowledge of minimum vehicle heights.

A further criterion applied to sagging curves is to limit the vertical acceleration that the driver experiences to 0.5 ms^{-2}, to reduce driver discomfort, although drivers will tolerate vertical accelerations of up to 5 ms^{-2} (Section 27.2.2). The acceleration is given by:

$$\text{vertical acceleration} = v^2/r = v^2(\text{vertical curvature}) = v^2 d^2y/dx^2 < 0.5 \text{ ms}^{-2}$$

where r is now the vertical radius. Thus the limiting minimum r in m would be $v_k^2/26$ which is about 400 m at 100 km/h. Another advantage of the parabolic transition curve is that, as $d^2y/dx^2 = 2K$, it gives constant vertical acceleration for constant v.

It is sometimes useful to use a *hilliness index*, which is usually the sum of the products of all the absolute values of the gradients times the length over which they occur, divided by the total length of the road. This can be seen to be equivalent to the absolute rise and fall rate of the road and increases rapidly with grade.

A typical desirable limiting grade to permit heavy vehicles to maintain speed is 5 percent (Section 18.2.5a). Eight percent is usually regarded as the maximum in this respect, although steeper grades are manageable over very short lengths of road. High crash rates are associated with steep grades. One reason for this is that drivers are very poor judges of vertical curves and the sharpness of crests. They therefore tend to maintain their previous speed, despite a loss of sight distance in the vertical plane (Section 19.4). On the other hand, trucks are usually forced to travel very slowly on steep grades, due to limits on their engine power and braking capability. Techniques for managing steep grades are described in Sections 18.1.4 & 23.2.3(10)c.

Many designers also impose a limit of 2 km on any straight lengths of vertical grade, as drivers tend to subconsciously increase speed on the down-grade. Low-radius horizontal curves on downgrades are particularly dangerous for trucks, which may find it difficult to slow down sufficiently to negotiate the curve at a safe speed (Section 18.2.5).

When an alignment requires both horizontal and vertical curvature within the same road length, aesthetics is usually best served by having the two angle changes occur over the same length of road.

A different limit on vertical alignment occurs over short lengths of road with sharp vertical curvature, typically at:

* raised crossings and road humps (Section 18.1.4),
* railway level-crossings,
* gutter crossings (Section 7.4.3),
* sagging curves with radii under 30 m, and
* the bottom of steep, short grades.

In these cases even conventional vehicles can experience clearance problems and have their undersides snared on the pavement surface, causing vehicle damage and pavement scoring (Section 14.1C1). The problem is worsened with some special-purpose vehicles (Section 27.4.1) which can have ground clearances as low at 75 mm over distances of about 10 m between tyre groups. However, the main problem occurs with heavily-loaded vehicles overhanging beyond their rear axle.

Further discussion of grades is given in Sections 7.4.2, 18.2.5 & 19.4.3.

19.4 SIGHT DISTANCE

19.4.1 Definitions

Sight distance is formally defined as the distance at which an attentive driver with good visual acuity (Section 16.4.2) can see a specified object on the pavement ahead, given clear and well-lit conditions and the object centrally located in the driver's field of vision. With a straight alignment, this (optimum) sight distance is not an issue. For instance, Section 24.4.1 suggests that drivers in good conditions can detect other cars at least a kilometre ahead. However, sight distance — even on straight alignments — may be limited by:

* the driver's visual acuity (Section 16.4.2 discusses the decrements caused by low acuity and how they can be reduced by the use of corrective lenses, e.g. by wearing glasses).
* the available light (Section 24.4.1 suggests that headlights on unlit roads rarely provide visibility beyond 100 m ahead, so this is a very relevant limit on night-time sight distance).
* the object not being centrally located in driver's field of vision (Section 16.4.1 discusses how this factor can both reduce acuity and increase detection time).

On curved alignments, the following additional factors may reduce sight distance:

* vertical crests (Section 19.3),
* horizontal curves requiring drivers to 'see around the corner',
* the basic horizontal curve problem can be exacerbated by roadside objects on the inside of curves, such as:

safety barriers, signs, trees, bridge abutments, batter slopes, and the sides of cuttings (significant low-cost improvements to sight distance can often be achieved by managing the placement of these objects) and

parked vehicles.

* the blocking effect of any forward vehicles if the curve is to the traffic side (i.e. to the left if driving on the left).

The height of the driver's eye is commonly taken as between 1.05 and 1.20 m above the pavement, but depends on both driver physiology and vehicle characteristics. For example, it is about 2.5 m for trucks. Driver eye height is of critical importance for determining vertical sight distance but has little effect on horizontal sight distance.

The major factor influencing driver behaviour on both vertical and horizontal curves is that most drivers must see a hazard before modifying their driving. Commonly, the risk of encountering a potential hazard, currently hidden by a physical obstruction, is not high enough to modify their behaviour. Hence, as argued in Section 18.2.5, speeds on curves depend on perceived curvature and not on sight distance. Similar behaviour has been observed during overtaking (Troutbeck, 1981). In addition, attempts to lower crash risks by increasing sight distance via an improved road realignment can sometimes be diminished in effectiveness by the increase in free speed that accompanies an increase in horizontal curve radius.

Having seen an 'object', the driver must then react to that sighting. That reaction will depend on the circumstances and, for road design, each combination of circumstance and reaction will give rise to a particular design sight distance. There are seven main types of design sight distance and these are discussed in Sections 19.4.3–7. However, it is firstly necessary to use Section 19.4.2 to define vehicle stopping distance, which underlies some of these design sight distances.

Some of the 'reactions' referred to in the preceding paragraph are merely theoretical constructs, as field evidence and the headlight data in Section 24.4.1 suggest that modern drivers will drive with sight distances as low as 100 m. With a 100 m sight distance, a driver travelling at 130 km/h with a short 0.5 s response time and using the maximum comfortable deceleration of 3 ms^{-2} will still be travelling at 28 km/h when the object is reached at 100 m. If the driver is sufficiently concerned to use an uncomfortable deceleration of 4 ms^{-2}, the speed at the object will still be 25 km/h. Collisions at these speeds can be serious. Nevertheless, no specific links have been claimed between sight distance and crash rate (Section 28.5e).

Care must be taken to avoid introducing occasional sight distance restrictions in an otherwise high quality visual environment, as there will be a heightened chance of drivers ignoring the restricted sight distance.

19.4.2 Vehicle stopping distance

Vehicle stopping distance, D_s, is defined as the distance that a driver needs to stop and is determined as the sum of the time that the driver takes to respond (Section 16.4.2) to a message to stop, and the ability of the vehicle to then stop. Thus, it is actually the driver/vehicle stopping distance and is given by:

D_s = (response–time distance) + (braking distance)

Response-time distance in metre is given by $T_r v_k/3.6$. T_r is the driver/vehicle response time in seconds and is obtained from Table 16.1 and Section 16.5.3 in the following manner. Item 1 in the table (detection) does not apply as the driver is assumed to be looking at the pavement ahead; Item 2 (identifying) is not relevant; Items 3, 4 & 5.1 apply in most cases. Summing these modified values gives a basic mean value of 1.2 s (+ 1.8/–0.4 s). Codes typically use T_r = 1.5 s for urban areas and T_r = 2.5 s for rural areas. Although code writers would usually give other reasons to justify its use, it is best to consider the rural 2.5 s as covering the bulk of the population and as accounting for the higher 'unexpected' risks associated with tight rural alignments. v_k is the speed of the vehicle in km/h. It is important to determine v_k from the measured free speed of the road (Section 18.2.5), if this exceeds the original design speed.

Braking distance is found using Equation 27.1d and assuming a uniform deceleration, a (ms^{-2}), to give:

$$\text{braking distance} = (v_k/3.6)^2/2a$$

Typical values of *a* for cars and trucks are given in Section 27.2.2. As drivers use friction conservatively, an upper bound on the deceleration can be linked to a pavement coefficient of friction (see, for example, Section 12.5.3 where f_l = SFC, and Figure 27.9), by Newton's force–acceleration equation and the friction equation:

$$\text{resisting force} = [\text{coefficient of friction}] \times [\text{interface force}] = Ma$$

$$f_l Mg = Ma$$

where M is the vehicle mass and g is gravity acceleration. Thus, in this case:

$$a = gf_l$$

Maximum non-skid dry-pavement values of f_l are typically about 1.2 (Section 21.2.8), however a dry road can deliver up to 1.7 (Figure 27.9). The wet coefficient of friction described in Section 12.5.2 is usually between 0.35 and 0.65.

The effect of a grade with a slope of S (m/m, positive for uphill) can be included by adding the mass component down the slope to the force to give an effective acceleration of:

$$a_{eff} = a + gS$$

or, at the skidding limit:

$$a_{eff} = g(f_l + S) \qquad (19.3)$$

Hence time to stop, T_s, in seconds is given as:

$$T_s = T_r + v_k/3.6a_{eff} \qquad (19.4)$$

and the vehicle stopping distance, D_s in metres, is:

$$D_s = T_r v_k/3.6 + (v_k/3.6)^2/2a_{eff} \qquad (19.5)$$

With T_r = 2.5 and at the skidding limit, Equation 19.3 gives:

$$D_s = 0.7v_k + v_k^2/(254[f_l + S]) \qquad (19.6)$$

Equations 19.3 and 19.6 can be used to calculate appropriate local values of D_s. Table 19.1 lists some typical design values for cars in Column 7 and for trucks in Column 8.

Values of D_s need to be increased on curves where some of the available friction will be devoted to providing the centripetal force given in Equation 19.1. Vector analysis shows that the available pavement friction in this case would reduce from the measured value of f_l to $f_l\sqrt{(1 - [v_k^2/127rf_l]^2)}$.

Table 19.1 Various vehicle stopping distances, D_{sd}. Some assume that drivers use their full braking capacity. This is rarely the case. Thus the high decelerations in Column 5 indicate that the figures in Columns 3 and 4 should not be used for design purposes.

1	*2*	*3*	*4*	*5*	*6*	*7*	*8*	*9*
Speed v_k km/h	react $D_{2.5}$ m	brake D_{bd} m	stop D_{sd} m	acceln a_c ms^{-2}	time T_s s	Car D_{sdc} m	Truck D_{sdt} m	Design D_{sdd} m
40	28	6 (19)	33	10.2 /1.8	1/10	50	40 /60	45
60	42	13 (16)	54	10.6 /2.5	2/11	90	80 /120	90
90	62	26 (12)	88	12.0 /3.5	2/12	170	180 /250	175
100	69	33 (11)	102	11.7 /3.8	2/12	200	220 /300	210
110	76	40 (11)	116	11.7 /4.0	3/12	230	270 /360	260
120	83	47 (11)	130	12.0 /4.3	3/12	270	330 /420	310
130	90	55 (11)	145	13.0 /4.5	3/12	310	390 480	365
		Measured, hard-braking				Normal, moderate braking		

Key to columns

Column 1: v_k = initial speed

Column 2: $D_{2.5}$ = distance travelled in 2.5 s (Section 16.5.3)

Column 3: D_{bd} = Minimum non-skid braking distance measured under dry and favourable conditions (Samuels and Jarvis, 1978). They imply a friction coefficient, f_l, of about 1.2 (Section 19.4.3). The distances will approximately double on unpaved roads.
( ) = coefficient of variation

Column 4: D_{sd} = Minimum non-skid stopping distance, ∑(column 2 and column 3)

Column 5: a_c = Minimum non-skid constant deceleration implied in columns 3 and 4. The first number is based on $v_b^2/2x$ (Equation 27.1d) where x is from column 3. The number after the / is based on using x from column 4. Most drivers consider 4 ms^{-2} to be a maximum usable deceleration (Section 27.2.2).

Column 6: T_s = Minimum non-skid time to stop using deceleration in column 5.

Column 7: D_{sdc} = car stopping distance from Equations 19.5–6. [Column 2] + [Equations 19.5–6 with f_l = 0.3]. Column 2 is conservative. 0.3 is a practical minimum value for f_l (Sections 12.5.2 & 19.2.4).
The column thus gives conservative stopping distances which are much higher than the values in Column 4 for hard braking by alert drivers in good conditions.

Column 8: D_{sdt} = observed truck stopping distance (dry/wet). The associated stopping distances for trucks in different situations are obtained by multiplying the distances in Column 8 in Table 19.1 by the following factors:

Situation	Multiplying factor
Antilock brakes	0.9
Design assumption	1.0
Very good driver or rigid truck	1.1
Bus or empty truck	1.2
Prime mover or articulated truck	1.4
Poor driver	1.5

Column 9: D_{sdd} = typical stopping distance assumed in design. This column is for trucks and is from Austroads, 2003.

19.4.3 Stopping sight distance (sSD)

Available sight distance, aSD

The *available sight distance, aSD,* is defined as the distance at which a driver can first see a stationary object too high to drive over (commonly 200 mm high) in the lane ahead. Figure 19.6 demonstrates this definition for the vertical alignment case. aSD is established in design by placing the chosen object and eye heights on the alignment and using a straight-edge; in the field it is typically measured by using two sighting targets of the required height connected by a distance measuring device. In practice aSD tends to be more influenced by changes in object height than in eye height. At night, the available light (Section 24.4.1) will also limit aSD.

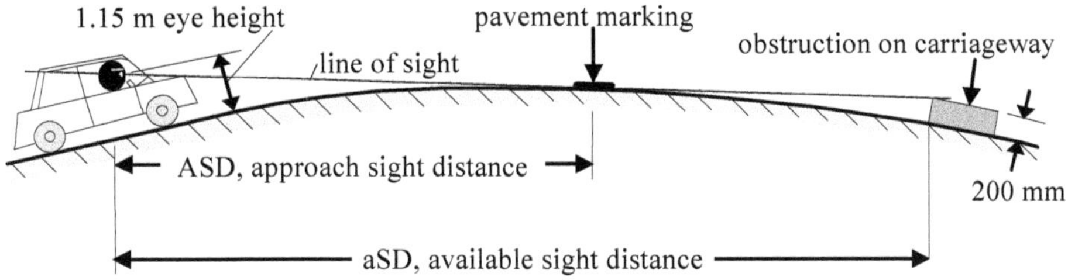

Figure 19.6 Available sight distance, aSD, definition.

The use of 200 mm as a standard object has been widely questioned. It is small enough to be driven over, although it may still cover events such as a body on the road, major subsidence or an oil slick. Crash statistics indicate that a low sSD due to vertical alignment has little impact on traffic safety.

Stopping sight distance, sSD

The *stopping sight distance, sSD,* gives a driver travelling at the design speed (Section 18.2.5) sufficient time to avoid hitting a standard (e.g. 200 mm high) object by coming to a complete stop. The stopping distance, D_s, was defined in Section 19.4.2. sSD is therefore achieved if:

$$\text{aSD} \geq D_S \textit{ or } 1.4D_s \qquad (19.7)$$

The 1.4 is a common conservative increase used in many design codes. Restrictions on speed and/or overtaking are imposed if the inequality (19.7) design criterion is not met. Some texts describe sSD, somewhat strangely, as the *non-passing sight distance.*

Manoeuvre sight distance, mSD

Drivers often do not need to stop on seeing an object. The alternative — which drivers would seem to prefer — is to steer around the object, with a resultant reduction in sight distance needs. It is only practical to take advantage of this liberalisation if there is room to manoeuvre past any potential obstruction. This requires, for instance, adequate carriageway width if the site has sharp vertical curves. A design case might involve an object the width of a vehicle in the centre of the carriageway. This would imply that an adequate pavement width would be about three lanes, or about 10 m.

The associated *manoeuvre sight distance, mSD,* can be calculated as the response distance, $T_r v_k/3.6$, from Equation 19.4, plus the distance required to simultaneously:

* move sideways, which diverging behaviour shows can easily be achieved at 1 ms^{-1} (Section 17.3.6) and will thus require at most 3 s. Even with the maximum comfortable longitudinal deceleration (3 ms^{-2}, Table 27.1), this lateral shift can be achieved within a

longitudinal speed drop of only 3x3x3.6 = 32 km/h. This assumption is not viable if there is also a tight horizontal curve (Cox, 2003).

* slow down from v_k to some acceptable speed whilst passing the stationary object. This distance can be calculated from Equation 27.1d, which shows that the distance to slow down to half speed would be $0.75D_s$.

As the response distance is the same for stopping or slowing down, the manoeuvre sight distance is only marginally less than the stopping sight distance, and so it is rarely used in design. Cox (2003) gives a detailed analysis of manoeuvre sight distance.

Approach sight distance, ASD

A related geometric value is the *approach sight distance* (*ASD*), (Figure 19.6), which is used for some intersection design (Section 20.3.2) and requires the driver to be able to see items such as pavement markings on the road surface, i.e. the object height is zero. It is usually used for unsignalised intersections. At a signalised intersection, the lowest object that need be seen is the rear light of another vehicle, which is often taken as 600 mm above the pavement surface, to give ASD_{sig}.

19.4.4 Intermediate sight distance (ISD)

The intermediate sight distance, ISD, conservatively assuming that both vehicles must stop rather than one manoeuvre aside, gives two vehicles coming in opposite directions in the same lane an opportunity to each halt before colliding. The case is illustrated for the typical horizontal alignment case in Figure 19.7. This gives the design criterion as:

$$ISD \geq 2D_s$$

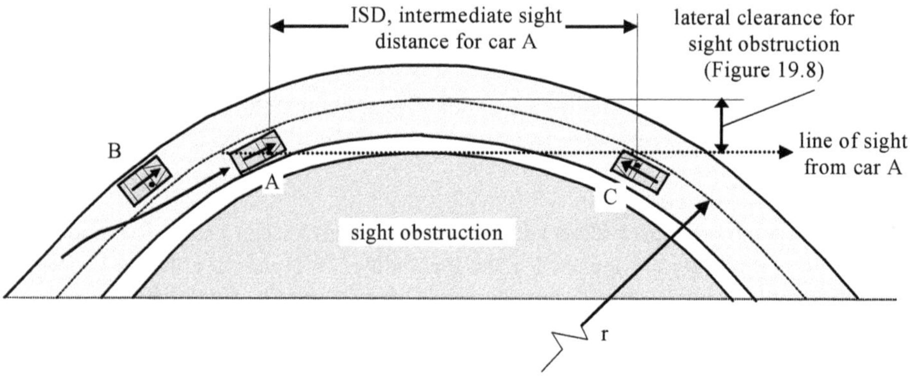

Figure 19.7 Intermediate sight distance, ISD, applied to horizontal alignment sight distance calculation. Drawn for left-side driving. The drawing is for the case where car A is overtaking car B and requires sight distance to car C coming in the lane car A is temporarily occupying. For formal calculations the drivers' eyes are usually assumed to be located on the appropriate longitudinal lane line.

When ISD is applied in the vertical alignment case a common conservative assumption is to use a 200 mm high object rather than one as high as a typical oncoming vehicle. If the height of the oncoming vehicle is used, the resulting sight distance is sometimes called the *full overtaking sight distance* and is less than ISD (another more appropriate use of the term *full overtaking sight distance* is given in Section 19.3.4).

For a design speed of v_k = 110 km/h, Figure 19.8 shows the lateral clearances that would be needed to achieve ISD for various curve radii. Many of these would be impractical to achieve on conventional two-way roads. Another dilemma here is that an obvious solution to low ISD values is to increase the curve radius. However, Section 18.2.5 has shown how this will lead to an increase in vehicle curve speeds and hence to greater ISD values which, from Equation 19.6, increase with speed squared.

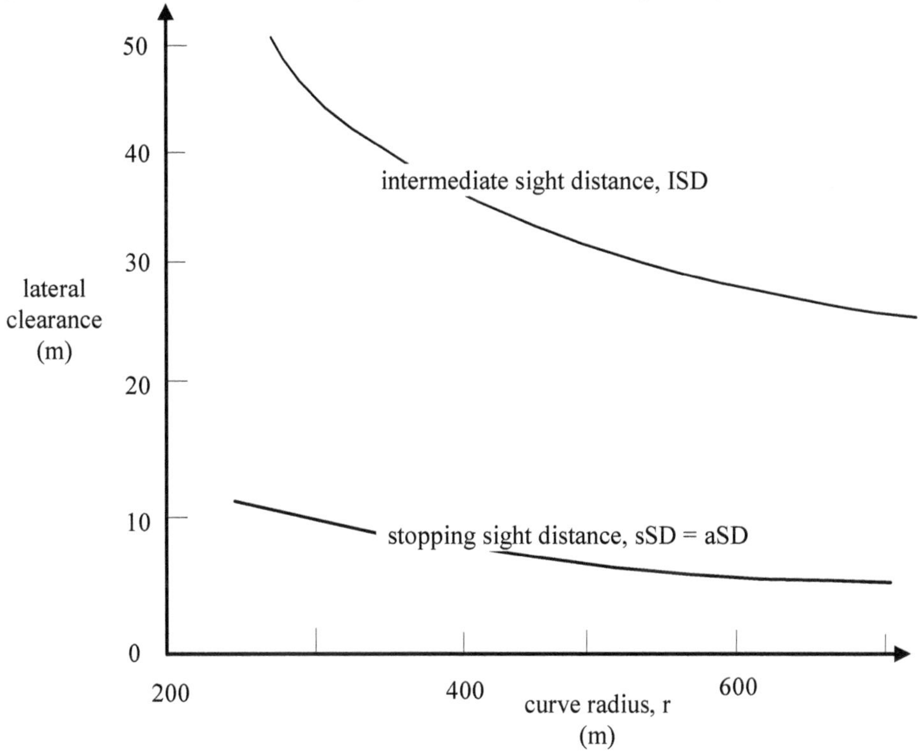

Figure 19.8 Lateral clearance requirements for intermediate sight distance (ISD) and a design speed of v_k = 110 km/h. Lateral clearance is defined in Figure 19.7.

On divided highways where the chances of meeting an oncoming vehicle should be minimal, it could be expected that ISD values would not apply and that horizontal alignments could be checked using sSD values.

19.4.5 Continuation (or abort) sight distance (CSD)

The overall overtaking manoeuvre was discussed in Section 18.4.1 and is illustrated further in Figure 19.9. The continuation sight distance, CSD, allows a driver to safely abort or complete an overtaking manoeuvre (Figure 19.9b). Until this point, aborting the manoeuvre is always less demanding than completing it. Thus, continuing to completion is used to define a 'point of no return', which becomes the point at which the driver of vehicle A has travelled a distance D_1 consisting of the distance:

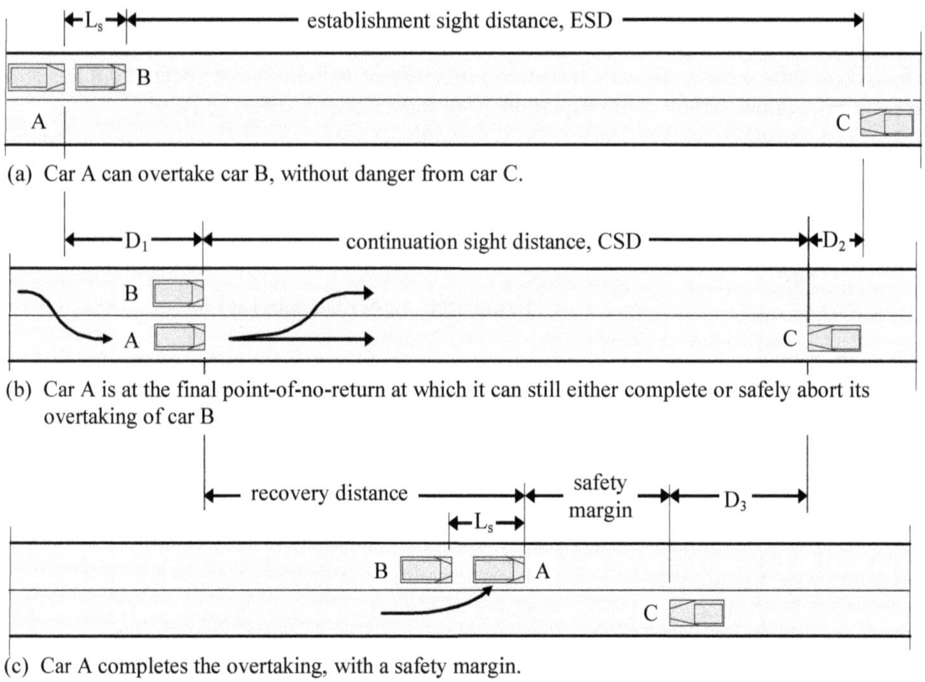
Ls
establishment sight distance, ESD
B
A
C
(a) Car A can overtake car B, without danger from car C.
D1
continuation sight distance, CSD
D2
B
A
C
(b) Car A is at the final point-of-no-return at which it can still either complete or safely abort its overtaking of car B
recovery distance
safety margin
D3
Ls
B
A
C
(c) Car A completes the overtaking, with a safety margin.

Figure 19.9 Various components of sight distances used for overtaking. See also Figure 22.5. Note that ESD (Section 19.4.6) and CSD are only at their 'design' values if the recovery distance and safety margin in (c) are both at their minimum values. Drawn for left-side driving.

* that elapses before that driver perceives that an overtaking opportunity exists, and
* taken to draw alongside vehicle B and be in a position to continue on and just complete the manoeuvre before meeting the oncoming vehicle C (Figure 19.9c). From Section 18.4.2, it will be seen that this distance will depend on whether the driver is making a flying or accelerative overtaking. The latter conservative assumption is usually adopted initially, although some traffic codes have found that this has led to unacceptably large values for CSD and have sought an uneasy compromise between the two values.

To make a positive decision at the point of no return, the driver of vehicle A must be able to estimate the distance needed to safely regain the correct lane. This is the distance CSD that is thus the sum of:

* the recovery distance needed for vehicle A to regain its lane,
* the distance travelled forward by the oncoming vehicle C during this time (D_3, Figure 19.9c), and
* a safety margin between vehicles A and C allowed for by the driver of vehicle A (Section 18.4.2 & Figure 19.9c).

To estimate the recovery distance, Figure 19.9a shows that the initial spacing of the vehicles A and B is L_s. The time to reach the point of no return (Figure 19.9b) is T_{ab}, and so:

$$L_s + v_B T_{ab} = v_A T_{ab}$$

where v_A and v_B are the speeds of vehicles A and B, giving

$$T_{ab} = L_s/(v_A - v_B)$$

The recovery distance has proved hard to estimate and a number of alternative assumptions have been made at this stage. If it is assumed simply that the recovery will take the same time, then:

recovery distance = $T_{ab}v_A = L_s v_A/(v_A - v_B)$

Another assumption is that the recovery distance is about $2D_1$, which implies that the spacing before overtaking (L_{sa}) is half the spacing at lane return (L_{sb}). As a third assumption, some codes also assume that $(v_A - v_B) = 15$ km/h.

Calculations and observations show that the CSD values are generally more severe than the related ISD ones (Section 19.4.4), mainly because they assume no drop in the speed of the oncoming vehicle (Troutbeck, 1981). Typical values of ISD and CSD for various speeds are shown in Figure 19.10. This issue is pursued further in Section 22.5.

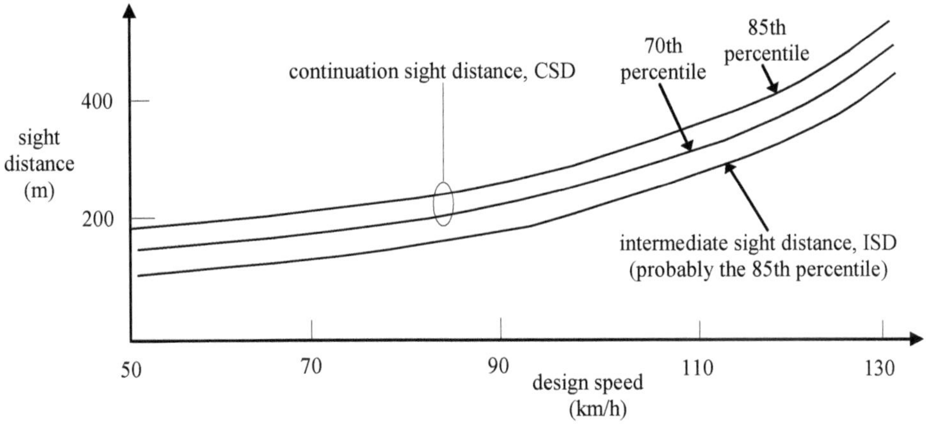

Figure 19.10 Continuation and intermediate sight distances for overtaking (Troutbeck, 1981).

19.4.6 Establishment sight distance (ESD)

The *establishment sight distance*, ESD, allows a driver to safely initiate and complete an overtaking (Figure 19.9a). It is sometimes also referred to as the *full overtaking, overtaking sight distance* (another use of the term is given in Section 19.3.2) , or *passing sight distance.* ESD can be calculated from the times and distances of each stage of the simple passing manoeuvre illustrated in Figure 19.9. It thus comprises the sum of:

* D_1, Section 19.4.5,
* CSD, Section 19.4.5,
* D_2, the distance travelled forward by the oncoming vehicle while the overtaking vehicle travels D_1, and
* a safety margin between the two vehicles.

The definition also requires the driver to act with restraint under the two conservative assumptions that:

(a) hidden from current view just beyond the ESD is a potentially threatening on-coming vehicle, and

(b) if such a vehicle does finally come into view — and this will occur relatively infrequently — the driver will be within the recovery distance and will not abort the overtaking manoeuvre safely.

Not surprisingly, drivers often treat the consequent conservative ESD values with some scepticism, being prepared to occasionally abort an overtaking rather than forego many potentially safe opportunities. This issue is pursued further in Section 22.4. Alternatively, ESD can be calculated from the establishment critical gap concept referred to in Section 18.4.2. These two ESD estimates are shown in Figure 19.11.

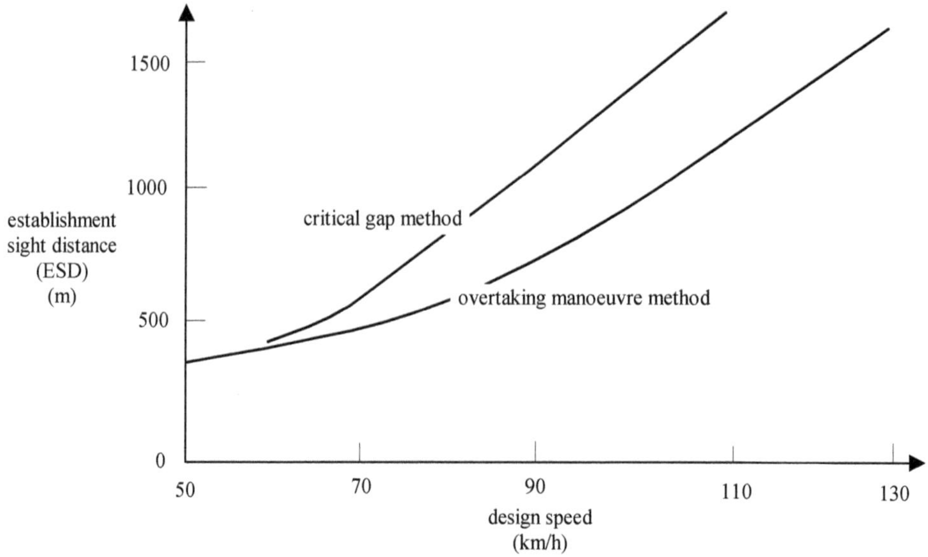

Figure 19.11 The establishment sight distance, ESD, for overtaking (Troutbeck, 1981).

19.4.7 Headlight, intersection, and entering sight distance

Headlight sight distance is the distance that can be seen in a vehicle's headlights and has been mentioned in Section 19.2.3 and is discussed in Section 24.4.1. In assessing this distance, it is necessary to consider whether a driver will be affected by the glare of the headlights of oncoming vehicles (Section 24.2.2). Median plantings (Sections 6.4 & 22.4.2) can often be used to reduce the influence of this factor.

Intersection sight distance provides vehicles approaching an intersection with sufficient sight distance for their drivers to decide whether to stop or to cross the intersecting road legally and safely. It is further defined and fully discussed in Section 20.3.7. Intersection sight distances are greater than stopping sight distances (Section 19.4.3) and will therefore govern design in areas where uncontrolled intersections occur.

Entering sight distance gives a driver entering a road sufficient sight distance to be able to join the road from a stop and accelerate to the design speed without being overtaken by a vehicle travelling at the design speed on the entered road. It is usually measured from driver's eye to driver's eye. The definition implies that:

(distance to accelerate to design speed) + (entering sight distance) =
(design speed)(time to accelerate to design speed) + (safe headway)

Acceleration distances and times are given in Section 27.2.2 and headways in Sections 17.2.2 & 17.2.4. Typical practical values for entering sight distance are:

design speed, km/h	entering sight distance, m	intersection sight distance (cars), m
40	120	80
60	180	120
80	340	170
100	500	230–300
120	500	300–350

19.5 ALIGNMENT ECONOMICS

Design codes usually specify geometric standards for road design in terms of a design traffic volume (Section 6.2.3) and a design speed (Section 18.2.6). The latter is the major external determinant of alignment characteristics. Cars gain major benefit from horizontal alignment improvement whereas trucks benefit most from vertical alignment improvements.

In undulating or mountainous terrain, it may not be easy to achieve curves with large radii and low slopes. The selection of a design speed will therefore have a major bearing on earthwork costs. For instance, a 20 km/h increase will approximately double the cost of the earthworks for a heavily trafficked (AADT > 1000 veh/d) road. For lightly trafficked roads with design speeds between 50 and 90 km/h, case studies indicate that construction costs increase by about 9 percent for each 10 km/h increment in design speed.

However, higher design speeds give benefits to road users in terms of improvements in travel time, convenience, and safety that may offset the higher construction costs. This can be assessed via a benefit–cost analysis (Section 33.5.8). Unfortunately, inadequate total funds can often prevent the adoption of schemes with quite favourable benefit/cost ratios.

A more detailed discussion of this topic from a road location viewpoint is given in Section 6.2.2, including a discussion of the need to control earthwork costs. Minimising such costs is only appropriate if total user costs (Section 5.2.1) are also minimised.

19.6 BIKEWAYS

19.6.1 Path types

Bicycles and their use as a transport mode are discussed in Section 27.6. This section discusses traffic facilities for bicycles. Every street will be used by cyclists and they therefore require consideration at all stages.

A key initial task is to estimate the demand for bicycle facilities. Travel demand in general is discussed in Section 31.3, but bicycles form a special and very different subset. Thus, in addition to the tools in Section 31.3, it can be useful to also predict usage by observing usage in similar existing applications. Key variables to look out for are local bicycle ownership rates and the hilliness of the terrain. Structured interviews with samples of the local community can also be used.

Routes for bicycle travel are called bikeways (or cycleways) and can include roadside paths, designated lanes on a road carriageway, bicycle crossings and overpasses, and trails (or tracks) through open urban spaces and countryside.

It is a misconception to believe that total physical separation of bicycles from cars is a necessity. Even if such a network of separate off-road facilities existed, cyclists' requirements are such that they would still need to use the existing road system for at least part of most journeys, mainly because many origins and destinations are located on roads and streets. In one typical study area, 35 percent of school bicycle usage and 80 percent of adult usage occurred on main roads.

In addition, many cyclists prefer to use main roads because they often provide a better cycling surface with respect to smoothness, maintenance, and the absence of surface debris such as loose gravel. Surface condition affects comfort, as bicycles have no suspension and relatively hard tyres, and safety (surface condition contributed to 20 percent of bicycle crashes in one survey). In the absence of bicycle lanes, cyclists usually travel in the metre of lane adjacent to the kerb and so special attention must be given to the surface in this area.

The main types of bikeways are illustrated in Figure 19.12. They are:

(a) *Exclusive bikeways* (or bicycle paths in urban areas and bicycle trails outside the urban zone)

Exclusive bikeways (Figure 19.12a) require the construction of special facilities in extra right-of-way and so are relatively expensive. Although the cyclist is completely segregated from the road system, conflict may still occur at intersections, or with pedestrians using the same path. Some cyclists may feel insecure if their passage is not visible to others.

(b) *Shared* (or dual-use) *paths*

There are two main types of shared path (Figure 19.12b) which allow cyclists and pedestrians to legally share the use of designated sections of roadside or independent path. They thus provide a joint facility of lower overall cost. It is usual for the cyclist to be required to give way to pedestrians.

b1. Unlined footpaths. In many jurisdictions, it is illegal for cyclists to use roadside footpaths, although the practice is widespread. In some areas cycling is only permitted on such footpaths at specific essential locations.

b2. A segregated path (Figure 19.12c) is a shared path, which is marked by a line to permit cyclists to use one side and pedestrians the other side of the same pavement.

(c) *Shared* (or dual-use) *road lanes*

Shared lanes (Figure 19.12d) use line marking and signs to provide cyclists with an on-road facility. The shared lane can be either a traffic lane or a parking lane. The options work best on streets with low speed limits. Lane widths are given in Section 19.6.3.

Ideal sites for shared routes are residential streets with low traffic volumes and located adjacent to arterial roads. This latter system is particularly effective when used in conjunction with street closures (Section 7.3) which still permit the passage of cycles.

Particular attention needs to be given to 'squeeze points' within the shared-lane system, such as intersections and bridges, where a reduction in the 'free' carriageway width forces cyclists into the main traffic stream.

A typical warrant would suggest the implementation of shared lanes when the one-way bicycle AADT exceeds 50/day and the traffic AADT exceeds 1500/day. Shared lanes are often not effective when the vehicle flow is above 300 veh/h.

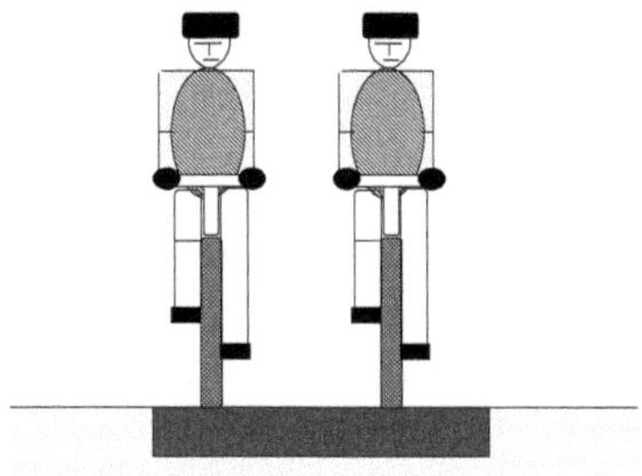

(a) Exclusive bikeway: space allows complete separation from other traffic.

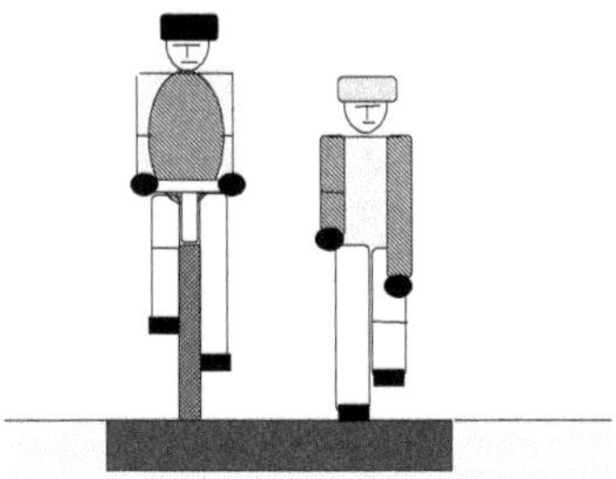

(b) Shared path; requires special signing.

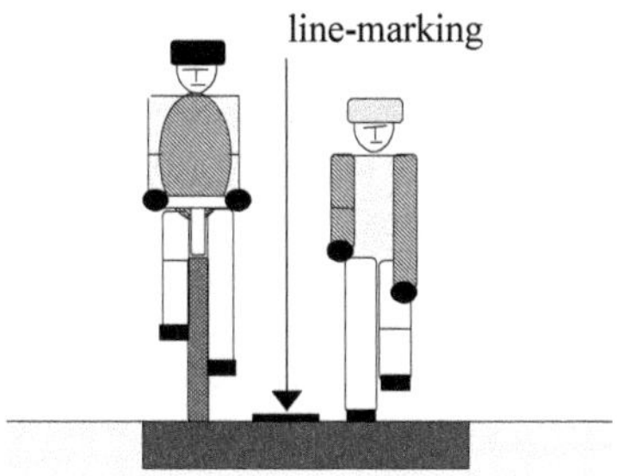

(c) Segregated path.

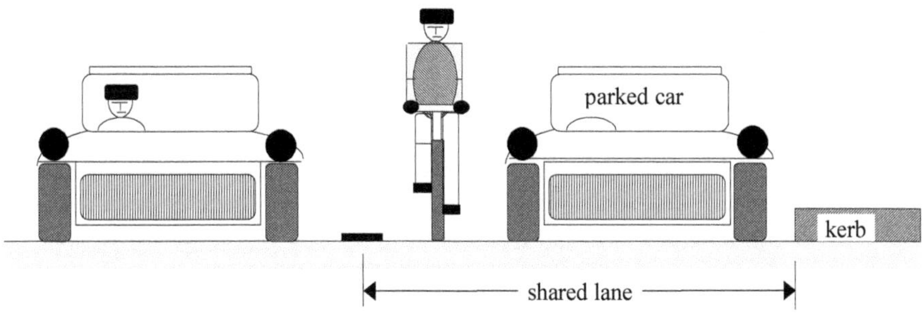

(d) Shared lane: requires special signing.

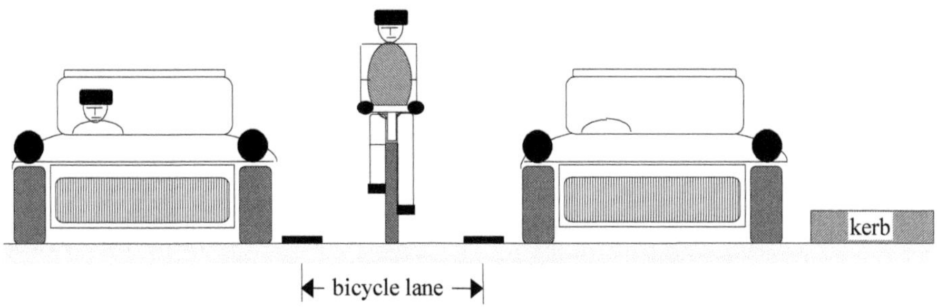

(e) Bicycle lanes: may require raised lane markers to deter traffic.

Figure 19.12 Types of bikeway

(d) *Bicycle lanes* (or restricted bikeways)
Bicycle lanes (Figure 19.12e) use signs, pavement markings, coloured surfacings, or physical barriers to denote an area of road for the exclusive use of bicycles. They are

used where vehicle speeds and/or volumes are high. Lanes on the travelled way are rarely justified when the total AADT is below 1000 veh/d and the bicycle AADT is below 400 veh/d.

Bicycle lanes can be located in the median, between the footpath and kerb, on the sealed shoulder of rural roads, or on the travelled way. They are typically 1.5 m wide. When bicycle lanes are installed it is essential that steps be taken to prevent parked cars from using these lanes. This problem can be avoided by using indented parking bays. Using dashed line-marking where motor vehicle intrusion is possible provides cyclists with some warning of a potential hazard.

19.6.2 Location

The origins of bicycle trips are usually relatively dispersed and trip demand is sometimes difficult to predict. It has been estimated that one optimum bikeway to serve a typical school would draw no more than 5 percent of the bicycle trips undertaken by the pupils of that school. Because of such factors, the construction of trial sections of bikeway is recommended before installing full systems.

In locating a route it is necessary to consider route convenience, directness, comfort and smoothness, grade, intersections, safety, and effects of the route on neighbouring properties and pedestrians. A smooth riding surface is a fundamental need. Normal footpath construction (75 mm of crushed rock and 25 mm of asphalt, often using only 7 mm size aggregate, or 125 mm of reinforced concrete on a 50 mm sub-base) is usually acceptable. Most paths must be capable of carrying small maintenance trucks. Surface differentials should be less than 5 mm. Longitudinal grates on stormwater inlets should be avoided as they tend to snare bicycle tyres (Section 13.4.1). Once completed, the paths must be kept clean of rubbish and loose stones.

It is also desirable that bicycle routes avoid areas where there is a risk of vandalism or a perceived threat of anti-social behaviour towards children (such as child molesting). Similarly, some cyclists will avoid large areas of open space or zones where the path is shielded by trees. Locations should thus be such as to ensure ready visibility by the cyclists of the surrounds and by others of the bicycle track.

The successful operation of bicycles as part of the public transport system, either individually or as part of a dual-mode trip, usually requires secure storage for bicycles and specialised clothing at the work-end of the cycle trip, i.e. either at work or at a public transport station or stop.

19.6.3 Design

For the purpose of designing facilities for their use, cyclists can usefully be placed in three categories. These are:

* advanced cyclists, who utilise a range of skills and experience, and gain most from on-road measures,
* basic cyclists, who lack many skills, experience and/or confidence, and
* children, whose attitudes and dimensions differ markedly from those of adults.

The following are typical design standards that have been adopted in some jurisdictions.

A design speed of 15 km/h on the flat and 30 km/h downhill is common, however experienced commuter cyclists on a good flat path have an 85th percentile speed closer to

20 km/h. The maximum flow of around 10 000 cycle/h occurs at about 10 km/h. Grades are of major importance to cyclists (Section 27.6). A maximum acceptable continuous grade is 3 percent, with grades of up to 6 percent rideable for lengths of 40 m or less. A grade of 10 percent should never be exceeded. Horizontal radii of no less than 4 m on the flat and 8 m on grades are suggested. A stopping sight distance of 40 m for sign placement is common. Lighting levels should be at least 0.3 cd/m^2 (Section 24.3). Roughness levels should not exceed 3.5 IRI (Section 14.4.4) and there should not be vertical 'steps' of more than 15 mm, although surfaces should not deviate vertically by more than 5 mm from a 3 m straight-edge.

Typical design dimensions for cyclists are (in mm):

1 length of:	
1.1 rider and cycle	1750
1.2 rider and cycle operating space	3000
2 vertical clearance of:	
2.1 rider and cycle	2400
2.2 pedal from path	150
3 rider eye-height	1350
4 width of:	
4.1 a wheel track	500
4.2 a stationary rider and cycle	800
4.3 an operating rider and cycle	1200
5 one-way lane width for:	
5.1 on-road lanes beside a car-parking lane	1500
5.2 < 2000 cycle/d	1800
5.3 > 2000 cycle/d, cycles only	2200
5.4 maximum rider comfort and to prevent cars squeezing past a cyclist	3000
5.5 shared bicycle and car parking lanes	3750
5.6 shared lanes if car parking is frequent	4000
5.7 shared lanes wide enough to permit cars to attempt to squeeze past parked cars — to be avoided	>4500
6 two-way lane width for:	
6.1 < 2000 cycle/d	2600
6.2 > 2000 cycle/d	3000
In Holland where cycling is a dominant mode, 2.5 m paths carry up to 9000 cycle/h.	
6.3 curves with R < 30 m	3500
7 horizontal clearance to:	
7.1 obstructions	250
7.2 other cyclists or pedestrians	400
7.3 vehicles travelling at 60 km/h	500
7.4 vehicles travelling at 80 km/h	800
7.5 vehicles travelling at 100 km/h	1000
7.4 vehicles travelling at over 100 km/h	2000

The vehicle clearances are used to avoid the dangerous aerodynamic effects caused to cyclists by such vehicles. Wheelchairs can be accommodated within the cycle dimensions: indeed, the 1800 mm path width usually permits a pair of wheelchairs to pass each other.

Path-to-path *intersections* should be designed so that the paths meet at right angles and approaching cyclists automatically slow down.

Intersections with roads and streets can be both a major crash hazard and cause unsafe cycling behaviour. Particular attention should therefore be paid to such features of intersection design as ensuring that cyclists can:

* easily wait,
* readily activate traffic signals, and
* are not forced to cross roads at midblock where traffic speeds are highest.

Desirably, the path crossing should be located well before (e.g. 2 to 5 m) the vehicular intersection in order to both prevent drivers giving visual priority to the search for other cars, and to give them an opportunity to see any cyclists at a reasonable stopping sight distance (Section 19.4.3). The crossing path should be clearly marked, possibly by using coloured and/or raised paving.

Turning drivers are a particular hazard as their attention is often devoted to the search for gaps in the cross traffic. The problem is worsened with two-way bicycle paths, as the cyclist may be coming from an unexpected direction. A 6 m clearance from the parallel kerb gives turning drivers a chance to see cyclists before any risk of a collision.

A useful measure at traffic signals is the provision of a cyclists' stop line in advance of the normal stop line (Section 20.3.7). This technique:

* provides a holding area for waiting cyclists,
* ensures that they are very visible to drivers, and
* allows them to clear the intersection in advance of the motor vehicles.

CHAPTER TWENTY

Intersections

20.1 INTRODUCTION

20.1.1 Performance indicators

An intersection (or *junction*) is a location where at least two different traffic streams cross or merge and is an inevitable part of any conventional road system. The traffic to be handled may include cars, trucks, bicycles, motorcycles and pedestrians — all with their quite different response characteristics. Intersections between two greatly different types of traffic or class of road should be handled with great care.

Within an intersection there are locations where the individual vehicles in the various streams have the potential to occupy the same pavement area at the same time, thus decreasing capacity and increasing the crash risk. Intersections are therefore major determinants of both the traffic flow patterns, capacity, and the safety of the entire road system. In this context, the desire to optimise traffic flow and minimise crash risk can produce conflicting objectives. An example discussed in Section 23.2.4 is the determination of the length of the yellow interval at a set of traffic signals. Three key performance parameters for an intersection are thus:

(a) its *capacity* to manage the traffic flows wishing to pass through it (capacity calculations are given in Section 17.3.3 for basic cases, in Section 20.3.6 for roundabouts, and in Section 23.6.3 for traffic signals),

(b) its *safety* record, which is discussed in Section 20.2.3, and

(c) the inevitable *delay* it causes to vehicles. A total delay per vehicle of about 15 s usually corresponds to good conditions whereas most drivers would consider 45 s as indicative of poor conditions. However, Section 7.3 discussed how some intersections in local precincts may be designed to increase delay and thus discourage through traffic.

Delay is the sum of traffic and geometric delay.

Traffic (or *congestion*, *queuing* or *operational*) *delay* is the delay a vehicle encounters in crossing another traffic stream. The fundamentals of traffic delay for low traffic volumes are given in Section 17.3.3 and Equation 17.24. Traffic delay for signalised intersections is discussed in Section 23.6.5.

Geometric (or *fixed*) *delay* occurs when the geometry of an intersection requires vehicles to:

(1) travel further to pass through the intersection, and/or

(2) slow down to negotiate it.

It applies particularly to roundabouts (Section 20.3.6). Methods of calculating geometric delay for a variety of intersections are given in McDonald *et al.* (1984).

The following intersection types occur within an hierarchy of intersections:

(1) the simplest intersections are uncontrolled and at-grade (Section 20.1.3),

(2) the next stage commonly sees the introduction of traffic signs (Section 20.3.8) to assign traffic priority
(3) a further stage uses channelisation to separate conflicting traffic (Section 20.3.5),
(4) when performance drops further, control by a roundabout (Section 20.3.6) is considered
(5) if that is impractical or inappropriate, control by traffic signals (Section 23.1) is adopted,
(6) traffic is rerouted to avoid traffic conflict and finally, when all other measures are exhausted,
(7) grade-separation by underpasses and/or overpasses is introduced.

Types (1) to (5) are at-grade intersections and are discussed in Section 20.1.3. Type (5) is discussed in Chapter 23 and types (6 & 7) in Section 20.1.2. Intersections may contain a mix of these types.

20.1.2 Grade-separated intersections

Grade-separated intersections solve the problem of the multi-objective criterion alluded to in Section 20.1.1, but at considerable cost in terms of money and land. Traffic movements that would otherwise come into conflict are separated vertically by bridges and tunnels. The costs of these structures and of the large land requirements are usually only outweighed by the benefits of conflict removal, in the cases of freeways and intersections with consistently high traffic flows. Grade-separation is rarely cost-effective if the crossing flows are below 1 000 veh/d, usually deserve consideration when these flows exceed 3000 veh/d, and are usually economically justified for crossing flows in excess of 6000 veh/d.

Grade-separated intersections on freeways are commonly called *interchanges*. Minimum and typical spacings for interchanges are about 1.5 and 3 km in urban areas and 3 and 6 km or more in rural areas. The segments of the interchange that link directly to the freeway are known as entry and exit *ramps*. Ramps and speed change lanes are designed in terms of length and radius to provide a transition between the design conditions at either end of the ramp (e.g. between freeway design speeds and the much lower operating speeds on a local arterial). Special merging/diverging features of entry and exit ramp design are discussed in Section 17.3.6. As an exit ramp cannot begin until a safe distance after the previous entry-ramp has finished and its incoming traffic has safely merged — typically 500 m, mainly determined by the need to accommodate weaving manoeuvres (Section 20.2.1) — the length of these exit ramp tapers and the associated speed change lanes — typically at least 500 m — are a key control on minimum interchange spacings of 1.5 km and 3 km mentioned above.

There are a variety of types of interchange. The commonest are the Y, diamond, partial cloverleaf, trumpet, roundabout, directional and cloverleaf — each is illustrated in Figure 20.1. Some connect freeways to lesser routes (Figures 20.1a–f,j–k) and others provide freeway-to-freeway connections (Figures 20.1g–i,l). In each case, common design practice is to give preference to the movement having the highest traffic flow, and occasionally to the movement giving route continuity.

The simplest interchange is the Y (Figure 20.1a) — or fork or flyover interchange — which is common at the beginning and end of bypasses and other bifurcations.

Mirror image Figures 20.1a–l for right-hand driving.

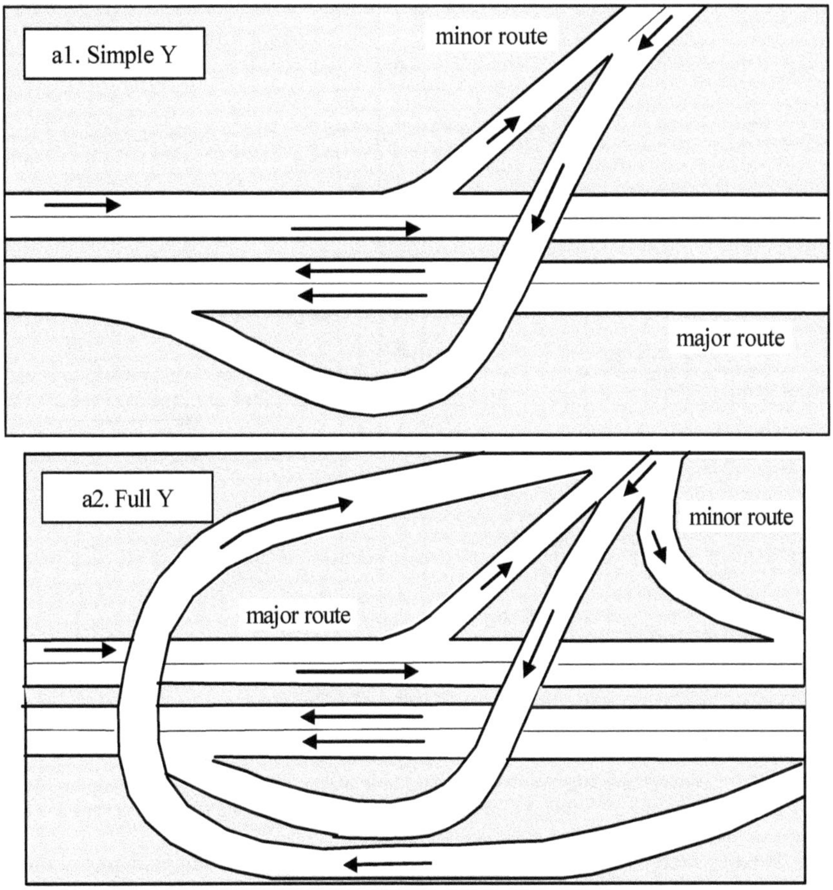

Figure 20.1a Y interchange, which accommodates the beginning and end of another road. In more constrained situations, the Full Y may need to be a three-level interchange.

Of the interchanges catering for movements on two intersecting roads, the simplest is the *diamond* and its six commonest forms are shown in Figure 20.1b–g, approximately in order of increasing cost. They differ in the way in which they treat the less important (or minor) routes. (In the at-grade form these are called crossroads (or *X*- or *four-way intersections*).

The *conventional* (or *spread*) *diamond* (Figure 20.1b) is usually cost-effective for crossing flows in excess of 1500 veh/d and is frequently used to connect freeways to rural roads in areas where the space consumed is not critical. It requires traffic control devices on the minor road and these are usually the factor that limits the capacity of the interchange.

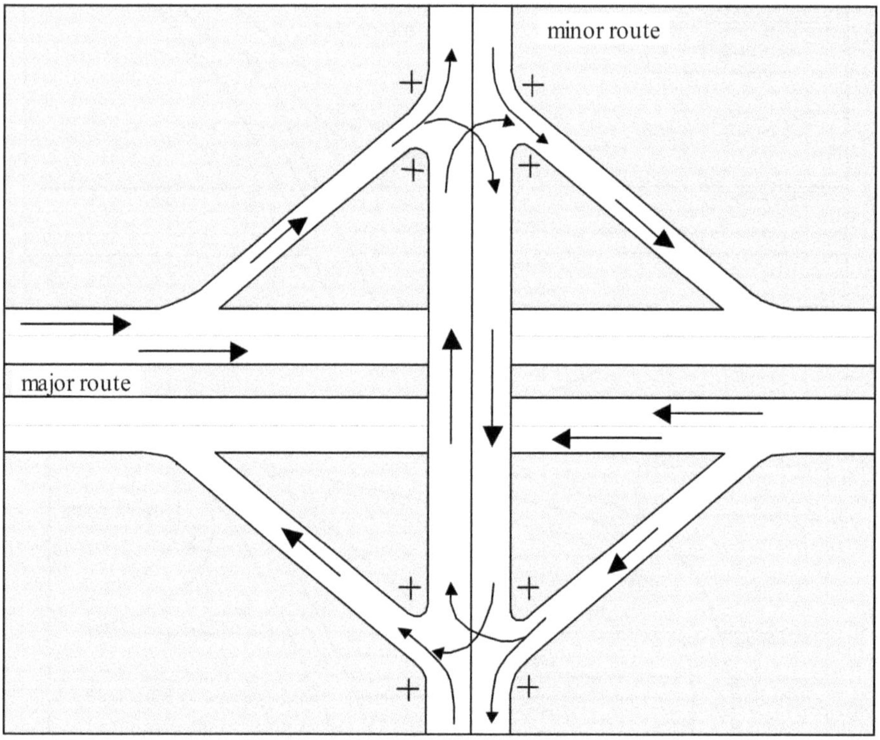

Figure 20.1b A conventional diamond interchange. Note that in this and subsequent figures, (a) the + indicates a spot where a traffic control device is needed, and (b) ramp shapes and lengths are diagrammatic and not realistic.

As demand increases, the conventional diamond can be translated into a cloverleaf interchange, provided sufficient land is available (see Figure 20.1j below). The minor road is commonly on an overpass and so the ramps meet the minor road at the bottom of a vertical curve on the minor road. It is important to ensure good sight distances at these terminals.

In areas of more intense land-use, keeping the ramps closer to the major route (Figure 20.1c) creates the urban diamond. To fit within a narrow urban reservation, the distance, w, is held to a minimum and the diamond is sometimes called *compressed* (or *compact*). The ramps commonly meet the minor route at a pair of closely-spaced, at-grade intersections at the end of a vertical curve in the ramp. In the common urban case illustrated in Figure 20.1c, the minor road carriageways are separated and four individual sets of traffic control devices are used to manage these intersections. The arrangement becomes a *split diamond* when there is significant separation, s, between the two minor route carriageways. Sometimes s becomes so large that the intersection becomes two separate half-diamonds, each serving one major-route direction and the two minor-road carriageways operating as separate two-way roads. In a *dumb-bell interchange*, roundabouts (Section 20.3.6) rather than signals are used at the intersections with the local street.

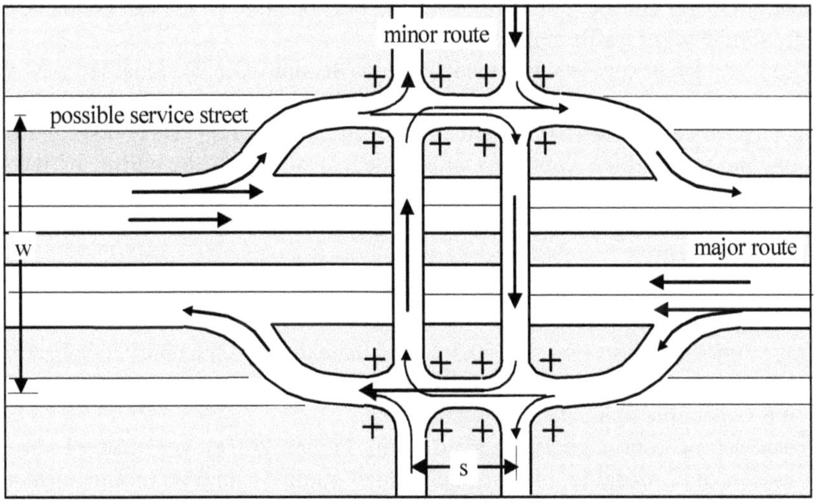

Figure 20.1c Urban diamond interchange, with the dashed lines indicating service roads. The connection between service road and ramp needs careful design to avoid creating a traffic hazard. The intersection becomes a split diamond if the separation of the two carriageways on the minor route becomes large.

The *tight* (or *closed*) *diamond* shown in Figure 20.1d is a version of the compressed diamond in which the four intersections with the minor route are reduced to one large set of linked traffic signals involving many overlapped turning-movements (Section 23.6.1).

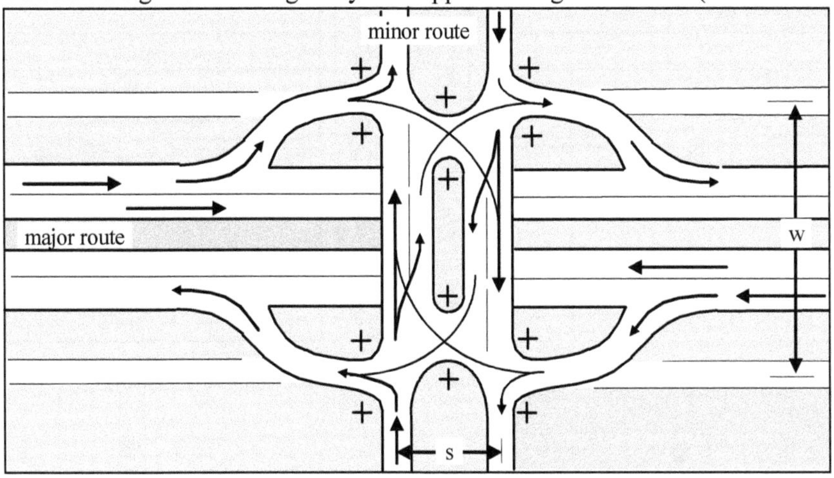

Figure 20.1d Tight diamond interchange.

In a *single-point diamond* (Figure 20.1e, or *single-point urban, urban, or single-signal*) interchange (SPI), the arrangement is further simplified by eliminating the overlapped turning movements by using pronounced channelisation (Section 20.3.5). If there is little separation, s, between the two carriageways of the minor route, the physical channelisation using islands (Section 22.4.1) disappears and reliance is placed solely on line markings. The separation, s, must be managed to provide space between opposing centre-turners. The *single point* in the name refers to the turning point in the middle of the pavement; all major conflict points occur in this one location, which can be managed by a relatively simple set of traffic signals.

There are three key sets of signal phases (Section 23.6.2). One set provides for traffic leaving the major route, the second for traffic joining the major route and the third for minor route traffic passing straight through the intersection. Of course, the through traffic on the major route continues unimpeded at all times, by virtue of the grade-separation.

The key advantages of the single-point over the urban and tight diamond interchanges are that:

* it consumes relatively little space outside the linear reservation,
* there is a single conflict area,
* its non-locked turns (Section 20.3.3) eliminate many of the conflicts inherent in the locked turns used in the tight diamond,
* it is better able to handle traffic variations,
* one set of four signals (marked + in Figure 20.1e) can control the whole intersection, working on a simple three-phase (exit, entry, and cross-traffic) operation, and
* vehicles need only stop once in negotiating the intersection.

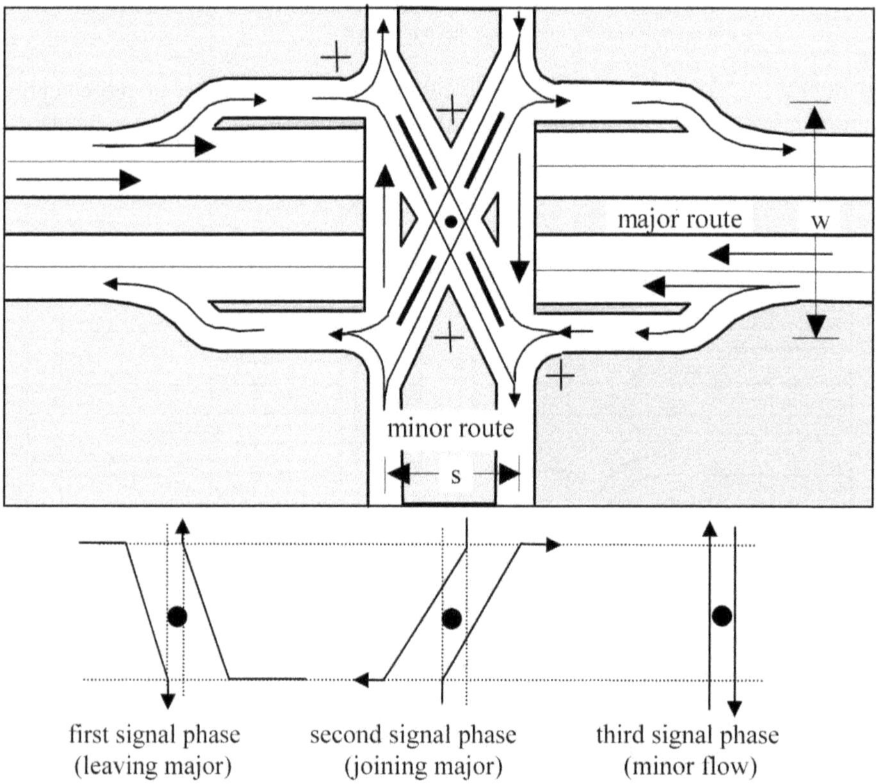

Figure 20.1e Single-point diamond interchange. The + indicates a traffic signal.

On the other hand, the disadvantages of the single-point intersection are that:

* it can cost more to build than the conventional diamond as it needs a wider bridge structure,
* it makes it difficult to widen the minor route at some future time,

* it requires long all-red periods (Section 23.6.2) in the signal phasing to clear the large intersection area,
* there can be constraints on sight distance and on other geometric requirements, particularly due to small ramp radii,
* it can be difficult for pedestrians and cyclists,
* the high-speed 'opposing' turn movements (Figure 20.1e) can be seen as unconventional and can disturb drivers who find the opposing drivers on their left, and
* any small turn-radii can create problems for large trucks.

The single-point concept can also be applied to at-grade intersections, by using signals on the major route approaches.

The two arrangements in Figures 20.1f&g are developments of the diamond which allow for more control and separation of the traffic flows and are therefore more suitable for high traffic flow situations. These grade-separated *roundabouts* also require more space and so are inappropriate in tight urban situations.

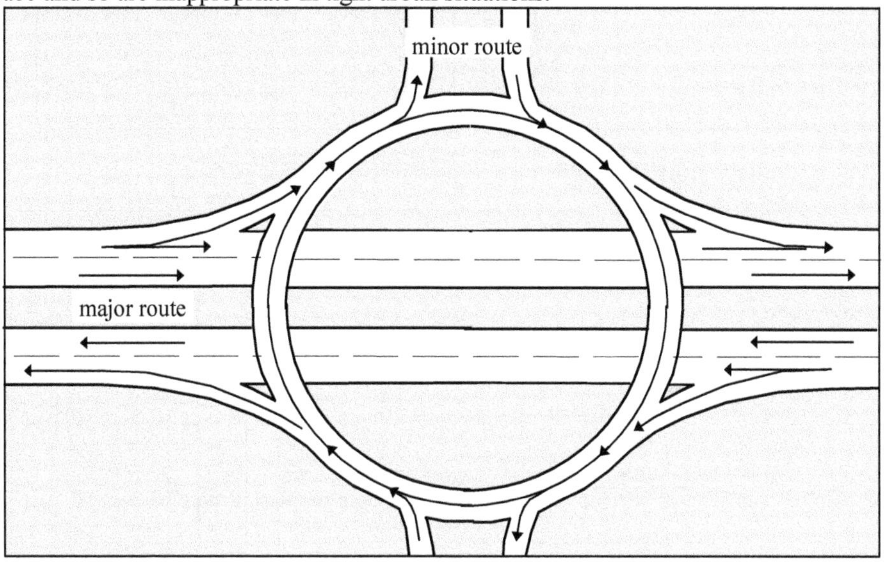

Figure 20.1f Bridged roundabout: a diamond interchange which uses roundabout control on the minor route.

The three-level grade-separated roundabout in Figure 20.1g covers an intersection between two heavily trafficked roads of equal importance and is usually only justified when all lanes are consistently running at capacity. Note that the remaining merging conflicts (Section 20.2.1) can be further reduced, if sufficient space is available, by using the additional lanes shown by dotted lines in the upper left quadrant. These two arrangements can operate with traffic conflict on the roundabouts controlled by either signals or specific rules for merging priority (Section 20.2.2).

The (single) *trumpet interchange* shown in Figure 20.1h is used for major T-intersections; the double trumpet version works for major crossroad intersections and permits a single tollbooth for flows between arterials.

The *cloverleaf* interchange illustrated in Figure 20.1i is the simplest of the interchanges between two major routes. However, it can be very space consuming as the loop radii must be very large if drivers are to maintain normal freeway speeds (Section

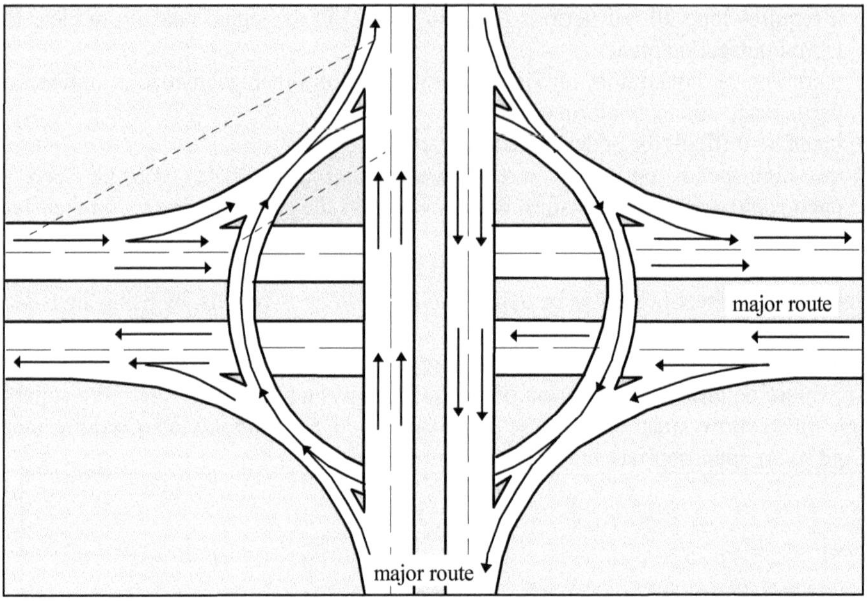

Figure 20.1g Three-level diamond interchange.

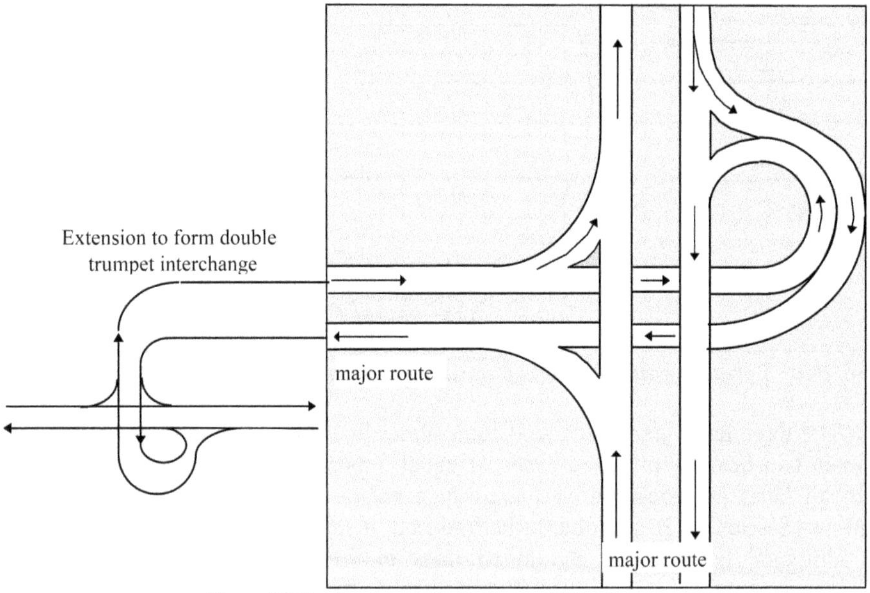

Figure 20.1h Single and double trumpet interchange.

18.2.5) and if ramp clearances between entry and exit ramps are to be maintained (Section 17.3.6). The use of lower design speeds for the loops can alleviate this problem and lessen the distance that vehicles must travel to make a centre turn — however, the lower operating speeds will noticeably reduce capacity. The loops are usually circular in order to discourage changes in speed.

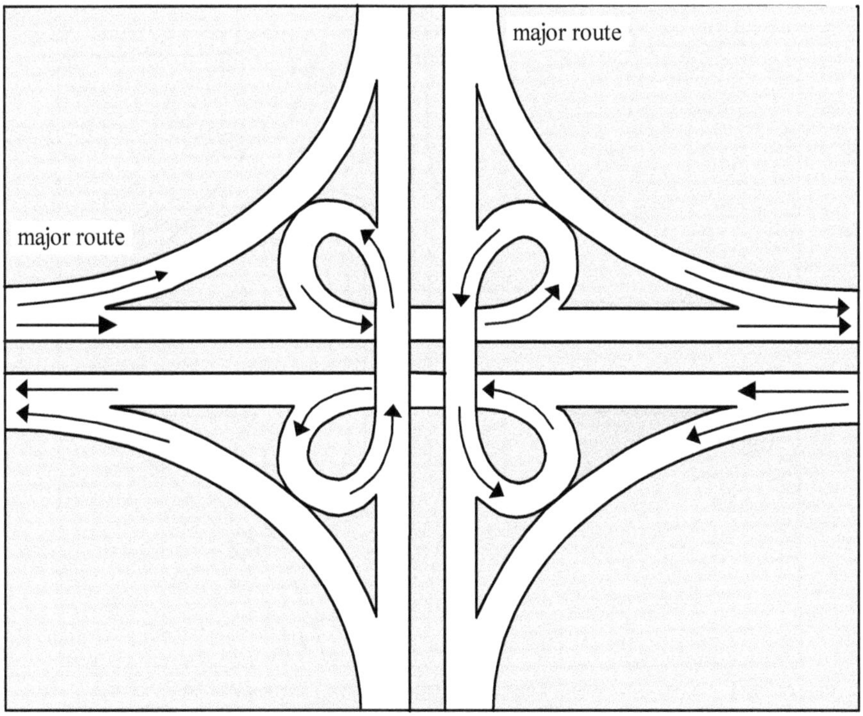

Figure 20.1i Cloverleaf interchange.

The space problem of the cloverleaf can also be alleviated by using the *partial cloverleaf* (or *half cloverleaf* or *parclo*) in Figure 20.1j. However, it requires traffic control devices along the minor route. If slip lanes in the unused quadrant are possible, the devices are only needed on one of the minor road carriageways.

The simplest form of the partial cloverleaf in Figure 20.1k caters for light traffic on the minor road and requires two sets of traffic control devices on that road.

The *directional interchange* — illustrated by the point-symmetric 'full-stack' form in Figure 20.1l — provides the most direct set of unimpeded movements but requires four levels of roadway and so is very expensive and space consuming. It can be reduced to three levels by replacing one level by cloverleaf loops. Some of the space can be saved in the *turbine interchange* explained in the inset to Figure 20.1l. However, this solution requires very high and long approach structures.

At the other extreme, in stage construction two roads intersecting at-grade may be initially upgraded by a simple overpass for one road and the arrangement shown in Figure 20.1m used instead of one of the more elaborate arrangements in Figures 20.1b–l.

In an *at-grade intersection* the various traffic movements operate at the same level and are thus in potential conflict. Such intersections can be divided into the following three categories:

(a) *Uncontrolled intersections*, where no regulatory devices are used and normal priority rules apply (Section 20.2.2). Drivers in the non-priority flow must wait for a super-critical gap (typically > 6 s, Section 17.3.4) in order to cross or merge with another traffic stream. Traffic passing through such an intersection is usually assumed to have a Poisson distribution. If the priority flow is q_m then Equation 17.24 gives the consequential capacity of the non-priority stream.

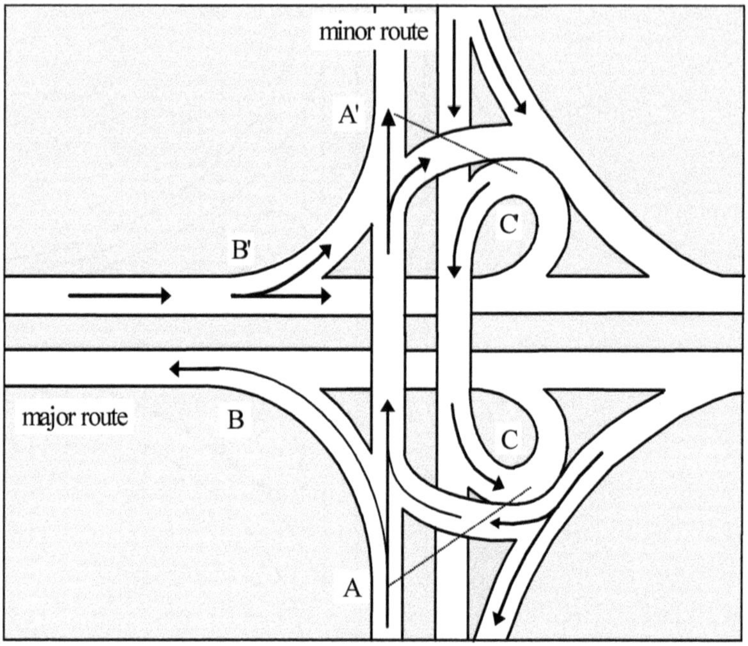

Figure 20.1j Partial cloverleaf (or parclo) interchanges. The loops may be on either side of the minor route, and may be adjacent or diagonally opposed. The links AB and A'B' are simpler alternatives to AC and A'C' shown as dashed lines but intrude into the land in the 'unused' (left-side) segments of the intersection. The use of AC and A'C' would require more complex signaling, as illustrated in Figure 20.1k below.

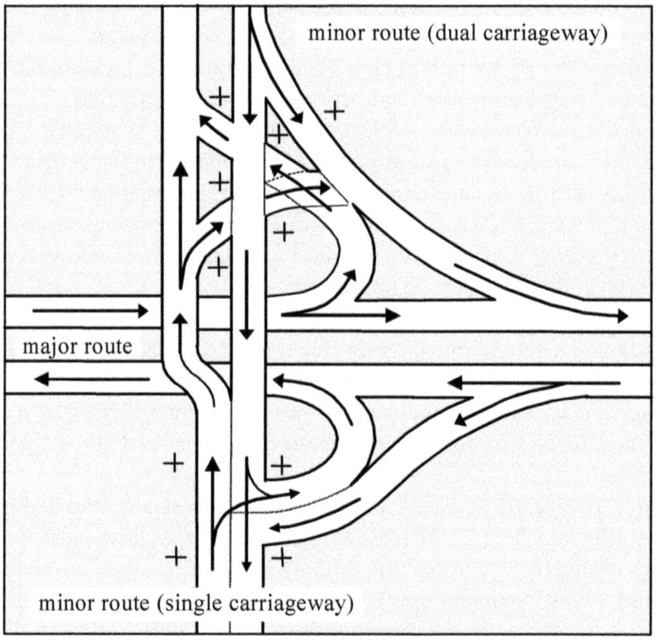

Figure 20.1k Simple parclo interchange.

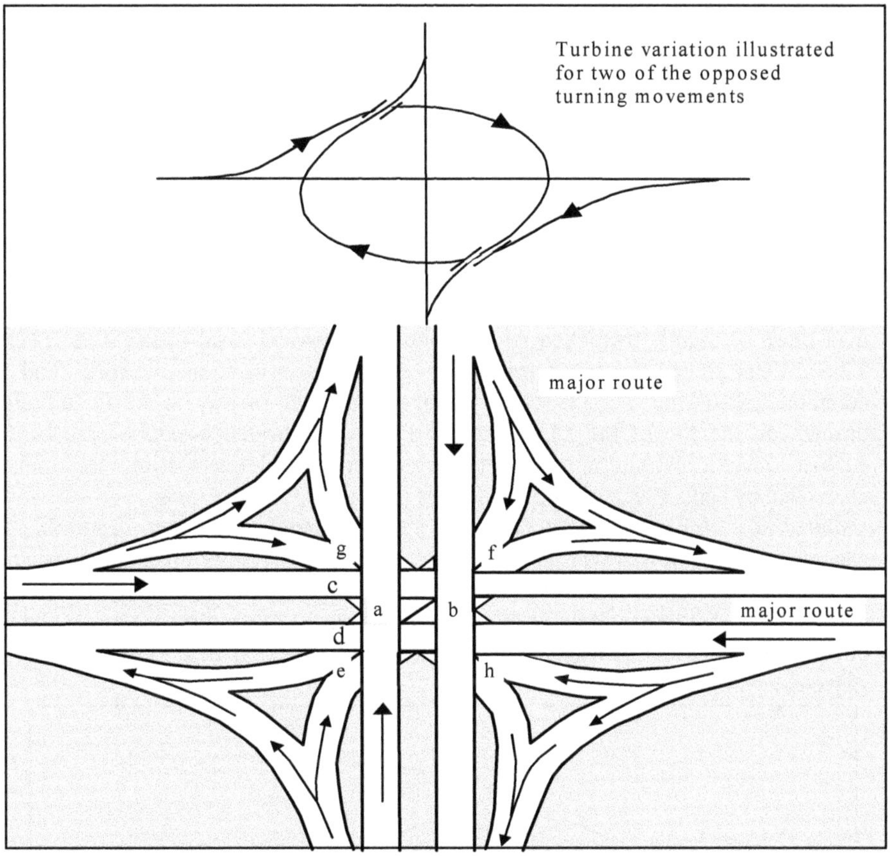

Figure 20.1l Directional interchange. There are many forms of this interchange. The one shown is a 'full stack' and requires four levels of roadway. The levels in order are a&b, c&d, e&f and g&h. Another form is known as the 'double Y'.

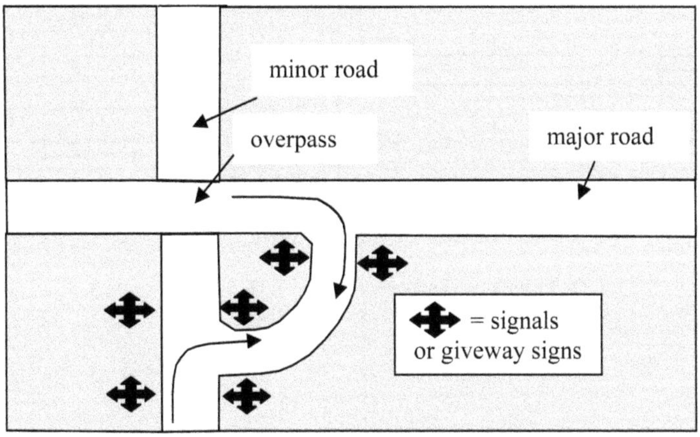

Figure 20.1m At-grade intersection converted to grade-separation with overpass but minimal interchange provision.

(b) *Controlled intersections*, where regulatory controls, such as Give Way signs, Stop signs, traffic signals, and police are to be used, but where there is no channelisation of flows. The installation of signs at intersections can be very cost-effective in terms of crash reduction. Give Way and Stop signs are discussed in Section 20.3.8 and traffic signals in Chapter 23. Police control can be useful when it is necessary to be sensitive to the length of a particular queue or to permit very long cycle times (Section 23.6.2). However, in other respects, police are usually unable to out-perform a set of traffic signals. Priorities in these control systems are usually related to both the intended function of the intersection and the actual traffic volumes on each leg. Design features of these intersections are discussed in Section 20.3.3.

The tendency has been to provide some form of control at all intersections other than those on lightly trafficked streets such as loops and local distributors (Section 7.2.1). Even there, the trend towards *T-intersections* in these networks and the common use of the T-intersection priority rule (right-of-way to traffic travelling through on the top of the T) has reduced the uncontrolled category even further. Railway *level crossings* are another form of controlled intersection: they can be controlled by signs (Section 21.3.2), signals (Section 23.2.3), or gates.

(c) *Controlled, channelised intersections*, where both regulatory controls and channelisation of flows (Section 20.3.5) are used. A variation of this is the jug-handle intersection where centre-turning traffic is taken away from the primary intersection and undertakes its turns at another location (Figure 20.2). In most cases it would be better to improve the phasing of the traffic signals (Section 23.2.5), than to use the space-consuming jug-handle intersection.

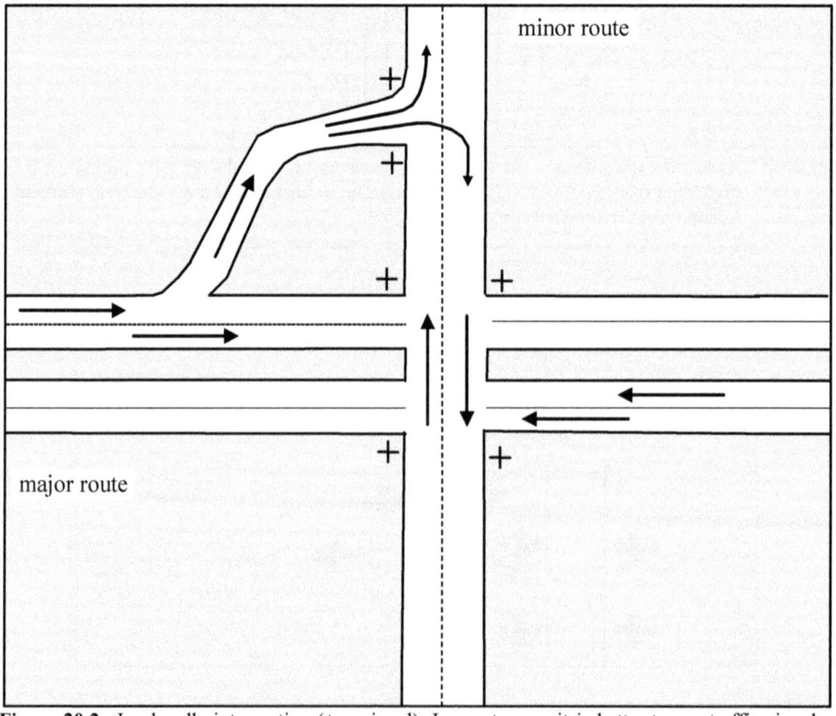

Figure 20.2 Jug-handle intersection (+ = signal). In most cases it is better to use traffic signals with carefully phased turn arrows (Section 23.2.5). Refer also to a wider version in Figure 20.10.

An at-grade intersection is a potential source of major traffic conflict and congestion. It requires drivers to be able to judge the time they will take to cross the intersection, the speed and distance of any crossing traffic, and to estimate its time of arrival at the intersection (Section 20.3.8). These are not easy tasks. The at-grade intersection therefore demands special design attention and is the subject of Section 20.3. Moreover, each intersection will have its own characteristics which will need to be individually studied and treated. Before discussing these design details, it is necessary to further understand how drivers behave at intersections.

20.2 DRIVER BEHAVIOUR AT INTERSECTIONS

20.2.1 Traffic conflicts

The basic *traffic manoeuvres* that occur whilst driving in traffic lanes are shown in Figure 20.3. The components of the manoeuvres are *merging*, *weaving*, *diverging*, and *crossing*. The associated theory is discussed in Sections 17.3.4–6 where it is shown that traffic will not enter or cross a traffic stream unless it contains a gap in excess of some critical size. Whilst each component presents the driver with some increased risks, the crossing component is particularly hazardous. This is seen, not only in relatively high long-term crash levels, but more commonly in traffic *incidents* in which drivers take some form of evasive action to avoid a crash (Section 28.1.3). Such incidents are called traffic conflicts and the locations at which they occur are known as *conflict points*, which could be expected to be points of relatively high crash risk.

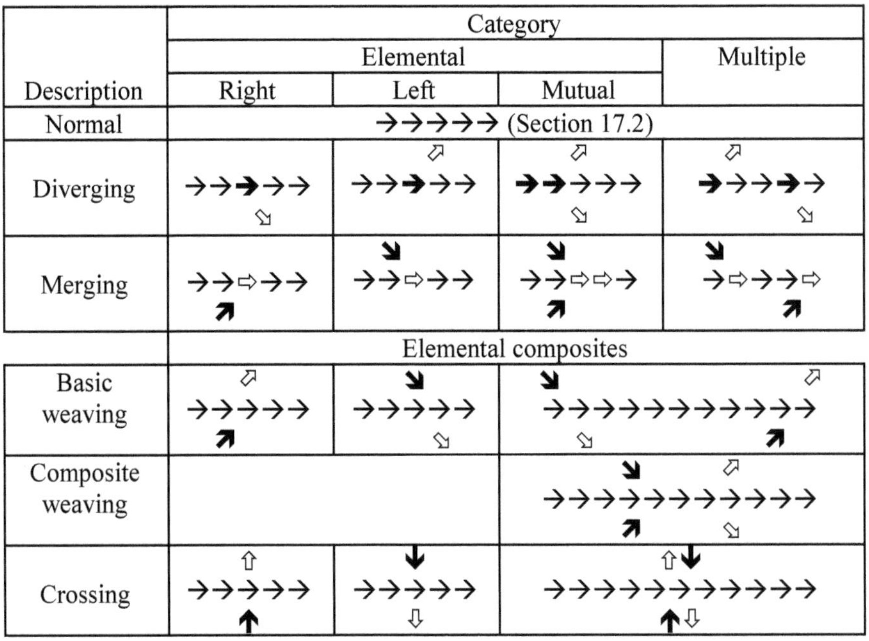

Category
Elemental
Multiple
Description
Right
Left
Mutual
Normal
(Section 17.2)
Diverging
Merging
Elemental composites
Basic weaving
Composite weaving
Crossing

Figure 20.3 Traffic manoeuvre types. → is normal traffic and ➔ ⇨is the path of the manoeuvring traffic.

A crash will only happen at a conflict point if two vehicles arrive there simultaneously. The chances of this happening can be reduced by:

* planning measures, such as totally separating some traffic flows by re-routing;
* using control devices such as traffic signals to prohibit the simultaneous operation of crossing movements;
* traffic engineering measures such as:
  (a) channeling traffic within an intersection to direct and separate the various traffic flows, thus reducing driver stress;
  (b) replacing complex multiple manoeuvres by simple elemental ones;
  (c) increasing the separation between decision and/or conflict points, giving drivers a sufficient distance in which to sight the other driver and thus ensure that the drivers' response time is less than the elapsed time before the possible conflict occurs; and
  (d) making conflicting streams cross at larger intersection angles (i.e. at closer to 90°).

The right-of-way rule to be discussed in Section 20.2.2 must be used to resolve all conflict situations remaining after such measures have been applied.

There are five basic types of intersection conflict, as illustrated in Figure 20.4. Those that are particularly hazardous are noted on the figure. The conflict points where two-lane roads meet at a crossroads (or X) intersection are illustrated in Figure 20.5 where it is seen that 32 conflict points exist. A similar intersection with five approach-legs has 72 conflict points. If a number of vehicles are present on each leg, there are 118 conflict points at even the four-leg intersection. The large number of possible conflict points highlights the potential risks associated with at-grade intersections. In general, the number of potential conflict points depends on the:

(1) number of approaches to the intersection,
(2) number of lanes on each approach,
(3) use of measures (a) to (d) above, and
(4) type of priority control.

The observation of conflicts and the subsequent prediction of conflict points is a useful traffic engineering technique. Such points not only indicate potential crash sites, but also other system inefficiencies such as increases in driver stress, travel time, and fuel consumption. However, no direct link between conflict numbers and crash rates has been shown to exist, nor can it be expected as conflicts are not necessarily directly linked to crashes (Andreassen and Cairney, 1985). In many instances a simple link between traffic flow and crashes will be found to be more useful (Sections 28.1.2 and 28.5).

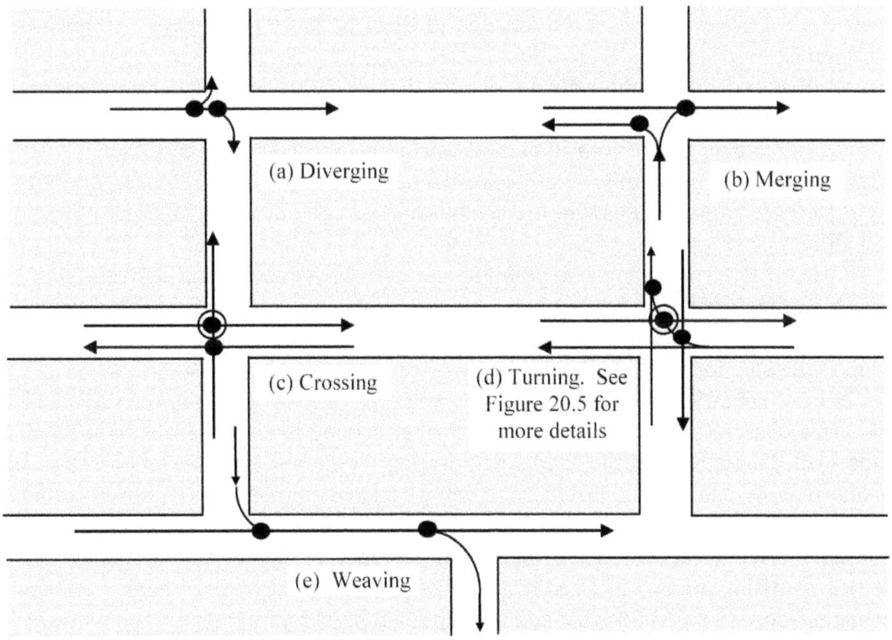

Figure 20.4 Basic types of intersection conflict. The conflict points are marked with a •. The O around the • indicates particularly hazardous conflicts. Mirror image for right-hand driving.

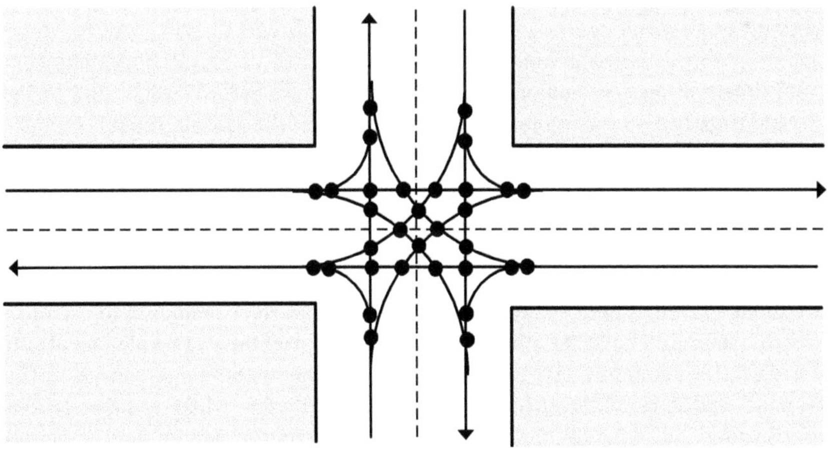

Figure 20.5 Conflict points at a crossroads intersection, assuming only one lane in each direction. The points are shown by a •. Mirror image for right-hand driving.

20.2.2 Traffic priority

Points of conflict between two traffic streams are resolved in the first instance by granting priority to one stream. Priority rules should be free from ambiguity and the need for interpretation, as drivers may need to make correct decisions in fractions of a second. *Priority to proceed,* or its negative alternative, *giving way,* is thus an important concept in driver control. *Right-of-way* is a term used to describe these procedures. A typical legal definition is:

> Right-of-way is the right of one road user to proceed in a lawful manner in preference to another road user approaching under such circumstances as to give rise to danger of a collision unless one grants precedence to the other.

In law, proceeding unlawfully (e.g. above the speed limit) usually removes the right-of-way preference from the unlawful driver, but does not give it to the other road user, who must still proceed with due care.

Two other terms that need defining are *off-side* which is the driver's side of a vehicle, i.e. the right side for left-hand drive and the left side for right-hand drive, and *near-side* which is the passenger's side of a vehicle, i.e. the left side for left-hand drive and the right side for right-hand drive.

In most countries the right-of-way rule is to *give way* (or *yield*) *to the vehicle on the right*. This observation is independent of whether driving is on the left or right-hand side of the road. Giving way to the right when driving on the left (or vice versa) is known as *off-side priority* and applies in left-hand drive countries, whereas giving way to the left whilst driving on the left (or vice versa) is known as *near-side priority* and applies in right-hand drive countries. There are some advantages to nearside priority (Quayle, 1980), but offside priority is a distinct advantage with roundabouts, where it favours the circulating flow (Section 20.3.6), and with divided highways, where it favours traffic on the highway.

Other options exist as alternatives to the offside and nearside priority rules. For example, there is the *rule of prior arrival*, giving priority to the first vehicle into an intersection. Another rule grants priority to proceed to drivers on particular routes, e.g. on designated *priority roads* (the *major/minor* road system), or on the horizontal (through) leg of a T-intersection. The major/minor system allocates priority according to the road network and usually also incorporates specific signing of the priority roads.

Give-way rules imply a negative prohibition for the driver whereas the priority-to-proceed and right-of-way rules represent a positive granting of privilege to one set of drivers (Quayle, 1980). Studies of human behaviour (Section 21.2.7) indicate that people respond better to positive instruction and that drivers expect one road at an intersection to have priority.

At an uncontrolled intersection (Section 20.1.3), a driver intending to travel straight across will generally have to observe the primary right-of-way rule. Centre-turners (Section 17.4.3h) commonly give way to vehicles in the oncoming stream of traffic. The turners must therefore filter through the opposing stream when a super-critical gap appears (Section 17.3.4). The provision of special *turning lanes* to store these queuing vehicles without impeding through traffic is discussed in Section 20.3.3 and will be shown in Figure 20.12. Any turning driver must also observe various secondary laws. For instance, centre-turners must usually give way to opposing vehicles turning from the kerb of the same road. As discussed in Section 20.1.3, vehicles in a T-intersection must usually give way if they are in the (vertical) stem of the T.

There are some conflicts that cannot be simply resolved. For instance, if the simple give-way rule applies and vehicles are simultaneously passing through all legs of an intersection, there are two situations where an impasse will occur (Figure 20.6). Each of the vehicles shown must give way to one of the others. Who gives way to whom? These are called *intransitive situations*.

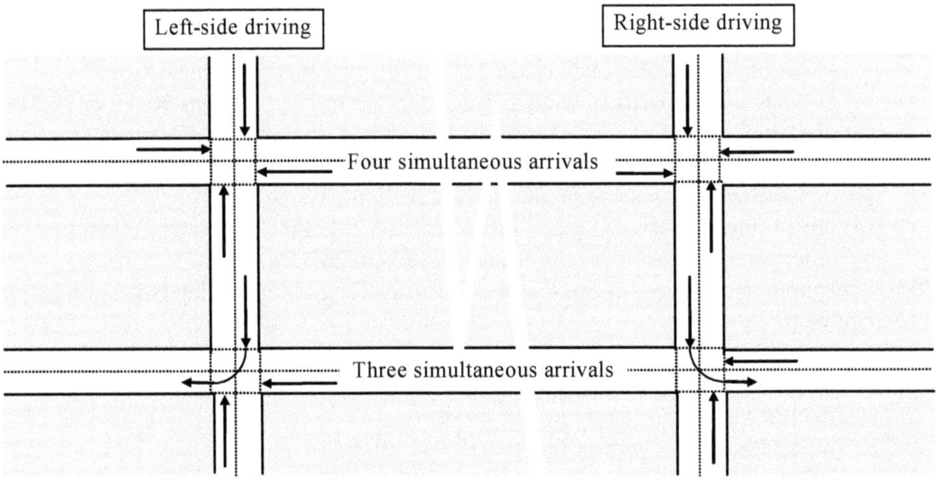

Figure 20.6 Unresolved conflicts with give-way-to-the-right (from Quayle, 1980).

20.2.3 Safety

Intersection safety is discussed in an overall context in Sections 28.1.2 and 28.5d. Typical intersection crash rates per 10^9 vehicle involved are as follows (van Every, 1982);

Traffic situation	Crash rate per 10^9 vehicle
Cross-road intersections	600
T-intersections	350
diverging	20
merging	15

A particular long-standing safety problem with uncontrolled local intersections is that some drivers approach them at too fast a speed to safely stop if an unexpected hazard does occur on the intersecting street. The approach speeds increase as the traffic volume on the orthogonal street decreases, suggesting that most drivers indulge in deliberate risk-taking at uncontrolled intersections and rely on evasive action to avoid some problems. The most direct solution to this problem is to use Give Way or Stop signs to give priority to one street (Section 20.3.8).

Once a vehicle is within an intersection, the use of 'protected' turn slots (Section 20.3.3) can reduce crash rates by up to 50 percent (McLean, 1997).

20.3 INTERSECTION DESIGN

20.3.1 Planning

The planning of an intersection should consider the following factors:

(a) the cost of the works, including land acquisition costs,
(b) the surrounding topography and environment and any landscaping needs,
(c) the pavement cross-section and lane configuration of the approach roads,

(d) the expected traffic flows, movements, composition (Section 17.4.3f), and temporal patterns including:
 * the future rate of growth of each of these variables (Section 31.5), and
 * the role of pedestrian, bicycle, truck and bus traffic (Sections 30.4.2 & 30.5),
(e) the alignment, grade, and visibility restrictions on the approach roads (Sections 6.2.1, 7.3.2 and 19.4),
(f) vehicle speeds on the approach roads (Section 18.3), and
(g) given their relatively poor safety record, whether the intersection can be eliminated or sited in a safer location (Section 20.3.2).

The latter three observations mean that a driver travelling within an intersection must be provided with:

1. *Good vision*
 (1.1)advance warning of the intersection,
 (1.2)a clear view of the entire intersection,
 (1.3)adequate sight distance (Sections 19.4 & 20.2.3),
 (1.4)easy identification and tracking of other vehicle movements and thus an awareness of other relevant, potentially-conflicting vehicles within and approaching the intersection;
2. *Helpful design*
 (2.1)a minimum number of conflict points,
 (2.2)a manageable level of demanding tasks (Section 20.2.1),
 (2.3)sufficient time to respond to information inputs (Section 16.5.3),
 (2.4)clear directions as to the required manoeuvres (Section 20.2.1), leading to confidence in the course required to negotiate the intersection, and
 (2.5)consistent traffic engineering treatments.
 Factor (2.1) will be largely covered by the geometrical arrangement of the intersection (Section 20.3.2) and factors (2.2–5) by traffic control devices such as signs (Sections 20.2.3 & 20.3.8) and signals (Section 23.1).
3. *Reasonable expectations*
 (3.1) no need for prior knowledge of the details of the intersection,
 (3.2) no requirement to perform unusual driving manoeuvres,
 (3.3) a clear order of precedence and priority within the intersection, and
 (3.4) the design must particularly assume that drivers will:
 * act according to habit,
 * tend to follow natural, short paths, and
 * often become confused when surprised.

The geometric design of an intersection based on these factors is discussed in the following section.

20.3.2 Layout

From the discussion in Section 20.3.1, it will be seen that any general layout of an intersection should consider the following features:

A. General layout issues
 (a.1) property boundaries and building lines;
 (a.2) property access (Section 6.2.4);
 (a.3) grade-separation of some or all traffic movements (Section 20.1.2),

(a.4) the horizontal and vertical geometry of the intersection and its approaches (e.g. swept paths for horizontal curves, Section 27.3.8; and dips and crests for short vertical curves, Section 19.3);

(a.5) special facilities for trucks and other commercial vehicles. Turning requirements are often critical and require a knowledge of truck turning circles (Section 27.3.7);

(a.6) channelisation within the intersection and medians on the approaches (Section 22.4.2).

B. Specialist features

(b.1) drainage, including managing large flat areas (Sections 7.4.3, 13.3 & 4).

(b.2) traffic signs and signals;

(b.3) safety aspects of trees and traffic signal, lighting, and utility poles (Sections 6.4, 23.2.1 & 28.6.1);

(b.4) parking provisions (Section 30.6);

(b.5) lighting effectiveness (Section 24.3);

(b.6) overhead obstructions; and

(b.7) public utilities (Section 7.4.3).

The layout of grade-separated intersections was discussed in Section 20.1.2. This section concentrates on at-grade intersections which are important in their own right and which form key parts of most grade-separated intersections. They may be defined as T, Y, crossroads, or multi-way depending on the number and relative orientation of the legs.

T intersections are relatively safe, with traffic on the non-through leg usually being forced to slow down to safe speeds by both traffic priority rules (Section 20.2.2) and inherent driver response. In rural areas their safety record is enhanced by the provision of good sight distance, whereas a sight distance restriction can be an effective speed deterrent in an urban situation.

The Y-interchange was illustrated in Figure 20.1a. In the simple Y-intersection shown in Figure 20.7, the crossing of two traffic streams at included angles of less than 75° can lead to relatively severe high-speed, head-on crashes, particularly if traffic flows are high. The problems associated with this arrangement are:

(1) poor visibility for the driver looking out of the vehicle towards the rear in order to check merging vehicles,

(2) right-of-way confusions,

(3) large areas of potential conflict, and

(4) long crossing distances for slow-moving vehicles and pedestrians.

A technique for converting these Y-intersections into the safer T-intersections is shown in Figure 20.7.

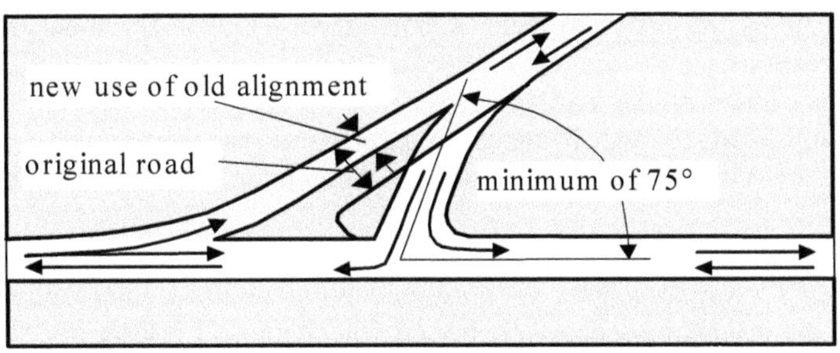

Figure 20.7 Conversion of a Y-intersection into a T-intersection.

Figure 20.8 shows similar techniques for treating oblique crossroad intersections to reduce the four problems listed above. The methods involve creating a more orthogonal intersection form (Figures 20.8a&b), or dividing the intersection into two staggered T intersections (Figure 20.8c). Each of these solutions can produce their own new problems, as discussed below.

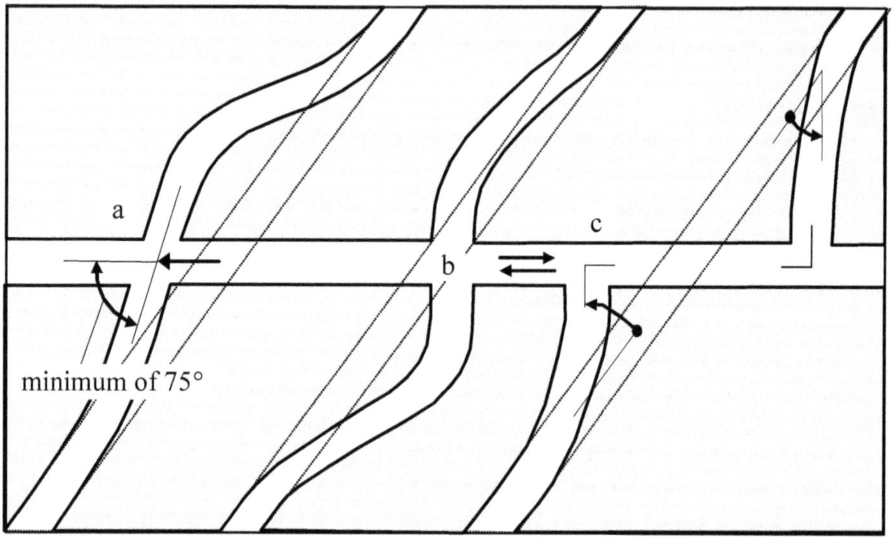

Figure 20.8 Three methods for converting oblique cross-roads into T-intersections. The original alignment is shown dashed. (a) and (b) may be constructed at intersection angles of as low as 75°.

Simple crossroad intersections can also create crash hazards and have a poor crash record at sites with low traffic volumes. Solution (a) should only be used when a protected centre-turn lane exists. This centre-turn effect can be minimised by a suitable orientation of the stagger, as shown in Figure 20.9.

As suggested in Figure 20.8c, crossroads can be converted to pairs of T-intersections, called staggered T's. There is now some doubt as to whether this treatment has a net safety benefit. The method is most effective when the traffic flows on the various legs are unequal. Roundabouts (Section 20.3.6) are a preferred solution when the flows are about equal. Both these treatments make the presence of main road traffic more obvious to the crossroad traffic and also automatically slow down any crossroad traffic. A problem with the staggered T arrangement in Figure 20.8c is that it can create hazardous centre turns soon after the driver has joined the main road stream. However, even this alternative requires the driver to make a simultaneous judgement about traffic in both main road directions.

Multi-way intersections involve at least five approach roads meeting at one point. They can confuse drivers and, if unavoidable, are best managed by using a roundabout (Section 20.3.6). Other methods for treating these intersections are shown in Figure 20.10.

Intersections on a major road should be kept at least 5 seconds of travel apart to relieve a driver's information load. As illustrated in Figure 20.11, intersections should not be placed on the inside of a sharp curve because the alignment restricts the sight distance. Better locations are on tangents or at sites such as A and B on the outside of curves. Even A and B may be inadvisable due to the superelevation of the main road tilting its pavement out of the line of sight of vehicles on the side road.

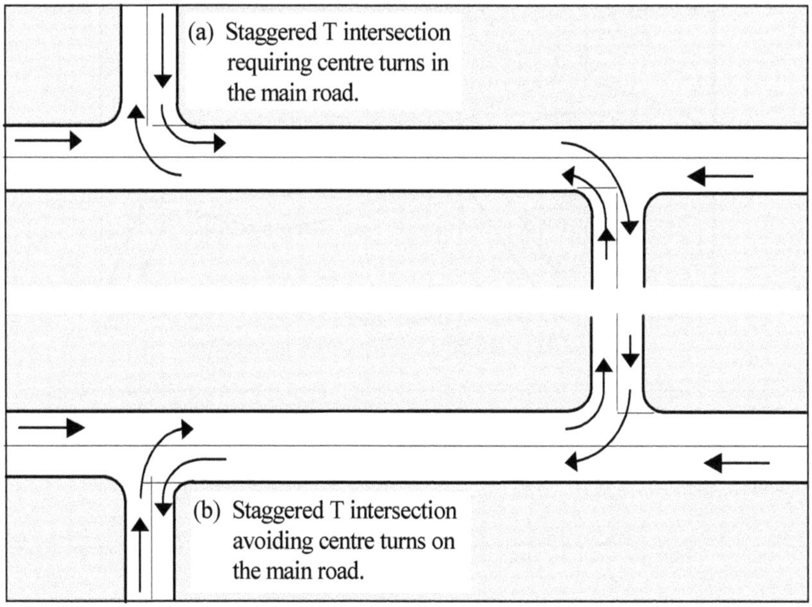

(a) Staggered T intersection requiring centre turns in the main road.
(b) Staggered T intersection avoiding centre turns on the main road.

Figure 20.9 Centre turns with staggered T intersections. Reverse drawing for right-hand driving.

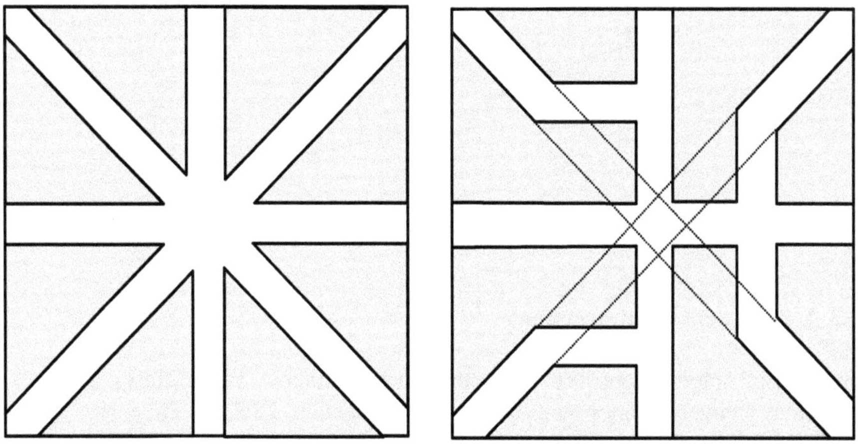

Figure 20.10 Methods for converting a multi-way intersection (left) into a crossroads intersection (right). The original alignments are shown dashed. Refer also to the jug-handle solution in Figure 20.2.

Similarly, Figure 20.12 shows how intersections on crests of hills could be avoided, due both to poor sight distance and to reduced acceleration response providing poor intersection capacity.

The types of at-grade intersection, as defined by treatment type, are discussed in Sections 20.3.3 to 6. Specific treatments for residential street intersections are described in Section 7.3.

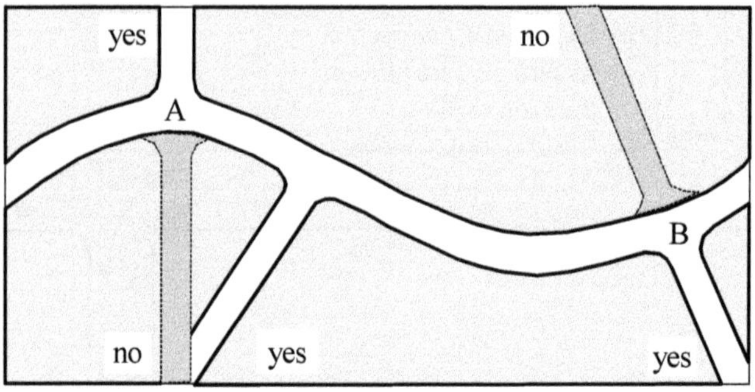

Figure 2.11 Horizontal location of new intersections.

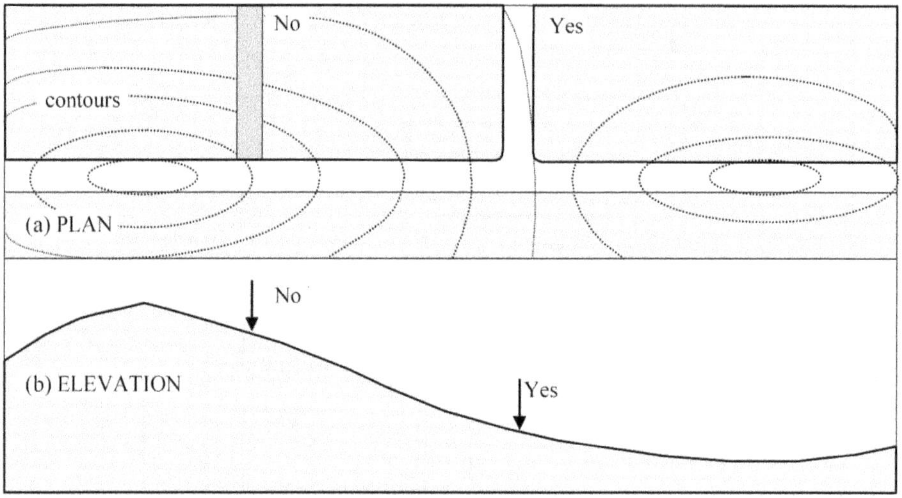

Figure 20.12 Vertical location of new intersections.

20.3.3 Unchannelised intersections

Unchannelised intersections are used for streets with low traffic flows and at some intersections between minor and major roads (Section 20.2.2). They are frequently controlled by Stop or Give Way signs (Section 20.3.8). It is common to improve the performance of such intersections by marking the lanes over at least the last 100 m or so prior to the intersection. For low turning flows, the turning vehicles use the same lane as the through vehicles: this is called a *shared lane*. A *trap lane* is a conventional through lane that becomes a no-option turning lane.

The need for a special *turning lane* (or *slot*) to provide for turning vehicles and thus reduce the impedance that they cause to through traffic will depend on the:

* traffic flows in both directions — one traffic stream creates the turners and the other, opposing, stream causes the delays to both the turners and to any intended through traffic trapped behind them, and
* the flow on the cross street which provides the gaps into which the turners must merge.

Typically, every 150 veh/h of opposing traffic causes 1 s of delay to a centre-turner. Similarly, the introduction of turn lanes will increase through-capacity in almost a direct link to the amount of turning traffic. The tcu of 3 for a centre turn in Table 17.2 suggests a turning lane capacity of 1800/3 = 600 veh/h, whilst experience suggests that turn-lanes are required at closer to 300 veh/h. When turn demand increases two, or even three, centre-turn lanes may be used. This provision of multi-turn lanes both increases capacity and decreases the time needed to serve the turning traffic.

When through lanes are not running at or near capacity, it is often desirable in terms of both total throughput and crash reduction to convert a through lane into a dedicated centre-turn lane.

The turning lane can frequently be created without resort to property acquisition or new construction. Recall from Section 6.2.5 that if 3.7 m lane widths are recommended on the 'open road', then a four-lane road — nominally 14.8 m wide — can be subdivided into a five-lane road using three 2.8 m lanes and two 3.2 m lanes (Figure 20.13). The narrower lanes should not pose a problem to drivers as it is assumed that the drivers using them will have slowed down (Section 16.5.1).

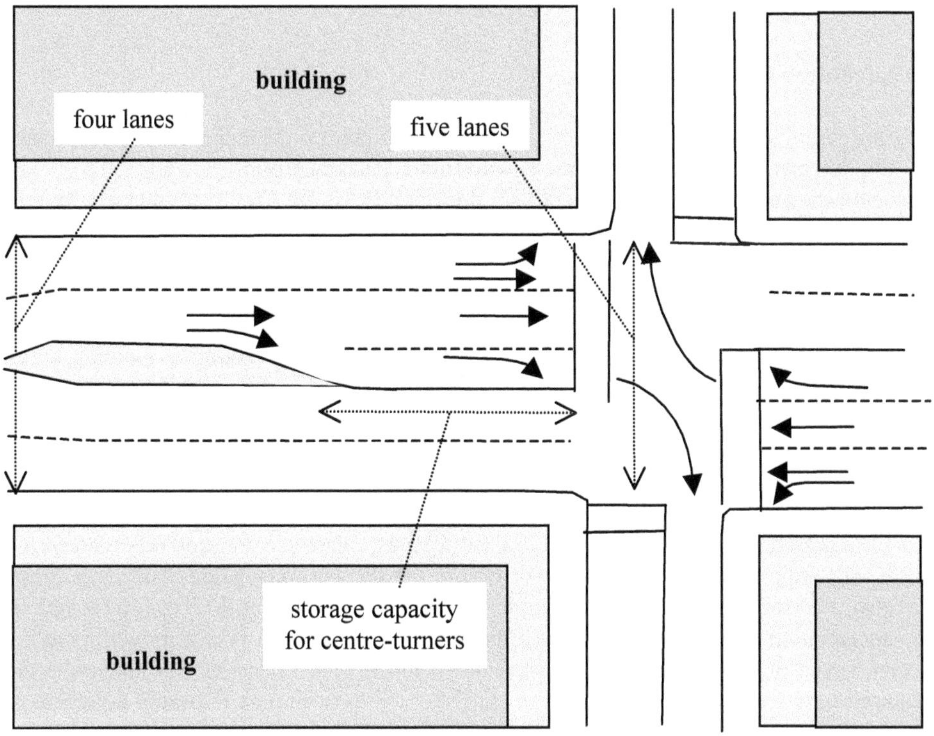

Figure 20.13 Typical intersection improvements using pavement markings to create a centre-turn lane. Mirror image for right hand driving.

Centre turns are commonly line-marked as *non-locked* (or near-side, Section 20.2.2) *turns*, as indicated in Figure 20.14. This is usually about 50 percent safer than the *locked* (off-side) alternative, but requires attention to ensure that pedestrians are protected from the turners. Opposing centre-turn lanes on either side of an intersection can result in reduced sight-distance for a turner if the opposing lane is occupied. Thus there is a risk of increased crashes between turners and opposing traffic.

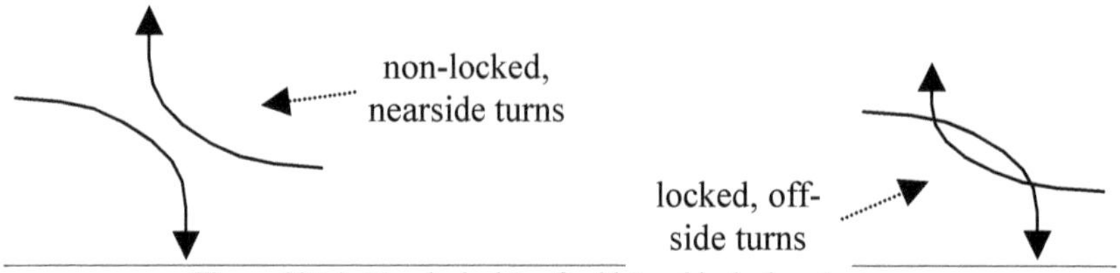

Figure 20.14 Non-locked (preferable) and locked centre turns.

When there are frequent needs for centre turns into side streets or abutting property, the centre lane is sometimes continuously devoted to centre turns from either direction. This technique gives turners unrestricted access but may cause some conflict with other turners. The arrangement has a poor crash record when there are more than two through lanes in each direction. The alternative of turning lanes in medians (above, and Section 22.4.2) provides more control but less flexibility.

20.3.4 Flared intersections

Flaring refers to the use of tapers or short additional lanes outside the normal boundaries of the intersection. It may be used with both unchannelised (Section 20.3.3) and channelised (Section 20.3.5) intersections. If turning flows are low, a simple taper may be adequate. Otherwise, widening the pavement outside the line of the through carriageway provides additional storage and/or speed-change lanes (for the turning vehicles) or passing lanes (for the through vehicles). This prevents slowing, turning vehicles from delaying through traffic by moving them from the through lanes, thus improving safety, and also provides extra storage capacity for these vehicles by acting as holding areas. Two related issues sometimes arise:

* vehicles are prevented from entering the turning lane by the queue in the through lane, thus delaying turning traffic, or
* the queue in the short turning lane backs up into the through lane, thus delaying through traffic.

Longer versions of these lanes are described as either *acceleration* or *deceleration lanes*, depending on their usage after or before the turning manoeuvre. The same terms are also used for merging and diverging lanes on freeways (Section 17.3.6). The length of the deceleration lane must at least equal the length required for a vehicle travelling at the design speed of the through road to slow down to the speed necessary to undertake the required turn. Similarly, an acceleration lane must be long enough to permit a vehicle to accelerate from the turn speed up to the design speed of the road to be joined. Data on vehicle acceleration and deceleration characteristics are given in Section 27.2.2. The distances involved can be quite large. For example, design codes typically require about 300 m to permit a vehicle to accelerate from 40 km/h to 100 km/h. Deceleration lanes less than the desired length can present safety hazards to trucks, particularly when coupled with tight curves or steep down-grades.

20.3.5 Channelised intersections

Channelisation provides a defined and separate path through an intersection for each traffic movement, removing a major source of confusion from the driver's mind and also making the driver's behaviour more predictable by other drivers. Channelisation is usually achieved by pavement markings (Section 22.2), delineators (Section 24.4.3), and such physical separators as small, raised-medians (Section 22.4.2) which positively protect the lane from other traffic movements.

Channelised kerb-turning acceleration/deceleration lanes are called *slip lanes.* They are typically justified when there are more than 40 kerb-turners per hour. Figure 20.15 shows a channelised T intersection of a minor road with a distributor. Figure 20.16 illustrates a solution (the *seagull*) for the case where the second road is a major divided-arterial with more room for, and a greater need for, channelisation.

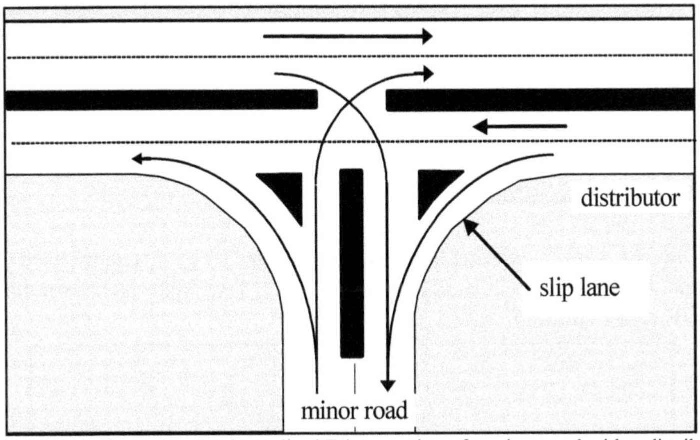

Figure 20.15 Example of a channelised T intersection of a minor road with a distributor road. Mirror image for right hand driving.

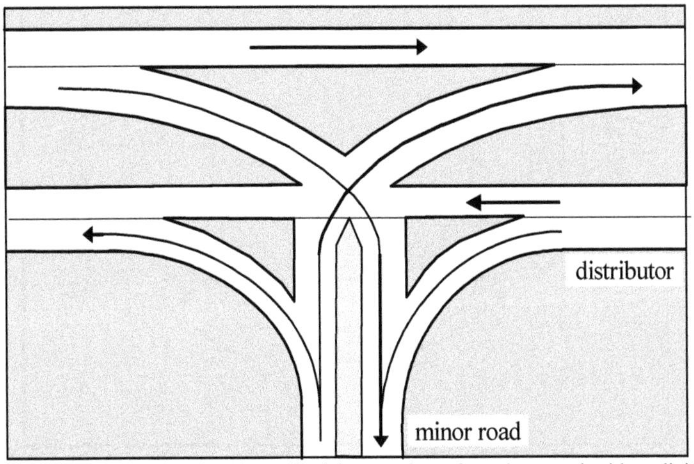

Figure 20.16 Example of a channelised intersection of a minor road with a divided arterial road (the seagull intersection). Mirror image for right hand driving.

The slip lane usually requires a generous radius of curvature and a small raised island (Section 22.4.1) to separate it from the other lanes and provide a pedestrian refuge. A common design basis is that entry to a slip lane should not require a lateral movement in excess of 1 m/s. An alternative to providing a generous curve radius is to use a much longer entry lane on the approach road and a tighter radius in the actual curve.

These features, and the operation of give-way rules and signs, may avoid the need for traffic signals. By reducing conflict areas, drivers can negotiate smaller gaps in other flows and pedestrians can cross with greater safety. Experience indicates that channelisation can reduce crashes at problem intersections by up to 40 percent. However, channelisation can often be a problem for large vehicles due to turning path problems (Section 27.3.7).

Slip lanes used in conjunction with a '(kerb) turn any time with care' sign can significantly improve the capacity of the intersection (Sections 17.4.3 & 23.6.3). One advantage of the technique over '(kerb) turn on red' (Section 23.2.5) is that the turning vehicle does not need to stop in light traffic. Its main disadvantage is the extra cost of land acquisition.

A comprehensive review of channelised intersections is given in Neuman (1985).

20.3.6 Roundabout operation

A *roundabout* (or *rotary intersection*, or *traffic circle*) is a channelised intersection serving at least three concurrent roads (Figure 20.17). Two interchange roundabouts were illustrated in Figures 20.1f&g.

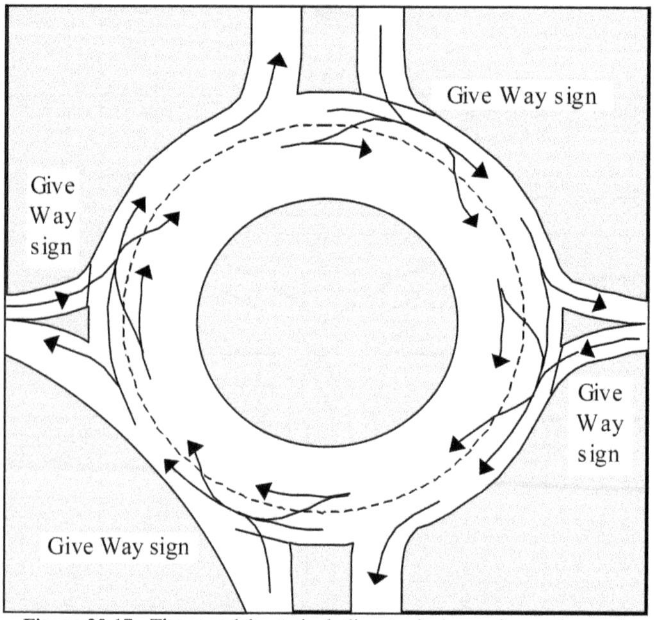

Figure 20.17 The roundabout, including typical merging and weaving movements required of drivers.

A roundabout can be so small that the central island is little more than a 1 m diameter painted circle or raised spherical segment. This is called a *traffic dome*. A roundabout is usually considered 'mini' if the diameter of the central circle is below 4 m, and 'small' if it is between 4 and 25 m. On the other hand, a roundabout can be so large that the layout resembles a circular road with a series of T-intersections or mini-roundabouts at each entering leg. The last case is called a *ring intersection.* It is also possible to use multiple roundabouts, sharing common lengths of circulating flow, in order to accommodate a series of close intersections.

The roundabout simplifies most intersection manoeuvres. Entry into the roundabout depends on the give-way rule (Section 20.2.2) which must apply to each entry leg to ensure that traffic already circulating has priority over traffic wishing to enter. This is particularly the case in right-hand drive countries where the give-way-to-the-right rule would otherwise give entering traffic priority over traffic already in the roundabout, and thus quickly clog the intersection. Standard give-way signing is therefore used at each entry point (Section 20.3.8). Thus, a driver enters the roundabout by using gap acceptance (Section 17.3.4) and the critical gaps discussed below are between those associated with crossing a traffic stream (6 s) and with merging into it (1.5 s). The entries to the roundabout should therefore be flared to allow drivers to merge without stopping.

The priority rule means that a fundamental difference between a roundabout and a conventional intersection is that in a roundabout a turner receives priority over through traffic. The reverse, of course, is true in a conventional intersection.

The key to the internal operation of a roundabout is that the traffic travels clockwise around a central island (Section 22.4.1) when driving on the left, and anti-clockwise when driving on the right. This process reduces and sequences the number of conflict points that the driver encounters. Within the roundabout, drivers have the advantage of being able to merge, weave, and diverge at a relatively uniform speed. The speed will also be below the approach road speed, thus lowering the severity of any impacts.

Roundabouts have good safety records and typical studies show statistically significant reductions of at least 60 percent in total crashes after the introduction of roundabouts on main roads and 30 percent on local streets. As with signal installations (Section 23.1), there is often a slight increase in rear-end crashes and a major decrease in right-angle crashes. Crashes involving turning vehicles also decrease. Crash rates increase, approximately, with the square root of the number of vehicles using the roundabout.

Typically, a majority of the severe crashes will involve vehicles striking motorcyclists, cyclists, or pedestrians from behind. A common bicycle crash is for an entering motor vehicle to hit a bicycle already circulating in the roundabout. Multilane roundabouts are relatively dangerous for these road users. The crash risk is reduced by lowering the entry speed for motor vehicles.

Replacing a conventional intersection with a roundabout will usually lower net fuel consumption.

The advantages of roundabouts relative to traffic signals have not always been recognised. Roundabouts are particularly suitable for intersections:

(a) where it is desired to give minor street traffic more priority than they would receive from Stop or Give Way signs (Section 20.3.8),
(b) on streets, distributors and rural roads with poor crash records but where the flows are too low to justify a more elaborate treatment,
(c) where traffic signals would cause excessive delays, e.g. where the intersection would require more than three signal phases,
(d) which are of the multi-way, T- or Y types (Section 20.3.2),
(e) with high centre-turning flows,

(f) where it is important to give good service to low-volume off-peak flows, and
(g) where signal maintenance would be a problem.

Roundabouts can be difficult where:

(h) there is not sufficient space for an adequate geometric layout,
(i) the traffic flows on the various legs are unbalanced to the extent that there is not sufficient circulatory traffic to break up the predominant movements and permit traffic on other legs to enter the roundabout,
(j) vehicular, cycle or pedestrian flows are high,
(k) the intersection is within an area of traffic control or linked arterial signal system (Section 23.4–5),
(l) over-dimensional vehicles are common,
(m) all intersecting streets are multilane, and/or
(n) drivers do not slow down on approaching the roundabout.

20.3.7 Roundabout design

Design objectives for roundabouts usually fall into one or more of the following three categories:

(1) *Amenity*: roundabouts can reduce vehicle speeds and discourage high-speed cars, large vehicles and through traffic. Small roundabouts aimed at doing this are discussed in Section 7.3.
(2) *Safety*: See Section 20.3.6. Motorcycle, cycle and pedestrian crashes may increase unless special consideration is given to their needs. For instance, splitter islands can be provided on any hazardous leg to create a mid-crossing refuge for pedestrians and cyclists. Cyclists can also be aided by providing separate cycle paths outside of the roundabout.
(3) *Reduction of delays* at uncontrolled intersections. Turning movements are helped, but efficient overall operation requires reasonably balanced flows on the various approach arms.

A prime design requirement is that a roundabout should be 'readable' to the approaching driver wishing to navigate through it. This requires special attention to visibility.

A related key design requirement for a roundabout intended to meet objective (1) or (2) is that it should deflect the path of a vehicle passing through an intersection. This deflection 'naturally' reduces speeds and causes geometric delay (Section 20.1.1). It is achieved by:

* the use of a sufficiently large central island,
* aiming the entry lanes at the centre island, thus forcing the traffic to deflect (and slow),
* the use of 'splitter' islands in the joining arms, and
* staggering the alignment of entries and exits.

In small roundabouts where there is no imposed lane discipline, the drivers' paths can best be described by the clothoid (Section 19.2.7). In terms of the driving task (Section 16.5.1), drivers will initially make a navigation decision to decide on a course through the roundabout. They will then use tracking behaviour based on a sequence of guide points within the roundabout in order to pass through it. It is therefore essential to provide drivers with both navigational cues and a sequence of guide points.

As discussed in Section 20.1.1, the delays at a roundabout will comprise the above geometric delays plus traffic delays, as each vehicle is required to wait for an acceptable gap. Indeed, as roundabouts operate on the basis of gap acceptance, their capacity can be calculated using gap-acceptance techniques (Sections 17.3.3), treating each entry leg as a separate T intersection. Critical gaps and the following minimum entering-headways (Section 17.4.3c) of 3.5 & 2.0 s have been observed, leading to Equation 17.32 for entry capacity becoming:

$$(\text{entry capacity})/q_{cf} = e^{-3.5q_{cf}}/[1 - e^{-2.0q_{cf}}] \quad (20.1)$$

where q_{cf} is the circulating flow within the roundabout in veh/s. The follow-headway (Section 17.3.4) decreases as the diameter of the roundabout increases. It is typically about 60 percent of the critical gap. Practical capacity is usually taken as 80 percent of the above prediction.

The SIDRA program (Section 23.6.4) is effective in assessing roundabout performance. It works on the above gap acceptance basis and assumes that the headways of the circulating traffic are represented by a bunched negative-exponential model (Section 17.3.8). However, this is not the approach adopted in the U.K. where the use of roundabouts is the most widespread. Their approach is to approximate Equation 20.1 with:

$$(\text{entry capacity})/q_{cf} = [(\text{entry capacity at } q_{cf} = 0)/q_{cf}] - f_c$$

where f_c is effectively a tcu (Section 17.4.3h) giving the entry vehicle equivalent of a circulating vehicle. The unknowns on the right hand side of the equation are determined by regression as functions of the site geometry. For example, they give f_c as:

$$f_c = 0.212(1 + 0.2e_e)t_d$$

and

$$(\text{entry capacity at } q_{cf} = 0) \leq 303e_e$$

in veh/h, where e_e is the effective entry width in m and depends on the approach width, the actual entry width and the length over which the entry place is developed, and t_d is a function of the inscribed circle diameter of the roundabout (Kimber, 1980). During high circulating flows this method usually predicts higher entry flows than experience would suggest are likely to occur.

Capacities, queues and delays at roundabouts based on this approach can be calculated using the ARCADY program, first released in 1981 (Semmens, 1985). ARCADY 6 was recently released. ARCADY lumps the traffic in individual lanes into a single approach flow, and this, together with its empirical basis, limits its effectiveness.

When the circulating flow or the geometry of the site demands that the circular carriageway be multilane, the question arises as to the lane marking to use. Many jurisdictions only begin lane marking when the carriageway is three lanes wide, and experience no operating problems. At the other extreme, a few have experimented with forms of spiral lane marking, such that the kerb lane is always required to leave the roundabout at the next exit. Without helpful line markings, many drivers are reluctant to use the inner circular lanes.

Traffic signals are sometimes used on the approach legs of busy roundabouts to:

* cover peak hour conditions,
* ensure that all entry legs receive sufficient priority, thus balancing the delays caused by unbalanced flows,
* control queue lengths,
* regulate traffic patterns (particularly weaving), and
* provide safe crossing for pedestrians and cyclists.

20.3.8 Give Way and Stop signs

Signs are discussed generally in Chapter 21 and intersection priority in Section 20.2.2. This section discusses the use of Give Way (or Yield) signs and Stop signs to define priority at particular intersections. When such signs are used on the minor road, it is common to place *Stop lines* (or *holding lines*) transversely across the minor road on the line of the outside edge of the major road carriageway (Sections 22.3 & Figures 22.2c&d). (Advanced stop lines for cyclists are discussed in Section 22.3.) Drivers must be able to see the associated signage sufficiently well in advance to permit them to stop if necessary at the Stop line. Thus there must be adequate sight distance (at least ASD, Section 19.4.3) for the signs on the approach leg.

A driver approaching a Give Way sign is only obliged to stop at the Stop line if required by the give-way rules (Section 20.2.2) or if proceeding would be unsafe. Thus the Give Way sign is used where good intersection and entering sight distances (Section 19.4.7) mean that a complete stop is not essential. 'Good' usually means that the driver can see any possible conflicting cross-traffic well before reaching the Stop line and this distance is defined in some detail below.

A driver at a Stop sign is required to stop at the Stop line, check the traffic on the crossroad, and then only enter or cross the crossroad if it is safe to do so. Stop signs are used if:

* either the intersection sight distance or entering sight distance is unsatisfactory and cannot be improved,
* for safety, a complete stop is necessary before entering the intersection,
* the speed environment is being unexpectedly lowered, or
* Give Way signs have been tried, but with a poor crash outcome.

Good practice is to consider the Give Way sign first and to use the Stop sign as a fallback option.

Sight distance was discussed generally in Section 19.4. The basis for intersection sight distance calculation is shown in Figure 20.18. The two triangles ABC and FED are commonly known as *intervisibility triangles*. The first step is to decide which vehicles may not stop on approaching the intersection. Commonly, these will be vehicles p and q on the priority road, which has a free speed of v_P (Section 18.2.5).

The vehicles that may need to stop are vehicles m and n travelling on the minor (non-priority) road. Their stopping distance, D_{sm}, and time to stop, T_{sm}, are functions of their speeds, v_m and v_n, and can be obtained from Table 19.1 or Equations 19.4 and 19.6. If A_i is the distance from the driver's eye to the front of the car, then the distances:

for legality, $AG = D_{sm} + A_i$

for safety, AC and $DF = D_{sm} + A_i$

define the last points, A & F, at which drivers in vehicles m and n respectively have an option to stop. The available sight lines, AB and FE, are usually measured from driver eye height of vehicles m and n (typically 1.1 m above the pavement, Section 19.4.1) to a similar height on vehicle p or q at the centre of their lane (although the figure shows a slightly more realistic condition). The visibility distances along the priority road are the distances BC and ED defined by the option points A and F and the available sight lines AB and EF.

Before passing A and G, vehicles m and n therefore need a clear view that ensures that no priority road vehicle, p or q, will:

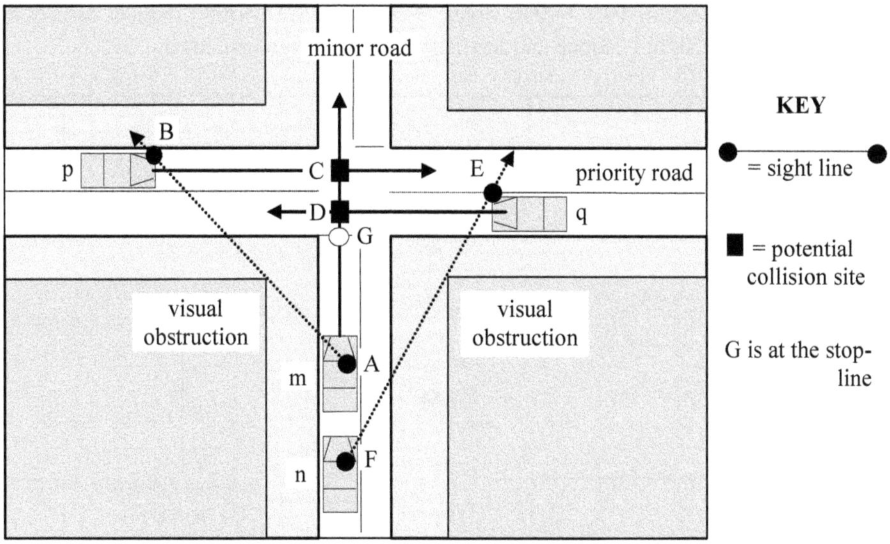

Figure 20.18 Intersection sight distances. Mirror image for right-hand driving.

(a) impact on their path if they proceed straight across the intersection (intersection sight distance), or

(b) run into their rear if they do a right or left turn into the priority road (entering sight distance).

Both cases require a minimum clearance gap between vehicles (Section 17.3.4) to provide a safety margin.

Case (a) is covered by the requirements calculated above. Once drivers are within A and/or F they must slow down so that they can stop at G rather than impede any vehicle that looms into view on the priority road. Case (b) also requires the entering vehicle to have sufficient time to accelerate up to the speed of the priority road.

Driver m must be able to see vehicle p at least {$v_P T_{sm}$ + the minimum clearance gap} away for a Give Way sign to be satisfactory (Figures 20.18 & 19), i.e. a Give Way sign requires:

$$BC > v_P T_{sm} + \text{the minimum clearance gap}$$
$$DE > v_P T_{sm} + \text{the minimum clearance gap}$$

It is also necessary to similarly check that the sight distance from G gives vehicle m sufficient time to cross the road from a standing start at the stop line (case a above) without being hit by a vehicle on the priority road.

Note from Table 19.1, that vehicle times to stop are about 10 s, whereas a vehicle travelling at 80 km/h will have travelled over 20 m in this, so requirements based on time to stop will usually lead to very safe outcomes for the average driver. Some jurisdictions simply merely assume that A and F are located about 3 m prior to the edge of the carriageway being crossed.

If the intersection is uncontrolled (Section 20.1.3) and it is not clear that one of the streams of traffic will have priority, then it is also necessary to reverse the above process and check CA, CF, BC, and DE against the stopping distance for vehicles p and q.

Intersection sight distance can be improved by controlling the location of the visual obstructions which are sight distance limiters in Figures 20.18 & 19. In residential areas, this is usually done by controlling the height of corner fences so that they are below driver eye height (1.1 m, Section 19.4.1). However, it may not always be practical to provide adequate intersection sight distance and in such cases Stop signs are used.

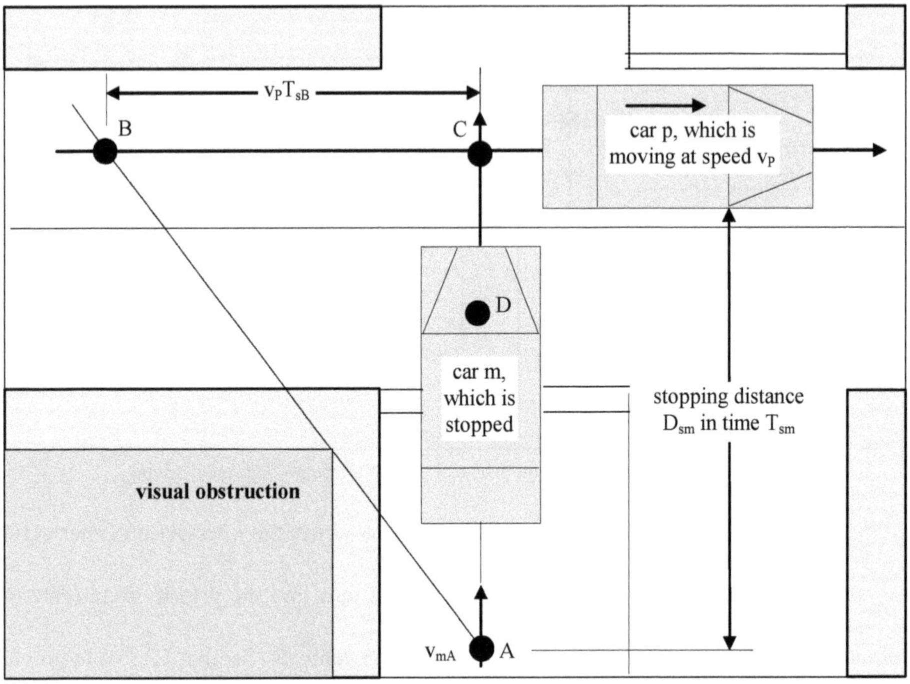

Figure 20.19 Sight distance geometry. ABC is the intervisibility triangle.

A typical set of practical guidelines for when a Stop sign is required by regulation in lieu of a Give Way sign is given in Table 20.1.

Table 20.1 Typical set of practical guidelines for when a Stop sign is required.

Operating speed of major road, V_P (km/h)	Assumed value of AD if not given by text or local conditions, (m)	A Stop sign considered in lieu of Give Way sign if sight distances BC or DE are less than these values (m)
60	10	35
75	10	45
80	16	70
90	16	80
100	16	90

The sign placement systems used in a controlled intersection are:

(a) use of the appropriate signs on the minor streets in the road hierarchy only, allowing traffic on the major streets to have automatic priority, visible to them via transverse line markings on the entries of the minor streets. This is consistent with overall traffic planning methods (Sections 7.2.2 & 20.2.2).

(b) use of the appropriate signs on all entries. This is not recommended over (a) where intersection sight distance is satisfactory, as it will give redundant information. In difficult cases, if one route is controlled by Stop signs with all vehicles on it coming to a halt at the Stop line (Section 22.3), there will always be adequate sight distance to at least justify Give Way signs on the other legs: or

(c) use of Stop signs on all entries, irrespective of sight distance. This is the sometimes controversial *four-way stop*, which is widely used in North America, particularly when the minor street traffic flows are at least 35 percent of the total flow through the intersection. The main objections to the technique are:
  * the unnecessary delays that it causes,
  * that it diminishes respect for the Stop sign, and
  * that it uses the Stop sign in places where a stop may not be essential (e.g. where sight distance is satisfactory).

20.4 PEDESTRIAN CROSSINGS

Pedestrian safety is discussed generally in Section 28.7. Most road laws require that pedestrians intending to cross a road give right-of-way to motor vehicles, except where the pedestrian has specific priority, as at many intersections and at designated pedestrian crossings. This rule is sometimes reversed in residential precincts (Section 7.2.4).

Pedestrian crossings are special features used to help pedestrians to cross a road. They should only be installed where:
  * visibility is adequate,
  * the 85th percentile vehicle speed is not too high, and
  * no more than four lanes need to be crossed at any one stage.

The kerbside requirements for a pedestrian crossing are discussed in Section 7.4.4. The main types of pedestrian crossings are as follows:

(a) No crossing at all. In this category, the pedestrian has no right-of-way and must cross by using gaps in the traffic stream (Section 17.3.1). Citizen concern at this arrangement occurs when the traffic flows exceed about 300 veh/h (Section 7.4.1).

(b) *Safety zones* (Section 22.4.4), which are pedestrian refuges in the centre of the road, either at pedestrian crossings or as crossable median strips (Section 22.4.2). These allow the pedestrian to cross in two stages and hence worry about only one direction of traffic at a time. They usually produce a significant reduction in pedestrian crashes.

(c1) Marked crossings, possibly with warning devices, but without protective signs or signals. These can be relatively unsafe and are not recommended.

(c2) *Zebra crossings*, so named because of the striped road marking defining where the pedestrian should cross. In the U.S. they are called *marked crosswalks*. Pedestrians on the zebra marking have priority over motor vehicles. Zebra crossings are only used midblock, and never at intersections.

There is commonly a 'pedestrian crossing' sign on the roadway in advance of the crossing. In addition, there is often some form of distinguishing signal in the form of illuminated signs or pedestrian symbols or flashing yellow or amber lamps. In the U.K. this lamp is called a *Belisha Beacon* and there are zigzag markings on the road approaching the crossing.

A problem with zebra crossings is that drivers do not always notice pedestrians crossing mid-block and sometimes do not feel obliged to comply with the requirement to give-way to mid-block pedestrians. Zebra crossings are generally regarded as unsatisfactory when the 85th percentile vehicle speed exceeds 75 km/h or when more than four traffic-lanes must be crossed. In such circumstances, and when flows are high, drivers are even more reluctant to concede priority to any pedestrians. Another objection to the zebra crossing is that it cannot be made part of a linked traffic signal system (Sections 23.4–5).

Crossing safety can be enhanced by having pedestrian refuges at each kerb and at the centreline of the road.

(d) Conventional vehicular traffic signals where the pedestrian must follow the same phasing and use the same circular signal discs as the vehicular traffic. The pedestrian is thus required to walk with the green turning traffic. The situation presents risks to the pedestrian which are usually alleviated a little by over-riding local priority laws to favour the pedestrian.

(e) Pedestrian-operated signal crossings.

(e1) Conventional pedestrian-operated signal crossings. These are based on conventional vehicular traffic signals but will show green to the vehicular traffic unless activated by a pedestrian. The response to the pedestrian's request can be delayed by the needs of the overall signal system (Sections 23.4–5). The signals are additionally equipped with special pedestrian features including activation controls that are easily accessible to all pedestrians and audible crossing signals for the visually impaired (Section 23.2.3). Pedestrians are further helped if the system gives a countdown indication of the remaining crossing time available (usually in seconds).

As this system will give pedestrians apparent right-of-way, it is necessary to ensure that the signals and any pedestrians are very visible to heavy conflicting traffic flows.

The vehicle stop-line is placed well upstream from the pedestrian crossing lines.

The flashing signal (Section 23.2.3(12)) has been successfully used in some jurisdictions.

Pedestrian-operated signals are relatively safe, but can cause four times as much vehicle delay as zebra crossings and twice as much delay as manned school crossings (Maclean and Howie, 1980).

(e2) The *pelican crossing* is a related mid-block device in which the yellow traffic-phase corresponds to a flashing-green phase for pedestrians, allowing them to complete but not initiate a crossing and permitting vehicles to pass through the crossing if they would not endanger pedestrians. It therefore reduces vehicle delays.

(e3) The *toucan* crossing shares the facility between pedestrians and cyclists.

(e4) The *puffin crossing* uses pedestrian sensors to:

* detect pedestrians waiting to cross at the site,
* extend the crossing green-phase in order to permit pedestrians to complete the crossing, and
* terminate the crossing green-phase if no pedestrians are waiting.

It can thus avoid:

* a crossing phase being introduced when no pedestrians are present, and
* the crossing signals remaining activated when the waiting pedestrian has already crossed against the red by using a gap in the traffic flow.

(f) Special crossings such as manned school crossings. Manned installations are usually highly effective.

(g) Overpass footbridges (Section 15.4.7) or underpass tunnels.

It is not possible to rank these types in order of desirability. Each proposed installation must be independently examined in terms of benefits and costs to pedestrians and to through traffic. Typical warrants for the installation and operation of some of the above devices, based on delay calculations for vehicles and pedestrians, are shown in Table 20.2.

The average reaction time of a pedestrian to a signal change is about 2 s (see Section 16.5.3 for background). Pedestrian walking speeds are given in Table 20.3. With these times and speeds, Equation 17.27 can be used to calculate the proportion of the time that a pedestrian would encounter delays when wishing to cross the flow at the warranted level of 600 veh/h. This data is shown in Table 20.4.

Table 20.2 Typical pedestrian crossing installation warrants.

Device	pedestrian/h, q_p, crossing within 20 m of site exceeds:	Traffic flow, q veh/h, exceeds[c]:	q_pq >
c2 Zebra crossing[a]	60[b]	600[b]	90,000
e1 Pedestrian-operated	350 (over 3 h)[d]	600[b]	–
e1 Ped-operat'd at school	50	600[b]	30,000
f School crossing	20	50	–

a But not used where it would be unsafe or interfere with traffic flow.
b In some jurisdictions the warrants is only half these values.
c Measured between kerb and median, or kerb and kerb if no median.
d Drops to 175 over 8 h.

Table 20.3 Free-flow pedestrian crossing-speeds[e, f, g] in m/s.

Pedestrian type	running	50th percentile	15th percentile
Peak value		1.8	
Children	6 (12 y.o.)	1.6	1.3
Adults	8	1.4[h]	1.1
Elderly		1.4	0.9
congested		0.8	0.8

e Grades of up to 5 percent have little effect on these speeds.
f Most pedestrians are able to increase their speeds by up to 40 percent under pressure (Moriarty, 1980).
g The data has a standard deviation of 0.3 m/s.
h This is associated with densities of about 0.8 person/m^2 and space headways of about 1 m.

Table 20.4 Factors influencing the effectiveness of a pedestrian crossing.

Factor	Width to be crossed (m)			
	3.5	7.0	10.5	14.0
Time to cross, t s	4.5	7.0	9.5	12.0
Vehicles in this time, qt in Equation 17.27	0.75	1.17	1.58	2.00
Proportion of times	0.13	0.33	0.47	0.58
Number of crossing gap/h, from Equation 17.24	280	190	120	80
Distance (m) covered in t s by a vehicle at 70 km/h	88	136	185	233

Given the 2 s average reaction time of a pedestrian, the accepted traffic gaps for lane crossing without signals, are 1.6 s (15th percentile), 2.6 (50th percentile) and 4.3 s (85th percentile). Table 20.4 suggests that the warrants in Table 20.2 might only apply when pedestrians begin to encounter significant delay. The last row also indicates that elderly pedestrians using a wide street might have difficulty seeing an approaching gap (Sections 16.4.2 & 24.4.1 suggest that a good young eye will detect a car at 1000 m).

Speeds for pedestrians crossing roads at intersections are given in Table 20.3. This implies lateral spacings of over 1 m, which equates to a peak pedestrian flow of 1.1 person/s/m-width of path. It is normally assumed that pedestrians occupy a minimum area given by a 600 x 450 mm ellipse, or about 2 person/m^2. This gives flow rates of about 1.3 person/s/m. People walking will usually detour around areas with more than this density. An ideal walking density on a footpath is as low as 0.1 person/m^2. However, pedestrians accept a density of 4 person/m^2 when waiting for a signal change. On the other hand, high proportions of elderly pedestrians or people with shopping or baby carriages will push these densities in the opposite direction.

If the planned facility is located near an establishment catering for disabled or old people, then the walking speeds and visibility distances used should be modified to take into account the needs of this user population.

It has been observed that most pedestrians tend to slow down when on a crossing and threatened by a vehicle. When a conflict does appear imminent it is, naturally enough, usually the pedestrian who accedes to the motorist.

Pedestrians are notoriously unwilling to use formal crossing facilities covered by (b) to (g) above, unless they are directly on the pedestrian's desired route. For this reason, Authorities sometimes instal kerbside barrier fences to prevent pedestrians from crossing a road at other than a designated crossing. They are often forced to do this by traffic injury statistics (Section 28.7) showing that most pedestrians are injured near, but not on, a pedestrian crossing. This statistic, and the need to provide drivers with adequate advance warning of a pedestrian's movements, also lead to the common practice of banning kerbside parking adjacent to and upstream of a pedestrian crossing.

Pedestrians can also be aided by a number of the physical treatments listed in Section 7.3 — such as footpath widening at a crossing point — which act to shorten crossing distances, increase visibility, and reduce vehicle speeds. Similarly, medians (Section 22.4.2) and traffic islands (Section 22.4.1) can very effectively reduce hazardous crossing distances. Kerb treatments at crossings are discussed in Section 7.4.3. Plateau-topped road humps can be used to provide an effective crossing point (Section 18.1.4).

CHAPTER TWENTY-ONE

Traffic Signs

21.1 TRAFFIC CONTROL DEVICES

21.1.1 The role of traffic control devices

Traffic control devices are provided to aid in providing safe, predictable, and orderly movement of traffic. The messages that they are intended to convey to drivers and pedestrians should be consistent with the other features of the road. Their specific use is usually governed by regulations and warrants developed by the relevant traffic Authority.

Traffic control devices include signs, signals (Chapter 23), pavement markings (Chapter 22), kerbing, traffic islands (Section 22.3.1), medians (Section 22.3.2) and other installations provided for road users for one or more of the following purposes:

(a) *To instruct or direct road users*, i.e. to provide instructions or regulations which are required by law to be obeyed and which might otherwise be overlooked. Such devices are called *regulatory devices* and failure to comply with them is an offence. An example would be a sign defining priority at an intersection (Section 20.2.2). There are two sorts of regulatory traffic control device:

  * *Prohibitory* (or *negative*) devices indicate a forbidden action (e.g. 'no entry') or restriction (e.g. '5 t load limit'). They are usually white discs with a red annular border and the forbidden action described inside the annulus. Prohibitory messages lead to a slower response time and higher error rate than do mandatory (positive) messages.
  * *Mandatory* and *permissive* (or *positive*) devices indicate an essential action (e.g. 'turn left'), an instruction to proceed (e.g. 'through traffic'), or an exclusive action (e.g. 'bikes only'). They are usually coloured discs with white symbols.

(b) *To identify relevant features*, i.e. to mark or warn of hazards ahead which would not otherwise be self-evident or expected. Such signs are called *warning devices*. Typical examples are signs that warn of sharp curves, poor surfaces, intersections (Section 21.3.2), advisory speeds (Section 18.1.3), reduced clearances, and temporary roadworks. Warning signs are often characterised by white triangles with a red border. They can have a high benefit/cost ratio in terms of crash reduction.

(c) *To inform road users*, i.e. convey useful information. Such devices may be *permanent* information signs such as those giving navigation or locational data, informing and advising road users of such items as directions, distances, destinations, routes, points of interest, and location of services. They are called *guide* (or *information*) signs. They also include *temporary* information signs which give data on local traffic conditions, roadworks, road closures, diversions and the like.

Traffic signs can be easily misused. For example, three major risks with traffic control devices are that they may:

* provide inappropriate or inadequate information (Section 21.2.1),
* cause a physical safety hazard (Section 12.3.3), and/or
* cause visual blight (Sections 6.4–5),

Examining the first of these risks in more detail, inappropriate traffic signs can:

* cause information overload (Chapter 21.2.6),
* be contrary to driver expectations,
* readily confuse drivers when they are making decisions under pressure, particularly if the sign was unexpected,
* attempt to achieve more than is feasible,
* be difficult to interpret and lead to drivers disrespecting their message, particularly if the signs are poorly maintained,
* be lost in a forest of roadside signs,
* be obscured by environmental factors such as trees, dust and smoke, and
* be hidden by the road alignment.

The configuration, placement and maintenance of traffic signs therefore deserve careful attention.

21.1.2 Traffic sign definitions

The purpose of this chapter is to deal with *traffic signs*, which are one form of traffic control device. The distinction between a traffic *sign* and a traffic *signal* (Chapter 23) has traditionally been that a sign passively conveys a static, unchanging message whereas the message on a signal varies over time. Also, 'passive' implies that the communication is only from sign to road user, and that there is no interaction from road users to signs. There are now many devices in which this distinction is blurred and so this discussion will treat the traffic sign as a traffic control device that conveys an external visual signal to a driver. A general review of traffic signs is given in Castro *et al.* (2004).

It is essential that traffic signs be of uniform design, application and location and convey a uniform message. This is needed:

* to assist rather than confuse drivers,
* to increase decision-making quality,
* to reduce decision-making time in unfamiliar situations, and
* be able to withstand legal scrutiny (as traffic signs often perform a role in enforcing and prescribing driver behaviour).

Decision-making time is critical as, for reasons that will become more obvious later in the chapter, a driver should be able to interpret the message on a sign correctly in one glance.

However, despite this need for uniformity, there is no worldwide uniformity of traffic signing, with the most widely followed document being an international Convention (or Protocol) on Road Signs and Signals produced by the United Nations in Geneva in 1968. It almost exclusively uses icons based on symbolic codes and a few textual messages. It generally forms the basis for Section 21.3.

21.1.3 Retroreflective sheet

Signs that convey passive messages are now almost exclusively made of retroreflective sheet on a supporting backboard. The sheet has high specular (i. e. mirror-like) reflection properties which are usually achieved by inserting glass beads (Section 22.1.3(4)) or other reflective microstructures into a transparent medium covering the sheet surface.

Lighting theory in general is discussed in Section 24.2.2. Retroreflective sheet is usually specified by the *coefficient of luminance intensity* (CIL) approach developed for corner-cube reflectors (Section 24.4.3). CIL is the ratio of the reflected luminance intensity to the illumination falling on the device, per unit of sheet area. This is sometimes called the *specific intensity per area* (SIA) and the units are cd/lx.m^2 = cd/lm. SIA values depend on the angle at which the observation is made. Typical values for new sheet viewed almost (0.2°) normal to the surface, and with a close (4.0°) to normal light source, are:

	Specific intensity per area (SIA), or CIL per area of sheet, cd/lx.m^2						
	sheet colour						
sheet quality	silver-white	white	yellow	red	standard green	blue	brown
high	220	250	170	45	24	20	12
average	85	55	50	16	10	7	4

Altering the viewing angle from 0.2° to 0.33° cuts the CIL for white, for instance, from 250 to 180 cd/lx.m^2; similarly, altering the light source angle from 4° to 30° cuts the CIL for white from 250 to 150. It is commonly assumed that the maximum angle at which a retroreflective sign must be seen is 30° from the direction of travel. The table also indicates the losses incurred using colours other than white or yellow (see also Section 21.3.1). Section 21.2.2 will indicate that:

* sign luminance levels should be between about 30 and 100 cd/m^2, with an optimum of around 80 cd/m^2,
* a minimum luminance of 3 cd/m^2 is needed to ensure the night-time legibility of signs and
* significant decrements occurred below 30 cd/m^2.

From the above data, retroreflective sheets can be seen to meet these needs. For example, the illumination provided by headlights is between 0.01 and 0.25 lux (Section 24.4), so a CIL of about 200 cd/lx.m^2 would result in a sign luminance of between 2 and 50 cd/m^2.

Retroreflective sheets deteriorate with time and usually need replacement when they have lost 80 percent of their retroreflectivity. This typically happens after 8 to 15 years. In particular, a high class white sheet loses about 4 cd/lx.m^2 per year. Red sheets deteriorate relatively rapidly. Cleaning the sheets can be useful at any time, adding about 10 cd/lx.m^2, and is particularly effective towards the end of their lives.

The legibility distance of a retroreflective sign (Chapter 21.2.5) increases logarithmically as its CIL increases. More generally, if a light source has a luminance intensity of LI, the illumination on the reflector will be b/D^2 where b is a constant and D the distance between the light and the reflector. This produces a luminance at the reflector of CIL.A.b/D^2 where A is the area of the reflector. A person standing a distance D_f away — ignoring angular effects — would then perceive an illumination of CIL.A(b/DD$_f$)^2.

Reflectorised signs are placed skew to the observer to minimise the amount of unnecessary light reflected back to the driver.

21.1.4 Variable message signs (VMS)

Part of the blurring of definitions referred to in Section 21.1.2 is that active systems where the sign message may vary have become more common as information technology has improved. *Variable message signs* (*VMS*) were first used in the early 1970s. They can have messages added or deleted manually, remotely by a manually-initiated command, or remotely and automatically as a consequence of traffic conditions. They are particularly common in situations where the traffic flow varies greatly — for example in tidal flow, congestion, or incident-prone conditions, or where unexpected local weather events can impact on driving conditions.

The commonest technologies used for VMS are:

* flip faces and rotating prisms, containing an entire character, word or message, and
* a dot matrix, fed by incandescent bulbs, optical fibres, and/or light-emitting diodes (LEDs).

LEDs are based on the ability of some crystals to emit a narrow colour band of light when an electric current is passed through them. The crystals are embedded in solid plastic, which makes the device very robust. Thus LEDs need less maintenance (typically they last at least 10 years), are more reliable, are easily dimmable, and use less energy than incandescent globes (they can often be supplied by local solar power). Nevertheless, only fibre optics can produce the white preferred for many traffic signs. In addition, LED light intensity (Section 24.2.2) decreases as the temperature rises.

It is necessary to check LED colours for people who are colour blind (Section 16.4.4) or wearing sunglasses. In 2005 ITE issued a Specification for LED signals.

21.2 TRAFFIC SIGN THEORY

21.2.1 General

The basic requirement for a traffic sign is that it must be capable of fulfilling an established need for traffic information. Conveying such information from signs relies on the use of either:

* *legends*, i.e. words conveying literal messages (e.g. 'keep left'), or
* *pictorial elements*, such as graphic symbols, shapes, and colours.

Standard shapes and colours are discussed in Section 21.3.1. Symbols (Section 21.2.7) may be abstract (e.g. a speed derestriction sign) or indicative (e.g. a cross for a crossroads sign). Common symbolic signs include the class of prohibitory signs (Section 21.1) which use a red annulus around the actual symbol and a red diagonal slash through the symbol. The advantages of symbolic signs are their:

* increased legibility and conspicuity due to the use of larger sign elements (this can double the legibility distance of the sign, Section 21.2.5),
* ability to be rapidly and easily comprehended, even when their colour is obscured, and
* potential for overcoming problems associated with illiterate drivers.

Some signs are hybrids (Section 21.2.1); e.g. the words 'bus only' in a green annulus to denote a route reserved for buses.

The effectiveness of a sign can be checked by considering that the potential user of a sign must successfully pass through the following four stages, often as the consequence of a single glance:

(a) *detect* the sign. Is it:
  (a1) visible (Section 21.2.2)?
  (a2) conspicuous (Section 21.2.3)?
(b) *read* the sign. Is it legible (Section 21.2.4):
  (b1) at an adequate distance (Section 21.2.7)?
  (b2) in the time available (Section 21.2.8)?

(c) *understand* the sign (Section 21.2.7). Is it:
  (c1) comprehensible?
  (c2) unambiguous?
  (c3) precise?
(d) *act* on the sign in the intended fashion (Section 21.2.8). Is its message:
  (d1) credible?
  (d2) correct?
  (d3) appropriate?

The requirements associated with these stages are shown as questions in the above list and are now each examined in some detail.

21.2.2 Visibility

Item (a1) in Section 21.2.1. The prime requirement for a sign to be detected is that it is visible, i.e. that it can be usefully seen. This requires firstly that its position and attitude make it possible for it to be seen by a driver. To produce a light signal, a sign must be illuminated — this may be done by external light and/or from an internal source (sometimes called trans-illumination).

As the second part of the task, the light signal delivered from the illuminated sign to the driver must allow the driver to recognise the sign's visual signal separately from all the other visual signals being received from the sign's luminous background. Visibility is enhanced as the contrast in *luminance* (Section 24.2.2 — note that the subjective equivalent of luminance is *brightness*) between the sign and its background increases. The ratio of sign-to-background luminance is measured by the *luminance contrast ratio* (*lcr*).

Thirdly, as discussed in Section 16.4.1, the driver's eye must have the capacity to resolve the detail in the light signal. This capacity sets the criteria for the physical attributes of the sign. For example, the luminance contrast ratio has a series of criterion values based on Section 16.4.3. These are:

* a *threshold* value of lcr = 2, at which a sign detail will just be visible. This ability degrades at luminances of 30 cd/m^2 or less and the decrement becomes dramatic below 3 cd/m^2 when the eye becomes very sensitive to light but suffers a corresponding drop in resolving ability (Section 24.2.2).
* a value of lcr = 6 to ensure *useful* visibility, and
* an *optimal* value of about lcr = 10 which is the level found in the white-on-red design used for Stop and Give Way signs.

It is necessary to provide both the sign legend and its background with retroreflectivity in order to convey properly any colour-coded message at night. To some extent this will lower the legibility of the sign (Section 21.2.4) by lowering the luminance contrast ratio between legend and its immediate background.

Luminance levels range from −6 to −1.5 log cd/m^2 at night, −1.5 to +1.5 in twilight, and +1.5 to +6 log cd/m^2 in daylight. Observation shows that for a typical road user, the following sign luminances are relevant:

* at least 0.3 cd/m^2 (−0.5 in log units) is needed to ensure that the sign colour can be detected,
* at least 3 cd/m^2 (+0.5 in log units) is needed to ensure that sign detail can be resolved. This is thus the threshold performance level.

* at least 5 cd/m^2 for new signs to cover performance degradation in service. This is not easy to achieve at night, as Section 24.4 shows that street lighting will supply 2 cd/m^2 or less and vehicle headlighting 4 cd/m^2 or less,
* at least 30 cd/m2 (+1.5 in log units) to avoid significant visual degradation,
* at least 40 cd/m2 for older drivers,
* about 80 cd/m2, which many authorities adopt as an optimum sign luminance level,
* over 100 cd/m2 (+2 in log units) brings little additional visual benefit, probably because of the negating influence of irradiation and glare.
* values as high as 1.7 kcd/m^2 are needed to enhance conspicuity in brightly lit urban areas, and even higher levels are needed in strong sunlight.

For night-time visibility, the luminance of unlit signs depends on retroreflectivity of the sign face (Section 24.4.5), on the position and performance of the driver's headlights (Section 24.4.1), and on the placement of the sign, which must be low enough to be illuminated by the light from the driver's headlights. In urban areas, it is also necessary to place signs sufficiently high for them to be seen over parked vehicles — this requirement will counter the above headlight requirement. When illumination by headlights is unlikely, it will be necessary to ensure permanent self-illumination, either by fixed external illumination or by internal illumination of signs.

Sign shape is best conveyed by illuminating the border, rather than the whole sign, which would blur the edges via *halation* (Section 16.4.2). Halation also means that increasing overall sign luminance can reduce the legibility of a sign.

21.2.3 Conspicuity

Item (a2) in Section 21.2.1. In addition to being visible, a sign must also be conspicuous, i.e. it must attract the driver's attention with certainty and within a short observation time, regardless of the location of the sign relative to the driver's initial line of sight (Cole and Jenkins, 1980). Conspicuity thus deals with sensory effects — that is, with those features of a sign that force a viewer to give it perceptual prominence — and can be subdivided further into attention conspicuity and search conspicuity.

Attention conspicuity is the ability of the sign to attract attention when the driver is not prepared for its occurrence. The attention conspicuity of a sign is related to its:

* Luminance and luminance contrast ratio (Section 21.2.2). The probability of detecting a visual signal will increase with an increase in its luminous contrast ratio and, to a lesser extent, with its absolute luminance. The luminances needed for conspicuity are many times greater than those needed for visibility (Section 21.2.2) and values as high as 1.7 kcd/m^2 are needed in brightly lit urban areas, although even this level would not be enough to enhance conspicuity in strong sunlight (Bryant, 1980).
* Size is a particularly powerful determinant of conspicuity as the eye tends to favour nearby objects which provide a large visual angle (Cole and Jenkins, 1982).
* Large and sharply-defined edge-contours (or borders) are particularly effective.
* Other features such as surface highlights, shape, graphical boldness, location relative to the driver's line of sight and relevance to the driver in the driver's current cognitive state (Section 16.3.2) can each have a visual impact (Cole and

Jenkins, 1982). The impacts may be different for older drivers or drivers new to an area.

* Drivers often fail to perceive signs carrying known or redundant messages.
* Colour does not contribute much to the conspicuity of a sign, possibly because the colour property does little more than compensate for the luminance loss associated with using colour rather than white (Section 21.3.1).

A sign with good attention conspicuity will always have good *search conspicuity*, but the reverse is not always the case (Hughes and Cole, 1984). Search conspicuity is needed when a sought object lacks attention conspicuity and a visual search is initiated. With experience, drivers learn the need to initiate such searches in particular traffic circumstances (Castro *et al.*, 2004).

Sections 16.4.5 & 16.5.2 show that a driver will take about 300 ms to have one glance (or look) at the scene ahead, and about 500 ms to actually notice a feature in the scene. Hence visual searching for a sign can consume significant portions of a driver's available time. A measure of search conspicuity is the maximum angle from the eye's line of sight at which an object can be detected with a 90 percent probability within 250 ms (Cole and Jenkins, 1980, 1982). This can be readily translated into traffic engineering practice with respect to sign location and type. For example, common practice is that signs must be placed within 8° to 10° of the line of sight to ensure that they are in a visual zone providing reasonable visual acuity and according to driver expectations (Section 16.4.2). Thus the search conspicuity measure should be set at 8° or less.

Nevertheless, Section 16.4.1 shows that a driver's peripheral vision usually operates to about 90° on either side of the line of sight. Detection of something moving or visually interesting in the peripheral field will usually lead to an appropriate eye movement. Section 16.5.3 shows that this eye movement process will take at least 500 ms and hence peripheral signs will not evince a rapid response. Further, the peripheral cone of effective vision decreases with travel speed to become about 50° at 30 km/h, 40° at 60 km/h, and only 20° at 100 km/h.

Many current traffic control devices are not particularly conspicuous. In one test series observers only located 50 percent of the devices they were instructed to search for, although they did detect 96 percent of the key regulation signs such as Stop, Give Way and Speed Limit (Hughes and Cole, 1984). There are also marked differences in the conspicuity of common signs. For example, the Give Way sign has relatively poor conspicuity and is much less conspicuous than the Stop sign. One reason for this is that the word STOP provides a much bolder legend whilst the Give Way sign subtends a smaller visual angle. It has also been suggested that its white background results in low contrast relative to a bright sky (Cole and Jenkins, 1982). Other signs that failed to attract attention included parking, street name, and tourist signs (Hughes and Cole, 1984).

Usually only one sign will be noticed in a single glance. Thus, the use of more than one sign in a driver's effective visual field will result in ineffective signing. Furthermore, giving a driver more observation time will not necessarily increase the number of signs noticed because many drivers will spend the extra time available taking more information from the first sign noticed (Cole and Jenkins, 1982). A driver's search capability increases with experience and declines with age and with the visual clutter of the scene being searched (Castro *et al.*, 2004).

Roadside advertising signs cause a small but statistically significant distraction to occur. The general effect is not of great magnitude at about one crash every 6 years per sign per kilometre per 10 kveh/d. Drivers appear to have defences against most visual distraction. However strongly illuminated signs and signs involving novel, sensuous, flashing (Section 23.2.3) or moving displays of high information content or calling for attention via peripheral vision (Section 16.4.1) should not be located near roadsides. Aesthetic factors will probably play a greater role than safety factors in decisions on advertising sign location.

21.2.4 Detecting detail

Item (b) in Section 21.2.1. In addition to making a sign visible, its message must also be made legible, i.e. sufficient detail within the sign must be visible at a given distance and in a given time.

Some signs, such as the Stop and Give Way signs, can convey their message by their shape and colour alone, with shape being the far more important determinant of message legibility. Hence there is a strong emphasis on the border of such signs. However, if the message within the sign must be read — as with a direction sign — the legibility of that message becomes of paramount importance.

The first step is to determine which is the critical detail. It is usually the smallest detail to be resolved within the sign. Once this is established, the next step is to determine the maximum distance at which this detail can be seen. This is called the *legibility distance. Visibility distance* is based on the same concept as the legibility distance, but implies that no detail within the object being detected needs to be observed, i.e. that the detail and the object are the same.

The application of the minimum angle of resolution data in Section 16.4.2 shows that the legibility distance for 90 percent of people with normal vision is 1/0.290 = 3.4 m for every millimetre of detail dimension. Young people will be about 10 percent better, and old people about 10 percent worse than the population average. One corollary is that drivers with sub-normal visual acuity (Section 16.4.2) will have less time in which to read the message on a sign than would a visually normal driver.

Using the above data and Figure 16.3a, the legibility distance of a letter of the alphabet, or other complex symbol, would be based on the dimension H being the *stroke width* of the component line elements within the letter or symbol. However, the strokes in a letter have both width and length. It is found that the optimal letter legibility is achieved with a stroke length to width ratio of between five and ten, with a preference for about six. The exact ratio depends on the shape and style of the letter, on the separation from other letters, on the shape of the word, on the sign colours, and on such external factors as background luminance, luminance contrast ratio, and viewing distance. With a near-optimum stroke length to width ratio, a letter or symbol can be recognised as a whole without all its fine detail being resolved. This gives a minimum angle of resolution of six times the basic value of 290 μrad (Section 16.4.2), i.e. 1.75 mrad, for well-designed letters and symbols (Bryant, 1982). The associated legibility distance is 1/0.00175 = 600 mm for every millimetre of letter height.

The spacing of letters is also relevant as *visual interaction* can occur between the contours of the individual letters. An inter-letter spacing of about 0.3 times letter height appears optimal for legibility. However, the effect is usually relatively insignificant and in practical cases letter spacing can usually be determined on aesthetic considerations. Letter width is ideally about 0.9 times letter height, so a letter and its surrounding space should occupy about 1.25 times the letter height. Some numerals (such as 5 and 8) have the potential to be easily confused.

The fonts used for letters and numerals in traffic signs are therefore very carefully chosen from laboratory, prototype and in-service field tests to ensure that they provide maximum legibility. Many countries use the letter set originally produced in the U.S. by the old Bureau of Public Roads. The modified E set is particularly popular. Helvetica variants had been adopted for many European road signs but has recently been replaced by a Euroface font, said to be 40 percent more legible than Helvetica. In the U.S. the need to accommodate older drivers led in 2005 to the replacement of the Series E by a font called Clearview, said to be 40 percent more legible than Series E. One of the main changes in Clearview is to improve the internal openings in letters such as a, b, d, and e and thus limit the halation (irradiation) effect discussed in Section 16.4.2 (Castro *et al.*, 2004).

Section 21.2.3 mentioned that a large edge-contour enhances the conspicuity of a sign. However, visual interaction between this contour and the detail within a sign can cause a drop in legibility. Thus, the need for conspicuity and bold graphics must be balanced against the need for legend legibility. In addition, legibility is best with straight borders.

In strong and uniform daylight, most observers prefer dark letters on a light background (negative contrast). However, when ambient light is poor and signs rely on illumination, light lettering on a dark background (positive contrast) is usually more legible, as the smaller light regions with their greater illumination have less chance to overwhelm their borders by irradiation of the letters (Section 16.4.2).

21.2.5 Detecting whole words

The effective legibility distance is the distance at which the intended message can be read. It is not always necessary to depend on the details of individual letters being seen, as legibility is also a function of the shape of the word or symbol and its familiarity to the reader. There is a redundancy in written words which means that drivers can often recognise a word without distinguishing every detail of each letter in the word. This is particularly so for familiar messages. Thus their legibility distance is much greater than would be predicted from knowledge of the stroke widths of the letters and the driver's visual acuity. This effect is much more pronounced if upper case (i.e. capital) letters are avoided and letters in lower case are used wherever possible. This is because the lower case letters give words a varying contour, whereas upper case words are all rectangular.

Clearly, legibility distances are best determined by tests on the specific message or sign in question. Given this qualification, typical legibility distances in good light are 60 m for a finger-sized detail, 80 m for complex signs, 125 m for legends and 250 m for symbols and fibre-optic signs (Bryant, 1982). However, these distances can be reduced by such factors as:

* distraction due to roadside activity,
* drivers not being in a 'normal, alert' condition (e.g. intoxicated, drug-affected or drowsy),
* low luminance levels at night or due to rain, snow, or fog (Section 21.2.2) dropping legibility distances by up to 50 percent,
* headlight seeing distances being as low as 30 m and rarely above 90 m (Section 24.4.1), although retroreflective sheeting (Section 24.4.5) can restore much of this night-time loss, and
* dirty and ageing signs causing legibility distance drops of about 30 percent.

21.2.6 Glance legibility

The discussion so far has concerned the distance at which a sign can be read. Another form of legibility relates to reading the sign during quite brief periods of exposure. This is called *glance legibility*. Clearly, the limited time that a driver has to read a sign will restrict the length and complexity of the message that can be extracted. From peripheral vision studies, the maximum legend length, L, for certain resolution in a single 500 ms glance (Section 16.5.1), is given by:

$$L = 54W - 0.024D$$

where W is the stroke width, D is the observation distance and consistent units are used (Cole and Jacobs, 1978). When this is rewritten as:

$$W/D \geq 0.000440 + (L/54D)$$

it is seen to be much more demanding than the expression from Figure 16.3 for a single detail, viz.:

$$W/D \geq 0.000290$$

In a single glance a driver can resolve and read about one new word, or six characters (Cole and Jacobs, 1978). The number of words that can be read over a longer period is given approximately by:

$$\text{time available (s)} = 0.32(\text{number of words}) - 0.2$$

which is about three words per second and is consistent with the glance times in Table 16.1 (Jacobs and Cole, 1978a). Taking 300 ms per word (the driver would have travelled over 5 m) is a good reason for keeping sign messages terse. However, the resolution of a few words may be sufficient to enable the whole message to be 'read', particularly if it is a familiar one. Thus, if the sign has a familiar message or a relevant context, then an extra two words could be assumed to be read, i.e. about five words per second. The short-term memory (Section 16.3.3) can retain about seven words or similar chunks of information.

The reading process will be enhanced if as much of the sign as possible is within a driver's field of central retinal vision. Indeed, Section 16.4.1 showed that a driver's zone of maximum visual acuity is usually only a degree or so on either side of the line of sight. Obviously, increasing the size of a sign will improve its legibility distance and give a driver more opportunity to observe it. However, there will be some loss of data read at each observation (or glance) as more of the larger sign will now be seen by peripheral parts of the retina, where acuity drops and contour interaction susceptibility increases (Anderton and Cole, 1982). Large signs can thus be visually inefficient.

The reading process will also be degraded if the driver's attention is diverted by the demands of driving in heavy or threatening traffic.

21.2.7 Understanding the sign

Item (c) in Section 21.2.1. The discussion to date has concerned the sensory perception of a sign. Comprehension is the next critical stage and relates to the driver's response to the sensory perception. The basic process is one of pattern recognition. The pattern received has to relate to a pattern coded in the driver's memory.

The key questions related to comprehension are whether the perceived message is:

* credible,

* meaningful (e.g. to a driver from outside the area),
* correctly understood,
* interpreted unambiguously, and
* responded to in the intended manner.

In assessing whether a sign message will be properly comprehended, it is important to understand:

* the exact nature of the message that the sign is intended to convey,
* the context of its use, and
* the population to whom it is directed.

Given the answers to these three points, it is then necessary to ask:

* whether the message will be seen as relevant, and
* whether the message and response combination is readily learnable.

Comprehension can be measured in terms of both the speed and effectiveness of this response. In cases where an immediate response is not required, the best test is often the retention of the message in the short-term memory. These issues are particularly important for symbolic signs.

21.2.8 Acting on the sign

Item (d) in Section 21.2.1. The previous sections have discussed whether a driver could 'read' the message on a sign. That message will command attention if it has either meaning or novelty, such that the reception of its sensory stimulus in the driver's mind will lead to it being given cognitive priority. That priority will particularly depend on the relevance and functional importance that the driver assigns to the sign. That relevance and importance, in turn, usually relates to whether:

(a) the driver has to act on the sign,
(b) the message will affect the driver's own well being, and/or
(c) previous signs have been perceived to be useful.

On the other hand, a driver noticing an irrelevant conspicuous sign will give it little or no cognitive priority. Indeed, there is considerable evidence that drivers take little notice of many signs (Bryant, 1980). This particularly applies to signs that are only obeyed when there is a high risk of prosecution. Studies of VMS signs (Section 21.1.4) giving route diversion data indicate that only about a third of the drivers in a position to use the information, actually do so.

If a driver is to act on the message received from a noticed sign, the sign must be located sufficiently in advance to allow the driver to properly detect, read, understand, and then respond to its message. The response may range from no action to a complete stop.

Figure 21.1 illustrates how the correct location is chosen, using the example of a sign warning of a hazard ahead requiring a driver to stop. Working backwards, an approaching driver must be able to stop before the hazard. The vehicle stopping distances given by Equation 19.5 in Section 19.4.2 are listed in Table 19.1. They are used to locate the last point, C, at which the braking process can safely begin upstream of the hazard. Section 16.5.3 gives a basic driver response time of 1.8 s (1 to 4) and a vehicle braking response time of 0.2 s (5), giving a total of about 2 s.

However, before using this 2 s the driver must have detected, read and understood the sign. From Sections 21.2.3&6 this could take about 500 ms for a simple, pertinent and well-located sign such as the roadside hazard warning sign in this example. Thus we have a total response time of about 2.5 s which is the number used in many design codes. This is not a conservative assumption as if the sign is not well-located, lengthy and complex, the response could be at least 4 s. On the other hand, an increase in a sign's luminance contrast ratio (Section 21.2.2) will decrease the time a driver takes to react to it.

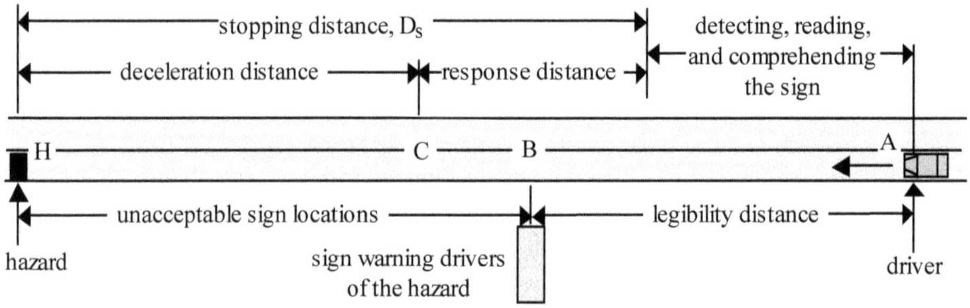

Figure 21.1 Sign location assumptions based on permitting driver action to occur.

Thus, assume that AC is the distance travelled in 2.5 s. For the driver to have begun the visual process leading up to acting on the sign message, the sign must be located at B which is within the driver's legibility distance from A (Section 21.2.4). Thus the sign must be at least BH from the hazard.

As another example, using Table 19.1, it is possible to calculate direction sign distances which are consistent with a detecting, reading, comprehending time of 3 s for complex signs, and an action time which depends on the complexity of any change of direction required. Taking a total response time of 3.6 s (rather than the standard 2.5 s) and a low 60 m legibility distance (although this may not allow sufficient time for a complex sign to be read), a driver will have reached the sign before braking commences. This requires a large BH distance.

However, an upper limit on BH is based on the assumption that a driver should not be required to remember a sign message for more than 15 s (said to relate to short term memory). At 100 km/h, this suggests that BH should not exceed 15 x 100 / 3.6, or about 400 m. Thus, signage on roads with speeds over 130 km/h may tax a driver's short-term memory.

21.3 TRAFFIC SIGN REQUIREMENTS

21.3.1 Shape and colour

The visual significance of sign shape has been discussed in Sections 21.2.3–5. The actual shapes of signs commonly have the meanings given in Table 21.1.

Colour on signs commonly has the meanings given in Table 21.2. The specific colours employed for traffic signs are tightly specified. They are described technically by their *chromaticity*, which is based on chromaticity co-ordinates, which are defined using the CIE 1931 Standard colorimetric system (CIE, 1978). Each chromaticity is represented by a single point on this planar diagram, which is commonly called the *CIE Chromaticity Diagram*. Colours need to be used with restraint as they lead to a loss of visibility over white, mainly due to a loss of luminance (Section 24.2.2). The relative visibilities of some common colours are:

white	1.00
yellow	0.95
green	0.7
red	0.1
blue	0.05
violet	0.0005

On the other hand, for the same luminance levels, people perceive other colours to be brighter than yellow or white.

Table 21.1 Common meanings of sign shapes.

Shape name	Shape	Usage
Octagon (red)	⬡	Stop
Disc (red)	●	Regulation associated with speed & pedestrians
Equilateral triangle (pointing up, yellow)	Δ	Warning
Diamond (yellow)	◆	Warning
Equilateral triangle (pointing down, red)	∇	Give Way (or yield)
Rectangle (vertical)	▤	Generally used for regulation
Rectangle (horizontal, green)	◘	Guide signs
Rectangle (horizontal, yellow)		Warning of roadwork
Shield	❶	Route marking
Diagonal cross (or crossbuck)	X	Railway level crossings*

* However, there are advantages in the alternative use of a symbolic steam train symbol (Cole and Jacobs, 1981).

Table 21.2 Meaning of colour used on signs.

Colour	Usage
red	Stop, Give Way, Wrong Way, and speed reduction signs, etc. Thus red normally means 'extreme' hazard or prohibition.
black, white	Regulatory signs and backgrounds and legends on other signs.
yellow	Warning, advisory, and temporary signs. Thus yellow is normally used for warning of hazards.
orange	Warning of roadworks.
green	Permissive, guide, and direction signs.
blue	Guide, direction, mandatory, and service signs.
brown	Tourist and recreational information signs.

The retroreflective materials used for traffic signs are discussed in Section 24.4.5.

The legibility of the individual letters, numerals, and symbols used on signs is discussed in Sections 21.2.4–6. Blank signs can arise with VMS (Section 21.1). Surveys suggest that this confuses drivers and so it may be wise to carry some message on a VMS.

21.3.2 Direction signs

Direction signs are guide signs used to:

(a) indicate or confirm directions and destinations that may be reached by using one or more of the lanes available to the driver,

(b) help locate the driver, and

(c) give some guidance through any channelisation (Section 20.3.5).

Direction signs are often supported by:

* advance signs placed before an intersection, particularly where the complexity of the intersection or the level of the traffic flow requires drivers to be in the correct lane well in advance of the intersection, and
* reassurance direction signs placed after the intersection.

A distinction should also be made between advance direction signs prior to an interchange and the simpler signs needed for at-grade intersections where speeds will be lower and early lane choice will be less critical.

Advance direction signs can be made relatively complex as they are scanned rather than read in full; on the other hand, reassurance signs contain minimal information. There are two basic forms:

* the *diagrammatic sign* has destinations positioned on a map-like representation of the intersection and directions represented by the position of the town name on the diagram. Such a sign would typically contain between two and ten destinations.
* the *stack sign* has destinations in a vertical list (or stack) and directions represented by a small adjacent arrow. The search time for finding a name in a stack will depend on both the location of the name and the length of the stack. Typically, the time needed to search a stack of N names is [0.25N – 0.17] seconds (Jacobs and Cole, 1978b). Thus, stacks of up to seven names can be searched in 1.5 s with 95 percent success. However, design codes often recommend a maximum of three destination names or five lines of information, with no more than two names for each direction of travel.

For a given sign area, stack signs are superior to diagrammatic signs in terms of legibility, because the destination names on stack signs can be made relatively larger. It is commonly believed that diagrammatic signs have a superior ability to convey complex messages. Diagrammatic signs are preferable where:

(1) the necessary direction violates driver expectancies, or

(2) a simple graphic design can be found to represent the situation.

Major routes are marked by special signs that are intended to supplement direction signs. They permit a route to be traversed by following the appropriate numbers and shapes and therefore must usually be used in conjunction with a good map and some advance planning or knowledge.

21.3.3 Placement

Signs located in positions where they might be exposed to traffic should use either *frangible* or *breakaway* poles to avoid serious effects resulting from a driver's impact with a sign (Section 28.6.1).

On two-way roads, signs are normally only placed on the side of the carriageway associated with the driver's direction. For safety reasons, signs should only be placed in the centre median of a divided road if they have special relevance to traffic in the median lane.

Following Section 21.2.2 and the limited reading times available to a driver, there should only be one sign of a particular type on any single post. The location and use of intersection signs such as Stop or Give Way are discussed in Section 20.3.8 and some comments on their conspicuity are given in Section 21.2.3.

From Section 21.2.8, a common driver sign response time is T_r = 2.5 s and so a driver travelling at 80 km/h would travel about 60 m before beginning an effective response to a sign. There is little to be gained by placing signs closer than this.

Despite these facts, it is common for a large number of traffic signs to be installed in a busy location, thus diminishing their effectiveness, increasing crash risks, and creating visual blight.

CHAPTER TWENTY-TWO

Pavement Markings

22.1 GENERAL

22.1.1 Purpose

Messages applied to the pavement surface to regulate, control, warn, partition or guide traffic are known as *pavement markings* and are a form of traffic control device. They may be used to supplement traffic signs (Chapter 21) or signals (Chapter 23) or to act in a stand-alone mode. Pavement markings have a major advantage in that they convey continuous information within the driver's direct field of vision. Their installation typically has a benefit–cost ratio of over 50.

As with the traffic signs discussed in Chapter 21, the basic requirement for a pavement marking is that it is sufficiently visible for drivers to be able to interpret its meaning in adequate time to properly react to its message. Thus the conditions in Section 21.2.1 still apply.

Pavement markings rely on their shape to provide information and are usually either longitudinal lines, transverse lines, arrows, letters, or numerals and are discussed in Section 22.2 under these headings. Lines are typically between 100 and 200 mm wide; arrows and letters have specific dimensional requirements to be discussed in Sections 22.1.2 & 22.2–4.

Pavement markings are usually white or yellow. Other colours are far less visible. White is preferred technically as in most night-time conditions yellow is indistinguishable from white and is less visible (Section 21.3.1). However, yellow is used extensively in some regions (particularly the U.S.) to separate traffic flows in opposing directions and to mark the outer edge of the travelled way.

Constraints on visibility are discussed in Section 22.1.2. Other disadvantages of markings are that they:

(a) can lower skid resistance,
(b) cannot be applied to unsurfaced roads, and
(c) carry less informative messages than do signs.

22.1.2 Visibility

A line marking letter of height H seen at a distance D from a driver with an eye height of a, will subtend a visual angle of aH/D^2 at the driver's eye. A driver's minimum angle of resolution (Section 16.4.2) is about 500 μrad, which is about the same as that for complex symbols and a little larger than that for the same legend on a sign (290 μrad, Section 21.2.4). Given this result, the legibility distance of the letter is:

$D = 45\sqrt{(aH)}$

which is seen to only increase with the square root of the letter height. The need to make markings as large as possible to achieve adequate visibility is counteracted by their increased cost and pressures under (c) and (d) below to minimise their total area.

The visibility of pavement markings in most cases is determined by the luminance contrast (Section 24.2.2) between the marking and the adjacent portion of the road surface. The luminance contrast ratio should be at least two. Both the pavement surface and the marking will be equally illuminated and therefore the contrast ratio is exclusively determined by the differences in their reflective properties, as defined by their individual luminance coefficients (Sections 24.2.2 & 24.4.4). Minimum acceptable coefficients for pavement markings are about 100 mcd/lx.m^2, whereas desirable levels are at least 400 mcd/lx.m^2.

At night-time, and particularly where headlights provide the prime illumination, the visual effectiveness of a linemarking will depend on its retroreflectivity (Section 24.4.3). This is measured by CIL values (Section 24.4.3) rather than by luminance coefficients. White lines without glass beads and white retroreflective sheet typically have CIL values of 50 mcd/lux.m^2. Using glass beads as specified in Section 22.1.3 raises this to between 200 to 500 mcd/lux.m^2.

Apart from the basic size and lighting issues discussed above, the visibility disadvantages of pavement markings are that:

(a) their visibility is reduced in dusty or wet conditions, in bad weather, or at night. In heavy rain, the linemarking may be effectively submerged and therefore undetectable.

(b) at night vehicle *headlights* (Section 24.4.1) will be almost parallel to the pavement marking and so will provide relatively low illumination of the marking. {Indeed, unreflectorised paints are practically invisible in headlighting. In this respect rough surfaces will be better than smooth, reflectorised paints (Section 22.1.3) will raise effectiveness, but the greatest benefit is with the use of glass beads. Hence their widespread use, as discussed in Section 22.1.3 (4).}

(c) they wear under traffic and require frequent maintenance (methods for determining the wear-resistance of paving paints under traffic are specified in ASTM D913 and BS 3262),

(d) they can be obscured by traffic, and

(e) they lower pavement skid resistance. {This is most critical for longitudinal linemarking. Typically, linemarkings should have a sideways force coefficient, SFC, of at least 0.40 (Section 12.5.4).}

22.1.3 Materials

The materials used for pavement markings are as follows.

(1) *Paints for spraying, brushing, roller coating, and glass-bead application.*
The paint used is specified with respect to colour, consistency, luminance coefficient, speed of drying, and service life (wear resistance). Colour and consistency are routine paint specifications. Luminance is discussed in Section 22.1.2. Installation of pavement markings is greatly aided by using paint with a short drying time. Typical times are:

* under 30 s (instant),
* 30–120 s (quick-dry),
* 2–7 min (fast-dry), and

* > 7 min (conventional).

The paint used is usually either high solvent or water-borne.

* High solvent paints are usually composed of alkyd resins contained in hydrocarbon solvents constituting over 50 percent of the product. This material is cheap and easy to apply, but has a relatively short life and the solvents cause environmental problems.
* Water-borne paints based on acrylic or PVA resins have usually had long drying times and poor durability. However, the newer water-borne paints are acrylic latex-based and are much more effective and suitable for most applications. They have superior retention of large glass beads (see 4 below). Their retroreflectivity degradation rate is about 0.15 percent/day, or 50 percent in a year.

The dry film thickness of the applied paint is about 300 μm, although 200 μm might be suitable for lightly trafficked lines such as edge and centrelines.

Paints may have trouble bonding to dirty or moist surfaces.

Epoxies and polyesters are also used.

(2) *Thermoplastics.*

Thermoplastics such as methylacrylate are usually initially dearer than paints, but have a longer service life. They are suitable for most applications but are particularly cost-effective for heavily trafficked areas. They are commonly applied at a temperature of about 215 C. Overheating can cause discolouration and/or embrittlement. Glass beads (see 4 below) can be dropped or sprayed on to the thermoplastic to provide reflective properties.

(2a) *Ribbed thermoplastics.*

There is a growing use of thermoplastics with raised transverse ribs to both improve visibility and to give an audible sound in any vehicle crossing the line (e.g. to alert drivers straying from their intended path). The process is sometimes called *profile linemarking* and the product may be called ribbed markings or rumble strips or audio-tactile profiled (ATP) road markings. Similar devices for speed control are discussed in Section 18.1.4. The ribs are typically at 250 to 500 mm spacing and 6 to 10 mm high. They are intended to raise noise levels within a vehicle by at least 6 dB(A) (Section 32.2) and so provide aural and tactile simulation in addition to the visual messages from conventional linemarking. Lower ribs are used when there is likely to be pedestrian or cyclist usage.

The strips can also be applied beside the conventional edge line. Such strips are typically supplied rolled, milled or corrugated. The rolled strips consist of a series of transverse pavement indents about 750 mm long, 25 mm wide and 10 mm deep. Milled strips are much narrower but with wider indents, at 400 x 200 x 10 mm. Service experience in the U.S. tends to favour the milled strips. Corrugated strips are usually used on concrete pavements.

In at-risk areas, edge lines using ribbed strips can reduce run-off the road crashes by about 20 percent and have benefit/cost ratios as high as 45.

As such line markings are relatively expensive, they are sometimes reserved for black spots, for sites with a high run-off the road propensity and an AADT above 2 000, and for most relevant situations where the AADT exceeds 10 000.

(2b) *Pimpled thermoplastics.*
In this method the thermoplastic is cast onto a substrate as droplets rather than a uniform layer. This improves the wet weather visibility of the lines as the droplet peaks (*c.* 4mm) protrude through a thin water film.

(3) *Cold applied plastic.*
Cold-applied plastic is typically 2 mm to 3 mm thick pre-cut sheet, with glass beads (see 4 below) either pre-mixed or dropped on to the sheet.

(4) *Glass beads.*
The reflective luminance of a painted or thermoplastic marking is greatly enhanced by the use of small uncoated glass beads with a diameter between 400 and 1000 μm. They are applied to the surface at a rate of about 300 gm/m^2 and become immersed in the paint to about 60 percent of their diameter — thus 400 μm beads could be used with 250 to 300 μm paint thicknesses. Beads must be carefully applied to achieve the 60 percent immersion. Use of larger beads can assist in meeting this objective. Beads lose some of their value if wet and all of their effectiveness if covered by rainwater.

Optically, the beads function as spherical reflectors in the manner shown in Figure 22.1 and particularly increase retroreflectivity under headlight conditions (Section 22.1.2). They are discussed further in Section 24.4.4.

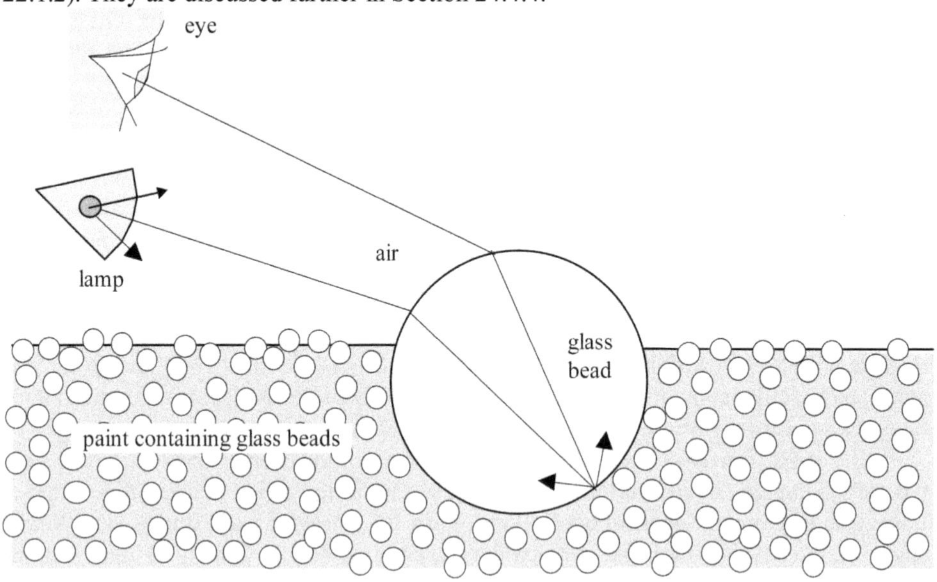

Figure 22.1 Retroreflective behaviour of a glass bead in a paint film.

(5) *Raised pavement markers* (RPMs) which may be retroreflective (RRPM) or non-retroreflective. These are discussed in Section 24.4.3.

22.2 LONGITUDINAL LINES

Longitudinal lines provide delineation to either:

(a) separate opposing flows of traffic,

(b) divide traffic in the same direction into lanes, or
(c) define the outer edge of the trafficked carriageway.

These functions are discussed below under the headings of broken and unbroken lines.

Given minimum through lane widths of 3.0 m (Section 6.2.5), continuous lane markings are not commonly used on pavements under 6.0 m in width, except at intersections (Section 20.1.3).

22.2.1 Broken lines

A broken line is a series of dashes and is commonly used to indicate that the line formed by the dashes may be crossed in some circumstances. Its main use is to assist the positioning vehicles travelling along a carriageway or through an intersection. Broken lines fall into the following five types.

(1) *Separation lines* (Figure 22.2) separate opposing traffic-lanes on two-way roads (Section 6.2.5) when overtaking is feasible (Section 19.4.5) and when the pavement width exceeds a minimum value (typically 6 m). For such an important traffic control device, they perform poorly at night and in wet conditions (Section 24.4). Overall separation lines reduce crashes by about 3 percent, rising to 15 percent at hazardous sites, and 40 percent on curves and tangents.

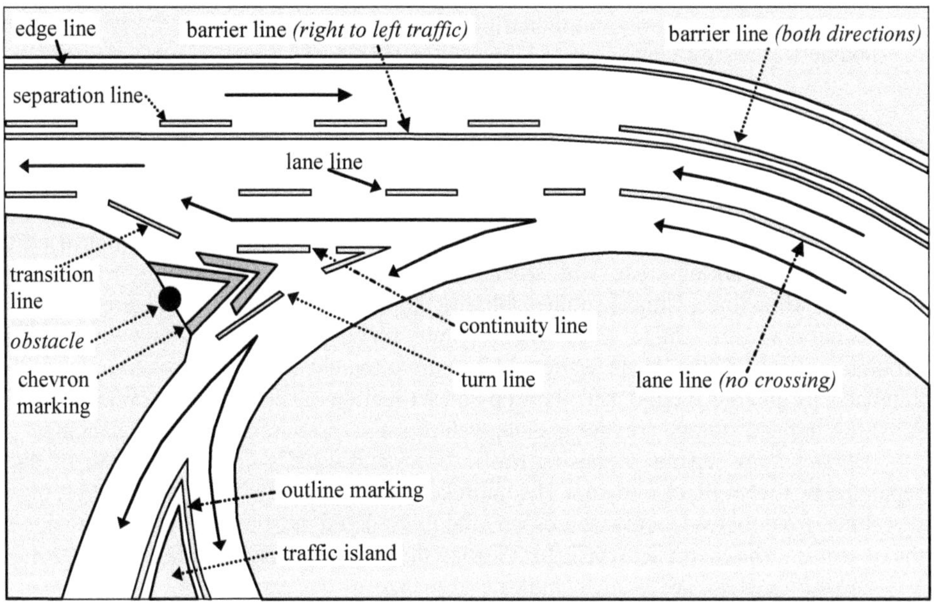

Figure 22.2 Pavement marking types between intersections. This figure is drawn for left-hand driving. A mirror-image will give right-hand driving situations.

(2) *Lane lines* (Figure 22.2) separate traffic flows in the same direction, but permit lane-changing. They may be made narrower than separation lines to aid the discrimination between the two. Indeed, one of the major confusions in current marking practice is that there is usually no certain way that a driver can tell whether a line is a separation or a lane line, although the distinction is vital. Thus, lane marking must be undertaken with great care. Problem areas arise where a lane ends abruptly — a lane drop

(Section 6.2.5) — and where freeway entry and exit lanes come too close together (Section 17.3.6).

(3) *Continuity lines* (Figure 22.2) indicate the edge of the portion of carriageway assigned to a particular traffic stream, or guide drivers past roadside hazards (Section 28.6).

(4) *Turn lines* (or *turning markings*) (Figure 22.2) indicate the intended course for turning vehicles within an intersection, and usually relate to centre turns (Section 20.3.3). Turning lanes are discussed in Sections 20.3.3 & 22.4.2.

(5) *Edge lines* (Figure 22.2) provide short-range delineation of the outer edge of the travelled way and are used:
* commonly, on pavements over 7 m in width and on high-volume, sinuous roads,
* occasionally, on pavements below 7 m, and
* rarely, on pavements under about 5.5 m in width, for fear that they might induce drivers to travel too near to the centre of the road.

It is the practice in some regions to only use edge lines where the alignment is good.

Edge lines improve vehicle lateral placement, discourage traffic from using shoulders, reduce shoulder maintenance, increase driver comfort, and reduce overall crash rates by about 10 percent and rates for relevant crash types by about 25 percent, although the data is inconclusive (Willis *et al.*, 1984). Edge lines are much appreciated by drivers and are noticed more often than signs. They appear to function together with other lane markings to provide drivers with a corridor of good visual guidance. Wide edge-lines also reduce the incidence of extreme lateral placement of vehicles on a carriageway, particularly in the case of alcohol-affected drivers and the shoulders of narrow carriageways (Johnston, 1983).

22.2.2 Unbroken lines

There is a strong trend for unbroken lines to indicate lines that should not be crossed, although a number of exceptions exist. For example, unbroken lines are often used to define kerbside parking areas, with some confusion avoided by commonly using yellow rather than white lines. Unbroken lines fall into the following four types (Figure 22.2).

Lane lines are single lines that separate traffic flows when lane changing is prohibited. Broken lane lines (Section 22.2.1) are often changed to unbroken lines on the approach to an intersection and its stop lines (Section 20.3.7). These *stand-up lanes* provide a defined storage area for queuing vehicles and discourage late lane changing.

Barrier lines separate opposing traffic flows and usually consist of a pair of lines separated by the width of one line. The unbroken line in the pair indicates that crossing is prohibited from that side and that a driver who has crossed must return to the usual side of the pavement. The theoretical basis for locating these lines is discussed in Section 22.5.

Transition lines are used to indicate a change in the carriageway marking, e.g. a drop in the number of lanes.

Edge lines. See the general comment for broken edge lines in Section 22.2.1.

22.3 TRANSVERSE LINES

Transverse lines are lines marked across the traffic stream and are generally associated with traffic control signs. They are either:

* S*top lines* which indicate the point behind which vehicles must stop when required by a Give Way sign, Stop sign (Figure 22.3 & Section 20.3.7), or red traffic signal (Section 23.2.2). Occasionally, advanced Stop lines are provided for cyclists to enable them to be more visible to drivers and to move off ahead and clear the intersection before the motorised vehicles. Stop lines must be placed sufficiently far back from the intersection corner to accommodate pedestrian movements and to allow transverse vehicles to clear the intersection (see also Section 23.2.4).
* Markings for pedestrian crossings and the various crossing variations discussed in Section 20.4.

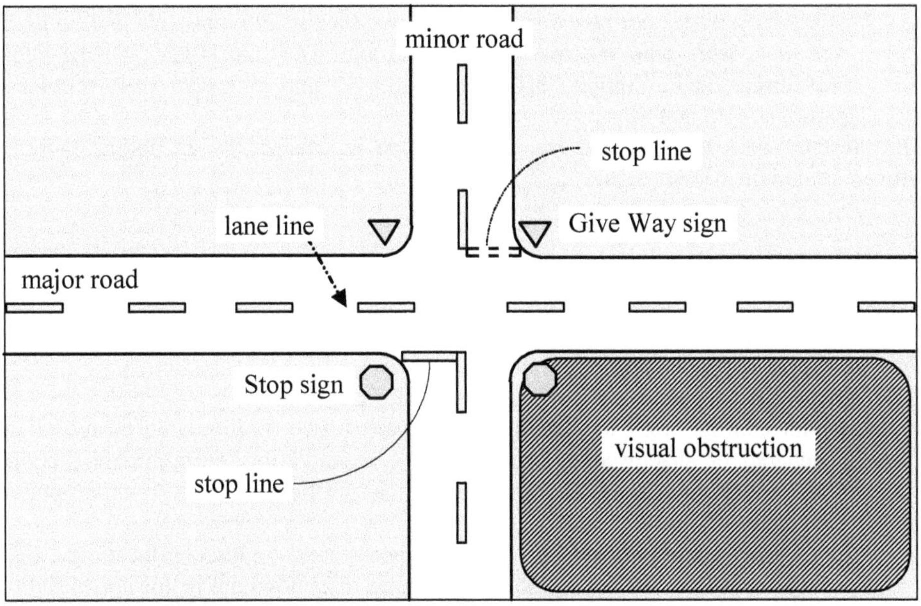

Figure 22.3 Pavement marking types at intersections. Note that Stop line patterns (Section 22.3) will vary from region to region. This figure is drawn for left-hand driving. A mirror image will give right-hand driving situations.

22.4 OTHER MARKINGS

This category covers traffic islands and medians created by diagonal, chevron or solid markings, direction arrows, pavement messages, parking and loading areas, and kerb markings. Some of the main devices are as follows.

22.4.1 Traffic islands

A traffic island is a small area of pavement that is not available for routine traffic use. It is located between lanes and within the carriageway. The purposes of a traffic island are to:

(a) minimise potential conflict between traffic in different traffic lanes, commonly by separating the traffic streams (see also Section 22.4.2),

(b) direct traffic streams,

(c) prevent undesirable traffic movements,

(d) protect pedestrians,
(e) protect traffic control signs, and
(f) avoid creating a new traffic hazard.

Islands are formed using either:

* diagonal, chevron, or completely infilled painted markings,
* unsurfaced areas,
* coloured or roughened pavement surfacing,
* raised areas. These are often edged with mountable kerbs (Section 7.4.3) to avoid creating a safety hazard. However, when the area is to be used as a pedestrian refuge (Section 20.4), semi-mountable kerbs are often preferable (Section 7.4.3).
* safety bars. These are raised blocks that are typically 200 mm wide and up to 50 mm high, with minimum spacings of 1500 mm. They are larger than the jiggle bars discussed in Section 18.1.3. Bars over 30 mm high can cause problems, particularly to motorcyclists.

The presence of a traffic island may be signalled in advance by an outline marking painted around the island (Figure 22.2).

22.4.2 Medians

Medians are long traffic islands and are usually formed as longitudinal strips between carriageways. Their functions are to:

(a) physically separate opposing streams of traffic;
(b) limit conflict areas for turning traffic, particularly by providing protected (or sheltered or shadowed or stand-up) *turning lanes* for centre-turners (Section 20.3.5 & Figure 22.4). These serve alternate traffic directions on either side of the intersection.

When turning movements are not concentrated at particular locations, when traffic speeds are below 100 km/h, and when total flows are below about 20 000 veh/d, the median lane may be used to provide a continuous two-way centre-turn slot for traffic from both directions.

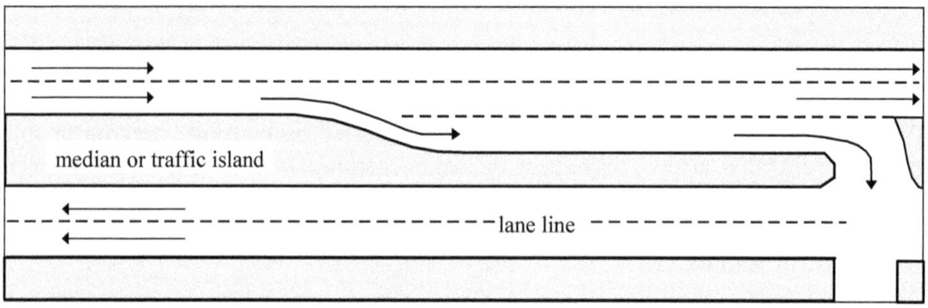

Figure 22.4 Protected turn lane in a central median. This figure is drawn for left-hand driving. A mirror image will give right-hand driving situations.

(c) provide space to shelter crossing traffic at an unsignalised intersection, allowing a pair of carriageways to be crossed in two manoeuvres (Chapter 20);
(d) reduce headlight glare (Sections 24.2.2 and 24.4.1);
(e) provide a recovery area for out-of-control vehicles (Section 28.5);

(f) provide an emergency stopping area;
(g) provide a pedestrian refuge (Section 20.4), often called a *safety zone;* {It may be a median, traffic island, or painted extension to a pedestrian area.}
(h) accommodate level differences between carriageways;
(i) provide scope for landscaping; and
(j) provide space for roadside traffic control signs and other traffic furniture.

A *flush median* is created by using pavement markings to delineate its edges and to crosshatch its interior. Medians can be made more obvious and less crossable by using mountable kerbs (Section 7.4.3) to contain a raised interior surface producing a *raised median*. The additional advantages of this more-costly arrangement are that it:

(1) physically separates oncoming traffic, preventing most unsafe accidental encroachments, and thus increasing traffic safety (typically, raising a median reduces crashes by about 20 percent), and
(2) prevents improper turning movements.

Medians are continuous on freeways. On major roads, medians should ideally be continuous between major intersections. However, the provision of an opening at a cross or side street is commonly considered when at least three such streets occur between the major intersections. On lesser roads, an opening is usually provided at each side street, although spacings at less than 100 m (or 150 m for wide medians) tend to destroy the acceptance of the median. One factor that controls this minimum length is the need to provide an adequate length for any slot for centre-turners. Such slots should be provided wherever possible. Median openings are not usually provided for less than 20 veh/d. Of course, in locating medians it is also necessary to consider local street and traffic patterns and any visibility restrictions.

The following guides suggest median widths of at least:

* 600 mm to separate traffic lanes,
* 1 m to provide a continuous median and/or shelter a sign,
* 1.5 m to provide a pedestrian refuge,
* 2 m to accommodate groups of pedestrians, wheelchairs, and baby carriages,
* 3 m to separate local and through traffic,
* 4 m to indent to provide separate turning lanes,
* 5 m to protect crossing traffic,
* 7.5 m to completely separate traffic, achieve very low crash rates, and/or permit effective tree and shrub planting,
* 15 m to meet some freeway standards for run-off protection (Section 28.6.1), however on lesser roads the use of such wide medians can adversely affect the performance of signalised intersections due to large crossing times.

Median traffic barriers (such as guardrail and New Jersey barriers) are discussed in Section 28.6.2.

22.4.3 Pavement arrows

Pavement arrows are used to ensure correct lane usage and are the common way of designating turning lanes (Section 22.4.2). The minimum length in the direction of travel of an arrow is commonly 5 m. At the 100 km/h stopping distance of 200 m (Table 19.1), a person whose eye is 1.2 m above the pavement (Section 19.4.3) will see this arrow with a visual angle of about 150 μrad. This is at the limit of detail that can be resolved, even by a good eye (Section 16.4.2). Hence, the shape of the flat arrow must be distorted to present a meaningful picture to the driver. In particular, the arrow shapes must be elongated to counteract the longitudinal compression that occurs as the driver projects the arrow into a vertical plane. Arrows or arrow components that are neither parallel to nor perpendicular to the driver's line of sight are particularly prone to distortion. The shapes used commonly account for this. Similar distortion techniques must be used for painted verbal or numerical messages.

22.5 THEORETICAL BASIS FOR THE LOCATION OF BARRIER LINES

The barrier lines described in Section 22.2.2 are used on two-way roads to separate opposing traffic flows where there is a sight distance restriction (Section 19.4) which would make overtaking unsafe. The provisions are generally designed to cater for the needs of the 85th percentile driver. It is thus implied that the remaining 15 percent must wait for a set of above-minimum conditions to occur. The use of barrier lines is a fundamentally negative approach to the difficult passing manoeuvre. Drivers are told when they cannot pass, they are not told when passing might be feasible.

The sight distance restrictions in Section 19.4 are conventionally based on the establishment sight distance, ED, which is a conservative lower limit based on when an overtaking manoeuvre can be safety initiated and completed (Section 19.4.6). However, most drivers are not prepared to be so conservative and field observations indicate that they are prepared to begin an overtaking if it can subsequently be safely aborted (Troutbeck, 1981). That is, for most drivers the continuation sight distance, CD, which is based on the *point of no return* concept, is a better final locator of the solid barrier line (Section 19.4.5 & Figure 19.8). The use of CD to define sight distance is indeed the basis of the *short-zone* concept used in many modern codes of practice and illustrated in Figure 22.5.

Consider the driver of car A travelling from left to right across the page in Figure 22.5 and wishing to overtake car B. At point G driver A cannot see an oncoming vehicle C until it is within the distance ED. Thus, in the *long-zone* concept a sign at G would warn drivers not to commence an overtaking after G, until the sight distance had improved. A special line marking between G and P would reinforce this. When drivers reach the *point of no return*, P, the sight distance with respect to an oncoming vehicle has dropped to the CD limit. Driver A in the process of overtaking must now return to the normal side of the road, on the assumption that an oncoming vehicle is just about to appear at P'. The 'Return to your side' roadside sign shown in Figure 22.5 is hypothetical, but the use of semi-circular return arrows to supplement a driver's view of an oncoming solid barrier line is common practice in Europe. The collision point, F, indicates where driver A will just avoid impacting the previously obscured oncoming vehicle C.

Thus solid barrier line marking begins a distance equal to the recovery distance (Section 19.4.5) beyond P. A driver seeing the solid line must plan to return to the correct side before reaching the solid line — in effect the beginning of the line marks where the obscured vehicle C would reach in the worst case. The solid barrier line operates on the principle of total non-violation. The *short-zone* concept thus defines where the driver must begin to terminate the overtaking manoeuvre, usually by the driver looking ahead and seeing an oncoming solid barrier line which will soon require a return to the normal side of the road. Thus, the compliant driver begins returning to the normal side of the road

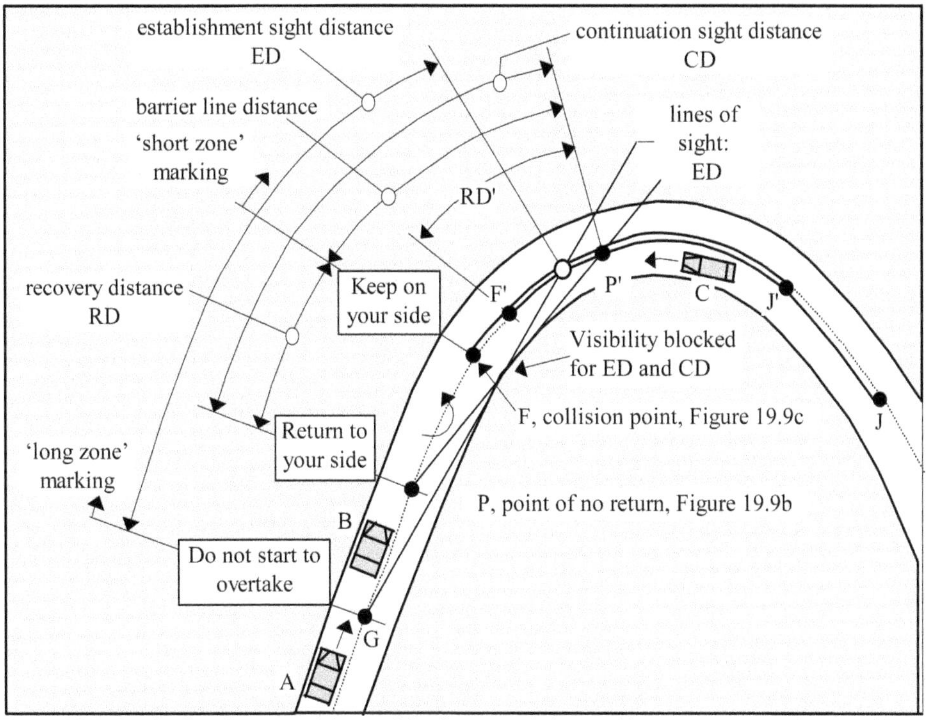

Figure 22.5 Barrier line distance, minimum sight distance and short and long zone concepts for overtaking, with three levels of signing. Drawn for driving on the left-hand side — mirror image for right-hand side driving.

and achieves this before F. To summarise, the concept assumes that the driver at point P and beyond will react to the sight of an approaching solid barrier line at F as if to the sight of an oncoming vehicle.

The procedure for establishing a solid barrier line is, ideally, that an observer at a possible point P indicates to an associate where to stand to be just seen in the oncoming lane. If this distance is greater than CD, then no barrier line is as yet needed. The observer advances and again measures the available sight distance. If it drops to equal CD, then the points P and P' are established by the locations of observer and associate. Points F and F' at the end of the barrier line are then found by measuring forward from P and P' by the recovery distance (Figure 22.5). The point J is the same as the equivalent point F for setting out the barrier line for the opposing flow.

The establishment sight distance, ED, is also used to determine whether or not barrier lines need to be considered at all. A road with its sight distance always greater than ED is sometimes called an overtaking road. In some countries ED (rather than CD) is used to decide when to terminate barrier lines, in order to eliminate excessive stopping and starting of the solid lines.

CHAPTER TWENTY-THREE

Traffic Signals

23.1 GENERAL

23.3.1 Definitions

Traffic signals are traffic control devices that regulate the passage of traffic through a traffic device by changing the colour and/or the symbol displayed in a signal face in the device. The types of signal in use are:

(1) crossing signals assigning priority to potentially conflicting traffic at:
  * crossing at intersections,
  * midblock pedestrian-crossings (Section 20.4),
  * facilities for emergency services, and
  * construction sites;

(2) lane signals indicating the operating direction for a lane; e.g. for tidal flow systems where the direction of travel in a lane is reversed as the peak flow reverses; and

(3) merge signals, managing the rate at which traffic joins a road via an entry ramp (ramp metering).

Signals in type (1) are the appropriate traffic control devices to alleviate:

(a) excessive delays experienced by:
  – traffic at Stop or Give Way signs or at an uncontrolled intersection (Sections 20.1.3 & 20.3.8),
  – turning traffic,
  – pedestrians.

(b) crashes involving:
  – crossing traffic,
  – turning traffic, and
  – pedestrians.

(c) traffic inefficiencies leading to excessive environmental degradation:
  – noise (Section 32.2.3),
  – fuel consumption (Section 29.2.3), and
  – air pollution (Section 32.3.1).

23.1.2 Performance measures

Suggested performance measures for crossing signals based on the three purposes listed as (a), (b) and (c) in Section 23.1.1 are given in Table 23.1. The listing indicates the range of objectives which traffic signals are intended to fulfil.

Traffic signals are frequently very effective traffic management measures with a typical benefit/cost ratio of 2.5 for signal installation. They will also have negative

Table 23.1 Performance measures for traffic signals (Richardson and Graham, 1980).

Performance measure	Explanation
a. Excessive delays	
a1. Vehicle delay (Section 23.6.5)	Delay is the difference between actual trip time and hypothetical free flow times (Section 17.2.6) and directly influences motoring costs (Section 29.2.6) and indirectly influences driver annoyance. Delay variability can also be important. Public transport can be given specific assistance (e.g. by providing bus priority, Section 23.7).
a2. Pedestrian and cyclist delay (Section 23.6.4)	Pedestrian and cyclist delay (Section 7.4.1) is a separate measure to motor vehicle delay. Indeed, decreasing one may increase the other.
a3. Queue lengths (Section 23.6.6)	Queue length indicates local traffic problems and measures overall system failures (Section 17.4.4).
a4. Degree of saturation (Section 23.6.3)	Degree of saturation is an indirect measure of level of service (Section 17.4.4).
a5. Capacity (Section 23.6.3)	Capacity is the vehicle throughput in terms of the number of vehicles able to pass through the intersection per hour. (Section 17.4.1)
b. Crashes (Section 23.1.3)	This measure must include both vehicular and pedestrian crashes. Methods of relating it to traffic volume are given in Section 28.5.
c. Traffic impacts	
c1. Energy consumption	See Section 29.2.1.
c2. Vehicle stops (Sections 23.6.5, 29.2.4 & 32.3.1)	The number of vehicle stops is not necessarily the same measure as vehicle delay and may have a different effect on fuel consumption. Vehicle stops are important when factors such as vehicle operating costs, air pollution, driver annoyance, and safety are considered. The negative contribution of stops to fuel consumption and air pollution is particularly significant. Hence, the joint minimisation of delays and stops is a common design objective (e.g. the stop penalty, Section 23.5.2).
c3. Air & noise pollution (Sections 32.3 1 & 32.2)	This measure and measure c1 are mainly affected by the need to stop and accelerate and by the time spent idling. They can thus be regarded as a composite of measures a1 and a2.
c4. Environmental capacity (Sections 7.4.1 & 17.4.1)	Techniques to improve measures such as a1, a2, a4, c1 and c2 may heighten vehicular intrusion into residential areas. The alternative is to use signals to prevent unwanted traffic or to favour particular routes.
d. System performance	It is conceptually possible to develop a model that will produce an optimal community solution for all the above variables and for the entire local transport system.

influences. For instance, signals may (1) reduce the overall traffic capacity of an intersection, (2) increase some types of crash, and (3) lead to the unwanted diversion of traffic to adjacent unsignalised routes (Huddart, 1980). Care must therefore be taken to avoid over-signalisation, particularly in light traffic. One of the causes of over-

signalisation is the demand by people in residential areas for reasonable entry into or passage across the major arterial roads bounding their area.

The traffic signal objectives listed above have led some commentators to the following typical guidelines for traffic signal installation:

(a) When concurrently over a 4 h period, one road carries at least 600 veh/h and the intersecting road at least 200 veh/h or 150 ped/h. The use of central medians relaxes this pedestrian requirement (Sections 20.4 and 22.4.3).

(b) When the less heavily trafficked road carries at least 100 veh/h over a 4 h period, and when there are obvious intersection problems that can only be solved by signalisation.

(c) When at least 80 percent of the above warrants are reached and there are at least three reported crashes per year.

(d) When there are at least three casualty crashes a year and no other crash counter-measure appears likely to be effective.

23.1.3 Crashes at traffic signals

Traffic signal installation is not a panacea for crash reduction, mainly because signals change the pattern of crashes rather than automatically cause an absolute reduction. Installation will commonly have the following mixed effect on crashes:

* Total and major crashes will drop (e.g. by between 15 and 30 percent, Section 20.2.3), although a rise may occur at lightly trafficked intersections. The percentage reduction is largest at sites which were previously experiencing at least three casualty crashes per year. Signal co-ordination (Section 23.5) usually produces a small (5 percent) reduction in total crashes.
* Crash severity will drop.
* Right-angle crashes will be reduced by over 50 percent.
* Crashes between turning and oncoming vehicles will increase by about 50 percent. However, the use of turn arrows (Section 23.2.5) will reduce this crash type.
* Rear-end crashes will reduce in heavy traffic and increase by up to 50 percent in light traffic. As the rear enders are less costly than right-angle crashes (Table 28.5), the net benefits in light traffic are usually still positive.

Crashes at signalised intersections typically increase with traffic flow to a power of about 0.3, although rear-end crashes increase close to linearly with flow.

A special category of crashes is associated with failing to obey the red signal. The behaviour is commonly called red light running or red-signal violation, and is mainly associated with right angle crashes or crashes between turning and through vehicles. Red light running can be caused by:

* high traffic flows,
* large intersections,
* intentional violation of the signal,
* driver error, or
* poor intersection design (e.g. hiding signal faces from the driver's view, requiring the driver to cross too many lanes, or poor signal phasing (Section 23.2.4).

Some U.S. observations have suggested that there is one red-signal violation for every three cycle changes (Porter and England, 2000). Typically, the consequences of red-light violation represents about 20 percent of all crashes at signalised intersections. These crash types typically result in a higher proportion of injury crashes (Section 28.1.5) than in other urban crashes. The traffic engineering aspects of red light running are discussed further in Section 23.2.4.

23.2 SIGNAL HARDWARE DESIGN

23.2.1 Basic elements

The basic design requirements for traffic signal hardware are illustrated in Figure 23.1. The key element is the *aspect*, which is a single optical system capable of being illuminated at any given time. In lay terms, an aspect is an individual 'light'. Traditionally the light source providing the illumination has been an incandescent bulb, but this has recently been replaced by LEDs (Section 21.1). LEDs require less maintenance (e.g. they can have a 5 year life) and less power (e.g. 10 w compared with 150 w) and provide a more uniform signal than those lit with the more conventional incandescent lamps. LEDs do have a higher capital cost. The aspect's optical system includes a light source, a rear reflector, and a forward lens (or *refractor*).

The diameter of the exposed circular forward face, which may be a lens or an LED array, is commonly either 200 or 300 mm. The larger face is used for higher traffic speeds, but is coming into common use for all speeds. In lamp-based systems the lens takes the luminous flux (Section 24.2.2) from the light source, distributes it in the required direction, and filters it to give the desired signal colour. The three signal colours used are red, yellow and green. The information contained in a traffic signal is therefore conveyed primarily by a single mode, that of the colour of the aspect, although the relative position of the aspect does provide a clue to its message. In some cases, the circular lens is partly masked so that the shape of the illuminated aspect (e.g. an arrow head) also conveys a message.

A group of aspects applying to traffic approaching from one direction is called a signal face. Ideally, there should be one signal face for each traffic lane, at the stop line and across the intersection (e.g. to provide a signal for the lead vehicles in each stationary queue at the stop line).

A traffic signal *lantern* is an assembly of aspects for all relevant traffic directions, including the mounting frame, signal surrounds, and power supply connections. The lantern is carried on poles and pedestals or hung from cables. All components are commonly painted yellow or grey or left in the galvanised state, except that the visors and target boards attached to the signal face (Section 23.2.2) are usually painted black. This assembly is called a traffic signal set.

The objectives of systems comprised of traffic signal sets are listed in Table 23.1. They are achieved in practice by the signal timing plan, which defines the overall sequencing (Section 23.2.4) and timing (Section 23.6.1) of the signal aspects. The plan is executed by the traffic signal controller, which is basically composed of:

* a logic unit,
* a switching unit, and
* an electrical output to the light sources in each signal aspect.

The five main types of control systems are described in Sections 23.3–5.

Signal control for an individual intersection can be at four levels, as pursued further in Section 23.3:

(a) plan control, which allows signal timing plans to be manipulated (Section 23.3.1),

(b) stage control, which allows different pre-determined aspect sequences to be selected within the signal timing plan (called a multi-plan system, Section 23.3.1),
(c) step control, which allows some manipulation of the content of the pre-determined aspect sequences (Section 23.3.2), and
(d) phase control, which allows the proportion of time devoted to each aspect to be adjusted, usually by vehicle actuation (Section 23.3.2).

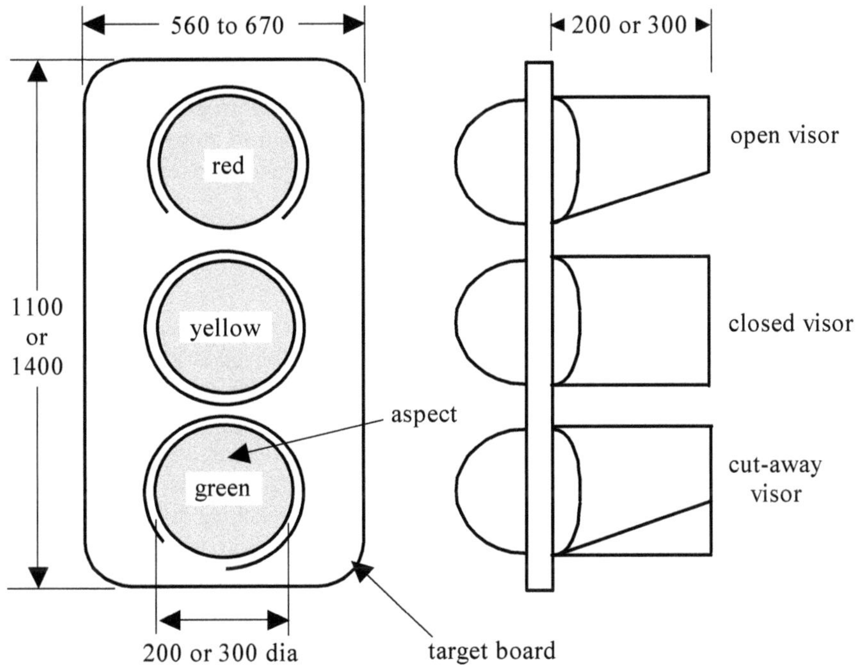

Figure 23.1 Traffic signal lantern assembly with typical dimensions (in mm). The left view is the signal face. The arrangement of visor types is illustrative rather than typical.

Traffic signal maintenance is discussed in Section 26.4 and has been greatly improved by the development of automatic fault monitoring systems linked directly to maintence teams. There is no clear provision for indicating when signals are not operating, although the flashing yellow described in Section 23.2.3(10) could be used in conjunction with the fall-back use of static priority-signing (Section 20.2.2). Flashing signals alone provide no priority guidance and run the risk of being ignored by impatient drivers.

23.2.2 Optical properties

The red, yellow, and green signal colours used are defined in terms of the *CIE standard colorimetric system* using the colour co-ordinates of the *CIE chromaticity diagram* (CIE, 1980 & Section 21.3.1). The definitions were first issued by CIE in 1959 and revised in 1975. To assist people with colour-vision abnormalities (or colour blindness, Section 16.4.4), the colours used are closely specified. As people with abnormal colour-vision commonly confuse red and green, the red used tends to have a yellow component and the green has a blue rather than a yellow bias. In addition, the red and green aspects are consistently placed top and bottom for vertical signals and right and left for horizontal (lane control) signals, so that their position is unique to their colour.

A key need is to maintain contrast between the signal and its background. Typically, a general-purpose 200 mm diameter signal aspect has minimum luminous intensities for red, green and yellow of 200, 265 and 600 cd (Section 24.2.2). These intensities:

* ensure that aspects will be easily seen at distances of up to 100 m, even when the luminance of the sky is as high as 10 kcd/m^2 — a value only found within 5° of the sun on a clear bright day.

  High-intensity, extended-range aspects typically have luminous intensities of 900 (red), 1200 (green) and 2700 cd (yellow), increasing the seeing distance to 240 m. These intensities are still too low to cause discomfort glare at night (Section 24.2.2). The 100 m and 240 m distances correspond to stopping from speeds of about 65 km/h and 110 km/h (Tables 19.1 & 23.2) and thus respectively represent approach speeds for urban and arterial/rural driving.
* ensure, by their inter-colour differences, that the three colours have equal visual impact (Section 24.2.2), and
* accommodate
  - the ageing process in the human eye,
  - abnormal colour vision, e.g. the intensity chosen for red is four times the normal value to ensure that it will be adequate for red-blind drivers,
  - dirt on the aspect face, and
  - the colour-filtering effect of some sunglasses and windscreens.

The luminous intensities quoted above apply along the axis of the light beam emanating from the aspect. This is usually aimed at an approaching driver with eyes 1.5 m above the pavement, at the relevant stopping range, and in an appropriate vehicle location.

To avoid confusion, each aspect should be of uniform appearance when viewed within about 25° of the beam axis and should appear to be black when unlit.

Sun phantom refers to the reflection of sunlight and skylight from the optical surfaces of the forward lens of lamp-lit signal sets. There are two types of sun phantom. Uncoloured (or white or reflex) sun phantom is a reflection from the external face of the lens. It washes out (desaturates) the colour of the signal and is diminished by using a convex lens face. Coloured (or mirror or true) sun phantom is produced by light entering the optical system and being reflected back from its inside surfaces, thus giving rise to a spurious signal indication. It is obviously a problem in extreme latitudes, when the sun is low for long periods. Coloured sun phantom can be controlled by using long angled-down visors (or cowls), louvres fitted to the visors, a red signal luminous intensity of at least 200 cd, and a lens which inhibits internal reflection.

Signals illuminated by LEDs (Section 21.1) avoid the sun phantom problem. However, they can occasionally create a problem for red-green colour-blind people (Section 16.4.4) when viewing LED green signals under direct sunlight. Thus, the relative placement of red and green signals remains of paramount importance.

Visors also reduce the possibility of aspects being seen by traffic for which they are not intended. For this reason, vertical louvres may be placed across the visor opening. A low reflectance target board (Figure 23.1) is often used to surround a signal face to improve over-all visibility and conspicuity, particularly with respect to background luminance (Section 21.2.2). The 200 and 600 cd intensities described above are based on their use.

In most cases a signal lantern will remain visible to an approaching driver provided its mounting height does not exceed 6 m and it is not laterally offset from the driver's line of sight by more than 6 m. Common practice is to provide a primary signal at the required stopping location, as marked by a Stop line (Section 22.3) and a secondary signal where it can be readily seen by a driver parked at that Stop line and looking at the road ahead.

23.2.3 Aspect usage

Each separate line of traffic leading to an intersection and characterised by its direction, lane usage, and right-of-way provision is called a *traffic movement* (or 'stream' in the U.K.). A movement which receives right-of-way during more than one phase is called an *overlap movement.* A traffic signal aspect gives priority to one or more movements. The main features for each aspect are its type (i.e. whether the lit aspect is steady or flashing), colour, and shape.

A signal *phase* (or 'stage' in the U.K. where 'phase' often means a sequence of this book's phases) is a single, static set (or state) of all the aspects simultaneously displayed at an intersection. The start and finish of a phase are therefore identified by consecutive changes in aspect signals resulting in movements gaining and/or losing right-of-way. A phase will thus contain at least two signal aspects — a green aspect giving one movement right-of-way and a red aspect for all legally-conflicting movements.

Steady aspects and their meanings are as follows.

(1) *Green disc* (disc is often erroneously called a 'circle'). A green disc means that drivers may enter the controlled area of the intersection, through an entrance that is usually defined by the stop lines at the associated set of signals. Drivers may then proceed straight ahead or turn left or right. However, the driver of a turning vehicle:
  * can ignore a red disc displayed on the carriageway that the vehicle is about to enter,
  * must give way to higher priority traffic, such as pedestrians,
  * must either *filter-turn* through gaps in any opposing flows or utilise the intergreen time (Section 23.6.2) at the end of the current green movement for the opposing flow. Turning capacities in this manoeuvre thus depend on gap-acceptance behaviour, as given by Equation 17.24 and discussed in Sections 17.3.3 & 20.2.2. If waiting drivers are alert, the gaps will only arise once the platoon of previously-queued opposing vehicles has been discharged and the opposing flow is then comprised of a random stream of unstopped vehicles. Therefore, as the opposing flow increases less filter-turning will be possible (Section 23.6.3).

(2) *Green arrow.* A green arrow means that drivers may proceed in the direction indicated by the arrow without the expectation of opposition from any opposing flow. Further discussion of this aspect is given in Section 23.2.5.

(3) *Yellow disc.* A yellow disc means that drivers should not enter the controlled area of the intersection if they can safely stop beforehand. See the discussion in Section 23.2.4 below. The yellow used is often erroneously called 'amber'.

(4) *Yellow arrow.* See (2) and (3).

(5) *Red disc.* A red disc means that drivers should not enter the controlled area of an intersection. There is a tendency for a few drivers to disobey the initial red signal at an intersection and this action is the cause of about 25 percent of intersection casualties (Hulscher, 1984).

(6) *Red arrow*. A red arrow means that drivers should not proceed in the direction indicated by the arrow. See (2) and (5) and Section 23.2.4 for further discussion.

(7) *Red cross*. A red cross means that the lane signified is not available to drivers.

(8) *Green 'walk'* (or *walking man*). This signal means that pedestrians may cross a carriageway of an intersection leg or at a pedestrian crossing. Special signals may be used for cyclists.

It is commonplace to also provide non-visual signals — such as sound and vibration emitters — to allow visually-handicapped pedestrians to locate the push-button used to request the pedestrian phase, to indicate a relevant phase to the pedestrian, and to announce any relevant phase change.

Another measure is to provide a display of the remaining time in seconds left to undertake the crossing process. This particularly aids pedestrians with restricted walking speeds. This technique has been associated with a halving of pedestrian crashes.

The use of a phase providing a pedestrian movement that is fully protected from crossing traffic is rarely an efficient solution outside areas with high pedestrian flows. It is particularly inefficient when the intersection handles major through-traffic. Thus it is usually employed in conjunction with some green traffic movements and pedestrian priority rules. This tactic may be relatively unsafe if used in conjunction with a green arrow permitting traffic to move in potential conflict with the pedestrians.

(9) *Red 'don't walk'* (or *stopped man*, or *hand*). This signal means that pedestrians should not leave the footpath.

Flashing lights are used in various circumstances. Their intermittent signal attracts attention and the eye basically sees them as moving objects in the peripheral vision (Section 16.4.1 & 6). This gives them high priority for attention within the brain. Rapid but sharply delineated flashing heightens the need for attention. Flashing signal aspects are:

(10) *Amber or yellow discs*. Generally, these mean that drivers must exercise caution.

a. Some jurisdictions use flashing yellow in a conventional signal to indicate that it is in need of maintenance (Section 26.4). Any priority at the now uncontrolled intersection would either rely on additional Stop or Give Way signing or be that given by the local underlying priority rule (Section 20.2.2).

b. Extending this concept further, some jurisdictions change all, or some, signal faces to flashing yellow (or red) for flows below about 600 veh/h and rely on Give Way signing. However, U.S. data indicates a significant increase in crash rates during the flashing mode. The technique is also made unnecessary by the use of vehicle-actuation (Section 23.3.2).

c. Some jurisdictions use a flashing yellow disc as a warning signal. This should be done sparingly, to preserve the effectiveness of the flashing mode. It is used in the following cases:

c1. On a high speed road (typically with a speed limit of at least 80 km/h), to warn of a forthcoming signalised intersection, rail crossing or school crossing.

c2. On a road approaching a signalised intersection with a high speed road.

c3. Where there is a steep grade ahead requiring a change of gears and/or where braking at a signalised intersection would require special effort (see also Sections 18.1.4 & 19.3).

c4. Where there is limited sight distance (Section 19.4) to a forthcoming signalised intersection, railway crossing, school crossing or to the back of a queue (Section 23.6.6).

c5. On the approach to a known crash black spot (Sections 28.3.1 & 28.5).

(11) *Twin red discs with alternating flashes* mean that drivers must stop.

(12) *Red or yellow 'don't walk'* (or *stopped man* or *hand*) pedestrian signals (Section 20.4(d)). In the flashing mode, this signal means that pedestrians should not leave the footpath and that those on the carriageway should hurry to their destination footpath. Many pedestrians do not understand this signal and its use was discontinued in the U.S. in 1986.

23.2.4 The yellow aspect

The yellow signal warns of a change of phase. However, the sequence used for signal aspects is usually green–yellow–red–green as it is no longer practice to give a yellow warning of the impending change from green to red, as modern vehicles do not need lengthy advance notice of the need to get under-way. The yellow signal is therefore now reserved for a warning of imminent change from green to red. There is sometimes an all-red clearance period before the green disc is displayed to other, potentially conflicting, movements (see below). In this period, all potentially conflicting directions are shown a red disc. The yellow and all-red phases thus perform an *intergreen* function (Section 23.6.2), separating the movement phases on two potentially conflicting movements by an intergreen time from the end of the green period for one movement to the beginning of the next green for a conflicting movement.

Although the yellow aspect is sometimes called a *stopping yellow*, its intended use does not require all vehicles to stop. The most rational interpretation of the yellow aspect — and, fortunately, the usual legal requirement — is that a driver confronted with a yellow aspect must stop, if this can be done with safety. If stopping would not be safe, the driver may enter the controlled area of the intersection with the yellow aspect showing. If the intersection is large, a pair of all-red aspects may be needed to then allow the driver to clear the intersection safely. This avoids requiring the driver to estimate the length of the yellow aspect.

The situation is illustrated in Figure 23.2 where the decision point, A, is located at the driver's preferred stopping-distance prior to the Stop line. The basis for this distance is given in Table 19.1 as the minimum non-skid stopping distance, D_s. The first item to determine is the driver response time (Section 16.5.3). Some design codes recommend 1.0 s and this is supported by measurements in Singapore (Wong and Goh, 2000) giving 1.0 s $\pm$ 0.3 s (15th percentiles). As the driver will be alert to the potential for a signal change, component 1 in Table 16.1 will be zero and components 2 and 3 (1.1s) will apply. The preferred stopping-distance will therefore relate largely to the driver's deceleration level, a, which Section 27.2.2 examines in some detail.

Stopping distances based on Table 19.1 and assuming a 1.0 s response time and constant deceleration at both the hard braking a = 12 ms^{-2} and the common a = 3 ms^{-2} deceleration rate (Section 27.2.2) are given in Table 23.2.

The preferred stopping distance will vary between drivers and will depend on road and traffic conditions. For instance, drivers deciding to brake must avoid being struck in the rear by a following vehicle. Nevertheless, relating the alternatives in the table to common observation suggests that many drivers determine their decision point at the hard braking, high deceleration end of the range. The decision point A is therefore not well defined, either theoretically or in the minds of drivers.

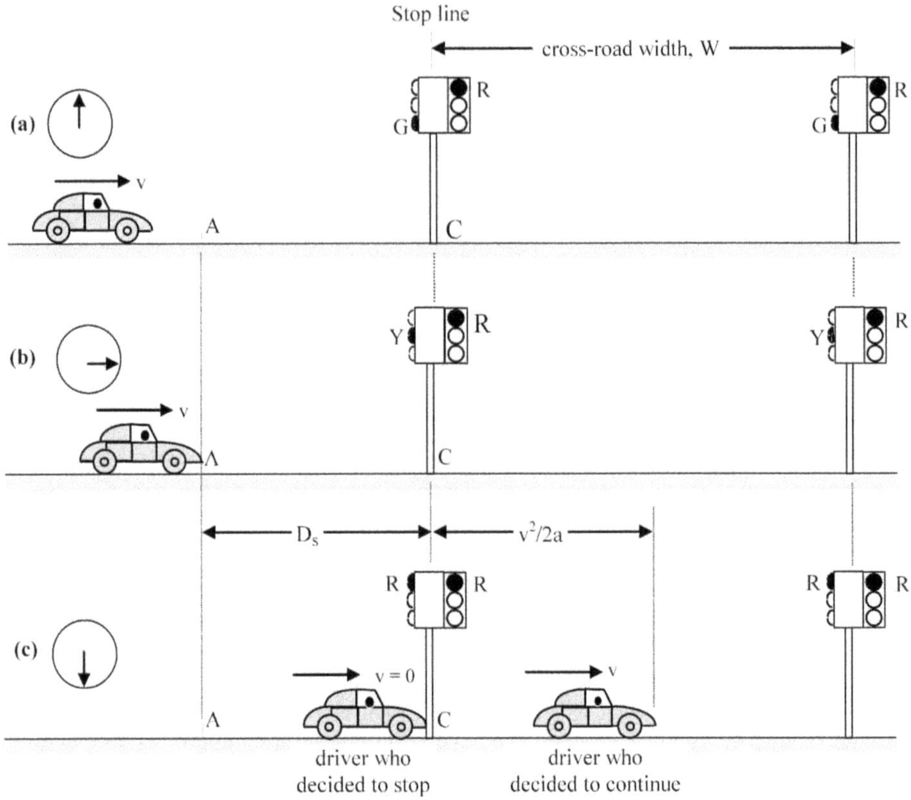

Figure 23.2 The operation of a phase which includes a yellow signal. 'A' is the decision point. Note that D_s is the minimum non-skid stopping distance and includes an allowance for response time (Section 19.4.3 & Table 19.1). Typical values of D_s for hard braking are 13 m at 60 km/h and 33 m at 100 km/h.

Thus, from the onset of the yellow signal there must be sufficient time for a driver to either stop before the Stop line or to traverse this just inadequate stopping distance plus the width of the intersection at constant speed (equal to the driver's approach speed, conservatively assuming that the driver does not accelerate to a faster speed), before the crossing traffic is permitted to proceed. On seeing the yellow aspect, drivers before point A will come to a stop and drivers beyond point A will proceed through the intersection. Drivers actually at point A will have to decide between the two choices illustrated in Figure 23.2c. By 'coincidence', the distance that the driver who does not stop has travelled past the stop line when a conservative counterpart would have just stopped at that line, is equal to $v^2/2a$ and close to D_s for conservative driving.

The actual times required for a vehicle to stop are also given in Table 23.2 and plotted in Figure 23.3, together with the distance a vehicle would travel in the yellow time if it did not decelerate. It can be seen that a yellow time of about 2 s will — for the factors assumed — permit drivers to exercise the 'stop' option by hard braking and 4 s will allow many to gently brake to a stop. Even a 2 s yellow requires only 33 m of travel with moderate braking from 60 km/h. These predictions are based on a deterministic analysis. When practical variabilities in deceleration rates and response times are taken into account, longer yellow times are required. Typical yellow and all-red times are 4 s and 2 s respectively. The table indicates that high speed drivers may need to consider hard braking.

Table 23.2 Illustrative data on driver stopping behaviour at signals. Data based on a 1.0 s response time and a deceleration of a = 12 or 3 ms^{-2}, as indicated by (12) or (3). n.a. = not applicable.

	Minimum stopping:				Distance travelled through stop line in 3 s			
	distance		time					
Speed km/h	m (12)	m (3)	s (12)	s (3)	m (12)	m (3)	lanes (12)	lanes (3)
40	11	26	2	5	22	7	5	1
60	19	55	2	7	31	n.a.	8	n.a.
90	38	120	3	9	37	n.a.	9	n.a.
100	46	140	4	10	37	n.a.	9	n.a.
110	54	170	4	11	37	n.a.	9	n.a.
120	63	200	4	12	37	n.a.	9	n.a.
130	72	240	4	13	37	n.a.	9	n.a.

Assume as a variant of Figure 23.2b that the driver first sees the yellow aspect at a point B before A (Figure 23.4). As the insert in Figure 23.4 (from Figure 23.3) shows, there is an *option zone*, BA, in which the driver who sees a yellow signal can stop before C and can therefore decide whether to stop or to break the law. On the other hand, if braking is soft or the yellow time t* is short, the braking distance will be inadequate. In this case, if a driver passes decision point A before seeing the yellow aspect at B, then the driver will not be able to stop in time and must enter the intersection during the yellow aspect. AB is called the *dilemma zone* for the driver. Observation supported by the data in Table 23.2 suggest that, with a reasonable yellow time, drivers avoid creating a dilemma zone by using the hard-braking option, i.e. A is always after B for practical intersection widths. This applies whenever:

v(intergreen time) – (vehicle length) – (stopping distance) > (intersection width)

This possibility of drivers making different decisions, including the probability of a hard-braking decision, means that there is a heightened potential for rear-end crashes at the approaches to traffic signals (Section 23.1.3). There will also be a great deal of variability in the response of individual drivers and the distance at which they would stop for a yellow signal could vary by ± 20 m. The subjective factors which influence a driver are speed when distant from the signal and distance to the signal when close to the signal.

If there is a possibility that drivers may not have cleared the intersection when the stopping-yellow aspect is finished (Figure 23.2c), then it is necessary to use an all-red phase to allow those drivers to clear the intersection. This might mean that the rear of the vehicle is just clear of conflict with cross-traffic (i.e. at the far kerb line) or that it is clear of pedestrians in a crossing zone which will be on the far side of the kerb line. Some jurisdictions would require the new entering drivers to give way to the clearing driver, thus indicating that a clearing-red interval is unnecessary. However, experience indicates that it is usually unwise to rely on such counter-intuitive behaviour on the part of entering drivers.

Such an all-red phase would rarely need to be more than 2 s long. For instance, the distances travelled in 3 s by a driver who does not stop at the decision point are also shown in Table 23.2 and demonstrate that most intersections are readily cleared. Use of an all-red phase requires the law to permit a driver to be travelling within the intersection whilst facing a red aspect. One alternative is to lengthen the yellow time and run the risk of more drivers entering the intersection on the yellow. Long yellow times are inefficient in terms of traffic flow but are favoured by many nervous or elderly drivers. A better alternative is to use detectors (Section 23.7) to determine if a vehicle that has preceded rather than stopped requires protection within the intersection.

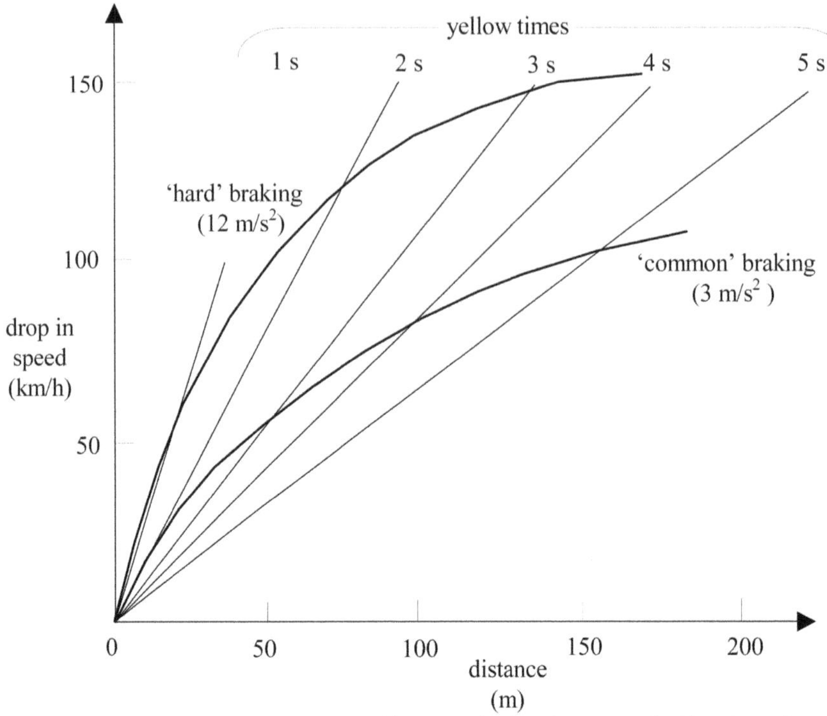

Figure 23.3 Graphs of distances travelled by drivers for hard and common braking (**bold** lines) and if travelling at constant speed during the yellow time.

This approach does not cater directly for cyclists, who will take much longer to clear an intersection, and can be a cause of bicycle–car crashes. This is particularly the case for cyclists using a minor street with a relatively short green time, and crossing in front of stopped cars on a major road, eager to start early in their next green time. In such cases, it may be necessary to extend the clearance time to include the bicycle crossing time.

The yellow time also gives centre-turners, trapped in an intersection by a lack of super-critical gaps in the oncoming traffic, more chance of clearing the intersection before they begin impeding the cross-traffic.

23.2.5 Arrow aspects

The best arrow shapes are ones that minimise the area of the arrow located in the region of high luminous-intensity at the centre of the disc (see irradiation effect, Section 16.4.2). Arrow aspects are often mounted at the same levels as the aspect disc of the same colour, with left arrows on the left of the disc aspect and right arrows on the right.

Green arrows (Section 23.2.3(2)) are used in the following situations:

– for kerb-turners when:
  * kerb turn volumes are high (e.g. in excess of 300 veh/h for kerb-turners),

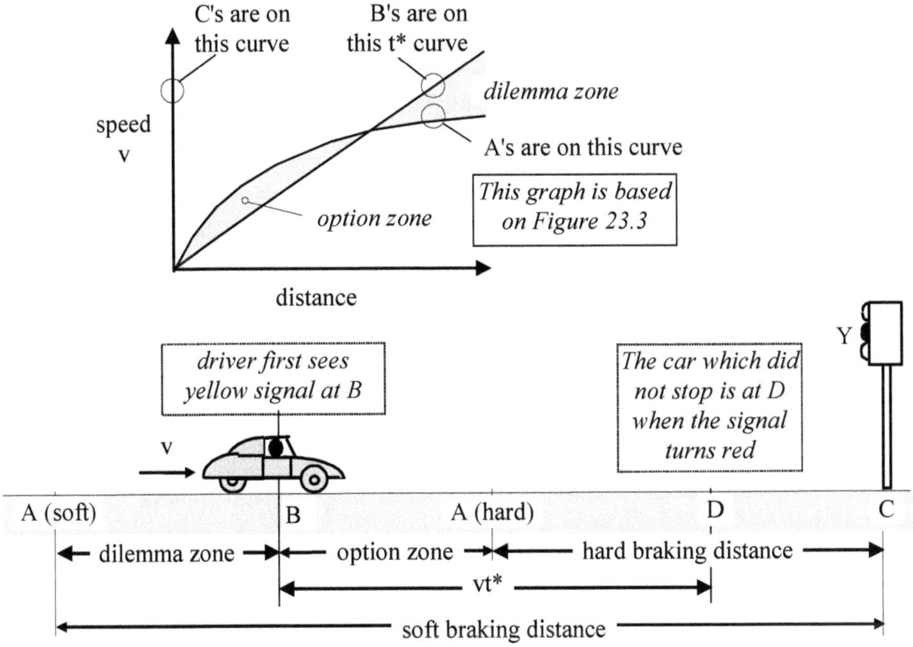

Figure 23.4 Dilemma zones and option zones associated with phases containing a yellow aspect. t* is the yellow time. A is the driver's decision point. For design it is usually assumed that point D is at C.

  * pedestrians are constrained from crossing (particular care must be taken to ensure that this is actually the case), and/or
  * there are no other movements permitted into the cross-road.

– for centre-turners when:
  * there are many centre-turn crashes,
  * opposing traffic volumes are high,
  * opposing traffic speeds are high,
  * super-critical gaps in the opposing flow are infrequent or hard to judge,
  * sight distances to opposing traffic are poor, as a consequence of either alignment, the use of wide medians, or an opposing centre-turn lane heavily used by large trucks,
  * turners experience high delays, waiting for more signal cycles than through traffic,
  * queuing turners block other traffic, and/or
  * turning movements are heavy.

Green arrows reduce the capacity of the intersection to handle through movements, but usually have a net safety benefit, reducing turn-against crashes and increasing rear-end crashes. However, they can confuse unfamiliar drivers in the opposing stream who assume that they also have freedom to move. Thus, green arrows usually require an associated red-disc aspect for opposing traffic in order to protect the turners from oncoming traffic. A simultaneous green arrow can be given to the opposing traffic if the two centre turns do not interlock. Green arrows are best used in conjunction with a turn slot (Figure 20.11) where vehicles waiting for the arrow can queue without blocking through traffic. Green arrows are most commonly employed in four common aspect sequences:

(a) *leading turn*. The green arrow is introduced before the two through-movements have been activated via green discs.

(b) *early start* (or *late release*). The green arrow occurs at the beginning of the associated through movement, whilst the oncoming traffic's green signal is delayed. This arrangement is difficult to achieve with some signal controllers (Section 23.3).

(c) *early cut-off*. The green arrow occurs near the end of the through movement, with the oncoming traffic's green movement prematurely halted.

(d) *trailing* (or *lagging*) *turn*. The green arrow is activated after the two through-movements have been deactivated via red discs. This sequence usually creates greater delays than a leading turn and runs the risk of collisions with late through traffic.

(e) *turn prohibition*. A red arrow is used to prohibit turns during a portion of the through movement.

Methods (a) and (b) are usually more appropriate than methods (c) and (d), for reasons suggested at the end of (d) above. Combinations of (a) and (d), and of (b) and (c) are also used. Note that these six arrangements are all *protected* (or *fully controlled*) modes, in that the turner is protected from oncoming traffic. They usually operate in sequence with a major green movement during which no red arrow prevents turners from filtering through any gaps in the opposing traffic. This is called *permitted* (or *partially controlled*) turning, and the full sequence is known as a *protected-permitted* system.

Red turn-arrows used with a green disc provide a *non-permitted* system that is used to:

* protect pedestrians given priority by a green *walk-signal*,
* prevent vehicles making hazardous filter-turns. This commonly occurs when:
  – sight distance is restricted,
  – there are two lanes of turning traffic,
  – the approach speeds of oncoming vehicles are high,
  – there are high oncoming flows,
  – after a leading green turn arrow when unsuccessful turners can be tempted to use dangerously small gaps in the oncoming traffic, or
  – the crash record is poor.
* avoid conflict with other turning movements,
* permit separate control of individual traffic-lanes,
* prevent side-traffic from entering an already-congested arterial, and
* give priority to public transport vehicles.

It is common to mix arrows and discs with, for instance, a green arrow being used to give priority to one lane in an approach but a red disc then prohibiting all other movements in that approach. The general rule is that:

(1) arrows have precedence over discs, and

(2) a green arrow implies that that movement will encounter no conflicting movements and has absolute priority.

When a green arrow is followed by the potential for a filter turn (Section 23.2.3), it is common to terminate the green arrow with at least 5 s of red arrow, to help drivers to make realistic filtering decisions based on actual levels of opposing through traffic.

The practice of permitting a kerb turn whilst a red signal is showing has become almost universal in the U.S., where it is called *right turn on red* (*RTOR*). The turning vehicle must still stop, and then give way to any pedestrians and vehicles travelling with the green signal on the cross-road. The advantages are time savings, increased capacity and reduced fuel consumption. The disadvantage is reduced safety and diminished respect for red signals. Although most drivers execute the manoeuvre without creating conflicts with pedestrians or green-signal through traffic, observation suggests that about 5 percent of turns do cause pedestrian conflict and that about 25 percent of the drivers in such conflicts incorrectly take priority away from the pedestrian. Many U.S. states also permit a vehicle to kerb turn from one one-way street to another, even when the red signal is showing.

23.3 ISOLATED SIGNAL CONTROL

Traffic signal control systems were described generally in Section 23.2.1. There are five types of systems. The first three are described as *isolated signal controllers* and are discussed in this section. The other two are described in Sections 23.4–5.

23.3.1 Plan systems

In a plan system, the signals are pretimed using estimated future traffic conditions and patterns, often based on prior traffic counts. A *single-plan* system provides traffic-control by using a single, fixed-plan for the timing of a defined sequence of signal phases. It is the most primitive form of signal control and, in traffic that differs from that on which the plan was based, can result in such inappropriate behaviour as the use of unnecessary phases in light traffic.

A *multiple-plan* system uses a sequence of fixed-plans, selected by time of day. It is common to calculate timing plans for four major traffic patterns: morning peak, afternoon peak, business hour pattern (medium demand), and non-peak (low demand). Multiple-plan systems can thus avoid some of the worst features of a single-plan system. However, they remain unresponsive to the actual traffic conditions. An improved variant is the vehicle-actuated fixed-plan system described as (3.3) in Section 23.3.2.

23.3.2 Vehicle-actuated systems

The use of vehicle detectors (Section 23.7) permits the use of vehicle-actuated (VA) signals in which the signal timing plan depends on the actual traffic at the signal location. Vehicle actuation thus provides a servo-system and is most effective in low traffic flows where the signal system is able to respond to the needs of individual vehicles. VA systems tend to favour the minor traffic flows and so may be inherently inefficient. Adjustments to the timing of particular aspects will usually also affect the time to cycle through all the aspects (called the *cycle time*, Section 23.4.1).

In heavy traffic, VA signal controllers regress to fixed-plan operations of the type discussed in Section 23.3.1 and the monitoring of traffic flows at the intersection becomes less relevant. As VA systems only respond to traffic at the signal location, they will be unlikely to achieve system optimisation. To overcome this drawback, techniques are used which also look at conditions at more than one intersection and these are discussed in Sections 23.4 & 5.

The three basic types of VA signal controllers are as follows (Huddart, 1980):

1. *Systems based on selecting a minimum green-time* (Section 23.6.2) and extending it under certain traffic conditions:
   1.1 *Primitive minimum green-time controller.* A minimum green-time (typically between 4 s and 10 s) is chosen in advance to:
   * meet driver expectations, and
   * provide sufficient time to clear any traffic waiting between the detector and the Stop line.

   This minimum green time is extended if vehicles are detected using the movement. However, a predetermined maximum green time is chosen in accord with driver expectations.

   1.2 *Delay controller.* The predetermined minimum green time is extended to ensure that it is sufficient to clear all vehicles that had been delayed before movement began.
2. *Systems with traffic conditions in the green movement used to limit the green time.* These extend System 1 by delaying the commencement of green time (*late start*), or terminating its operation (*early cut-off*). The intergreen time (Section 23.6.2) may be accommodated within the early cut-off time.
   2.1 *Variable green-time controller.* The maximum green time in Type 1.1 is reduced if there is a gap (Section 17.3.8) in the green movement traffic which exceeds some predetermined size (typically 3 to 8 s). If the detector is in advance of the stop line, this gap must be large enough to permit the last vehicle detected to legally pass through the intersection.
   2.2 *Saturation-flow controller.* The gap size used in Type 2.1 is reduced to the gap at saturation flow; i.e. the condition occurs when the associated flow drops below saturation (Section 23.6.3).
   2.3 *Waste-time controller.* The maximum green time in Type 1.1 is reduced when the accumulation of gap times in excess of a given value reaches a predetermined sum called the *waste* time (typically 10 s) (Akcelik, 1981).
3. *Systems with traffic conditions in the red movements also used to limit the current green time.* These are based on and extend Systems 1 and 2.
   3.1 *Minor-flow (or semi-actuated) controller.* The maximum green time for System 2 is only invoked if there is a demand from minor-flow traffic currently facing a red phase.
   3.2 *Queue-sensitive controller.* The predetermined gap size used for Type 2.1 becomes lower as the number of vehicles waiting on the red increases.
   3.3 *Minimum delay controller.* The maximum green time provision applies when the next extension of green time, δT, would lead to:

$$N(\delta T + I) - \sum_{\text{all lanes}} (\delta T / T_g) c > 0$$

N is the number of vehicles currently held by any red signals, as counted by the detectors. I is the time it will take for the delayed vehicles to receive the green signal after the red has occurred on the other movement (the intergreen time, Figure 23.3). T_g is the time gap (Section 17.3.1) for traffic in the green movements, so $\Sigma(\delta T/T_g)$ is the number of vehicles passing through in δT. c is the time by which they would be delayed, possibly the cycle time. The expression becomes more positive when more vehicles join the red movement queues (N becomes larger) or as gaps increase in the green-movement traffic (T_g becomes larger).

3.4 *Optimisation controller*. The maximum green time is based on optimising an objective function dependent on a number of variables. For example, total fuel consumption can be minimised by considering both delays and stops (Section 29.2.4).

3.5 *Fixed-plan (or plan-selection) controller*. The maximum green time is chosen using the measured vehicle flows to either:
* select a predetermined fixed-plan, or
* generate a special fixed-plan using TRANSYT-type calculations (Section 23.5.2). This is called a *plan-generation* system.

An extension of this method is fixed-plan ATC, described in Section 23.5.1.

3.6 *Equal degree of saturation controller*. This method is due to Webster and is explained in Section 23.6.3. It is also the method used within SCATS (Section 23.5.3). A site and simulation review by Akcelik *et al.* (1998) showed that the SCATS version of the method produced lower delays and shorter queues than all the other VA methods.

4. *Systems optimising traffic flow by using phasings based on real-time calculations and measured current traffic conditions*. This approach has become more widespread as computing capabilities have advanced.

With pedestrians the situation is a little simpler, as push-button demand-responsive phases can be routinely expected to avoid the need for special, fixed sequences of phases in signals used by pedestrians. In addition, pedestrian demand may be programmed to over-ride the requirements of the VA system.

In the absence of any traffic most VA systems revert to an all-red display.

23.4 RESTRICTED SIGNAL CO-ORDINATION

Having discussed the three isolated methods of traffic signal control in which each signal set at an intersection acts independently of all other sets at other intersections, the following two sections discuss two systems in which there is interaction and co-ordination between signal sets. First, methods with limited co-ordination are discussed.

23.4.1 Adjacent signal sets

Traffic signal co-ordination can apply to as few as two sets of traffic signals. The basic need is to determine the correct offset between the commencement of a phase at the upstream signal for the connecting leg and the subsequent commencement of the equivalent phase for the same leg at the downstream intersection. Section 17.3.8 showed that traffic platoons from an upstream signal can persist for 1.5 km and separation distances of about a kilometre have been found to permit excellent signal linking. Signals interacting in this manner are called 'paired'.

A decision as to whether to link (or co-ordinate) two paired signals is usually made on the basis of their separation distance, the traffic flows, and the running speed (Section 18.2.11) on the road joining them. The traffic flow in veh/h at which co-ordination is warranted has been suggested to be:

130[signal spacing (m)]/[running speed (km/h)]

23.4.2 Linked arterial (or serial) systems

These systems are sometimes called progressive control systems. They address the management of traffic passing in one direction through a sequence of signals along a route. Pre-determined timing offsets between adjacent signal sets and mid-link detectors (Section 23.7) can be used to give a succession of green aspects to a traffic platoon. The phase offsets along the route are maintained using accurate clocks, cable connections, or radio links. With vehicle-actuated fixed-plan systems (Section 23.3.2), the linking can also ensure that the plans used along the route change in a co-ordinated fashion.

23.5 AREA TRAFFIC CONTROL (ATC)

23.5.1 General

Area traffic control (ATC) permits the central control of all the signals within a network of intersections and thus allows the co-ordination of traffic in different streams and passing through many intersections. Benefit–cost ratios of over 20 can be achieved by introducing this approach. An ATC system can adjust the:

(a) proportion of time devoted to each aspect (or phase) within an overall cycle,

(b) time to cycle through all the phases (the cycle time, c, Section 23.6.2), although it is common for many adjacent intersections to use the same cycle time.

(c) signal timing offsets between intersections (Section 23.6.7), although these are usually fixed,

ATC systems use a master central-controller linked either directly to local signal controllers or to a set of sub-area computers and hence to the local controllers. The master controller receives information and sends command signals back. Often the local controls are still able to act alone and display some local intelligence, although over-ridden for area-wide reasons by control systems higher up in the hierarchy.

Section 23.1.2 indicated that signalisation may reduce the capacity of an individual intersection. It will therefore be appreciated that an ATC system may further reduce an individual intersection's capacity to meet the demands placed on it. ATC must therefore be justified on an area rather than a local basis. In this respect, a careful comparison between the operation of ATC and isolated signals was made in Parramatta, Australia (Luk *et al.*, 1983). In congested urban traffic it was found that there was no significant difference between the performance of the two control modes with respect to average speed and fuel consumption, although stops were reduced by 20 percent. This was because any control system that merely reallocates road space cannot be expected to produce significant effects in congested conditions. However, on the less congested arterial network, ATC resulted in improvements of 20 percent in average speed, 50 percent in stops and 13 percent in fuel consumption, highlighting the advantages of signal co-ordination. Overall, ATC reduced journey times by 8 percent and stops by 18 percent.

The two major ATC modes are *fixed-plan*, which operates at level (a) above, and *dynamic* (Akcelik, 1981), which operates at all three levels. In fixed-plan ATC, sets of desired signal timing plans for various patterns of traffic flow within the network are calculated off-line and in advance of the on-line operation of the system (Section 23.5.2). The plans for each signal are then changed in accordance with measured vehicle flows and time-of-day. A single plan may be in operation for between 15 minutes and 2 hours. To smooth short-term demands, the systems usually have response times of at least 3 minutes (i.e. at least two or three signal cycles). The system is similar to System 3.5 in VA operations (Section 23.3.2), but uses a broader range of traffic data. The aim is to produce system optimisation in terms of Table 23.1 for each traffic condition. However, evidence is that these systems do not produce significant benefits over the simpler isolated, fixed-plan signals described in Section 23.3.1 (Hunt *et al.*, 1981).

In dynamic ATC, which is closer to pure ATC than fixed-plan ATC, the offsets, cycle times, and signal timing plans for each intersection are varied on-line by a computer-based algorithm responding to measured traffic patterns. Such a system is adaptable and flexible, but demanding on computer power, data gathering (particularly from vehicle detectors, Section 23.7), and data transfer. There are two major dynamic ATC systems — SCATS and SCOOT — which are discussed in Sections 23.5.3–4. The Italian SPOT system has similar attributes to SCATS. A detailed review of ATC systems was given in Wood (1993).

Before discussing these two systems, it is necessary to describe how signal cycle lengths and offsets are calculated. Signal timing plan calculation is described in Section 23.6.4.

23.5.2 TRANSYT

The commonest method of cycle length and offset calculation is based on *TRANSYT*, TRL's traffic flow model for arterial roads. TRANSYT was first introduced in 1967 and version 13 was released in 2008. It is in three parts:

(1) a routine to determine a timing plan for an individual set of signals (Section 23.6.1–4),
(2) a macroscopic traffic flow model for sharing flows between links, as in the examples in Section 31.3.7, and for modelling the flow along the link between two intersections, and
(3) an optimisation routine (usually based on hill-climbing) which seeks to minimise the following performance index:

$$(\text{total delay in veh–h/h}) + K_s(\text{total number of stops})$$

The following values are commonly used for K_s, which is termed the *stop penalty*:

Minimisation objective	Stop penalty, K_s
queue length	–30
delay	0
user cost	20
fuel consumption	40
stops	∞

TRANSYT can handle platooned flows from up to four upstream links entering the design link and can handle traffic entering or leaving in mid-link. It is most effective for consistent, heavy traffic and leads to plans favouring traffic on main roads at the expense of minor road traffic. This bias is shared by the main alternative to TRANSYT — the green-wave methods of Section 23.6.7 in which the operator actually nominates the routes that are to be given priority.

In running TRANSYT, the timing offset of one signal from the other is altered until a local optimum of the performance index is reached. This iteration process continues for other offsets at that intersection and is then repeated at each other intersection in turn until no further improvements in the performance index are observed. Group cycle times are also iterated, and any change will require the whole offset set to be rechecked. Pedestrian and public transport flows can be accommodated. The process is clearly demanding of both time and computer space.

23.5.3 SCATS

The SCATS system is a prime example of a dynamic ATC system. In SCATS, each signalised intersection has its own microprocessor-based signal controller capable of operating progressively in one of four modes.

* An *isolated* vehicle-actuated intersection controller. This mode is common during heavy congestion or system failure (Section 23.3.2). The use of local signal controllers dramatically reduces the demand on the regional computer by:
  - performing locally all the repetitive manipulations of data collected from its own detectors (Section 23.7),
  - assessing and using relevant data, and
  - at decision points at intervals of between 1 and 120 s, transferring relevant pre-processed data, such as smoothed vehicle flows, to the regional computer.
* A *linked* intersection controller (Section 23.4) with a store of fixed-time plans. The fixed-time plans give:
  - all the phase times (Section 23.6.2) as percentages of the cycle time,
  - limits on the amount by which a regional computer can alter cycle and phase times, and
  - provisions for using local data to fine-tune a selected plan by such measures as skipping a phase.
* A *slave-controller* operating under instructions from a *regional* computer. The regional computer manages groups of many slave intersections sharing a common cycle time. It receives data from the local controllers and then selects:
  - the appropriate timing plan for each of its intersections, using the equal degree of saturation method (Section 23.6.3),
  - the timing offsets between each pair of adjacent intersections, and
  - the common cycle time for its region, which is updated every cycle in steps of up to 6 s, using saturation measures. Lengthening cycle times upwards is called *stretching.*
* A *central* computer. This monitors and reports on the operations of the regional sub-systems and undertakes fault diagnoses in real time.

23.5.4 SCOOT

An alternative ATC method developed by TRRL is called SCOOT (Split Cycle and Offset Optimising Technique). It differs from SCAT in that its central controller uses measured traffic flows and an on-line traffic model based heavily on the TRANSYT approach (Section 23.5.2). SCOOT constructs a time-profile of the traffic flow from its data in order to predict future traffic, particularly future queues and delays at downstream intersections. The effect of alterations to the current signal timing plans are calculated once per signal cycle for a number of variations from a set of fixed plans. The software then selects the variation which optimises delays and stops, based on the TRANSYT performance index in Section 23.5.2. Changes of up to ±4 seconds in one second multiples are then made to green times, offsets, and cycle times (although cycle times are kept common within defined regions). This central traffic-modelling approach requires a more dispersed vehicle-detector arrangement than does SCAT. SCOOT is now widely used within Britain. It is described in detail in Hunt *et al.* (1981) and Wood (1993) and its background is given in Robertson (1986). Version 4 was released in 1998.

The French *PRODYN* system uses a similar approach but does most of the calculations within the local controllers.

23.6 SIGNAL TIMING

23.6.1 Basic model

The key steps in the design of traffic signals for an individual intersection are to:

* identify all the traffic movements occurring at the intersection (Section 23.2.3),
* determine a physical intersection arrangement to accommodate these movements (Section 20.3),
* determine movements that can share a set of signal aspects,
* estimate the flows associated with each movement (Section 31.5), and
* determine the green times for each movement.

This last step is now discussed.

As discussed in Section 23.2.3, aspect changes determine the sequence and duration of each signal phase. One complete sequence of signal phases is called a *signal cycle* and the associated time is known as the *cycle time* (or cycle length), c. The duration of a green aspect is called the *displayed green time* (or running interval), G, for the associated movement and the remainder of c is called the *stopped interval* for that movement. The set of this data for each traffic signal is embodied in its *signal timing plan.*

The basic sub-model on which the analysis and design of a signalised intersection is based is a graph of the traffic flow for a movement as a function of time, as shown in Figure 23.5.

23.6.2 Signal phasing

To illustrate the application of the above definitions, a simple phasing diagram for a T-intersection is shown in Figure 23.6. All movements are represented by an arrow and require their own green/red aspects. Movement 1 is an overlap movement (Section 23.2.3) and all the others are single-phase movements. The lines with bars at the end indicate vehicles stopped at a red aspect. Such a phasing system can be described by a *phase-movement* matrix, as illustrated in Table 23.3 for Figure 23.6. The phases themselves may be subdivided to accommodate the yellow and red aspects occurring during the intergreen time (Section 23.2.4). The *cycle diagram* shown in Figure 23.7 corresponds to the phasing system in Figure 23.5. Phase changes occur at the end of each green period (e.g. at A, B, and C in Figure 23.7). The *phase split* is the proportion of cycle time devoted to each phase. The *green split* represents the amount of green time given to each movement.

The separation time, I, is the intergreen time (sometimes called the intergreen period, clearing interval, clearance interval or change interval). As shown in Figure 23.7,

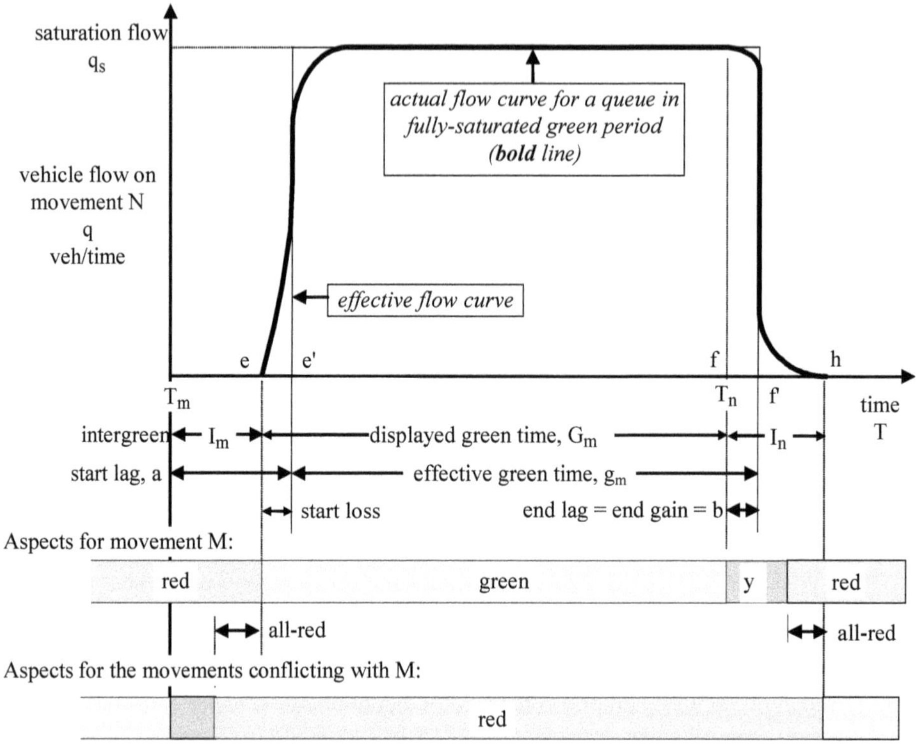

Figure 23.5 Basic intersection timing sub-model and associated definitions (Akcelik, 1981). Saturation flows are defined in Section 23.6.3.

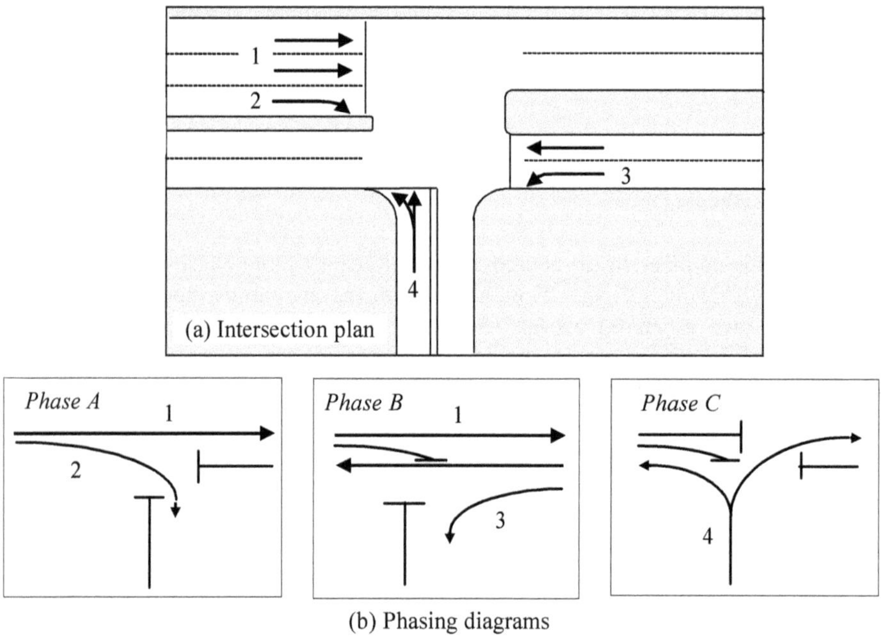

Figure 23.6 An example of a signal-phasing diagram and intersection plan, drawn for left-side driving. (Akcelik, 1981). Note that it might be possible to continue movement 1 during Phase C.

it consists of the yellow and all-red periods discussed in Section 23.2.4 and is a part of the *stopped interval* (Section 23.6.1). It is taken as the initial part of the movement's phase sequence. The time for a phase is the sum of its intergreen and green periods (I + G).

Table 23.3 Phase-movement matrix for situation in Figure 23.6.

Movement	Starting phase	Terminating phase
1	A	C
2	A	B
3	B	C
4	C	A

When a signal aspect changes to green, it is assumed that the flow across the Stop line (Section 20.3.8) increases rapidly to a maximum value known as the *saturation flow*, q_s. Following Section 17.4.4, q_s, is defined as the maximum departure rate possible in a single lane when a waiting queue is provided with continuous green time. This limiting, maximum flow remains constant until the queue is exhausted or the green period ends. Figure 23.5 shows a fully-saturated green period when the queue persists until at least the end of the green period.

This actual flow–time curve is shown by the *bold* curve in Figure 23.5. The idealised behaviour is called the *effective flow curve* and is shown by the light curve whose area $q_s g$ equals the actual number of vehicles passed by the green phase. This defines the magnitude of the *effective green time*, g. The time between the end of the intergreen time and the beginning of the effective green time is known as the *start loss*. The intergreen time plus the start loss is called the *start lag* (*a* in Figure 23.5). Likewise, there is an *end lag* (or *gain*, *b* in Figure 23.5) during which traffic still flows after the displayed green time, G, is over. In these terms, the effective green time, g, for a movement is given by:

$$g = (\text{number of vehicles passed in green phase})/q_s = I + G - (a - b)$$

The lags *a* and *b* and their net difference (a – b) are relatively indeterminate, however the number of vehicles passed can be readily measured by the VA detectors and q_s can be measured for each situation.

To illustrate the approach further, Figure 23.8 is a typical signal timing diagram based on the example in Figures 23.6&7 and including the start and end lags for each of the four movements. All timings are seen to be related back to the major phase-change times, T, which thus provide clear benchmarks within the system for use outside the system, as in an ATC operation (Section 23.5).

Phasing for turning vehicles was discussed in Section 23.2.5. The turners must either filter through the oncoming traffic when both through movements have a green disc or take advantage of a special green arrow, as in phase A for movement 2 in Figure 23.6.

For the phasing example shown in Figure 23.6, there are two possible maximum time paths through a signal cycle. They are movement 1 followed by movement 4, and movement 2, followed by movement 3, followed by movement 4. The maximum time path determines the capacity and timing of the intersection and the movements on this path are called *critical movements*. If there are a number of movements in each phase, but no overlap movements, the critical movement for the phase is the one that requires the longest time.

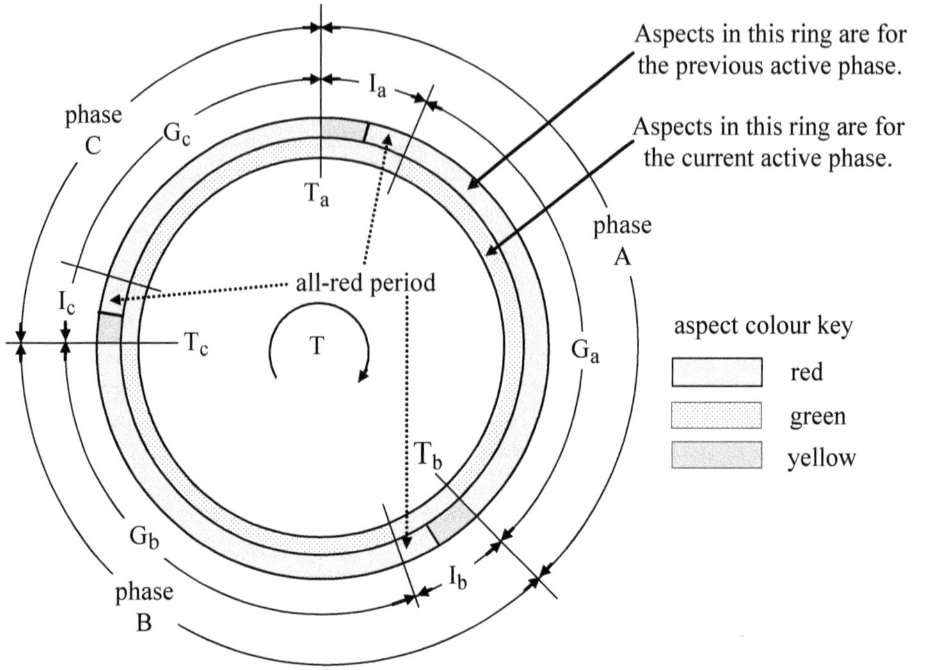

Figure 23.7 Signal cycle diagram for the example in Figure 23.5 (Akcelik, 1981). T is the time, with the cycle time, c, given by $(I_a + G_a + I_b + G_b + I_c + G_c)$.

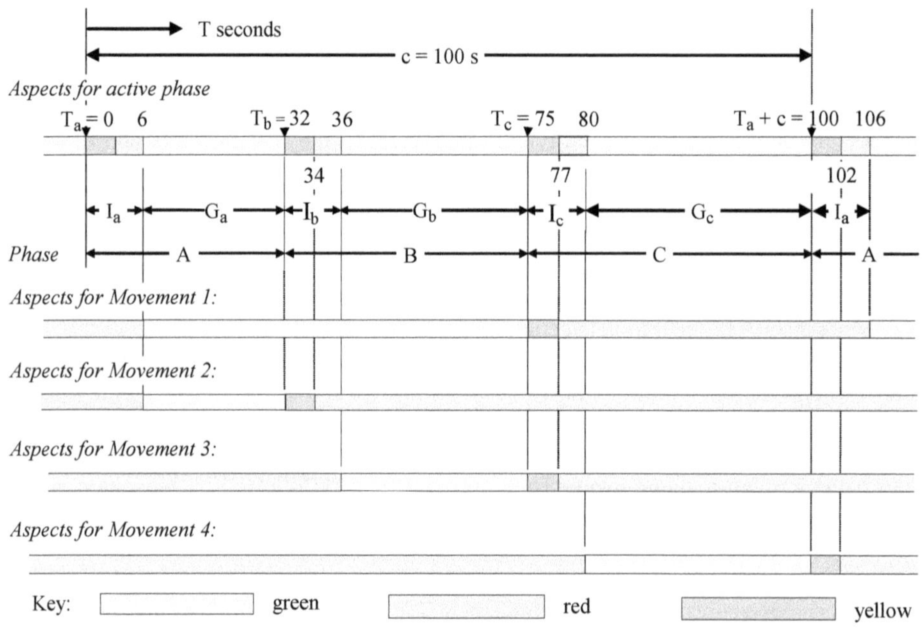

Figure 23.8 Typical signal timing diagram for the example in Figures 23.6&7 (Akcelik, 1981). The numbers above the bars give time in seconds. The diagram is an unwrapped version of Figure 23.7.

Practical considerations often mean that turning lanes and flares (Sections 20.3.3&4) have a limited storage length, as in the case illustrated in Figure 20.10. In a development called the *continuous flow intersection*, Phase A is eliminated by storing the waiting centre-turners in Phase C (movement 2) on the other — lower — side of the waiting through movement 3. Hence, the turning movement 2 can take place in Phase B.

23.6.3 Movement flows

Saturation flows, q_s. were defined in the previous section. A typical value is $q_s = 1800$ veh/h with the actual value depending heavily on factors such as the delay in the reaction time of many drivers — called the *entering headway* (Section 17.4.3c) — after the green is displayed. Measured values of q_s have a coefficient of variation of about 20 percent. Commonly, a single, isolated lane will have a lower than average q_s. In a multilane situation, the central through lane will have the highest q_s.

If q_m is the traffic flow on a movement over a number of cycles, the *flow ratio*, y, of the movement describes to its capacity if green time is unlimited, i.e.:

$$y = q_m/q_s \tag{23.1}$$

Note that, as q_s is not available for a full cycle, $y > 0.5$ suggests potential problems.

The *flow capacity* of a movement, q_{cm}, in the practical case when green time is limited can be obtained by dividing the maximum number of vehicle departures, gq_s, by the full time available, c, to give:

$$q_{cm} = gq_s/c$$

The *movement degree of saturation*, x, is the ratio q_m/q_{cm} and thus:

$$x = q_m/q_{cm} = q_m/(gq_s/c) = q_mc/gq_s$$

It is therefore a measure of how well the available green time, g, is being used by the traffic. Recall from Section 17.2.5 and Equation 17.7 that there will be a space between moving vehicles of at least $L_{sm} + c_4V - L_v$, which in terms of time is $(L_{sm} + c_4V - L_v)/V$ per vehicle. The qg vehicles that pass through during the green time will have a length of qgL_v and will therefore occupy the stop line for a total time of qgL_v/V giving a time efficiency of qL_v/V. If drivers are closely spaced and Equation 17.6 applies, this becomes:

$$\text{best possible time efficiency} = L_v/(L_{ts} + c_4V)$$

where L_{ts} is the territorial surround of each vehicle (Figure 17.2) and c_4V represents the extra space that drivers require as they increase their speed.

Webster proposed the *equal degree of saturation method design method* in which the same x was produced for each critical movement. It is the basis of the signal control method used in SCATS (Section 23.5.3).

The *intersection degree of saturation*, X, is the largest value of x of all the critical movements. One common design approach is to select the green times for the critical movements to minimise X. Ohno and Mine (1973) showed that this occurred when the green times of the critical movement were in the same proportion as their y values. If g/y is a constant for each movement, then $g/y = gq_s/q_m$ is also a constant. Thus another constant for each movement is $gq_s/cq_m = q_{cm}/q_m = x$. Therefore the 'X minimisation' method is also an equal degree of saturation method.

Defining the *green time ratio*, u, as:

$$u = g/c \tag{23.2}$$

leads to:

$$x = y/u \tag{23.3}$$

An intersection will be adequate for the traffic it serves if its capacity exceeds the actual arrival flow for each movement. If it does not, increasing queue lengths and congested conditions may develop. This leads to the basic design requirement that:

$$x < 1 \tag{23.4}$$

for each movement. x values can be misleading where pedestrian phases force the use of inefficient traffic phases. In addition, experience indicates that the condition is actually:

$$x < x_p$$

where x_p is the *practical degree of saturation* and is sometimes taken as 0.95, although particular movements may often have a lower value. Indeed, many capacity-based design methods attempt to ensure practicality by requiring $x_p = 0.9$ for all movements. Akcelik's (1981) extension of the Webster equal-saturation method emphasises the role of x_p. Substituting Equation 23.3 into the inequality leads to:

$$u > y/x_p$$

The *intersection flow ratio*, Y, is given as:

$$Y = \Sigma y = \Sigma(q_m/q_s) = (\Sigma q_m)/q_s$$

For an equal degree of saturation design x = X and so:

$$g = cq_m/Xq_s$$

and

$$\Sigma g = c - T_L = (c/q_sX)\Sigma q_m = cY/X$$

hence:

$$Y = X(c - T_L)/c$$

where T_L is the *lost time,* which is the total time during a cycle in which no green is available to any movement. It is given by:

$$T_L = \Sigma I - \Sigma(\text{yellow})$$

Unless $Y < 1$, it will be impossible to find a signal setting which provides adequate capacity. Indeed, empirical data indicate that Y values should be kept below 0.75 for effective operation. A further basic design requirement is for the capacity to be adequate for the demand on all approaches to an intersection. If inequality (23.4) applies to each movement, then it will be possible to find a set of signal phases for all movements that will provide sub-saturation flows on all approaches, provided that:

$$\Sigma y < \Sigma u$$

i.e.

$$Y < \Sigma u$$

Using Equation 23.1 gives:

$$Y < (\Sigma g)/c$$

and so this design requirement for the intersection flow ratio is:

$$Y < (c - T_L)/c$$

Given that T_L at a site will be relatively constant, increasing c will increase the intersection flow ratio and thus the intersection efficiency. However, a large c will penalise pedestrians and minor traffic.

23.6.4 Operational design

The operational design of traffic signals involves the selection of a phasing system and the calculation of optimal signal settings in terms of cycle times and green times. Following Table 23.1, the total delay of all vehicles passing through the intersection is the most widely used performance measure to be minimised. Preliminary estimates of this parameter can be made from the following three simple measures:

(a) Intersection flow ratio, Y. The use of Y alone can lead to misleading results as it does not properly account for the role of c and T_L.

(b) Intersection degree of saturation, X. This provides a much better preliminary measure of performance than does Y. A value of X = 0.90 is a common maximum value and corresponds to a capacity level of service of about D (Section 17.4.2).

(c) Cycle length, c. Cycle length typically ranges from 60 s in light traffic to 120 s in heavy congestion. There are a number of bases on which c can be selected.

* For fixed-plan signals (Section 23.3.1), there exists a single c and associated green time split that will minimise total delay. This c value is usually close to the average c needed to minimise delay at vehicle-actuated controllers with equal approach volumes on each leg (Section 23.3.2).
* For clearing queues, the appropriate optimum value is given by *Miller's formula*, as:

$$c_0 = (T_L + 2\sqrt{[T_L/q_{sl}]})/(1 - Y)$$

where q_{sl} is the lowest of the saturation flows, in veh/s, for any of the critical movements.
* The use of a high c minimises the lost time ratio, T_L/c and hence is favoured in delay optimisation settings. It tends to produce large platoons of fast flowing traffic.
* c may be chosen to produce a desired operating speed, for example to minimise pollutants (Section 32.3) or fuel usage (Section 29.2).
* c = 120 s is sometimes recommended as a maximum acceptable value for both system reasons and because longer cycle lengths are often taken by drivers to indicate a malfunctioning signal.
* The use of c < 60 s makes it difficult to provide for pedestrian signalisation. For example, Section 20.4 shows that pedestrians with a 2 s response time would take about 14 s to cross a four-lane road, which would occupy half of one phase with a c = 60 s.

A number of performance measures other than total intersection delay can be used. These flow from Table 23.1 and include minimising stops, queue lengths, fuel consumption and total user costs, and maximising flows. To provide a general objective function a TRANSYT approach (Section 23.5.2) can be used in which the total delay in veh-h/h is added to K_S times the number of stops per hour. For this case the appropriate cycle time, c_0, is (Akcelik, 1981):

$$c_0 = [(1.4 + 0.01K_s)T_L + 6]/(1 - Y)$$

where K_s is the stop penalty (Section 23.5.2). The equation is an average expression that does not explicitly allow for minimum saturation flows, for any critical movement, or for the relative values of the critical movement flow ratios.

Concentrating solely on one intersection can cause optimisation problems. Signal co-ordination — as discussed in Sections 23.4–5 — can alleviate some of these. A more subtle effect occurs with delay minimisation, which will usually give priority to the more heavily trafficked arms of an intersection as this will attract further traffic to those arms, giving them even more priority, and so on. This may increase overall system delays and subvert planning strategies for the region.

The *SIDRA Intersection* (version 3, 2006) computer program available from Akcelik and Associates allows the computer-aided design of individual intersection timing plans, particularly for complex phasing systems and intersection layouts. It works on a lane-by-lane basis and calculates a practical cycle time based on Akcelik (1981). It can handle short lane lengths and shared lanes (e.g. Figure 20.12). SIDRA uses TRANSYT-type input and thus traffic demand can be represented by traffic flow data. Alternative programs are *SOAP*, produced by the University of Florida; and *OSCADY PRO* (2006) produced by TRL.

23.6.5 Delays and stops

The preceding section and Table 23.1 showed that delays and stops were key variables in assessing the performance of a signal system. To examine them further requires the integration of the flow–time curve in Figure 23.5 to give numbers of vehicles as a function of time, as shown in Figure 23.9. When the intersection is saturated j is at f '.

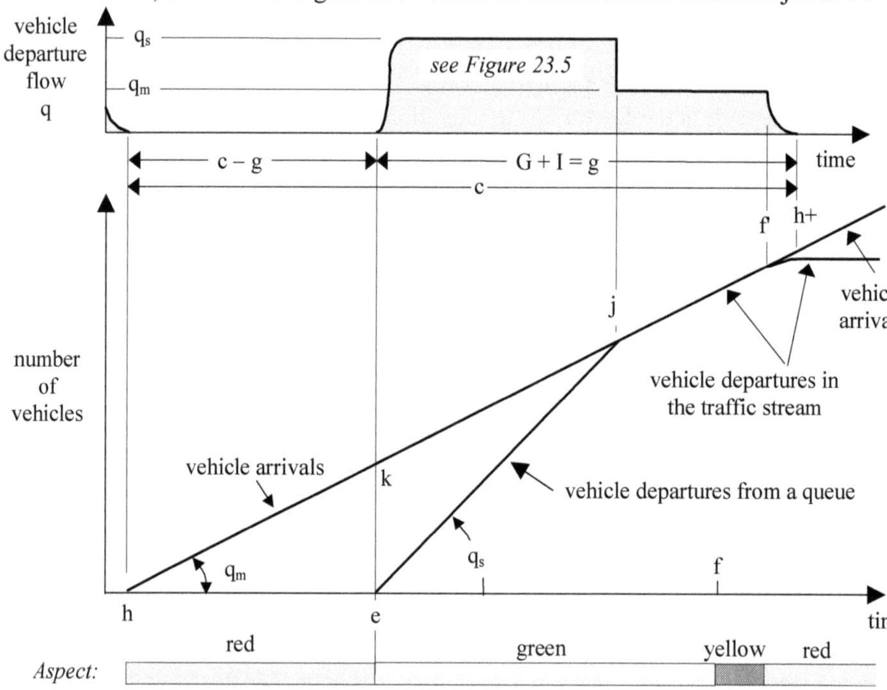

Figure 23.9 Vehicle–time curves based on Figure 23.5.

The main case to consider is when the arrival pattern for a traffic movement is as shown by the curve hj and the departure pattern by ej. There is a uniform arrival flow of q_m. In this case the queue is cleared by point j and the subsequent arrival and departure curves are identical over jh+. The new queue begins to form between f and h_{next}.

The delay caused by the red signal is given by the area ehj in Figure 23.9 in veh-time units and is called *uniform delay*, d_u. It affects the $q_m(c - g)$ vehicles waiting during the movement's red phase. The delay is based on the above simple model of traffic arriving uniformly at q_m during the entire cycle, including the red phase, and then departing uniformly at saturation flow q_s (Section 23.6.3) during its green phase until the queue is discharged (subsaturation, j < f'). The maximum number of vehicles queued is ek and the maximum waiting time is he.

From geometry and using Equation 23.2, the number of delayed vehicles at j is $(c - g)q_m/(1 - y)$ and multiplying by $(c - g)/2$ gives the area hej and thus the total time delay in veh-h. The average delay per vehicle is obtained by dividing this area by the total number of vehicles, cq_m, to give the uniform delay per vehicle in a uniform traffic stream as:

$$d_u = (c - g)^2/2c(1 - y)$$

This delay applies for $x < 1$ and is seen to increase as the square of red time and diminish as cycle time is increased. More complex equations are available for cases when the traffic is not uniform or when the intersection is over-saturated.

If saturation occurs with $x > 1$, then the queue will not be cleared by the end of the movement's green phase, i.e. j occurs after f, resulting in some vehicles being stopped more than once at one intersection. This important multiple stop aspect is ignored by some methods, which calculate stops directly from the proportion of vehicles stopped. With saturation, a more complex, unstable situation arises. Part of the queue will overflow into the next cycle. For the first cycle with no pre-existing queue, the delay area is $(c^2q_m/2 - g^2q_s/2)$ and so the uniform delay per vehicle in this case is:

$$d_u = (c - g/x)/2$$

These overflows or random peaks in the flow during a movement's red phase may not be able to be fully discharged during its next green phase. This will create *random* (or *overflow*) *delays*, d_r, in excess of those predicted by the uniform delay model. If this begins to affect upstream intersections, then congestion occurs (Section 17.4.4). Random delay is predicted by two 'classic' equations due to Webster and Miller and two more recent ones due to Akcelik. The first two are derived from assumed Poisson distributions of the traffic parameters. However, Webster adjusted the second term in his solution to account for simulation and field results whereas Miller's is the unadjusted result of a more complex and theoretically-correct analysis. For each movement, the equations predict the average random delay per vehicle in seconds, d_r, as:

Webster:

$$d_r = x^2/2(1 - x)q - 0.65(c/q^2)^{1/3}x^{2 + 5g/c} \quad (23.5)$$

Miller:

$$d_r = (c - g)e^{-1.33\phi}/2cq(1 - y)(1 - x) \quad (23.6)$$

where

$$\phi = (1 - x)\sqrt{(q_s g)}/x$$

or

$$d_r = (1 - x)\sqrt{(qc)}/x^{1.5}$$

Akcelik: for $x < 1$:

$$d_r = k(x - x_0)/q_{cm}(1 - x) \quad (23.7)$$

where x_0 is the x at which random delays begin to occur, and k is obtained by calibration and is typically between 0.5 and 1.5.

Akcelik: for $x > 1$:

$$d_r = 900T_{con}[(x - 1) + \sqrt{\{(x - 1)^2 + 8k(x - x_0)/cT_{con}\}}] \quad (23.8)$$

where T_{con} is the congestion time in h and k is an empirical value lying between 0.5 and 1.5.

Aspects of Equation 23.8 have previously been discussed in Section 17.2.6 as Equation 17.12i.

The prediction of vehicle stops will depend on the average number of stops per vehicle (called the *stop rate*). This will have both a uniform and a random component.

Thus, delay, queue length and stop rate are inter-related and each has a uniform and a random component. Total delay, d, therefore comes in two parts, uniform and random:

$$d = d_u + d_r$$

Webster suggested that the average total queue at the beginning of the green phase is the larger of $q_m(c - g)$ and $q_md + q_m(c - g)/2$. This leads to the overflow queue being predicted by either zero or $q_md - q_m(c - g)/2$. The equivalent Miller overflow prediction comes from his delay formula above as $e^{-1.33\phi}/2(1 - x)$. The overflow queue length is a major contributor to the random component of delay in the two equations above. A simpler approach is to assume overflow problems for degrees of saturation above 0.80.

23.6.6 Vehicle trajectories

It is useful to examine delays and queues by looking at the flow of individual vehicles through an intersection, thus extending the vehicle trajectory concept introduced in Section 18.4.1 and Figure 18.5. Such trajectories are shown in Figure 23.10. The non-linear aspects of the graphs are best represented by the actual speed reaching the desired speed via a negative exponential relationship (Akcelik *et al.*, 1999).

ABCD is the trajectory of the platoon leader, vehicle 1, which is stopped for time BC whilst waiting for a green aspect. EFGH is the trajectory of a typical vehicle 2, which receives some impedance in passing through the intersection. FG is its stopped delay time and its queue length is shown in the figure. LMN is the trajectory of vehicle 3, the first vehicle to pass through this green aspect without being required to stop. If the intersection is operating below capacity, subsequent vehicles will pass through without the need to slow down. However, no vehicle will do better than vehicle 3 if the intersection flow is at or above capacity. Triangle BCM represents the amount of queuing taking place in an intersection just at capacity. The over-capacity/overflow-queue case was discussed in Section 23.6.5 and involves some vehicles stopping and starting more than once before passing through the intersection. This case for vehicle 2 is illustrated by the dashed line IJK.

It is sometimes suggested that less-than-perfect driver behaviour in queues adds a third component to total delay. The minimisation of delays associated with the progression of traffic from one signal to the next are further discussed in Section 23.6.8.

23.6.7 Offsets

If a group of signals are required to operate in an ATC mode (Section 23.5), it is usual for all such signals to have a common cycle time calculated using the performance optimising approaches discussed in Section 23.6.5. Co-ordination between signals is primarily achieved by using an offset that staggers adjacent green phases. Offsets relate to a selected *platoon* (Section 17.3.8) of traffic released by one green phase in a signal set and received at the next downstream signal set by the green phase for the relevant movement. Offsets are usually calculated from a point in the upstream green phase to the same point on the downstream green phase. Where the two green phases are of different lengths, it is common to calculate from the midpoints of the phases.

The aim of offset selection is usually to have a dispersing platoon passing through the first intersection also pass unimpeded through the second. The offset to achieve such a progression of green phases will thus depend on the average running times on the

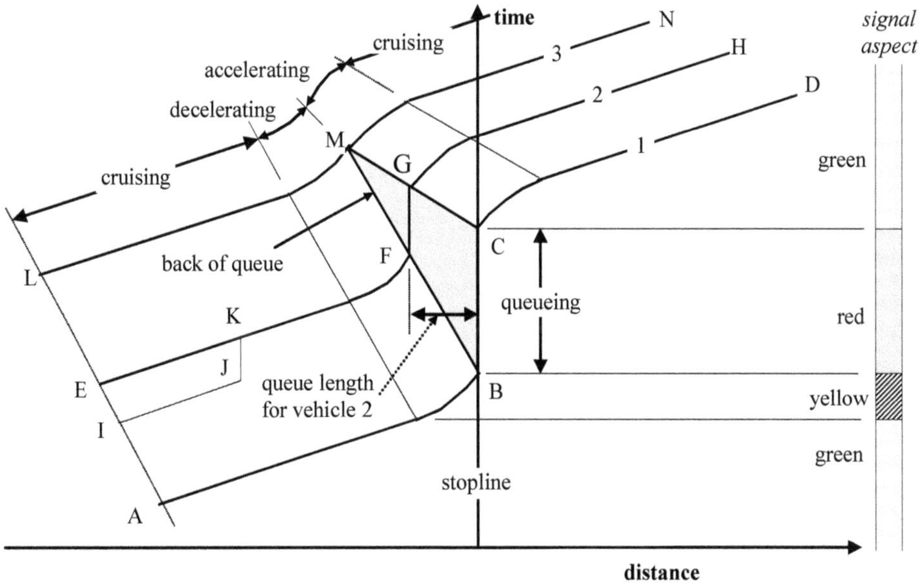

Figure 23.10 Vehicle-trajectory diagrams for a signalised intersection.

connecting link between the two sets of signals (Sections 17.2.5 & 18.2.11) and the platoon dispersion characteristics (Section 17.3.8) of vehicles using the link. As a first approximation, the offset is given by:

(average cruising time) + [(upstream green time) – (downstream green time)]/2

The situation is usually represented pictorially on a space–time trajectory graph such as Figure 23.11. A platoon of vehicles leaving intersection A can be represented by a band of trajectories. The inverse slope of the band represents the assumed (or design) speed of the vehicles. The *bandwidth* is time between the first and last vehicle travelling at the design speed. In reality the platoon will be dispersed with some vehicles travelling above, and others below, the design speed. The objective is to ensure that the bandwidth can pass through a downstream intersection by receiving a green phase for the movement at the right time and of the right length.

An effective bandwidth progression produces a *green wave* travelling in the flow direction. In the case shown in Figure 23.11, the bandwidth does not spread with platooning and keeping the two green times equal permits perfect progression. In practice, the bandwidth spreads with time and it is necessary to decide whether to give progression to the front of the platoon [offset = running time] or to the rear of the platoon [offset = (running time) + (dispersion)].

Increasing the size of the bandwidth will capture the dispersed vehicles but will not minimise total vehicle stops and delays or even maximise total flows. Instead, bandwidth size can be calculated using TRANSYT (Section 23.5.2) or one of two U.S. bandwidth maximisation programs — *MAXBAND* and *PASSER*. PASSER IV is a macroscopic, deterministic, optimisation model that evolved from MAXBAND. Its output includes optimised values of cycle time and offsets.

Whilst the approach in Figure 23.11 is visually useful, it falls down in congested areas or where there is more than one major flow direction in the area. Even for two adjacent intersections with different green times and comparable flows in each direction, it will be difficult to find an offset satisfying flow in both directions. A key point is that,

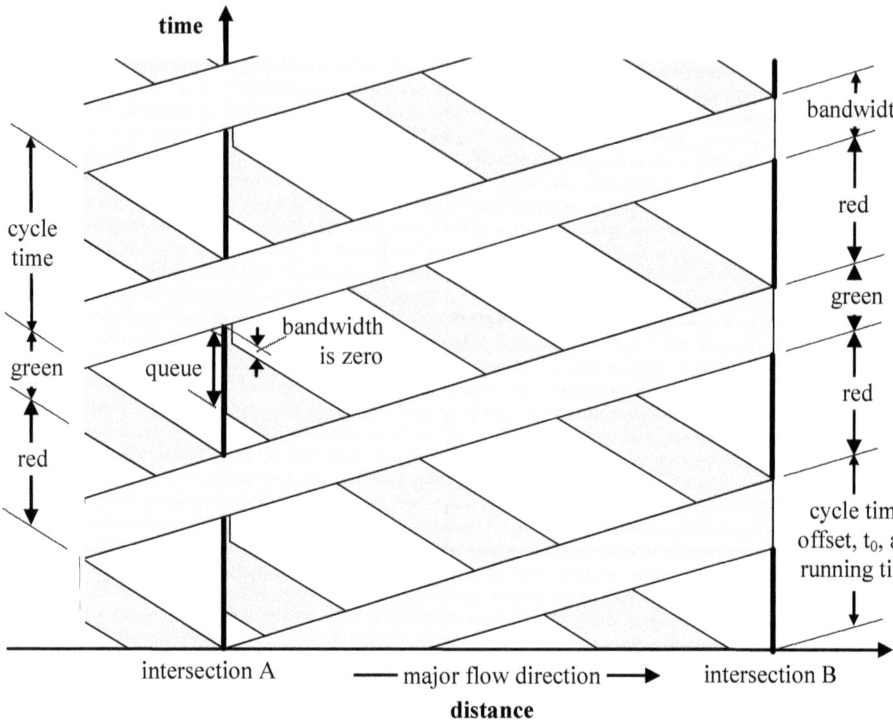

Figure 23.11 Linked signals on a time–distance (space–time) diagram. Trajectories shown are limits based on average running times. The diagram assumes that there are common red phases for the movements in both directions — this will not be so in many cases where leading and/or trailing green arrows are used (Section 23.2.5).

normally, the running time between the two intersections A and B is not a multiple of the cycle time. Assume, for instance, that the running time between the intersections is t_0. For otherwise uniform conditions, counter-flow traffic released on intersection B's green will then arrive at A $2t_0$ after it has changed to green. This will only be the beginning of a green phase for the movement at A in the unlikely event that $2t_0$ equals a multiple of the cycle time at A. In the example in Figure 23.11, traffic in the low volume direction is seen to encounter almost inevitable delays at intersection A. Flow on cross-streets may also suffer as a result of the priority given to the traffic on the co-ordinated link.

One of the differences between signals acting in real isolation and signals working in near proximity to other signals is that the former receive traffic with relatively randomly-distributed headways. In the latter case, there will be a greater tendency for the arrival traffic to be platooned. This influences the type of optimisation formula that should be used. 'Packaged' platoons of traffic can obviously be more efficiently handled than randomly dispersed traffic and signals have been installed on links simply to maintain platoons in their undispersed state. Likewise, flow entering a green-wave link from a side street will normally do so when there is no platoon on the link, and therefore will usually encounter a red signal at the first signalised intersection.

23.7 VEHICLE DETECTORS

23.7.1 Detector technology

A key item in both isolated and ATC systems is the need to detect the presence of a vehicle in a lane. Initially, vehicle detectors were a troublesome item but the technology has improved rapidly. The main methods of vehicle detection are the inductance loop, radar-type reflection devices based on infrared lasers or ultrasonics, and the image-processing software associated with video pictures. These techniques continue to grow in sophistication and usefulness. In particular, video pictures offer a wide range of potential and present applications. Relevant detector operating requirements include:

(a) high sensitivity to vehicles inside the detection zone,
(b) low sensitivity to vehicles outside the detection zone,
(c) low road damage,
(d) short installation time,
(e) low cost, and
(f) low maintenance.

There are two basic types of loop detector:

(1) *passage* detectors which produce a short output simply indicating the passage of a vehicle over the loop;
(2) *presence* which produce an output signal which lasts for as long as the detector senses a vehicle within its detection zone. This duration is called the *occupancy time*.

Detectors can also be classified as:

* small area detectors which detect a vehicle over a length equal to or less than a typical vehicle length of 2 m. They are usually passage detectors located in the traffic flow area.
* large area detectors which detect vehicles within a length of up to 20 m. They usually operate as presence detectors prior to the Stop line.

The most common current vehicle detector is the *inductive wire loop.* Loops are placed about 50 mm below the surface the pavement surface (although loops have worked at depths of up to 500 mm) in rectangular shapes about 2 m by 2 m. A power supply provides the loop with an alternating current having a frequency of about 100 kHz. This produces an alternating magnetic field around the wire. For cars and trucks the vertical flux field above the loop induces an electromotive force (EMF) in the metal of the vehicle chassis passing horizontally through the field, causing eddy currents to flow. These create their own secondary magnetic fields, which in turn induce opposing EMFs back into the loop, thereby causing a change in the relationship between the current flowing in the loop and the voltage across its terminals. The change in inductance depends on the sensitivity of the loop. The process involved is not particularly well understood and little is known about the interaction between a loop and its surrounding environment. The major limitations of the loop detector are its:

* *inherent sensitivity.* The maximum loop sensitivity is at a distance above the plane of the loop, which depends on the loop dimensions. Hence, different loop dimensions give optimum response for cars and trucks, as trucks have a much higher chassis.
* *low sensitivity.* Loops can give unreliable detection of light vehicles such as small motorcycles and cycles. These vehicles are better detected by skewed and 'figure 8' loops rather than by rectangular loops.

* *out-of-lane sensitivity* (or cross talk). This is particularly relevant as the relative change in inductance depends not only on the size and shape of a vehicle but also on its position in a lane. The careful choice of loop dimensions and the use of two or more loops in combination can do much to reduce sensitivity to vehicles in other lanes. Cross-talk can also occur if two loops are too close together (e.g. less than 1.2 m apart), if loop cables share common conduits, saw-cuts or cabinet space (e.g. a separation of under 50 mm), or if cable shielding is inadequate.
* *performance.* This can vary from site to site, depending on road geometry, pavement structure, and the presence of other utilities.
* *external influences.* There can be effects from extraneous nearby conductors, coupling between loops in adjacent lanes, and pick-up of electromagnetic noise from external sources.
* *lead-in cables*. The lead-in cables from the control box to a loop may vary the apparent behaviour of the loop, as the lead-in cable possesses inductance and capacitance of its own. Long lead-in cables can produce an inherently unstable electrical condition, which can be alleviated by operation at lower frequencies. Ideally, the lead-in characteristic impedance should equal the loop reactance at its operating frequency. The insulation must also have stable dielectric properties and water must be excluded.

If a detector fails, modern systems default to the detector being either permanently occupied or unoccupied, depending on which is the safer option.

23.7.2 Detector applications

A detector located in a lane leading to a Stop line is used to give data related to the passage and presence of vehicles, their headways, and possibly their speeds and platooning characteristics. The manner in which this data is used by the signal controller was discussed in Section 23.3.2. The detector location is a matter of some argument as its output must satisfy three diverse needs, viz.:

* detecting vehicle position,
* determining the characteristics of approaching platoons, and
* measuring queue lengths at a red aspect.

A Stop line detector can only determine whether at least one vehicle is waiting. A single detector some vehicle lengths upstream can only detect if at least N vehicles are queued. The use of more than one detector in the traffic lane partly overcomes this dilemma.

Whilst individual vehicle presence can be difficult to measure reliably, counts of vehicles detected over time are relatively stable and reliable. Section 23.3.2 showed that this fact is often used in VA signals. In addition, detectors are also directly used for traffic counting (Section 31.5).

There has been a growing trend in recent years to give priority at signals to *high* occupancy vehicles such as buses and car-poolers and signals are usually set to minimise total delay in terms of people rather than vehicles (Section 30.5.1). The techniques can be broadly classified as passive or active. Passive systems rely on prior estimates of priority vehicle flows at various times (i.e. a fixed-plan system), whereas active systems detect the presence of a priority vehicle (i.e. 'vehicle actuated' system). The main passive approaches are to:

* give transport vehicles priority within a network by adjusting signal timings to favour their routes, and

* exempt transport vehicles from turn restrictions.

Active systems can also, when a transport vehicle is detected,

* extend the relevant movement's green phase,
* call-up a special transport phase,
* advance a bus phase in a cycle,
* provide special turn arrows,
* in the case of trams travelling in the centre of a road, clear turning vehicles blocking the tram's path,
* split the signal phases to give transport vehicles more than one green phase per cycle.

CHAPTER TWENTY-FOUR

Lighting

24.1 INTRODUCTION

24.1.1 Design principles

The visual information revealed to a driver during night-time driving is obtained from light provided by either:

* road lighting,
* vehicle headlights, or
* self-luminous sources.

As the eye functions better at high ambient light levels (Section 16.4.1), road lighting is an essential part of road design. The general design principle for road lighting is that the road environment and any objects in it need to be made sufficiently visible to:

(a) enhance traffic flow, by aiding the driver in the tracking and navigating task (Section 16.5.1). This is done by displaying the pavement ahead, particularly the pavement centreline and edges, as these provide navigational guidance by permitting detection of:
   * the current position of the vehicle,
   * the future direction of the road, and
   * any channelised intersections, lane markings, kerbs, and safety barriers.

(b) enhance traffic safety, mainly by:
   * displaying the pavement ahead, to reveal unexpected objects on the travelled way, other vehicles using the pavement, parked cars, and changes in road condition;
   * revealing the roadside surrounds in order to permit the driver to detect the presence, position, and movement of other road users. Objects to be detected include pedestrians, traffic control devices, and vehicles approaching from joining streets.

(c) enhance pedestrian and cyclist activity by providing them with greater security from aggressive acts;

(d) promote the utility of the roadside, mainly through better appearance and greater business activity;

(e) reduce roadside crime.

Many of the arguments advanced for road lighting relate to design principle (b) — traffic safety — and, in particular, to a reduction in both the rate and severity of night-time crashes. On an exposure basis, a disproportionate number of crashes occur at night (Section 28.1.3) and such crashes are, on the average, more severe than daytime crashes. The severity increase is mainly due to the relative preponderance of multi-vehicle and pedestrian crashes. Broadly, multi-vehicle crashes are twice as likely to be fatal at night than by day and pedestrian crashes are four times more likely to be fatal at night.

At night the driver's visual environment will be severely degraded and the driver will have diminished visual sensitivity (Section 16.4.3). This degradation will be worsened by the glare of oncoming headlights (Section 24.4.2). Therefore, part of the poor crash record can be attributed to the degraded visual conditions at night. However, social habits mean that driver alcohol levels (Section 28.2.2) and fatigue propensity (Section 16.5.4) will also be higher at night — thus making driving markedly more hazardous in a manner that cannot be aided by enhanced lighting levels. In addition the lighting poles themselves can constitute a significant traffic hazard (Section 28.6.1).

Nevertheless, there is general belief that road lighting can reduce fatal crashes by about 70 percent, casualty crashes by at least 25 percent, and property-damage crashes by 15 percent. For example, the application of lighting caused the following typical reductions in reported crashes (sources include CIE, 1992):

road type	percent decrease in crashes
simple intersections	15
urban freeways	15
urban roads	35
rural roads	40
interchanges	40
complex intersections	50

The financial savings resulting from these reductions more than offset the cost of the lighting and benefit–cost ratios as high as four have been suggested. Of course, the relationship between increasing light levels and crash rates will follow a line of diminishing returns.

The ideal design of a road lighting scheme aimed at satisfying principles (a) & (b) would start with an examination of the relevant aspects of the driving task (Chapter 16), leading to a determination of its important visual components and an understanding of how these are influenced by road lighting. In practice, however, only simple visual components can be analysed and even these are difficult to relate to lighting levels. For instance, attempts to relate lighting levels to car-following behaviour as a performance measure were not successful (Armour, 1980). It is therefore necessary to examine the role of lighting in a more pragmatic fashion and focus on the lighting system itself rather than on its interactions with other parts of the road system. This focus will be pursued in the following sections.

24.1.2 Light sources

The light sources (*lamps*) commonly used for road lighting are usually in one of the following six categories.

(a) *Incandescent* (or *tungsten filament*) *lamps*: These lamps provide good optical control and colour rendering and are cheap to purchase. However, they are inefficient to operate and have a short operating life.

(b) *Colour-corrected, high-pressure, mercury-vapour lamps* (*CCMV*): These are commonly used for major routes because of their reliability, efficiency and good colour rendering.

(c) Low-pressure, tubular, *fluorescent lamps*: In many ways these lamps are similar to CCMV lamps in (b), but they have a lower light output. This restricts their use to applications where only low light outputs are needed. They are also difficult to maintain and are temperature-sensitive.

(d) Low-pressure, *sodium-vapour lamps* (*SOX*): Although these are the most efficient light source, their monochromatic yellow gives poor colour rendering.

(e) High-pressure *sodium-vapour lamps* (*SON*): These give better colour rendering than SON lamps (d) and — although not perfect — can challenge CCMV lamps (b).
(f) *Metal-halide lamps*: These are similar to CCMV lamps (b), and have excellent colour rendering. However, they have a shorter life and higher cost.

The assembly housing the lamps, reflector, refractor, diffuser, other covers and enclosures, shields, and electrical fittings is called a *lantern.* The lantern and its supporting pole form a lighting installation. Lantern spacing is determined by methods discussed in Section 24.3. It is typically about five times the lantern height. The spacing is usually decreased on curves, particularly if the lamps are on the inside of the curve.

24.2 LIGHTING THEORY

24.2.1 Design method

Light makes the surface of an object visible by being reflected from the surface and into the eye of the observer (Section 16.4.1). The degree to which the object is visible is discussed in Section 21.2.2 and shown to depend mainly on the object's size, shape, and contrast. Contrast is influenced by lighting and is measured by the *luminance contrast* ratio defined in Section 24.2.2.

The properties of the pavement surface are an important component of road lighting. In operation, a pavement surface will lie somewhere between the extremes of a perfect diffuser providing uniform reflected light and a perfect mirror reflecting the light sources. Dry, well-textured pavements provide a good diffuser and are rarely a cause for concern. However, light will be reflected from a film of water on a wet pavement in a mirror-like (*specular*) manner, which is not conducive to good visibility. This specularity will be reduced and the reflections made more diffuse if the water film is broken up by irregularities and surface asperities.

Light sources for road lighting were discussed in Section 24.1. Road lighting will produce illuminated patches on the pavement surface and illuminate objects at the edge of the road. If the lighting levels are sub-optimal, these patches will appear to the driver as a T, with the stem extending towards the observer. If lighting levels are even lower, the illuminated patches contract into bright streaks between dark patches.

Under wet conditions the stem of the light patch becomes a disconcerting ribbon of reflected light. As the distorted specular image is as bright as the light source, the detection of objects, reflective pavement markers, and traffic markings on the pavement becomes difficult, if not impossible. If the pavement surface is smooth (Section 12.5.1), such problems will arise at quite low rainfall levels. They will be compounded by the accompanying reduction in windscreen visibility and in surface friction. Hence, the relevance of porous friction-courses and coarsely-textured pavement surfaces, as these reduce water film thickness and change the T patches to a more useful elliptical shape providing diffused reflection (Sections 12.2.2 & 12.5.4).

The underlying design approach used for road lighting is based on the light reflected from the pavement surface being beamed into the direction of traffic. Dark objects on the pavement are thus seen by drivers as a dark contrast against a bright background. Vehicle lighting, on the other hand, illuminates objects against a dark background. The road and vehicle lighting provide the illumination and the lit surfaces of the pavements and the objects respectively provide the reflected light (or luminance, Section 24.2.2). It is desirable to minimise lighting overlap between these two light sources in order to avoid both the object and the pavement being illuminated.

The technique is called *silhouette vision* (or silhouette lighting). Silhouette visibility begins at the contrast increment at which an object can just be detected separately from its background (Section 16.4.3). Lighting needs to raise objects well beyond this threshold level because:

* the presence of glare from oncoming vehicles will cause the viewer's threshold detection ability to be degraded (Section 24.2.2), and
* in the driving task it is necessary to detect relative motion between objects, as well as to detect the objects themselves.

Silhouette vision is used for the following reasons:

* the reflection characteristics of the pavement surface permit relatively high levels of reflected light (i.e. high luminances),
* the illuminated pavement carries key pieces of information for the driver,
* the reflection characteristics of objects on the pavement can be poor, and
* at the low levels of light encountered at night, discrimination of colour and detail are impaired (Section 16.4.4) and objects will be seen mainly by contrast (indeed, the use of colours other than white at night may critically decrease the available contrast).

However, silhouette vision has the following disadvantages:

* dark objects on crests and curves may not have an illuminated pavement in their background,
* traffic islands may obscure the illuminated pavement: e.g. separate lighting may be needed to illuminate pedestrians at median crossings,
* a dark object on the pavement may also be lit, thus reducing, or even reversing, the contrast of its silhouette (this lighting can come from vehicle headlights or arise because most road lighting is bi-directional and also shines in the direction of traffic),
* objects may have an inherently low contrast relative to the pavement,
* silhouettes are poor when the pavement is wet, and
* in heavy traffic, other vehicles obscure the pavement surface.

24.2.2 Lighting terminology

In order to specify the properties of the lighting to be used, it is necessary to introduce some lighting terminology. The more important terms used in lighting design are defined below and illustrated in Figures 24.1&2.

Light (or *luminous energy*) is the radiant energy in the visible spectrum (Section 16.4.4) emitted by a radiation source such as a lamp (Section 24.1.2). Light output can be measured in terms of the energy emitted per time, i.e. the power of the radiation. Such a power-based unit is *luminous flux* which is the light emitted by a lantern or received by a surface (e.g. by a pavement). It is measured in *lumen* (lm) and a typical value for a lantern is about 25 klm.

The *luminous efficiency* (or *efficacy*) of a lantern is the lumens produced per watt of power supplied to the lantern. For a typical lantern, this efficiency is usually below 30 percent.

Luminous intensity, LI, is the light (or luminous flux) emitted by a point source (e.g. by a lantern) per solid angle (steradian). It is thus a power-density term and the unit

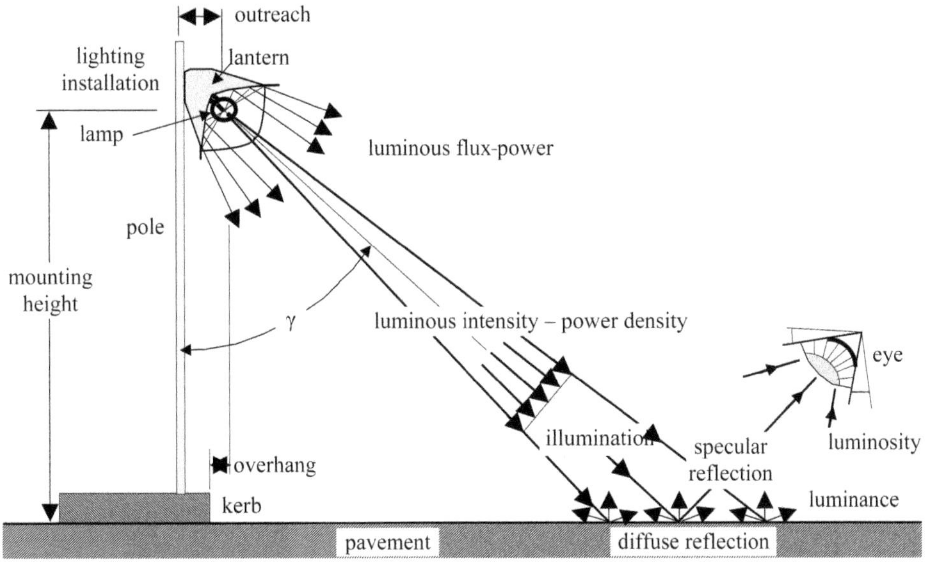

Figure 24.1 Terms used in the design of road lighting.

is the *candela* (cd) which is equivalent to a lumen/steradian (lm/sr). From solid geometry, a light source with an *LI* of a candela emits a total luminous flux of 4π lumen. The candela, rather than the lumen, is one of the seven basic units of the SI system. Originally based on the *LI* of a candle, it is now defined as the *LI* perpendicular to a black body surface that has an area of 1.6 mm^2 and a temperature equivalent to the temperature of solidification of platinum at 101.325 kPa. The *LI* of a lantern usually varies with direction. A set of *LI* contour lines drawn on a sphere centred on a light source is called the iso-candela diagram for the light source.

Illumination (or *illuminance*), E, is the most useful measure of the light arriving at an object and is the appropriate lighting measure for a pavement. It is defined as the luminous intensity incident on a surface per unit area (e.g. of pavement). Its unit is the *lux* (lx) and is equivalent to lm/m^2, or $cd.sr/m^2$. It is dependent on the luminous intensity of the light source, the lighting direction, and the area lit, but does not depend on the properties of the lit surface. From geometry, it can be shown that the lit area of the solid angle cone of luminous intensity decreases inversely with the square of the distance from the luminous source. Thus, the resulting illumination will decrease at the same rate. From Figure 24.2, this distance is H/cosγ and so:

$$E \propto LI/(H/\cos\gamma)^2$$

As the surface area of the pavement is increased by 1/cosγ, then:

$$E = LI\cos^3\gamma/H^2$$

Luminance (or *emitted light* or *photometric brightness*), L, is the most useful measure of the light leaving an object and perceived by the driver's visual system. It is defined as the luminous intensity per unit area reflected in a given direction from a point on a lit surface. Its units are therefore candela per unit area, i.e. cd/m^2, and it measures the light reflected from a surface. Luminance ranges from –6 to –1.5 log cd/m^2 at night, –1.5 to +1.5 in twilight, and +1.5 to +6 log cd/m^2 in daylight. Section 21.2.2 showed that:

* significant visual degradation occurs below 30 cd/m^2 (+1.5 in log units)

* sign luminance should be at least 3 cd/m^2 (+0.5 in log units) to allow detail to be resolved, and
* sign luminances over 100 cd/m^2 (+2 in log units) bring little additional benefit.

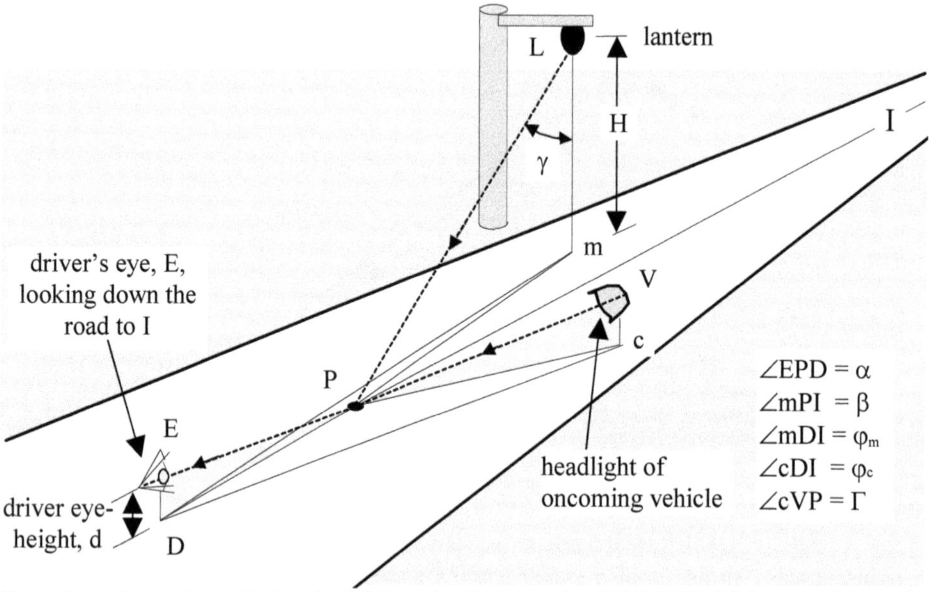

Figure 24.2 Geometric terminology for lighting using the international CIE (1976) lighting co-ordinates. The angles α and ϕ usually have little influence.

It has been suggested that road lighting should provide a new road surface with a luminance of between 0.5 and 2 cd/m^2, depending on the importance of the road, with 1 cd/m^2 being a common value for an 'average' road. Over time and as the lighting degrades, the average value should be kept above 0.7 cd/m^2.

Luminosity (or *brightness*) is the visual sensation related to a surface appearing to emit or reflect light. Luminosity is usually taken as the subjective equivalent of luminance.

The *luminance coefficient* (or *reflectance*) of a surface, q, is the ratio of the luminance, L, to the illumination of the surface, E, and the units are cd/m^{2}lx = 1/sr. Thus:

$$L = qE = q\cos^3\gamma/H^2$$

q is usually measured horizontally but more generally is primarily a function of H, ß and γ defined in Figure 24.2.

The *average luminance coefficient*, q_o, is calculated from q over the range of applicable ß and γ values. The *reduced luminance coefficient*, q_r is the luminance coefficient, q, for given values of the angles ß and γ. It is usually given by:

$$q_r = q\cos^3\gamma$$

Thus, with c as a constant:

$$L = cq_r(LI)/H^2$$

The *luminance contrast ratio* is the ratio of the luminance difference between an object and its background, and the background luminance. Given the discussion in Section 24.2.1, objects therefore fall in one of the following two luminance contrast ranges:

$$lcr = -1 < \text{(object darker than its background)} < lcr = 0$$

$$lcr = 0 < \text{(object brighter than its background)} < lcr = \infty$$

The definition assumes that the observer is adapted to the background luminance. This issue and an observer's threshold luminance contrast ratio are discussed in Sections 16.4.3 & 21.2.2.

The *visibility level* of an object is its luminance contrast ratio divided by the observer's threshold contrast ratio.

*Diffuse reflection (*Section 24.2.1) r, similarly to the luminance coefficient, it changes the illumination on a surface into the luminance of the surface, where r = 0 for a perfect diffuser and r = 1 for a mirror.

*Specular reflection (*Section 24.2.1). Two specular reflection factors are used. S_1 compares the luminance of the pavement, as seen by an approaching driver, to the value directly under the lantern. S_2 relates the mean luminance of the whole of the relevant pavement surface to that directly under the lantern. High values of S indicate difficult, specular conditions. Typical values for S_1 are between 0.1 to 2 and for S_2 are between 1 to 4.

Glare is formally defined as luminance greater than that to which the eye is accustomed. Glare from headlights (Section 24.4.1) and strong roadside lights is of particular concern in road lighting, as it dramatically reduces visibility and is worsened by:

* drivers having little ability to adapt to glare,
* drivers often being forced to look towards the glare source, and
* the pulsating nature of many glare sources.

Glare is accentuated if the average luminance of the visual field is low compared with the luminance of both fixed and vehicle light sources. Glare is usually subdivided into disability and discomfort glare.

Disability (or *veiling*) *glare* affects visibility but may pass unnoticed. It is caused by a scattering of light within the eye, which effectively adds to the background luminance. Its effect on visibility is analogous to placing a luminous veil between the eye and any observed object and is measured in terms of its equivalent veiling luminance in cd/m^2. The threshold value of the luminance contrast ratio is degraded (i.e. increased, Sections 16.4.3 & 21.2.2) for some drivers, particularly older people. The stronger the glare source and the closer it is to the viewer's line of sight, the worse the effect. The combination of these two factors means that the glare from the lights of an oncoming car peaks twice, when it is:

* over 150 m distant and on the driver's line of sight, and then
* within 8 m and therefore of high effective illumination.

Disability glare is worsened by reflections from the surfaces of spectacle lens, but the effect is minimised by the use of modern lens materials.

Discomfort glare relates to higher glare levels at which the driver experiences a physical reaction. It is usually measured on a nine-point CIE mark-scale from '1 = unbearable, severe' to '9 = just noticeable, full glare control'. The middle fifth point is 'just acceptable'. Despite the name, subjects usually relate more to the loss of visibility than to any physical effects.

24.3 LIGHTING SPECIFICATIONS

Following Section 24.1, the performance criteria used in lighting specifications are commonly that the lighting supplied must meet the following objectives.

(a) *Indicate the way ahead.*

This requires an orderly array of lanterns of adequate power and appropriately positioned in relation to the course of the roadway. The lighting of freeways is often specified in terms of warrants for five general levels of lighting:

(1) continuous lighting (usually used when flows exceed 50 kveh/d),
(2) full lighting of interchanges only,
(3) partial lighting of interchanges only,
(4) lighting of terminating roads only, and
(5) no lighting.

Levels (2) to (4) apply with generally decreasing traffic volumes on urban freeways and level (5) applies to rural freeways where vehicle headlights and delineation treatments are often effective. (Section 24.4.4).

(b) *Reveal objects on the carriageway and its shoulders and verges.*

A driver's identification of a distant object must occur sufficiently early for the driver to be able to take action to avoid the object. This visibility distance is given by Section 21.2.4 as:

(minimum size of object to be resolved)/
(minimum angle of resolution of the eye in the circumstances)

and should exceed the vehicle action distance. This distance is conservatively taken as its safe stopping distance — typically 200 m for travel at 100 km/h (Table 19.1). Section 21.2.4 defines a *legibility distance* that is a more severe requirement as it requires discrimination of detail within the object.

The roadway lighting must provide the needed carriageway luminance, and is usually calculated independently of any side lighting from shops or similar adjacent installations.

(c) *Illuminate the roadside and eliminate dark areas in residential streets.*

The lighting must provide a level of illumination over the area between property lines that will:

(c1) enable safe and comfortable pedestrian and cyclist movements, enhance business activity, permit identification of people, discourage illegal acts and identify premises;
(c2) identify street features, such as street names, kerbs and changes in level; and
(c3) indicate street alignment and intersections.

When residents object to the spill-over of such street lighting into their properties, the objections are usually in terms of the incidence of the light, per se, rather than to the level of illumination.

A performance-oriented specification for road lighting would be based on these criteria and written in terms of visibility requirements, whereas a recipe (Section 25.3) lighting specification would be in terms of lighting equipment. Of course, even if the performance approach is followed, the specification of the actual lighting installations must finally be in terms of equipment and geometric layout.

In the absence of specific visibility requirements linked to the performance criteria, the following four lighting parameters usually ensure adequate road user performance, comfort, and satisfaction (CIE, 1977):

(1) *Direct illumination.* This is specified where it is important to ensure the visibility of fixed objects on a carriageway, shoulder, or verge. For instance, the lighting system must be placed to provide sufficient visual guidance to allow a driver to navigate and track the way ahead.

(2) *General illumination level.* The specification of a luminance level (Section 24.2.2) is used to ensure that the silhouettes of objects are seen with sufficient contrast to provide visibility (Section 24.2.1).
(3) *Uniformity of luminance.* Specifying uniformity ensures that no dark areas exist capable of concealing objects bigger than 300 x 300 mm or moving vehicles more than momentarily. The latter requirement usually means that no dark area should extend more than 25 m longitudinally when seen from 75 m. It is desirable that the ratio of the minimum to the average luminance be at least 0.33 overall, and that the minimum/maximum ratios along centrelines be at least 0.25. Minimum to average ratios for longitudinal sections should not be less than 0.5.
(4) *Control of glare.* Glare from the lanterns provided for road lighting is minimised by specifying the light distribution characteristics of the lanterns, using either cut-off or semi-cut-off distributions to ensure that the only light reaching the driver's eye directly is at an angle, γ (Figure 24.1), of under 20°. Glare control thus conflicts with a desire to illuminate as much road as possible with one lantern.

24.4 UNLIT ROADS

24.4.1 Headlights

Even on a straight piece of road, in order to manoeuvre around a detected object, a driver with a response time of 2.5 s (Section 16.5.3) travelling at about 100 km/h will need information on the roadway 70 m ahead of the vehicle. In fact, Section 24.3 suggested that twice this distance is often needed. Table 19.1 suggests that braking distances will be of the order of 200 to 300 m. These requirements are easily satisfied in good daylight, when a driver would experience visibility distances (Section 21.2.4) of the order of 1000 m. From Section 16.4.2 an average young driver would detect detail as large as $(290/10^6)1000 = 300$ mm at this distance and would, for instance, be able to detect an oncoming car.

A quite different situation exists on unlit roads at night. As it will usually prove uneconomic to illuminate long lengths of low-volume road, drivers in rural conditions will particularly depend on vehicle headlighting and other delineation alternatives to road lighting. However, vehicle headlights provide a relatively poor visual environment extending, at the best, 100 m ahead of the vehicle. This is well short of most braking distances. Furthermore, conventional headlight technology appears to have been fully exploited as, for instance, doubling current headlight intensities would only increase visibility distances by about 20 percent. Refer also to the discussion of headlight sight distance in Section 19.4.7.

The luminance in cd/m^2 of an object lit by a headlight is proportional to the luminance coefficient of the object and the luminous intensity of the headlight (in cd) and inversely proportional to the square of their distance apart (Section 24.2.2), further illustrating the problem with headlights. In addition, the effectiveness of headlights can be dramatically reduced by the presence of fog, which can cause major reductions in the luminance contrast (Section 24.2.2) presented to the driver. Although drivers usually make adjustments when visibility distances fall below 250 m, these adjustments are commonly inadequate.

Conventional headlight systems produce four types of beam:

(1) An upper (or driving or main) beam for driving on the open road in the absence of road lighting and of oncoming vehicles. Such beams have intensities of up to 200 kcd. Measurements at signs suggests that at 100 m this is equivalent to an illumination of the sign of a little under 0.05 lux, rising to about 0.25 lux near the sign. Most beams will permit people in dark clothing to be seen at 90 m, implying a luminance of about 3 cd/m^2 (sign data suggests about 4π x 0.06 = 7 cd/m^2.) Even at legal speeds, the associated visibility distance (Section 24.3) is often less than the vehicle's stopping distance (Table 21.1). This means that a driver would need to drive around, rather than stop before, some unexpected night-time object detected in the vehicle's headlights.

Glare-free light (Section 24.2.2) from approaching vehicles can more than double visibility distances by lighting the road surface and hence allowing objects on it to be seen in pure silhouette rather than by the driver relying on reflective differences. It is usual for most approaching drivers at night to 'dip' their upper beam lights, using comfort from reduced glare rather than visibility as the criterion for their action.

(2) A lower (or meeting or dipped) beam used when meeting oncoming traffic. It usually relies on parabolic reflectors and is directed towards the kerbside to further reduce glare and to facilitate the relative positioning of oncoming vehicles. Intensity levels are about 10 kcd and visibility distances between 30 and 90 m.

(3) A town (or running) beam which is an even lower-intensity beam and is used primarily to provide vehicle conspicuity in well-lit residential streets. It does not produce any glare.

(4) A set of presence (or parking) lights, which have much lower intensity than the town beams and therefore have little conspicuity and no value in moving traffic. Vehicle tail (or rear) lights serve to indicate the presence of the vehicle, that the brakes are being used, that the vehicle is turning, and/or that the vehicle is reversing. Typical lighting intensities are about 5 cd for parking, 20 cd for presence, 300 cd for braking, and 700 cd for turn indicators.

There is also some argument for using vehicle headlights in the daytime to increase vehicle conspicuity (Section 21.2.3), particularly for motorcycles where specially-designed *running lights* have found some favour (Section 27.5). Reductions of about 3 percent in multivehicle crashes have been claimed. The light levels need only be sufficient to provide conspicuity in the ambient daytime conditions. The objections to the proposal are usually cost and glare annoyance to other drivers.

24.4.2 Night driving

Many of the visual aspects of night driving were discussed in Sections 16.4.2&3 and 21.2.2. In night driving, objects are generally seen with foveal vision at the centre of the retina of the eye, rather than with peripheral vision (Section 16.4.1). At low light levels the eye loses acuity and peripheral sensitivity. Problems begin at luminances below 30 cd/m^2 and are serious at 3 cd/m^2. Yet Sections 24.2.2 & 24.4.1 have shown that frequently street lighting and headlights do not provide 3 cd/m^2. The objects seen in the available light fall into four categories:

(a) large objects, whose light images are much larger than the full summation area of the retina. In this case the border is the most detectable feature of the object. This has relevance to traffic signs and their edge contours (Section 21.2.3).

(b) intermediate objects;

(c) small objects, where the light from the object falls fully within the retina of the eye (Section 16.4.1). This category contains many objects of concern in night driving, such as tail lights, reflectors, and delineators. A graph for determining the increment in luminance, δL, needed for an expected small object to be just visible against a given background luminance is given in Figure 24.3.

(d) just-visible objects. Figure 24.4 shows how the visual area needed for an object to be just visible increases as the luminance decreases. The objects on which these data were based were 'expected'. The needed visibility area is quadrupled and the visibility distance is halved when the object is unexpected. This is equivalent to reducing the luminance by 6 cd/m^2.

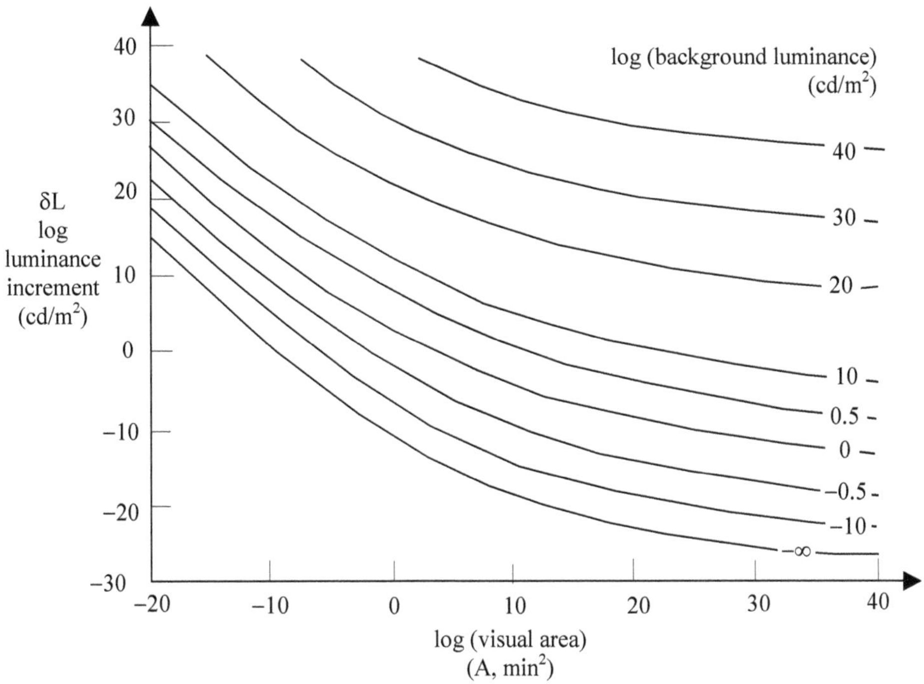

Figure 24.3 Luminance increment (δL) of an object over its background as a function of its visual area. The graph can be used for predicting 'just visible' distances of disc or square objects under night driving conditions when the observer is expecting the object.

24.4.3 Corner-cube reflectors

Corner-cube (or prismatic) reflectors are usually formed of plastic with many reflective, metalised, orthogonal cube-corners moulded into their surface to produce a series of orthogonal mirrors so placed that incident light is reflected two or three times and then re-emitted in the direction from whence it came. This process is called *retroreflection* and is shown in Figure 24.5a.

Section 16.4.2 showed that the eye normally cannot resolve detail less than about 300 μrad. Hence, even a 100 mm delineator will become a point source at distances over about 300 m. Indeed, most corner-cube reflectors are intended to be effective at up to 300 m in light from a car's high beam. The optical performance of delineators is usually measured by their *coefficient of reflex luminance intensity* (*CIL*) (or coefficient of retroreflection) which is the output/input ratio of the reflected luminous intensity (usually 0.2 degrees off-axis, corresponding to a driver's eye height) divided by the illumination falling on the device (usually along its axis). The units of CIL are thus cd.m^2/lm = cd/lx. CIL can also be defined on an area, rather than point source, basis as luminance (rather than luminous intensity) divided by illumination, which is thus CIL/m^2 in units of cd/lx.m^2 = cd/lm. This latter definition gives the SIA measure used in Section 24.4.5. It can be seen as the ratio of the reflected luminance of a sheet in cd/m^2 to its illumination in lm/m^2. Visibility distance increases approximately with the logarithm of the CIL level.

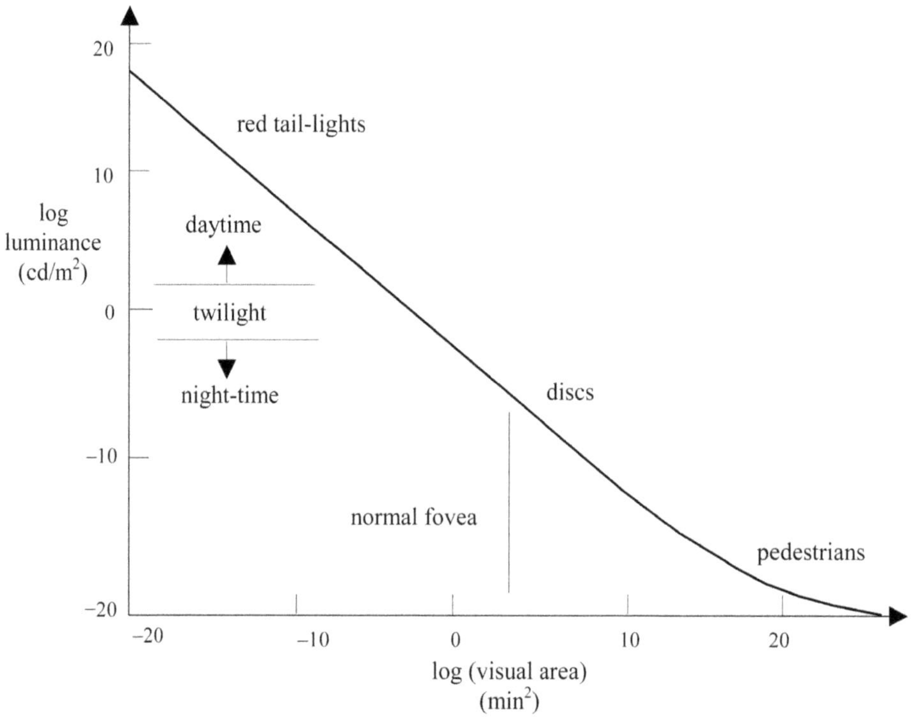

Figure 24.4 Relationship between luminance and visual area required for an object to be 'just visible' under open-road, night-driving conditions. The observer was expecting the object.

Typical CIL values for corner-cube reflectors range from 300 to 10 000 mcd/lx. Values of over 1000 are necessary if delineators are to be effective in the face of glare from oncoming traffic.

Specifying CIL levels is only part of the answer to the problem of ensuring that retroreflective devices are effective at night. As shown in Section 16.4.3, the eye also sees by contrast and so the light from the delineator must also be compared with that of its background.

24.4.4 Delineation

Delineation refers to the sources of visual information which drivers use to decide on the vehicle trajectory and speed that is most appropriate for the geometry of the oncoming road. Drivers receive delineation guidance from both formal and informal sources. Informal sources such as the pavement surface, pavement markings, pavement edges, shoulders, verges, and fences often only provide low-quality guidance. Formal and reliable guidance comes from signs (Chapter 21), edge lines (Section 22.2.2), illuminated pavement markings (Chapter 22), chevron boards, reflective devices, and guide posts.

Pavement delineation generally works by enabling drivers to negotiate a road by short-range tracking. The more remote guidance devices such as chevron boards and guide posts provide for the strategic searching, i.e. the long range tracking, used in navigating the course ahead (Section 16.5.1). Chevron boards are arrowhead signs placed on the roadside on the outside of a curve to indicate the direction of a curved alignment. They have been found to be one of the best forms of long range delineation (Johnston, 1983).

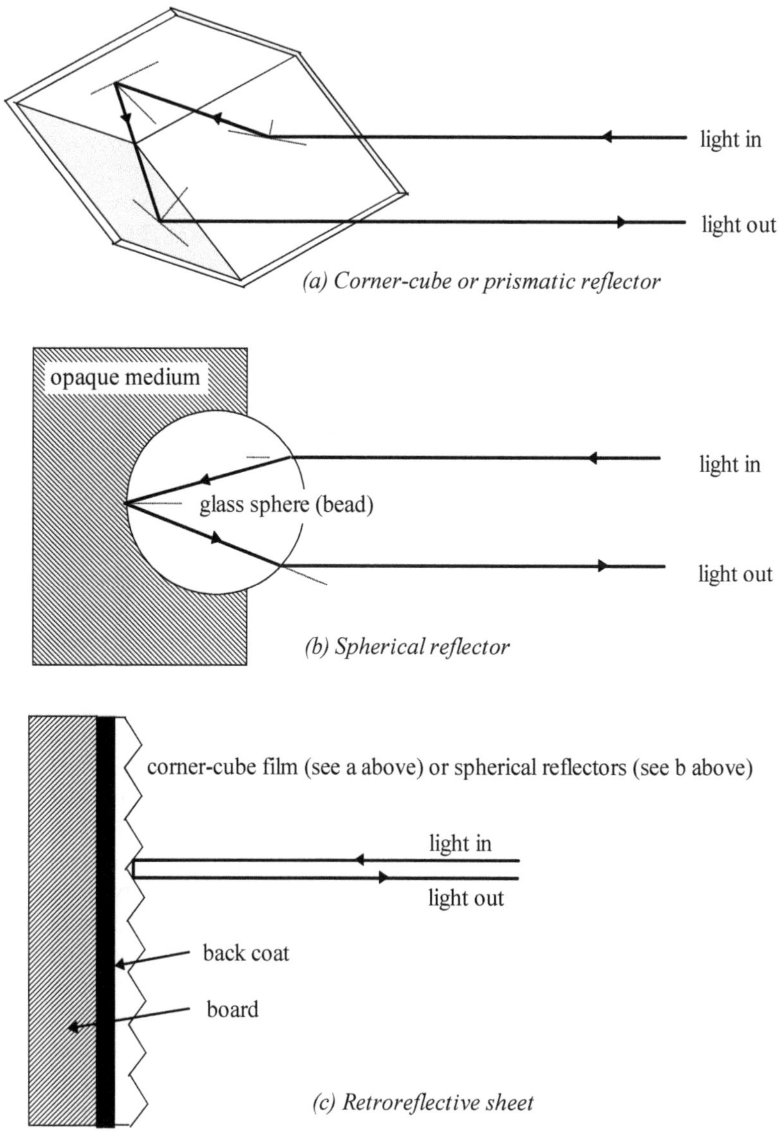

Figure 24.5 Three types of retroreflectivity.

Section 24.4.1 demonstrated that, for night-time driving, drivers will often somewhat recklessly drive with their visibility distance less than their stopping distance. In critical locations this problem is circumvented by using small devices known as delineators. These may be non-reflective devices known as *raised pavement markers* (*RPM*) or *road studs* which are usually made of white or yellow ceramic material and attached to the pavement surface by adhesive. A typical spacing is 35 m. RPMs can provide good wet-weather visibility by raising the visible elements above the level of the water on the pavement surface. They are also considerably more effective than painted lines in influencing the variance in the lateral placement of vehicles in poor light. Both the better visibility and the rumble effect of the RPMs dissuade drivers from encroaching on edges and centrelines. RPMs are also useful in fog, sand, and light snow, but much less effective in heavy snow conditions where they may be covered by snow and/or damaged by snow-removal equipment.

A major advance over the RPM is the *reflective RPM* (or *RRPM* or *cat's eye*) based on the corner-cube reflector (Section 24.4.3). Studies suggest that RRPM are more effective at night than white lines. For instance, by reducing a diver's inadvertent crossing of the centreline, they can reduce head-on and single vehicle crashes by up to 20 percent, and improve driver performance in curves.

The optical performance of RRPMs is measured in terms of their CIL value (Section 24.4.3). After installation there is a rapid drop of CIL from about 300 mcd/lx when new to about 20 mcd/lx after six months in service. RRPM lose their effectiveness over time due to:

(a) dirt accumulating on the reflective elements,
(b) traffic abrading the reflective elements,
(c) breakage of the reflective elements, and
(d) failure of the attachment between RRPM and pavement.

Even when degraded, RRPM give better in-service performance and are about 30 times brighter than retroreflective sheet or paint on a dry day, and 1000 times brighter than paint on a wet day. To accommodate the above decline in properties, it is common to base acceptance criteria in CIL values for RRPM which have been subjected to eight weeks of vehicle exposure. The tests are conducted in both wet and dry conditions.

RRPMs are substantially more expensive than the painted lines discussed in Chapter 22, and so their use has not been as widespread as their relative effectiveness would at first indicate. A significant advance over the RRPM is the intelligent road stud which uses its own power source (usually from solar power) to direct a light signal towards an oncoming motorist. They are particularly effective on curved roads and can achieve close to daylight visibility distances (*c*. 900 m).

Corner-cube reflectors or rectangles of reflective sheet are also attached to guide (or dolly) posts as *post-mounted delineators* (PMD). The posts are usually 1 m high and of flexible white plastic or of 100 x 50 mm timber cross-section with their wider face to the traffic to reduce damage to vehicles. At night PMDs have a visibility distance (Section 24.3) under dipped lights of about 175 m, with a coefficient of variation of 20 percent. Guide posts and their PMDs are used to mark the edge of the roadway and to cater for strategic guidance by indicating the alignment of the oncoming road. In the latter role they thus aid a driver's strategic navigational searching, especially in the presence of extraneous horizontal and vertical lines. Their height above the road surface also allows them to provide some advance guidance on road direction over the crest of a hill. They have little effect on the mean lateral placement of vehicles, particularly if the road already has good pavement markings and RPMs, but they do reduce lateral placement variance on straight roads and on the outside of curves. Reflectors are sometimes placed on the top of New Jersey safety barriers (Section 28.6.2)

The PMD spacing range is from 10 m to 150 m on straight lengths of road, with spacings in the range 100 to 150 m being applied for long range tracking (150 m represents 4.5 s of driving at 120 km/h). The shorter spacings imply a reliance on the PMD to supply short-range delineation, despite the fact that this need is better met by devices closer to the pavement surface and thus to the driver's visual field such as RRPMs and edgelining. On curves, PMD can provide some useful short-range delineation and spacings are, typically, 50 m for 1000 m radius curves down to 5 m for 20 m radius curves. A common additional requirement is that at least three PMDs should be visible at all times, this giving a shape to the curve.

One advantage of both RPM and PMD is that they usually provide a perspective that is below, rather than at, the level of the driver's eye. Hence, drivers obtain both a better view of the way ahead and are more sensitive to variations in their vehicle's position relative to the delineation.

24.4.5 Retroreflective sheet

See Section 21.1.3.

CHAPTER TWENTY-FIVE

Construction

25.1 INTRODUCTION

The planning, location and design of roads and bridges have been discussed in preceding chapters. Once these tasks have been completed, the overall proposal accepted (Chapter 5), the funds obtained, and the land surveyed and acquired (Sections 6.1–3), the construction phase can begin.

The first step in the construction phase is to divide the approved proposal into a number of separate construction projects. Their extent will depend on the work to be done and the financial and construction resources available. Each project will have a well-defined physical scope and a carefully estimated total cost (Section 5.2.3). The scope will be embodied in the specification describing the work to be performed (Section 25.3.1b).

This project work can be divided into the eight broad categories shown in Table 25.1. The categories are approximately listed in a construction sequence. The list is output-oriented and does not directly give such key inputs as the selection and assessment of material (Sections 8.5&6), equipment management, material handling, and the measurement of quantities.

The management of any environmental impacts follows the methods described in Sections 6.4, 13.2.1, 13.3.1, 25.8 & 32.2–3 and minimises the effect of the work on adjoining areas, air-borne dust, and water-borne mud. Dust and mud originate from exposed surfaces and these should therefore be kept to a minimum and stabilised at an early stage by grassing or covering with geofabric (Section 10.7). Dust can be suppressed by watering. Filter traps, sediment basins and spillage traps can be used to capture water-borne material or hazardous liquids before they leave the site.

This chapter will only deal with those aspects of these categories that are not covered elsewhere, and which deserve special attention or do not follow conventional civil engineering procedures. Before proceeding to discuss all these processes, it is first necessary to describe how they are managed because, for a project to be successful, it must be adequately managed.

25.2 BUDGETING AND PROGRAMMING

Section 5.1 showed that the construction cost of a road is typically about 0.1 c/passenger-km or 0.25 c/tonne-km, or about 20 M$/km for a heavily-used 6-lane road. Such expenditures require careful management.

A discussion of budgeting and programming (or scheduling) at program level is given in Section 5.3. For a construction project, budgeting and programming procedures usually involve:

(a) estimating the physical and financial resources needed to undertake the project;
(b) planning the activities required to complete the project;

Table 25.1 List of project tasks. For key to the category numbers, refer also to Tables 14.2 & 26.1.

D Design (Section 6.2)	
D1 road design (Chs 6, 18 & 19)	D2 pavement design (Chs 11, 13)
D3 bridge design (Ch 15)	D4 traffic design (Chs 17, 20, 24)
D5 risk management	
S Surveying (Section 6.1–3)	
S1 surveying	S2 mapping (Section 6.1.2)
S3 site investigation (Section 6.2)	S4 setting out
75 permanent survey marks and reference points	
R Right-of-way (Section 6.3)	
R1 environmental approval (Ch 32)	R2 acquisition (Section 6.3)
R3 permits and approvals	R4 clearing buildings
R5 clearing vegetation (Section 6.4)	R6 clearing top soil
45 fencing	84 site establishment
46 services and utilities (Section 7.4.3)	
0 Formation (Chapter 8 & Sections 11.8 & 25.4)	
011 earthworks in rippable material	012 earthworks in unrippable material
02 enhancing and/or manufacturing material (Sections 8.3–6 & Chapter 10)	
03 embankment and cutting construction	04 trenching
051 formation trimming	052 formation finishing
100 Pavement (Chapters 11 & 12; Table 26.1, Categories 100 & 200)	
111 providing the sub-base	112 providing various basecourses
122 preparation for surfacing	123 providing surfacing
130 edging	140 forming unpaved shoulders
150 providing paths	160 ensuring property access
3 Drainage provision Chapter 13 and Table 26.1, Category 300)	
31 surface drains	32 drainage pits
33 culverts	332 culvert end-walls
34 underground drains (see 04)	
4 Roadside (Section 6.4; Table 26.1, Category 400)	
410 topsoiling and seeding	416 providing trees and shrubs
421 providing rest areas	423 providing parking facilities
424 providing roadside features	43 tidying
44 erosion protection	48 environmental management
5 Traffic facilities (Chapters 21 to 24; Table 26.1, Category 500)	
51 signs	521 guardrail
522 delineation	53 road marking
54 signals	55 roadside services
56 lighting	57 communications
6 Structures (Chapter 15; Table 26.1, Category 600)	
61 bridge substructure	63 bridge superstructure
64 bridge deck	661 floodways
664 retaining walls	
7 Job safety (Section 25.9)	
71 traffic control	72 bad weather
73 worksite safety	74 public safety
8 Overheads (Chapter 25; Table 26.1, Category 800)	
81 project administration	811 contract administration

Table 25.1 continued

812 costing and budgeting	813 programming and scheduling
82 documentation	83 insurance
84 inspection	86 commissioning
85 supervision and surveillance	87 handover of project to operators

(c) forecasting the costs, resources and times associated with those activities;
(d) programming the imminent work and associated resource flows;
(e) recording actual costs, outputs and times, and comparing them with (d);
(f) replanning and reprogramming activities, given variances detected in (e) and changing needs;
(g) documenting the costs incurred, resources used and time taken on the completed project and providing feedback for future projects, and
(h) producing an 'as built' record of the project.

A typical standard list of activities needing construction is contained in Parts 0 to 8 of Table 25.1. The conventional way of costing (or estimating) these activities is to estimate the required quantity of each individual job activity — e.g. from lists such as Table 25.1 — and multiply each by the cost per quantity (the cost rate) for that activity. For example, it might be estimated from the design drawings that there are 6000 m^3 of a basecourse to place (item 112) and it is known from previous jobs that this will cost \$10/m^3 to deliver to the site and \$3/m^3 to lay and compact. The cost of that basecourse is then 6000(10 + 3) = \$78 000. The sum of all these products represents the total cost of the job, although the estimation of the unexpected or intangible may confuse such predictions. For example, weather, site difficulties, and contractor performance can affect the accurate forecasting of expenditure flows. A very similar and related process is described under the Schedule of Rates discussion in Section 25.3.3.

The main costs incurred by the constructing authority will be progress payments to contractors and wages, material and plant-hire costs, and costs associated with the constructing authority's own staff. The sum of all these costs, together with an allowance for unforeseen contingencies, forms the basis for the total estimated cost of the project and must be less than the project budget allocated according to Section 5.3.

The project tasks in Table 25.1 must not only be costed but the time required for each must also be carefully estimated. These estimates of cost and time are only predictions, and will have a variability associated with them. Financial and time allowances, called contingencies, must be provided to cover the more likely events. Contingencies must be applied to each specific event, and to the project outcome as a whole. Clearly, a great deal of judgement and experience must be applied to this process.

The project tasks must also be organised as a logical and efficient sequence of events. The process of allocating and programming construction resources is usually undertaken by some form of network planning, treating the project as a network of individual activities whose optimum sequencing is initially obscure. The resource needs, interrelationships, and starting and finishing dates of each activity are carefully defined as parts of the project network. Even producing the data to this stage is often useful in itself. Once available, the data is then operated on by one of the techniques described below.

Routing is the identification of a path through the activity network that minimises some objective such as total project time or cost: *sequencing* is specifically focussed on minimising time. The construction program is then built around minimising the objective function for the critical route. Common routing methods are described below and are based on either:

* simple bar (or Gantt) charts,
* program evaluation and review (PERT) methods, or
* the critical path method (CPM) approach.

A bar chart shows activities versus time. The activities that comprise the job are listed vertically and the progress of the job in time is plotted horizontally. Each bar runs horizontally from the start of the activity until its completion. The bars do not show the interaction between activities nor which activities are critical. An extension of the bar chart is the *line of balance method*, which does include consideration of the interaction between activities. In particular, it considers the production rates of various dependent processes, which are plotted in terms of time and construction stage.

PERT and CPM address these issues by seeing the project as a network of interconnected activities. The PERT is more applicable to development as it uses a range of possible completion times, whereas CPM is more applicable to construction control as it uses a single completion time. CPM detects that sequence of activities within the project that will govern its completion date. It highlights activities that must be completed before other activities can start, and indicates any spare time (float) between activities in sequence. The activities are then scheduled, and resources allocated.

The selected schedule and associated cost-estimates permit short-term budgeting and programming to be undertaken. The programming discipline ensures that:

* at an early date, problems will be spotlighted and action taken to resolve them before construction begins, and
* progress is monitored from the commencement of design to the completion of construction. Deviations from the program can be detected and action taken to bring the work back on schedule.

Programming and scheduling are thus important management tools, which can also be coupled to ordering and payment procedures. Important features of an effective programming system are that it is based on the best and most up-to-date information and is kept under continuous review.

The money during a project will not relate directly to the outcomes achieved by the project, and other means of monitoring physical progress are required. A simple measure of the value of work satisfactorily completed is *earned value*, which is the product of the proportion of the work physically complete and the total estimated cost of that work. Thus, the total earned value is the sum for all units of work of the product of each unit of work completed times its estimated cost. At the end of the job, the earned value will equal the cost of the job.

The estimate of proportion of work physically complete may require some judgement, and is most simply given as the ratio of units of work completed to the estimated total number of units of work required. Section 25.3 indicates the need for careful documentation of events and of effort and for measuring and reporting procedures to be defined and followed. To achieve this, the job site must be well organised, both administratively and technically.

Stage construction is a process by which a road is open to traffic throughout the various stages of its construction history. It is often thought of as a series of construction stages separated by relatively long periods of time — often years (Section 12.1.1). However, it can also cover stages of only a week or two in length, provided adequate traffic levels and pavement moisture contents are present over those few weeks. One major technical advantage of stage construction is that the holding periods can be used for trial and observation of the efficacy of the work to date. Arguments of this type indicate some benefits of stage construction over the completion of major single jobs (Berry and

Both, 1980), at least for sub-freeway roads. However, in general, each financial investment in a stage will provide a lower marginal rate of return than previous stages.

25.3 CONTRACTS

25.3.1 Definitions

Once physical and financial timings are determined and designs are completed, a project may be undertaken by either an organisation's own work force (sometimes called *force account* or *day labour*) or purchased by contract.

A contract is a voluntary but legally-enforceable agreement between a purchaser (or customer, or consumer, or principal, or owner) and a supplier (or contractor) and is an essential part of a process in which:

* the potential *purchaser defines* the project to be undertaken and *invites* suppliers to consider undertaking the project,
* the potential *supplier* (or tenderer) *offers* definitely and with intent and capacity, *to supply* the purchaser in accordance with requirements of the intended agreement, in return for a significant financial reward, and
* the *purchaser* unconditionally and with intent and financial capacity, *accepts* the offer. If the potential supplier introduces a condition, its offer becomes a counter offer, and hence requires another cycle of the process.

The legal elements of a contract are the offer, the reward, and the acceptance (Antill and Farmer, 1991). The contract must be unambiguous, self-consistent, and capable of being the basis of constructive legal arguments. Its objectives must be legal and achievable, and the two parties to it must both have legal standing. An oral agreement can be treated as a contract.

As the contract is a two-way agreement, both parties have rights and responsibilities and both stand equally before the law. If one of the parties does not act in accord with the agreement, the other (innocent) party can legally remedy the situation, usually by claiming compensation for the costs that the default has caused the innocent party.

The four key elements of the contractual agreement determining the work to be done are as follows:

(1) *General conditions of contract.* These define the broad rights and responsibilities of both parties with respect to the terms and conditions under which the supplier will carry out the required work. They are usually identical to those used in other contracts for similar projects.

(2) *Specification.* In general there is a pyramid of specification types (Korteweg, 2002):

1. *Systems* specifications that stem from societal needs (Chapters 30 and 31), and which specify the outcomes required of the project.
2. *User* specifications that are solution-free and that relate to the road transport network and specify measures such as accessibility (Section 31.2.4), travel time (Section 31.2.3) and safety (Chapter 28).
3. *Functional* specifications of the functional needs relating to the particular road component but which are design-free, and specify outputs such as roughness (Section 14.4), skid resistance and noise generation.
4. *Construction* specifications that specify such specific items as pavements, bridges and landscaping.
5. *Component* specifications that adequately specify the items in #4.

6. *Material* specifications that specify the materials from which the items in #4 are made.

Specifications in levels 1 to 3 above typically relate to contracts between a road agency and its government (Section 4.2). Specifications in levels 4 to 6 are typically those between a road agency and its suppliers.

The specification is a written description of the actual work associated with the project which is the subject of the contract. It particularly sets out:

* the required form, quantity and quality of, and standards for, the project, its components and its outcomes,
* any obligatory procedures,
* how the work performed under the contract will be assessed and paid for, and
* the consequences of any non-compliance with its terms (Section 25.7).

Where applicable, it also includes a schedule of the quantities of and payments for the required work.

Some parts of the specification (and some drawings) will be common to many contracts and other parts will be unique to the particular project. The common parts will normally relate to standard or routine procedures and practices. There is considerable advantage in using existing in-house, national or industry documents for these common activities.

Specifications may be structured in three ways, which are discussed further in Section 25.7.3:

(a) *End-result* (or *performance*) specifications in which the customer specifies the outcome of the process — in particular, how the product will perform over time. Such specifications require a knowledge of the link between properties measured during construction, and operating performance (e.g. the models in Section 14.4.5).
Their advantages are that they:
help ensure that the product achieves its design intent,
provide the contractor with scope to produce cheaper products, and
permit the use of alternative materials.

(b) *Performance-based* specifications which are in method format, but specify properties related to product performance (e.g. fatigue characteristics for pavements Sections 11.4–5).

(c) *Method* (or *recipe* or *prescription*) specifications, in which the customer specifies what is to be done and how.

The following table links the specification purposes, types and structures discussed above.

Parties	Specification type	Specification structure
Government and Road Agency	1. Systems	a. End-result
	2. Users	a. End-result
	3. Functions	a. End-result or b. Performance
Road Agency and Supplier	4. Construction	b. Performance or c. Method
	5. Components	b. Performance or c. Method
	6. Materials	c. Method

There is a temptation for customers to adopt various formats simultaneously but a customer who specifies a process can have little ground to disagree with the outcome of that process. For example, if a customer fully specifies how compaction is to be performed, it would be inconsistent to separately specify the resulting density. Likewise a customer who specifies only the density to be achieved, could not demand further compaction once that density was achieved.

Specifications can rarely completely cover every aspect of a job and the quality of unspecified workmanship is usually assumed to be determined by recourse to current procedures and practice.

(3) *Drawings*. These are graphic descriptions of both the detail and the extent of the works that have been referred to in the specification. Modifications to these drawings, after the contract is signed, should follow a carefully defined set of procedures.

(4) *Engineer/Reviewer*. Some General Conditions call for an Engineer with a well-defined role to be appointed by the Customer. In large Contracts the appointment may be made jointly by the Customer, Supplier and other interested groups (such as lending agencies). In this latter case, the Engineer would be independent of all parties. The Engineer's responsibilities are basically to administer the contract within the practices and code of ethics of the engineering profession. Primarily, these require that the Engineer — despite appointment by the Customer — act impartially when interpreting the contract. The Engineer must also ensure that:

(a) the work is of the required quality,
(b) the contract provisions are followed,
(c) both parties are treated justly in any conflict or dispute, and
(d) the Supplier is paid fairly and expeditiously for work completed.

25.3.2 Contract procedures

The contractual process is usually initiated by the potential Customer inviting potential suppliers to tender (i.e. bid) a price to produce a required project outcome, as defined in the Customer's *tender documents.* The invitation may be a public advertisement, or limited to a list of tenderers who have prequalified for work of the magnitude envisaged, mainly by their performance on previous contracts. For complex jobs, tenderers may be invited to preregister or submit an expression of interest for tendering for the work. The Customer then sends tenders to a select group chosen from the field of preregistrants.

Following the approach in Section 25.3.1, the Customer's tender documents will include the intended contract conditions, a written specification of the project, drawings of the project, and conditions for submitting tenders. These conditions may include a requirement that the tenderers inspect the site before tendering. The tenderers bid for the work on the basis of the tender documents, but must also assume the need to abide by any applicable government rules and regulations. A tenderer may add its own conditions to its bid, but this will usually count against the bid or require a new bid.

The Customer then selects a Supplier from the tenderers, usually on the basis of the favoured tenderer's:

(a) price,
(b) technical, managerial, physical and financial capacity,
(c) other current commitments,
(d) standing in the industry,
(e) record of past performance, and
(f) ability to complete the project satisfactorily.

The Customer then enters a contract with the favoured tenderer to undertake the work in accordance with the selected bid. Although it is thwart with difficulties, the Customer and the favoured tenderer may negotiate terms within the final contract which differ in some way from the terms in the tender document and the offered price.

The Supplier may let contracts to do part of the work to another set of suppliers (or contractors) that are known as sub-contractors. It is usual to then refer to the contract discussed earlier as the Main Contract and to the associated supplier as the Main Contractor.

Progress payments during the course of a contract are usually made either following a schedule in the contract or in accordance with work undertaken. These payments may be adjusted for items such as inflation following a 'rise and fall' clause in the contract, and as defined external charges (such as wage levels) vary. Final contract payments may also include variations for work either added to or deleted from the original contract, either following contract procedures or by mutual agreement. Another area for variation relates to variation in the agreed completion dates. There may be penalties in the contract for late completion and bonuses for early completion.

When a contract is completed, there is commonly a review of the contract and its outcomes. This review will include assessing the adequacy of the design, the quality of the project management, and the performance of the construction contractor. One purpose of the review is to provide feedback to future preregistration and prequalification processes.

Disputes within a contract may arise over such matters as the cost and scope of the works, delays, unexpected site conditions, or performance. If a dispute is not resolved it may be escalated into a formal claim for, for instance, a contract adjustment, extra payments, or extra work. Contracts will usually include a provision for resolving any such differences between the two parties to the contract. This may be by using the Engineer provision (Section 25.3.1(4)), an agreed independent arbitrator, or by appeal to a court of law. The claimant may also use the provisions of the appropriate legal system.

There may also be a warranty covering the successful operation of the road for some years after completion. Warranties relate to the achievement and maintenance of contractually-defined performance levels (e.g. maximum roughness, no potholes) over a time following contract completion (the warranty period). If the performance level is not achieved the contractor may be required to make a cash refund (e.g. 5 percent of the contract sum), or to repair or replace the deficient component. A warranty is thus a form of guarantee, acceptance of liability for defects, or maintenance commitment. The warranty period is sometimes called the defects liability period. Warranties are relatively simple to apply in end-result specifications and very difficult to apply in method specifications.

25.3.3 Forms of contract

Contracts may be defined with respect to the type of activity supplied and are broadly categorised into those related to the:

* conduct of a piece of work,
* supply and delivery of material and goods, or
* commonly, some combination of the first two.

Most current road and bridge contracts are in the last category. Contracts can also be divided with respect to the manner in which the prices are bid. There are three major forms:

(a) *Price–cost* contracts in which the Customer pays the actual cost of all requested work, plus some fixed or percentage fee. This form is not commonly used in roadbuilding.

(b) *Schedule of rates* contracts in which the tenderer bids a rate (e.g. dollars per tonne) to perform the various defined tasks which make up the project and which are listed in a schedule in the specification. Typical quantities to which the rates in such a schedule are applied are the volume of earth to be moved or placed, the area of land to be cleared, the area of subgrade to be prepared, and the length of drains to be installed. The tendered sum is calculated by multiplication and summation from the tendered schedule of rates and the estimated quantities in the specification. The contract is normally for payment on the basis of the actual measured, rather than the estimated, amount of work.

Section 25.7.3 discusses a *modified payments* scheme where the rates are varied depending on the quality delivered. This approach requires a performance specification (Section 25.3.1).

Earthwork measurements for a schedule of rates contract are usually on the basis of one of three separate volumes:

* original earthworks volumes for excavation. This 'in situ' or 'in place' volume is the volume of the material in its natural pre-construction condition.
* loose volumes. 'Loose' volume is the volume after the material has been won but before it has been placed in its new position. It is thus used for the supply and delivery of filling material. The estimation of loose volumes is usually based on the capacity of the haulage trucks used (Section 25.5): and
* final compacted volume of the fill.

Typical ratios of these volumes to the in situ volume are:

Material	in situ	loose	compacted
sand	1.00	1.10	0.95
clay	1.00	1.45	0.90
other soil	1.00	1.25	0.90
rock	1.00	1.65	1.40

(c) *Lump-sum* contracts in which the tenderer bids a total amount, called the 'contract sum', to complete the entire project. To aid tenderers, a lump-sum specification may also contain a schedule of the individual items of work to be performed and the estimated quantities of that work, although these quantities would be excluded from the contract. The tenderer might also be asked to submit a schedule of the rates the tenderer would charge for particular work items, for use in any future negotiations over contract variations beyond the original scope of work and its associated lump-sum price.

An advanced form of lump-sum contract is the *design and construct* contract in which the work to be done includes both designing and building. This allows the tenderer to be innovative and to increase the 'buildability' of the tendered design. On the other hand, the Customer's specification of the work must be less specific and concentrate on performance requirements. The tenderer's bid will include not only a price but also a design, which will need to be assessed on appropriateness as well as on cost. Unless specific provisions are made, losing tenderers are unable to recover the costs they expended on their designs.

Clearly, the most appropriate form of design and construct contract is a contract written in terms of well-defined performance indicators for the product. Because some — probably most — of these indicators will relate to performance over time, the contract payment process must also have a strong time-based component. Some intermediate forms of contract adventurously attempt to predict performance from properties measured at the completion of construction.

Undertaking a contract requires some degree of risk to be taken by each party. It is sometimes naively assumed that the contractor should carry all the risk. Whilst this is theoretically possible, it would be reflected in indefensibly high tender prices. To obtain an optimum use of resources, contracts need to include some risk sharing between the parties (Section 25.7). The underlying principle is that risks are best allocated to the party best able to manage that risk. In addition, unforeseeable risks need to be treated equitably.

The three contract forms listed above are in increasing order of risk to the contractor and so will contain increasing provisions to cover that risk. For example, in a schedule of rates contract the Customer shares more of the risk than in a lump-sum contract and the unpaid pre-tender work done by each contractor is reduced. However, innovative tenders are discouraged.

Two associated forms of contract are *partnering* and *alliancing*. Partnering utilises current contract approaches with informal commitments to cooperate whereas alliancing requires the commitments to be part of separate formal agreements. These are best operated in a quality management context (Section 25.7.1). A contract based on quality management:

* accepts that the two parties to the contract have separate but complementary objectives,
* provides for positive relationships between the parties,
* provides mechanisms for teamwork and cooperation, and
* reduces mistrust and provides mechanisms for conflict resolution.

25.4 CONSTRUCTION PLANT

The first step in planning the use of construction plant follows from the programming processes described in Section 25.2, which will have led to the determination of the type, size and sequence of all construction operations. From this base, the next step is to select the type and number of plant needed to meet these task objectives. Such calculations must take into account the availability of both funds and particular pieces of equipment, and the likelihood of plant breakdowns and severe weather.

In considering the type of plant to select, an optimum balance will exist between using large, expensive and inflexible plant to complete a job quickly, and the alternative of using small, cheap and flexible plant at the sacrifice of short completion times. Another feature of plant operation is that most equipment must be seen as part of an overall equipment system; for example, satisfactory completion of the earthworks will require the capacity of the earthmoving plant to match the capacity of both the compaction and earthmining plant. Whilst it is normal to assume that most plant will operate at some percentage (say 75 percent) of its theoretical capacity, there must still be the ability in the system to handle any transient peaks in supply. Finally, in selecting suitable equipment, heed should be paid to the old construction dictum that 'there is not good and bad construction equipment — just good and bad construction engineers'.

Typical roadmaking plant may be placed in the following categories:

(a) *Plant for rock drilling.* Drills may be either hand-held or machine-mounted, and usually require a compressed air source (an air compressor). Drilling is usually followed by blasting (see b below).

(b) *Plant for initial preparation of the formation*: commonly bulldozers, although explosives are also used.

(b1) *Bulldozers* (or *tractor dozers* or *crawler tractors*) are tractors used for pushing and ripping material, and towing other equipment. The term *dozer* is widely used to cover all types. A bulldozer has a horizontal front-mounted blade which is transverse to the line of the body and which can be raised or lowered in a vertical plane. If the blade can be tilted vertically, the machine is a tilt-dozer; if it can be angled horizontally it is an angle-dozer.

(b2) *Explosives* are used for blasting operations during roadmaking, in building bridge foundations, and for quarries producing roadmaking materials (Section 8.5.9). The original explosive used in civil engineering was blasting powder (gunpowder), which is a mixture of potassium nitrate (saltpetre) or sodium nitrate with charcoal and sulfur. However, most modern explosives are based on nitroglycerine, which is a liquid. This move to nitroglycerine began with dynamite, which is a mixture of nitroglycerine and siliceous earth. It was then replaced with blasting gelatin (gelignite) in which the nitroglycerine is dissolved in colloidal cotton to form a relatively insensitive gelatinous mass. Another important group of explosives includes ammonium nitrate mixtures such as ANFO (ammonium nitrate and fuel oil) and slurries of ammonium nitrate and nitroglycerine or trinitrotoluene (TNT).

(c) *Plant for ripping* (see Section 8.5.9). This equipment usually consists of tines mounted on bulldozers or graders and avoids the more expensive drilling and blasting process discussed in (a) and (b).

(d) *Plant for earthmoving.* Earthmoving may be undertaken by bulldozers (see b) for distances up to about 75 m or where ground conditions are difficult. For distances of up to about one kilometre, material is usually moved by scrapers (see f) and, for longer distances, by loading onto trucks via front-end loaders or elevating graders. The trucks are usually specially designed dump trucks. Their productivity depends on the time they take to load, travel, discharge, and return. Haul times are usually the most critical, and depend on the rolling resistance and grade of the haul path.

(e) *Specialised earthmining plant* such as mechanical (power) shovels for working above grade; back hoes (or backacters) for trench digging; and draglines, grabs, and clamshells for digging below grade.

(f) *Plant for formation construction* (commonly graders, scrapers or bulldozers, see b1).

(f1) A *grader* is a major piece of road-construction plant. It is self-propelled and has a blade positioned between the front and rear axles and able to be raised, lowered, tilted, moved sideways, and revolved horizontally and vertically. The blade typically has a curved plate face (or mulboard or mouldboard) which allows the blade to turn-over the material as well as move it horizontally. The two front-wheels can be leant from the vertical to assist steering and to resist lateral forces produced by earth pressure on the skewed blade. The grader is used to:

* cut, move and spread material (this earthmoving role overlaps with those of the bulldozer, scoop and loader/truck),
* prepare and later maintain the formation, usually to grade requirements,
* prepare and later maintain shoulders and side drains, and
* (sometimes) rip and scarify soil, using tines carried ahead of the blade.

The process of using a grader is known as grading (or blading). A small bulldozer blade can also be fitted.

(f2) A *scraper* is a self-propelled machine with a cutting edge between front and rear axles and which loads, transports, and spreads material. A tractor may be used to assist in pushing the scraper. An older form of scraper is the *drag*, which is a heavy piece of timber dragged along the surface.

(g) *Plant for material preparation.* Before placing imported material it may be necessary to mix materials together to improve the grading of their size distribution (Section 8.3.2). This can be done in-place using graders (see f1), although the efficiency drops away as the material strength increases. In such cases, and on large jobs, batch or continuous mixing plants are usually used. The material is first graded into individual sizes by passing it across screen tables containing openings of a defined size. A grizzly uses a fixed, sloping screen-table and a power screen uses rotating or vibrating screen tables. After this separation, the material is then blended together to give the desired distribution of particle sizes.

(h) *Plant for spreading material.* Generally, material is placed in layers of uniform thickness. Otherwise the reduction in thickness during compaction will be greater in the areas with a relatively thick layer. For example, if care is not taken, a correction layer on top of a rutted pavement may see the rut reproduced on the surface of the new pavement. Tipping and spreading material on the formation must be done accurately and without causing segregation (Section 8.3.2). Tipping can be done directly from the back of a truck or via special spreading devices. It is often followed by watering to achieve the moisture content leading to maximum density (Section 8.2.4). The material is usually placed by:

(h1) A *grader* (see f1),

(h2) A *spreader*, which is a wide-bladed device able to spread dumped material to a defined thickness, or

(h3) An *autograder*, which is a grader capable of automatically maintaining its wide — up to 8 m or two lanes — blade along established survey lines. The paving material can either be placed on the formation in advance or delivered from a truck via a hopper and spread laterally by an auger. This avoids stopping and starting and hence gives a more uniform end-product. Compaction then follows and any non-uniform compaction may require a further trimming pass to be undertaken. This last pass can be avoided by ensuring that most compaction occurs during the screeding process. Autograders and other similar devices commonly use floating-screed pavers in which the screeding template floats on the material being spread and is only pulled horizontally by the tractor unit. A windrow in front of the screed provides some surge capacity to allow the paver to run at constant speed. For a given screed geometry, the vertical position of the screed will depend on the amount of material in the windrow, the towing speed, and the asphalt temperature. These should be kept constant if a uniform output is required. Controls on the screed geometry permit deliberate level changes to be introduced via observation, string-line sensing, or laser-based reference levels. The screed is vibrated and loaded to achieve better placement and surface compaction.

(h4) Asphalt is usually placed by purpose-built *paving machines* — similar to the floating-screed autograder. A machine with a 3 m wide board — one lane-width — travelling at about 0.2 m/s can readily place a 50 mm asphalt layer, needing 260 t/h of asphalt. The longitudinal joints between adjacent pavement layers are discussed in Section 12.4.

(h5) In constructing a rigid concrete pavement over a sub-base, the concrete is placed either within conventional side-forms or by slipforming using mechanical plant. The conventional side-forms often incorporate rails to support and guide the paving machines. The basic equipment consists of a concrete spreader, vibrating and poker compactors, a finishing beam, joint-forming equipment, surface texturing devices, and curing facilities. With slipforming, most of these operations occur within a single machine equipped with travelling side-forms. In addition, some extrusion machines use a top-form as well.

(i) *Plant for quarrying and rock crushing*, see Section 8.5.9.
(j) *Compaction plant*: see Section 25.5.
(k) *Asphalt plant*: see Section 25.6.
(l) *Maintenance plant*: see Chapter 26.
(m) *Trucks, prime-movers and other motor vehicles*: see Chapter 27.
(n) *Field accommodation.*

Another distinction is between whether the plant is tracked or rubber-tyred. Tracked vehicles can operate in difficult construction conditions but will always require special transport facilities for use on roads. Thus, a related distinction is between whether the equipment is only suited for off-road work.

The cost of operating construction plant can be usefully placed in the following categories:

* *capital costs*: costs of purchase or lease, modifications, and depreciation or replacement.
* *operating costs*: costs of insurance, legal registration, delivery, setting-up time, labour, fuels, lubricants, daily service and cleaning, repair of abnormal damage, and dismantling.
* *maintenance costs*: costs of maintenance labour, materials, and replacement parts.

25.5 COMPACTION PLANT

The need for and benefits of soil compaction with respect to pavement layers are described in Section 8.2. Compaction is also used to produce a tight and smooth running surface on unpaved roads (Section 12.1.1). Moisture also helps this process, partly by bringing fine particles to the surface interstices.

The common methods of compaction are ramming, rolling, and vibrating. Maximum compaction is achieved when the material is moistened to optimum moisture content (Section 8.2.4). Liquid additives are occasionally used to reduce interparticle friction. Hand ramming using wooden stumps was the traditional compaction method, but it is now little used although its mechanical equivalents are frequently employed for compacting in small areas or against bridge abutments (Lay, 1992). Modern large-scale ramming uses pugmills developed for brick making. A detailed review of the performance of many types of compaction equipment is given in Parsons (1992). Asphalt compaction is also discussed in Section 12.4. The rollers and vibrators used for the compacting process are as follows:

(a) *Smooth steel-wheeled roller*. This roller relies solely on its mass to achieve compaction of soil or asphalt. The most common form is the three-wheeled self-propelled roller, although tandem rollers weighing between 6 and 12 tonne are often used on asphaltic concrete and 2 to 3 t rollers are used for asphalt footpath construction. To avoid bow wave effects, rollers should be self-driven or pulled and a common option has the roller towed behind a tractor. The individual rollers are typically two metres wide, requiring two passes of a tandem roller for one traverse of a typical traffic lane. Large roller diameters are preferred as they minimise horizontal displacement of soil under the roller and encounter lower horizontal rolling resistance. Although smooth rollers are very versatile, they may cause problems in initially uneven ground by overcompacting the high spots and undercompacting the low spots. This will lead to differential settlement in service.

Asphalt when placed is typically at about 80 percent of its intended density. To achieve the final 20 percent and to produce a smooth surface, it is often compacted using one or two passes of a smooth roller, followed by a pneumatic-tyred roller (*c* below). The process requires care as thin asphalt layers stiffen rapidly as they cool.

(b) *Grid roller*. This roller is used to break soft rock, compact gravels, and force large pieces of rock below the surface (Sections 8.5.2&9). It consists of two drums of open-mesh grid, with a self-cleaning device to prevent clogging of the grid openings. It is pulled at quite a high towing speed of 25 km/h to raise the impact forces and thus help crushing. When used for compaction, the drums work by both direct pressure and a kneading action.

(c) *Pneumatic-tyred roller*. This roller typically has seven or nine wheels each carrying 3 t. It is useful for asphalt (Section 12.2), spray and chip seals (Section 12.1.5) and for materials with low cohesion (Section 8.4.3) such as sands. It produces a more uniform compaction than does the smooth steel-wheeled wheel. The tyre pressure is kept as high as is practical, without bogging the roller or destroying traction. The kneading effect of the tyre partly compensates for contact stresses that are lower than those achieved with smooth steel-wheeled rollers. Due to the stiffness of the tyre wall, the tyre inflation pressure only approximately equals the tyre contact pressure (Section 27.7).

(d) *Sheep's-foot* (or *tamping-foot*) *roller*. This roller has a steel drum fitted with feet and its history was discussed in Section 3.3.6. It produces high contact pressures with low mass, and is useful for compacting clay. The roller avoids the creation of a deceptive surface layer that can give a misleading indication of the compaction achieved. It can also be used successfully on rough or steep areas inaccessible to other rollers. A grid roller (b) can be used to smooth the surface left by the sheep's-foot roller.

(e) *Vibrating roller*. These smooth rollers are about as effective as similar non-vibrating rollers of twice their mass, as the vibration raises the effective compactive effort (Section 8.2.3), particularly by causing momentary drops in internal friction (Section 8.4.3). They are thus better for non-cohesive materials than for clays. They can have an influence to a depth of a metre. Low-frequency, high-amplitude rollers are used for unbound courses and high-frequency, low-amplitude rollers are used for asphalt, mainly because of the better surface finish. Care must be taken to limit the vertical force used on mixes with a high bitumen content. Oscillating rollers vibrate horizontally as well as vertically.

(f) *Impact roller*. This square wheeled roller was developed in South Africa for use on silty sands.

Heavier rollers are usually better than light ones, other things being equal. One caution is that a roller may not necessarily compact a material, i.e. increase the material density. For example, it may merely relocate material and — because there is little restraint on vertical movement near the roller — may cause rutting and ridging. Care must also be taken to ensure that the increased interparticle stresses do not produce excessive breakdown of large particles, as this rise in the fineness of the particle sizes may cause a deterioration in other properties (Section 8.3.3). This particularly applies to soft material, such as chalk, which may therefore need to be compacted with a light roller.

In defining compaction procedures, Sections 8.2.4–5 showed that there is a maximum density obtainable with a particular combination of roller, contact pressure, layer thickness, sub-layer stiffness, and number of passes. This is usually determined by calibration during preliminary field trials (e.g. Figure 8.4). Normally between eight and 15 passes of a roller are needed to adequately compact a clayey soil. It will always be an economic limit on the number of passes to be applied.

The *optimum moisture content* for compaction (Section 8.2.4) will be lowered if either the contact pressure or the number of passes is increased. Although the rate of compaction decreases with speed of rolling, overall economics usually favour rolling at as high a speed as is possible. The maximum depth of soil that can be compacted in a single layer is typically:

* 150 mm for clay,
* 300 mm for clay in non-critical locations, and
* 500 mm for cohesionless soils.

If the surface of the receiving layer is too smooth to provide the necessary shear transfer (Section 11.1.1), it will need to be roughened or scarified before the new layer is placed.

The effectiveness of a compaction process is assessed by measuring the achieved densities. Compaction testing and density measurement procedures are discussed in Section 8.2.2. The nuclear gauge is the usual first choice for field testing.

Each placed layer resulting from a single day's production should be separately tested. Tests should not be conducted within 200 mm of the edge of the work or within 5 m of a lateral construction joint. Within a layer, the material tested should be homogenous with respect to material type and appearance, response to rolling and moisture content. Appropriate provision must also be made for any material particles which are larger than can be handled by the chosen compaction test.

Impact Soil Testers provide quick non-destructive ways of determining the adequacy of the compaction of bound and unbound courses, particularly as a means of construction control and as a determinant of construction uniformity. A typical device (e.g. Clegg) is essentially a laboratory compaction hammer (Section 8.2.5) with a base, a series of vent holes, and a carrying handle added to the guide tube. The hammer is dropped on to the basecourse. Peak accelerations of the hammer on impact then give a measure of dynamic soil properties. For the top 75 mm of compacted material, such a tester is a useful means of:

(a) detecting relatively poorly-compacted areas, and
(b) determining when rolling is no longer producing additional compaction.

25.6 ASPHALT PLANT

Asphalt is discussed as a material in Section 12.2, its placement and compaction is discussed in Sections 12.4 & 25.5, and this section examines how it is produced. The basic functions of an asphalt plant are as follows:

(a) delivering aggregate, which may be wet and cold, from storage bins;
(b) drying and heating the aggregate;
(c) mixing specified proportions and grading of aggregate, filler, and binder; and
(d) discharging the mix to suitable delivery vehicles.

These are the key quantities to control as input. The two main types of plant in common use are:

(1) *Batch plants*: These are flexible plants able to operate intermittently and change readily from one mix type to another. Batch sizes range from 0.3 to 6 t, and outputs of up to 120 t/h are possible. Bitumen must be initially heated to at least 160 C to achieve a suitable wetting and mixing viscosity of under 200 mPa.s (Sections 8.7.4–5). Aggregates are dried by heating to temperatures of between 150 and 190 C, a process which reduces the moisture content to about 0.3 percent. All the ingredients are separately proportioned on scales and mixing occurs in a pugmill chamber. The mixing time must be sufficient to coat all the particles of aggregate with bitumen. However, it must be kept as short and at as low a temperature as possible, as the exposure of the thin films of bitumen to heat and air will result in early oxidation and hardening of the bitumen (Section 8.7.7). Mixing normally occurs in the 135 to 165 C temperature range, although exceptional mixes extend the range to 120 to 190 C.

(2) *Continuous plants*: Drum-mixing plants represent the most recent development in this category, with the aggregate being dried, heated and mixed in one operation. Large quantities of an unchanged mix are preferred and a typical output is 200 t/h. The resulting product is commonly called *plant mix* (Section 12.2.1). The ingredients are proportioned by volume through calibrated gates, belt scales, and metered pumps. There is often automatic temperature control. The heating and mixing occur in a relatively inert atmosphere to minimise oxidation and hardening. The mixing can take place at lower temperatures (115 – 130 C) but higher moisture contents (1 percent) than with batch plants. The gas flow in the drum draws off the fine particles in the aggregate and sand. Cyclones are used to help recirculate these fines or to add more fine filler material to compensate for their loss.

Plant can also be categorised as:

* stationary,
* portable, i.e. easily relocatable, or
* fully mobile, i.e. on wheels.

For effective mixing of asphaltic concrete, there are upper and lower limits on the mixing temperature. It must be sufficiently high to:

* keep the bitumen binder sufficiently fluid to adequately coat the aggregate, and
* dry the aggregate (Section 12.3.2);

and sufficiently low to avoid causing:

* the bitumen to drain away from the aggregate,
* unnecessary hardening of the bitumen (Section 8.7.7), and
* extra fumes, which can cause irritations (see below).

Asphalt can be stored in retention bins for periods of up to 12 hours before special measures, such as the use of an inert atmosphere or silicone additives are needed to prevent excessive hardening of the bitumen. Asphalt is usually transported in covered trucks to retain heat and prevent the ingress of moisture. Insulated trucks are used for long hauls or in cold climates. The minimum transported lot is about 6 t. Minimum delivery temperatures are commonly in the range 115 to 155 C, extending to 95 to 175 C, depending on mix type.

The fumes produced during asphalt mixing and placing have caused eye, nose and throat irritation and nausea. They can be minimised by not overheating the bitumen and by using fume extractors. However, the major contributor to fume generation is often remnants of diesel oil sometimes used to clean or lubricate the equipment.

As output, the key measures to control are the density and/or void content of the mix (Section 12.2.3). Usually, these cannot be determined until the mix has been placed and compacted (Section 25.5). Sometimes the compaction process is modified to accommodate variations in the properties of the product provided by the mixing plant.

25.7 QUALITY MANAGEMENT

25.7.1 Total quality management

Following Section 25.3.2, construction is undertaken by a Contractor for a Customer. *Total quality management* (TQM) is used to ensure that the process delivers value and requires that:

(a) the Customer accurately determines its needs and defines fitness for purpose, leading to the Specification containing a clear statement of required Contract outcomes,
(b) co-operative Contractor–Customer relationships are established and maintained,
(c) the Customer accepts the Contractor's quality plan,
(d) Customer and Contractor deliver the Specification outcomes by planning and managing well-defined work processes and procedures (process control, Section 25.7.2),
(e) the chosen processes and procedures are managed to deliver their intended outcomes (quality assurance, Section 25.7.3),
(f) where appropriate, processes are altered during production to eliminate further rework and ensure continuous process improvement,
(g) any non-conformances in the processes and products are reported and rectified, and
(h) objective evidence is kept of the proper execution of each of the six actions listed above.

The combination of process and acceptance control in (d) and (e) is the key component of the quality system. Indeed, for the Contractor, process control is a form of quality assurance. For both parties:

* there is little merit in detecting unsatisfactory work after a job is completed, and
* it is better to improve quality by determining and correcting the causes of poor quality, rather than by increased inspection and testing.

Process control methods:

* check for changes in a process, so that its output can be kept within a desired quality range,
* give early warning signals that may lead to a useful increase in sampling frequency for control testing,
* facilitate any necessary changes in the process,

Some Customers also use process control, but their primary focus should be on the outcome of the process.

Process control uses extensive baseline information, supervisory observations, and test results. A core tool is the process control chart which plots the observed data in a time sequence relative to both the baseline data, and upper and lower control limits. The charts give visual evidence of the need to act and can avoid over-reacting to an unexceptional event.

Quality management and control can be applied prior to construction (e.g. to the materials used), during construction (e.g. to the equipment usage), and after construction (e.g. to the pavement profile). However, applications after construction are usually forms of acceptance control.

25.7.2 Variation and risk

The processes lead to products. The products of all construction processes will be of variable quality, despite the best of human intentions and desires. Overall variation will arise from variation inherent in the:

* materials used,
* production processes, and
* sampling and testing methods.

Indeed, if one of these is much larger than the others, then there is little to be gained in strenuous efforts to reduce the variation in the others.

Techniques allowing a product to be accepted or rejected in accordance with a specification are known as acceptance control (or compliance judgement) methods. They compare test results and observations with absolute levels set by the specification. The same process information can be used for both process and acceptance control, however the two will often use different tests and sampling frequencies.

In non-quality managed contracts, the decision-maker would be the Customer. In a contract based on quality management, the Contractor makes the acceptance decision, reporting non-conformances to the Customer. The Customer can audit test the process, although such audit testing would be less frequent than control testing. During construction, *hold points* may be introduced to allow work to be checked before it is lost to view.

Variation means that, unless all the product is tested, there will be a risk that in taking and testing a sample from a population (Section 33.4.1), a wrong decision will be made as to the quality of the product. In particular:

* *Customer's risk* is the risk that an unusually good sample will lead to accepting a generally bad product. More precisely, it is the risk that a population of unacceptable quality will be accepted by the consumer under a given sampling scheme, on the evidence of a valid but optimistic set of samples.
* *Contractor's risk*, on the other hand, is the risk that an unusually bad sample will lead to rejecting the Contractor's generally good product. More precisely, it is the risk that a population of acceptable quality will be rejected under a given sampling scheme, on the evidence of a valid but pessimistic set of samples. A key insight is that an occasional poor result does not necessarily mean a poor product.

The probability of acceptance of a sample will depend on the:

* Customer's and Contractor's risks, and
* distribution of the test results,

and is discussed further in the next section.

When a sample is taken and tested, there are various ways in which the test results can be expressed and subsequently used by the Specification. The average value of the test results is not, on its own, a useful property as it gives no indication of the variation of data. Technically, variation is measured by the standard deviation of the results and is combined with the average to produce the more useful *characteristic value* (Section 33.4.3). For example, the 95th percentile characteristic value is that exceeded by 95 percent of the product. A production lot is accepted if the sample characteristic value exceeds a pre-set value, L. The method operates at maximum efficiency if the standard deviation of the population is known (Gray and Robinson, 1980).

A common specification-related measure of variation is the *proportion defective,* or the *acceptable quality level,* A. This is the proportion of results outside the specified level, L, in a sample of tests taken from the lot being examined. It is effectively a count of defects. However, the term *defect* comes from quality control in the manufacturing industry and is too strong a word to use to properly describe roadbuilding material that is outside a specification limit (Gray and Robinson, 1980). Both L and the acceptable proportion defective should be determined at the design stage. In this context, quality control should not aim for 'best quality material' as:

* there will be an increasing cost associated with increasing quality, at a rate which will tend to snowball, and
* the benefits arising from increased quality will tend to diminish.

Thus there will be some point at which quality and benefit/cost ratios are optimised and a sufficient price is paid for a sufficient quality.

25.7.3 Specification aspects

Because variation exists in all elements of a system, it must be recognised within a specification, in process control, and in quality assurance. These systems must therefore have a strong statistical basis and the risks associated with any decision must be clearly understood and the consequences unambiguously allocated to either Customer or Contractor. A specification where quality and risk are unclear will not only result in uncertain outcomes, but will also attract higher tender prices as tenderers attempt to cover themselves against the obvious uncertainties.

As a consequence of process control and quality management, the Contractor offers the Customer a product that the Contractor believes complies with the contract's specification of process and product (Section 25.3.3). The Customer accepts or rejects the product. The ways in which the two parties can manage this process in a logical and equitable manner, in the face of the above variability, hinges on:

* the Customer clearly specifying its requirements for the production process and the product. In doing this, the Customer should — as far as possible — avoid specifying process (a method specification, Section 25.3.1), and concentrate on specifying the end product, leaving the Contractor to select the most appropriate process. Only where the end product is difficult to measure, should it be necessary to specify the process to be used. A Contractor-nominated process can be made part of the contract.
* the Contractor being sure it correctly understands the Customer's requirements,
* the Contractor applying process control (Section 25.7.1) to its production processes,
* the contract providing quality assurance (Section 25.7.1) by specifying a reasonable regime of sampling, testing, and statistical assessment, making allowance for the variability of both the product and the sampling, and the criticality of each product property. Testing is expensive but the frequency at which samples are taken and tested is often the easiest and best variable to increase in times of uncertainty. It is therefore necessary to specify precisely:
  + the sampling process (Section 25.7.2),

+ the testing procedure (Chapter 8), and
+ the acceptance plan to be used for acting on the test results (see below)
* the Customer auditing the Contractor's system.

The tests may be conducted as part of the Contractor's testing for process control or the Customer's audit testing.

These processes will provide a basis for the Contractor's decision to release its product and the Customer's decision to accept it.

The acceptance plan for a Contract can be rationally represented by a plot of the probability of acceptance of a product as a function of its proportion defective (Section 25.7.2). This is called the *operating characteristic* of the specification and the Contractor's and Customer's risks will be two extreme points on this plot. For example, the target quality for pavement layer thickness could be defined by a 25 percent proportion defective relative to design thickness with probabilities of lot acceptance of 90 percent for asphalt surfaces, 99 percent for intermediate courses and 95 percent for subgrades (Auff, 1982).

A useful variation on this process is the modified payments approach. Following definitions in Section 25.7.2, test levels A (the acceptable quality level) which result in unequivocal acceptance, or in unnegotiable rejection, R (the rejectable quality level), are separately defined in the performance specification relative to the design intent or specification level, L ($R < L < A$). Any product of quality, Q, where $R < Q < L$ is accepted, but at an agreed reduced payment based on a function of $(L - Q)/(L - R)$. Charges might be levied for replacing product where $Q < R$. Similarly, bonus payments might be made for cases where $Q > L$. The selection of the desired levels of the property values ($R < L < A$) will stem from the various design considerations discussed, for example, in Chapters 11, 12, 14 and 15.

When detected by the Contractor, deviations from the agreed test results and processes are non-conformances. When detected by the Customer, they are *non-compliances*. Each should be resolved to the satisfaction of both parties.

25.7.4 Physical properties

Improved and more cost-effective control of the properties achieved in road construction can usually be achieved by sampling more often, by random sampling, and by using better equipment. The first two of these techniques were discussed in Section 25.7.3 and this section discusses the role of equipment selection.

Measurements of pavement level, thickness (see below), and profile (Section 14.4) indicate that these features are normally distributed (Section 33.4.6 and Auff, 1982). It is common to assume that most other properties are also normally distributed. A typical variability achieved on pavement levels, at 95 percent probability, ranges between ±5 mm and ±75 mm, depending on the method used to control levels. Four common categories of level control for pavement construction depend firstly on the equipment used (Section 25.4) and are (Auff, 1982):

(a) spreader or autograder, with final trimming by grader and dumpy level — average standard deviation = 12 mm, although values as low as 3 mm have been achieved;
(b) grader, with control less than dumpy level but better than by eye — average standard deviation = 15 mm;
(c) grader, with control by eye to taped, closely-spaced pegs — average standard deviation = 25 mm; and

(d) grader, with control by eye to taped, widely-spaced pegs — average standard deviation = 30 mm; and
(e) non-professional staff — average standard deviation = 45 mm.

The survey control and the spacing of construction pegs are both of vital importance in reducing level variability.

It is suggested (Auff, 1982) that there are three distinct standards of pavement level control, viz.: Standard I, Standard II, and Standard III, as defined by the following data:

	Specification standard deviation (mm)	
Standard*	Target	Limiting
Ia	9	10
Ib	10	14
IIa	15	17
IIb	18	21
IIIc	21	24
IIId	25	29
IIIe	37	43

*a, b and c refer to the control categories given above.

Thickness is commonly measured by digging a pit or using an auger to provide core samples. Ground-penetrating radar (50 kHz) can also be used, as some radar is reflected back to the surface at each interface.

25.8 VIBRATION AND NOISE

Construction operations can cause ground vibrations which may lead to cracking in adjacent structures due to the movements produced and to differential settlement due to enhanced compaction (Sections 8.2 & 25.5). As the strain caused in the ground by the vibration is proportional to the particle speed, the resulting damage correlates most closely with particle speed. Damage is independent of the frequency of vibration, unless resonance occurs at a natural frequency of the affected structure. Particle speed can be readily measured and limits are commonly set to prevent damage. Common limits in mm/s are:

* 10 to minimise complaints or if the vibration is continuous, such as from vibrating rollers (Section 25.5) and pile drivers (Section 15.2.4),
* 20 to control architectural damage,
* 50 to prevent major damage, and
* 100 in strong, coherent rock formations

Ground vibrations are also caused by operating traffic and British experience is that as many people are bothered by traffic vibration as are by traffic noise. Although people are usually concerned that such vibrations will damage their homes, there is no evidence to support this fear (Watts, 1984). An ISO standard provides specific guidance on building vibrations (ISO, 1990).

The effects of noise due to operating traffic are discussed in Section 32.2.3. Noise generated during construction can affect people in the following three ways:

(a) Noise can mask normal auditory warning signals and therefore increase the hazardous nature of the construction environment.
(b) Noise can interfere with speech and thus cause communications to deteriorate. No voice communication is possible at over 90 dB(A) at distances above 500

mm. Many earthmoving devices produce noise levels above 90 dB(A) and so it would be difficult, if not impossible, to make contact with an unprotected operator.

(c) Extreme noise (typically, > 90 dB(A)) can cause a temporary or permanent loss of hearing acuity (*sociocusis*), raising of the hearing threshold level. The effect is approximately proportional to the time of exposure to the noise. Hearing loss associated with this is termed a noise-induced, permanent, threshold shift. Hearing loss also occurs with age (oresbycusis) and it is often difficult to distinguish it from sociocusis.

Noise alleviation generally is discussed in Section 32.2.5. The four common noise-alleviation methods at construction sites are:

(1) avoiding or limiting noisy activities and monitoring noise levels (Section 32.2.2);
(2) suppression of the noise at source, e.g. quietening an air compressor and maintaining engine exhausts;
(3) changes to the noise propagation path, e.g. with cabins, enclosures and barriers;
(4) personal protection via earmuffs and ear plugs. This approach is usually very cost-effective, but intended wearers of devices can become lax in their use and some devices prevent all noise reaching the wearer, thus making communication difficult.

25.9 TRAFFIC AT CONSTRUCTION SITES

Construction and maintenance work can often be carried out with little disruption to existing traffic, by the use of such measures as detours, by-passes, lane diversions, movable traffic barriers (Section 28.6.2), or contra-flow lanes (Section 30.4.2). The *work zone* is defined as length of road in which drivers are restricted by the construction work.

When traffic and roadworkers are required to share the same piece of road, direct traffic control must be employed to minimise traffic delays and crashes inflicted on motorists, danger to construction workers, and disruption to the construction job. Even when careful measures are taken, it will commonly be found that both crashes and delays increase:

* crash rates during construction are about 60 to 120 percent greater than the preconstruction rate at the site. The approaches to the site are usually the areas of greatest danger.
* delays are caused by reductions in effective lane capacity. For example, closing a lane will commonly halve the capacity of the adjacent 'free' lane. This is mainly due to two effects in Section 17.4.3:
  (a) the clearance factor, and
  (b) the gawking factor (passing drivers slowing to observe operations).

The prime need is for an effective traffic-control plan to be developed for each work site, followed by the nomination of a person at the site to be responsible for the implementation and management of the plan. The contents of the plan might include the overall strategy being adopted, the staging of the work, modifications to be made to the worksite, and the signs, delineators, and related devices which must be installed and policed. The main objectives of the devices are to:

(a) give advance warning of the hazard caused by the construction work (e.g. 'roadwork ahead' signs and roadside lamps),

(b) indicate the extent and nature of the hazard (e.g. 'soft edges' and 'loose stones'),
(c) guide drivers and pedestrians around or through the hazard (e.g. impose a work-site speed limit or sign-post detours),
(d) protect the work-site (e.g. portable barriers, trucks with energy-absorbers attached to their rear and parked upstream of the work site, and high-visibility jackets worn by all staff on the work site, and
(e) provide any necessary traffic control.

Such devices must be erected before work commences, checked regularly during the job, and removed as soon as is practical after the job is completed. Policing may also be important, as poor observance of signs by drivers can jeopardise both their own safety and that of the construction workers.

The common minimum luminous intensity (Section 24.2.2) for lamps intended to warn of a traffic hazard is 0.75 cd. These are intended to give an effective night-time warning at distances of at least 100 m and it will be seen from Table 19.1 that this will be adequate for drivers travelling at 60 km/h or less. These minimum-standard lamps will thus be adequate on many residential streets if located at the hazard, but for rural roads and many arterials they will need to be either placed in advance of the hazard or replaced by higher intensity units. Indeed, rotating or halogen strobe lights are far more effective than static lamps.

High-intensity warning signals typically use flashing amber or yellow lamps with a minimum luminous intensity of 50 cd. The lamps usually have a highly directional output and so must be correctly aligned with respect to the traffic flow. This lighting is adequate for most daytime use, giving an effective warning at 150 m for travel at 90 km/h or less (Table 19.1).

Studies of driver behaviour at construction sites show that the sight of the operation itself exerts a major influence on vehicle speeds past the work (Cordingley and Jarvis, 1982). The presence of some signs causes a further speed reduction. However, the only control treatments which caused major additional speed reductions were those which contained an element of driver surveillance — such as police officers or *flaggers* (or *flagmen* or *batmen*) from the construction team. Flaggers typically use red flags or stop–go signs on the end of a pole to signal to approaching drivers to stop or slow down. They are heeded, at least partially, by most drivers.

There are definite limits to the delays that drivers will tolerate at a construction site and traffic management breaks down when these limits are exceeded. One of the commonest difficult situations occurs on two-way two-lane roads in which the closure of one lane means that traffic in two directions must share the other lane. This is called a *shuttle lane*. Flaggers have usually been used to control this situation but their role is becoming industrially and economically less acceptable. A common replacement for flaggers is the use of portable traffic signals using either conventional red–yellow–green aspects (Section 23.2.1) or stop–go signs. With respect to the actual signal settings used, the capacity of a shuttle lane relates primarily to the duration of the all-red period and thus to the length of the shuttle lane, which must be cleared before the opposing traffic can be given a green light.

For traffic guidance on wide, multi-lane carriageways it may be necessary to use conspicuous dynamic displays and changeable message signs. These 'automatic' control systems can be operated by either cable, radio, clocks, or vehicle detectors (Section 23.7). Clock (or fixed time) operation is not recommended as it produces excessive delays and encourages non-observance.

Worksite traffic barriers (Section 28.6.2) should be capable of safely resisting the impact of traffic travelling at highway speeds. If the worksite contains workers, fragile temporary works (e.g. scaffolding), or would be hazardous to vehicles, it is not sufficient to use barriers that only operate as guides, separators or delineators. It is also important to ensure that the barriers themselves are very visible, can restrain traffic and do not themselves create a hazard. Portable concrete or water-filled traffic barriers (Section 28.6.2) have proved very effective in this regard.

Mud transferred from the work site to any existing pavement must be cleared as soon as possible to avoid creating a safety hazard and causing environmental pollution.

CHAPTER TWENTY-SIX

Maintenance

26.1 MAINTENANCE DEFINITIONS

26.1.1 Road management

The road is a capital investment that deteriorates in the same way as other assets, whether used or not. The technical background to this deterioration and techniques used for recording its extent and progression and inhibiting and reversing its progress are given in Sections 12.1.4, 12.2.6 & 14.3. The financial implications for users with respect to their routine vehicle operating costs are discussed in Section 29.1.2. Unexpected road deterioration can also create a safety hazard.

Organised effort is necessary to lessen the rate of deterioration of a road and to restore service levels to earlier values and thus to reduce vehicle operating costs and increase safety. Section 26.2.2 will show that the specific effort needed is usefully developed from a comparison of the existing condition of a road with standards established with respect to the current and future levels of service to be provided by that road. The effort needed will fall into one of three categories.

(a) *Reconstruction* completely replaces the existing pavement and often also alters the pavement width and/or alignment to accommodate increased traffic volumes and/or loads in excess of the original design levels. Such work is discussed in Chapter 25.

(b) *Rehabilitation* (or *remedial maintenance*) is carried out on the existing road alignment and carriageway width to regain original conditions. It restores, reverses or repairs, rather than retards and stops. Resheeting (Section 12.2.5) is a typical rehabilitation treatment.

(c) *Maintenance* (or *preventative maintenance*) is a process which retards or stops, rather than reverses, deterioration. Much of this maintenance need will be independent of road usage (e.g. landscape maintenance), however typically about 30 percent of routine pavement maintenance costs are due to traffic usage.

To illustrate the commitment to these three categories, a typical subdivision of the work done within a Road Authority is shown below, recognising that the data will vary greatly with circumstances:

Item (definitions in text)	Percent (to nearest 5)	Range
New facilities (Chapter 25)	30	10–60
Pavement:		
a. Reconstruction	20	5–40
b. Rehabilitation	10	0–20
c. Maintenance	20	10–50
Bridge maintenance	5	0–10
Traffic management	5	0–10
Organisational management	10	5–20

26.1.2 Approaches to maintenance

This chapter concentrates on Category (c), Maintenance, in the list in Section 26.1.1. In this context, maintenance can be defined as all work that is required to reduce to a practical minimum the deterioration of the quality, efficiency, safety, and appearance (many users see the quality of a road in visual terms) of a road below the levels which pertained immediately after construction. Maintenance is usually cheaper than rehabilitation and so overall costs can usually be minimised by increasing the attention to maintenance at the expense of rehabilitation. There are thus two broad approaches to maintenance (Jordan *et al.*, 1980):

(1) A basic approach in which the aim is to avoid both over-maintenance and rapid deterioration of the road network. In attempting to find the balance between maintenance and rehabilitation, this method requires a great deal of day-to-day judgement; and

(2) A systems approach which assesses needs and then predicts funding levels from an analysis of the behaviour of the road network under past expenditure patterns. Section 26.2.2 provides a basis for such an approach and the wider and embracing field of pavement management systems is discussed in Section 14.6.

Many factors contribute to the complexity of managing road maintenance; they include:

* despite the comments above, administrations are often reluctant to spend adequate funds on maintenance,
* maintenance scheduling still relies heavily on judgement and experience,
* many of the activities are small-scale and labour-intensive,
* productivity is very dependent on the use of appropriate plant and personnel, and
* users react strongly to poorly maintained roads.

With respect to the last point, Road Authorities often enjoy an immunity from legal liability for any consequences of less than perfect road maintenance, even with respect to known dangers, and — despite the economic imperatives — this commonly negates any general legal duty to keep a road in repair. Therefore, legal actions often cannot be taken against a Road Authority for the consequences to a particular road user of road deterioration. However, possible immunity from legal liability does not cover misfeasance (Section 18.1.4). Thus, it may be necessary to demonstrate that the deterioration of the road was not the result of a negligent act.

26.2 MAINTENANCE MANAGEMENT

26.2.1 Maintenance types

Maintenance was defined as work category (c) in Section 26.1.1. It can be divided into three subcategories according to the mode of management used (Jordan *et al.*, 1980):

(c1) *Periodic* (or *programmed*) maintenance — work which can be predicted and is therefore able to be rigorously planned. Some maintenance is relatively minor, such as mowing grass, cleaning signs, painting road markings, and grading shoulders. Other periodic works, such as joint resealing, may need to be treated as a minor reconstruction job (Category (a) in Section 26.1.1). Some periodic maintenance is preventative — e.g. resealing (Section 12.1.4) — and may be better considered as rehabilitation (Category (b) in Section 26.1.1).

(c2) *Routine* maintenance — work which is not amenable to detailed planning but which is assessed in advance with reasonable accuracy on the basis of past experience and accepted levels of service. It is usually functionally associated with stopping the deterioration of the pavement by repairing potholes, cracks and other surface defects. Prompt action is effective and has a high benefit–cost ratio as it prevents the ingress of moisture into the pavement structure (Section 9.3). In net present value terms, routine maintenance is usually less costly than periodic maintenance.

(c3) *Service restoration* — work which restores service as a consequence of unexpected events beyond the control of the Road Authority. Typical examples are clearing snow and debris, de-icing pavements, and repairing flood and wind (fallen tree) damage.

Work in categories (c2) & (c3) and the routine component of work in (c1) is usually organised and executed by permanent *gangs* (or *patrols*) of maintenance workers, whereas the specialised tasks in (c1) are normally done by gangs specialising in the type of work involved. Usually the permanent maintenance gangs consist of three or four men and are equipped with hand tools and their own truck and operate over defined lengths of road. They usually cover about 100 km of roads carrying over 500 veh/d. Although a supervisor is required for every ten or so gangs, the system is usually self-managing, relying on the pride of the gang in the condition of its own stretch of road.

An alternative, somewhat older, approach is the *lengthman* system. Lengthmen are individual workers assigned to a particular length of road, usually about 10 km long. They carry out most of their tasks with hand tools and without a motor vehicle. Thus, a permanent gang is usually needed to undertake any larger tasks. The lengthman system suffers from low productivity and inflexible operations and has been largely superseded by vehicle-based gangs.

26.2.2 Maintenance planning

Maintenance planning ensures consistent and adequate levels of service, standards, methods, and procedures. It requires a means of objectively preparing maintenance programs related to needs and constraints. A maintenance management flow chart to achieve this is shown in Figure 26.1 and assumes that the following procedural steps are followed (Jordan *et al.*, 1980):

(0) Receive policy and funding guidance from management (step xiii), based on estimates supplied to them of the cost of delivering various service standards.

(1) Establish standards: The standards chosen will:

* determine whether a road is receiving too little or too much maintenance,
* promote consistent maintenance action, and
* guide maintenance staff.

The standards fall into two categories:

(a) *service* standards defining levels of road defects at which maintenance action is justified. There are certain defects such as potholes, broken edges, loose gravel, slick surfaces, defective bridge decks, damaged signs, and accumulated litter which the public expects will be rapidly repaired, at least on major roads. Data allowing most of these standards to be determined will be found in the relevant section elsewhere in the book.

(b) *resource* standards listing typical gang sizes, plant and materials; and then relating these resources to work methods, outputs and costs.

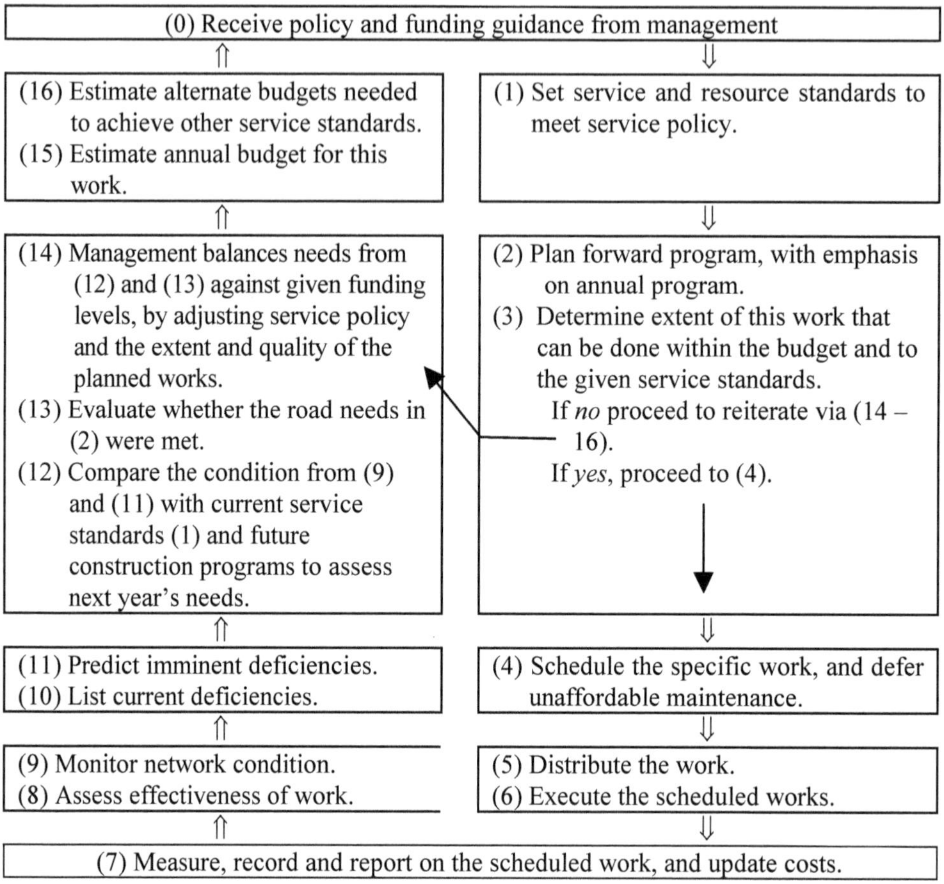

Figure 26.1 Maintenance management flow-chart for an annual program (based on Jordan *et al.*, 1980).

(2) Plan a forward program based on the known funding levels (0), extent of work and service standards (1). The annual program will cover:
   (2.1) periodic maintenance from known work load,
   (2.2) routine maintenance from estimates of needs, and
   (2.3) restoration from previous experience.

(3) Determine extent of this work that can be done within the budget and to the given service standards. If *yes*, proceed to step (4), if *no* proceed to reiterate via steps (14 – 16). This iteration process is the same approach used generally for all roadworks and described in Section 5.2.3 (Figure 5.1).

(4) Based on the overall forward plan from (2), prepare a works schedule — field staff commonly select specific works for their area, but first priority is usually given to remedying deficiencies which affect road safety.

(5) Order the necessary resources, provide schedules assigning gangs to specific tasks, and recommend when inspections and performance reviews should occur. Note that routine maintenance operations are relatively simple tasks but the allocation of priorities to the tasks is usually complex.

Applying the process in reverse allows next year's budget to be estimated.

(6) Execute the works and manage the operations within the agreed plan and allocated budget. The management and execution of maintenance operations assumes a local management structure of the type shown in Figure 26.2, although the Road Authority will often have a central maintenance engineer keeping a general eye on maintenance practices throughout the Authority. The district engineer is usually responsible for works in a district, and delegates the planning and organisation of all maintenance works to a district maintenance engineer. A district maintenance supervisor manages actual works within the annual maintenance program and supervises the foremen who organise and direct the maintenance gangs. This work may be done by either the Authority's own workforce or by contract. In all the possible scenarios, only the district engineer would always be an employee of the Authority.

A complete list of maintenance activities is given in Table 26.1. In operation, such a table would be just one — basic — part of a multidimensional representation of activities. For instance, for each activity, the maintenance engineer would list the resources needed and used, by type, quantity and price in accordance with the process outlined in Figure 26.1.

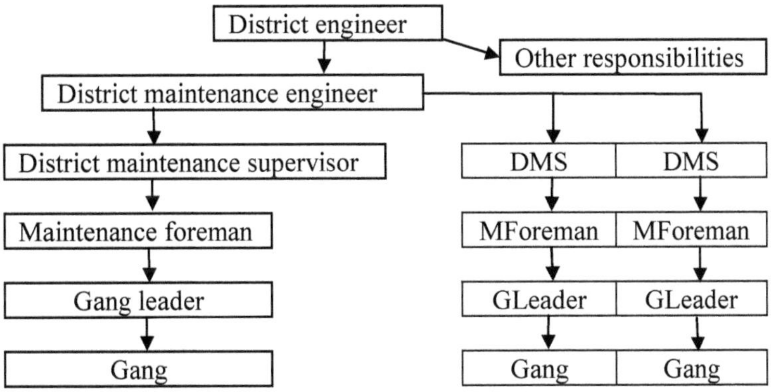

Figure 26.2 Typical organisational structure for district management.

(7) Measure, record, report, and cost the works to provide data for future work planning, expenditure monitoring, and the reordering of equipment and supplies (Section 14.5.1). This should be performed using the Authority's financial system, and a defined reference and location system (Section 26.5.1). The actual reporting system must capture data on:
   (a) when and where (usually by road segment) the maintenance was performed;
   (b) what activities were undertaken; for example, using the activity definitions in Table 26.1; and
   (c) what quantities (material volumes, man-hours, plant-hours) were used and at what cost. This maintenance activity recording system has two foci in that it must:
      * service the local cyclic pattern of maintenance planning, organising, directing, monitoring (this element), cost-controlling, evaluating, planning, and forward management, and
      * be compatible with the systems the Authority uses for ordering, payroll, equipment management, administration, and accounting. For example, the reporting system should produce expenditure summaries, location details and

Table 26.1 (pp. 654–7) Maintenance activities (after Jordan *et al.*, 1980). Refer also to Tables 5.3 & 14.2.)

Categories	Task definition	–*Type*
100	**Paved carriageway**	
110	***Pavement structure***	
111	Repair localised subgrade failures (Section 14.1A)	*– service, rehabilitation**
112	Repair localised pavement failures (Section 14.1A)	*– service, rehabilitation**
113	Reconstruct short sections of up to 500 m^2 in area	*– rehabilitation*
114	Clean joints in slabs and fill and seal cracks in pavements	*– service*
115	Lift and fix displaced concrete slabs (Section 14.1A.13)	*– rehabilitation*
120	***Pavement surface***	
121	Repair potholes (Section 14.1C9)	*– service**
122	Repair surface	*– service*
122.1	Emergency repairs	*– service*
122.2	Fill (or seal) surface cracks and repair concrete joints with injected sealants (Section 14.1B)	*– service*
122.3	Correct glazed or bleeding areas (Section 14.1C6&8)	*– service*
122.4	Level minor surface defects, make minor adjustments of concrete slabs (see Item 114 for major work)	*– service*
122.5	Patch and repair other surface failures	*– service*
122.6	Regulate surface shape (Sections 12.1.5–6, 12.2.6)	*– rehabilitation*
123	Apply to the surface:	*– rehabilitation***
123.1	Surface enrichment (Section 12.1.4)	
123.2	Reseal (Section 12.1.4)	
123.3	Surface dressing (Section 12.1.3–4)	
123.4	Thin overlay (Section 12.1.5–6)	
124	Sweep, shovel and otherwise remove debris, snow and ice	*– operations*
130	***Pavement edges***	
131	Repair by removing to transverse steps (Section 14.1A10)	*– service**
132	Repair by returning to longitudinal line	*– service**
140	***Unpaved shoulders***	
141	Grade to maintain shape	*– operations*
142	Remove vegetation	*– operations*
143	Resheet by adding material	*– rehabilitation*
144	Repair scour	*– operations*
150	***Paths:*** Return surfaces of all paths to their original condition	*– service*
160	***Access***	*– rehabilitation*
161	Maintain entries to crossroads	
162	Maintain entries to private property	
200	**Unpaved carriageways (Section 12.1.1)**	
210	***Existing surface***	*– operations**
211	Grade to restore shape (dry maintenance)	
212	Provide dust suppressants (Section 12.1.1)	
213	Rip excess material	
214	Remove loose material	
215	Patch soft spots	
216	Add new material and (OMC) water	

	217	Compact with roller (wet maintenance)	
	218	Spray weeds with herbicide	
220	***New surface:*** Resheet, replace surface losses (Section 12.1.1)		*– rehabilitation*
300	**Drainage**		
310	***Surface drains***		
	311	Clean and clear	*– service*
	312	Spray with herbicide	*– service*
	313	Repair	*– rehabilitation*
	314	Improve or add (Section 13.3.1)	*– improvement*
	315	Maintain kerb and channel (Section 13.3.2)	*– rehabilitation*
320	***Pits and sediment traps***		
	321	Clean, empty and clear	*– service*
	322	Repair lids and grates	*– maintain*
	323	Repair pits	*– maintain*
330	***Culverts and bridges with spans less than 6 m***		
	331	Clean and clear	*– service*
	332	Repair culverts	*– rehabilitation*
	333	Repair culvert end-walls	
	334	Repair bridges (with a span of less than 6 m)	
340	***Underground drains (Section 13.4.2)***		
	341	Check that drains are functioning	*– service*
	342	Remove blockages, clean and clear	
	343	Clean inlets	
	344	Clean outlets	
	345	Repair drains, inlets and outlets	*– rehabilitation**
350	***Roofwater outlets in road reservation (Section 13.3.1)***		
	351	Check for illegal outlets	*– operations*
	352	Repair and clean legal outlets	*– rehabilitation*
400	**Roadside (Section 6.4)**		
410	***Vegetation within road boundary (check 480)***		
	411	Mow	*– operations**
	412	Control by spraying with herbicides	*– operations*
	413	Prune vegetation for visibility and safety	*– operations**
	414	Remove weeds and unwanted vegetation	*– operations*
	415	Remove or burn fire hazards	*– operations**
	416	Reseed and replant	*– operations*
	417	Upkeep other landscaping	*– operations*
	418	Control local soil erosion	*– operations*
420	***Motorists' services***		
	421	Rest areas	*– service*
	421.1	Provide water, firewood and other supplies	
	421.2	Clean	
	421.3	Repair buildings and furniture	
	421.4	Control vegetation	
	421.5	Maintain access to roadside	*– operations*
	423	Maintain car parks, bus stops and taxi ranks	*– operations*
430	***Litter control***		***– service***
	431	Collect and dispose of litter from bins	

	432 Collect and dispose of litter from roadside	
	433 Maintain rubbish receptacles	
440	***Earthworks***	
	441 Remove unnecessary soil, rock and sand	*– operations*
	442 Protect batter slopes	*– rehabilitation**
	443 Provide windbreaks for sand dune protection	*– rehabilitation**
	444 Maintain other roadside earthworks	*– maintain*
	445 Maintain retaining walls	*– service, rehabilitation*
450	***Property-boundary fences and gates***: maintain and repair	
460	***Other service agencies within the road boundary:*** check facilities	
470	***Roadside hazards***	
	471 Check for traffic hazards	
	472 Remove any transient roadside hazards	
	473 Remove designated major roadside hazards	
480	***Manage sites of environmental significance***	
	481 Check signs indicating significance of site	*– operations*
	482 Use agreed environmental management plan	*– operations*
500	**Traffic facilities**	
510	***Signs and markers (Chapter 21)***	
	511 Provide, supply and erect new road signs	*– improvement, operations**
	512 Clean, paint, repair or replace	*– rehabilitation**
520	***Road furniture***	
	521 Provide, supply and erect new road furniture	*– improvement*
	522 Repair or replace guardrail and New Jersey barriers	*– rehabilitation**
	523 Repair/replace impact absorbing devices	*– rehabilitation**
	524 Repair or replace guideposts and delineators (Section 24.4.3)	*– rehab'n**
	525 Repair/replace internal fencing	*– rehabilitation*
	526 Clean, paint or repair other road furniture	*– rehabilitation*
	527 Maintain traffic islands	*– operations*
530	***Road markings (Chapter 22)***	
	531 Sweep, spot, paint or replace	*– operations*
	532 Repair or replace raised pavement markers (RPM)	*– rehabilitation*
540	***Traffic signals (Chapter 23)***	
	541 Service detectors, controllers and lights	*– service**
	542 Repair detectors, controllers and lights	*– repair**
	543 Ensure power	*– operations**
	544 Check operations	*– operations**
	545 Check communications	*– operations**
550	***Roadside telephones:*** service and repair	*– service, repair**
560	***Lighting of roads, bridges, features, crossings, signs etc. (Chapter 24)***	
	561 Service lighting	*– service*
	562 Repair lighting	*– repair**
	563 Ensure power to lighting	*– operations**
	564 Ensure links with power supply authority	*– operations**
	565 Ensure accounts are paid	*– operations**
570	***Communications links***	
600	**Structures (Chapter 15)**	
610	***Bridge substructure***	

	611	Service abutments, cribwalls, wingwalls, fenders	*– rehabilitation**
	612	Service, following 631–634	*– rehabilitation*
	613	Service bearings	*– service*
620	***Maintain free-flowing waterways***		*– service**
630	***Bridge superstructure***		
	631	Spray to control termites and vegetation	*– service*
	632	Waterproof	*– service*
	633	Paint	*– service*
	634	Tighten bolts	*– service*
	635	Check steel and prestressing	*– service*
	636	Repair masonry, spalled concrete and corroded steel	*– maintain*
	637	Repair appurtenances and catwalks	*– maintain*
640	***Bridge deck***		
	641	Repair deck surface	*– repair**
	642	Repair footways	*– repair**
	643	Repair kerbs	*– repair**
	644	Repair railings	*– repair**
	645	Clean drains and channels	*– operations*
650	***Miscellaneous***		
	651	Maintain floodways	*– service, rehabilitation*
	652	Maintain cattle grids	*– service, rehabilitation*
	653	Rail crossings	*– service**
	653.1	Maintain rail crossings	
	653.2	Repair pavements near rails	
700	**Restoration**		
710	***Traffic accidents***		*– restoration**
	711	Control traffic	
	712	Clear debris from carriageway	
	713	Repair associated damage to pavements, structures, traffic facilities etc.	
720	***Storm and flood damage***		*– restoration**
	721	Clear debris from carriageway	
	722	Clear water, snow and ice from roadway	
	723	Repair associated damage to pavements, structures, traffic facilities etc.	
730	***Fire damage***		*– rehabilitation**
	731	Clear debris from carriageway	
	732	Repair damage to pavements, structures, traffic facilities etc.	
740	***Road openings due to utilities operating in the road reserve***		*– operations**
	741	Protect and restore the roadway (Section 7.4.3)	
	742	Protect traffic	
750	***Permanent reference points (Section 26.5.1)***		
	751	Clean and repair	*– service, rehabilitation*
800	**Overheads**		
810	***Administer maintenance activities***		
820	***Inspect roadway***		
830	***Supervise and assure work***		
840	***Operate depots and camps***		

* Indicates tasks which are driven by needs. Many of the other tasks can be scheduled to meet resource availability.

** Classed as rehabilitation rather than maintenance if the work is greater than about 500 m^2 in area.

job data suitable for foremen to schedule gangs, for field management to reallocate resources and consider alternatives, for district management to monitor performance, and for central management to review progress in accord with policy.

(8) & (9) Monitor and assess the condition of the existing road network by annual inspections. Procedures for the monitoring of pavement condition are discussed in Section 14.5.

(10) & (11) Predict imminent deficiencies and list current deficiencies.

(12) Compare the condition from (9) and (11) with current service standards (1) and future construction programs to assess next year's needs.

(13) Evaluate whether the road needs in (1) & (2) were met and compare the condition with current service standards and future construction programs to assess next year's needs.

(14) Management balances needs from (12) and (13) with funding levels by adjusting service policy and the extent and quality of the planned works. Section 5.2.2 discusses the interaction between funding levels, needs, and service standards. In this context, it is necessary for management to balance needs and funding levels by adjusting the extent and quality of the planned works.

(15) Estimate annual budget for this work.

(16) Estimate alternate budgets needed to achieve other service standards.

26.2.3 Unpaved roads

Unpaved roads were discussed in Section 12.1.1 and their maintenance represents a special case. For example, the first objective must be to keep the cross-section of the road of such a shape that it will quickly shed any water. Typically, this requires a surface graded laterally at about 6 percent on both sides of, and away from, a centreline crown. This is usually done by regular dragging or grading (Section 25.4), a process which is also used to restore any displaced surfacing, correct unevenness, remove ruts, fill potholes, and smooth corrugations. Grading is sometimes followed by roller compaction of the surface.

It may also be necessary to remove loose material, patch soft spots, and replace any permanently-lost surfacing material. This addition of new material is called *regravelling* and is discussed in Section 12.1.1.

Primarily because of the lower standards adopted and lower traffic flows, the maintenance costs of unpaved roads per kilometre usually average only about 20 percent of the same costs for paved roads. As volumes and standards increase, the costs for the two types of road begin to equalise.

26.2.4 Roadsides

Maintenance of the roadside must recognise that it serves as a:

(a) safety area for errant vehicles (Section 28.6.1),

(b) emergency stopping area,

(c) location for public utilities (Section 7.4.3),

(d) place for people to use,

(e) home for flora and fauna, including the opportunity for regeneration of local species (Section 6.4) [Indeed, many roadsides have becomes sites of environmental significance and require careful management.],
(f) site for snow from snow clearance, and
(g) potential fire hazard with accompanying requirements for fire prevention and control. Firebreaks may be needed in areas with a prior history of fires, containing potential fire sources, or acting as part of a larger fire control strategy. However, it is rarely necessary to use the whole of the reserve as a firebreak.

26.3 RESIDENTIAL STREET MAINTENANCE

Although the problems may be of a different scale and type, residential street maintenance must follow the same principles as were discussed in Section 26.2.2, i.e.:
(a) determine the current situation,
(b) assess the needs,
(c) consider solutions, and
(d) determine an optimum program consistent with budgetary constraints.

Most of the items of work that need to be performed for residential street maintenance can be selected from the list of activities in Table 26.1. One change from the situation with major roads is that, within the condition 'index' derived from Table 14.4, the role of high-speed factors such as road roughness will diminish, whereas appearance measures will gain in importance.

An aspect of residential street maintenance that distinguishes it from other road maintenance tasks is the much greater occurrence of utility services within the road reserve (Section 7.4.3). This raises a number of issues. In the first place, their location is sometimes a problem, as it is not always possible to obtain plans showing service locations. Thus exploratory digging and metal detectors are often required. An associated problem is that excavation adjacent to a service, particularly by backhoes and other trenching machines (Section 25.4e), is the most common cause of service damage.

The trenches are a major aftermath of the provision or maintenance of utility services. Trench backfilling and pavement reinstatement need to be performed with care and skill. Desirably, utility agencies provide advance notice that work is to be undertaken, and provision is made in mutual agreements for the utility agency to meet the costs of any reinstatement. However, the actual procedure for pavement openings for service trenches and the subsequent pavement restoration rarely follows a common, formal, or adequate procedure. It seems almost inevitable that liaison and organisation difficulties occur in this area and so close co-operation is required between large organisations on what would appear to many to be a minor issue. The most important factors in trench reinstatement are:

* *careful observation of the existing material during excavation.* This will guide the selection of replacement material.
* *the excavation.* This should be wide enough for compaction equipment and should not have reverse (or belled) slopes that would make compaction difficult.
* *the quality of the backfill.* Good practice is to use well-graded sand (Section 8.3.2) or cement-stabilised material (Section 10.2) as a backfill. A well-graded mix prevents fines from the adjacent material migrating into the backfill. The material around the actual utility service should be non-cohesive (Section 8.3.2) to permit better placement in confined spaces. Organic material should not be used.

* *the degree of compaction achieved.* Adequate compaction is rarely achieved by unsupervised non-road groups working without clear specifications. The common compaction techniques in use are hand-held compactors, rolling with the wheel of a tractor or truck and, less commonly, mechanical compaction (Section 25.5). Particular attention should be paid to the selection and compaction of the bedding material used by the utility.
* *the inspection of the procedures used.* The lack of inspection of completed work probably explains the poor service record of many reinstatements.

26.4 TRAFFIC FACILITY MAINTENANCE

All traffic facilities require regular maintenance. This is particularly the case with respect to traffic signals (Chapter 23) where the need may not always be apparent, but where the effects can be significant. In one survey prior to automatic fault detection, 86 percent of the signalised intersections investigated had timing faults (90 percent of which related to incorrect intergreen times, Section 23.6.2), 68 percent had detector faults (Section 23.7), and 30 percent had lantern faults (Section 23.2.2). Each intersection wasted an average of 70 L of fuel a day due to poor maintenance (Daley and Ogden, 1980).

The acceptable probability of a failed signal being hazardous is about 0.01, based on both implied public opinion and generally acceptable hazard levels. In addition, the crash rate is about 20 times greater at a blank signal. When a traffic signal fails, it is therefore desirable that it does so in a safe mode, preferably also alerting users to the fact that it is no longer operating normally. If a red signal continues to display due to a fault, drivers will usually tolerate a delay of between 1 and 2 minutes, depending on whether the traffic is light or heavy. The worst case from a safety viewpoint would be for all signal faces to show green discs or 'walk' symbols. Flashing yellow discs all around are commonly seen as the most desirable 'failure' mode, although this method suffers the major drawback of not allocating priority to particular intersection legs. This is important as the accidents occurring at blank signals are predominantly priority-related (Moore and Hulscher, 1985). One technique sometimes used is to have small Give Way or Stop signs (Section 20.2.2) permanently displayed. These apply when the signals are not operating.

Section 17.4.4 showed how traffic incidents can have a significant impact on the level of traffic service provided and can be the major (e.g. 60 percent) cause of urban congestion. There are typically between 10 and 100 stopping incidents per Gm of vehicle travel — compare these rates with a fatality rate of 0.01 (Section 28.1.3). Thus road service is best maintained by an incident management plan which:

* promptly *detects* disruptive incidents such as crashes, disabled vehicles, illegally parked vehicles, spilled loads, and faulty signals. The basic data will be received from loops (Section 23.7), remote-controlled cameras, roadside emergency phones, road patrols, and telephone calls from bystanders. It will be processed and acted upon by the traffic control system (Section 23.3–5).
* *institutes* emergency traffic management,
* *dispatches* work gangs to the site to remove the incident. These gangs should be well rehearsed in the procedures to follow in a variety of incidents. Typically, they will clear disabled vehicles, debris, etc. from the roadway.
* *informs* other drivers of the unexpected delay via variable message signing and broadcasts.

* as relevant, *informs* police, fire brigades, ambulances, emergency services, tow-truck services, public transport, and utility services.

* *reports* on the removal of the incident so that traffic can readjust.

Such incident management systems also improve driver security, help public relations, avoid secondary crashes, and minimise the need to call on police resources.

26.5 REFERENCE SYSTEMS AND MAINTENANCE MODELS

26.5.1 Reference systems

Road maintenance management requires a method for defining the location of road features, pavement segments, and deteriorated sections of road. Most systems are built around a series of *permanent reference points*. These fixed physical markers may be fairly sparsely located, such as at the beginning and end of the road, or on specific features along the road, such as culverts, bridges and intersections. At the other extreme, special labelled posts may be placed along the roads at intervals as close as every 100 m (Schacke, 1984). As an intermediate stage, routine kilometre posts can be used. One of the problems with using roadside distance posts is that, if a part of the road is shortened, it can have repercussions throughout a distance-based system. This can be resolved by using complex adjusting systems (Schacke, 1984) or by simply storing both the before and after change data. As *global positioning systems* (*GPS*) are now widespread, it is also possible to locate points purely by their map coordinates.

When roadside features are relied upon, the reference point must be specifically marked as most items of maintenance interest will need to be located to accuracies of better than ±5 m. These interest areas can be located relevant to the permanent reference points by odometer readings and compass and/or road directions.

The reference systems provide a means of storing readily accessible material on road geometry, intersections, abutting properties, pavement structure, bridges, traffic management hardware, traffic flows, crashes, roadside features, drainage, and pavement conditions. The geometrical features of each road can be combined to define the road network and this can be interconnected with national mapping features (Section 6.1.2). Thus a comprehensive road inventory can be established. Table 14.3 illustrated the need for and use of such a reference system.

26.5.2 Network models

Chapter 14 discussed the performance of a single piece of road. Detailed management systems for that piece of pavement are discussed in Section 14.6. At a larger scale, a number of macroscopic models have been developed to analyse a network of roads. These take advantage of the road inventories described in Section 26.5.1 and perform analyses of the extent and consequences of the condition of the road network. The best-known is the *Highway Development Model* (*HDM*) developed by the World Bank and now managed by PIARC. It has largely supplanted such earlier models as the *Road Transport Investment Model* (*RTIM*) of TRRL and Australia's *NIMPAC* model. The three models have basically the same approach, as shown in Table 26.2. The basic roles of the models are to (Potter and Hudson, 1981):

(a) identify and rank pavements needing improvement,

Table 26.2 Road investment/maintenance models.

1	**Specify or determine**
1.1	Constraints (Chapters 5 & 6)
1.1.1	Funding levels (Section 5.3)
1.1.2	Design standards (Section 6.2)
1.1.3	Assessment standards (Section 14.5.3)
1.2	Givens (Chapters 14 & 31)
1.2.1	Initial road network (Section 6.1.2)
1.2.2	Initial conditions – structure (Chapter 11) and service (Chapter 14)
1.2.3	Traffic forecasts and conditions (Section 31.5 & Chapter 17)
1.3	Infrastructure costs
1.3.1	Construction (Chapter 25)
1.3.2	Maintenance (Chapter 26)
1.4	Pavement (and bridge) deterioration models (Section 14.5)
1.5	Operating costs
1.5.1	Vehicle operations (Section 29.1.2)
1.5.2	Crashes (Section 28.8)
1.5.3	Travel time (Section 31.2.3)
1.5.4	Social and community impacts (Chapter 32)
2	**Annually**
2.1	Calculate
2.1.1	Traffic flow and conditions
2.1.2	Resulting travel costs
2.1.3	Travel costs, assuming no expenditure
2.1.4	Resulting road conditions
2.2	Compare predictions with assessment standards
2.3	Estimate necessary rehabilitation and reconstruction
2.4	Calculate
2.4.1	Resulting road expenditure
2.4.2	Adjustment to future expenditure to match budget
2.4.3	Adjustment to work program to match budget
2.4.4	Discounted value of this expenditure
2.4.5	Total community costs (#2.1.2 + #2.4.2)
2.4.6	Net benefits (#2.1.3 – #2.1.2)
2.4.7	Discounted value of the net benefits
2.4.8	Incremental benefit–cost ratio (#2.4.6/#2.4.2)
3	**Increment**
3.1	The road network, on the basis of #2.4.3
3.2	Return to #2 for the next year
3.3	If the design life is reached, go to #4
4	**Calculate for the design life**
4.1	NPV (Section 33.5.7) of the costs by summing all #2.4.4's
4.2	NPV of the benefits by summing all #2.4.7's
4.3	An overall benefit–cost ratio (#4.2/#4.1)

(b) produce budgets at the level of the road network, and
(c) forecast future network needs.

The World Bank began its development in 1968, working with both TRRL — leading to its RTIM — and with MIT, issuing the Highway Cost Model (HCM) in 1971 and eventually producing HDM in 1977. HDM was issued in second and third versions in 1979 and 1987. Like RTIM, HDM was largely based on results from TRRL's work in Kenya. A subsequent version of the model replaced the Kenya equations with equations from a study in Brazil (Geipot, 1982c). A description of the resulting HDM-III and the related MICR model was given in Watanatada *et al.* (1987). The newer HDM-4 model is described in Roper (2001). PIARC took over the development HDM-4 in 1998. HDM-4, version 2, was in service in 2004 (PIARC, 2004). It has a much improved road-user effects model, a facility to permit a sensitivity analysis of the key design parameters, and allows an unlimited number of budget scenarios to be pursued.

HDM calculates total life-cycle costs — construction, road user, maintenance — using selected design and maintenance strategies. It does this by simulating physical, economic, traffic and surface conditions for a road network over an analysis period of up to 30 years. The model provides a range of options as its output. The pavement deterioration models that HDM uses must be calibrated for local conditions. Despite the use of a range of maintenance data, one of the drawbacks with HDM is that the subtlety of the data required means that the influence of different maintenance practices cannot always be explored with confidence.

A variation on these methods is the *Arizona* approach which takes into account the probabilistic nature of road deterioration, using a Markovian approach based on accumulated experience (Section 14.6.1).

26.6 VALUE OF ROAD AND BRIDGE ASSETS

Accounting practices often require monetary values to be placed on existing road and bridge assets. Fundamentally, the value of the asset will depend on its ability to deliver future service benefits. Specifically, the valuation will depend on the asset's:

* initial cost,
* current condition (Chapter 14), reflecting the depreciation of the asset,
* performance of its intended task,
* net income generation (e.g. via tolls, Section 29.4),
* future usefulness and functional adequacy (e.g. capacity constraints, Chapter 6),
* replacement cost,
* remaining life, and
* ongoing maintenance cost (this chapter).

Normally, market value would also be a consideration, but there is rarely a market for fixed road assets.

Commonly, the valuation is based on the replacement cost, reduced by depreciation, and modified by considering performance, net income, and usefulness. Whilst depreciation directly reflects current condition, it is often simply estimated from the age of the asset. A more sophisticated method is based on remaining life. As this chapter has highlighted, remaining life will depend on:

* the physical condition of the asset, which will relate to past usage and to annual maintenance and rehabilitation activities, and
* the need for the asset.

The various condition indices developed in Chapter 14 are useful in this context. Related techniques for vehicle depreciation are discussed in Section 29.1.2. In summary,

Asset value = (replacement cost) –
(value lost by ageing, deterioration, obsolescence, and/or traffic irrelevance)

CHAPTER TWENTY-SEVEN

Road Vehicles

27.1 DEFINITIONS AND DATA

27.1.1 The vehicle fleet

A study of traffic behaviour requires a knowledge of the characteristics of the vehicles commonly found in the traffic stream, particularly their:

* overall dimensions,
  Basic vehicle dimensions are given in Section 27.3.2 and, in so far as they affect lane widths, are also examined in Section 7.4.2 (Table 7.3) and Section 16.5.1. Dimensions also influence parking design (Section 30.6.2). Vertical underbody clearances affect the geometry of gutter crossings (Section 7.4.3), road humps (Section 18.1.4), and short, steep grades (Section 19.3).
* visibility from the vehicle (Section 27.2.1),
* visibility of the vehicle (Sections 16.6.1, 28.2.1–2 & 28.3),
* manoeuvrability (Sections 16.5.1–2),
* maximum speed (Section 18.1.1),
* acceleration (Section 27.2.2),
* deceleration due to in-gear performance (travelling in gear but without pressure on the accelerator pedal) and/or braking (see Section 27.2.2; vehicle braking performance is utilised in Section 17.2.3 and Table 19.1.),
* hill-climbing ability (the basis of the grade discussions in Sections 7.4.2 and 19.3),
* steering (Section 16.5.2),
* cornering (cornering is examined in Section 19.2.2 and seen to significantly influence road alignment design),
* lighting (Section 24.4.1),
* fuel consumption (Section 29.2.1),
* emissions (Section 32.3),
* safety and occupant protection,
* wheel and axle loads (pavement design is very dependent on a knowledge of these loads, Section 11.4.1) and
* suspension characteristics (Section 27.3.5).

As roads are an old, well-established, and slowly changing technology, it can be assumed that in most cases vehicles will be designed to fit the road, rather than vice versa (Lay, 1992). This will provide some relief for the road engineer in the short term, but in the long term all aspects of roads will change as vehicle technology changes.

In the developed world, road design is mainly concerned with motorised (i.e. self-powered) vehicles, bicycles (Section 27.6), and pedestrians (Section 30.5.3). This chapter concentrates on motorised vehicles and cycles. Notable absences from this book are the various animal-powered vehicles in common use in many parts of the less-developed world (Sections 3.2 & 3.5.1). The types of motor vehicles likely to be encountered on public roads are:

* cars and their derivatives (Section 27.2.1),
* bicycles (Section 27.6),
* motorcycles and mopeds (Section 27.5),
* cars towing trailers, caravans or boats,
* light vans (Section 27.3.1, *light* commonly means under 4 t in unladen mass and having less than three axles),
* heavy vehicles such as trucks and buses (Section 27.3.1),
* road trains (Section 27.4.2), and
* tractors, particularly in developing countries.

Vans and other light commercial vehicles are largely involved in door-to-door deliveries and usually account for the largest proportion of freight vehicles on urban roads. They are usually very simple mechanically and have large doors to facilitate entry and exit. The term *commercial vehicle* usually includes these light commercial vehicles and most of the trucks and other heavy vehicles discussed in Section 27.3.1.

Each Road Authority (Section 4.2) usually has specific but slightly artificial definitions of each vehicle type for the purpose of legal classification and regulation. A typical classification is:

Vehicle type	Book Section	Typical GVM (t)	Typical Power (kW)	Number of axles
toy vehicles				
bicycles	27.6			2
mopeds	27.5			2
motorcycles	27.5			2
tractor				
small car	27.2.1	1.0	65	2
medium car	27.2.1	1.25	80	2
station (or ranch) wagon	27.2.1	1.5	80	2
sports vehicles	27.2.1	1.5	90	2
large car	27.2.1	1.5	110	2
cars towing		1.25+	80+	3+
utility (or pick-up)		2.5	100	2
mini-bus (< ten passengers)	27.2.1			
light vans	27.3.1	< 4		2
small bus	27.3.1			2
light truck		2.7		2
large bus	27.3.1	14	200	8
rigid truck	27.3.1	8	120	6
articulated truck	27.3.1	Many types available, Section 27.3.1		
B-double	27.3.1	45	320	15
road train	27.4.2	54	320	22
oversize vehicle	27.4.1	Many types available, Section 27.4.1		

GVM = gross vehicle mass (Section 27.3.2)

One set of regulations based on vehicle classification will define whether a particular vehicle type can use public roads subject to such local restrictions as roadworks, weakened bridges and deteriorated pavements. These permitting-to-use regulations may cover such matters as dimensions, loads, emissions, fuel consumption, visibility, braking, lighting, safety and occupant protection, instrumentation (Section 27.7) and noise.

Vehicles occasionally permitted to use the roads whilst outside such ordinary limits are discussed in Section 27.4.1.

Data on the vehicle fleet are important for:

* estimating the size and characteristics of the current vehicle fleet (Section 30.5.1),
* predicting future road usage, and
* predicting the rate at which new vehicle features will propagate through the vehicle fleet.

The vehicle fleet can be defined in a number of ways. It can refer to the number of vehicles:

* legally registered as permitted-to-use the roads (vehicle registration records are usually an important data source),
* actually using the roads (Section 31.5),
* using the roads in a specific area or time zone, and/or
* of a particular vehicle type.

As the differences between these categories can be great, the term 'vehicle fleet' should be used with care. The current fleet of registered vehicles results from the combined effect of the:

* existing list of registered vehicles,
* new vehicles added to the fleet, as reflected in annual numbers of new vehicles registered, and
* scrapping of older vehicles, although the associated data are usually difficult to obtain.

The scrapping rate of vehicles from the fleet is often expressed in terms of either the typical life expectancy of a vehicle in the fleet, or the *vehicle survival functions* for the fleet. Life expectancy can be misleading — for instance, because older cars are scrapped at a faster rate, the average age of a car fleet is about half its median age (Tanner, 1984). The median life of cars usually exceeds a decade (see also Table 4.1). If the relative price of cars increases, so too do their average lives.

Vehicle survival functions indicate the proportion, n, of vehicles registered in a given year and that can be expected to remain in service after a specified time, y years. They are usually logistic curves (Section 33.2) of the form:

$$n = 1/(a + [1 - a]e^{by})$$

where a and b are constants. The scrapping rate is given by 1 – n. Typically:

$$n = 1/(1 + 0.01e^{0.35y}) = 1/(1 + e^{-4.60 + 0.35y})$$

and

$$1 - n = 1/(1 + 100e^{-0.35y}) = 1/(1 + e^{4.60 - 0.35y})$$

A car may pass through a number of owners during its life. British and U.S. data indicate that only a small proportion of cars are resold in the first year, the maximum proportion being sold in either the second or third year, with 50 percent being resold within three to four years. A large proportion of cars in most countries comprises company cars owned by organisations rather than by individuals, although they are used as if they were privately owned. In addition these cars tend to be newer, larger, travel further, and be used more frequently than the average car. Car ownership issues are pursued further in Section 30.5.1. Depreciation costs are discussed in Section 29.1.2.

27.1.2 Mechanical engineering

The forces that directly determine the performance of an operating vehicle can be divided into three categories:

* rolling resistance,
* air resistance, and
* inertia.

Rolling resistance forces, R_r, are present in any moving vehicle. They are caused by energy losses in:

the pavement (a flexible pavement — Section 14.5.2 — can increase rolling resistance by 60 percent),
the tyre–pavement interface (Sections 12.5.2 and 27.7.1),
the tyre (Section 27.7.1),
the suspension (Section 27.3.5),
devices attached to the engine (e.g. alternators, pumps and cams),
the moving parts of the engine and power transmission systems (e.g. friction losses in pistons and bearings), and
the operation of peripheral devices directed at passenger comfort (e.g. accessories such as cooling fans, power steering, and air conditioning).

R_r is approximately proportional to vehicle mass, M, and thus:

$$R_r = cMg$$

where c is the *rolling resistance coefficient* which is typically given by c = 0.01 for speeds up to 100 km/h. However, Section 29.2.3 suggests that a more accurate representation might be:

$$R_r = (c_0 + c_1V)Mg$$

where V is the speed of the vehicle.

The second category of forces covers those due to air resistance, which increase from zero with the square of vehicle speed and exceed rolling resistance above about 60 km/h.

The third category are the inertial forces which arise on an operating vehicle due to:

* the effects of gravity due to a sloping vertical road alignment (Section 19.3),
* the need to alter a vehicle's speed (e.g. braking or accelerating) or direction (e.g. cornering forces), and
* motions induced by pavement roughness (Section 14.4) and texture (Section 12.5).

The first two of these inertial forces can be calculated directly from Newton's Laws.

For the vehicle to move at a given speed it must overcome these three sets of forces over both distance and time. The *power* that a vehicle requires to do this is the product of its speed and the vector sum of the forces opposing or assisting motion. The power needed by most motor vehicles is supplied by spark-fired internal combustion engines operating on the *Rankine cycle* and fuelled by petrol (gasoline) or, less commonly, liquefied petroleum gas (LPG), alcohol (from sugar or biomass), or compressed natural gas (CNG) (see also Sections 27.8 & 29.2.1).

The major alternative to the internal combustion engine is the *compression ignition* (or *diesel*) engine, which has gained wide acceptance in commercial vehicles due to its lower operating costs and longer life. Alternatives to the Rankine cycle such as the *Brayton (gas) cycle* and *Stirling cycle* are rarely used. Electric vehicles currently have a minor role, as their limited storage capacity results in a short operating range and thus limits them to urban operations.

The rate at which the engine consumes the energy of its fuel, and the engine's mechanical efficiency, are major determinants of the power provided. The effect of this power usage on vehicle operating costs is discussed in Section 29.2.1.

27.2 PASSENGER CARS

27.2.1 General

As discussed in Section 16.1, the road transport system is a series of interactions between its component parts of vehicle, road, and driver. Nowhere is that interaction more evident than in the operation of the passenger car. The operation of a car in traffic has three main sets of objectives, viz.:

(1) performance objectives related to travelling from A to B in a particular style.

(1.1) At an instantaneous level, the performance objectives translate into vehicle performance objectives for speed (Chapter 18), acceleration (Chapters 17 and 19), deceleration (Sections 21.2.8 and 27.7.1), handleability and steerability (Sections 16.5.2 and 19.2.2).

(1.2) At a trip level, they translate into objectives for reliability, ease of operation, comfort, safety and security.

(2) fuel consumption and operating cost objectives (Sections 29.1–2); and

(3) greenhouse/exhaust emission objectives (Section 32.3).

Cars can be broadly classified as small, medium, or large; or as two- or four-wheel drives; or as single vehicles or cars towing other vehicles; or as sedans (typically defined as an enclosed car with a single compartment for at least four passengers), station (or ranch) wagons, sporting and specialised use vehicles, vehicles carrying less than ten passengers, or utilities (or pick-ups). They can also be characterised with respect to performance from 'standard' car trial data published by the motoring press, by engine capacity and power output, by mass, by power-to-mass ratio, and by wheelbase. Fuel consumption characteristics (Section 29.2.1) can be related to engine capacity and mass and acceleration performance to wheelbase.

The typical car has a mass of about 1.25 t. Incentives to reduce vehicle mass come from lower operating costs (Section 29.1.2) and higher payloads, and have resulted in increasing substitution of aluminium and plastic for steel. Most car bodies are of *monocoque* construction in which the body shell of the car is used to provide much of the overall structural strength. The vertical pillars between the roof and the floor are key components in this structure. The front, centre, and rear pillars are called the A, B, and C pillars respectively. These pillars can have a marked effect on a driver's ability to see a full 360° from the driving position. Rearwards vision is assisted by the provision of driver's-side, central, and passenger's-side mirrors. It has been estimated that about 8 percent of crashes involve poor rearward vision cars now have camera-assisted vision.

A general review of the role of passenger car design in road safety is given in Section 28.2.1. There it is suggested that vehicle defects contribute to about 8 percent of crashes. About 85 percent of these are readily detectable defects in the brakes and tyres (Section 27.7) and their primary cause is inadequate inspection and maintenance by the driver. Thus regular compulsory vehicle inspection is not a cost-effective safety measure. The risk of this sort of defect causing a crash increases with vehicle age, being four times as high in a seven-year-old car as in a two-year-old one. Vehicles in a crash are over twice as likely to have a defect as are vehicles in a random traffic sample, possibly reflecting largely on the attitudes of their owners (Grandel, 1985).

About 40 percent of fatal crashes result from frontal collisions and the remainder are equally divided between right- and left-side collisions and overturning. Rear-end collisions play a minor role in fatalities, although truck under-run crashes remain a cause for concern. Fire is a relatively minor problem. Increasingly, vehicles are being tested by independent NCAP programs to determine whether they adequately protect their occupants in a crash.

27.2.2 Acceleration and deceleration

The acceleration/deceleration performance of cars is of importance in a number of traffic engineering situations — such as traffic signal timing, the placement of signs and signals, and the layout of lane merges.

The accelerations observed in traffic are more dependent on the driver than they are on the type of vehicle, whereas speeds and decelerations are more vehicle-dependent. A key reason for the difference is that acceleration has only one input — the depression of the accelerator pedal by the driver. However, deceleration can occur (a) when the driver reduces pressure on the accelerator, (b) when the driver removes any pressure from the accelerator and the vehicle coasts – a process which is very vehicle dependent, (c) when the driver uses the brake pedal.

For ordinary driving, field data from various countries (e.g. Jarvis, 1982 and McDonald *et al.*, 1984) indicates *mean*:

* accelerations of 0.5 ms^{-2} to 1.5 ms^{-2} (trucks) or 2.0 ms^{-2} (cars). The low values are critical in determining the time large vehicles will take to clear intersections and crossings (e.g. Section 23.2.4).
* decelerations of around 1.0 ms^{-2}, increasing with initial speed.

Many factors cause variations from these means. For example acceleration/deceleration:

* is not constant throughout the manoeuvre. If required to stop in a given distance, drivers braking from a high speed will tend to reach maximum deceleration early and drivers from a low speed will reach it late in the braking manoeuvre, as the driver varies the force applied to the brake pedal.
* increase as the required speed changes increase.

An analysis of the standard U.S. urban *drive-cycle* (Section 29.2.1) shows that its range of accelerations (+) and decelerations (–) lie between ±1.5 ms^{-2}. British data suggests an extreme range of ±6.0 ms^{-2}, but with 99 percent of drivers falling within ±3 ms^{-2}. This latter range covers moderately quick response situations in urban driving (Watson and Milkins, 1985). For example, drivers required to halt when a signal changes will generally do so if it involves decelerations of 3 ms^{-2} or less. Other studies place the limit at 4 rather than 3. Passengers generally find up to 2 ms^{-2} to be normal, 3 ms^{-2} comfortable, over 4 ms^{-2} to be uncomfortable, and over 5 ms^{-2} to be severe.

Given the enormous personal and regional differences involved, the situation can be as summarised in Table 27.1. The table demonstrates that drivers rarely use the peak acceleration and deceleration levels quoted in car trial results (Jarvis, 1982) and the above measured data do not even reflect the fact that most cars can decelerate far more quickly than they can accelerate. However, one practical limit to deceleration is the locked-wheel situation (Section 27.7.1) which occurs when the wheels stop turning — the vehicle may then slew out of control and the driver lose the ability to steer the vehicle. Fear of this effect is a key reason why drivers are reluctant to use their full deceleration capacity.

Table 27.1 Typical range of speed-change accelerations and decelerations used by drivers. A ± indicates acceleration or deceleration; a – indicates deceleration only.

Acceleration or deceleration case	Speed change (ms^{-2})
No change of speed	0
Preferred maximum acceleration	1
Maximum acceleration likely in an urban trip	1.5
Maximum acceleration contemplated	2
Maximum used with comfort whilst under pressure (covers at least 80 percent of drivers)	±3
Maximum deceleration whilst still feeling in control	–3.5
Uncomfortable acceleration/deceleration; maximum deceleration in poor conditions	±4
Deceleration used in an emergency	–4.5
Acceleration/deceleration considered severe	±5
Deceleration consciously used by only a few drivers and in good (dry) conditions	–7
Deceleration used for minimum vehicle stopping distance in column 3 in Table 19.1	–12
Deceleration achievable in some emergency situations	–14

The equations for accelerating and braking can be developed using the following variables:

a is the constant acceleration or deceleration (always positive),
v is the actual speed and v_b is the speed after accelerating or before decelerating,
x is the distance travelled and x_b is the total distance travelled, and
t is the time and t_b is the total time travelled.

1. In the simplest case, if it is assumed that a vehicle has constant acceleration, a_c, then its equations of motion are:

Accelerating — *Braking*

$$a = a_c \qquad a = -a_c \qquad (27.1a)$$

$$v = a_c t \qquad v = v_b - a_c t \qquad (27.1b)$$

$$x = a_c t^2/2 \qquad x = v_b t - (a_c t^2/2) \qquad (27.1c)$$

$$= v^2/2a_c \qquad = (v_b^2 - v^2)/2a_c \qquad (27.1d)$$

$$x_b = v_b^2/2a_c \qquad x_{stop} = v_b^2/2a_c \qquad (27.1e)$$

$$t_b = v_b/a_c \qquad t_b = v_b/a_c$$

2. A somewhat more complicated analysis assumes that acceleration and deceleration are linearly related to speed and a_o, the maximum acceleration, occurs at t = 0:

Accelerating to $v = v_b$ — *Braking from $v = v_b$*

$$a = a_o(1 - v/v_b) \qquad a = -a_o v/v_b$$

$$v/v_b = 1 - e^{-a_o t/v_b} \qquad v/v_b = e^{-a_o t/v_b}$$

$$xa_o/v_b^2 = a_o t/v_b - 1 + e^{-a_o t/v_b} \qquad xa_o/v_b^2 = 1 - e^{-a_o t/v_b}$$

$$= -v/v_b - \ln(1 - v/v_b) \qquad = 1 - v/v_b$$

$$x_b = \infty \qquad x_b = v_b^2/a_o$$

The speed equations in this set are shown in Figure 27.1a. They are useful for broad studies of vehicles in traffic.

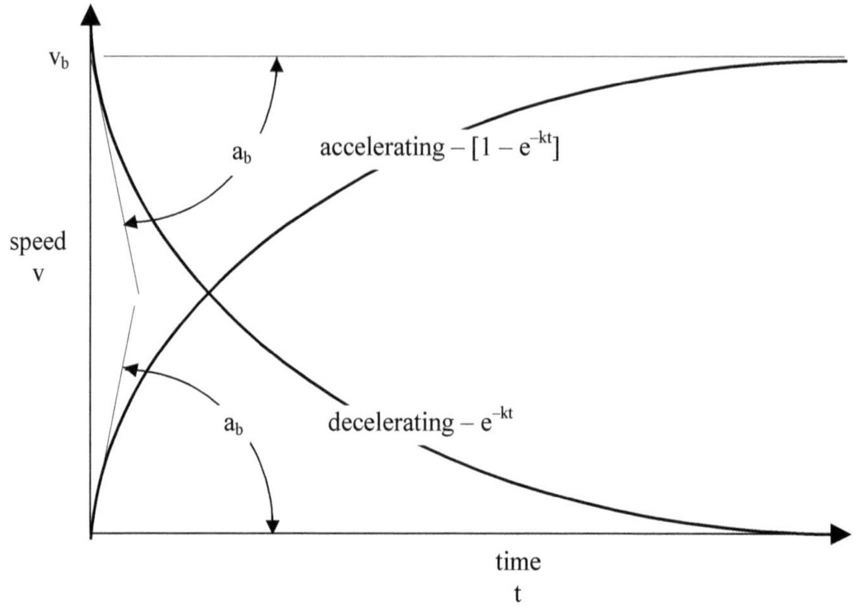

Figure 27.1a Speed–time curves based on a linear link between acceleration and speed. Note that $k = a_o/v_b$.

3. Another alternative model which has some relevance for trucks uses acceleration linearly related to time, and a_o, the maximum acceleration, occurring at $t = 0$.

Accelerating to $v = v_b$	*Braking from* $v = v_b$
$a = a_o(1 - t/t_b)$	$a = -a_o(1 - t/t_b)$
$v/a_ot_b = t/t_b - 0.5(t/t_b)^2$	$(v_b - v)/a_ot_b = t/t_b - 0.5(t/t_b)^2$
$t_b = 2v_b/a_o$	$t_b = 2v_b/a_o$
$2x/a_ot_b^2 = (t/t_b)^2 - (1/3)(t/t_b)^3$	$2(x - v_bt)/a_ot_b^2 = (t/t_b)^2 - (1/3)(t/t_b)^3$
$x_b = 4v_b^2/3a_o$	$x_b = 2v_b^2/3a_o$

4. Empirical studies suggest that the x–v relationship (e.g. Equation 27.1d) might be better represented by an equation of the form;

$$v^3 = v_b^3 + c_1x + c_2x^2$$

Idealised, but more realistic, acceleration–time and speed–time curves for use in studying individual vehicle manoeuvres are shown in Figures 27.1b & c. The model assumes zero acceleration and zero rate of change of acceleration (*jerk*) at the beginning and end of the manoeuvre. Actual curves will vary markedly in their shape, particularly with respect to the position and value of the maximum acceleration point, A. Accelerations typically last between 5 and 50 seconds, the time increasing as the final speed increases.

In crash investigations, it is sometimes necessary to estimate vehicle speed from skid marks. The marks indicate that the brakes were applied with sufficient force for the vehicle's wheels to *lock* (Section 27.7.1). The braking force will be (SN/100)Mg, where SN/100 is the *longitudinal locked-wheel coefficient of friction* (Section 12.5.3), and M is the mass of the vehicle. Deceleration is therefore given by:

$$-a = (\text{braking force})/\text{mass} = (SN/100)g$$

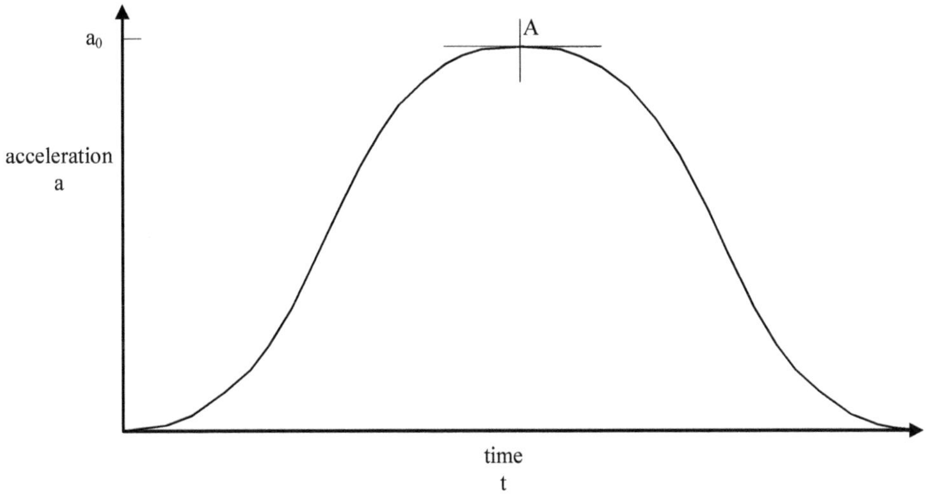

Figure 27.1b Idealised acceleration–time curve.

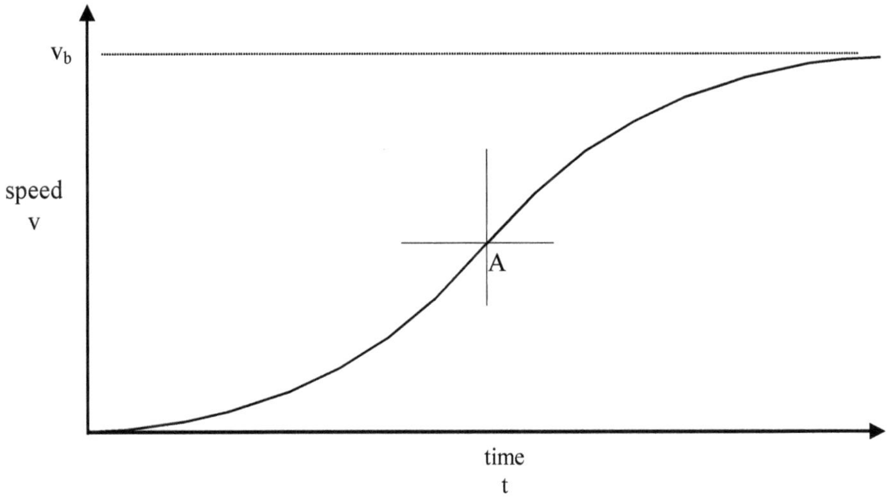

Figure 27.1c Idealised speed–time curve.

Table 27.1 suggests that a good, dry surface would produce a peak deceleration of 14 ms^{-2}, or 1.4g which leads to an SN of at least 140 to achieve this case. Using Equation 27.1d and assuming that the vehicle comes to a stop (v = 0) under constant deceleration, gives the locked-wheel stopping distance, D_s, as:

$$D_s = v_b^2/2a_c = v_{bk}^2/(254[SN/100])$$

where D_s (in m) and SN can be measured on site, thus allowing evaluation of v_b (or v_{bk} in km/h), the actual speed before braking.

Grades can significantly reduce a vehicle's acceleration capacity. From physics, the absolute reduction would be 10S ms^{-2} where S is the grade in radian. Measurements indicate that the reductions are closer to 5S ms^{-2}.

27.3 HEAVY VEHICLES

27.3.1 Types of heavy vehicle

Heavy vehicles are usually defined as having more than two axles and/or an unladen mass of over 4 t. A *truck* (or *lorry*) is a heavy vehicle designed to move freight (Section 30.5.2) and is commonly taken as having an *unladen* (*tare*) mass of at least 3 t. The unladen mass is the mass of the truck when it is not carrying any payload. The *kerb mass* of a vehicle is the unladen mass, and specifically includes the mass of all standard equipment and accessories and a maximum content of fuel, oil and coolant. The most common truck types are the *rigid* (or *single-body* or *straight*) truck (Figure 27.2a) and the *combination* truck (Figures 27.2b&c). A *tip truck* (or *tipper*) is a truck in which the load-carrying tray can tilt to discharge freight.

The rigid truck is the basic truck configuration. Its single body has between two and five axles, all attached to an integral chassis that is effectively rigid in a horizontal plane. The axles may be used as singles, in pairs (*tandems*), or in threes (*triaxles*). Groups of axles are known as *bogies*. One or two of the axles may be *driving* axles, receiving power from the engine via the transmission, and one or two may be *turning* axles, controlled by the steering wheel. When axle pairs are used in this way they are respectively called *twin-drive* and *twin-steer* axles.

Combination trucks consist of a number of separable mobile components. For instance, a *trailer* is a mobile freight-carrier without motive power of its own and designed to be attached to another powered vehicle. A *truck-trailer* (Figure 27.2b) is a rigid truck towing a trailer via either a drawbar or pin connection. An *articulated* vehicle (or *tractor-trailer*, Figure 27.2c) differs from a truck-trailer in that the rear portion of the articulated vehicle, called a *semi-trailer*, is pivoted (pinned) to and superimposed on the forward, powered portion of the combination. It is therefore not a full trailer — hence, the term *semi*. The powered forward portion is called the *prime mover* (or *tractor*, thus the U.S. term, *tractor-trailer*). If a draw-bar is used between the prime mover and the trailer, the vehicle is not articulated but is a form of truck-trailer and the trailer in this case must be self-supporting and not a semi-trailer.

The prime mover provides the motive power for the truck combination. This power must be sufficient to permit the truck to:

* start on the maximum expected parking grade,
* maintain its forward momentum on the maximum expected operating grade, and
* achieve sufficient acceleration to pass other vehicles travelling just below the legal speed limit on a flat grade.

The prime mover may have two or three axles. A prime mover where the driving cabin is located ahead of, or over, the turning axle is called a *forward control* vehicle.

The *fifth wheel* shown in Figure 27.2c is a horizontal plate that allows a trailer connecting point to bear vertically on a prime mover and rotate horizontally. The *kingpin* is the vertical post through the centre of the connecting point. It transmits the horizontal force between prime mover and trailer.

The road train (Figure 27.2d) is discussed in Section 27.4.2. The *B train* (or *B double*, *turner double* or *Western double*; Figure 27.2e) fills the gap between articulated vehicles and road trains. It is produced by carrying the rear of an *A train* semi-trailer (Section 27.4.2) on a suspension — called a *B dolly* — that protrudes sufficiently further to the rear to include a fifth wheel which supports the dog trailer. There is no articulation under the semi-trailer and this increases steering stability (e.g. in lane changing) and makes reversing easier.

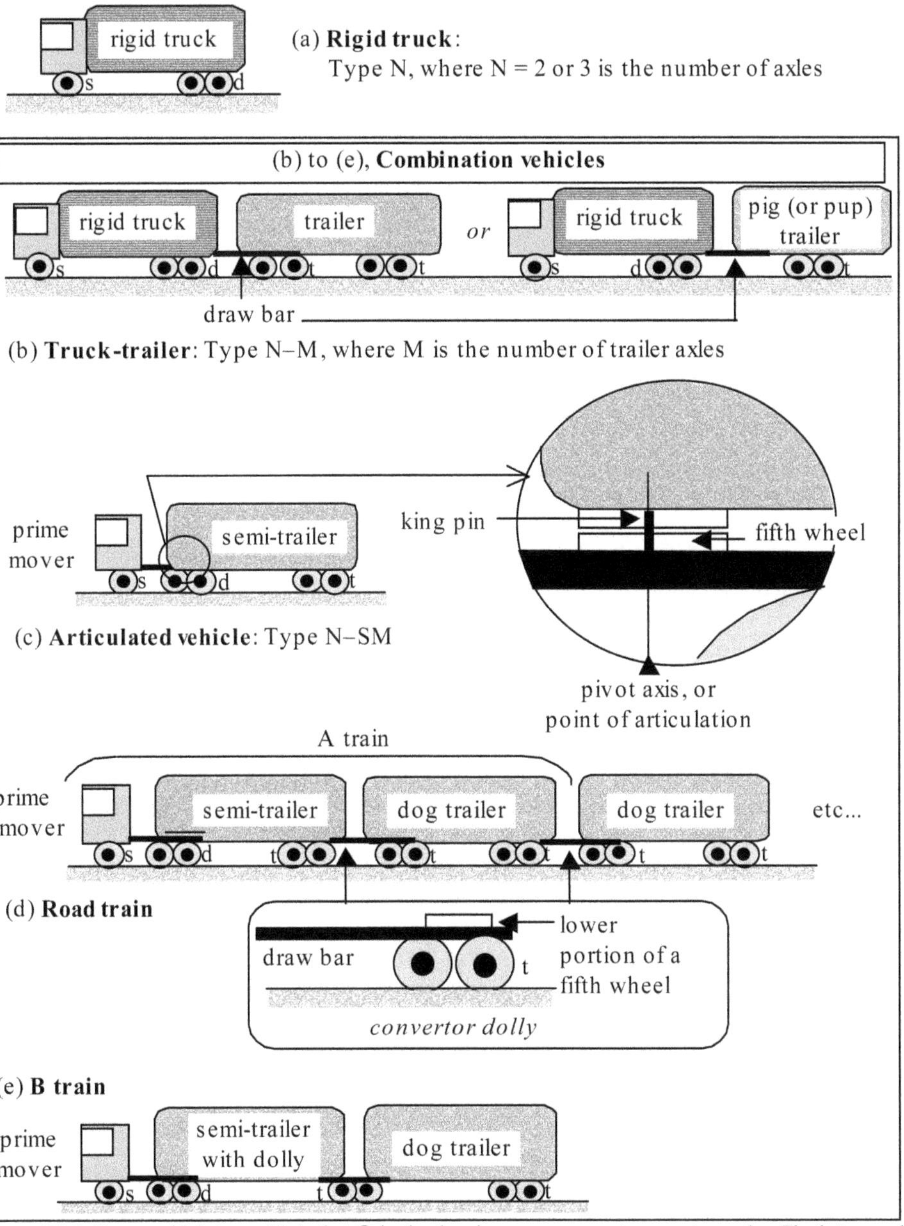

Figure 27.2 Heavy vehicle types. Note that ⊙ in the drawings can represent one or two axles. The letters s, d, and t indicate whether the axle (or axle group) arrangement is for steering (i.e. turning), driving (i.e. powering), or trailing (i.e. following unpowered).

The more common truck and trailer body types in use are:

(a) *flat top* – a flat load-carrying platform that may be either the rear of a rigid truck or a trailer,

(a1) open — standard version,
(a2) covered — has a temporary cover, which may be no more than a tarpaulin,
(a3) timber carrier (or *jinker*) — has side posts to retain the timber,
(a4) livestock carrier — has side walls to retain the livestock,
(a5) car carrier — built to carry cars,
(a6) container carrier — built to carry container bodies (b4),
(a7) mobile load platform (Section 27.4.1).

(b) a fully and permanently enclosed freight area,
(b1) *van* — standard version,
(b2) *pantechnicon* (or furniture-removal van),
(b3) refrigerated van,
(b4) *container body* — a standard freight shipping container, usually carried on (a6),

(c) a container for carrying liquids,
(c1) *tanker* — standard version,
(c2) tanker with hazardous freight (Section 27.3.9),
(c3) rigid truck carrying pre-mixed concrete,

27.3.2 Dimensional and mass limits

The unladen mass of a truck was discussed in Section 27.3.1. A key road-related factor in the design and regulation of trucks is the mass that they carry. *Gross vehicle mass* (*GVM*) is the vehicle's total laden mass and is comprised of the unladen mass and the payload. Typically, the maximum GVM is about three times the unladen mass. There is a universal legislative concern with gross vehicle mass as:

(a) Pavement and bridge deterioration will be related to GVM (Chapters 14 & 15). Note, however, that fatigue-like deterioration will depend on the number of passes of the load, as well as on its magnitude (Section 14.3.2).
(b) The momentum of the vehicle is linearly related to GVM and hence its impact and collision effects will increase steadily as GVM increases.
(c) The size of the vehicle will, by and large, increase as its GVM increases. Generally the amount of annoyance, fear, noise, vibration, visual intrusion, and traffic disruption that a large truck produces in urban areas will increase as its size increases (e.g. Sections 30.5.2 and 32.2.3).

The administrative convenience of gross vehicle mass is that a single set of measurements can be used to attempt to control these factors.

A typical limit on the gross mass of a vehicle is about 40 t but many countries allow vehicles with a much larger mass to operate under a variety of route-specific conditions. This situation has arisen because, if the factors listed in (a) to (c) above are managed, there is no inherent reason for any limit on GVM.

The stresses in pavements are much more related to wheel loads (Section 11.4.1) and in bridges to axle loads (Section 15.4.2) rather than to GVM. Maximum wheel loads are naturally limited by the maximum loads on individual tyres, as discussed in Section 27.7.3. The unladen mass per axle typically ranges between 1 t for a light truck to 4 t for an articulated vehicle. The maximum axle load will occur under GVM conditions and with an axle with dual tyres at each end, as illustrated in Figure 27.3a. In practice, the critical road-related limits on vehicle mass are usually those imposed on:

* axle load,
* axle configuration, and

* the spacing of adjacent axles.

To provide a reasonable service life, such 'static' limits understandably result in wheel loads well below the load capacity of normal tyres.

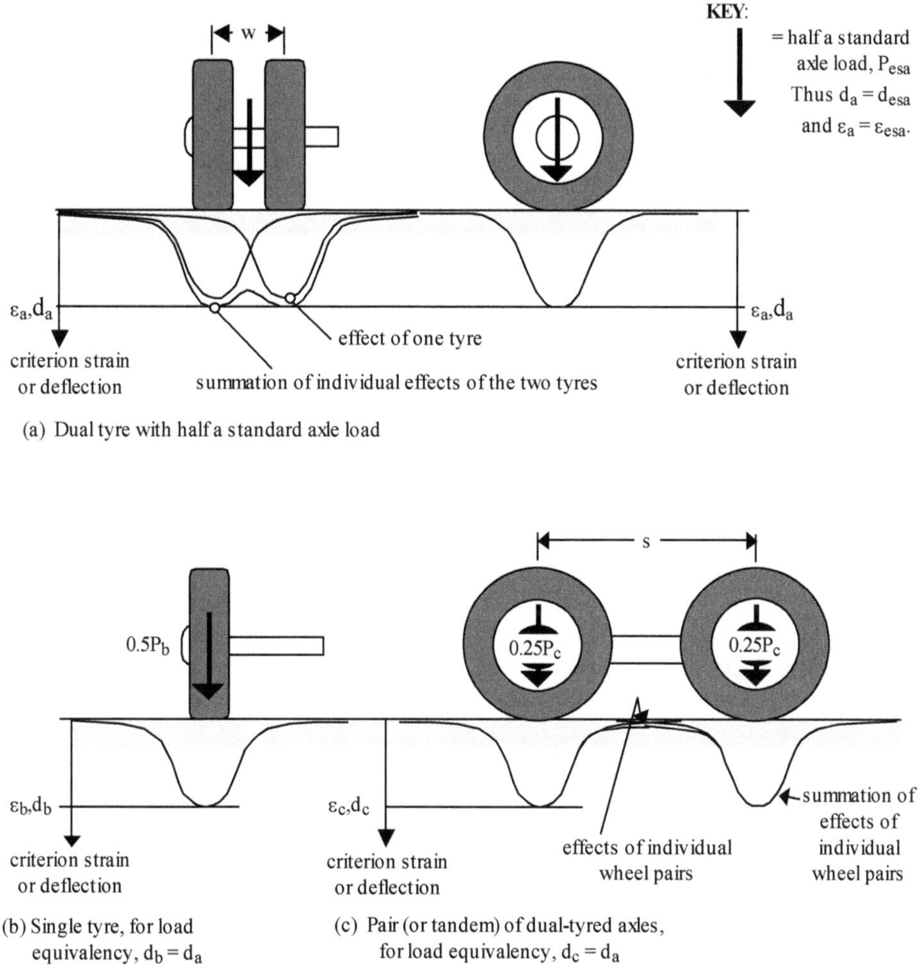

Figure 27.3 A technique for comparing different axle configurations and determining the equivalent load on each group.

Axle load as a force normal to the pavement–tyre interface (F_V in Figure 27.7) is due to four factors:

(a) the distribution of the gross mass of the stationary, horizontal vehicle to the individual axles, in accordance with the principles of structural mechanics,

(b) the changes in the normal component of the gross mass, due to the grade and cross-fall of the pavement (e.g. Chapter 19),

(c) the dynamic forces produced by the normal (vertical) accelerations of the various components of the mass of the moving, suspended vehicle (Section 27.3.5), and

(d) the vertical force couples produced by the horizontal accelerations arising from cornering, braking, and speed increases.

Legal limits on axle load usually apply only to (a), although there are growing attempts to also cover (c).

There is a great deal of Road Authority concern with limits on axle loads. This is because an Authority's existing pavement and bridge infrastructure will have been built to carry design loads based on the then current limits, supplemented with some extrapolation to cover increases in the load limit during the infrastructure's design life. The limits are therefore usually enforced and defended. Any increase in the limits will cause an increase in the rate of deterioration of the existing infrastructure (Section 14.3.3) and, in the extreme, may cause failure during a single pass of the new, increased load. Nevertheless, an economic systems-analysis of networks with high tonnages of freight may well show an overall benefit from axle load increases, provided adequate funds are made available for pavement upgrading. The best current tool for doing such an analysis is the *HDM* model (Section 26.5.2)

Not all trucks will be loaded to the legal limit. On the one hand, some will be unable to obtain a full load for their return trip and, for others, the amount of freight that can be hauled will be limited by dimensional rather than by mass requirements, i.e. volume rather than mass will determine the payload. On the other hand, operating costs per tonne will drop as the mass carried increases, at least in the short run. For this and other reasons it is possible to encounter vehicles on the road which are loaded above their legal limit. In developed countries, up to 10 percent of trucks may be in this overload category. The damaging consequences of overloading are discussed throughout Chapters 11, 14 & 15 (see Index entries). They increase disproportionately with the amount of overload.

If axle loads are kept below an acceptable level, then any GVM can be carried by a pavement by adding more acceptable axles and hence redistributing the GVM over a longer length. For example, one truck of GVM = M, can be designed to do no more pavement damage than two trucks of GVM = M/2. For bridges, the distribution of the mass within a truck length is also often controlled by a *bridge formula* (Section 15.4.2) controlling the overall axle configuration, as there are more load interactions over distance in bridges than there are in pavements. In addition, the design of long-span bridges often assumes that the trucks on the bridge do not fill each lane with bumper-to-bumper traffic. Unusual loads are discussed in Section 27.4.1.

The spacing of adjacent axles becomes critical if the stresses induced by individual axles can significantly interact within the pavement to produce stress levels greater than that from a single axle. Of course, the actual interactions are between individual tyres in the same wheel-path. In the case where there are two adjacent axles (the *tandem* system) shown in Figure 27.3c, there is some interaction between axles and so the permissible sum of the two individual axle loads with a close axle spacing could be less than for two single axles. Typically a 10 percent reduction might occur for a $s = 1$ m axle spacing. If $s \leq 1.0$ m the arrangement is called a *close-coupled* axle group, and regulatory bodies often require the group to be considered as a single axle. A similar close-coupled philosophy applies to groups of three axles, with a reduction in permissible axle load for each axle within a metre of the central axle. Methods for determining the precise reductions and the equivalent loads applied to different axle groups are described in Section 27.3.4.

Permissible dimensions and axle loads and GVMs for trucks vary from country to country, although considerable standardisation exists within the European Union. Some typical values are given below. A comprehensive collection of data from many countries is maintained by the International Road Transport Union (http:www.iru.org). Truck turning-paths are discussed in Section 27.3.8.

Quantity	Maximum range	Notes
Container height	2.4–2.6 m	
Truck height	2.9–4.8 m	The usual limit is 4.0 m.
Container width	2.44 m	
Truck width	2.1–2.9 m	The usual limit is 2.5 m.
Container length	3.0–12.2 m	Or 10 ft, 20 ft and 30 ft.
Truck length	6–40* m	Lengths of 6 to 12 are considered *medium*, and over 12 m is considered *long*.
Axle load	5–14 t	The load capacity of a light truck-tyre is about 2.7 t (Section 27.7.3).
Gross vehicle mass	10–100* t	

* Can be increased by a specific local permit (Section 27.4).

27.3.3 Weighing methods

Various load-sensitive devices are used for weighing trucks and other heavy vehicles. They may weigh the truck when stationary, or when in motion.

Lever-based *weighbridges* which weigh the entire stationary truck provide the 'primary' weighing standards. They can achieve accuracies of 0.02 percent, i.e. ±10 kg in a 50 t truck. However, they suffer the disadvantage of high costs and the problems posed by needing to divert traffic to them and by drivers avoiding them. Mobile phones aggravate this last problem as news of the presence of an operating weighbridge is soon broadcast.

Some major system errors are possible in vehicle weighing. Any device which weighs less than a whole truck will introduce errors into the mass estimate due to the need to estimate the sharing of weight between the two parts of the truck.

Portable devices — commonly known as *loadometers* — are often used to overcome the disadvantages of weighbridges, but can suffer from systematic errors as a single loadometer weighs one wheel at a time. To minimise error, it is desirable to either have a loadometer under each wheel of the truck being weighed and/or to have all the wheels raised to the same horizontal plane. Even then, hysteretic non-linearities in the load-deflection response of the truck suspension system (Section 27.3.5) mean that different wheel loads can be recorded, depending on how the vehicle to be weighed is brought to rest on the weighing platform. These non-linearities will depend on the design and maintenance of the suspension system, with the latter capable of causing wheel loads to vary by up to 1 t. Even a linearly-sprung suspension will cause measured loads to vary if the tyre–pavement contact areas are at even-marginally different levels, or if there are strong, gusty winds. Typically, each 1 mm of difference changes the recorded mass by 100 kg.

Techniques which can weigh a vehicle in motion (*weigh-in-motion* or *WIM* systems) fall into three main categories:

(a) instrumented bridges,
(b) devices on the pavement, and
(c) devices in the pavement.

The instrumented bridge uses the bridge structure as a transducer, with strain gauges placed on selected components. One such device is the ARRB CULWAY system which uses simple box culverts rather than bridges (Brown and Peters, 1988). Axle sensors are usually also needed to help convert the structural actions back into truck actions. Initial structural calculations may be involved and each structure may require its own calibration.

Because of the effects of suspension dynamics to be discussed in Section 27.3.5, the measured dynamic loads from single devices on or in the pavement can readily differ by ±30 percent from the static load. However, with careful attention to site and approach preparation, accuracies of ±10 percent are possible with most weigh-in-motion devices. A leading supplier in this field, International Road Dynamics, quotes ±6 percent in traffic and ±1 percent at slow speeds. Because of these inherent inaccuracies, WIM devices are sometimes used to select potentially-overloaded vehicles in a traffic stream for static weighing. WIM accuracies are improved by using a series of closely-spaced WIM units to pick up the truck's loading waveform. For example, a three-sensor series of WIMs can improve accuracies to ±5 percent. Devices in the pavement include a beam supported on load cells, plates supported on load cells or strain-gauged, and load-sensitive materials such as capacitance-sensitive strips, piezoelectric ceramics and fibre-optic cables.

Devices on the pavement are usually either strain-gauged plates or capacitance-sensitive plates (e.g. *PAT*). For accurate weighing, thicker devices need ramp approaches and compensating levelling plates.

27.3.4 Equivalent standard axles

Figures 27.3b&c show a technique for comparing different axle configurations and determining the equivalent load on each group. For a single wheel, this is called the *equivalent single wheel load* (*ESWL*). For a single axle, it is called the *equivalent single axle (ESA)* or the *ESA load* (*ESAL*). A comprehensive review of the load equivalents of various axle and tyre groupings is given in Vuong and Jameson (2003).

In comparing different types and configurations of axles and tyres for the relative pavement damage that they cause, the developments in Sections 14.3.3 and 14.5 are used. There are four sets of criteria which could be used for making the comparison:

(a) *the damage caused* — whilst this is the logical measure, the relevant data is difficult to obtain experimentally in the short term (Section 14.3.1), however the 1993 AASHTO design guide uses such a relationship derived statistically from the AASHTO road test conducted in the late 1950s.

(b) *damage criteria* based on Sections 14.1A–C, such as a chosen pavement roughness or extent of cracking (Section 14.5.1). These criteria can be more readily measured than (a) but will usually depend on the pavement type and condition and remain difficult or impossible to obtain in the short term.

(c) *the deflections produced* (Section 14.5.2 & Figure 27.3c), given that the surface deflection is the aggregation of the effect of many individual and localised damage components. Deflections (and strains) can be measured instantaneously but may not relate well to long-term damage (Section 14.3.2). The use of deflection as the criterion (see Figure 27.3) was introduced by Boyd and Foster in 1950.

(d) *the strains produced,* as these strains accumulate to produce the surface deflection. It is common to concentrate on the major contributors to damage, which are usually the maximum compressive (shear) strain in the subgrade (Section 11.4.3) or the maximum tensile strain in the basecourse (Section

11.4.4). This approach allows single, discrete strain-readings to be used and is illustrated in Figure 27.3.

For short term assessments, deflections are usually preferred over strains as they are easier to measure and give a more aggregated picture. Deflections are therefore commonly used for comparing two configurations with the same axle load, with any available strain measurements employed as a useful cross-check (Figure 27.3).

Taking the single axle with dual tyres as the base case the equivalence ratios for the other common cases, with an assumed longitudinal axle spacing of 1.32 m, is shown below (Vuong and Jameson, 2003):

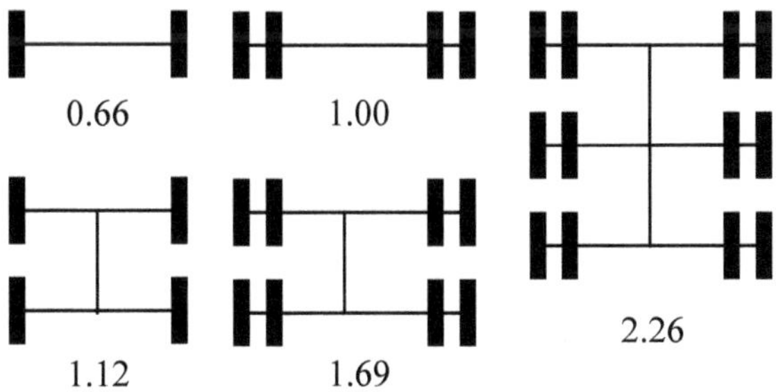

Section 14.3.3 showed that the effects of different axle loads may be predicted by using the fourth power law (Equation 14.7) to calculate their *load equivalencies* relative to a *standard axle load* (or *equivalent design axle*), P_{esa}, which is usually taken as the legal limit for a single dual-tyred axle. The *number of equivalent standard axles*, N_{esa} for any particular load and axle configuration is defined as the number of standard axle loads which would cause the same damage as a single pass of the load and axle configuration in question. It is also variously called the *axle factor* [AF], *equivalent axle load number* [EAL], *load equivalence factor* [LEF, AASHTO], *axle load equivalency factor*, *wheel load equivalency factor*, *pavement load equivalency factor*, *pavement damage factor*, or *truck factor*.

To calculate N_{esa} for a particular axle configuration, it is first converted to a standard axle, using one of the above damage criteria, commonly deflection or strain, as shown for $d_{b\text{ or }c} = d_{esa}$ in Figure 27.3b&c. For the example of the load P_c on the configuration in Figure 27.3c and which causes a deflection of d_c, the conversion process selects the load P_a on the standard axle which also causes a deflection of d_c. Thus, P_a on the standard axle produces the same deflection as the load P_c on the original configuration. As the load on the equivalent standard axle is P_{esa} and not P_a, Equation 14.7 is used to give N_{esa} as:

$$N_{esa}(P_{esa})^4 = 1(P_a)^4$$

or

$$N_{esa} = (P_a/P_{esa})^4 \tag{27.2}$$

The use of some other failure criterion may produce an exponent other than 4: e.g. Sections 14.3.2 & 3 suggest about 5 for strain and 2.5 for rutting.

Assuming a linear load–deflection response, which is reasonable for moderate loads on stiff pavements (Section 14.3.2), gives:

$$P_a = P_{esa}d_c/d_{esa}$$

and so:

$$N_{esa} = (d_c/d_{esa})^4$$

Once N_{esa} is known for a particular axle configuration and loading and, if it is also known that N_s passes of the standard axle load will cause failure (Figures 14.9 and 14.10), then it can be deduced that N_s/N_{esa} passes of that axle configuration and loading will cause failure. Hence, its relative impact on the life of the pavement can be calculated.

It must be emphasised that N_{esa} refers to the number of passes of a standard axle load to cause equivalent damage, and not to a particular load value. Confusion is often created by the process shown in Figure 27.3 for comparing axle configurations and which produces load equivalents between configurations, such as P_a. The selection of P_a is not based on the fourth power law. The fourth power law (Equation 14.8) is invoked only when the relative effects of P_a and P_{esa} on pavement life are being calculated.

To illustrate the issue further, consider a pavement being loaded by a standard axle, P_{esa}. At N_o passes of this axle, it would take another N_s passes to cause failure of the pavement (Figure 27.4). Consider what would happen if at N_o passes the load is increased to P_a. As the axle configuration has not changed, the fourth power law (Equation 14.8) can be used directly to predict that it will now take only $(P_{esa}/P_a)^4 N_S$ passes to cause pavement failure. From Equation 27.2, this can be written as N_s/N_{esa}, which is consistent with the definition of N_{esa} (i.e. N_{esa} passes of P_{esa} for one pass of P_a).

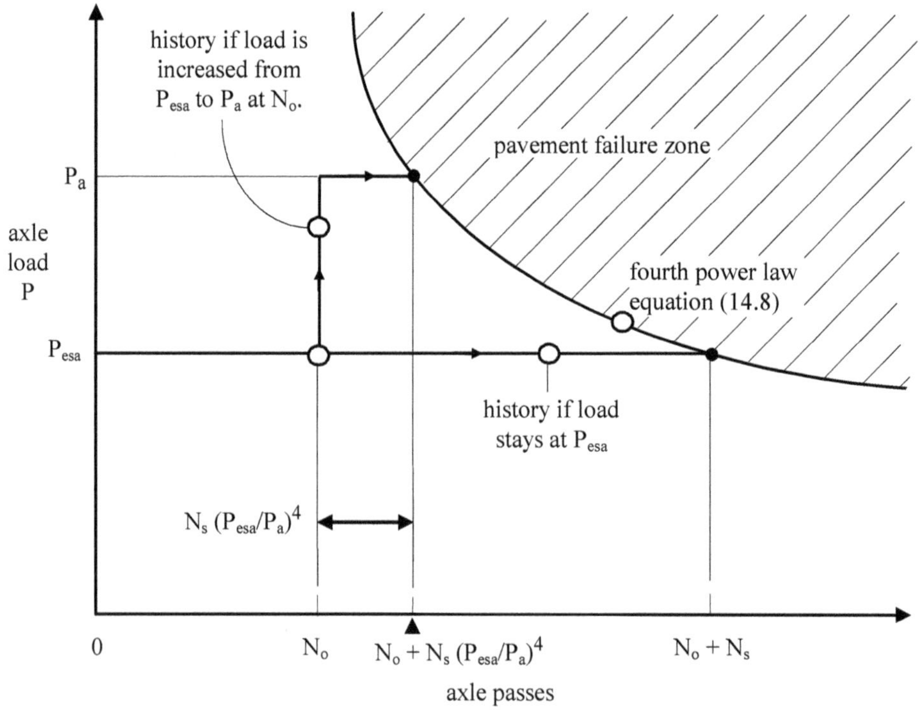

Figure 27.4 Illustration of the influence of axle loads and axle passes. Pavement failure is indicated by •. The ordinate axis could be a criterion strain or deflection, rather than a load. Note that $N_o + N_S = N_t$ where N_t is defined in Figure 14.10.

A vehicle will usually be composed of a number of distinct axle groupings (or configurations, in the terms of the above example). Each will have its own N_{esa} and the N_{esa} for the vehicle will be the sum of these individual N_{esa}s. This N_{esa} per vehicle reflects the relative damage of a single pass of the vehicle relative to a single pass of the standard axle. Typical measured N_{esa} per vehicle for U.S. roads are given in the following table. These are often lower than 1.0, as many vehicles are not fully laden on all axles.

Road type (Section 7.2.1)	N_{esa} for each truck type (Section 27.3.1)		
	rigid	semi-trailer	truck-trailer
Collector	0.8	0.7	n.a.
Urban sub-arterial	0.4	0.4	1.3
Rural sub-arterial	0.4	1.1	0.7
Urban arterial	1.8	1.4	3.3
Rural arterial	0.5	1.2	1.4
Urban freeway	0.3	0.8	1.7
Rural freeway	0.4	1.4	2.6

27.3.5 Suspension behaviour

A moving truck is a dynamic mechanical system, moving horizontally and vertically. The parts of the system are interconnected by *springs* (the suspension and parts of the tyres) and *damping* devices (the *shock absorbers* and parts of the tyres). The system includes some unsprung mass (mainly the tyres, axles, and parts of the suspension and transmission), but is mainly a sprung mass carried by the suspension. The interconnecting springs do not respond linearly, provide increasing stiffness as they are deformed, and display considerable hysteresis (Section 33.3.3). It is usually necessary to define separate loading and unloading curves. Thus any analyses are complex and the non-linear stiffness causes many of the problems referred to in Section 27.3.3 when parts of vehicles are weighed on non-level sites. With many suspension types, there is also a dual mode of behaviour with a separate, stiffer system coming into action at higher loads.

The vertical movements affect the pavement loads (see item (3) in Section 27.3.2) and are due to:

For the sprung mass:

- * body bounce (uniform vertical movement),
- * body pitch (longitudinal fore-and-aft rocking), and
- * body roll (transverse rocking).

For the unsprung mass:

- * axle bounce,
- * pitching between adjacent axles, and
- * axle tramping (transverse rocking along the axle length).

In normal operations, the sprung mass of a truck body bounces at between 1 and 3 Hz and pitches at 6 to 7 Hz, the chassis flexes at between 3 and 10 Hz, and the suspension bounces and pitches at about 10 Hz.

For a vehicle travelling at about 100 km/h, this produces longitudinal wavelengths of between 3 m and 30 m. Section 14.4.2 showed that these road surface wavelengths are covered by those that can be measured by laser profilometry.

The vehicle suspension softens and damps the dynamic load effects caused by these movements. Without a suspension, there would be high momentum transfer and rigid-body impacts between the entire vehicle and the road, with magnified forces imposed on both freight and pavement. In addition, the jerky nature of the vertical movement would result in a relatively uncomfortable and damaging ride (Sweatman, 1983).

There have always been market pressures on vehicle manufacturers to develop good suspensions for passenger vehicles, for the driver cabins of trucks, and for carriers of fragile freight. Until recent years, few such pressures have existed to force the development of better suspensions for heavy trucks carrying typical coherent, robust cargo. However, the relatively-high stiffness of the road pavement means that these are the very vehicles likely to produce the highest impact loads and thus cause the most pavement damage (Section 14.3.2). Typically, the stiffnesses of pavement, tyre, and suspension are 60 000 kN/m (Figure 11.5), 1 000 kN/m, and 300 kN/m respectively.

The effect will also vary between wheels. The rougher outer wheel-path lanes with narrow shoulders (Section 9.2.3) will worsen the response in these wheel-paths. The wheel-path difference may also exaggerate forces induced by body roll. Drive axles carrying greater unsprung mass will also produce higher loading effects.

The forces that a truck applies to the pavement thus depend on both the static load on each wheel, the suspension type, the tyre pressure and stiffness, and the chassis design. In addition, they are functions of the roughness of the road (Section 14.4) and the speed at which the truck is travelling (Sweatman, 1983). On smooth roads, the increase over the static load will be minor or even non-existent. Typically, however, the dynamic forces applied by the wheels of a moving truck have an RMS amplitude of between 10 and 30 percent of the static load. The ratio of this RMS value to the static load is called the *dynamic load coefficient* of the vehicle suspension system. The damaging effect of these dynamic loads on the road rises from 10 to 30 percent to 2 to 14 times if — due to the preceding pattern of pavement roughness — the dynamic forces from all trucks occur at the same points on the pavement surface.

Some of the dynamic effects are avoided by the use of *load-sharing* suspensions within an axle group. The mechanism of load-sharing is explained by reference to Figure 27.5a which shows a rigid truck with a single steer-axle and a tandem rear-axle group. When standing on a bump, independent axles in the tandem group would behave as in Figure 27.5b, with increased loads on one axle, whereas the load-sharing suspension in Figure 27.5c would ensure that the pavement forces are not increased by any unevenness.

Poor load-sharing can increase road damage by a factor of up to 3. Even good load-sharing is not the total answer as, in operation on the road in normal conditions, a load-sharing suspension provides time-averaged axle loads on all axles in the group. This means that, contrary to the static analysis in Figure 27.5, instantaneous axle loads will tend to be unequal due to dynamic forces generated over the road profile. The simple load-sharing mechanism cannot handle all these impact forces. Indeed, some load-sharing, tandem, drive suspensions can produce relatively high dynamic forces (Sweatman, 1980).

27.3.6 Suspension types

Individual suspensions use six main techniques for providing the required stiffness properties:

(a) *Coil* springs which are conventional spiral springs.

(b) *Leaf* springs (Figure 27.6a) which function largely as steel beams and are so-named because of their arrangement as overlapping or laminated leaves, although they are also called semi-elliptical springs because of their shape, or steel springs because of their material. The main leaf is connected to the vehicle body at one end by a pin and at the other by a hanging bracket capable of swinging longitudinally. The other leaves follow its contour and support it when the wheel drops downwards or when the chassis rises. Clips are used to hold the leaves together.

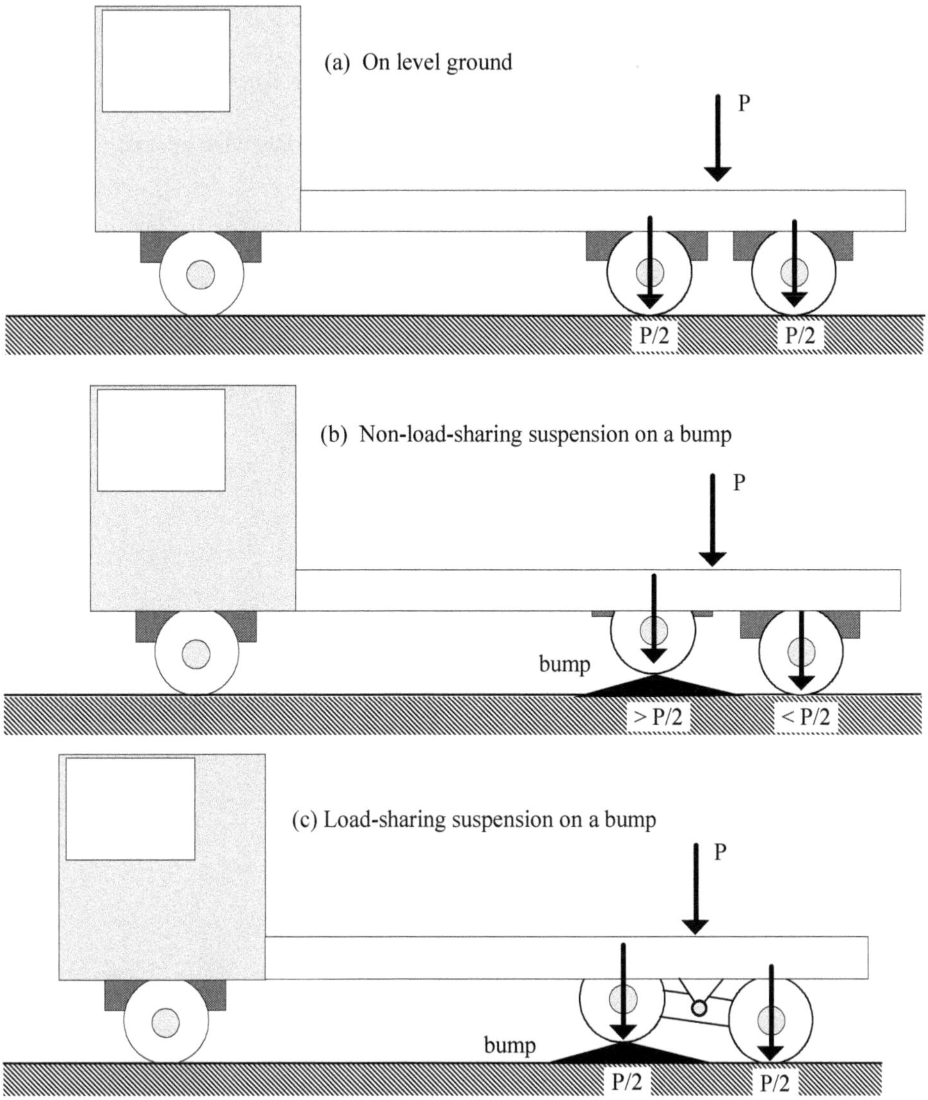

Figure 27.5 Load-sharing suspensions. Figures 27.5a&b assume independent leaf-spring suspensions (see Figure 27.6a). The mechanism illustrated in Figure 27.5c is a variation of the walking beam suspension shown in more detail in Figure 27.6d and discussed below.

(c) *Air bags or air springs* (Figure 27.6b). These usually consist of compressed air contained in flexible rubber bags, with levelling bleed valves and inter-connected air supplies for load-sharing. These usually have good load-sharing capabilities.

(d) Rubber in compression. These units are often alternate laminated layers of rubber and steel (see discussion of elastomeric bridge bearings in Section 15.2.5). Rubber in shear or torsion is also used in some suspensions.

(e) *Torsion bar* (Figure 27.6c) which uses the torsional stiffness of a bar as the springing device.

(f) *Walking beam* (Figure 27.6d) which is an extension of the leaf spring, adding a central pivot to provide static load-sharing (Section 27.3.5). It has a natural frequency of about 10 Hz and is only lightly damped. It has poor dynamic properties and is probably the worst of the common suspension types.

As well as carrying vertical gravity forces, all suspensions must also be able to transmit horizontal forces due to braking, accelerating, and cornering.

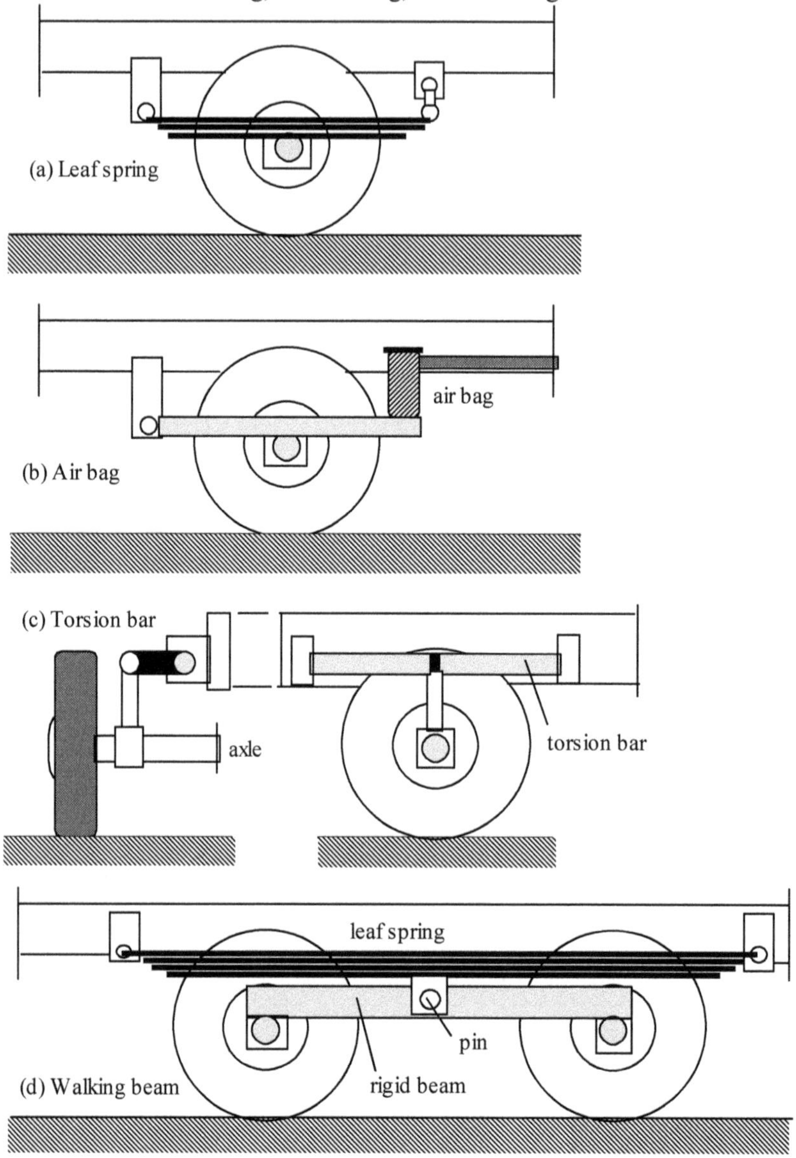

Figure 27.6 Suspension types.

Lazy axles are undriven axles in a driven tandem-group. When mounted ahead of the drive axle they are known as *pusher* axles and when behind the drive axle they are called *lag* axles. A *retractable* axle is one which can be vertically raised or lowered to vary the axle load that it carries. A major issue with retractable axles is to ensure that they are able to properly share their load with other axles. Inadequate design can readily lead to axle overloading.

27.3.7 Braking

The truck braking systems commonly used are as follows:

(a) Hydraulic brakes with vacuum booster (vacuum-hydraulic systems) which comprise hydraulically-activated systems in which the driver's effort is augmented by that of a vacuum unit.

(b) Hydraulic brakes in which the normal hydraulic system receives energy from a compressed-air booster cylinder (air-over-hydraulic systems).

(c) Mechanical brakes individually operated by separate air-powered chambers supplied from a central compressed-air supply (full air systems).

(d) Engine and exhaust brakes. Engine brakes convert the engine from a producer to a consumer of power, commonly by altering the exhaust-valve timing to use the engine as a compressor. The charge of compressed air is released to the atmosphere.

An exhaust brake closes a valve in the exhaust pipe between the manifold and the silencer. Air in the engine is compressed against this valve, thus retarding the engine. Electric engine retarding systems convert the energy to electric power and are usually used to regulate vehicle speed down hills.

The role that tyres play in braking is discussed in Section 27.7.4.

The progression of brake application is from the rear forwards, rather than vice versa, to avoid jack-knifing. This creates longitudinal forces in the truck. Such forces can also occur as a result of trailer inertia if the prime mover is either over-braked or loses traction. Braking tends to compress the articulated vehicle about the vertical pivot axes between its chain of separable units (Figure 27.2c). Forms of yaw instability can then occur when these destabilising longitudinal forces cause parts of the truck to deflect out of the longitudinal line. In *jack-knifing* one pivot deflects, in *trailer swing* at least two, and in *fishtailing* the back of the rear unit deflects out of line. The problems are compounded by the fact that the driver at the front of the truck is often unaware that these dangerous instabilities are developing.

Truck stopping-distances are given in Table 19.1. These distances should be increased by 50 percent for curved roads.

27.3.8 Turning paths

The need for a knowledge of the turning paths of large vehicles is particularly evident in intersection design (Section 20.3.1). For a rigid truck (Section 27.3.1), the radius of the turning path increases as the wheelbase increases and the *steering lock* (i.e. the maximum angle through which the turning wheels can turn) decreases. Most rigid trucks can turn within a 12 m radius. The turning path is called the *swept path*. Articulated vehicles, by their nature, can usually turn in much smaller radii (e.g. under 8 m), but cover a larger area in making the turn. This is caused by the *off-tracking* of the trailing wheels which follow a turning path inside of that of the forward wheels at low speeds and outside of it at high speeds or in sudden manoeuvres. The latter effect is sometimes called rearward amplification. The degree to which the path of the rear of the vehicle follows the front is called its *trailing fidelity*. It can be influenced by road roughness as well as vehicle speed.

The *swept width* of the vehicle is its rear width plus its off-tracking. The swept path and width can both be critical intersection design factors.

Turning-path templates, mechanical models, formulas, and computer programs for the local truck fleet are often available to aid the layout of intersections and road curves. An additional allowance of 600 mm is often made to cover assumptions in such models, driver error, and illegal vehicles. *Vehicle overhang* predictions for motorcycles and trucks must be used to offset roadside furniture away from the kerb face.

Overturning caused by the lean of high vehicles must also be considered on curves with extreme cross-slopes and tendencies for drivers to travel at excessive speeds (see also Section 19.2.5).

27.3.9 Safety

Vehicle crashworthiness can be measured as the rate of serious injury or death per crash. It varies significantly between vehicles. Crashworthiness plays a large part in determining how seriously a passenger is injured in a crash, although point of impact, seating position, and restraint use also play significant roles.

Crash rates for cars, in terms of distance travelled, are discussed in Section 28.1.1. As Chapter 28 concentrates on car-related crashes, this section discusses the specific issues associated with truck crashes. Total crash rates for trucks range from about:

* 0.5 to 1 crash/Gm on freeways,
* 3 crash/Gm on undivided, multilane, rural roads, to
* 10 crash/Gm on undivided multilane urban roads.

The rates increase as the percentage of trucks in the traffic stream increases. It can be seen from Section 28.1.1 that these rates are about half the car rate. Typical fatality rates are:

* 13 fatality/Tm for all trucks,
* 17 fatality/Tm for rigid trucks, and
* 31 fatality/Tm for articulated vehicles (Australia, 2003).

As crashes involving trucks are usually much more severe than those only involving cars, the truck fatality rate per distance travelled is commonly twice that of cars. Typically, fatalities involving trucks constitute about 15 percent of road fatalities, but, of those fatalities, only about 13 percent are occupants of the involved truck (OECD, 1983c). Most of the damage occurs to the car and its occupants. This imbalance is caused by the gross differences between truck and car in terms of their relative mass, stiffness, geometry, momentum, and kinetic energy. Hence a number of measures have been taken in recent years to improve truck–car safety. These include:

* better rear visibility for trucks,
* side and rear bars on trucks to prevent cars running under the truck body,
* anti-jack-knifing and other brake-management devices (Section 27.3.7),
* roll-over warning devices (see below),
* speed limiters (Section 28.2.5), and
* controls on drivers' hours (Section 16.5.4).

Generally, fatality rates per distance increase as the laden mass of the truck increases, although fatality rates per kilometre per tonne carried do not increase with increasing mass — suggesting that larger trucks are not inherently less safe.

Truck roll-over crashes (Section 19.2.2) usually occur on curves when the horizontal centripetal acceleration of the truck produces a tendency for the truck to roll over about its longitudinal axis (Figure 19.2). A typical roll-stiffness of a truck on its suspension is 2 MN/rad and limiting lateral accelerations are about 0.37g (Section 19.2.3). It will be clear from Section 19.2.3 that this is well below the lateral acceleration for sideways skidding, and so roll-over will usually be the dominant lateral failure mode in trucks. Furthermore, a driver in the isolated cabin of an articulated vehicle may receive little or no warning that the truck trailer is beginning to roll over. Roll-over resistance can be raised by:

(a) lowering the centre of gravity of the load,
(b) widening the wheelbase and spring base, and
(c) reducing accelerations due to lateral cornering, pavement cross-fall, and wind gusts (Mai and Sweatman, 1984). Super-elevation can assist here (Section 19.2.1).

A practical measure of roll-over resistance is the load transfer ratio, which measures the load transferred from one side of the vehicle to the other. In situations where the road geometry and/or cross-fall exaggerate the tendency to roll over, some success has been achieved using roadside signs to warn drivers of the hazard and to suggest safe speeds.

One of the problems with many truck crashes, and with roll-over crashes in particular, is that the freight being carried can often be quite dangerous if spilt. Common hazardous freights are flammable liquids such as petroleum products, and corrosive liquids. Spillages occur in about 1 in 30 hazardous freight crashes and are most likely in collisions with trains, run-off-the-road incidents, and overturnings. Measures to manage this problem include requiring special licensing for such movements, restricted routes, and a clear identification of the hazardous freight being carried and the counter-actions recommended in the case of an incident. Risk assessment also plays a major role. This is usually based on the following four factors:

(a) the magnitude and location of the flows of the hazardous freight through the transport network;
(b) possible crash scenarios (this requires a good crash database to supply the appropriate risk levels, Section 28.3).
(c) exposure of people and environmental elements to the hazardous freight, and
(d) effects of the release of the hazardous freight into the environment (this requires the hazardous freight to be classified and estimates to be made of the probable consequences of its release, its release rate, and the area likely to be affected).

Route selection for hazardous freight is based on minimising the assessed risk. This means avoiding sites with high crash risks and high exposure, and — where possible — minimising travel distance and travel time. An excellent review of hazardous freight transport is given in OECD (1988b).

27.3.10 Environmental impacts

Most people react adversely to the presence of a truck. The specific environmental impacts of the truck relate to:

* fear created in bystanders and other travellers (the larger the truck, the more fearsome it is),
* noise (Section 32.2),
* air pollution (Section 32.3),
* air-borne and ground-borne vibration (Section 25.8),
* parking obstruction,

* wet-weather spray (Section 27.7.1),
* traffic obstruction (Section 17.4.3),
* crash risk (Section 27.3.9), and
* visual intrusion.

27.4 EXCEPTIONAL VEHICLES

27.4.1 Oversized vehicles

Section 27.3.2 deals with heavy vehicles which meet the conventional mass and dimensional limits. Section 27.4.2 discusses road trains, a class of exceptional vehicles which meets a fairly well-ordered set of conditions. For various reasons, a random set of oversized or over-mass vehicles also use the roads from time to time. These vehicles operate under permit from the appropriate Authority and within stringent operating conditions. Usually the permits are granted only for indivisible loads or vehicles.

Indivisible loads are ones which cannot be divided for transport purposes without disproportionate effort, expense, or risk of damage. Typical examples are power-station equipment and manufacturing plant. In most countries, requests to carry such loads are routine and provision for them is a controlling load for many bridges (Section 15.4.2).

Indivisible vehicles may be mobile cranes, earthmoving equipment, drilling rigs, or large flat-top platforms known as *low loaders* (or *flat-bed transporters*) used to carry indivisible loads (Section 27.3.1). These platforms are commonly pulled by a prime mover and carried by up to ten axles with four or eight tyres per axle. Often the wheels will be unsprung, but connected together hydraulically in order to equalise group and individual loads and to enable the platform to be tilted. Tilting may be needed, for example, when the platform encounters significant pavement superelevation (Section 19.2.1). The method by which the platform is towed is also of importance. A draw-bar (Figure 27.2c) keeps all the payload on the flat-top platform but gives the driver less control and 'feel', whereas a king pin connection (Section 27.3.1) to the prime mover gives the driver better control but places some load on the prime mover (Figure 27.2c).

27.4.2 Road trains

As illustrated in Figure 27.2, combination vehicles consist of rigid trucks or prime movers towing one or more trailers (Figures 27.2b–e). A *large* (or *long*) *combination vehicle* (*LCV*) is any combination vehicle which exceeds routine dimensional limits. One such vehicle is the low loader discussed in Section 27.4.1.

Another LCV is the road train which is a freight vehicle consisting of a prime mover and two or more interconnected but independent trailers. The common types of road train are shown in Figure 27.2d, which also lists the various terms used. A typical road train is an articulated vehicle hauling one trailer, i.e. a prime mover and two trailers. In some countries road trains are known as *doubles*, *trebles*, *triples*, etc. depending on how many independent articulated elements are used in the combination (Section 27.3.1).

To produce a road train from a conventional articulated vehicle, the additional trailers must be free-standing (Figure 27.2d). They are called *dog trailers* and can be formed by using either:

* a trailer and draw bar, as in Figure 27.2b, or

* a semi-trailer seated on the fifth wheel (Section 27.3.1) of a *converter dolly* to which a draw bar is attached (to produce an *A dolly*).

The *A train* (or *twin trailer*, *double trailer combination*, *double bottom*, or *double*) is a road train with only two semi-trailers, as shown in Figure 27.2d. *B trains* are illustrated in Figure 27.2e and the range has recently been extended by the use of A and AB triples which can increase payloads whilst giving better tracking performance.

A key facet of most road trains is that they are assembled by coupling together vehicles which themselves comply with the regulatory requirements for general road transport operations. This enables road trains to be made up from or disassembled into vehicles able to be used on all roads. To provide this flexibility, it is essential that the vehicles and their components be interchangeable.

The selection of routes for road trains relates largely to the disadvantages of road trains to other road users and it is therefore necessary to consider traffic flows, dust hazards, road structures, turning geometries, overtaking opportunities and road geometry in selecting road train routes. Nevertheless, operating experience indicates that the problems associated with the operation of road trains are largely imaginary.

27.5 MOTORCYCLES AND MOPEDS

The single-track vehicle (e.g. the motorcycle, moped, and bicycle) is a device which is statically unstable, but which gains stability in dynamic operation due to the rotational inertia of its wheels (see 'self-aligning torque' in Figure 27.7). It has three potential collapse modes:

(a) capsize mode, in which the vehicle falls on to its side about a longitudinal horizontal axis (also called roll-over). This usually occurs at low speed when the rotational inertia is low. It can be controlled by a steering response and a subsequent increase in speed.

(b) wobble mode, in which the front fork and the turning wheel of the vehicle oscillate about an inclined but near-vertical axis. The mode is controlled by increasing the torsional stiffness of the front fork.

(c) weave mode, in which the vehicle oscillates about a vertical axis, resulting in a weaving track. It is controlled by increasing the frame stiffness and, to a lesser extent, the fork stiffness.

One of the major differences between driving a car and a single-track vehicle is the need to consider these modes and be sensitive to the effect of tyre stiffness and damping.

In addition, single-track riders can use a range of procedures to achieve the same vehicle manoeuvre. For example, a motorcycle steers in quite a different way to a car. For slight turns, the rider leans towards the turn. For sharp turns, the rider first deflects the front wheel outwards and away from the direction of the intended turn in order to quickly develop the necessary lean inwards. After about 500 ms, the rider then turns the front wheel into the turn. At the end of the turn the rider uses the same practice to bring the bike back to vertical, this time turning into the turn. Many riders are not aware of their use of these two somewhat counter-intuitive manoeuvres. The use of leaning in order to turn produces vehicle overhang and means that good lateral clearances must be used on the inside of curves used by motorcyclists (Section 27.3.8).

With respect to braking, motorcyclists rarely utilise the full capacity of their machines. Partly this is due to a fear of loss of control following excessive front-wheel braking, coupled with a lack of feedback as to when the wheels are about to lock (Section 27.7.1).

Motorcycles offer little protection to riders involved in crashes and the rider in a crash frequently makes direct contact with the road surface and with other vehicles. In addition, a motorcyclist's right-of-way is frequently violated by other vehicles whose drivers have not perceived the presence of the motorcycle. This lack of motorcycle conspicuity (Section 21.2.3) is the dominant factor in a large proportion of car/motorcycle crashes. Methods of enhancing conspicuity include fluorescent jackets, driving with the headlights on, and the use of specially-designed running lights. For all the above reasons, motorcyclists are about 20 times more at risk per kilometre travelled than are car occupants. This relatively high risk scenario underlines the importance of motorcyclists wearing of helmets as this measure reduces motorcycle fatalities by about 30 percent (see also Section 28.3).

Following Section 16.5.5, there is a good deal of evidence that young (under 25 years), inexperienced (less than two years driving experience) motorcyclists have a relatively high crash rate, particularly when riding large (over 250 mL) motorcycles. This is partly due to the complex vehicle-handling skills required. It has been suggested that motorcyclists should learn on small sub-250 mL machines to improve riding skills and to ensure reasonable experience before driving large machines.

The *moped* (*mo*torised, but with *ped*als) has been regarded traditionally as a power-assisted bicycle and thus fits between the motorcycle and the bicycle and has the potential to satisfy a mobility need currently largely filled by families owning two or more cars. Mopeds range in type from scooters to motorcycles. Many have more than one gear, but there is rarely provision for the manual selection of gears. Engine size is the basis of most current moped definitions. Generally, a moped has an engine of less than 50 mL, a design speed of less than 50 km/h and an unladen mass below 65 kg. Operating speeds are usually 40 km/h or less. An earlier requirement that mopeds should also be capable of being pedalled by muscular power has gradually disappeared, thus making their name somewhat of a misnomer.

27.6 BICYCLES

The bicycle (or bike) is a human-powered two-wheeled vehicle subject to the same 'single-track' performance characteristics mentioned in the opening of the last section. There are three main types of bicycle:

* urban or town bikes (e.g. step-through frame, parcel rack, horizontal handle-bars, splash-guards, lights),
* mountain bikes (good brakes, wide range of gears, robust frame), and
* sporting and touring bikes (dropped handle-bars, light frame).

A bicycle's main resistance to motion comes from aerodynamic drag on the machine and its rider, which is more significant than rolling resistance over the road surface. For example, at 20 km/h the power to overcome air drag is about twice the power needed to overcome tyre rolling-resistance. The bicycle and rider convert energy at a peak efficiency of about 30 percent, producing up to 400 W of tractive power at a peak speed of 45 km/h, and 150 W at the normal maximum speed of 20 km/h. However, even fit people have problems producing 100 W during a 30 minute trip, and travelling for longer periods sees the output drop to about 40 W.

Normal cycling speeds are 16 to 20 km/h but are sensitive to gradient and wind. For instance, the power needed for riding on a horizontal surface has to be increased by 75 percent for a 5 percent grade and by 300 percent for a 10 percent grade. A pavement surface with a poor longitudinal profile may produce excessive vehicle impacts and vibrations and lead to skidding (Whitt and Wilson, 1982).

The advantages of the bicycle include that it:

(a) demands only moderate effort (*c.* 100 W, see above),
(b) is relatively energy-efficient — a cyclist at 20 km/h uses the same power as a walker travelling at a normal walking speed of about 7 km/h.
(c) emits few fumes,
(d) makes little noise,
(e) has low capital and operating costs,
(f) offers door-to-door mobility without timetabling restrictions,
(g) takes 1/16th of the parking space of a car,
(h) is quick in congested traffic,
(i) is safe when isolated from other, faster modes, and
(j) can aid the user's health, although breathing traffic fumes is not advantageous.

Its main disadvantages are that it has:

(a) a lack of safety in conjunction with other traffic (see below).
(b) a security problem when parked,
(c) a lack of weather protection (U.K. data suggests a 15 percent drop in ridership levels due to rainfall, Emmerson *et al.* (1998)),
(d) possible user discomfort such as warm weather and lack of shower facilities for riders (U.K. data suggests a 3 percent change in ridership numbers for an unfavourable 1 C change in temperature, Emmerson *et al.* (1998)).
(e) a limited distance range,
(f) limitations caused by grades and extreme weather,
(g) limited carrying capacity, and
(h) slow travel speeds.

From a safety viewpoint, per kilometre travelled, cyclists are between about 5 and 20 times more at risk than car occupants (a complication here is that many cycle crashes are unreported). Crash rates vary dramatically between countries. In the U.S. and Australia, cyclist fatalities are about 2 percent of traffic fatalities, in Holland — where cycling is very popular — they are about 20 percent. Most cyclist fatalities occur in cycle–vehicle crashes and involve turning and crossing. Hence, cycle safety is aided by raising the conspicuity of cycle and rider; flashing tail-lights are particularly good in this respect. When all cycle crashes are considered, only 20 percent are cycle–vehicle crashes and 50 percent are falls involving only the cyclist.

About 80 percent of cyclist fatalities are due to head injuries, and so helmet wearing is an effective counter-measure. Typically, wearing a helmet reduces a cyclist's risk of sustaining a head injury in a crash by 70 percent and, if a head injury is sustained, reduces the chance of death by 95 percent.

Many cyclists are relatively undisciplined in traffic and this contributes to their crash record. For example, cyclists often travel against the direction of traffic and cross intersections with insufficient sight distance, despite the high risk associated with these behaviours.

The main variables affecting cyclist safety are:

* cyclist conspicuity,
* use of protective equipment,

* cyclist behaviour,
* traffic flow, particularly in kerbside lanes,
* traffic speed,
* traffic type,
* amount of separation from traffic,
* parking adjacent to a bicycle route,
* property access across a bicycle route,
* bicycle flow,
* pedestrian flow on or across the bicycle route, and
* pavement condition.

'Bicycle route' here refers to the area used by bicycles, and not necessarily to a separate path (Section 19.6.1).

The main applications of bicycles are neighbourhood riding, short-distance commuting, recreation, and sport. Track and traffic facilities for bicycles and various bikeway issues are discussed in Section 19.6. The bicycle can be merged into the public transport system by providing:

* secure storage at public transport stops,
* racks or trailers permitting bicycles to be carried on the outside of buses, and
* facilities for bicycles to be taken inside public transport vehicles.

27.7 TYRES

27.7.1 Tyre components and types

As key parts of a road vehicle, tyres have a number of roles to play. They must:

(a) use the tyre–pavement interface to provide horizontal forces for traction, cornering and braking (Chapter 19),
(b) distribute vertical loads from the vehicle until they cause an acceptably-low contact stress on the pavement (Section 11.2.3),
(c) contribute to the vertical dynamic stiffness of the vehicle, providing both a softer ride and lower impact forces on the pavement (Section 27.3.5),
(d) minimise rolling resistance forces (Sections 27.1.2 & 29.2.1–2) (which is lower with higher inflation pressures),
(e) provide adequate lateral stiffness and strength for the vehicle to corner safely and comfortably (Section 19.2.1),
(f) minimise noise generation (Section 32.2.3), and
(g) minimise wear of both the tyre and the pavement surface. This role is in some conflict with role (a). The link between tyre wear and pavement surface is discussed in Section 12.5.1.

The tyre structure (or carcass or casing) is a thick, flexible shell made of reinforced rubber (Figure 27.7). It carries the external loads applied to the tyre. The internal pneumatic pressures which give the tyre its unique properties may be carried by a tube or, in the case of a tubeless tyre, by the inner liner of the tyre structure sealed against the rim. The tread is that part of the outer cover of the tyre structure which bears on the pavement surface, and contributes directly to roles (a), (f) and (g) above.

Section 27.7.2 discusses the friction properties necessary for a tyre tread to supply adequate horizontal forces required by role (a) above. To meet its other needs, the principal requirements for a material used in tyre-making are that it be strong, elastic (or flexible), durable, insulating, wear-resistant, hard to cut, tear-resistant, fatigue-resistant, impact-softening, rough, and able to bond well with steel or fabric reinforcement. Many of these needs run counter to the friction needs, making tyre design a challenging task.

The above demanding list of requirements for the tyre material is met by natural or synthetic compounds of rubber (Section 8.7.8) mixed with extending oils, fillers, and hardeners such as carbon black, softeners, protective chemicals such as antioxidants, and vulcanisation products such as sulfur. Rubber is a *visco-elastic* material (Section 33.3.3) and its viscous properties (Section 27.7.1) are sensitive to temperature and strain rate. The *hysteretic* effects referred to in Section 27.7.2 come from a braking of the bonds between the long rubber polymers and the particles of carbon. Synthetic rubber is preferred for passenger cars and natural rubber for trucks. The most common synthetic rubbers used in tyre manufacture are *SBR* (*styrene-butadiene-styrene*) and *BR* (*polybutadiene)*.

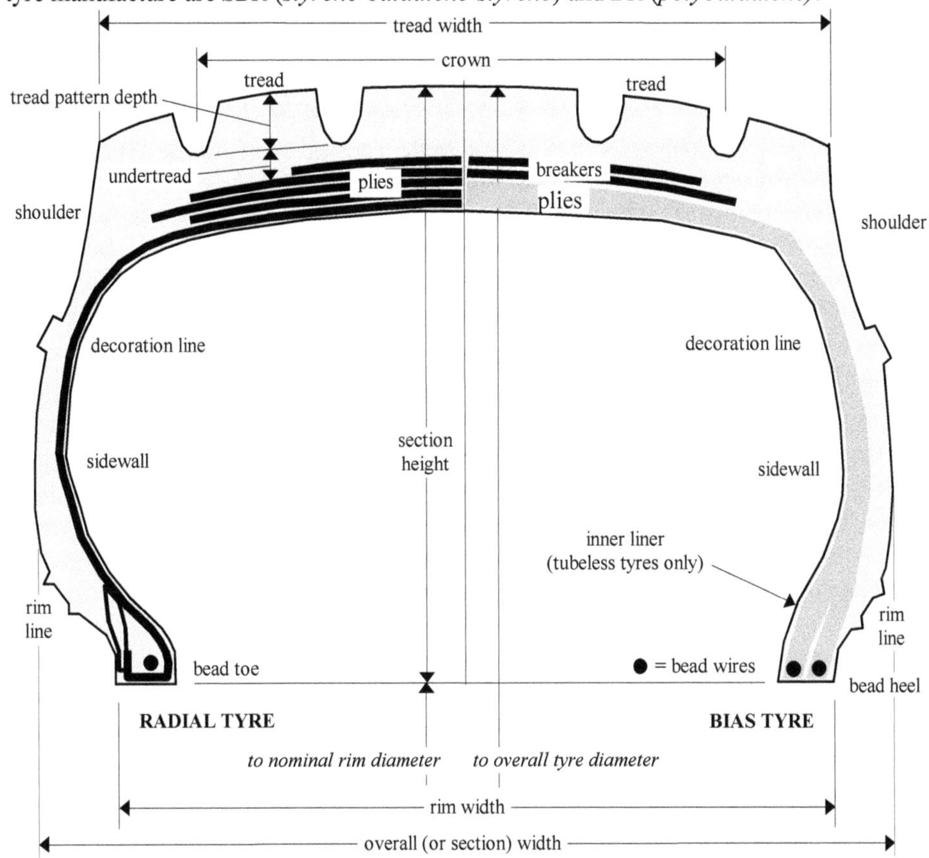

Figure 27.7 Tyre structure. The *aspect ratio* is the section height divided by the overall width.

The tyre structure is reinforced with plies which provide it with increased strength and stiffness. There are three main ply types:

* the *bias-ply* (or *cross-ply*) tyre is the older tyre type. Its plies are belts of cord (Figure 27.7, rightside) which cross the tread circumferential centreline at an angle (or bias) of between 30° and 40°. This angle is very much a design compromise between steering-related handling characteristics and such other requirements as braking performance and wear resistance.

* the newer *radial-ply* tyre (Figure 27.7, leftside) has one or more plies comprised of textile or steel belts at or near 90° to the tread circumferential centreline. The radial tyre provides a considerable improvement in tyre performance, avoiding some of the bias-ply compromises as it provides independent mechanisms by which the conflicting demands can be satisfied. The advantages of radial tyres are:
  (a) longer life,
  (b) better vehicle handling,
  (c) lower rolling resistance, and
  (d) better braking and skid resistance.

  However, they require more care in operation and maintenance than do bias tyres.
* the *belted-bias* tyre is a compromise between bias- and radial-ply tyres, with a radial belt forming the tread section.

The plies are anchored to the rim beads (Figure 27.7), which are bundles of high-strength steel wires which also help to hold the tyre on the rim.

The key elements of a tyre which have not been discussed previously are (Figure 27.7):

* *shoulders* (or *buttresses*) which protect the carcass from damage. They contain voids and/or grooves to help dissipate heat.
* *breakers* which are often used in bias-ply tyres to absorb shocks to the carcass and to prevent separation of tread and carcass.
* *sidewalls* (or *flex areas*) which are external parts of the tyre that protect the carcass but do not contact the pavement. They are subject to more flexure than other parts of the tyre.

Splash and spray from trucks travelling on wet pavements can be a major traffic hazard. Splash results from the impact action of the tyre and comes in drops greater than 1 mm in diameter. Spray is comprised of smaller drops lifted and atomised by the turbulent aerodynamic action of the truck itself and by suction caused by the tyre treads. The effects can be reduced by the use of porous road surfaces (Section 12.2.2), by spray-suppressing devices mounted on the truck, and by better aerodynamic design of the truck.

Tyres play a minor role in road crashes. Commonly, tyre failures contribute to less than 1 percent of fatal crashes and about 1 percent of all recorded crashes.

27.7.2 Tyre mechanics

The three forces and three moments to which the tyre is subjected are shown in Figure 27.8. Each of the six has an associated tyre–vehicle stiffness and will contribute to the dynamic operation of the vehicle. Specific comments on each are:

* *vertical force*, F_V. Its four components are discussed in Section 27.3.2. The lateral displacement of the associated reaction, R_V, reflects the pressure distribution under the tyre and is a major contributor to the *rolling resistance moment* (Section 27.3.3). The vertical stiffness of a truck tyre is typically 1 MN/m.
* *self-aligning torque*. See the discussion of rotational inertia in Section 27.5.
* *roll-over moment*, see Sections 19.2.2, 27.3.9 & 27.5.
* *longitudinal force*, F_H. This includes the rolling resistance force (Section 27.1.2) and the momentum-change forces due to accelerating and braking. These are discussed below.

* *cornering force*, F_L. This causes a change in vehicle direction and arises when there is an angle — called the *slip angle* — between the direction of movement of the vehicle and the plane of the wheel (Section 19.2.2 and Figure 19.2).

These forces and moments develop via vertical pressures and horizontal friction at the tyre–pavement interface. Figure 27.9 shows that the coefficient of friction given by $F_{L \text{ or } H}/F_V$ can range from (about) 0.1 to 1.5. The properties of the pavement surface are discussed in Sections 12.5.1–2 and obviously have a major role to play in determining the magnitude and variation of this coefficient.

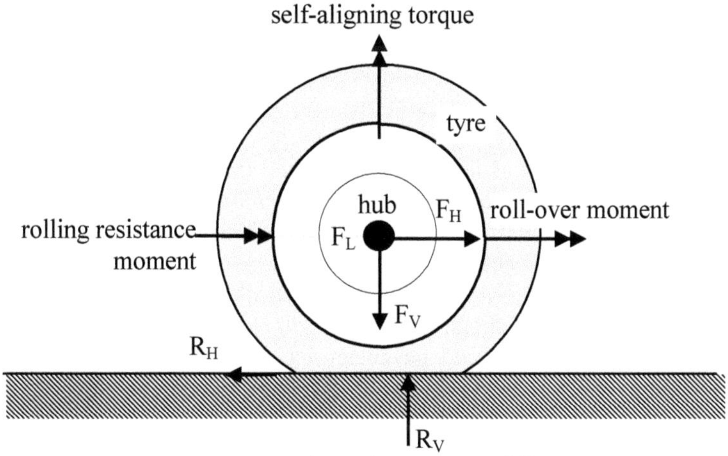

Figure 27.8 Forces (⟶) and moments (⟶⟶) on a tyre.

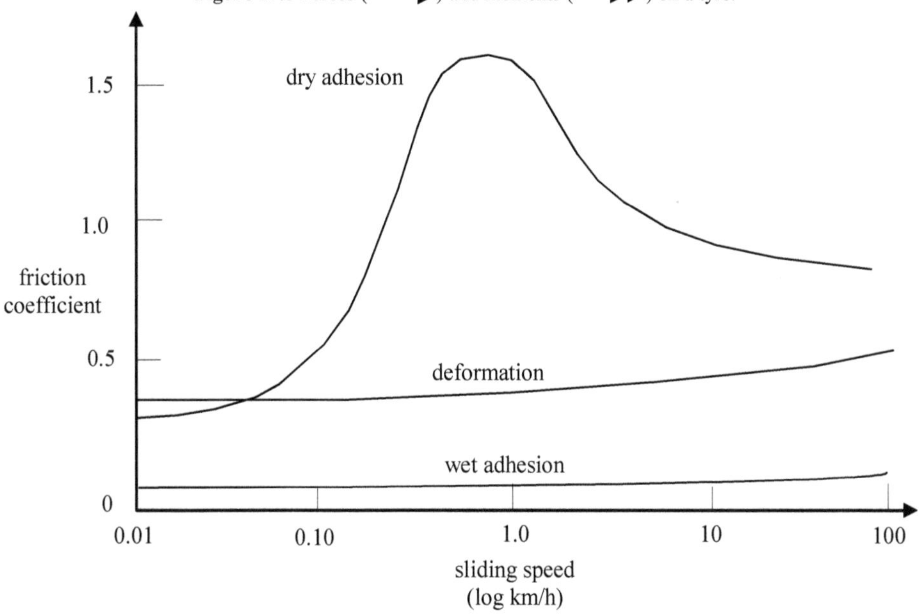

Figure 27.9 Typical adhesion and deformational friction coefficients for tyre rubber sliding on a pebble-like surface.

The tyre develops its frictional contact forces at the contact surface by three modes:

* *sticking to it*. This is called *adhesion* and arises from the shearing of molecular bonds formed when the tyre rubber is pressed into intimate contact with the pavement surface.
* *rubbing over it*. The tread design contributes to this mode.
* *deforming around pavement asperities*. This depends on the temperature-sensitive stiffness and hysteresis of the rubber (Section 33.3.3), and the tread design. Of course, it also depends on the macrotexture of the pavement (Section 12.5.1). The deformational energy is transformed into heat in the tyre.

The horizontal force developed at the interface is the sum of these components. Figure 27.9 shows the relative contribution of adhesion and of deformation (and rubbing).

The tyre clearly requires energy to develop each of these modes. In addition, the tyre must distort at the macro-level to form the oval-shaped tyre–pavement contact area (Section 27.7.3). This combination leads to hysteretic-energy losses in the rubber (Section 33.3.3), which are dissipated as heat in the tread and the breakers (see Figure 27.7). As vehicle speed increases, the losses are fairly constant up to about 70 km/h and then slowly increase. Losses increase linearly with tyre mass. Thus there is friction and rolling resistance, even if the pavement is nominally smooth.

As suggested in Figure 27.8, the rolling radius of a tyre is less than its free radius. This effective change in tread circumferential length means that further local longitudinal stresses are developed at the tread–pavement interface. The effect can also produce damaging standing waves in the tyre sidewall if the two radii are significantly different — as is the case with under-inflated tyres.

27.7.3 Tyre pressures and loads

Tyre *contact areas* (or *footprints*) are an important property for pavement design (Section 11.2.3). Section 27.3.2 indicated that maximum permissible axle loads lay between 5 and 14 t, implying maximum wheel loads of between 1.2 and 3.5 t. Assuming a typical maximum tyre (or wheel) loading of 2.5 t and a truck tyre pressure of 825 kPa, the theoretical contact area at full load is $2.5 \times 10^7/825 = 30{,}000\ mm^2$.

The contact width is relatively constant. For truck tyres it is about 80 percent of the rim width (Figure 27.7), but this may be significantly influenced by the tyre type. Thus, the change in contact area resulting from a change in load or pressure occurs largely as a change in the contact length and the contact shape tends to change from a circle to an ellipse to a rectangle as the load increases or the tyre pressure decreases. Measured contact areas are about 33 percent above those calculated from load/pressure, as a result of the bending stiffness of the tyre wall reducing the actual interface pressure.

Thus, the average tyre–pavement contact stress is usually less than the tyre inflation pressure, but locally high pressures will occur. A decrease in tyre pressure changes the local peak pressure point from near the centre of the contact area for over-inflated tyres, to under the tyre walls for under-inflated tyres. The peak stresses — which occur near mid-length and at the side walls — are commonly 1.5 times inflation pressure and can be over three times inflation pressure. By concentrating the force, high tyre-pressures can have a relatively serious effect on pavement life. To help manage these effects, some jurisdictions limit the allowable mass per tyre width (typically 10 kg/mm).

For a rolling tyre, the vertical reaction force R_V (Figure 27.8) will usually be ahead of the hub to produce a moment on the hub to counter that produced by the horizontal force F_H acting above the surface. Consequently, the rolling-tyre contact area is larger in the forward portion and the treads there are being forced radially inwards. The tyre exerts radial, inward-acting, horizontal shear-stresses on the pavement surface as it resists being spread over a wider surface area. These shear stresses can exceed the inflation pressure. Longitudinal shear stresses are also produced. These peak near the front of the tyre at about the tyre pressure. They then twice change size along a line towards the rear of the tyre. Thus quite high shear stresses, with a potential to cause pavement damage, can occur near the tyre edge. In the rear portion of the contact area the pressures are lower and the treads are moving radially outwards.

With dual tyres, the load on the outer wheel is usually less than that on the inner wheel, particularly for high tyre-pressures. Dual tyres are sometimes replaced by a wide, single tyre carrying the same load. They are said to improve fuel consumption, ride, and handling. However, they increase road damage by a factor of between 2 and 10.

27.7.4 Braking, accelerating and turning

The overall braking process is discussed in Sections 12.5.3 and 27.3.7. The braking force developed by a tyre depends on its slip angle (Section 27.7.2) and *slip ratio.* The slip ratio is the ratio of the speed of the tyre at the pavement contact surface to the road speed of the vehicle. Hence, a slip ratio of zero indicates no slip between tyre and pavement; unity indicates total slip in a state when the wheels are not rotating and which is known as the *locked wheel* condition. Under normal operations, the braking force increases linearly with slip ratio from zero slip ratio to a peak force at a slip ratio of about 0.2. Beyond 0.2 the slip ratio increases rapidly in a process known as *spin down* until the locked-wheel condition occurs. The force at locked wheel is called the *slide force* and is about two thirds of the maximum braking force. Hence, if a car's wheels do spin down and lock, good driving practice is to release the brakes a little until the wheels are rolling again, as this counter-intuitive behaviour will give the car a 50 percent increase in braking capacity. *Anti-lock brakes* (*ABS*) sense when spin down is about to occur and correct for it by releasing brake pressure and thus avoiding wheel lock.

When the wheels of a vehicle do lock, skidding occurs in that the wheels operate as blocks of material in contact with the road surface and lose their ability to provide a steering response. The vehicle may therefore rotate or slide sideways. When only the rear wheels lock, the back of the car will swing around. At low speeds, this can be corrected by a steering manoeuvre known as 'steering out of a skid'. In the case where the front wheels lock, the vehicle continues on its original course and control is regained by releasing the brakes.

It can be seen from Figure 27.9 that, whereas water does not influence deformational friction significantly, it plays a major role in determining the available level of adhesive friction and is thus the major determinant of peak braking capacity. Dry adhesion is potentially the major contributor to friction, but only contributes fully if full contact is maintained between the tyre and the pavement.

As the amount of available water on the pavement surface increases, a water film will develop over the pavement surface. When a tyre travels along the pavement, its face sinks into the water film. The treads in the tyre and the texture of the pavement surface serve an important secondary role here in providing drainage channels through which water can be dispersed by the rolling tyre. Vehicle speed also plays a role as increasing speed decreases the time for the water to disperse. In these circumstances fluid pressures in the film produce an upwards force which balances the gravity force in the tyre and keeps the tyre completely separated from the pavement surface. As with pore pressure in soil (Section 9.2.3), the pressured water film has no shear strength. The tyre therefore cannot generate any force, F_H, for braking or accelerating or F_L for turning. The powered wheels spin freely and the non-driven wheels tend to stop rotating, particularly if braked. The process is known as *dynamic hydroplaning* (or *aquaplaning*). It requires the water film to have a thickness in excess of 6 mm and speeds over 80 km/h (or 40 km/h for worn, cross-ply tyres) and is therefore relatively rare. Partial hydroplaning can occur at much lower speeds, although its effect is not usually dramatic. The biggest drop in the available braking force occurs as the water film thickness increases from 0 to 4 mm. *Viscous* hydroplaning occurs with bald tyres on smooth pavements.

On the other hand, in dry weather Figure 27.9 shows that the greatest friction will be obtained by maximising the contact area — hence the wide, smooth tyres used in motor racing. Thus optimum tread patterns vary according to application and situation. There is also suction formed within the tread channels, which can lift water and dust from the pavement (Sections 12.1.1 & 27.7.2).

27.8 FUEL

Following Figure 8.13, petroleum (or crude oil) is refined to produce many products including, for automotive purposes, petrol (or gasoline or motor spirit, Section 33.1), and the heavier diesel fuel.

As discussed in Section 8.7.1, petrol is a volatile mixture of hydrocarbons and contains alkenes, olefines and aromatics. Its boiling point is between 30 and 210 C. Combustion of petrol with oxygen gives:

$$C_nH_m + (n + m/4)O_2 > nCO_2 + (m/2)H_2O$$

The grade of petrol, which also represents its *anti-knock number* is expressed as the nominal *Research Octane Number* (*RON*). It places the anti-knock performance of the fuel in a scale from 0 to 100, where 100 was the value assigned to the good performance in early engines of a pure hydrocarbon called iso-octane, and 0 was the value assigned to the very poor performance of normal heptane.

Petroleum fuels serve their market well, and are formidable opponents for any proposed alternative fuel, because of their:

(a) transportability as a fuel,
(b) widespread availability,
(c) high energy/volume ratio,
(d) ease of energy conversion,
(e) low cost/energy ratio, and
(f) lack of a viable and competitive alternative.

Indeed, because of transport's total dependence on petroleum, it may be one of the last users to utilise substitute fuels.

Internationally, road transport consumes about 80 percent of transport energy. Energy is also consumed indirectly by the transport sector via such activities as vehicle manufacture and maintenance. About 50 EJ of energy is used to produce a car and it is estimated that this is equivalent to about 25 percent of the direct propulsion-related consumption.

The overall efficiency of usage of petroleum fuels, i.e. useful energy obtained as a fraction of energy input, is about 15 percent, with a range between 10 and 25 percent. Data on the energy-intensiveness of various vehicle types and on the efficiencies of various transport modes are given in Tables 30.1 and 30.2.

In order to meet energy conservation, sustainability and cost reduction goals, automotive fuel consumption in transport can be reduced by three major ways:

(1) reducing the amount of travel by:
 * the transport demand management measures discussed in Section 30.3.
 * technological change. The major 'technological change' which has long been seen as imminent is in telecommunications and information technology. It has often been argued that these developments will substitute telecommunication for transport and hence reduce the need for physical travel (Section 30.3). However, such changes are not yet noticeable and detailed reviews suggest a complementarity between telecommunication and transport, which diminishes some of the substitution effects.

(2) increasing the efficiency of travel through:
 * smaller vehicles for personal travel and larger vehicles for freight; e.g. demonstration vehicles have operated at about 1800 km/L, and motorcycles run easily at 100 km/L, compared with about 10 km/L for a typical car,
 * more efficient engines (Sections 27.1.2 & 29.2.2); this remains the measure with greatest potential returns in terms of both effectiveness and ease of implementation. However, the effect of vehicle-based fuel conservation measures is gradual as it takes about five years for 40 percent of the vehicles on the road to possess a new feature (Section 27.1.1).
 * more efficient transport modes (Chapter 30),
 * better traffic flow (Section 29.2.4),
 * better driving practices, and
 * better operating procedures (Section 29.2.6).

(3) fuel substitution of petrol with such alternatives as:
 * compressed natural gas (methanol) (Section 27.1.2),
 * liquefied petroleum gas (Section 27.1.2),
 * ethanol (from plants),
 * hydrogen,
 * alcohol (Section 27.1.2), or
 * electricity (from batteries or cable).

 To date, these alternatives are rarely cost-effective.

Data suggest a short-term elasticity (Sections 30.5.1 & 33.5.4) of fuel consumption use with respect to fuel cost per kilometre of about –0.2 and a long-term value of about 0.5. That is, the long-term percentage drop in fuel consumption is about half the percentage increase in fuel price. Of this, –0.3 is due to changes in trip length and –0.2 to the short-term changes.

CHAPTER TWENTY-EIGHT

Safety

28.1 ROAD CRASHES

28.1.1 Overview

Safe roads are one of the prime objectives of road design and operation (Section 5.1). Consequently there are safety implications throughout the other chapters of this book. Rather than repeat those implications, this chapter will explore some of the wider aspects of road safety.

Road safety is achieved by reducing road crashes (or collisions). *Accident* is the colloquial term sometimes misused to describe an event that would more properly be called a *crash*. There is little that is accidental in most crashes, as this chapter will demonstrate.

To illustrate the extent of the road safety problem, some international data for traffic fatalities are given in Table 28.1. In the seven countries listed — all of which have good road systems — over 60 000 people were killed on the roads in one year. This is an unacceptably high number.

Table 28.1 International data on road safety, from various websites, for 2005. Note that vehicle numbers are given in Table 4.1 and car ownership levels are discussed in Section 31.3.4.2.

Country	People killed (k)	People killed per Tm of travel
Australia	1.6	8
Canada	2.9	9
U.K.	3.3	6
Japan	7.9	10
New Zealand	0.4	10
U.S.	43.4	9
Germany	5.4	8

The collection of crash data is described in Section 28.3. It is a common observation that crash numbers vary positively with economic changes such as those measured by changes in a country's gross national expenditure (GNE). This is partly because such changes cause a similar positive change in overall traffic flows. In addition, the probability of a crash is higher in the type of recreational trip that occurs as discretionary income (Section 33.5.3) rises.

Crash numbers at various similar sites commonly follow a Poisson distribution (Section 33.4.5) although when crash numbers are high, a normal distribution often becomes more appropriate.

Traffic safety measures are often discussed in terms of the 'three E's' of enforcement, education and engineering. A discussion of traffic law and its enforcement is given in Section 16.7. Driver education is discussed in Section 16.6.1 and is related specifically to safety via Section 28.2 which covers the characteristics of the road user which contribute to road crashes, and Section 28.3 which examines crash control techniques (Section 28.3). Engineering-type measures are the province of most of the remainder of the chapter. For instance, seat belts are discussed in Section 28.4, traffic engineering countermeasures in Section 28.5, and roadside protection in Section 28.6.

Other chapters include discussions of crash countermeasures with respect to speeds (Section 18.1.1), road alignment (Chapter 19), intersections (Section 20.2.3), parking (Section 30.6), signs (Section 21.3.3), pavement markings (Chapter 22), signals (Section 23.1), lighting (Section 24.1), delineation (Section 24.4), construction (Section 25.9), maintenance (Section 26.4), and vehicles (Chapter 27). These cross-references are listed here to reinforce the overlap between the road environment, the vehicle, and driver behaviour.

28.1.2 Crash exposure

Crash exposure is the frequency of encountering events that might cause a crash, i.e. crash opportunities. It thus measures the intensity of potentially unsafe use. As illustrated in Table 28.1, one prime predictor of the number of road crashes that will occur is the distances that vehicles are driven, known as the vehicle-kilometres of travel. Very few jurisdictions require all drivers to record the distance they travel and so direct data is not readily available. In aggregate, the distance can be estimated as a first approximation from fuel sales or traffic flows. The total quantity of fuel consumed reflects overall vehicle use and so this readily-obtained value is sometimes used as a national exposure measure (e.g. Table 32.1), or converted via fuel consumption rates (Sections 29.2.6–7) to an estimated vehicle-kilometres of travel. Measured traffic flows (Sections 17.1.1 & 31.5) on a particular route or in a given network can be multiplied by the route or network length and the time period to also estimate distance travelled. Somewhat more precisely, changes in traffic flow from year to year reflect changes in distance travelled.

Traffic flow is of itself a good overall measure of crash exposure, as illustrated by the data in Section 28.5. One reason for this is that flows reflect the actual number of vehicles 'met' by a particular driver.

Other direct exposure measures include time spent on the road, the size of the driver population, the number of vehicles registered to use the road, the length of roads, and the number of conflict points along the road (Figure 20.3). An intersection exposure measure is discussed in Section 28.5.

There are also a number of indirect exposure measures which some analyses have found play a significant role. These include the number of people employed, the number unemployed (see Section 28.2.3), discretionary income (Section 28.1.1), GNE (Section 28.1.1), and retail sales. These generally enter the debate via their links to various direct exposure measures. For instance, a rise in discretionary income increases leisure travel and night-time travel.

Whilst exposure measures will give good general guidance, to use them in a specific study usually requires segmentation with respect to (Perry and Callaghan, 1980):

(a) driver characteristics. For instance, U.S. data shows that (Evans, 1991):
  * between the ages of 15 and 45 a woman in a crash is 25 percent more likely to be killed than a male,
  * a 70-year-old in a crash has three times the probability of dying as a 20-year-old.

(b) road characteristics (e.g. urban, rural, freeway — Section 28.5);

(c) environmental characteristics (e.g. time of day and week). For example, rain increases rates by between 10 and 20 percent, and can double some specific risks. The effect is due in part to decreased visibility (Chapter 24) and in part to a drop in pavement skid resistance (Section 12.5.2).

(d) vehicle characteristics. From U.S. data, in a crash between cars with masses of 900 kg and 1800 kg, the driver of the lighter car has a thirteen times higher fatality risk (Evans, 1991).

Exposure data is often used in association with data on crash numbers. For example, *Smeed's* famous analysis considers the number of crashes in terms of the number of vehicles or people able to use the road system. This is reflected in his well-known empirical formula based on aggregate national data from a range of countries (Smeed, 1949):

$$\text{fatality/veh} = 0.0003(\text{person/veh})^{2/3} = (\text{person/[200 000 veh]})^{2/3}$$

The equation suggests that the fatal crash rate falls as a country becomes more motorised and the person per vehicle ratio drops. It has since been realised that Smeed's formula applies to the conditions in only one country in one year. A wider data analysis has suggested that the best that can be said is (Andreassen, 1985):

$$\text{fatality/veh} = \text{constant (veh)}^{b}$$

where $-1 < b < 0$. That is, the fatality rate per vehicle declines as the vehicle population increases. Similarly, other studies have indicated that the fatality rate per vehicle-kilometre declines exponentially with time, i.e.:

$$\ln(\text{fatality/veh-km}) = \text{constant} + c(\text{time})$$

where $c < 0$. These data all suggest that, over time, communities become more sensible in their use of the car and the truck. Table 28.1 also suggests that the provision of a good highway system can reduce crash rates.

28.1.3 Crash risk

Using exposure data, the actual number of crashes can also be seen as the product of the *exposure* and the *risk* of that exposure. Risk has a number of related meanings. In this book, it is defined as the probability of a crash in a given situation, i.e. the number of crashes divided by the number of potential crashes. *Liability* is usually taken to mean the expected number of crashes, which is also a risk prediction. Risk analysis looks at the consequences of risk and exposure and calculating risk from exposure and crash data gives a performance measure for a particular situation, by allowing comparisons between road elements with different levels of use. Care must be taken here, as the link between exposure and crash numbers may not be linear. Other related definitions of risk more subjectively define it as crash potential, propensity, or conditional probability. From an economic viewpoint, risk can also be defined as the expected loss.

The *relative risk* (or *liability index* or *hazard index*) of a given situation is the ratio of the risk of a crash in that situation divided by the overall crash risk. It measures whether a crash would be more or less likely to happen in a given situation, rather than on the road in general. It can also indicate the presence of factors likely to contribute to crashes. A related site-specific exposure measure is the number of crash-avoidance manoeuvres (or incidents or near-misses) which occur at the site. Such painstakingly-recorded avoidance data form the basis of the traffic *conflicts* assessment method discussed in Section 20.2.1.

A valuable approach to the indirect estimation of crash exposure was developed by Thorpe (1964), who postulated that a driver in a two-car crash was either:

* responsible for the crash, with a probability which was the same as that driver's probability of being involved in a single-vehicle crash; or
* not responsible for the crash, with a probability equal to the proportion of that driver category in the driving population.

The proportions of driver category 'a' found in the driving population, in single-vehicle crashes and in two-car crashes respectively are represented by ${}_aP_p$, ${}_aP_{sv}$ and ${}_aP_{2c}$ respectively. The proportion of driver category 'a' found in two-car crashes will be $2{}_aP_{2c}$ and, from Postulate 1, the proportion of driver category 'a' responsible for the crash will be ${}_aP_{sv}$. This leaves the proportion of drivers who were not responsible for the two-car crash in which they were involved as $2{}_aP_{2c} - {}_aP_{sv}$. Thorpe's Postulate 2 is that this is the same as the proportion of those drivers on the road, thus:

$$2{}_aP_{2c} - {}_aP_{sv} = {}_aP_p$$

This equation allows the exposure ratio of category 'a' drivers, ${}_aP_p$, to be estimated from the crash data which gives ${}_aP_{2c}$ and ${}_aP_{sv}$. The approach can also be used to estimate the relative risk of a driver category:

$$\text{relative risk for category 'a'} = {}_aP_{sv}/{}_aP_p$$
$$= 1/(2[{}_aP_{2c}/{}_aP_{sv}] - 1)$$

This approach can be used to rapidly construct graphs of the type shown later in Figure 28.2.

Everyone in the community is subjected to some degree of risk. In risk analysis it is commonly argued that the risk levels in Table 28.2 apply in developed countries. Note that people will tolerate much higher risks when they are self-imposed (e.g. driving) than externally imposed (e.g. public transport).

Table 28.2 Commonly accepted risk levels in terms of fatality/person-year expressed as 10^{-N}.

Most people for this risk type:	N
* class the event as rare, and therefore accept the risk, e.g. of being killed by a lightning strike whilst walking. This is called the *public health risk*.	6
* are willing to spend money to reduce the risk, e.g. buying protective clothing to wear whilst involved in the activity (e.g. motorcycle riding).	4
* find the risk unacceptable, e.g. driving under the influence of alcohol (Figure 28.1).	3

That implication in this listing is that traffic risks are bordering on the unacceptable and this is borne out by public reaction in most countries when the rates begin to rise.

28.1.4 Crash rates

Crash rates per distance travelled give the risk of using the transport system. For example, fatalities per veh-Tm of travel are listed for a number of countries in Table 28.1, although the data is rarely reported and the values are often estimates. Over the years, this rate has been somewhere between 5 and 20 in developed countries, but is usually much higher in developing countries. As communities become more adept at using the car and particularly as road systems and vehicle safety both improve, this risk measure has fallen. Across time and across countries, people learn to live with the car. As an example of another distance-related rate, the rate for injuries among young drivers is usually between 1000 and 4000 injury-crashes/veh-Tm.

For developed countries, the car usage data in Table 4.2 shows that the average vehicle in many countries travels about 15 Mm/y, suggesting an average annual fatality rate per vehicle of 10^{-4} (Table 28.2). For about 50 years of driving, this gives a lifetime risk of a fatal traffic crash as about 7 x 10^{-3}, or 1 in 150. Thus, at about three trips per vehicle per day, or 1000 trips per year, the risk of a fatality due to a crash during a trip is about 10^{-5} per trip. The annual risk per person of a serious injury in a crash is about 10^{-3}, which is about seven times the fatality risk, and the annual risk of a property-damage-only crash is about 0.2. Assuming that a driver makes a decision every 10 m of driving, leads to about one decision in every 5 000 000 leading to a crash. The annual pedestrian fatality risk is typically about 25 x 10^{-6}, but for people over 65 years of age the value rises to about 60 x 10^{-6}.

The rates given above are all average values and should not be extrapolated too far. As two salutary examples, data suggests that:

* the number of crashes experienced by an individual driver does not increase linearly with distance travelled, but with (distance travelled)$^{0.3}$, and
* crash rates are at least twice as high at night as they are during the day; e.g. the average of 10 fatality/Tm becomes, on U.S. figures, 5 fatality/Tm in daylight and 20 fatality/Tm at night.

Major differences exist in the risk of using particular road-based modes of travel, as illustrated by the crash data in Table 28.3.

Table 28.3 Casualties per occupant-Tm in the U.K. in 1977 and 1987, based on Sabey and Taylor (1980).

Mode	Fatality	Serious injury	All injuries
motorcycle (Section 27.5)	200–130	2750	9650
pedestrian (Section 28.7)	70	600	
bicycle (Section 27.6)	60–65	1000	5300
car (Table 28.1)	6–10	90	420
truck (Section 27.3.8)	3[a]	–	–
all motor vehicle users	8[b]	120	520[c]

a 1989 data. For further truck safety data, see Section 27.3.9. Given the relative safety of a truck occupant, the fatality rate for crashes involving trucks is usually much higher at about 30/Tm.

b Lower than Table 28.1 as it excludes pedestrians and is for occupant rather than vehicle travel.

c Refer also to the data in Section 28.5.

Risk-taking is discussed in Section 28.2.4 where it is shown that individual drivers will make both objective and subjective judgements about both assessing a risk level and then deciding whether to accept it.

28.1.5 Crash categories

Crashes may be categorised in various ways. For example, the *pattern of vehicular behaviour* that led to the crash can be used to provide four broad categories of crash types. This approach builds on the traffic conflict work in Sections 20.2.1–2. A simple version is shown in Table 28.4, far more complex definitions are used for reporting actual crashes.

The proportion of crashes in each of these crash-based categories can vary remarkably between locations. For example, in 1982 category c comprised 24 percent of urban injury crashes requiring hospitalisation in London and 80 percent in Addis Ababa (Jacobs and Sayer, 1984). Similarly, the major contributors to:
* minor urban crashes are in category a.2
* serious urban crashes are in categories a.3–5, and
* serious rural crashes are usually in category b.

Table 28.4 Categorisation of crash types, based on vehicular behaviour leading to the crash. For off-side and near-side definitions, see Section 20.2.2.

	Category	Example
a	Multiple-vehicle collisions: a.1 head-on a.2 rear-end a.3 right-angle, front to off-side a.4 right-angle, front to near-side a.5 turn across oncoming vehicle	 a.1 misjudged overtaking a.2 inattention approaching stop-line a.3&4 failure to stop at cross-road control device a.5 centre turn without gap
b	Single-vehicle collisions	Collision with roadside object after running off the pavement
c	Collisions with other road users such as: c.1 pedestrians c.2 cyclists c.3 motor-cyclists	
d	'other'	Object falls on vehicle

Crash location is also a critical categorisation factor. In most of the (densely settled) countries of Europe, urban area crashes predominate with Great Britain being markedly high at 75 percent. However, in more sparsely settled countries such as Australia, the reverse is true and rural crashes predominate. Urban crashes also tend to predominate in developing countries and cities in those countries have about 15 times more crashes per year per vehicle than do the cities of the developed world (Jacobs and Sayer, 1984). Rural crashes are generally more severe than urban ones, due to the higher speeds involved, generally poorer roads, and a greater tendency for the victims not to wear seat belts (Section 28.4).

Crashes can also be categorised according to their *primary cause*, which generally may be divided into three categories, viz., those caused by the:
* driver
* vehicle, or
* road.

The well-known *Haddon countermeasure matrix* shown in Table 28.5 illustrates the interactions involved. The countermeasures listed can be seen to be the 'reverse' of a crash 'cause'. The three numbered columns indicate how the *stage in the crash* also forms another set of categories.

Crashes can also be categorised according to their consequences, most commonly by their *severity*. *Severity level* is usually based on the consequences of the crash, typically using the following six categories:

A. Involving an injury, and assessed using the most severely affected person in the crash:

(1) fatal,

Table 28.5 Haddon countermeasure matrix (Haddon, 1980). Data for Stage 1 is available from the broad studies described in Sections 28.1.2–3. Data in Stages 2 & 3 usually comes from the in-depth studies described in Section 28.3. S = Section.

Crash	Safety stages in the application of countermeasures		
Causative category	1. Before-crash or primary	2. In-crash or secondary	3. After-crash or tertiary
Influences	crash frequency (S28.1.2–3)	crash severity (S28.1.5)	
D. Driver	D1.1 Training (S16.6.1) D1.2 Education (S16.6.1) D1.3 No alcohol or drugs usage (S28.2.2) D1.4 Driver attitudes attuned to safety (S16.3.2) D1.5 Driver inexperience covered (S16.5.5 & 28.2.6) D1.6 Pedestrians, cyclists wear conspicuous clothing D1.7 Clear priority rules at intersections (S20.2.2)	D2.1 Use of passenger restraints (S28.4) D2.2 Motorcycle and cycle helmets used (S 27.5)	D3. Timely and effective emergency services
V. Vehicle	V1.1 Effective and assisted braking (S27.2.2) V1.2 Good tyres (S27.7) V1.3 Good steering (S27.1.2 & 16.5.2) V1.4 Roadworthiness maintained V1.5 Good visibility from within vehicle V1.6 Speed control devices in vehicle V1.7 Vehicle use reduced in high-risk situations (exposure control)	V2.1 Passenger restraints fitted (S28.4) V2.2&3.2 Vehicle crashworthiness and impact protection (S27.3.9) V2.3 Vehicle interior protects occupants (S27.1.2) V2.4 Vehicle exterior protects pedestrians (S27.1.2)	V3.1 Well-equipped incident teams
R. Road	R1.1 Good road geometry (Chapter 19) R1.2 Good delineation (S24.4) R1.3 Good road lighting (S24.2) R1.4 & 2.3 Good traffic engineering (S7.2,20.3&23) R1.5 & 2.4 Effective signs and markings (S20-22) R1.6 Good pavement surface conditions (S12.5.2&26) R1.7 Vehicles and pedestrians separated (S7.2&20.4) R1.8 Supervision of child and elderly pedestrians (S20.4)	R2 Roadside hazards: R2.1 – reduced (S28.6.1) R2.2 – protected (S26.6.2)	R3.1 Damaged road and traffic devices quickly restored R3.2 Emergency telephone call facilities available

(2) admitted to hospital but not fatal,
(3) medical treatment required but not admitted to hospital,
(4) injuries, but no medical treatment,

B. Property damage only:
(5) crash reported,
(6) crash not reported.

For example, high-speed collisions with fixed objects would have a high crash-severity whereas collisions with frangible objects would have a low crash-severity. When applied to an individual, the categories in A can also be used to define the *injury class*.

A road *fatality* — A(1) — is usually defined as any death which can be attributed to a road crash and which occurs within 30 days of that crash. Fatalities represent only a small proportion of total injury crashes (Table 28.1) but, because of the human suffering they cause, their reduction receives a disproportionate amount of attention. This issue is pursued further in Section 28.8. Injuries A(2–4) are sometimes divided into;

* incapacitating,
* non-incapacitating, and immediately apparent, and
* non-incapacitating, and not immediately apparent.

Because of their frequency, unreported crashes — B(6) — can often be a significant contributor to the total cost of road crashes, although their individual cost is relatively low. This is illustrated in Table 28.9 below. A distinction is sometimes made between crashes reported to the police and/or road agency and the much larger number reported to insurance companies.

28.2 DRIVER BEHAVIOUR AND CRASHES

Driver behaviour was examined at length in Chapter 16. As driving is a relatively simple task, a driver's physical characteristics play an insignificant role in causing road crashes. Even vision, which services about 90 percent of the driving task, is not important as drivers with poor vision usually correct for the deficit, often by not driving at all. Thus crash studies show only a weak link between crash record and visual capability, with the link being strongest for crashes in poor light.

28.2.1 Crash responsibility

Road crashes are not 'accidents' but are the outcomes of a far from accidental chain of events linked to the crash categories given in Section 28.1.4. They represent failures of the system as a whole rather than of its isolated components. A number of attempts have been made to detect the prime causative factor in this system failure and the driver is found to provide somewhere between 65 to 95 percent of the prime causes, with the vehicle and the road providing the remainder. A breakdown of the causes is given in Table 28.6.

The role of vehicle factors is examined further in Sections 27.2.1 & 27.3.9. Driver-related causes predominate in Table 28.6. All these causes relate to poor judgement and none relate to an inherent lack of skill. Of course, such human errors made in driving do not necessarily result in collisions, or even attract the attention of other drivers, and the wider term *unsafe driving action* is used to describe any driver action which has safety implications — that is, which leads to a traffic incident (Section 20.2.1) or increases the risk of a crash. See also the related discussion of traffic conflicts in Section 20.2.1.

The predominant unsafe driver actions are:

A. Unsafe actions due to poor judgement before the crash:

A1 inattention,
A2 failing to see or perceive other vehicles,
A3 not looking for or anticipating other vehicles,

Table 28.6 Primary causative factors in road crashes (Sabey, 1980).

Cause of crash	Percentage	Example
Driver factors alone	65	Run off straight road
Driver and road factors	25	Collide with roadside post
Driver and vehicle factors	5	Wheels lock during braking
Road factors alone	2	Skid on slippery road
A4 not adjusting for visual restriction,		
A5 driving too fast for the conditions,		
A6 following too closely.		
Vehicle factors alone	2	Brakes fail
Driver, road and vehicle factors	1	Right-angle crash at night
Total	100	

B. Operationally unsafe actions due to poor judgement:

B1 failing to yield to vehicles with right-of-way (Section 20.2.2),
B2 turning too closely in front of another vehicle,
B3 diverging too near to the front of another vehicle, and
B4 merging too closely into another traffic stream.

C. Operationally unsafe actions due to poor execution of a known driving task:

C1 generally, inadequately controlling the vehicle being driven, and
C2 specifically, poor positioning when turning.

It follows from Table 28.6 that unsafe driver actions account for about 80 percent of all crashes. Japanese studies indicate that the unsafe actions in category A are the major cause of 75 percent of crashes and it is known that about one-third had taken no evasive driving action prior to their collision. Indeed, various studies have suggested that unsafe Action A1 in its various forms is a contributor to over 80 percent of all crashes and Action A2 alone may be the major cause of 50 percent of crashes (see also Section 16.6.1).

These factors indicate that the problem is one of a lack of judgement and consequent erroneous decisions, rather than a lack of skill. The prime causes of this lack of judgement falls into two categories of:

* risk-taking (Section 28.2.4), and
* poor decision-making.

Drivers in traffic typically commit an unsafe driving action every 700 m or so, or once every 2 minutes. One decision in 50 is potentially unsafe (Quimby, 1988). About 15 percent of unsafe driving actions are potentially serious and 5 percent cause a response from other drivers. From the crash risk data in Section 28.1.3, it can be deduced that about 1 in 100 000 unsafe driving actions will result in a crash.

As shown above, these unsafe actions by drivers are the major primary cause of road crashes, i.e. they are the major link in the chain of circumstances which leads to the crash. Of course, this does not mean that the driver is the only cause of the crash, as even the most errant and dangerous driver usually needs to encounter some other object in order for damage to occur. Therefore, people responsible for vehicle and road design cannot use the above conclusion to wash their hands of any responsibility for road crash prevention. They must accept the near-certainty of irresponsible, errant driver behaviour leading to unsafe driving actions and thus design the road system to tolerate this less-than-ideal response. Indeed, the design professions have an obligation to design for human error and not to piously condemn the next round of crash victims. Furthermore, society expects its traffic Authorities to exercise a road safety responsibility.

28.2.2 Alcohol

One major contributor to dangerous driver behaviour is the prior consumption of alcohol (or drugs). Any alcohol in a driver degrades driving ability, but the extent of the effect will vary between individuals (Johnston, 1983). Alcohol causes this deterioration mainly by depressing functions within the central nervous system. In this sense it acts as an anaesthetic. This depressant role reduces inhibitions, releases aggression, and increases risk-taking and hence raises the chance of being involved in a crash. Furthermore, drivers are rarely aware of the effect that alcohol has on their driving ability. Subtle physical consequences include reduced arousal and impaired information processing (Sections 16.4.3&5 and 16.5.2&3). The visual degradation explains why alcohol-affected drivers are over-involved in 'failure to see' crashes (Section 28.2.1), such as running into parked cars.

Alcohol-affected drivers (colloquially called *drinking drivers*) are over-represented in fatalities and other severe crashes and in crashes occurring at times when alcohol consumption is high, e.g. at night and at weekends. Typically, about 30 percent of traffic fatalities — both drivers and pedestrians — are above the local legal alcohol limit for drivers. About two-thirds of these victims will have a record of persistently driving whilst alcohol-affected. Similarly, a tenth of injured people possess alcohol levels far above those encountered in drivers not involved in crashes. Of course, alcohol-affected drivers will injure many sober drivers. Alcohol also usually figures disproportionately highly in pedestrian fatalities.

The effect of alcohol on the relative risk of a fatal collision is illustrated in Figure 28.1. A person's alcohol level is measured by their *blood alcohol concentration* (*BAC*) which is the mass of alcohol in gram found in a litre of blood. [In pure SI units this would be measured as kg/m^3 which is the same as gm/l. However convention is often to use mgm/100 ml as a percentage. Thus 5 gm/l BAC becomes 0.05 percent BAC.] Typical traffic laws set the legal BAC limit somewhere between 0.5 to 1.0.

Empirical data suggest, approximately, that the risk of a crash trebles as BAC goes from zero to 0.5, and is 20 times at a BAC of 1.0. The most common BAC for fatalities is usually about 2.0. At that level, the chances of a crash being fatal are about three times higher than they are for a driver with zero BAC. The actual situation is worse than that portrayed in Figure 28.1 as the relative risk can be almost 250 times greater at particular times of the day. The data in the figure are also conservative — that is, real risks are actually higher — as the presentation underestimates the exposure factors.

Unfortunately, driving whilst affected by alcohol is often commonplace. A British study (Clayton *et al.*, 1984) suggested that 60 percent of people attached little social stigma to drinking an average of 1.7 L of beer on each of two weekly occasions, and then driving home. In many countries, 5 to 10 percent of late-night drivers are above the legal limit.

Typically, alcohol-affected drivers are male, single, and under middle-age. On the average, apprehended alcohol-affected drivers have drinking habits that differ significantly from those of the community at large. For instance, their alcohol consumption pattern and blood alcohol content is excessive by normal community standards and the majority of drivers who attract police attention are well in excess of the legal limit rather than fractionally above it. In addition, drink-drivers have more frequent and serious traffic convictions and more criminal convictions. For instance, serious offences such as dangerous driving, failing to stop after a crash, and driving whilst disqualified are relatively more frequent amongst such drivers. Thus this small population of excessive drinkers contributes disproportionately to two major health problems — alcoholism and road crashes. Indeed, drink-driving might be best regarded as a problem in community health.

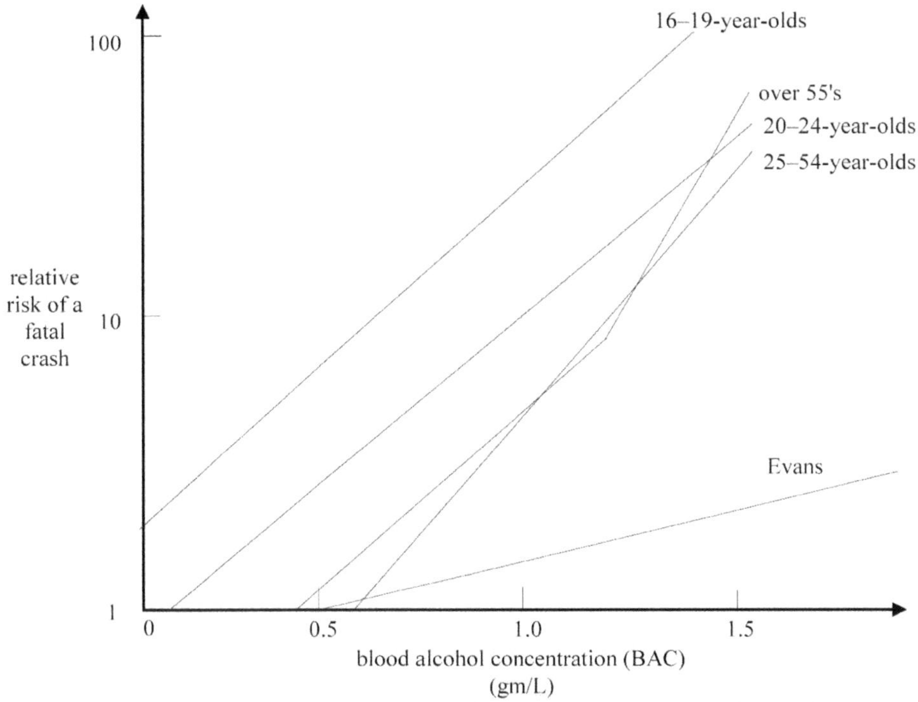

Figure 28.1 Relative risk of a fatal crash as a function of BAC and age. From Mayhew *et al.* (1986).

Common impediments to measures aimed at minimising the alcohol–driving problem are:

* permissive community attitudes to drinking alcohol and then driving,
* peer group pressures on drivers to both drink and drive, and
* the fact that young male drivers often couple their initial driving experience with a high level of vehicle usage (i.e. high exposure) and frequent high alcohol intakes.

The last point, and Section 16.5.5, indicate that there are good reasons for setting the driving age below the legal drinking age. Raising the legal age for alcohol consumption has been successful in the U.S. A counterview is that granting a driving licence should be delayed as long as possible, thus reducing exposure and therefore total crash numbers. Applying the test of *reductio ad absurdum* to this counterview, no one would be licensed to drive. Both view and counterview have some logic, but the counterview fails because crash avoidance is not the prime objective of the road transport system.

Nevertheless, evidence does indicate that patterns of alcohol consumption can be changed, causing a drop in alcohol-related crashes. This change does not need to be associated with a drop in the amount of alcohol consumed. The change requires a suite of measures including enforcement, publicity, low-alcohol drinks, and appropriate treatment of detected offenders. The treatment of detected offenders tends to reduce the incidence of drink-driving, rather than reduce drinking levels — i.e. drivers better manage their problem, rather than eliminate its cause.

Drugs play a similar role to alcohol, but have a wide range of effects and are much harder to detect. In many communities up to 50 percent of the population take some form of prescription drug over a two-day period. It has been variously estimated that in many countries prescription and illegal drugs contribute to at least 10 percent of road fatalities. A 1990 American study of truck crashes found that two-thirds of the drivers tested positive for some form of drug, one-third for a drug of abuse, and only one-sixth for alcohol. A 1994 study of Australian crash fatalities found that 22 percent of the drivers had drugs other than alcohol in their blood (Drummer, 1994). The data for the presence of specific drugs in road fatalities were:

* cannabinoids such as marijuana and cannabis, 11 percent
* stimulants, 4 percent
* benzodiazepines, 3 percent
* opiates (e.g. codeine), 3 percent.

Marijuana on its own can, on the one hand, cause a driver to behave more carefully and, on the other hand, reduce driving skills. When used in combination with alcohol, the reduction in driving skills is marked and there is a significant increase in crash risk.

28.2.3 Sociological characteristics

A significant number of drivers in serious road crashes are in some way socially maladjusted, and the population of drivers involved in road crashes is noticeably different in sociological terms from the general population of drivers (Section 16.3.1). The factors commonly over-represented in drivers associated with serious road crashes are:

* personally inherent, being:
  - male,
  - youthful, and
  - on a low income, or unemployed (however, overall crash rates fall as unemployment increases, as the associated recession reduces the amount of driving done by high-risk young drivers).
* socially inherent, being from:
  - a socially deprived area (Abdallah *et al.*, 1997),
  - an unstable family background.
* psychological, being (Section 16.3.2):
  - readily distracted,
  - unable to exercise self-control,
  - unable to manage hostility
  - prone to anti-social behaviour, and
  - prone to uncontrolled aggression.
* self-imposed:
  - taking risks (Section 28.2.4),
  - drinking alcohol or taking drugs (Section 28.2.2),
  - smoking.
  - suffering from transitory emotional stress (Section 16.3.2),
  - being fatigued (Section 16.5.4),
  - being distracted, and
  - chronically violating the law.

This character description of the crash offender is not meant to support the concept of accident proneness. Far too few people in the categories mentioned are involved in road crashes for any one group to be labelled *accident prone*. It is more correct to discuss groups of drivers as being *over-represented* in road crashes, e.g. more people with a high alcohol content are found in a sample of road crash victims than would be found in a random sample taken from the population at large.

Indeed, there is no method to accurately predict the future crash involvement of individual drivers. Even a driver's past crash-record has little predictive power with respect to their future crash-record, mainly because of the relative rarity of a crash occurring. If all the people who had had crashes were removed from the driving population, there would be almost no effect on the subsequent crash data. Typical developed country data in this respect are:

(a) 80 percent of all crashes in a year involve drivers who have not had crashes during the previous two years;
(b) about half of all crashes in a year involve drivers who have not had moving violations in the previous two years; and
(c) less than 3 percent of drivers have had more than four convictions or crashes in three years.

The type of socially irresponsible behaviour that is over-represented in road crashes cannot be cured by a stroke of the legislative pen or a change in a road-design parameter. Indeed, road crashes can be said to be just a rather observable symptom of some larger problem. For instance, drivers prone to take high risks will not have their personalities altered by the lowering of speed limits. A more appropriate measure, therefore, might be to raise the risk of apprehension by higher enforcement levels. In the longer term, it is necessary to influence attitudes (Section 16.3.2).

28.2.4 Risk-taking

Section 28.1.3 indicated that one of the problems in road safety is that the risk of an individual driver having a road crash is very small. However, the annual risk of receiving a serious injury in a road crash of about 1 in 1000 was close to the risk level that the community found unacceptable (Table 28.2) and is confirmed by the fact that most communities are disturbed by current crash levels. However, the risk of being apprehended whilst committing a serious traffic offence is often less than 1 in 1000, which is too low to alter the driving behaviour of many individual drivers.

The risks just discussed are objective risks, calculated from established data. The risks that drivers assign as a result of their own views and inclinations are subjective risks. Troubles arise when subjective risks are lower than objective risks. A number of factors can create this common situation (Williams and Haworth, 2007):

(a) Research consistently indicates that most people believe that their driving skills are above average. A common finding is that about 75 percent of drivers consider that they have above average driving ability.
(b) Drivers usually assume that crashes are due to the faults of other drivers. They also believe that they can control their crash involvement. For instance, Section 28.1.4 showed that the annual risk of a crash was about 0.2, whereas Williams and Haworth (2007) report typical driver assessments of this risk at about 0.01.
(c) More specifically, drivers often underestimate risk as is evidenced by the fact that professional drivers tend to rate many common road situations as riskier than do other drivers (Cairney, 1982). Inexperienced, new drivers commonly overestimate the risks of having a crash in general, but underestimate the risk of particular driving manoeuvres, such as following closely behind another vehicle.

(d) Factor (c) is heightened by a human tendency to underestimate the frequency of events that are hard to imagine or recall.
(e) A related common tendency is for a driver to assume that he or she is less likely to encounter a serious crash than is the 'average' driver. And if they are in a crash, they put the probability that they might be responsible as about 0.05. Many drivers use such an assumption to allay fears that they would otherwise have about driving.
(f) People generally under-estimate the risks from actions (such as driving) which are under their personal control.
(g) Further, some drivers — particularly young ones — find pleasure and peer approval in risk-taking (Section 16.3.2). Some deliberate risk-taking is apparent in all drivers but varies significantly between them. It usually results from a genuine — if ill-informed — underestimation of the risks involved, as outlined above.

The attitudes listed above are reinforced by the very low risk of a crash in any one trip (Section 28.1.4), compared with the daily crash reports that are presented to drivers through the media.

The controversial *risk compensation hypothesis* states that drivers tend to adjust their behaviour to a riskier level to compensate for any lowering of risks by safety improvements. The potential crashes that are avoided at the improved site may thus still occur, but at another site. The process is known as *crash* (or *accident*) *migration*. Other explanations for it are that improvements lower a driver's assessment of the risks in an area, and/or that drivers unwisely compensate for delays at an 'improvement'. *Risk homeostasis* occurs when the compensation is total and so no overall change in crash numbers occurs after a safety measure is introduced.

Most researchers (e.g. O'Neill *et al.*, 1985) believe that even the compensation hypothesis is weak, and that no more than 30 percent of 'saved' crashes are so affected. A key reason for this view is that there are very many examples of safety improvements that have reduced crash numbers.

28.2.5 Behavioural modifications

In attempting to reduce the effect of road crashes, it is not surprising that people are tempted to first suggest methods that involve altering driver behaviour. There are a variety of forms that such proposed modifications could follow:

(a) training in driving skills (Section 16.6.1). It is often argued that drivers should be trained to improve their driving skills and hence reduce their crash propensity. However, there is little or no evidence that skill training will reduce crashes. The main reason for this is that the most effective approach is not to teach the physical skills needed in a crash typically resulting from an unwise decision, but rather to teach how to avoid making such unwise decisions (Section 16.6.1).
For instance, drivers should learn how to avoid being in a skid rather than how to 'steer out' of it. Indeed, giving drivers improved driving skills commonly enhances their risk-taking propensities and raises their crash rate (Section 28.2.4).
(b) training in appropriate behaviour via the use of physical restraints (Section 18.1.4), e.g. via devices placed in the carriageway or speed governors (or *limiters*) on vehicles. These methods do appear to influence overall driving patterns.
(c) imposing punishment and penalties (or fines) for inappropriate behaviour (Section 16.7). The effect of punishment and penalties is usually of only a short-term nature and modifies behaviour only in circumstances seen by the driver to carry appreciable

risk of detection. Indeed the risk of detection is more important than the punishment level. Furthermore, the punishment needs to be reasonably immediate and appropriate for the driver to later associate it with the event for which the punishment was imposed. Thus, automatic methods of detecting offences and quickly distributing penalties — such as by *speed cameras* and *owner-onus* — have had measurable effects on driving behaviour. ['Owner onus' is a system where the owner of a vehicle is guilty of the related traffic offence, unless the owner can nominate the actual driver. It means that only the vehicle's ID need be detected in order to make a prosecution.] A major problem with using many more conventional law-enforcement processes to produce behavioural change is that they can be very costly and have produced little evidence of cost-effectiveness.

(d) creating an awareness of the consequences, possibly to the extent of causing fear and apprehension. Graphic advertisements showing the dramatic human consequences of inappropriate driving behaviour have had a measurable effect on the number of road crashes.

(e) appeals to drivers' sensibilities. A famous case in road safety history was reported as follows (Wigglesworth, 1978):

> In April 1978, on a motion introduced by the Opposition, the Victorian Legislative Council unanimously carried a resolution calling on all drivers to make the next weekend 'death free'. That resolution was supported by the Premier, by the Chief Commissioner of Police (who promised massive police support), and by a joint statement by the Anglican and Roman Catholic Archbishops.

The number of road fatalities on the same weekend in the preceding two years had been six and five. In the 'death free' weekend there were nine fatalities.

There is no firm evidence that any measures aimed at altering driver behaviour will be successful in the long term and considerable evidence to the contrary. However, attitudinal changes — whilst more gradual — can be effective and long-lasting. It could be argued that seat belts do provide evidence of a successful modification of driver behaviour. In the author's view, the behavioural changes required were so minor as not to qualify in this category. Seat belts are discussed further in Section 28.4.

28.2.6 Effects of age

In developed communities, motor vehicle crashes are the most significant cause of death for teenagers and young adults (15–24 year age group), most commonly when they are drivers rather than passengers. For young drivers, risks peak at about 18 or 19 years when their risk of being a crash fatality is about 2.5 times the population average. Their pattern of crashes is also atypical: for instance, they are over-involved in out-of-control crashes and single-vehicle crashes. Young males are over-involved in crashes resulting from excessive speed.

Driving risk diminishes as the driver grows older (Figure 28.2) with a noticeable safety improvement in the third year of driving. The phenomenon is not new and was first reported in the U.S. in 1930. The question as to whether the link with a driver's youthfulness reflects immaturity or inexperience, particularly as age and driving experience are identical for nearly all drivers, was explored sociologically in Section 16.3.2 and more widely in Section 16.5.5. It can be argued that immaturity probably accounts for the high initial crash-rate and experience for its subsequent decline with age.

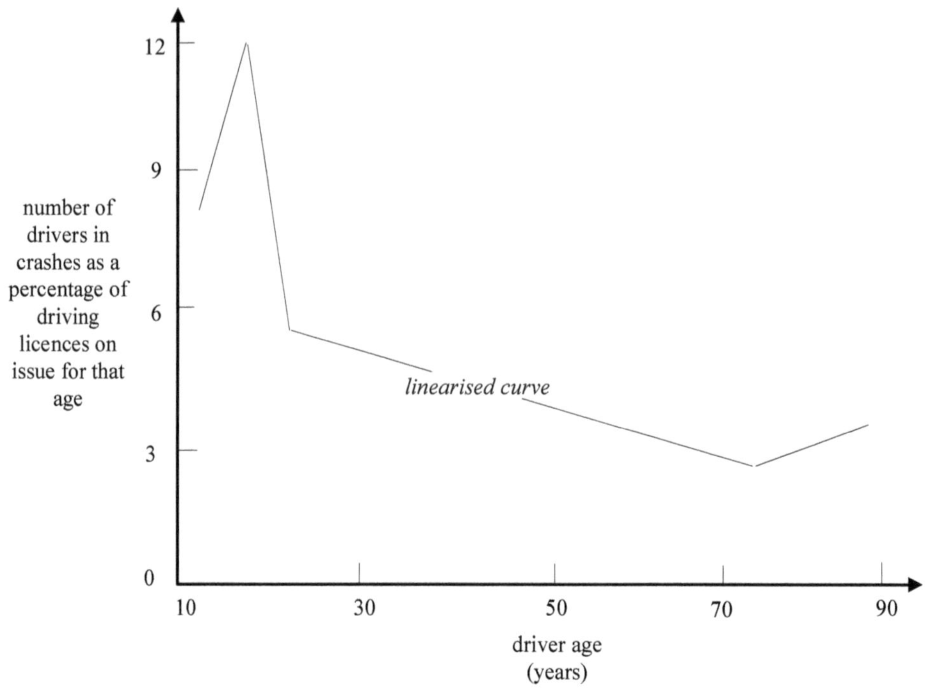

Figure 28.2 Drivers involved in crashes by age (Herbert, 1980).

The *elderly driver* represents the other end of the spectrum. Crash statistics do not suggest that age alone should be used as a barrier to prevent people driving. Many elderly people compensate for their reduced abilities, or drive only when the risks are low, thus automatically limiting their exposure by excluding high-risk situations. Indeed, Figure 28.2 shows the steady decline of crash involvement with age in terms of the proportion of the driving population involved. However, more recent data analysis suggests an up-turn in:

* crashes per distance travelled for the over-50s,
* crash responsibility for the over-60s, and
* most measures of crash rate (Section 28.1.3) for the over-65s.

Due to higher exposure, elderly males have a worse absolute record than females but their crash per distance record is very similar and elderly females appear to have higher crash culpability levels. Older drivers tend to be over-represented in low-speed, multi-vehicle and intersection crashes and most of their crashes occur in daylight conditions in their own local urban area. The behavioural effects of increasing age on drivers are discussed in Section 16.5.5.

28.3 CRASH CONTROL

28.3.1 A systems approach

All the factors in Section 28.2 mean that the road engineer must accept the road-using public for what it is — a risk-taking, sometimes aggressive, sometimes drunken, often inexperienced, and usually poorly-trained set of car-driving operatives.

As argued in Section 28.2.5, there is therefore no alternative but to build safety into the road and the vehicle. This view is put with some over-statement to deter the occasional tendency of the community to argue that errant drivers deserve what they get and that road engineers should not waste money on providing for their safety. This latter view has little to recommend it, given the above facts and our professed respect for human life.

The two major options available to reduce the road toll are to reduce:

(1) the total number of crashes, and/or

(2) the severity of individual crashes.

The factors that can be used to control each are given in the Haddon matrix (Table 28.5).

By medical analogy, the crash reduction stage can be subdivided into three sub-stages as follows:

(a) *Observe the symptoms and identify the problem.*
This sub-stage should be based on a statistical analysis of the data and on-site studies. It has been noted of U.K. experience that 'accident problem areas highlighted by the lay public are far less serious than the locations requiring action as identified by a proper, systematic analysis of recorded accident statistics' (Sabey, 1985).

(b) *Diagnose the cause.*
The process again requires the systematic data analysis used in (a). This will indicate proportions of the cause due to each of the nine elements in the Haddon matrix (Table 28.5).

(c) *Select and implement the most appropriate treatment.*
Taking the road-related Haddon categories R/B and R/I as the example, four methods of prioritisation are used:

(c1) sites are pinpointed where previous crashes have clustered or have over-occurred. As crashes are relatively rare, care should be taken with such statistical analyses to ensure that a selected site is not a 'good' site having an exceptional run of random crashes. The analysis is done in terms of:

(c1.1) absolute crash numbers,

(c1.2) absolute crash rates (Sections 28.1.1–3),

(c1.3) crash rates relative to other similar facilities, or

(c1.4) benefit/cost ratio estimates (Section 33.5.8), including a measure of the severity of the crashes, and a prediction of the subsequent crash reduction due to the measure.

The sites so located are called *black spots* — as crashes were once marked with a black spot on a map (Section 28.3.2);

(c2) an area-wide analysis is undertaken of the frequency and location of a particular feature common in a variety of previous crash situations;

(c3) estimates of crash rate are used by combining α times the measured crash rate for the site from (c1.2) with $(1 - \alpha)$ times the crash rate for the feature used at the site given by (c2). Typically, $\alpha = 0.3$.

(c4) the *traffic-conflict* method (Section 20.2.1) is used in conjunction with traffic flow observations.

It is usually necessary to check public acceptability of the proposed treatment. The operation of a treatment program is described in Section 28.5.

Because crashes are relatively rare events (Section 28.1.3), any driver-related countermeasures will only be cost-effective if they are very cheap and technically efficacious. There can be major variations in the cost-effectiveness of particular measures.

For instance, the cost to save a life is in the proportion 1:10:20 for the three compulsory measures of motorcycle helmets, seat belts and daytime motorcycle headlights (Section 27.5).

28.3.2 Crash data

In order to plan and design crash-control measures it is firstly necessary to understand those crashes that have already occurred. Broad-brush crash data were given in Section 28.1.1. Such broad data and detailed data describing crash circumstance, cause, and effect can only be assembled from the analysis of databanks of the results of crash investigation studies which should include data on:

1. *basic crash description*:
   time, location, weather, those involved (see #4 below), vehicle type (see #5 below), consequences (Section 28.1.4 — crash severity, property damage, casualties), the associated road-user movements, speeds of involved vehicles.
2. *road features*:
   land use, road type, road geometry, lighting, speed limit, pavement surface type and condition (Section 12.5.2), roadside furniture (Section 28.6), intersection type, traffic control devices. This data set is usually best obtained by linking the crash data to a separate file on road features, produced by the road inventory system (Section 26.5.1).
3. *traffic features*:
   vehicle flows, composition, common movements, and typical speeds. This data set is usually best obtained by linking the crash data to separate files on traffic flows and speeds and on traffic features, produced by the traffic monitoring system (Section 31.5) and road inventory system (Section 26.5.1).
4. *driver features*:
   personal data (Section 28.2.3 — age, sex, occupation, marital status, drug or alcohol influence, trip purpose), injury class. This data set is usually best obtained by linking the crash data to a separate file on drivers, produced by the driver licence system (Section 16.6.2).
5. *vehicle features*:
   type, age, condition. This data set is usually best obtained by linking the crash data to a separate file on vehicles, produced by the vehicle registration system (Section 27.1.1).
6. *hypotheses as to crash cause*:
   These data can be obtained by the routine investigation of each serious crash by the police or similar bodies, or by the in-depth study of a sample of crashes by specialised teams of experts. These latter studies are, of course, much more expensive and demanding and so occur far less frequently than routine collections. However, the data in conventional crash databanks will lack depth and have an accuracy of about 95 percent for fatal crashes, 80 percent for hospitalisations, and 50 percent for all injuries. It will suffer from significant under-reporting, particularly for non-injury crashes (here insurance and hospital records may prove more accurate than official data), and for crashes involving uninsured vehicles (such as cycles). Errors are also likely when reports are by non-experts: for example, police data on alignment, pavement surface, and crash severity are often unreliable.

As suggested by item 6 above, it is possible to make a number of deductions from the data collected at a crash site. These usually involve applying the conservation of momentum principle. Speeds at impact can be calculated from a knowledge of the length of skid-marks, and the friction coefficient, f, of the road surface (Sections 12.5.2 & 27.2.2). The latter gives the deceleration force as fMg, where M is the skidding mass, and hence the deceleration as fg. This gives a_c in Equation 27.1d, the skidding distance is x and, as v = 0, the speed at impact can be calculated from Equation 27.1d as v_b. Speeds before impact are calculated from momentum conservation.

In the Haddon countermeasure context (Table 28.5), such databases can be used for the R1 crash-prevention stage by influencing future decision-making and for the R2 crash-reduction stage, by rectifying current problem areas.

28.4 SEAT BELTS

Lap seat belts derive from aviation practice and were occasionally installed in cars prior to the 1950s. In 1955 seat belts were first offered as optional equipment in some U.S. Fords. The measure was not very successful, but — by the early 1960s — anchorages for installing seat belts became a routine new-car feature. In 1964, South Australia introduced a legal requirement that anchorages be fitted in the front driver's seats of all new cars. In the same year, a number of U.S. states introduced laws requiring lap seat belts to be installed in the front seats of new cars (Campbell and Campbell, 1986). Subsequent seat belt wearing rates in South Australia are given in Table 28.7 and showed that it was not enough to have provisions for seat belts.

Table 28.7 South Australian seat belt wearing rates in the years after belts were made compulsory in cars.

	Vehicles fitted in	Wearing rate (percent)	
Year	the driver's position	Drivers	Passengers
1964	15	64	48
1966	28	46	42
1968	46	36	22
1970	60	28	23
1972	68	81	68
1974	78	72	58
1976	85	70	56
1977	95	91	71

Perhaps the strongest technical support for seat belt wearing came from Australia's Snowy Mountains Authority between 1960 and 1968. A review of its work demonstrated the value of seat belts in vehicles being operated by the Authority in often extreme conditions of terrain and weather. It concluded, *inter alia*, that the lap belt should be used only if a more effective belt was not available, that dynamic testing of belts was necessary, and that wearing should be compulsory. As a result, the first legislation making seat belt wearing compulsory was introduced in Victoria, Australia, in 1970. By 1972 compulsory wearing applied throughout Australia. The effect in South Australia is evident in Table 28.7.

It has been estimated that fatalities and injuries are reduced by 40–60 percent through seat belt wearing (Evans, 1986). Further, a restrained occupant has ten times less chance of being ejected in a crash than does an unrestrained occupant. The greatest benefit of a seat belt is in frontal collisions and roll-over crashes. Crash victims wearing seat belts have a lower probability of a severe injury and of head, face or chest injuries than do unbelted victims. When seat belt wearing was made compulsory in Britain in

1983, there were significant reductions in car-crash hospitalisations, particularly for head injuries (Rutherford, 1987). There is little or no evidence that properly designed and worn seat belts cause serious injury. Indeed, there is no doubt that a properly-worn seat belt is the single most effective measure in protecting vehicle occupants from death and injury in road crashes.

Seat belts have reduced the human damage caused by road crashes, but it must be realised that they do not reduce road crashes. On the other hand, there is no evidence that the wearing of seat belts encourages people to drive less safely (Hampson, 1982). In addition, because of the sociological aspects described in Section 28.2.3, unbelted motorists are more likely to be involved in crashes than are their belted-up counterparts. Typically, many drivers killed in road crashes are not wearing seat belts, despite high overall wearing rates. The road engineer then has the obligation to protect the inherent risk-takers from themselves.

The *air bag* is a relatively sophisticated and expensive extension of the seat belt. It works by expanding within about 40 ms of impact to fill the space between driver and steering wheel (or side panel) and protect the driver from forward (or side) impact. By 200 ms it has deflated sufficiently to permit the driver to leave the seat. The air bag supplements rather than replaces the seat belt. In terms of reducing the risk of a fatality, the air bag alone is only half as effective as a seat belt. When used in conjunction with a seat belt, the air bag increases its effectiveness by about 10 percent. Air bags alone are particularly valuable for side impacts.

More recently, automatic stability control had had a major positive effect on vehicle safety. The system ensures that the car travels in the direction that the driver intends and so is particularly valuable in skidding or emergency avoidance manoeuvres. It works by detecting the vehicle's actual heading and then applying braking or accelerating efforts to individual wheels at millisecond intervals to reduce any discrepancy between actual heading and steering wheel intention.

28.5 TRAFFIC ENGINEERING MEASURES

It has been estimated that about 60 percent of crashes resulting in injury could be avoided using proven remedies and measures (some suggestion of this has been given in Table 28.6). Of this reduction, 15 percent would come from measures applied to the road, 20 percent from measures applied to the vehicle, and 25 percent from measures applied to the driver. On the other hand, the primary purpose of the road system is to provide accessibility, with safety as a secondary objective. Thus the benefits of any safety treatment must be measured against its implementation cost and its effect on accessibility.

Of relevance to this section are the measures that can be applied to the road, and these can be categorised as changes to the following fifteen factors:

(a) **Traffic flow**.

The total number of crashes occurring on a piece of road will be monotonically related to the number of vehicles using the road (i.e. to the exposure, Section 28.1.2), and so the number of crashes will increase with the traffic flow and the road length. Typically, changes in exposure levels account for about 70 percent of the annual variation in crashes in a 'stable' region. The key variable is the amount of travel in veh-km. Management of traffic demand (Section 30.3) can therefore be classed as a crash-control measure.

For a vehicle flow of q, typical crash–flow relationships (based on the author's judgemental amalgamation of data from various sources, locations and times) predict crashes as proportional to q^N, where N is given by:

Circumstance	N
Pedestrians	0.5
Cyclists, where q is the pedestrian or cyclist flow	0.7
All crashes	0.7
Head-on, out-of-control, single-vehicle crashes, and all crashes on divided roads with $V < V_c$	0.8 (Section 17.2.2)
Rear-end, side-swipe, and multi-vehicle crashes	1.2
Undivided roads with q < 3000 veh/d AADT	> 1.2

Similar relationships for intersections are discussed in (d) below.

(b) **Road type**.

The number of crashes per veh-km for a particular road type is approximately constant. Typical values for serious injury crashes on links of 500 m or more in an urban environment, and for fatalities generally, are given in Table 28.8. In comparing this data with Tables 28.1 and 28.2, a large deduction must be made for intersection crashes. The data indicate how the type of road provided has a major influence on crash numbers.

(c) **Access control** (Section 7.2.2).

This particularly relates to the reduction of conflict points through such measures as minimising the number of intersections, median openings (Section 22.4.2), driveways, and other access points. As implied in Table 28.8, there is strong evidence to indicate that crashes per veh-km increase as the intensity of the abutting land-use developments increases. Access control can produce dramatic crash reductions. Broadly, reducing access points by one per kilometre reduces crash rates by about 2 percent (private access) or 10 percent (commercial access), or by about 40 crashes per veh-Tm. Where well-used access points are provided, they should be designed as small intersections.

(d) **Intersections**.

The approximate risk of crashing at an intersection, per intersection passed through is estimated as (Migletz *et al.*, 1985):

* for turning traffic, 2×10^{-6},
* for at-grade cross-traffic, 10^{-6},
* for grade-separated intersections, 0.5×10^{-6},
* for a minor intersection, 0.3×10^{-6}.

Thus, crashes per veh-km increase as the number of intersections per kilometre increases. In terms of Table 28.8, the crash rates per veh-Tm should be increased by 300I where I is the number of intersections per km.

For an individual intersection, when there are two crossing flows, q_1 and q_2, simple gap acceptance considerations (Section 17.3.5) lead to the prediction that crashes are proportional to $(q_1q_2)^a$ where 'a' is between 0.5 and 0.7. Some empirical studies suggest that total crash numbers are simply proportional to $q_1 + q_2$ and that crossing crashes are proportional to q_2. Peak crash numbers therefore occur when $q_1 = q_2$. However, the different exposures experienced by the two flows generally leads to $q_1^b q_2^c$, typically, b = 0.65 and c = 0.25, where q_2 is the flow on the minor road. However, when one of the cross-roads is a minor road controlled by stop signs, the role of b and c is reversed, indicating that the minor road flows dictate the number of crossing accidents in these cases. Minimum crash numbers also occur when $0.6 < q/q_c < 0.8$, where q_c is the capacity flow (Section 17.2.2), as low flows encourage risk-taking and high flows increase the potential for conflict.

Table 28.8 Crash rates as a function of road type. From McLean 1984, Grayling 1980 and many more recent sources. The data presented below are based on the author's judgemental amalgamation of data from such sources. The upper range of the values quoted usually occurs with high levels of abutting development and poor access control (see c below). See Table 28.1 for fatality data from various countries.

Road type	Serious injury crashes per veh-Tm	Fatalities per veh-Tm
Single-lane roads	8 000–12 000	
Unsurfaced two-way roads	2 000–3 000	30
Narrow two-way roads	1 500–2 500	
Two-way road without shoulders	400–1 250	26
Undivided arterial roads	300–600	
Sub-arterial roads	300–600	22
Freeways between 1200 and 0600 h	400	
Residential collector roads	350	
Unsurfaced roads	300	
Divided arterial roads,	100–300	14
Roads without side traffic		
Old freeways	150	
Freeways and expressways without chance of unexpected side traffic	100	10
New freeways	50	7
Road with peak safety provisions	13	2
Average for all roads	300–500	23

Measures for improving intersection safety are also discussed in Sections 20.2.3 (general) & 20.3.6 (roundabouts).

(e) **Geometric design** (Chapter 19).

This issue particularly applies to curves and grades. Driver work-load is high on sharp curves. The risk of a serious injury on a typical road curve is about 0.04×10^{-6} and increases exponentially with curvature (radius^{-1}). It is relatively high for radii of less than 450 m and gradients of over 4 percent, particularly in combination (Sections 19.2.6 & 19.3), and for all horizontal curves with a radius of under 150 m. Increasing the radius of a curve from 70 m to 140 m can reduce the crash rate by 40 percent. Crash reductions are greatest for single vehicle crashes.

Curves are also dangerous when they occur at the end of long lengths of straight road and so the provision of a consistent alignment standard is critical (Section 18.2.6). Straight lengths of road are also a problem as drivers lose concentration if their task is too easy. In geometric design this implies avoiding monotonous sections of road where the driving task demand is low (Section 16.2). Substantial sight distance (Section 19.4) improvements have a noticeable safety benefit, provided a consistent standard is maintained along the road length.

(f) **Cross-section** (Section 6.2.5).

The base rate for serious-injury crashes between intersections is given in Table 28.8 as about 600/Tm. Broadly, the rate drops by about 20 percent for each 1 m increase in pavement width, and by 10 percent for each 1 m increase in shoulder width. More specifically, the rate changes little for drops in pavement width from 13 m (2 lanes and 3 m shoulders) to 10 m (2 lanes and 1.5 m shoulders), but increases to around 1200/Tm when the carriageway drops 7 m (2 lanes and no shoulders). Any drop in pavement width below 5 m has a major effect on multi-vehicle crashes. British work suggests that an extra 1 m of pavement width at intersections reduces crashes by 5 percent. Table 28.8 also gives rates for other cross-sections. On the other hand, there is a suggestion that crash severity increases as pavement width increases beyond the width needed to manage the existing traffic (Section 6.2.5).

Lane width selection is discussed in Section 6.2.5 and the influence of lane width on driver behaviour and crash rates is discussed in Section 16.5.1. The effect on crash rates is seen to be marginal or non-existent. Unless driver discipline is good, the provision of three-lane roads, where use of the centre lane is uncontrolled, can be hazardous. The use of specific overtaking lanes is usually the preferred option (Section 18.4.4).

Section 11.7 suggests shoulders of at least 600 mm in width if the traffic exceeds 200 veh/d. A dramatic reduction of about 75 percent occurs in the fatal crash rate for two-way rural roads when their shoulders are surfaced with a bound layer (Section 12.1.3 and Armour, 1984). In urban areas where speeds are lower, the reduction would be a much lower 10 percent. The improved shoulders provide extra room for emergency use. For instance, the typical crash caused by poor shoulders occurs when a car strays off the pavement surface, the driver realises that a wheel is no longer on a good surface, over-corrects the steering to return the vehicle to the pavement, and thus causes the car to yaw. The crash then occurs when the car either becomes out-of-control in the loose shoulder material, or regains the pavement surface in an over-reaction and heads across the lane into the face of oncoming traffic. The situation is significantly worsened when the vertical edge drop-off exceeds 50 mm, due to poor maintenance (Section 14.1A10).

Drivers who run-off the paved surface must contend with unexpected changes in the *slope* (Section 6.2.5) and surface friction of the verge. Such drivers are aided by verge slopes that are flatter than:

Verge slope	Purpose. Slopes of the given value or flatter:
10 to 1	prevent a vehicle that is out-of-control from becoming even less manageable.
6 to 1	provide a manageable driving situation and thus permit the verge to function as a full part of the safe clear zone (Section 28.6.1) beside the pavement. Guardrail is used on steeper slopes on high-speed roads.
4 to 1	prevent some out-of-control vehicles from becoming even less manageable (the number of single-vehicle run-off crashes will be 50 percent above the '6 to 1' rate). Guardrail is usually used on steeper slopes.
1.5 to 1	require the use of guardrail.

(g) **Roadside obstacles**.

See Sections 28.6 & 27.3.8.

(h) **Structures.**

Following Section 15.2.5, particularly avoid the use of:

* bridge components prone to vehicular impact,
* narrow bridge decks (see f above), and
* overpass spans, which place piers in dangerous locations.

(i) **Wide, clear medians and median barriers.**

See Sections 22.4.2 & 28.6.1&2.

(j) **Skid resistance and drainage of the pavement surface.**

See Sections 12.5.2&4 and 13.3.1. Section 12.5.2 shows that skid resistance is lower when a pavement surface is wet or covered with detritus and Section 12.5.4 shows that crash rates increase as skid resistance drops. Thus, skidding accidents predominate in wet weather or after a summer shower. Crashes can also result when vehicles hydroplane on poorly drained or badly rutted pavements (Section 27.7.2).

(k) **Roughness** (Section 14.4).

Roughness as measured by IRI is discussed in Sections 14.4.4–5. It has been suggested that changing pavement roughness from poor (IRI 5) to good (IRI 1.5) will reduce crash rates by at least 10 percent. This is a net effect as a larger decrease in wet-weather crashes will be counterbalanced by the increase in dry-weather speeds raising crash numbers.

(l) **Traffic management.**

See Sections 7.2.3 & 31.6.

(m) **Speed control.**

See Section 18.1.1. Crash frequency depends on $(\text{speed})^{2.5}$ and so control of traffic speed can be an effective safety measure.

(n) **Traffic guidance, warning and control devices** (Chapters 21, 22, 23 and 24).

These devices must be visible to approaching drivers at day and at night. Night driving issues are discussed in Section 24.4.2.

(o) **Lighting and delineation.**

Crash reductions due to lighting are described in Section 24.1. Upgrading delineation from poor to excellent can reduce crash rates by about 30 percent. Rumble strips to alert drivers that they are running off the road can reduce such crashes by 20 percent (Section 22.1.3).

Discussion of the safety benefits arising from many of the facilities listed above will be found mainly in the chapters and sections quoted. For new works, the above measures should be considered during the preliminary and detailed design stages (Sections 6.2.2–3) and also in a post-construction/pre-opening safety review (or audit) of the job.

For roads that are already in service, the application of the above measures to a black-spot site (Section 28.3.1) where crashes have been occurring is called a *black-spot program*. Typically, black-spot programs reduce crashes at a site by about 30 percent.

A black spot implies that the road system is of variable safety, at least partly as a result of the actions of the Road Authority.

Common law is law developed as a result of accumulated court decisions, rather than by legislative action. Common law actions against a Road Authority for *negligence* and/or to establish legal *liability* can arise with respect to the alleged contribution to a crash of a piece of road with less than normal safety. To succeed in this case, the claimant must usually prove that:

(a) damages were sustained;
(b) the Road Authority owed a duty of care to the claimant;
(c) its conduct was negligent, falling short of that duty of care; and
(d) this short-fall was the cause of the damages sustained.

Points (a) and (d) are technical issues. Point (b) and the associated role of *misfeasance* are discussed in Sections 18.1.4 & 28.6.1. Point (c) should be judged against local:

* funding levels for roadworks, as limited resources could diminish liability for reducing known dangers, and
* standards of conduct, which set an accepted level of reasonable care, skill and diligence.

28.6 THE ROADSIDE

28.6.1 Roadside objects

Roadside objects found in or adjacent to the verge can be major safety hazards, contributing to the overturning or abrupt deceleration of errant vehicles. About 5 percent of all crashes, 20 percent of all serious injury crashes, and 30 percent of all fatal crashes arise from collisions with fixed objects. The fatality rate for these crashes is twice the average of all crash types and three-quarters of the roadside fatalities result from head injuries (Lozzi, 1981). The most severe type of crash occurs when the object is struck laterally by a (slewing) car. In this case, the fatality rate per crash is seven times the overall average, with the passenger being particularly vulnerable. Rural fixed-object crashes are 70 percent more severe than urban ones.

The crash rate for run-off crashes per vehicle-kilometre of travel decreases as the traffic flow increases. Depending on the density of pre-existing hazards, their removal can reduce crash rates by up to 50 percent. Examples of hazardous roadside objects are:

* bridge abutments and parapets (Section 28.5h),
* roadside furniture (e.g. Chapters 21 and 23),
* *utility poles* are the major hazard in urban areas. A detailed examination of crashes with such objects led to the following conclusions:
  (a) poles do not significantly protect pedestrians or property from errant vehicles, despite nonsensical claims to the contrary made by some utilities;
  (b) improving skid resistance is often an effective countermeasure; and
  (c) poles near the kerb are over three times more likely to be involved in a crash than are those more than 3 m away. About two-thirds of all such crashes occur within 6 m of the carriageway.

  One problem is that Authorities often locate poles with no road safety consideration.
* poorly-designed guardrail (Section 26.2.2),

* trees with a trunk diameter of over 100 mm (Section 6.4) are the major problem in rural areas
* boundary fences,
* boulders,
* steep roadside slopes (Section 28.5f),
* abrupt drops of over 1 m,
* drainage ditches (sloping drainage openings larger than 750 mm can be covered by a grating; smaller openings can match the slope of the surrounding ground, the slope on the approach to a ditch can be 1 in 5 or flatter), and
* the ends of drains and culverts (unshielded culvert ends can match the slope of the surrounding ground; culvert end-walls can protrude less than 100 mm above the ground slope).

Hazardous locations for roadside objects are:

* the outside of bends that are likely to be taken at high speed or with high traffic flows,
* the inside of the exit end of a sharp bend where the driver's recovery manoeuvre may cause over-correction,
* the closed side of a T intersection,
* intersection corners, and
* where centre-turners may force through traffic in the same carriageway to leave the carriageway.

Clearly, new roadside features should be carefully designed and positioned to avoid creating new hazards. Indeed, it is usually a simple matter to avoid the creation of, or to reduce the hazardous nature of, roadside hazards. Nevertheless, despite the prospect of a charge of misfeasance as a result of negligent action (Section 18.1.4), some Authorities have been slow to remove known roadside hazards, hiding behind an assumed immunity from liability for their creation and continuance (Section 28.5). This attitude has not aided the provision of safe roadsides.

U.S. observations in the early 1970s indicated that about 80 percent of high-speed (> 90 km/h) vehicles which ran off the carriageway came under control or to rest within 10 m laterally. An idealised clear zone where roadside obstacles are undesirable was defined as extending at least the following typical distances from the edge of the carriageway:

* 10 m for high-speed (> 90 km/h) roads,
* 6 m for moderate-speed roads, and
* 3 m for low (< 50 km/h) roads in urban areas.

If the aim was to allow the vehicle to also recover and continue on its original course, these distances were increased by a further 2 m. The introduction of a clear zone will typically halve the number of single-vehicle run-off crashes. For similar reasons, medians at least 10 m wide have been recommended on freeways.

However, it is difficult to sustain the above 'safe' width figures from first principles. For instance, Table 21.1 gives the stopping distance of vehicles travelling at 130 km/h as about 350 m. If such a vehicle deviated from the road at even a modest 15° it would have travelled about 90 m laterally before stopping. Applying the equations in Section 27.2.2 to a divided highway, it can be shown that when an errant vehicle has travelled laterally the 15 m (median plus inner shoulders) distance needed to bring it into contact with oncoming vehicles, it will still be travelling at 120 km/h. Thus, it is not surprising that more recent Californian data suggests that the clear zone for no effect is closer to 25 m than 10 m (Corben *et al.*, 2003). Furthermore, it has recently been argued (McLean, 2002) that the data on which the 10 (or 9) m was based can be reinterpreted to show that

85 percent of the benefits from a clear zone occur within the first 6 m. Hence, trade-offs are possible in the 6 m to 25 m zone.

There are actually two issues. One concerns encountering roadside objects within the roadside and the other concerns striking oncoming traffic on the other side of the median. With respect to the first issue, it is often not practical to provide sufficient clearance to roadside hazards. An effective solution for both issues is to install re-directing traffic barriers, as discussed in Section 28.6.2. The safety advantages of shrubs in medians are discussed in Section 6.4.

A second approach to the first issue is to only use objects in the clear zone that will bend or break on impact, rather than rigidly resisting the impacting vehicle. The effect of a collision is therefore dramatically reduced. When applying this approach to roadside poles, which do not support overhead power-lines, there are two main methods:

* A weak zone is introduced at the base of the pole to produce a *break-away* pole. For timber poles, the weakening is often achieved by sawing one or two vertical slots between pairs of drilled holes — if two slots are used they are placed in orthogonal vertical planes. For metal poles, break-away is commonly achieved by using a base that can slip horizontally. The slip base is usually about 75 mm above ground level. These poles are not suitable in areas with high pedestrian or parking activity.
* *Frangible* (i.e. impact-absorbent) poles are used. These tend to harmlessly wrap themselves around any vehicle that strikes them. This method is usually used when pedestrian activity is high or where the poles are very close to moving traffic.

If the poles support overhead power-lines, break-away poles are used but a second weak zone is introduced below the cross-heads supporting the power lines. This ensures that the power lines remain intact after breakaway as, during impact, the pole between the weakened zones 'pops-out', leaving the conductors spanning between adjacent poles and supporting the popped out cross-head.

The second issue of head-on traffic requires wider medians (as discussed above) or continuous barriers within the median (see Section 28.6.2 following).

28.6.2 Traffic barriers

Traffic barriers are put in place to protect:

(a) pedestrians and road workers from being struck by errant vehicles.

(b) vehicles from contact with oncoming errant vehicles.
On narrow medians, barriers are located in the median to prevent head-on collisions when the traffic flow at least exceeds 5000 AADT (Section 22.4.2). These are sometimes also called median barriers.

(c) vehicles from contact with any of the various fixed (i.e. not frangible or able to breakaway) roadside objects listed in Section 28.6.1. Barriers are located:

(c1) near roadside hazards (Section 28.6.1) and within the clear zone defined in Section 28.6.1. Observation suggests that most drivers will have regained control of an errant vehicle after 120 m of unimpeded travel. From Table 21.1, this is close to design stopping distances at 100 km/h. Hence, barriers commonly do not begin until about 120 m before a hazard.

(c2) before the edge of embankments, where the slope or drop exceeds the values in Sections 28.5f & 28.6.1.

(c3) where the road pavement narrows abruptly, as at some bridges and culverts (Section 15.2.5).

(c4) on the outside of substandard curves (Section 19.2.6) using, as a guide, a speed deficiency of:

[85th percentile speed] – [advisory speed] > 15 km/h

In cases (b) and (c), barriers are used when economic or practical constraints prohibit the provision of a very wide, clear median.

Barriers do not prevent crashes, but diminish their effects. For example, fatal crashes can be reduced by up to 75 percent with a benefit/cost ratio of about 6. No matter how well designed, a barrier is still a fixed object and is itself a hazard. Thus, it should only be used when the benefits of its protection outweigh both the costs and the risks of any installation. The benefits are predicted from the product of the cost of a crash times its probability of occurrence. This means that any warrant for barrier use will depend on traffic flows as well as on roadside condition. The costs will include both installation and maintenance costs under both routine and post-crash situations. Barriers are also chosen on the basis of their appearance and ease of maintenance. There are four main design considerations:

* the barrier type,
* the barrier design,
* the transverse placement of the barrier, and
* the longitudinal placement of the barrier.

There are two main types of barriers:

1. *Crash attenuating systems* ahead of rigid objects. These are cushions and other energy-absorbing (or -attenuating) devices that absorb energy via dampeners, the deformation of metal, rigid foam, and sand- or water-filled barrels,
2. *Longitudinal barriers* beside a carriageway. A key component of any longitudinal barrier is its terminations and transitions with other barriers.

The performance of longitudinal barrier systems is measured by their ability during impact to:

* maintain their structural integrity — the errant vehicle must not break through, penetrate, vault over, or wedge under the barrier;
* cause vehicle decelerations which are within human tolerance and which minimise vehicle damage; and
* redirect the vehicle with an acceptable post-impact trajectory, particularly a small exit angle and a small rebound distance, so that the risk to other traffic is minimised.

The main vehicle variables that determine the effect of a collision with a barrier are the vehicle's type, speed and impact angle. Barriers operate most successfully when redirecting vehicles hitting with a glancing impact back into the traffic rather than when either preventing right-angle impact or when required to completely arrest a vehicle. Indeed, typical barriers are only designed to accommodate glancing impacts at 15° or less. About 6 percent of barrier crashes lead to fatalities. Most barriers (and guardrails) have been developed for four-wheeled vehicles of under 2 t, travelling at 110 km/h or less, and with an impact angle of 25° or less. In this situation, the critical design parameters are:

* post spacing,
* post strength,
* guardrail or cable height, and

* guardrail or cable coefficient of friction;

and, to a lesser extent:

* guardrail or cable tension, and
* soil stiffness.

The most demanding situation is usually to restrain a fully laden articulated truck, with a mass closer to 40 t rather than 2 t, and a high effective centre of that mass.

The types of longitudinal traffic barrier in use are:

2.1 *Kerbs*. Section 7.4.3 shows that kerbs under 150 m high will have little value as a barrier.

2.2 *Guardrails and other non-rigid barriers* (such as cable systems). Guardrail is usually comprised of steel W- or box-sections mounted on timber or steel posts, often offset from the post by a spacer to prevent impacting vehicles from snagging on the post. Nevertheless, such posts can still be a hazard to motorcyclists.

2.3 *Rigid barriers*. These are constructed of concrete and have negligible impact deflection, with any energy dissipation taking place in the vehicle and by rubbing along the barrier. The barrier cross-section is shaped as in Figure 28.3 to allow a vehicle in a glancing impact to only slightly climb the barrier face before running along the barrier and then, possibly, safely re-enter the traffic stream. Rigid barriers usually require less transverse space than guardrails. The three common types of rigid barrier are:

2.3.1 *New Jersey barriers*. In the past, rigid barriers have typically been *New Jersey* (or *concrete safety-shaped*) barriers (Figure 28.3a). However field experience is that the New Jersey barrier may be a little too flat, permitting smaller vehicles to roll over, and a little too low for trucks with a high centre of mass (Section 19.2.2). Most national standards (e.g. AS/NZS 3845 with its Type F barriers) now specify alternative or modified New Jersey barriers.

2.3.2 *Vertical concrete barriers*. These typically have a 50 by 180 mm kerb at the base of a vertical face. This nib avoids rolling impacting vehicles in the way possible with the original New Jersey shape. The vertical shape also tends to capture vehicles better, encouraging them to run parallel to the barrier.

2.3.3 *STEP system*. The alternative STEP system introduced by the Dutch and approved in the U.K. and Belgium is shown in Figure 28.3b. Its height is sufficient to eliminate most glare from the lights of oncoming traffic. However, it is a compromise between the demands of truck centres of mass and the need to prevent drivers feeling closed in by being above normal driver eye-height of at least 1.05 m (Section 19.4.1) (van der Drift, 1998).

2.3.4 The next step is to use barriers with a vertical face, perhaps faced with a small 50 by 150 mm kerb. These have proved effective in high speed situations, although the decelerations immediately on impact may be greater. Due to the tendency of vehicles to roll over such barriers if they are under 1.2 m high, any poles should be set back at least 300 mm behind the barrier.

2.4 *Wire-rope safety fences*. The posts are offset back from the cables which present a continuous face to the traffic. In a typical design, after impact the cables stay intact and a small number of posts near the impact fail (e.g. Schmidt, 2004). Vehicles are 'caught' in the loop of the cables and energy is absorbed by rubbing along the cables.

2.5 *Bridge barriers (or railings) and parapets* (Section 15.2.5).

2.6 *Temporary devices*. These can be portable inter-locking water-filled barriers for use at worksites. They are commonly made from impact-resistant polyethylene and measure about 2000 (long) x 900 (high) x 600 (wide) mm. These barriers usually provide a safe end-treatment without the need for special provisions and are also energy-absorbing. They are far superior to the use of cones and barricades. An alternative solution is to use pre-cast concrete barriers.

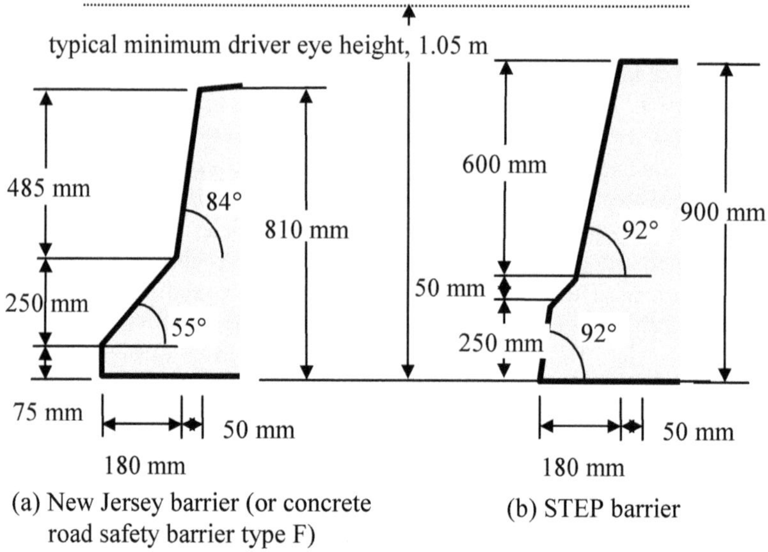

Figure 28.3 Typical profile of New Jersey and Dutch STEP traffic barriers. Precise details are given in local standards, e.g. AS/NZS 3845 in Australia and New Zealand.

28.6.3 Barrier terminals

Barriers must also present safe ends (or *terminals*) to errant vehicles. Poorly designed terminals, particularly when located in the clear zone (Section 28.6.1), can be very hazardous with the potential to overturn or impale vehicles. In addition, the placement of kerbs in front of a barrier can reduce its effectiveness by causing the vehicle to bounce as it impacts the barrier. A guardrail terminal must meet three main criteria:

* for non-rigid barriers, it must anchor the guardrail so that it can develop the tensile strength needed to catch and redirect vehicles,
* it must direct vehicles away from hazardous fixed objects, and
* it must not act like a spear, ram, or overturning device during any possible end-impact. An impacting vehicle must be able to vault the barrier.

A favoured type of end treatment for W-section guardrail is the *break-away cable-terminal* in which the tensile forces are resisted by a cable anchored to the bottom of the last post. The nose of the guardrail terminal consists of a steel plate eccentric to the line of the guardrail and bent into a 300 mm radius return piece. The first and second posts are weakened to fail on impact, preventing appreciable compressive forces from developing. The large nose increases the vehicle contact area. Over about the last 10 m, the terminal sectors are also gradually flared back by about 1.2 m. On severe impact, a vehicle will pass through the break-away barrier and so there should be a clear flat area behind the break-away barrier, extending 25 m along the barrier and 6 m behind the barrier.

Even with well-designed terminals, sections of barrier less than 30 m long would often present more hazards than would justify their use. Considerable attention must also be given to the transition between one guardrail type and another.

28.7 PEDESTRIANS

28.7.1 Pedestrian behaviour

The behaviour of pedestrians in traffic is described in Section 20.4 and as transport users in Section 30.5.3. Pedestrians (and cyclists, Section 27.6) are the most physically vulnerable of all road users and so, not surprisingly, they constitute over half the road fatalities in many densely-populated urban areas. Crash severity (Section 28.1.4) depends mainly on vehicle speed — most pedestrians struck by vehicles travelling at over 70 km/h will die; most struck by cars travelling below 30 km/h will live. Thus rural pedestrian crashes have a high fatality rate. Only people with gunshot wounds have a higher mortality rate than pedestrians injured in car crashes, and only motorcyclists have a longer stay in hospital. Adult pedestrians usually predominate in night-time and arterial road crashes and child pedestrians predominate in crashes in local streets. Only 20 percent of pedestrian fatalities occur at intersections — many occur near but not at pedestrian crossing facilities (see 1.1 below). Children and elderly pedestrians are usually over-represented in pedestrian–vehicle crashes and are the *prima facie* cause of the majority of them. The behaviour of child pedestrians is discussed further in OECD (1983a). The causes of pedestrian crashes can be listed under a number of headings.

1. All pedestrians

1.1 a reluctance to use the traffic-control devices provided;
1.2 ill-considered road-crossing manoeuvres;
1.3 over-optimistic estimates of a driver's ability to see and avoid them: this problem is markedly reduced if pedestrians wear conspicuous, reflective clothing at night.

2. Adult pedestrians

2.1 intoxication (in Australia 40 percent of adults involved in pedestrian crashes had been drinking prior to the crash).

3. Children and elderly pedestrians

3.1 frequent road crossings, leading to high exposure;
3.2 random and unpredictable road-crossings;
3.3 poor judgement of traffic behaviour;
3.4 uncertainty about the operation of traffic-control devices.

4. Children

(Note: Children are not merely 'little adults', but have quite different behaviour patterns. The group at greatest risk is usually 8- to 12-year-old urban males. Road crashes are usually the greatest killer of children in the 0 to 16 year age-group).

4.1 child-like, rather than adult behaviour;
4.2 sudden emergence from behind or between stationary vehicles and into a driver's line of travel;
4.3 inability to see or be seen over parked cars;
4.4 under-developed visual and hearing capabilities until 16 years (Section 16.4.1);
4.5 poor judgement of speed;
4.6 inability to use peripheral information as warning signals; and

4.7 drivers often lack height cues and so may misjudge a child's height and assume adult behaviour.

5. Elderly pedestrians (Section 16.5.5)

5.1 poor response times (Section 16.5.3);

5.2 poor vision (Section 16.4.2);

5.3 slow walking-speed (Section 30.5.3).

6. Drivers

6.1 human error (see Chapter 16 and Section 28.2.1);

6.2 a tendency to leave most of the crash-avoidance manoeuvres to the pedestrian.

These factors are not listed to argue that pedestrians should be chastised or ignored, but rather to indicate that the traffic engineer has a duty to protect pedestrians, particularly children and the elderly, from their own inevitable inadequacies, misadventures, misjudgements, and errors on the road.

28.7.2 Pedestrian countermeasures

Traffic engineering measures to reduce pedestrian crashes include:

(a) good footpaths;

(b) pedestrian malls (Section 7.2.1);

(c) overbridges and underpasses (Section 15.4.5);

(d) fences to channel pedestrians to marked crossways (aimed particularly at factor 1.1 in Section 28.7.1);

(e) special pedestrian facilities at traffic signals (Sections 20.4 and 23.1);

(f) median strips, raised islands, and safety zones to provide pedestrian refuges from traffic and hence reduce the crossing task (Sections 20.4 & 22.4.2);

(g) supervised crossings for children (Section 20.4);

(h) parking prohibitions near pedestrian crossings (aimed particularly at factors 4.1–3 in Section 28.7.1, Section 20.4);

(i) special safety provisions at bus, taxi, and train stops;

(j) good peripheral lighting (Section 24.3);

Factors to consider when deciding which measures to implement are the:

* number of pedestrians,
* type of pedestrians (e.g. children, elderly, intoxicants),
* times when the crossing is used (e.g. night or day),
* number of vehicles passing,
* speed of the vehicles passing,
* road geometry, particularly sight distances, and
* extent of car parking and the associated driver behaviour.

Animals on the road can constitute a major traffic hazard, particularly at night. The problem can be reduced by using post-mounted reflectors (Section 24.4.3) which shine the car-headlight beam into the roadside and hence deter animals from crossing the road.

28.8 THE COST OF ROAD CRASHES

28.8.1 Cost components

Estimates of the cost of crashes are needed in order to:

(a) indicate the magnitude of the crash problem to policymakers (it has been estimated that crashes can consume about 2 percent of a nation's GDP, Himanen *et al.*, 1992), and

(b) conduct benefit–cost analyses to prioritise planned highway improvements or safety measures (Section 5.2.3).

The costable consequences of a road crash fall into one or more of the following categories. The italicised words are comments on the cost implications of the item.

1. Property damage — *readily costable*:
    1.1 vehicle damage, towing and temporary replacement,
    1.2 public property damage,
    1.3 private property damage.
2. Personal injury — *estimated from relevant and recent court and public-sector decisions*:
    2.1 loss of salary or other output equivalent – see below,
    2.2 medical (ambulance, physician, hospital, funeral, etc.),
    2.3 compensation for disablement other than in #2.1,
    2.4 rehabilitation and welfare,
    2.5 pain and suffering,
    2.6 loss of life — see Section 28.8.2.
3. Effects on close associates (indirect costs) — *estimated from relevant and recent court and public-sector decisions*:
    3.1 bereavement, suffering and anxiety — see Section 28.8.2,
    3.2 loss of financial support — see #2.1 and below.
4. Public costs (in addition to #1.2) — *often difficult to trace*:
    4.1 police,
    4.2 legal,
    4.3 insurance,
    4.4 medical support and subsidies — see #2.2,
    4.5 traffic delays caused by crash,
    4.6 emergency services,
    4.7 reduced output and consumption — see #2.1 and below,
    4.8 reduction in the probability of a large simultaneous or public loss of life. Communities are usually prepared to pay a disproportionately large price for avoiding the simultaneous loss of many lives. For example, compare the sums spent per life at risk on air safety relative to road safety, or compare concern over a single child lost in the countryside relative to the lesser concern for any child pedestrian.

Items 2.1 and 4.7 are not only significant, particularly in fatal crashes, but also subject to some controversy. In economic terms, lost salary is best seen as lost consumption by the incapacitated person. This can be estimated by dividing the region's gross consumption (e.g. the Gross Domestic Expenditure, GDE) over the period of incapacitation by the 'working' population. To particularise in any more detail creates problems. For example, if the victim were not a member of the working population it would be necessary to use either shadow pricing, opportunity costs, or weights applied according to community values in order to estimate the victim's expected consumption. The annual amounts can be converted to a net present value (Section 33.5.7) of the victim's loss as a result of the crash.

The calculations are done in terms of gross rather than net consumption, based on the argument that a person's reason for existence includes the consumption of goods

(Atkins, 1982). In the past, net rather than gross consumption was used and gave unrealistically low values. The gross consumption figure may need to be adjusted to include the victim's:

* special quality-of-life issues,
* expected remaining working-life, if the incapacitation is severe,
* atypical gross consumption,
* dependents' drop in the consumption caused by the loss of the victim's income, and
* the contribution of the victim's associates to the victim's care.

28.8.2 Value of human life

Not surprisingly, the economic value of human life is difficult to estimate. A minimum value could be obtained from the technique given for Items 2.1 & 4.7 in Section 28.8.1, estimating the cost of a person's incapacity and summing this over the rest of the person's life. Techniques of this nature calculate the cost of a life on the basis of the consequences of the fatality and the tangible costs subsequently occurred. For these reasons they are called *ex post* (or *human capital*) techniques. They are particularly appropriate for non-fatal crash costing where Item 3.1 does not dominate but will produce minimum cost-estimates for fatal crashes.

A maximum value of life would be the infinite value assigned by many moral and emotional arguments. In addition, communities usually place higher than average values on the lives of children and the elderly.

Between these two extremes, it can also be observed that communities are usually prepared to accept that there is a limit on the amount of public money spent protecting citizens — particularly risk-taking citizens — from low-risk hazards.

The numerate way around the problem of valuing life is to follow this lead and base costs on what the community would be prepared to pay to avoid a crash, i.e. to reduce the crash risk such that the number of crashes is reduced by one. This marginal before-the-event costing is known as *ex ante* (or *willingness-to-pay*) costing, and is based on welfare economics (Section 33.5.1). It is relevant for policy-making estimates and is more publicly defensible than *ex post* methods as a means of valuing human life. It can be gauged, for example, by retrospectively examining how much the community has been prepared to spend on safety measures expected to reduce fatal crashes by a certain amount.

Typically, the *ex ante* value of a life might be three times the *ex post* cost, depending on the size of the 'pain and suffering' provision (Item 2.5) in the *ex post* costing. The 'three times' value suggests that Item 2.1 alone is about double the more tangible costs. *Ex post* and *ex ante* costing are not opposing views but are at either end of the spectrum of possible valuations of human life. *Ex post* can be thought of as giving the cost of a human life, whereas *ex ante* gives the value of that life.

28.8.3 Cost comparisons

Table 28.9 shows typical contributions of the major costable components in the list in Section 28.8.1 to the *ex post* costing of a crash. Note the importance of gross consumption (Item 2.1).

Table 28.9 Contributions of crash categories as a percentage of total crash costs (after Atkins, 1981).

Cost category (Section 28.8.1)	Crash type			
	Fatal	Injury	Property damage	All
1.1 Vehicle damage	5	5	20	30
2.1 Loss of salary	20	5	0	25
2.2&3 Injury	5	10	0	15
4.1-3 Insurance & legal	2	3	10	15
4.5 Traffic delays	10	5	0	15
Range	20–45*	20–30	25–45*	
All categories	40	30	30	100

* These limits apply when all unreported property-damage crashes are included.

The average proportional cost of individual crash types is shown below, with the cost of a property-damage-only crash taken as unity.

Crash category	Average cost
Property-damage-only	1
Only minor injury	4
Serious but non-fatal injury	30
Fatal crashes	150

Despite the skew towards fatal crashes in this data, the distribution of crash numbers is very skewed towards the other (property damage) end of this spectrum. This produces the result shown in Table 28.9 where each of the major crash categories — fatal, injury and property damage — contributes about an equal amount to total crash costs. However, as illustrated in Table 28.10 the costs are far from uniform across a particular crash type.

Table 28.10 Crash type contributions.

Crash component cost as a percentage of total crash cost		
Crash type	Property damage	Personal injury
Vehicle–vehicle	70	30
Vehicle–fixed object	70	30
Vehicle alone (e.g. roll over)	40	60
Vehicle–pedestrian	5	95

The cost of crashes varies between the different crash categories, as illustrated in Table 28.11.

Table 28.11 Relative cost of a particular crash type, with a head-on crash taken as of unit cost (Andreassen, 1992 and others).

Crash type	Relative cost	
	Urban	Rural
Vehicle–train	1.3	1.6
Vehicle–pedestrian	1.1	0.7
Vehicle–vehicle, head-on	1.0	1.0
Vehicle leaves road on curve, hits object	0.8	0.7
Vehicle leaves road on straight, hits object	0.6	0.6
Vehicle–vehicle, opposing directions, turning	0.6	0.5
Vehicle leaves road on curve	0.5	0.5
Vehicle–vehicle, adjacent intersection approaches	0.5	0.4
Vehicle–fixed object	0.5	0.3
Vehicle–cycle	0.4	0.4
Vehicle–vehicle, angle	0.3	0.3
Vehicle–vehicle, rear	0.3	0.3
Vehicle–vehicle, lane change	0.2	0.3

CHAPTER TWENTY-NINE

Road User Costs and Charges

29.1 ROAD COSTS

29.1.1 The cost of the road system

An investment in a road is intended to provide a facility with a desired level of operating characteristics and a desired useful life. Commonly, operating characteristics dominate decision-making for urban roads and road durability dominates rural road decisions.

The total cost of a road system can be divided into the five major categories A to E listed below and in Section 29.1.2. In principle and with the exception of the costs in category C, their assessment should require no more than a set of accurate and detailed accounting records. However, such costs are frequently unavailable.

A. Capital cost of the infrastructure

A1. This covers construction expenditure needed to create a new road (Section 25.2) which will operate over a period of time (its useful life). It is sometimes called the *sunk cost*, as these costs have now sunk beyond recovery. Once such costs have sunk from sight, attention moves to the costs defined in A2–5.

A2. Reconstruction expenditure (Section 26.1) is expenditure to extend an existing road's useful life. It would also include Type 3 congestion costs (Section 29.4.5) which relate to adding capacity to a road.

A1–2. Because road building is such a 'lumpy' investment over time, in the *public sector* costs A1 and A2 are sometimes converted to equivalent annual costs, e.g. to provide a basis for usage pricing (Section 29.4). This can be done by using interest charges to amortise the capital costs (Section 33.5.6). The interest rates can be established by at least hypothetically treating the construction cost as a loan or as the opportunity cost (Section 35.5.2) of the capital used.

In the *private sector* the annual cost of capital is the interest paid or return required on the loan of the sum of the construction costs and any interest charges during construction, less any capital repayments made out of operating profits.

A3. Periodic expenditure related to any leased, rented or hired land or other facilities.

A4. An 'expenditure' which is the negative of the residual value of the road at the end of its useful life (Section 33.5.8).

A5. Depreciation expenditure, which represents the write-down in the value of the road asset (A1, A2 + A4) over time.

Funds accumulated from prices based on such capital costs need not be re-invested in roads. If the expenditure is to replace a road asset, it may first be appropriate to subject the planned replacement to a new road needs analysis (Section 5.2). If the funds are to be reinvested in a road, the amount will need to be based on the *replacement* expenditure which is the amount required to provide a future new road, when the current road has reached the end of its useful life. Commonly, such work will also enhance the level of service of the road, by adding to its traffic and/or load capacity. Hence, it will often exceed A1 + A4. See also the discussion of long-run marginal costs in Section 29.4.1.

B. Recurrent internal cost of operating the infrastructure
An *internal cost* is one borne by the users of the system. These system operating costs are typically only about 4 percent of the vehicle operating costs discussed in Section 29.1.2 below. Internal costs associated with operating the road infrastructure are:

B1. Expenditure to operate the traffic management (Chapters 7 & 17) system and maintain the roadside (including the replacement of short-life items, Chapter 6).
B2. Maintenance expenditure to minimise the deterioration of the road — particularly the pavement (Section 26.1).
B3. Rehabilitation expenditure to restore the road — particularly the pavement — to its original level of service.
B4. Profit based on the skills applied, risks taken and otherwise unrewarded effort devoted to the project.

The important distinction between maintenance and rehabilitation is explained in Section 26.1.

C. Recurrent social costs caused by the operation of the system
Social costs are costs caused by the operation of the transport system and imposed across the community. Most are recurrent costs. They become external costs when they do not involve expenditures by the road builder or road user. Social costs can be based on the damage caused, the cost of prevention or on willingness to pay (Section 33.5.3). In terms of Sections 29.4.1 and 33.5.2, the costs in this category might be either short-run or long-run marginal costs, depending on the circumstances and needs. Social costs include:

C1. Environmental costs, e.g. those due to noise and air pollution (Sections 32.2–3). These are difficult to estimate but will be of the same order as the vehicle fuel costs (Section 29.2) and are estimated in some detail in Tsolakis and Houghton, 2003.
C2. Type 2 congestion costs (Section 29.4.5).
C3. Some crash costs (Section 28.8) — see also E5.
C4. Unaccounted for social costs (for issues, see Section 32.1).

29.1.2 The cost of operating the vehicles

As indicated in Section 5.1, vehicle operating costs are typically about ten times the infrastructure costs in Categories A and B in Section 29.1.1. They are sometimes called *private, running,* or *road user costs*. They fall into two categories:

D. Time-dependent and distance-independent (or fixed) vehicle operating costs

D1. Drivers' wages usually are the single biggest cost component, typically at about 40 percent of total costs, and increase linearly with time. The actual costs of wages and personal time are discussed in Section 31.2.3 and include Type 1 congestion costs (Section 29.4.5). For private travel, these costs in some circumstances can be taken as zero, however studies of driver travel behaviour suggest that drivers do place a value on their private travel time (see Section 31.2.3).
D2. Vehicle depreciation costs refer to the part of the capital value of a vehicle that is used up in the course of its operation and by obsolescence. The capital value may be

indicated by the vehicle purchase price. Depreciation costs are rarely clear cut and can be estimated in one of three ways:

* assuming that value is lost linearly over time,
* assuming that value is lost at a constant proportion, i.e. that it decays with the negative exponential of time, or
* by using market values.

Depreciation cost is rarely used for private travel estimates, as it does not influence most drivers' use of their vehicles.

Note that all expenditure on vehicle purchase effectively ends up in purchases of new vehicles, as other transactions on second-hand vehicles are transfer payments within the car market (Mogridge, 1983).

D3. Vehicle insurance.

D4. Annual vehicle taxes (Section 29.3.1&2). Their correct handling depends on the purpose of the analysis. They are direct costs for the private user and a legitimate influence on vehicle depreciation, but they are not relevant in governmental project evaluation where social benefits and costs (Section 31.1) and net gains to society are under consideration (e.g. Section 5.2). Resource costs are total costs less any taxation costs (Section 29.4.2).

D5. Overheads, such as fleet management costs. They are not applicable to assessing costs for private cars but are charged for commercial vehicles.

D6. Parking and garaging costs.

D7. Road pricing (Section 29.4.6) based on time in the system.

E. Usage-dependent (or variable or direct running) vehicle operating costs.

These will all include Type 1 congestion costs (Section 29.4.5).

E1. Fuel costs (Section 29.2). These costs will include a fuel tax, which in many jurisdictions is much larger than the tax-free fuel price. This situation arises partly as governments attempt to discourage excessive fuel use, but mainly because fuel tax has proved to be a convenient and simple way of raising large amounts of taxation revenue.

E2. Lubricating oil costs.

E3. Tyre costs, which are usually related to tread wear.

E4. Maintenance and repair costs. These increase linearly with distance travelled and with road roughness (see below).

E5. Crash costs (Section 28.8).

E6. Usage-dependent depreciation, tax, insurance and overheads. These costs replace the time variable in D2–D5 with a usage variable, such as distance travelled. This is usually a more valid approach for commercial vehicles. However, it is also a time-dependent cost as a reduction in trip time for a constant freight task would reduce its size.

E7. Road pricing (Section 29.4.6) based on distance travelled or points passed.

D + E. Total vehicle operating costs.

Total vehicle operating costs can be estimated as the sum of the individual costs in categories D and E. Alternatively, PIARC's HDM-4 (Section 26.5.2) and the U.K. *COBA* (Section 33.5.11) models are often used to predict these costs. The increase in total vehicle operating costs due to a change in road roughness (Section 14.4) is usually central to these estimates. Total vehicle operating costs increase with roughness (IRI, Section 14.4.2) for IRI values over 3, as illustrated by:

percent increase in total vehicle operating costs = $3.125([5IRI/16] - 1)^2$ (29.1)

In summarising extensive Brazilian data, Paterson (1985) suggested that total vehicle operating costs rise by 10 percent as the condition of a paved road degrades from good to fair and a further 20 percent as it degrades from fair to poor. Unpaved roads are equivalent to the 'poor' category. This is about a 2.5 percent change for each unit increase in IRI (other estimates put the percentage as high as 5), implying that total vehicle-operating-costs increase exponentially with roughness. This increase in operating cost is far more than the road-maintenance cost (Chapter 26) needed to reduce roughness and restore the pavement to a good condition (Table 14.2).

Automobile clubs regularly issue data on the total vehicle costs for cars. Drivers usually perceive their costs to be only a fraction of these total costs.

The following data illustrates the total vehicle operating costs for a six-axle articulated owner–driver truck in 1990–1 (Thoreson, 1993 and Section 27.3.1):

Type of operating cost	Category	Percentage of cost
Dependent on time and independent of distance	D2. Depreciation (etc)	29
	D3&4 Insurance and tax	8
	D Total fixed costs	**37**
Dependent on usage and distance	E1&2. Fuel, oil	41
	E4. Maintenance, etc	14
	E3. Tyres	8
	E Total variable costs	**63**

These percentages have been confirmed by 1997–8 data (32:68). Typical vehicle operating costs from the World Bank's Brazil Study are given in Table 29.1.

Table 29.1 Typical commercial vehicle operating costs as a percentage of the total cost, from the Brazil Study (Geipot, 1982f).

Cost component	Paved roads	Unpaved roads
E1. Fuel	35	26
E4. Maintenance, oil	25	28
E3. Tyres	11	11
D2. Capital	15	17
D1. Salary	13	18
Total	100	100

Total operating costs for trucks per mass or per volume-carried drop as the mass or the volume, respectively, is increased.

The sum of all these costs, other than those in Category C, is sometimes called the road track-cost. Vehicle operating costs dominate all the other costs by a ratio of about 25:1.

29.1.3 Cost allocation to road user classes

When the costs in Sections 29.1.1&2 are known, the next task is often to allocate all costs amongst the various classes of road user. The first step is to see infrastructure costs (categories A to C) as composed of avoidable (or separable, attributable or variable) costs and unavoidable (or joint costs), viz.:

infrastructure costs =

$$\text{(Category A costs)} + \text{(Category B costs)} + \text{(some Category C costs)}$$
$$= \text{(joint costs)} + \sum_{\text{for all users}}\text{(avoidable costs)} \qquad (29.2)$$

Avoidable costs are those which can be allocated directly to a particular road user as a consequence of that user's use of the road, and are avoidable because they would not be incurred if the usage did not occur. These are comprised of the costs of:

* operating the traffic management facilities (Category B1 in Section 29.1.1),
* maintaining the road to stop it deteriorating further (Category B2), and
* compensating for the external effects of the traffic (Category C), such as the social costs due to congestion (see Types 1 & 2 congestion pricing, Section 29.4.5), noise (Section 30.3), or through traffic (Section 29.4.6).

For roads, typical avoidable costs are the extra pavement thickness needed to carry truck loads and the cost consequences of the damage caused by those loads. An avoidable cost will include some costs shared with other users and some unique to one user group. In terms of economic theory, avoidable costs are also marginal costs (Section 33.5.1).

Typically, avoidable costs are between 20 and 70 percent of total costs, with the percentage increasing as the traffic flow increases. For example, about 60 ± 20 percent of pavement costs can be related to vehicle usage in terms of tonnes carried or equivalent standard axles (Section 27.3.4). Typically, avoidable costs could be allocated amongst the various truck types in the following manner:

Comparison of truck avoidable costs, with a 2-axle rigid truck as unity		
Type of freight vehicle (see Figure 27.2)	Basis for distributing the costs	
	truck-km	tonne-km
2 axle rigid	1.0	1.0
3 axle rigid	1.4	1.0
4 axle rigid	1.9	1.1
3 axle articulated	2.1	1.3
4 axle articulated	2.7	1.3
5 axle articulated	3.2	1.3
6 axle articulated	1.0	1.5

Joint costs are those that would be incurred, even if no traffic used the road. They include the costs of such non-traffic items as drainage, fencing, and time-related degradation of the pavement. They are often difficult to assess, as in the case of the joint cost of a pavement. For example, the pavement joint cost is sometimes assumed to be the cost associated with a pavement built solely for cars. That is:

$$\text{total cost} = \text{(cost of a pavement built solely to carry cars)}$$
$$+ \text{(extra costs if the pavement is also to carry trucks)}$$
$$= \text{(joint cost)} + \text{(avoidable cost due to trucks)} \qquad (29.3)$$

Common cost is the component of a joint cost that would disappear if the traffic disappeared. For example, the common cost of a pavement would be the cost of the formation, minimum basecourse thickness, sealing coat and drainage, but not the cost of any basecourse above the minimum and required by traffic (e.g. Figure 11.13). The proportion of a common cost assigned to each vehicle class can usually be assessed: in the example, by using the N_{esa} (Section 27.3.4) contributed by each vehicle class. Drainage costs represent a typical remaining, non-common, component of joint cost. When this distinction is made, total cost is the sum of common joint cost, remaining joint costs avoidable. Thus, Equation 29.3 becomes:

total cost = (common joint cost) + (remaining joint cost)
+ (avoidable cost due to trucks) (29.4)

The distinction between these three costs is often difficult to make. Each of the examples above, for instance, could be attacked on theoretical grounds. This is a major complication in equity pricing (Sections 29.4.1) which requires using the costs associated with each road-user class as the basis for assigning a proportion of road-user costs to each road-user class. Nevertheless, a fairly common finding is that the charges levied on cars more than meet their avoidable costs, whereas those on large trucks do not.

Typical costs allocated to trucks are given in the following table using a number of different measures.

Typical percentages of joint and total costs allocated as charges to trucks, on the basis of the various pricing techniques in Section 29.4. The last column makes the common assumption that 40 percent of avoidable costs are due to trucks.

Allocation of costs based on:	Joint cost	Total cost
Market-will-bear (Section 29.4.3)	80	90
Benefits received (Section 29.4.4)	50	70
Usage in tonne-km	25	55
Usage in pcu-km (Section 17.4.2)	20	50
Usage in veh-km	15	50
Usage in veh (i.e. number of vehicles)	10	45

When the total costs are all allocated, by whatever means, the costs and any prices based on them are said to be *fully allocated.*

29.2 FUEL CONSUMPTION

29.2.1 Background

The cost of the fuel is a major component of vehicle operating costs (Categories D & F in Section 29.1.2). The fuel used by a vehicle is also a useful surrogate for its contribution to pollution levels associated with gaseous emissions and noise (Chapter 32) and to global warming. Fuel is consumed by a vehicle's engine to produce the power needed to overcome (Section 27.1.2):

* rolling resistance,
* air resistance,
* inertia, and
* accessories.

Factors of importance in determining the fuel consumption of a vehicle are those which influence these four resistances. The factors can be placed in six categories:

1. *vehicle-related*: mass, size, shape, axle ratio, accessory use.

   Mass is the most important of these factors, as it has an approximately linear link to both rolling resistance (Section 27.1.2) and air resistance (proportional to cross-sectional area), and hence provides a good estimate of relative fuel consumption. Commonly, fuel consumption increases by about 100 nL/m for every kilogram increase in mass from one vehicle to the next. Vehicle accessories such as loaded roof racks, air conditioning, power-assisted devices, heaters and lighting, can raise fuel consumption by 4 to 50 percent.

2. *engine-related*: power, capacity, condition (manufacturing tolerance and maintenance can both have a 5 percent influence), compression ratio, starting temperature, idling fuel rate (Section 29.2.4), lubricant performance (3 percent), transmission type.

   There is an approximately linear link between fuel consumption and engine capacity or number of cylinders (Tanner, 1984), but it is not as strong as the link to vehicle mass.
3. *transmission-related*: Except for very low speeds, automatic transmissions require more fuel than manuals — as shown later in Figure 29.2. Transmission losses account for about 27 percent of engine output for automatic transmissions and 15 percent for manuals.
4. *suspension-related*: tyre type, tyre pressure, wheel alignment (2 percent), brake condition (2 percent). Energy losses in tyres falls from about 80 percent of all losses at very low speeds to 20 percent at about 100 km/h (Section 27.7.1). The selection of tyre size and type can affect fuel consumption by 4 percent and tyre pressure can affect it by 2 percent.
5. *environment-related*: wind speed, air temperature, air pressure, and humidity. For example, fuel consumption changes by 1 percent for each 5 C change in ambient temperature.
6. *traffic-related*: A major determinant of fuel consumption is the speed-time behaviour of the vehicle in traffic. In an early study, Pelensky (1970) determined fuel consumption rates by direct measurement in urban conditions. He found, for instance, that 1 km of travel at 50 km/h was equivalent to 3 minutes of additional running time, 4 additional stops, 2.5 minutes of stopped time, and 4 sharp turns. The work underlined the importance of traffic impedance measures in predicting urban fuel consumption. Similarly, pioneering work by Hermann showed that the best single measure to predict fuel consumption by an individual car on a specific trip was its trip speed (Section 18.2.12), i.e. trip length divided by the trip time. Such an expression is developed below in Equation 29.18.

Fuel consumption of a vehicle is typically measured formally in a laboratory equipped with dynamometers which are able to operate in accordance with locally-defined *drive cycles* that attempt to reproduce local on-road conditions. The fuel consumption determined in this manner is usually lower than would be obtained by real drivers in real traffic (OECD, 1982). A number of governments conduct such tests and publish the data. Another data source for fuel consumption is in the publication by car clubs of the results of road tests and car trials; however, this data is usually too ill-defined or influenced by too many variables to be useful analytically.

29.2.2 Deterministic fuel consumption models

It is now appropriate to discuss the range of deterministic models used to predict the fuel consumption and emissions produced by a single vehicle. There are six general types of model and this discussion will proceed from the most disaggregate to the most general, providing a logical sequence of development, but by-and-large the historical development has been in the opposite direction. A comprehensive guide to the use of these models for predicting fuel consumption in urban traffic is given for cars in Bowyer *et al.* (1985) and for trucks in Biggs (1988).

Section 29.2.1 listed all the facets of car and engine design and maintenance that can contribute to its fuel consumption performance. Fuel consumption predictions at engine level are handled by using a map of torque vs engine vs power vs efficiency vs gearing, an example of which is shown in Figure 29.1. Such maps are basic vehicle design tools, and are derived from a series of steady-state measurements of output torque (using a dynamometer), engine speed, and instantaneous fuel consumption. 'Steady state' in this instance means prediction intervals of 1 s or more. For smaller intervals, transient engine maps are needed. Torque vs engine speed maps allow the driving pattern of a specific vehicle to be traced, as in the solid line abcdef in Figure 29.1. Note how peak efficiency is not reached until full power in top gear. Driving at a constant and low speed corresponds to point a and is seen to be much less efficient than accelerating. Even worse, deceleration can be at zero efficiency.

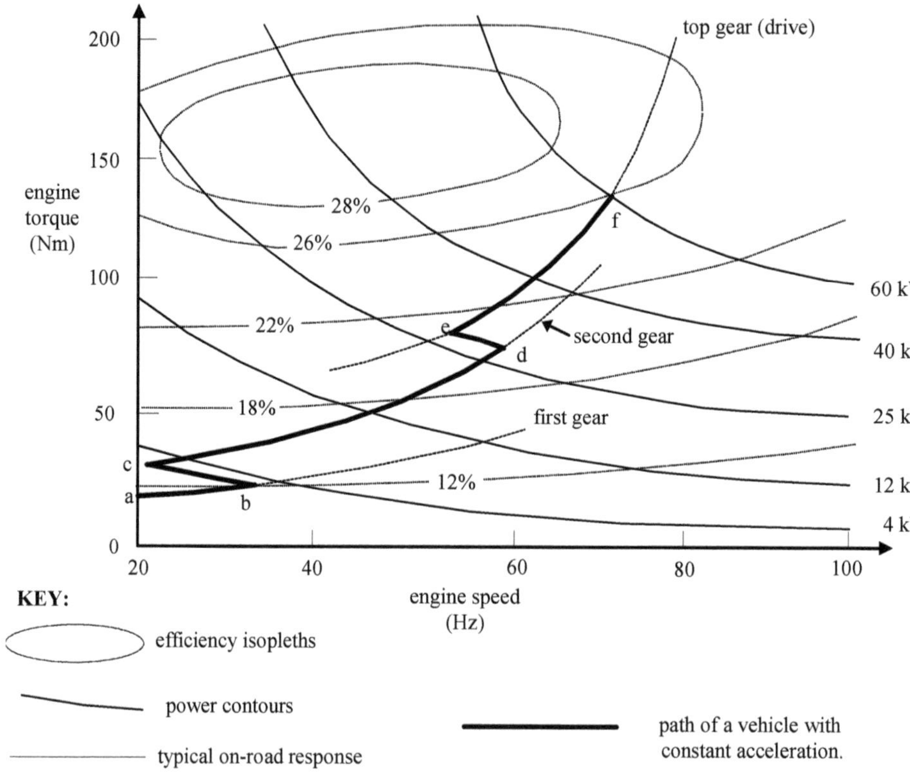

Figure 29.1 Typical torque vs engine-speed map, also showing power contours and thermal efficiency isopleths up to 28 percent. Efficiency is defined in Section 33.5.2. The line abcdef shows the path of a vehicle at constant acceleration.

An extension of this approach is to develop vehicle maps that plot fuel consumption (or emissions) as contours on graphs of vehicle acceleration against vehicle speed. Because they ignore the effects of gear changes (Figure 29.2), the contours have a coefficient of variation of about 20 percent.

Engine and vehicle maps are not particularly useful in road studies. As fuel is directly converted to power, engine fuel consumption, C, could be expected to be a linear function of the power produced. This assumption leads to models that are of more use. Following Section 27.1.2, the rate of fuel consumption with time, C_t, is given by:

$$C_t = (\text{constant})(\text{total power needed})/\eta_e \quad (29.5)$$

The engine *efficiency*, η_e, is usually independent of engine speed, but does increase slightly as the demand for power increases. From Section 29.2.1:

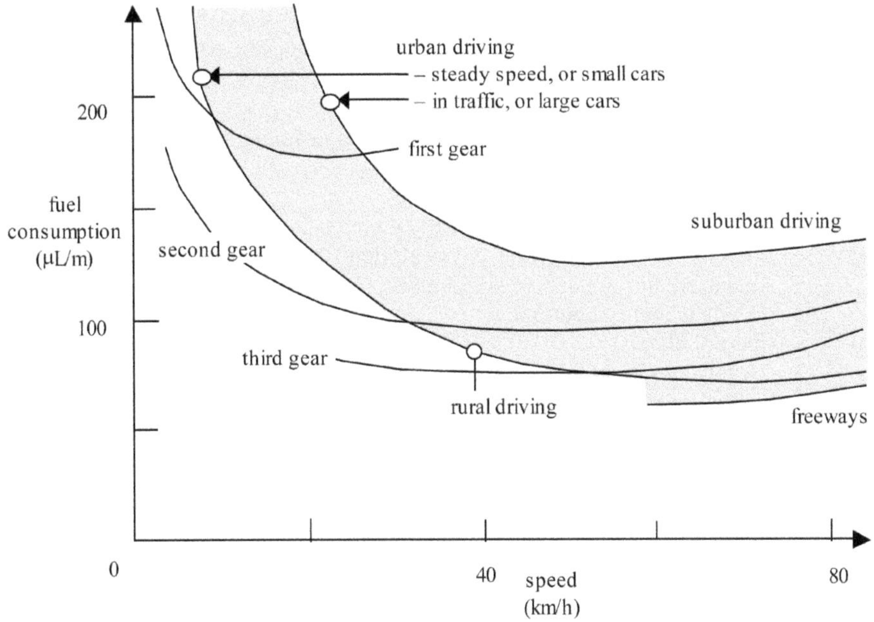

Figure 29.2 Variation with speed of the fuel consumption of common cars. The data for various gears is for level pavements and at constant speed (Biggs and Akcelik, 1985).

total power needed = power needed to overcome Σ{(engine resistance) + (demands of accessories) + (rolling resistance) + (air resistance) + (inertia)} = (power needed when engine is idling) + (tractive power, p_{dt}) (29.6)

This expression can be substituted into Equation 29.5 to give the fuel consumed. The procedure can be simplified by realising that, when the vehicle is stationary, Equations 29.5&6 provide the time rate of fuel consumption when idling, C_{ti}, which satisfies the demands of engine friction and accessories.

When tractive power is needed, it is also necessary to consider the efficiency, η_d, of the transmission and related devices (the *drive train* efficiency). This gives the total tractive power needed from the engine as:

tractive power needed from the engine = p_{dt}/η_d (29.7)

where η_d is the efficiency of the drive-train. This then allows Equations 29.5&6 to be rewritten as:

$$C_t = C_{ti} + (\text{constant})p_{dt}/\eta_d\eta_{dt} \qquad (29.8)$$

which is useful, as C_{ti} is relatively easy to measure and is approximately proportional to engine rotational speed (Hz).

29.2.3 Instantaneous fuel consumption models

To move from the vehicle-based emphasis of Section 29.2.2, to an on-road situation requires the conversion of the vehicle's power need, p_{dt}, in Equation 29.8 into on-road terms. From physics, mechanical engineering, and an empirical examination of test data, for a vehicle travelling in a straight line on a smooth and level pavement, it can be predicted that:

$$(\text{constant})p_{dt} = d_1v + d_2v^2 + d_3v^3 + d_4a + d_5a^2 + d_6av + d_7av^2 + d_8a^2v + d_9a^2v^2$$

where a is acceleration, v is instantaneous speed, and d_i are constants. The individual terms cover the following aspects:

d_1v — rolling resistance (Section 29.2.1),
d_2v^2 — drive-train power losses. This is usually a minor factor as its effect and is largely covered empirically by d_3v^3,
d_3v^3 — air resistance (Section 27.1.2),
d_4a — some inertial effects, but is usually a minor factor,
d_5a^2 — inertial term of uncertain physical significance needed to empirically cover fuel used during positive acceleration (Bowyer *et al.*, 1984). Zero for negative 'a'. d8 is more important.
d_6av — inertial effects,
d_7av^2 — largely empirical factor which is usually insignificant,
d_8a^2v — inertial term of uncertain physical significance needed to empirically cover fuel used during positive acceleration (Bowyer *et al.*, 1984). Zero for negative 'a'.
$d_9a^2v^2$ — largely empirical factor which is usually insignificant.

Hence Equation 29.8 becomes:

$$C_t = C_{ti} + (d_1v + d_2v^2 + d_3v^3 + d_4a + d_5a^2 + d_6av + d_7av^2 + d_8a^2v + d_9a^2v^2)/\eta_e\eta_d \quad (29.9)$$

If the power required is negative, the fuel used is given solely by C_{ti}.

Equation 29.9 is thus a power-based, instantaneous (or incremental), deterministic fuel consumption model. Such models are less sophisticated than the engine maps in Section 29.2.2 and assume that the drive train and engine can be treated as a single unit that is insensitive to gear selection. They are useful for predicting trips of 60 s or more, in both steady state and accelerative manoeuvres (Bowyer *et al.*, 1984). Their major traffic use is to derive the more aggregated models to be discussed below. For instance, there is little practical value to a traffic engineer in a model that deals in the instantaneous speed and acceleration of a single vehicle. Indeed, an issue to be watched in the subsequent work is whether the models developed are dealing with the properties of a single vehicle or those representative in some way of the vehicle fleet.

By taking the efficiency term, $\eta_e\eta_d$, into the coefficients, i.e. $c = d/\eta_e\eta_d$, where c_i are constants, Equation 29.9 becomes:

$$C_t = C_{ti} + c_1v + c_2v^2 + c_3v^3 + c_4a + c_5a^2 + c_6av + c_7av^2 + c_8a^2v + c_9a^2v^2 \quad (29.10)$$

Given the earlier comments on the d terms, Equation 29.10 can be further simplified, with little loss of accuracy, to:

$$Ct = C_{ti} + c_1v + c_3v^3 + c_6av + \{c_8a^2v\} \quad (29.11)$$

where { } = 0 for a < 0. Using test data due to Watson, Equation 29.11 becomes:

$$C_t\,(\text{mL/s}) = 0.70 + 0.44(v_k/100) + 2.2(v_k/100)^3 + 0.76a(v_k/100) + \{0.089a_k^2(v_k/100)\}$$

here v_k is v in km/h and a_k is a in km/h/s.

The corresponding fuel consumption per distance, C_d, is:

$$C_d = C_t/v$$
$$= C_{ti}/v + c_1 + c_3v^2 + c_6a + \{c_8a^2\} \quad (29.12)$$

For driving at a constant speed (a = 0), Equation 29.10 becomes:

$$C_d = C_{ti}/v + c_1 + c_3v^2 \quad (29.13)$$

which is graphed in Figure 29.3. Typical numerical data gives:

$$C_d\,(\mu\text{L/m}) = 2200/v_k + 24 + 84(v_k/100)^2 \quad (29.14)$$

The speed for optimum fuel consumption is given by differentiating Equation 29.13 to obtain:

$$v_{op} = (C_{ti}/2c_3)^{1/3}$$

Thus, if the corresponding fuel consumption/distance, $C_{d/op}$, is known, then C_{ti} and c_3 are readily determined from simple trials as:

$$C_{ti} = 2v_{op}(C_{d/op} - c_1)/3$$

$$c_3 = (C_{d/op} - c_1)/3v_{op}^2$$

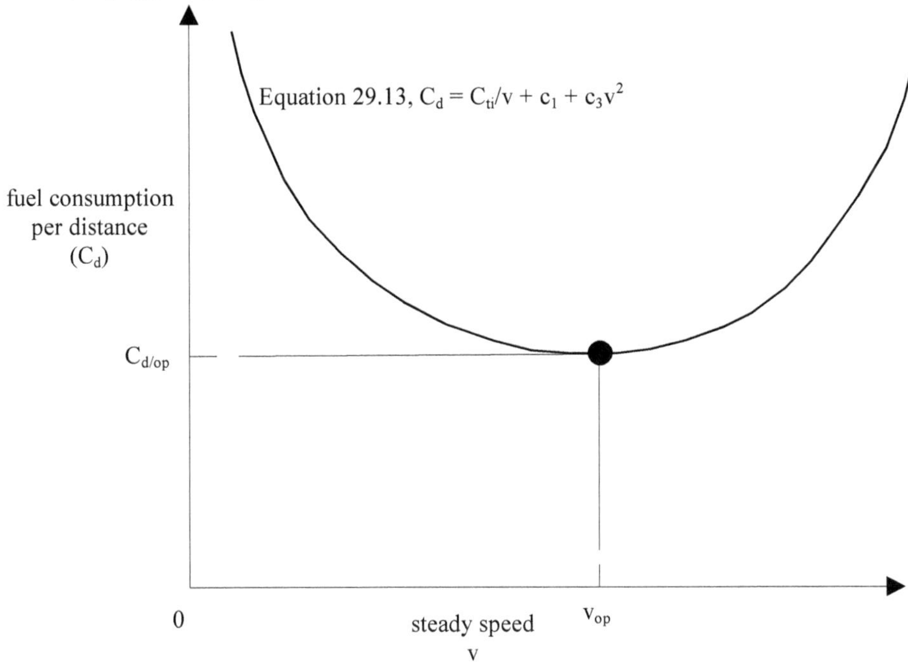

Figure 29.3 Fuel consumption per distance at constant (a = 0) speed, showing the presence of a minimum fuel consumption point, $C_{d/op}$, at a speed, v_{op}.

The indication of an optimum fuel consumption at a particular speed must be discounted by the fact that the cost of running time will decrease monotonically with speed, as illustrated in Figure 31.1b.

Instantaneous fuel consumption models can be used directly in association with microscopic traffic simulation models (Section 17.3.10) or when individual vehicle speed–time traces are available. They can also be used to derive the simpler models discussed below.

29.2.4 Traffic-oriented fuel consumption models

A useful model for urban traffic-engineering purposes called the *elemental model* (Akcelik, 1980), gives the average fuel consumption per vehicle, C:

$$C = C_{dr}D + C_{ti}T_{sd} + e_3h \qquad (29.15)$$

where:

* C_{dr} is C_d at running speed, v_r, is defined in Section 18.2.11 and is the speed the vehicle adopts between traffic control devices and for the prevailing traffic flow and road conditions. C_{dr} is thus a function of v_r (e.g. from Equation 29.13 with $v = v_r$) but must also take into account the speed fluctuations around v_r caused

by such factors as the effect of other vehicles and the need to manoeuvre and navigate. It can thus be seen as comprised of components due to v_r and to Δv_r. Values of C_{dr} as a function of running speed are given in Bowyer *et al.* (1985).

* D is the distance travelled,
* C_{ti} was defined in Section 29.2.1,
* T_{sd} is the time stopped (when the engine is *idling*),
* e_3 is a 'stop–start' parameter giving the fuel used in a v_r-brake-stop-start-v_r sequence. It is thus a function of v_r (Akcelik *et al.*, 1983). Accuracy can be greatly enhanced by replacing e_3 by $(e_{3b} + e_{3a})$ where e_{3b} depends on v_r before stopping and e_{3a} on v_r after accelerating. Values of these two e_3 coefficients are given in Bowyer *et al.* (1985). They can also be deduced by numerically integrating Equation 29.11 over the speed-acceleration-time profiles given in Section 27.2.2. Alternatively, empirical data suggest that both e_3's are functions of v_r and v_r^2. The link with the *stop penalty* is discussed below.
* h is the number of stops per vehicle.

The constants are usually chosen to be average parameters for the vehicle population. Typical values used are given below for D in m and T_{sd} in seconds. The C_{dr} value is reasonable for 20 km/h < v_r < 90 km/h (Akcelik *et al.*, 1983).

constant	cars	heavy vehicles	units
C_{dr}	0.08–0.10	0.12	mL/m-veh
C_{ti}	0.33–0.60	0.70	mL/s-veh
e_3	0.02–0.035	0.035–0.075	L/stop

The model assumes that the three elements of distance, D, delay, T_{sd}, and stops, h, are independent of each other. A common running speed is implied, and the operation of traffic control devices is handled via T_{sd} and h.

In terms of fuel used per distance travelled, C_d, the equivalent expression to Equation 29.15 is:

$$C_d = C_{dr} + C_{ti}T_{sd}/D + e_3h/D$$

The coefficient e_3 can be related to the stop penalty, K, which is part of the TRANSYT performance measure for traffic signals (Section 23.5.2):

(total delays) + K(total stops)

In fuel consumption terms and in the context of Equation 29.15, K is proportional to:

$$(C_{ti}T_{sd} + e_3h) \alpha (T_{sd} + e_3h/C_{ti})$$

which implies $K = e_3/C_{ti}$, however, given its use of different units and definitions for T_{sd}, TRANSYT's K is:

$$K = 3600[e_3/C_{ti} - \text{(time delay at a stop due to deceleration and acceleration)}]$$

The value of the elemental model is that it allows the effect of changes in traffic systems to be explored. It is less useful for predicting absolute values, because of the assumptions listed above.

29.2.5 Lumped fuel consumption models

Lumped fuel consumption models are due to Watson and predict about 80 percent of the variance in his test data for fuel consumption and emissions in terms of vehicle speed and speed change. The most effective of these in extending Equation 29.13 are:

$$C_d = k_0/v + k_1 + k_2v + k_{10}(\Sigma\{[v_a^2 - v_b^2]/D_d\}) \qquad (29.16)$$

where the last term gives the positive kinetic energy (PKE) change per distance and where D_v is the distance over which the speed changes from v_b to v_a. Watson's coefficients for a 1980 4 L Ford Cortina are in Table 29.2. Such coefficients are usually obtained from tests on chassis dynamometers.

Table 29.2 Coefficients for Watson's lumped model. HC, CO and NO_X are discussed in Section 32.3.

Item being predicted	k_0	k_1	k_2	k_{10}
Fuel consumption (mL/m)	2.90	−0.031	0.094	0.0012
HC (mg/m)	16	−0.24	1.5	0.006
CO (mg/m)	18	−2.2	2.3	0.036
NO_X (mg/m)	114	−0.7	1.9	0.016

To provide greater consistency with the traffic models, Equation 29.16 has been modified to (Akcelik *et al.*, 1982):

$$C_d = k_0/v_r + k_1 + C_{ti}T_{sd}/D + k_2v_r + k_{10}(PKE) \tag{29.17}$$

The modification allows the model to predict the effect of changes in delays due to stops. The k coefficients are directly related to the e coefficients in Equation 29.15. A detailed account of the coefficients to use in Equations 29.16&17 is given in Watson *et al.* (1982). Although the model has the ability to explain a particular set of experimental data, it is generally less useful than the traffic model in Section 29.2.4 as it cannot be confidently transferred to new traffic situations.

29.2.6 Urban fuel consumption models

Urban (or average travel speed) models predict the fuel consumption of individual vehicles in terms of such broad urban parameters as trip speed and trip time. At the widest scale, models can predict total transport energy usage from knowledge of the area of the city.

For individual vehicles, Figure 29.2 showed how the running speed of traffic along a road can be linked to fuel consumption, and the different curves indicated how speed fluctuations influence the results. Hence, the traffic models in Sections 29.2.4 & 5 were developed to accommodate these variations. An urban model can be developed by omitting the v^2 term in Equation 29.13 — which has a minor influence for low speed travel — and using $v = v_s$, where v_s is the speed of a traffic stream rather than of an individual vehicle. This gives:

$$C_d = j_{ti}/v_s + j_1 \tag{29.18}$$

or

$$C = j_{ti}T_{sd} + j_1D \tag{29.19}$$

$$= j_{ti}T_{sd} + j_1v_sT_r \tag{29.20}$$

which are suitable for broad urban planning purposes. In these equations T_{sd} is the time stopped, T_r is the running time and D is the distance travelled. Thus trip time T_t is:

$$T_t = T_r + T_{sd}$$

Of the other terms in Equation 29.20:

J_{ti} is directly linked via c_{ti} (Section 29.2.3) to engine idling, i.e. to the energy needed to maintain the engines in the fleet in operation.

j_1 is linked via the definition of c_1 in Section 29.2.3 to the fuel needed to provide tractive force to the vehicle fleet and therefore depends on vehicle characteristics, grades, and the driving environment.

For fuel consumption in mL/m and v_s in km/h, typical values of j_1 and j_{ti} are 0.074 and 1.10 for a fleet of 1.2 t cars, and it is suggested that j_1 is given by:

$$j_1 = 0.021 + 0.044M$$

where M is the mass of the average car in tonnes (Biggs and Akcelik, 1985). Fuel consumption per mass has continued to drop as engine designs improve and so a 2008 version of the equation would be closer to:

$$j_1 = 0.016 + 0.030M$$

For even broader urban studies, typical C_d values for vehicles on the road in the U.S. in 1992 were:

$$C_d\ (\mu L/m) = 110\ \text{(cars) or } 160\ \text{(light trucks)}$$

For heavy *trucks*, the following equation has been suggested:

$$C_d\ (\mu L/m) = N([547/v_s]^2 + 105 + [v_s/7.75]^2)$$

where N = 0.90 for rigid trucks and 2.20 for articulated trucks. v_s can be eliminated to give (Thoreson, 1993):

$$C_d\ (\mu L/m) = 260 + 6(GVM)$$

where GVM is the gross vehicle mass in t, as defined in Section 27.3.2.

Finally, at the broadest scale, data from 45 cities around the world suggests that the total transport use in a city can be estimated from the area of the city (Hughes *et al.*, 2004). The relationship is linear and explains about 90 percent of the variance. Thus the prime measure to contain overall use of transport energy in a city is a restraint on the growth of the area of the city. This powerful but simple relationship explains other models, such as a hyperbolic link between population density and energy use per capita. In addition, the distance travelled measures (C_d) developed earlier in this chapter will also correlate well with urban area.

ARRB's ARFCOM model (Biggs, 1988) was an excellent general tool for predicting truck fuel consumption, but may now be out-dated. Models for specific cases are now available, however, care must be taken in using any truck fuel consumption model due to the vast differences between vehicles and the rapid changes in truck technology (e.g. major variations in truck unladen mass, tyre design, engine efficiency, and body aerodynamics).

29.2.7 Rural fuel consumption models

Because there is less speed fluctuation in rural driving, the models in Section 29.2.3 can often be extended to the rural case. For example, the up-turn in fuel consumption that occurs in cars at higher rural driving speeds is provided by the subscript 3 terms covering air-resistance in Equations 29.9–15.

Rural fuel consumption models have also been developed directly from a regression analysis of field data from Kenya, the Caribbean, India and Brazil (Geipot, 1982d&f). The results are typically of the form:

$$C_d = e^{b_1 + b_2R + b_3CI}$$

where R = roughness is defined in Section 14.4 and CI = curviness index in Section 19.2.1. Regression equations of this type were used in the RTIM and early HDM road-network management models described in Section 26.5.2. They cannot be relied upon without first checking if they can be extrapolated to new circumstances or applied to specific local conditions. Typically, C_d increases by 5 percent for each increase of 1 IRI in roughness. The effect of roughness changes on fuel consumption can be further estimated by the following equations:

$\delta C/C = (R - 50)/500$

or $= 0.063(IRI) - 0.10$

The fuel consumption increases are mainly caused by roughness profiles in the 50 to 500 mm megatexture wavelength range (Section 12.5.1) and is closely linked to the conditions that generate road noise (Section 32.2.3).

Provided the pavement has some form of bound surface, the actual surface type is not of great significance in this respect. However, for unpaved roads, fuel consumption increases can range from 5 to 40 percent, depending on conditions (Geipot, 1982a; Savenhed, 1985). Another important variable affecting fuel consumption is the grade of the road (Section 19.3) which effectively increases the vehicle's mass. Its influence depends on operating speed and is typically:

Up-grade	Percent increase in fuel used for:	
%	speeds over 25 km/h	speed = 80 km/h
4	5–10	10
6	15–30	20
8	25–40	30
10	45–65	50

More precise means for estimating all these effects are given in Bowyer *et al.* (1985).

29.3 ROAD FUNDING

29.3.1 Funding sources

Roads produce the costs listed in Categories A to E in Sections 29.1.1–2. Costs in Categories D and E are carried directly by the road user. However, costs in Categories A and B are incurred by the Road Authority and in Category C by the community. The five mechanisms used to provide the funding needed to sustain A & B and to compensate for C are discussed below. Of course, these mechanisms may often be preceded by the borrowing of a lump sum to undertake the work, in which case the methods then relate to the repayment of that loan.

(a) *Taxes on the benefits caused by the road.*
For example, many suburban streets are funded by either levies on the developer of the housing scheme or industrial park or by property taxes on abutting land-owners. For arterial roads, taxes can be applied to the increases that the enhanced accessibility provided by the new road produces in (Section 31.2.4):

* land values,
* local employment, and/or
* economic output.

Such taxes play an important role in letting market forces and economic rationality influence land development, thus avoiding cross-subsidisation.

(b) *Direct tolls on road users.*
Tolls are charges placed directly on each road user. They may be applied for one or more of the following reasons:

* to finance the maintenance of the facility,
* to repay the cost of constructing the facility, e.g. a toll bridge,

* to promote economic efficiency (Section 33.5.2) in road provision,
* to help efficiently allocate resources within the transport sector,
* as a rent for the use of road space that is in high demand (e.g. congestion pricing, Section 29.4.5, & value pricing, Section 29.4.4), and
* to compensate for social costs of operation (C in Section 29.1.1, e.g. environmental costs).

Raising finance via direct tolls rather than via other means may be undertaken:

* as part of a national privatisation philosophy,
* to bring private-sector attitudes to road management,
* to circumvent constraints on Government fund-raising, and/or
* to employ otherwise under-utilised construction capacity, e.g. constructing a road may be cheaper than paying interest on the cost of unused plant.

Road tolls meet frequent opposition from road users who commonly believe that the fuel taxes (Section 29.1.2E(1)) that they pay are more than enough to fund the road system.

Methods for determining the amount to actually charge a road user are the subject of wide philosophical debate and are reviewed in Section 29.1.3 for charges relative to other users and in Sections 29.4.1–5 for the absolute level of the charges. Toll elasticities are commonly between –0.1 and –0.5.

Tolls on roads and bridges have a long history (Section 3.4.2). Most tolled facilities have been privately owned, and this tradition continues today. Toll bridges usually present a straightforward case, as they commonly possess a route monopoly. However, a toll road (or tollway) competes with free public roads and so to be successful requires either a captive user-market or an improved level of service. Tolls are difficult to apply universally to the road system, as toll collection is not easy on roads that are in public ownership and have few limits on access (Section 4.2). However, road pricing attempts to overcome these problems (Section 29.4.6) and GPS-based tolls will make universal tolling simpler.

There is a range of possible tolling schemes:

* point tolling charges a fee every time a point is passed (e.g. on a toll bridge),
* cordon tolling charges a fee every time a cordon enclosing an area is crossed,
* distance tolling charges a fee which depends on the distance travelled,
* time tolling charges a fee which depends on the trip duration, and
* parking fees.

In each of these, the toll may be varied with time of day (peak charging), traffic conditions (congestion charging) and/or vehicle occupancy. Tolls may be collected by:

1. manual methods which require vehicles to completely stop at a toll booth,
2. semi-manual methods (e.g. coins thrown in a basket) where vehicles must slow down to about 5 km/h,
3. automatic methods which require vehicles to slow down (typically to about 30 km/h),
4. automatic methods which can operate at highway speeds (typically at up to 200 km/h).

Methods 1 to 3 require facilities to be built between lanes. Methods 1 and 2 further require that the facilities can accommodate human operators, and so can be the equivalent of a lane wide, thus doubling the road space needed before allowing for roadside structures. In addition, the need for vehicles to effectively stop can halve capacity as vehicles stopping from 40 ms^{-1}, paying and starting with accelerations of about 1 ms^{-2} (Table 27.1) will require 40 + 10 + 40 = 90 s and a longitudinal length of about a kilometre to undertake the toll process (see Table 21.1). The same non-stop vehicles only take 1000/40 = 25s. As a consequence, experience is that manual tolling can require the road space needed to be five times as wide as elsewhere on the highway. This widened pavement tapers over a kilometre or so. On the other hand, method 4 requires no increases in road space.

In a *shadow toll* the road builder is paid a fee from other revenue sources (e.g. 1 and 3) on the basis of the volume of traffic using the road. The method is used when direct tolls would create unwanted route-diversion or be difficult to collect.

(c) *Usage-related charges.*
Various charges are levied on vehicle owners and operators and on road users. Pedestrians are the only road users generally immune, as even cyclists will pay some form of sales tax. The particular usage charges on vehicles fall into three categories:

* acquisition, e.g. sales tax,
* ownership, e.g. annual vehicle (registration) and driver licence fees, and
* operation, e.g. fuel tax and tax on vehicle maintenance.

(d) *Impact fees.*
These are fees levied on property developers and reflect the transport costs of the extra traffic caused by the impact of their new development on the external transport system that serves their development.

(e) *Direct grants from government.*
In this system, the government allocates grants from its general revenue to the Road Authority, often on the basis that a benefit–cost analysis (Section 5.2.3) has indicated that the project will return various benefits to the community which will sum to many times the cost of grant. Such a charging system can be economically efficient if the road can be used by everyone, has excess capacity, and has no unfunded social costs (see C in Section 29.1.1).

29.3.2 Funding levels

The funds collected under the various schemes in Section 29.3.1 may more than off-set the cost of road expenditure (Section 29.1). For example, governments often use fuel taxes as a way of raising general revenue. A common and vexed question then is whether an adequate proportion of the funds is returned to pay for road construction and maintenance. For example, it is reasonable for road users to regard fuel taxes as a price placed on motoring and to express concern when that revenue is not returned to the road system. In addition, the road user is not the only beneficiary of the provision and maintenance of roads and could well argue that the user's contributions should not be set to meet the total costs required.

However, there is no inherent reason why funds collected under the schemes in Section 29.3.1 should be used for road construction. For instance, Table 4.1 suggests that a motor-vehicle tax (e.g. c in Section 29.3.1) may well be intended by legislators to be a form of *consumption* tax, analogous to a tax on alcohol or tobacco, and intended to restrain use and/or make a contribution to Government revenue. Some governments have tried to link the user taxes directly to road expenditure (as in the PAYGO system, Section 29.4.2). Whilst this method has appeal to road-builders, there is again no inherent reason why user taxes should be linked to non-marginal road expenditures. However, it is rational to expect user taxes to pay for the marginal costs of the damage they cause (Section 33.5.2). In a rational world, funds spent on roads would be determined on the basis of needs, warrants, and economic assessments (Chapter 5), rather than on the revenue collected. However, in an even more rational world, a component of that revenue would be related to road needs via some form of usage or pricing mechanism (Sections 29.4.6 & 33.5.2).

29.4 CHARGING FOR ROAD USE

A *minimum objective* for any road charge is that the revenue collected should at least meet the cost of operating the road system. This ensures that the value of the trip at least matches the cost of providing and making the trip. To be effective, the charge must be allocated reasonably amongst the road user classes, in the manner described in Section 29.1.3. A *consequent objective* is to internalise any external costs created by a road user, so that that user's transport decision-making reflects the real costs of the contemplated journey.

Pricing theory is discussed in Section 33.5.2. In the context of that theory, prices (or charges) are set to meet the *minimum objective* and achieve one of the following four *optional objectives*, which are further discussed in Sections 29.4.1–4 below:

1. maximising reasonable usage of the road,
2. achieving financial balance for the operator,
3. charging what the market will bear, and
4. receiving a proportion of the benefits received.

Whichever optional objective is chosen, a practical constraint is that the charging method must be easy to administer and operate. Following Section 29.3.1(b), charging methods are either:

* *open*, collecting the charges as a vehicle passes specific points along a road, or
* *closed*, collecting the charges as a vehicle enters and/or leaves a cordon around an area.

29.4.1 Option 1 — Charging to maximise reasonable usage

A sensible *optional objective* for public sector roads is to ensure that road charges produce reasonable road usage. If existing roads have been funded by direct and non-recoverable government grants (Section 29.3.1e), then 'reasonable usage' could be defined as that level of usage which has the roads delivering their outcomes with maximum efficiency. In terms of *economic efficiency*, this is achieved by applying *marginal pricing* (Section 33.5.2). There are three forms of marginal pricing.

(a) *Short-run marginal pricing*
Short-run marginal pricing in effect says 'the government has given us these roads, let us now ensure that they are used as much as possible, whilst earning just enough income for us to operate them and keep them in good repair.'

Thus the prime purpose of the short-run marginal price is to ensure that the road is maintained, rather than to influence travel behaviour. To do this, the short-run marginal price is based on the *avoidable costs* associated with operating the road (Section 29.1.3) and is sometimes called *occasioned cost equity pricing* or *first-best pricing*. The cost may also be extended to include the external social costs associated with the use of the road, typically calculated for congested conditions (Category C in Section 29.1.2).

Usage charges can also be based on charging users a *rent* for occupying particular road space — perhaps via a time-based toll or a parking fee — based on the marginal costs of providing that space.

(b) *Long-run marginal pricing*
The *long-run marginal price* would include the short-run costs in (a) and also cover, at least, the cost of reconstructing the road at some future time to restore it to its original condition, and thus negate the result of progressive irreversible deterioration (Section 14.6 & Category B3 in Section 29.1.1). It may also include the amortised capital costs (Category A in Section 29.1.1), but there is little technical merit in this latter inclusion.

There is an optimal volume/capacity ratio for each road (Section 17.4.4) and it is sometimes argued that the long-run price should cover the cost of reconstructing and improving the road to achieve this optimal ratio.

Thus, *short-run prices* are relevant to decisions related to immediate road usage and *long-run prices* apply to decisions related to long-term road usage. Which of these prices to use is often a source of theoretical and administrative controversy. If a road operator decides not to include a reconstruction component in its usage charge it:

(1) should preserve its future right to close the road rather than reconstruct it, and
(2) must have another source of funds available (see Section 33.5.2 for the obvious source) if it wishes to keep the road operating beyond the reconstruction period.

(c) *Prices in excess of marginal prices*
Prices based on but in excess of marginal prices, may be charged in order to:

* avoid sending distorted price signals to users who are not aware of their *operating and social costs*, as marginal road costs are a small percentage of operating and social costs (Section 5.1 & Categories D & E in Section 29.1.2).
* recover *joint costs* incapable of rational distribution to individual users (Section 29.1.3),
* achieve financial balance (Section 29.4.2), and
* raise funds required for other purposes.

The level of the excess price is often set using market-will-bear pricing (Section 29.4.3).

29.4.2 Option 2 — Charging to achieve financial balance

Marginal pricing in accordance with Option 1 (Section 29.4.1) will lead to system efficiency, but suffers the disadvantages of:

* not recovering the funds expended on the creation and operation of the system, and
* possibly being seen as inequitable by those not directly benefiting from that expenditure.

Financial-balance (or *cost-recovery*) pricing as an *optional objective* therefore extends *marginal pricing* to also include the amortised annual cost of the long-run capital investment in the road system (Category A in Section 29.1.1). This greater charge will balance the total road costs (Categories A to C), leaving users to pay their own operating costs (Categories D & E). When the financial-balance price is extended to include vehicle operating costs, the associated charge is known as the *resource price*. Strictly, resource costs are the full financial costs of an activity, less any taxation charges. The price is sometimes called a *long-run financial-balance price*.

Financial-balance pricing provides a return on the funds invested in the construction of the system. It is therefore appropriate for private toll roads (Section 29.3.1b) and for a public road system intended to fairly compete with a private sector system. There are three problems with financial-balance pricing.

* As it does not relate to marginal costs, it will not lead to an economically efficient allocation of resources.
* Most road transport investments are large and spasmodic. Even if these are amortised (Category A.1 in Section 29.1.1), it is difficult to assign them without being unfair to a user who happens to be present soon after one such investment has been made.
* The allocation of the relatively large joint construction costs to particular road users remains unresolved (Section 29.1.3). In practice, allocations are often based on road use, e.g. on vehicle-kilometres of travel.

If the charge for using a public road is set at a proportion of the financial-balance price, a government could meet any revenue shortfall via general taxation. Experience indicates that this approach is both ineffective and inefficient, relative to collecting the revenue directly from road users.

Many jurisdictions omit social costs (Category C) from the financial-balance calculation — often for no better reason than that these costs are difficult to assess. This decision is often coupled with an administratively simple approach which replaces the amortised annual costs by the proposed *sunk costs* (Section 33.5.9) in the annual road budget. This method focuses on the year in which the expenditure was incurred. The lumpiness referred to in Section 29.1.1 is assumed to be smoothed out by the ongoing stream of projects in large road administrations. This simplified form of financial-balance pricing is known as *pay-as-you-go* (or *PAYGO*) pricing. It runs the risk of creating a self-perpetuating road-building budget.

In commenting on the policy dilemma between reasonable usage (or efficiency) pricing (Section 29.4.1) and financial balance (or cost recovery) pricing discussed in this sub-section, Roth (1996) has remarked 'efficiency in resource use has sometimes to be sacrificed for efficiency in investment'.

29.4.3 Option 3 — Charging based on what the market will bear

The financial balance approach in Option 2 (Section 29.4.2) can be extended to a *market-will-bear* (or *Ramsey* or *willingness to pay*, Section 33.5.3) price. If the number of travellers is N and the price charged is P, then P is chosen to maximise NP — given that N will drop as P increases. Market-will-bear pricing is therefore based on the *elasticity* of the demand for the product (Section 33.5.4); the less elastic the demand (i.e. the more captive the traveller), the higher the price that can be charged. The price thus depends inversely on the elasticity. The demand elasticity for fuel is about –0.3 in the short run, and –1 in the long run when longer-term travel adjustments have occurred (Section 34.5.4).

For a road to be efficiently used, its market-will-bear prices must be below its marginal price; for it to be financially viable, the market-will-bear price must exceed the financial balance price.

A market-will-bear price is appropriate where *optional objective* of the charge is to:

* maximise the revenue raised. It is thus the charge one would expect if the road were operated by a focussed private-sector group operating without price regulation.
* determine the minimum price needed to ration road space or suppress travel demand (Section 29.4.6).
* allocate the super-marginal costs (see Section 29.4.1). The least deviation from economic efficiency would be achieved by using *market-will-bear* pricing.
* avoid distortions between competing modes, which would occur unless the ratio of market-will-bear price to the marginal price is the same for each mode. This is called a *second-best option* (relative to marginal pricing, Option 1).

Many transport economists favour the *market-will-bear* method to 'solve' the technically intractable problem of choosing the right pricing mechanism.

29.4.4 Option 4 — Charging based on benefits received by the user

In this approach the *optional objective* is to achieve equity in the pricing, as is usually also the case with short-run marginal costing (Option 1a, Section 29.4.1(a)). Benefits-based charges can be used to redistribute the benefits of a project or to tax those who receive the greatest benefits.

(a) *Benefit pricing*

Users can be charged on the basis of the benefits they receive from their use of the road (Section 5.1) and is sometimes called *received benefit equity pricing*. Section 34.5.4 shows that this approach reduces the user's consumer surplus.

The direct benefits considered relate to travel time and vehicle operating cost savings associated with the use of the road relative to some other facility (Sections 29.1 & 2). In addition, Richardson (2004) estimates that 30 percent of travellers would pay a toll for the comfort of using a road of noticeably higher quality than the available alternatives.

The benefits considered could also be related directly to the benefit components of the benefit–cost analysis previously used to justify the road as a *public good* (Section 33.5.8). Given the variety of the benefits (Table 5.1), some of the benefits-based charges could be collected through taxes unrelated to road use — for example, through taxes on land value and business profits. If the benefit–cost ratio of the original road proposal significantly exceeded one, and the proposed charge exceeded the marginal price defined in Option 1 (Section 29.4.1), then the benefits-based charge could be used to raise revenue.

In a rational world, benefits-based price would be expected to have some relation to the market-will-bear price of Option 3 (Section 29.4.3). However, the latter relates more to the direct benefits the user receives from the trip and not to the broader issues canvassed in a benefit–cost analysis.

(b) *Value pricing*

Value pricing is one form of pricing that comes close to satisfying both market-will-bear and benefits-received objectives. In value pricing the user is charged a higher price when the road is lightly used, thus permitting higher travel speeds and increased driving comfort. The benefits received are reduced travel times.

The price must be at or above the market-will-bear price in order to discourage other drivers from destroying the idyllic traffic conditions. The process also runs contrary to the common market process of lowering the price of under-used services.

(c) *Social equity pricing*
The value prices may be based on *social* (or *equity* or *distributional*) *objectives* (Section 5.2.3) if a government wishes to use market-will-bear prices applied to most users to *subsidise* the prices charged to particular users. This could be used to provide benefits to children, the poor, or those in remote communities. Many of these subsidies will be only vaguely defined and some may be so obscure as to be largely unnoticed, except in so far as they will move the pricing system away from the economic efficiency provided by Option 1 (Section 29.4.1).

29.4.5 Congestion pricing

Congestion pricing is a pricing option that imposes higher charges on vehicles using a congested system (Section 17.2.2) and thus relates to the operating characteristics aspect of the road investment (Section 29.1.1). It should assist in reducing congestion. Because congestion varies with time, this form of pricing usually also varies with time. It is commonly applied to vehicles entering the system. Congestion is a cost that is external to the motorist. There are three major types:

(1) the cost that the vehicle causes (mainly by increased travel times) to other road users already in the system,
(2) the cost of congestion to activities along the roadside (consider, for example, the Section 7.2.2 amenity objectives which would be impacted by congestion), and
(3) the cost of adding capacity to the system to alleviate the problem.

The congestion price is usually based on funding some combination of these three costs.

Type (1) congestion cost is calculated on the basis that the number of vehicles involved is q/V per distance (Equation 17.2) where q is their flow and V their average speed. Assuming that the vehicles have an average running cost per time of C_t \$/time is given by Equation 29.18 with

$$C_t = C_d V = j_{ti} + j_1 V$$

Recall from Section 29.2.6 that j_{ti} relates to how the engine idles. The total cost rate of the flow is:

$$qC_t/V \text{ \$/[distance-time]} = qj_{ti}/V + qj_1$$

Differentiating to find the effect of altering q gives:

$$d(\text{total cost rate})/dq = (j_{ti}/V) - (qj_1/V^2)(dV/dq) + j_1$$

But from Equation 29.18 $(j_{ti}/V) + j_1$ is the 'base' cost of the trip per distance travelled, of which the new driver is already aware as the trip distance does not alter. Hence, the congestion price per distance travelled is:

$$-(qj_1/V^2)(dV/dq)$$

which is positive when dV/dq is negative, i.e. when the extra vehicle slows down the traffic. In terms of trip time (T = 1/V) this congestion price per distance travelled becomes (Luk, 1996):

$$qj_1 dT/dq$$

Recall that j_1 relates to fleet idling characteristics. Relations between T and q are discussed in Section 17.2.6. For example, the simple Normann equation (17.12b) would lead to a congestion price per distance travelled of $(q/q_{pc})(V_{uf}/V)(c/V)$.

Within a system, the cost of congestion is taken as the difference between the cost of travelling in the congested system and the cost of travelling in an uncongested system. As there will be more cars in the congested system, the calculations for the two circumstances are commonly done for the same transport task. Some major studies have mistakenly assumed that vehicles will not need to stop in the uncongested case. However, traffic control devices still operate in light traffic.

29.4.6 Road pricing

There is a strong need for better pricing of road transport services. It has been commented that 'a relatively light share of intermediate inputs in the cost of producing transport services tends to insulate transport from movements in the prices of traded products in the economy and thus from the effect of competitive forces or of technical progress acting on those prices. Transport is also less sensitive than other industries to the cascading effect of turnover taxes' (Bennathan and Johnson, 1990).

In conventional road systems there is a limited degree of road pricing via fuel prices, parking fees and other usage-based charges. Nevertheless, the pricing of roads is a somewhat esoteric activity as:

* road usage is not sold on the open market and so there is no market to readily indicate the price that might be charged for using a road,
* roads frequently are not constructed in order to return a profit (e.g. roads built to provide accessibility to a remote community, Section 5.2.3), and
* road space in high demand is conventionally rationed, not by price, but by congestion (Section 29.4.5).

Road pricing theory attempts to provide a way around these dilemmas.

The core concept of road pricing is that users should be charged directly for the use of each segment of road. For example, car drivers would be charged a toll for using a road or crossing a barrier. It is thus a universal form of the road tolls discussed in Section 29.3.1(b). Three main factors have historically counted against road pricing:

* the lack of a suitable charging mechanism,
* political concern over voter dislike for a new tax, and
* unless the tolls are earmarked for roads, those paying the tolls and those discouraged from driving by the tolls will usually be technically worse off than in an untolled system.

These factors have meant that — despite wide support from transport economists — road pricing remains a theoretical concept. The only working applications of this concept in urban areas are in Singapore, Bergen, Stockholm and London.

Road pricing usually implies charging road users the marginal cost (Sections 29.4.1 & 33.5.2) of their trip or their parking, to ensure that:

* they are aware of the cost to the road system of their trip, and
* their travel is economically rational (Section 29.4.1).

In the short term, this will influence the route drivers choose. In the longer term, it will influence their demand for further road space and their use of roads rather than some other transport mode. Road pricing on a *market-will-not-bear* (Section 29.4.3) basis has been suggested as a means of deterring vehicles from using certain socially unacceptable routes. Pricing at between marginal and market-will-bear levels would provide a key plank in any transport-demand-management strategy (Section 30.3).

If road pricing is only applied on some roads, it will force some users off the priced roads and onto the unpriced ones. This will normally result in significant transport inefficiencies, particularly as the unpriced roads will typically be of a lower standard than the priced ones. It will also lower the market-will-bear price of the priced road, possibly to less than the appropriate minimum marginal price.

CHAPTER THIRTY

Road Transport

30.1 DEFINITIONS

Transport exists to supply people and products with the mobility they need to meet their travel demands. It can thus be categorised according to trip purpose. The principal purposes are:

* travelling to and from work and school (job travel),
* travelling between businesses (business travel),
* travelling to and from shops and personal facilities (support travel),
* travelling to and from entertainment or to enjoy travel (pleasure travel), and
* moving freight.

In some countries it is common to use the word *transportation* rather than *transport* to describe these tasks. Transport as an activity can be considered to be composed of four basic stages:

(a) moving (e.g. walking) from the trip origin to a transport pick-up point;
(b) waiting at the pick-up point;
(c) transport from pick-up point to transport delivery point;
(d) moving from the transport delivery point to the trip destination.

The mobility at stage (c) may be supplied by various methods of transport, each called a *transport mode*. Cars, walking, bicycles, motorcycles, trucks, buses, taxis, trams, trains, planes, and ships are the common transport modes. Of course, a number of changes of mode can occur within stage (c), and each can be decomposed into the same four sub-stages. For example, bus routes often operate as feeder services to railway stations. *Traffic* (Chapter 17) is the movement that occurs in a particular transport mode. The three major operational components of a transport system are:

* the fixed running-surface and terminals (the *infrastructure*),
* the vehicles (or rolling stock), and
* the operating scheme.

Transport can also be categorised as *public transport* (or *mass transit*) and *private transport.*

* Public transport ranges from road-based systems like buses and taxis, to systems like railways which operate on private right-of-ways.
* Private passenger transport covers personal travel by private means; typically by car, although walking and cycling are also examples of private transport. Private freight-transport is almost entirely by road-based trucks.

The balance struck between the modes is jointly a matter of governmental transport policy and market forces.

Transport supply is influenced by the community's ability to construct and operate a transport infrastructure. It is best measured by the various performance measures discussed in Section 30.2. Table 30.1 gives the approximate travel *capacities* of some of the modes of surface transport. In practice, it will usually be found that capacity is the link capacity reduced by the performance of the transfer points (stations or stops) at each end of a link. The column headed 'most effective' indicates the mode commonly found to be most effective at the traveller flows noted in the last column.

Table 30.1 Traveller capacity of various modes of surface transport in 1000s of travellers per hour = ktraveller/h = k*tr*/h (based partly on Meyer *et al.*, 1981 and Hensher, 1999).

Most effective	Transport mode	Capacity in k*tr*/h
On foot	file of pedestrians (Table 30.5)	2
	lane of private cars (Sections 17.4.1 & 30.5.1)	2.3 x 1.2 = 3
car	street	1
	freeway	3
	lane of cyclists (Section 19.6.3)	10
	bus on street	2 to 12
bus or	bus lane with special provisions (Section 30.4.2)	10 to 35
light rail	busway (Section 30.4.4)	10 to 40*
	light rail (Section 30.4.2)	5 to 20**
rail	under-used railway	8 to 50
	well-used railway	50 to 80

* maximum actually achieved is 11 (Ottawa West)
** maximum actually achieved is 10 (Boston Green)

One of the characteristics of road transport is that the components of this system are individually managed. Governments provide most of the fixed running-surface and fixed operating system, a variety of groups provide the terminals, and millions of operators own the motor vehicles and movable operating systems. The operators have a wide choice of vehicle types and operating schemes, although Governments commonly pass laws to constrain their choice and methods of operation (Chapter 27).

30.2 PERFORMANCE MEASURES

In order to evaluate or compare transport systems, it is necessary to consider:
* the demographics of the area served (Section 31.3.2),
* the type of service being provided, and
* the performance of that service.

The first of these factors is usually fixed, and so the management of a transport system usually focuses on the service and its performance. This may be examined in terms of the four measures of:

1. capital invested,
2. effectiveness, via
   a. quantitative measures (e.g. patronage or usage), and
   b. qualitative measures (e.g. customer perceptions),
3. operating efficiency, and
4. equity.

These are each discussed below.

30.2.1 Capital invested and patronage

In economic terms, the supply (or availability) and consumption of a transport service will require the expenditure of scarce resources. Capital invested measures the financial resources applied to a transport system and is subsequently an indicator of transport supply. It covers both the construction of new facilities and the reconstruction of and marginal improvements to existing facilities. Thus consumer demand theory (Section 33.5.1) provides a conceptual framework for analysing transport.

Patronage is a key transport demand indicator. The demand unit may be the number of passengers, the equivalent number of passengers paying the full fare, the number of freight deliveries, the tonnes of freight delivered (see Sections 30.4.2 & 30.5.2), or the value of the freight delivered. Typical patronage measures for the number of passengers moved are:

(a) *passengers in total*. This is a measure of throughput.

(b) *passengers per kilometre of track*. This is one of the better measures, both generally and for specific routes.

(c) *passengers per vehicle-kilometre of travel (VKT)*. This measure can easily be read from vehicle odometers and indicates both the extent to which the available service is being utilised and route coverage.

(d) *passengers per hour*. This indicates the amount of service supplied.

(e) *passengers as a proportion of the number potentially available to be moved*. This market measure is useful for comparisons with other systems.

(f) *passengers as a proportion of the maximum possible (capacity)*. This usage/capacity ratio is an *occupancy* or *utilisation* measure. Simple examples compare occupancy per vehicle with vehicle passenger capacity or passenger-kilometres to seat-kilometres. The discussion elsewhere in this chapter (e.g. Sections 30.2.3, 30.3 & 30.4) shows that occupancy is a critical factor in the determination of the transport and energy *efficiency* of a transport mode.

One of the on-going difficulties with public transport is that the capital available is usually insufficient to meet passengers' performance requirements, as listed in Section 30.2.2 below.

30.2.2 Effectiveness

Measures of effectiveness in general record the relationship between an actual output and the desired outcome. In the transport case, they measure how a service meets individual, community, and government expectations and needs. They may be applied to either the system or to its components. Management of a system to achieve maximum effectiveness is known as transport systems management (TSM). Traffic management, on the other hand, relates to the regulation of the flow of traffic by local measures, and is described in Sections 7.2.3 and 30.7. Transport demand management (Section 30.1.3) is a subset of TSM.

The expectations of customers, owners and stakeholders result in a set of transport needs which must be translated by management into key operating objectives. Ten common objective categories are listed below.

(a) *accessibility* (Section 31.2.4). A good measure of accessibility is the number of people or facilities within walking distance of a transport access point — i.e. within its catchment or watershed.

Mean acceptable walking distances to common transport access points are commonly about 400 m (Taylor, 1980), with only 10 percent of travellers prepared to walk 800 m, unless their destination has relatively high levels of service. However, data for Perth (Ker and Ginn, 2003) suggests that the mean acceptable distance to a quality rail station is closer to 1 km, with some travellers walking 3 km. Brisbane data (Burke and Brown, 2007) give a median distance of 600 m and an 85th percentile of 1300 m. If the conditions for walking are unattractive or if population densities are high, these distances will be significantly reduced.

At the typical walking speed of 1.4 m/s (Section 20.4), 500 m represents a 6 minute walk. A 500 m limit also means that low-density suburban areas will have difficulty generating more than around 1000 total trips per kilometre of transport route (assuming four trips per household). Clearly, residential population density and origin and destination locations are key variables determining accessibility levels.

Closeness to work is another accessibility factor. Some studies have suggested only 50 percent satisfaction with zero trip times (and also with a 40 minute trip time) and that times of 10 to 20 minutes give peak (80 percent) satisfaction (Young and Morris, 1980).

(b) *service frequency* (or timeliness). Vehicle headways or passenger waiting-times are good measures. The minimum number of vehicle-hours to be supplied in a given time is a less direct measure, which is equal to the number of vehicles in operation. Similarly, the *VKT* patronage measure (Section 30.2.1c) gives a good indication of service frequency and also gives an estimate of the total traffic load on the system. Vehicle-hours of travel are more useful for estimating operating costs.

(c) *service reliability.* A useful reliability measure is the percentage of vehicles departing and arriving within x minutes of their scheduled time. This is a critical consumer performance measure as unforeseen delays, unscheduled waiting times (Section 31.2.2), and variable trip times are all poorly regarded by travellers.

(d) *service quality.* This measure applies particularly in such areas as convenience, comfort, security, cleanliness and condition, and affordability. Interview teams are usually needed to obtain the indicators related to these factors.

(e) *service safety*, as measured by the number of *crashes* and other unplanned incidents (see Chapter 28). This is a key consumer issue and various indicators are discussed in Section 28.1.2. Events per distance travelled, or period of operating time or per period of passenger time are the most widespread measures.

(f) *travel time* and *travel cost*, which must be seen as 'reasonable', e.g. with respect to any alternatives.

(g) *service information.* This has grown out of the ability of modern systems to provide with real-time information about vehicle departure and arrival times. Such data is now sought both before and during trips, from special facilities, from websites or from dedicated channels on personal devices. The information provided must:

* be in a form relevant to a travellers' needs, particularly at interchanges,
* cover routine real-time operating data,
* give rapid advice on operating incidents, emergencies and construction and maintenance impacts.

(h) *service alternatives.* This reflects the desire by travellers to explore the options that suit their needs at a particular instance and for these options to create genuine competitive demands on transport operators. The options will relate to the other factors in this list and to the location and route of the service. The data should span all systems and not just those managed by a particular operator. Trip planning aids should also be provided. Park-and-ride facilities (Section 30.6.3) should be part of the integrated service.

(i) implementing policy objectives. Table 30.2 shows how particular transport policy expressions are translated into travel management objectives.

Table 30.2 Translating policy expressions into objectives for transport management.

Transport policy	Related management objectives
1. Better use of road-space	1.1 Increased throughput of travellers (e.g. by increased car occupancy, Section 30.4.1) 1.2 Improved traffic conditions (Section 17.4) 1.3 Spreading of the peak demand (Section 30.3) 1.4 Reduced private vehicle demand
2. Improved road-based public transport (Section 30.3)	2.1 Reduced bus delays and trip times 2.2 Improved bus reliability 2.3 Improved 'image' of bus transport 2.4 Increased bus patronage (Section 30.4.2) 2.5 Reduced operating costs (Section 29.1.2)
3. More shared use of all transport modes	3.1 Better modal interchanges 3.2 Information provided about other modes
4. Increased environmental and energy awareness	4.1 Smoother traffic flow (Section 29.2.4) 4.2 Reduced transport demand (Section 30.3) 4.3 Lower pollution emissions (Section 32.3)
5. Increased economic development	5.1 Transport provision encourages development in vicinity (Table 5.1, Part C)

30.2.3 Efficiency and equity

In resource terms, *efficiency* is concerned with delivering the maximum service for the minimum consumption of resources (Section 33.5.2). Direct service efficiency measures are obtained by dividing one of the patronage measures listed as (a) to (f) in Section 30.2.1 by the total cost of the service. For example, using measure (a) for passenger travel would give (passenger trips per dollar spent) as the efficiency measure. The marginal efficiency of a new addition to the system is estimated by dividing the extra patronage by the total cost of the new facility. If the efficiency measure is based on the net cost of the service (i.e. total costs less revenue collected), the measure relates more to the community than to the transport system. In financial terms, efficiency is also concerned with maximising the net revenue collected and minimising the cost of collecting that revenue.

The energy and space efficiency of various modes is measured by such ratios as passengers per joule and passengers per square metre of floor space. Typical measures are shown in Table 30.3 and a more vehicle-specific set has been given in Section 27.2.1. The table illustrates that it is important for public transport vehicles to operate at near their capacity — otherwise they may be very inefficient.

Equity relates to the manner in which transport resources and services are distributed among the particular markets and market segments being served.

Table 30.3 Energy and space efficiency of various modes. Note that the actual conventional measures quoted (e.g. J/m) are actually the inverse of efficiency, being input/output.

Vehicle / Item	Train	Tram	Bus	Car	Truck	Light truck	Motor-cycle
Number of seats	600	50	50	4–5	2	3	1–2
Seat area (m^2) needed at 50 km/h	1.4	4.0	4.6	23	n.a.	n.a.	1.5
Energy per distance travelled (kJ/veh-m)	190	38	12–24	5	12	6	3
Energy/passenger/distance (kJ/pass-m)[1]							
a. all seats occupied	0.3	0.5	0.5	1	6	2	1
b. 20% occupied	1.6	2.7	2.4	5.5	n.a.	n.a.	n.a.
c. peak periods[2]	0.5	0.9	1.0	4.3	n.a.	n.a.	n.a.
d. off-peak[3]	1–10	1–10	1–10	3.6	n.a.	n.a.	n.a.
e. daily average[4]	1.6	1.3	2.4	4	n.a.	n.a.	2.2
Energy per tonne of freight per distance (kJ/t-m)							
a. urban	0.8	n.a.	n.a.	n.a.	4	23–36	n.a.
b. non-urban	–	n.a.	n.a.	n.a.	2	–	n.a.
c. intrastate	–	n.a.	n.a.	n.a.	3	n.a.	n.a.
d. interstate	–	n.a.	n.a.	n.a.	0.7–1.3	n.a.	n.a.

Notes:

1. Standing passengers would reduce these figures. The equivalent energy efficiency figures in kJ/m for walking and cycling are 0.00018 and 0.00006 respectively.
2. Allows for lower loadings in the counter-peak direction.
3. Loadings in off-peak, and hence energy efficiencies, are usually variable.
4. Based on average occupancies throughout the day of 120 persons for trains, 20 for trams and 10 for buses.

30.3 TRANSPORT DEMAND MANAGEMENT

Items 1.3 and 4.2 in the list of transport management objectives in Table 30.2 in Section 30.2.2 highlight the need to consider measures to reduce transport demand in order to optimise the use of the entire transport system. These measures may prove more effective than the alternative strategy of increasing transport supply or capacity. *Transport demand management* (or *travel demand management* or *environmental traffic management*) is the introduction of measures to reduce the amount of travel undertaken, totally or by a particular mode. It is thus a subset of transport systems management (Section 30.2.2) and can be applied generally, even in the absence of congestion. Congestion management methods are discussed in Section 17.4.4. Measures that can be applied to achieve transport demand management are:

(a) *Management of initial travel demand.* Measures such as those listed below are employed to reduce the need for and extent of travel:
  * staggered and flexible working hours, spreading of the peak demand time,
  * using telecommunications rather than travelling (telecommuting and teleconferencing, Section 27.8),
  * incentives and inducements to reduce travel,
  * regulations inhibiting travel, and

* land-use management and zoning via town planning (can be used in the long term, Section 31.3.2).

(b) *Development of alternative modes* (e.g. Section 27.8). Measures include:
* improving the performance of competing modes (as in Table 30.2),
* subsidising the fares on competing modes and publicising their availability. However, Section 30.4.1 shows that fare subsidies can be counterproductive. Similarly, bonuses can be granted to users of a preferred mode.
* imposing restrictions on the use of the target mode (as in d and e below), and
* making it easier for passengers and freight to transfer from the target mode to some other mode.

(c) *Uninhibited congestion.* No effort is made to increase supply in the face of excess demand and, consequently, potential travellers are discouraged. The inefficiencies associated with this approach are described in Section 17.4.4.

(d) *Traffic management.* Traffic on a road is deliberately limited or prohibited by various measures. Such physical control of the capacity of the system will have negative effects on travel efficiency. The measures to prevent travellers from using particular routes can be:
* physical devices (Sections 7.3.1 and 30.7),
* regulation, relying on signs, area permits, and enforcement, and
* changes in social attitude.

(e) *Tolls, taxes, and levies.* Increasing the price of the transport mode is a classical method for managing transport demand (Section 29.4). General experience and the elasticity figures for car travel in Section 27.8 (at best, –0.2) both indicate that car usage charges have to be greatly increased — to many times running costs — in order to significantly influence car travel.
* A common method is to increase fuel charges or taxes (Section 29.3.1). This is the neatest and simplest solution for road travel but suffers from its universal impact on all road travel. A way around this problem is to use a reduction of other taxes (e.g. income tax or local rates) to return some of the revenue collected to those unreasonably disadvantaged. Note, however, that decreasing general taxes will increase travel by increasing funds available for travel.
* More sophisticated methods of tolling or road pricing are described in Section 29.4.

(f) *Parking control* (Section 30.6). This method has the merit of ease of application — particularly via increased pricing — but is likely to meet considerable local resistance, engendered by the reliance of most business and commercial centres on road-based customers. Furthermore, common experience is that increased parking charges have more effect — usually undesired — on land use than they do on car usage. Parking controls also do not inhibit through traffic.

(g) *Better utilisation and efficiency of the mode suffering from excess demand.* Key methods are improving travel efficiency and vehicle occupancy through such measures as ride-sharing (Section 30.5.1) and dedicated lanes for high-occupancy vehicles (HOVs) such as buses and taxis (Section 30.4.4).

30.4 PUBLIC TRANSPORT

30.4.1 General

The objectives of public transport are to:

1. *assist the community to:*
   1.1 manage large travel demands,
   1.2 provide as much of its transport demand as is feasible,
   1.3 ensure a healthy, pleasant, sustainable and safe environment,
   1.4 minimise the economic and social costs of transport,
   1.5 make the best use of the existing infrastructure, and
   1.6 preserve the desired form of existing areas;
2. *with respect to individual travellers:*
   2.1 provide attractive standards of accessibility (Section 30.2.2a),
   2.2 provide attractive service levels (Section 30.2.2b),
   2.3 assist those without access to a car, and
   2.4 provide an alternative to the car.

For public transport to be attractive, it must be available, affordable and competitive.

Availability is a matter of supply and is addressed in Section 30.2.1. Its challenge is exacerbated by the fact that public transport is often used for trips to work and to school. Hence, demand may be relatively high during short (peak-hour) periods of the day and relatively low for the remainder of the day.

Affordability is often achieved by subsidising fares. It is common international experience that public transport systems require significant subsidies — on the average of about 50 percent but in places up to 70 percent — in order to at least meet marginal operating costs. Given the likely continued dominance of the car in many sectors (Section 30.5.1), these subsidies are increasingly seen as a normal part of the operation of the system and possibly captured by the increased land values that they create. Many travellers become captive to public transport, as demonstrated by a fare elasticity of about −0.3, with a range between −0.10 and −1.30 (Section 33.5.4). Recall that an elasticity of −0.30 means that a 10 percent fare increase would cause a 3 percent drop in patronage.

Competitiveness with respect to the car is a major factor and using subsidies to further reduce fares commonly generates new travellers or takes them from other favoured modes (Section 30.3), rather than transferring them from private cars (OECD, 1980). For example, reducing public transport fares in densely populated areas often takes passengers from such modes as walking and cycling and leaves car travel little affected (Section 30.5.1). The influence of the car is directly linked to levels of car ownership (Section 30.5.1).

Two additional factors which particularly influence the demand for public transport are listed below.

Captive users, such as the very young and the very old and the very poor, have no alternative but to use public transport (Section 30.5.1).

Community attitudes in car-based societies often favour public transport collectively, but disfavour it when making individual decisions. For example, typical traveller attitudes towards public transport unrealistically require a high level of service at minimum fare, a 'no need to look at a timetable' peak schedule, off-peak schedules adequate for any possible demands, and a network linking all conceivable origins and destinations.

In comparing or predicting the use of particular transport modes, it is necessary to distinguish between participation rates and activity rates. The participation rate describes the number of people using a particular mode as a fraction of the population. The activity rate is the number of trips per day made by those participating in a mode. That is:

(total mode trips)/day = [number of participants in the mode][trips/day/participant]

The usefulness of this approach is that the activity rate in trip/day/participant remains quite constant across locations and groups. The key variable is the number of participants. This can often be quite accurately estimated by considering the facilities provided, their market penetration, and the appropriate levels of market segmentation. Trip rates are a little over 2/person-day for most modes, but participation rates can vary from 90 percent for cars to one percent for motorcycles (Wigan, 1983).

30.4.2 Road-based public transport

Roads differ from other transport infrastructures in that they are all-pervasive, with most local laws usually requiring each piece of land to have access to a road. Thus road-based transport can potentially provide all users with door-to-door service.

Public transport use of the road system is via high-occupancy vehicles such as taxis, buses, trams (streetcars), and light-rail vehicles operating in road medians. Buses are discussed in Sections 30.4.3–4 and some data on trams is given in Tables 30.3&5. A critical review of light-rail relative to buses is given in Hensher (1999). Taxis (or taxi cabs) represent a significant transport component, providing many regular and emergency services. They are often an important substitute for buses. A significant number of the users of taxi cabs may be economically disadvantaged and taxis are usually not the often-assumed preserve of the upper income groups. A degree of government regulation of private buses and taxis has often been found necessary. This relates to the licensing of the vehicles, the drivers, and the routes over which they operate.

A range of options known as paratransit exists between these relatively well-established systems and the private car (Section 30.5.1). A summary of their service characteristics is given in Table 30.4.

Thus in an economic and transport efficiency sense, road-based transport modes have a major role to play as people movers, due to their door-to-door facility. They also have a dominant role in the provision of essential mobility to people in low travel-demand situations — situations that may arise due to the locational, personal, or social disabilities of the intending traveller relative to the bulk of travellers. Understandably, such needs normally surface at local community levels, as this is where the individual restrictions on personal mobility are first felt through social isolation or deprivation. The groups most affected are discussed as captive travellers in Section 30.5.1.

The desired destinations are usually also distinguished by their local nature, e.g. to shopping, medical, and recreational facilities. Thus the major, additional service needs of this low-travel group are for non-arterial trips, off-peak trips, and provisions for disabled people. Rarely can the satisfaction of these needs be supplied on economic grounds, as any service provided will have high organisational demands associated with low patronage. Therefore, such facilities must be planned with the prime objective of satisfying a defined and agreed need, rather than of operating an economically efficient facility.

30.4.3 Bus transport

The major road-based public transport mode is the bus. In many cities, buses are the major public transport mode. Common bus types are listed in Table 30.5. The larger buses are normally diesel-powered and often can also run on LPG or from electric wires

Table 30.4 Paratransit service, service characteristics.

Mode	Vehicle from:	Driver from:	Timetable fixed by:	Route fixed by:	Fare fixed by:
Supply service:					
Normal bus	Operator	Operator	Inflexible or operator	Inflexible or operator	Operator
Demand services					
Dial-a-bus	Operator	Operator	Flexible or passenger	Flexible or passenger	Operator
Route deviation	Operator	Operator	Inflexible, operator or passenger	Part flexible, operator or passenger	Operator
Exclusive taxi	Operator	Operator	Passenger	Driver or passenger	Operator
Shared taxi	Operator	Operator	User group	User group	Operator
Jitney[1]	Operator	Operator	Operator or passenger	Operator	Operator
Subscription services					
Dedicated bus	Operator	Operator[2]	User group	User group	Operator and user group
Social-service bus	Group/ public	Group[2]	Group	Group	Group
Van pool[3]	Employer	Group[2]	Group	Group	Employer
Car pool[3]	Group	Group	Group	Group	Group

1. The jitney is a vehicle used on a service operating over a fixed route, but with a flexible schedule and picking up and depositing passengers on demand.
2. The driver may be paid or a volunteer, but the responsibility for the provision of the driver usually remains with the operator.
3. See Section 30.5.1.

Table 30.5 Common bus types and their characteristics.

Bus type	Approximate values for:		
	passenger capacity	relative cost	relative life
paratransit (Section 30.4.2)	3–10	0.3	0.7
minibus	6–20	0.4	0.7
midibus	30	0.7	0.8
standard	60	1.0	1.0
large	90	1.4	1.3
double-decker	110	2.0	1.3
articulated	110	2.8	1.3
tram (Section 30.4.1–2)*	150	5.0	1.7
super double-decker	170	2.5	1.3

* tram track costs will be at least double bus way track costs.

(trolleys). Standard and double-decker bus types are categorised by whether their engine is forward of the front axle, under the floor or at the rear of the bus.

The major transport advantages of the bus are:

(a) if fully loaded, it carries far more people per unit area than a car;
(b) if fully loaded, it is more energy-efficient than a car (Table 30.3);
(c) it has a better crash record per passenger-kilometre than a car;
(d) it has a lower capital cost per passenger than trams and trains;
(e) it has flexibility with respect to re-routing and re-scheduling;
(f) it has the potential to supply both a door-to-door and an arterial service, which no other public transport mode can do;
(g) the size of the unit operated can be quickly changed to meet immediate demand;
(h) it can use existing 'free' road space and right of way; and
(i) it can effectively service population densities as low as 8 people/ha or 8 houses/ha, and passenger demands as low as 20 passenger/bus-hour.

However, the bus has some serious disadvantages:

(j) door-to-door service can only be supplied at great cost (it is equivalent to a taxi service). For instance, bus routes at an 800 m spacing would be needed to bring buses within the 400 m walking distance just tolerated by many people (Section 30.2.2b);
(k) bus stops (Section 30.4.4) are often dirty, uncomfortable and inhospitable and the bus schedule information provided at them is often inadequate;
(l) bus services sometimes have a reputation for unreliability and poor time-keeping (common operating objectives for buses are to reduce both trip times and their variability) and published bus schedules are sometimes found to be confusing and unhelpful. This problem can be alleviated by the use of bus locating systems, feeding current estimates of bus arrival times into VMS signs at bus stops and into other customer networks and personal devices;
(m) the bus will have longer trip times than a car, unless given very significant traffic priority (Section 30.4.4). Typically, bus travel times are 20 to 40 percent longer than car travel times. A key component of this is the 20 to 30 s that a bus typically spends at each bus stop;
(n) the bus is usually less comfortable than a car and may be unpleasantly crowded;
(o) to the user, out-of-pocket expenses for a bus trip often appear higher than for a car trip;
(p) buses can be involved in crashes. The most common cause of serious bus crashes is vehicle roll-over (Section 19.2.3). Injury severity is typically heightened by a lack of occupant restraint via seat belts (Section 28.4) and poor side protection for occupants. Passengers in crashes are often completely or partially ejected through openings such as windows and doors, with heightened risk of serious injury;
(q) for scheduling reasons (e.g. to and from depots), trips may have to be made with no passengers (known as *dead running time*).

In a typical city or town, buses would rarely account for more than 10 percent of person-movements although a few cities — such as Ottawa and Brisbane — have developed extensive bus systems. For those bus travellers who are not captive to the bus, the choice between car and bus is an example of modal choice (Section 31.3.6). The traveller's decision is based on:

* the direct cost of the bus fare,
* the cost of the fare relative to the cost of car travel, which is often perceived to be free (Section 30.5.1),
* trip time,
* associated time losses which are predominantly comprised of:

– walk time (Section 30.2.2b),
– waiting time, which averages somewhat less than half the bus headway ('Less than half' allows for intelligently-random arrivals by passengers. When service headways exceed a certain level, passengers begin planning their arrival at the bus stop.) and
– transfer time between modes.

* service quality issues such as the reliability of the service.

By placing a dollar value on trip time and waiting time, it is possible to add them to fare and vehicle costs to develop a *generalised cost* (or *disutility*, Section 31.2.2) for each mode, to then allow the total perceived costs of all modes to be compared. This approach can be extended by using the methods in Sections 31.2.2 and 31.3.6, but will still not cover the effect of the quality of the service on patronage.

Long-distance bus operations (i.e. between urban areas rather than within them) play a significant transport role for inter-urban commuters (e.g. regular, daily customers) and for irregular travellers (e.g. tourists).

30.4.4 Provisions for buses

It was seen in Section 30.2.2a that people will usually walk up to 400 m to a bus stop. However, the stops must be at less than a $2 \times 400/\sqrt{2} = 570$ m spacing to cater for people living off the route and experience suggests that everyone in an area will be adequately served if the stops on a square grid of bus routes are at a 500 m spacing. More specifically, bus stops are located at points which:

* are conveniently located,
* are secure and comfortable places to wait and queue,
* minimise the impact of waiting passengers on any sensitive land-uses,
* permit passengers to board and alight and cross the road with safety,
* take advantage of convenient traffic locations (e.g. at traffic signals), and
* minimise the delays that the stopped bus will cause to other traffic using the road. This can be reduced by the use of full-width bus bays when traffic flows are over 800 veh/h/lane.

The basic bus stop is simply a sign-posted kerbside pick-up and set-down point. It will usually require vigorous policing of parking restrictions at the stop to be effective. Seats and a shelter increase the amenity of the stop. There must be a good walking surface between the bus, the footpath, and any seats and shelter. Some consumer problems with bus stops are listed in (k) and (l) in Section 30.4.3. *Nub* (or bulb or bulge or kerb extension) stops take this a step further by extending the raised stop into the parking lane. There is usually no problem with parked vehicles, but traffic delays increase as the bus remains in a traffic lane whilst loading and unloading passengers. The nub stop also provides further space for amenities at the bus stop.

Because buses and other high-occupancy vehicles (HOVs) such as trams and taxis share roads with private motor vehicles, they are subject to the same traffic delays, particularly at peak hours. However, if the objective is passenger-moving efficiency and reliability, then attention must be given to the special needs of buses and to giving them priority within the traffic system. This can be via:

A. adjustments to lane operations

(a.1) adequate lane width (Section 6.2.5),

(a.2) parking restrictions in lanes used by buses and other HOVs,

(a.3) stopping bays for buses (see above),

B. adjustments to intersection operations

(b.1) set-back bus-only kerb-side lanes (Figure 30.1) at busy intersections which are:

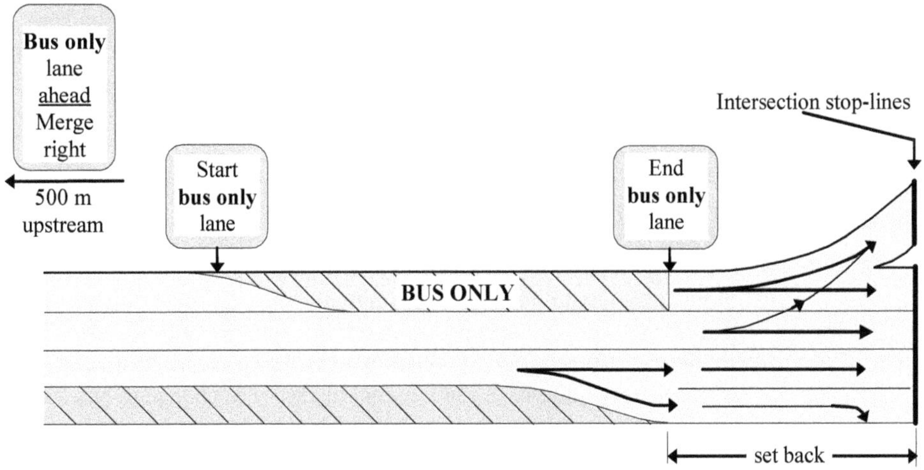

Figure 30.1 Typical set-back bus lane.

* long enough to permit buses to manoeuvre into them,
* close enough to a signalised intersection to ensure that a bus waiting in them will be in a queue which will clear the intersection during one green phase, and
* set as far back from the actual stop-line as possible to preserve as much as possible of the capacity of the intersection (Section 17.4.3),

(b.2) turning-radii at intersections which are suitable for at least low-speed turns by buses (Section 27.3.8),

(b.3) traffic signals to accommodate buses via special turning arrows, dedicated green-phases and extended clearance intervals (Section 23.2.3 & 23.7),

(b.4) bans on conflicting turning movements by cars, and

(b.5) priority at exits and entries to freeways,

C. special facilities for buses

(c.1) queue bypasses — usually at toll booths, metered on-ramps, and bridge and tunnel approaches,

(c.2) through lanes reserved for use by buses and other HOVs. There are three types, depending on the method of operation. When operating in the:

* same direction as the traffic, they are called *with-flow* (or *concurrent-flow*) lanes. They are usually located on the kerb-side lane, with a lane width of at least 4.5 m to provide a buffer from other traffic.
* same direction as the current peak flow, they are called *tidal-flow* lanes. They are typically in a freeway median and separated from general lanes by a permanent barrier or wide (4 m) buffer zone. They are at least 6.0 m wide to permit emergency passing.
* opposite direction to the normal traffic, they are called *contra-flow* lanes. They typically take advantage of unused central road space in the low-flow

direction. Contra-flow lanes can cause traffic safety problems when pedestrians or turning movements are high.

If well-utilised, conversion to a reserved bus-lane can increase the total passenger-throughput of a traffic lane. In normal traffic, cars provide about 2000 passenger/h/lane (1800x1.2) and buses about 10 000 (± 5 000), taking lane capacity as 1800 car/h. In a lane dedicated to buses, if a bus has a pcu of 3 (Table 17.1), this gives 600 bus/h and with 50 passenger/bus corresponds to 30 000 passenger/h/lane. Thus the capacity increase car:bus:bus-lane is in the ratio 1:5:15. At low utilisation levels, bus lanes still provide higher travel speeds, but they are usually not cost-effective. Experience suggests that HOV lanes will only be practically effective if:

(1) the city has a population of over one million,
(2) the urban centre has over 100,000 jobs,
(3) substantial traffic congestion exists,
(4) there is a demand for at least 25 bus/h in peak hours,
(5) there are separate passenger loading/unloading facilities, such as park-and-ride (Section 30.6.3) provisions at bus stations, and
(6) there are sufficient car travellers who will experience extra difficulty with car travel, and have the freedom of choice to change to bus travel.

(c.3) Special bus facilities (*transitways* or *busways*), providing separate carriageways for buses and other HOVs. They may be short lengths of roads in urban shopping districts or major transport arteries. In the latter case, the capital cost of the busway would be somewhere between a tenth and a hundredth of the cost of a fixed-rail system. An excellent review of the viability of busways is given in Hensher (1999).

A number of bus-priority schemes have failed because the disbenefits to the rest of the travelling public exceeded the benefits to the priority vehicles. For example, although a bus or tram will carry many more passengers than a car, if it is relatively slow-moving and bulky it will delay a large number of individual cars. The total of these motorists' delays may exceed the benefits gained by the bus passengers, particularly if the buses are not fully laden.

30.5 PRIVATE TRANSPORT

30.5.1 Travel by car

The common advantages of private car transport in comparison with public transport are that it:

(a) provides a universal mode for trips of all lengths and to all origins and destinations,
(b) provides door-to-door service,
(c) requires no mode changes,
(d) is always available,
(e) has a high operating speed,
(f) gives assured seating,
(g) provides weather protection,
(h) gives assured privacy
(i) produces pride of ownership,

(j) is in good repute as a service, and
(k) has a low perceived cost (many travellers perceptually divorce the running cost of their car from their travel cost).

The disadvantages of the private car relative to other modes are that it:
(l) has a high real cost,
(m)has a high social cost (Chapter 32),
(n) is relatively inefficient in terms of fuel and space,
(o) has a high crash potential per passenger carried, and
(p) has variable journey times (the coefficient of variation of a typical urban trip is around 15 percent).

For the average household, the net result is that the private car increases the efficiency with which family members can participate in various activities. Therefore, it is not surprising that in many cases the car is the dominant form of private transport, although other important modes are walking (Section 30.5.3), cycling (Section 27.6), and motorcycling (Section 27.5). In many situations the car is also the dominant total transport mode. Even in compact cities and towns, the car can account for 90 percent of all person-movements.

Travel by car is very responsive to car ownership levels. An international review showed that, on the average, a person living in a household owning one car makes only 60 percent of the public transport trips of a similar person in a no-car household (Webster *et al.*, 1984). If the household has more than one car, the proportion drops to 40 percent. The percentage drop is also greater in areas where public transport use was (previously) high. Broadly, adding a car to a household loses one person's public transport trips. Typically, a 1 percent increase in cars owned per person causes a 0.2 percent drop in public transport trips. Thus car ownership deserves further consideration. Some discussion of car ownership has already been given in Section 27.1.1 and the issue is examined further in Section 32.5.

Home interview surveys (Section 31.3.1) provide useful sources of car ownership data to supplement data from conventional vehicle registration records. This data is usually expressed in terms of cars per person or per household, commonly using a logistic curve (see Figure 33.1) of the form:

$$\text{car/person} = C_1/(1 + C_2e^{-C_3t})$$

where t is time and the C's are constants. C_1 is the *market saturation* level, and the role of C_2 and C_3 can be seen from Figure 33.1. In most countries, if every eligible driver had a car, C_1 would be about 0.7. Predicted saturation levels in cars per person range from — for the countries studied in Section 4.3 — 0.2 in Japan to 0.8 in the U.S. (Tanner, 1983).

Market saturation is sometimes called the *epidemic diffusion* technique and was introduced by Tanner of TRRL, who extended it to include the effect of the future rate of economic growth and future fuel prices on the path to saturation, in simple terms, by adjusting C_2 and C_3. It has more recently proved necessary to replace the logistic curve by a more general power curve. The alternative approach (Mogridge, 1983) is to use the sort of data shown in Table 32.3 to extrapolate forward on the basis of estimates of future household incomes and types. An alternative version of this approach is to estimate on the basis of comparisons with other sectors or regions.

At a more disaggregate level (and following Section 33.5.1), car ownership is affected by such factors as:

* household income,
* the occupation and age of the household head,

* household size and location,
* the number of licensed drivers in the household,
* public transport availability,
* the availability of subsidised or company cars (Section 27.1.1), and
* the prevalence of materialistic attitudes.

Household income usually dominates, with disposal income the strongest of the income variables. The best car ownership link has been with the income per person 20 years earlier, perhaps suggesting a lag in adjustments to land use and the transport system (Tanner, 1983). Lesser emphasis is placed on the role of commercial vehicles in such analyses, both because they are fewer in number and because their movements can best be predicted from a knowledge of freight flows.

Thus the emphasis in analysis is usually placed on the household rather than the person or the adult as the decision-making unit. The characteristics of the household are varied with location and with time. Commonly, the number of cars per household (or per person) increases with income to either some power or logarithmically, typically with an income elasticity of about 0.5 (Section 33.5.4). A similar elasticity applies to car usage. The combination of the two means that car travel also increases as incomes increase, with an elasticity of close to 1.0.

A useful dependent variable is often the proportion of households, P(N), with N = 0, 1, 2, ... cars (Tanner, 1983; Wigan, 1987). Common P(N) vs income relationships are illustrated in Figure 30.2. The P(0) and P(1) forms are often respectively taken to be negative exponential and gamma distributions (Section 33.4.4), i.e.:

$$P(0) = ae^{-bx}$$

and

$$P(1) = cx^{d}e^{-fx}$$

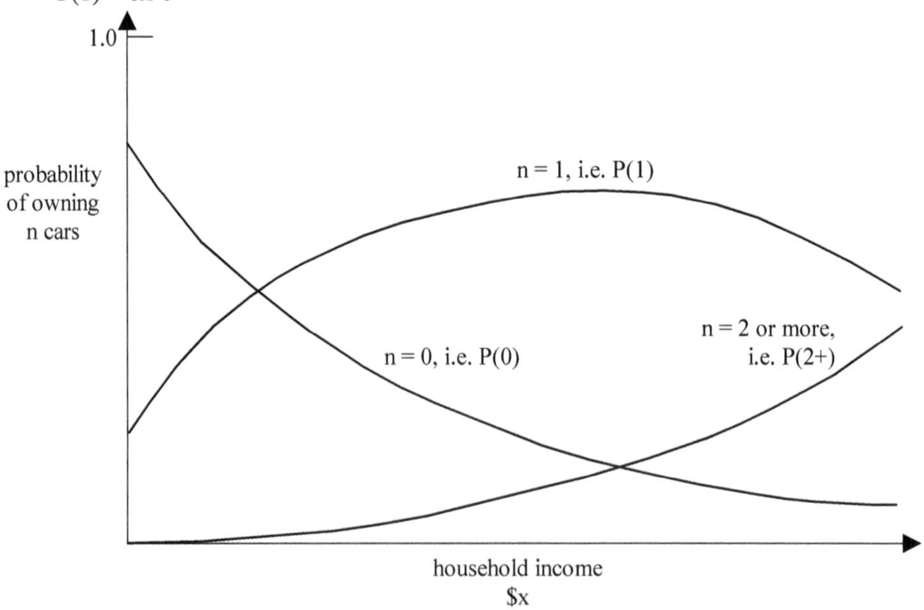

Figure 30.2 Typical car-ownership model.

Note that:

$$P(2+) = 1 - P(0) - P(1)$$

The average *occupancy* of the urban car in developed countries is only about 1.2 people; hence it is not surprising that there is a marked interest in providing for and encouraging higher vehicle occupancy as a means of reducing traffic demand. One technique for doing this is to encourage ride-sharing. *Ride-sharing* is a generic term covering:

* *car-pooling,* drivers rotate the use of their cars within their group,
* *car-sharing,* one group member supplies the car,
* *van-pooling,* travellers share a passenger van, often supplied by their employer, and
* *bus-pooling,* travellers share a bus rather than a car.

Ride-sharing is aided by:

(a) the availability of high-occupancy vehicle lanes (Section 30.4.4),
(b) employer subsidies, e.g. by van purchase,
(c) preferential or concessional parking facilities,
(d) ride-matching services, and
(e) guaranteed ride programs (to cover sharing failures).

Particular interest has arisen in car-pooling, but attempts to encourage it have not always been successful. Some of the reasons for this lack of success are as follows:

(1) A substantial proportion of the occupants of multi-occupant vehicles come from the same household and family group. Thus, the low success rate with respect to genuine car-poolers is even lower than might be expected from casual observation.
(2) The car freed by car-pooling is usually employed for other travel.
(3) Pooling usually means longer trips for the driver.
(4) Pooling is usually only practical for people who work together. Thus, van-pooling is usually the most effective means of encouraging multi-occupancy travel.
(5) Many people regard car-pooling as inflexible and distrust their fellow travellers.
(6) The lack of a suitably large and accurate database of potential car-poolers.

Another approach is *car-sharing*, where one vehicle is used separately by a variety of drivers. Commercially, this area is dominated by the car rental industry. Usually, for co-operative schemes the most difficult part of such schemes is ensuring that a participant can easily obtain a vehicle when they need one.

One of the problems with private transport is that there are a number of groups in the community who do not have use of a car except, perhaps, as a passenger. If these people have access to and a need for public transport, then they are *captive* users of the public transport system. The major groups involved as captive travellers are:

(A) people without a car available to them, e.g. the poor, the aged, and the ill;
(B) people who do not possess a driver's licence, e.g. young people below the legal age for a licence, disqualified drivers, disabled people, and people eligible but without a licence.

There are also *car captives* whose trip origins and/or destinations are only accessible by car.

30.5.2 Freight

The transport of freight by road is a vital and major industry. The figures in Table 4.1 for the percentage of tonnes carried by road indicate the dominance of road freight around the world. The smaller percentages for tonne-kilometres in Australia and the U.S. indicate the effectiveness of railways and shipping in moving bulk products, such as the output of mines, over long distances.

The freight industry typically represents about 5 percent of the GDP of a region. It contributes about 25 percent of the cost of many low value commodities (e.g. stone) and, typically, 4 percent of the cost of high value commodities (e.g. instruments). Freight involves both:

* *vehicular operations*, such as the manufacture, driving and operating of trucks. These aspects are discussed in Section 27.3.1, and also
* *logistics management*, such as fleet administration, freight-handling operations at freight terminals, inventory control and warehousing, freight management and freight forwarding. Freight forwarding is providing a total freight transport service to a client, and is not associated solely with one transport mode or method. These aspects are increasingly linked to the operations and cost structures of the organisations supplying and receiving the freight.

As with personal travel, the demand for the transport of freight arises from the:

(a) level of consumer demand for freight, which itself arises from economic activity. There is an almost linear link between percent changes in the GNP and percent changes in road freight (OECD, 1986a). The linearity usually increases if non-farm GDP is used. Alternatively, a linear link has also been observed between the logarithms of the GDP and the freight tonnage. The elasticities (Section 33.5.4) commonly exceed unity, with the freight task growing more quickly than the GDP.

(b) spatial separation of the points of production and consumption and of various intermediate transfer terminals, warehouses, retail outlets, and waste-disposal facilities.

The tonnage of freight to be moved is estimated from predictions of the freight generated in a particular area, as described in Section 31.3.4. Market segmentation with respect to freight is best achieved using 'trip purpose' defined by reference to trip origin and destination as the explanatory variable. Note that the firm (or business) is the freight equivalent of the household discussed in Section 30.5.1. These freight flows are then allocated to the various transport modes by some form of modal split, using the methods in Section 31.3.6. Depending on the styles of distribution and production involved, the final freight tonnage delivered may have given rise to several times that amount delivered at various intermediate facilities. This will distort both tonne and tonne-km indicators of the freight task.

The road component of the modal split is largely carried by trucks and is used to predict truck flows. Roads are used for both the long-distance movement and also for local distribution of freight. Indeed, all freight will utilise the road system at one or more stages during its manufacture or distribution. Long-distance haulage will usually involve bulk movement whereas freight distribution is usually characterised by small shipments of a range of products from a range of origins to a range of destinations.

Trucks are discussed in Section 27.3. Other vehicles to participate in the road freight task are vans and light commercial vehicles (Section 27.3.1), although these may also be used by service and trades people carrying their tools of trade, rather than freight. They often prove difficult to classify and consider in planning studies.

Trucks operating in an urban distribution system are not operating in a particularly efficient manner. If trucks are delayed, the consequences will be:

* increased freight costs,
* reduced service reliability, and

* increased damage to goods.

Particular road-related measures to improve truck operations overall include:

1. Planning

1.1 designated truck routes with minimal stops,

1.2 good access to loading and put-down points, and

1.3 good provisions for truck parking.

2. Road design

2.1 flat grades,

2.2 minimum cross-fall,

2.3 wide lanes

2.4 helpful lanemarkings,

2.5 improved roadside and overhead clearance,

2.6 wide radius corners,

2.7 intersection layouts which accommodate large, turning vehicles (Section 20.3.1), and

2.8 signal phasing which does not assume high accelerations.

3. Enforcement

3.1 more consideration of truck accelerating and braking capabilities by car drivers.

4. Management

4.1 planning of pick-up and delivery operations. In many cases, the average urban truck is actually on the road for about 3 hour/day and only travels about 100 km/day. The remainder of the time is spent at pick-up and delivery points (Ogden *et al.*, 1981).

Almost invariably, the road system provides the last means of freight distribution, bringing goods to their final point of sale and use. As a consequence, there is often an inherent incompatibility between such distributional freight movements on urban roads and the aspirations of urban dwellers. The problem is worsened when long-distance freight movements also pass through urban areas, or when large trucks are used to move freight between urban hubs (e.g. between a rail terminal and a product distribution centre). Thus trucks can create significant detrimental impacts on the rest of the community (Chapter 32), as indicated below:

Truck impacts on:	Impact type
General	* safety threat to smaller objects (Section 27.3.8).
Pedestrians	* reduced kerbside visibility when parked.
	* inhibited crossing of the road.
Car drivers	* other vehicles slowed down.
	* considerable road and parking space taken.
Residents	* reduced local amenity.
	* lower land values.
	* increased noise and air pollution.
	* noticeable ground vibration (Section 25.8).
Road Authority	* road deterioration increased (Section 14.3.3).

To manage many of these aspects, it is common for Authorities to introduce regulations to control some aspects of truck operations. These may include licences for operators and drivers (Section 16.6.2), control of vehicle mass and dimension (Section 27.3.2), environmental controls (Chapter 32), truck route control (see above), and controls on the movement of dangerous or hazardous goods (Section 27.3.8).

30.5.3 Pedestrians

Cycling and walking are known as *soft* transport modes. Cycling is discussed in Section 27.6 and walking is covered in this section. Important operational characteristics of pedestrians are their vulnerability, variety, lack of constraint and unpredictability. Their motivations are very different to those of motorists.

Walking is, of course, the fundamental human transport mode and is a component of almost all trips. Pedestrian trips both begin and end at conventional traffic generators (Section 31.3.4) and at the 'stops' associated with the traditional transport modes.

Nevertheless, pedestrians and cyclists are often discriminated against in transport planning by trip estimates (Section 31.3.4) which are too low for the following reasons:

(a) as these modes are unregulated and free, very little relevant data exist,

(b) walk trips are often ignored in travel surveys, although they may represent about 5 percent of work trips.

(c) the value of a pedestrian's time savings and the cost of a pedestrian injury (Section 28.8.3) are difficult to quantify.

Car ownership levels significantly influence walk trips. When a household buys its first car it loses the equivalent of half of one person's walk trips. Purchase of the second car loses an entire person's walk trips. Each 10 percent increase in cars owned causes a 1.5 percent drop in walk trips (Webster *et al.*, 1984).

Key behavioural aspects of pedestrian trips are that:

(a) trip distance is the primary factor influencing an individual to make a trip on foot. Section 30.2.2 showed that 400 m is the average distance at which a walk trip for transport purposes becomes unacceptable. Of course, people may walk larger distances for recreational or health reasons.

(b) pedestrians attempt to minimise travel distance, effort and excess time, even when this involves some traffic risk;

(c) safety is just one of a number of considerations determining pedestrian behaviour. Convenience and access will often cause safety devices to be ignored.

(d) compared with vehicular traffic, pedestrian behaviour is relatively anarchic and subject to rapid flow variations.

The pedestrian's concerns are somewhat more direct and personal than are those of the motorist. Pedestrians are concerned with:

the footpath itself:

* the condition of the surface,
* the absence of obstructions,
* how crowded it is, and
* the continuity, directness and coherence of the footpath system,

the surroundings:

* their attractiveness and interest, particularly whether design elements have been kept to a human scale,
* the presence of adjacent facilities,
* their quietness,
* continual cleansing and rubbish removal, as litter is both prevalent and visible in many pedestrian areas,
* protection from extreme temperature, wind, and precipitation, and
* freedom from crime,

the journey:

* the provision of easy access to cars, car parks and/or public transport,
* the convenience of the path,
* minimisation of walking distances,
* tactile paving devices and strong visual contrasts between road and pedestrian areas to aid pedestrians with sight impairment,
* freedom from even minor annoyance from cyclists and other pedestrians,
* adequate crossing opportunities and kerb ramps,
* minimised kerb delays prior to being able to use a crossing,
* traffic safety,
* adequate rest areas.

Within the footpath there is a need to establish an hierarchy of path users. Commonly, people with disabilities are given highest priority. Next, the lower the speed of movement, the higher to priority. For example, bicyclists should give way to scooters and scooters should give way to joggers and joggers should give way to pedestrians. Speed limits on footpaths would typically lie between 10 and 20 km/h.

There is a basic incompatibility between cars and pedestrians using the same conventional road space. Thus, it is also necessary to decide whether there will be a complete or partial prohibition on vehicular access. Partial prohibition may apply either in time, with vehicles restricted during some hours or days, or in space, with vehicles restricted from some parts of a paved area. At least in the short term, any vehicles denied passage through the pedestrian area will be diverted onto adjacent roads and this extra traffic flow will require consideration. In addition, there will be reduced direct access for car travellers.

Pedestrian planning should see pedestrian facilities as more than footpaths by the side of the road and may well include independent pedestrian networks and pedestrian-only areas such as malls (Section 7.2.3). Malls usually strike initial trade resistance but later prove to be assets for a city, particularly if they are serviced by an effective transport system. There is also a need for a means of servicing the activities linked to the pedestrian-only area.

Data on walking speeds and flows is given in Section 20.4. Commonly encountered pedestrian densities and their equivalent level of service (Section 17.4.1) are given in Table 30.6.

Table 30.6 Transport characteristics of walking (based in part on May *et al.*, 1985)

Situation	ped density (ped/m^2)	level of service	mean speed (m/s)	mean flow (ped/s/m)
Open areas such as squares, concourses:				
free flow (avoid conflicts)	0.3	A	1.3	0.3
Closed areas such as public buildings, terminals, shopping malls, office foyers:				
Walk at normal speed	0.5	B	1.3	0.4
Walking is impeded	0.7	C	1.2	0.6
Walking speed is restricted.	1.0	D	1.1	0.8
All peds are forced too slow	1.4	E	0.9	1.0
Congested causes stoppages	2.0	F	0.7	1.2
Frequent forced stoppages	2.6			1.2
At traffic signals (Section 20.4):				
Crossing road	4.0		1.3 (± 0.5)	1.4

ped = pedestrian

Given the above general densities, the minimum width needed to accommodate a pedestrian is usually taken as 600 mm (i.e. 1.3 person/m^2 at a 1 m headway) corresponding to Level of Service E. *Disabled* people require greater widths, ranging from 750 mm for a walking-stick user to 850 mm for a wheelchair. Minimum path widths are given in Table 7.3.

Other pedestrian issues are discussed in various other parts of this book. For example, Section 20.4 deals with pedestrians at intersections and with pedestrian crossings of roads, Section 28.7 with aspects of pedestrian safety, Section 24.3 with lighting for pedestrian visibility, Chapter 7 with residential streets, Chapters 21 and 22 with signs and markings for pedestrian activities, and Chapter 32 with environmental needs.

30.6 PARKING

30.6.1 Parking needs

The need to consider parking provisions seriously can be gauged from the fact that a car that is typically used for about 15 000 km/year at an average speed of 40 km/h will be in motion on the road for about 400 hours. For the remaining 8400 hours of the year — i.e. 96 percent of the time — the car will be parked.

Parking can be provided either:

* on-road, using unused road space at the roadside or in the median, or
* off-road, either within the premises being used by the driver, or in separate off-street parking facilities (ground level facilities are sometimes called parking lots, and multilevel facilities called parking stations).

The public expects a parking area to have a good, well-drained surface to drive and walk on, clear line-marking, adequate lighting, good traffic engineering for entrance, circulation and exit, and a pleasing appearance. The normal planning approach to the provision of parking space is to relate parking needs to the associated land uses and their propensity to generate trips (Section 31.3.4). Standards have frequently been set for particular land-use developments, requiring a minimum number of spaces to be provided before planning approval will be given. Typical car parking space needs per residential building are:

Number of bedrooms in residential building	Car-parking spaces needed if public transport is: available	absent
1	0.25	1.0
2	0.35	1.1
3	0.50	1.3
n > 10	0.25n	1.0n

A typical desirable space need for shops is one parking space for every 12 m^2 of retail floor space. Special parking provisions, such as greater widths for door-opening, are needed for disabled travellers.

A planning approach based on the above table needs to be applied with some caution. As the figures illustrate, the availability of other transport modes to cater for the trips generated by the land use should be taken into account. The provision of inadequate parking spaces can cause local traffic chaos but, equally, the over-provision of car parking facilities can draw travellers from other modes, decrease the efficiency of a city's transport system, and create urban ugliness. A parking restraint philosophy is discussed below.

Parking studies can aid the development of a balanced approach to parking by indicating:

(a) existing parking spaces, particularly their number, location, and type;
(b) existing parking practices, including usage of available spaces, parking duration and fees, peak parking times, and illegal parking;
(c) the need for parking time-limits;
(d) the adequacy of enforcement measures; and
(e) parking demand, including spatial and time distribution, demand generators, duration, trip purpose, and origin.

Before a parking scheme is introduced, it is necessary to consider its effects on:

(1) road capacity (Section 17.4.3),
(2) safety — parking and parked vehicles on roads are associated with about 10 percent of urban crashes (roads with no parking are usually safer than roads with parking),
(3) existing traffic management measures,
(4) service to adjacent properties (see the land-use discussion above),
(5) local traders, who will usually over-react to any proposed changes,
(6) adjacent parking areas suffering from diverted parking,
(7) the visual appearance of the area,
(8) the balance between public and private transport, and
(9) the directions *of urban development.*

Cities are so dependent on vehicular travel that such an obvious restraint as inadequate parking can have rapid and unfavourable influences. On the other hand, control of the amount of parking space and the setting of parking fees and service levels are effective means for managing traffic demand. Therefore, the implementation of parking policies must of necessity be used in conjunction with other traffic, transport, and urban development measures to ensure that:

* adequate alternatives are provided for the person dissuaded from driving, and
* retail trade is not unduly damaged.

As a policy example, the removal of inner-city parking spaces can be matched by the provision of equivalent parking spaces at an outer-urban public transport station and by improved service levels on the connecting transport facility. As a complication in any application of parking policies, many employers provide staff with free on-site parking.

Parking policies therefore should not be developed in isolation from a consideration of other issues. If new parking policies are being introduced to rectify a parking problem, it must also be recognised that the problem will inevitably have arisen from some other inadequacies. The most probable source of those inadequacies will be in land-use planning and control procedures (Section 31.3.4) and the second most probable source will be inadequate public transport.

30.6.2 Parking provisions

For the driver, parking involves a number of stages, each of which should be considered in the planning of a parking system. The first stage involves the conventional trip between origin and destination (Section 31.3.5). Thus, the need discussed in Section 30.6.1 to consider land uses in designing parking facilities. The second stage involves searching for an empty parking space. Studies have shown that this 'random' searching is a significant portion of the traffic in many urban areas. Hence, it is good practice to pay particular attention to signs indicating the location of, and current availability of space in, particular parking areas. The third stage involves parking and paying any parking fees. The fourth and final stage involves travelling from the parking space to the traveller's final destination. The travel generated by this activity can cause local problems if not catered for adequately.

Parking provisions can be either informal, with no designated parking-bays, or formal, with marked parking-bays. Section 30.6.1 indicated that the provisions could be either on-road (i.e. along the roadside) or off-road in special facilities.

On-road parking should only be used after considering the:

(1) road width (at least 9 m between kerbs for one-side parking and 12 m for parking on both sides),
(2) volume and type of traffic,
(3) parking turnover,
(4) nature of neighbourhood (local residents may need to be given special concessions via windscreen stickers, etc.), and
(5) type of road.

On-road parking is usually prohibited:

* near intersections,
* at bus stops,
* near pedestrian crossings,
* near fire hydrants, and
* at points of poor visibility.

Angle parking is often more convenient than parking parallel to the kerb and always accommodates more parked vehicles than parallel parking, but it is at least 20 percent less safe, more difficult for commercial vehicles, and has a greater effect on road capacity. Part of the increase in crashes will simply be associated with the increased number of parking spaces generating more parking movements. Vehicles usually park with their front to the kerb and this makes loading the vehicle and leaving the parking place hazardous manoeuvres. Angle parking is therefore rarely recommended.

Off-road (or *off-street*) *parking* facilities need to be considered in relation to:

(1) parking demand (Section 30.6.1),
(2) effects
    (2.1) their role as traffic generators (Section 31.3.4),
    (2.2) the effect of the extra traffic on the local street system,
    (2.3) their influence on public transport usage,
    (2.4) their influence on the community's economic well-being,
    (2.5) space consumed, and
    (2.6) their effect on urban appearance,
(3) design
    (3.1) walking distances, with few drivers prepared to walk more than 250 m,
    (3.2) whether the entrances and exits are adequate (see next paragraph).

The internal design of off-road parking facilities is based primarily on the space needed to manoeuvre large cars. Typical car dimensions are given in Table 7.3. These lead to typical parking-bay widths of 2.7 m, although 2.5 m is often used in private and special-purpose car-parks. Bay lengths are equal to the nominal car-length (< 5 m) for 90°

parking and equal to nominal car-length plus one metre for kerbside parking. Wider bays are needed if the location is to be used by disabled motorists who need space to open-doors, load, unload, and manoeuvre their vehicles.

Special attention to parking layout is required in large car parks where internal traffic flows are also important. The three key variables in this case are the inflow capacity, the parking capacity, and the outflow capacity. The inflow and outflow values increase as the area per car parked increases, whereas the parking capacity decreases. Hence, a conflict exists at design stage. A more complex and often more relevant concept is the turnover capacity, which is the maximum rate at which cars can exit, whilst the same rate of cars is entering. As a first approximation it can be assumed that each of the activities of searching, parking, and unparking by car will inhibit all other cars by 5 seconds (Ellson, 1984).

30.6.3 Parking operations

Parking operations can be categorised in three ways:

1. time-based:
   (a) without time restraint,
   (b) with time constraints on long-duration parking, or
   (c) with significant time restraint.
   Time limits will range from 0.25 h — e.g. near post offices — to 8 h outside work-places. They will require regular enforcement. An important aspect of parking policy covered by this category is the split between long-term and short-term usage. A typical conflict that can be resolved is that which arises over the use of parking spaces by shop employees rather than by shoppers. Long-term parkers tend to arrive earlier and be work-trip commuters. Short-term parkers are usually shoppers or on personal business.
2. fee-based:
   (a) free,
   (b) with charges for entry (i.e. based only on the need to park),
   (c) with charges for parking duration (thus increasing the turnover of available parking spaces),
   (d) with charges based on time of day (e.g. favouring off-peak travel), and/or
   (e) with charges for special vehicles (e.g. taxis, buses, car-poolers and commercial vehicles).
   The charges levied are usually monitored and enforced by the use of parking officers in association with parking meters or prepaid parking permits. Parking charges commonly have an elasticity (Section 33.5.4) of about –0.3.
3. user-based:
   (a) preferential parking for special drivers (e.g. local residents),
   (b) park-and-ride facilities, providing parking for commuters using public transport.

30.7 TRAFFIC MANAGEMENT

30.7.1 Objectives and issues

Transport systems management is defined in Section 30.2.2 and *transport demand* (and *supply*) *management* in Section 30.3. They are both concerned with the performance of the whole transport system and are usually policy- or strategy-oriented. On the other hand, *traffic management*:

* is directed at making the best short-term use of the existing road system,
* has a problem-solving orientation, and
* rarely causes a change in the actual amount of travel.

Indeed, traffic management has very local objectives that can be categorised as:

* *traffic on the road network*. Typical measures include the installation of traffic signals, traffic signal co-ordination (Chapter 23), improved intersections (Chapter 20), incident management (Section 26.4), the use of road priority (Section 20.2.2) and road hierarchies (see below). These measures lead to lower fuel consumption (Section 29.2), reduced congestion, increased capacity, less delay, and easier driving.
* *traffic in residential areas*. Typical *local area traffic management* measures include low-speed zones, street closures and pavement narrowing (Section 7.2.3).
* *safety* (Chapter 28). Typical measures include pavement marking (Chapter 22), street lighting (Chapter 24), and local area traffic management.
* *provisions for pedestrians and cyclists*. Typical measures include pedestrian malls (Section 30.5.3), and cycle lanes and paths (Section 19.6).
* *access to business, shops and recreation*. Typical measures include turn provisions at signalised intersections (Sections 20.3.3 & 22.4.2).
* *parking*. Typical measures relate to parking policy (Section 30.6.2).

In Section 7.2.4 it was argued that *road hierarchies* and *classifications* had limited value in a residential system. However, the reverse situation applies with respect to the overall road network. The advantages of a road classification system in a road network are:

* increased consistency in resolving traffic management conflicts;
* recognition of the conflicting roles of traffic service and local amenity, i.e. the need to cater for particular traffic flows which will cause noise, vibration or pollution impacts on fronting properties and reduce the chances of locals crossing the route;
* a clear division of administrative responsibility and accountability;
* a basis for compensation or alleviation; and
* a basis for controls on land use.

Traffic management schemes must always take account of the following issues:

* Clear-cut solutions rarely exist.
* Schemes can divert and re-route traffic, but will rarely diminish it. If traffic is prevented from using one street, it will immediately use some other street. As most traffic management measures are isolated efforts, this issue must be given frequent consideration.
* Schemes which increase capacity may not increase speed as unsatisfied demand may be attracted to the improved system, resulting in increased throughput but little level of service improvement.

* Schemes can be in obvious conflict with other objectives and so trade-offs are usually essential. For example, a street closure may lessen through traffic passing a person's house but increase the distance that person must travel on many trips.
* Schemes often raise the issue of *equity*, as the introduction of any scheme will benefit some members and disadvantage others. These distributional inequities are difficult to handle on a local scale (Section 5.2.3), although at a community level the application of the Pareto principal discussed in Section 33.5.8 ensures that a method exists for ensuring overall benefit to the community.

The sensitive nature of these issues makes it important to ensure that any traffic management scheme is accompanied by careful before-and-after studies so that any changes are documented and quantified. Key parameters to measure are traffic flows, running times, changes in access, traffic intrusion, crash history, noise, and pollution levels.

30.7.2 Simulation models

The traffic simulation models discussed in Section 17.3.10, and the transport network analyses described in Section 31.3, may need to be used to predict the effect of traffic management measures. Potentially useful traffic models for this task are TRANSYT/13 (Section 23.5.2), CONTRAM/8 (TRL), SATURN/10 (Atkins), EMME/3 (Inro) and TRIPS for road networks (Section 31.3.6). SCAT and SCOOT (Section 23.5.3–4) could also be seen as models handling wider road network analyses.

CHAPTER THIRTY-ONE

Transport Planning

31.1 GENERAL

To understand transport planning, it is first necessary to understand the meaning of the two words that comprise the term. Section 30.1 defines *transport* as the movement of people and goods. *Planning* is a task undertaken to:

* *predict* future needs and the consequences of possible future courses of action, and thus to
* *ensure* that resources available for development are efficiently (Section 33.5.2) and effectively (Section 30.2.2) allocated.

The *efficient* allocation of resources can be pursued on purely technical grounds, but their *effective* allocation — particularly by offering the community options for the future and thus assisting it to satisfy its broad socio-economic *goals* and *objectives* — requires the community to define those goals and objectives and the criteria by which it will assess whether they have been achieved. Such public *goal formation* is a value-based *political process* at a broad and abstract level and therefore is difficult and often inappropriate for planners to initiate and manage. On the other hand, planning can aid the process by informing the decision-makers of relevant facts, showing the consequences of particular goals, and itemising conflicts and inequities that their achievement might create.

A major difficulty is that the community — particularly its political component — has historically been slow to state its goals and objectives in terms useful to transport planning and, having done so, quick to subsequently change those goals and objectives. Any planning must therefore be able to change as community objectives change.

Thus, transport planning can be summarised as following a number of hierarchical stages, which also relate to the discussion in Chapters 5, 6 and 30:

1. *External factors*:

1.1 Political stage — covers planning at a conceptual and policy level, and leads to changes in regional and urban development and in transport balance and co-ordination. It therefore considers broad social and economic goals which relate to:

a. world issues such as peace and sustainability,
b. regional issues such as the desired form and structure of communities and regions,
c. liveability issues such as the environment (Chapter 32), public health, individual well-being and the quality of life,
d. economic issues such as wealth creation, productivity and competitiveness, and
e. transport issues such as performance (Chapter 30), safety (Chapter 28), accessibility and equity (this chapter).

1.2 Development stage — covers strategic and systems planning related to:

a. land use,
b. transport policy,
c. transport networks,
d. corridors for transport facilities (Section 5.2.1), and

e. the management of transport demand (Section 30.3).

2. *Internal factors*:
2.1 Transport stage — covers:
  a. predictions of transport demand by location and magnitude, and
  b. consideration and selection of options to satisfy the demands within each transport corridor. This stage is discussed in the remainder of this chapter.
2.2 Roads stage — covers the planning and design of road facilities to satisfy in an efficient manner:
  a. the needs defined by stage 2.1b, and
  b. the goals related to closely-linked issues such as road safety (Chapter 28), environmental protection (Chapter 32), and local and regional equity (Tables 5.1 & 2).

  The appropriate processes are discussed in Chapter 6.

It will usually be found that road transport, because it is so all-pervasive (Section 30.4.2), will be a powerful and leading factor in the achievement of the transport stage — rather than merely a minor contributor. For example, road schemes usually:
* have relatively high benefit–cost ratios (Section 5.2.3),
* make large contributions to national economies (Section 4.3),
* significantly increase mobility and accessibility (Section 31.2.4), and
* deliver major social benefits in such areas as enhanced freedom to choose where to live, work, play, shop, be educated, and be entertained.

On a negative note, road transport changes can have a range of undesired impacts on the environment and on different community groups, at all levels of aggregation (Section 32.5).

After defining some essential concepts in Section 31.2, Section 31.3 will examine the road-related aspects of the four planning stages (1.1, 1.2, 2.1, 2.2) in some considerable detail.

31.2 TRANSPORT THEORY

31.2.1 Mobility

As discussed in Section 30.1, transport is rarely a commodity desired in its own right. Rather it is a *demand* derived from the need for passengers or cargo at an origin at location A to transfer to a new destination at location B. It is thus the activities located at A and B, their physical separation, and the ease with which that separation can be traversed, which largely determine the direction, intensity, pattern, timing, and demand for travel between A and B. Only recreational travel sometimes falls outside this A and B category.

Mobility describes the ease with which a traveller can move from A to B. It has the dimensions of:
* speed and time (see below),
* cost (Section 31.2.2),
* flexibility (Section 30.5.1),
* reliability (Sections 30.2.2 & 30.4.2),
* safety and security (Chapter 28), and

* externalities (Sections 32.2–4).

The journey from A to B is called a *trip,* which is defined as a one-way movement between an origin and a destination. A *tour* is an origin-to-origin movement. Tourist or pleasure tours may involve no other destination but most tours will include a stop at at least one non-origin destination and therefore comprise at least two trips.

Following the widespread availability of the car (Section 3.5.4), the mobility it has provided has increasingly allowed greater separation of A and B. This separation is such that it is now reflected in the physical structure of the built environment as, for example, in low-density modern suburbs. Communities now often find themselves reliant for their very functionality on the mobility of the motor vehicle. Moreover, people have come to clearly enjoy and value the mobility that the private vehicle offers. It is one of the new rights created during the 20th century (Lay, 1992).

A major constraint on the location of A and B is measured by the time and/or cost of making the trip from one to the other. The common factor for all travellers is that each has a time budget of only 24 hours to devote to a day's activities and so trip time is a primary determinant of the quality of travel. Internationally, trip times for an average journey to and from work are usually in the range 0.7 to 2.0 h/day. This led to the suggestion by Zahavi in 1974 that there is an average constant daily trip-time budget for most urban dwellers of about 1 hour (e.g. two 30 minute trips to and from work), independent of transport mode. For example, cars would dominate the hour in low density cities and walking in high density ones. The constancy has held but the absolute time has slowly increased, and in 2008 the budget appears to be about 70±25 minutes.

Allocations of a traveller's time budget provide fruitful evidence of travel behaviour, with the prime factors determining personal travel being stage in life-cycle, role in family, vehicle availability, and location of origin and destination (see also Section 32.5). It has been observed that the time that a household spends travelling rises as its income rises. Most of this rise relates to the increase in car ownership that occurs with an increase in income (Section 30.5.1). External determining factors are ease of travel, working hours, eating and sleeping hours, and external timetables such as those set by school age children.

In developed urban areas, people typically make 3 to 4 trips per day (see Section 31.3.3 for detailed data). For freight movement, time constraints can be set by industry, by unions, by community regulations, and by the need to maximise the number of trips.

31.2.2 Travel costs

Section 31.2.1 showed that travel cost was a major explanatory variable in transport. In order to understand its role, it is often necessary to use *generalised travel costs.* These are the sum of the actual cost of any road toll or public transport fare, plus the *perceived costs* of all other factors associated with the trip, such as private vehicle operations, walking time, waiting time, convenience, and comfort (refer also to Sections 29.1.2 & 30.4.2). Naturally, the dollar values of these perceived and unquantified costs is a matter of some contention. A traveller's perceptions of cost will be influenced objectively by the presence of taxes and hidden subsidies and subjectively by the traveller's biases towards a particular transport mode. Indeed, the ratio of perceived cost to actual cost can be regarded as a form of user *preference rating*. The relative size of these preference ratings indicates the extent to which a user is prepared to trade-off one against the other, e.g. the time lost against the fare saved.

Estimates of the value of time are discussed in Section 31.2.3. Some typical values for generalised travel costs, normalised to give a unit coefficient for the car-driver-alone case are:

	Cost components			
Traveller	Car time, T_c (min)	Distance*, D (km)	Out-of-vehicle time, T_0 (min)	Fare, F ($)
Car driver alone	T_c	+ 1.2D	+ $24T_0$	
Car driver with passenger	$0.7T_c$	+ 0.8D	+ $24T_0$	
Car passenger	$0.3T_c$	+ 0.4D		
Public transport passenger	T_c	+ $4.0T_0$	+ $2T_0$	+ 0.5F

* A measure of vehicle operating costs (Section 29.2.4).

Note how much more highly the unproductive out-of-vehicle time is perceived to be valued.

Generalised costs can be regarded as a measure of the economic *disutility* (Section 33.5.1) of a particular transport mode, reflecting both the quality and the quantity of travel. One use to which they can be put is the construction of trip time versus trip cost curves of the type needed for trip-assignment models (Section 31.3.6). This is done in the following manner. It is known from Section 29.2.6 that vehicle operating cost versus travel–time curves are as shown in Figure 31.1a. (Travel time is time spent in motion, and is thus a component of trip time.) Next, generalised costs can be used to predict the costs of the user's running time (Section 18.2.11), as in Figure 31.1b. Combining Figures 31.1a and 2b gives the total travel cost for one traveller, as in Figure 31.1c. Finally, Figures 17.5 and 17.8 are used to produce the total costs for a traffic stream, as in Figure 31.1d.

31.2.3 Value of trip time

The value of trip time to be used in transport planning studies can be seen as either:

(1) a marginal value – what would the traveller do with the extra time, and — therefore — what would the driver be willing to pay for the extra time? This view sees time as a commodity. Such values are useful in benefit–cost analyses. Above a threshold time saving, this value commonly increases with the square of the time saved. It also increases with journey length.

(2) an average value — what is the traveller's wage rate? Such values are useful in calculating resource usage. The travel time saving will be only some proportion of the traveller's total income (direct salary increased to cover such overheads as leave and pensions). The next table gives the value of trip time as a percentage of direct salary. In some cases it may be better to use only the discretionary component of the direct salary.

(3) as an implied value — what value was used in the actual decision-making? Such values are useful in calibrating behavioural models (Section 31.3.5).

These values depend to varying degrees on the trip purpose, the person travelling, and the trip type. A major concern with using savings in trip time is that all time savings are conventionally valued equally in aggregate. For example, 60 1-second savings are taken as equivalent to a single 1-minute saving, although people attribute lower values to small time savings or to savings associated with short trips. A related problem is seen from the example in Section 31.2.2 where travellers apply different values of time to each trip component — walk, wait, and travel. A summary of common findings is:

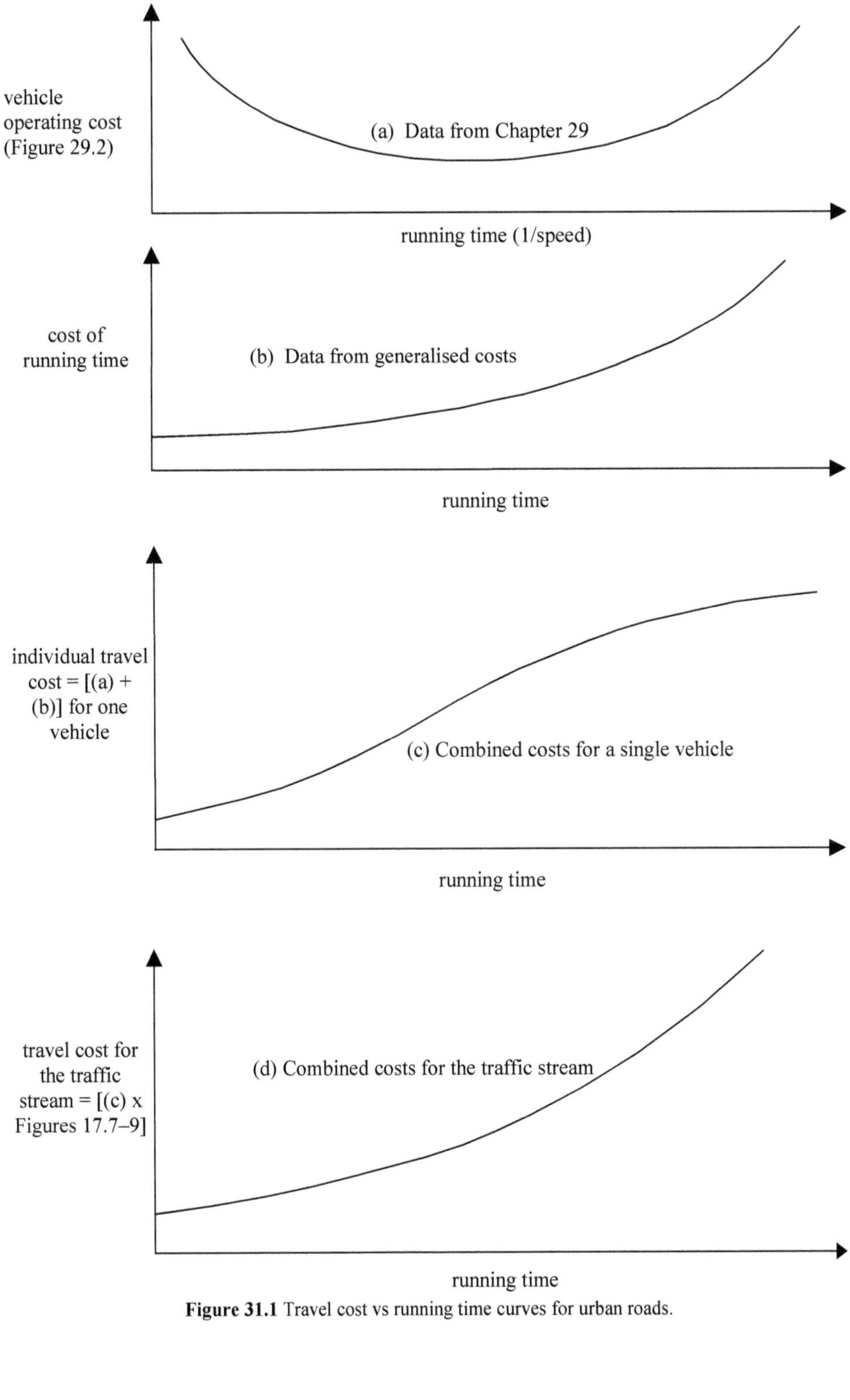

Figure 31.1 Travel cost vs running time curves for urban roads.

Trip type	Value of time saved as a percentage of direct salary	Reasons for reduction of value below 100
Work	70	Many business trips made outside of working hours.
To and from work	45, car and long public transport 40, public transport 20, part-time work	The traveller may not use the time saved in a way that would justify the use of the full salary: e.g. the traveller may use it to travel further or to sleep in longer.
Private	30	From behavioural studies.
Family/ leisure	20	Allows for children and other unsalaried people in the travelling population.

People also value travel time reliability and typically value the variable component of their trip time on a 'known' trip (such as the journey to work) at about three times the value of their average trip time (Cohen and Southworth, 1999).

In valuing time for freight movements, it is necessary to consider not only the value of the operator's time, as defined above, but also the cost of the freighted goods being unavailable to potential users, or the benefits of early or reliable delivery. Many delivery points for freight only give a truck a limited loading time window, and so variability of travel times become even more important.

Obviously, time savings and their associated dollar values are difficult to calculate with certainty. Nevertheless, they can be particularly important. For example, Table 5.1 shows that they can constitute about 60 percent of the benefits of an urban road improvement scheme.

31.2.4 Accessibility

The *accessibility* of destination B relative to origin A is a function of:

* the desirability of making the trip from A to B,
* the route-based separation of A and B,
* the mobility available for the trip (Section 31.2.1), and
* the quality of the outcome.

Whereas mobility depends only on transport, accessibility also depends on *land use*. It is commonly used on an aggregate basis to measure the ease with which a variety of desired activities at locations B_j can be reached from point A. Because it depends on estimates of the desirability and quality of activities, accessibility is not an easy quantity to measure objectively.

A simple accessibility measure for point A was proposed by Hansen as the number of desired activities available within a fixed trip time from A. The basic measure of aggregate accessibility is then given by:

$$\text{aggregate accessibility of A} = \sum_j s_j f(C_{ABj})$$

where s_j is a measure of desirable activities at location B_j such as the number of trips it generates (Section 31.3.3), C_{ABj} is the cost of travelling from A to B_j and f is an *impedance function* translating that cost into a behavioural response. The impedance function is usually of the form e^{-aC} where a is a constant, and reflects a traveller's lessening sensitivity to cost increases. It is thus an expression of the well-known economic concept of *diminishing marginal utility* (Section 33.5.1). Thus:

$$\text{aggregate accessibility of A} = \sum_j s_j e^{-aC_{ABj}}$$

and the ratio $\Sigma s_j e^{-aC_{ABj}}/\Sigma s_j$ is sometimes used to give a comparative measure of the accessibility of different areas. If s is based on employment generation, rather than on trip generation, then the summation measures employment accessibility, rather than transport accessibility.

Accessibility is an economic good that can be traded off against such other goods as land value and amenity. Land values in particular will reflect the accessibility of a location (Section 33.5.8) and therefore neither can be used independently to measure social welfare. That is, there is no inherent reason why transport should correct accessibility inequities, as the property market will already have adjusted to that inequity.

Clearly, an understanding of why origin A and destination B are located where they are is an essential part of understanding the demand for travel. Different households and different industries will have different abilities and desires to adjust their trip origins and destinations and there will be a clear market-based link between the type and quantity of travel consumed and the qualities of those origins and destinations. One immediate consequence of such factors is that origins and destinations are constantly changing, particularly as a result of a strong two-way interaction between land use and transport. These issues are pursued further in Section 31.3.2.

Within this fluid system, each individual decision-maker will reach an equilibrium between the transport demands for A and B to be accessible and all the other factors (e.g. land values, a desire for space, access to other destinations) leading to a separation of A and B. Overall, the decision-maker's utility (Section 33.5.1) will be maximised.

31.3 TRANSPORT SYSTEMS ANALYSIS

The following methods of transport systems (or network) analysis have evolved in response to the needs described in Section 31.1. It will be seen that they are based on seven consecutive analytical stages corresponding to the following seven sub-sections. Specific methods for each stage are named in the text. Transport studies are aimed at a wide spectrum of schemes, from long-range transport plans to studies with shorter and closer targets. With long-range planning, the study objectives will probably need to narrow as the study proceeds. Shortened planning horizons often result in marginal adjustments to existing schemes.

31.3.1 Data collection

The first step in transport planning is to obtain adequate and sufficient data. Transport data have commonly been concerned with general conditions at a given time (i.e. *cross-sectional* data) rather than with specific conditions over a period of time (i.e. *time-series* or *historic* data). This is largely because most data must still be obtained via specific surveys, which are expensive and time-consuming, with only a small but increasing amount collected automatically. Essential population, employment, and income forecasts are based on:

* extrapolations of current trends,

* general statistical predictions from planning and forecasting agencies,
* a consideration of various public policy alternatives, and
* an iterative application of the land-use plan produced for the design year using the models in Section 31.3.2.

Data sources for general infrastructure and land use are the various maps commonly available (Section 6.1.2). It may also be necessary to use data from government agencies on a region's topography, resource base, community services, and economic base (Section 6.1.3). Data sources for transport infrastructure are the maps and reports from transport agencies. Many Road Authorities maintain databases of the geometric characteristics of their road network, often in a GIS context (Sections 6.1.2 & 26.5.1).

An understanding must also be gained of the current and future plans and policies of decision-makers and of likely future economic and cultural trends. From this basis, the goals of the area and the appropriate evaluation criteria can be understood (Section 31.1).

Major data sources for traffic movements are census records, traffic and commodity flow measurements (Section 31.5), vehicle and cordon surveys (Section 31.3.6), and household surveys (Section 31.3.1). Sources of car ownership data are discussed in Section 30.5.1.

Data on car usage, the use of other transport modes, routes taken, and time of travel are often obtained from *household* surveys tailored to the particular interest area. Survey questions are directed at the three levels of:

* the household (e.g. house type, number of residents, number of cars),
* the person (e.g. sex, age, number and type of trips), and
* the trip (e.g. origin, destination, route, time of day, purpose).

The phrasing of the questions requires care and attention. An extensive review of the uses that can be made of home interview survey data is given in Wigan (1987).

Questioning may be done by an interviewer, as in a *home interview* survey, or by a self-administered questionnaire. The latter is simpler and cheaper, but less reliable and evinces a lower response rate. Sampling of the homes to be included in the survey needs to be treated carefully. Accurate estimates require a 20 percent sample for populations of 50 000 or less, i.e. about 2000 household interviews, down to 4 percent for populations of 1 000 000 or more, i.e. about 10 000 household interviews. The results of the surveys need to be approached with some caution, particularly with respect to measurement errors (e.g. misunderstood questions), trip under-reporting (e.g. forgotten trips), off-peak travel, use of minor modes, and time spent on activities (Barnard, 1985).

31.3.2 Land use

The movement of people and goods is intimately related to land use: the need for the movement is generated at one land use and consummated at another. Government policies on national and regional growth and on the size and form of cities thus have a decisive influence on transport requirements. For example, the demand for urban transport is influenced by decisions as to whether or not towns should be:

* unplanned, or grow under restraint or incentive;
* developed according to market force, linear, vertical, satellite, or dormitory models;
* centrally oriented or multi-nodal.

Therefore, transport planning must recognise that it is an integral part of the regional planning process and must fit community objectives (Sections 31.1, 31.3.1–3, & 32.1).

Consequently, the second step in transport planning is to use demographic and policy studies to produce land-use predictions and plans for the chosen design year, which may be 20 years ahead. There are three approaches to this task:

(a) extrapolation of existing trends;

(b) deduction from assumed levels of economic activity and policy application; and

(c) use of the intervening opportunities model.

Each must be used within the limits of known plans and policies for the years leading up to the design year. For example, planning policies exert a major influence on residential densities and industry location.

Approaches (a) and (b) are relatively straightforward: however, (a) depends on stable continuous change and (b) depends on the accuracy of the prediction of the levels of economic activity and policy application. Method (c) is of value for predicting land use.

The basic assumption of the intervening (or competing) opportunities model is that there is a finite number of opportunities for locating a particular land use, and that these opportunities can be ranked in some order. The probability that a suitable opportunity is accepted monotonically decreases as the number of intervening opportunities between the one encountered and the highest-ranked opportunity increases. The traveller is assumed to consider each opportunity in turn in terms of some increasing parameter like trip time, cost, or distance. Each one has a stated probability of being accepted. Input to the model includes:

(1) the pattern of land use at the present and at at least one past point,

(2) the available opportunities at each location,

(3) a mechanism for ranking opportunities by their desirability, and

(4) a forecast of the overall growth rate of the region. The growth rate of a sub-region (or zone) is proportional to this overall regional growth rate by a factor determined by the model.

As land use depends on the transport facilities provided, and as these facilities will themselves depend on the land use, the process input will depend on the process output and a proper solution of the transport/land-use system will need a number of iterations. As a first step, some future set of transport facilities must be assumed.

For practical reasons, it is also necessary to aggregate individual land uses (e.g. individual properties) in the region being planned into zones which:

* are of similar land use, property cost, job generation, and topography,
* are compatible with zones used for other related purposes — e.g. local government, census or postal zones, and
* rarely exceed 2 km^2 in area.

The selection of zone size is a balance between the need to use small zones to achieve internal homogeneity and the data-handling costs associated with the use of many zones.

The Lowry model provides an alternative technique for linking transport to employment and residential location. The model input includes the labour participation rate, local travel characteristics, and zonal attractiveness for residential and service needs. Employment is categorised as either basic (i.e. providing the impetus for urban development) or service (e.g. shops and schools). Distance travelled to work is an inverse function of travel time and trip cost (Section 31.2.2). It uses the gravity model (Section 31.3.4) to estimate trip numbers.

The model first predicts desired residential locations for workers from an assumed set of heavy (or basic) industry locations. It then locates service (or secondary) industries consistent with observed travel behaviour. Workers in service industries are then allocated to residential locations. An iterative process occurs as the needs of the service industry workers are then satisfied. Thus the model readily considers location demands, however household characteristics are difficult to consider and supply factors such as transport must be included via the zonal attractiveness weightings and travel costs. A further application of the model is given in Section 31.3.4.

The Lowry/gravity model approach therefore places more emphasis on trip distance than the accessibility measures in the intervening opportunities model. A review of the strengths and weaknesses of the two approaches to transport modelling is given in Cheung and Black (2008).

Two further alternative residential location models are one due to Alonso, which is based on the economic concept of a competitive market, and one that uses the multinomial logit model (Section 31.3.5).

31.3.3 Estimates of trip numbers

The third stage in the planning exercise is to predict the number of trips, t_{ij}, made for each trip purpose. This is based on the forecasts and land-use patterns in Sections 31.3.1–2, and on using past data to provide regression equations relating trip production and attraction to particular zonal attributes such as:

* population, population density, and population characteristics,
* housing numbers,
* household numbers, car ownership levels (Section 30.5.1), family income, income distribution,
* employment levels, job distribution and type, and
* retail, educational and recreational opportunities.

Trips can be categorised as:

personal

* home-based (e.g. journeys to and from work, education, shopping, and social/ recreational facilities, employer's business),
* local but not home-based (e.g. on employer's business, shopping from work, journey to next attractor),
* from outside the region (e.g. tourists), and

freight

* commercial vehicle (Section 30.5.2 and below).

Typical examples of predictive equations for such trips are given in Table 31.1. In a typical city or town, the journey to/from work will represent about a sixth of all trips, but a third of all trips by distance travelled.

For further refinement, *category analysis* is often used, classifying the population according to characteristics that explain their trip-making behaviour. For example, women may be more likely to work near their home than are men. Another common classification is of households by size, employed persons, dependent persons, income, and number of cars. Car ownership is a particularly important predictive variable and is considered in some detail in Section 30.5.1.

Trip generation refers to the trips or tonnes of freight produced or attracted by a particular land use (trip origins or destinations). The daily sum of these trips is called the *generated traffic* and is important in system studies, such as those described in this section, and in examining the impact of individual proposals.

If a transport improvement is made between the trip origin and destination, some new traffic will be:

Table 31.1 Estimate of the number of home-based trips made from a zone each day. All data in the equations is for the zone. d = number of dwellings, p = population, e = number of employees, v = number of vehicles.

Trip type	Cause	Equation	Equation no.
work	production	0.52p + 0.41d	31.1a
	attraction	1.60e	31.1b
shopping	production	0.35v	31.2a
	attraction	1.10(e for shops)	31.2b
social/ pleasure	production	0.54v	31.3a
	attraction	0.06p + 0.14d + 0.21(e for service)	31.3b
'other'	production	0.58v	31.4a
	attraction	0.12d + 0.03(e for manufacturing)	31.4b
all trips[a]	production	2.5p + 0.9e	31.5a
	attraction	1.4p + 3.6e	31.5b

a. The aggregate number of trips in this category may be much greater than the sums of the individual producers and attractors.

* *induced* by this improvement (e.g. by encouraging new activities, land use changes and/or by causing shifts in origins and destinations), or
* *diverted* from other routes, times or modes to the improved route.

For example, a major new road may generate about 20(+60/–20) percent more traffic than existed prior to the completion of the project. Experience suggests that reducing travel times by x percent will lead to an increases in volume of x/2 percent in the short term and x percent in the long term. On an area basis, the increase is about half these values. Reduced travel times will often encourage longer trips as people manage their time budget (Section 31.2.1). Induced and diverted traffic and associated land use changes must be carefully considered, as they will normally reduce the benefits otherwise ascribed to the project.

Table 31.2 gives typical equations which estimate the trips generated by a unit within a zone. It therefore usually gives higher rates than the aggregate zonal data in Table 31.1. The major factors which can affect the traffic generated by a land development are the:

* type of development,
* intensity of the activity undertaken,
* access to the development,
* parking provisions, and
* availability of other transport modes.

Examples of how these factors are considered are illustrated in the equations in Table 31.2. In addition to the factors in Equations 31.6, the trip rate increases significantly as household income rises (Section 32.5).

The number of car trips can be estimated from a knowledge of the trips generated (Table 31.2), the proportion of workers likely to use cars (the modal split, Section 31.3.6), and the car occupancy levels (Section 30.5.1).

A general review of the generation of freight traffic is given in Section 30.5.2. The key independent variable determining freight traffic generation is the (number of workers)/(site area) and for the freight traffic attraction it is the population density. Other key variables are type of land use, freight type and time of day. Wholesale trips are explained in terms of retail jobs per hectare.

Table 31.2 Estimates of the number of trips generated by a particular land use. An extensive collection of trip generation data is provided in ITE (2004). h = people/household, w = (people working external to the household)/household, a = dependents/household.

Land use	Typical equation	Equation no.
House	0.1h + 1.9(vehicle/household) + 0.75w	31.6a
Notes a & b	0.52h + 1.47(vehicle/household) + 0.43	31.6b
Note c	1.6h + 1.4w + 0.3a	31.6c
Office block	64 + 0.037(gross floor area in m^2)	31.7a
	[21 + 0.037(gross floor area in m^2)]/(peak hour)	31.7b
Motel	[1 + 0.015(gross floor area in m^2)]/(peak hour)	31.8a
	[0.009(gross floor area in m^2)]/(am peak hour)	31.8b
Vehicle fuel and repair	[32 + 0.003(site area in m^2)]/hour	31.9a
	[0.015(site area in m^2)]/(peak hour)	31.9b
Shop	0.01(shop area in m^2)	31.10
Shopping centre	0.07(shop area in m^2), Note d	31.11
Factory	0.06(factory area in m^2)	31.12
School: primary	1.0(number of students)	31.13a
secondary	1.5(number of students)	31.13b
tertiary	2.0(number of students)	31.13c

a. Residential dwellings are commonly assumed to generate between two and twelve trips per day for separate dwellings and from three to eight for apartments. The minimum of two trips corresponds to the common one trip in each of the morning and afternoon peaks.

b. This equation is the summation of Equations 31.1–4 in Table 31.1. The form is similar to Equation 31.6 and the differences come from the different factors regressed in the zonal and local equations.

c. This equation is produced by statistically relating levels of car ownership to population (Section 30.5.1). The three terms respectively represent 'other' trips, work trips, and shopping trips.

d. Coefficients as high as 2 have been noted.

In the case of an existing system, estimated trip numbers should be checked against measured numbers of actual trips in the base year, and any significant discrepancies eliminated.

31.3.4 Trip distributions

Having established the spatial arrangement of land uses (Section 31.3.2) and predicted their traffic attracting and generating consequences (Section 31.3.3), the fourth stage of the planning process is to predict the numbers of trips between these origins and destinations (or interchanges between origin-destination pairs). Origins are zones where trips are commenced (or produced) and destinations are zones where trips are finished (or attracted). The number of trips is estimated using some form of model, assumed inter-zonal trip times, and the data in Section 31.3.3 on trip productions and attractions.

The number of trips, n, between zones are labelled using the method illustrated in the origin–destination matrix, or trip table, shown in Table 31.3. The quantity n_{ij} is the number of trips produced in zone i by the attraction of zone j, I_i is the total number of trips produced by zone i, and J_j is the total number of trips attracted by zone j. I_i and J_j are defined algebraically as:

$$\sum_j n_{ij} = I_i \qquad (31.14)$$

Table 31.3 General origin–destination matrix, or 'trip table'.

Destination zones	Origin zones								
	1	2	3	–	–	i	–	m	Σ
1	n_{11}	n_{21}	n_{31}	–	–	n_{i1}	–	n_{m1}	J_1
2	n_{12}	n_{22}	n_{32}	–	–	n_{i2}	–	n_{m2}	J_2
3	n_{13}	n_{23}	n_{33}	–	–	n_{i3}	–	n_{m3}	J_3
–	–	–	–	–	–	–	–	–	–
–	–	–	–	–	–	–	–	–	–
j	n_{1j}	n_{2j}	n_{3j}	–	–	n_{ij}	–	n_{mj}	J_j
–	–	–	–	–	–	–	–	–	–
n	n_{1n}	n_{2n}	n_{3n}	–	–	n_{in}	–	n_{mn}	J_n
Total	I_1	I_2	I_3	–	–	I_i	–	I_m	

$$\sum_i n_{ij} = J_j \quad (31.15)$$

Typically, each origin is also a destination, and vice versa, and this symmetry produces a square trip table. When only total trips ($n_{ij} + n_{ji}$), or maximum trips irrespective of direction, are plotted then only half of the matrix is filled and the result is called a triangular trip table.

Three different model types can be used to predict the number of trips made between origins and destinations. They are the Fratar, gravity, and intervening opportunities models. The *Fratar* model assumes that the future number of trips is related to the present number and to the estimated growth, and that the distribution of trips is proportional to the present distribution modified by growth factors. The calculations are done iteratively, beginning with the current situation.

The *gravity* model assumes that the number of trips between an origin and a destination is related to the trip production and attracting properties of each and to the difficulty or cost of travelling from one to the other. In its historical form, if the demand generated at a zone i is n_{gi}, if the demand attracted to a zone j is n_{aj}, and if the cost of travelling from i to j is c_{ij}, the trips from i to j, n_{ij}, can be estimated as:

$$n_{ij} = kn_{gi}n_{aj}f(c_{ij}) \quad (31.16)$$

where k is a constant (Figure 30.2). In the simplest application of the model the demand factors, n_{gi} and n_{aj}, are taken to be the actual populations at i and j. The data in Section 31.3.3 suggests that demand is proportional to the populations and so the joint proportionality constant can be included in k, thus permitting n_{gi} and n_{aj} to be total populations.

The number of trips will decrease as the costs increase and it is usual to assume that they are proportional to e^{-cost}. The exponential form is consistent with entropy maximisation theory (Section 31.3.6) and avoids negative trip rates. Thus:

$$n_{ij} = kn_{gi}n_{aj}e^{-c_{ij}} \quad (31.17)$$

In the original 'gravity' model $f(x) = x_{ij}^{-2}$ where x_{ij} is the distance between i and j. An exponent of –1.5 has often been found to apply for travel between separated towns.

Trip cost is usually expressed in terms of generalised cost (Section 31.2.2) and the trip times used to estimate it are based on assumed operating conditions. The traditional models provide no means of adjusting these times as a result of subsequent predictions,

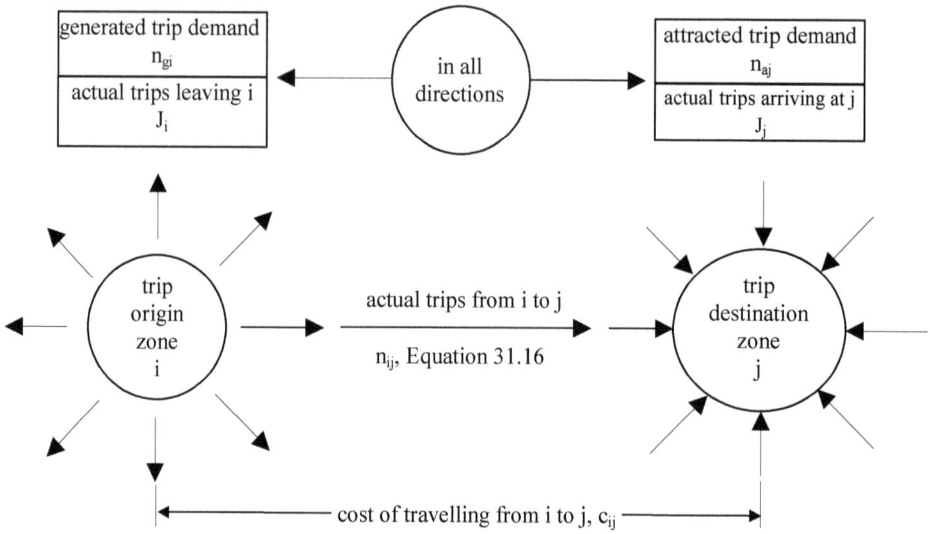

Figure 31.2 Gravity model.

e.g. if the link becomes congested. The iterative feedback necessary to resolve this issue is discussed in Section 31.3.8.

So far the discussion has been based on an *unconstrained* gravity model where the constant k can be used to bring the model in line with survey data. The *constrained* gravity model is one in which the number of trips made:

* from each origin (the sums of the individual trips out of an origin to various destinations), and
* to each destination (the sums of the individual trips into a destination from various trips generated by the origin and attracted to the destination),

are constrained to equal the estimated values. It is thus the conventional form of the model.

In the common *singly-constrained* gravity model only the trips from the origin, i, are constrained, as there is only one constant k to determine. An example of such a model is TRL's *LUTE*. Trips are distributed from each origin i to the destinations j in proportion to specified attractivenesses of each destination, following Equation 31.17. LUTE can additionally place budget constraints on average trip times and numbers of trips per day (Bland, 1982). The *fully (doubly) constrained*:

$$n_{ij} = p_i q_j I_i J_j e^{-c_{ij}} \qquad (31.18)$$

contains two constants, p and q, evaluated using the constraint Equations 31.14&15 to give:

$$p_i^{-1} = \sum_j q_j J_j e^{-c_{ij}}$$

$$q_j^{-1} = \sum_j p_i I_i e^{-c_{ij}}$$

The strengths of the gravity model are:

(a) in the prediction of the longer-term consequences of future growth, rather than in short-term changes, and

(b) in its relatively simple computational needs.

The weaknesses of the gravity model are that it:

(a) cannot include measures which influence individual destination choice,
(b) uses aggregate rather than individual data,
(c) cannot readily accommodate changes between transport modes, and
(d) has a fairly poor predictive record, stemming from both its underlying assumptions and the data that it uses.

The *intervening opportunities* model, previously discussed in Section 31.3.2, can also be used at this stage. It assumes that the number of trips to a destination is proportional to the number of opportunities there and inversely proportional to the number of intervening opportunities. Thus, the number of trips between i and j depends on the number of origins at i, I_i, and a probability term relating to intervening and final destinations. This term is usually taken as the difference between the negative exponentials of the probability density for accepting destinations up to but excluding j and the same term up to and including j. This can be written as

$$n_{ij} = I_i[e^{-miNj-1} - e^{-miNj}]$$

where m_i is a constant, N_j is the number of all the ranked trips up to and including those to the zone j in question and $N_{j-1} = N_j - J_j$. As was shown above, I_i and J_j can be determined from transport survey data and N_j can be calculated from the same data. m_i must be determined by calibration (Section 31.3.6).

The models all operate by dividing an area into zones incorporating common origin–destination behaviour and representing well-defined units (using census, postal, or natural boundaries). Each zone has a centroid that represents its approximate travel centre. The analyses assume that all trips begin and end at a centroid. The centroids are connected together by travel links. The number of trips from one centroid to another provides the entry in each cell of the origin–destination matrix (Table 31.3). The results reflect the coarseness of these zones. In addition, the matrices are notoriously hard to maintain and update.

Short-term and mode-specific trip predictions are usually obtained from disaggregate models of the type discussed in the following section. As with the case of trip numbers in Section 31.3.3, for an existing system, these predictions can be usefully checked against measured distributions in a base year.

31.3.5 The choice of transport mode

The fifth stage in the modelling process is to split the calculated number of trips between the available transport modes. This allocation is known as *modal split.* For *captive travellers* (Section 30.5.1), each trip can only be made by one mode. These trips are thus assigned first.

For the remaining travellers, modal split is the result of many individual travel decisions. An individual's choice of transport mode is influenced by the traveller's perception of four main sets of characteristics:

(a) *personal and household characteristics*, such as age, sex, income, car ownership (e.g. Sections 32.5 & 33.5.1);
(b) *trip characteristics*, such as its purpose, urgency and length;
(c) *transport system characteristics*, such as travel costs, times, comfort, convenience, prestige (see Chapter 30, particularly Section 30.4.1); and
(d) *trip-end characteristics*, such as home and work-place types and density. See also Section 31.3.2.

The remaining trips are distributed by a *choice model* which:

* considers some or all of the above characteristics,
* sees the trip-maker as a classic consumer, and
* assumes that the trip-maker seeks to maximise the *utility* of each trip, where utility is a function of the cost, and time, and perceived quality of the trip (Section 33.5.1).

The basis of most choice models is that the probability of a user selecting a particular travel alternative is a function of the ratio:

$$U_k/[\Sigma_m U_m]$$

where U_k is the utility of mode k and Σ_m denotes the sum for all modes. The ratio is more conventionally expressed in terms of *disutilities* (the negative of U) which are proportional to the *generalised costs*, c, described in Section 31.2.2.

As the effects of changes in the attributes of a particular travel mode are less than linear, relatively large changes are required to affect behaviour. Exponentials of (–c) are therefore used and the probability is estimated as the ratio of negative exponentials, i.e.

$$\text{probability of using mode k} = [e^{-ck}]/[\Sigma_m e^{-cm}]$$

For example, if there are just two choices, and one costs δc more than the other does, then the probability, p_{kc}, of using the cheaper one is:

$$p_{kc} = e^{-c}/[e^{-c} + e^{-c-\delta c}] = 1/[1 + e^{-\delta c}] \qquad (31.19)$$

This is the *logistic curve* (Equation 33.1) and explains why *logit* modelling techniques are common in this area (below & Section 33.2).

When considering the effect of a change in conditions, it is usually more appropriate to use a marginal approach. This assumes that a traveller changes transport mode because the perceived increment in benefits received from the change to the new mode is greater than the increased (generalised) costs caused by the change.

Choice models based on applying behavioural theory to the trip-making decisions of individual travellers are called *disaggregate* (or *behavioural* or *individual choice* or *discrete choice*) models. They supplement the more aggregated evaluations of the older models discussed above and in Section 31.3.4. Using regression or maximum likelihood techniques (Section 33.4.2&3), disaggregate models are calibrated against either:

* individual trip data for an existing situation,
* *revealed* preferences (or choices) obtained by observing travel decisions,
* *stated* preferences (or choices) of interviewees for items in a list of proposed situations, or
* some mix of revealed and stated preferences.

The travellers make choices between distinctly different options (i.e. their choices are mathematically discrete). Despite their name, disaggregate models may still involve aggregating the data according to chosen socio-economic characteristics and zonal averages. *Activity-based* models also incorporate the type of time-constraint data discussed in Section 31.2.1.

The most favoured of the individual choice models is the *multinomial logit model* (Section 33.2) based on Equations 31.14–17. One such model is *BLOGIT*, which also provides the probabilities of selecting alternatives and direct cross-elasticities. An advantage of logit methods is that they can often use existing *home-interview-survey* data (Section 31.3.1).

One of the major problems in applying choice-based methods to travel prediction is that a model which simply predicts that all urban travellers will travel by car will have a quite high, but useless, predictive power when applied to the totality of many urban trips. As was demonstrated in Sections 30.4 & 30.5.1, for many people the car is the only travel choice — or at least the best choice. Only major changes, such as a relocation of origin and destination, will usually alter this, and then only temporarily. Another disadvantage of the models is that their basis in personal preference makes them better suited to short-term (operational) rather than long-term (planning) decisions.

Disaggregate and aggregate models can be compared as follows:

Disaggregate models	Aggregate models
Use a lot of data	Use less data, by working on aggregated data
Get more out of the data	Get less out of the data
Avoid bias due to correlated data	Are influenced by correlated data
Have a behavioural basis	Have little behavioural basis
Use a complex model	Use a simple model
Examine policy options	Do not allow policy options to be explored
Are robust over time*	May not be robust over time
Are robust across regionals*	Cannot be transferred between regions

* Provided the analysis accounts for the underlying socio-economic and attitudinal behaviour of the commuters.

31.3.6 Trip assignment

(a) Estimating actual traffic flows

In the sixth and final stage of the model, the trips that have been forecast to be made by a particular mode are then assigned to that mode's links within the transport network, thus giving the estimated flow on each link.

The simplest assignment technique used is the *all-or-nothing* method in which all trips are assigned to the path best meeting some criterion, such as providing the smallest trip time, shortest travel distance, or lowest generalised cost (Section 31.2.2). The technique therefore assumes that the capacity of each link is infinite, so that the amount of traffic on it affects neither the time nor the cost of travel. The method allows a problem to be solved in one iteration, but is clearly unrealistic for congested conditions and is often very sensitive to demand changes.

Various capacity restraints have been developed to modify this technique and most require the knowledge of a *speed–flow* (or *time–flow*) relationship for each link (see Figures 17.8 and 31.1d). In this context it is instructive to see the gravity model of Equation 31.16 as representing a *demand curve* (Figure 33.6), and the time–flow curve of Figure 31.1d as representing the supply curve, as in Figure 31.3. Trip time is replaced by trip cost.

Given the link flows, the link speeds and thus link travel times can be calculated. More detailed assignment methods directly consider the influence on total trip time of such factors as intersection queues and delays and the upstream and downstream effect of long queues. This approach is called *dynamic assignment* and is used by the CONTRAM model described in Section 30.7.2. The equilibrium aspects of the iterative approach are discussed further in Section 31.3.7. As intelligent transport systems have been deployed, the need has developed to do these calculations taking into account actual conditions — for instance by including the effects of such ephemeral factors as construction delays, traffic incidents, illegal parking, emergency vehicles, weather, major events and parking availability.

Models such as CONTRAM and SATURN that undertake this traffic-oriented task have been discussed in Section 30.7.2. Rather than use dynamic assignment, SATURN

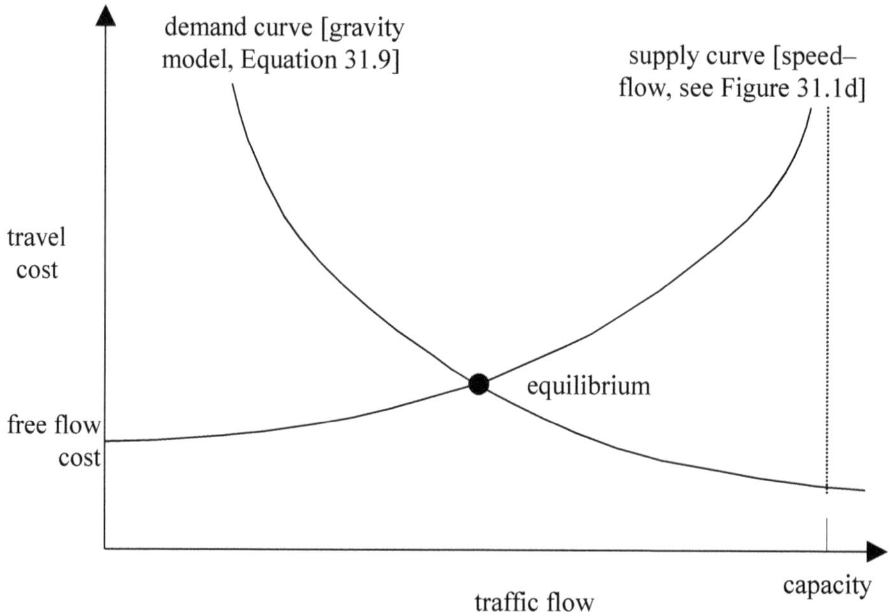

Figure 31.3 Transport demand and supply (compare with Figure 33.6).

uses 'cyclic flow profiles' — a concept derived from TRANSYT (Section 23.5.2) — which assumes that flows within one signal cycle are repeated from cycle to cycle.

A further complexity is that the time of travel may be a choice for some travellers who may be able to avoid travel at times when travel costs are high. If this group is significant, a further set of iterations of the predictive model will be needed.

At this stage, the route and mode that each trip will take are now known. These groups of forecast trips are known as *desire line* patterns and are often shown on a map of trip flows. Such traffic flow maps are also used to plot data measured in operating networks (Section 31.5).

(b) Checking the predictions

The predictions of the modelling process can now be checked against field data giving trips on all available routes and modes, from known origins and destinations. Such flow data is readily obtained from traffic counting (Section 31.5). *Origin-destination* (or *travel pattern*) surveys are of seven types:

(a) *cordon* (or *numberplate*) surveys:

A cordon (or screenline) is drawn around the survey area and the identification (e.g. numberplates) and times of vehicles entering and leaving are recorded. Numberplates must later be matched to produce the data. The method gives little data on origins and destinations within the cordon.

(b) roadside interviews:

This method gives accurate and reliable data but at considerable relative cost and driver delays. Suitable interview sites and police co-operation are needed.

(c) reply-paid postcards:

This technique can be used to reduce drivers' delays at interviews by handing the postcard to the driver for subsequent completion.

(d) home interview surveys:

See Section 31.3.1.

(e) floating car:

See Section 18.1.5.

(f) records of trips on public transport:

See Section 30.4.1.

(g) traffic flows:

The measured flows (Section 31.5) are compared with the flows in an Origin-Destination matrix (Table 31.3) generated from an assumed set of origins and destinations. The origins and destinations are then incrementally adjusted to make the calculated flows match the measured flows. A typical method using this process is embodied in the ME2 program.

An alternative method is occasionally possible. Using the symbols defined in Table 31.3 and Equation 31.19, the number of trips on a link a is given by T_a where:

$$T_a = \Sigma_i \Sigma_j n_{ij} p_{ij}$$

These equations can be solved for n_{ij} (the O–D matrix) provided there is at least as many T_a as n_{ij}. A further complication with this alternative is that often $p_{ij} = f(T_a)$.

(c) Calibration or projection?

Unfortunately, many past applications of transport modelling have used stage (b) of the model process to calibrate rather than to check models. This unwarranted process can give the models a misleading sense of accuracy. In summary, the two alternative approaches to transport model development are (Figure 31.4):

(a) calibration procedures, which — if used alone — merely reproduce the data, and

(b) projection procedures intended to respond to system changes.

As the figure illustrates, these two approaches produce different outcomes.

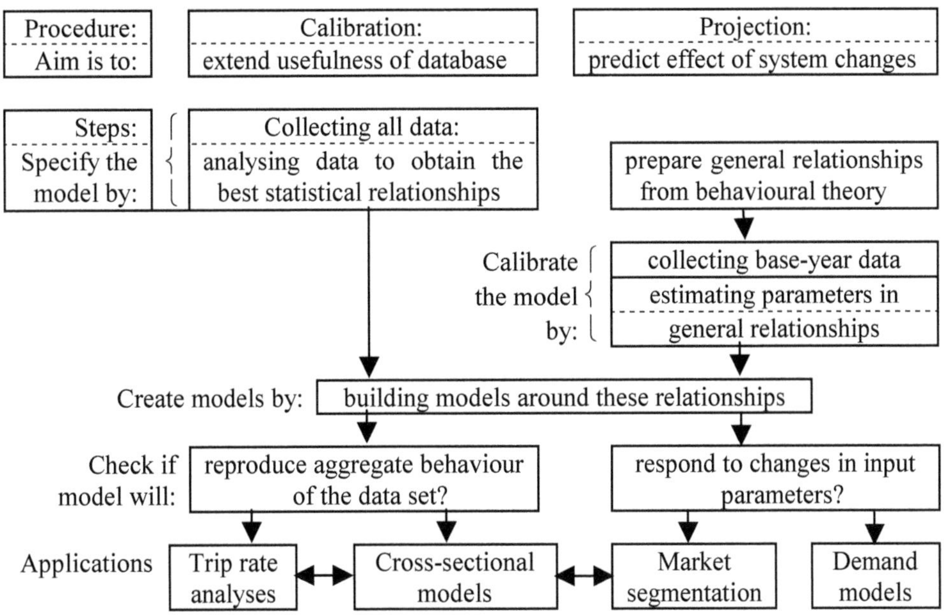

Figure 31.4 Process of model development.

The resulting model set defined in Sections 31.3.1–6 often proves cumbersome and expensive to operate.

31.3.7 The demand–supply equilibrium

The traditional transport planning method described in the sequence of stages in Sections 31.3.1 to 31.3.6 is often referred to as the *four-step* method. The four sequential steps represent the demand for and then the supply of transport (Section 30.1) and are:

* *demand modelling:*
    step 1 – data collection for trip production and attraction (Sections 31.3.1–3),
    step 2 – trip distribution (Section 31.3.4),
* *supply modelling*:
    step 3 – modal split (Section 31.3.5), and
    step 4 – trip assignment (Section 31.3.6).

In terms of the discussion in Section 31.3.6, demand-side models are commonly projection models, and supply-side models are commonly calibration models.

The inter-relationship of demand and supply modelling is shown in Figure 31.5 where the distinction between the two is seen to be between system and user.

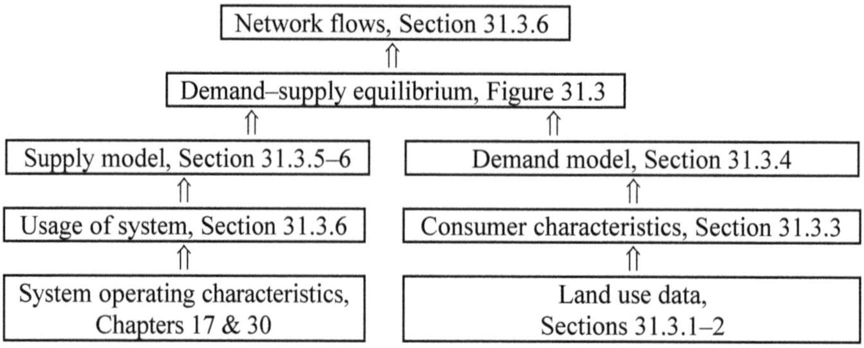

Figure 31.5 Relationship between transport demand and supply modelling.

As mentioned in Section 31.3.6, many models initially exclude the possibility that the supply characteristics of the transport system might influence demand, or introduce the effect very late in the analysis. This unrealistically implies that the amount of trip making is independent of the transport system. In such models no change in such important factors as trip times, fares, parking charges, or congestion would affect the number of trips made. Hence, the models over-predict transport usage. In reality, demand is usually ahead of supply and transport facilities commonly reach capacity constraints. Thus, many operating systems include some *suppressed demand* (Section 31.5). [A recognition that developed out of the earlier models from the 1960s was that there was no realistic, absolute solution to transport problems. Some proposed solutions of the time were both economically unachievable and socially undesirable.]

There are a number of ways of causing demand and supply to interact within model. Section 31.3.6 mentioned some cases where continuous demand–supply relationships can be used: for example, a travel time prediction for a link that is a function of the flow on the link. Such non-linear relations mean that the model must be run iteratively to obtain an accurate prediction. The iteration of the models continues until a stable prediction is obtained. In practice, this iteration series is an expensive process and often only the first projective stage has been run.

Beginning with a reasonable estimate of the final traffic flows can reduce the iterations. A useful alternative approach begins the iteration with limits (e.g. 1.3) on high flow/capacity ratios. This gives a first equilibrium near the 'capacity-unconstrained' case. The ratio is then reduced towards 1.0 in small steps as the incremental changes in the equilibrium situation are tracked.

The more sophisticated alternative is the *incremental* method that analyses a proportion of the total trips at each iteration, in a manner analogous to the build up of traffic within the system. When a link reaches capacity before demand is satisfied, trips must be re-routed, either during the current calculations or in the next iteration. It may also be necessary to consider the longer term responses, with trips being made at other times thus spreading the congested period, by other modes, or with changes in employment and land use (i.e. a revisit of Sections 31.3.1–3).

A major way around the iterative aspect of the problem is to use the fact that steady-state traffic occurs when there is equilibrium between demand and supply. The associated *(user) equilibrium* methods rely on Wardrop's (1952) two fundamental principles that are, for a given set of origins and destinations:

(1) Trip times (or costs) for an individual vehicle on the route used are equal to or less than those on any unused route. The principle is thus based on 'selfish', *user-optimised equilibrium*. It is assumed that drivers have access to perfect information and are able to select optimum routes. Estimates of the difference in travel cost between a driver's guess at the optimum route and the actual optimum vary between 25 percent in a dense network to 5 percent in an open one.

(2) Average trip times over the network are a minimum. This principle is thus based on *system-optimised equilibrium*. It will produce lower total system time than the first principle.

Thus traffic control is a 'game' between the individual users seeking to optimise their own trips and the system managers seeking system optimisation. The speed–flow relations in Section 17.2.5 and Figure 31.1 can be used to examine the consequences of this game.

One key equilibrium model was the *TRAFIC* program of Nguyen and James of the University of Montreal, which was extended by Montreal to the *EMME/3* model (Inro).

Drivers are not as deterministic and rational as the equilibrium models would suggest, and further reality can be introduced by representing the various choice processes in Sections 31.3.3–6 by probability distributions (e.g. of perceived travel costs), rather than by deterministic outputs. The distributions are usually either logit- or probit-based. The resulting models are called *stochastic user equilibrium* models. [Stochastic, logit and probit are defined in Section 33.2.]

31.4 APPLICATION OF TRANSPORT MODELLING

Consideration of the output of the models discussed in Sections 31.3.6&7 should lead to broad strategies and detailed proposals for both land use and transport infrastructure. These will certainly require careful evaluation, as their subsequent adoption will involve trade-offs between the demand for accessibility and other competing economic, social and environmental demands of society. This is a core task and plays a central role in Chapters

5 (Section 5.2) and 32 (Section 32.1), in the earlier parts of this chapter (Section 31.1), and in the discussion of benefit–cost analysis in Section 33.5.8. For the evaluation to be effective, the model output of the intended plan must be expressed clearly, specifically and understandably. Procedures that should be avoided include:

Within the model:

(a) targeting at a planning year, rather than at staged development within realistic annual budgets (Section 31.3),
(b) over-collecting and under-analysing data (Section 31.3.1),
(c) imputing excessive accuracy to the input data (Section 31.3.1),
(d) imputing of excessive accuracy to the output of models.
(e) over-emphasising complex models and analyses (Sections 31.3.6&7),
(f) not testing plans at intermediate stages,
(g) using inconsistent evaluation criteria,

Beyond the model:

(h) extrapolating existing levels of service (Section 30.2),
(i) assuming fixed community preferences over time (Section 31.3.2),
(j) limiting the analysis to predetermined solutions,
(k) prematurely imposing constraints, mainly by discarding options too early,
(l) ignoring community, financial, social, political and environmental constraints (Chapter 5),
(m) avoiding public participation (Section 32.1).

In the context of this chapter, it is appropriate to make point (d) as planning and transport forecasting models have a poor record as predictive tools. This point was well made in the past by Atkins (e.g. Atkins, 1986) and has recently been remade by Flyvberg *et al.* (2005) who considered 183 road projects from various countries built between 1969 and 1998. In half of these projects, the actual transport usage (or patronage) was at least 20 percent away from the forecast, and in a quarter of the projects, the error was at least 40 percent.

If the plan is adopted, the completed project will also require its performance to be evaluated, particularly against the criteria adopted at the assessment stage (Section 30.2). The output of this post-project evaluation should be used in the course of the future planning evaluations of new projects.

The specific way that this section fits into the rest of the book is shown in Figure 31.6. Given point (m) above, particular attention should be given to the final evaluation, checking, measuring and assessing points in Figure 31.6, not so much to review past performance but to feed into future decision-making. The discussion in this sub-section therefore brings us to the topic discussed in Chapter 5 and extended in Chapter 6 and thus completes the circle, in so far as the arrangement of this book is concerned (Figure 1.1). The remainder of the text discusses traffic data needed by road engineers.

31.5 TRAFFIC COUNTING AND FORECASTING

31.5.1 Counting traffic

Traffic counting primarily measures the traffic flow, which is the number of vehicles passing a given point in a given period of time (Section 17.1). Traffic forecasting predicts these flows at some future time. Counting and forecasting serve a variety of purposes and are an essential part of:

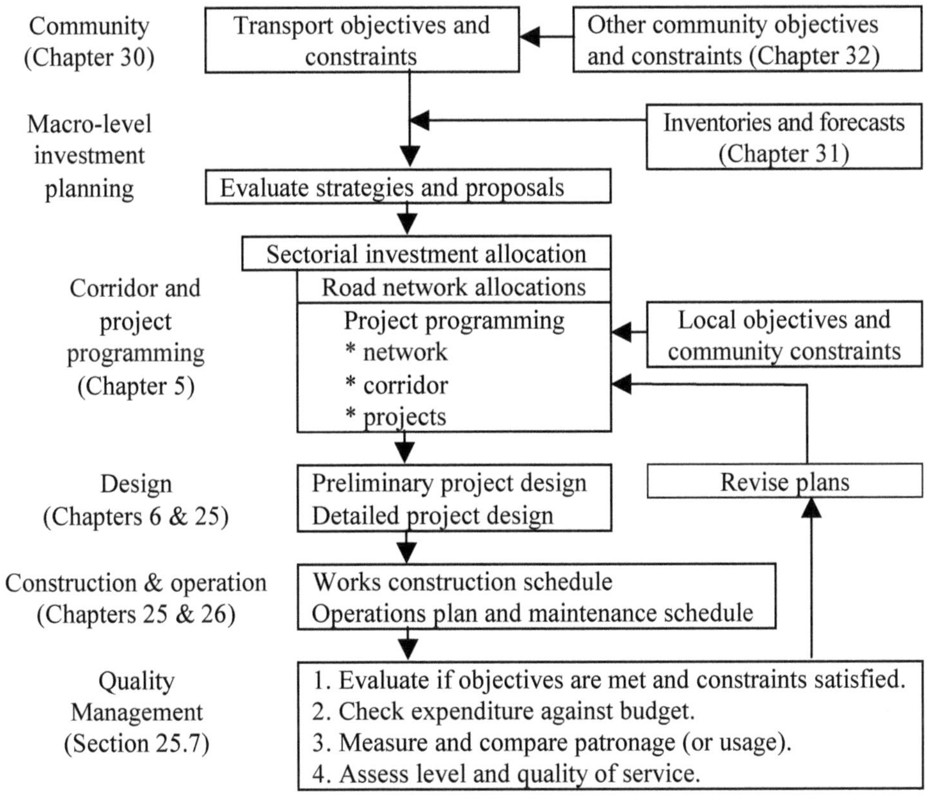

Figure 31.6 Major stages in the planning hierarchy for road investment proposals.

* transport planning (Section 31.3.6), by providing basic input to transport models in terms of existing flows on particular routes,
* studies of travel patterns (Section 31.3.6), particularly in areas not covered by surveys of consumers and users,
* origin–destination surveys (Section 31.3.6), by providing cross-checks of survey data related to existing travel patterns,
* evaluations of land-use proposals (Section 31.3.2&3),
* road design (Section 6.2),
* intersection design (Chapter 20),
* pavement design (Section 11.4), which needs the number, N_{esa}, of equivalent standard axles (Section 27.3.4) in one lane during the pavement design life, n (Sections 11.4.1 & 14.6.2),
* bridge design (Section 15.4.2) has a similar need to pavement design,
* planning of traffic management schemes (Section 30.7),
* measures of system efficiency and effectiveness (Section 30.2),
* indicators of areas needing closer examination, when compared with capacity calculations (Section 17.4.3), and
* exposure calculations for road crashes (Section 28.1.2).

Manual counting is done by roadside observers, aided by automatic notepads, field sheets, clip boards, tally counters, handheld calculators and organisers, and portable computers. Manual methods have an accuracy of about 10 percent and can readily be extended to obtain data on vehicle type, occupancy, and direction. However, origin and destination data must be obtained by surveys of the type described in Section 31.3.6. Automatic counting is done by:

* *pneumatic tubes*, which work by activating a pressure-sensitive air switch,
* *treadle switches*, in which two metal strips — separated at intervals by rubber pads — are pushed into electrical contact by the passage of a tyre,
* pressure-sensitive (e.g. piezo-electric) cables,
* pressure-sensitive pads and other weigh-in-motion devices (Section 27.3.3),
* *loop detectors* (Section 23.7), either as stand-alone devices or as part of a traffic signal installation, and
* video techniques.

The limiting factors for most of these devices are cost and the availability of power. Their use for speed detection is discussed in Section 18.1.5.

Except for the loop detectors and video, the automatic devices only detect axle passes and their output is converted to vehicle passes by dividing by an estimate of the average vehicle/axle ratio of the traffic stream. The main independent variable here is the percentage of trucks in the stream. This percentage is also important for estimating traffic capacity (Section 17.4.3) and the number of truck axles is critical for pavement design (Sections 11.4.1 and 27.3.4). If the counter can also detect the speed of the crossing vehicle, then it is possible to determine the spacing of the vehicle axles and hence automatically classify the type of vehicle being counted. Weigh-in-motion systems of the type discussed in Section 27.3.1 also permit vehicle gross and axle mass to be recorded.

A basic problem with traffic counts is the variations that occur in traffic flows. Time variations relate generally to the fact that traffic is constantly on the move in a variety of directions, and specifically to the following factors:

(a) morning and evening weekday travel demands causing peak flows, particularly with respect to flow to and from the CBD. A typical urban hourly pattern for a weekday is shown in Figures 31.7 and 31.8;
(b) variations between week days;
(c) daily recreational peaks, e.g. to sporting venues;
(d) seasonal recreational peaks, e.g. to holiday resorts;
(e) seasonal freight flows, e.g. crop harvests;
(f) differences between usage on urban arterial and residential streets, and
(g) changes in demand and supply, e.g. changing patterns of land use.

Thus a single census-like snapshot of the traffic is not easily produced and alternative procedures are usually needed.

An ideal solution is to have permanent, automatic counting devices installed on each road length. If this is neither economically nor technically feasible, the major alternative is to have permanent devices at a limited number of sites forming a representative but 'random' sample. Counts from these sites are used to provide data on the variation in traffic flow over time. Once-a-year traffic counts at 'temporary' (or 'sample') sites on each length of road to give a spatial distribution of traffic. These wide-coverage counts can be adjusted by the time variation factors obtained from the permanent sites.

The minimum counting periods at a temporary site is often 15 minutes. The counting period may be repeated on a weekly basis over four weeks. Typical coefficients of variation for 24 h counts predicted from less than 24 h counting periods are:

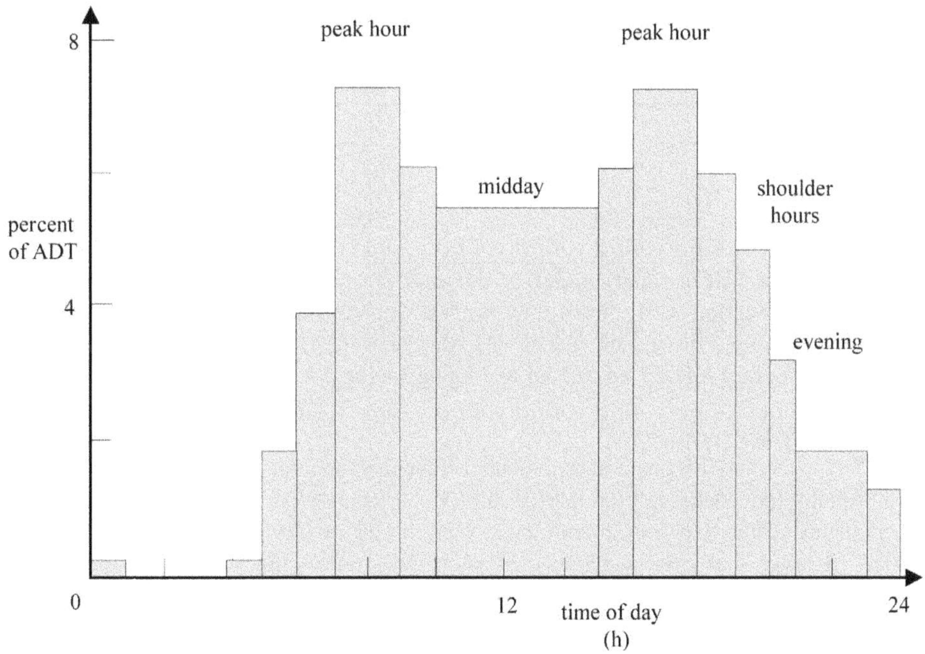

Figure 31.7 Typical weekday hourly traffic pattern.

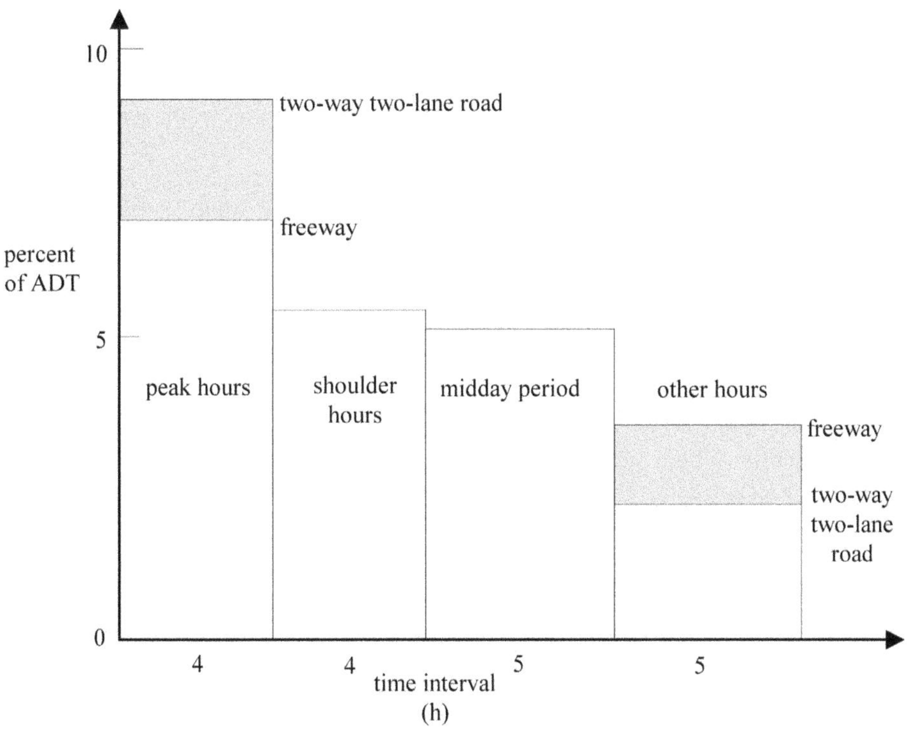

Figure 31.8 Typical weekday calculated hourly traffic pattern.

Length of count (h)	Coefficient of variation (%)
16	2.5
12	5
6	7.5
4	10
2	15

The commonest traffic-count output is the *annual average daily traffic* (or AADT, Section 17.1). Section 17.1 shows how *highest hourly flows* are often used in design. Following the above discussion, the AADT may be obtained by extrapolating one day's data at a temporary site by an annual averaging factor obtained from a permanent site. Thus:

AADT at any location =

(24 h count at that location)(averaging factor from a permanent station)

It is usually not necessary to use a full 24 h count. For instance, U.K. practice recommends using two 6 h counts each year, based on the evening peak, to predict AADT. It notes that even a single 6 h count on a representative day can provide an estimate of AADT within 10 percent of the actual. In the U.K. procedure, the 24 h count is estimated at 2.2 times the 6 h count and the AADT equation becomes (Phillips *et al.*, 1984):

$$\text{AADT} = 2.2(6\text{ h count})(354\text{ to }432)$$

↑

depends on location, road type, and vehicle mix (=365.25 for perfect uniformity).

Figure 31.9 illustrates a typical measured distribution of flows over a year. It also demonstrates the common situation in which the plot is based on a bimodal distribution. The two modes are a:

* Poisson (P) representing random trip-making, and
* normal (N) or lognormal around some 'central' flow value (Section 33.4.4) at about 5 percent of AADT, representing commuter trip-making.

The second component is often insignificant on rural roads.

31.5.2 Forecasting traffic from count data

Traffic forecasting can use the tools discussed in Sections 31.3 and 31.4 or can assume that the future will be a steady extension of the past and estimate future traffic counts from current count data. This latter approach is usually done on the basis of a steady annual growth in traffic at a rate of r. Thus Equation 33.14 gives the new flow as $(1 + r)^{n-1}$.

For pavement design, the cumulation of flows is needed to give the number of equivalent standard axles, N_{esa}, and so Equation 33.16 is used to give:

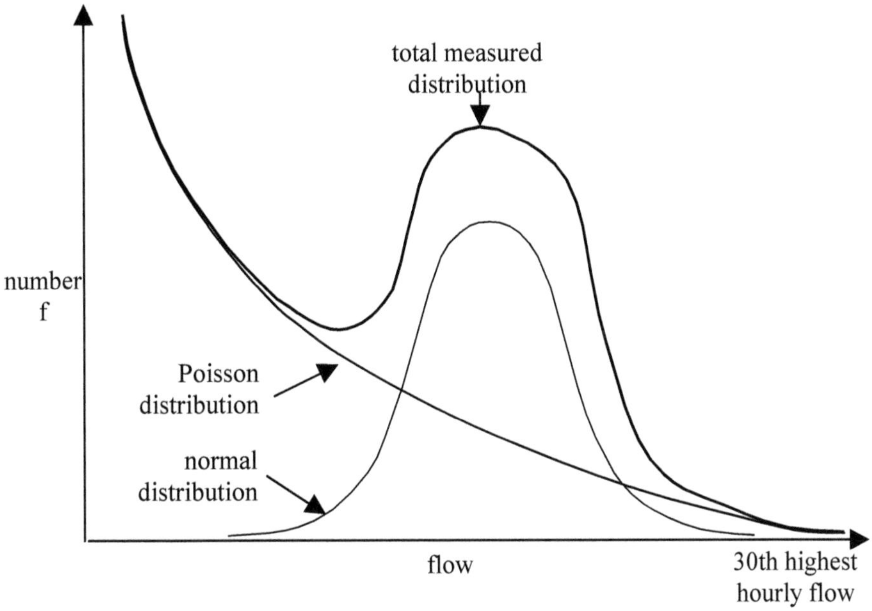

Figure 31.9 Hourly flow distribution.

$$N_{esa} = c(AADT)([1 + r]^{n+1} - [1 + r])/r$$

The coefficient c is often taken as 40, based on the assumptions that 10 percent of vehicles are trucks and that each truck is equivalent to 1.1 ESAs, i.e. 0.1 x 1.1 x 365 = 40.

More generally, traffic forecasts are made on the basis — not of a constant r — but by the extrapolation of existing trends, as established from traffic counts, judgement, and the predictions of transport planning models (Section 31.3.4). The final traffic flow forecast from any model — be it computer-based or a concept in the designer's mind — will be in five parts:

(a) existing traffic (from traffic counts),
(b) traffic change resulting from changes in the economy,
(c) traffic diverted by 'local' road changes,
(d) traffic generated (suppressed or latent demand) by road improvements on the route (or deterioration on other routes),
(e) traffic generated by other transport and land use changes (Section 31.3.4).

In the absence of road pricing (Section 29.4.5), suppressed demand can be a powerful factor and is often under-estimated (e.g. Section 31.3.9).

Traffic flow data collected over, or forecast for, an area are often represented using a traffic flow map (or *desire-line diagram*), showing the width of each traffic route in proportion to the traffic flow using the route over some specified period (Section 31.3.7).

The estimation of vehicle kilometres of travel (*VKT*) is another key transport data need that is often fulfilled by traffic counting and forecasting, followed by the estimation of trip lengths. It can also be estimated from knowledge of total fuel usage and fleet fuel consumption rates (Section 29.2.1). In most countries, VKT has been increasing steadily with time.

CHAPTER THIRTY-TWO

Environmental Factors

32.1 EVALUATION AND ASSESSMENT

32.1.1 Definitions

Chapter 5 discussed the wide range of factors that must be considered in assessing the worth of a planned road project. In particular, Section 5.2.3 introduced the concept of a social audit to consider the economic, strategic, distributional, and environmental impacts of a project. This chapter discusses the environmental component, although it is rarely possible to place the impacts of a road project in neat, self-contained categories.

Although the environment embraces all our surroundings, there are three aspects of interest to this chapter:

(a) the environmental framework, which is the physical stock that has evolved as a result of past activity. It has two components;
  (a1) the *ecosystem* of land form, minerals, rivers, air, flora, and fauna (Section 6.4),
  (a2) the *built environment* of buildings and machinery;
(b) the current *activity systems* operating within and drawing upon this framework, with a primary emphasis on the human activity system.

These aspects are interactive. For instance, to meet human needs, human activity draws upon the environmental framework in four ways:

(1) for *life support* (e.g. air and food), thus creating physical consequences (e.g. air pollution and imported flora);
(2) for *amenity* (e.g. transport and scenery), thus creating social consequences (e.g. accessibility restrictions and landscape changes);
(3) for *material resources* (e.g. mining), thus creating economic consequences (e.g. land development); and
(4) for *waste reception* (e.g. sullage flows), thus creating ecological consequences (e.g. water pollution).

The concept of sustainability requires that such activities be managed to ensure that it does not comprise the ability of future generations to meet their own needs. This can be achieved by both minimising the impacts of any activity and by ensuring that that activity is both efficient and effective (Sections 30.2.2&3).

32.1.2 Environmental impacts

It is better to avoid an environmental impact, rather than plan to mitigate the impact. Nevertheless, each human transport activity will have some impact on the environment. The full range of environmental impacts caused by a road project is listed in Tables 5.1 & 32.1 and covers the planning, economic, social, and engineering aspects of the project (Section 31.1). Any course of action will impact to some extent on the environment and on community goals and objectives, and the community will normally wish to guard and improve rather than destroy the environment. Thus, if a project is to proceed, environmental impacts will usually need to be clearly seen in the context of the various benefits delivered — these are also given in Table 5.1.

Table 32.1 Environmental factors which may be positively or negatively impacted by road projects. Factors impacted during planning and design are indicated by a P, during construction by a C, when the road is operating by an O, and at all stages by an A.

Environmental factors	Table 5.1	Impact type	Book Section
1. Factors relating to the community as a whole			
1.1 Diversion of resources to construction	C2	C	30.1
1.2 Effects of the construction process	C3	C,O	
1.3 Damage to or destruction of:			
a. community facilities and services	C6–7		
b. unique or historical features	C7	P,C	25
c. public open space and parklands	C7	C,O	6.4
d. landscape and townscape	C10	C,O	6.4
2. Factors relating to the social environment			
2.1 Apprehension about compensation	C1	P,C	6.3
2.2 Depletion of low-cost housing	C6	A	32.5
2.3 Relocation of residents	C6	A	32.5*
2.4 Relocation of job opportunities	C6	A	32.5
2.5 Social concerns due to disruption of:			7.2,
a. local patterns of social linkage	C7	A	32.5
b. local social stability	C7	A	32.5
c. the quality of local life	C7	A	32.5
d. social life and social patterns	C7	A	32.5
2.6 Decrease in local road safety	T3	O	28.5
3. Factors relating to the physical environment	C12		
3.1 Noise		C,O	32.2
3.2 Air pollution		C,O	32.3
3.3 Dirt and dust		C,O	32.3
3.4 Water pollution		C,O	32.4
3.5 Vibration		C,O	13.4
3.6 Poor drainage, more sediment transfer and soil erosion		C,O	25.8
3.7 Other ecological impacts		C,O	–

The levels beyond which these environmental impacts become unacceptable are called environmental capacities. *Environmental impact assessment* means the identification of all environmental capacities, the prediction of any likely environmental impacts, and an assessment of whether each impact will reach or exceed the associated environmental capacity. The approach must:

* be broadly based to ensure adequate consideration of the effects of all possible courses of action (or non-action),
* predict the spatial and temporal distribution of those effects, and
* range from being subjective and informal, through to applying structured, numerate analyses.

Its major advantage of an environmental impact assessment is that it often signals projects or issues requiring special attention, rather than grades particular facets.

Not all impacts have an environmental focus and it is easy to overlook wider community impacts. In particular, the four major, negative community impacts of a road scheme were listed in Table 5.1. They are:

(a) *Community severance.* This occurs if a new road restricts existing local access and thus has a divisive effect on the community. This can be a consequence of the physical structure of the road (e.g. access-control fences and embankments — sometimes called static severance), or of the lack of crossing gaps in the traffic using the road (e.g. measured by crossing delays, Section 17.3.5 — sometimes called dynamic severance). The effects may be merely perceived, or they may result in changes in behaviour, particularly in trip making.

(b) *Displacement.* This occurs if property acquisition eliminates homes or jobs.

(c) *Migration.* This occurs when the changes caused by the road scheme alter the population in the area.

(d) *Disintegration.* The accumulated changes in (a) to (c) can destroy the existing community fabric in a way that is particularly hard on those unable to leave the area.

Of course, the road planner and builder must minimise these impacts.

Environmental risk management is based on conventional risk management practice. It identifies and analyses events which have some risk of impacting on the environment. It then evaluates the probability and severity of their impacts. This allows the acceptability of the risk to be assessed and a management strategy to be proposed. Environmental risk management also requires regular system reviews and continued stakeholder consultation.

32.1.3 Environmental evaluation

An environmental evaluation interprets the environmental impact assessments. The process takes place at two levels:

(a) At a broad level, occurring as soon as practical after a possible project appears on a forward program. It is best seen as one part of the systematic evaluation of proposals, as described in Section 5.1 and further defined in Sections 5.2.3 and 31.3.8. Its purpose is to eliminate inappropriate proposals and to identify those requiring further consideration.

(b) At detailed level, considering specific aspects of projects selected under (a), and following Sections 32.1.2&3.

The evaluation particularly assesses the importance, significance, and consequence of all situations where the environmental impact exceeds the environmental capacity. It will also check whether proposals will:

* meet regional planning objectives,
* comply with the relevant environmental laws and regulations,
* protect the environment, mainly by keeping most impacts below their associated environmental capacity, and
* be environmentally acceptable to the community at large and to political groups in particular. This usually includes clarifying unresolved issues in such a way that informed debate and decision-making can occur, using as full an information base as is possible and feasible.

Most projects will require only a relatively simple environmental evaluation, although the process will often have a subjective component, as in attempts to put a monetary value on a noise impact (Section 32.2.4). In these circumstances, the degree of objectivity of the evaluation is not a measure of its significance and objective measures should not be allowed to overshadow subjective ones.

It will often be necessary to find a means of combining the results of a number of different environmental impact assessments. The six main methods for doing this are given below (see also Section 5.2.3). They should all be proceeded by the production of a *comprehensive list of environmental impacts*, such as the components of Tables 5.1 & 32.1, needing investigation.

(a) Ad hoc methods relying on local practice and personal experience. For example, items in the impact list can be given numerical ratings. However, it may be necessary to use the political process to provide the value judgements on which to base the weightings.
(b) *Overlays* — this method places the items on the impact list on to overlays of a regional map, in order to produce a composite picture that can then be subjectively assessed.
(c) *Matrices* — these add a second dimension to the impact list by adding an orthogonal list of the activities likely to cause the impacts.
(d) *Networks* — these trace the impacts via cause and condition to effects and are thus a third dimension added onto the matrix method.
(e) *Benefit–cost analysis* (BCA) — this is the conventional economic approach used in recent years (Sections 5.2 & 33.5.8).
(f) *Planning balance sheet* — this extends the BCA by attempting to incorporate social and distributional effects into BCA via a matrix showing the type of cost or benefit (Section 5.2.3).

Methods (a), (e) and (f) permit a single value, or score, to be calculated and preferred options selected.

32.1.4 Community participation

It is important that the community be involved in the process of impact assessment and evaluation. A major role of community participation is to bring under consideration information that only the community can provide. This role is concerned with the delineation of viewpoints rather than with decision-making. The process of establishing these viewpoints is essentially political. Even after participation has occurred, it may be necessary to impute community values in order to obtain numerate assessments.

There will be differing interests and priorities within the community, and each proposal will impact each community segment in a different way and cause its own response. A typical community segmentation is into:

(a) people adversely affected by the project,
(b) people interested in and/or beneficially affected by the project,
(c) people with relevant expertise,
(d) pre-existing local groups such as residents' associations, and
(e) single issue groups such as some environmentalists, public transport enthusiasts, and political factions.

Another segmentation is with respect to affected locations, viz.:

* at home,

* at work,
* at recreation,
* in special groups (e.g. the elderly),
* at special functions (e.g. hospitals),
* while driving, walking, cycling, etc.,
* while observing (e.g. bird-watching), and
* while touring.

When assessing the views of a community segment, the number of people involved in it, their abilities, and their political influence should be irrelevant. It is content of their ideas and views that must be considered. However, the participation of most people will be voluntary and those who do become involved will usually be a minority of those affected and will often be predominantly from categories (a) and (e) (Evans and Bush, 1980). Thus opinions gained will at the best be a random sample and may well be ill founded or of a strident segment, rather than of the whole community. Thus, inarticulate groups may need to be assisted and over-articulate groups may need to be muted.

The interactions between the Road Authority and the community will vary from manipulation by the Authority at one extreme to citizen power at the other, with consultation and compromise being a median position.

32.1.5 Environmental impact statements

In many regions the environmental impact assessment discussed in Section 32.1.1 has been formalised into a defined legal requirement known as an *environmental impact statement* (*EIS*). Such a report is mandatory for many road projects.

The job of preparing an EIS is often of such magnitude and diversity that it requires the formation of a study team specifically dedicated to the task. It is important that the team has a well-defined scope and task statement. The activities that need to be undertaken by the study team are:

* prepare a work statement, work program, and funding schedule;
* define the decision-making process and present it in a way that will be clear to all participants;
* involve all relevant agencies and Authorities and ascertain and define their roles: prepare inter-agency agreements and establish joint groups;
* establish a community involvement program and identify relevant individuals, community organisations and other appropriate interest groups;
* make contact with all involved parties, attempting to strike an appropriate balance between interests (see also Section 32.1.3);
* undertake the work program,
* ensure that the team's work is carried over to future implementation.

Such a study team would:

* use members from an appropriate range of disciplines,
* maintain a flexible attitude,
* mould its methods to suit the circumstances,
* keep its work continuous and iterative, and
* accept that predictions of the future are inherently uncertain.

The study team can exist:

(1) within the bureaucratic framework of the Road Authority;
(2) as an autonomous team under a Road Authority manager; or

(3) as an autonomous team operating under an independent manager.

However, the Review Panel shown in Figure 32.1 should be independent of all interested groups.

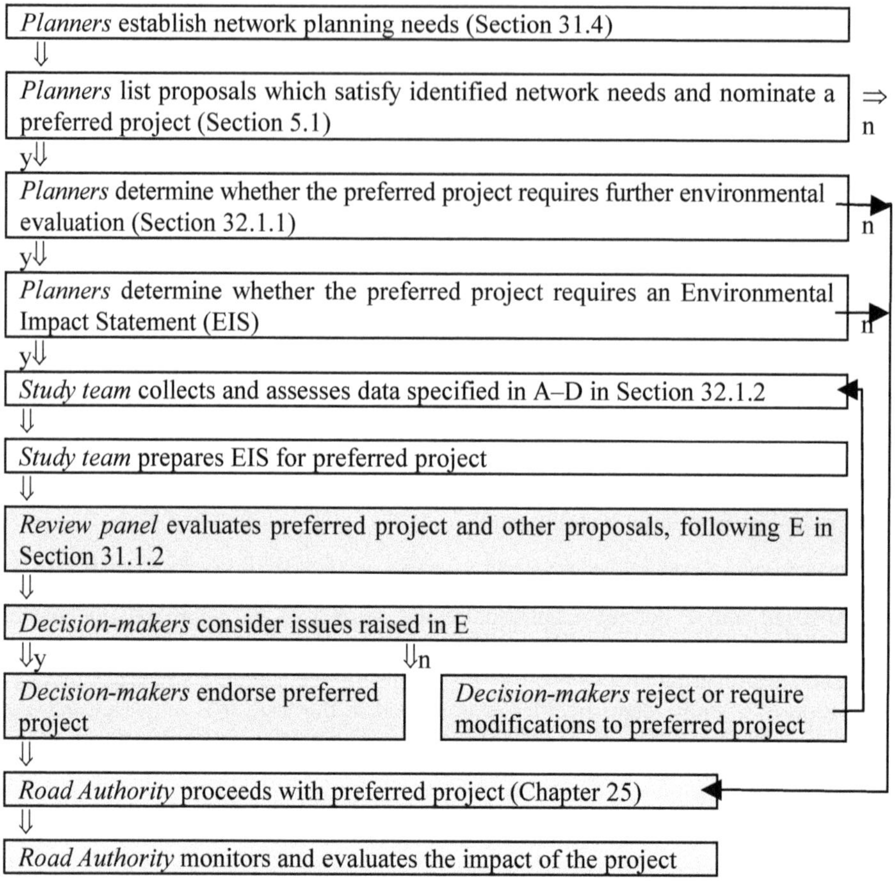

Figure 32.1 Typical framework for an environmental impact statement. Refer also to text in Section 32.1.2.

The EIS prepared with the aid of the study team will usually contain material under the following headings, although parts A to D may be produced by the proponents for a proposal, and part E by an independent review body. The process is shown further in Figure 32.1. A comprehensive review of environmental impact assessment processes is given in OECD (1994).

Components of an EIS

A. *Listing of prior community-wide decisions:*

A1. the community's goals and objectives (Section 31.1),

A2. the agreed needs of the transport network (Section 31.4), and

A3. circumstances when an EIS is needed.

B. *Listing of overview material* (for the range of alternative proposals addressing A1 and A2, including the nominated project and the do-nothing and do-minimum options):

B1. adequate description of all the alternative proposals,

B2. relevant laws and regulations,
B3. the existing environmental framework (Section 32.1.1), and
B4. relevant environmental capacities (Section 32.1.1).

C. *Relevant specific material* (for at least the nominated project and the do-minimum option):
C1. specific objectives, and
C2. detailed descriptions.

D. *Listing of prior arguments*:
D1. justification for selecting the nominated project (Section 33.5.8),
D2. likely environmental impacts for all proposals studied in-depth (Section 32.1.1),
D3. environmental impact assessment for all proposals studied in-depth (Section 32.1.1), and
D4. environmental evaluation of all proposals studied in-depth (Section 32.1.1).

E. *Decisions coming from the EIS process*:
E1. key issues associated with the nominated project,
E2. modifications that should be made to the nominated project,
E3. critical trade-offs that should be made,
E4. safeguards that should be applied,
E5. totally unacceptable actions, and
E6. areas where further immediate data or ongoing feedback is needed.

32.2 NOISE

32.2.1 Theory

Sound is a pressure wave that results from periodic wave-like mechanical disturbances of a medium (usually air), within the range of frequencies and amplitudes to which the human hearing system responds. The transmission of sound is affected by the physical characteristics of the medium and the obstacles in its path. In air, the periodic disturbances travel as pressure fluctuations above and below atmospheric pressure, moving at about 340 m/s at 20 C. Their main measurable characteristics are wavelength (the distance between successive pressure fluctuations), frequency (the time rate at which the fluctuations occur), and pressure (the root mean square pressure, p_r, or peak pressure variation from atmospheric). Sound *intensity* is the power transmitted per area by this process and is proportional to the square of sound pressure. It is thus a measure of the *loudness* of a sound. An important consequence of the waveform of sound is that the total sound from two independent sources is not the arithmetic sum of their individual pressure levels (see below).

Sound spreads out in all directions from a point source. The radiating sound waves from traffic usually take about 2 to 3 m to 'settle down' to a well-defined periodic form. Sound energy is dissipated (i.e. *attenuated*) with distance, and sound pressure decreases following an inverse square law. The area influenced by sound increases with distance squared and this spreading-out effect dominates sound attenuation for distances under 300 m. Appreciable attenuation effects occur within about 10 m. The higher frequencies are more rapidly attenuated and so, at long distances, all sounds are changed towards a rumble.

32.2.2 Hearing noises

Noise is unwanted sound, or sound that produces an adverse response in the hearer. Thus sound can be subjectively assessed as noise. In an objective sense, noise is usually found to be either a loud sound or a sound containing either many inharmonic components or excessive amplitudes.

The loudness of a sound was defined in Section 32.2.1. It is assessed in the following manner. For the case of a sound with a frequency of 1 kHz (the soprano's high C), the softest sound that can be heard by good, young ears has a root mean square pressure of $p_{r0} = 20$ μPa. At the other extreme, the threshold of ear pain occurs when p_r is above 60 Pa, i.e. some 10^6 times as great. Not surprisingly, the ear does not appear to respond linearly to pressure changes over this range, and so a logarithmic scale based on $\log[p_r/p_{r0}]$ is used to provide a better measure of human response. This scale uses p_{r0} as its reference pressure. Multiplying the log[pressure ratio] by 20 gives the so-called *sound pressure level*, spl, which is therefore defined by:

$$\text{spl} = 20\log[p_r/p_{r0}] \qquad (32.1)$$

This means that 'the softest sound that can be heard by good, young ears' has an spl = 0 and the 'threshold of ear pain' has an spl = 130. With $p_{r0} = 20$ μPa and p_r in Pa, the equation becomes:

$$\text{spl} = 120 + 20\log[p_r/20]$$

Given the definitions of sound intensity and loudness in Section 32.2.1 and the fact that intensity is proportional to the square of pressure, Equation 32.1 can be written to quantify the loudness of a sound as:

relative sound intensity = 10log[(sound intensity)/(reference sound intensity)] dB (32.2)

This loudness measure gives the well-known *decibel* (dB) scale. The *bel* is the log[intensity ratio], and the *deci* relates to the use of the coefficient *10* in Equation 32.2. Equations 32.1&2 show that the 'softest sound', spl = 0, corresponds to zero dB and pain to 130 dB.

Each increase of 10 dB is equivalent to a 3.2 times increase in p_r and a 10 times increase in power output. This is usually subjectively considered to be the equivalent of a doubling of the loudness level. The numerical and physical significance of the decibel scale are illustrated in Table 32.2. Where the table uses the term 'tolerated', this applies to newly encountered noise — there is evidence that people do grow accustomed to noise levels that were initially disturbing.

The human ear responds to various sound frequencies in an uneven fashion. For example, between 1 and 5 kHz the audibility threshold in dB is relatively constant near the reference value of 0 dB at 1 kHz. However, the threshold rises to about 50 dB for a 50 Hz sound. An 'A' weighting scale has been produced to model this response to relatively-quiet noise. Sound measured with this weighting — which is usually done electronically by filters within the measuring instrument — is designated *dB(A)*. Changes in dB(A) correlate well with changes in subjective annoyance. The two other filters in some use are B and C. Their attributes are summarised below:

A: relates to threshold hearing and thus to quiet circumstances,

B: relates to speaking, and

C: relates strong, loud noises.

The C filter has been used to explain responses to sounds in excess of 50 dB produced by trucks.

Table 32.2 Relationship between the RMS pressure, p_r, of various sounds and their loudness, spl, in decibels.

Description of the sound *Non-road examples of type of sound*	p_r (xPa)	x (xPa)	spl (dB)
Just-audible 1 kHz (high C)	20		0
Background sound in a recording studio	63		10
Wind in trees; caves	200	μPa	20
Just-tolerated as a continuous noise in a bedroom: *whispers, rustling leaves*	630		30
Unacceptable as a continuous noise in a bedroom	1.1		35
Quiet conversation; *private office, refrigerator humming*; lightly trafficked street; *bird sounds; quiet countryside*	2.0		40
Unacceptable as a continuous noise anywhere indoors or as a peak noise in a bedroom or where relaxation is intended; interferes with speech, speech impossible beyond 10 m	3.6		45
Living room, shared office, *quiet stream*	6.3		50
Unacceptable to 10 percent of population (noise-sensitive people); annoying to 25 percent of population; acceptable to 25 percent of population: street carrying moderate traffic; *female voice*	11.0	mPa	55
Unacceptable to people in intellectual pursuits and in white-collar work places: busy street; *open office*; *voices unconsciously raised*; *normal conversation*	20.0		60
Unacceptable to people; comfortable discussion impossible; interferes with radio, TV and relaxation: *male voice face-to-face*	36.0		65
Avoided by people: *busy pedestrian mall*; *loud conversation*	63.0		70
Very busy road with many trucks; considered to be loud; *vacuum cleaner*; *train station*	200		80
Ear damage threshold: car horn, truck accelerating in low gear; considered to be extreme: *workshop*; *dog barking*	630		90
Factory; *chain saw*; *near airport*; *pop group*	2.0		100
Shouted speech inaudible: *drill*	6.3		110
Deafening: *inside helicopter*	20.0	Pa	120
Causes pain:	63.0		130
Jet aircraft, pneumatic drill	200		140
Jet aircraft at take-off	630		150

The *noise exceedence*, L_n, scale is used to measure the noise exceeded during some defined percentage, n, of the day. Rather than use the entire 24 hours, better estimates of annoyance are obtained from L_n values calculated over the 18 h period from 0600 h to 2400 h (Figure 32.2). These values are called $L_n(18)$ values. For instance, $L_{10}(18)$, is the sound level in dB(A) exceeded for 10 percent of the time between 0600 h and 2400 h (these 0.1 x 18 = 1.8 h need not be continuous). For normal traffic the L_{10} value for a 3 hour period, $L_{10}(3)$, is about 1 dB(A) greater than $L_{10}(18)$. Some jurisdictions calculate $L_{10}(18)$ as the average of the 18 $L_{10}(1)$ values that occur in the 18 h period.

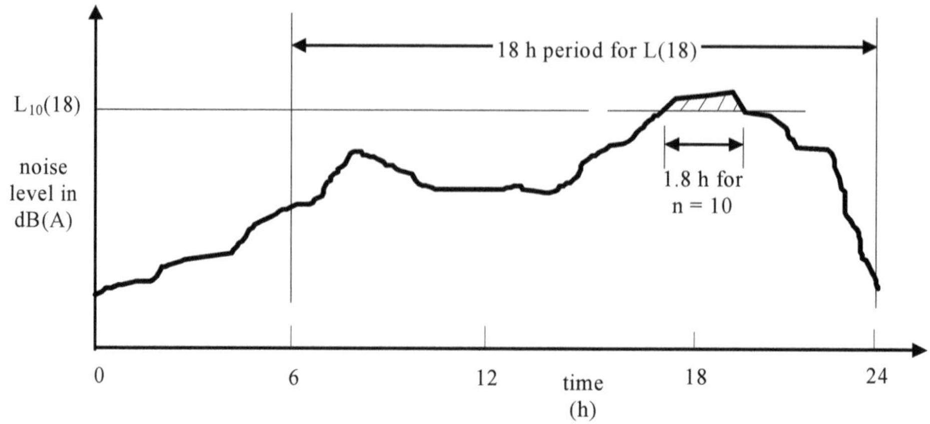

Figure 32.2 Definition of $L_n(18)$.

An extension of this approach is the use of an *equivalent continuous sound level*, L_{eq}. This is the steady sound level over some defined time period, with the same energy equivalent as the actual, varying, sound level over the same time period. It is given by:

$10^{L_{eq}/10}$ = mean of all the $10^{dB(A)/10}$ (32.3)

$L_{eq}(T) = 10\log$[mean of all the $10^{dB(A)/10}$ over time T]

An empirical relationship between L_{eq} and L_{10} based on urban data is:

$L_{10}(18) = L_{eq}(24) + 3.5$

Another term sometimes used is the *noise pollution level*, L_{np}, which is:

$L_{np} = L_{eq} + 2.56$[standard deviation of dB(A)] (32.4)

In studies of freeway noise, a very high correlation has been found between most of these sound scales indicating that, at least for freeways, the scale chosen is not particularly important.

Sound is usually transmitted from its source to the hearer by air, and will therefore depend on the air temperature, pressure, and humidity. For example, sound waves spreading out horizontally are refracted by temperature gradients in the air. When the temperature decreases with height e.g. on a warm day the waves are refracted upwards and noise effects are diminished. When the weather is cold and the sky is clear, temperature can increase with height. In this case, sound waves are refracted downwards and distant noises are more readily heard. Similarly, waves will be bent downwards when travelling downwind and upwards when travelling into the wind. As a consequence of such effects, 300 m from a noise source, weather conditions can cause noise variations of ±10 dB(A).

Impervious objects (such as mounds and barriers) will prevent the direct (line-of-sight) transmission of sound, but the sound waves can bend past such objects by diffraction around edges. Most vegetation will only gently dissipate sound and cannot be relied on for sound absorption.

For determining levels of traffic noise, the next section will show that dominant roles are played by traffic flow and speed, vehicle type, steep road grades requiring gear changes and braking, and the texture of the road surface.

32.2.3 Traffic noise

Traffic noise is the summation of the noise produced by each individual vehicle in a given traffic situation. At low speeds, such as those encountered on urban roads, most noise produced by an individual vehicle is caused by vehicle air-intakes, engines, fans, transmissions, exhausts, brakes, and bodywork. Vehicles are generally at their quietest when travelling at about 30 km/h. Poor driving practice and stop-start conditions, such as those caused by congestion and traffic signals, increase noise levels.

As vehicle speed rises, the noise attributable to the tyre–road contact and to air disturbance increases and tends to dominate. It is the major contributor above 40 km/h for all vehicles, other than motorcycles. Typical relationships are:

for cars, $p_s = 30\log v_k + 10$ dB(A)
for trucks, $p_s = 20\log v_k + 40$ dB(A)

The wavelength of the longitudinal surface profile (Sections 12.5.1 & 14.4.2) that contributes most to tyre–road contact noise is between 10 and 500 mm. This was described in Section 12.5.1 as the *macrotexture* and is measured by the sand patch method. Its effect can be estimated from (Samuels, 1982):

$$p_s = [-16 + 30\log v_k + 3.9\log(d_M{}^2A + V/d_M)] \text{ dB(A)} \tag{32.5}$$

where:

v_k is the vehicle speed in km/h,
d_M is the mean macrotexture depth of the surface in mm,
A is the tyre contact area in mm^2, and
V is the maximum volume of air enclosed by the road surface in the tyre tread and available for *pumping* from the tyre–pavement interface (in mm^3).

The noise produced thus increases logarithmically with speed, macrotexture, and the coarseness of the tread. Reducing the mean of the macrotexture depth to below 1.6 mm does not cause further reductions in the noise levels produced.

Taking a dense-graded asphalt as the datum, a 15 mm spray and chip seal (Section 12.1.4) can raise noise levels by 4 dB(A), whereas a porous friction course (Section 12.2.2f) can lower the level by between 1 and 6 dB(A). Overall, the range between a good and bad surface can be as much as 10 dB(A) and so the effect is substantial and should be managed. A typical poor noise surface would be deep-grooved concrete (Section 12.5.5), and a typical good surface would be the porous friction course.

The noise generated by grooved concrete pavements can be reduced by keeping transverse grooves below 10 mm in width and minimising longitudinal variations. Indeed, 2 mm square grooves made by dragged devices (either laterally or longitudinally) are often the most effective way of reducing the noise generated by concrete surfaces. This can also help reduce the tonal qualities of concrete road noise which can sometimes distress people.

The statistical pass-by index (SPBI) was developed in Europe in the 1990s to handle the effects of different vehicle types. It uses the model of Equation 32.3 to combine their effects logarithmically and in proportion to their presence in the vehicle stream:

$$10^{SPBI/10} = \Sigma(\text{vehicle A proportion})10^{LA/10}$$

Tyre–road contact noise is actually composed of noise produced by the pavement by pumping of air entrapped in the pavement macrotexture, and by the tyre tread patterns. The noise produced by tyres occurs in the 800–1000 Hz frequency range and by road surfaces in the 200–500 Hz range. The A-filter scale attenuates the road surface component and thus emphasises the tyre component. The tyre noise does not increase as the macrotexture increases. The stiffer the tyre, the more noise it produces. Modern tread designs often randomise the spatial elements to eliminate any pronounced noise frequency peaks, which are detected by the ear as annoying whines.

Most jurisdictions have requirements controlling the noise levels produced by individual vehicles. The measurement of these noise levels requires careful specification of the road surface and a clear definition of the position and operating behaviour of the vehicle and the microphone. The use of a fixed microphone beside the road (the *pass-by* method) is relatively simple and cheap, gives good statistical data, and a reasonable representation of the noise heard by people at the roadside. However, the detection of the magnitude of the noise components from the various noise sources (e.g. the tyre and pavement) usually requires trailer-mounted devices travelling near the noise source, although roadside readings as vehicles coast-by can be used to eliminate the engine-noise component. The procedures mostly derive from ISO Recommendation R362 issued in 1964 (now ISO 5130) and ISO 1189 (1997).

Different noise standards are usually imposed for cars, motorcycles, trucks and buses. Typical dB(A) values in various standards are:

	Microphone location at		
vehicle type	15 m	9 m	7.5 m
cars	72	95[a]	78
trucks and buses	80	90[a]	84
motorcycles		90[a]	80

Note: These values are high by some common current world standards.

Noise generated by *heavy vehicles* deserves particular attention as often it is used also as a surrogate measure for many of the other nuisances created by such vehicles. These include interference with other traffic, smell, vibration, and a fear of large machinery. Studies have shown that perceptions of traffic noise are strongly related to the noisiest three trucks in the previous hour.

So far the discussion has concentrated on the noise from a single vehicle. The noise generated by a traffic stream requires the aggregation of the noise produced by the individual vehicles in the stream. The level of traffic noise generated at any roadside point will fluctuate with time between the extremes of a background noise and the passage of peak 'noisy' vehicles. This statistical variation is handled by the L_n and L_{eq} single number scales described in Section 32.2.2. In modelling, the aggregation is often taken as coming from a line source which is a function of vehicle and traffic characteristics.

The simplest model is that the increase in noise in dB(A) over a background level is given by 10logq, where q is the flow in veh/h. This implies that noise increases at only about a twentieth of the rate at which the traffic causing it increases. Thus, most people will be insensitive to noise changes due to changes in traffic volumes on a heavily trafficked road. Their initial experience will usually be continued into the future. The effect of traffic speed is also largely logarithmic and approximately given by 30logV, as suggested by Equation 32.5.

The most widely used prediction model is the U.K. *CORTEN* (Calculation Of Road Traffic Noise) model which requires an estimate of traffic flows, speeds, heavy vehicle percentages, road geometry and surroundings, to predict L_{10}(18h) and L_{10}(1h) values at a point location. The model has been shown to usefully predict measured L_{10} noise levels with a standard deviation of about ±2 dB(A) (Samuels and Saunders, 1982). CORTEN has recently been developed into CRTN and in this form can handle the new traffic noise indices defined in the EU Directive on Environmental Noise (2002/49/EC).

A useful model is the U.S. FHWA's Traffic Noise Model (TFM/2) which was released in 1998 (Batstone *et al.*, 2001) and updated in 2003. TFN has a database of some

6000 measured traffic noise events that form the basis of its predictions. It handles the sound path geometry particularly well and so is useful for assessing noise barriers (Section 32.2.5).

Measurements of traffic noise are usually taken:

* 1 m away from the façade of a building facing the roadway, to allow readings to be taken readily and in a manner that is repeatable and reproducible. If it is not possible to get within 1 m of the façade, measurements can be taken in an equivalent position where there is no sound-reflecting surface within 15 m of the instrument. In the case of L_{10} readings, for example, 2.5 dB(A) would need to be added to the readings to account for façade reflection.
* not closer than 5 m from the nearest lane of traffic to avoid distortions caused by the traffic stream.

Measurements are taken when the road surface is dry and when either the average wind speed at 1.2 m is below 2 m/s or where the wind direction is more towards the instrument than along the road. Traffic noise measurements usually have a standard deviation of about 5 dB(A). Measurements taken within rooms require the occupants' specific operating conditions to be duplicated.

The reduction in traffic noise from the outside to the inside of a house will typically be about:

* 25 dB if the windows and doors are shut and there are no wall openings,
* 20 dB if the windows and doors are shut and there are wall openings (e.g. air vents),
* 10 dB if the windows and doors are open.

To achieve 25 dB would therefore require some form of forced ventilation.

Most current noise standards attempt to achieve a measured noise level not exceeding 60 to 65 dB(A). It can be seen from Table 32.2 that this will cause affected people to be concerned by the noise, and is often a compromise been the noise attenuation that is practically possible, and peoples' tolerance to noise over time. A standard of 55 dB(A) would be desirable, but rarely achievable. Although people are also annoyed by isolated peak noises, these are not covered in the above measures and this is a deficiency in most commonly applied roadside noise standards. It is usually argued that it is better eliminating the causes of the noise peaks, rather than attempt to attenuate them.

32.2.4 Noise annoyance

Sections 32.2.2–3 have mentioned the role that human annoyance plays in determining procedures for noise measurement. Many annoyance indicators are listed in Table 32.2. A conceptual model of the relationship between traffic noise and its effect on people in their homes is shown in Figure 32.3.

The annoyance that people feel when they are exposed to traffic noise is related to both the physics of the noise (Section 32.2.2) and to the following sociological and psychological factors:

* whether the noise interferes with valued activities such as sleep, telephone conversations, family discussions, the audibility of radio and television programs, and business conferences. These interferences form the predominant complaints against traffic noise as it is reasonably assumed that the noise environment in a house should be such that it is possible and easy to talk at any time, and to sleep or

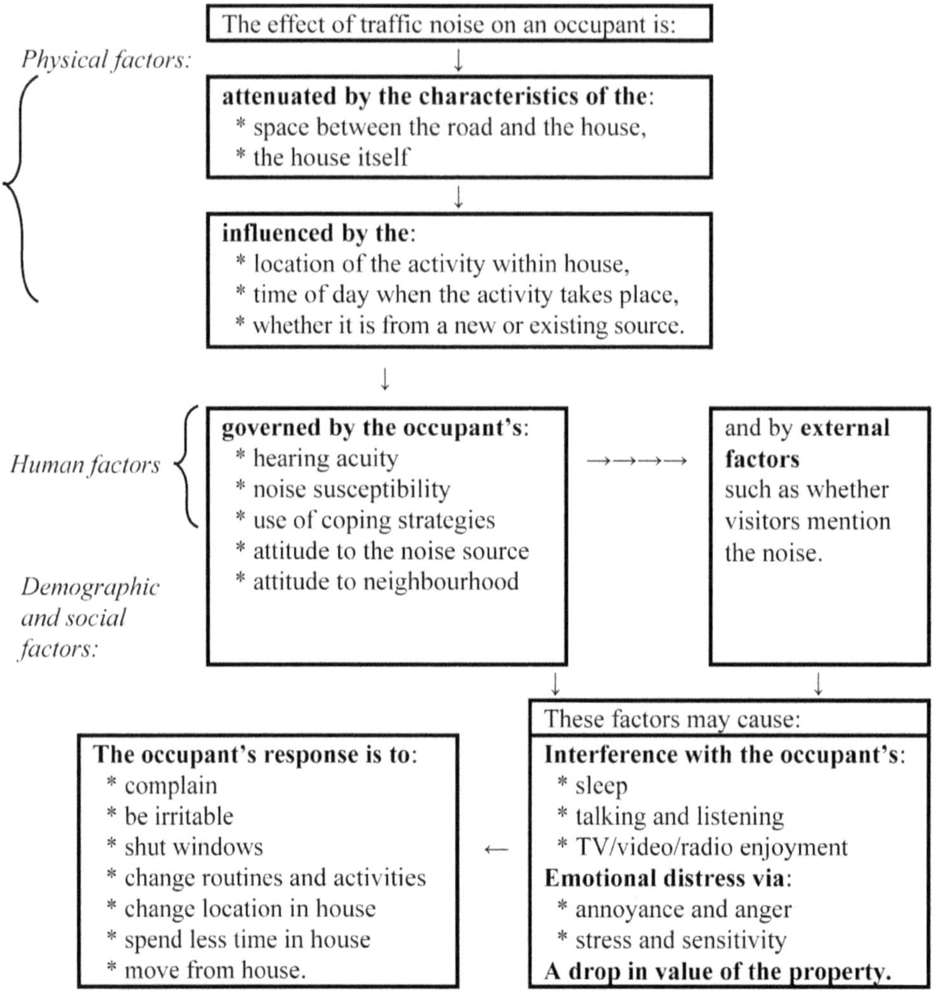

Figure 32.3 Conceptual model of the relationship between road traffic noise and its effect on people in their homes.

work when required without disturbance. When noise is intolerable, residents will completely avoid using some parts of their property,

* whether visitors to a house comment on the noise (this is one of the first indicators of noise annoyance),
* the hearer's sensitivity and adaptability to noise,
* whether or not the hearer thinks the noise is reasonable and unavoidable,
* the general neighbourhood reaction to the noise,
* whether the hearer is also troubled by other traffic issues, e.g. pedestrian danger and delay, vehicle fumes, and dust and dirt outside their house,
* the effect of the noise on land values. House values typically drop by about 0.5 percent for each 1 dB(A) increase in traffic noise,
* overall satisfaction with the neighbourhood,
* the hearer's occupation and socio-economic status,

and, more quantitatively,

* when the imposed noise levels are more than 5 dB(A) above the background noise level, and
* when there has been a recent permanent noise change of at least 2 dB(A).

Once these factors have been accounted for, the common physical method for assessing noise annoyance in residential areas is to compare the measured noise level with a tolerable (or acceptable) background noise. The tolerable noise is determined from past experience, from the noise levels listed in Table 32.2, or from measurements in the absence of the annoying noise.

The maximum tolerable level of outside noise, measured at the external façade of a house, can be estimated as follows. A typical tolerable in-door noise level is 45 dB(A) by day and 35 dB(A) by night (Table 32.2). A typical noise level change between the outside and the inside of a house (Section 32.2.3) is 20 dB(A). Thus an $L_{10}(18)$ of about 65 dB(A) by day and an $L_{10}(6)$ of 55 dB(A) during sleeping hours could be acceptable outside the façade. Clinical methods may also be possible. For example, studies in severe noise environments have found links between measured noise levels and noise annoyance, blood pressure, stress-related adrenaline levels, and memory levels in children (Bond, 1996). Despite this 'logic', the complexity of the noise annoyance mechanism outlined in this section means that there is not a robust noise-acceptability relationship at the relatively-low noise levels typical of many traffic situations.

It has been estimated that the social costs of traffic noise are about 0.2 percent of GDP (Himanen *et al.*, 1992), or 0.5 percent of vehicle operating costs. At a vehicle level, this is about 0.5 c/km. European data for year 2000 puts the cost of noise at about $30/person/dB(A)/y (Mackie and Nellthorp, 2001). Somewhat higher estimates and a more comprehensive review are given in Tsolakis and Houghton (2003).

Low-frequency noise can also cause structures to vibrate and objects to shake and rattle. Indications are that the various 18 hour noise exposure measures given in Section 32.2.2 are closely related subjective vibration nuisance ratings, with L_{eq} being the most useful.

32.2.5 Noise alleviation

Noise problems can be avoided by:

* planning strategies, e.g. keeping traffic routes away from noise-sensitive land-uses,
* zoning controls, e.g. preventing noise-sensitive uses from being located near traffic routes, and
* building regulations, e.g. requiring buildings in noise-sensitive uses to be appropriately designed and insulated.

From Section 32.2.3, the at-source factors that can most reduce noise problems are:

(a) smoother traffic flow,
(b) less traffic (e.g. by re-routing trucks or imposing truck curfews),
(c) slower traffic,
(d) fewer 'noisy vehicles', particularly noisy trucks, and
(e) a smoother road surface.

The techniques needed under (a) to (c) are usually those associated with traffic management (Section 31.6).

Once noise has been generated, it can be reduced (i.e. attenuated) by treatments such as those listed below. However, their effects will usually be marginal rather than comprehensive.

(1) Noise barriers. Barriers may be earth mounds, the faces of cuttings, crib walls, rock walls, concrete walls, or timber fences. Earth mounds constructed at the same time as the major earthworks are usually the most economical form of noise treatment, and can often use material leftover from the pavement excavation. A typical noise barrier will drop noise levels by about 10 dB(A) — i.e. halve the noise — rising to 20 dB(A) if the noise path is increased by 3 m or more. Barriers should be located to at least prevent any line-of-sight between noise source and receiver. The problem is complicated in urban areas where large, solid buildings can reflect or scatter sound in a complex manner. The barriers themselves usually reflect any sound that strikes them upwards and/or back across the road. However, absorptive barriers may be required if a reflective barrier could worsen conditions on the other side of the carriageway. Barriers need a mass of about 3 kg/m^2 to prevent excessive transmitted sound.
(2) Attention to house design and layout, e.g. avoiding having large windows facing the noise source and locating the house wisely (Sections 7.2.2&5).
(3) House insulation, e.g. by double-glazing, the use of alternatives to open window ventilation, and plugging openings around doors. Reductions of about 10 dB(A) in noise levels are possible and so this is usually a cost-effective means of reducing noise impacts. Noise insulation occurs almost automatically in cold climates as a result of thermal insulation measures, but can be expensive in hot climates where natural ventilation via open windows is a key cooling device.
(4) Distance. As a guide, quiet buildings need to be at least 10 m from a road carrying 200 veh/day, 50 m from one carrying 1000 veh/day and a hundred metres or more from heavy traffic. Buffer zones of this size are rarely feasible in many existing urban areas but can be achieved in new developments with large reservations and building setbacks. In addition, the use of adsorptive barriers can almost halve the necessary separation distances.
(5) Introducing noise-tolerant land-uses. This may be expensive as it will usually involve purchasing noise-affected properties and selling them to new occupants who are less concerned with the noise level. In the longer term, a similar effect can be achieved by planning controls discussed at the beginning of the section.

Noise alleviation at construction sites is discussed in Section 25.8.

32.3 AIR POLLUTION

32.3.1 General

The motor vehicle — via its crankcase, fuel system, exhaust, and brakes — is the source of a number of forms of air pollution: oxides of carbon (CO_x) and nitrogen (NO_x), brake-lining dust, hydrocarbons (HC), smoke, aldehydes, lead salts and particles, rubber, gaseous petrol, and carbon particles. These are discussed in the following sections. Crankcase and fuel system emissions are mainly HC. Air pollution due to vehicle emissions is of concern as it can:

* affect health via psychological suffering, discomfort, and/or illness,
* cause a nuisance through odours and reduced visibility,
* create ugly visual impacts,
* damage buildings via dirt and corrosion, and
* impact on other aspects of the environment.

The social costs of air pollution caused by transport have been estimated at about 0.3 percent of GDP (Himanen *et al.*, 1992), or about 2 percent of vehicle operating costs. In vehicle operating terms, this is about 2 c/km or, for trucks, 3 c/t-km. Somewhat higher estimates and a more comprehensive review are given in Tsolakis and Houghton (2003).

Air pollutants are classed as either primary or secondary pollutants. Primary pollutants are present in the atmosphere in the form in which they were emitted whereas secondary pollutants are formed from reactions of primary pollutants in the atmosphere.

Once emitted into the atmosphere, pollutants are dispersed by the wind. Thus wind speed, wind direction, local air currents, and atmospheric stability will determine the amount of dispersion and dilution that occurs once the pollutants are emitted. Air pollution is exacerbated by some weather conditions and alleviated by others. The worst cases occur in an inversion layer of still, warm air associated with a high-pressure system. The stillness prevents any horizontal dispersion of pollutants and the inversion layer prevents any vertical dispersion. Mountain ranges can also create large, still pockets of air. Temperature and relative humidity will influence the formation of secondary pollutants. These meteorological effects can swamp the role of traffic volume in determining local pollution levels. There are three primary measures of airborne pollutants:

(a) the total amount emitted in an area in a year, which provides a general indicator,
(b) the amount affecting a particular site, which will depend on the local traffic, topography and meteorology, and
(c) the amount within the vehicle cabin, which will affect the driver and passengers.

It was noted in Section 29.2.1 that the models developed there for fuel consumption also had particular relevance for emission studies. The type of vehicle, the way it is driven, and the traffic conditions will all have a major influence. For example, intersections — where vehicles brake, stop, idle, start, and then accelerate — will be major point sources of emissions. In heavy traffic, the major emission sources are the slow-moving lines of queuing vehicles with their stop–go driving (Section 29.2.1g). Thus the modelling of emissions is very difficult, and the above factors mean that no emissions model can be confidently recommended.

32.3.2 Oxides of carbon

Carbon monoxide (CO) is produced during incomplete fuel combustion and is emitted through the vehicle exhaust. The amount emitted depends almost entirely on the air–fuel mixture used and can be dramatically reduced by the use of leaner (lower fuel content) mixtures and by catalytic converters. However, many converters increase hydrogen sulfide emissions. Older vehicles are usually worse than newer vehicles. For urban travel, the CO output of a typical private car decreases with speed and ranges from 5 mg/m on a freeway to 40 mg/m in congested traffic. A typical relationship would be that the CO output is $(8 + [400/V_k])$ mg/m, where V_k is the speed in km/h. The output is noticeably less from diesel-engine vehicles, but this is partly offset by their added output of NO_x, sulfur, and solid particles. In most developed countries motor vehicles produce from 60 to 80 percent of the non-natural CO emissions.

CO is a colourless, odourless gas which is non-reactive in air but is toxic as it is absorbed by the lungs and has a greater affinity than oxygen for the haemoglobin in the blood with which it reacts to form carboxyl-haemoglobin. It thus interferes with the absorption of oxygen by red blood cells. The consequential physiological effects include slow reflexes, reduced visual and other mental activity, headaches, vomiting, and — at very high levels — an increased propensity for angina pectoris and other coronary conditions. In the absence of oxygen, CO will lead to asphyxiation. CO does not appear to affect materials or vegetation. It oxidises to CO_2 and thus also contributes to the greenhouse effect (see below).

A desirable CO level in the atmosphere is 8 parts per million or less, an acceptable CO level is 30 parts per million, and a detrimental level is 60 parts per million, measured over one hour.

Carbon dioxide (CO_2) is produced during combustion, no matter how efficient the combustion process might be. It is produced in greater quantities than carbon monoxide with typical rates being 200 mg/m for cars and 200 mg/t-m for trucks. These rates typically cause about 20 percent of the CO_2 emissions in developed countries. CO_2 is an inert gas and its prime physiological effect is to cause asphyxiation by displacing oxygen. Environmentally, it is a major contributor to the greenhouse effect. European data for 1998 puts its community cost at approximately \$50/t (Mackie and Nellthorp, 2001). Later estimates are given in Tsolakis and Houghton (2003).

32.3.3 Hydrocarbons

Hydrocarbon-based (HC) pollutants, other than methane, are the result of fuel evaporation in the fuel tank and carburettor (15 percent), crankcase blowby (20 percent), and incomplete fuel combustion (65 percent). They are reduced by the use of leaner mixtures, reduced use of the choke and brake, lower compression ratios, a retarded spark, higher operating temperatures (they are high when engine is cold), a well-designed combustion chamber, and catalytic converters. For urban travel, the HC output of a typical private car decreases with speed and ranges from 1 mg/m on freeways to 3 mg/m in congestion. A typical relationship is that the HC emission in mg/m is given by $(1 + [60/V_k])$. The outputs are lower for diesel engines. In most countries motor vehicles produce about 40 percent of total HC emissions.

HCs are sometimes categorised as *volatile organic compounds* or *reactive organic gases*. They produce the major *smell* associated with car exhaust fumes. A few HCs can affect human and plant health but most are non-toxic. At very high levels, some are mutagenic. At practical levels, they are mainly of concern in so far as they help to produce smog and are a precursor to tropospheric *ozone* (O_3), as will be explained in Section 32.3.4.

A high level of *benzene*, a HC (C_6H_6) found in most petroleum-based fuels, is carcinogenic and can cause anaemia and leukaemia. Benzene is present in fuel and can also form during combustion.

Formaldehyde is another volatile HC compound found in vehicle emissions. It is produced during combustion.

32.3.4 NO_x

NO_x emissions are formed at high temperatures and peak at a particular air–fuel ratio near the stoichiometric value. They can therefore be reduced by controlling this ratio. Spark retardation, recirculating the exhaust gas, and catalytic converters are also effective. In most countries about 50 percent of NO_x emissions come from motor vehicle exhausts as a consequence of incomplete combustion and oxidation. They are higher with diesel engines, due to the higher peak combustion temperatures. NO_x emissions are about 1 to 3 mg/m for a typical car, irrespective of its speed.

NO_x emissions are mainly nitric oxide (NO) and nitrogen dioxide (NO_2) and are colourless or yellow-brown. Nitric oxide is harmless but can convert into nitrogen dioxide as a secondary pollutant. NO_2 is more reactive than NO but far less reactive than ozone. It is a minor health risk, creating physiological effects by decreasing gaseous exchanges in the blood and lower lungs. This can irritate the lungs and reduce immunity to respiratory infections. Concentrations above 0.30 parts/million can cause adverse pulmonary effects in people with chronic asthma or obstructive airways disease. Its one-hour acceptable and detrimental atmospheric levels are 0.10 and 0.25 parts/million. NO_2 also causes colour fading and reduces visibility.

The irradiation of a mixture of HCs and NO_x by sunlight produces *photochemical smog* as a secondary pollutant (or *oxidant*, a term based on the chemical method originally used for air pollution analysis) and thus creates a major air pollution problem and contributes to the greenhouse effect. Generally, smog causes throat and eye irritation, vegetation and material damage, and low visibility. Its one-hour acceptable and detrimental levels in the atmosphere are 0.12 and 0.15 parts/million.

Photochemical smog is characterised by high ozone levels. Ozone is relatively reactive and is a powerful oxidising agent. It irritates the mucous membranes and causes noticeable respiratory symptoms after about two hours exposure. In the longer term, it diminishes the performance of the respiratory system, has a deleterious effect on most objects and vegetation, and converts relatively harmless NO to the more toxic NO_2.

32.3.5 Lead

Lead is added to petrol to control engine knock and increase lubrication. It can be eliminated from petrol, but with an energy penalty at the refinery. However, the removal of lead permits the use of catalyst technology, which significantly reduces exhaust emissions and vehicle fuel consumption.

From a vehicle viewpoint, lead creates problems as it tends to be emitted erratically from the vehicle exhaust, mainly because the lead compounds involved change phase with temperature. Basically, low temperature lead compounds are sticky solids that adhere to the surface of the exhaust. At high temperatures, these compounds are vaporised and emitted. The key variable is probably the percentage of the lead intake that is subsequently emitted. As cars age, this percentage tends to reach between 40 and 60 percent. The mass of lead emitted per distance travelled also tends to increase as a vehicle ages. About 60 percent of the lead emitted is airborne. In most countries the motor vehicle contributes about 80 percent of airborne lead.

Even relatively low lead levels in the blood may impair physiological functions such as those of the kidney, liver, reproductive system, and blood, produce nervous and neurological disorders, and cause mental retardation. Children are particularly susceptible. The detrimental atmospheric level for lead is commonly taken as 1.5 $\mu g/m^3$.

32.3.6 Fine particles

Dust from unpaved roads and shoulders represents a source of vehicular air pollution. Dust can affect the value of pasture land, lower the value of wool, interfere with crop development, cause livestock losses during road transport, and raise vehicle operating costs. Rates of dust loss are discussed in Section 12.1.1.

In developed countries motor vehicles contribute about a third of the suspended particulates in the air. Particulates from the combustion process are reduced by reducing the sulfur level in the fuel. Particles under 10 μm in size (called PM10) are able to penetrate the nose and mouth and enter the respiratory tract, where they may be carcinogenic. They are mainly droplets of unburnt fuel emitted by diesel engines. The remainder are largely from brake pads and tyres. European data for 1998 puts their community costs at between $50 and $750/kg/y (Mackie and Nellthorp, 2001).

32.3.7 Predictive models

Predictive models for air pollution due to road traffic can be based on the production data given in Sections 32.3.2–6 coupled with traffic, terrain and weather data. It is important for any such model to properly model the dispersion due to the vehicle, the turbulence due to the traffic stream and to individual large vehicles in that stream, the prevailing wind conditions, the topography, adjacent buildings and large trees.

One such model for a line of traffic is CALINE, produced by the Californian Department of Highways. More involved versions of the model are required when the consequences of traffic impediments such as congested intersections are to be predicted. A development associated with increased computer power replaces such steady-state models with puffs of emissions from individual polluting events.

The second stage of the predictive process requires a dispersion model to estimate the wider effects of the traffic line model. Most of these are based on the U.S. EPA's Industrial Source Complex (ISC) model.

Data input to such models is not easy and their output should be used with some caution and preferably checked against field monitoring.

32.4 WATER POLLUTION

The common components of the waterborne pollution load in road drainage are:

(a) *fine particles* (commonly < 2.5 μm) from engine fuel and lubricant leakage and fuel combustion. These petroleum-based pollutants are significant as they tend to float on the water and degrade very slowly.

(b) *solid particles* from:

- (b1) unpaved roads (Section 12.1.1). Unpaved roads produce very high sediment levels (typically 500 mg/L). These are a problem as they may far exceed most standards set for discharge to receiving waters and can stay in suspension for lengthy periods. The effect can only be usefully reduced by using a bound pavement surface.
- (b2) bound pavements (Section 12.2),
- (b3) safety barriers and road maintenance operations,
- (b4) debris from cuttings and embankments,

(b5) vehicles. These particles include the following heavy metals:
  * iron, chromium, cadmium, and aluminium from vehicle body parts,
  * copper from vehicle bearings,
  * lead from fuel and lubricants (Section 32.3.5),
  * zinc from tyres and lubricants, and
  * various compounds from brakes, paint and body metal.

(b6) dust – particularly nearby construction sites,

(c) *waste* from:
  (c1) vehicle occupants (litter),
  (c2) vehicles in adjacent areas (debris and drippings),

(d) *biological material* from:
  (d1) roadside soil,
  (d2) roadside herbicides and fertiliser (Section 6.4),
  (d3) bacteria, and

(e) *de-icing chemicals*. Their use is discussed in Section 9.5. They are predominantly sodium chloride, but may also contain iron, nickel, lead, zinc, chromium and/or cyanide.

The quantity of contaminants on pavement surfaces has been estimated as:

* residential areas:	340 g/kerb-m
* industrial and commercial areas:	800 g/kerb-m

Sixty percent of the solids are over 250 μm in particle size although the finer material has a disproportionately large polluting effect. The actual extent of these contaminants will depend on seasonal factors, local weather, land use and local events, as well as on traffic levels and composition. The contaminants also build up in a fairly linear manner with time since the last flushing of the surface (usually a rainstorm).

Most of these pollutants will be contained in the first flush of run-off after a storm and will be present in the run-off water as suspended solids. Their slow settlement rates mean that their effects may occur over large areas of receiving water.

To protect freshwater systems, suspended solids should be below 25 mg/L, however typical measured values are closer to 50 mg/L. This suggests that particle removal devices are usually a desirable component of road drainage systems. Reducing the extent of water-borne pollutants will both improve the environment and also reduce the need to clear blocked pits, sediment traps, and pipes (Sections 13.4.2 & 13.5). Where there are table drains (Section 7.4.5), pits and/or traps, these should be maintained regularly.

32.5 THE COMMUNITY

The basic needs served by transport are described in Sections 30.1&2 which, *inter alia*, discussed broad community objectives and accessibility levels. Section 31.1 pursues this question of broad objectives and definitions from a planning viewpoint with, again, a discussion in part on accessibility. Chapter 5 describes the impact of transport facilities on the community and this chapter *in toto* deals with road transport impacts on the community and its subsequent reactions. Having in mind these particular discussions of community roles, this section brings together some additional community-related aspects of transport.

32.5.1 Community characteristics

The principle personal characteristics that determine an individual's level of service or disservice from a transport facility and the preferences or values placed on those services and disservices, are as follows:

(1) *Command of resources*. Income and education level are the best available measures of command of resources. A high command of resources (i.e. wealth) ensures high mobility and favoured locations for residence and recreation. A poor command of resources (i.e. poverty) means little choice in residential location.

(2) *Social class*. Class definitions are elusive. To some extent, they may be characterised by the length of time over which a family has been in command of resources. High-class populations may be less dependent on the physical proximity of relatives, friends, services, and neighbourhood facilities.

(3) *Age, stage in life cycle*. The demands of an individual will clearly alter with time. This category therefore affects demand for access to such age-related services as baby-care centres, pre-schools, schools, universities, work, and old people's centres. The presence of children in a household will have a major effect on lifestyle options. Strong spatial concentrations of particular age and life-cycle groups are common.

(4) *Role in household*. The type of dwelling needed varies with household role as does, for example, the influence of daytime disruptions on child-rearing parents. Men are less affected than women in this regard.

(5) *Ethnic background*. Particular ethnic groups require access to their own ethnic services such as clubs and restaurants. Frequently these groups concentrate in particular areas.

(6) *Lifestyle orientation*. Such factors as career emphasis, social life, working mothers, and family style (nuclear, group living) can influence accessibility demands. Social mobility, or a striving for a perceived change in 'position', is also an important determinant. To some extent, lifestyle orientation is determined by the 'command of resources' category. In many communities, car ownership is an essential part of the life style of a majority of families.

(7) *Length of residence*. People who have been in an area for a long time may become more resistant to changes to that area.

Family expenditure surveys provide valuable insights into many of the basic constraints influencing transport use by various community sectors. Their basic purposes are to:

* quantify all the main dimensions of finance and expenditure in the personal or household sector,
* provide a picture of the pattern of consumer finance and expenditure and its inter-relationship with household characteristics,
* allow transport importance to be quantified vis-a-vis other household commitments, and
* explore the social impact of transport changes.

The basic survey unit is the household, which is defined as a group of persons living together with common housekeeping arrangements. Expenditure data are collected on the basis of payments actually made and are primarily collected by personal interview and personal diaries. Useful related information can also be obtained from home interview surveys (Section 31.3.1).

Such data usually shows that the competition between different needs conditions the behaviour of different households so tightly that in some cases the family moves into deficit funding over quite long periods of time by drawing upon capital to sustain housing, food and movement needs. This shows up more with low-income households, but young families in urban fringes who are housing-rich but financially stretched indicates that similar situations arise early in many family life-cycles. Other data of the type that are collected in expenditure surveys is show in Table 32.3. It highlights both the frequent dominance of the car in household travel and the major role that transport (and thus the car) plays in household activities — only food and possibly housing take larger shares.

Table 32.3 Household expenditure shares on transport (data from 1980 to 2000).

Item	Britain	Australia	U.S.
Transport expenditure as a percent of household expenditure:			
* private transport	12	11	13
* public transport	2	1	1
* total transport	14	12	14
Private transport expenditure as a percent of total transport expenditure:			
* car overhead (including purchase)	33	28	36
* car operation and maintenance	46	51	56
* other vehicles	4	–	–
* total private transport	83	79	92

32.5.2 Transport choices

Given the above constraints and restraints, personal transport can be seen as an hierarchy of choices:

* residential location (Section 32.5.1),
* workplace location (Section 31.3.5),
* number and type of vehicles owned/available (Section 30.5.1),
* frequency and purpose of trips (Section 31.3.5),
* time of day for trip,
* trip mode, and
* trip route.

Although there is a great deal of mutual dependence, a choice at one level in this list influences decisions made at levels below it. In making choices at each of these levels, the individual must select between alternatives possessing a number of different attributes. Each decision will depend on any external constraints and on the traveller's perception, knowledge, and ranking of each attribute. For example, in choosing a residential location, the inner-urban resident rates closeness to work, entertainment, and public transport more highly than does an outer-suburban resident. Inner-suburbanites place a lower rating on closeness to open country, ethnically similar neighbours, and air cleanliness than do their outer-suburban counterparts. To some extent these attitudes are a reinforcement of the person's current situation. There are some constants — most people require access to shops and families with children require access to schools.

CHAPTER THIRTY-THREE

Peripheral Technologies

33.1 INTRODUCTION

This chapter discusses a number of topics essential to an understanding of parts of the book, but which may not be well understood by the typical reader with an engineering education.

33.2 SOME TERMS FROM MATHEMATICS

A *model* is a quantitative description of the behaviour of a system, developed to predict its quantitative performance in new situations and/or to illuminate the workings of the system under existing conditions. Controlled experiments, physical analogues and mathematical functions are common devices used as models to represent reality. *Macroscopic* models use data aggregated across at least two non-homogeneous categories within the system. *Microscopic* models only aggregate across homogeneous categories. An *endogenous* variable is one determined within the model under consideration, whereas an *exogenous* variable is one determined outside the model. A *stochastic* variable is one varying randomly with time and a stochastic model is one containing stochastic variables. A *deterministic* model is one that does not include random variables.

The *log*istic curve is an equation involving *log*arithms and, therefore, involving exponentials. The commonest logistic curve in the study of populations (e.g. in traffic and transport) is:

$$y = 1/(1 + ae^{-bx}) \tag{33.1}$$

which produces the S-shaped (sigmoidal) curve in Figure 33.1 and which has some resemblance to the normal cumulative distribution function shown later in Figure 33.5b. A different, statistical, definition of population is given in Section 33.4.1. The logistic curve represents many natural situations where growth is initially unconstrained and exponential and then tends towards a negative-exponential response (Section 33.4.5) as constrained, saturation conditions are reached, viz.

$$y' = abe^{bx}/(a + e^{bx})^2$$
$$= by(1 - y)$$

The latter form illustrates one of the characteristics of the logistic curve in that its growth rate is proportional to both the size of the population, y, and how close conditions approach saturation, (1 – y). At the limits of y':

$$e^{bx} << a, \ y' \rightarrow (b/a)e^{bx}$$
$$e^{bx} >> a, \ y' \rightarrow abe^{bx}$$

The intercept 1/(1+a) represents the initial conditions.

Models based on the logistic curve defined in Equation 33.1 are called *logit* models. A *probit* model similarly uses the normal cumulative distribution function, F (see Figure 33.5b below). It is used where the prediction is a yes–no situation rather than a

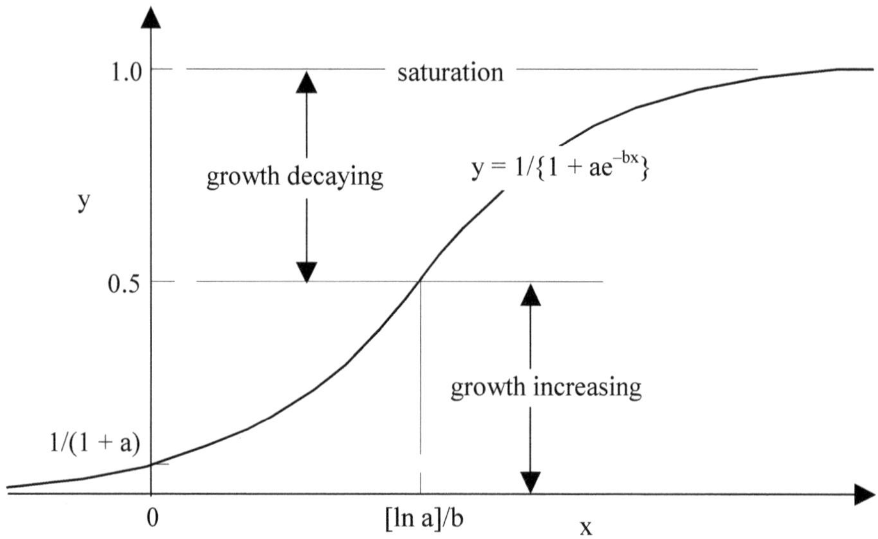

Figure 33.1 Common logistic curve.

continuous dependent variable. The predictions are therefore in terms of the probability, P, of a 'yes' which is modelled as:

$$P = 1/(1 + e^{-y})$$

or,

$$y = \ln(P/[1 - P])$$

The convenience of the y function is that it ranges from $-\infty$ to $+\infty$ for P ranging from 0 to 1. Hence, any predicted value of y is converted into a 'meaningful' value of P. Refer also to the discussion of the binomial distribution in Section 33.4.4.

33.3 MECHANICS OF MATERIALS

33.3.1 Stresses and strains

Mohr's circle is a useful graph in which the stress condition on the faces of a two-dimensional element is represented by a plot of the normal stresses along one axis and the shear stresses on the other. A circle drawn through these points represents the stress conditions on all orientations of the element (Lay, 1982). Some examples from the later text are shown in Figure 33.2.

The *mean normal stress*, s, is the average of the orthogonal normal stresses, σ_x, σ_y, and σ_z, acting on an element in the x, y, z plane:

$$s = (\sigma_x + \sigma_y + \sigma_z)/3$$

It is sometimes also called the *bulk stress* or the *first stress invariant*. The mean normal strain, e, is similarly given by:

$$e = (\varepsilon_x + \varepsilon_y + \varepsilon_z)/3$$

where ε_x, etc. are orthogonal normal strains. The *deviator stress, s,* (or *stress deviation*) is the normal stress applied to an element, minus the mean normal stress, i.e.:

$$s_x = s - \sigma_x, \text{ etc,}$$

$$e_x = e - \varepsilon_x$$

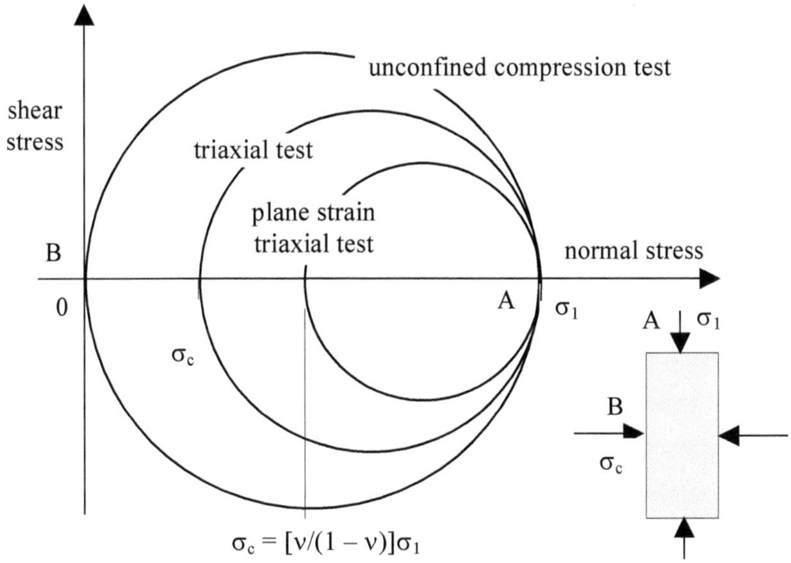

$\sigma_c = [\nu/(1-\nu)]\sigma_1$

Figure 33.2 Mohr's circle

The usefulness of the mean normal and deviator stresses stems from the fact that the mean normal stress does not affect the shear failure mode, which dictates the strength of most materials. The deviator stress, on the other hand, directly influences shear failure.

Experimentally, three-dimensional stresses and strains are applied by the triaxial test. Forms of the triaxial test are given in Sections 8.6r&s. In soil mechanics, this test is usually conducted on a cylindrical specimen (Figure 33.3) with the independent loading stress, σ_1,applied to the flat surfaces and the confining stress, σ_c, to the curved surface.

Plane stress and plane strain are discussed in Lay (1982). For a plane strain triaxial test, σ_c is adjusted as σ_1 increases to ensure that no lateral expansion of the specimen occurs. This represents the situation in a wide pavement where the rest of the pavement resists any local expansion (Figure 33.3, $\varepsilon_c = 0$). The maximum value of σ_1 is determined by the test. The Mohr's circles for the conventional and the plane strain triaxial tests are shown in Figure 33.2.

33.3.2 Elasticity

Phenomenologically, elasticity is the property of a material that permits it, after loading, to return to its original unloaded condition. Most elastic material behaviour is also linear and it is common to regard the term 'elasticity' as implying 'linear elasticity'. In terms of the theory of elasticity, elasticity means that there is a one-to-one correspondence between any stress state and any strain state, with stress being proportional to strain via the appropriate elastic constant. Stiffness generally is the ratio of an action (e.g. a force) applied to a body divided by the consequent deformation.

An *elastomer* is an elastic material containing large, polymer molecules which cross-link to provide enhanced elasticity. Similarly, a *plastomer* is a polymer that gains enhanced plasticity (see Section 33.3.3) under high strains via polymeric action. In an elastomer, the cross-linking typically produces long chains of molecules whereas the molecular network in polymers is more three-dimensional.

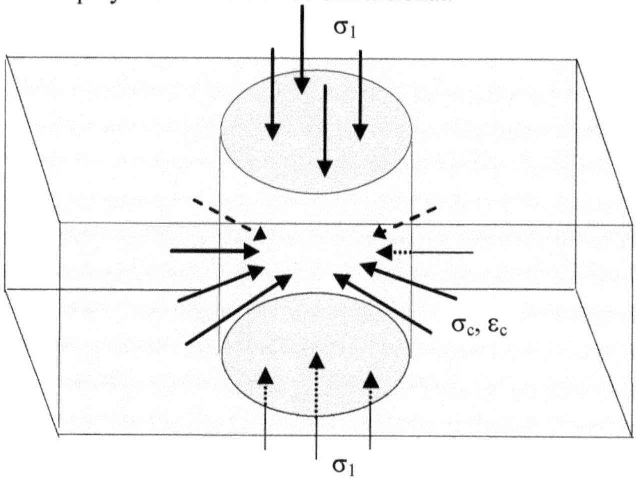

Figure 33.3 The triaxial test in a pavement context. The plane strain case occurs when $\varepsilon_c = 0$.

The term *elasticity* is also used in economics (Section 33.5.4). The relationship between the two usages of 'elasticity' is that an elastic material is one that responds 'perfectly' to an applied stress with a unique change in shape and an elastic economic system is one which has a one-for-one response of usage change to price change. That is, an x percent increase in price will cause an x percent decrease in patronage.

In their commonest precise form for an isotropic (uniform) elastic body (Sokolnikoff, 1956), the six elastic equations are:

$$\varepsilon_x = (\sigma_x - \nu[\sigma_y + \sigma_z])/E \quad (33.2)$$

$$\Gamma_{yz}/2 = (1 + \nu)\tau_{yz}/E$$

where Γ_{yz} is the angular change between the y and z faces, τ_{yz} is the shear stress on face y in direction z, E is the elastic (or *Young's*) stiffness modulus and ν is Poisson's ratio. ν is a measure of volumetric change in a loaded material. Γ_{yz} is related to the conventional shear strains ε_{yz} etc., where ε_{yz} is the shear strain on face y in direction z, by:

$$\Gamma_{yz} = \varepsilon_{yz} + \varepsilon_{zy}$$

Only two constants (e.g. E and ν) are required to describe an isotropic elastic material. For example, the shear modulus, G, defined by:

$$\tau_{yz} = G\Gamma_{yz} \quad (33.3)$$

is given by:

$$G = E/2(1 + \nu)$$

Adding Equations 33.2 for x, y, and z together shows that the mean normal stress (Section 33.3.1) is related to the mean normal strain by:

$$s = 3Ke \quad (33.4)$$

where K is given by:

$$K = E/3(1 - 2\nu)$$

and is called the *bulk modulus.* Likewise, for shear:

$$s_x = 2Ge_x \quad (33.5)$$

Equations 33.4&5 provide an alternative way of considering the three stress–strain relations in Equation 33.2.

Section 11.2.2 indicates that many pavements are *cross-anisotropic* (there is one vertical and one horizontal stiffness) and such materials require five rather than two constants for a full characterisation (e.g. E_v, E_h, ν_{vv}, ν_{vh}, and G_{vh}). A general *anisotropic* body requires 21 constants.

The *work* done (or *energy* absorbed) during stressing of a body of stressed area dA and length dl is $\int \sigma dA d\varepsilon dl$ and thus the area under the stress–strain diagram, $\int \sigma d\varepsilon$, is equal to the work done per volume of material.

33.3.3 Viscosity

Ductility is the ability of a material to deform in tension without cracking or separating. *Plasticity* is the ability of a material to deform permanently without losing its strength — indeed, it usually occurs at constant stress. Elastic materials such as steel become plastic once some threshold stress level (the yield stress) is exceeded. *Thermoplasticity* is the ability of a material to become plastic as its temperature is raised. Plastically deformed materials unload elastically and the area between the loading and unloading stress–strain curves therefore represents the energy/volume permanently absorbed during plastic deformation. This process of permanent deformation and energy losses occurring during a loading/unloading cycle is known as *hysteresis* (or *hysteretic damping*). The deformation remaining after plasticity has occurred is called *permanent* (or *irreversible* or *residual*) deformation.

Viscosity is the property of a material that causes it to resist flow and other relative internal motion at all stress levels. It is thus the fluid equivalent of stiffness in a solid. When a viscous material is loaded, the amount of flow deformation will depend on the loading time, the load level, the temperature, and the viscosity of the material. Viscous deformation in very viscous material (e.g. concrete, timber, or plastic steel) is sometimes called *creep*. Whereas elastic energy is recoverable, the energy used in viscous flow is dissipated.

A material is considered purely viscous and the flow is called *Newtonian* if, for two parallel layers a short distance apart, the following proportionalities apply:

viscous force between the two layers = C_v(velocity gradient between the two layers)

and, in the limit:

$$\tau_v = C_v d\Gamma_v/dt$$

where C_v is a constant called the *coefficient of viscosity*, τ_v is the viscous shear strain, and $d\Gamma_v/dt$ the viscous strain rate. Compare this last equation with Equation 33.2. C_v usually drops as the temperature of the material increases. Many materials are not purely viscous and their flow response is better represented by:

$$\tau_v = C_v(d\Gamma_v/dt)^c \qquad (33.6)$$

where c is a constant and is a measure of shear susceptibility. If $c = 1$, the material is Newtonian and if $c < 1$, the material is considered to be shear-susceptible. Viscosity measurement is discussed in Section 8.7.4.

In visco-elastic behaviour, the strain present is the sum of the elastic strain, Γ_e, and the viscous (or creep) strain, Γ_v:

$$\Gamma = \Gamma_e + \Gamma_v \qquad (33.7)$$

These two aspects of behaviour can be seen as:

* an *in-phase elastic* response, in which the deformations stop increasing when an applied force is a maximum, and

* a 90° *out-of-phase viscous* response, in which the deformations stop increasing (i.e. are at a maximum) when the applied force is zero.

The *phase angle* represents the relative contributions of these two phases.

Thus the shear modulus in Equation 33.3 can be represented as a complex number to accommodate the phase difference, with its absolute value given by the ratio of the maximum shear stress to the maximum shear strain (τ_{max}/Γ_{max}). This is called the *complex shear modulus, G**. A graph relating phase angle, δ, and complex shear modulus to loading rate, at a given temperature, is called a *master curve*. An example of such a curve is found in Figure 8.15.

Stokes' law governs the setting rate of small particles in a fluid. It states that the terminal settlement speed is proportional to the diameter of the particle, for diameters of 40 μm or less. It thus helps to explain the geological sedimentation process and can be used experimentally to estimate particle size from time measurements.

33.3.4 Strength and stability

The *strength* of a material is the maximum stress that it can sustain and depends on many factors, including stress condition. *The unconfined compressive strength* is the strength obtained when a specimen of the material is tested in uniaxial compression without any external restraints being imposed on orthogonal expansion; that is, when tested under biaxial plane stress. The testing of a cylinder by loads applied only to its flat ends is the commonest example, as in Figure 33.3a with $\sigma_c = 0$. The corresponding Mohr's circle is shown in Figure 33.2 (Section 33.3.1).

Generally, material *hardness* describes the resistance of a material to surface deformation (or indentation). That is, a material is hard when its surface cannot be easily marked. In this sense, hardness is a measure of material strength, but not necessarily of fracture toughness. However, for aggregates (Section 8.5) some particular hardness concepts are evaluated by tests such as those in Section 8.6b,e&j. These definitions relate as much to how difficult it is to crush and compact the material, as they do to its in-service performance. In quarrying, hardness thus relates largely to the strength and fracture toughness of the material.

Modulus of rupture is an unfortunate term that can be defined as the maximum stress in the component at failure (rupture), calculated on the original dimensions of the member and on the assumption of elastic behaviour up to failure. The term is misleading and has little to recommend it.

Stability is a precisely defined technical term that refers to a property of systems which causes them, upon being disturbed, to return towards their pre-disturbed state. It is often somewhat confusingly used in pavement engineering to refer to a material's ability to resist permanent deformation. See also the discussion of the CBR test in Section 8.4.4.

The *coefficient of friction* is the ratio of the friction force at and along an interface to the force normal to the interface.

33.4 STATISTICS

33.4.1 Populations and samples

In statistics, a *population* is a collection of items having at least one common characteristic. A *lot* is a portion of the population of items that is of homogenous quality. A *sample* is a set of items which have all been taken from the same lot.

The values of the characteristics of each item are established by measurement. *Repeatability* is the precision with which a measurer and associated measurement procedure can repeat the measurement. *Reproducibility* is the precision of the measurement procedure when used by different measurers. Quantitatively, each is usually taken as the difference between measurements that is exceeded in only one case in 20.

Consider a population of n discrete items or events, each of which has a common characteristic with a value x. The measurements show that M items have a particular value of $x \pm \delta x/2$, where δx typically represents the measurement accuracy (Figure 33.4a). Dividing the M corresponding to an x by the total number of items, n, gives the probability density, y, for that δx interval, i.e. y is the probability of an event in that δx range of x. y will depend on the size of δx. For continuous values of x, δx becomes dx, and y then becomes f(x), the *probability density function* (pdf) for x. Such pdf's define the rate of change of probability and are important as they describe the way in which the values of an item are distributed.

As the pdf, f(x), is the probability of an item having a value in the range $x \pm dx/2$, it is the rate at which the probability of the item having a value of x alters as x changes. This leads to the probability of the item's value being less than or equal to the value of x being the integral of this rate. This integral is called the *cumulative distribution function* (cdf), F, and is (Figure 33.4b):

$$F = \int_{-\infty}^{x} f(x)dx \qquad (33.8)$$

or

$$f = dF/dx$$

The graph is called the *cumulative frequency diagram* (or *ogive*).

33.4.2 Means and averages

The arithmetic mean (or [most] *expected value*), x_m, of x for a population of n is the *average* of the x values of all the n items. That is:

$$x_m = \left(\sum_{1}^{n} x\right)/n$$

For continuous values of x, x_m is given by (Figure 33.4a):

$$x_m = \int_{-\infty}^{\infty} xf(x)dx$$

It is thus the first moment of the pdf of x. The mode is the most common value of x, i.e. the value that occurs most frequently.

Regression towards the mean is the statistical tendency of high values to reduce with time and low values to increase with time. That is, the best prediction of a future value is that it will be closer to the mean than the previous one. In crash studies, this means that

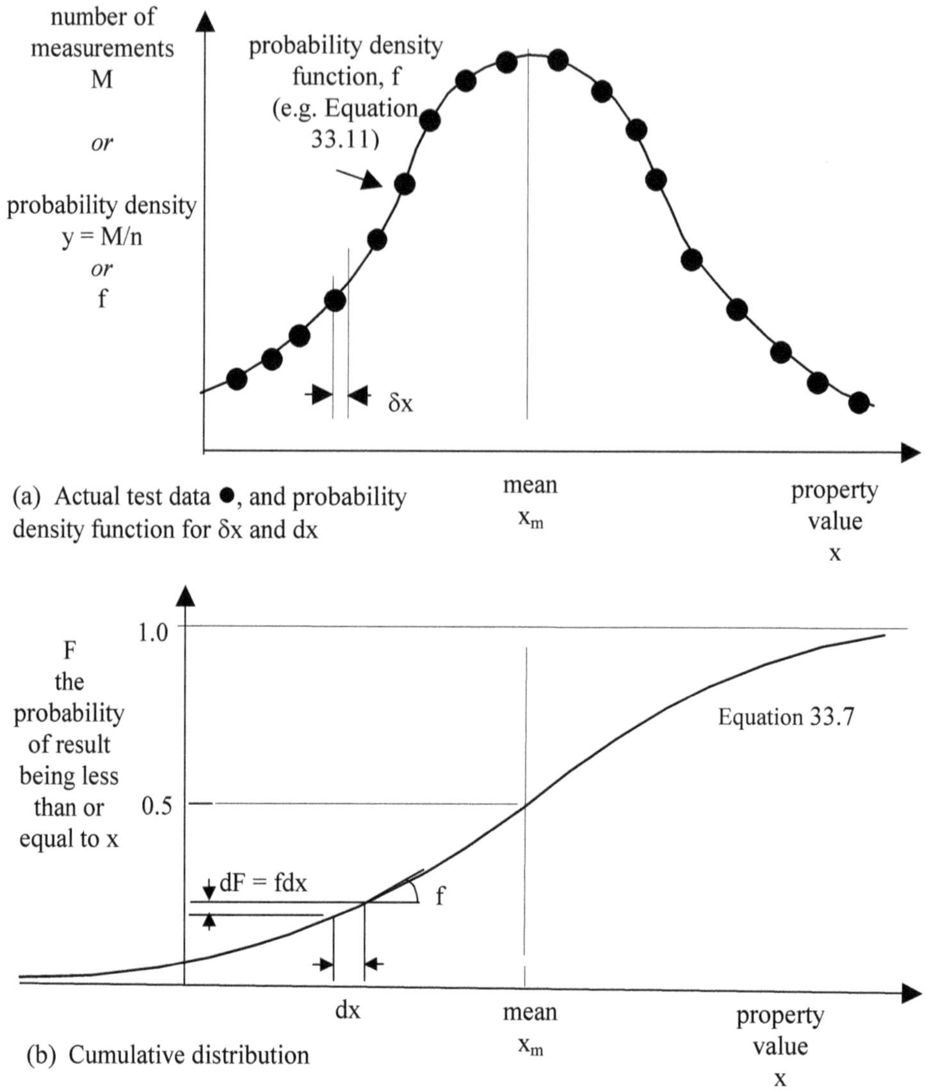

Figure 33.4 Definitions of distribution functions.

an unusually large number of crashes is usually followed by a crash reduction of around 40 percent, even if no countermeasures are employed.

A practical example of the application of regression towards the mean is in before-and-after studies at a high crash-rate site which has been subjected to some remedial treatment. As the pretreatment crash rate was higher than average, regression towards the mean predicts that the crash rate after treatment will drop, even if the treatment is ineffective.

As another example, assume that all the components of a traffic system were grouped according to whether they were associated with either 0,1,2,n,... crashes in a period of time. Then, because crashes are very rare events, in the next time period of the same length the average number of crashes occurring at those elements components had previously had n crashes, would be less than n. The only exception would be in the largest category — that which had previously had no crashes ($n = 0$). The average number of traffic crashes in elements in this category would increase from zero to greater than zero.

The *geometric mean* of the values of n discrete items is their product to the power (1/n). It is thus:

$$x_{mg} = \prod_{1}^{n} x^{1/n}$$

The *harmonic mean*, x_{mh}, is the reciprocal of the mean of the reciprocals of a quantity. Thus:

$$1/x_{mh} = \int_{-\infty}^{\infty} (1/x) f(x) dx$$

The *mode* (or modal value) is the peak (or most common) value in a pdf.

33.4.3 Deviations

The population *standard deviation*, σ, is a measure of the dispersion of data in a population of items. It is based on the root mean square concept for averaging absolute values. For n values of x with a mean of x_m, σ is given by:

$$\sigma = \sqrt{[(\sum_{n}(x - x_m)^2)/n]} \qquad (33.9)$$

The square of the standard deviation, σ^2, is the *variance* of the data. For a continuous variable the variance is:

$$\sigma^2 = \int_{-\infty}^{\infty} (x - x_m)^2 f(x) dx$$

where f is the pdf. Variance is thus the second moment of the distribution of x, just as x_m is the first moment.

The sample standard deviation, s, is the value for the population sample tested. There is a strong bias in small samples, which is rectified by *Bessel's correction*, viz.:

$$\sigma/s = \sqrt{(n/[n - 1])}$$

The *coefficient of variation* is σ/x_m as a percentage.

A *characteristic value* is one with some stated probability, F, of being exceeded, with F expressed as a percentage. The definition is illustrated in Figure 33.5a, which is an expansion of Figure 33.4b. The *median* of a distribution is the value above and below which lie fifty percent of the values and is thus the 50th percentile characteristic value. However, characteristic values are most frequently used to represent the extremes of a distribution. For example, in structures, it is common to use a 5th percentile value for loads and a 95th percentile value for material strength (Section 15.3.2). Values for pavements are typically based on 15th and 85th percentiles. Speed studies commonly use 85th percentile values (Section 18.2.4). See also the discussion of extreme values in Section 33.4.6.

Many properties follow the *normal distribution* described in Section 33.4.4 and defined by Equation 33.11. This equation shows that the 85th percentile for a normal distribution will occur at 1.04σ from the mean. Indeed, one justification for selecting 85 percent is that the inflection point in Equation 33.11 occurs at:

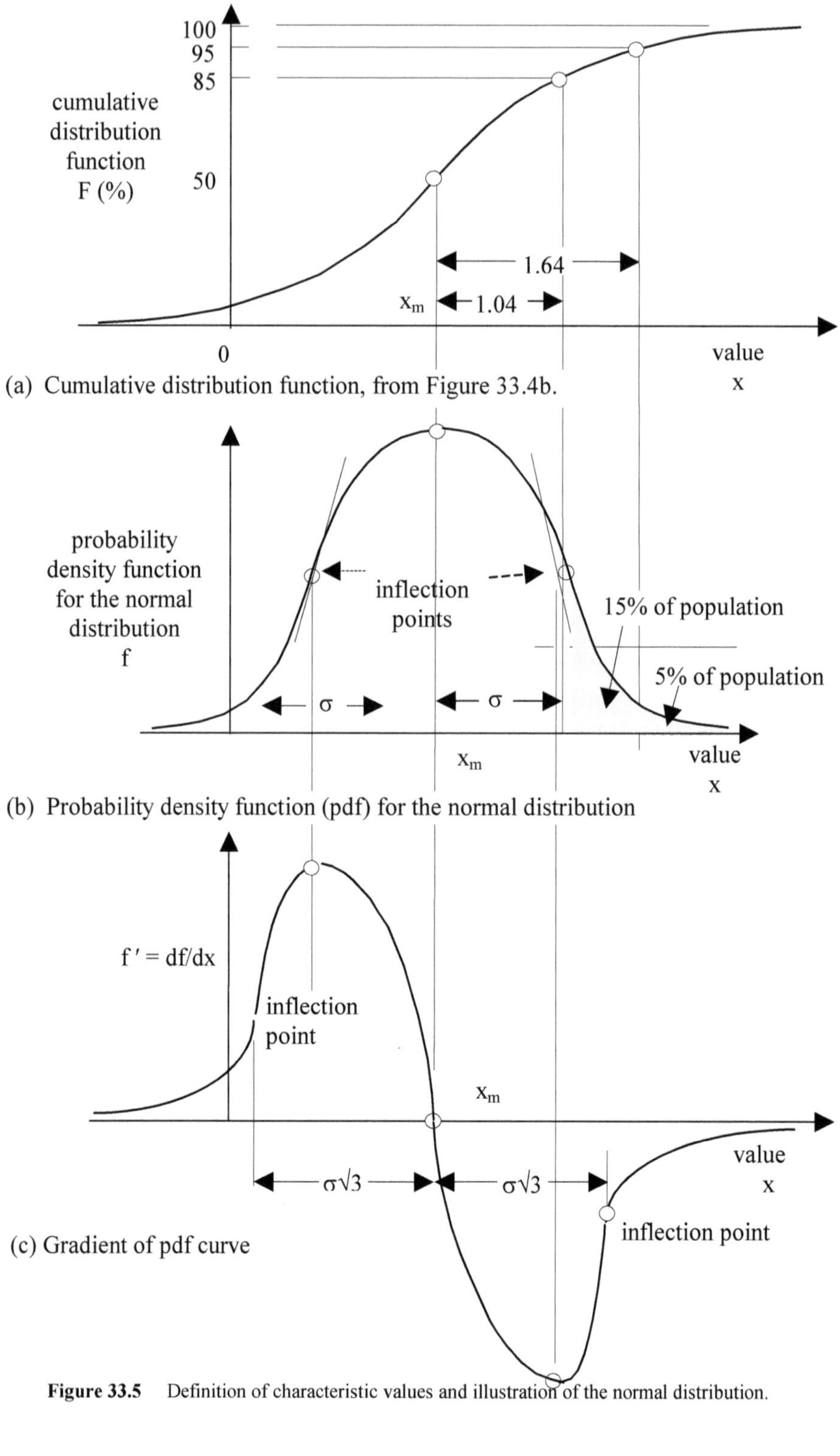

(a) Cumulative distribution function, from Figure 33.4b.

(b) Probability density function (pdf) for the normal distribution

(c) Gradient of pdf curve

Figure 33.5 Definition of characteristic values and illustration of the normal distribution.

$x = x_m \pm \sigma$

Thus the range $(x_m - \sigma)$ to $(x_m + \sigma)$ is a good mathematical definition of the range of major change in characteristics. The 95th percentile characteristic values occur at:

$x = x_m + 1.64\sigma$

which is very close to the f' inflection point at (Figure 33.5c):

$x = x_m + \sqrt{3}\sigma$

33.4.4 Significance and regression

It is often necessary to test whether particular results differ from the general population. For instance, if data are collected after a safety measure is installed it is necessary to know whether its values represented only a chance variation, or a significant change, from the rest of the data. Statistical convention is to accept a change as significant if:

(a) it has less than a 5 percent chance of having occurred by chance. This is described as 'probably significant' and usually denoted as 5 percent, $p = 0.05$ or by *.

(b) it has less than a 1 percent chance of having occurred by chance. This is described as 'significant' and usually denoted as 1 percent, $p = 0.01$, or by **.

(c) it has less than a 0.1 percent chance of having occurred by chance. This is described as 'highly significant' and usually denoted by 0.1 percent, $p = 0.001$ or by ***.

For engineering purposes, even the $p = 0.05$ test can be a very severe one.

A common approach to establishing p values is to use the *Student's t test* which checks whether the difference, δ, between the mean of the sample under test and that of the general population is significant. It assumes that the properties are normally distributed (Section 33.4.4). The value t is calculated as:

$t = \delta\sqrt{(n - 1)}/s = \delta\sqrt{n}/\sigma$

where n is size of the sample, s is the sample standard deviation and σ the population standard deviation. t is tabulated in texts as a function of n, but for large samples ($n > 20$), the t values for the above p levels can be approximated by:

Case	Probability of event being a chance effect	Student's t	Common symbol
probably significant	0.05	2.0	*
significant	0.01	2.5	**
highly significant	0.001	3.5	***

A number of other tests are also in use. For example, the *F test* checks whether the variances are significantly different. This test also has the advantage of allowing several samples to be compared simultaneously to see if they come from the same population. It operates by comparing the variability within and between samples and the F-ratio is the ratio of between-groups to within-groups variance. F-ratio tables are then used to give the appropriate p (0.05, 0.01, 0.001) level.

A third significance test applies when the issue is the frequency of a category (e.g. yes or no) rather than its size. The standard error techniques described below can be used when just two categories are involved, but in cases where more categories exist, the *chi-squared* (χ^2) test is used. Following a similar pattern to before, a variable (χ^2) is calculated from the sum of the squares of the differences between observed and expected values, divided by the expected value. This is compared with tabulated values to give the appropriate p level for whether or not the differences between samples are really significant or possibly random. The test can be a severe one.

For a population of pairs of data, a *regression* equation estimates the dependent value in each pair from the independent value, by selecting coefficients by the *least squares method* which minimises the variances between each real dependent value and the associated value predicted by the equation. If the total variance of the actual dependent values from their mean is V, the total variance explained by the regression equation is termed r^2V, i.e.:

$$V = r^2V + (\text{unexplained variance})$$

r is called the *correlation coefficient* and thus measures how well the regression equation fits (or correlates) the data. A value of $r = +1$ indicates a perfect fit (or full positive correlation), $r = 0$ indicates no fit, and $r = -1$ indicates perfect negative correlation (a completely perverse fit).

The *standard error of estimate* (s_{ee}) is closely related to the standard deviation (σ) and the correlation coefficient. It can be thought of as the standard deviation of data calculated with respect to a regression line, rather than to the mean of the data. s_{ee} is thus given by:

$$s_{ee} = \sqrt{(\{\sum_n [y - y_r]^2\}/\{n - k - 1\})}$$

where y is the data, y_r the regression line prediction, n the number of data points, and k the number of dependent variables (x) in the regression equation. This expression can be compared with Equations 25.1 and 33.8. The link with the correlation coefficient is given by:

$$r^2 + s_{ee}^2/V = 1$$

Because of its link with the deviation of data from a line, the standard error also has a link with *confidence limits*. Approximately, for a straight line, the 95 percent confidence limit is $\pm 2\sqrt{s_{ee}}$.

The *maximum likelihood* method extends the regression method by making a specific assumption about the pdf of the individual variances. It then maximises the likelihood (or probability) of the predicted values being the actual values. If the pdf of the individual terms is normal (Section 33.4.4), the maximum likelihood equation is identical to a least-squares solution. For crash studies the pdf may need to be based on the Poisson or negative binomial distributions discussed in the next Section.

33.4.5 Counting distributions

When events are discontinuous, the associated distributions are sometimes called counting distributions. A common simple discontinuous situation involves n independent events, each of which has only two possible outcomes (e.g. yes or no). The probability (P) of these individual outcomes being 'yes' is known and does not change from trial to trial. The resulting distribution of the outcomes is called a *binomial distribution*.

The distribution function for the probability of x yes's in n events can be deduced to be:

$$f(x) = [n!/x!(n - x)!]P^x(1 - P)^{n-x}$$

The mean number of yes's to be expected is, of course, nP and the variance of x can be calculated as $nP(1 - P)$. The *binomial cumulative distribution function* gives the probability F(x) of x or less yes's in the n events as:

$$F(x) = \int_0^x f(x)$$

The *Poisson distribution* extends the binomial distribution to the case where there is a very large number of discontinuous events and where one of the two outcomes (e.g. 'yes') is of very low individual probability. In the limit, $n = \infty$ and $P = 0$, although the overall expected number of 'yes' outcomes, $m = nP$, remains constant. The probability density function for obtaining x outcomes where x may not equal m, is $p_t(x)$ where:

$$p_t(x) = m^x e^{-m}/x! \qquad (33.9)$$

In a stationary Poisson process, m does not vary with time. From Equation 33.7, the Poisson cumulative distribution function — i.e. the probability, P, of obtaining x outcomes where x may not equal m is:

$$P_t(x) = \sum_0^x p_t(x)$$

The gamma distribution is a *Pearson type III* counting distribution based on the gamma, Γ, (or *factorial*) function:

$$K! = \Gamma(K + 1) = K\Gamma K = \int_0^\infty e^{-m} m^K dm$$

for $K > 0$. It is commonly applied to traffic volumes, where m is the number of vehicles, q is the vehicle flow, and t the time over which the observation occurs. The presence of a vehicle in a small time interval is assumed to be a relatively improbable event (Section 17.3.2). Thus $m = qt$ and the associated pdf is:

$$f(t) = [qe^{-qt}(qt)^{K-1}]/\Gamma K \qquad (33.10)$$

The distribution has a mean of K/q and a variance of K/q^2.

If $K = 1$, the gamma distribution becomes the *negative-exponential* distribution represented by:

$$f(t) = qe^{-qt}$$

This distribution has a mean of $1/q$ and a variance of $1/q^2$. The mean number of vehicles in time t is $m = t/q$. The distribution is frequently used to describe traffic flow headways (Section 17.3.2).

If K is a positive integer greater than 1, the distribution is called an *Erlang* distribution; e.g. for $K = 2$, the distribution is:

$$f(t) = q^2 te^{-qt}$$

The variance in this case is m.

33.4.6 Other distributions

The distributions discussed in the preceding section covered discontinuous processes, but tending in the last example towards continuity. In many cases an event can take on a continuous range of property values. When these values vary randomly and symmetrically from a single average value, with the likelihood of a value decreasing steadily as its distance from the most expected value increases, we possibly have a *normal* distribution. It may be thought of as the simplest distribution meeting the above criteria or as a distribution found from experience to fit most symmetrically distributed random events. It is a limiting form of the binomial distribution.

The pdf for the normal distribution (e.g. Figure 33.5c) is called the *standard normal curve* and its equation is:

$$f = e^{-[x - x_m]^2/\sigma^2}\sqrt{(2\pi)} \qquad (33.11)$$

where x_m and σ are the population mean and standard deviation. Explicit equivalent expressions for F do not exist.

Often a property cannot be symmetrically distributed; e.g. pavement thickness cannot be less than zero but could be very large and hence we would expect a skewed distribution with a mean nearer to zero than infinity. In such cases the data can often be made to look normal by using the logarithms of the values rather than the values themselves. A normal distribution of the logarithms of a value is called a *log-normal* distribution.

The *gumbel* distribution relates to the extreme values of a quantity (such as loads). It assumes that the distribution of all the values of the quantity, x, has an upper tail of the form (Figures 33.7&8):

$$F = 1 - e^{-g(x)}$$

where g is a monotonic function of x. This can be applied to the tails of most distributions (such as those discussed above). Assuming that:

$$g = e^{-a(x - b)}$$

the gumbel distribution of the values that exceed some amount X is given by:

$$1 - F = F(>X) = e^{-e - a(x - b)}$$

where a and b are constants depending on the distribution of all the values.

The *recurrence interval* (or return period or average exceedence probability) is the average time (T) between events. Such an event has a 1/T probability of occurring in any one year, when T is in years. The probability of such an event occurring in L years is $[1 - (1 - [1/T])^L]$ which is approximately L/T. Thus F(>X) can be taken as L/T. For example, the chance of a 50 year flood flow occurring during the 50 year design life of a bridge is 36 percent. It is not wise to confidently predict recurrence intervals that are greater than the record length. However, if estimates must be made, the Gumbel plot can be used as a predictive tool for extreme events beyond current records. This gives the link between an annual extreme value, X, and its return period, as:

$$X = c + d.\ln(\ln(T))$$

where c and d are constants.

The power, P, contained in a vibration is proportional to the square of the root mean square of its amplitude. To handle a set of random vibrations, as arise with road roughness (Section 14.4), it is necessary to consider the power over the entire frequency spectrum of the vibrations present. By analogy with probability and probability density (F and f), a *power spectral density* (*PSD*) is defined where PSD is the power per unit frequency over the frequency range.

33.5 TRANSPORT ECONOMICS

33.5.1 Theory

Welfare economics is the study of the extent to which the community's needs and objectives are being fulfilled. It therefore focuses on the alternative ways in which a community's resources can be used to attempt to fulfil those needs and objectives. The evaluation of road proposals is clearly part of welfare economics.

Utility is a measure of the ability of a service to satisfy those needs and objectives and thus reflects the value and satisfaction users gain from the service. It is therefore a key tool in welfare economics. Utility depends on the desired properties of the service and it is usually assumed that utility is the sum of terms, which are themselves products of a service property (e.g. quantity) and a user factor (e.g. quality). Utility can be either cardinal (i.e. measurable) or ordinal (i.e. rankable). Marginal utility is the additional utility supplied by an additional unit of service. In terms of economic theory, the aim of any rational activity is to maximise utility, subject to the external constraints imposed on the system. In the context of road transport, the constraints would be such items as limits on noise and air pollution. The opposite (or negative) of utility is *disutility*, which thus relates to the cost the user experiences in obtaining the service. Disutility is closely related to generalised cost (Sections 31.2.2).

An economic *externality* exists when one group's activities affect another group's welfare, without any payment or compensation being made. It is thus independent of buyer–seller operations and represents a market failure. Road-related externalities are usually of the consumer-upon-consumer type, typically where traffic affects the welfare of residents (Section 5.1).

Once objectives have been established and utilities defined, transport economics is often about the selection of the best alternative to satisfy those objectives. The labour, land, and capital resources available are inevitably in limited supply and exhaustible, but can be maintained or even improved by investment. This is the field of *microeconomics*, which is concerned with the allocation of such scarce resources to needs that are competing and relatively unlimited.

In the transport context, many services have been offered by the public sector. In these cases, market economics have traditionally played a secondary role. The pattern has been to 'satisfy community transport objectives whilst attempting to operate efficiently' rather than to 'operate profitably whilst satisfying some community objectives'. This reversal of the classical private enterprise role complicates rather than simplifies the economic analysis of public sector transport systems.

Economic costs are the total costs born by the community, and apply whether or not there are financial flows. Transfer costs are discussed in Section 33.5.9.

Marginal costs are those additional costs incurred in adding a small (marginal) extra to a facility. Roads are usually a *decreasing cost facility*, i.e. one where the marginal cost (Section 29.4.1) falls as the usage increases, and is less than the average cost (Section 29.4.2). *Public goods* are ones that are available to everyone, not able to be appropriated, and whose use by one person does not prevent their use by another. A road is a slightly impure public good as:

* it only serves people in a particular area,
* it may become congested, and
* there is a marginal cost associated with its use, although it is evident from Table 5.1 and Section 29.1.2 that this cost is a small part of total vehicle operating costs.

33.5.2 Pricing

The role of the price of a commodity in a market economy is to provide a measure by which:

* investments for producing the commodity can be planned,
* resources for producing the commodity can be assigned,

* the commodity produced can be allocated,
* costs of production can be recovered, and
* the effort of those involved can be rewarded.

Hence there is no single, correct price for any commodity. Rather, it is necessary to establish the objectives of the pricing process and then find the best price to achieve those objectives.

Pricing is perhaps the key issue in road transport economics. Road pricing methods are discussed in Section 29.4.5.

A range of pricing possibilities is discussed in Section 29.4, each price giving quite different answers. Of those prices, the one that needs more explanation is marginal pricing. Marginal (or cost-occasioned) pricing is based on the increase in costs caused by the extra work needed to supply the unit of service being sold, i.e. it is based on avoidable costs (Sections 29.4.1 & 33.5.1). Marginal prices can be short run or long run. The *short-run* price covers only the immediate costs of providing the service, and the *long-run* price also covers the costs of staying in business. Marginal pricing will not occur in an *imperfect market*, as in the cases where the product is produced by a monopoly, there is government intervention, or there is no market.

Efficiency is the ratio of results produced to resources consumed. An efficient operation maximises this ratio. If everything in an economy is optimally organised, marginal pricing of an existing service will lead to *economic efficiency* as it will attract the maximum use of the service, whilst raising enough revenue to support its continued operation. Marginal pricing is not related to the economic merits of a particular investment in a service, as is a benefit–cost analysis (Sections 5.2.3 & 33.5.8). Rather, it relates to the operation of the service once it has been constructed. Marginal pricing regards the past resources used to create the facility providing the service as irrelevant historical facts. It assumes that the service has already passed a benefit–cost ratio filter and is therefore pumping adequate benefits into the community via a number of avenues in addition to the revenue stream created by pricing.

Marginal pricing is thus appropriate for public investments such as roads whose construction was justified by a benefit–cost analysis. Once constructed, any such roads with a benefit–cost ratio greater than one should be used to the maximum extent feasible. However, as these benefits are widely dispersed, they will not be fairly reflected in usage charges. As the marginal price is the minimum price that could be charged without making a short-run operating loss, it also tends to prevail in competitive industries, although it will not ensure long-term commercial viability.

If utility is determined solely by price, then the selection of transport alternatives will be in terms of the price of a particular service compared with the price of other services and with the *opportunity cost* of the service selected. Opportunity cost is value of whatever is foregone by using the selected service. A relevant example is the opportunity cost of capital, which is the foregone interest that would have been earnt if the capital had not been spent on transport infrastructure, but had been left in a secure investment.

A major reliance on the price mechanism for resource allocation in transport is thwart with difficulty, for the following reasons.

(a) The other factors influencing utility are ignored (Section 33.5.1). The complexity of the constraints, subsidies and interactions at work (such as taxi licensing, public transport deficits, and cross-subsidisation between routes).
(b) Many of the services supplied — such as the provision of roads — do not have a market value. Nevertheless, they will be strongly linked to the community's

objectives and hence, although unquantifiable financially, will be important in measuring the effectiveness of any action.

(c) Many of the services will operate as virtual monopolies — canals in the 18th century, railways in the 19th century and roads in the 20th (Section 3.4). As a true market cannot work in these instances, governments have often stepped in with controlling regulations and/or ownership. The inevitable resulting inefficiencies have often not counterbalanced the original ill-effects of the monopoly — although they remain part of the community's objectives.

33.5.3 Demand and supply

As the price of an item increases, the supply from producers increases, and demand from consumers then decreases. If the demand and supply curves for a particular set of socio-economic conditions are as shown in Figure 33.6, the *equilibrium* price and quantity sold of an item are determined by the intersection, A, of the two curves. 'A' is called the equilibrium point.

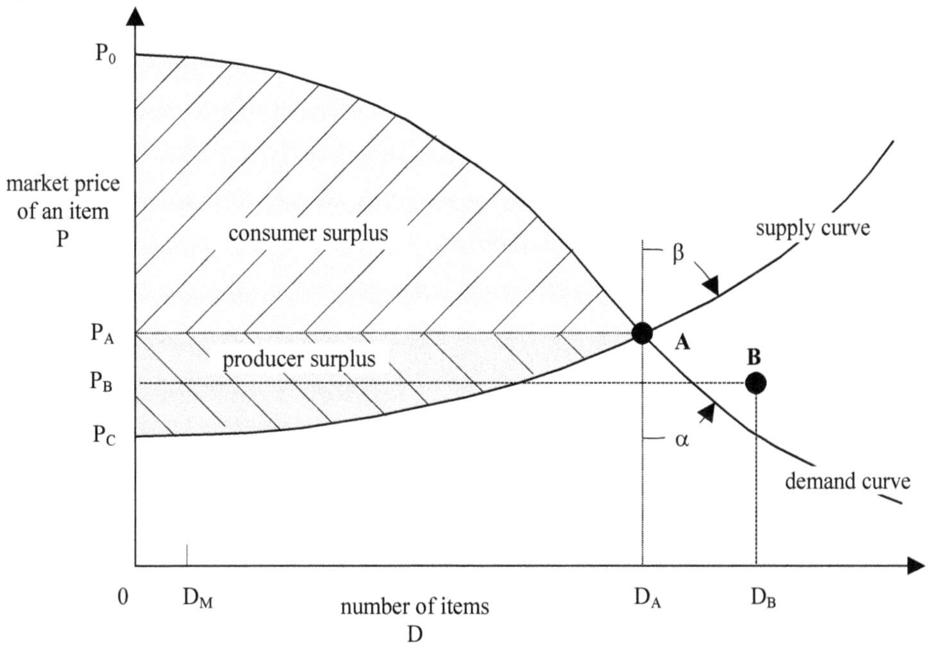

Figure 33.6 Demand and supply curves. Note from Equation 33.12 that the elasticities of demand and supply, E, at A are related to the graph by $\alpha = (D_A/P_A)E_S$ & $\beta = (D_A/P_A)E_D$.

The consideration of demand and supply concepts is important in transport where users' travel needs can be taken as the demand side and the provision of transport infrastructure as the supply side. The flat ends of the demand and supply curves near D = 0 are typical of transport situations and, for supply, indicate that there is a minimum number, D_M, of the items that can be produced. For example, if a bus is operated, there will always be more than one seat provided, even at minimum service levels. The demand curve represents the maximum price consumers would pay for the item and it is assumed that this represents the value of the item for each consumer. This value is often linked to the benefit the consumer receives from the item.

Therefore, the area under the demand curve up to the quantity consumed, OP_0AD_A, represents the *gross benefit* that consumers gain from D_A items. It can also be described as the total worth to consumers of the items purchased and as a measure of the consumers' *willingness to pay* for those items. However, the price of the items is only OP_AAD_A. The difference between these two areas is called the *consumer surplus* and is discussed in Section 33.5.5.

The area under the supply curve, OP_CAD_A, represents the total costs of production. The difference between this area and the cost area is called the *producer surplus*, and is discussed in Section 33.5.5. If the price were raised above the equilibrium price to P_A, the demand would drop below D_A but the supply would rise above D_A.

The demand for travel is measured in units (on the abscissa of Figure 33.6) of trips, tonnes (t), trip-km, or t-km. Establishing curves such as P_0A is not easy and few examples exist. It is difficult to play experimentally with the real fares of real travellers. In addition, the demand estimated must be the amount travellers would actually choose under the specified conditions, rather than the amount they would like to have, and so questionnaires must be used with caution. The conditions may also change. One of the most sensitive external conditions is *discretionary income* (income remaining after all essentials are paid for) — travel demand rises as discretionary income rises.

A characteristic of transport demand and supply is that the market responses are far from instantaneous. If transport prices rise, some users will need to take time to rearrange their affairs, e.g. change home location, in order to respond rationally to the rising prices. Various other factors will also influence decision-making, such as the attractiveness of the existing home location to other family members.

It has been a common past mistake in transport planning to see the only response to unsatisfied demand to be the supply of further infrastructure. However, there are two alternative responses:

* the demand can be managed (Section 30.3), and
* market forces can be allowed to work with the excess or unsatisfied demand either tolerating the congestion or using some alternative strategy, such as changing travelling hours.

In a properly run business, the cost of transport to the producer will be below the P vs D supply curve in Figure 33.6. It may fall as D increases. The producer's profit is between this cost curve and the P_AA line and will include the producer surplus.

33.5.4 Elasticity

The definitions and explanations in Section 33.5.3 leads to the concept of (economic) elasticity. If the price of a product P, is increased by δP, then the use of that product, D, will change by δD. The ratio

$$E = (\delta D/D)/(\delta P/P) = (\delta D/\delta P)(P/D) \qquad (33.12)$$

is known as the point elasticity of the demand as δD and δP approach zero. The word point is included as the definition is only used at a point (i.e. for small changes or for linear systems). Other elasticity definitions are:

* *shrinkage ratio*: in Equation 33.12, δD and δP are finite and D and P are taken as their initial values;

* *arc elasticity*: in Equation 33.12, δD and δP are finite and D and P are taken as their average values during the change;

* *log-arc elasticity*:

$$E_{l\text{-}a} = (\log[D + \delta D] - \log D)/(\log[P + \delta P] - \log P)$$

All three measures give similar results for small δD and δP.

Figure 33.6 showed that the demand elasticity, E_d, and supply elasticity, E_S, are related to the curve gradients at A. Elasticities can be positive, as with the increase in car usage that occurs as disposable income rises. However, demand elasticities normally begin at zero and become increasingly negative. An elasticity of 0 is described as inelastic (price changes would have no effect on demand) and an elasticity of –1 is described as elastic. Generally, an elasticity of –E means that a p percent price increase would reduce demand by Ep percent. Typical demand elasticities for car travel are:

peak hour	–0.10 to –0.70
off peak	–0.20 to –1.10
all day	–0.10 to –1.10
long term	–0.20 to –1

For truck freight, the elasticity is in the range –0.70 to –1.30.

In transport, demand elasticities usually take some time (e.g. 12 months) to stabilise, and thus tend to be long-run elasticities. The slow demand response times are reflected in the long-term elasticities that are measured when all changes have eventually occurred. Short-term elasticities reflect the immediate response of the system and may be smaller than long-term ones (as in the above case) or longer, when some of the initial response is motivated by curiosity or when some disenchantment occurs.

Cross-elasticity occurs when demand for one item (e.g. use of one travel mode) is influenced by the price of another item (e.g. the price of another travel mode). It is defined by analogy as $(\delta D_1/D_1)/(\delta P_2/P_2)$ where the subscripts refer to the two items. For competitive situations, cross-elasticities are positive. The ordinary elasticities are sometimes called own-elasticities to distinguish them from cross-elasticities.

33.5.5 Surpluses

In a rational world, the market price a consumer pays for an item would be equal to or more than its cost and less than the value (or utility, Section 33.5.1) that the consumer places on the item. A useful concept is that of *consumer surplus* (or *total consumer surplus* or *net benefit*) which is the difference between value and price. In terms of Figure 33.6, consumers are paying P_A for D_A items, giving sales of value given by the area OP_AAD_A, sometimes called the *total worth* of the system. The consumer surplus is therefore the area P_AP_0A. The term *surplus* is also used in a wider sense, with *social surplus* (or *social return*) being used to describe the net benefits of a project to society.

Note that some definitions use both the demand–price curve and a maximum value 'demand' curve. This text defines the curve P_0A as the lower of the two (the difference is not critical here). Similarly, it is possible to use both the supply–price curve and a true cost 'supply' curve.

The *producer surplus* (or *value added*) is the difference between price and cost and so is the area P_CP_AA. The *total surplus* (or *social welfare*) produced by the use of the system is the area P_CP_0A.

If conditions change and the number of travellers increases to D_B, then new travellers given by $a = (D_B - D_A)$ must value the service at approximately $b = (P_A + P_B)/2$ and so the total increase in value is ab. The cost of achieving this is approximately:

$$c = (P_B D_B - P_A D_A)$$

Thus, the increase in consumer surplus for small changes will be:

$$ab - c = (D_A + D_B)(P_A - P_B)/2$$

which equals the average number of users times the drop in price. It is thus represented by the area $P_B P_A AB$. This increase in consumer surplus is directly linked to the benefit–cost method to be discussed in Section 33.5.8.

33.5.6 Growth rates

If a quantity Q_1 in year (or any other time unit) one is increasing in size at a growth rate of r per year — i.e. its annual increment in size is Qr — then its size in any year, Q_n, with year 1 as the base year, is given by:

$$Q_n = Q_1 \sum_{n=1}^{n-1} (1 + r_{n-1}) \qquad (33.13)$$

For a continuous process, i.e. the time increment dt is small:

$$dQ/dt = rQ \qquad (33.14)$$

This process is known as *compounding*. For variable growth rates, Equations 33.13 or 14 must be integrated for $r = r(t)$. For a constant growth rate, r:

$$Q_n = (1 + r)^{n-1} Q_1 \qquad (33.15)$$

where $(1 + r)^{n-1}$ is called the *compound amount factor* or the future amount factor. For a continuous process, integration of Equation 33.14 gives:

$$Q_n = e^{rt} Q_1$$

A good example of compounding is money invested at an *interest rate* of r per annum, usually expressed in percent as 100r.

Accumulation is the process where the quantity grows as in compounding but where, additionally, each year's interest increment is added to the existing quantity. A good example of accumulation is the total number of vehicles using a piece of road over its life. The total quantity accumulated, $Q_{\Sigma n}$, beginning with Q_1 in year one, is:

$$Q_{\Sigma n} = Q_1 \sum_{n=1}^{n} (1 + r)^{n-1}$$

$$= Q_1[(1 + r)^n - 1]/r \qquad (33.16)$$

The ratio $Q_{\Sigma n}/Q_1$ is sometimes called the *cumulative growth factor*, which is thus $[(1 + r)^n - 1]/r$

If the level in the preceding year was Q_b, then the cumulation for year 1 onwards (i.e. in terms of Q_b but excluding Q_b from the accumulation) is given by Equation 33.16 as:

$$Q_{\Sigma n} = Q_b[(1 + r)^{n+1} - (1 + r)]/r \qquad (33.17)$$

This is the expression used in Section 31.5 where the base year traffic Q_b is not part of the design traffic — for example, Q_b could represent the traffic prior to reconstruction which is the basis for the traffic growth but not for the new pavement design.

In comparing the costs of an initial lump-sum investment of Q_0 in a scheme, as an alternative to a series of annual investments of Q_{1E}, then Equation 33.16 still applies. The two schemes are equivalent over n years when:

$$Q_0(1 + r)^{n-1} = Q_{1E}[(1 + r)^n - 1]/r$$

using Equations 33.15 and 33.16. Thus the equivalent annual investment is:

$$Q_{1E} = Q_0 r/[1 - e^{-rt}] \qquad (33.18)$$

This process is known as *amortising*. For a continuous process, the equivalent investment rate is $[e^{rt} - 1]/r$

N = n–1 is used to represent the number of income-earning years. In effect, n represents accounting practice with events beginning at year 1 and N represents mathematical practice with events beginning at year 0. If Q_0 was borrowed, then Q_{1E} is the amount that would need to be earned each year to pay off Q_0. The coefficient of Q_0 in Equation 33.18 is therefore called the *capital recovery factor*.

33.5.7 Present values and discounting

The *present value* of a future amount of money is the money that would need to be invested today in a deposit cumulating at a nominated interest rate, to be equal in value to the future amount at the time in question. It will thus be less than the future amount. Equation 33.15 is used to predict that our current money, Q_1, when invested at interest rate r will have a future value of;

$$Q_N = (1 + r)^N Q_1 \qquad (33.19)$$

Hence, the present value, Q_1, that would need to be invested for N years to get the required future amount is:

$$Q_1 = Q_N/(1 + r)^N \qquad (33.20)$$

where $(1+r)^{-N}$ is called the *present worth factor* or the *present value factor*. If future resources become available each year for N years then their total present value Q_{1T} is:

$$Q_{1T} = \sum_{N=0}^{N} Q_N/(1 + r)^N \qquad (33.21)$$

For a given year, the *net value* (NV) is:

$$NV = (\text{value of the benefits}) - (\text{value of the costs}) \qquad (33.22)$$

Thus, Equation 33.21 leads to the *net present value* (NPV) being;

$$NPV = \sum_{n=1}^{n}[(\text{net benefits}) - (\text{net costs})]/(1 + r)^n \qquad (33.23)$$

$$NPV = \sum_{n=1}^{n}(\text{net benefits})/(1 + r)^n - \sum_{n=1}^{n}(\text{net costs})/(1 + r)^n \qquad (33.24)$$

Because the NPV approach discounts earnings in year n by dividing them by $(1+r)^N$, the interest rate r is called a *discount rate*, the process is called *discounting*, and the discounted annual net benefits is called the *discounted cash flow*.

33.5.8 The benefit–cost method

Section 33.5.1 described how welfare economics analyses the allocation of resources within a community. Section 33.5.5 showed how the concept of consumer surplus leads to the idea of a project producing net benefits. This measure provides the basis by which projects can be numerically assessed within the context of welfare economics.

The specific tool developed for this purpose is the benefit–cost method of analysis (BCA; sometimes called cost–benefit analysis, CBA) which is a tool allowing the present value and net worth of projects to be evaluated. BCA is now widely applied to assess the optimal allocation of resources amongst projects with competing priorities and thus to prioritise a list of public sector schemes in order of economic desirability. BCA is generally accepted as the best of the available methods for schemes that do not have a marketable output.

Scarcity of key data, uncertainty about the future, and social concerns are the three main factors that reduce its reliability. Social concerns with BCA are discussed in Section 5.2.3 and can be very real. Indeed, BCA should not be used for projects with a high social or citizen welfare component. On the other hand, one of the assets of BCA is that it does take a society-wide view. Thus, it is particularly appropriate for public sector projects and less appropriate for private sector ones where only benefits accruing directly to the investor are to be considered. Two of the essentials of BCA are:

(1) to include all costs and benefits, irrespective of who pays or gains, and

(2) to quantify each of these financially.

Knowing the annual benefits and costs of a scheme and the appropriate discount rate (Section 33.5.9), the net present value (NPV) of the scheme can be calculated from Equations 33.21&23. A scheme is optimised when its NPV is at a maximum. If only Equation 33.21 is used, the NPVs of the benefits and the costs of schemes can be calculated separately and a benefit–cost ratio (BCR) for each scheme can be calculated by dividing one by the other, i.e.:

BCR = (NPV of the benefits)/(NPV of the costs)

From Equation 33.22, this can be rewritten as:

BCR = 1 + {NPV/(NPV of the costs)}

Thus an NPV < 0 or a BCR < 1 both indicate a scheme with no economic merit. It is assumed that all costs and benefits are the maximum amounts the recipient would pay to obtain the benefit or avoid the cost.

The advantage of BCR is that its use as a ranking device leads to the selection of the optimum economic solution from a range of projects that are not mutually exclusive. It provides a sound way of comparing two similar schemes. An incremental BCR is used when additions to a scheme are being considered. The extra (or marginal, Section 33.5.1) cost of an investment in additional capability must be less than the marginal benefits of that addition in capability. In particular, the increase in the NPV of the relative benefits divided by the increase in the NPV of the relative costs gives the incremental BCR, which should exceed one.

Theoretically, BCA relies on the identification of potential Pareto improvements over a clearly defined base-case. This means that any improvements resulting from a scheme are potentially able to compensate those who suffer from the scheme, such that one condition can be found in which no one is worse off and at least one person is better off. The point at which no further Pareto improvements are possible is called the point of *social efficiency*. This process requires transfers between individuals that may not occur in reality or which creates winners and losers, and hence community divisiveness. These consequences expose a distributional weakness in BCA discussed previously in Section 5.2.3 and which presented alternatives that have arisen to overcome these weaknesses. Individuals assess these separate compensating effects in terms of trade-offs based on their (economic) willingness to pay. When the effects involved in the trade-offs do not have money values, these must be imputed using either:

(a) opportunity cost (Section 33.5.2); or

(b) revealed or stated preferences, i.e. examining behavioural choices and trade-offs (Section 31.3.6).

33.5.9 Specific benefits and costs

The particular benefits and costs of road schemes are listed in Table 5.1. Many are *externalities* in that they are experienced by groups not participating in, or having any rights over the key travel activity. The costs used are total economic costs and may differ from market prices, which may not reflect the real cost of an item to the economy, due to such factors as insufficient competition, taxes, subsidies, social values, and the impact of this public investment on competing private investment opportunities. Adjustments made to market prices to account for these effects are called *shadow prices*. In terms of Figure 33.6, the supply curve must be based on the sum of the producer's cost and the external cost.

Transfer payments between two groups within the system are not included in costs as they have no overall effect and cannot change social welfare. For this reason, for instance, internal taxes are subtracted from any benefit or cost calculation as the transfer of taxes is only a transfer of funds and not a resource user or producer. Taxes levied within a society do not affect that society's net gain from a project.

Likewise, changes in land values due to the project are not included if these are a consequence of transport changes and improved accessibility caused by the project and thus already counted in the BCA (Section 31.2.4). Absolute land values are a basic factor as land used for roads is lost to the community for use for any other purpose. Its value will reflect both planning and environmental constraints and the land's potential for future development. Market values may not reflect the real (or social) value of the land for the following reasons;

(a) planning blight (Section 6.3),

(b) government subsidies to land-intensive industries (e.g. protection to farmers), and

(c) externalities of the existing land-use, e.g. there may be community benefit in removing an eyesore.

Because the value of any item must reflect its full worth to society, it is best measured in doubtful cases as the value that would be obtained for the item if it were used in the best possible alternate way. This prevents biasing results towards projects which use material or land already owned or staff already on the payroll.

As a BCA is normally conducted to decide on a course of action, it is common not to include any previous expenditure in the cost calculations. Such expenditures — *sunk costs* — are already committed and will not be influenced by subsequent decisions that are directed towards future action.

Capital expenditure is treated as a resource cost recorded in the year in which the capital item was acquired. It is not amortised, depreciated, or inflated other than through the discounting rate, NPVs, and the residual values (see below). Items are thus shown at their full cost when they enter the project. Nevertheless, any non-capitalised money borrowed for a project will be part of the cost of that project. Inflation is only considered if the price of an item is known to be going to inflate at a significantly different-from-average rate.

The *residual value* of a project is the net realisable value of the project at the end of its investment life. Roads and bridges rarely have any resale value and so their residual value will usually be the value of the land, earthworks, and drainage in their inaccessible state. This can still be a significant amount. The residual value must also be discounted back to its NPV. Projects can be taken to have infinite life, unless some very specific life span does exist, as the present value approach automatically takes account of life effects. Maintenance costs represent another key project cost and are treated in the same NPV manner discussed above for capital costs.

33.5.10 Discount rates in a benefit–cost analysis

In order to add costs incurred and benefits gained in different years, the values are then discounted to present values in the manner described in Section 33.5.7. The *discount rate* to be used is the subject of considerable significance and controversy. A high discount rate will favour projects that result in early benefits and early obsolescence, while a low rate will favour long-term projects. It is sometimes argued that people inherently prefer projects with high early pay-offs and thus the rate should be high. The values chosen for public sector projects are basically judgmental and are called social rates of discount. Even if the discount rate is set at the lending interest rate, it is necessary to perilously assume values for future interest rates.

A commonly recommended BCA discount-rate for public projects is a value which equalises the marginal return to the community from investment in the private and public sectors, said to be the real (inflation-free and before-tax) return on private sector investment, adjusted for the lower risk associated with public investment.

Because a wide range of discount rates may be chosen by various analysts, it is wise practice to examine the sensitivity of any BCA conclusions to variations from the chosen discount rate.

The *internal rate of return* of a project is that discount rate which will make the NPV zero. It is used as a measure of the relative worth of a project and projects with a value less than the interest rate would have a negative ranking. It is probably the best neutral measure for public sector assessments but cannot handle interest rates varying with time. An advantage of using both internal rate of return and BCRs is that they need not be expressed in money terms.

Projects may need to be deferred if the first year rate of return does not match the market-place investment-rate, i.e. if the money could be better invested for a year elsewhere.

33.5.11 Applications of benefit–cost analysis

The theory of benefit–cost analysis (BCA) was described in Section 33.5.8. BCA has been extensively used for road program assessment in a number of countries and within the World Bank. It was first applied to major road studies in the U.K. in the 1960s and a BCA program called *COBA* was released early in the 1970s. It is now one of the leading BCA procedures. COBA manuals are issued by TRL in the U.K. with Version 11 being released in 2001.

COBA operates by calculating user costs on the improved network over a 30 year life and then subtracting the user costs over the existing network over the same period to obtain the net user benefits. From these are subtracted the associated construction and maintenance costs, to give the net value of the project. Of course, all these costs are discounted in the manner described in Section 33.5.7.

To operate COBA requires estimates of the construction and maintenance costs, which are relatively straightforward. The user costs, however, require a description of the road network involved and of the traffic flows likely to occur on it over the 30 year life. This requires use of the traffic forecasting methods described in Section 31.5. Once the flows are known, the operating characteristics of each link in the network can then be estimated (Section 31.2.2). From these data it is possible to predict vehicle operating costs (Section 29.1.2), value of time costs (Section 31.2.3) and accident costs (Section 28.8) to obtain total user costs.

One critical assumption in COBA is that, when looking at the new network, it considers that drivers may well change their routes (Section 31.3.7) but does not assume that the distribution of the trips will be altered by the new scheme and uses a fixed trip matrix. It may therefore be necessary to conduct a separate study of trip distributions, following the principles in Section 31.3.5.

References

...s9.3 = Location within the book

AASHTO, see AMERICAN ASSOCIATION OF STATE HIGHWAY AND TRANSPORTATION OFFICIALS

ABBOTT, P., TYLER, J. & LAYFIELD, R. 1995. Traffic calming: vehicle noise emissions alongside speed control cushions and road humps. TRL Report 180...*s18.1*

ABDALLAH, I. M., BARKER, D. & RAESIDE, R. 1997. Road accident characteristics and socio-economic deprivation. *Traffic Engrg & Control* 38(12):672–6, Dec...*s28.2*

AKCELIK, R. 1978. On Davidson's flow rate/travel time relationship. Discussion. *Aust. Road Res.* 8(1):41–4...*s17.2*

-- 1980. Objectives in traffic system management. Program and Papers, Joint SAE/ARRB Seminar on 'Can traffic management reduce fuel consumption and emissions and affect vehicle design requirements?'...*s29.2*

-- 1981. Traffic signals: capacity and timing analysis. Aust. Road Res. Brd. Res. Rep., ARR No. 123...*s23.3,5&6*

-- & BESLEY, M. 1996. *SIDRA 5 User's Guide*. ARRB...*s17.2*

-- , RICHARDSON, A. & WATSON, H. 1982. Relation between two fuel consumption models. Papers. Joint SAE-A/ARRB 2nd Conf. on Traffic, Energy and Emissions, Paper 7, May...*s29.2*

-- Editor with WATSON, H., RICHARDSON, A. & BAYLEY, C. 1983. Progress in fuel consumption modelling for urban traffic management. Aust. Road Res. Brd. Res. Rep., ARR No. 124...*s29.2*

-- , -- & CHUNG, E. 1998. An evaluation of SCATS master isolated control. *Proc. 19th ARRB Conf.* 3–24...*s23.3*

-- , -- & ROPER, R. 1999. Fundamental relationships for traffic flows at signalised intersections. Aust. Road Res. Brd. Res. Rep., ARR No. 340...*s23.6*

AITCHISON, G. D. & RICHARDS, B. G. 1965. A broad-scale study of moisture conditions in pavement subgrades throughout Australia. In *Moisture equilibria and moisture changes beneath covered areas. A symposium-in-print*, 184–232. Butterworths: Australia....*s9.4*

ALDERSON, A. 2006. The collection and discharge of stormwater from the road infrastructure. *ARRB Res. Rep.* 368, May...*s13.1,13.2,13.4,13.5*

AMERICAN ASSOCIATION OF STATE HIGHWAY AND TRANSPORTATION OFFICIALS 1993. Guide for design of pavement structures. (with 1998 supplement), GDPS-4-M, AASHTO: Washington...*s11.4,11.5*

-- 2004. *Policy on geometric design of highways and streets.* 5th Edn. AASHTO: Washington...*s17.3*

ANDERTON, P. J. & COLE, B. L. 1982. Contour separation and sign legibility. *Aust. Road Res.* 12(2):103–9...*s21.2*

ANDREASSEN(D), D. C. 1985. Linking deaths with vehicles and population. *Traffic Engrg & Control* 26(11):547–9...*s28.1*

-- 1992. Preliminary costs for accident types. Aust. Road Res. Brd. Res. Rep. ARR 217, Feb...*s28.8*

-- & CAIRNEY, P. T. 1985. An old chestnut – T and cross-intersections *and* International calibration study of traffic conflict techniques. *Aust. Road Res.* 15(1):58–62...*s20.2*

ANTILL, J. M. & FARMER, B. E. 1991. *Antill's Engineering Management.* McGraw-Hill: Sydney...*s25.3*

ARGUE, J. R. 1979. Urban surface drainage design; some hydrologic/hydraulic interactions. *Program and Papers. 17th ARRB Regional Symp.* Perth, 1–12...*s13.2*

-- 1986. Storm drainage design in small urban catchments. Aust. Road Res. Brd Special Rep. SR 34, Dec...*s13.2,13.4*

ARMOUR, M. 1980. The effect of lighting conditions on car following behaviour. *Proc. 10th ARRB Conf.* 10(4):43–53...*s24.1*

-- 1982. Vehicle speeds on residential streets. *Proc. 11th ARRB Conf.* 11(4):190–205...*s18.3*

-- 1984. The relationship between shoulder design and accident rates on rural highways. *Proc. 12th ARRB Conf.* 12(5):49–62...*s28.5*

-- 1985. Effect of road cross-section on vehicle lateral placement. *Aust. Road Res.* 15(1):30–40...*s16.5*

-- and McLEAN, J. R. 1983. The effect of shoulder width and type on rural traffic safety and operations. *Aust. Road Res.* 13(4):259–70...*s11.7*

ARRB, see AUSTRALIAN ROAD RESEARCH BOARD

ASCHAUER, D. 1989. Is public expenditure productive? *J Monetary Economics*, 23:177–200...*s4.3*

ASHTON, N. R., BUCKLEY, D. J. & MILLER, A. J. 1968. Some aspects of capacity and queueing in the vicinity of slow vehicles on a rural two-lane road. *Proc. 4th ARRB Conf.* 4(1):595–615...*s18.4*

-- & BRINDLE, R. E. 1982. An attempt at evaluating local area safety improvements in an Australian study. Papers. Seminar on Short-Term and Area Wide Evaluation of Safety Measures. OECD Road Research Programme. SWOV. April, 236–45...*s7.3*
ATKINS, A. S. 1981. The economic and social costs of road accidents in Australia. Office of Road Safety, Dept Trspt Rep. CR 21, June...*s28.8*
-- 1982. The economic costs of road accidents in Australia: some issues in estimation: concept and application. *Proc. 11th ARRB Conf.* 11(5):206–20...*s28.8*
-- , S. 1986. Transportation planning models — what the papers say, *Traffic Engrg & Control*, Sep:460–7..*s31.4*
AUFF, A. A. 1982. Quality control of dimensions in road construction. Aust. Road Res. Brd. Special Rep., SR No. 25...*s25.7*
-- 1984. Analysis of some asphaltic concrete data for quality control purposes. *Proc. 12th ARRB Conf.* 12(2):146–58...*s12.2*
AUSTRALIAN ROAD RESEARCH BOARD 1982. Proceedings. NAASRA/ARRB workshop on the use of relative compaction in the control of roadworks. ARRB, Dec...*s8.5*
-- 1985. Accelerated loading facility seminar. Gosford: NSW...*s8.6*
-- 1993. *Unsealed roads manual*. ARRB, May...*s12.1*
AUSTROADS 1992. *Pavement design*. AUSTROADS: Sydney...*s11.4,14.5*
-- 2000. *Austroads guide to the selection of road surfaces....s12.5*
-- 2003. *Rural road design — a guide to the geometric design of rural roads*. AP-G1/03 Austroads: Sydney...*s18.2&7,21.2*
-- 2004. *Pavement design*. AP-G17/04...*s11.4, 14.6*
AUSTRALIA 2003. *Articulated truck fatalities*. Dept Infrastructure, Transport etc., Monograph 15...*s27.3*

BAIL, J. 1981. Design for kerb ramps. *Roy. Aust. Plan. Inst. J.* 19(1):19–21...*s7.4*
BANKS, J. H. 1991. The two-capacity phenomenon: some theoretical issues. *Trspt Res. Rec.* 1320: 234–241...*s17.2*
BARNARD, P. O. 1985. An examination of home interview survey trip under-reporting as documented in Australian transport studies. *Aust. Road Res.* 15(1):41–5...*s31.3*
BARRETT, J. R. & SMITH, D. M. 1976. Stress history effects in basecourse materials. *Proc. 8th ARRB Conf.* 8(3):Session 7, 30–42...*s11.3,11.4*
BATSTONE, M., HUYBREGTS, C., SAMUELS, S. & WEST, P. 2001. Evaluating the American FHWA traffic noise model in Australia. *Proc. 20th ARRB Conf.* CD... *s32.3*
BEAVIS, H. M 1984. Observations on sections of rural road at three sites in Western Australia and South Australia 1967–82. *Aust. Road Res.* 14(2):78–81...*s9.3*
BENNATHAN, E. & JOHNSON, M. 1990. Transport in the input–output system. *Trspt Res. Rec.* 1274:104–116...*s29.1&4*
BERRY, D. & BOTH, G. J. 1980. Stage development of traffic capacity improvements to major rural roads in Victoria. *Proc. Workshop on Economics of Road Design Standards.* BTE and AGPS, Vol. 2:58–78...*s25.2*
BIGGS, D. C. & AKCELIK, R. 1985. An interpretation of the parameters in the simple average travel speed model of fuel consumption. *Aust. Road Res.* 15(1):46–9...*s29.2*
-- 1988. ARFCOM — models for estimating light to heavy vehicle fuel consumption. Aust. Road Res. Brd Res. Rep. ARR 152, Sept...*s29.2*
BLAND, B. H. 1982. The LUTE land-use and transportation model. TRL SR716, Berkshire, England...*s31.3*
BOND, M. 1996. Plagued by noise. *New Sci.*, 16 Nov, pp14–15...*s32.2*
BOWERING, R. H. 1970. Properties and behaviour of foamed bitumen mixtures for road building. *Proc. 5th ARRB Conf.* 5(6):38–57...*s10.4*
BOWYER, D. P., AKCELIK, R., BIGGS, D. C. & BAYLEY, C. 1984. Fuel consumption savings from traffic management: findings from a scientific audit. *Aust. Road Res.* 14(2):95–6...*s29.2*
-- , -- & BIGGS, D.C. 1985. Guide to fuel consumption analysis for urban traffic management. Aust. Road Res. Brd. Special Rep., SR No. 32...*s29.2*
BOYCE, L. & DAX, E. C. 1981. The illiterate on the road. Aust. Road Res. Brd. Special Rep., SR No. 24...*s16.6*
BOYD, M. J. 1979. Accuracy of design flood estimates for medium sized catchments in Eastern New South Wales. *Aust. Road Res.* 9(3):22–9...*s13.2*
-- , W. & FOSTER, C. 1950. Design curves for very heavy multiple wheel assemblies. *Trans ASCE* 115:534–6...*s27.3*
BRIAUD, J-L., JAMES, R. W. & HOFFMAN, S. B. 1997. Settlement of bridge approaches (the bump at the end of the bridge). Transportation Research Board, Synthesis of highway practice #234... *s15.2*

BRINDLE, R. E. 1982. Town planning and road safety in smaller urban areas. *Program and Papers. 19th ARRB Reg. Symp.* Wagga Wagga, May, 1–28...*s7.2*

BRODIE, H. D. 1970. An investigation into flexible pavements in arid regions. *Proc. 5th ARRB Conf.* 5(4):5–43...*s8.3*

BROWN, I. D., GROEGRER, J. A. & BIEHL, B. 1987 Is driver training contributing enough towards road safety? Chapt 8 in J. A. Rottengatter *et al.* (Eds) *Road users and traffic safety.* Netherlands: van Gorcum...*s16.6*

BROWN, J. I. & PETERS, R. J. 1988. Development and performance of CULWAY: a culvert-based weigh-in-motion system. *Proc. 14th ARRB Conf.* 14(6):88–105...*s27.3*

BRYANT, J. F. 1980. Signs of high brightness. *Proc. 10th ARRB Conf.* 10(4):252–62...*s21.2*

-- 1982. The design of symbolic signs to ensure legibility. *Proc. 11th ARRB Conf.* 11(5):161–71...*s21.2*

BULLEN, F. 1984. Use of coralline materials in pavements. *Proc. 12th ARRB Conf.* 12(2):63–70...*s8.5.5*

BURKE, M. & BROWN, L. 2007. Distances people walk for transport. *Road & Trspt Res.* 16(3):16–29…*s30.2*

CAIRNEY, P. T. 1982. An exploratory study of risk estimates of driving situations. *Proc. 11th ARRB Conf.* 11(5):233–40...*s28.2*

CAMPBELL, B. J. & CAMPBELL, F. A. 1986. Seat belt law experience in four foreign countries compared to the United States. Univ of N. Carolina, HSRC and AAA Fndn for Traffic Safety. Dec...*s28.4*

CARSE, A. & BEHAN, J. 1980. Static chord modulus of elasticity of high strength concrete in uniform compression and flexure. *Proc. 10th ARRB Conf.* 10(3):46–56...*s8.9*

CASTRO, C. & HORBERRY, T. (Eds) 2004. *The human factors of transport signs.* Florida: CRC...*s21.1&2*

CATCHPOLE, J. 2005. Learning to take risks; the influence of age and experience on risky driving and young drivers: who takes risks and why? ARRB Research Reports ARR 362 & 364, July…*s16.5*

CHEUNG, C. & BLACK, J. 2008. A reappraisal of the intervening opportunities model of commuter behaviour. *Road & Trspt Res.* 17(2):3–18, June…*s31.3*

CHRISTIE, R. 2001. The effectiveness of driver training as a road safety measure: a review of the literature. RACV Report 01/03, Nov

CIE: see COMMISSION INTERNATIONALE DE L'ECLAIRAGE

CLAESSEN, A. EDWARDS, J., SOMMER, P. & UGE, P. 1977. Asphalt pavement design — the Shell method. *Proc. 4th Int. Conf. Structural Design of Asphalt Pavements.* 1: 39–74...*s11.2,11.4*

CLAYTON, A. B., McCARTHY, P. E. & BREEN, J. M. 1984. Drinking and driving habits: attitudes: and behaviour of male motorists. Trspt Road Res. Lab. U.K. TRRL Su Rep. SR 826...*s28.2*

CLEGG, B. 1983. Design compatible control of basecourse construction. *Aust. Road Res.* 13(2):112–22...*s8.4,8.6,11.3,25.5*

COHEN, H. & SOUTHWORTH, F. 1999. On the measurement and valuation of travel time variability due to incidents on freeways. *J. Trspn Stat.* 2(2):123–31, Dec ..*s31.2*

COLE, B. L. & JACOBS, R. J. 1978. A resolution limited model for the prediction of information retrieval from extended alphanumeric messages. *Proc. 9th ARRB Conf.* 9(5):383–89...*s21.2*

-- & -- 1981. A comparison of alternative symbolic warning devices for railway level crossings. *Aust. Road Res.* 11(4):37–45...*s21.3*

-- & JENKINS, S. E. 1980. The nature and measurement of conspicuity. *Proc. 10th ARRB Conf.* 10(4):99–107...*s21.2*

-- & -- 1982. Conspicuity of traffic control devices. *Aust. Road Res.* 12(4):223–38...*s16.4,21.2*

COLE, W. F. & CERAM, F. I. 1981. Influence of Washington Degradation Factor and secondary mineral content on acceptance limits for basalt aggregate for road purposes. *Aust. Road Res.* 11(1):58–60...*s8.6*

-- & SANDY, M. J. 1980. A proposed secondary mineral rating for basalt road aggregate durability. *Aust. Road Res.* 10(3):27–37...*s8.5*

COMMISSION INTERNATIONALE DE L'ECLAIRAGE 1976. Glare and uniformity in road lighting installations. CIE Pub. No. 31. CIE: Vienna...*s24.2*

-- 1977. Road lighting lantern and installation data. CIE Pub. No. 34. CIE: Vienna...*s24.3*

-- 1978. Light as a true visual quantity: principles of measurement. CIE Pub. No. 41 TC-1.4. CIE: Vienna...*s21.3*

-- 1980. Light signals for road traffic control. CIE Pub. No. 48 TC-1.6. CIE: Vienna...*s23.2*

-- 1992. Road lighting as an accident countermeasure. CIE Pub. No. 93 TC 4.02. CIE: Vienna...*s24.1*

CORBEN, B., TINGVALL, A., FITZHARRIS, M., NEWSTEAD, S. & JOHNSTON, I.. 2003. An analysis of crashes involving median encroachments on high speed roads. Oroc 23 ARRB Conf, Session 28…*s28.6*

CORDINGLEY, R. C. & JARVIS, J. R. 1982. The effect of various traffic control treatments employed at roadwork sites. *Proc. 11th ARRB Conf.* 11(5):21–29...*s25.9*

COUNTRY ROADS BOARD, VICTORIA 1980a. The design of flexible pavements. *Tech. Bull.* 31, Sep...*s6.2,8.4*

-- 1980b. Bridges. *CRB News* 44: 9–13, Oct....*s15.2*

COX, R. 2003. Reduced sight distance on existing rural roads. How can we defend it? Proc. 21st ARRB Conf. Session S16...*s19.4*

CRONEY, D. & CRONEY, P. 1991. *The design and performance of road pavements*, 2nd Edn. McGraw-Hill; London...*s11.2,12.2,14.1*

DAFF, M. & SIGGINS, I. D. 1982. On road trials of some new types of slow points. *Proc. 11th ARRB Conf.* 11(4):214–37...*s7.3*

DALEY, K. F. & OGDEN, K. W. 1980. A review of traffic signal performance: the case for improved maintenance. *Proc. 10th ARRB Conf.* 10(4):75–84...*s26.4*

DAVIDSON, K. B. 1978. The theoretical basis of a flow–travel time relationship for use in transportation planning. *Aust. Road Res.* 8(1):32–35...*s17.2*

DAVIES, A. L. & EADES, G. W. 1980. The applicability of terrain evaluation to route location and feasibility surveys for highways in Queensland. *Proc. 10th ARRB Conf.* 10(2):45–57...*s6.1*

DEPARTMENT OF MAIN ROADS, NEW SOUTH WALES 1980. Pavement thickness design. MR Form 76, Jan...*s8.4*

DICKINSON, E. J. 1978a. The design of bituminous plant mixes for road and airfield paving. *Aust. Road Res.* 8(2):32–38...*s12.2*

-- 1978b. The cooling of asphalt layers during the compaction operation. *Proc. 9th ARRB Conf.* 9(4):247–59...*s12.4*

-- 1981b. The flow and deformation behaviour of bituminous binders under pavement service conditions. *Aust. Road Res.* 11(3):3–10...*s8.7*

-- 1982. The performance of thin bituminous pavement surfacings in Australia. *Proc. 11th ARRB Conf.* 11(3):35–51...*s8.7,12.1*

-- 1984. *Bituminous roads in Australia.* Aust. Road Res. Brd...*s8.7,8.8,12.1,12.3*

-- 1989. The effect of climate on the seasonal variation of pavement skid resistance. *Aust. Road Res.* 19(2):129–44, June...*s12.5*

DORMAN, G. & METCALF, C. 1964. Design curves for flexible pavements based on layer system theory. *HRB TRR Record* 71:69–84..*s11.4*

DOWD, B. P., IOAKIM, R. & ARGUE, J. R. 1980. The simulation of gutter pavement flows on South Australian urban roads. *Proc. 10th ARRB Conf.* 10(2):145–52...*s13.3*

DRUMMER, O. 1994. Drugs in drivers killed in Australian road traffic accidents. Victorian Institute of Forensic Pathology Report, Department of Forensic Medicine, Monash Uni...*s28.2*

DUNLOP, R. J. 1980. A review of the design and performance of roads incorporating lime and cement stabilised pavement layers. *Aust. Road Res.* 10(3):12–25...*s10.1,10.2,10.5*

ELLSON, P. B. 1984. Parking: turnover capacities of car parks. Trspt Road Res. Lab. U.K. TRRL Lab. Rep. LR1126...*s30.6*

ELVIK, R. 2005. Speed and road safety: synthesis of evidence from evaluation studies. Transportation Research Record, *J. Transportation Research Board,* No 1908:59–69....*s18.1*

EMMERSON, P., RYLEY, T. J. & DAVIES, D. G. 1998. The impact of weather on cycle flows. *Traffic Engrg & Control* 39(4):238–243...*s27.6*

EVANS, L. 1986. The effectiveness of safety belts in preventing fatalities. *Acc. Anal. Prev.* 18(3):229–41...*s28.4*

-- 1991. *Traffic safety and the driver.* Van Nostrand-Reinhold; New York...*s28.1*

EVANS, R. G. & BUSH, J. 1980. Participatory road studies — a review of techniques. *Proc. 10th ARRB Conf.* 10(5):89–99...*s32.1*

EVANS, R. P., HOLDEN, J. C. & McMANUS, K. J. 1996. Application of a new vertical moisture barrier construction method for highway pavements, *Road & Trspt. Res.* 5(3):4–12...*s9.4*

EVANS, R. S. & HAUSTORFER, I. J. 1982. Groundwater investigation methods for freeway design and construction. *Proc. 11th ARRB Conf.* 11(2):66–76...*s9.2*

FARAH, H., POLUS, A. & COHEN, M, 2007. Multivariate analyses for infrastructure-based crash-prediction models for rural highways. *Road & Trspt Res.* 16(4):26–41, Dec.

FIELDING, B. J. 1980a. The Washington degradation test — mechanism and use. *Proc. 10th ARRB Conf.* 10(3):94–105...*s8.6*

-- 1980b. A test to assess the durability of crusher fines. *Proc. 10th ARRB Conf.* 10(3):205–9...*s8.6*

FILDES, B. N. & TRIGGS, T. J. 1982. The effects of road curve geometry and approach distance on judgements of curve exit angle. *Proc. 11th ARRB Conf.* 11(4):135–44...*s16.5*

FLYVBERG, B., HOLM, M. & BUHL, S. 2005. How (in)accurate are demand forecasts in public works projects. *J. Am. Planning Assn* 71(2):131–146, Spring...*s31.4*

FORMAN, R. T. *et al. Road ecology: science and solutions.* Washington: Island Press...*s6.4*

FREEMAN, R., MANN, T. & GONZALEZ, C. 2003. Design and assessment of unsurfaced and aggregate-surfaced roadways in the U.S. Army Corps of Engineers. *Proc. 23rd ARRB Conf.* Session S15...*s8.4*

GALIN, D. 1981. Speeds on two-lane rural roads — a multiple regression analysis. *Traffic Engrg & Control* 22(8/9):453–60...*s18.2*

GEHL, J. 1980. The residential street environment. *Built Environ.* 6(1):51–61...*s7.4*

GEIPOT, 1982a. *Study of vehicle behaviour and performance.* Vol. 6 of *Research on the Inter-Relationships Between Costs of Highway Construction, Maintenance and Utilisation, Brasilia.* Republican Federation de Brasil...*s18.2,29.3*

-- 1982b. *Study of pavement maintenance and deterioration. Study of vehicle behaviour and performance.* Vols. 6 & 7, loc cit...*s14.4*

-- 1982c. *Highway costs model MICR.* Vol. 8. loc. cit...*s26.5*

-- 1982d. *Model of time and fuel consumption.* Vol. 9. loc. cit...*s29.2*

-- 1982e. *Study of road user costs.* Vol. 5. loc. cit...*s29.1&2*

GERKE, R. J. 1987. Subsurface drainage of road structures. Aust. Road Res. Brd. Special Rep. 35, April...*s9.3,12.5*

GERRARD, C. M. & WARDLE, L. J. 1976. Tables of stresses, strains and displacements in three-layer elastic systems under various traffic loads. Aust. Road Res. Brd. Special Rep. SR 4...*s11.2*

-- 1980. Rational design of surface pavement layers. *Aust. Road Res.* 10(2):3–15...*s11.4*

GERRARD, C. M., McINNES, D. V. & HARRISON, W. J. 1972. Basecourse selection — a comparison of two-layer theory and Texas triaxial test data. *Aust. Road Res.* 4(9):26–40...*s8.6*

GIFFEN, J. C., YOUDALE, G. P. & WALTER, P. D. 1978. Use of non-standard gravels as bases. *Proc. 9th ARRB Conf.* 9(4):54–61...*s8.6*

GILLESPIE. T. D. & SAYERS, M. 1983. Measuring road roughness and its effects on used costs and comfort. *ASTM Spec. Tech. Publn* 884...*s14.4*

GIPPS, P. G. 1981. A behavioural car-following model for computer simulation. *Trspt Res.* B. 15B:105–11...*s17.2*

-- 1992. ALIGN_3D: a package to optimise route alignment. *Road & Trspt Res.* 1(2): 50–9, June...*s6.1*

GRAHAME, R. & GOLDSBOROUGH, R. 1980. Developments in pavement construction processes and equipment in Queensland. *Proc. 10th ARRB Conf.* 10(2):175–89....*s10.3,10.6*

GRANDEL, J. 1985. Investigation of the technical defects causing motor vehicle accidents. Field Accidents 'Data collection: analysis: methodologies and crash injury reconstructions'. SAE Publn 159:301–20, SAE paper 850434...*s27.2*

GRAY, W. J. & BARAN, E. 1979. An interpretation of surface deflection measurements using linear elastic theory for a full depth asphaltic concrete pavement. Aust. Road Res. Brd. Internal Rep. AIR 1034–1...*s11.3*

-- . & ROBINSON, G. K. 1980. Basic ideas of acceptance control in road construction. *Proc. 10th ARRB Conf.* 10(2):65–79...*s25.7*

GRAYLING, B. W. 1980. Prediction of casualty accident rates on urban arterials in Melbourne. *Proc. 10th ARRB Conf.* 10(4):54–61...*s28.5*

GRIEBE, P., HERRSTEDT, L. & NILSSON, P. 2000. Speed management in urban areas. *Routes/Roads,* 306(II):23–30, April...*s7.2,18.1*

HAAS, R. C. 1973. A method for designing asphalt pavements to minimise low-temperature shrinkage cracking. *Asphalt Inst. Res. Rep.* 73–1, Jan...*s11.4*

-- , HUDSON, W. R. & ZANIEWSKI, J. 1994. *Modern pavement management.* Krieger: Florida...*s11.2,14.6*

HADDON, W. 1980. Advances in the epidemiology of injuries as a basis for public policy. *Public Health Reports* 95(5):411–21...*s28.1*

HALL, F. L. 1987. An interpretation of speed–flow–concentration relationship using catastrophe theory. *Trspt Res.* A.:21A(3):191–201...*s17.2*

HALLETT, S. 1990. Drivers' attitudes to driving: cars and traffic. 'Transport and Society.' Rees Jeffreys Discussion Paper 14, Feb...*s16.3*

HAMORY, G. I. 1980. Experience with the use and operation of nuclear moisture/density meters. *Proc. 10th ARRB Conf.* 10(3):164–75...*s8.2*

-- & McINNES, D. B. 1972. Compressive strength assessment of road materials. *Proc. 6th ARRB Conf.* 6(5):143–65...*s8.6*

HAMPSON, G. 1982. The theory of accident compensation and the introduction of compulsory seat belt legislation in New South Wales. *Proc. 11th ARRB Conf.* 11(5):135–40...*s28.4*

HEATON, B. S. & BULLEN, F. 1980. Properties of air-cooled slag road base for rational pavement design. *Proc. 10th ARRB Conf.* 10(3):198–204...*s8.5.6,11.3*

-- 1982. Properties of stabilised blast furnace slag roadbase. *Proc. 11th ARRB Conf.* 11(3):168–75...*s8.5*

-- , EMERY, J. & KAMEL, N. 1978. Prediction of pavement skid resistance performance. *Proc. 9th ARRB Conf.* 9(3):121–6...*s12.5*

-- , FRANCIS, C. L., JAMES, W. & CAO, T. 1996. The use of BOS steel slags in road pavements. *Proc. 18th ARRB Conf.* 18(3):49–62...*s8.5*

HENSHER, D. 1999. A bus-based transitway or light rail? *Road & Trspt Res.* 8(3):3–21...*s30.1&4*

HERBERT, D. C. 1980. Road safety in the seventies: lessons for the eighties. Traffic Accident Research Unit, DMT, N.S.W. Res. Rep. 4/80, Mar...*s28.2, 28.3*

HERMAN, R. & POTTS, R. B. 1961. Single lane traffic theory and experiment. *Theory of Traffic Flow Proc. of Symp.*, Elsevier...*s17.2*

HIGHWAYS AGENCY 2008. *Design manual for roads and bridges, Vol 7, Pavement design and maintenance*. U.K...*s11.4*

HIGHWAY RESEARCH BOARD 1962. Factors influencing compaction test results. HRB Bull. 319...*s8.2*

HILF, J. W. 1957. A rapid method of construction control for embankments of cohesive soil. *ASTM Spec. Pubn* STP 232...*s8.2*

HIMANEN, V., NIJKAMP, P. & PADJEN, J. 1992. Environmental quality and transport policy in Europe. *Trspt Res.* A 26A(2):147–157...*s28.8,32.2&3*

HOBAN, C. J. 1982. The two and a half lane rural road. *Proc. 11th ARRB Conf.* 11(4):59-70...*s18.4*

-- , 1984. Measuring quality of service on two-lane rural roads. *Proc. 12th ARRB Conf.* 12(5):117–31...*s17.4*

HOLDSWORTH, J. & SINGLETON, D. J. 1980. Environmental capacity as a basis for traffic management at local government level. *Proc. 10th ARRB Conf.* 10(5):165–74...*s7.4*

HOLLAND, J. E. & RICHARDS, J. 1982. Road pavements on expansive clays. *Aust. Road Res.* 12(3):173–79...*s8.4*

HOLT, C. & FREER-HEWISH, R. 2000. The significance of recent research on specifications and standards for soil-lome pavement layers – the need to rationalise the design procedure. *Road & Trspt Res.* 9(2):14–32, June..*s10.6*

HUDDART, K. W. 1980. Australian traffic signalling. Aust. Road Res. Brd. Res. Rep., ARR No. 113...*s23.1,23.3*

HUGHES, B., CHAMBERS, L., LANSDELL, H. & WHITE, R. 2004. Cities, area and transport energy. *Road & Trspt Res.* 13(2):72–84, June...*s29.2*

HUGHES, P. K. & COLE, B. L. 1984. Search and attention conspicuity of road traffic control devices. *Aust. Road Res.* 14(1):1–9...*s21.2*

HULSCHER, F. R. 1984. The problem of stopping drivers after the termination of the green signal at traffic lights. *Traffic Engrg & Control* 25(3):110–16...*s23.2*

HUNT, P. B., ROBERTSON, D. I., BRETHERTON, R. D. & WINTON, R.I. 1981. SCOOT — a traffic-responsive system of co-ordinating traffic lights. TRRL, U.K., TRRL Lab Rep LR 1014...*s23.5*

HUSCHEK, S. 1990. The influence of road surface roughness on tire noise generation in the Federal Republic of Germany. *ASTM Special Publication, Surface characteristics of roadways*, STP 1031, 430–41...*s12.5*

INGLES, O. G & METCALF, J. B. 1972. *Soil stabilisation: principles and practice*. Butterworths: Sydney...*s8.4,9.2,10.1*

-- & NOBLE, C. 1975. The evaluation of basecourse materials. *Proc. Symp. on Soil Mechanics: Recent Developments*. S. Valliappan Ed. Univ. N.S.W...*s8.2, 8.3, 8.6*

INSTITUTION OF TRANSPORTATION ENGINEERS 2004. *Trip generation handbook*, 2nd edition, ITE: Washington. ...*s31.3*

INTERNATIONAL ROAD FEDERATION 2007. *World Road Statistics* 2001–2005. IRF: Geneva...*s4.1,4.3*

INTERNATIONAL STANDARDS ORGANISATION 1982. *Measurement of noise emitted by stationary road vehicles survey method*, ISO 5130...*s32.2*

-- 1990. *Mechanical vibration and shock — vibration of buildings — guidelines for the measurement of vibrations and the evaluation of their effects.* Amended 1994&6. ISO Geneva, ISO 4866...*s25.8*

-- 1992. *SI units and recommendations for the use of their multiples and of certain other units.* ISO 1000-1992, 3rd Ed...*s2.3*

-- 1997. *Acoustics — methods of measuring the influence of road surfaces on traffic noise*, ISO 11819...*s32.2*

ITRD (INTERNATIONAL TRANSPORT RESEARCH DOCUMENTATION) 2005. *ITRD English alphabetical thesaurus*. Version 2. TRL, Berkshire, England for ITRD...*s2.2*

JACOBS, G. D. & SAYER, I. A. 1984. Road accidents in developing countries — urban problems and remedial measures. Trspt Road Res. Lab. U.K. TRRL Suppl. Rep. SR 839...*s28.1*

JACOBS, R. J. & COLE, B. L. 1978a. Acquisition of information from alphanumeric road signs. *Proc. 9th ARRB Conf.* 9(5):390–95...*s16.4,21.2*

-- & -- 1978b. Searching vertical stack direction signs. *Proc. 9th ARRB Conf.* 9(5):396–400...*s21.3*

JAMESON, G. W. 1984. Variability of soils classification tests. *Proc. 12th ARRB Conf.* 12(2):131–45...*s8.2*

-- 1985. A field evaluation of nuclear gauge calibration methods. Part 1: Manufacturers' calibration. Part 2: Field and blocks — field calibration. *Aust. Road Res.* 15 1, 20–29 and 15 2, 83-89...*s8.2*

JARVIS, J. R. 1980a. The off-road testing of road humps for use under Australian conditions. *Proc. 10th ARRB Conf.* 10(4):293–305...*s18.1*

-- 1980b. Legal aspects of road humps on public roads. Aust. Road Res. Brd. Res. Rep., ARR No. 109...*s18.1*

-- 1982. Driving patterns and driver characteristics on the road. *Program and Papers. Joint SAE-A/ARRB 2nd Conf. on Traffic, Energy and Emissions. Session 6, Paper 21*, May...*s27.2*

JENKINS, S. E. & SHARP, K. G. 1985. Some laboratory tests to investigate the colour differences between concrete paving blocks and asphaltic concrete. *Aust. Road Res.* 15(3):187–89...*s11.6*

JEWELL, R. J. 1969. An evaluation of criteria for selection of pavement basecourse materials in Western Australia condensed by D.B. McInnes. Bull. 5. Aust. Road Res. Brd. October. *Proc. Symp. on Selection and Construction of Base Materials in Roads...s8.3*

JOHNSTON, I. R. 1982. Modifying driver behaviour on rural road curves — a review of recent research. *Proc. 11th ARRB Conf.* 11(4),(11)5–24...*s19.2*

-- 1983. The effects of roadway delineation on curve negotiation by both sober and drinking drivers. Aust. Road Res. Brd. Res. Rep., ARR 128...*s16.5,22.2,24.4,28.2*

JORDAN, J. R., LOGUE, G. & ANDERSON, G. M. 1980a. Review of the NAASRA study of road maintenance standards: costing and management. *Proc. 10th ARRB Conf.* 10(2):232–49...*s26.1,26.2*

JORDAN, P. G. 1984. Measurement and assessment of unevenness on major roads. Trspt Road Res. Lab. U.K. TRRL Lab. Rep. LR 1125...*s14.4*

KENIS, W. J., SHERWOOD, J. A. & McMAHON, T. F. 1982. Verification and application of the VESYS structural sub-system. *Proc. 5th Int. Conf. Struct. Des. Asphalt Pave.* Delft, 1: 333–48...*s11.2*

KENNEDY, C. K. 1985. Analytical flexible pavement design. *Proc. Inst. Civ. Engrgs*, Part 1, 78: 897–917, August...*s11.3,11.4,14.3*

KER, I. & GINN, S. 2003. Myths and realities in walkable catchments: the case of walking and transit. *Road & Trspt Res.* 12(2):69–79...*s30.2*

KIMBER, R. M. 1980. The traffic capacity of roundabouts. Trspt Road Res. Lab. U.K. TRRL Lab Rep. LR 942...*s20.3*

KORTEWEG, A. L. 2002. Functional specifications in contracting. *Routes/Roads* 315 (III):24–34...*s25.3*

LAMM, R., PSARIANOS, B. & MAILAENDER, T. 1999. *Highway design and traffic safety handbook.* NY: McGraw-Hill...*s6.2*

LAVE, C. & ELIAS, P. 1994. Did the 65 mph speed limit save lives? *Acc. Anal. Prev.* 26 1:49–62...*s18.1*

LAWRANCE, C. J., BYARD, R. J. & BEAVEN, P. J. 1993. *Terrain evaluation manual.* HMSO: London...*s6.1*

LAY, M. G. 1978. *Collected ARRB research, Vol. 1.* Aust. Road Res. Brd. Administrative Manual, AAM3...*s8.4,14.5,16.5,28.1*

-- 1980. Load combinations — a calibration oriented approach to more rational design rules. Aust. Road Res. Brd. Res. Rep. ARR 110...*s15.4*

-- 1982. *Structural steel fundamentals.* Aust. Road Res. Brd...*s8.6,11.2,11.3,11.4,14.3,15.2*

-- 1985a. *Source book for Australian roads.* 3rd Ed. Aust. Road Res. Brd...*s1*

-- 1985b. Pavement maintenance management systems in Australia. Papers. *PTRC 13th Summer Annu. Meet. Seminar P271*, 69–78...*s14.6*

-- 1992. *Ways of the world.* Rutgers University Press: New Jersey...*s2.2,3.1,8.7,10.1,14.4,15.3,31.2*

-- 1993. Modelling pavement behaviour. *Road & Trspt Res.* 2(2):16–27, June...*s11.3,14.3*

LEONHARDT, F. 1991. Developing guidelines for aesthetic design. In *Bridge aesthetics around the world*, pp. 32–57. Trnsptn Res. Brd: Washington...*s15.3*

LILLEY, A. A. & CLARK, A. J. 1980. *Concrete block paving for lightly trafficked roads and paved areas.* Cement and Concrete Assoc. U.K. 2nd Ed. Pub. 46.024...*s11.5*

LINN, M. & SYMONS, M. 1981. Stabilisation of fine-grained soils for pavements using lime and flyash. *Papers. First Nat. Conf. on Local Govt Eng. I.E.* Aust. Adelaide. 272–79...*s10.3*

LOCKWOOD, S. 2005: Systems management and operations: a culture shock. *ITE J* 75(5), May, 43–7...*s17.4*

LOGIE, G. 1980. *Glossary of transport*. Elsevier Scientific: Amsterdam...*s2.1,2.2*

LOZZI, A. 1981. Motorcar lateral impacts and occupant injuries. *Int. J. Veh. Design* 2(4):470–79...*s28.6*

LUK, J. 1996. Congestion pricing and distribution of toll revenue. *Proc. 18th ARRB Conf.* 18(7):325–342...*s29.4*

-- , SIMS, A. & LOWRIE, P. R. 1983. The Parramatta experiment — evaluating from methods of area traffic control. Aust. Road Res. Brd. Res. Rep., ARR No. 132...*s23.5*

MACCARRONE, S., KY, A. & GNANASEELAN, G. 1996. Rutting and fatigue properties of high performance asphalt, *Proc. 18th ARRB Conf.* 19(2):133–147...*s12.2*

MACKENZIE, K. R. & FLETCHER, N. G. 1980. Structural design of asphalt pavements in Australia or the use and misuse of equivalencies. *Proc. 10th ARRB Conf.* 10(2):219–31...*s11.3,11.4*

MACKIE, P. & NELLTHORP, J. 2001. Cost–benefit analysis in transport. In *Handbook of transport systems and control*. K. Button & D. Hemsher (Eds). Permagon: Amsterdam...*s32.2&3*

MACLEAN, A. S. & HOWIE, D. J. 1980. Survey of the performance of pedestrian crossing facilities. *Aust. Road Res.* 10(3):38–45...*s20.4*

MAI, L. & SWEATMAN, P. F. 1984. A study of heavy vehicle stability. *Proc 12th ARRB Conf* 12(7):47–57...*s27.3*

MANINGIAN, S. 1996. Towards a national strategy to minimise wildlife mortality. *Proc. 18th ARRB Conf.* 18(6):143–155...*s6.4*

MARTIN, T. 1994. Pavement behaviour prediction for life-cycle costing. ARRB Research Report 255, Dec...*s14.4*

-- 1996. A review of existing pavement performance relationships. Aust. Road Res. Brd. Res. Rep. ARR 282, June...*s14.4*

-- 2005. Structural deterioration of sealed granular pavements in Australia, *Road & Trspt Res.* 14(2):3–15...*s11.4*

-- & ROPER, R. 1997. A parametric study of the influence of maintenance and rehabilitation strategies on network life-cycle costs. Aust. Road Res. Brd. Res. Rep. ARR 306...*s14.62*

MAY, A. D. 1990. *Traffic flow fundamentals*. Englewood Cliffs: Prentice Hall...*s17.2*

-- , TURVEY, I. G. & HOPKINSON, P. G. 1985. Studies of pedestrian amenity. Inst. Trspt Stud., Univ. Leeds, Working Paper WP 204...*s30.5*

MAYHEW, D. R., DONELSON, A. C., BIERNESS, D. J. & SIMPSON, H. M. 1986. Youth: alcohol and relative risk of crash involvement. *Acc. Anal. Prev.* 18(4):273–87...*s28.2*

McDONALD, M., HOUNSELL, N. B. & KIMBER, R. M. 1984. Geometric delay at non-signalised intersections. Trspt Road Res. Lab. U.K. TRRL Su Rep. SR 810...*s18.3,20.1,27.2*

McGOVERN, E. W. & ALDERTON, K. L. 1972. Development of durable road tar binders from high temperature coke oven tar. *Proc. 6th ARRB Conf.* 6(4):39–72...*s8.8*

McKELVEY, G. M. & THOMAS, I. G. 1984. The roundabout that isn't. *Aust. Road Res.* 14(1):37–39...*s7.3*

McLEAN, J. R. 1978. Observed speed distribution and rural road traffic operations. *Proc. 9th ARRB Conf.* 9(5):235–44...*s18.2,18.4*

-- 1979. An alternative to the design speed concept for low speed alignment design. *Proc. 2nd Int. Conf. Low Volume Roads. Trsptn Res. Rec.* 702:55–63...*s18.2*

-- 1984. The principles of geometric road design. *Program and Papers. 21st ARRB Reg. Symp.*, Darwin:1–21...*s28.5*

-- 1989. *Two-lane highway traffic operations: theory and practice*. Gordon & Breach: London & New York...*s17.2,17.3*

-- 1997. Practical relationships for the assessment of road feature treatments — summary report, Aust. Road Res. Brd. Res. Rep. ARR315, Dec...*s15.2,20.2, 28.5*

-- 2002. Review of the development of U.S. roadside design standards. *Road & Trspt Res.* 11(2):29–41, June...*s28.6*

METCALF, J. B. 1977. Principles and application of cement and lime stabilisation. Aust. Road Res. Brd. Res. Rep. ARR 49...*s10.2*

-- 1978. The use of local materials for low cost roads in Australia. Aust. Road Res. Brd. Res. Rep. ARR 50...*s8.2,9.4,11.7*

-- 1979a. Introductory Lecture Notes: 1. Compaction; 2. Principles and application of stabilisation; 3. Pavement design. Aust. Road Res. Brd. Res. Rep. ARR 47A. Revised and reprinted 1981...*10.2,10.4*

-- 1979b. Overtesting in pavement construction. *Aust. Road Res.* 9(1):56–58...*s8.4,14.5*

MEYER, J. R. & GOMEZ-IBANEZ, J. A. 1981. *Autos, transit and cities*. Harvard Univ. Press: Boston...*s30.1*

MIGLETZ, D. J., GLAUZ, W. D. & BAUER, K. M. 1985. Relationships between traffic conflicts and accidents. FHWA/RD-84/041, July...*s28.5*
MIHAI, F. 2008. The effect of civil liability legislation on negligence claims arising from road maintenance activities. *Proc. 23rd ARRB Conf....s18.1*
MILLAR, L. & GENEROWICZ, B. 1980. A study to assess the effectiveness of varying levels and types of enforcement on driver behaviour and urban signalised traffic intersections. *Proc. 10th ARRB Conf.* 10(4):141–57...*s28.1*
MINTY, E. J., PRATT, D. N. & BRETT, A. J. 1980. Aggregate durability tests compared. *Proc. 10th ARRB Conf.* 10(3):10–20...*s8.6*
MOGRIDGE, M. J. 1983. *The car market.* Dion: London...*s29.1,30.5*
MOORE, S. E. & HULSCHER, F. R. 1985. Improving the intrinsic safety of blacked-out traffic signal installations in New South Wales. *Aust. Road Res.* 5(1):9–19...*s26.4*
MOORS, H. T. 1972. The Washington degradation test used and abused — a review. *Proc. 6th ARRB Conf.* 6(5):166–78...*s8.6*
MORIARTY, P. 1980. The potential for non-motorised transport in Melbourne. *Proc. 10th ARRB Conf.* 10(5):43–50...*s20.4*
MORRIS, P. O. 1975. Compaction: a review. Aust. Road Res. Brd. Res. Rep. ARR 35...*s8.2*
-- & GRAY, W. J. 1976. Moisture conditions under roads in the Australian environment. Aust. Road Res. Brd. Res. Rep. ARR 69...*s9.2,9.3*
--, -- & COWAN, D. G. 1968. Strength, density, moisture content and soil suction relationships for a grey brown soil of heavy texture. *Proc. 4th ARRB Conf.* 4(2):1064–82...*s9.2,9.4*
MULHOLLAND, P. J. 1989. Into a new age of pavement design: a structural design guide for flexible street pavements. ARRB Spec. Rep. SR 41, April...*s11.4,12.1*

NEESON, J. A. 1973. Letter to the Editor. *Aust. Road Res.* 5(1):78–79...*s8.2*
NEUMAN, T. R. 1985. Intersection channelisation design guide. TRB NCHRP Rep. 279 *20.3*
NYOEGER, E. 1964. Petrological investigation into the secondary minerals of an older basalt flow north of Melbourne. *Proc. 2nd ARRB Conf.* 2(3):997–1007...*s8.6*

OGDEN, K. W., RITCHIE, S. G. YOUNG, W. & DUMBLE, P. L. 1981. Analysis of appraisal of freight generation characteristics. Aust. Road Res. Brd. Res. Rep., ARR No. 117...*s30.5*
OHNO, K. & MINE, H. 1973. Optimal signal settings. *Trnsptn Res.* 7(3):243–292...*s23.6*
OLIVER, J. W. 1978. Calibration and use of the British pendulum tester for the measurement and prediction of pavement skid resistance. *Proc. 9th ARRB Conf.* 9(3):127–32...*s8.6,12.5*
-- 1984. An interim model for predicting bitumen hardening in Australian sprayed seals. *Proc. 12th ARRB Conf.* 12(2):112–20...*s8.7*
O'NEILL, B., LUND, A. K., ZADOR, P. & ASHTON, S. 1985. Mandatory belt use and driver risk taking: an empirical evaluation of the risk-compensation hypothesis. In *Human behaviour and traffic safety.* Eds L. Evans and R.C. Schwing. Plenum: New York...*s28.2*
ORCHARD, D. F. 1964. Factors influencing the wear of coarse aggregate in the Los Angeles Test. *Proc. 2nd ARRB Conf.* 2(2):963–80...*s8.6*
ORGANISATION FOR ECONOMIC CO-OPERATION AND DEVELOPMENT 1979. Evaluation of load carrying capacity of bridges. OECD Road Research Programme: Paris...*s15.3,15.5*
-- 1980. Transport choices for urban passengers — measures and models. OECD Road Research Programme: Paris...*s30.4*
-- 1981a. Traffic control in saturated conditions. OECD Road Research Programme: Paris...*s17.4*
-- 1981b. Guidelines for driver instruction. OECD Road Research Programme: Paris...*s16.6*
-- 1982. Automobile fuel consumption in realistic traffic conditions. OECD Road Research Programme: Paris...*s29.2*
-- 1983a. Traffic safety of children. OECD Road Transport Research Programme: Paris...*s28.7*
-- 1983b. Traffic capacity of major routes. OECD Road Research Programme: Paris...*s17.4*
-- 1983c. Impacts of heavy freight vehicles. OECD Road Research Programme: Paris...*s27.3*
-- 1983d. Bridge rehabilitation and strengthening. OECD Road Research Programme: Paris...*s15.5*
-- 1984. Road surface characteristics. OECD Road Research Programme: Paris...*s12.5*
-- 1986a. Technico-economic analysis of the role of road freight transport. OECD Road Transport Research Programme: Paris...*s30.5*
-- 1986b. Guidelines for improving the safety of elderly road users. OECD Road Transport Research Programme: Paris...*s16.5*
-- 1988a. Pavement damage due to heavy freight vehicles and climate. OECD Road Transport Research Programme: Paris...*s11.4*

-- 1988b. Transporting hazardous goods by road. OECD Road Transport Research Programme: Paris...*s27.3*
-- 1991. OECD full-scale pavement test. OECD Road Transport Research Programme: Paris...*s14.3*
-- 1994. Environmental impact assessment of roads. OECD Road Transport Research Programme: Paris...*s32.1*
-- 2005. *IRRD thesaurus*, English alphabetical, Version 2.... *s2.2*
OXFORD 2007. *The shorter Oxford English dictionary* based on Historical Principles, 6th Edition (text also draws on 1993 4th Edition). Clarendon Press: Oxford...*s2.1*

PAPAGIANNAKIS, A. & MASAD, E. 2008. *Pavement design and materials*. New Jersey: Wiley...*s8.6*
PARSLEY, L. L. & ROBINSON, R. 1982. The TRRL road investment model for developing countries RTIM2. Trspt Road Res. Lab. U.K. TRRL Lab. Rep. LR 1057...*s14.4*
PARSONS, A. W. 1992. *Compaction of soils and granular materials*. HMSO: London...*s25.5*
PATERSON, W. D. 1985. *Prediction of road deterioration and maintenance effects: theory and quantification*, Vol III, Trnspt Dept World Bank, Washington...*s11.4,12.1,14.3,14.4,29.3*
PATRICK, S., TARANTO, V., BLANKSBY, C., LUK, J., RITZINGER, A, & FRASER, S. 2006. Review of passenger car equivalency factors for heavy vehicles. *Proc. 23rd ARRB Conf...s17.4,*
PAUL, R. 2003. Deformation resistance of Victorian asphalt mixes...Proc. 21st ARRB Conf. Session S11...*s12.2*
PEARCE, P. & PROMNITZ, J. 1982. Highway rest areas: studies from two Australian states. *Aust. Road Res.* 12(1):29–40...*s6.4*
PERRY, D. R. & CALLAGHAN, W. M. 1980. Some comparisons of direct and indirect calculations of exposure and accident involvement rates. *Proc. 10th ARRB Conf.* 10(4):263–74...*s28.1*
PERMANENT INTERNATIONAL ASSOCIATION OF ROAD CONGRESSES 2004. *Highway development and maintenance systems*, Paris: PIARC...s26.5
PHILLIPS, G., BLAKE, P. & REESON, R. 1984. Estimation of annual flow from short period traffic counts. Trspt Road Res. Lab. U.K. TRRL Su Rep. SR 802...*s31.5*
PIARC, see PERMANENT INTERNATIONAL ASSOCIATION OF ROAD CONGRESSES
PIDWERBESKY, B. 1997. Predicting rutting in unbound granular basecourses from Loadman and other in situ non-destructive tests. *Road & Trspt Res.* 6(3):16–25...*s14.5*
PILGRIM, D. H. (Ed) 2001. *Australian rainfall and runoff.* Canberra: Institution of Engineers, Australia...*s13.2*
POLUS, A. & MATTAR-HABIB, C. 2004. A new consistency model for rural highways and its relationship to road safety. *J. Trspn Engrg* (ASCE) 130(3):286–93...*s18.2*
PORTER, B. E. & ENGLAND, K. J. 2000. Predicting red-light running behavior; a traffic safety study in three urban settings. *J. Safety Res.* 31(1):1–8...*s23.1*
PORTER, K. F., MORRIS, P. O. & ARMSTRONG, P. J. 1980. Weighting formulations for the ARRB road condition rating system. Aust. Road Res. Brd. Internal Rep. AIR 262-7...*s14.5*
POSTANS, R. L. & WILSON, W. T. 1983. Close-following on the motorway. *Ergonomics*, 26:317–27...*s17.3*
POTTER, D. W. 1977. *Manual of operation for ARRB wave propagation equipment for the measurement of elastic moduli*. Aust. Road Res. Brd. Technical Manual, ATM 2...*s11.3*
-- & HUDSON, W. R. 1981. Optimisation of highway maintenance using the Highway Design Model. *Aust. Road Res.* 11(1):3–15...*s12.1,14.1,14.6*
POWELL, W. D., POTTER, J. F., MAYHEW, H. C. & NUNN, M. E. 1984. The structural design of bituminous roads. Trspt Road Res. Lab. U.K. TRRL Lab. Rep. LR 1132...*s11.3,11.4,14.6*
PREM, H., RAMSAY, E., FLETCHER, C., GEORGE, R. & GLEESON, B. 1999. Estimation of lane width requirements for heavy vehicles on straight paths. ARRB Res Rep ARR 342, Sept...*s6.2*
PRETTY, R. L. 1980. A further treatment of opposed turning movements in traffic signal calculations. *Proc. 10th ARRB Conf.* 10(4):1–7...*s17.4*

QUAYLE, G. M. 1980. Intersection priority: Australia at the crossroads. *Proc. 10th ARRB Conf.* 10(4):233–44...*s20.2*
QUIMBY, A. R. 1988. In-car observation of unsafe driving actions. ARRB Res. Rep. ARR 153...*s28.2*

RAWLINSON, W. R. 1987. Peak side friction and yaw rate during survey of Hume Highway. *Aust. Road Res.* 17(2), June:129-31...*s19.2*
RICHARDS, B. G. 1980. The analysis of the total stress–strain response of an expansive clay subgrade at Macalister, Queensland. *Proc. 10th ARRB Conf.* 10(2):1–8...*s11.2,11.3*
-- & CHAN, C. Y. 1971. Theoretical analyses of sub-grade moisture under Australian environmental conditions and their practical implications. *Aust. Road Res.* 4(6):32–49...*s9.3*

RICHARDSON, A. J. 2004. Estimating individual values of time in stated preference surveys. *Road & Trspt Res.* 15(1), March: 44–53…*s29.4*
-- & GRAHAM, N. 1980. Objectives and performance measures in urban traffic systems. Program and Papers. Joint SAE Aust./ARRB Conf. on 'Can Traffic Management Reduce Vehicle Fuel Consumption and Emissions and Affect Vehicle Design Requirements?' ...*s23.1*
ROADS AND TRANSPORTATION ASSOCIATION OF CANADA 1977. *Pavement management guide*. RTAC: Ottawa...*s11.4,12.2*
ROBERTSON, D. I. 1986. Research on the TRANSYT and SCOOT methods of signal co-ordination. *Inst. Trnsptn Engrgs J.* 56(1), Jan:36–40...*s23.5*
ROE, P. G. & WEBSTER, D. C. 1984. Specification for the TRRL frost heave test. Trspt Road Res. Lab. U.K. TRRL Su Rep. SR 829...*s9.5*
ROPER, R. 2001. Adaptation of HDM-IV user effects models for application in Australia. *Proc. 20th ARRB Conf.* CD...*s26.5*
ROTH, G. 1996. *Roads in a market economy*. Aldershot: Ashgate…*s29.4*
RTAC, see ROADS AND TRANSPORTATION ASSOCIATION OF CANADA
RUFFORD, P. G. 1977. A pavement analysis and structural design procedure based on deflection. *Proc. 4th Int. Conf. Struct. Des. Asph. Pave.* 1: 710–21...*s8.4*
RUTHERFORD, W. H. 1987. Compulsory wearing of seat belts in the United Kingdom — the effect on patients and on fatalities. *Trspt Rev.* 7(3):245–57...*s28.4*

SA & SAA, see STANDARDS (ASSOCIATION OF) AUSTRALIA
SABEY, B. 1980. Road safety and value for money. Trspt Road Res. Lab. U.K. TRRL Su Rep. SR 581...*s28.2*
-- 1985. Accident reduction by low cost engineering. Presented to Conf. N.Z. Inst. of Prof. Eng., Feb...*s28.3*
-- & TAYLOR, H. 1980. The known risks we run: the highway. Trspt Road Res. Lab. U.K. TRRL Su Rep. SR 567...*s28.1*
SALT, G. 1977. Research on skid-resistance at the Transport and Road Research Laboratory. *Trsptn Res. Rec.* 622: 26–38...*s12.5*
SAMUELS, S. E. 1982. Tyre/road noise generation. *Proc. 11th ARRB Conf.* 11(6):18–29...*s32.2*
-- & JARVIS, J. R. 1978. Acceleration and deceleration of modern vehicles. *Proc. 9th ARRB Conf.* 9(5):254–61...*s21.2,27.2*
-- & SAUNDERS, R. E. 1982. The Australian performance of the U.K. DoE traffic noise prediction methods. *Proc. 11th ARRB Conf.* 11(6):30–40...*s32.2*
SARRE, R. 2003. Liability on negligence and the High Court decisions in Brodie and Ghantous. *Road & Trspt Res.* 12(4):3–12, Dec…*s18.1*
SAVENHED, H. 1985. Vehicle fuel consumption on different types of wearing course. *PTRC 13th Summer Annu. Meet.*, July, Seminar P:155–65...*s29.2*
SAYERS, M., 1995. On the calculation of International Roughness Index from longitudinal road profile. *TRB Res. Rec.* 1501, 1–12..*s14.4*
-- , GILLESPIE, T. D. & QUEIROZ, C. A. 1986. International experiment to establish correlation and standard calibration methods for road roughness measurements. Report to World Bank by Univ. Michigan, Trsptn Res. Inst. May. UMTRI-85-15-1. Later published as 'The international road roughness experiment', World Bank technical paper 45, 1986, and 'Guidelines for conducting and calibrating road roughness measurements', loc cit 46...*s14.4*
SCALA, A. J. 1978. A systems approach to flexible pavement design in Australia. *Proc. 9th ARRB Conf.* 9(4):24–37...*s11.3,14.6*
-- 1979. An analysis of the deflection bowls in pavements measured by the Benkelman beam test. Papers. Proc. N.Z. Road. Symp. Road Research Unit, Paper G2...*s11.3*
SCHACKE, I. 1984. A pavement maintenance system. *Proc. 12th ARRB Conf.* 12(1):47–65...*s14.6,26.5*
SCHAFER, A. & VICTOR, D. 1997. The past and future of global mobility. *Sci. Amer.* 277(4):36–9, Oct…*s4.3*
SCHMIDT, A. 2004. The Brifen wire rope safety fence — the evolution continues. *Hwy Engrg in Aus.*, 36(5):5–14, Nov…*s28.6*
SCOTT, J. 1978. Adhesion and disbonding mechanisms of asphalt used in highway construction and maintenance. *Proc. Assoc. Asphalt Pave. Technol.* 1: 19–48...*s12.3*
SEDDON, P. A. 1980. The behaviour of interlocking concrete block paving at the Canterbury test track. *Proc. 10th ARRB Conf.* 10(2):58–64...*s11.6*
SEMMENS, M. C. 1985. ARCADY2: an enhanced program to model capacities: queues and delays at roundabouts. *TRRL Res. Rep.* 35...*s20.3*
SHACKEL, B. 1973. Repeated loading of soils — a review. *Aust. Road Res.* 5(3):22–49...*s8.2,11.3*

-- , MAKINCHI, K. & DERBYSHIRE, J. R. 1974. The response of a foamed bitumen stabilised soil to repeated load triaxial loading. *Proc. 7th ARRB Conf.* 7(7):74–89...*s10.4*

SHARP, K. G. & ARMSTRONG, P. J. 1985. Interlocking concrete block pavement. Aust. Road Res. Brd. Special Rep. SR 31...*s11.5,11.6*

SHELL 1978. *Shell pavement design manual.* Shell Int. Petrol Ltd: London...*s11.2,11.4*

SHINAR, D. 1978. *Psychology on the road.* Wiley: New York...*s16.5,28.2*

SHIPWAY, C. H. 1964. A study of the aggregate crushing test. *Proc. 2nd ARRB Conf.* 2(2):981–96...*s8.6*

SMEED, R. J. 1949. Some statistical aspects of road safety. *J. Roy. Stat. Soc.*, CXII, Part 1, Series(4):1–24...*s28.1*

SMELT, J. M. 1984. Platoon dispersion data collection and analysis. *Proc. 12th ARRB Conf.* 12(5):71–87...*s17.3*

SMITH, R. B. 1980. A laboratory and field study on mechanical breakdown of gravel during construction. *Proc. 10th ARRB Conf.* 10(3):1–9...*s8.5*

-- & CREWS, K. I. 1983. Compaction control testing of an experimental pavement. *Papers. N.Z. Road Symp.* 2: 83–90...*s8.2*

-- & PRATT, D. N. 1983. A field study of the in situ California Bearing Ratio and Dynamic Cone Penetrometer testing for road subgrade investigations. *Aust. Road Res.* 13(4):285–94...*s8.4*

SOKOLNIKOFF, I. 1956. *Mathematical theory of elasticity.* 2nd Ed. McGraw-Hill: New York...*s33.3*

SPARKS, G. H. 1984. The compaction of coarse fills. *Aust. Road Res.* 13(4):295–99...*s11.8*

-- & HAMORY, G.I. 1980. The performance of a stabilised limestone base material. *Proc. 10th ARRB Conf.* 10(2):80–96...*s10.4*

STANDARDS ASSOCIATION OF AUSTRALIA 1980. Determination of durability of bitumen. AS 2341.13. Part of AS 2341. Methods of testing bitumen and related roadmaking products...*s8.7*

STANDARDS AUSTRALIA 2002. Road and traffic engineering — glossary of terms. AS 1348...*s2.1,17.1*

STANLEY, J. K., HOGG, T. M. & DELANEY, D. J. 1973. The theory of benefit and cost measurement, with reference to residential disruption costs of urban road improvements. *Aust. Road Res.* 5(2):23–35...*s5.2*

SWEATMAN, P. F. 1980. Effect of heavy vehicle suspensions on dynamic road loading. *Proc. 10th ARRB Conf.* 10(5):281–89...*s27.3*

-- 1983. A study of dynamic wheel forces in axle group suspensions of heavy vehicles. Aust. Road Res. Brd. Special Rep., SR 27...*s27.3*

TANNER, J. C. 1962. A theoretical analysis of delays at an uncontrolled intersection. *Biometrika* 49 1 & 2:163–70...*s17.3*

-- 1983. International comparisons of cars and car usage. Trspt Road Res. Lab. U.K. TRRL Lab. Rep. LR 1070...*s4.3,30.5*

-- 1984. Modelling the car market. Trspt Road Res. Lab. U.K. TRRL Lab. Rep. LR1116...*s27.1,29.3*

TATE, F. & TURNER, S, 2007. Road geometry and drivers' speed choice. *Road & Trspt Res.*, 16(4):53–64, Dec...*s18.2*

TAYLOR, M. A. 1980. Soft: slow and going places: pedestrians and cyclists in urban transport. *Proc. 10th ARRB Conf.* 10(5):100–110...*s28.7,30.2,30.5*

ten BRUMMELAAR, T. 1983. The reversal point in the perspective road picture. *Aust. Road Res.* 13 2:123–7...*s16.5*

THORESON, T. 1993. Survey of freight vehicle costs. Aust. Road Res. Brd Res. Rep. ARR 239, Jan...*s29.1,29.2*

THORPE, J. D. 1964. Calculating relative involvement rates in accidents without determining exposure. *Aust. Road Res.* 2(1):25–36...*s28.1*

TIMOSHENKO, S. 1941. *Strength of materials.* Vol. II. Von Nostrand: New York...*s11.2*

TRANSPORTATION RESEARCH BOARD 2000. *Highway capacity manual.* Spec. Rep. 209. TRB: Washington...*s17.2,17.4*

TRB, see TRANSPORTATION RESEARCH BOARD

TRIGGS, T. J. & HARRIS, W. G. 1982. Reaction time of drivers to road stimuli. *Human Factors Rep.* HFR-12, Monash Univ., Melbourne, June...*s16.5*

TROUTBECK, R., 1980. Overtaking sight distances for rural road design. *Proc. 10th ARRB Conf.* 10(4):84–96...*s18.4*

-- 1981. Overtaking behaviour on Australian two-lane rural highways. Aust. Road Res. Brd. Special Rep., SR No. 20...*s18.4,19.4,22.5*

-- 1984. Overtaking behaviour on narrow two-lane two-way rural roads. *Proc. 12th ARRB Conf.* 12(5):105–16...*s16.5*

TSOLAKIS, D. & HOUGHTON, N. 2003. Valuing environmental externalities. *Proc. 21st ARRB Conf.* 21(S12)...*s29.1,32.2&3*

TYRRELL, A. P., LAKE, L. M. & PARSONS, A. W. 1983. An investigation of the extra costs arising on highway contracts. Trspt Road Res. Lab. U.K. TRRL Su Rep. SR 814...*s6.2*

ULLIDTZ, P. 1987. *Pavement analysis*. Elsevier: Amsterdam...*s11.3, 14.5*
UZAN, J., LIVNEH, M. & ISHAI, I. 1980. Thickness design of flexible pavements with different layer structures. *Aust. Road Res.* 10(1):8–20...*s8.4,11.3*

van DIJK, W., MOREAUD, H., QUEDEVILLE, A. & UGE, P. 1972. The fatigue of bitumen and bituminous mixes. *Proc. 3rd Int Conf. on Structural Design of Asphalt Pavements*, 1: 354–66, London....*s11.4*
van der DRIFT, R., 1998. The STEP barrier — a better road barrier. *Traffic Engrg & Control*, 39(2):86–9, Feb....*s28.6*
van EVERY, B. E. 1982. A guide to the economic justification of rural grade separations. *Aust. Road Res.* 12(3):147–54...*s20.2*
VICROADS 2002. Using polish resistant aggregates. Geopave Technical Note 42, May...*s12.5*
VINER, M. E., SINMAL, R. & PARRY, A. R. 2005. Linking road traffic accidents with skid resistance — recent developments. TRL Paper PA/INF 4520/05...*s12.5*
VINES, F. R. & FALCONER, G. D. 1980. Experience with coral and volcanic road construction material in Western Samoa. *Aust. Road Res.* 10(1):32–38...*s8.5*
VUONG, B. 2008. Validation of a three-dimensional nonlinear finite element model for predicting response and performance of granular pavements with thin bituminous surfaces under heavy vehicles. *Road & Trspt Res.* 17(1):3–31, March...*s11.2&3*
VUONG, B. & JAMESON, G. 2003. Equivalent load for a quad axle. *Proc. 23rd ARRB Conf.* Session 32...*s27.3*

WALLACE, K. B. 1981. Some important aspects of design of pavements for high rainfall areas. *Program and Papers. 18th ARRB Reg. Symp.* Traralgon, 147–54...*s8.3*
-- & LEONARDI, F. 1978b. Seasonal changes in the strength of rural road pavements in coastal North Queensland. Aust. Road Res. Brd. Special Rep. SR 16...*s9.1,9.4*
WALLACE, K. B. & MONISMITH, C. L. 1980. A simplified thickness design procedure for pavements containing asphalt treated layers. *Proc. 10th ARRB Conf.* 10(2):190–204...*s11.3,11.4*
WARDROP, J. G. 1952. Some theoretical aspects of road traffic research. *Proc. Inst. Civ. Eng.* 2(1):325–78...*s17.1,18.2,31.3*
WATANATADA, T., HARRAL, C., PATERSON, W. D., DHARESHWAR, A. M., BHANDARI, A. & TSUNOKAWA, K. 1987. *The highway design and maintenance standards model.* Published for the World Bank by John Hopkins Univ. Press; Baltimore...*s26.5*
WATSON, H. C. & MILKINS, E. F. 1982. Further application of the PKE average speed fuel consumption and emissions model. Papers. Joint SAE-A/ARRB 2nd Conf. on Traffic Energy and Emissions, May, Paper 9...*s29.3*
-- , -- , PRESTON, M. O., BEARDSLEY, P. & CHITTLEBOROUGH, C. W. 1985. The effect of operating conditions on vehicle fuel consumption and emissions. *Proc. 3rd Int. Pac. Conf. on Auto. Engrg*, Jakarta, Nov, Vol 2:473–87, SAE 85 2230...*s27.2*
WATTS, G. R. 1973. Road humps for the control of vehicle speeds. Trspt Road Res. Lab. U.K. TRRL Lab. Rep. LR 597...*s18.1*
-- 1984. Vibration nuisance from road traffic — results of a 50 site survey. Trspt Road Res. Lab. U.K. TRRL Lab. Rep. LR1119...*s25.8*
WEBSTER, D. C. 1993. Road humps for controlling vehicle speeds. Trspt Res. Lab. Project Rep. 18...*s18.1*
WEBSTER, F.V., BLY, P. H., JOHNSTON, R. H., PAULLEY, N. & DASGUPTA, M. 1984. Changing patterns of urban travel. ECMT Paper, CM 84 26, Oct...*s30.4,30.5*
WHEELER, J. G. 1980. Pedestrian-induced vibrations in footbridges. *Proc. 10th ARRB Conf.* 10(3):21–35...*s15.4*
WHITT, F. R. & WILSON, D. G. 1982. *Bicycling science*. MIT Press: Mass...*s27.6*
WIDDUP, J. D. 1980. The use of rolling wheel deflections to predict flexible pavement life: a review. *Proc. 10th ARRB Conf.* 10(2):115–28...*s14.3*
WIGAN, M. R. 1983. Bicycle ownership: use and exposure in Melbourne 1978–79. Aust. Road Res. Brd. Res. Rep., ARR 130, May...*s30.4*
-- 1987. Australian personal travel characteristics. ARRB Spec. Rep. 38, Nov...*s31.3*
WIGGLESWORTH, E. J. 1978. Compensation — an un-affordable handicap to the prevention of trauma. *Proc. 15th Annu. Conf. Ergonomics Soc. of Aust. & N.Z.*, Melbourne. 'Human Factors and Contemporary Society': 64–74...*s28.2*

WILLIAMS, A. & HAWORTH, N. 2007. Overcoming barriers to creating a well-functioning safety culture: a comparison of Australia and the United States. *ITE J* 77(6):24–31, June…*s28.2*
WILLIS, P. A., SCOTT, P. P. & BARNES, J. W. 1984. Road edgelining and accidents: an experiment in south-west England. Trspt Road Res. Lab. U.K. TRRL Lab. Rep. LR1117...*s22.2*
WITT, H. P. 1976. Determination of the long term effect of air on a thin film of bitumen heated at 100 degrees C ARRB durability test. Aust. Road Res. Brd. Technical Manual. ATM 3...*s8.7*
WONG, Y. D. & GOH, P. K. 2000. Driver perception response time for braking action during signal change interval. *Road & Trspt Res.* 9(3):17–26, Sept...*s23.2*
WOOD, K. 1993. Urban traffic control: systems review. Trsprt Res. Lab. U.K. Project Rep 41...*s23.5*
WORLD BANK 1994. *Infrastructure for development*. Washington…*s4.3*
WORLD HEALTH ORGANISATION 2004. Asphalt (bitumen). Concise International Chemical Assessment document 59, www.inchem.org/documents/cicads...*s8.7*
WYLDE, L. J. 1982. Mineralogical factors in the durability of basaltic roadbase materials. Aust. Road Res. Brd. Special Rep. SR 22...*s8.5*

YAGAR, S. 1983. Capacities for two-lane highways. *Aust. Road Res.* 13(1):3–9...*s17.4*
YODER, E. J. & WITCZAK, M. 1975. *Principles of pavement design*. 2nd Ed. Wiley: New York...*s11.1,11.4,12.2*
YOUDALE, G. P. 1984. The design of pavements surfaced with asphalt for particular temperature environments. *Proc. 12th ARRB Conf.* 12(3):70–85...*s11.4*
YOUNG, W. & MORRIS, J. M. 1980. Perception and evaluation of the journey to work. *Papers 6th Aust. Trspt Res. Forum*, Brisbane:609–30...*s30.2*

ZAHAVI, Y. 1974. Travel time budgets and mobility in urban areas. Rep FHWA 8183...*s31.2*

Index

To aid cross-referencing, the Index refers to sections rather than to pages. This will also help the reader familiar with the book to detect relevant sections.

B

C

D

E

F

G

H

J

K

L

M

N

O

P

Q

R

S

T

U

V

W

X

Y

Z

Greek, etc.

Milton Keynes UK
Ingram Content Group UK Ltd.
UKHW051850071024
449327UK00025B/1903

9 780367 864873